Zoology

For Degree Students

Non-Chordates II | Cell Biology

About the Author

Dr V K Agarwal, MSc, PhD, has rich teaching experience of more than four decades at Meerut College, Meerut. He has taught subjects such as Cell Biology, Genetics, Molecular Biology, Chordate Embryology (Developmental Biology), Invertebrate Zoology, Chordate Zoology, Evolution, Ecology, Animal Behaviour (Ethology) to undergraduate and postgraduate students.

Dr Agarwal has to his credit the authorship of several widely acclaimed books such as *Cell Biology*, *Genetics*, *Chordate Embryology* (*Developmental Biology*), *Ecology* (*Environmental Biology*), *Ethology* (*Animal Behaviour*), *Zoology for Degree Students-BSc First Year* (includes *Invertebrate Zoology*), *Zoology for Degree Students–Semester I*, *Zoology for Degree Students–Semester III*, *Molecular Biology, Genetic Engineering,* and a multi-coloured edition of *Cell Biology, Genetics, Molecular Biology, Evolution and Ecology.* The author has also co-authored *CBSE – IX Biology.* He has also published few research papers on *Insect Physiology* and *Toxicology*.

Zoology

For Degree Students

Non-Chordates II | Cell Biology

B.Sc. (Honours) Semester - II

As per UGC Choice Based Credit System (CBCS)

V K Agarwal
MSc, PhD
Former Head of the Zoology Department
Meerut College, Meerut, Uttar Pradesh

S. CHAND
PUBLISHING

S Chand And Company Limited

(ISO 9001 Certified Company)

S Chand And Company Limited

(ISO 9001 Certified Company)

Head Office: Block B-1, House No. D-1, Ground Floor, Mohan Co-operative Industrial Estate, New Delhi – 110 044 | Phone: 011-66672000

Registered Office: A-27, 2nd Floor, Mohan Co-operative Industrial Estate, New Delhi – 110 044 Phone: 011-49731800

www.**schandpublishing.com**; e-mail: **info@schandpublishing.com**

Branches

Chennai	:	Ph: 23632120; chennai@schandpublishing.com
Guwahati	:	Ph: 2738811, 2735640; guwahati@schandpublishing.com
Hyderabad	:	Ph: 40186018; hyderabad@schandpublishing.com
Jalandhar	:	Ph: 4645630; jalandhar@schandpublishing.com
Kolkata	:	Ph: 23357458, 23353914; kolkata@schandpublishing.com
Lucknow	:	Ph: 4003633; lucknow@schandpublishing.com
Mumbai	:	Ph: 25000297; mumbai@schandpublishing.com
Patna	:	Ph: 2260011; patna@schandpublishing.com

First Edition 2018
Reprints 2019 (Twice), 2020

Reprint 2021 (Twice)

ISBN : 978-93-525-3411-1

Product Code : H6ZOO68ZOOL10ENAA18O

PRINTED IN INDIA

By Vikas Publishing House Private Limited, Plot 20/4, Site-IV, Industrial Area Sahibabad, Ghaziabad – 201 010 and Published by S Chand And Company Limited, A-27, 2nd Floor, Mohan Co-operative Industrial Estate, New Delhi – 110 044.

Preface

It gives me immense pleasure and satisfaction to present this textbook, **Zoology for Degree Students - Semester II**, covering subject **Non-chordates - Coelomates** and **Cell Biology**. This comprehensively written textbook provides up-to-date information on the subject in an informative and systematic manner.

This textbook has been designed as per the UGC Choice Based Credit System (CBCS) syllabus of Zoology, BSc (Honours) - Semester II. It includes the following two courses:

1. **CC–III Non-Chordates II : Coelomates**
2. **CC–IV Cell Biology**

The central concept of **Invertebrate Zoology** is the evolution of animal form and function. For example, the evolution of epithelia allowed eumetazoans to physiologically regulate internal, extracellular compartments such as the gut and coelom. This regulation not only improved the performance of digestive and other functions, but also liberated eumetazoans to some extent from the vagaries of environmental variations, thus enabling them to sustain relatively high levels of activity. The adoption of bilateral symmetry by eumetazoans resulted in a directionally polarized body equipped with sensory organs used for pursuing food and makes or escaping from enemies. The arthropods, especially the 20 million-plus species of insects, with their mobile appendages and protective, supportive exoskeletons, found the keys to the colonization of land and air (Ruppert, Fox and Barnes, 2004).

Cell Biology is an experimental science and cell biologists believe students should gain some knowledge of how we know what we know. The experimental pathways illustrate the stepwise nature of scientific discovery, showing how the result of one study raises questions that provide the basis for subsequent studies. To a large degree, cells are like tiny black boxes. We have developed many ways to probe the boxes, but we are always groping in an area that cannot be fully illuminated. A discovery is made or a new technique is developed and a new thin beam of light penetrates the box (Karp, 2010).

Some of the special features of this textbook are:

1. It includes Non-Chordates or Invertebrate Zoology having descriptions of Coelomate Phyla such as Annelida, Arthropoda, Mollusca and Echinodermata.
2. Non-chordate Zoology also includes comparative anatomical account of certain organ systems of non-chordate animals.
3. Cell Biology discusses important topics such as Viruses, Prokaryotic Cells (Archaebacteria, True Bacteria and Cyanobacteria and Mycoplasma) and Eukaryotic Cells (Plants and Animals).
4. Cell Biology further discusses topics such as Biomolecules and structure & function of various Cellular Organelles.
5. Modern topics such as Vesicular Trafficking, Cellular Movement, Mechanics of Cell Division, Cell Signalling and Cancer have also been discussed.

The advances in Non-Chordate Zoology and Cell Biology are ever growing due to exponential growth of the subject. Thus, I have tried my best to incorporate the latest advances in the subject.

Undergraduate students will find this book immensely useful as it lucidly explains the concepts with the help of well-illustrated diagrams and information-packed boxes.

To friends and family, I remain indebted and thank them for their timely support during development of this textbook. Special thanks to my wife Usha Gupta for her unending inspiration during crafting of this textbook. I wish to thank my daughters Dr Surabhi, Dr Shubhra and Er. Anubha they need special mention for restricting me to remain a healthy, cool and normal person. I also wish to thank Trilok Jain who has illustrated the figures of this book.

Many thanks to the editorial team at S Chand, especially, Surabhi Khare, Shalini Negi, Himani Ramola, Priyanka Sharan, Arun Kumar, Ketan Mithrani, Sanjeev Arora and Mohit Bharti for their efforts in bringing out this book in shortest possible time. Their zeal to excel is really heart-warming.

Suggestions and feedback from the readers will be gratefully acknowledged.

V K Agarwal
Villa–20
Green Wood City
Haridwar Bypass Road
Meerut - 250 005
Uttar Pradesh (India)

Syllabus

BSc (Hons) Zoology – Semester II

CORE COURSE III

NON-CHORDATES II: COELOMATES

Unit 1:	**Introduction to Coelomates** Evolution of Coelom and Metamerism	**2**
Unit 2:	**Annelida** General characteristics and Classification up to classes Excretion in Annelida	**10**
Unit 3:	**Arthropoda** General characteristics and Classification up to classes Vision and Respiration in Arthropoda Metamorphosis in Insects Social Life in bees and termites	**17**
Unit 4:	**Onychophora** General Characteristics and Evolutionary significance	**4**
Unit 5:	**Mollusca** General characteristics and Classification up to classes Respiration in Mollusca Torsion and detorsion in Gastropoda Pearl formation in bivalves Evolutionary significance of trochophore larva	**10**
Unit 6:	**Echinodermata** General characteristics and Classification up to classes Water-vascular system in Asteroidea Larval forms in Echinodermata Affinities with Chordates	**12**

CORE COURSE IV

CELL BIOLOGY

Contents

NON-CHORDATES II : COELOMATES (CC–III)

CELL BIOLOGY (CC–IV)

(CC–III)
NON-CHORDATES II : COELOMATES

1 Introduction to Coelomates (Cladistic methods, Evolution of Metamerism and Coelom)

1.1 INTRODUCTION OF SOME MODERN CLADISTIC METHODS

One of the remarkable achievements of biological research has been the establishment of a **natural system**, the recognition of species and the arrangement of those species in a hierarchic pattern of relationships. This pattern of evolutionary or kinship relationships is called a **phylogeny**, or a **"tree of life"**. The science that deals with the discovery and depiction of phylogeny is known as **phylogenetic systematics** or **cladistics**. The goal of cladistics is to discover and portray the kinship relationships among species, ultimately in a **phylogenetic tree**, or **cladogram**.

In practice, one establishes a kinship relationship between two named groups, or **taxa** (the singular is **taxon**), by detecting a trait or **character**, that is expressed in these taxa alone. Such a uniquely shared character is called a **synapomorphy** (= shared dervied character), and taxa united by one or more synapomorphies are called **sister taxa**. Because a synapomorphy is shared by the sister taxa and no other taxon, it must have evolved in the immediate **ancestor** of the sister taxa. Thus the synapomorphy observed in the descendants actually originated in the ancestor as an evolutionary novelty, now called an **autapomorphy** (= self derived character). As the sole descendants of that immediate ancestor, the sister taxa constitute a **monophyletic** (= one origin) **taxon**. The ultimate goal of cladistics is to reconstruct a detailed tree of life (phylogeny) exclusively based on monophyletic taxa.

These concepts can be explained on the basis of arthropoda (Fig. 1.1). For taxa, we can consider the crustaceans (such as shrimps, crabs, and lobsters) and tracheates (such as insects, millipedes, and centipedes). Representatives of these two taxa show that all share the character "mandibles" (jaws derived from appendages on the third head segment), which occurs nowhere else and thus is a synapomorphy of crustaceans and tracheates. The synapomorphy shows that the sister taxa (Crustacea and Tracheata) constitute a monophyletic taxon, the ancestor of which first evolved mandibles as an evolutionary novelty or autapomorphy. Once the monophyletic taxon has been identified, it is given a formal name – in this case **Mandibulata**.

Further, let's suppose that, instead of crustaceans and tracheates, the study began with crustaceans and chelicerates (including horseshoe crabs, scropions and spiders) and found that they shared the character "jointed appendages", which was then hypothesized to be a synapomorphy of two taxa (Fig. 1.1). The character "jointed appendages", however is not shared by crustaceans and chelicerates alone, but also by the tracheates. This means that the "jointed appendages" of chelicerates and crustaceans is not a shared derived character but rather a shared ancestral trait, or **symplesiomorphy**, that evolved in the stem species (ancestor) of all arthropods. The erroneous union of crustaceans and chelicertaes based on a symplesiomorphy is called a **paraphyletic taxon** (Fig. 1.1). A paraphyletic taxon contains some, but not all, of the descendants of the stem species. A major challenge of modern phylogenetic research is to identify and weed out paraphyletic taxa. In this example, "jointed appendages" is actually an autapomorphy of the monophyletic taxon Arthropoda, which includes the sister taxa chelicertata and mandibulata (*i.e.*, Crustacea + Tracheata, Fig. 1.1).

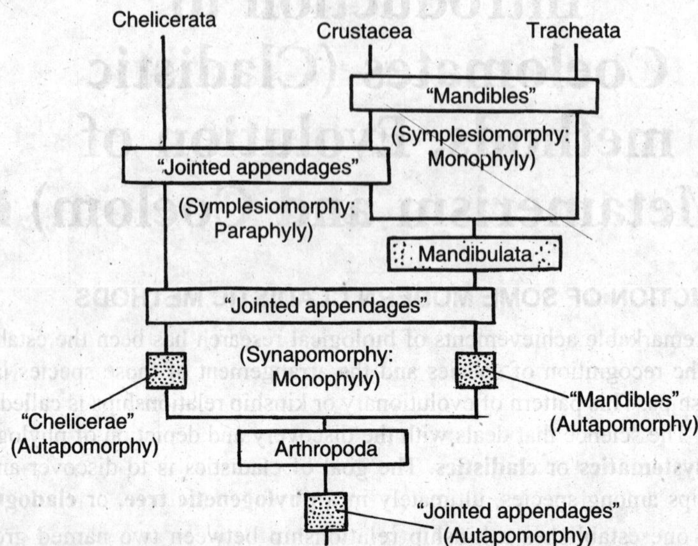

Fig. 1.1. Cladistics (or phylogenetic systematics). Example of the use of cladistics in phylogenetic reconstruction. The observation of a uniquely shared character (synapomorphy), "mandibles", in Crustacea and Tracheata unites them as sister taxa in a monophyletic taxon, Mandibulata. Thus the stem psecies, or ancestor, of Mandibulata must have first expressed "mandibles" as an evolutionary novelty (autapomorphy). Similarly, the observation that "jointed appendages" occur only in chelicerates and mandibulates and nowhere else indicates that the character "jointed appendages" is a synapomorphy of the two sister taxa, which form the monophyletic taxon Arthropoda. The ancestral arthropod therefore, first acquired "jointed appendages" as an evolutionary innovation, or autapomorphy. If an observer, for lack of complete knowledge, unites chelicerates and crustaceans into what he/she believes in a monophyletic taxon because both share "jointed appendages", the union is false because "jointed appendages" are not unique to chelicerates and crustaceans alone, but to all arthropods. Thus the character "jointed appendages" did not evolve in the immediate ancestor to chelicertates and crustaceans alone, but in a deeper ancestry that also gave rise to the tracheates. The chelicertates and crustaceans express this deeply ancestral trait as a shared primitive character, or **symplesiomorphy**. The false union based on a symplesiomorphy is called a **paraphyletic taxon**. The autapomorphy "chelicerae" is a pair of pincer like appendages on the second segment of the head (after **Ruppert** *et al*., 2004).

In fact, a paraphyletic taxon fails to include all descendants of one ancestor, but a **polyphyletic taxon** includes the descendants of more than one ancestor. This mistake occurs when a shared similarity results from evolutionary **convergence** rather than common ancestry. Similarity attributable to common genetic inheritance is called **homology**, whereas the superficial similarity that arises from convergence is known as **homoplasy** (or **analogy**). Only homologous structures are useful in phylogenetic reconstruction based on monophyletic groups. An example of a polyphyletic taxon would be one that united birds, bats and insects together because all share "wings". Bird wings, bat wings and insect wings, however are unrelated homoplasous structures and the members of this polyphyletic taxon descended from three separate ancestors, each of which independently evolved its own unique wings.

The exercise of the cladistic method to species, groups of species, and groups of groups of species leads to the dichotomously branching hierarchic structure that typifies a phylogenetic tree. **Hierarchic structure** means that species are nested in larger, more inclusive taxa, which are contained in still more inclusive groups. So a species of shrimp is a kind of crustacean, which is a kind of mandibulate, which is a kind of arthropod, which is a kind of animal. The hierarchy shown in Fig. 1.1 can also be summarized in tabular form in a **phylogenetic hierarchy:**

Arthropoda
- Chelicerata
- Mandibulata
  - Crustacea
  - Tracheata

A properly construcuted phylogenetic tree is a deeply informative picture. In it, the kinship relationships of species and higher taxa are shown graphically and the synapomorphies are indicated for each pair of sister taxa. The autopomorphies associated with stem species, moreover, offer insight into the form of the ancestros which then may be compared with fossil evidence, if such exists.

1. Reassessment of the Linnean Categories

Modern cladistic is the method of choice for establishing kinship relationships among taxa. Once monophyletic taxa are identified and named on a phylogenetic tree or in a phylogenetic hierarch, the hierarchical relationship among the taxa are obvious (Fig. 1.1). The question then is: "what is gained in assigning a Linnean category, *i.e.*, phylum, class, order and so on, to the taxon names already present?" Many systematics today have abandoned the Linnean categories because they are unnecessary, because they were established by Linnaeus for what he considered to be the immutable levels of creation, and because new cladistic ranks of many taxa do not coincide well with the old Linnean categories (see **Ruppert** *et al.*, 2004). For some systematists, liberation from the Linnean categories will shift emphasis away from pigeonholing ("To what class does the animal belong?") to science ("what are the postulated synapomorphies of Crustacea and Tracheata and how can they be tested?"). For other systematists, however, the Linnean hierarchy of categories will continue to provide a familiar and constant frame of reference.

In seventh Edition of book "**Invertebrate Zoology**" authored by **E.E.Ruppert, R.S. Fox**, and **R.D. Barnes** (2004), authors have embraced the new cladistic approach to systematics and largely manages without Linnean categories. According to them, the relative ranks of taxa can be obtained from the phylogenetic trees. Thus, to soften the transition from the classical to modern approach, Linnean categories are assigned to taxa as superscript abbreviations when a taxon name is used as a text heading. The standard abbreviations are : P for Phylum, C for Class, O for Order, and F for Family. An uppercase S prefixed to a category abbreviation identifies a supertaxon, a lowercase *s* identifies a subtaxon and a lowercase *i* indicates and infrataxon. Super–, sub– and infrataxon are ranked as follows:

Supertaxon
- Taxon (for example phylum, class, order)
  - Subtaxon
    - Infrataxon

In the phylogenetic hierarchy given earlier, the superscript would be : Arthropoda^{p}, Chelicerata^{sp}, mandibulata^{sp}, Crustacea^{ip} and Tracheata^{ip}.

2. Phylogeny of Protozoa (and Origin of Metazoa)

The phylogenetic relationships among the protozoan taxa have been notoriously difficult to the application of microanatomical and molecular techniques (Fig. 1.2). In general, the amoeboid taxa (and some flagellates) are thought to represent primitive taxa, although their interrelationships are presently unclear. At least five monophyletic taxa have been proposed. One of these is the **Euglenozoa**, which includes the euglenoids and kinetoplastids. Another is the **Chlorophyta**, which includes the Volvocida, Green algae and Multicellular green plants (Metaphyta). Ciliates, dinoflagellates, and apicomplexans form another monophyletic taxon, called **Alveolata**, based on their common possession of alveoli and mitochondria with tubular cristae. From the perspective of animal evolution, the most interesting monophyletic taxon is the **Opisthokonta**, characterized by a posterior flagellum on motile cells as well as gene–sequence similarities. Included in this taxon are the sister taxa Choanoflagellata and Metazoa (multicellular animals) as well as Fungi and perhaps Microsporidia. The only other Opisthokont taxon is the former protozoan group Myxosporidia which is now included among the metazoans as Myxozoa (Box 1.1).

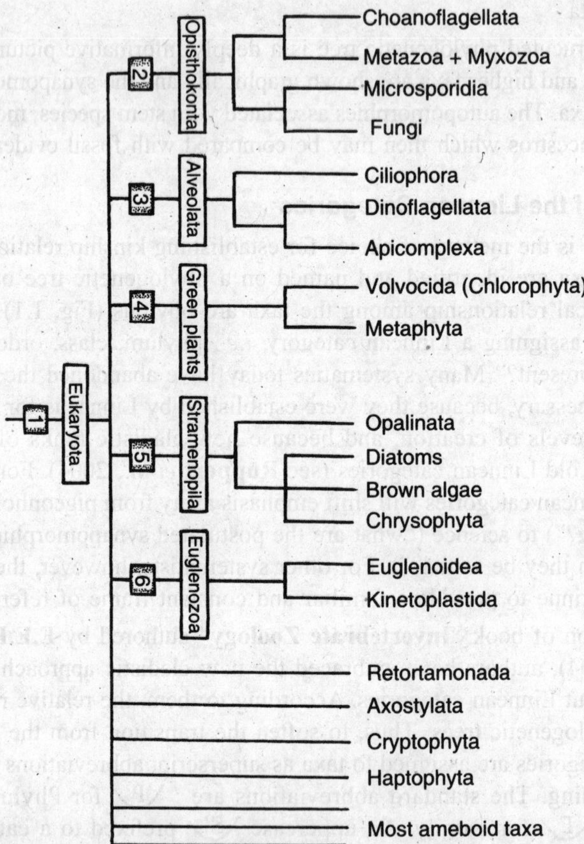

Fig. 1.2. Phylogeny of Protozoa and Metazoa based on gene–sequence data and morphology. Morphological characters: **1. Eukaryota.** Nucleus is enclosed in a nuclear membrane. **2. Opisthokonta.** Locomotory flagellum is posterior on motile cells. **3. Alveolata.** Alveoli are below the plasma membrane, mitochondrial cristae are tubular. **4. Green plants.** Chloroplasts are enclosed in two membranes and include chlorophylls a and b, starch is produced as a storage product, and a cellulose cell wall is present. **5. Stramenopila.** Flagella have three–part, tubular mastigonemes. **6. Euglenozoa.** One or two anterior flagella emerge from a pit, a para–flagellar rod is present, and mitochondrial cristae are discoid (after Ruppert *et al.*, 2004).

Box 1.1 Characters of Myxozoa

Cell and tissue parasites of connective tissue, muscle, and internal organs of invertebrates, fishes; within host, parasite is motile, multinucleated syncytium; periodically certain nuclei (called **germinal nuclei**) in syncytium divide, become cellular, and form infective spores, which are ingested with food by next host; formerly considered a protozoan taxa related to Apicomplexa, myxozoan spores attach to host tissues and cells with one or more eversible "polar filaments" now known to be nematocysts (atrichous isorhizas), each in its own nematocyte; spores are multicellular include two or four nematocytes, one or two cells that will become a new parasitc syncytium and two test (valvogenic) cells forming spore walls; test cells joined together by septate and adherens–type junctions typical of metazoans, unknown in protozoans; extra–cellular matrix containing collagen, also typical of metazoans, between some cells; 1200 species. **Examples.** *Myxobolus, Buddenbrockia, Tetracapsuloides, etc.*

Table 1.1 Approximate Numbers of Species Described Extant Metazoa (Ruppert, Fox and Barnes 2004).

Serial Number	*Taxon (Phylum)*	*Species*
1.	Porifera	8,000
2.	Placozoa	1
3.	Cnidaria	10,000
4.	Ctenophora	80
5.	Platyhelminthes	20,000
6.	Orthonectida	20
7.	Dicemida	75
8.	Nemertea	1,150
9.	Mollusca	1,00,000
10.	Annelida	12,000
11.	Echiura	150
12.	Sipuncula	150
13.	Onychophora	110
14.	Tardigrada	800
15.	Arthropoda	1,113,000
16.	Gastrotricha	500
17.	Nematoda	22,000
18.	Nematomorpha	325
19.	Priapulida	18
20.	Loricifera	10
21.	Kinorhyncha	150
22.	Gnathostomulida	80
23.	Rotifera	2,000
24.	Seisonida	2
25.	Acanthocephala	1,150
26.	Kamptozoa	150
27.	Cycliophora	1
28.	Phoronida	14
29.	Brachiopoda	350
30.	Bryozoa	5,000
31.	Chaetognatha	150
32.	Hemichordata	92
33.	Echinodermata	6,000
34.	Chordata	52,000

1.2 SOME BASICS OF MODERN SYSTEMATICS AND EVOLUTIONARY TREE

1. Europeans tend to use the terms "systematics" and "biosystematics" for the field of classification as a whole, whereas North Americans tend to use "taxonomy" more frequently.

2. Ernst Mayr (1991) said that the field of systematics can be thought of as a **continuum**, from the routine naming and describing of species through the compilation of large faunal compendia and monographs to more synthetic studies, such as the fitting of these species into classifications that depict evolutionary relationships, biogeographic analyses, studies of population biology and genetics and evolutionary and speciation studies. Mayr designated three stages of study within this continuum, which he called **alpha, beta** and **gamma**, corresponding to the three general levels of complexity he perceived in systematics. When a group of organisms is first discovered or is in a poorly known state, work on that group is necessarily at the alpha level (*e.g.*, the describing of new species). It is only when most, or at least many, species in a taxon become known that the systematist is able to work at the beta or gamma levels within that group (*e.g.*, to perform evolutionary studies). Some biologists choose to refer to those people working at the alpha level as **taxonomists**, reserving the term **systematists** for those engaging in studies at the beta or gamma level (**Brusca** and **Brusca** 2002).

3. Monophyletic, paraphyletic and polyphyletic group. A **monophyletic group** (or taxon) is a group of species that includes an ancestral species and all of its descendants – that is, a **natural group.**

A groups whose member species are all descendants of a common ancestor but that does not contain **all** the species descended from that ancestor, is called a **paraphyletic group.** Paraphyly implies that for some reason (*e.g.*, lack of knowledge, purposeful manipulation of the classification) some members of a natural group have been placed in different group. Examples of paraphyletic taxa include class Reptilia (which exclude birds); Class Polychaeta and Class Oligochaeta.

A third possible kind of taxon is a **polyphyletic group** – a group comprising species that arose from two or more different immediate ancestors. Such composite taxa have been established primarily because of insufficient knowledge concerning the species in question. One of the principal goals or systematists is to discover such polyphyletic or "artificial" taxa and through careful study, reclassify their members into appropriate monophyletic taxa.

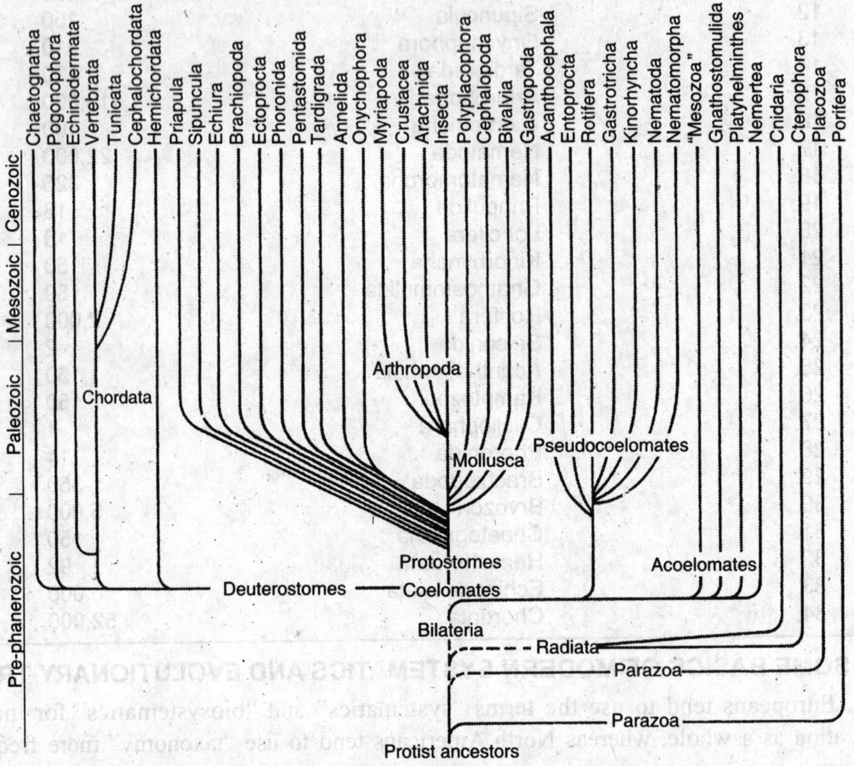

Fig. 1.3. A traditional evolutionary tree (not cladogram) that depict phylogeny among the Metazoa (Brusca and Brusca, 2002).

4. Recently several **biochemical** and **molecular biological** techniques for measuring similarity (homology) among organisms have been developed over the past 40–50 years; these include DNA hybridization, starch gel electrophoresis of proteins and amino acids, immunological similarity indices, and most recently nucleotide sequencing of genes. All of these kinds of "characters" have been used to create **phylogenetic tree.**

5. A **dendrogram** is a branching diagram or tree depicting the relationships among groups of organisms. It is a graphical means of expressing relationships among species or other taxa. Most dendrograms are intended to depict evolutionary relationships, with the base representing the oldest (earliest) ancestors and the higher branches indicating successively more recent divisions of evolutionary lineages (Fig. 1.3).

6. Clades and grades. As depicted in Fig. 1.4, a **clade** is a monophyletic group or branch of a tree which may undergo very little or a great deal of diversification. A clade, in other words, is a group of species related by direct descent. A **grade**, on the other hand is a group of

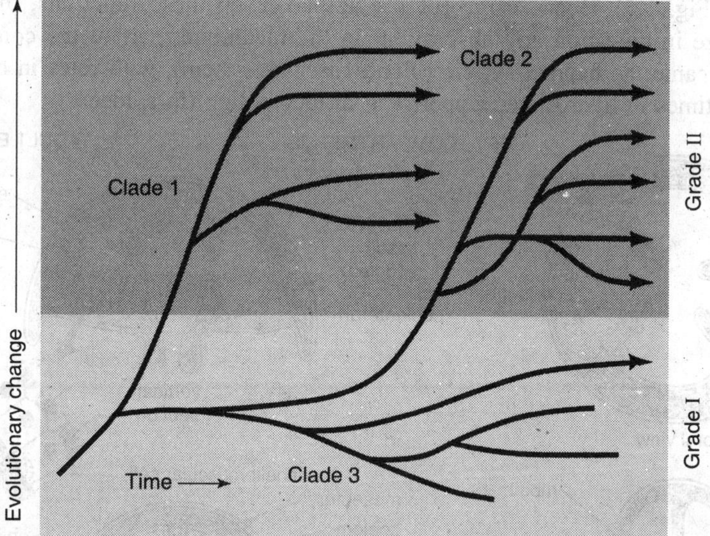

Fig. 1.4. Clades and grades. Clades are monophyletic branches that may undergo various degrees of diversification. Grades are groups of animals classified together on the basis of levels of functional or morphological complexity. Grades may be monophyletic, paraphyletic or polyphyletic. In this figure, grade I is monophyletic, encompassing only a single clade (clade 3); grade II is polyphyletic, because the associated level of complexity has been achieved independently by two separate lineages, clade 1 and clade 2 (after Brusca and Brusca, 2002).

species (or higher taxa) defined by a particular level of functional or morphological complexity. Thus, a grade can be polyphyletic, paraphyletic or monophyletic. A good example of a grade is the large group of gastropod taxa that have achieved shellessness. These "slugs", however, do not constitute a clade, because shell loss has occurred independently in several different lineages; thus "slugs" are a polyphyletic group. An example of monophyletic grade is the subphylum Vertebrata (animals with backbone).

7. **Phylogenetic systematics (cladistic) and Evolutionary tree**. Most workers today use a method known as **phylogenetic systematics**, or **cladistics**, when construing biological dendrograms and their resultant classifications. Phylogenetic systematics had its origin in 1950 in a textbook by the German biologist **Willi Hennig**; the English translation (with revisions) appeared in 1966. Its popularity has grown steadily since that time. Through the years, cladistics

has evolved well beyond the framework Henning originally proposed. The goal of phylogenetic systematics is to produce explicit and testable hypotheses of genealogical relationships among monophyletic groups of organisms. As a systematic methodology, cladistics is based entirely on **recency of common descent** (*i.e.*, genealogy). The dendrograms used by phylogenetic systematics are called **cladograms**, and they are constructed to depict only genealogy, or ancestro–descendant relationships. The term **cladogenesis** refers to splitting of one species (or lineage) into two or more species (or lineages). It is the splitting process that produces genealogical (ancestor–descendant) relationships (**Brusca** and **Brusca** 2002).

1.3 COMPARTMENTALIZATION (Physiological regulation and Specializations)

Large–bodied bilaterians evolved a new internal compartment separate from the gut that is called **coelom**. It functions as a hydrostatic skeleton and has roles in circulation, reproduction, and excretion (Fig. 1.5). Because the coelom has taken up these functions, the gut became free to specialize in digestion and absorption. In the meantime, part of the connective tissue compartment became the haemal system (blood vascular system), with roles in circulation and excretion, sometimes in hydrostatic support and in locomotion (in spiders).

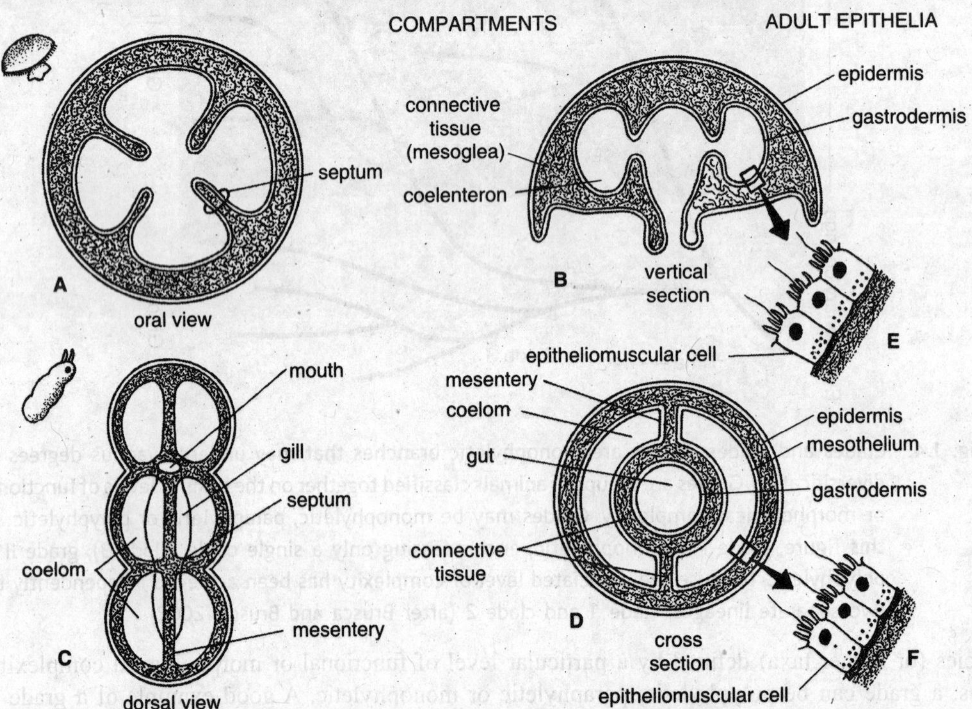

Fig. 1.5. Body compartmentalization in bilateria. The body of radial animals (A and B) consists of two epithelia, the outer epidermis and inner gastrodermis that enclose the coelenteron and mesoglea. This arrangement of epithelia allows for physiological regulation and cephalization of these two extracellular body compartments. In contrast to cnidarians and ctenophores, coelomate bilaterians specialize and regulate three extracellular compartments, the connective tissue, the gut lumen and the coelom. The additional compartment is made possible by a body design that consists of three rather than two, epithelia. These three epithelia are epidermis, gastrodermis and mesothelium (C and D). The fact that the epithelial musculature of cnidarians is identical to the coelomic epithelial musculature of bilaterians (E and F) suggest that the bilaterian muscles in general evolved from a coelomic epithelium (after Ruppert *et al.*, 2004).

1.4 COELOM IN LARGE BILATERIA

A coelom is a fluid–filled cavity or canal lined by an epitherlium known as **mesothelium** (or middle epithelium) which is derived from the mesoderm (Fig. 1.3D). The coelom is unique to Bilateria and provides a new third internal compartment for physiological regulation and specialization. It also functions as a static skeleton, especially in soft–bodied worms, and in internal transport excretion, and reproduction. the coelom occupies some of the connective–tissue space between epidermis and gastrodermis (Fig. 1.5D), but during its development, it displaces the connective tissue peripherally. Its cavity is filled with **coelomic fluid**, which is circulated by mesothelial cilia or contraction of body–wall muscles of mesothelial origin.

Animals with a coelom are known as **coelomates**. The simplest and probably most primitive coelomic cavities are lined by a monolayered mesothelium composed chiefly of monociliated,

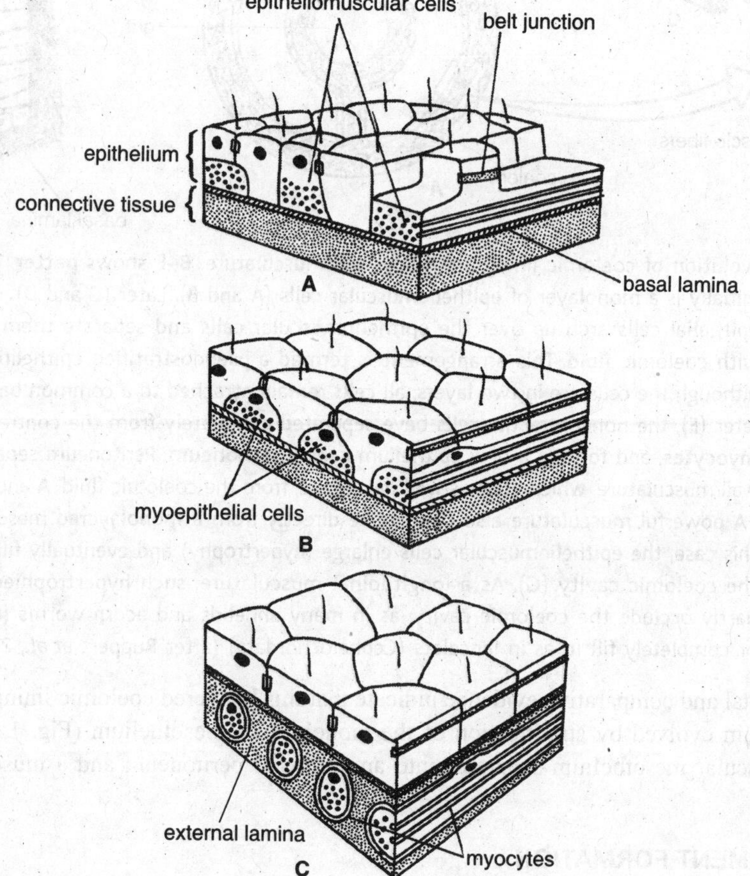

Fig. 1.6. Muscle tissue: three evolutionary grades of organization. Epitheliomuscular cells (A) and myoepithelial cells (B) are both part of a multifunctional epithelium. In C, myocytes, or true muscle cells, are below the epithelium in the connective tissue compartment and form a distinct muscle tissue separate from the overlying epithelium (after Ruppert *et al.*, 2004).

contractile epitheliomuscular cells (Fig. 1.4A). This single epithelium is simultaneously the body musculature and the coelomic lining (Fig. 1.5D, F and 1.7A, B). A more complex, bilayered coelomic lining occurs in some invertebrate bilaterians and in the vertebrates. These coelomic cavities are lined by a thin noncontractile epithelium called **peritoneum** (Fig. 1.7E), which separates the coelomic fluid from the retroperitoneal (*i.e.*, behind the peritoneum) muscle and connective tissue.

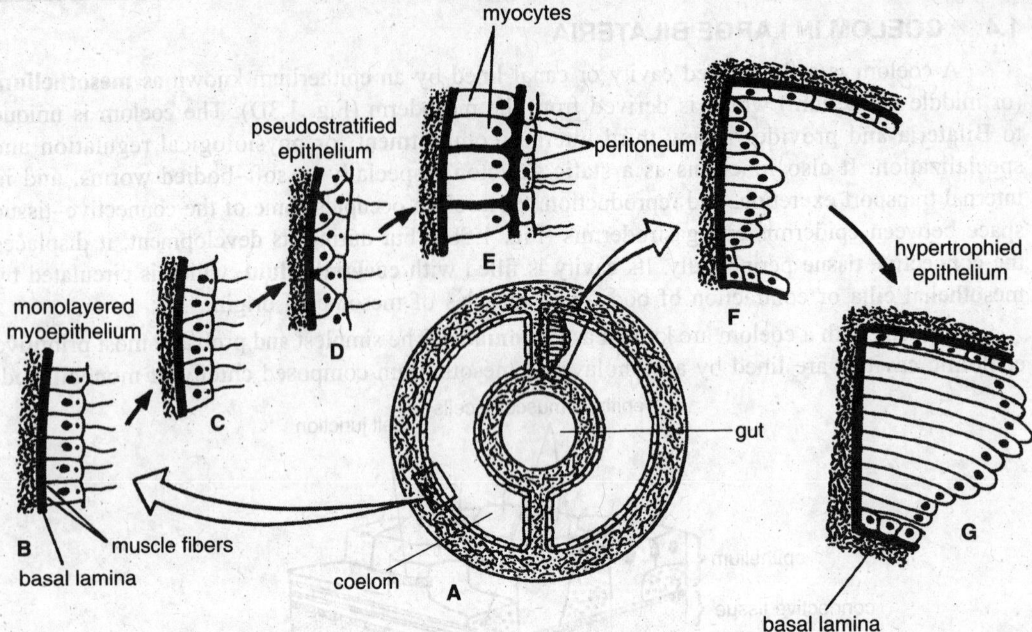

Fig. 1.7. Evolution of coelomic lining and body-wall musculature. **B–E** shows **patter 1**: Mesothelium initially is a monolayer of epitheliomuscular cells (A and B). Later (C and D), non–contractile epithelial cells arch up over the epitheliomuscular cells and separate them direct contact with coelomic fluid. This arrangement is termed a pseudostratified epithelium (D) because although the cells are in two layers, all cells remain attached to a common basal lamina. Still later (E), the noncontractile cells have separated completely from the contractile cells, now **myocytes**, and formed a new epithelium called **peritoneum**. Peritoneum separates the body wall musculature, which often is well developed, from the coelomic fluid. **A** and **F–G, pattern2** : A powerful musculature also can evolve directly from a monolayered mesothelium (A). In this case, the epitheliomuscular cells enlarge (hypertrophy) and eventually fill and obliterate the coelomic cavity (G). As a longitudinal musculature, such hypertrophied muscles may partly occlude the coelomic cavity, as in many annelids and acorn worms (Hemichordata), or completely fill it, as in lancelets (Cephalochordata) (after Ruppert *et al.*, 2004).

Developmental and comparative evidence indicate that this bilayered coelomic lining that includes the peritoneum evolved by stratification of the monolayered mesothelium (Fig. 1.5B to E). The epitheliomuscular mesothelium separates into an epithelial peritoneum and a muscular myocyte layer.

1.5 SEGMENT FORMATION

Bilaterial pairs of coelomic cavities contribute to the formation of **segments** or **somites**. which are a series of similar modules along the length of the body (Fig. 1.8B). Typically, there is one pair of coelomic cavities per segment, as in each of many segments of an earthworm. Left and right cavities surround and approach each other above and below the gut. At the point of contact, the abutting walls of the two cavities form a longitudinal partition in the midsagital plane called a **mesentery** (Fig. 1.8C, D). A **dorsal mesentery** is positioned above the gut, a **ventral mesentery** below. A **septum** (Fig. 1.5C) is a transverse partition, resembling a bulkhead

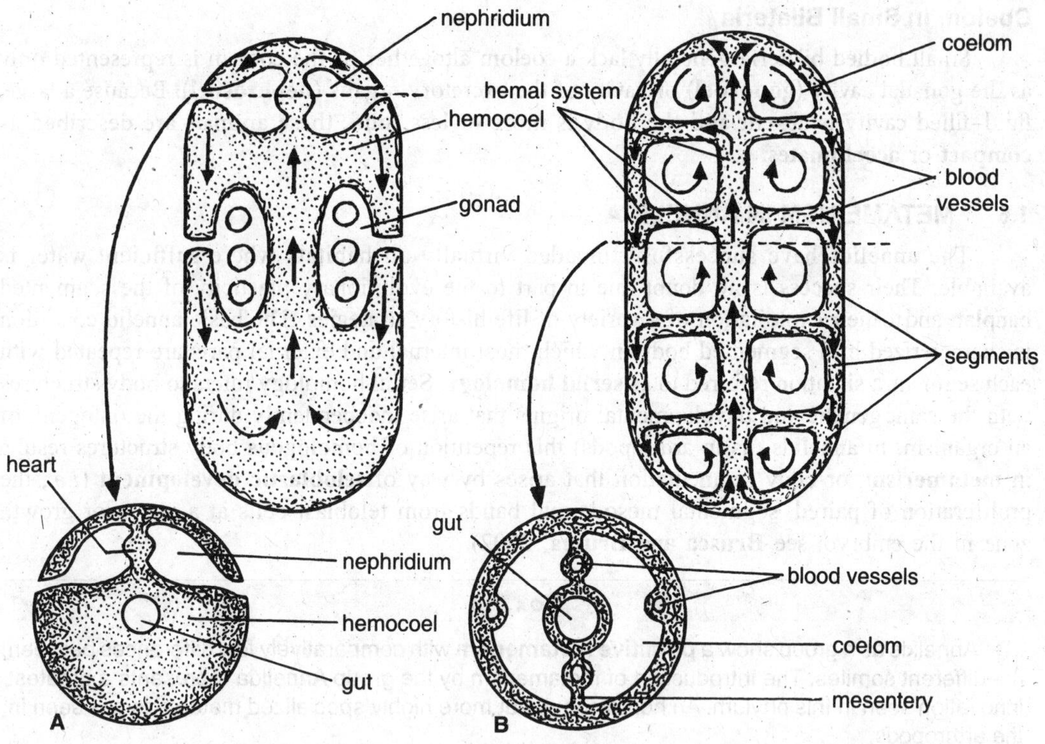

Fig. 1.8. Internal fluid transport system of Coelomate Bilateria. A–When a haemocoel (haemal system) is present, it constitutes the main body cavity and the coelomic cavities are small and specialized as gonadal or nephridial sacs. B–When coelomic cavities are well developed, they constitute the body cavity and the haemal system is confined to vessels and small sinuses (after Ruppert *et al.*, 2004).

(*i.e.*, water–tight division or dividing wall) in a ship or an aeroplane, between successive pairs of cavities. Septa and mesentries divide the segmented coelom into separate fluid–filled compartments (Fig. 1.8B), each of which may be regulated, by quality of its lining epithelium, for a particular specialized function. Specialized coelomic cavities, for example, often enclose gonads as genital coeloms (gonocoels) or enclose a heart as a pericardial cavity. The pleural, pericardial and abdominal cavities of mammals are also specialized coelomic cavities. **Pleural cavity** is space included between the parietal and visceral layers of the **pleura** which are serous membrane covering the lung and lining of thoracic cavity. A **serous membrane** is a delicate, watery membrane which lines the closed cavities of the body.

Coelomic fluid has not been as well analyzed as blood, in general, it contains water, ions, low–molecular–weight solutes, and cells. Overall, it differs from blood in having a low, rather than a high, protein concentration. In some coelomates, germ cells are released from the gonad into the coelomic fluid. There they take up nutrients, grow, and complete their differentiation into gametes. The mesothelium is responsible for regulating the composition of the coelomic fluid for the growth and maintenance of the gametes. Other free cells of various functions, collectively called **coelomocytes**, originate from the mesothelium and circulate in the coelomic fluid. Phagocytic coelomocytes police the coelomic fluid for bacteria and other foreign or unwanted material. Respiratory proteins such as haemoglobin often are found in the coelom, but in contrast with their occurrence in invertebrate blood, they are invariably intracellular in coelomocytes.

Coelom in Small Bilateria

Small bodied bilateria generally lack a coelom altogether or the coelom is represented only as the gonadal cavity (**gonocoel**) or cavity of the excretory organs (**nephrocoel**). Because a large, fluid–filled cavity is absent and the body is more or less solid, these animals are described as **compact** or **acoelomates.**

1.6 METAMERISM IN ANNELIDA

The annelids have successfully invaded virtually all habitats where sufficient water is available. Their success is no doubt due in part to the evolutionary plasticity of the segmented bauplan and to their exploitation of a variety of life history strategies. The basic annelid condition is characterized by a segmented body in which most internal and external parts are repeated with each segment, a situation referred to as **serial homology**. Serial homology refers to body structures with the same genetic and developmental origins that arise independently during the ontogeny of an organism. In annelids (as in arthropods) this repetition of homologous body structures results in **metamerism**, or body segmentation that arises by way of **teloblastic development** (*i.e.*, the proliferation of paired, segmental mesodermal bands from teloblast cells at a posterior growth zone in the embryo; see **Brusca** and **Brusca**, 2002).

Box 1.2

Annelids as a group show a **primitive metamerism** with comparatively few differences between the different somites. The introduction of metamerism by the group Annelida represents a greatest innovation seen in this phylum. An homologous but more highly specialized metamerism is seen in the arthropods.

Metamerism is encountered for the first time in the annelids, and forms the most important distinguishing character of the phylum (Box 1.2). The **prostomium**, *i.e.*, the anterior part of the body infront of mouth, and the **pygidium,** *i.e.,* the terminal part of the body carrying the anus, are not segments. New segments are formed just infront of pygidium. The oldest segments are, therefore, anterior and the youngest, posterior. Metamerism is not confined to external structures but affects the internal organs as well. Muscles, small blood vessels, nerves, excretory organs (nephridia), and gonads are repeated in each segment. Only the alimentary canal, nerve cord, and the main blood vessels are unaffected, although they too extend through each segment.

Metamerism is usually more complete or uniform throughout the body of the embryo than of the adult, in which it may become obscured due to the formation of head and other structures such as limbs. Internal organs may also become "concentrated" or restricted to certain segments. Modification of the embryonic metamerism is especially noticeable in the vertebrates.

Metamerism is of great interest, because, besides the annelids, it is also found in the most successful groups of the animal kingdoms, namely, Arthropoda and Vertebrata (Chordata). Evidently, it has originated independently in the annelids and the vertebrates. How it developed (= originated) is not clear however.

Theories Regarding Origin of Metamerism

1. Transverse fission theory. Some zoologists hold that the body of a metameric animal has evolved from that of an unsegmented animal by transverse fission into individuals, which remained joined end to end and gradually became integrated both morphologically and physiologically.

2. Division of ancestral organ theory. Some other zoologists maintain that the repetition of organs, such as blood vessels, nephridia and gonads, in various segments has resulted by the division of a single ancestral organ, and not by formation of new organs as the fission theory suggests.

Comparison of Pseudometamerism and Metamerism

The true metamerism seen in the annelids should not be confused with pseudometamerism found in the tapeworms (phylum Platyhelminthes). The differences between the two are given in Table 1.2.

Table 1.2 Differences between proglottids of tapeworm and true segments of Annelida.

Proglottids of tapeworm	*True segments of earthworms*
1. Number is very large and indefinite.	1. Number is small and almost definite, varying within small limits only.
2. New proglottids arise at the anterior end so that the oldest proglottid is the most posterior.	2. New segments arise from the posterior end so that the oldest segment is the most anterior.
3. Proglottids are not preformed in the embryo.	3. Segments are preformed in the embryonic somites.
4. Proglottids are voluntarily cast off.	4. Segments are not voluntarily cast off.
5. All proglottids are identical in structure and function.	5. Though fundamentally alike, some segments are very specialized in structure as well in function.

There is thus a superficial resemblance between the proglottids of a tapeworm and segments of an earthworm.

Every mature proglottids contains a complete set of male and female reproductive organs and in this respect resembles a bisexual animal, such as liver fluke or planaria. This resemblance has led to the speculation that the tapeworm is a colony of bisexual individuals produced asexually by budding. However, the tapeworm is a single animal. The scolex contains the brain and forms its head while the strobila contains other viscera and constitutes its body. The power of increasing the number of proglottids by budding and of forming sex organs in all of them is a special adaptation to parasitic life for ensuring the continuity of the race.

Significance of Metamerism

Although a spacious, fluid-filled coelom provided an efficient hydrostatic skeleton for burrowing, precise control of body movements was not possible in the earliest coelomates. The force of muscle contraction in one area was carried throughout the body by the fluid in the undivided coelom. This defect was remedied when a series of partitions (septa) evolved in the common ancestor of annelids and arthropods. When the septa divided the coelom into series of compartments of most other body systems, such as circulatory, nervous and excretory, were repeated in each segment (*i.e.*, metamerism).

The evolutionary advent of metamerism was highly significant, because it made possible development of much greater complexity in structure and function. Metamerism not only increased the efficiency of burrowing but also made possible independent and separate movements by the separate segments. The need for fine control of movements led, in turn, to the evolution of a more sophisticated nervous system. Moreover, repetition of body parts gave the organisms a built-in redundancy, as in some human-made systems. This provided a safety factor: if one segment should fail, the others could still function. Thus an injury to one part would not necessarily be fatal.

The evolutionary potential of the metameric body plan is amply demonstrated by the large and diverse phylum Arthropoda. Metamerism also arose independently in the deuterostome line, which includes the numerous and adaptively diverse chordates (**Brusca** and **Brusca**, 2002).

1.7 EVOLATION OF COELOM IN ANNELIDA

Coelom is a cavity present between the body wall and internal organs (*i.e.*, visceral organs) and is lined with a single-layered mesodermal epithelium called the **peritoneum**. The peritoneum which lines the body wall is called the **parietal peritoneum**, and that which covers the internal or visceral organs is called the **visceral peritoneum**. The coelom contains the **coelomic fluid**, which is secreted by the peritoneum and has free mesenchyme cells, the **colomocytes** or **coelomic corpuscles**. The coelomic fluid receives materials absorbed by the alimentary canal and carries it to the tissues. Excretory substance finds its way into the coelomic fluid and thence out of body through the nephridia. Sex cells arise from the coelomic epithelium.

Status of Coelom in Bilateria

The bilateral animals can be grouped according to their body cavity type or lack of body cavity (*i.e.*, acoelomata, pseudocoelomata and eucoelomata). In the **eucoelomata bilateria** (*i.e.*, higher animals) exists a fluid-filled space that surrounds the gut (and other visceral organs). The evolution of the coelom, like bilateral symmetry, was a major advancement in animal architecture because it provides a "tube-within-tube" arrangement that allows much greater body flexibility than is possible in animals lacking an internal body cavity. The coelom also provides a space for the visceral organs and permits greater size and complexity by exposing more cells to surface exchange. The fluid-filled coelom additionally serves as a **hydrostatic skeleton** in some forms, especially many worms, aiding in such activities as movement and burrowing.

Schizocoelous and Enterocoelous Coelom

Eucoelomate bilateria possess a **true coelom** lined with mesodermal peritoneum. The true coelom arises within the mesoderm itself and may be formed by one of two methods: schizocoelous and enterocoelous formations (Fig. 1:7). Both of these terms are descriptive, for *schizo* comes from the Greek *schizein*, to split; *entero* is derived from the Greek *enteron*, meaning gut; and *coelous* comes from the Greek *koilos*, meaning hollow or cavity. In schizocoelous formation, the coelom arises, as the word implies, from the splitting of mesodermal bands that originate from cells in the blastopore region (Mesoderm is one of the three primary germ layer that appear very early in the development of bilateral animals, lying between the inner most endoderm and outermost ectoderm). In enterocoelous formation the coelom comes from pouches of the archenteron or primitive gut.

Once development is complete, the results of schizocoelous and enterocoelous formation are indistinguishable. Both processes give rise to a true coelom lined with a mesodermal peritoneum (Gr. *peritonations*, stretched around) and having mesenteries in which the visceral organs are suspended.

1. Coelom in Polychaetes and Oligochaetes. Polychaetes (*e.g., Nereis*) and oligochaetes (*e.g., Pheretima*) possess a spacious coelom extending from the anterior to the posterior end of the body. It is divided into a series of small chambers, the **coelomic compartments**, by transverse partitions, the **septa** or **coelosepta**, that extend from the annuli of the body wall to the alimentary canal. The septa are composed of a double layer of peritoneum with a thin sheet of musculo-connective tissue in between. They are usually perforated by sphinctered aperture or apertures to allow flow of coelomic fluid from one segment to another.

In polychaetes, the gut is suspended by dorsal and ventral longitudinal **mesenteries** formed of a double layer of peritoneum. With the result, each coelomic compartment is divided into right and left halves. Actually, however, such perfect compartmentation is rarely seen because the

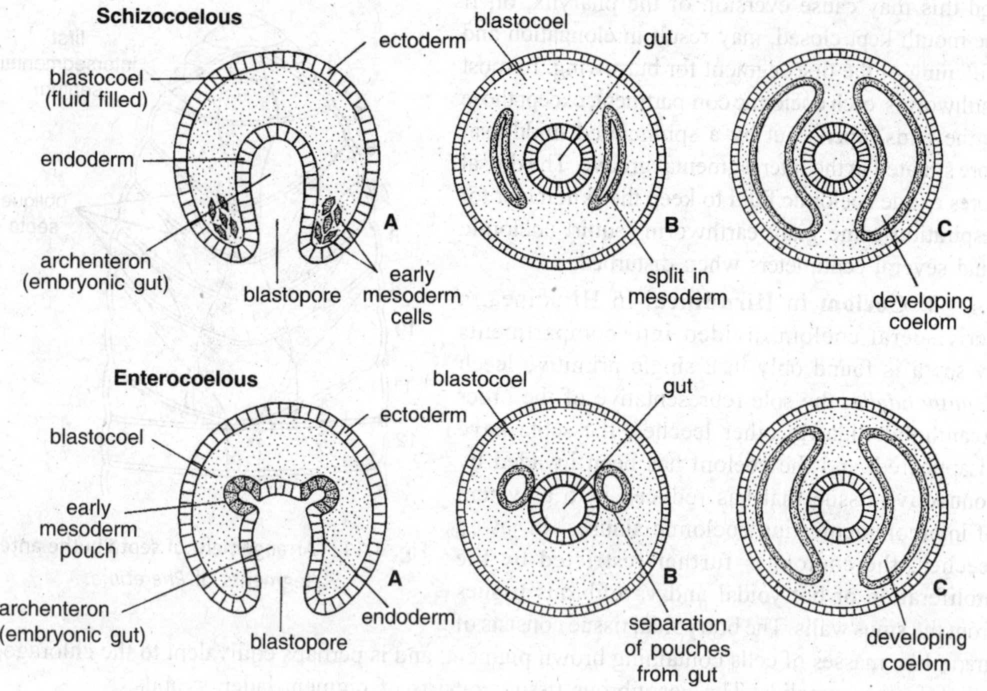

Fig. 1.9. Type of mesoderm and coelom formation. In schizocoelous formation, the mesoderm originates from the wall of archenteron near the blastopore and proliferates into a band of tissue that splits to form the coelom. In enterocoelous formation, most mesoderm originates as a series of pouches from the archenteron; these pinch off and enlarge to form the coelom. In both formations, the coeloms expand to obliterate blastocoel (after Hickman *et. al.*, 1997).

mesenteries are almost always incomplete and the septa are weak or have partly or fully disappeared in many polychaetes. *Arenicola,* for instance, retains only three septa in the anterior region and a few in the posterior regions, those of the middle region having been lost. In *Aphrodite, Glycera* and some other annelids, the peritoneum is ciliated, and the movement of cilia brings about an active circulation of coelomic fluid in the coelom. This distributes materials to the tissues so efficiently that the blood vascular system is reduced.

In oligochaetes, the septa are relatively complete, except in the anterior and posterior parts of the body. Certain anterior septa in large earthworms are cone-shaped and very muscular (Fig. 1.10). Contraction of these septa increases the pressure of coelomic fluid,

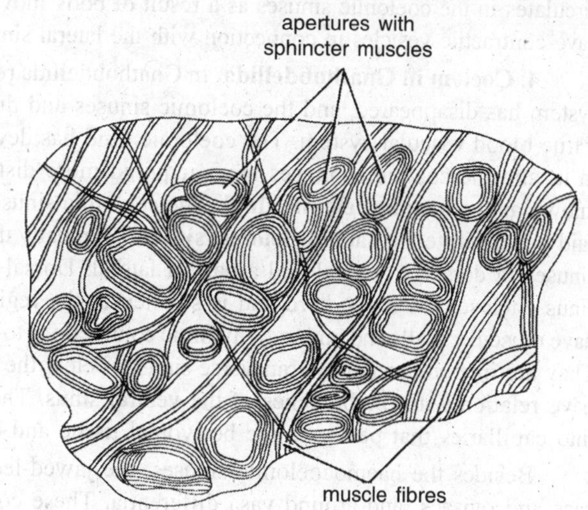

Fig. 1.10. A part of intersegmental septum of *Pheretima* (earthworm).

and this may cause eversion of the pharynx, or, if the mouth kept closed, may result in elongation and stiffening of the first segment for burrowing. In most earthworms, each coelomic compartment, except those at the ends, opens out by a sphinctered middorsal pore situated in the intersegmental groove. The dorsal pores exude coelomic fluid to keep the skin moist for respiration some giant earthworms squirt coelomic fluid several centimeters when disturbed.

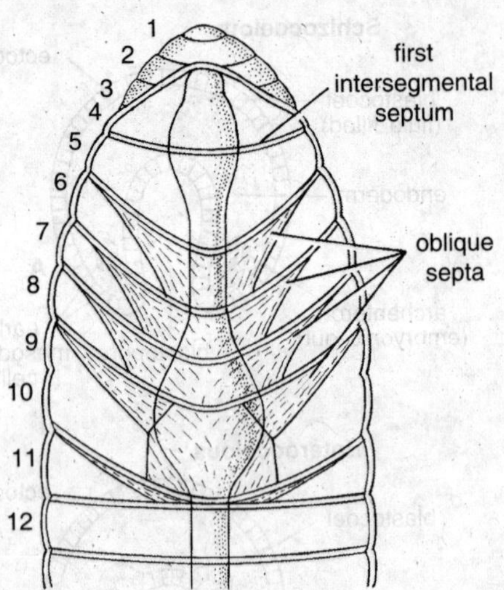

Fig. 1.11. Arrangement of septa in the anterior segments of *Pheretima*.

2. Coelom in Hirudinea. In Hirudinea, a perivisceral coelom divided into compartments by septa is found only in a single primitive leech *Acanthobdella*, the sole representative of the order Acanthobdellida. In other leeches, the septa have disappeared, and the coelom has been invaded by connective tissue that has reduced it to a system of intercommunicating coelomic sinuses. In jawed leeches, the coelom is further restricted by the proliferation of botryoidal and vasofibrous tissues from the sinus walls. The botryoidal tissue consists of grape-like masses of cells containing brown pigment, and is perhaps equivalent to the chloragogen cells of other annelids. The vasofibrous tissue consists of pigment-laden strands.

3. Coelom in Rhynchobdellida. In Rhynchobdellida (*e.g.*, *Pontobdella*), the coelomic sinuses coexist with the blood vascular system and act as an auxiliary circulatory system. There are four longitudinal sinuses: a dorsal sinus enclosing the dorsal blood vessel, a ventral sinus having the ventral blood vessel and the ventral nerve cord, and a pair of lateral sinuses. The lateral sinuses are connected with the median sinuses by irregular intermediate sinuses in each segment. A slender subepidermal sinus occurs between the epidermis and the muscles of the body wall, and communicates with the lateral and intermediate sinuses. Coelomic fluid with floating coelomocytes irregularly circulates in the coelomic sinuses as a result of body movements. Certain forms, such as *Piscicola,* have contractile vesicles in connection with the lateral sinuses for the circulation of coelomic fluid.

4. Coelom in Gnathobdellida. In Gnathobdellida (*e.g., Hirudinaria*), the original circulatory system has disappeared, and the coelomic sinuses and fluid have become modified to function as a true blood vascular system. The coelomic fluid has developed haemoglobin, which is dissolved in it, and the coelomic sinuses have acquired more distinct walls. Hence, the coelomic fluid is known as the **haemocoelomic fluid**. The coelomic sinuses with their branches and tributaries are said to constitute the **haemocoelomic system**. Typically there are four longitudinal haemocoelomic sinuses: a dorsal, a ventral and a pair of lateral. Dorsal sinus is not always present. The ventral sinus encloses the nerve cord and is swollen in the region of each ganglion. The lateral sinuses have muscular walls and undergo rhythmic contractions to keep the haemocoelomic fluid circulating. They are interconnected by transverse sinuses below the ventral sinus in each segment. Nephridia have relation with the branches of the ventral sinus. The branches of the main sinuses break up into capillaries that penetrate the botryoidal tissue and body wall.

Besides the haemocoelomic sinuses, the jawed-leeches have coelomic spaces in the testis sacs and ovisacs, and around vasa differentia. These coelomic spaces have coelomic fluid.

5. Coelom in Archiannelida. The Archiannelida (*e.g., Polygordius*) have a spacious and compartmented coelom.

QUESTIONS

Long Answer Questions

1. Give an account of metamerism in animals.
2. What is coelom? Differentiate between different methods of coelom formation. Give an account of coelom in various annelids.
3. Write an essay on Modern cladistic methods.

Short Answer Questions

1. Tabulate differences between the true segments of earthworm and the proglottids of tapeworm.
2. Define the following terms:

   (*i*) Monophylitic; (*ii*) Paraphyletic; (*iii*) Polyphyletic.

Very Short Answer Questions

1. What is synapomorphy?
2. Define autapomorphy.
3. Give an example of symplesiomorphy.
4. Give examples of paraphyletic taxa.
5. Name three stages of study of systematics continuum.
6. What is dendrogram?
7. What is clade.
8. Name the organs not affected by segmentation.
9. What is homonomous segmentation?
10. Which animals show pseudometamerism?
11. In which phylum for the first time did true coelom appear?
12. Define true coelom.

Fill in the Blanks

1. Division of the body into an anteroposterior row of similar parts is called ....
2. New segments arise just in front of the ....
3. In annelids, the oldest segments are at the .... end.
4. Metamerism is more complete in the .... than in the adult.
5. Metamerism is also found in .... and ....
6. True coelom is originated from ....
7. Coelom is always lined by mesodermal .... and has .... fluid.

Multiple Choice Questions

1. Segmentation is found in

   (*a*) Annelida (*b*) Arthropoda

   (*c*) Chordata (*d*) all of these
2. Besides Annelida and Arthropoda the metamerism is exhibited by

   (*a*) Cestoda (*b*) Mollusca

   (*c*) Chordata (*d*) Acanthocephala

ANSWERS

Very Short Answer Questions

1. Synapomorphy is a shared derived character, *e.g.*, mandibles of crustaceans and tracheates of class Mandibulata, Phylum Arthropoda.

2. Autapomorphy is a self derived character of a taxon, *e.g.*, presence of mandible in Mandibulata and presence of chelicerae in Chelicerata.

3. Symplesiomorphy is a shared primitive or ancentral trait, *e.g.*, trait of jointed appendages that evolved in the stem species (ancestor) of all arthropoda.

4. Class Reptilia (which excludes birds), Polychaeta and Oligochaeta.

5. Mayr (1991) has recognised the following three levels: alpha, beta and gamma.

6. Dendrogram is a branching diagram or tree, depicting the relationships among groups of organisms.

7. Clade means a monophyletic group.

8. Gut, nerve cord and main blood vessels.

9. Segmentation with exactly similar segments.

10. Tapeworms.

11. Annelida.

12. The body cavity in triploblastic animals, lined with mesodermal peritoneum.

Fill in the Blanks

1. Segmentation (metamerism) **2.** Pygidium **3.** Anterior **4.** Embryo

5. Arthropoda, Vertebrata **6.** Mesoderm **7.** Peritoneum, coelomic fluid.

Multiple Choice Questions

1. (*d*); **2.** (*c*).

2

Phylum Annelida (Characters, Classification and Types)

Phylum Annelida (L., *Annelus* = a ring) includes about 12,000 species of segmented coelomate worms. They range from a deep-sea species measuring less than 1 mm in length to giant tropical earthworms (of Australia) which measure upto 4 m in length. Most annelids are aquatic found in the sea as swimmers, crawlers, burrowers and tube dwellers or in freshwater where they usually burrow. Terrestrial annelids are mostly burrowers and some are parasites. Phylum Annelida comprises three main classes: the **Polychaeta** (marine), the **Oligochaeta** (terrestrial and freshwater) and the **Hirudinea** (terrestrial, marine and freshwater leeches).

Annelids are soft-bodied, elongated, cylindrical, bilaterally symmetrical, metamerically segmented coelomate worms having a thin covering of cuticle, a dermomuscular body wall and often chitinous setae.

2.1 GENERAL CHARACTERS

1. Annelids are triploblastic and bilaterally symmetrical coelomates.
2. Body is soft, elongated, vermiform (worm-like) and metamerically segmented (*i.e.*, body divided into similar parts or segments which are arranged in a linear series along the antero-posterior axis). The segments are marked externally by transverse grooves and internally by transverse **septa**. Each septum comprises of a layer of connective tissue sandwitched in between two layers of peritoneum.
3. The head or **acron** is represented by **prostomium** and contains the brain. It is pre-oral body part. Terminal part of body is called **pygidium** and contains **anus**. **Mouth** exists in **peristomium** (the first true segment, see Box 2.1).
4. Body wall consists of a fibrous collagenous **cuticle**, a **glandular epidermis** in which the nerve fibres are situated and a connective tissue **dermis** of varying thickness. Muscles of body wall exist in outer continuous layer of **circular muscles** (just inner to dermis) and inner layer of **longitudinal muscles**. The latter is lined by **parietal peritoneum** (coelomic epithelium). Cuticle is secreted by ectoderm (or epidermis).
5. The paired appendages if present are not jointed.
6. Locomotory organs are segmentally arranged, paired, chitinous lateral bristles called **setae** or **chaetae**. They are absent in leeches.
7. Body cavity is a true **coelom** or **perivisceral coelom** lying between the two layers of mesoderm. Coelom is a **schizocoel** (Box 2.1) and divided into segmental compartments by septa.

Box 2.1

1. Prostomium and pygidium are not true segments since they do not arise from the segmental growth zone. The formation of new segments always takes place just infront of pygidium. The oldest body segments are therefore anterior and the youngest are posterior.
2. **Schizocoel.** A coelom which appears as a space in the embryonic mesoderm.

8. **Alimentary canal** is a straight tube running from the anterior mouth to the posterior anus. Digestion is **extracellular**.
9. Respiration occurs through general body surface or gills.
10. Circulatory system or blood vascular system is well developed and **closed type**. Blood is usually red and contains **haemoglobin** (a respiratory pigment). Respiratory pigment is often dissolved in the plasma.
11. Excretory organs are segmental, ectodermal in origin, ciliated, paired and called **nephridia** (metanephridia and protonephridia).
12. The nervous system consists of an anterior, dorsal ganglionic mass or **brain**; a pair of anterior **connectives** surrounding the gut and a long double or single solid **ventral nerve cord** with ganglionic swellings and lateral nerves in each segment.

    A **giant fibre system** is used for rapid conduction and evokes fast startle contraction. *Aphrodite* and *Chaetopterus* have no giant fibre system but other annelids including *Lumbricus* have giant fibre system.
13. Gonads develop from coelomic epithelium.
14. Sexes may be united (monoecious or hermaphrodite) or separated (dioecious).
15. Development is **direct** (without any larval stage) in monoecious annelids and is **indirect** (with **trochophore larva**) in dioecious and aquatic annelids.
16. Cleavage is spiral and determinate.
17. Asexual reproduction occurs in some annelids. Power of regeneration of their lost parts occurs in many annelids.
18. Aquatic or terrestrial.

2.2 OUTLINE CLASSIFICATION

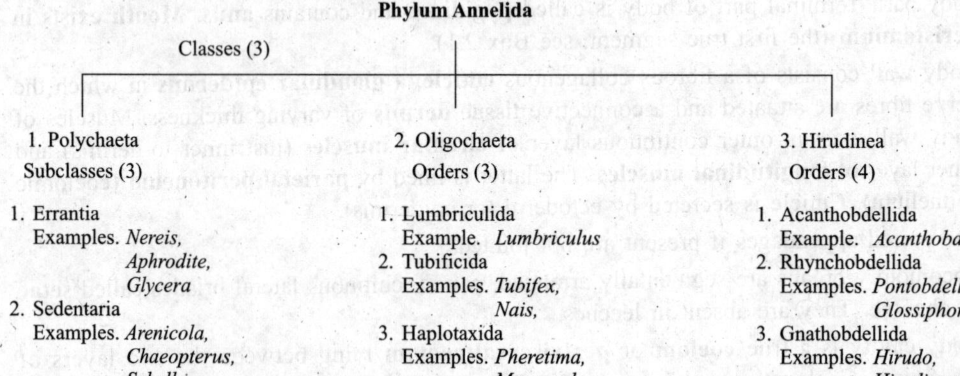

2.3 DETAILED CLASSIFICATION

Phylum Annelida is divided into **three** classes:

Class 1. Polychaeta

1. Class Polychaeta (Gr., *poly* = many; *chaete* = bristles) includes more than 8000 species having great diversity (classified into 86 families; see Box 2.4 and 2.5).
2. Exclusively marine; mostly carnivorous (predaceous or scavengers), some are herbivorous or deposit feeders.
3. Polychaetes may be errant (free-moving) or sedentary.
4. Polychaetes have varied habitats: surface-dwelling (crawling), gallery-dwelling, burrowing and tubicolous.
5. The majority are less than 10 cm long; some are as small as 1 mm and others as large as 70 cm.
6. Colour of the body may be red, pink or green.
7. Body is elongated and segmented. Each segment carries a pair of lateral, fleshy, paddle-like appendages or feet called **parapodia**. Parapodia bear many large setae or chaetae (in bundles).
8. Head is well developed (*i.e.*, cephalization well marked). It contains eyes, tentacles, cirri and palps.
9. Clitellum absent.
10. Cirri or branchiae (gills) or both may be present for respiration. Some part of parapodia is also used as a gas exchange surface.
11. Internal transport occurs through blood-vascular system or coelomic fluid or both. Gas transport uses three types of iron-containing respiratory pigments: **haemoglobin** (red in colour and most common), **chlorocruorin** (green in colour and occurs in *Serpula*, etc.) and **haemerythrin** [a nonhaem (lacking porphyrin) red protein pigment of *Magelona*].
12. Main sensory organs include **eyes, nuchal organs** or ciliated pits (chemoreceptors) and **statocysts** (*e.g.*, *Arenicola*).
13. Sexes are separate. Gonads are localized but extending throughout whole body.
14. Fertilization external (see Box 2.2); free-swimming larval stage is trochophore larva. No cocoon is formed.
15. Power of **regeneration** is well developed. For example, *Chaetopterus* can be regenerated from a single segment but this power is restricted in anterior most 14 segment.
16. Asexual reproduction occurs by lateral **budding**.

Box 2.2 Epitoky and Swarming

Copulation is rare in polychaetes. Instead synchronous emission of sperm and eggs occurs. Epitoky and swarming bring a dispersed benthic population (*e.g.*, *Neries*) together for a brief pelagic existence, when gametes are shed and the likelihood of fertilization is increased. **Epitoky** is the formation of pelagic reproductive individuals or **epitokes** (*e.g.*, *Heteroneries* and Samoan palolo worm) that are adapted for leaving bottom burrows, tubes and other habitats. **Swarming** is a synchronous behaviour in which sexually mature individuals congregates in a relatively short time to increase the likelihood of fertilization.

Class 2. Oligochaeta

1. Class Oligochaeta (Gr., *oligo* = few; *chaete* = bristles) includes about 3100 species. Oligochaetes approximate the polychaetes in size.

2. Mostly freshwater or terrestrial (*i.e.*, damp earth) forms.
3. Body segmented. Parapodia absent. Setae are usually arranged segmentally. Setae are short and embedded in the integument.
4. Head is indistinct and eyes and cephalic appendages such as tentacles and cirri are absent.
5. Clitellum is present. It secretes a cocoon for the eggs.
6. Most oligochaetes are scavengers and feed on dead organic matter, particularly vegetation. Digestive tract is straight and relatively simple containing mouth, pharynx, crop, gizzard, intestine and anus. Intestine has **typhlosole** (A dorsal longitudinal invagination of the intestine wall to increase the absorptive surface).
7. Intestine is surrounded by a layer of yellowish peritoneal cells, called **chlorogogen cells**. They have a role similar to liver of vertebrates (Box 2.3).

Box 2.3 Chlorogogen Tissue

Chlorogogen tissue of earthworms plays a vital role in intermediate metabolism. It is the chief centre of glycogen and fat synthesis and storage. Storage and detoxification of toxins, haemoglobin synthesis, protein catabolism and formation of ammonia and synthesis of urea also take place in these cells. In terrestrial species silicates obtained from food material and the soil are removed from the body and deposited in the chlorogogen cells as waste concretions.

8. Oligochaetes contain two types of metanephridia—**holonephridia** and **enteronephric nephridia**–to excrete their nitrogenous wastes such as ammonia and urea.
9. Circulatory system is basically similar to that of polychaetes.
10. Oligochaetes move by peristaltic contractions. Circular muscle contraction and the consequent elongation of segments are most important in crawling and always generate a coelomic fluid pressure pulse. Longitudinal muscle contraction is more important in burrowing, dilating the burrow or anchoring the segments against the burrow wall. Setae are extended during the longitudinal muscle contraction and retracted during circular muscle contraction.
11. Oligochaetes are **hermaphrodites** with well developed reproductive systems limited to a few segments (with testes anterior to ovaries and with only one or two pairs of male and female gonads).
12. There is **copulation** and reciprocal transfer of sperm. **Fertilization** and **direct development** occur within a cocoon.

Class 3. Hirudinea

1. Class Hirudinea (L., *hirudo* = a leech) includes about 500 species of leeches.
2. Most are freshwater; some are marine or terrestrial. Some are ectoparasites.
3. Body is elongated and dorso-ventrally flattened.
4. Body consists of definite number of segments which may be further divided externally into 2 to 4 annuli. Head is indistinct.
5. **Setae** (except one genus), **parapodia** and **cephalic appendages** are absent.
6. Body is provided with a small suctorial **anterior sucker** and a large **posterior sucker** used for adhesion or locomotion. Both suckers are situated ventrally.
7. Clitellum present.
8. Coelom is greatly reduced. It is represented by a specialized circulatory system, called **haemocoelomic system.** True blood vessels are also present.
9. Coelom is filled up with characteristic **botryoidal tissue.**
10. The septa are absent. Metamerism is shown by nephridia and nervous system (*i.e.*, by gangila of the ventral nerve cord).

11. **Hermaphrodite**; fertilization is internal (cross fertilization due to copulation).
12. Development is **direct** (*i.e.*, no larval stage); eggs are laid in cocoons.
13. Vertebrates form the principal hosts of ectoparasitic, or blood-sucking leeches. Digestion of blood, which is slow, depends on exopeptidases (enzymes) produced by the leech and a symbiotic bacterial flora.

Class 1. Polychaeta

Class **Polychaeta** is divided into following **three** subclasses.

Subclass 1. Errantia

1. Subclass Errantia (L., *errans* = wandering) includes free-swimming, crawling, burrowing and tube-dwelling worms.
2. Body segments are numerous and similar except at anterior and posterior ends.
3. Each body segment carries a pair of well-developed parapodia. Parapodia contain **acicula** (stiff basal setae) and **setae.**
4. Head is well-developed, being distinguished into **prostomium** carrying eyes, tentacles and palpi. **Peristomium** with mouth.
5. Pharynx is usually protrusible and armed with chitinous jaws and teeth to capture prey.

Examples. *Nereis, Aphrodite, Glycera, Polynoe, Syllis.*

Box. 2.4 Families of Errantia

According to **Fauchald** (1977), **Parker** (1982) and **Barnes** *et al.*, (1992) subclass Errantia includes the following **17** families:

Family 1. **Aphroditidae. Example.** *Aphrodite.*
Family 2. **Polynoidae. Example.** *Polynoe.*
Family 3. **Sigalionidae. Example.** *Sigalion.*
Family 4. **Polyodontidae. Example.** *Polyodontes.*
Family 5. **Phyllodocidae. Example.** *Phyllodoce.*
Family 6. **Amphinomidae. Example.** *Amphinome.*
Family 7. **Alciopidae. Example.** *Alciopa.*
Family 8. **Tomopteridae. Example.** *Tomopteris.*
Family 9. **Hesionidae. Example.** *Podarke.*
Family 10. **Syllide. Example.** *Syllis.*
Family 11. **Nereididae. Example.** *Nereis.*
Family 12. **Nephtyidae. Example.** *Nephtys.*
Family 13. **Glyceridae. Example.** *Glycera.*
Family 14. **Eunicea** (Considered as superfamily). **Examples.** *Eunice, Palola, Arabella.*
Family 15. **Histriobdellidae. Example.** *Histriobdella.*
Family 16. **Ichthyotomidae. Example.** *Ichthyotomus.*
Family 17. **Myzostomidae. Example.** *Myzostoma.*

Subclass 2. Sedentaria

1. Sedentary, tubicolous worms living in calcareous or chitinous tubes secreted by epidermal glands.
2. Body is divisible into two or more regions.
3. Segments are not of same type; each segment with a pair of reduced parapodia lacking acicula. Parapodia often modified to perform specific functions.

4. Head is commonly with palps, tentacles and other structures for feeding. They feed on plankton or detritus.
5. Pharynx is often non-protrusible and devoid of jaws or teeth.
6. Gills (or branchiae), when present, localized to anterior segments.

Examples. *Chaetopterus, Sabella, Arenicola, Serpula, Terebella, Spirorbis.*

Box 2.5 Families of Sedentaria

According to **Fauchald** (1977), **Parker** (1982) and **Barnes** *et al.*, (1992) subclass Sedentaria is divided into following 17 families:

Family 1. Orbiniidae. Example. *Orbinia.*
Family 2. Spionidae. Example. *Spio, Polydora.*
Family 3. Megelonidae. Example. *Megelona.*
Family 4. Chaetopteridae. Example *Chaetopterus.*
Family 5. Cirratulidae. Example. *Cirratulus.*
Family 6. Flabelligeridae. Example. *Flabelligera.*
Family 7. Opheliidae. Example. *Ophelia.*
Family 8. Capitellida. Example. *Capitella.*
Family 9. Arenicolidae. Example. *Arenicola.*
Family 10. Maldanidae. Examples. *Maldane, Clymenella.*
Family 11. Oweniidae. Example. *Owenia.*
Family 12. Sabellariidae. Example. *Sabellaria.*
Family 13. Pectinarridae. Example. *Pectinaria.*
Family 14. Ampharetidae. Example. *Ampharete.*
Family 15. Terebellidae. Example. *Terebella.*
Family 16. Sabellidae. Example. *Sabella, Myxicola.*
Family 17. Serpulidae. Example. *Serpula.*

Subclass 3. Archiannelida

1. Subclass Archiannelida (Gr., *archi* = first) includes small-sized animals which are aquatic—marine or freshwater.
2. Body elongated, thread-like.
3. Segmentation chiefly internal. There is no external segmentation.
4. Parapodia and setae are reduced or absent.
5. Tentacles occur on prostomium.
6. Nephridia are either simple (protonephridia) or complex (metanephridia).
7. Nervous system is primitive, *i.e.*, it is continuous with epidermis.
8. Hermaphrodite or unisexual (sexes are separate in *Nerilla*).
9. Larva is typical trochophore.

Subclass Archiannelida includes the following **three** families:

Family 1. Nerillidae. Examples. *Nerilla, Nerillidium.*

Family 2. Polygordiidae. Examples. *Polygordius, Protodrilus.*

Family 3. Dinophilidae. Examples. *Dinophilus, Diurodrilus.*

(**Note.** Archiannelida was previously considered as a separate class of phylum Annelida. Now it is regarded as a heterogenous group of polychaetes (**Fauchald,** 1975).

Class 2. Oligochaeta

Class Oligochaeta is divided into following **three** orders on the basis of differences in the reproductive organs (**Jamieson 1978**).

Order 1. Lumbriculida

1. Male gonopores and at least one pair of testes in the same segment.
2. Each body segment consists of four pairs of setae.
3. Clitellum is one cell thick and includes male and female gonopores.
4. All are freshwater animals.

Example. *Lumbriculus*.

Order 2. Tubificida

1. Male gonopores situated in the segment immediately anterior or posterior to the segment containing testes.
2. One pair of testes followed by one pair of ovaries in adjacent segments.
3. Clitellum is one-cell thick including male and female gonopores.
4. Setae hair-like or modified. Each segment contains setae bundles-each containing two or more setae.
5. Marine, freshwater and a few terrestrial forms.

Examples. *Tubifex*, *Nais*, *Dero*, *Chaetogaster*, *Enchytraeus*.

Order 3. Haplotaxida

1. Male gonopores are situated atleast one segment posterior to last pair of testes.
2. Basically have two segments with testes followed by two segments with ovaries. One pair of testes or ovaries or often both absent. If there is one pair of testes, these are separated from the ovaries by one or two segments.
3. Clitellum one-cell or more than one cell in thickness. Male gonopore is not situated in the clitellum.

Examples. *Lumbricus*, *Megascolex*, *Pheretima*, *Drawida*, *Eutypheus*.

Class 3. Hirudinea

Class Hirudinea is divided into following **four** orders:

Order 1. Acanthobdellida

1. This order includes only one species which is parasitic on the fins of salmon fishes.
2. Body comprises thirty segments only. Each segment contains four annuli.
3. Anterior sucker is absent but posterior sucker is well developed and composed of four segments.
4. Anterior five segments are provided with two pairs of setae.
5. First five segments show internal segmentation (septa are present).
6. Proboscis is short; jaws absent; anus dorsally situated.
7. Acanthobdellida forms a connecting link between Oligochaeta and Hirudinea.

Example. *Acanthobdella peledina*.

Order 2. Rhyncobdellida

1. Blood-sucking ectoparasites of snail, fishes, frogs, tortoise and turtles. May be freshwater or marine.
2. Each typical segment consists of 3, 6, or 12 annuli.

3. Proboscis is protrusible and without jaws and teeth.
4. Setae are absent.
5. Both anterior and posterior suckers are present.
6. Coelom is reduced to sinuses but botryoidal tissue absent.
7. Circulatory system is separated from the coelomic sinuses. Blood is colourless.
8. Eyes absent.

Examples. *Pontobdella, Glossiphonia, Helobdella, Placobdella, Piscicola.*

Order 3. Gnathobdellida (or Arnchynchobdellida)

1. Most are ectoparasites on insect larva, worms, snails, amphibia, cattles and human beings; some are predators.
2. Freshwater and terrestrial forms.
3. Each typical body segment consists of five annuli.
4. Body with anterior and posterior suckers.
5. Proboscis is absent.
6. They contain three jaws with strong cuticular teeth in their buccal cavity.
7. They contain five pairs of eyes. Host finding occurs by means of olfactory sense (smell).
8. Setae are absent.
9. Coelom is reduced to sinuses and body cavity filled with botryoidal tissue.
10. Blood is red coloured.

Examples. *Hirudo, Hirudinaria, Haemopis, Macrobdella, Xerobdella.*

Order 4. Pharyngobdellida

1. Freshwater or amphibious carnivorous (predaceous) leeches.
2. Proboscis nonprotrusible. Jaws absent but they may contain one or two piercing stylets.

Examples. *Erypobdella, Dina.*

2.4 TYPE STUDIES

1. *Aphrodite*

Classification

Phylum	–	**Annelida**
Class	–	**Polychaeta**
Subclass	–	**Errantia**
Genus	–	***Aphrodite***

Habit and Habitat

Aphrodite, is a type of **scale worm** which is commonly called **'sea-mouse'**. It is a small marine burrowing form found in mud in deep waters. It occurs in Europe and on the north west and north east coasts of United States. *Aphrodite* lives almost half buried in the mud; its posterior end of body is exposed at the surface. It occasionally creeps on the sea floor.

Structure

Aphrodite is about 12 cm in length. The body is short, broad and oval in shape and dorso-ventrally flattened. Body contains 30 to 35 segments. Its dorsal surface is convex and covered with

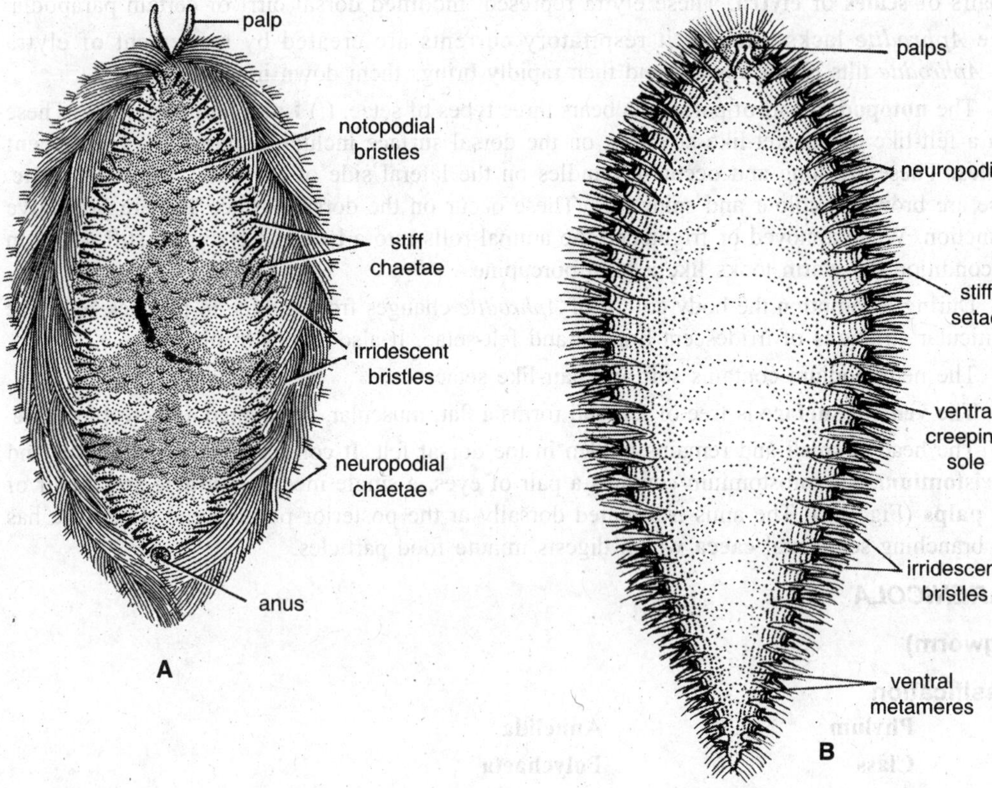

Fig. 2.1. *Aphrodite*. A—Dorsal view; B—Ventral view.

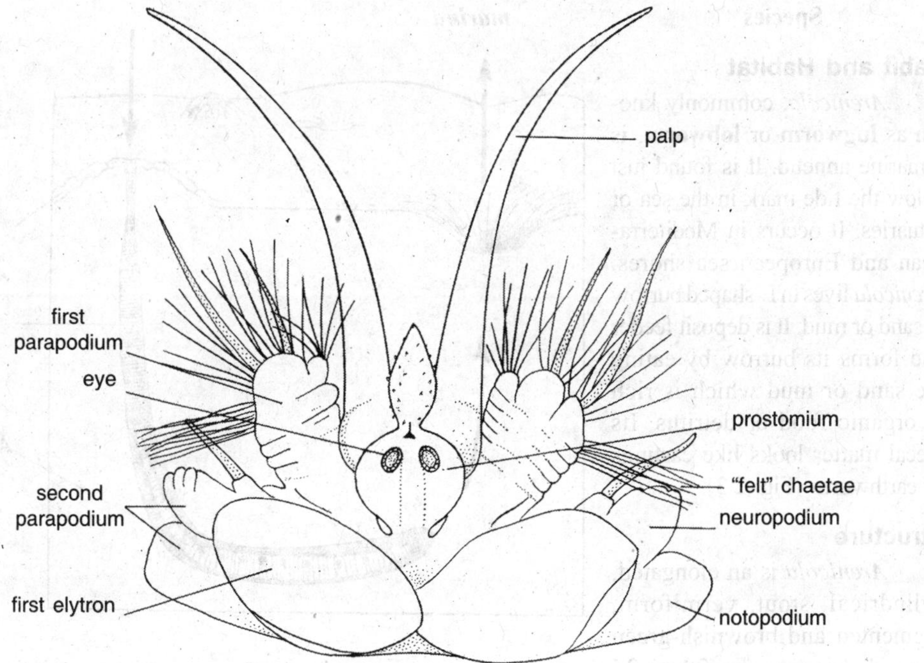

Fig. 2.2. *Aphrodite*. Anterior end, including the first pair of dorsal scales or elytra.

15 pairs of **scales** or **elytra**. These elytra represent modified dorsal cirri of certain parapodia. Since *Aphrodite* lacks cilia, in it respiratory currents are created by movement of elytra (*i.e.*, *Aphrodite* tilts elytra upward and then rapidly brings them down in sequence).

The **notopodial** part of parapodia bears three types of setae: (*i*) **Long hair-like setae.** These form a felt-like or blanket-like covering on the dorsal surface including elytra. (*ii*) **Irridescent bristles.** These are long and occur in bundles on the lateral side of the body. **(*iii*) Stiff setae.** These are brown in colour and spine-like. These occur on the dorsal surface and are protective in function. When annoyed or frightened the animal rolls into a ball and erects its stiff setae. In this condition the worm looks like a mini porcupine.

During locomotion the body colour of *Aphrodite* changes from golden to peacock blue due to cuticular striations of irridescent bristles and felt-setae. It also shines brilliantly.

The neuropodium contains soft and hair-like setae.

The **ventral surface** is free of felt and forms a flat, muscular, segmented and creeping **sole**.

The head is small and remains hidden in the dorsal felt. It consists of a **prostomium** and a **peristomium**. The prostomium contains a pair of eyes, a single median **tentacle** and a pair of long **palps** (Fig. 2.2). The anus is situated dorsally at the posterior pointed end. **Intestine** has long branching segmental **caeca** which digests minute food particles.

2. *ARENICOLA*

(Lugworm)

Classification

Phylum	–	**Annelida**
Class	–	**Polychaeta**
Subclass	–	**Sedentaria**
Genus	–	***Arenicola***
Species	–	***marina***

Habit and Habitat

Arenicola, commonly known as **lugworm** or **lobworm**, is a marine annelid. It is found just below the tide mark in the sea or estuaries. It occurs in Mediterranean and European sea shores. *Arenicola* lives in L-shaped burrow in sand or mud. It is deposit feeder and forms its burrow by eating the sand or mud which is rich in organic food or detritus. Its faecal matter looks like castings of earthworm (Fig. 2.3).

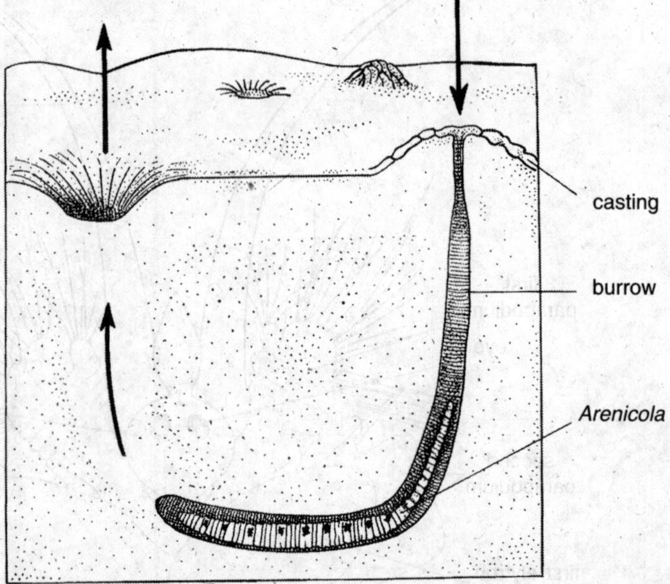

Fig. 2.3. *Arenicola*. L-shaped burrow and position of the annelid inside the burrow.

Structure

Arenicola is an elongated, cylindrical, stout, vermiform, segmented and brownish-green annelid. It measures 20 to 25 cm in length. Its body is divisible into three regions (Fig. 2.4):

1. Anterior region or prebranchial region. It comprises small trilobed **prostomium** with no eyes or tentacles, a **peristomium,** an achaetous (= without setae) segment and six segments bearing setae and parapodia. The parapodia are rudimentary.

2. Middle region or branchial region. It contains thirteen segments. Each segment contains rudimentary parapodia and branched **gills** above the parapodia (Fig. 2.5). The gill are modified dorsal cirri of parapodia.

3. Posterior region or postbranchial region. It is much thinner body part than others and contains variable number of segments. It is without gills, parapodia or setae.

Each body segment bears **annuli. Mouth** lies ventral to the prostomium. The buccal region and pharynx protrude as **proboscis** which is covered by chitinized papillae (*Arenicola* uses the eversible pharynx to punch and anchor in the sediment. It pick up particles on the adhesive glandular surface). Jaws are absent. The **anus** opens through the last segment. There are six pairs of nephridia in segments 4 to9.

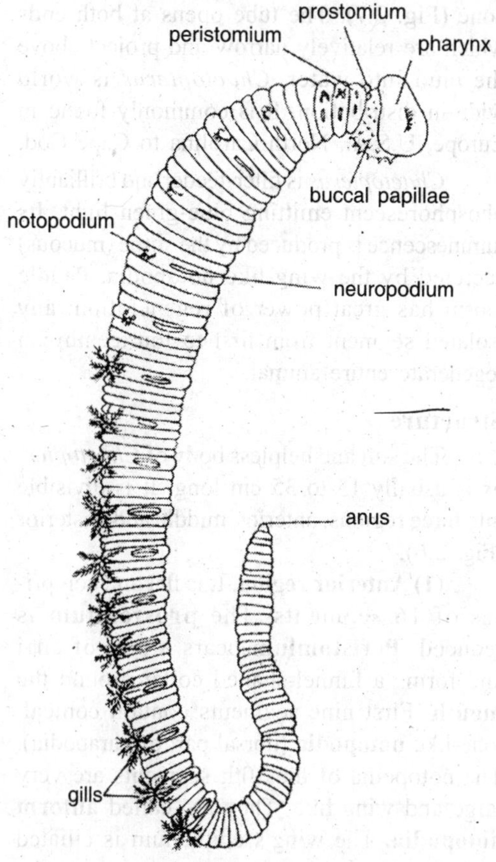

Fig. 2.4. *Arenicola.* Lateral view.

Arenicola contains two types of sensory organs— a pair of **nuchal organs** and a pair of **statocysts**. Nuchal organs consist of a pair of ciliated pits or slits, they are often eversible and situated in head region. The **statocysts** or **gravity receptors** of *Arenicola* are situated within the body wall of the head. Each statocyst is open pit or canal that opens to exterior (at lateral body surface). It contains various foreign objects such as spicules, diatom shells and sand or qurts grains all covered with chitinoid material. *Arenicola* always burrows head downward and if an aquarium containing a worm is tilted 90 degrees, the worm makes a compensating 90 degree turn in burrowing. If statocysts are destroyed this compensating ability is lost.

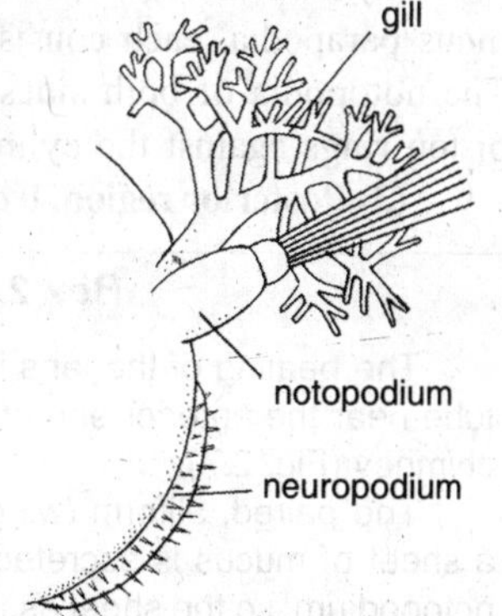

Fig. 2.5. *Arenicola.* Parapodia bearing a gill.

3. *CHAETOPTERUS*

Classification

Phylum	–	**Annelida**
Class	–	**Polychaeta**
Order	–	**Sedentaria**
Type	–	***Chaetopterus***

Habit and Habitat

Chaetopterus is commonly called **paddle worm**. It is a marine tubicolous worm. It lives permanently in U-shaped parchment-like tube formed in the sand or mud in the intertidal zone (Fig. 2.7). The tube opens at both ends which are relatively narrow and project above the mud into water. *Chaetopterus* is world wide in distribution. It is commonly found in Europe, U.S.A., North Carolina to Cape Cod.

Chaetopterus is filter-feeder and brilliantly phosphorescent emitting blue-green light. Its luminescence is produced by the slime (mucous) secreted by the wing-like notopodia. Paddle worm has great power of regeneration; any isolated segment from first to fourteenth can regenerate entire animal.

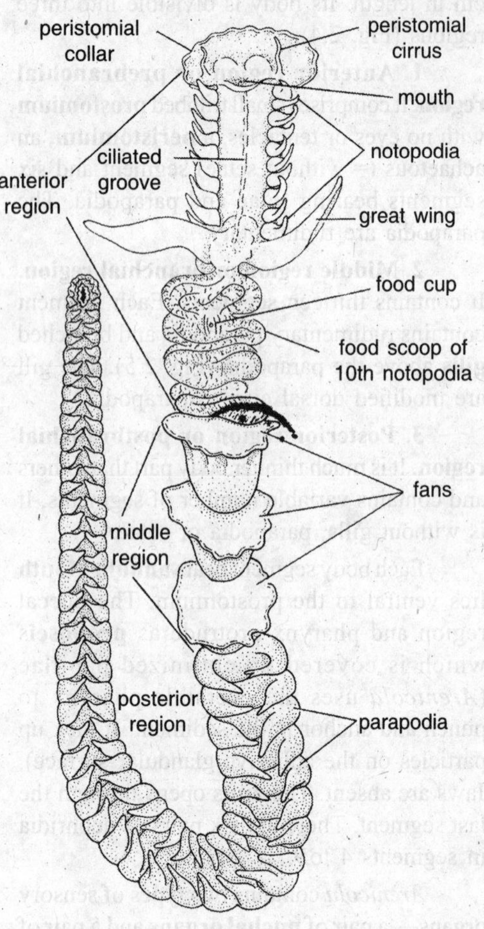

Fig. 2.6. *Chaetopterus*. Structure.

Structure

The soft and helpless body of *Chaetopterus* is usually 15 to 35 cm long. It is divisible into three regions; anterior, middle and posterior (Fig. 2.6).

(1) Anterior region. It is flat and comprises of 16 segments. The **prostomium** is reduced. **Peristomium** bears a pair of cirri and forms a funnel-shaped collar around the **mouth**. First nine segments contain conical, lobe-like **notopodia** (dorsal part of parapodia). The notopodia of the 10th segments are very large and wing-like. They are called **aliform notopodia.** The wing's epithelium is ciliated and con-tains numerous mucous glands. In the next three segments parapodia are reduced. Their **neuropodia** (ventral part of parapodia) form **anchors** or **suckers** which attach the worm within the tube. A **ciliated groove** runs on the mid-dorsal surface from the mouth to a ciliated **food-cup** or **dorsal cupule.**

(2) Middle region. It comprises of three segments only. Each segment bears a pair of biramous parapodia, each consisting of a very small **neuropodium** and a very large **notopodium.** The notopodia of both sides of each segment fuse to form semicircular **fans** that project like piston rings against the cylindrical wall of the tube.

(3) Posterior region. It consists of several segments and each segment carries typical parapodia.

Box 2.6 Feeding Mechanism of *Chaetopterus*

The beating of the fans produces a current of water that enters the chimney of the U-shaped tube near the anterior end of the form, flows through the tube, and then flows out of the opposite chimney (Fig. 2.7).

The paired, aliform (wing-like) notopodia are stretched out around the walls of the tube and a sheet of mucus is secreted between them. The mucus film is continuously secreted from each notopodium, so the sheet assumes the shape of a bag. The posterior end of the bag is grasped by a ciliated cup on the middorsal side of the worm a short distance behind the aliform notopodia. Water is brought into the tube by the rhythmic beating of the fan parapodia, passes through the mucous bag which strains suspended detritus and plankton.

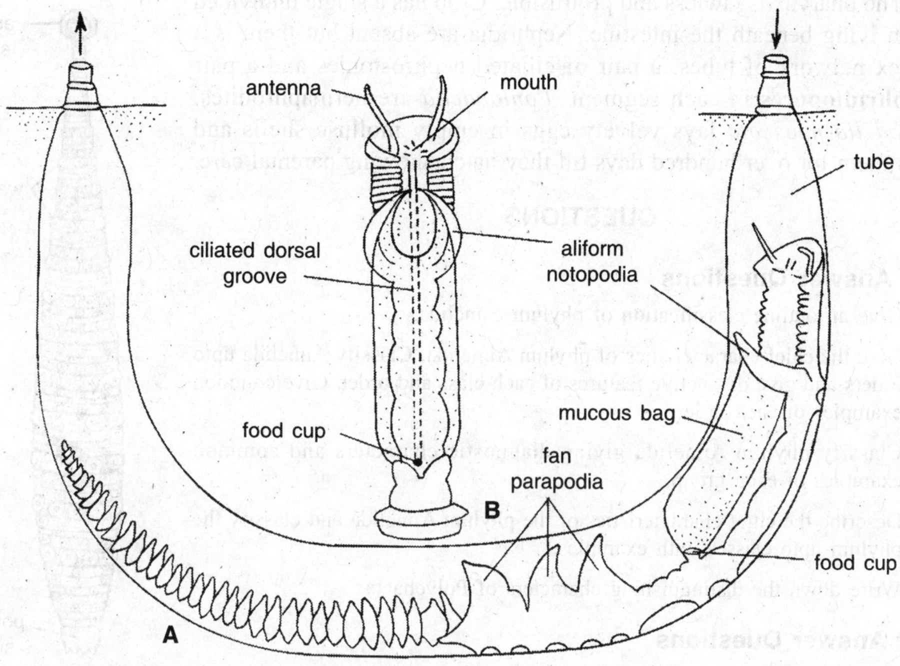

Fig. 2.7. *Chaetopterus*. Feeding mechanism. A—Anterior part of the body (dorsal view); B—Worm in tube (lateral view).

Large objects brought into the tube by the water current are detected by peristomial cilia; the aliform notopodia then are pulled back to let the large objects pass by. The food-laden mucous bag is continuously being rolled up into a ball by the dorsal cupule. When the ball reaches a certain size, the bag is cut loose from the notopodia and rolled up with the ball. The cupule then projects forward and deposits the mucous food ball into a ciliated middorsal groove which extends to the anterior of the worm and the ball is carried to the mouth. A 18 to 24 cm long specimen of *Chaetopterus* may produce mucous film for the bag at the rate approximately 1 mm/sec, with food balls averaging 3 mm in diameter.

4. *Pontobdella*

Classification

Phylum	–	**Annelida**
Class	–	**Hirudinea**
Order	–	**Rhynchobdella**
Genus	–	***Pontobdella***

Habit and Habitat

Pontobdella is a marine fish leech. It is ectoparasitic on sharks, skates and rays. *Pontobdella* is found in Europe and USA.

Structure

The body is elongated, cylindrical and club-shaped measuring 20 cm in length (Fig. 2.8). It consists of an anterior and a posterior **sucker.** The anterior sucker is saucer-shaped. It can be firmly applied to the tissue of the host. The posterior sucker is deep and cup-shaped. Body surface is rough and leathery. It is covered with many green tubercles or warts. Metameres are not very distinct. Eyes and gills are absent.

The pharynx is jawless and protrusible. Crop has a single undivided caecum lying beneath the intestine. Nephridia are absent but there is a complex network of tubes, a pair of ciliated nephrostomes and a pair of **nephridiopores** in each segment. *Pontobdella* are hermaphrodites. *Pontobdella muricata* lays velvety eggs in empty mollusc shells and guards them for over hundred days till they hatch showing parental care.

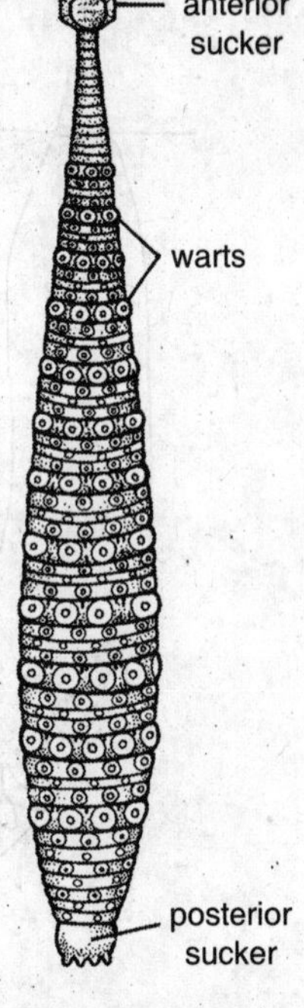

Fig. 2.8. *Pontobdella.* External structure.

QUESTIONS

Long Answer Questions

1. Give an outline classification of phylum Annelida.
2. Give the chief characteristics of phylum Annelida. Classify Annelida upto orders and give distinctive features of each class and order. Give common examples of each order.
3. Classify phylum Annelida giving diagnostic characters and common examples of each group.
4. Describe the chief characteristics of the phylum Annelida and classify the phylum upto classes with examples.
5. Write down the distinguishing characters of Polychaeta.

Short Answer Questions

1. Assign the following animals to their systematic position giving their distinguishing features.

   (*i*) *Aphrodite* (*ii*) *Arenicola* (lugworm)

   (*iii*) *Chaetopterus* (*iv*) *Pontobdella*

Very Short Answer Questions

1. Name the subclass to which *Polygordius* belongs.
2. From which germ layer is the peritoneum derived.
3. What is pygidium?
4. Is prostominum a metamere?
5. Where are new segments formed in earthworm?
6. What is schizocoel?
7. In earthworm, haemoglobin is dissolved in the plasma. Is it so?
8. Parapodia are absent in annelids of classes.

Fill in the Blanks

1. Segmentation in Annelida is called ....
2. The most distinguishing characteristic of the phylum Annelida is ....
3. The cuticle in Annelida .... rather than chitinoid.
4. Body cavity in annelids is ....
5. Body of an annelid may be divided into three regions: .... , trunk and ....
6. Annelids have .... circulatory system.
7. Excretion in Annelida occurs by ....
8. Larva of annelids is ....
9. Coelom of annelids is lined by peritoneum of .... origin.
10. Currently phylum Annelida is divided into .... classes.

11. Skin of annelids serves for ....

12. In annelids, respiratory pigment is present in the ....

13. In metanephridia the nephridium opens into coelom by ciliated funnel-shaped opening, called ....

14. Fans of *Chaetopterus* are modified ....

15. .... is a tubicolous polychaeta.

16. Giant fiber system is absent in ....

17. In *Arenicola*, gills are present in the .... region of body.

18. Clitellum is absent in annelids of class ....

Multiple Choice Questions

1. Members of phylum Annelida show following characters except one which is wrong. Mark it

(*a*) segmented and triploblastic (*b*) excretory organs and nephridia
(*c*) a true coelom is absent (*d*) blood vascular system is closed type

2. Segmentation in Annelida is

(*a*) homonomous (*b*) heteronomous
(*c*) both (a) and (b) (*d*) neither (a) nor (b)

3. Segmentation in Annelida affects

(*a*) external features only (*b*) internal organs only
(*c*) both (a) and (b) (*d*) none of these

4. Body cavity of annelids is

(*a*) coelom (*b*) pseudocoel
(*c*) haemocoel (*d*) none of these

5. The coelom in Annelida is

(*a*) pseudocoelic (*b*) enterocoelic
(*c*) schizocoelic (*d*) acoelous

6. *Nereis* belongs to class .... of Annelida

(*a*) Oligochaeta (*b*) Polychaeta
(*c*) Hirudinea (*d*) None of these

7. *Aphrodite* is popularly called

(*a*) scale worm (*b*) palolo worm
(*c*) lugworm (*d*) sea mouse

8. The excretory organs of annelids are

(*a*) metanephridia (*b*) malpighian tubules
(*c*) flame cells (*d*) renette cells

9. Parapodia are locomotory organs in

(*a*) Oligochaeta (*b*) Chilopoda
(*c*) Polychaeta (*d*) Diplopoda

10. Larvae form of polychaeta is known as

(*a*) tornaria (*b*) glochidium
(*c*) trochophore (*d*) bipinnaria

11. Zoological name of paddle worm is

(*a*) *Arenicola* (*b*) *Chaetopterus*
(*c*) *Polynoe* (*d*) *Pheretima*

12. *Tubifex* belongs to class .... of Annelida.
(*a*) Polychaeta (*b*) Oligochaeta
(*c*) Echiurida (*d*) Hirudinea

13. Branchiae are respiratory organs in
(*a*) *Arenicola* (*b*) *Nereis*
(*c*) *Aphrodite* (*d*) *Polynoe*

14. Locomotory structures of earthworms are
(*a*) bristles (*b*) cilia
(*c*) parapodia (*d*) setae

15. *Arenicona* is commonly known as
(*a*) lugworm (*b*) paddle worm
(*c*) sea pen (*d*) palalo worm

16. True coelom is a space that lies between
(*a*) body wall and alimentary canal (*b*) ectoderm and endoderm
(*c*) two layers of endoderm (*d*) visceral and parietal mesoderm

17. Segmentation in archiannelids is
(*a*) external only (*b*) internal only
(*c*) both external and internal (*d*) not found

18. Special cells that help in excretion in earthworm are
(*a*) chloragen cell (*b*) chlorocytes
(*c*) choanocytes (*d*) flame cells

ANSWERS

Very Short Answer Questions

1. Archinnelida.
2. Mesoderm.
3. Terminal part of body of an annelid bearing anus.
4. No.
5. Just infront of pigidium.
6. Coelom formed by split in mesodermal bands.
7. Yes.
8. Oligochaeta and Hirudinea.

Fill in the Blanks

1. Metamerism **2.** Metamerism **3.** Collagenous **4.** True coelom
5. Prostomium, Pygidium **6.** Closed **7.** Nephridia **8.** Trchophore
9. Mesodermal **10.** Three **11.** Respiration **12.** Plasma
13. Nephrostome **14.** Parapodia **15.** *Sabella* **16.** *Chaetopterus*
17. Branchial **18.** Polychaeta.

Multiple Choice Questions

1. (*c*) **2.** (*c*) **3.** (*c*) **4.** (*a*) **5.** (*c*) **6.** (*b*)
7. (*d*) **8.** (*a*) **9.** (*c*) **10.** (*c*) **11.** (*b*) **12.** (*b*)
13. (*a*) **14.** (*d*) **15.** (*a*) **16.** (*d*) **17.** (*b*) **18.** (*a*).

3

Nereis (Neanthes) (Phylum Annelida)

3.1 CLASSIFICATION

Phylum	—	**Annelida**
Class	—	**Polychaeta**
Subclass	—	**Errantia**
Genus	—	***Nereis (Neanthes)***

3.2 HABIT AND HABITAT

Nereis is commonly known as **sandworm, rag worm** or **clam worm.** It is marine and found on the sea-shore in the shallow water (*i.e.*, tide levels) in rock crevices hidden under the stones or sea weeds. Mostly it lives in temporary U-shaped burrows in sand. The mucus secreted by its body wall binds the sand particles together. Jaws are used for preparing burrows.

Nereis is **carnivorous** and **predaceous**. It feeds on small insects, molluscs and worms. It is very active during night and passive during the day, *i.e.*, *Nereis* is **nocturnal**. During night, *Nereis* come out from their hidden places and start crawling over the sand. When breeding period approaches, *Nereis* leaves the burrow and comes at the water surface to lead a pelagic life then it is called *Heteronereis*.

Distribution. *Nereis* is cosmopolitan in distribution and found in coastal waters of Pacific and North Atlantic oceans, Europe and USA.

3.3 EXTERNAL STRUCTURE

Shape and size. The body of *Nereis* is elongated, cylindrical bristle worm. It is bilaterally symmetrical, metamerically segmented and dorso-ventrally flattened (Fig. 3.1). It has convex dorsal surface and flat ventral surface. The size of body may range from 30 to 40 cm in length and 2 to 6 mm in width.

Colouration. Different species of *Nereis* have different colours. For example, *N. pelagica* is reddish brown, *N. cultrifera* is greenish, *N. limnicola* is brownish in colour while *N. virens* is bluish-green in colour, having orange or red tinge on the appendages.

Segmentation. Body of *Nereis* is metamerically segmented. It is divided upto 200 or more **segments** or **metameres** which are marked externally by narrow grooves and internally by septa. All segments, apart from those most anterior and posterior, are very similar to each other.

Division of body. The body of *Nereis* is divisible into three distinct regions, *i.e.*, head, trunk and pygidium.

1. Head. The anterior end of body has a distinct **head** which shows a great degree of cephalization (Fig. 3.2). It consists of two main parts: (*i*) **prostomium** and (*ii*) **peristomium**.

(i) **Prostomium.** (Gr., *pro* = before; *stoma* = mouth). It is anterior triangular fleshy outgrowth. It is situated mid-dorsally in front of mouth (Fig. 3.2). Prostomium is not the true segment of the body as it is derived from preoral lobe of larva. Prostomium carries two pairs of simple, black and rounded eyes; a pair of short, cylindrical and tactile **prostomial tentacles**; a pair of stout (fleshy) longer and tactile **prostomial palps** and a pair of **ciliary pits** or **nuchal organs**.

(ii) **Peristomium.** (Gr., *peri* = around; *stoma* = mouth). It is posterior ring-like portion of head. Ventrally it bears a transverse **mouth** (Fig. 3.3). On each side, peristomium bears four pairs of long slender **tactile peristomial cirri**, two are dorsal and two are ventral.

Peristomium represents fused first and second body segments. Peristomial cirri correspond to the neuropodial and notopodial cirri of parapodia of rest of the body.

2. Trunk. Posterior to the head, the rest of the body is called **trunk.** It is formed of 80 to 200 identical **somites** or **metameres** depending on species. Each metamere is broader than long and bears a pair of **parapodia** on its lateral sides.

3. Pygidium. The posterior most segment is somewhat rounded and is called **tail somite, anal segment** or **pygidium**. It bears a pair of elongated **anal cirri** (Fig. 3.4). **Anus** is terminal.

Parapodium. (Gr., *para* = side; *podus* = foot). Each body segment except the peristomium and pygidium bears on either side a flat, fleshy and hollow flap-like **parapodium**. It is **biramous** and consists of an upper **notopodium** and lower **neuropodium** (Fig. 3.5). Dark coloured chitinous rods called **acicula** support both of these processes, and two tufts of **setae or chaetae** emanate from each aciculum forming fan-like bodies. Two additional outgrowths of the parapodia are noticeable: a dorsal and a ventral **cirrus**.

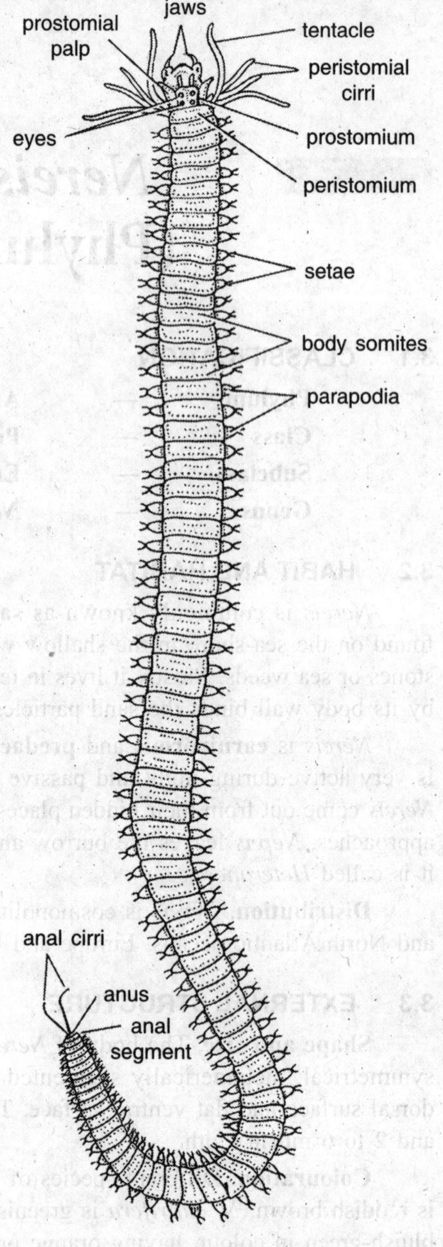

Fig. 3.1. *Nereis*. External features.

The parapodia are primarily the organs of **locomotion** used both in creeping and in swimming. Since parapodia are richly vascularized, they also serve the function of **respiration.**

Nuchal organ. These are present in the form of a pair of ciliated grooves one on either side of the prostomium. These are **chemoreceptors.** They also act as **olfactory sense organs.**

Nephridiopores. These are external openings of nephridia. They are present singly near the base of ventral cirri of parapodia.

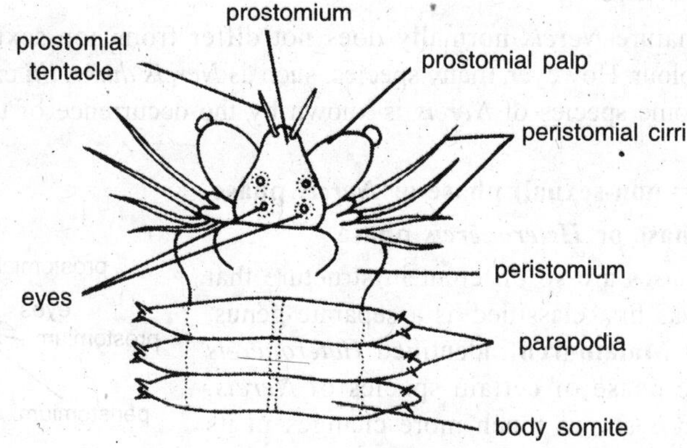

Fig. 3.2. *Nereis*. Head in dorsal view.

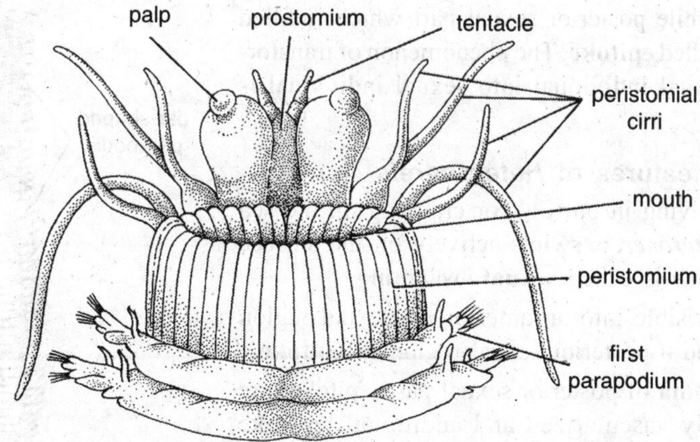

Fig. 3.3. *Nereis*. Head in ventral view.

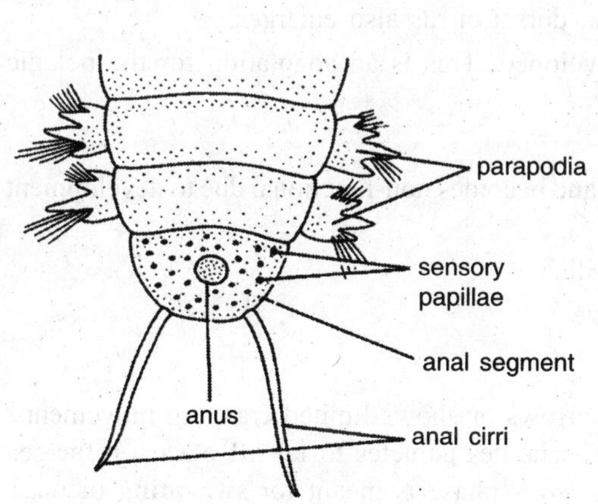

Fig. 3.4. *Nereis*. Anal segment or pygidium.

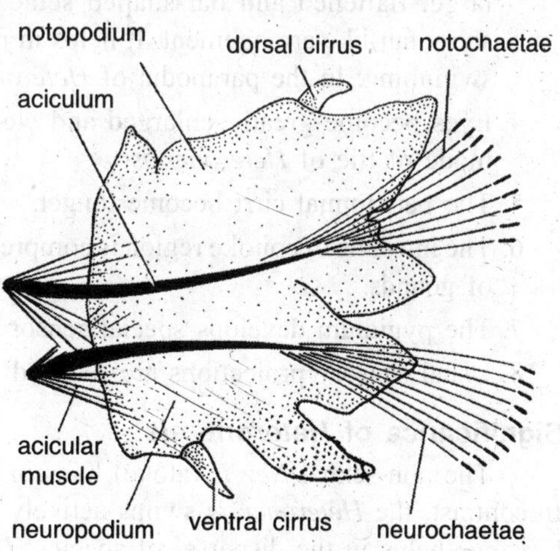

Fig. 3.5. *Nereis*. Parapodium.

3.4 HETERONEREIS

A sexually mature *Nereis* normally does not differ from non-sexual worm in external characters except colour. However, many species, such as *Nereis dumerilii* exhibit **polymorphism.** Polymorphism in some species of *Nereis* is known by the occurrence of two alternate forms or phases:

(*i*) Asexual (= non-sexual) phase or *Nereis* phase

(*ii*) Sexual phase or *Heteronereis* phase

These two phases are so different in structure that the sexual phase was first classified as a separate genus, *Heteronereis*. Later **Malamgren** indentified *Heteronereis* as sexually mature phase of certain species of *Nereis*. Sexually male *Nereis* shows much more changes in its body than the female. In the *Heteronereis* (Fig. 3.6), gonads (*i.e.*, testes and ovaries) develop only in the posterior part of body. Its anterior non-sexual part which lacks gametes is called **atoke** while posterior sexual part which is filled with gametes is called **epitoke**. The phenomenon of transformation of nonsexual individual into sexual individual is called **epitoky**.

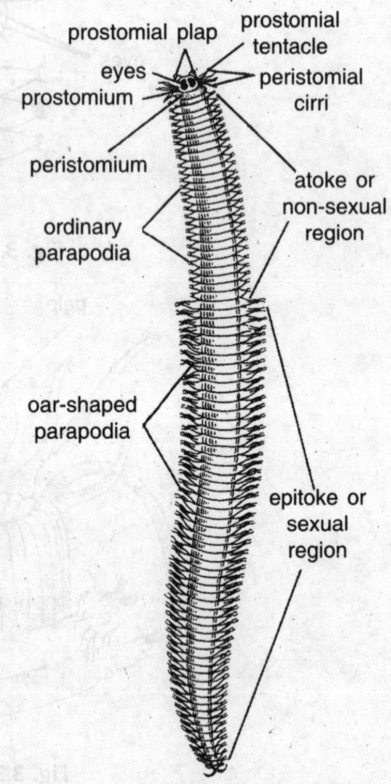

Fig. 3.6. *Heteronereis*. External features.

Characteristic features of *Heteronereis*

1. Instead of living in burrows or crawling on the sea bottom, *Heteronereis* swims actively on the surface of water. This is called **sexual swarming**.
2. Body is divisible into an anterior nonsexual region or **atoke** and a posterior sexual region or **epitoke**.
3. The **parapodia** of posterior sexual part (epitoke) are large, highly vascularized and additional leaf-like outgrowths for more rapid respiration (Fig. 3.7). The normal setae of these parapodia are replaced by new, large, flattened and oar-shaped setae. These setae show fan-like arrangement. It helps in more efficient swimming. In the parapodia of *Heteronereis*, dorsal cirrus also enlarges.
4. Eyes become greatly enlarged and well developed. This is an adaptation for the pelagic mode of life of *Heteronereis*.
5. The peristomial cirri become longer.
6. The intestine of epitoke region is compressed and becomes non-functional due to development of goands.
7. The pygidium develops special sensory papillae.
8. Usual sensory projections are reduced in size.

Significance of *Heteronereis*

The non-sexual *Nereis* (atoke) lives in the burrows or shows limited crawling movements. In contrast, the *Heteronereis* swims actively and discharges gametes to far off places in the sea and thus helps in the dispersal of species. *Heteronereis* phase is meant for **swarming** or mass **nuptial dance**.

Table 3.1 Differences between *Nereis* and *Heteronereis*.

Nereis	*Heteronereis*
1. Nereis is a non-sexual sluggish creature living in burrow or creeping at the bottom of sea.	1. *Heteronereis* is an active swimmer on the surface of water (pelagic).
2. Body is not differentiated into atoke and epitoke.	2. Body is divisible into anterior asexual **atoke** and posterior sexual **epitoke** regions.
3. The parapodia are thin, flap-like and less vascular.	3. The parapodia of posterior or sexual region are large, highly vascular and develop leaf-like outgrowths.
4. The setae on the parapodia are normal.	4. The setae are large, flattened and oar-shaped. These are arranged in fan-like shape.
5. Eyes are normal sized and the worm is less sensitive to light.	5. Eyes are large and conspicuous; worm is very sensitive to light.
6. Sensory projections are well developed.	6. Sensory projections become shrunken and anal segment develops special sensory papillae.
7. Dorsal cirri are less curved.	7. Dorsal cirri become shrunken.
8. The intestine is well formed.	8. The intestine is compressed and become functionless due to development of gonads.

Comparison of Parapodia of *Nereis* and *Heteronereis*

1. Parapodium of *Nereis*. Each trunk segment of *Nereis* bears laterally a pair of dorso-ventrally flattened, hollow flaps. For these fleshy body outgrowths **Huxley** used the term **parapodia.** The cavity of a parapodium is continuous with the body **coelom**. Each parapodium is a **biramous** structure consisting of two parts or lobes: an upper or dorsal lobe, the **notopodium** and a lower or ventral lobe the **neuropodium**. Each of these lobes is further divided into two smaller lobes, an upper and a lower. The dorsal margin of notopodium is produced into a short, cylindrical, tactile appendage, the **dorsal cirrus**. Ventral margin of neuropodium too produces such a tentacular process, called **ventral cirrus**. Of these two cirri, the dorsal one is much larger than the ventral (Fig. 3.5).

A bundle of bristle-like **setae** or **chaetae** fans out from each lobe of parapodium (*i.e.*, notopodium and neuropodium). In each case, these form a bundle and remain embedded in a **setal sac** or **setigerous sac**. The setae can be turned in various directions by strands of muscular fibres.

In the middle of each bundle of setae and deeply embedded in the parapodium is stout, straight, thick and dark coloured chitinous rod the **aciculum**. It serves as endoskeleton of parapodium and serves to support and provide attachment to the setal muscles.

Fig. 3.7. *Heteronereis*. Parapodium.

Setae or chaetae. The setae are fine, stiff and chitinous bristles which help in locomotion by acting as minute paddles; these being protective are used for offense and defence. Setae also provide hold on the smooth surface of the burrow. Setae are lodged in setal or **satigerous sac** and each seta originates from a single large **formative cell** present at the base of setigerous sac. The setae are continuously produced by the setal sac to replace the worn out old setae. Setae can be retracted and protruded out by special sets of muscles of parapodia.

Each seta consists of two movable articulated parts—the basal **shaft** or **stalk** and the distal **blade** which can be retracted into the shaft (Fig. 3.8). The setae of *Nereis* are of three types:

***(i)* Long blade setae** with a small slender shaft and the terminal long blade which is nearly straight, pointed and serrated at one edge.

***(ii)* Typical setae** which have a large stout shaft, short and slightly hooked blade, and

***(iii)* Oar-shaped setae** have a oar-shaped blade. These are found in *Heteronereis*.

The parapodia are highly muscular, well vascularized and glandular structures. The parapodia present in the middle of body are large and gradually become smaller towards two ends.

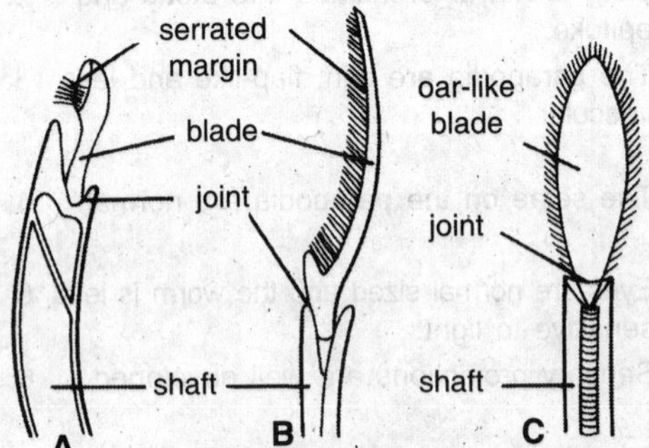

Fig. 3.8. *Nereis*. Various types of setae. A—Typical setae; B—Long bladed setae; C—Oar-shaped setae of *Heteronereis*.

2. Parapodia of *Heteronereis*. The structure of parapodia of *Nereis* and *Heteronereis* is basically identical but parapodia of *Heteronereis* shows the following peculiar features (Fig. 3.7)

1. These are bigger and more fleshy.
2. These have foliacious outgrowths.
3. The cirri are highly enlarged and vascular.
4. The setae are oar-like and arranged in fan-shaped manner.
5. Original parapodial muscles are replaced by new muscles.

The enlarged and modified parapodia enable *Heteronereis* to swim fast and lead a pelagic life.

Table 3.2 Comparison of parapodia of *Nereis* and *Heteronereis*.

Parapodium of *Nereis*	Parapodium of *Heteronereis*
A. Similarities	
1. Parapodia are flat, fleshy, leaf-like or flap-like outgrowths from the lateral side of body (Fig. 3.5).	1. Parapodia are large, fleshy, foliaceous outgrowths (Fig. 3.7).
2. Each parapodium is biramous structure consisting of a dorsal lobe, the **notopodium** and a ventral lobe, the **neuropodium**.	2. Same.
3. Each lobe consists at its base a cirrus.	3. It also contains a dorsal cirrus and a ventral cirrus.
4. Both in notopodium and neuropodium, there is a bundle of long, fine, stiff chaetae which are lodged in a setigerous sac.	4. Same.
5. The action of setae is controlled by special sets of muscles.	5. Same.
6. The centre of each bundle of setae is occupied by a thick chitinous rod.	6. Same.

Parapodium of *Nereis*	Parapodium of *Heteronereis*
B. Differences	
1. The parapodia are small, simple, non-vascular and without any outgrowth.	1. Parapodia are greatly enlarged in the posterior part of body. These become foliacious, highly vascular and develop additional leaf-like fleshy outgrowths.
2. The setae are simple.	2. The setae are flattened and oar-shaped.
3. Setae are arranged in the form of bundles.	3. Setae are arranged in a fan-like manner.
4. Less adapted for swimming.	4. Better adapted for swimming.

3.5 DIGESTIVE SYSTEM

Alimentary canal of *Nereis* is a straight tube having two opening to the exterior, an anterior **mouth** and a posterior **anus**. It is suspended in the body cavity or coelom by the dorsal mesentery and intersegmental septa. It is differentiated into following three regions (Fig. 3.9).

1. Foregut or stomodaeum which comprises buccal cavity and pharynx;

2. Midgut or meseneteron which comprises oesophagus and stomach-intestine;

3. Hindgut or proctodaeum which comprises only rectum.

Both foregut and hindgut are internally lined by **ectoderm** and **cuticle**. Midgut is lined by endoderm.

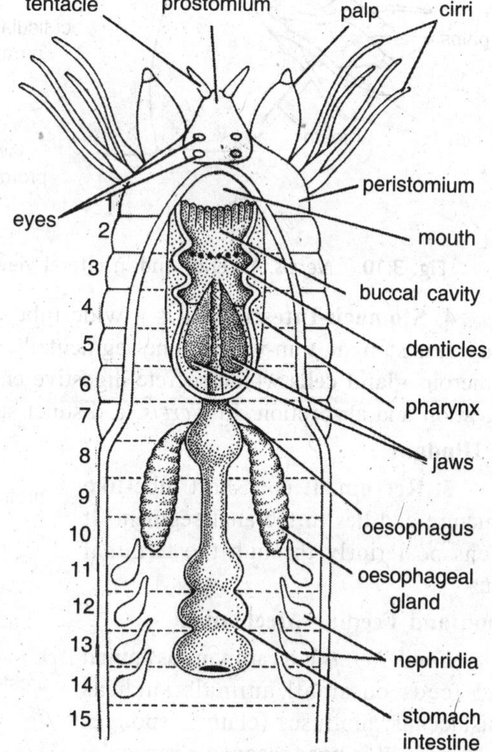

Fig. 3.9. *Nereis.* Anterior part of alimentary canal.

A. Foregut

1. Mouth. It is a transverse slit lying ventral to the prostomium. It is bordered laterally and ventrally by peristomium. The mouth opens posteriorly into buccal cavity.

2. Buccal cavity and pharynx. The buccal cavity is short but wide chamber situated inside the peristomium. The buccal cavity leads into a muscular protrusible **pharynx** which extends upto fourth segment. The buccal cavity and pharynx are bound together by a muscular sheath and internally they are lined with thick cuticle. The cuticle of buccal cavity is thickened at various points into dark brown **denticles** or **paragnaths**. In the posterior part of pharynx are found a pair of large, powerful, dark, movable, chitinous and laterally placed jaws with serrated edge. Each jaw has a rounded hollow base embedded in the muscles and an anterior pointed incurved apex, its inner edge is provided with **teeth**. Both pharynx and buccal cavity collectively form the feeding organs of *Nereis*.

The whole of buccal and pharyngeal region can be turned completely in-side out (Fig. 3.10). This is done by the help of two sets of muscles: **protractor muscles** which can evert out the pharynx and open the jaws as **introvert** or **proboscis;** and **retractor muscles** which can retract proboscis and cause closing of jaws (Fig. 3.10 and Fig. 3.11). Buccal cavity is everted due to pressure of coelomic fluid.

B. Mesenteron

3. Oesophagus. The pharynx opens posteriorly into a narrow tube, called **oesophagus** which runs upto ninth segment. A pair of large, unbranched, laterally placed glandular pouches, the **oesophageal caeca** open into oesophagus at the anterior end. The oesophagus is followed by stomach-intestine.

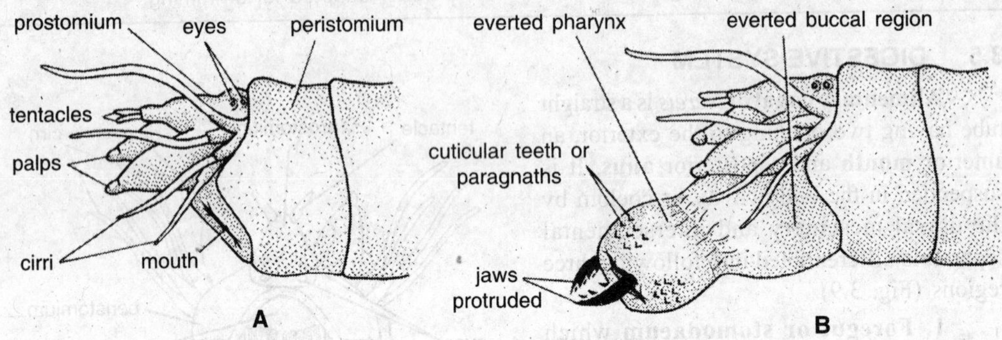

Fig. 3.10. *Nereis*. Anterior end in lateral view. A—Pharynx retracted; B—Pharynx everted.

4. Stomach-Intestine. It is a wide tube extending through the remaining length of the body. It is straight, thin-walled tube segmentally constricted by septa. Its epithelial lining contains numerous **gland cells** which secrete digestive enzymes. Stomach-intestine is the principle site of digestion and absorption. In *Nereis*, a distinct stomach is absent.

C. Hindgut

5. Rectum. It is a short tube-like structure and lies in the anal segment. It opens posteriorly through the terminal anus.

Food and Feeding Mechanism

Food. *Nereis* is a carnivorous animal and feeds on small animals such as crustaceans, molluscs (clams), sponges and larvae. It is **predaceous** or **raptorial** since its mode of feeding involves **seizing** and **devouring** living prey.

Feeding mechanism. Food is tested by the tactile **prostomial palps** and the **peristomial cirri**. The pharynx is rapidly everted, placing the jaws at the anterior end of the body and causing them to open. The food is seized by the jaws and the pharynx is retracted. The prey is, thus, dragged inside the buccal cavity during the inversion or retraction of buccopharynx. This is called **raptorial feeding** and it is performed when *Nereis* feeds on surface outside the burrow.

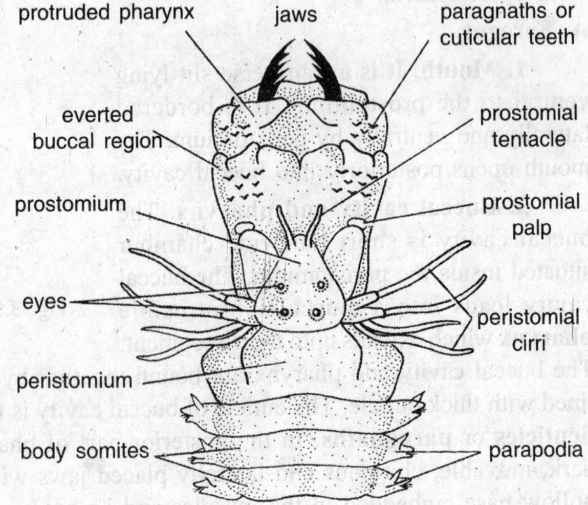

Fig. 3.11. *Nereis*. Head with everted buccal region (in dorsal view).

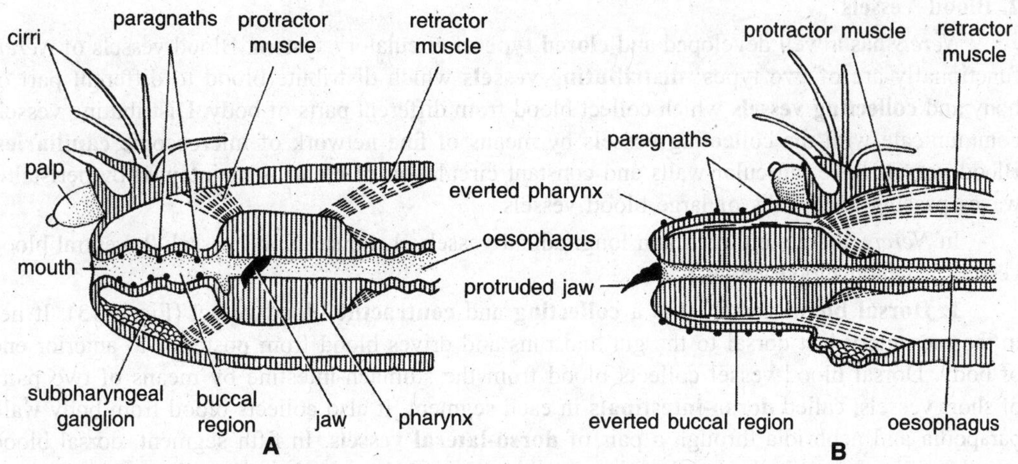

Fig. 3.12. *Nereis*. Diagrammatic representation of working of bucco-pharyngeal region. A—In inverted condition; B—In everted condition.

Nereis also feeds by many other methods (see Box 3.1). For example, in *Nereis*, another common method of feeding is **filter-feeding**. When *Nereis* lives in a U-shaped burrow, it feeds by ejecting out a cone of mucus in front of the mouth and creates a current of water by beating the parapodia. The currents of water enters the mouth end of the burrow and leave from other end. The food particles (detritus) coming along with the current of water are held in mucous cone which are later ingested by *Nereis*. This mode of feeding is called **filter-feeding**.

Box 3.1 Diversity of Feeding Habits of *Nereis*

Studies on *Nereis* by **H. Goerke** (1971) demonstrated the diversity of feeding habits that may exist even within a genus. Species of *Nereis* possess a muscular, eversible pharynx with a pair of heavy **jaws** (which are made of cross-linked tanned protein). Some species of *Nereis*, such as *Nereis pelagica*, *Nereis virens* and *Nereis diversicolor* are **omnivorous** and feed on algae, other invertebrates and even detritus. *Nereis succinea* and *Nereis longissima* feed primarily on detritus material in the substratum. *Nereis fucata*, which lives as a commensal in hermit crab shells, is **carnivorous**. *Nereis brandti*, which occurs on the Coast of USA and reaches 1.8 m in length, feeds primarily on green algae.

Digestion, absorption and egestion. The ingested food initially undergoes **mastication** in the bucco-pharyngeal region by the help of numerous denticles. The masticated food is pushed onwards inside the gut by peristaltic contraction of muscles present in the alimentary canal. Digestion is mainly **extracellular** and food is digested by enzymes secreted by the digestive glands and gland cells of the epithelial lining of stomach-intestine. Absorption of digested food occurs in stomach-intestine by the diffusion. The undigested residue is passed to rectum; from there it is ejected out through the **anus**.

3.6 BLOOD VASCULAR SYSTEM AND RESPIRATION

1. Blood

The blood of *Nereis* is bright red in colour. It consists of a fluid **plasma** and numerous amoeboid, nucleated, colourless and leucocyte-like **corpuscles**. The bright red colour of the blood is due to the presence of a respiratory pigment **erythrocruorin** which is dissolved in plasma. Blood flows through a system of blood vessels and helps in transportation of gases, food and excretory products.

2. Blood Vessels

Nereis has a well developed and **closed type** of circulatory system. Blood vessels of *Nereis* functionally are of two types: **distributing vessels** which distribute blood to different part of body and **collecting vessels** which collect blood from different parts of body. Distributing vessels communicate with the collecting vessels by means of fine network of microscopic **capillaries**. Blood vessels have muscular walls and constant circulation of blood is maintained by peristaltic waves passing over walls of large blood vessels.

In *Nereis*, there are three main longitudinal vessels: 1. dorsal blood vessel; 2. ventral blood vessel; 3. neural blood vessel.

1. Dorsal blood vessel. It is a **collecting** and **contractile** blood vessel (Fig. 3.13). It lies in the mesentery just dorsal to the gut and runs and drives blood from posterior to anterior end of body. Dorsal blood vessel collects blood from the stomach-intestine by means of two pairs of short vessels, called **dorso-intestinals** in each segment. It also collects blood from body wall, parapodia and nephridia through a pair of **dorso-lateral vessels**. In fifth segment, dorsal blood vessel bifurcates and both branches form a plexus to supply oesophageal wall.

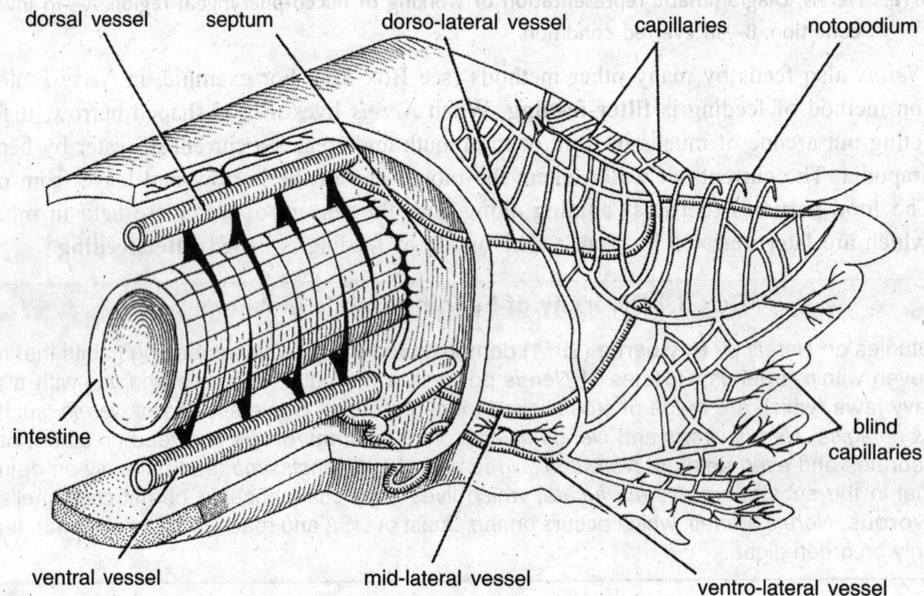

Fig. 3.13. *Nereis*. Blood vascular system.

2. Ventral blood vessel. It is mainly a **distributing** and **non-contractile** vessel. It lies mid-ventrally below the alimentary canal from fifth segment to the last. It carries blood from anterior end to the posterior end. In the fifth segment, it collects blood from the oesophageal wall by single median vessel. In the last segment, it communicates with the dorsal vessel by a **circumrectal ring**.

Blood vessels of a typical segment. In each segment except the first five segments, the dorsal and ventral blood vessels are connected on either side by a pair of loop-like **transverse vessels** or **lateral commissural vessels**. These vessels take the blood from the ventral to dorsal blood vessel but not directly. In each segment, these vessels divide into **afferent vessels** which supply blood to the body wall, nephridia and parapodia of their side. In these organs, the **afferent branches** break up into an extensive network of **capillaries**, then reunite to form the corresponding **efferent branches**. These unite to open into the dorsal vessel. In each segment, the ventral vessel also supplies blood to the stomach and intestinal wall by a pair of **afferent intestinal vessels, ventro-intestinal vessels.**

These form an extensive network of capillaries in the stomach-intestinal wall (Fig. 3.14). The capillaries reunite to form a pair of **efferent intestinal vessels,** a **dorso-intestinal vessels** and carry blood to dorsal blood vessel.

3. Neural blood vessel. It is a thin blood vessel which runs below the ventral nerve cord. It collects blood from the ventral body wall and returns it to the ventral blood vessel.

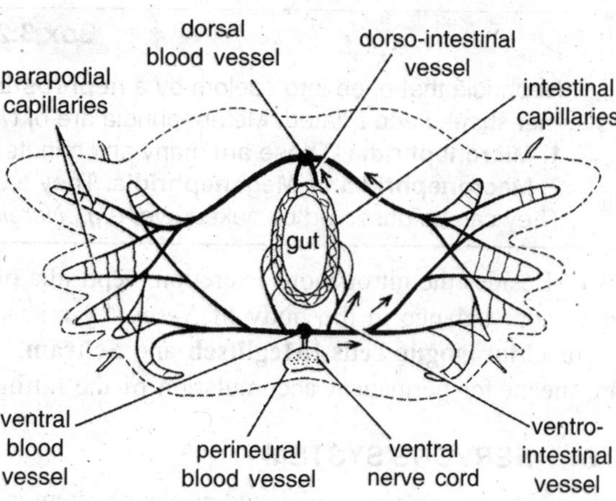

Fig. 3.14. *Nereis*. Diagrammatic representation of main blood vessels in a segment.

3. Respiration

Nereis lacks any special organs or gills for respiration. The parapodia with their blood supply and body wall (*i.e.*, general body surface) with its plexus of blood vessels serve as respiratory surfaces. Oxygen dissolved in surrounding water readily diffuses into the blood through the thin integument and parapodia. Similarly, from the blood carbon dioxide diffuses into the surrounding water. Blood plasma of *Nereis* contains the respiratory pigment, **erythrocruorin**, for transportation of respiratory gases.

3.7 EXCRETORY SYSTEM

In *Nereis*, the excretion is carried out by means of special type of coiled tube-like structures, the **nephridia**, one pair of which is found in each segment except the first and the last segment.

Structure

The nephridia of *Nereis* are macroscopic structures; they can be observed by naked eyes. Each nephridium consists of a coiled ciliated tube lying embedded in the syncytial mass of protoplasm. It is differentiated into two parts; a body and a neck. The main **body** of nephridium is an irregular, oval, compact and glandular mass which is placed transversely in the segment. It contains highly convoluted and mostly ciliated tubule. The nephridial tubule passes through the septum into the anterior segment in the form of **neck**, where it opens by a ciliated funnel or **nephrostome**. Posteriorly the tubule opens by a contractile (*i.e.*, sphinctered) **nephridiopore** located ventrally at the base of parapodium near the origin of ventral cirrus.

The nephridial tubule is an excavation in the mass of protoplasm. Such an open type of nephridium with a ciliated nephrostome is called **metanephridium**.

Function of nephridia. The outer surface of nephridium is richly supplied with blood capillaries. The gland cells present in the wall of ciliated tube remove waste materials, chiefly **ammonia** from blood. This liquid waste is expelled out of body through nephridiopores. The ciliated funnels (nephrostomes) opening into the coelom also removes dead coelomic corpuscles (coelomocytes or amoeboid corpuscles) which have eaten up bacteria, etc., and other foreign bodies.

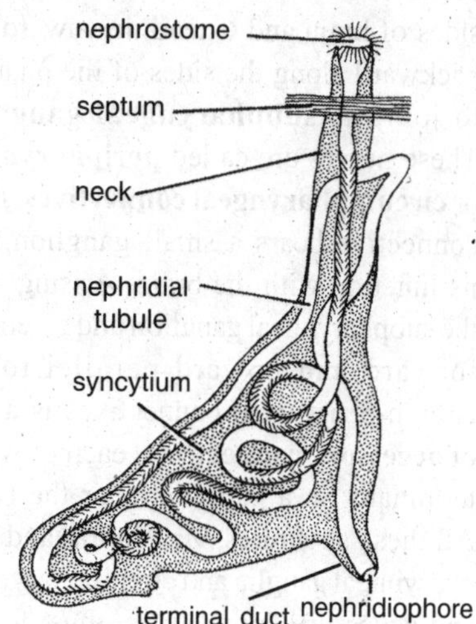

Fig. 3.15. *Nereis*. Nephridium.

Box 3.2

Nephridia that open into coelom by a **nephrostome** are called **metanephridia**. Nephrostome is funnel-shaped and ciliated. Metanephridia are of two types:

1. **Micronephridia.** These are many and minute or microscopic in size, *e.g.*, *Pheretima*.
2. **Macronephridia or Meganephridia.** They are large in size and few in number. They can be observed by naked eye, *e.g.*, *Nereis* and leech.

Besides the nitrogenous excretion, nephridia of *Nereis* do **osmoregulation;** it maintains salt and water balance in the body of *Nereis* by reabsorption of water and salts. Lastly, in *Nereis* occur **chlorogogue cells** (**Meglitsch** and **Schram**, 1991) in the visceral peritoneum; these cells are meant for permanent accumulation of the nitrogenous wastes.

3.8 NERVOUS SYSTEM

Nereis has a well developed nervous system in correlation to its active life. It is bilateral and metameric in its arrangement. It includes a central nervous system, peripheral nervous system and visceral nervous system (Fig. 3.16).

1. Central nervous sys-tem. There is a large bilobed mass of nervous tissue in the prosto-mium beneath the dorsal epider-mis. This is known as the **brain.** It is formed by the fusion of a pair of ganglia called the **suprapharyngeal ganglia** or **cerebral ganglia** (Box 3.3). A similar but smaller bilobed mass is situated in the third segment beneath the pharynx. It is called **sub-pharyngeal ganglion.** In fact, this gang-lion is formed due to fusion of two pairs of ganglia. A pair of slender cords of nervous tissue start from the sides of brain and extend downward and backward along the sides of the pharynx to join the **subpharyngeal ganglion.** These nerves are called **peripharyngeal** or **circumpharyngeal connectives.** Each connective bears a small ganglion near its junction with the brain. Arising from the subpharyngeal ganglion and extending forward and upward parallel to the circum-pharyngeal connectives is a pair of **accessory connectives**, each of which terminates in a ganglion near the brain. All these structures, *viz.*, supra- and sub-pharyngeal ganglia and circumpharyngeal and accessory connectives, thus, form a complete collar, called the **nerve ring,** round the pharynx.

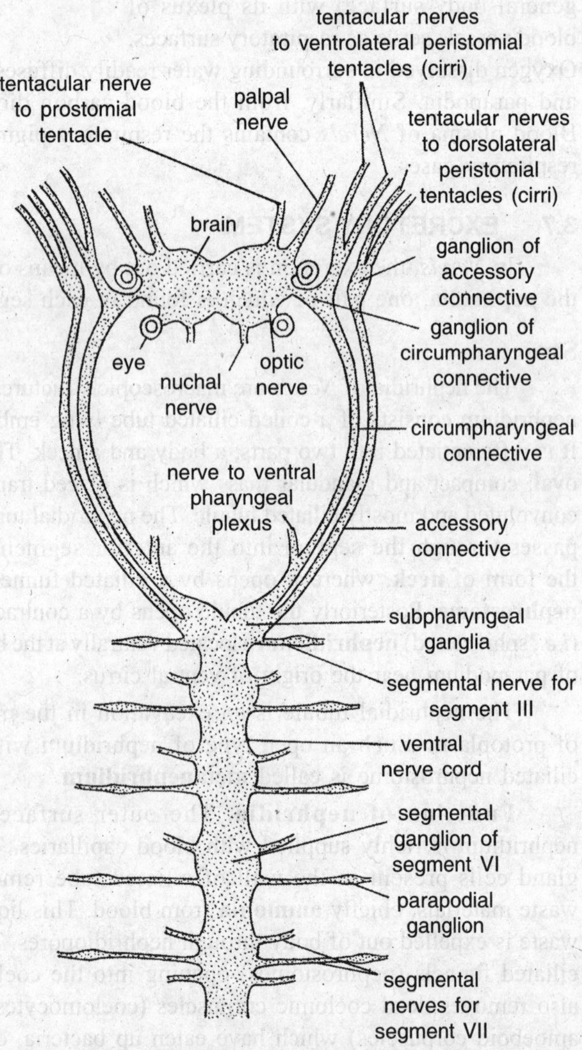

Fig. 3.16. *Nereis*. Nervous system.

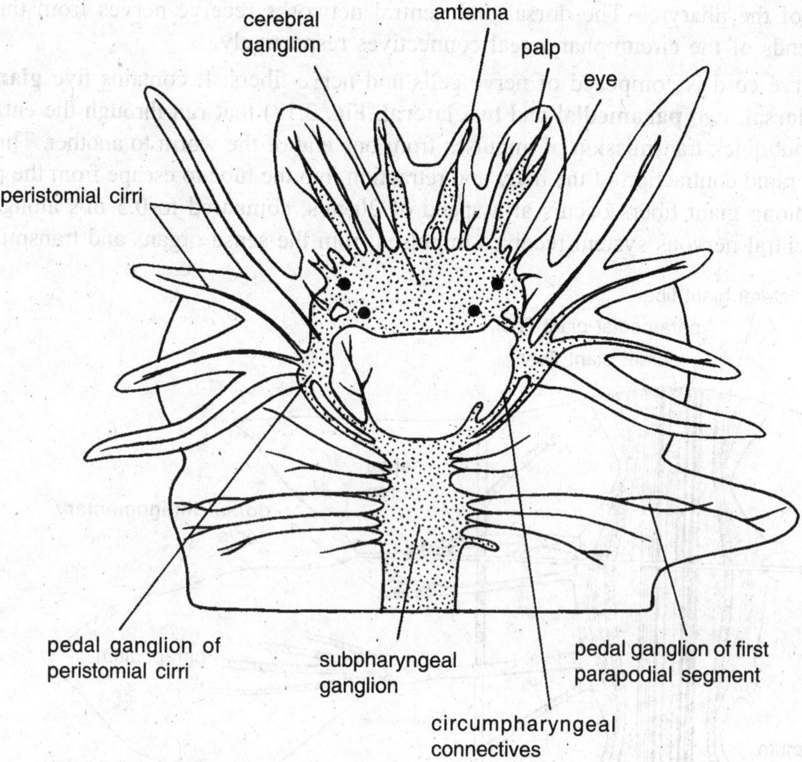

Fig. 3.17. *Nereis*. Central nervous system – nerve ring (after Brusca and Brusca, 2003).

Box 3.3

The brain of *Nereis* is often specialized into three regions, typically called the **forebrain**, **midbrain** and **hindbrain**. The forebrain innervates the prostomial palps, the midbrain the eyes and prostomial palps and the hindbrain the chemosensory nuchal organs. The **circumcentric connectives** (*i.e.*, circumpharyngeal plus accessory connectives) arise from forebrains and midbrains (Fig. 3.17).

A prominent cord of nervous tissue, called the **ventral nerve cord**, extends from the subpharyngeal ganglion to the end of body along the mid-ventral line of the body wall beneath the ventral blood vessel. This cord appears single but actually it consists of two solid cords fused together and surrounded by a common sheath of connective tissue. Ventral cord bears in each segment a swelling, the **segmental ganglion**, formed by the fusion of a pair of ganglia. An additional pairof **lateral inter segmental nerves** run parallel to nerve cord to connect the pedal ganglia (Fig. 3.18).

2. Peripheral nervous system. The peripheral nervous system mainly includes the nerves coming directly from the brain and ventral nerve cord to supply the different regions of the body. The brain gives out four short **optic nerves** to the eyes; two **tentacular nerves** to the prostomial tentacles; two **palpal nerves** to the palps. From the small ganglion of circumpharyngeal connectives two pairs of nerves go to the ventral peristomial cirri on each side. From the subpharyngeal ganglion a pair of nerves goes to the dorso-lateral peristomial tentacles. The nerves that arise from the segmental ganglia innervates the body wall musculature, and parapodia (via the **pedal ganglia**) and the digestive tract.

3. Visceral nervous system. Visceral nervous system innervates the musculature of the pharynx. It consists of a network of tiny nerves with occasional small ganglia on the dorsal and ventral side of the pharynx. The dorsal and ventral networks receive nerves from the ganglia and ventral ends of the circumpharyngeal connectives respectively.

The nerve cord is composed of nerve cells and nerve fibers. It contains five **giant fibers/ axons** (one **dorsal**, two **paramedial** and two **lateral**, Fig. 3.18) that run through the entire length and bring about quick transmission of impulses from one end of the worm to another. This enables the worm for rapid contraction of the body and retraction into the tube to escape from the predators. Conduction along giant fibers occurs at a speed of 20 m/s, compared to 0.5 m/s along ordinary fibers. The central nervous system receives impulses from the sense organs and transmit them to

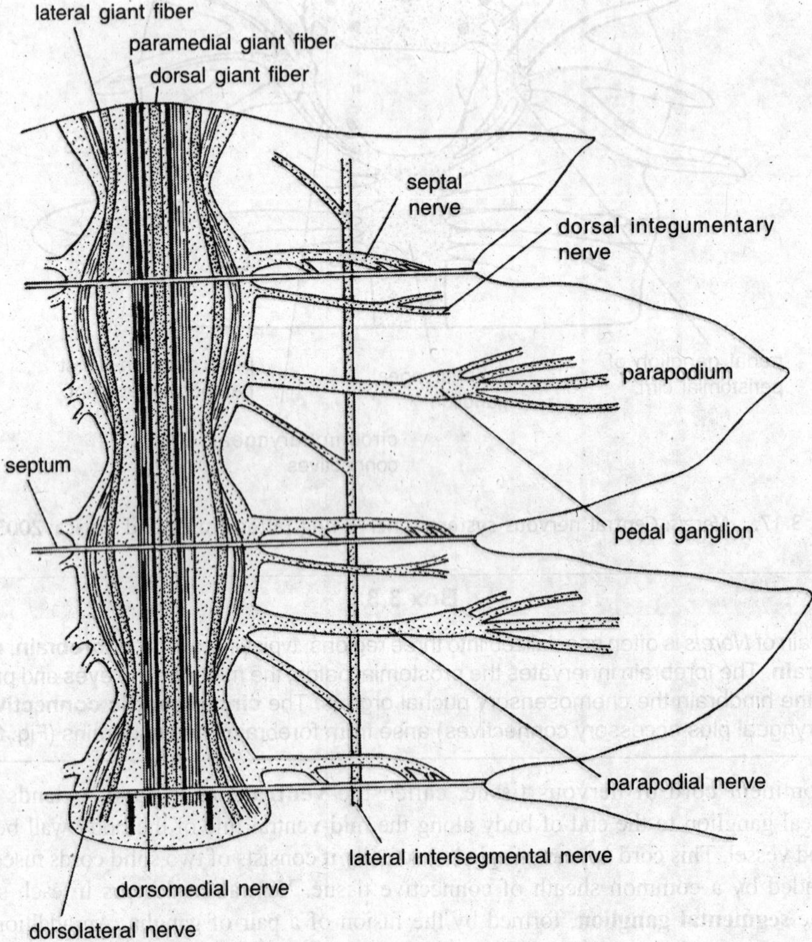

Fig. 3.18. *Nereis*. Peripheral nervous system (after Brusca and Brusca, 2003).

effectors, such as muscles of parapodia and body wall, via peripheral nervous system. The segmental ganglia act as reflex centers so that a clamworm devoid of head (*i.e.*, brain) continues normal movement.

3.9 SENSE ORGANS

In *Nereis* occur the following three types of specialized sense organs or receptors.

1. Tactile sense organs. The prostomial tentacles, prostomial palpi and peristomial cirri of head are the main **tactile** sense organs as they are sensitive to touch.

2. Chemoreceptors. Nuchal organs of *Nereis* act as chemoreceptors as well as olfactory sense organs as they are sensitive to **taste** and **smell** respectively. They are a pair of small pits in the posterior part of the prostomium on its dorsal side. They are lined by special ciliated epithelium with gland cells. They are lined by special ciliated epithelium with gland cells. They are connected with the posterior part of brain. They enable the clamworms to detect its prey present outside the burrow from the water current that enters the burrow with dissolved substances emanating from the prey.

3. Photoreceptors. There are four simple **eyes** (black spots) on the dorsal surface of the prostomium. Each eye has a cup made up of **pigmented retinal cells** produced inwards into clear rods, the opening of the cup is a **pupil** (Fig. 3.19). In fact, each retinal cell has three distinct parts: an outer nucleated part with nerve fibers, a highly pigmented main body and inner part of transparent cuticular rod. The external cuticle forms a transparent cornea. Inside the cup is a transparent, gelatinous, refractive **lens**. The retinal cells are joined to nerve fibers of the **optic nerve**. The eyes are photoreceptive; they are not related to the formation of image but they help in detecting changes in light intensity.

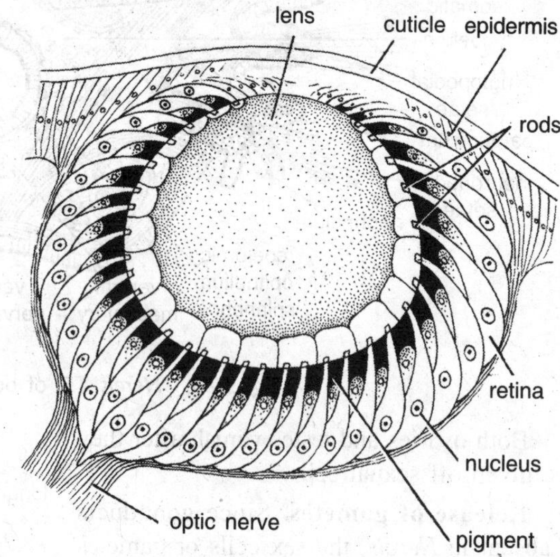

Fig. 3.19. *Nereis*. V.S. of eye.

3.10 REPRODUCTIVE SYSTEM AND DEVELOPMENT

Nereis is **dioecious**, *i.e.*, sexes are separate and **sexual dimorphism** occurs in only those species in which distinct non-sexual (*i.e.*, *Nereis*) and sexual (*i.e.*, *Heteronereis*) phases are found. The gonads are temporary or seasonal and develop only during the breeding season, *i.e.*, in summer season. They lack the **gonoducts**.

Gonads. These are formed by the proliferation of the cells of coelomic epithelium. They can be seen as groups of masses of germ cells on the ventral side in the coelomic space of each segment of body except few of anterior segments.

In male *Nereis dumerilli*, however, there is only one pair of **testes** lying in anyone segment between the 19th and 25th segments. But in other species such as *Nereis virens* and *Nereis diversicolor* the testes extend in many segments. During breeding season the germ cells which detach from the coelomic epithelium in male form the **sperm mother cells**. These cells float in the coelomic fluid, divide rapidly and undergo maturation division (meiosis) to form **spermatids** which transform into flagellated **sperms.** A mature sperm has a rod-shaped **head** and a long vibratile **flagellum** or **tail**.

In the female, the ovaries lie in many segments around blood vessels. The **oogonia** separate from coelomic epithelium, float in coelomic fluid and undergo **oogenesis** to form **ova** which are found floating in the coelomic fluid (Fig. 3.20). The ova are round in shape and filled with **yolk** granules. They are bounded by a **vitelline envelope**.

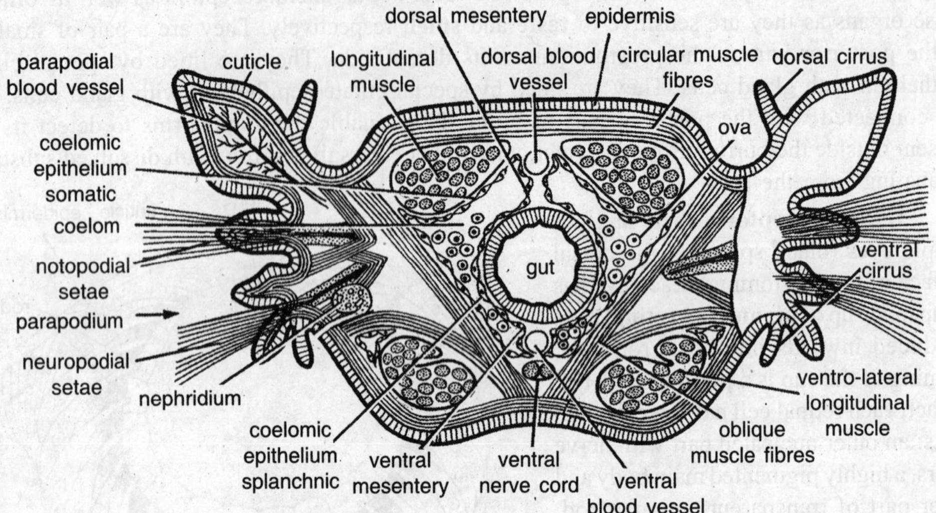

Fig. 3.20. *Nereis*. T.S. of body showing ova.

Both ovaries and testes vanish after they have given off sexual cells.

Release of gametes. Since gonoducts are absent in *Nereis*, the sex cells or gametes escape from body either through nephridia or through temporary rupture of the body wall. A pair of **dorsal ciliated organs** are found in each segment. These are in the form of short and much folded funnels of ciliated tracts opening into the body cavity by a wide opening and being closed to the exterior (Fig. 3.21). Ciliated organs resemble the coelomoducts and are said to act as **gonoducts** developing temporary openings during reproductive season.

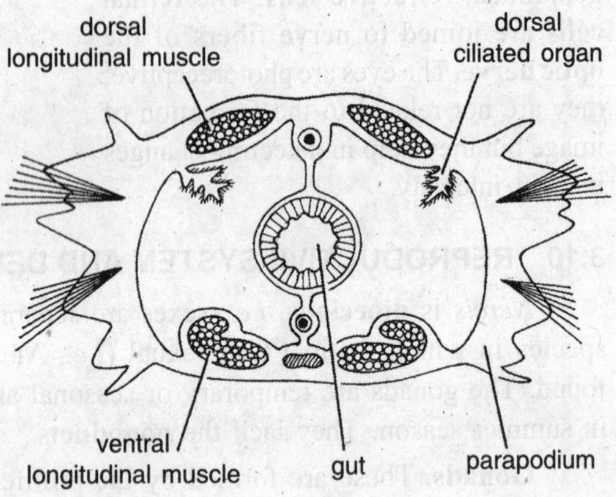

Fig. 3.21. *Nereis*. T.S. of a segment to show the position of ciliated organ.

Fertilization. The fully sexually mature clamworms leave the burrows in groups and come up at the surface of sea water to swim freely, it is called **swarming**. During swarming male and female *Nereis* (*i.e. Heteronereis*) simultaneously release gametes which fuse together during fertilization. Thus, in *Nereis* fertilization is **external** occurring in open seawater.

Development. Development of *Nereis* is **indirect** involving the formation of a trochophore larva which undergoes metamorphosis to become adult-like.

Structure of egg. The newly discharged planktonic egg of *Nereis* is spherical in shape and possesses two membranes: an outer thin **outer membrane** and a thick radially situated membrane, called **zona radiata** which encloses **ooplasm**. Ooplasm is rich in yolk having oil droplets and intact female nucleus (called **germinal vesicle**). A thick, gelatinous and transparent **vitelline envelope** surround entire ovum. However, immediately after fertilization several changes occur in egg: yolk granules aggregate at the lower end of egg to form the **vegetal pole**. The upper clear zone forms the **animal pole**. Such an egg is called **telolecithal egg**. The nucleus of ovum completes its meiosis and release **polar bodies** at the animal pole. The haploid nuclei of male and female gametes unite (during the process of **amphimixis** in fertilization) to form diploid nucleus of the **zygote.**

Cleavage and gastrulation. In *Nereis*, cleavage is **holoblastic**, **spiral** and **determinate** type. Blastula lacks a blastocoel so is called **stereoblastula.** Gastrulation involves **epiboly** and **invagination.**

The first two divisions are almost equal and vertical. The four cells produced lie in the same plane and are called A, B, C and D. Each cell gives rise to one of the quarter of the embryo. Cell D is larger than the others and forms the **dorsal** surface of the embryo; cell B forms the **ventral** surface and cell A and cell C form **lateral** side the embryo. The first quartet of cells undergoes three successive divisions (3rd, 4th and 5th). These divisions are **unequal** and at right angles to the first two. As a result three successive quartets of cells are separated; these contain comparatively smaller cells. The first four cells are called **macromeres** and the cells of three successive quartets are called **micromeres.** Since the cleavage divisions are spiral, the micromeres do not lie exactly above its macromeres, but somewhat displaced to the right in one quartet and to the left in the next quartet.

The micromeres of the first three quartets give rise to **ectoderm** of larva. The sixth cleavage separates the micromeres of fourth quartet. One of the cell (4d) of the fourth-quartet is known as **somatoblast** and gives rise to entire **mesoderm**. The macromeres form **endoderm**. The embryo at this stage is called **stereoblastula**, it is a solid blastula and lacks a blastocoel.

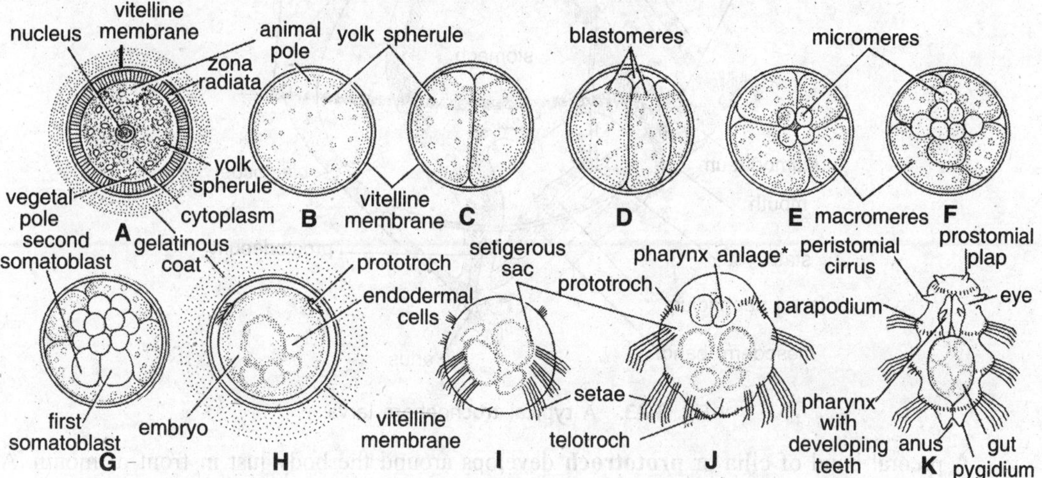

Fig. 3.22. *Nereis.* Stages in development. A—Unfertilized egg; B—Fertilized egg; C, D, E—2, 4, 8—celled stages; F—Formation of first somatoblast; G—Formation of second somatoblast; H—Young trochophore before hatching; I, J—Post-trochophore larva; K—Three segmented larva with parapodia and setae;* anlage = any group of embryonic cells when first identifiable as a future organ or body part; also called blastema.

Gastrulation. Gastrulation of *Nereis* is of **epibolic type** and is accomplished by the spreading (epibolic growth) of micromeres over the endoderm cells (macromeres) and somatoblasts. As the somatoblasts invaginate, the **blastopore** is formed as an uncovered area at the vegetal plate. It represents the ventral surface of the larva.

Development of trochophore larva. The gastrula develops into a **trochophore** or **trocho-phore larva.** At the top of presumptive head end, micromeres differentiate into **apical rosette** or **apical plate**, which differentiates into **cerebral ganglia** and a tuft of cilia. Another tuft of cilia develops slightly behind the apical plate; it is known as **prototroch**. **Stomodaeum** develops as ectodermal invagination and a similar invagination at the position of blastopore form the **proctodaeum**.

The cells derived from successive divisions of somatoblasts become arranged on the presumptive ventral side of the embryo and form the **ventral plate**. The ectodermal cells present along the ventral side are thickened into **ventral nerve cord**. The deeply located derivatives of somatoblast (*i.e.*, 4d) give rise to **mesodermal bands** which develop into muscles of body wall and of alimentary canal.

Structure of Trochophore Larva

The trochophore is a minute, transparent, unsegmented and pear-shaped larva. It has broad anterior **head** and narrow posterior **anal end**. It is bounded externally by a thin ectoderm. The head end is thickened into an apical plate. It bears a tuft of **apical cilia** (Fig. 3.23). A pair of **cerebral ganglia** and pigmented larval eyes are present on the **apical plate**. Larva contains a curved **gut** with a mouth, ectodermal **oesophagus**, an endodermal **stomach**, and an ectodermal **hind gut** opening by **anus**. It feeds on microorganisms.

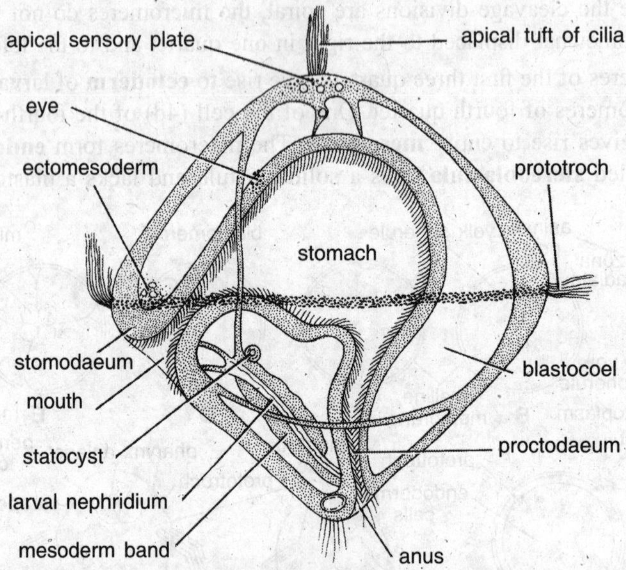

Fig. 3.23. A typical trochophore larva.

A preoral band of cilia or **prototroch** develops around the body just in front of mouth. A similar band of cilia occurs at the posterior end. This is post-oral band or **metatroch**. The larva lacks coelom but possesses a spacious **blastocoel** between the alimentary canal and ectoderm. A pair of protonephridia and larval muscles are found in the blastocoel. The mesoderm is represented by a pair of undifferentiated mass of cells. There is a statocyst near the nephridia.

A typical free-swimming trochophore larval stage occurs only in few polychaetes. In *Nereis*, larva is developed inside the egg membrane. The larva hatches out in the sea water by rupturing egg's vitelline envelope. It is called **post-trochophore** or **nectochaete**. It possesses three segments with bristles. Its larval structures are poorly developed.

Biological importance of trochophore larva. Annelid's trochophore larva is of great phylogenetic significance. It exhibits similarity with adult ctenophores and certain rotifers. It shows close affinity with the larva found in development of animals of some diverse groups such as polyclad Turbellaria (*i.e.*, Mueller's larva); Nemertinea (*i.e.*, pilibium larva), Bryozoa and Mollusca (*i.e.*, veliger's larva).

Visualizing these affinities, **Hatscheck** (1878) and others formulated the **trochophore theory**. According to which the present bilateral phyla are considered to have descended from some radially symmetrical group and trochophore represents a transitional stage in the line of emergence of group Bilateria (Rotifera) from radial group Ctenophroa.

Post-larval Development (Metamorphosis)

The trochophore larva of *Nereis* is pelagic, it drifts about in the sea swimming by its protoroch and is metamorphosed. During metamorphosis ciliated bands disappear and its pre-oral part form the prostomium having brain, tentacles and eyes. The post-oral part of larva elongates and forms segments of body by metameric segmentation (Fig. 3.24). The first segment forms the **peristomium** and last **pygidium**. The ciliary bands disappear and larva grows in length with the addition of new segments just infront of the pygidium.

Larval nephridia are replaced by permanent ones, larval setae are dropped, tentacles, palps and parapodia are formed. Young worm thus resulted settles at the bottom of the sea and starts forming its burrow and gradually attains adulthood.

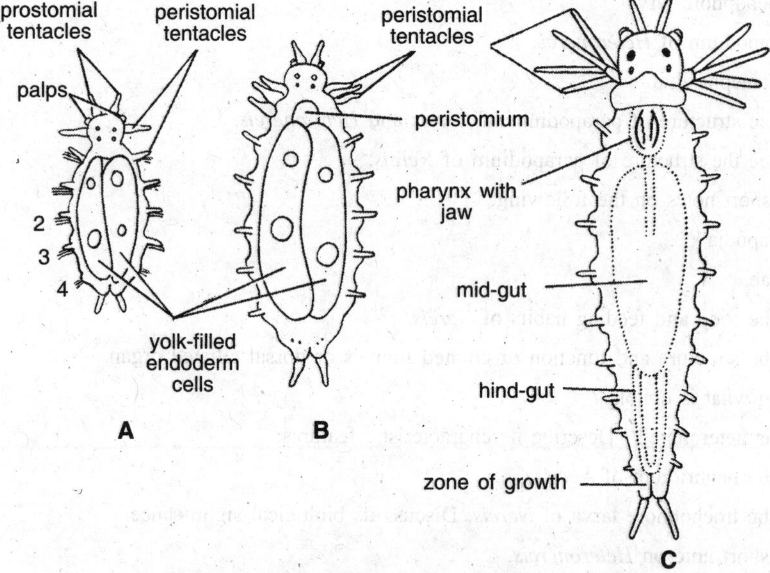

Fig. 3.24. *Nereis*. Metamorphosis and growth.

QUESTIONS

Long Answer Questions

1. Give an account of habit, habitat and external features of *Nereis*. Compare *Nereis* with *Heteronereis*.
2. What is heteronereis phase? How do the individuals of this phase differ from *Nereis*? Compare the structure of the parapodium with that of *Heteronereis*? Give labelled diagrams.
3. Describe the alimentary canal of *Nereis*. Add a note on the mechanism of feeding. What is the food of this animal?
4. Describe alimentary canal of *Nereis*. Discuss its adaptive features.
5. Describe the blood and blood vascular system of *Nereis*. How does respiration takes place in this animal?
6. Give an account of the organs of excretion in *Nereis*.
7. Describe structure of nephridium of *Nereis*. What do you understand by enteronephric nephridial system?
8. Describe the reproductive system of *Nereis* and state clearly how reproduction takes place in this animal.
9. With suitable diagrams illustrate and describe the development of *Nereis*.

10. Describe the trochophore larva of *Nereis*. Explain the trochophore formation in *Nereis*.

11. How are gametes discharged in *Nereis*? Describe the development of this animal.

Short Answer Questions

1. Write short notes on the following:

(*i*) Dorsal ciliated organ

(*ii*) Feeding in *Nereis*

(*iii*) Head region of *Nereis*

(*iv*) *Heteronereis*

(*v*) Parapodium

(*vi*) Trochophore larva

(*vii*) Parapodium of *Heteronereis*

(*viii*) Metanephridium

2. Describe structure of parapodium of *Nereis* and *Heteronereis.*

3. Describe the structure of parapodium of *Nereis*.

4. Write short notes on the following:

(*i*) Parapodia

(*ii*) Setae

5. Describe food and feeding habits of *Nereis*.

6. Describe structure and function of ciliated funnels or dorsal ciliated organ.

7. Explain what is epitoky?

8. What is heteronereis? Describe its characteristic features.

9. Describe nephridium of *Nereis*.

10. Describe trochophore larva of *Nereis*. Discuss its biological significance.

11. Write short note on *Heteronereis*.

12. What is the significance of *Heteronereis*? Compare the external features of *Heteronereis* and *Nereis*.

13. Nuchal organs of *Nereis*.

14. Describe feeding in *Nereis*.

15. Describe polymorphism in *Nereis*.

16. Distinguish between *Nereis* and *Heteronereis*.

17. Give difference between *Nereis* and *Heteronereis* phases.

18. Compare the parapodia of *Nereis* and *Heteronereis*.

19. Draw well labelled diagrams of the following:

(*i*) Head of *Nereis* with everted buccal region

(*ii*) Parapodium of *Nereis*

(*iii*) Parapodium of *Heteronereis*

(*iv*) T.S. *Nereis* through parapodia

(*v*) Nephridium of *Nereis*

(*vi*) *Heteronereis*

(*vii*) Alimentary canal of *Nereis*

(*viii*) Trochophore larva

Very Short Answer Questions

1. Give common name of *Nereis*?

2. Why is *Nereis* called clam worm?

3. What are paragnaths?

4. What are macronephridia?

5. Where are the eyes of *Nereis* located?

6. What process propells the food through gut of *Nereis*?

7. Name two modes of feeding found in *Nereis*.

8. Name the sense organs of *Nereis*.

9. Mention the principal digestive glands of clam worm.

10. What type of circulatory system occur in *Nereis*?

11. Name the process involved in urine formation in polychaetes.

12. What is *Heteronereis*?

13. What is epitoky?

14. What is swarming in *Nereis*? Describe in brief.

15. How many ciliary bands are found in trochophore larva? (*C.C.S.U., Meerut 2002*)

Fill in the Blanks

1. *Nereis* is commonly known as .... or ....

2. A fossorial animal such as *Nereis* lives in a .... burrow.

3. Head of *Nereis* consists of .... and peristomium.

4. A parapodium consists of two parts: dorsal notopodium and ventral ....

5. Pygidium of *Nereis* bears anal .... but lacks ....

6. Body cavity of *Nereis* is a true ....

7. Proboscis of *Nereis* consists of .... and ....

8. Digestion is .... in clam worm.

9. Wall of blood vessels in *Nereis* are made of ....

10. Larva of *Nereis* is called ....

Multiple Choice Questions

1. *Nereis* is commonly called

(*a*) ragworm (*b*) clamworm

(*c*) sandworm (*d*) all the above

2. *Nereis* is

(*a*) herbivorous (*b*) carnivorous

(*c*) omnivorous (*d*) parasite

3. Peristomium in *Nereis* is composed of

(*a*) one segment (*b*) two segments

(*c*) three segments (*d*) six segments

4. *Nereis* is

(*a*) raptorial (*b*) filter feeder

(*c*) herbivorous (*d*) both (*a*) and (*b*)

5. For excretion, clamworms have

(*a*) protonephridia (*b*) metanephridia

(*c*) malpighian tubules (*d*) podia

6. Blood of *Nereis* contains haemoglobin (= erythrocruorin) in

(*a*) plasma (*b*) amoeboid corpuscles

(*c*) red corpuscles (*d*) plasma as well as corpuscles

7. For Locomotion, *Nereis* uses

(*a*) parapodia (*b*) setae

(*c*) body wall musculature (*d*) all of these

8. Respiratory pigment of *Nereis* is

(*a*) haemoglobin (*b*) cyanin

(*c*) haemocyanin (*d*) erythrocruorin

9. Giant nerve fibers occur in the

(*a*) brain (*b*) nerve cord

(*c*) segmental nerve (*d*) ganglia

10. The chemoreceptor organs in *Nereis* is

(*a*) prostomial palp (*b*) prostomial tentacle

(*c*) nuchal organ (d) peristomial cirri

11. The name of the larva of *Nereis* is

(*a*) trochophore (*b*) bipinnaria

(*c*) zoea (*d*) cercaria

12. The clamworm having epitoke and atoke regions is called

(*a*) heteronereis (*b*) sexually mature *Nereis*

(*c*) heteronephrone (*d*) none

ANSWERS

Very Short Answer Questions

1. Clam worm.

2. *Nereis* feeds mainly on clams and is found near them. For this reason *Nereis* is called **clam worm**.

3. Paragnaths are dark brown denticles of cuticle of buccal cavity of *Nereis*. They help in mastication of food.

4. *Nereis* have large sized and few nephridia which are called macronephridia.

5. On prostomium.

6. Peristalsis.

7. Raptorial and filter-feeding.

8. Tentacles, palps, nuchal organs and eyes (*i.e.*, Tactile sense organs, chemoreceptors and photoreceptors).

9. Oesophageal gland/caeca.

10. Closed type.

11. Ultrafilteration, modification of primary urine to final urine by reabsorption and secretion.

12. It is the sexually mature phase of some species of *Nereis* (*e.g., Nereis dumerilii*). *Heteronereis* (or *Heteroneanthes*) also represent free-swimming, swarming, sexual phase of *Nereis*.

13. Sexually mature phase of *Nereis* is called *Heteronereis*. It has gonads (testes or ovaries) which develop in posterior part of the body. The parapodia of posterior part of body of *Heteronereis* are also modified, for swimming mode of life. *Heteronereis's* gonad containing posterior part of body bearing modified parapodia is called **epitoke**. Its non-sexual anterior part of body which is lacking in the gonads is called **atoke**. This phenomenon of transformation of non-sexual individuals into sexual ones is called **epitoky**.

14. Swarming means to come together, move or occur in great numbers. For breeding purpose, *Heteronereis* leave their burrows and come to the surface of the sea to swim about freely and to shed their gametes (*i.e.,* spawning). Such swarming behaviour often occur at night and during a particular phase of moon. Both sexes swarm together, sperms and eggs are released in huge numbers and this ensures fertilization.

Some clamworms (*e.g., Nereis limbata*) swarm on moonless nights during summer months soon after sunset. The female secretes a substance called **fertilium** to attract males and stimulate them to shed their sperms. This in turn exites the females and stimulates them to shed their eggs for fertilization. Swarming often continues for about an hour or so. The females die after liberating the eggs and the males perhaps also do not live long.

15. There are two ciliary bands found in trochophore larva of *Nereis*. These include a preoral ciliated band called **prototroch** and a post-oral ciliated band called **metatroch**. The former is just behind the equator and latter lies behind the mouth.

Fill in the Blanks

1. Clam worm, rag worm **2.** Self-made **3.** Prostomium **4.** Neuropodium
5. Cirri, parapodia **6.** Coelom **7.** Buccal cavity, pharynx
8. Extracellular **9.** Basal lamina **10.** Trochophore.

Multiple Choice Questions

1. (*d*) **2.** (*b*) **3.** (*a*) **4.** (*d*) **5.** (*b*) **6.** (*a*)
7. (*a*) **8.** (*d*) **9.** (*b*) **10.** (*c*) **11.** (*a*) **12.** (*a*).

4

Hirudinaria granulosa (Indian Cattle Leech) (Phylum Annelida)

4.1 CLASSIFICATION

Phylum	–	**Annelida**
Class	–	**Hirudinea**
Order	–	**Gnathobdellida**
Genus	–	***Hirudinaria***
Species	–	***granulosa***

Leeches are the most specialised annelids and have evolved from oligochaetes. The genus *Hirudinaria* includes following four Indian species: *H. granulosa, H. viridis, H. javanica* and *H. manillensis*. These species are also common in Burma (Myanmar), Pakistan, Bangladesh and Sri Lanka. *Haemadipsa* is a land leech of the hills of Southern India.

4.2 HABIT AND HABITAT

Hirudinaria granulosa is commonly called **'Indian cattle leech'.** It inhabits freshwater ponds and tanks, still waters near streams and rivers and water standing in rice-fields. It prefers shallow water where stones, logs, weeds, etc., provide shelter. *Hirudinaria* is an **ectoparasitic, sanguivorous** (blood sucking) animal which attaches to cattle and human beings when they visit the pond. It sucks the blood of fishes and frog also.

4.3 EXTERNAL FEATURES

Hirudinaria is an aquatic (freshwater) ectoparasitic sanguivorous (blood sucking) annelid. Its morphology and physiology is well adapted for its peculiar mode of habits and habitats.

External Structure

1. Size and Shape

Average leech measures 10 to 15 cm in length but it may grow upto a length of 35 cm. The body of *Hirudinaria* is soft, slimy, vermiform, elongated, bilaterally symmetrical and metamerically segmented. Body is dorso-ventrally flattened having flat ventral and convex dorsal surfaces. The body is broadest near the posterior end and narrowest at the anterior end. Body of leech is quite flexible whereas in contracted state it becomes cylindrical.

2. Colouration

Body of leech show protective colouration (camouflaging) being olive green dorsally, orange-yellow ventrally. Stripes of orange and black exist on the lateral sides.

3. Metamerism or Segmentation

Body of leech is divided into 33 segments, somites or **metameres** (*i.e.*, the number of segments is fixed in leeches). Except first two and last seven segments, external surface of each segment is superficially divided into variable number of **rings** or **annuli** (Fig. 4.1). All those segments which have five annuli are called **complete segments.** Segments having less than five annuli are called **incomplete segments.** Details of number of the annuli in different segments can be tabulated in the Table 4.1.

Table 4.1 Number of annuli in different body segments of *Hirudinaria*.

Segments	Number of Annuli	Designation of segment	Nature of segment
1, 2	One	Uniannulate	Incomplete
3	Two	Biannulate	Incomplete
4, 5, 6	Three	Triannulate	Incomplete
7 to 22	Five	Quinquannulate	Complete
23 to 26	Two	Biannulate	Incomplete
27 to 33	One	Uniannulate	Incomplete

Further, the surface of each annulus is divided by very fine longitudinal grooves (Fig. 4.3) into 18 rectangular areas on the dorsal and 18 on ventral side.

4. Division of Body

The body of leech can be divided into follow-ing six regions:

***(i)* Cephalic region.** First five segments which are incomplete form the head (Table 4.1). It contains **prostomium, anterior sucker, mouth** and **eyes.** In fact, a pair of **eyes** are situated on the dorsal side of first annulus in all the cephalic segments. These segments are, therefore, called **ocular segments.** There are no nephridiopores. On the ventral surface of cephalic region is present **anterior sucker** and **mouth.** The prostomium and first three segments form the **upper lip.**

***(ii)* Pre-clitellar region.** It comprises of three segments, *i.e.*, **6th** incomplete, **7th** and **8th** (com-plete). Each of these segments bear a pair of nephridiopores on its ventral surface.

***(iii)* Clitellar region.** It is composed of three complete segments, *i.e.*, **9th, 10th** and **11th.** A temporary girdle-like **clitellum** develops around this region during the breeding season. Male and female genital apertures are situated on the ventral surface of 10th and 11th segments respectively. All of these segments carry a pair of nephridiopores.

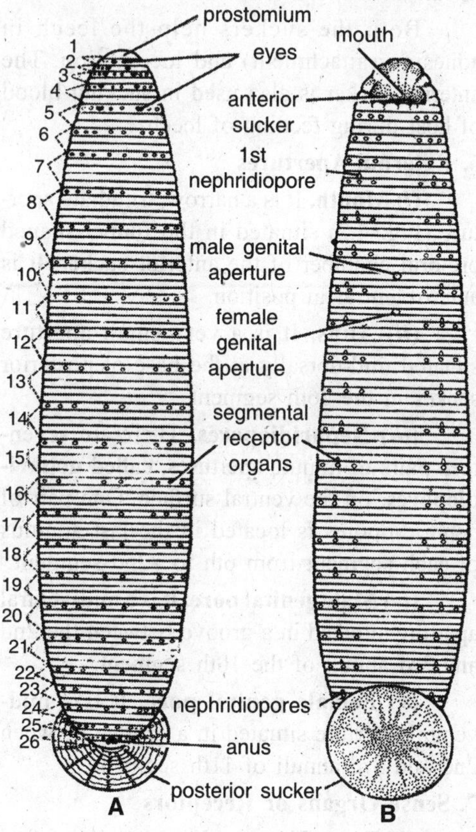

Fig. 4.1. *Hirudinaria granulosa*. External structure. A—Dorsal view; B—Ventral view.

(iv) Middle region. It is the largest region of the body and includes eleven complete segments (12th to 22nd). All segments contain paired nephridiopores ventrally.

(v) Caudal region. It is composed of last seven incomplete segments (27th to 33rd). These segments are arranged in concentric rings.

5. Suckers

Each end of the body of leech bears a muscular adhesive structure called **sucker.** The **anterior, cep-halic** or **oral sucker** is formed by the fusion of prostomium and first three segments. It is oval in shape and contains a cup-like depression, called **pre-oral chamber,** on the ventral side.

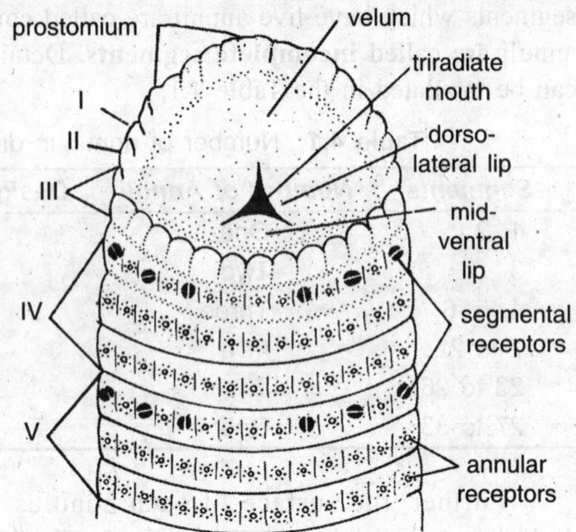

Fig. 4.2. *Hirudinaria*. Cephalic region.

The **posterior** or **anal sucker** is formed by fusion of last seven body segments. It is circular and disc-like. It is highly muscular. Posterior sucker is better developed and larger in size than the anterior sucker.

Both the suckers help the leech in adhesion (attachment) and locomotion. The anterior sucker is also used in sucking blood of host during feeding of leech.

6. External Apertures

(i) Mouth. It is a narrow, triradiate aperture (Fig. 4.2) situated in the funnel-shaped pre-oral chamber of the anterior sucker. It is antero-ventral in position.

(ii) Anus. It is a very small aperture situated mid-dorsally at the base of posterior sucker in the 26th segment.

(iii) Nephridiopores. There are seventeen pairs of minute apertures, called **nephridiopores,** on the ventral surface. One pair of nephridiopores is located in the last annulus of each segment from 6th to 22nd segment.

(iv) Male genital pore. It is a mid-ventral aperture situated in a groove between the 2nd and 3rd annuli of the 10th segment.

(v) Female genital pore. It is a mid-ventral aperture situated in a groove between 2nd and 3rd annuli of 11th segment.

7. Sense Organs or Receptors

(i) Eyes. The annulus of each of five segments carries a pair of eyes dorsally.

(ii) Segmental receptors. First annulus of each segment carries segmental receptors in the form of papillae. First pairs of them occur dorsally and three pairs ventrally.

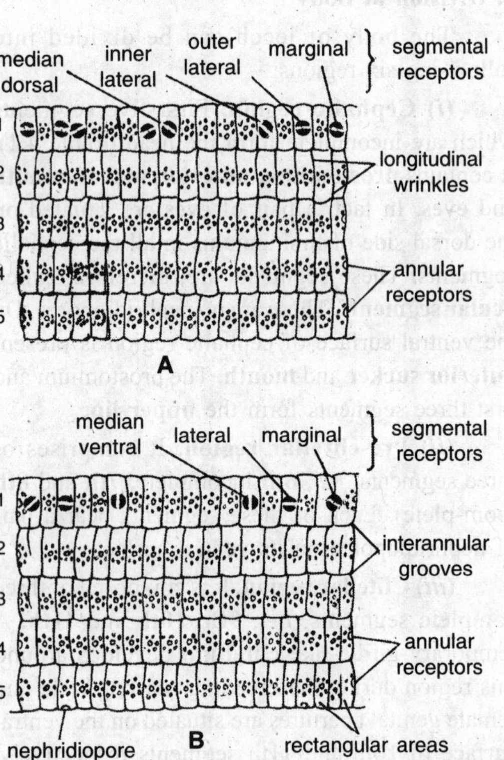

Fig. 4.3. *Hirudinaria*. A complete segment showing receptors. A—Dorsal view; B—Ventral view.

8. Annular receptors

There are thirty six **annular receptors** present in each annulus, of which eighteen are dorsal and eighteen are ventral in position. These occur in the form of minute elevated areas in 36 rectangular areas of each annulus (Fig. 4.3).

4.4 HISTOLOGY OF BODY WALL

The body wall of leech is glandular and dermo-muscular. It consists of five layers: cuticle, epidermis, dermis, muscle layers and botryoidal tissue (Fig. 4.4).

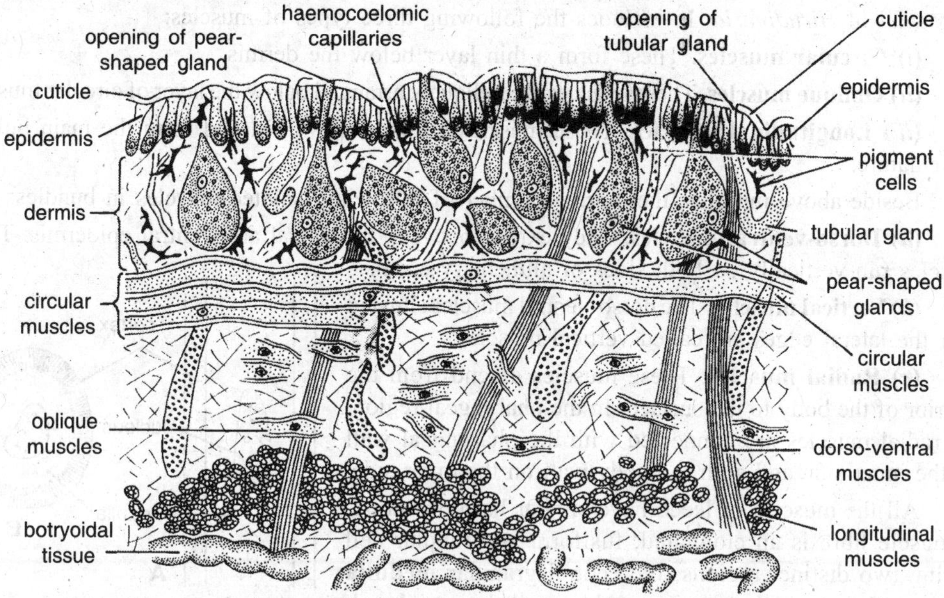

Fig. 4.4. *Hirudinaria*. T.S. of body wall.

1. Cuticle. It is the outermost, thin, transparent and elastic but firm protective layer. It is perforated by the openings of epidermal glands. Cuticle is secreted by the underlying epidermis and is cast off in patches periodically.

2. Epidermis. It consists of a single layer of hammer-shaped cells. The outer ends of these cells are closely fitted together while their narrower ends are nucleated. Some of the epidermal cells become modified to form unicellular pear-shaped glands and multicellular **receptor organs.**

Epidermal glands. These are of following types:

***(i)* Slime glands.** These glands are pear-shaped or tubular. These are distributed all over the body and secrete a slimy mucus that covers the whole body. The main body of the glands lie in the dermis and connective tissue and their long and narrow ducts open to the surface. Each gland contains a nucleus, a thin layer of cytoplasm and numerous secretory granules.

***(ii)* Sucker glands.** These glands are distributed only in the suckers and occur in clusters. These are pear-shaped in the anterior sucker and rounded in posterior sucker. The secretion of sucker glands serve to smoothen the surface for the attachment of suckers during locomotion.

***(iii)* Prostomial glands.** These are pear-shaped and occur in groups in the prostomium. Their secretion plugs the ends of cocoon (=ootheca or egg case).

***(iv)* Clitelar glands.** These glands are spherical in form and are found in the clitellar region (9th to 11th segment) and become active only during the breeding season. Clitellar glands are of two types: chitogenous and albumen glands. The **chitogenous glands** are located between the circular and oblique muscles and secrete a substance for the formation of wall of cocoon or ootheca. The **albumen glands** lie in the longitudinal muscles and secrete albumen which is filled up in the cocoon (it is used for the nourishment of developing embryos).

3. Dermis. It lies below the epidermis. It consists of fibrous connective tissue which is very thick and occupies much of the interior of body. Dermis consists of a network of connective tissue fibres, connective tissue cells, pigment cells, fat cells, muscle fibres and parts of haemocoelomic system. Epidermal glands also extend into this region.

4. Muscles. Musculature below the dermis is well developed forming largest part of the body wall of *Hirudinaria.* In includes the following three types of muscles:

***(i)* Circular muscles.** These form a thin layer below the dermis.

***(ii)* Oblique muscles.** These muscles are obliquely arranged below the layer of circular muscles.

***(iii)* Longitudinal muscles.** These muscles are quite powerful and form the main bulk of musculature.

Beside above muscles, there are a number of following isolated muscles in bundles:

***(a)* Dorso-ventral muscles.** They extend between the dorsal and ventral epidermis. These muscles run vertically on both sides of alimentary canal.

***(b)* Vertical muscles.** The dorso-ventral muscles situated near the lateral edges are called vertical muscles.

***(c)* Radial muscles.** These muscles extend from the interior of the body to the surface. In the pharyngeal region, the radial muscles are connected with the pharyngeal wall on the inner side and with the body wall on the outer side.

All the muscles of leech have similar basic structure. A muscle fibre is an elongated, fusiform and tubular cell having two distinct regions: the central granular **medulla** or **sarcoplasm** which contains nucleus and the peripheral, striated and contractile **cortex** or **myoplasm** (Fig. 4.5).

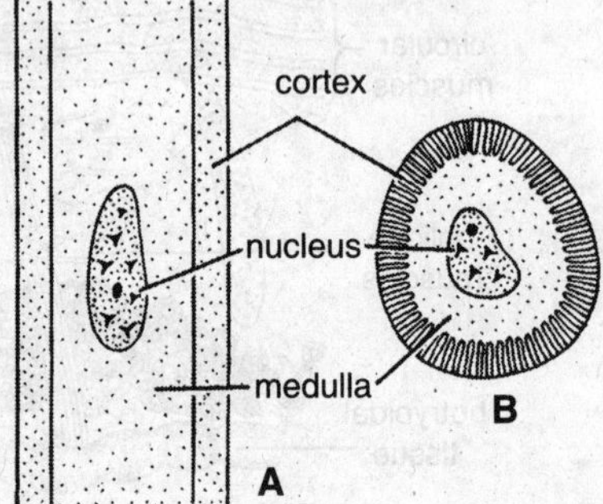

Fig. 4.5. *Hirudinaria.* A muscle fibre. A—L.S.; B—T.S.

5. Botryoidal Tissue

Botryoidal (*Gr. botryoeides* = cluster or grapes; *eidos* = shape) tissue is a characteristic tissue of jawed leeches and it resembles with bunches of grapes. It exists between the gut and muscles (longitudinal) almost completely filling the body cavity. Botryoidal tissue consists of a network of very fine capillary channels or sinuses of haemocoelomic system lined by small globular cells heavily laden with a **brown pigment** (which is a masked iron-compound probably derived from blood) partially corresponds to chloragogen tissue of oligochaetes. Its blood sinuses are roughly analogous to lymphatic system of vertebrates. Botryoidal tissue is a type of 'tissue of accumulation' which picks up particulate waste from blood and keep them forever.

Botryoidal tissue also contains another type of 'accumulation tissue' called **vaso-fibrous tissue** which contains an accumulated **green pigment** (another masked iron-compound probably derived from blood). Botryoidal tissue and vaso-fibrous tissues have excretory function.

Functions of body wall. The body wall of leech contains the following functions:

1. It protects internal organs from mechanical injury.
2. The cuticle and slimy mucous coat prevent loss of water through evaporation from the body.
3. The slimy mucous kills harmful organisms and protects the leech from infections.
4. Pigment cells of body wall produce camouflaging, protecting the leech from its predators.
5. It is richly supplied with haemocoelomic capillaries and makes gaseous exchange (respiration) possible through it.

6. It is sensory due to the presence of various types of receptors on it.
7. Clitellar glands secrete wall of cocoon and albumen.
8. Prostomial glands secrete plugs of the cocoon.
9. Body musculature helps in locomotion.

4.5 LOCOMOTION

The loss of septa, setae and coelomic compartments in leeches is correlated with a change from peristaltic burrowing to new modes of locomotion; leeches crawl or swim but never burrow using peristalsis. Leeches crawl on solid substratum by forming loops and swim in water.

Crawling

This type of locomotion is performed with the help of suckers and body musculature (Fig. 4.6). During this type of locomotion:

(1) Leech fixes its posterior sucker on the substratum and contracts its circular muscles and relaxes longitudinal muscles. This causes maximum lengthening of the body. The diagonal muscles may be postural, keeping the animal rigid while it is raised and attached posteriorly.

(2) Now the leech fixes its anterior sucker on the substratum and relaxes the circular muscles and contracts the longitudinal muscles. This releases the hold of posterior sucker and shortens the body.

(3) This results in forward movement of posterior sucker. Thus, posterior end of body come closer to the anterior end and the body of leech is raised up in the middle forming a loop or D. The diagonal muscles may also enable leeches to twist their raised bodies while attached posteriorly.

(4) Once again the posterior sucker is fixed, the anterior sucker is released and the whole process is repeated again and again and leech moves forward.

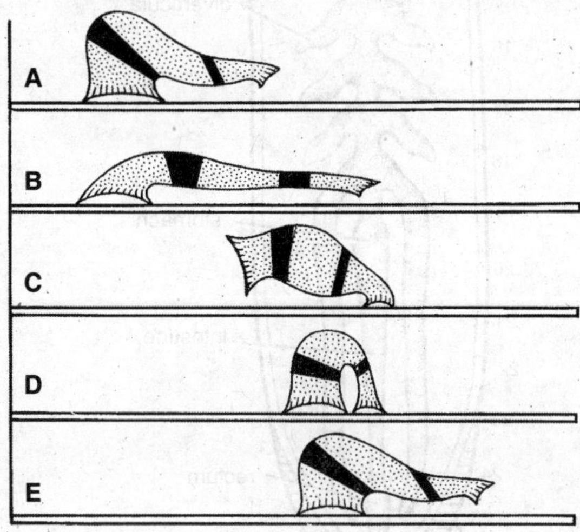

Fig. 4.6. *Hirudinaria.* Crawling.

Swimming. In the water the leech swims swiftly. During swimming the body becomes flat and the worm swims by successive undulating movements. The body is flattened by contraction of the dorsoventral musculature and waves of contraction along the longitudinal muscles produce vertical undulation.

4.6 DIGESTIVE SYSTEM

Digestive system consists of the alimentary canal and digestive glands associated with it.

1. Alimentary Canal

The alimentary canal of leech is a straight tube extending throughout the length of the body from the mouth to anus (Fig. 4.7). It is greatly modified due to its blood-sucking (sanguivorous) habit. Most part of the alimentary canal serves to store uncoagulated blood, while only a small portion serves for its digestion and absorption. The alimentary canal is differentiated into following three main parts:

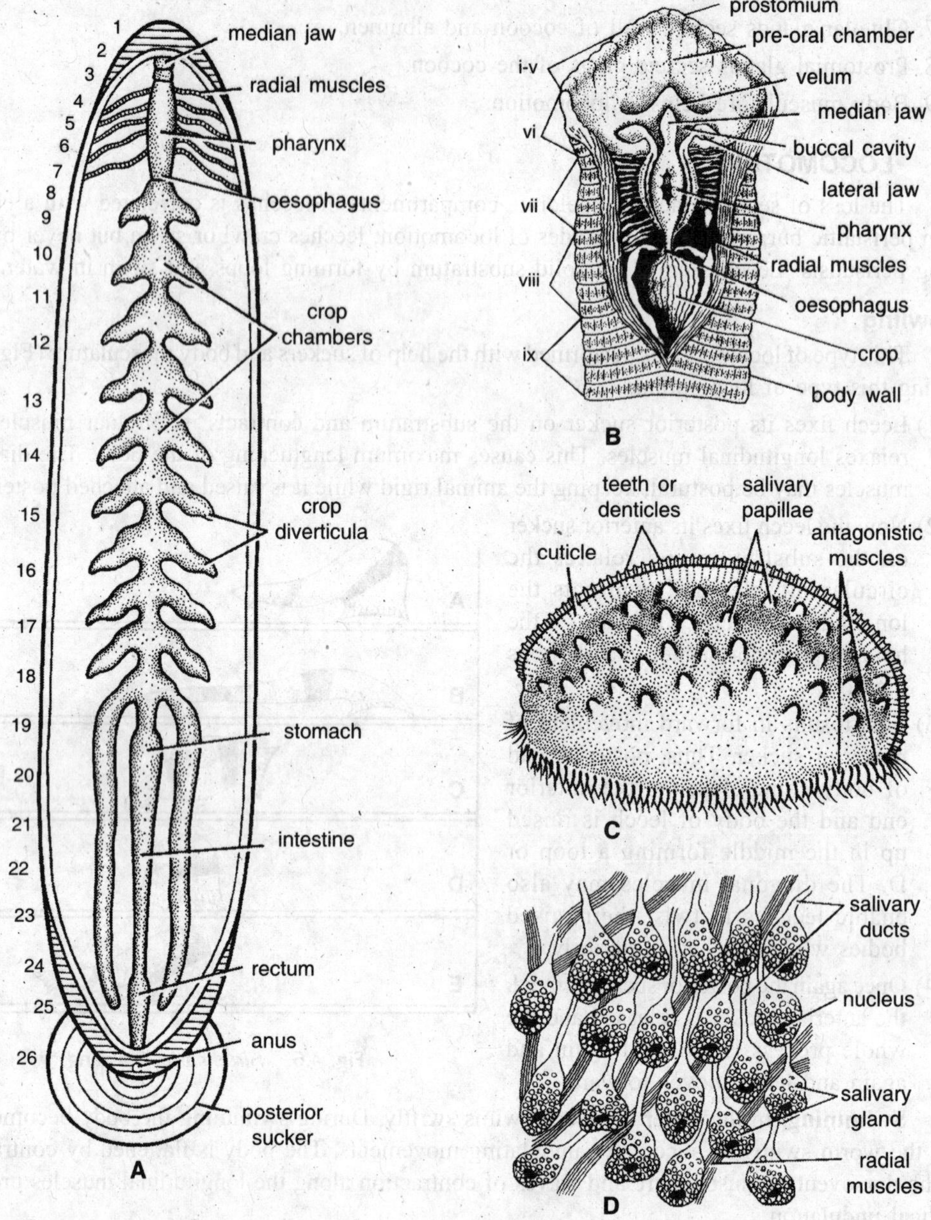

Fig. 4.7. *Hirudinaria*. A—Alimentary canal; B—Anterior part of leech dissected ventrally to show the buccal cavity and pharynx. C—A single jaw in lateral view; D—Salivary glands.

1. Stomodaeum (It is lined with ectodermal epithelium and is covered with cuticle).

2. Mesenteron (It is lined with endodermal epithelium and lacks any cuticular covering).

3. Proctodaeum (It is lined with ectodermal epithelium and covered by cuticle).

A. Stomodaeum. It includes the following three parts.

(1) Pre-oral chamber and mouth. The cup-shaped depression of the anterior sucker represents the **pre-oral chamber.** Its base is perforated by triradiate **mouth** which opens into the buccal cavity. The pre-oral chamber and buccal cavity are separated from one another by a membranous partition, the **velum.** Velum forms three lips (two dorso-lateral and one ventral) around the mouth. Pre-oral chamber helps in blood-sucking.

(2) Buccal cavity. It is a very short chamber. In the mucous membrane of buccal cavity are embedded three crescentic **jaws,** one mid-dorsal and the other two are ventro-lateral in position. Their position corresponds to the position of three slits of mouth. Each jaw is a laterally-compressed crescentic and muscular cushion-like structure covered with cuticle. The cuticle is thickened at the free edge to form a dentigerous ridge bearing a row of minute **teeth** (denticle). As these teeth form a single series on each jaw, the jaw is termed as **monostichodont.** The median jaw bears 103 to 128 teeth, while lateral jaws bear 85 to 115 teeth each. Each lateral side of jaw carries 42–45 button-shaped salivary **papillae.** These papillae bear openings of salivary glands. The jaws are moved forward and backward by special muscles. These work simultaneously as semicircular saws producing **triradiate Y-shaped wound** in the skin of host.

(3) Pharynx. The buccal cavity leads into an oval and highly muscular sac, the **pharynx.** It extends from 5th to 8th segments. Bundles of **radial muscles** extend from the wall of pharynx to the body wall. Their contraction widen the pharyngeal cavity and creates a suction force for sucking blood. The inner surface of pharynx bears longitudinal folds.

B. Mesenteron. It includes the following four parts:

(*i*) Oesophagus. It is a short and narrow tube connecting pharynx with crop. Its internal lining is greatly folded.

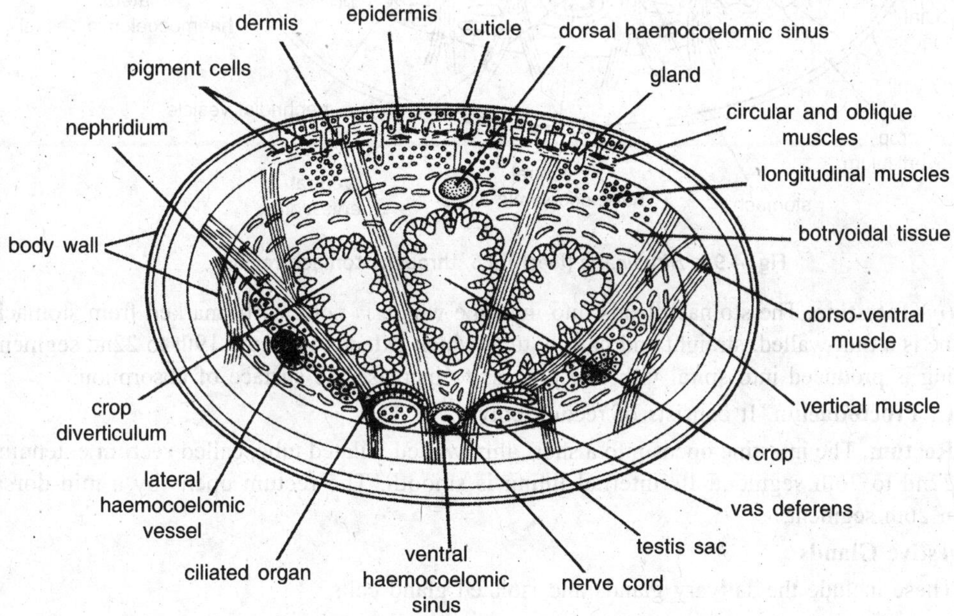

Fig. 4.8. *Hirudinaria*. T.S. through crop.

(*ii*) Crop. The crop forms the largest region of the alimentary canal extending from the ninth to the eighteenth segment. It has ten thin-walled **chambers,** one in each segment and they are connected with one another by more or less circular apertures surrounded by sphincters. Each chamber is differentiated into a small **anterior part** and a broad **posterior part.** The posterior part of each chamber is prolonged into a pair of lateral blind, backwardly directed outgrowths, the **diverticula**

or **caeca** (Fig. 4.8). The size of crop chambers and caeca go on gradually increasing towards the posterior side. The last chamber of the crop is the largest and its caeca are prolonged backwards upto twenty-second segment. Crop chambers and their caeca are highly elastic and used for **storing** blood meal. One cropful of blood lasts for several months and digested slowly.

***(iii)* Stomach.** The last chamber of crop extends posteriorly into a funnel-shaped narrow tube which communicates with the heart-shaped **stomach** through a sphinctered aperture. The stomach is a short tube lying in 9th segment. It has both **secretory** and **absorptive** epithelial cells and its inner lining is produced into many anastomosing transverse folds (Fig. 4.9).

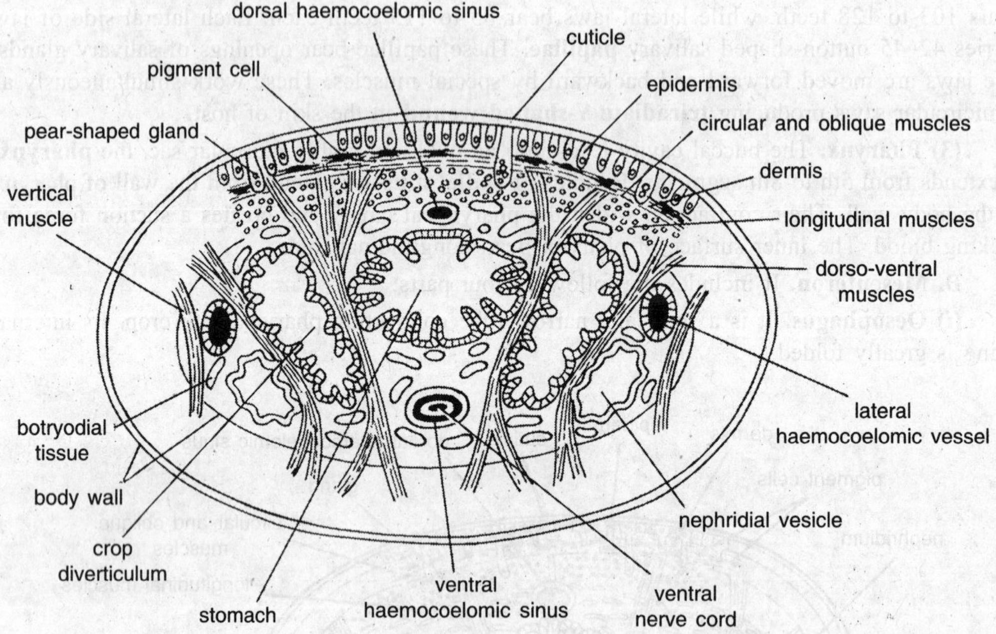

Fig. 4.9. *Hirudinaria* (leech). T.S. through stomach region.

***(iv)* Intestine.** The stomach leads into intestine which is not clearly marked from stomach. Intestine is a thin-walled, straight and narrow tube which extends from the 19th to 22nd segment. Its lining is produced into spiral villi-like process to increase the surface of absorption.

C. Proctodaeum. It consists of rectum only.

Rectum. The intestine opens into a short, thin-walled, dilated tube, called **rectum** extending from 22nd to 26th segment. Its internal lining is smooth. The rectum opens by a mid-dorsal anus in 26th segment.

2. Digestive Glands

These include the salivary glands and isolated gland cells.

***(i)* Salivary glands.** These are unicellular, pyriform, glandular mass of cells situated around and behind the pharynx (Fig. 4.7D). Each glandular cell has elongated fine ductule which finally enters the jaws to open in raised salivary papillae. The secretion of these glands is **saliva** which contains an anticoagulant substance called **anticoagulin** or **hirudin.** Hirudin prevents coagulation of blood while leech is feeding.

***(ii)* Gland cells.** These are in the form of **goblet cells,** found scattered throughout the lining (or mucous membrane) of crop. These secrete mucus in the crop.

Food and Feeding

Hirudinaria is sanguivorous in habit, feeding on blood of cattle and other mammals visiting the water source. In feeding, the leech applies its oral sucker very tightly on the skin of its prey, the jaws move towards and away from each other and they painlessly puncture the skin. The incision is anaesthetized by a substance of unknown origin. The oozing blood is sucked up by the suctorial pharynx. Pump-like action of the pharynx sucks in large quantities of blood to fill the crop. Secretion of **hirudin** prevents coagulation of blood.

At a single meal a leach usually sucks blood several times its own weight. It is sufficient for several months or even a year. After feeding, leech drops of the host and rests in concealed shelter.

Digestion

The sucked blood is stored in uncoagulated state in the crop-chambers where it is **haemolysed.** Red blood corpuscles are broken down. The haemoglobin from ruptured RBCs is separated and dissolved in plasma. Due to absorption of water from blood plasma by crop wall, the blood becomes thicker and jelly-like and dark red or blackish red in colour. From crop blood enters stomach drop by drop. This is controlled by sphincter that surrounds the opening of crop into stomach. Blood proteins (*e.g.*, globin) are digested in stomach by proteolytic enzymes.

Digestion in leech is peculiar in a number of respects. The gut secretes no amylases, lipases, or endopeptidases. The presence of only **exopeptidases** perhaps explains the fact that digestion in blood-sucking leeches is so slow. Also characteristic of the leech gut is a symbiotic bacteria flora that is important in nutrition. The bacteria (*e.g.*, *Pseudomonas hirudinicola* of *Hirudo medicinalis*) breaks down high-molecular-weight proteins, fats and carbohydrates. The bacteria may also produce vitamins and other compounds that are used by the leech host.

Rectum stores the undigested waste for sometime then egests it out through the **anus.**

4.7 RESPIRATION

Hirudinaria or leech is an aquatic animal. It lacks any special organ of respiration. The skin of leech serves as a respiratory organ. The capillaries containing haemocoelomic fluid extend in between the cells of epidermis through which exchange of respiratory gases takes place by **diffusion.** The carbon dioxide dissolved in haemocoelomic fluid passes out and the oxygen dissolved in water enters in the body. The skin is kept wet by the surrounding water, while the mucus secreted by the slime glands of the epidermis prevent it from drying.

4.8 HAEMOCOELOMIC SYSTEM

Coelom

The perivisceral coelom is absent in leech. It is obliterated and is filled with pigmented botryoidal tissue. The original coelom is replaced by a system of **longitudinal channels,** their branches and spaces or **sinuses.** It is called **haemocoelomic system** and filled with red **haemocoelomic fluid** containing colourless amoeboid corpuscles and extracellular haemoglobin dissolved in the fluid. The true coelom of leech is restricted to spaces around the testes sacs, ovisacs and gonoducts. It is filled with colourless coelomic fluid containing colourless corpuscles.

Haemocoelomic System

Leech lacks a blood vascular system but contains a haemocoelomic system (*i.e.*, modified coelomic circulatory system). Because the haemocoelomic system is evolved from the original coelom all of its vessels and channels are lined by an **endothelium,** the **peritoneum** along with connective tissue. The haemocoelomic fluid is propelled by the muscular contractions of the lateral longitudinal channels.

Haemocoelomic system of leech contains following four longitudinal channels:

1. A dorsal longitudinal haemocoelomic channel;
2. A ventral longitudinal haemocoelomic channel;
3. Two lateral longitudinal channels.

1. Dorsal Channel

The dorsal channel or dorsal sinus runs mid-dorsally below the body wall and above the alimentary canal. It is thin-walled, non-contrac-tile vessel which lacks valves. The haemocoelomic fluid flows in it from the posterior end to anterior end (from behind forward). Dorsal channel is an **distributing channel.** Anteriorly, dorsal channel ramifies in the sixth segment and supplies haemocoelomic fluid to first five segments. Posteriorly it bifurcates in the 22nd segment and two branches run downward around the rectum and join the posterior dilation of ventral sinus (Fig. 4.10).

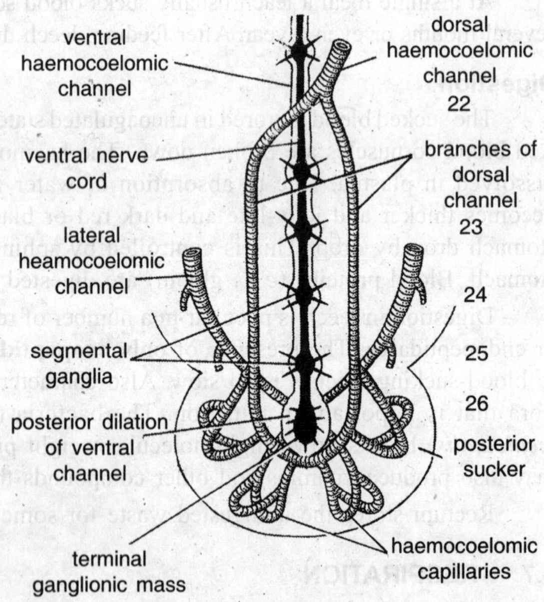

Fig. 4.10. *Hirudinaria*. Posterior end of body showing union of four haemocoelomic channels.

In each segment, the dorsal channel gives off two pairs of **dorsolaterals** taking the haemocoelomic fluid to the dorsal and dorso-lateral regions of body wall and several dorso intestinals going to the alimentary canal (Fig. 4.11).

2. Ventral Channel

The ventral channel is also a thin-walled, non-contractile distributing vessel running mid-ventrally below the alimentary canal from one end of body to the other. It is a broad vessel enclosing the entire central nervous system and lacks valves. In it, the haemocoelomic fluid flows from anterior to the posterior end. It gives off two pairs of branches in each segment.

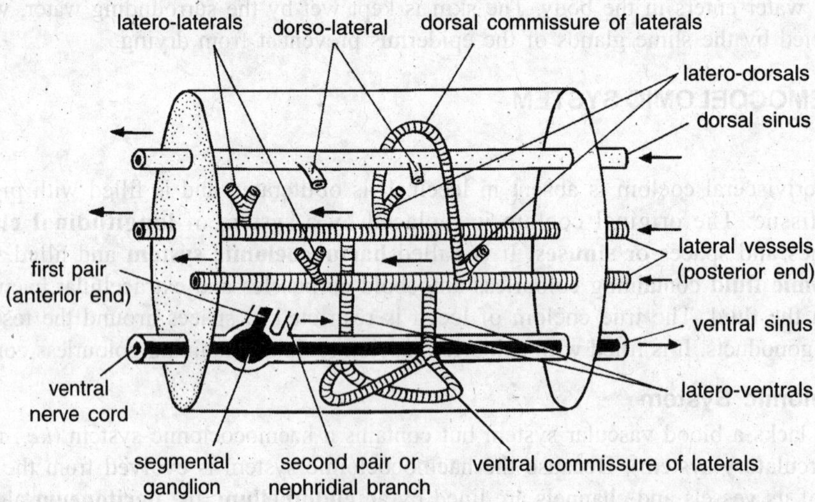

Fig. 4.11. *Hirudinaria*. Diagrammatic representation of main haemocoelomic channels of a segment.

(*i*) First pair or cutaneous branches. Two cutaneous branches are given off at the level of segmental nerve ganglion. Each branch immediately bifurcates into a **ventral branch** and an

abdomino-dorsal branch. The former forms a network in the ventro-lateral region of the body wall while the latter passes vertically upwards to dorsal body wall (Fig. 4.12).

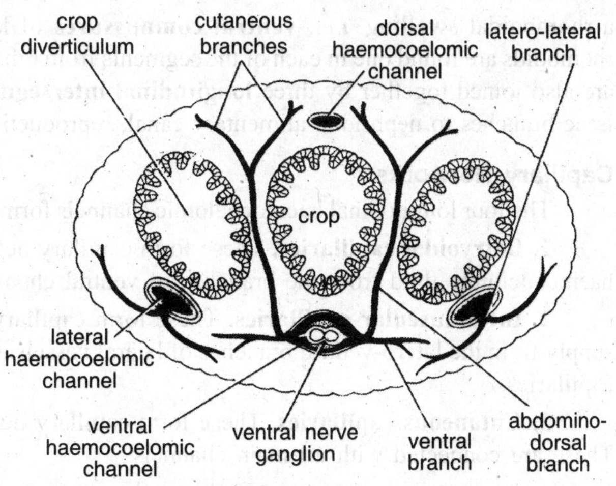

Fig. 4.12. *Hirudinaria*. T.S. body through dorso-abdominal branches.

***(ii)* Second pair or nephridial branches.** These branches are issued just behind the segmental nerve ganglion. Each runs outwards to supply the testis-sac and the nephridium, where it enlarges to a **perinephrostomial ampulla** which encloses the ciliated organ. There are only eleven pairs of nephridial branches or one pair in each ofthe segments from 12th to 22nd segment.

3. Lateral Channels

A pair of lateral channels extend one on either side of the alimentary canal from one end of body to the other. Each is a wide, thick-walled, muscular, contractile vessel with segmentally arranged valves. In it, hae-mocoelomic fluid flows from poste-rior to anterior end (from behind forward). The lateral channels break up in the sixth segment and form capillary network. Posteriorly they join the dilation of ventral sinus and thus, all four channels become continuous at this level (Fig. 4.10). Both lateral channels are **collecting** as well as **distributing** channels. In each segment, the lateral channel receives on its outer side two branches a latero-lateral and a latero-dorsal and gives off one branch, the latero-ventral (Fig. 4.13).

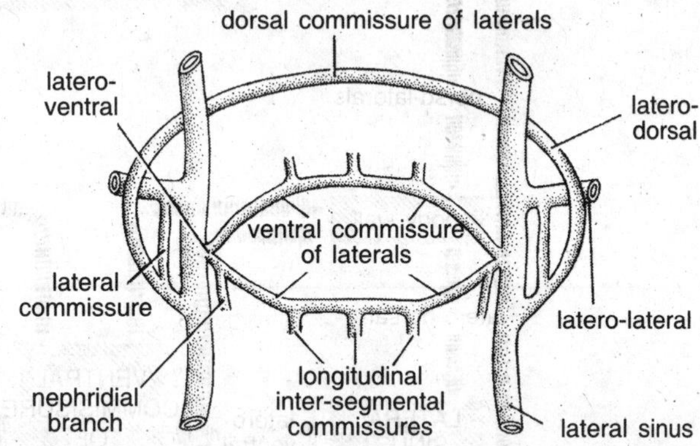

Fig. 4.13. *Hirudinaria*. Lateral vessels and their branches in a segment.

***(i)* Latero-lateral.** It is a small vessel. It collects haemocoelomic fluid from the lateral region of the body wall and nephridia and opens into the lateral vessels at the level of nephridial vesicle.

***(ii)* Latero-dorsal.** It is a large vessel opening in the lateral channel at the level of main lobe of the nephridium. It receives branches from the dorsal and dorso-lateral regions of the body wall, the gut wall and nephridium. In each segment the two latero-dorsals from the opposite side are connected together above the dorsal channel by a **transverse loop.** This is called **dorsal commissure** of lateral channels. There are 17 such dorsal commissures one in each segment from 6th to 22nd. A small branch, called **lateral commissure,** connects the latero-dorsal branch to the latero-lateral branch of its side.

The latero-laterals and the dorso-laterals are collecting vessels and their openings into the lateral channels are guarded with valves.

***(iii)* Latero-ventral.** Each lateral channel gives out a **latero-ventral** branch which bifurcates into an **anterior** and a **posterior** branch and each join its fellow of the opposite side mid-ventrally to form a rhomboidal swelling, *i.e.*, **ventral commissures** of laterals in each segment. Eighteen such rhomboids are found one in each of the segments from 6th to 23rd. Rhomboids of adjacent segments are also joined together by three **longitudinal intersegmental commissures.** The latero-ventrals issue branches to nephridia, alimentary canal, reproductive organs and ventral side of body wall.

Capillary Networks

The four longitudinal haemocoelomic channels form following three sets of capillary network:

1. Botryoidal capillaries. These form capillary networks around the viscera. These receive haemocoelomic fluid from the branches of ventral channel and return it to the lateral vessels.

2. Intermuscular capillaries. These form capillary networks in the muscles. They receive supply from the latero-ventral branches of lateral vessels and also communicate with the botryoidal capillaries.

3. Cutaneous capillaries. These form capillary networks which extend into the epidermis. These are connected with all main channels.

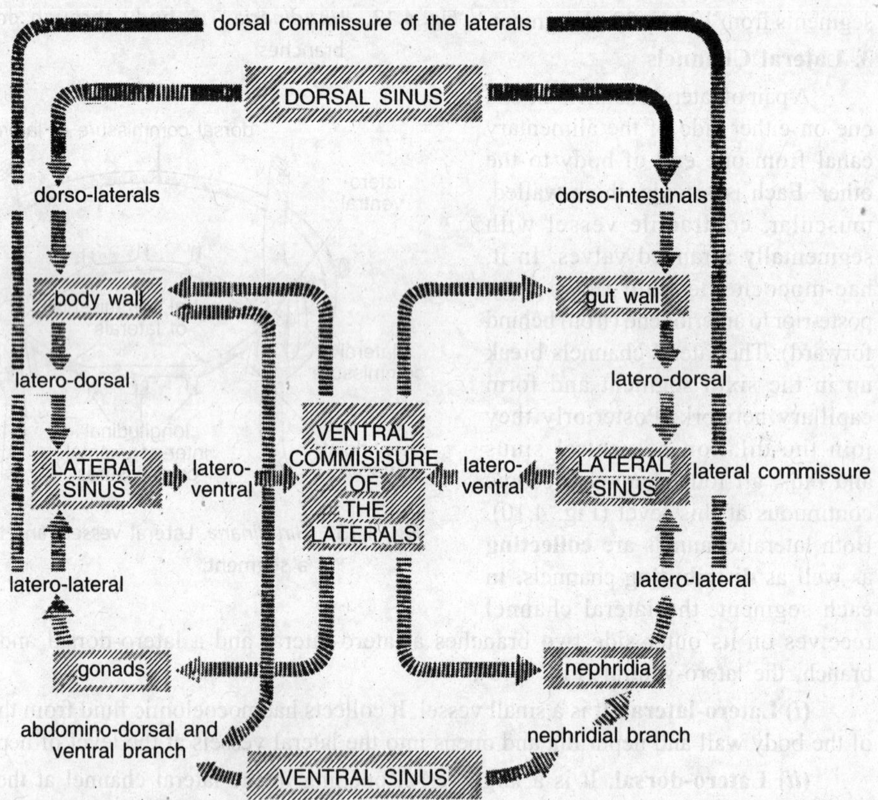

Fig. 4.14. *Hirudinaria.* Summary of course of circulation of haemocoelomic fluid.

Course of Circulation of Haemocoelomic System

The haemocoelomic fluid flows from posterior to anterior direction in the dorsal and lateral channels while it flows from anterior to posterior direction in the ventral channel. All the four longitudinal channels become continuous in the 26th segment. The dorsal and ventral channels distribute haemocoelomic fluid while the lateral channels distribute as well as collect the haemocoelomic fluid.

The **dorsal channel** supplies fluid to dorsal and dorso-lateral parts of body wall and the entire alimentary canal, from where it is collected by latero-dorsals which carry the fluid to the lateral channels.

The **ventral channel** supplies the fluid to the ventral, ventro-lateral and ventral parts of the body wall and the nephridia including ciliated organs. From all these parts, the fluid is collected by the latero-lateral branches of the lateral channels.

The **lateral channels** serve as tubular "coelomic hearts" and supply the fluid to the nephridia, reproductive organs, ventral gut-wall and the ventral body wall through latero-ventral branches. Haemocoelomic fluid from all these parts is collected by latero-lateral and latero-dorsal branches of the lateral vessels.

4.9 EXCRETORY SYSTEM

Nephridia are the excretory and osmoregulatory organs of leech. These are segmentally arranged, hollow tubes present one pair in each segment from 6th to 22nd segment. Thus, in total there are seventeen pairs of nephridia in *Hirudinaria*. These nephridia are of two types:

1. Pretesticular nephridia. First six pairs of nephridia in the segments 6–11 are without testis sacs and are called pretesticular nephridia.

2. Testicular nephridia. The remaining eleven pairs of nephridia are with testis sacs; they are called testicular nephridia. They are found in segments from 12th to 22nd.

1. Testicular Nephridia

A typical testicular nephridium is a horse-shoe shaped structure. It consists of six parts: (*i*) Main lobe; (*ii*) Vesicle and vesicle duct; (*iii*) Apical lobe; (*iv*) Inner lobe; (*v*) Initial lobe or testicular lobe; and (*vi*) Ciliated organ (Fig. 4.15).

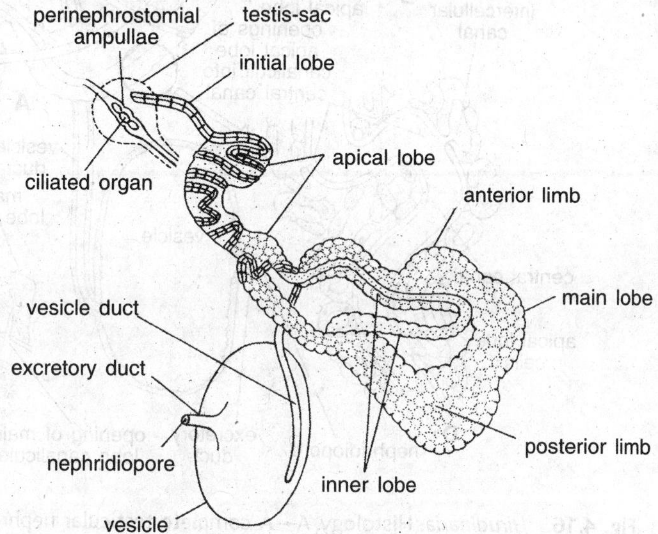

Fig. 4.15. *Hirudinaria*. A testicular nephridium.

(*i*) Main lobe. It forms horse-shoe proper of nephridium. It lies in a ventro-lateral position between the adjacent diverticula of crop. It consists of two unequal limbs: a longer **anterior limb** and a shorter **posterior limb.** Histologically main lobe comprises of large-sized polyhedral cells.

(*ii*) Vesicle and vesicle duct. From the ventral end of anterior limb of the main lobe arises a narrow **vesicle duct** which runs posteriorly to open into the vesicle. The **vesicle** or **bladder** is a large, oval, non-contractile thin-walled sac, lined internally by a non-ciliated epithelium. A short narrow **excretory duct** arises from the inner side of vesicle which opens to the exterior through **nephridiopore** surrounded by sphincter muscles.

(*iii*) Apical lobe. The posterior limb of the main lobe continues to form stout **apical lobe.** It lies in an antero-posterior position beneath the gut. The free end of apical lobe is bent on itself like the handle of a walking stick. The cells lining the apical lobe are large with regular intercellular canals.

4. Inner lobe. Running between the two ends of main lobe and situated in its concavity is found a short narrow **inner lobe.** It runs forward along the outer border of apical lobe for about half its length and then joins the apical lobe. Its cells are long and tubular. It carries a central lumen.

(*v*) Initial lobe. It is slender, much elongated transparent cord-like structure. It remains coiled around the apical lobe. Its posterior end joins the anterior limb of the main lobe. Anterior end of initial lobe is free and blind and extends over the testis sac to lie close to **perinephrostomial ampulla.** Histologically, initial lobe consists of a single row of elongated hollow cells placed end to end having the intracellular canal (Fig. 4.16) which gives off several diverticula in each cell.

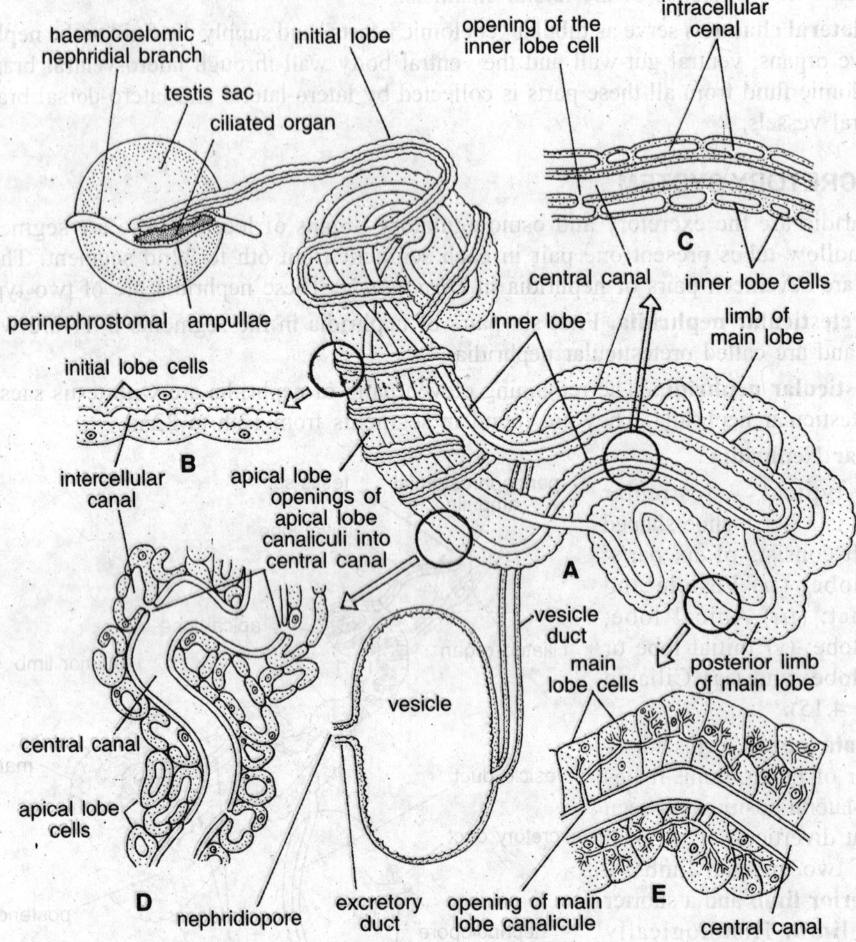

Fig. 4.16. *Hirudinaria*. Histology. A—A complete testicular nephridium showing central canal; B—A part of initial lobe in L.S.; C—A portion of inner lobe in L.S.; D—A portion of apical lobe in L.S.; E—A portion of main lobe in L.S.

(*vi*) Ciliated organ. A ciliated organ is completely enclosed within the perinephrostomial ampulla which is a dilation of the haemocoelomic system. A ciliated organ consists of a central **reservoir** which is perforated and a peripheral part containing closely-set **ciliated funnels.** The reservoir is spongy mass of connective tissue cells which manufacture **coelomic amoebic corpuscles.** The outer wall of ciliated funnel is perforated by numerous minute openings for the ciliated funnels. Each ciliated funnel is like an ear-lobe, with about one-fourth of its margin incomplete along one side and is covered with outwardly directed cilia on its margin and inner surface (Fig. 4.17).

Ciliated organ corresponds to coelomoduct of polychaeta and not to nephrostomes of nephridia. In the embryo, the ciliated organ remains connected with the nephridium and probably performs

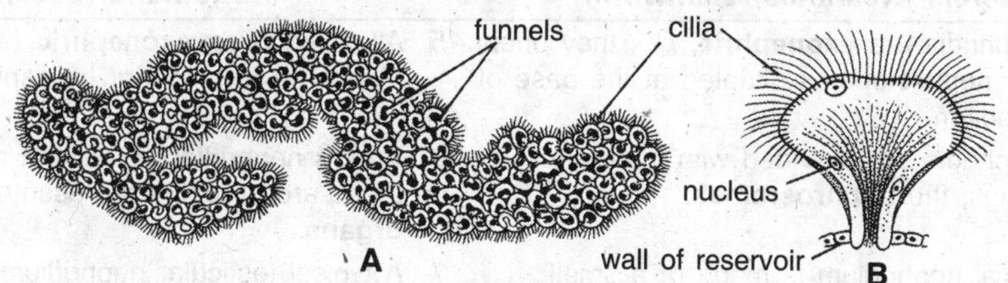

Fig. 4.17. *Hirudinaria*. A—Entire ciliated organ; B—A single funnel.

excretory function. In the adult, it loses excretory function and has no direct connection with the nephridium. Ciliated organ of adult, however, becomes associated with the haemocoelomic system and manufactures coelomic corpuscles.

There exists a central canal which originates in anterior part of apical lobe and traverses through the different lobes of nephridium and finally enters the vesicle duct. It forms a long continuous passageway across the nephridium and carries the excretory products secreted by the glandular cell masses of the nephridium and finally discharges into the vesicle through the vesicle duct.

2. Pre-testicular Nephridium

The first six pairs of nephridia are situated in segments 6th to 11th which lack testis sacs, perinephrostomial ampullae and ciliated organs. They are called pretesticular nephridia and resemble the testicular nephridia in structure. However, their initial lobes end loosely in the general connective tissue on either side of the ventral nerve cord (Fig 4.21).

Physiology

Nephridium of leech is both **osmoregulatory (Haupt 1974)** and **excretory** in function. Excess of water and nitrogenous wastes (chiefly **ammonia**) are excreted out by the nephridium. Nephridial gland cells separate the waste products from the haemocoelomic fluid. The excretory fluid is collected in the vesicle and voided through the nephridiopores. Leech excretre watery osmotic urine. The salts are reabsorbed from the urine by the nephridial tissues.

Waste particles are also picked up by botryoidal tissues (and by vasofibrous tissue) and are retained for life. Amoeboid corpuscles in the coelomic fluid are also phagocytic and play an important role in excretion.

COMPARISON OF NEPHRIDIA OF *NEREIS* AND *HIRUDINARIA*

Table 4.2 Comparison of excretory system of nephridia of *Nereis* and *Hirudinaria*.

	Nereis (*Neanthes; clamworm*)		*Hirudinaria* (*Leech*)
1.	Excretory system consists of **nephridia** only.	1.	Excretory system consists of **nephridia** and **botryoidal tissue.**
2.	**Nephridia** are segmentally arranged coiled tubes of very large size.	2.	**Nephridia** are segmentally arranged coiled tubes of large size.
3.	Paired nephridia are found in all segments except few anterior and posterior segments of the body.	3.	Paired nephridia are found only in the segments from sixth to twenty-second.
4.	All nephridia are alike.	4.	Nephridia are of two types: (*i*) **Pre-testicular nephridia,** which are six pairs in number and located from segments sixth to eleventh, one pair in each segment. (*ii*) **Testicular nephridia,** which are eleven pairs in number and distributed one pair in each segment from twelth to twenty second.

	Nereis (*Neanthes; clamworm*)		*Hirudinaria* (*Leech*)
5.	All nephridia are **exonephric,** *i.e.*, they open out by nephridiopores situated at the base of parapodium.	5.	All nephridia are **exonephric,** *i.e.*, they open directly to the exterior by nephridiopores.
6.	All nephridia are provided with internal openings; the **nephrostomes.**	6.	Adult's nephridia are without nephrostomes; these are probably represented by **ciliated organs.**
7.	A typical nephridium is made of a small narrow neck and an oval **main body** (Fig. 3.15).	7.	A typical testicular nephridium is made of a **main lobe,** a **vesicle** with duct, an **apical lobe,** an **initial lobe** and a **ciliated organ** (Fig. 4.15).
8.	Each nephridium is an oval and irregular in shape.	8.	Each nephridium is horse-shoe shaped.
9.	Sac-like vesicle is not found.	9.	Sac-like vesicle or bladder is present.
10.	Nephridia remove excretory wastes from blood and coelomic fluid.	10.	Nephridia remove excretory wastes from haemocoelomic fluid.
11.	No other excretory organ/tissue is found.	11.	The botryoidal tissue is also involved in excretion (removal and permanent storage).
12.	Ammonotelic.	12.	Ammonotelic.

Differences Between Testicular and Pretesticular Nephridia

Table 4.3 Differences in testicular and pretesticular nephridia of leech.

	Testicular nephridia		*Pretesticular nephridia*
1.	These are associated with testis-sacs.	1.	These are not associated with testis-sacs.
2.	These are eleven pairs in number and are present in only those segments that have testis-sacs.	2.	These are six pairs and are present in those segments which do no have testis-sacs.
3.	Testicular nephridium has six parts— main lobe, inner lobe, apical lobe, initial lobe, vesicle, ciliated organs.	3.	Ciliated organ is absent.

4.10 SENSE ORGANS OR RECEPTORS

The receptors or sense organs of *Hirudinaria* consist of specially modified epidermal cells. These are of following types; 1. Free nerve endings; 2. Annular receptors; 3. Segmental receptors and 4. Eyes.

1. Free nerve endings. These are found all over the body between the epidermal cells. The ganglia cells of these nerve endings lie beneath the epidermis. Free nerve endings are probably **chemoreceptors.**

2. Annular receptors. About 36 annular receptors are found to be equally distributed on the dorsal and ventral sides in each annulus (*i.e.*, 18 on dorsal surface and 18 on ventral surface). Each receptor is formed of a group of flattened overlapping cells and projects out of body wall like a minute papilla (Fig. 4.18). These are **tactile organs** or **tango-receptors.** Annular receptors get their nerve supply from the anterior and posterior lateral nerves.

3. Segmental receptors. First annulus of each body segment bears four pairs of segmental receptors on the dorsal surface and three pairs on the ventral surface. These are larger than the annular receptors and occur in the form of white patches elevated on the elliptical papillae. Each segmental receptors consists of two types of cells (Fig. 4.19).

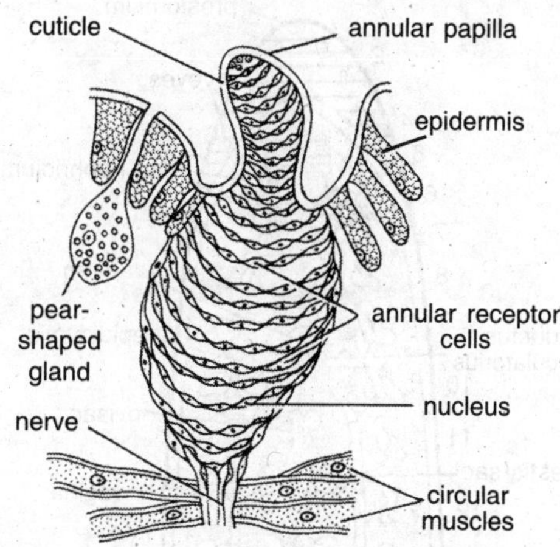

Fig. 4.18. *Hirudinaria*. V.S. of an annular receptor.

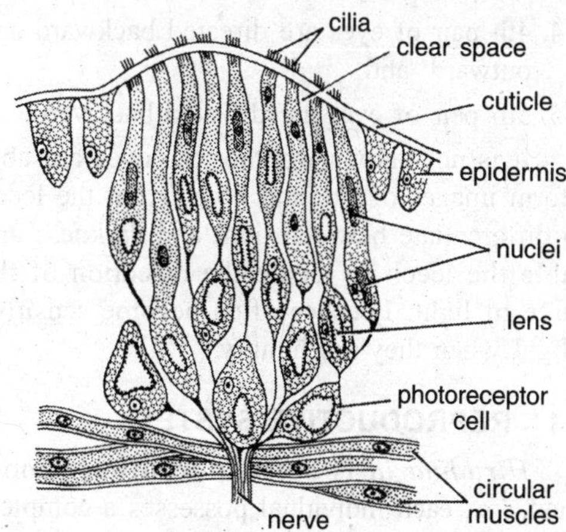

Fig. 4.19. *Hirudinaria*. V.S. of dorsal segmental receptor.

1. Tangoreceptors (or the tactile cells)
2. Photoreceptors (or the light perceiving cells)

(1) Tangoreceptors are about 5 to 10 long and slender cells which are separated from one another by clear spaces. These cells bear hair-like processes on their outer free ends. These are tactile in function (*i.e.*, sensitive to touch).

(2) Photoreceptors are found only in dorsal receptors. These consist of crescentic **optic organelle** or **lens** of hyaline substance which receives light rays.

4. Eyes. Leech possesses five pairs of simple eyes arranged as black spots in a semi-circle along the dorsal margin of the anterior suckér. A pair of eye is found in the first annulus of each of the first five segments. Each eye is cylindrical or cup-shaped with its long axis perpendicular to the body surface. Its wall is formed of black pigmented tissue and encloses numerous refractive or photoreceptor cells arranged in longitudinal rows. Each photoreceptor cell contains a large hyaline crescentic **lens** or **optic organelle** surrounded by a very thin peripheral layer of cytoplasm containing a small rounded nucleus. The distal free end of the eye is protected by a cap of transparent epidermal cells called **cornea** and a layer of cuticle. An optic nerve enters each eye at its base and runs along its median axis and gives off fine branches to all photoreceptor cells.

All the eyes are not of equal size; the first and second pairs are the largest while the remaining pairs are comparatively smaller in size, the fifth pair being the smallest. The eyes of leech are directed in different directions and each eye can receive light only from one direction:

1. 1st pair of eyes are directed forward;
2. 2nd pair of eyes are directed forward and outward;

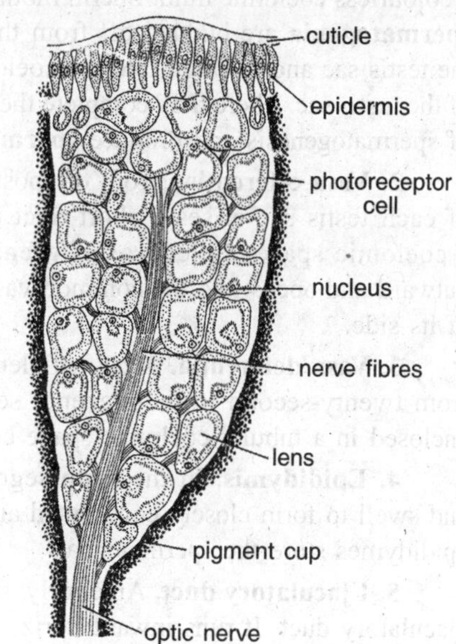

Fig. 4.20. *Hirudinaria*. V.S. of an eye.

3. 3rd pair of eyes are directed upward;
4. 4th pair of eyes are directed backward and outward; and
5. 5th pair of eyes are directed backward.

It is not known whether the eyes are able to form images of external objects but the leech can differentiate between light and darkness and enable the leech to locate the direction of the source of light. Leeches often become sensitive to light when they are hungry.

4.11 REPRODUCTIVE SYSTEM

Hirudinaria is **hermaphrodite** (monoecious), *i.e.*, each individual possesses a complete set of male and female reproductive organs (Fig. 4.21). In leech, copulation and cross-fertilization occur as a rule.

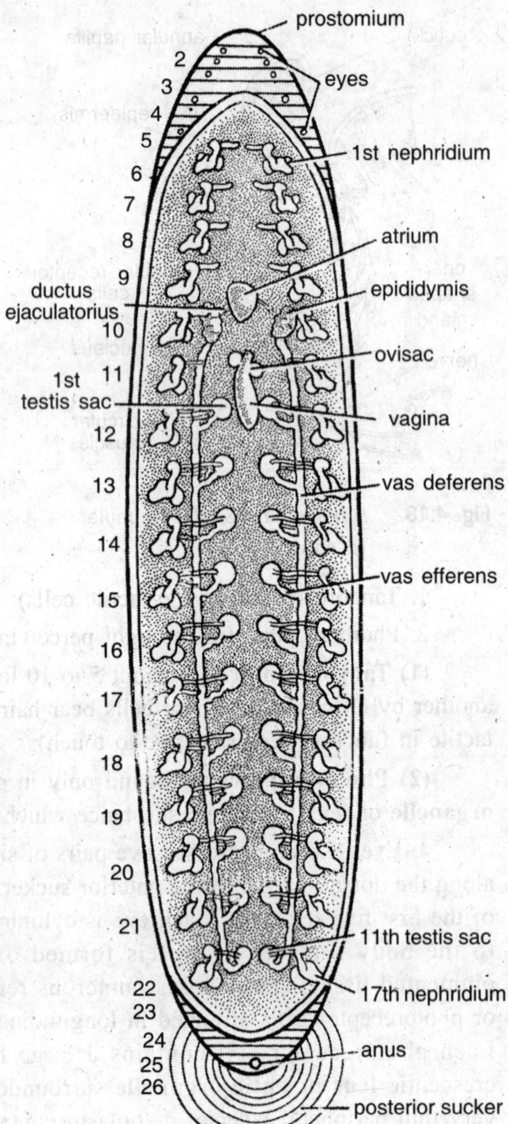

Male Reproductive Organs

The male sex organs are: 1. Testis sacs; 2. Vasa efferentia; 3. Vasa defferentia; 4. Epididymes; 5. Ejaculatory duct; 6. Atrium (Fig. 4.22).

1. Testis sacs. Each body segment from 12th to 22nd contains a pair of small, hollow and spherical **testis sacs** on either side of the ventral nerve cord. Thus, normally there are eleven pairs of testis sacs but sometimes 23rd and 24th segments also contain testis sacs. These testis sacs are true coelomic spaces and are filled with a colourless coelomic fluid. Sperm mother cells or **spermatogonia** are budded off from the wall of the testis sac and released into the coelomic fluid of the testis sac. There they complete their process of spermatogenesis and produce **spermatozoa.**

2. Vasa efferentia. From the posterior end of each testis sac arises a short duct enclosing a coelomic space, called **vas efferens.** It runs outward and opens into the common vas deferens of its side.

Fig. 4.21. *Hirudinaria.* Reproductive and nephridial (*i.e.*, urinogenital) system.

3. Vasa deferentia. A pair of slender longitudinal ducts, the **vasa deferentia,** run forward from twenty-second to the eleventh segment. It lies attached to the ventral body wall and is enclosed in a tubular coelomic space containing a fluid with amoeboid corpuscles.

4. Epididymis. In the tenth segment the vas deferens comes out of the coelomic space and swell to form closely convoluted and compact mass, the **sperm-vesicle** or **epididymis.** Both epididymes store the spermatozoa.

5. Ejaculatory duct. Anteriorly, from the inner side of the epididymis arises a short narrow ejaculatory duct. It runs inward horizontally and opens into a common atrium.

6. Atrium. It is a small pyriform sac situated midventrally in ninth and tenth segments (Fig. 4.22). It opens to outside in tenth segment by a mid-ventrally located **male genital pore.** The artium has following two distinct parts (Fig. 4.23):

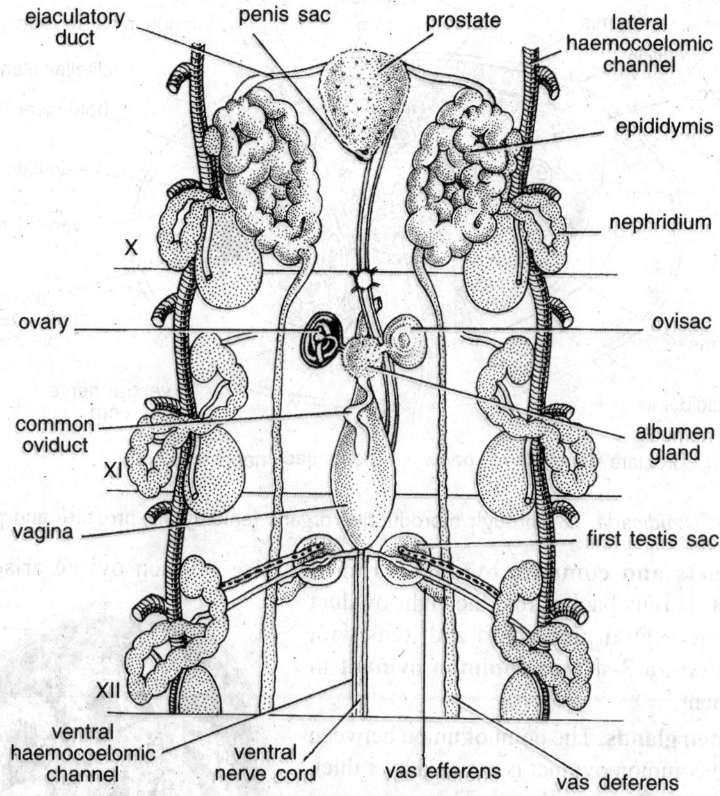

Fig. 4.22. *Hirudinaria*. Male and female reproductive organs.

(*i*) Prostate chamber. It is the anterior broad part of atrium. Its wall are thick and muscular and are surrounded by numerous unicellular **prostate glands.** Each gland opens into the prostate chamber by a slender duct. The spermatozoa are transferred into this chamber by the ejaculatory ducts. The sperms become glued together by the prostatic secretion into packets or bundles. These bundles are called **spermatophores.**

(*ii*) Penis sac. It is an elongated posterior part of atrium. It is highly muscular and contains a coiled filamentous tubular and eversible **penis.** The penis is often seen protruding through the male genital pore. It helps in copulation and transfer of spermatophores into the vagina of other leech.

Female Reproductive Organ

Female sex organs are: 1. Ovisacs; 2. Ovaries; 3. Oviducts; 4. Common oviduct; 5. Vagina; and 6. Albumen glands.

1. Ovisacs. There is a pair of ovisacs situated ventro-laterally in eleventh segment one on either side of the ventral nerve cord. Each ovisac is a hollow globular sac filled with colourless coelomic fluid. Each ovisac contains a single **ovary** (Fig. 4.23).

2. Ovaries. Leech contains a pair of ovaries, one in each ovisac. Each ovary is long, coiled, delicate, thread-like and contains a nucleated cytoplasm. The two ends of ovary are swollen and club-shaped. The ova are budded off from the ovary. The ovary lies floating in the coelomic fluid of ovisac.

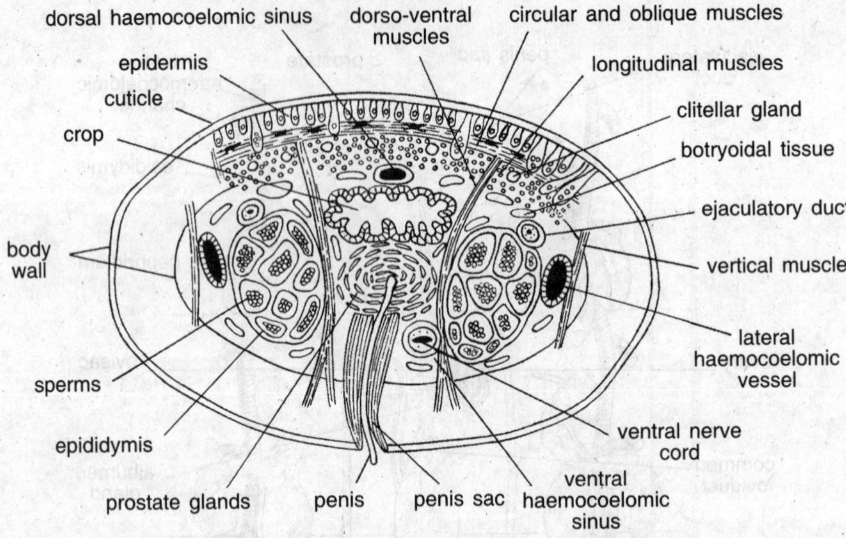

Fig. 4.23. *Hirudinaria*. T.S. through reproductive organs (epididymis, prostate and penis sac).

3. Oviducts and common oviduct. From the base of each ovisac arises a short and slender oviduct. It runs backwards, the right oviduct passes below the ventral nerve cord and joins with that of left to form a S-shaped **common oviduct** in the same segment.

4. Albumen glands. The point of union between the oviduct and common oviduct is covered by a thick layer of unicellular **albumen glands.** These open into the common oviduct.

5. Vagina. It is a large pear-shaped bag lying in the posterior part of eleventh segment (Fig. 4.22) During breeding season, vagina increases in size and becomes internally thrown into a large number of longitudinal folds. It opens to the exterior through a mid-ventral **female genital pore** in eleventh segment.

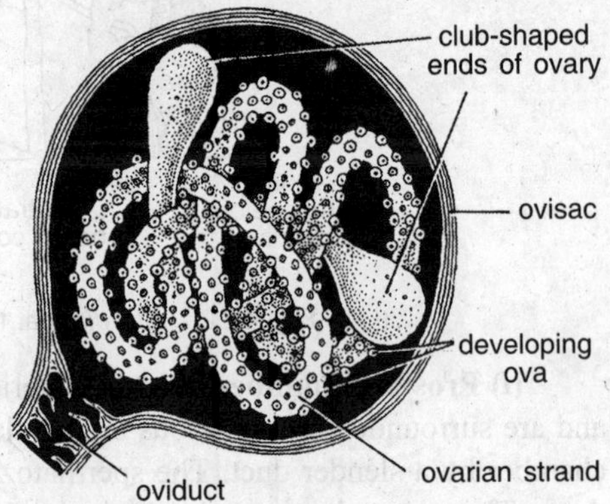

Fig. 4.24. *Hirudinaria*. An ovisac dissected to show the ovary.

The ova are budded off from the ovaries and pass through the oviducts into the vagina where they remain for some time.

Copulation and Fertilization

In leech, **copulation** takes place in the months of March and April. It occurs either on land or in water. During copulation two leeches come together by their ventral surfaces in a head-to-tail position (Fig. 4.25), so that the male genital pore of one lies against the female genital pore of the other. Then, the penis of one is inserted into the vagina of other and

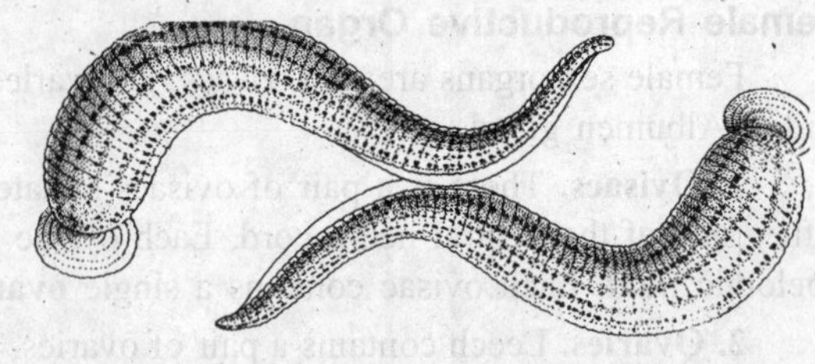

Fig. 4.25. *Hirudinaria*. Two leeches in copulation.

the spermatophores are deposited there. The **fertilization** of ova occurs in the vagina (*i.e.*, internal fertilization) and the zygotes are laid down into cocoon.

Cocoon Formation

Cocoon formation in *Hirudinaria* takes place in the months of April, May and June by the secretion of clitellum. In leech, a temporary clitellum is formed around segments ninth to eleventh during the breeding season. The clitellar glands secrete a frothy girdle which is snow white in colour (Fig. 4.26). It hardens to form a **cocoon.** The albuminous gland secretes **albumen** into the cocoon which is used as nourishment by the developing embryo. The fertilized eggs are extruded into the cocoon. The leech wriggles out backwards from the cocoon (*i.e.*, cocoon is passed off over the head of each individual). The two ends of cocoon are closed by **polar plugs** (*i.e.*, an anterior plug and a posterior plug which are secreted by the prostomial glands). The complete cocoon is formed within six hours and is laid down in some moist place but never in water.

Cocoon becomes hard, barrel-shaped and light yellow in colour. It is about 25 to 30 mm long and 12 to 15 mm wide. It is filled with jelly-like albumen and contains 1 to 24 developing embryos.

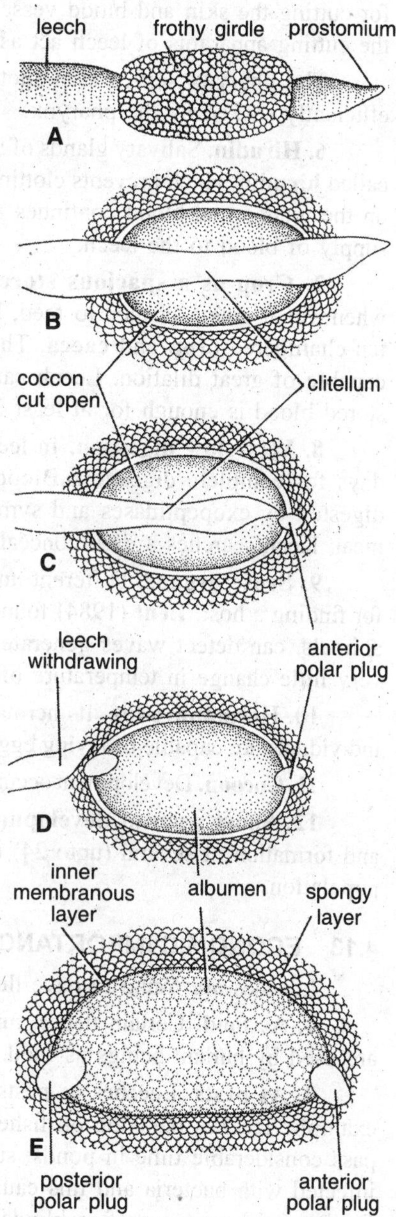

Fig. 4.26. *Hirudinaria*. Stages in cocoon formation A, B—Formation of frothy girdle. C—Formation of anterior plug; D—Formation of posterior plug; E—L.S. of a cocoon.

Development

Development in leech is **direct** involving no larva in its life cycle. It is completed in about 14 days. Young leeches after hatching out of the eggs swim in the albumen and feed on it. Soon the plugs drop off and young leeches come out of the cocoon.

4.12 PARASITIC ADAPTATIONS

Hirudinaria, the common Indian cattle leech, leads a parasitic mode of life. It is **semi-ectoparasitic** (*i.e.*, it normally leads an independent life but becomes attached to some mammal to feed) and **sanguivorous** (*i.e.*, it feeds by sucking blood of cattle and man). Its habit and habitat, morphology and physiology show the following important parasitic adaptations.

1. Habit and habitat. Leech lives in freshwater ponds, pools, slow-running streams and marshes which are regularly visited by cattles, human beings and other mammals for bathing and drinking the water. This ensures regular supply of host.

2. Body colour. The dark olive green colour helps leech in hiding undetected in weeds.

3. Shape. Its long, flattened, limbless and slimy body is well suited for its aquatic mode of life and for its swift swimming nature. Its **slimy body covering** protects it from desiccation, reduces friction during locomotion and also helps in the cutaneous respiration.

4. Suckers. Anterior and posterior **suckers** are the organs of attachment (with host as well as substratum); particularly posterior sucker is much more powerful and muscular.

5. Jaws and suctorial pharynx. The three **denticulated jaws** are so highly specialized for cutting the skin and blood vessels that host does not feel any pain. Certain secretions from the cutting apparatus of leech act as local **anesthetic.**

The pharynx serves as a suction pump. The blood flowing out of the host's wound is efficiently sucked by the pharynx.

6. Hirudin. Salivary glands of leech open on the surface of jaws and secrete an anticoagulant, called **hirudin** which prevents clotting of blood. As the jaws cut the skin of host, hirudin is poured on the wound and blood continues to ooze (without clotting). Thus, hirudin ensures continuous supply of blood to the leech.

7. Crop as a spacious store house. Getting a host is always a matter of chance. So whenever it gets a chance to feed, leech can suck and store enormous quantity of blood in its ten-chambered **crop** and **caeca.** The crop and its caeca are spacious, thin-walled, elastic and capable of great dilation. Leech can store blood two to five times its own body weight. This stored blood is enough for at least 200 days. Leech can also tolerate long period of fasting.

8. Very slow digestion. In leech, digestion is a very slow process. A full meal needs 200 days for complete digestion. Blood from the crop can enter the stomach drop by drop. It is digested by exopeptidases and symbiotic bacteria present in the intestine. After having a full meal, leech searches a dark concealed shelter and rests there for months.

9. Sense organs. Different annular and segmental receptors of leech are highly sensitive for finding a host. **Lent** (1984) found that *Hirudo*, another blood-sucking leech of warm-blooded animals, can detect waves generated by a host and is also attracted by the body secretion and very little change in temperature of water caused by the entry of a host in water.

10. Hermaphrodite. Its hermaphrodite sexuality has doubled the rate of reproduction as all individuals are capable of laying eggs in contrast to only 50 percent individuals of unisexual form.

11. Cocoon. Development occurs in cocoon which provides protection to developing young ones.

12. Short span of development. A short span of developmental period, *i.e.*, a fortnight, and formation of several (upto 24) individuals in a single cocoon ensures to maintain its regular population.

4.13 ECONOMIC IMPORTANCE

Leeches are found to have the following economic uses for human beings:

1. As food. Certain fishes, turtles, ducks and other birds, etc., eat leeches. These animals are used by human beings as food.

2. As pests. Leeches as pests cause great economic loss either directly or indirectly. For example, leeches parasitic on fishes cause great losses to fishery. Cattles and buffaloes which pass considerable time in ponds, suffer due to leeches. The wounds caused by leeches become infected with bacteria and this causes crippling effect and death of buffaloes and other cattles. Cattles and horses are often blinded by their attacks. The bite of Indian land leeches in forests is very painful. Due to their bites, human beings suffer serious loss of blood.

3. Medical Importance. Leeches owe medical importance from older time. They have been used in medicine for the following purposes.

(1) Leeches extract the blood of their victim in a painless manner. Due to this fact leeches were extensively used in **phelobotomy** (blood-letting) under the impression that these remove bad blood from the patient and cure many diseases. Tribals still use leeches for this purpose.

(2) Today, leeches are used to reduce **haematomas** (a mass of clotted blood that forms in a tissue, organ, or body space as a result of broken blood vessel) in areas of body that are difficult to treat surgically and to avoid leaving scars (see **Brusca** and **Brusca** 2003). Leeches are used as materia-medica in treatment of piles, tonsillitis and baldness.

(3) Hirudin as an anticoagulent remained in use in hospitals for a long time. However, it is not used now.

(4) In addition to the anticogulant, some leaches produce a number of other chemicals, including **anesthetics** and **vasodilators**, that are being studied for possible use in human medicine (**Conniff** 1987).

QUESTIONS

Long Answer Questions

1. Describe the external features of *Hirudinaria.*
2. Describe the structure and functions of body wall of leech and add a note on locomotion of leech.
3. Describe the alimentary canal of *Hirudinaria* and mention clearly the mode of sucking of the blood.
4. Describe the alimentary canal of *Hirudinaria* and mention clearly the mode of sucking blood in this animal and its peculiarities.
5. Give an account of the digestive organs and mechanism of feeding in leech. Explain in brief how these organs are adapted to its parasitic mode of life.
6. Give an account of the digestive system of *Hirudinaria.*
7. Describe the reproductive organs of leech. Give neat diagram.
8. Describe the urinogenital system of leech with labelled diagram stating its peculiarities and functions.
9. Describe the reproductive system and the process of reproduction and cocoon formation in leech.
10. Describe the reproductive system of leech and state clearly how reproduction takes place in the animal?
11. What is coelom? Give an illustrated account of the haemocoelomic system in leech.
12. Write an account of the coelom and the blood vessels of a leech and discuss the nature of the sinuses in this animal.
13. Describe the haemocoelomic system or circulatory system of *Hirudinaria.*
14. Give an account of the lateral haemocoelomic channels of leech.
15. What is nephridium? Describe the structure and function of excretory organs of *Hirudinaria.*
16. Describe the excretory system or nephridia of leech.
17. Describe lateral haemocoelomic channels and their branches in leech.
18. Giving labelled diagram, describe the excretory system of leech. Add a note on the physiology of excretion.
19. Describe a typical nephridium of *Hirudinaria* and compare it with that of *Nereis.*
20. Give a comparative account of the excretory organs of annelids studied by you.
21. Give an account of receptor organs of leech.
22. Describe parasitic adaptations in *Hirudinaria* and its economic importance.

Short Answer Questions

1. Give the histological structure of the body wall of leech.
2. Draw a labelled diagram of body wall of leech.
3. Describe the locomotion of leech.
4. Write a short note on the botryoidal tissue.

5. Describe locomotion in leech?

6. Write a note on copulation and cocoon formation in leech.

7. What is anticoagulin?

8. Discuss the role of botryoidal tissue?

9. What is haemocoel?

10. How leech is able to maintain a continuous flow of host blood while feeding?

11. Describe the role of radial muscles in leech.

12. Write short notes on the following:

(*i*) Cocoon

(*ii*) Crop of *Hirudinaria*

(*iii*) Feeding in leech

(*iv*) Haemocoel in leech

(*v*) Hirudin

(*vi*) Locomotion in leech

13. Write short notes on the following:

(*i*) Testicular nephridium of leech

(*ii*) Ciliated organ.

(*iii*) Ventral haemocoelomic system of leech.

(*iv*) Haemocoel.

14. How cocoon formation occurs in *Hirudinaria*?

15. Give an account of ciliated organs of leech.

16. Describe the process of respiration in leech.

17. What are basic differences between testicular and pretesticular nephridia?

18. Draw well labelled diagrams of the following:

(*i*) External features of *Hirudinaria*

or

Dorsal and ventral view of leech.

(*ii*) Alimentary canal of leech.

(*iii*) T.S. leech through stomach

(*iv*) T.S. leech through crop or through middle of body.

(*v*) Testicular nephridium of leech.

(*vi*) Reproductive organs of *Hirudinaria*. (*Kanpur 2001*)

(*vii*) T.S. leech passing through pharynx. (*C.C.S.U., Meerut 2002*)

(*viii*) Receptors of leech

(*ix*) Epidermal glands of leech

(*x*) Female reproductive organs of leech.

(*xi*) Urinogenital system of leech.

(*xii*) T.S. leech through epididymis and penis sac.

Very Short Answer Questions

1. Which segments of leech are triannulate?

2. What is haemolization?

3. Name the muscle layers in the body wall of leech.

4. Give the function of the chitogenous glands.

5. What is the location of albumen glands?

6. Name the glands which secrete hirudin?
7. What is the function of crop in leech?
8. Which segments of leech contain nephridia?
9. Mention the position of female genital aperture.
10. Where does fertilization occur in leech?
11. What is the significant function of clitellum?
12. Why does not blood coagulate when a leech sucks it?
13. Which anticoagulating substance is found in leech? (*C.C.S.U. Meerut 2001*)
14. How many segments form head in leech?
15. What are the parts of cephalic region of leech?
16. Why are the segments of cephalic region of leech called ocular segments?
17. How many segments form the caudal region in leech?
18. How will you collect leech?
19. Where is the botryoidal tissue found in leech?
20. How many pair of pretesticular nephridia are present in leech and in which segments?
21. Of how many segments the posterior sucker of leech is formed?

Fill in the Blanks

1. The zoological name of cattle leech is ....
2. Body of leech consists of .... segments.
3. Leech has two suckers: .... and posterior.
4. Cephalic region of the leech consists of .... segments.
5. Clitellar region of leech consists of three segments, from .... to ....
6. Posterior sucker of leech is formed of .... segments.
7. Leech has .... pairs of eyes.
8. In leech, segmental receptors occur on the .... annulus of each segment.
9. Male genital aperture of *Hirudinaria* lies in the groove between the .... and .... annuli of the segment number ....
10. The epidermal cells of the leech are .... shaped.
11. The body wall muscles are differentiated into .... layers.
12. Mouth of leech lies in a membrane called ....
13. Crop of leech consists of a row of .... chambers.
14. Triradiate wound in the skin of host is made by .... while feeding on host's blood.
15. Saliva of leech contains .... which prevents clotting of blood.
16. Coelom of leech is reduced by .... tissue.
17. Leech has .... pair of nephridia.
18. There are .... pairs of testes in *Hirudinaria*.
19. Fertilization in leech occurs in the ....
20. *Hirudinaria* belongs to the order .... of class Hirudinea.
21. Blood vascular system in leech is of .... type.
22. Body cavity of leech is filled with a network of connectives called .... tissue.
23. Coelomic corpuscles in leech are manufactured in ....

24. Ciliated organs lie in .... ampulla.

25. *Hirudinaria* is .... in nutrition.

Multiple Choice Questions

1. Common cattle leech belongs to the genus

(*a*) *Megascolex* (*b*) *Nereis*
(*c*) *Arenicola* (*d*) *Hirudinaria*

2. Locomotion in leech is brought about by

(*a*) parapodia (*b*) setae
(*c*) body musculature and suckers (*d*) suckers only

3. What is true about *Hirudinaria*?

(*a*) it has a cephalic sucker (*b*) mouth is triradiate
(*c*) segments are superficially subdivided (*d*) all of these

4. Which one of the following is an ectoparasitic annelid?

(*a*) *Hirudinaria* (*b*) *Lumbricus*
(*c*) *Pheretima* (*d*) *Ascaris*

5. Leech is a blood sucking animal nutritionally, therefore leech is

(*a*) herbivorous (*b*) carnivorous
(*c*) sanguivorous (*d*) omnivorous

6. A typical segment of leech is

(*a*) uniannulate (*b*) biannulate
(*c*) triannulate (*d*) quinquannulate

7. The anterior sucker of *Hirudinaria* is formed by the union of

(*a*) one segment (*b*) three segments
(*c*) seven segments (*d*) prostomium and three segments

8. The posterior sucker of *Hirudinaria* is formed by the union of

(*a*) one segment (*b*) seven segments
(*c*) twenty five segment (*d*) thirty three segments

9. In each segment of leech the sensillae lie on the

(*a*) first annulus (*b*) second annulus
(*c*) third annulus (*d*) fourth annulus

10. Number of annular receptors in each annulus

(*a*) 18 (*b*) 18 on dorsal and 18 on ventral
(*c*) 9 (*d*) none of the above

11. The material that fills the intervisceral spaces in leech is

(*a*) mesenchyme (*b*) mesogloea
(*c*) botryoidal (*d*) parenchyma

12. The largest organ of the alimentary canal of leech is

(*a*) buccal cavity (*b*) stomach
(*c*) intestine (*d*) crop

13. Hirudin of the saliva of leech prevents

(*a*) digestion of blood (*b*) clotting of blood
(*c*) putrefaction of blood (*d*) haemolysation of blood

14. Digestion of food in leech is brought about by

(*a*) proteolytic enzymes (*b*) bacteria
(*c*) both of these (*d*) none of these

15. Which segments possess nephridia in *Hirudinaria*?

(*a*) 7–10 segments (*b*) 6–22 segments
(*c*) 10–20 segments (*d*) 7–18 segments

16. Excretory matter of leech is mainly

(*a*) ammonia (*b*) urea
(*c*) uric acid (*d*) amino acid

17. Leech is

(*a*) unisexual (*b*) bisexual
(*c*) nonsexual (*d*) none of these

18. A temporary clitellum is formed in *Hirudinaria* on segments

(*a*) 7th, 8th and 9th (*b*) 9th to 11th
(*c*) 11th to 13th (*d*) none of these

19. In which segment do you find the ovisacs in *Hirudinaria*?

(*a*) 8th (*b*) 11th
(*c*) 15th (*d*) 13th

20. Number of testes in *Hirudinaria* is present in

(*a*) 10 pairs (*b*) 11 pairs
(*c*) 12 pairs (*d*) 13 pairs

21. Worm employed in medical practice in the past belongs to

(*a*) *Entamoeba histolytica* (*b*) *Hirudo medicinalis*
(*c*) *Pheretima posthuma* (*d*) *Ancylostoma duodenale*

ANSWERS

Very Short Answer Questions

1. Fourth, fifth and sixth.
2. Breakdown of erythrocytes to dissolve haemoglobin.
3. Circular, oblique and longitudinal.
4. Secretion of the material for the formation of egg case.
5. In the body wall of clitellum.
6. Salivary glands.
7. Storage of ingested blood.
8. Segments 6 to 22.
9. In the groove between the second and third annuli of the segment XI on the ventral side.
10. In vagina.
11. The clitellar glands secrete a snow-white frothy substance of which cocoon, ootheca or egg case is formed. Glands of clitellum also secrete an albuminous fluid to be deposited with the fertilized ova within the cocoon.
12. It is because the salivary glands of leech secrete an anticoagulant called **hirudin** besides the saliva on the wound which prevents clotting of blood while it is being sucked.
13. The anticoagulating substance, hirudin, found in the saliva of leech is secreted by salivary glands of the leech.
14. The head or cephalic region of leech is formed of first five segments. Segment one and two are uniannulate segement, third segment is biannulate whereas segment fourth and fifth are triannulate.

15. The head or cephalic region of leech includes prostomium, anterior sucker, mouth and eyes.

16. Since a pair of eyes are situated on the dorsal side of first annulus in all cephalic segments (*i.e.*, from segment one to five). These segments are, therefore, called **ocular segments.**

17. Caudal region of leech is made up of four incomplete segments 23rd to 26th. Segment 23rd is triannulate and the rest are biannulate.

18. Leeches are ectoparasites on vertebrates (cattles, human beings, etc.) and feed upon their blood. In nature, they are found mostly in freshwater ponds, tanks, lakes marshes and slow-moving streams. From these places leeches are collected in the old fashion. A person immerses his or her body in such muddy water where leeches are in abundance. Many leeches tend to adhere to the legs and later on are detached and collected. Sometimes, they are collected from buffaloes who love to wallow in mud and water.

19. Botryoidal tissue is situated beneath the longitudinal muscles surrounding the alimentary canal. It is filled in coelom reducing it into haemocoelomic channels and sinuses.

20. There are six pairs of pretesticular nephridia found in segments 6–11.

21. It is formed of seven segments.

Fill in the Blanks

1. *Hirudinaria granulosa*
2. Thirty three (33)
3. Anterior
4. Five
5. Segment nine, segment eleven
6. Seven
7. Five
8. First
9. Second, third, ten
10. Hammer or Malleate
11. Three
12. Velum
13. Ten
14. Leech jaws
15. Hirudin
16. Botryoidal
17. Seventeen
18. Eleven
19. Vagina
20. Gnathobdellida
21. Haemocoelomic
22. Botryoidal
23. Ciliated organ
24. Perinephrostomial
25. Sanguivorous.

Multiple Choice Questions

1. (*d*) **2.** (*c*) **3.** (*d*) **4.** (*a*) **5.** (*c*) **6.** (*d*)
7. (*c*) **8.** (*b*) **9.** (*a*) **10.** (*b*) **11.** (*c*) **12.** (*d*)
13. (*b*) **14.** (*b*) **15.** (*b*) **16.** (*a*) **17.** (*b*) **18.** (*b*)
19. (*b*) **20.** (*b*) **21.** (*b*).

5

Excretion in Annelida; Trochophore Larva and Vermicomposting

5.1 EXCRETION IN ANNELIDA

Most annelids possess some sort of **nephromixa**.

Nephromixia

In annelids, there is a frequent association of nephridia, especially **metanephridia**, with structures called **ceolomoducts**. Coelomoducts are tubular connections arising from the coelomic lining and extending to the outside via special pores in the body wall. Their inner ends are frequently funnel–like and ciliated, resembling the nephrostomes of metanephridia. Coelomoducts may have arisen evolutionarily as a means of allowing the escape of gametes to the outside; they are indeed considered homologous to the reproductive ducts of many invertebrates. Primitively, the coelomoducts and nephridia were separate units; however, through evolution they have in many cases fused in various fashions to become what are called **nephromixia** (**Brusca** and **Brusca** 2002). Nephromixia are of three types:

Protonephromixium. In it a coelomoduct is joined with a protonephridium and they share a common duct.

Metanephromixium or Mixionephridium. In them the coelomoduct is united with a metanephridium. However, the structural nature of this union is different in both of them.

Further, whereas coelomoducts originate from the coelomic lining, the nephridial components arise from the outer body wall, so nephromixia are a combination of mesodermally and ectodermally derived parts (**Bursca** and **Brusca** 2002).

Often nephromixia are serially arranged as one pair per segment, with the pore in the segment posterior to the nephrostome. However, variations on this theme are many and we may view the different conditions as having been derived from a basic primitive plan that arose with the evolution of the coelomate metameric bauplan. In fact, the success of an animal with segmentally arranged coelomic compartments depends on the physical and physiological maintenance of those separate segments. The removal of metabolic wastes (predominantly ammonia) and the regulation of osmotic and ionic balance must occur in each functionally isolated coelomic chamber.

1. Nephromixia in Polychaeta. The presumed primitive condition in polychaetes is a serially homonomous body with a pair of complete coelomic spaces in each segment (Box 5.1). Primitively, each coelom is served by two pairs of ducts that lead to exterior; each pair includes a coelomoduct and a nephrioduct. The inner end of each coelomoduct bears an open, ciliated funnel through which gametes escape, and the inner end of each nephridioduct bears protonephridia

Box 5.1

The bodies of some annelids are **homonomous**, bearing segments that are very much alike. Many others, have groups of segments specialized for different functions and are thus **heteronomous** (**Brusca** and **Brusca** 2002).

that function in excretion and osmoregulation. This primitive condition has been lost in all but a very few extant polychaetes, *e.g.*, it persists in *Vanadis* (family Alciopidae). In other few hundred species of polychaetes that possess protonephridia, the coelomoduct and nephridoduct are united to form a protonephridium (*e.g.*, various phyllodocids such as *Nereis*; Fig. 5.1.

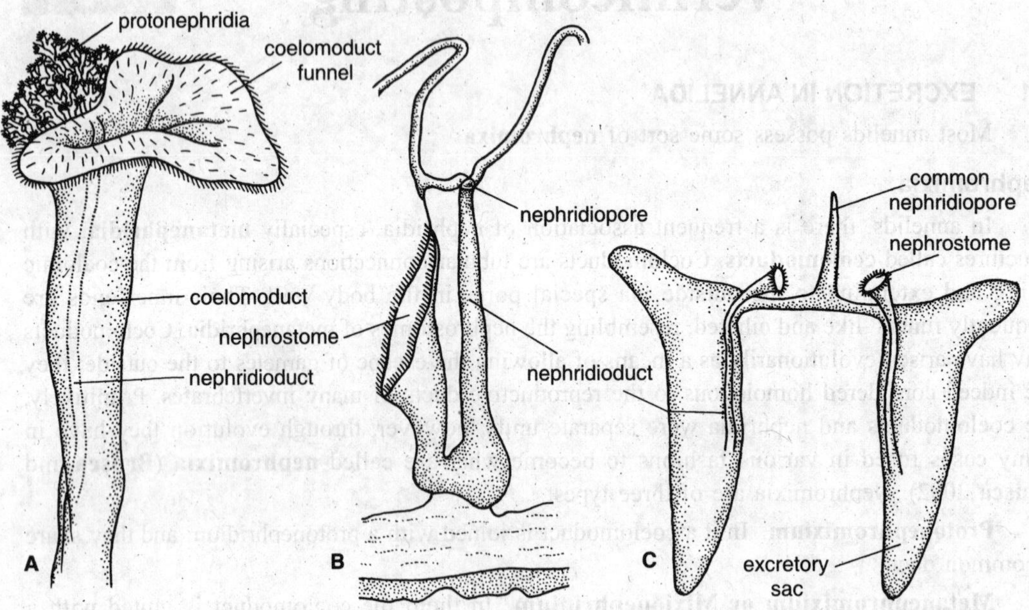

Fig. 5.1. Polychaete nephridia. A–A protonephromixium of a phyllodocid. Here a cluster of solenocytic protonephridia sits a top a nephridioduct. B–A mixonephridium of a spionid. C–A single pair of nephridia joined to a common duct in a surpulid (after Brusca and Brusca 2002).

In vast majority of polychaetes, however, possess metanephridia that open to coelom by a ciliated nephrostome. In some (*e.g.*, the capitellids such as *Arenicola*, *Capitella*, *Clymenella*), these metanephridia are entirely separate from the coelomoducts, but is most, either the coelomic and nephridial ducts are united to from a metanephromixium, or there is a single interior opening that leads to a singel duct as a mixonephridium (Fig. 5.1B). This last case, generally called simply a **metanephridium**, may represent the complete incorporation of coelomoduct and nephridium into a single organ.

In certain polychaete groups that posses incomplete septa, or that have lost septa between the coelomic spaces, the number of nephridia is reduced. In some sedentary polychaetes (*e.g.*, serpulids and sabellids) without complete intersegmental septa, there is typically only a single pair a nephridia. These are located at the anterior end of worm and lead to a single nephridioduct and common pore on the head (Fig. 5.1C).

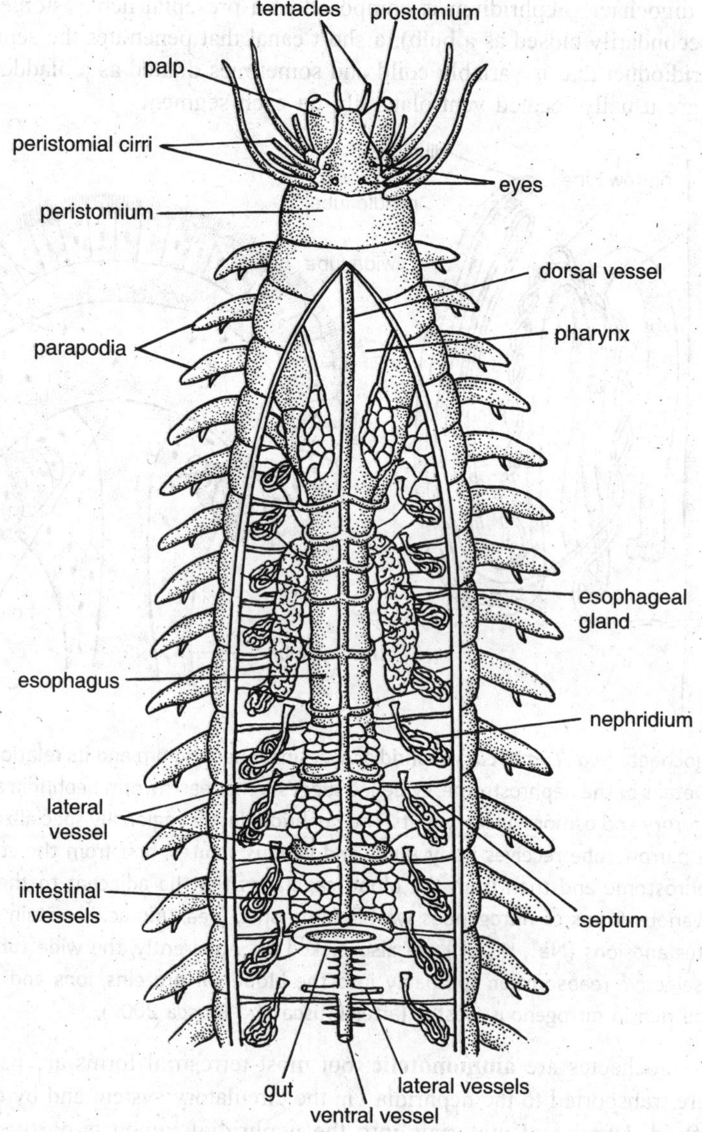

Fig. 5.2. A dissected nereid (dorsal view) showing position of nephridia in the body.

The open nephrostomes of metanephridia nonselectively pick up coelomic fluids. This action is followed by resorption of materials from the nephridium back into the body, either directly into the surrounding coelomic fluid, or into the blood in cases where extensive nephridial blood vessels are present (*e.g.*, in some nereids and the aphroditis). In either case the composition of the urine is quite different from that of the body fluids; the difference indicates a significant amount of physiological selectivity along the length of the nephridium.

2. Nephromixia of Oligochaeta. Oligochaetes bossess paired, segmentally arranged metanephridia usually in all but the extreme anterior and posterior segments. These nephridia are similar to the mixonephridia of polychaetes, but many show various secondary modifications or elaborations.

A typical digochaete nephridium is composed of a preseptal nephrostome (either open to the coelom or secondarily closed as a bulb), a short canal that penetrates the septum, and a post-segmental nephridioduct that is variably coild and sometimes dilated as a bladder (Fig. 5.3). The nephridiopores are usually located ventrolaterally on each segment.

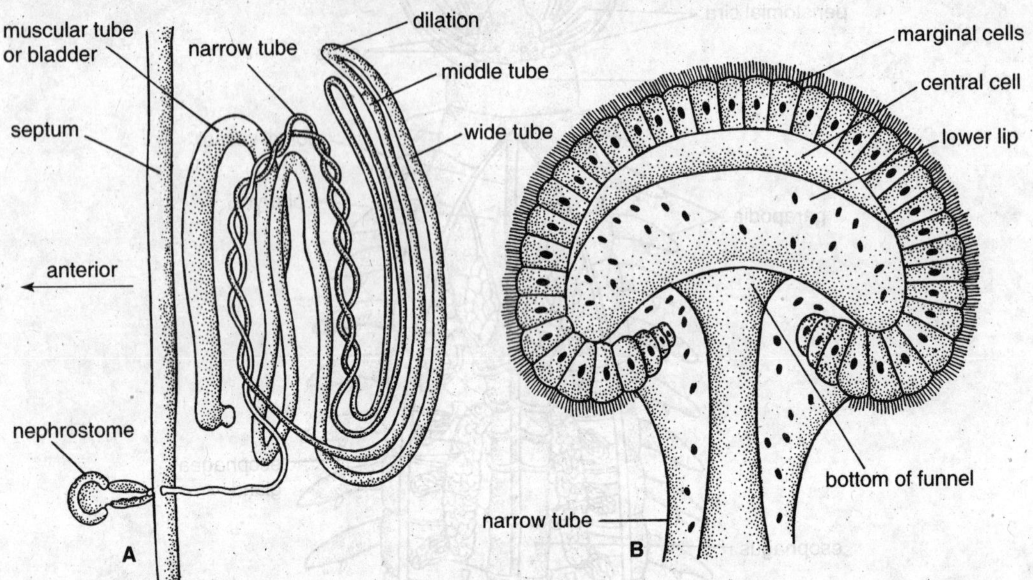

Fig. 5.3. Oligochaete (*e.g., Lumbricus*) nephridia. A–A single nephridium and its relationship to a septum. B–Details of the nephrostrome. Evidence suggests that earthworm nephridia are highly selective excretory and osmoregulatory units. The nephridioduct is regionally specialized along its length. The narrow tube receives **body** fluids and various solutes, first from the coelom through the nephrostome and then from the blood capillaries that lie adjacent to the tube. In addition to various forms of nitrogenous wastes (ammonia, urea, uric acid) certain coelomic proteins, water and ions (Na^+, K^+, Cl^-) are also picked up. Apparently, the wide tube serves as a site of selective reabsorption (probably into the blood) of proteins, ions and water, leaving the urine rich in nitrogenous wastes (after Brusca and Brusca 2002).

Aquatic oligochaetes are **ammonotelic**, but most terrestrial forms are partially **ureotelic**. These wastes are transported to the nephridia via the circulatory system and by diffusion through the coelomic fluid. Uptake of materials into the nephridial lumen is partly nonselective (in those worms with open nephrostomes) from the coelom, and partly selective across the walls of the nephridioduct from the afferent nephridial blood vessels. A significant amount of selective resorption occurs into the efferent blood flow along the distal portion of the nephridioduct, facilitating efficient excretion as well as ionic and osmoreguation (Fig. 5.3).

Chloragogen cells. The precise role of the chloragogne cells in excretion is not fully understood. While it is known that protein deamination and nitrogenous waste formation occur within these cells, the method of elimination of this waste is unknown. Individual chloragogen cells break free into the coelom and are probably engulfed by phagocytic amoebocytes that apparently accumulate wastes in a precipitated form. How these waste–filled cells are actually eliminated from the body remains an unanswered question?

3. Nephromixia in Hirudioidea. The excretory structures of hirudinoideans are structurally different from those of oligochaetes and polychaetes, but they are presumably derived from metanephridia. Leech nephridia are paired and segmentally arranged but are usually absent

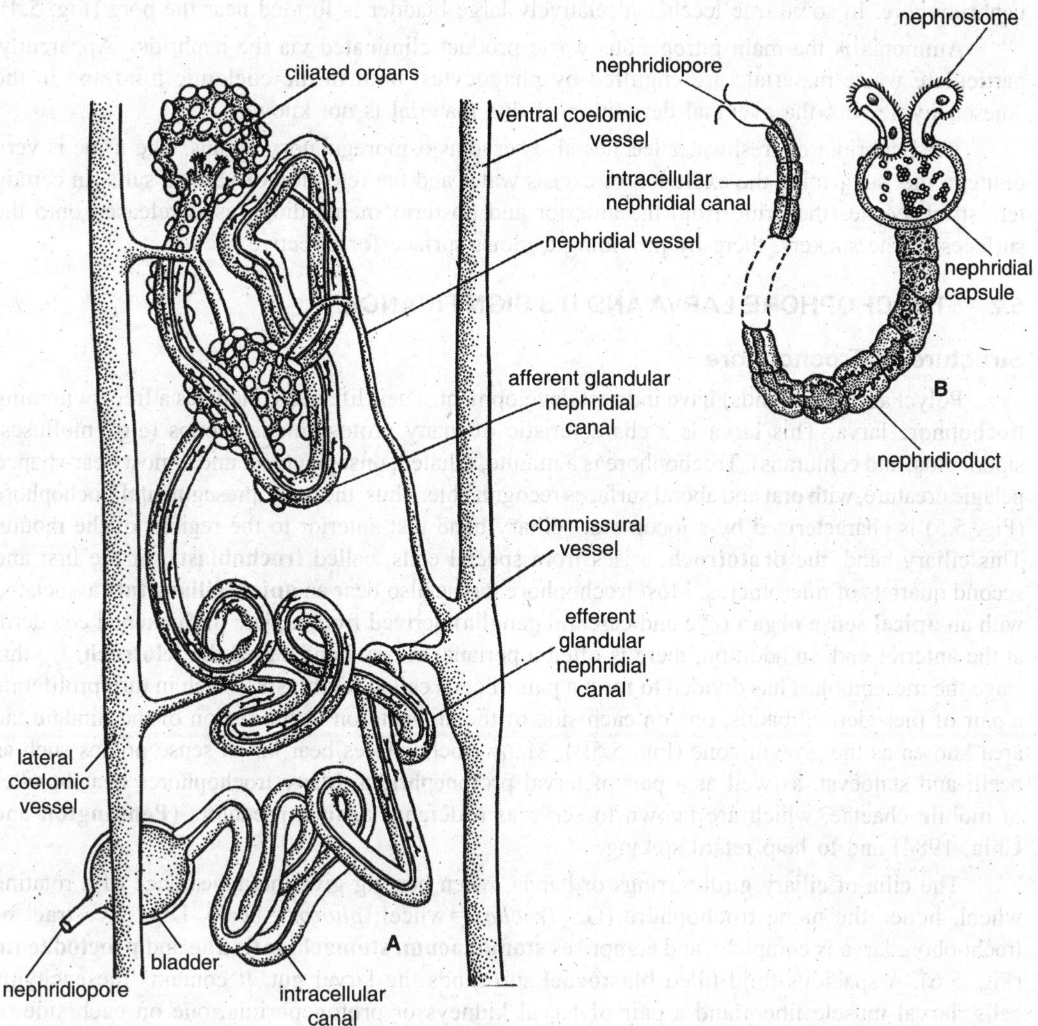

Fig. 5.4. Leech nephridia. A–A complex nephridium of *Erpobdella* and its association with the coelomic channels. B–Details of a nephridium of an arhynchobdelid leech.

from several anterior and posterior segments. The nephrostomes are ciliated funnels associated with coelomic circulatory vessels, an arrangement that probably evolved with the reduction in the main body coelom and the loss of septa. Some hirudineans possess clusters of nephrostomes called **ciliated organs**. Each nephridium leads ultimately to a ventrolateral nephridiopore. It is, however, the microscropic strucutre of each unit between the nephrostome and external pore that is so remarkably different from other metanephros (Fig. 5.4).

The nephrostome leads not into an open duct but into a blind chamber called the **nephridial capsule**. The capsule is connected to a nephridioduct, uniquely composed of a single row of cells through which an intracellular canal runs. This canal appears to be somewhat transitory, especially near the capsule, in that it forms from the coalescence of tiny intracellular tubules and vaculole–like chambers. Here selective filtration and resorption are suggested to occur. The intracellular nephridial canal connects with a short chamber derived from an ectodermal invagination at the nephridiopore. In some true leeches a relatively large bladder is formed near the pore (Fig. 5.4).

Ammonia is the main nitrogenous waste product eliminated via the nephridia. Apparently, particulate waste materials are engulfed by phagocytes, both in the coelomic fluid and in the "mesenchyme" but the eventual deposition of this material is not known.

The nephridia of freshwater leeches also serve as osmoregulatory organs. The urine is very dilute, a fact suggesting the excretion of excess water and the retention of various salts. In certain terrestrial leeches the urine from the anterior and posterior nephridiopores is released onto the surfaces of the suckers, there by providing a moist surface for effective suction.

5.2 TROCHOPHORE LARVA AND ITS SIGNIFICANCE

Structure of Trochophore

Polychaetes (Annelids) have indirect development. Their life cycle includes a free-swimming trochophore larva. This larva is a characteristic of many protostomous groups (*e.g.*, molluscs, sipunculus, and echiurans). Trochophore is a minute, ciliated, unsegmented, and almost pear-shaped pelagic creature, with oral and aboral surfaces recognizable. Thus, the early, presegmental trochophore (Fig. 5.5) is characterized by a locomotor ciliary band just anterior to the region of the mouth. This ciliary band, the **prototroch,** arises from special cells, called **trochoblasts**, of the first and second quartets of micromeres. Most trochophore larvae also bear an **apical ciliary tuft** associated with an **apical sense organ** (eye and cerebral ganglia) derived from a plate of thickened ectoderm at the anterior end. In addition, there is often a perianal ciliary band called the **telotroch**. By this stage the mesentoblast has divided to form a pair of cells called **teloblasts**, which in turn proliferate a pair of mesodermal bands, one on each side of the archenteron in the region of the hindgut, an area known as the growth zone (Fig. 5.5B). Many trochophores bear larval sense organs such as ocelli and statocyst, as well as a pair of larval protonephridia. Many trochophores bear bundles of mobile chaetae, which are known to serve as a defense against predators (**Pennington** and **Chia**, 1984) and to help retard sinking.

The cilia of ciliary girdles, rings or bands, when beating give the appearance of a rotating wheel, hence the name trochophore (Gr. *Trochos* = wheel; *phoros* = bear). Digestive tract of trochophore larva is complete, and comprises **stomodaeum, stomach, intestine** and **proctodaeum** (Fig. 5.6). A spacious fluid-filled **blastocoel** surrounds the larval gut. It contains mesenchyme cells, larval muscle fibers and a pair of larval kidneys or protonephridia, one on each side of the gut. The nephridia open by a common passage. Trochophore lacks segmentation and coelom.

The fully formed trochophore may be divided into three regions: 1. **pretrochal region** comprising area about and above the mouth; 2. **pygidium** consisting of telotroch and anal area; and 3. **growth zone** lying between the mouth and telotroch. In polychaetes, the growth zone forms all the trunk segments.

Trochophore may be **planktotrophs** (*i.e.*, feeding on plankton or planktonic organisms) and have long planktonic lives, or they may be **lecithotrophs** (*i.e.*, a non-feeding larva which utilises yolk as a source of nutrition) and have a short planktonic existence. The prototroch functions as the swimming organ, and in a feeding trochophore, it also collects the suspended food particles. The feeding trochophore also has a **food groove** between the preoral prototroch and the postoral metatroch. The larger cilia of prototroch drive water and suspended food particles backward. The short metatroch cilia beat in opposite direction and drive the food particles into the food groove. The short cilia of the food groove carry the food particles to the mouth.

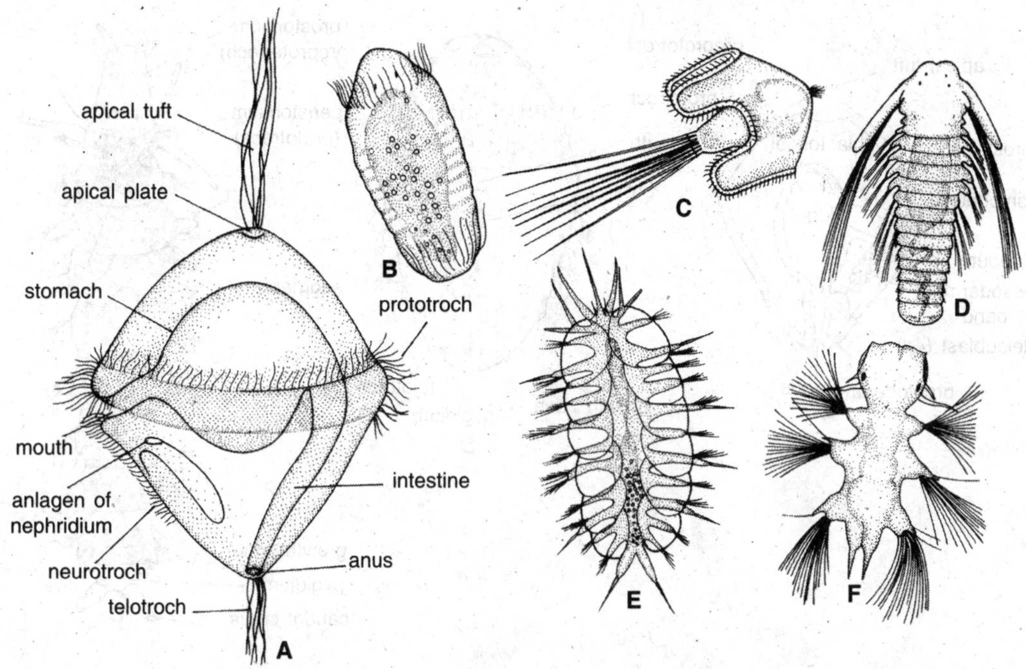

Fig. 5.5. Some polychaete larvae—trochophores and beyond. A—Anatomy of a generalised early trochophore larva; B—Young trochophore of a capillid; C—Trochophore of *Owenia*; D-Late spinoid larva; E-Late larval stage of a polynoid (Scale worm); F—Three-segmented nectochaeta "larva" of a nereid (after Brusca and Brusca, 2002).

Growth and Development of Trochophore

The trochophore larva grows and elongates by proliferation of tissue in the growth zone (Fig. 5.7), while segments are produced by the anterior proliferation of mesoderm from the teloblast derivatives on either side of the gut. These packets of mesoderm hollow (schizocoely) and expand as paired coelomic spaces, which eventually obliterate the blastocoel. Thus, the production of serially arranged coelomic compartments and the formation of segments are one and the same; the anterior and posterior walls of adjacent coelomic compartments form the intersegmental septa. Proliferation of segments by this process is called **teloblastic growth**. Externally, additional ciliary bands are added at each segment. These **metatrochal bands** aid in locomotion as the animal increase in size. Such segmental larvae are sometimes called **polytroch larvae**.

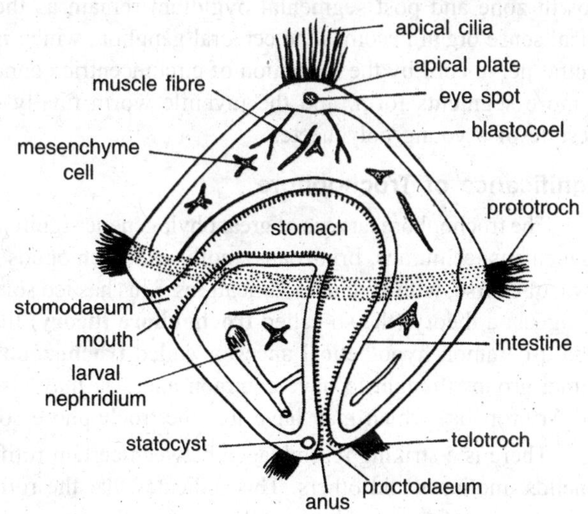

Fig. 5.6. Internal structure of a trochophore larva.

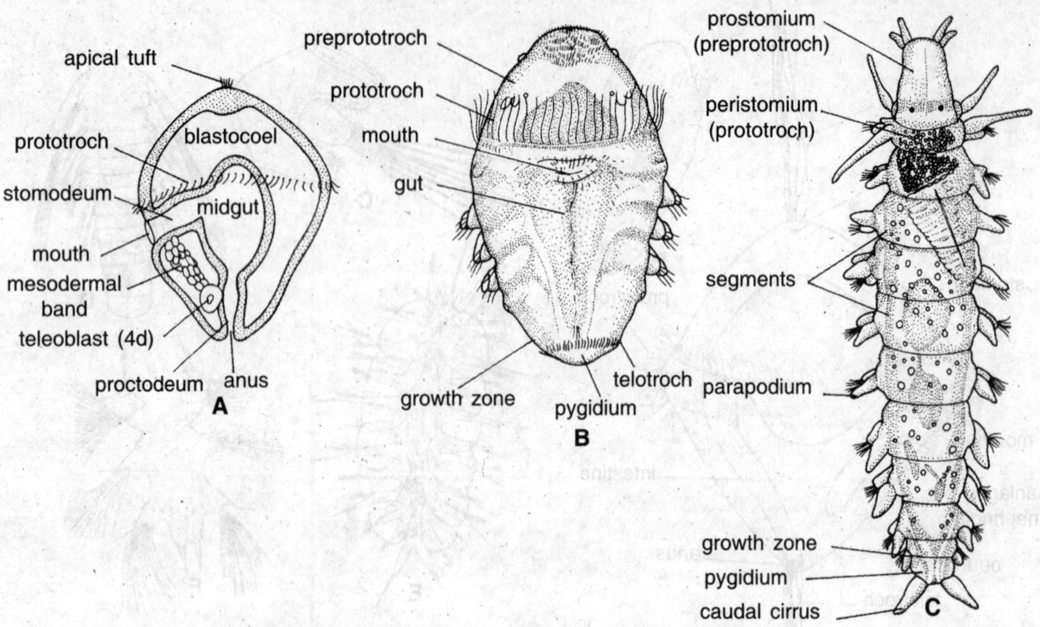

Fig. 5.7. Growth of trochophore larva. A—Trochophore larva of *Arenicola*. Note the teloblastic (4d) mesodermal bands destined to form the metameric coelomic spaces, B, C—Two stages in the development of *Eteone*; B—Early stage segmentation; C—Juvenile, showing the fates of the larval regions (after Brusca and Brusca, 2002).

The fates of the various regions of trochophore larva are now apparent (Fig. 5.7). The region anterior to the protrochal ring becomes the **prostomium**, while the prototrochal area forms the **peristomium.** Note that these two parts are not involved in the proliferation of segments and are thus presegmental. The segmental, metatrochal portion of the larva forms the trunk and the growth zone and post segmental pygidium remain as the corresponding adult body parts. The apical sense organ becomes the cerebral ganglion, which is eventually joined with the developing ventral nerve cord by the formation of circumcentric connectives. The body continues to elongate as more segments form, and the juvenile worm finally drops from plankton and assumes the lifestyle of a young polychaete.

Significance of Trochophore

The trochophore larva is of great phylogenetic significance. Many animals such as polychaetes, sipunculans, echiurans, bryozoans, molluscs, brachiopods and nemertines, develop a trochophore larva, of course, with certain modifications. This has led some embryologists (*e.g.,* **Hatschek,** 1878) to suggest a theory (the so-called **trochophore theory**) that these animal groups have descended from a common hypothetical ancestor called **trochozoon** having trochophore-like features. The animal groups thought to have common ancestry may also include other phyla (Platyhelminthes and Arthropoda) which may have lost the trochophore somewhere in their evolution.

There is a striking resemblance between certain rotifers and the trochophore larvae of some annelids, molluscs and others. This indicates that the rotifers and the animals closely related to the ancestors of the annelids, molluscs and other groups.

Trochophore larva (theory) accounts satisfactorily in many ways for a common ancestor of the coelomate prostomia but not for the acoelomate groups. Many authorities, however, believe that the resemblances of trochophore larvae may be coincidental, the result of adaptive radiation and not of evolutionary significance or actual relationship.

5.3 VERMICOMPOSTING

Vermiculture is the process of rearing of red worms (*Eisenia foetida*) for **vermicomposting** which is commonly called **indoor composting with earthworms**. **Composting** is a controlled process of decomposition used to transform organic materials such as kitchen scraps, yard wastes and paper products into **humus**. On the other hand, **vermicromposting** is a process of composting with earthworms. In a way, vermicomposting is the degradation of organic waste through the consumption by the earthworms. An earthworm, thus, is physically an aerator, crusher and mixer, chemically a degrader and biologically a stimulator of decomposition. Vermicomposting is one of the most modern exciting form of recycling of wastes.

Significance of Vermicomposting

Vermicomposting has the following advantages :

1. It helps in disposing of kitchen wastes.

2. It provides natural liquid fertilizer (called **worm tea**) and nutritious **worm castings** for the gardens and indoor plants. Worm castings are the faecal matter of earthworms and they act as soil conditioners.

3. It reestablishes natural worm populations in the yard (Yard means an enclosed or partly enclosed space near or round the building) and garden that were decimated by toxicity from chemical fertilizers and pesticides.

4. It helps to increase the number of free worms for use as pet food or fishing bait.

Typing of Earthworms Employed in Vermicomposting

A variety of earthworms commonly are used in vermicomposting. Examples of such earthworms are **red worm** (*Eisenia foetida*), red earthworm of China (*Lumbricus rubellus*), European night crawlers (*Eudrilus enginae*) and Blue worms (*Perionyx excavatus*, Box 5.2).

Box 5.2

In India, following species of earthworms are used in vermicomposting: *Dichogaster bolani, Drawida willsi, Perionyx excavatus* (Indian species) and *Eisenia foetida, Eudrilus eugeniae* (Exotic species) (**Verma** and **Agarwal** 2015).

The best kind of earthworms to be used in vermicomposting and vermiculture are red worms, *Eisenia foetida*. They are also known as **red wrigglers** or **manure worms**. These earthworms are reddish brown in colour. These are natives of Europe but have become naturalized (acclimatized) in a number of countries. Red worms thrive in decomposing organic matter such as leaf piles, compost heaps and old manure piles. These earthworms are good indicators of fertile soil because their presence indicates high organic matter content and lack of toxic substance in the soil. Red worms are most ideal for vermiculture and vermicomposting because

1. They are very efficient decomposers of organic wastes. They are capable of consuming organic wastes upto their own body weight daily.

2. Their worm castings may be used as soil amendment and are highly nutrient–rich.

3. These earthworms aerate the worm bin by continually forming tunnels in the food scraps.

4. Red worms are **hermaphroditic**, capable of producing an **egg capsule** or **cocoon** every 14–21 days, egg capsule containing over 1 dozen juvenile worms.

5. Red worms are highly prolific breeders, their development period is short and they become sexually mature in three months.

Another earthworm, the European night crawler (*Eisenia hortensis*) is also good in vermiculture systems. They are traditionally used as bait for anglers. They prefer to feed in deeper parts of the bin.

Strucutre of Vermibox

Vermibox or **Worm bin** is the apparatus used for vermiculture. A good worm bin has following characteristics:

1. A opaque container is required because red worms do not like light.

2. A deep bin is preferred because it allows more room for layering and burying fresh material.

3. Worm bin should have holes approximately 3″ apart in the sides and cover of the bin. the holes should be about ⅛″ wider and should begin about 4″ from the bottom of the bin.

4. Bedding material should be a high carbon material such as shredded leaves, shredded paper, ground cardboard or peat moss, or a combination of these materials. The bedding should fill the bin about ¾ full.

5. A small amount of **pulverized dolamite** (limestone) is added to help buffer the pH towards the worm–favoured neutral levels.

5. The bedding is dampened until the moisture content is 50% as red worms thrive in a very damp environment.

7. Vegetable wastes are buried underneath the bedding, which filter out any odours from the decomposing material, such as compost piles.

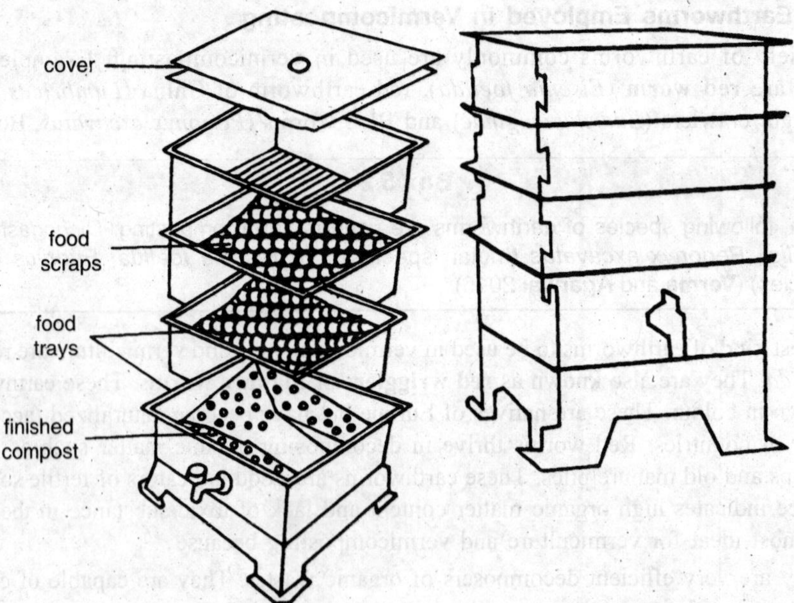

Fig. 5.8. Vermibox. Worm tower system.

Types of Vermiculture Systems

Following four types of vermiculture systems are used in vermicomposting:

1. Worm condo system. It is multilevel, 5–legged circular system in which the fisherman may use the unit to breed up to 20k bait worms which are easily harvested from the top.

This system can produce year–round nutritious worm–castings, easily removable almost free of worms but only top feeding varieties of red worms (*e.g., Eisenia foetida*) can be cultured.

2. Down under worm farm system. In this system, top feeding red worms live and breed in the top layer, while the **liquid compost tea** drains through the perforated base into the bottom container. This liquid is used as manure for the plants and valuable worm castings may be used in the gardens as a soil conditionar.

3. Ultimate home vermiculture system. It is a larger capacity unit with convenient working height for acess by all the family members. It has angled harvesting tray which allows easily drainage and liquid separation.

4. Expandable worm tower system (Fig. 5.8). It has one great advantage over other vemiculture systems as it allows automatic spearation of food scraps from finished compost. In this system, new food trays are added at the top while vermicompost of lower trays can be separated as the top feeding red worms move upward into trays with newly added waste material. In full operation, the worm tower allows a harvest a full tray of worm castings per month.

QUESTIONS

Long Answer Questions

1. Give a detailed account of excretion in Annelida.
2. Describe the trochophore larva and discuss its evolutionary significance.
3. Write an essay on Vermicomposting.

Short Answer Questions

1. Enlist the names of species of earthworm which are most commonly used in vermicomposting.
2. Write Short Notes on the following:
   (*i*) Nephromixia; (*ii*) Metanephridia; (*iii*) Chloragogen cells;
   (*iv*) Vermibox; (*v*) Red worm.
3. Draw a labelled diagram of a polychaete trochophore.

Very Short Answer Questions

1. What is prototroch?
2. Do you find segmentation in the trochophore larva?
3. Define planktotrophs.
4. What is lecithotroph larva?
5. What type of growth occurs in trochophore?
6. What is zoological name of red worm used in vermicomposting?
7. Who proposed trochophore theory?

Fill in the Blanks

1. Nephromixia of Annelida contain .... and ....
2. Metanephridia of Polychaeta open to coelom by a .... nephrostome.
3. In terms of excretion, aquatic oligochaetes are ....
4. Chloragogen cells are known to perform protein .... and .... formation.
5. The prototroch ciliary band of trochophore arise from .... cells and it helps in ....

Multiple Choice Question

1. Trochophore larva occurs in
   (*a*) annelida (*b*) brachiopoda
   (*c*) mollusca (*d*) all of these.

ANSWER

Very Short Answer Questions

1. A ring of cilia around the middle of the trochophore larva.

2. No.

3. An animal feeding on plankton, *e.g.*, trochophore larva annelids.

4. A non-feeding larva which utilizes yolk of embryonic cells as a source of nutrition.

5. Teloblastic growth.

6. *Eisenia foetida.*

7. Hatschek in 1878.

Fill in the Blanks

1. Coelomoducts, nephridia.

2. Ciliated.

3. Ammonotelic.

4. Deamination, nitrogenous waste.

5. Trochoblast, locomotion.

Multiple Choice Question

1. (*d*).

Phylum Arthropoda (Characters, Classification and Types)

The phylum **Arthropoda** (Gr., *arthros* = joint, *podos* = foot) includes nearly a million species. No other phylum of animals rival the arthropods in success: the arthropods contain the most showy group, the insects and the most variable, the crustaceans, of any other animal phyla. Much of their success can be traced to a remarkable flexibility of body plan.

Arthropods are bilaterally symmetrical and segmented animals that are equipped with jointed appendages attached to each body segment. Appendages are typically specialized for specific functions. The surface of body and appendages is covered by a continuous cuticle of complex structure, *i.e.*, **chitin** containing **exoskeleton** formed of a series of heavy skeletal plates or rings, connected by thinner, flexible articular membranes. Arthropods **lack true nephridia and cilia**. They undergo moulting periodically.

6.1 CHARACTERS

1. Body is **bilaterally symmetrical, triploblastic, metamerically segmented**. Few anterior segments (usually 5) have fused to form a distinct head.
2. They have a fixed number of segments.
3. Body is covered and protected with a thick chitinous **cuticle** forming an **exoskeleton**.
4. The exoskeleton is dead and cannot grow. Therefore, the old exoskeletoñ is thrown off periodically and new one is developed to take its place and to compensate the increase in size. The phenomenon is called **moulting** or **ecdysis**.
5. Body segments usually bear **paired** lateral and **jointed appendages**. Appendages are variously modified to act as organs of stimulus detection, food getting, mastication, defence, locomotion, egg carrying and copulation.
6. **Musculature** is not continuous but comprises of separate striped muscles.
7. Body cavity is a **haemocoel**. True coelom is represented by spaces within the gonads and excretory organs.
8. Digestive system is complete; **mouth** and **anus** lie at opposite ends of the body.
9. Respiration occurs through general **body surface, gills** (in aquatic forms), **trachea** or **book lungs** (in terrestrial forms).
10. Blood vascular system (or circulatory system) is of **open type**. It includes an ostia-containing **heart, arteries** but lacks capillaries. The arteries open into haemocoel.
11. True nephridia are absent. Excretion occurs through coelomoducts (*e.g.*, green glands, coxal glands) and **Malpighian tubules**.

12. Nervous system consists of a large bilobed **brain** and two ventral nerve cords.
13. Sense organs include **simple eyes, compound eyes** (with mosaic vision), chemoreceptors, tactile receptors (antennae and bristles), auditory organs and **statocysts.**
14. Cilia are entirely absent from all parts of the body.
15. Sexes are generally separate. Sexual dimorphism is common (*i.e.* male and female look different).
16. Fertilization is internal. They are either **oviparous, ovoviviparous** or **viviparous**.
17. Development is usually **indirect** involving one to several larval forms.
18. Parthenogenesis occurs in some crustaceans and insects.

6.2 OUTLINE CLASSIFICATION

An outline classification of phylum Arthropoda is as under.

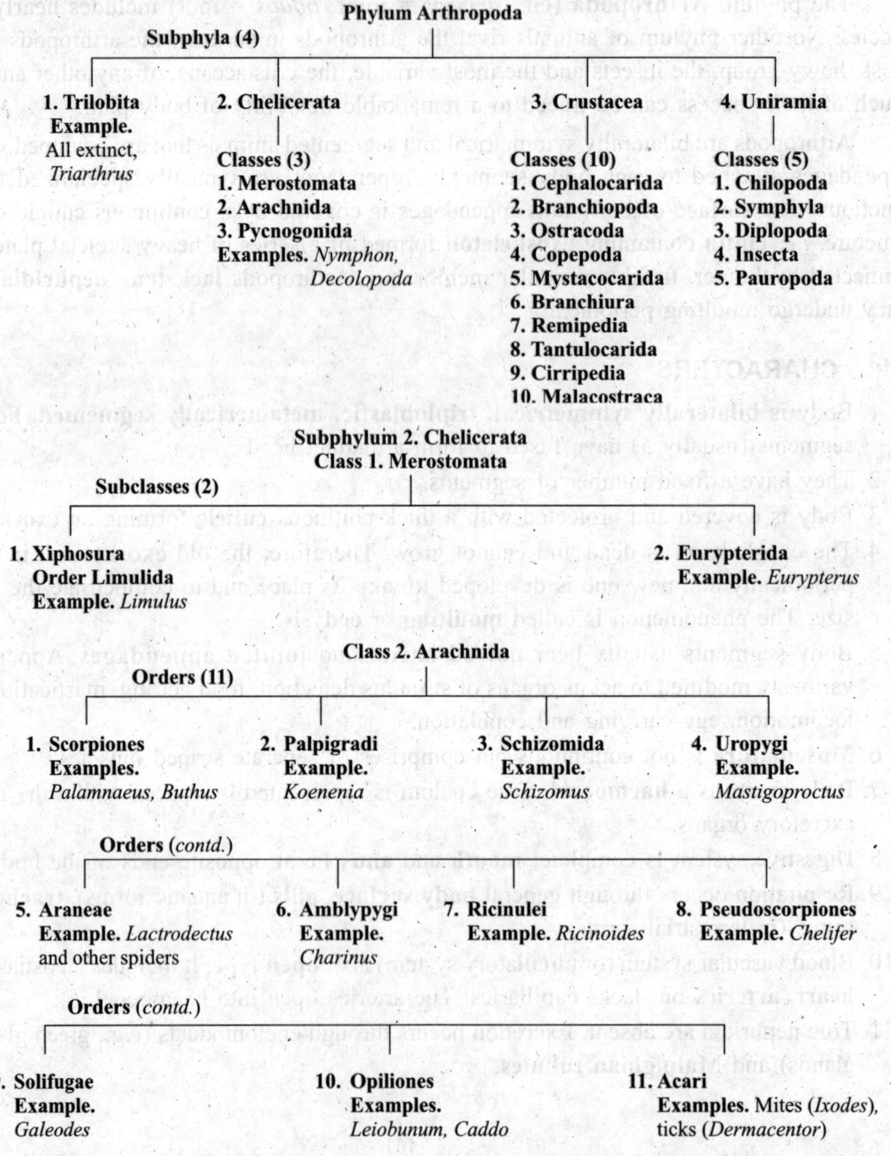

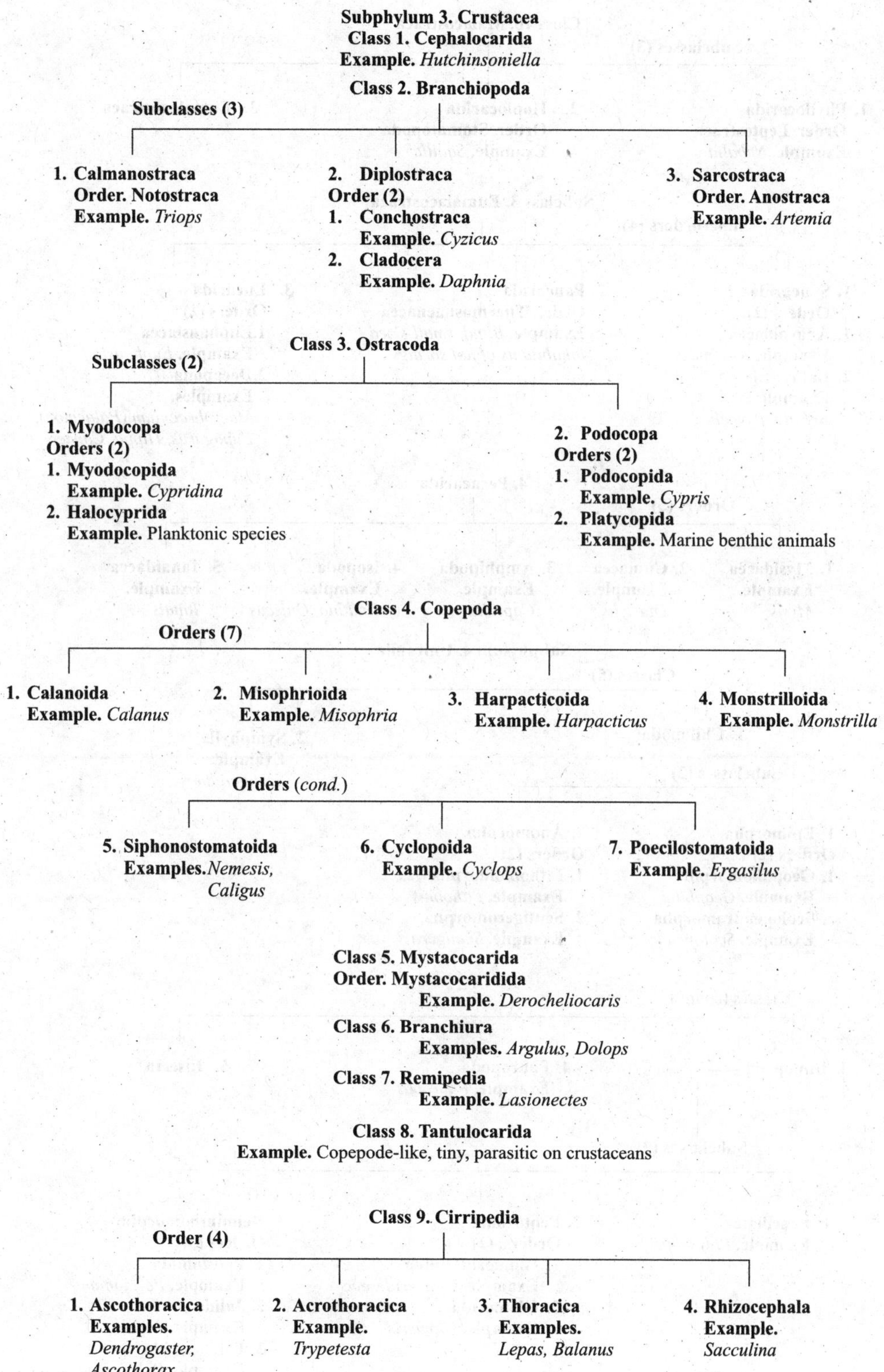
Subphylum 3. Crustacea
Class 1. Cephalocarida
Example. Hutchinsoniella
Class 2. Branchiopoda
Subclasses (3)
1. Calmanostraca
Order. Notostraca
Example. Triops
2. Diplostraca
Order (2)
1. Conchostraca
Example. Cyzicus
2. Cladocera
Example. Daphnia
3. Sarcostraca
Order. Anostraca
Example. Artemia
Class 3. Ostracoda
Subclasses (2)
1. Myodocopa
Orders (2)
1. Myodocopida
Example. Cypridina
2. Halocyprida
Example. Planktonic species
2. Podocopa
Orders (2)
1. Podocopida
Example. Cypris
2. Platycopida
Example. Marine benthic animals
Class 4. Copepoda
Orders (7)
1. Calanoida
Example. Calanus
2. Misophrioida
Example. Misophria
3. Harpacticoida
Example. Harpacticus
4. Monstrilloida
Example. Monstrilla
Orders (cond.)
5. Siphonostomatoida
Examples. Nemesis, Caligus
6. Cyclopoida
Example. Cyclops
7. Poecilostomatoida
Example. Ergasilus
Class 5. Mystacocarida
Order. Mystacocaridida
Example. Derocheliocaris
Class 6. Branchiura
Examples. Argulus, Dolops
Class 7. Remipedia
Example. Lasionectes
Class 8. Tantulocarida
Example. Copepode-like, tiny, parasitic on crustaceans
Class 9. Cirripedia
Order (4)
1. Ascothoracica
Examples. Dendrogaster, Ascothorax
2. Acrothoracica
Example. Trypetesta
3. Thoracica
Examples. Lepas, Balanus
4. Rhizocephala
Example. Sacculina

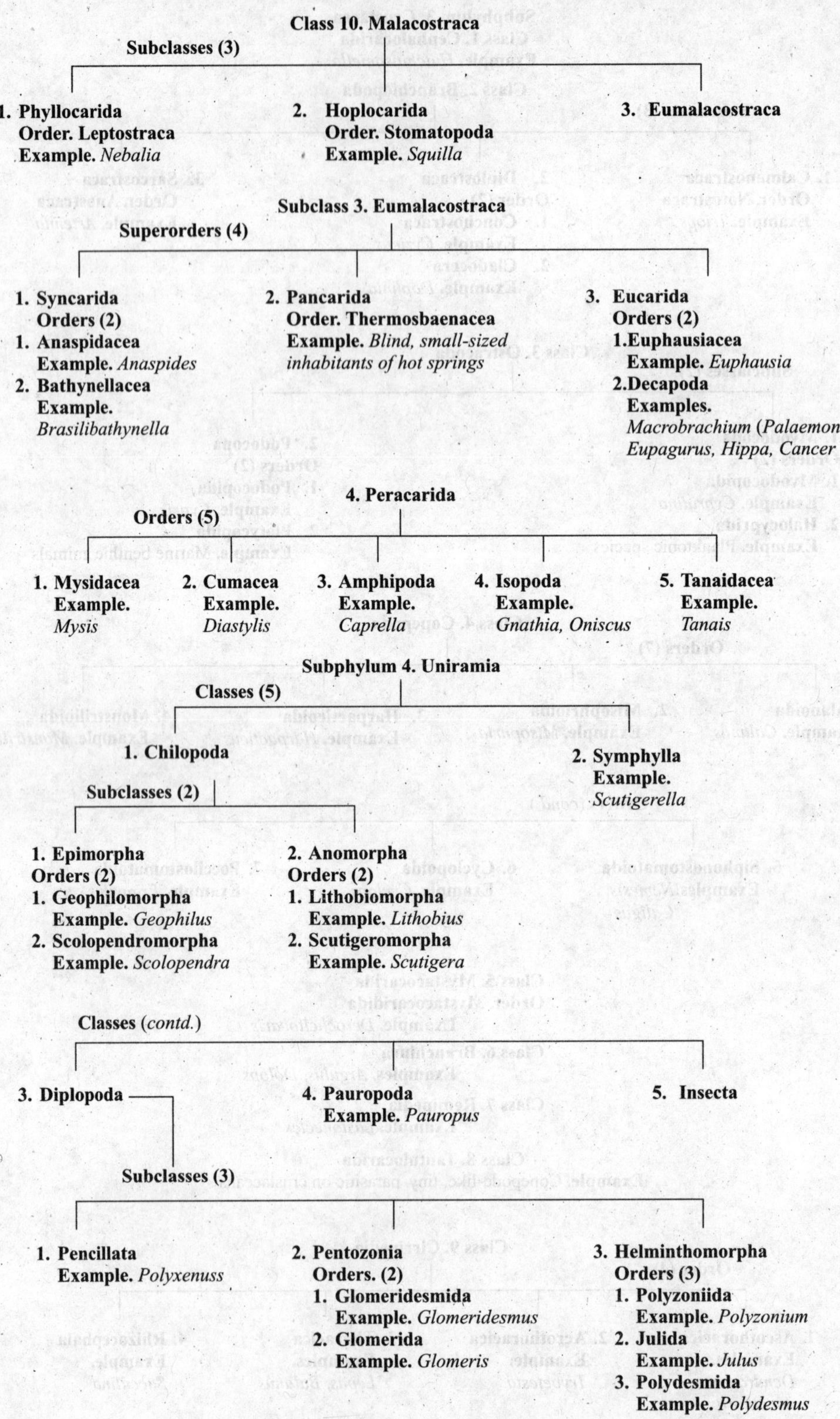
Class 10. Malacostraca
Subclasses (3)
1. Phyllocarida
Order. Leptostraca
Example. Nebalia
2. Hoplocarida
Order. Stomatopoda
Example. Squilla
3. Eumalacostraca
Subclass 3. Eumalacostraca
Superorders (4)
1. Syncarida
Orders (2)
1. Anaspidacea
Example. Anaspides
2. Bathynellacea
Example.
Brasilibathynella
2. Pancarida
Order. Thermosbaenacea
Example. Blind, small-sized inhabitants of hot springs
3. Eucarida
Orders (2)
1.Euphausiacea
Example. Euphausia
2.Decapoda
Examples.
Macrobrachium (Palaemon), Eupagurus, Hippa, Cancer
4. Peracarida
Orders (5)
1. Mysidacea
Example.
Mysis
2. Cumacea
Example.
Diastylis
3. Amphipoda
Example.
Caprella
4. Isopoda
Example.
Gnathia, Oniscus
5. Tanaidacea
Example.
Tanais
Subphylum 4. Uniramia
Classes (5)
1. Chilopoda
Subclasses (2)
1. Epimorpha
Orders (2)
1. Geophilomorpha
Example. Geophilus
2. Scolopendromorpha
Example. Scolopendra
2. Anomorpha
Orders (2)
1. Lithobiomorpha
Example. Lithobius
2. Scutigeromorpha
Example. Scutigera
2. Symphylla
Example.
Scutigerella
Classes (contd.)
3. Diplopoda
Subclasses (3)
1. Pencillata
Example. Polyxenuss
2. Pentozonia
Orders. (2)
1. Glomeridesmida
Example. Glomeridesmus
2. Glomerida
Example. Glomeris
3. Helminthomorpha
Orders (3)
1. Polyzoniida
Example. Polyzonium
2. Julida
Example. Julus
3. Polydesmida
Example. Polydesmus
4. Pauropoda
Example. Pauropus
5. Insecta

Class 5. Insecta

Subclasses (2)

1. Apterygota

Orders (5)

1. **Protura**
   Example. *Eosentomon*
2. **Aptera (Diplura)**
   Example. *Campodea*
3. **Collembola**
   Example. Springtails (*Entomobrya*)
4. **Microcoryphia**
   Example. Jumping springtails (*Machilis*)
5. **Thysanura**
   Example. Silver fish (*Lepisma*)

2. Pterygota

Orders (more than 17)

1. **Ephimerida**
   Example. Mayflies (*Ephemera*)
2. **Odonata**
   Example. Dragonflies (*Macromia*)
3. **Orthoptera**
   Examples. Grasshopper (*Poeciloceros*), locust (*Schistocerca*), cockroach (*Periplaneta*)
4. **Isoptera**
   Example. Termites (*Termes*)
5. **Dermaptera**
   Example. Earwig (*Forficula*)
6. **Plectoptera**
   Example. Stoneflies (*Taeniopteryx*)
7. **Mallophaga**
   Example. *Cuclogaster*
8. **Anopleura**
   Example. Human louse (*Pediculus*)
9. **Thysanoptera**
   Example. Thrips (*Taeniothrips*)
10. **Hemiptera**
    Example. Bedbug (*Cimex*)
11. **Homoptera**
    Example. *Cicada*
12. **Neuroptera**
    Example. *Mantispid*
13. **Coeloptera**
    Examples. Beetles (Lady bird beetle, *Adalia*), weevils (rice weevil, *Calandra oryzae*), fireflies (*Lamprohorous*)
14. **Lepidoptera**
    Examples. Butterflies (*Papilio*, *Kalima*), moth (*Bombyx mori*)
15. **Diptera**
    Examples. Mosquitoes (*Anopheles, Culex*), fruitflies (*Drosophila*), housefly (*Musca*)
16. **Hymenoptera**
    Examples. Wasps (*Polistes*, *Vespa*), honeybee (*Apis*)
17. **Siphonoptera**
    Example. Rat flea (*Xenopsylla*)

6.3 DETAILED CLASSIFICATION

Phylum Arthropoda is divided into **four** subphyla:

Subphylum 1. Trilobita (or Trilobitomorpha)

1. This subphylum is represented by fossil or extinct trilobites. They existed from Cambrian to Permian periods.
2. Body was divided into three lobes by two longitudinal furrows.
3. All body segments except the last contained **biramous appendages**.
4. Marine and benthozoic (bottom-dweller) arthropods.

   Examples. *Triarthrus, Dalmanites*.

Subphylum 2. Chelicerata

1. Body is divided into **cephalothorax** or **prosoma** and **abdomen** or **opisthosoma**.
2. **Antennae** and **true jaws** are absent.
3. Cephalothoracic or prosomic appendages comprises of one pair of **chelicerae**, one pair of **pedipalps** and four pairs of **walking legs**.
4. Chelicerae form first pair of appendages. They are preoral and feeding appendages.
5. Second opisthosomal segment bear gonopores.
6. Excretion occurs by **coxal glands.**
7. Mostly terrestrial and predaceous.

Subphylum 3. Crustacea

1. Body is divisible into **cephalothorax** (fused head and thorax) and **abdomen.**
2. Cephalothorax is covered dorsally by **carapace.**
3. Head bears a pair of **compound eyes** and five pairs of biramous appendages: 1. two pair of antennae (*i.e.*, **antennules** and **antennae**); 2. **mandibles**; 3. two pairs of maxillae (*i.e.* **maxillules** and **maxillae**).
4. Each segment of thorax and abdomen contains a pair of branched or biramous appendages.
5. Their skeleton is hard and calcified.
6. They respire by means of gills.
7. Excretion is by green glands.
8. All are aquatic—mostly marine, some are freshwater (*e.g.*, *Daphnia*) and a few are adapted for moist land (crabs and wood lice).
9. Crustaceans are so abundant in oceans that they have been called "the insects of the sea".

Subphylum 4. Uniramia

1. Appendages are **uniramous** or unbranched.
2. The head is composed of five pairs of appendages—1. a single pair of **antennae**; 2. a pair of **labrum**; 3. a pair of **Mandibles** and 4. first and second pair of **maxillae**, with latter forming a **labium**.
3. **Trachea** are used for respiration.
4. **Malpighian tubules** are employed as excretory organs. Excretory waste is ammonia.
5. Nearly all are terrestrial

   (**Note.** In old system of classification subphyla **Crustacea** and **Uniramia** were included in one subphylum **Mandibulata**. Four groups of uniramia, namely the **centipedes, millipedes, pauripods** and **symphyla**, form the common group **Myriapoda** including about 10,500 species).

Subphylum 2. Chelicerata

Subphylum Chelicerata is divided into **three** classes:

Class 1. Merostomata

1. Cephalothorax is covered by large, continuous dorsal shield.
2. Five or six pairs of abdominal appendages are modified as **gills**.
3. There is a well-developed spine or **telson** at the posterior end of body.
4. They contain a pair of compound eyes and a pair of simple eyes.
5. All are aquatic.

Class 2. Arachnida

1. This class includes about 10,000 species.
2. Body consists of two regions (or tagmata): **cephalothorax** (or **prosoma**) and **abdomen** (or **opisthosoma**).
3. The cephalothorax bears a pair of **chelicerae**, one pair of **pedipalps** and four pairs of **walking legs**. They lack **antennae** and **true jaws**.
4. Pedipalps are leg-like appendages that serve a sensory function, as in spiders, or are used for seizing the prey, as in scorpions.
5. Abdomen is without appendages.
6. Air-breathing mostly terrestrial arthropods.

Class 3. Pycnogonida (or Pentopoda)

1. This class includes about 1,000 species of sea **spiders**.
2. Arthropods with very small bodies and disproportionally **long legs**.
3. Body is narrow and comprises of **head** (cephalon), **trunk** (throax) and a short conical **abdomen.**
4. Head bears four **eyes** on a cen-tral tubercle and a cylindrical **proboscis**.
5. Trunk is formed of 4 to 6 cylindrical segments.
6. Appendages include a pair of **chelicerae** (called **chelifores)**, a pair of **palps**, a pair of **ovigerous legs** or **ovigers** and four to six pairs of **walking legs**.
7. Ovigers are peculiar and may be used in grooming and in males to carry the eggs.
8. There are no special organs for respiration and excretion.
9. Sexes are separate; the eggs are carried on the ovigers of males and give rise to larvae, called **protonymphon**, having three pairs of legs.
10. Most pycnogonids are marine, bottom dwellers and crawl about algae, hydroids and bryozoans. They are **carnivorous** and feed on hydroids, soft corals, anemones, bryozoans, small polychaetes and sponges.

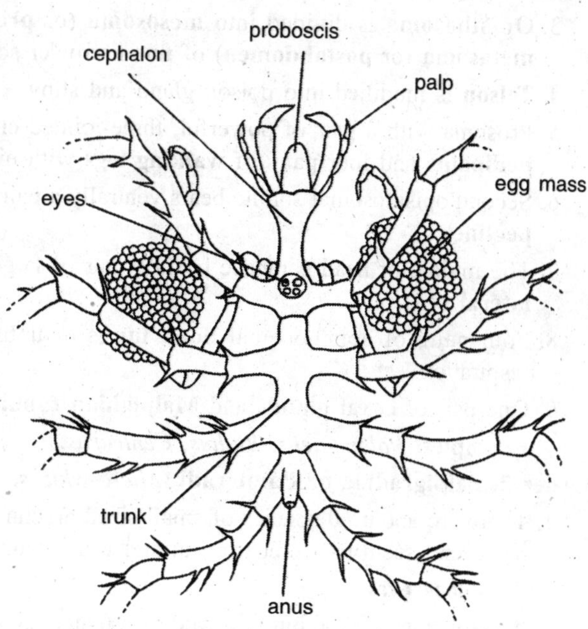

Fig. 6.1. *Nymphon rubrum*, a sea spider.

Examples. *Nymphon* (Fig. 6.1); *Pycnogonum*, *Pallene*, *Decolopoda*.

Class 1. Merostomata. Class Merostomata is divided into following **two** subclasses:

Subclass I. Xiphosura

1. Hemicircular prosoma or the cephalothorax is covered with a large horse-shoe shaped carapace.
2. Segments of abdomen (or opisthosoma) are fused and covered with a single plate.
3. Caudal spine or **telson** is as long as the body.
4. Cephalothorax with a pair of **chelicerae,** five pairs of walking legs and a pair of **chilaria.**
5. Walking legs are four to six-segmented and significant for burrowing and capturing the prey.
6. First four pairs of walking legs have projections, called **gnathobases** on the inner side of the basal coxae. They are used to fragment the prey.

   This subclass includes a **single** order.

Order Lemulida. Example. *Limulus* (king crab).

Subclass 2. Eurypterida

1. This subclass includes extinct giant arthropods that existed from the Cambrian and Permian period.
2. Cephalothorax is small and covered with a carapace.
3. Abdomen (opisthosoma) includes **mesosome** and **metasoma,** which lacks appendages.
4. Telson present.

   Examples. *Eurypterus*, *Pterygotus*.

Class 2. Arachnida

Class Arachnida is divided into following **eleven** orders.

Order 1. Scorpionida or Scorpiones (*Scorpions*)

1. There are about 2000 species of scorpions found in tropical and sub-tropical areas.
2. **Prosoma** is covered with a single shield or carapace.
3. **Opisthosoma** is divided into **mesosoma** (or **preabdomen**) of seven broad somites and **metasoma** (or **postabdomen**) of five narrower somites.
4. **Telson** is modified into poison gland and sting.
5. Prosoma with a pair of powerful, three-jointed chelicerae; a pair of large, distinct chelate **pedipalps** and four pairs of **walking legs** with nine podites.
6. Second opisthosomal somite bears ventrally a pair of comb-like sensory appendages, called **pectines.**
7. The middle of head carapace bears a pair of large **median eyes** and 1 to 5 pairs of small **lateral eyes.**
8. Four pairs of opisthosomal **book lungs** or unbranched **tracheal tubes** are present for respiration.
9. One pair of **coxal glands** and **Malpighian tubules** perform excretion.

   Examples. *Palamnaeus*, *Buthus*, *Centrurus*.

Order 2. Palpigradida or Palpigradi (*Micro-whip scorpions*)

1. There are about 60 species of small-sized arachnids which live in soil (in sand) and under rocks and seem to prefer tropical and warm temperate climates.
2. They lack **eyes.**
3. **Anus** is not terminal but followed by a many-segmented **post abdomen** or **flagellum** which is called **whip.**

4. Cephalothoracic carapace is formed of two pieces.
5. Chelicerae is large-sized and pedipalpi are leg-like.

Examples. *Koenenia*, *Eukoenenia*.

Order 3. Schizomida (*Schizomids*)

1. Cephalothoracic carapace is composed of three plates.
2. Pedipalps are leg-like.
3. First walking leg is antenna-like sensory limb.
4. Opisthosoma contains a large anterior portion and tiny post-abdominal region, "pygidium" of three segments with terminal flagellum.

Example. *Schizomus*.

Order 4. Uropygida or Uropygi (*Whip scorpion*)

1. This order includes about 85 species of whip scorpions.
2. Whip scorpions possess the well-developed, claw-like pedipalps of scorpions and the long, jointed-post abdomen of palpigrades.
3. **Chelicerae** is two jointed and ends in hook or fang.
4. Pedipalps are short but stout forming a pincer. They are raptorial.
5. Four pairs of eyes are usually present.
6. Abdomen (opisthosoma) is 12 segmented and bears a terminal flagellum or whip.
7. Respiration occurs by the help of two pairs of **book lungs**.
8. Excretion takes place by a pair of **coxal glands** and **Malpighian tubules**.
9. Abdomen contains a pair of large **anal glands** and when irritated the scorpion sprays the attacker with a fluid (containing acetic acid or formic acid) secreted from these glands. The secretion contains 84 per cent acetic acid and 5 per cent caprylic acid. The fluid can burn human skin. The repellent smell has given these scorpions another common name, **vinegarroon**.

Examples. *Mastigoproctus* (Fig. 6.2B), *Thelyphonellus*.

Order 5. Araneae (*Spiders*)

1. It includes about 32000 species of true spiders.
2. Prosomal segments are fused, opisthosoma is of 12 segments. Prosoma and opisthosoma are joined by a narrow **pedicel**.
3. Telson is absent.
4. Carapace of prosoma bears 8 eyes.
5. Each **chelicera** is two-jointed and it forms a hollow **fang** through which venom (poison) is injected into the prey.
6. **Pedipalps** are leg-like and are used for food handling. They are modified as copulatory organs in males.
7. The end of abdomen bears a group of modified appendages, the **spinning organs** or the **spinnerets**. These are used to make **silk webs**.
8. Spiders are extremely successful, small carnivorous animals. They play an important role in the ecological balance of their communities, serving as checks against some insect species.

Example. *Latrodectus* (black widow) and *Loxosceles reclusa* (brown recluse spider) both have extremely dangerous/fatal bites; *Lycosa* (wolf spider), *Synema* (crab spider), *Achaearanea* (house spider).

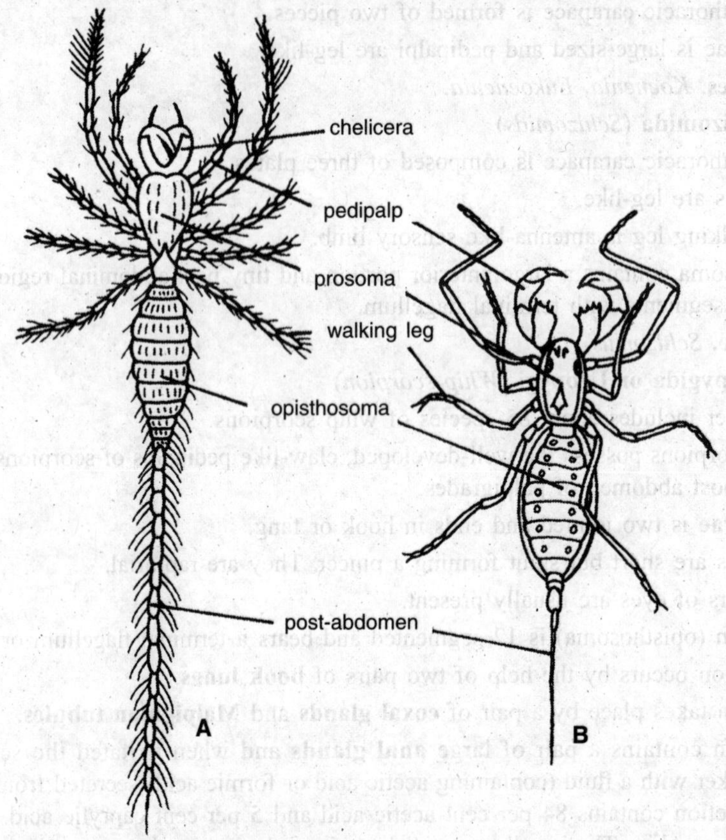

Fig. 6.2. A—*Nymphon hirsites*; B—*Mastigoproctus*.

Order 6. Amblypygida or Amblypygi (*Tailless whip scorpion*)

1. It includes 70 species of spider-like creatures.
2. Body is flattened; carapace of prosoma bears three pairs of reduced lateral eyes, usually a pair of median ocelli.
3. Abdomen is 12-segmented and connected to cephalothorax by a narrow **pedicel** as in spiders. Abdomen is without flagellum. Telson is reduced.
4. Chelicerae are moderate.
5. Pedipalps are powerful, spiny and raptorial.
6. First pair of legs are extremely long and slender, used as tactile appendages; remaining legs are cursorial.
7. They lack spinnerets.

   Examples. *Charinus, Terantula, Admetus.*

Order 7. Ricinuleida or Ricinulei (*Tick spiders*)

1. It includes 33 species of tick spiders which are found in Africa and from Brazil to southern United States.
2. Body is heavy and cuticle is very thick and often sculptured.
3. Carapace of prosoma bears a hood anteriorly, called **cucullus**, which covers chelicerae.
4. Prosoma and opisthosoma are separated by narrow pedicel; opisthosoma is wide and oval and formed of 12 united somites and ending in **anus** or short tubercle.

5. Chelicerae are two-segmented and pincer-like.
6. Pedipalps are leg-like and end in pincers.

Examples. *Ricinoides* (Africa), *Cryptocellus* (Southern USA).

Order 8. Pseudoscorpionida or Pseudoscorpiones (*False scorpions*)

1. This order includes about 2000 species of false scorpions.
2. Opisthosoma is broad and flat.
3. Carapace of prosoma is rectangular and bears one or two pairs of eyes. Eyes may be absent.
4. Chelicerae are small, two-jointed, chelate, forming pincers.
5. **Silk glands** open at the end of chelicerae. The glands are used to make a silk-lined chamber in which moulting occurs.
6. Pedipalps are powerful, chelate and contain **poison glands** that open at the tip.
7. Two pairs of spiracles lead to tracheae. Book lungs are absent.

Example. *Chelifer.*

Order 9. Solpugida or Solifugae (*Sun spiders*)

1. This order includes about 900 species of arachnids called **sun spiders** or **wind scorpions**.
2. These are not true spiders and are characteristics of hot deserts, where they hide in burrows or crevices until night, when they come out to feed. They are called 'sun spiders' because they are diurnal or 'wind scorpions' because they run very fast.
3. Prosoma is covered by shield of a single large anterior and three posterior plates. It contains a pair of closely placed eyes.
4. Abdomen (opisthosoma) is large and segmented containing 10 somites.
5. Chelicerae are two-jointed large and powerful; they form pincers.
6. Pedipalps are leg-like and contain an adhesive organ at the tip.
7. Respiration is by tracheal system.

Examples. *Galeodes* (*sun spider*).

Order 10. Phalangida or Opiliones (*Harvest* man or *Daddy longleg*)

1. This order includes about 4500 species of **daddy longlegs.** They derive their name from the fact that their legs are exceptionally long and thin.
2. Body is elliptical.
3. Undivided prosoma is broadly attached to nine-segmented opisthosoma.
4. Chelicerae are three-jointed, small and chelate.
5. Pedipalps are long and leg-like. Near the base of each pedipalp occurs a small pore that leads to **stink glands**. These give off an offensive secretion when animal is disturbed.
6. Carapace bears a pair of tubercles in the centre. Each tubercle bears an eye in centre.
7. Their reproductive system is **unique**. The male has a long tubular **penis**, and the female contains a protrusible **ovipositor**.

Examples. *Leiobunum, Caddo.*

Order 11. Acarina or Acari (*Ticks and mites*)

1. This order includes about 30,000 species of ticks (parasitic) and mites (free living).
2. Body is ovoid and compact; it is not divided into cephalothorax (prosoma) and abdomen (opisthosoma).

3. Chelicerae are typically modified for piercing.
4. Pedipalps are small but chelate.
5. Four pairs of legs with variable number of podites. They are used in creeping and climbing. In ticks each walking leg bears a pair of prehensile claws.

Examples. Mites (*Sarcoptes, Ixodes*) and tick (*Dermacentor*).

Subphylum 3. Crustacea

This subphylum includes about 38,000 aquatic species. It is divided into **ten** classes.

Class 1. Cephalocarida* (*Head shrimps*)

(*The root **carid** so widely used in the name of crustaceans taxa means "shrimp").

1. It includes about 9 species of most primitive tiny, eyeless crustaceans.
2. They are marine and found in soft bottom sediments. They feed on organic residues in bottom deposits.
3. The **head** is broadly rounded. The trunk is tapered, small and is divided into a **thoracic region** of eight segments, each bearing a pair of **biramous** appendages and **abdomen** which includes remaining 11 to 12 segments which lack appendages.
4. Antennae are short.
5. Appendages are flat phyllopods and their gnathobases-like endites help in detritus feeding.
6. They are hermaphrodites; development is **indirect** and includes metanauplius larval stage.

Examples. *Polyphemus*, *Hutchinsoneilla*.

Class 2. Branchiopoda (*Gill feet*)

1. It includes about 821 species which are mostly inhabitants of freshwater.
2. Small crustaceans with foliaceous trunk appendages, *i.e.*, appendages are flattened, leaf-like and act as organs of respiration and locomotion.
3. The first antennae and second maxillae are vestigial.
4. Last abdominal segment bears slender, paired processes (**cercopods**).
5. They are filter feeders.

Class Branchiopoda is divided into fullowing **three** subclasses.

Subclass 1. Calmanostraca

1. Possess carapace.

It includes **single** order.

Order 1. Notostraca (*Tadpole shrimps*)

1. The head and anterior half of trunk are covered with a large, shield-like dorsal carapace.
2. Body is composed of thorax with appendages and an abdomen without appendages.
3. They have unique external cuticular rings that do not correspond to the somites.
4. The **telson** bears a pair of annulate caudal rami (= cercopods).
5. Paired compound eyes are sessile (without eye stalk).

Tadpole shrimps prefer alkaline freshwater and are not disturbed by high turbidity.

Example. *Triops* (Fig. 6.3B).

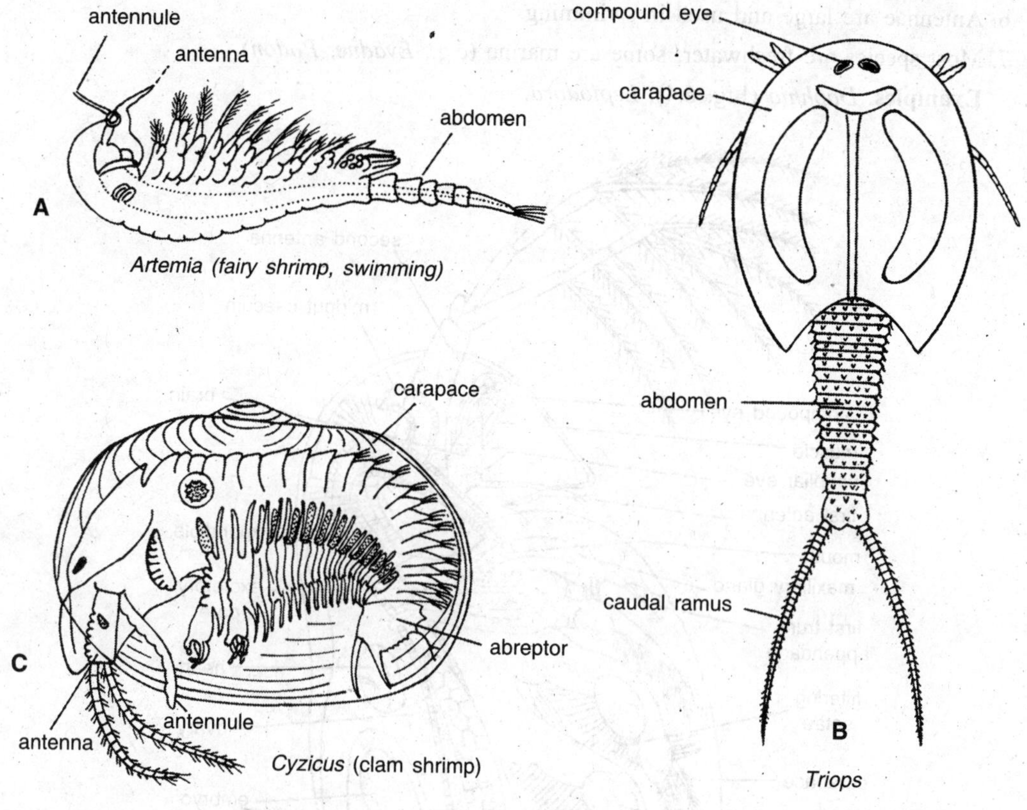

Fig. 6.3. Branchiopods A—*Artemia*; B—*Triops*; C—*Cyzicus*.

Subclass 2. Diplostraca

1. Body is enclosed within a laterally compressed bivalve carapace.

Subclass Diplostraca is divided into **two** orders:

Order 1. Conchostraca (*Clam shrimp*)

1. Entire body is completely enclosed within the carapace and animal looks strikingly like a little clam.
2. Ten to 32 trunk segments are present.
3. Second antennae are well developed, biramous and setose.
4. Compound eyes are sessile.

Examples. *Cyzicus* (Fig. 6.3C); *Lynceus*.

Order 2. Cladocera (*Water fleas*)

1. The bivalve carapace encloses the trunk but not the head and often terminates posteriorly in dorsal spine.
2. A space called **brood chamber** lies between the dorsal body surface and carapace.
3. Trunk has five or six pairs of appendages.
4. The tip end of the trunk, commonly called the **post-abdomen**, is turned ventrally and forward and bears special claws and spines for cleaning the carapace.
5. Head bears single median compound eye.

6. Antennae are large and used in swimming.
7. Most species are freshwater; some are marine (*e.g., Evadne, Podon*).

Examples. *Daphnia* (Fig. 6.4), *Leptodora*.

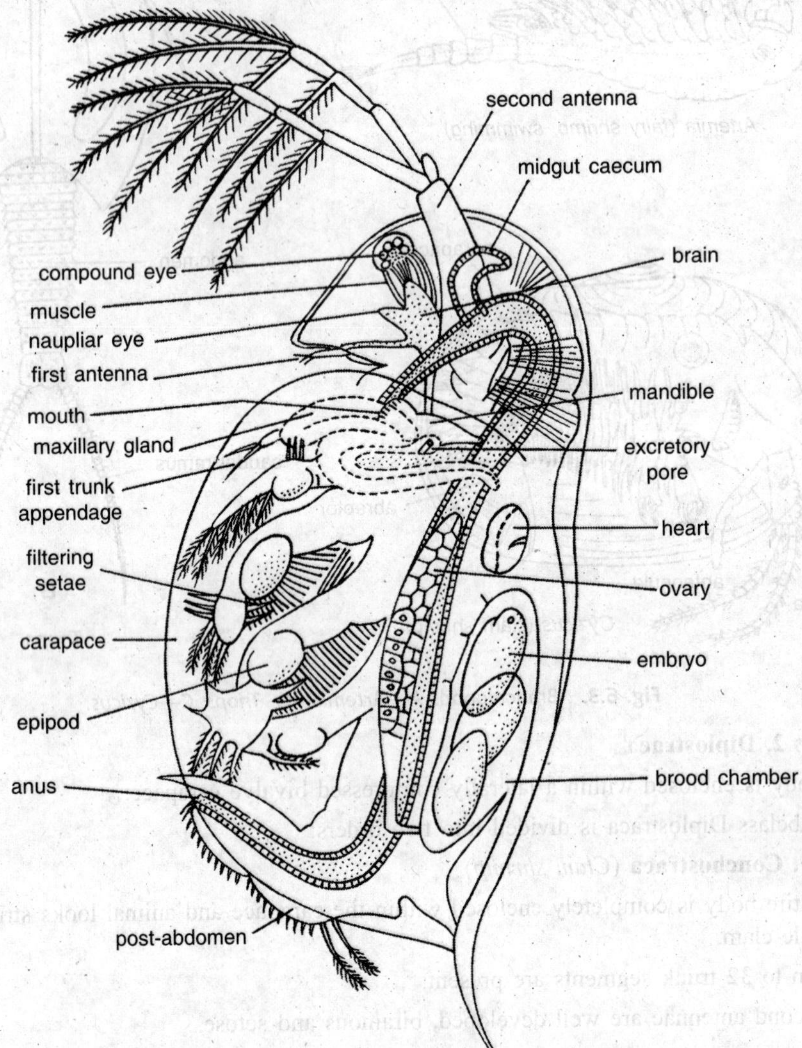

Fig. 6.4. *Daphnia pulex*. Female.

Subclass 3. Sarcostraca

1. Carapace is absent.

This subclass includes **single** order.

Order Anostraca (*Fairy shrimp*)

1. Carapace is absent.
2. Thorax contains 11 to 18 segments, each having a pair of appendages **(phyllopodia)**.

3. Abdominal segments are limbless. Anal somite is with a pair of caudal rami.
4. Have stalked compound eyes.

Examples. *Branchinecta*; *Artemia* (brine shrimp, Fig. 6.3A, an inhabitant of salt lakes and ponds).

Class 3. Ostracoda

1. Include about 5,650 species of mussel or seed shrimps.
2. They are marine and freshwater minute crustaceans. Some are free living (swimmers); others live on bottom.
3. Body is highly reduced and enclosed within hinged, bivalved carapace, which is commonly impregnated with calcium carbonate.
4. Head is large and trunk is short. All external trunk segmentation have disappeared.
5. Antennules and antennae are well developed.
6. There are only two pairs of legs in trunk. Maxillae are leg-like and are modified for swimming, feeding, clasping, etc.

Class Ostracoda is divided into **two** subclasses.

Subclass 1. Myodocopa

1. All are marine.
2. Shell valves contain **antennal notch**.
3. Second antennae is adapted for swimming.
4. Two pairs of trunk appendages.

Subclass Myodocopa is divided into **two** orders.

Order 1. Myodocopida

The long worm-like second pair of trunk appendages are adapted for cleaning the interior of valves.

Example. *Cypridina*, *Gigantocypris*, *Skogsbergia*.

Order 2. Halocyprida

1. All are marine.
2. First pair of trunk appendages are present or absent.
3. Second pair of appendages are absent or short and leg-like.

Example. Planktonic species.

Subclass 2. Podocopa

1. Carapace valves are without an antennal notch.
2. One or two pairs of trunk appendages are present.

Subclass Podocopa is divided into **two** orders.

Order 1. Podocopida

1. First pair of trunk appendages are leg-like.
2. This large order includes marine as well as freshwater species.

Examples. *Cypris*, *Pontocypris*, *Candona* (See Box 6.1), *Cypridopsis*, *Mesocypris*, *Darwinula*, *Cythere*.

Box 6.1 Zenker's organ

Candona suburbana is a very common freshwater podocopid ostracod. This nonswimming bottom dweller feeds on algae and decomposing vegetation. Zenker's organ, which is a part of the male reproductive system, ejects sperm to the penis.

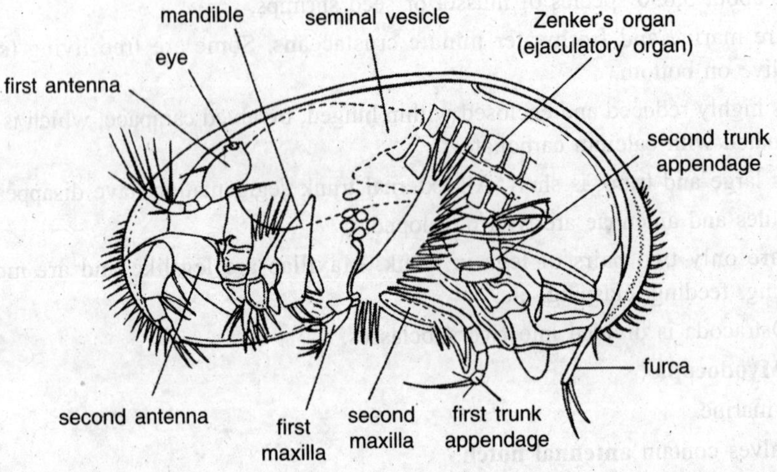

Fig. 6.5. *Candona suburbana*.

Order 2. Platycopida

1. Marine, benthic animals.
2. First pair of trunk appendages are not leg-like.

Class 4. Copepoda (*Copepods*)

1. It includes about 8405 species of small crustaceans.
2. Both marine and freshwater forms. Free living and parasitic.
3. Body is cylindrical and tapering and divided into head, thorax and abdomen.
4. Thorax bears five pairs of biramous appendages and abdomen is without appendages. Telson is with two caudal forks.
5. First pair of antennae are longer than the second pair and held outstretched.
6. Compound eyes are absent; naupliar eye is present.

Class Copepoda is divided into **seven** orders.

Order 1. Calanoida

1. Free living, planktonic forms.
2. First pair of antennae is very long having setae and 22 or more articles.

Examples. *Calanus*, *Calocalanus*, *Metridia*, *Acarita*.

Order 2. Misophrioida

1. Pelagic animals living on or above the bottom surface.
2. First pair of antennae are shorter, containing 11 to 16 articles.

Example. *Misophria*.

Order 3. Harpacticoida

1. Free living marine and freshwater forms. Bottom dwellers.
2. First pair of antennae are short having 10 articles.

Example. *Harpacticus*, *Canthocamptus*.

Order 4. Monstrilloida

1. Marine; larval stages are parasitic in polychaetes and gastropods.
2. Adults are nonfeeding, planktonic and lack second antennae and mouth parts.

   Example. *Monstrilla*, *Xenocoeloma*.

Order 5. Siphonostomatoida

1. Freshwater and marine forms.
2. Adults are parasitic on fish and invertebrates.

   Example. *Nemesis*, *Clevella*, *Caligus*, *Penella*.

Order 6. Cyclopoida

1. Marine, freshwater, planktonic and benthic forms.
2. Some forms are commensals and parasites.
3. First pair of antennae are short containing 10 to 16 articles.
4. Second antennae are uniramous (unbranched).

   Example. *Cyclops*, *Sapphirina*, *Oncaea*.

Order 7. Poecilostomatoida

1. Marine.
2. Adults are parasites on invertebrates and fish.

   Example. *Ergasilus*, *Chondracanthus*.

Class 5. Mystacocarida or Mystacarida (*Moustache shrimp*)

1. It includes 9 species of marine forms adapted for living between the intertidal sand grains.
2. Body is long and cylindrical and divided into head, thorax and abdomen.
3. Thorax contains less than six segments. Mandibular palp is present.
4. Naupliar eye is absent.
5. Both pairs of antennae are long and prominent.
6. Maxillipedes are present. Mouth appendages contain setae for collecting detritus.
7. Caudal rami work as pincers.
8. Sexes are separate. Larva is **nauplius**.

   This class includes **single** order.

Order Mystacocaridida

Example. *Derocheilocaris* (Fig. 6.6A)

Class 6. Branchiura (*Gill tails*)

1. This class includes about 150 species of marine and freshwater forms.
2. They are ectoparasitic on the skin and gill cavities of fishes.
3. Head and throax are covered by a shield-like carapace.
4. Abdomen is small, bilobed and unsegmented.
5. Both pairs of antennae are reduced and modified for attachment. First pair of antennae have a large claw for attachment to the host.
6. Compound eyes are present.
7. In some (*e.g.*, *Argulus*, Fig. 6.6B), first pair of maxillae contain two large suckers for attachment. Other appendages are vestigial.

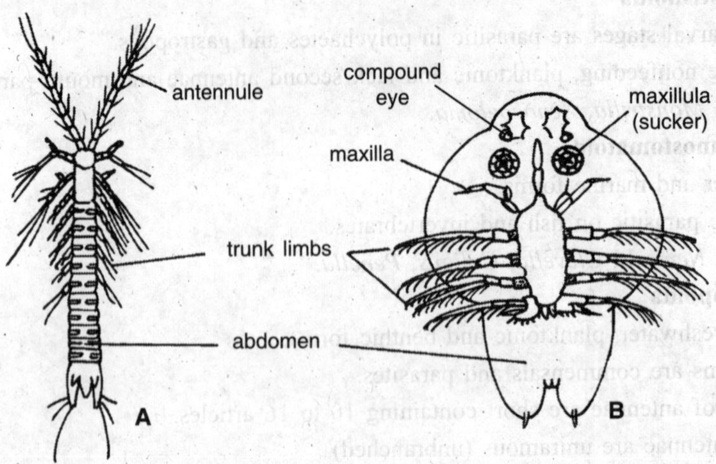

Fig. 6.6. A—*Derocheilocaris* (mystacocarid); B—*Argulus* (branchiuran).

Class 7. Remipedia

1. It includes only 2 species of crustaceans which are inhabitants of marine caves. They were discovered by **Yager** (1981).
2. Body is very long, segmented and worm-like.
3. Each body segment bears a pair of biramous appendages.

   Example. *Lasionectes.*

Class 8. Tantulocarida

1. It includes only 4 species.
2. Small marine crustaceans ectoparasitic on deepwater crustaceans.
3. Body is similar to copepods but lacking trunk segments.

   Example. *Copepode-like animals.*

Class 9. Cirripedia (*Curl foot barnacles*)

1. This class includes about 900 species of barnacles.
2. They are either free living or parasitic. All are marine.
3. Body is enclosed within a saccular bivalved **carapace** or mantle of lime. Carapace is divided into plates.
4. Both pairs of antennae are reduced or absent. **Mandibles** are blade-like.
5. Body is divided into head and trunk. External segmentation is very indistinct. Abdomen is absent.
6. Thorax bears six pairs of similar biramous appendages called **cirri**.
7. Most of them are **hermaphrodites**. Penis is long and tubular. Young ones hatches as **nauplius larva** that moults to produce **cypris larva**.

   Class Cirripedia is divided into **four** orders:

Order 1. Ascothoracica or Ascothoracida (*Naked barnacles*)

1. They are parasitic on echinoderms and corals.
2. Bivalve or saccular carapace; without calcareous plates on the mantle.
3. First antennae are prehensile. **Mandibular palp** is present.
4. Abdomen is present.

5. Female gonopore on first thoracic segment.

Examples. *Dendrogaster*, *Ascothorax*.

Order 2. Acrothoracica (*Naked boring barnacles*)

1. They live attached to some calcareous substratum, especially shells of molluscs and corals. They bore into substratum.
2. They lack calcareous plates on the mantle. A chitinous attachment disc is present.
3. Three to five pairs of cirri are present.

Examples. *Trypetesa*, *Kochlorine*, *Berndita*.

Order 3. Thoracica (*Goose barnacles and Acorn barnacles*)

1. They are free living and commensal forms; with or without stalk (peduncle).
2. Carapace or mantle is covered with calcareous plates.
3. Animal lies in the carapace upside down, *i.e.*, stands on its head.
4. Six pairs of well-developed cirri are present.
5. They are monoecious.

Examples. *Lepas*, *Scalpellum*, *Verruca*, *Balanus*, *Coronula*, *Xenobalanus*.

Order 4. Rhizocephala

1. Parasitic on decapod crustaceans and tunicates.
2. Mantle, appendages and alimentary canal are absent.
3. Body is reduced to ramifying stolon and gonad sac.
4. Stalk or peduncle forms root-like absorptive processes.

Examples. *Sacculina*, *Peltogasterella*, *Lernaeodiscus*.

Class 10. Malacostraca

1. It includes about 22,651 species which are marine or freshwater.
2. Body is divided into head, thorax and abdomen.
3. Trunk is typically composed of 14 segments and a *telson*. Its first 8 segments form the thorax and last 6 form the abdomen.
4. Thorax may or may not be covered with carapace.
5. All body segments bear appendages.
6. **Female gonopores** are always located on the 6th thoracic segment and the **male gonopores** are on the 8th thoracic segment.

Class Malacostraca is divided into **three** subclasses:

Subclass 1. Phyllocarida

Order Leptostraca or Nebaliacea

1. It includes about 20 species of most primitive living malacostracans (appeared in Cambrian period).
2. Small sized marine animals living in bottom mud.
3. They are suspension feeders.
4. They have **eight** segments in the abdomen.
5. Thorax is enclosed in a bivalve carapace and is provided with foliaceous appendages.

Examples. *Nebalia*, *Paranebolia*.

Subclass 2. Hoplocarida

Order Stomatopoda (*Mantis shrimp or armed shrimps*)

1. It includes about 300 species which are marine and specialized predators of fish, crabs, shrimps and molluscs.
2. Body is dorso-ventrally flattened. Carapace is small and shield-like but does not cover the last two thoracic segments.
3. Abdomen is large, broad and segmented.
4. First antennae (*i.e.*, **antennules**) with three flagella.
5. Second pair of thoracic appendages are large and chelate; it is developed for **raptorial feeding**.
6. Thoracic limb protopod (= protopodite) is composed of two segments. Endopods (= endopodite) comprise of four segments.
7. Compound eyes are stalked.
8. Pleopods are well developed; they contain gills.
9. Telson and uropods are large.

Examples. *Squilla*, *Lysiosquilla*, *Gonodactylus*.

Subclass 3. Eumalacostraca

1. Antennae are without three flagella.
2. Endopodite is composed of five segments.
3. Gills when present, they are restricted to thoracic limbs.

Class Eumalacostraca is divided into **four** superorders:

Superorder 1. Syncarida

1. Carapace is absent.
2. Eight thoracic segments are free or the first fused with head.
3. Thoracic appendages are biramous and similar. Chelate or subchelate legs are absent.
4. Freshwater (ground water)—inhabitants or cave-dwelling.

It includes **two** orders.

Order 1. Anaspidacea

1. It includes 10 species which are suspension feeders.
2. First thoracic somite is fused with head.
3. First thoracic legs are modified as maxillipedes and have distinct gnathobases.

Example. *Anaspides*.

Order 2. Bathynellacea

1. It includes about 25 species of minute malacostracans.
2. First thoracic segment is not fused with the head.

Example. *Brasilibathynella*.

Superorder 2. Pancarida

Order Thermosbaenacea

1. They are inhabitants of hot springs and ground water.
2. Small sized and blind animals having a reduced carapace which is fused with the first thoracic segment.
3. They carry eggs in dorsal brood pouch.

Example. Hot spring inhabitants.

Superorder 3. Eucarida

1. It includes large sized malacostracans.
2. Carapace is highly developed and fused with all the thoracic segments.

3. Gills are located near the base of thoracic appendages.
4. Compound eyes are stalked.

This superorder is divided into **two** orders.

Order 1. Euphausiacea (*Krills*)

1. It includes about 90 species of brightly shining krills.
2. All are marine; pelagic and shrimp-like, upto 3 cm long.
3. Sides of carapace do not enclose the gills tightly.
4. Thoracic appendages are biramous and anterior ones are not modified as maxillipeds.
5. Many contain light producing **photophores** and are biolumnescents.

Example. *Euphausia.*

Order 2. Decapoda (*Shrimps, crabs, lobsters* and *cray fishes*)

1. It includes about 10,000 species accounting for nearly 20 per cent of the known species of crustaceans.
2. They are marine, freshwater or terrestrial.
3. Their first three pairs of thoracic appendages are modified as **maxillipedes**. The remaining five pairs (*i.e.*, 10) of thoracic appendages are **uniramous walking legs** (hence the name decapoda).
4. The carapace encloses the gills in a branchial chamber.
5. **Statocyst** is present in the basal segment of first antennae (or antennule).

Examples. Shrimps—*Crangon* (sand shrimps), *Alpheus* (snapping shrimp), *Palaemon* or *Macrobrachium* (prawn), *Atya*; **crayfish**—*Procambarus*, *Astacus*, *Cambarus*; **lobsters**—*Nephrops*, *Homarus*; **marine burrowing shrimp**—*Thalassina*; **spiny lobsters**—*Panulirus*; **hermit crab**—*Eupagurus*, *Pagurus*; **coconut crab**—*Birgus*; **lobster like** — *Galathea*; **mole crabs**—*Hippa*, *Emerita*; *Brachyura*; **box crab**—*Calappa*, **Spider crab** *Maja*; **fiddler crab** — Uca; **mud crab**—*Xantho*; **rock crab**—*Cancer*; **blue crab**—*Callinectes.*

Superorder 4. Peracarida

1. This superorder carries 40 per cent of the known species of Crustacea.
2. They distinctly contain a **ventral brood chamber** or **marsupium** in the female. It is formed by large plate-like processes of certain thoracic coxae.
3. A carapace may be present or absent.
4. First thoracic somite is fused with the head.
5. There are eight pairs of thoracic appendages. First pair is usually a maxillipede and remaining seven are **legs**.
6. Development is direct.

Superorder Peracarida is divided into **five** orders:

Order 1. Mysidacea (*Opossum shrimps*)

1. Most are marine and some live in freshwater (*i.e*, in lakes and caves); pelagic and benthic; filter–feeder or scavengers.
2. They occur in large **swarms** forming an important food organism of economically important fish.
3. Ventral **marsupium** present *i.e.*, oostegites present from two to all of thoracic legs forming a **marsupium** due to which they are called **opposum shrimps**.
4. They have stalked compound eyes.
5. First and sometimes second pair of thoracic appendages are modified as maxillipedes.

6. Pleopods are present in males but reduced or absent in females.

Examples. *Mysis, Neomysis, Gnathophausia.*

Box 6.2

Mysis oculata lives only in deep lakes and undergoes remarkable **diurnal movement**. During the daylight hours they stay within a metre of the bottom, but during the night they might rise to surface.

Order 2. Cumacea

1. It includes 800 species which are marine, burrowing and filter feeding creatures. They live in burrows or mucous tubes in bottom mud or sand.
2. Swarmming occurs in males.
3. Head and thorax are greatly enlarged. Carapace covers the first three to four thoracic somites forming a gill chamber.
4. Abdomen is very narrow; it is usually without pleopods in females and bear two to five pleopods in males.
5. Uropods are slender, filiform structures and used in cleaning the anterior segments.
6. Antennae are vestigial in the female but very long in males.
7. Series of filamentous gills are present on first pair of maxillipedes.

Example. *Diastylis.*

Order 3. Amphipoda (*Both foot*)

1. It includes about 5,500 species which are marine freshwater or terrestrial. They may be free living and ectoparasitic species.
2. Some are pelagic, most are benthic, a few are filter feeders.
3. Body is laterally compressed and without carapace.
4. Compound eyes are sessile lacking eyestalks.
5. There are one pair of maxillipedes and seven pairs of uniramous thoracic legs having gills.
6. The abdominal appendages include three pairs of pleopods and three pairs of uropods. These are adapted for jumping.

Example. *Caprella* (skeleton shrimp), *Hyalella*; *Paracyamus* (whale lice), *Phronima* (parasite of jelly fish or tunicates); *Orchestoidea, Rhabdosoma.*

Order 4. Isopoda (*Equal foot*)

1. It includes about 10,000 species which are freshwater, marine or terrestrial. They are free living and parasitic.
2. They are not filter feeders. Mouth parts resemble with that of insects.
3. Body is dorso-ventrally flattened. Head is shield-shaped.
4. Carapace is absent; compound eyes are sessile.
5. There is one pair of maxillipedes and seven pairs of

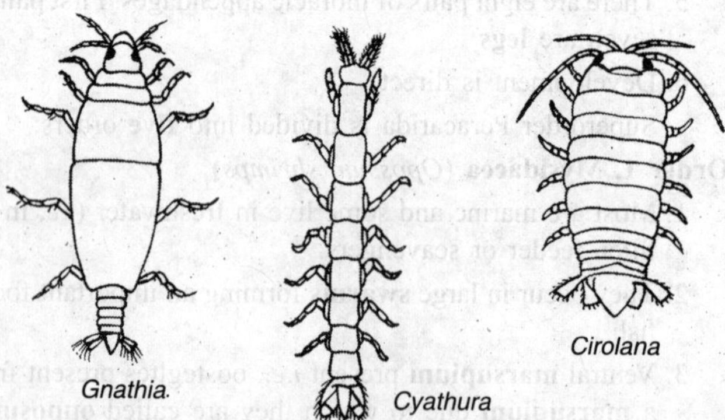

Fig. 6.7. Some isopods.

similar uniramous thoracic legs (Fig. 6.7). These are adapted for crawling (they creep like insects). Some aquatic isopods can also swim.

6. Some abdominal pleopods function as gills.
7. Uropods are fan-shaped.
8. Some isopods can roll up, armadilo-like when disturbed. Land isopods, often called **pill bugs, sow bugs** or **wood lice** are frequently found under logs in wooded areas.

   Example. *Gnathia, Cyathura, Limnoria, Cirolana, Astacilla, Asellus, Phreatoicus, Bopyrus, Ligia, Oniscus*.

Order 5. Tanaidacea

1. It includes about 550 species of small, benthic marine animals.
2. They burrow in mud or live in mucous tubes.
3. A small carapace covers anterior part of body; inner surface of carapace acts as gills.
4. Eyes are lobed, present on lateral immovable processes.
5. The first thoracic appendages are maxillipedes equipped with flattened epipodite.
6. Third thoracic appendages are adapted for burrowing.
7. Females have marsupium.
8. Each of them begin life as a female. After having served in this role, it molts to a male and serves in this capacity at least once.

   Examples. *Leptochelia, Tanais*.

Subphylum 4. Uniramia

Subphylum Uniramia is divided into **five** classes:

Class 1. Chilopoda (*Centipedes*)

1. It includes about 2500 species of centipedes.
2. They live in moist places such as soil and humus and beneath stones, bark and log. They are nocturnal and carnivorous animals feeding upon soft insects such as cockroach, plant lice, silver fish and earthworm and slugs.
3. First pair of trunk appendages (maxillepedes) are modified into **poison claw** or **forcipules**.
4. Body is divided into head and trunk. Head is convex in scutigeromorph but flattened in other centipedes.
5. Trunk is somewhat flattened and elongated and contains 15 to 170 segments. Each segment bears one pair of legs. All the legs are similar. The last pair of legs are elongated and sensory (or defensive) in function.
6. Last segment is limbless, called **genital segment**, and it bears genital opening.
7. Centipedes run very rapidly.

   Class Chilopoda is divided into **two** subclasses:

Subclass 1. Epimorpha

1. Development is **epimorphic**, *i.e.*, young contains all segments when they hatch.
2. The female broods its eggs by winding itself around the egg mass.
3. Adults contain 21 or more pairs of legs.

   This subclass is divided into **two** orders:

Order 1. Geophilomorpha

1. Elongated, blind burrowing form with 31 to 180 pairs of legs.
2. Each trunk somite with dorsal tergite and intertergite and ventral sternite and intersternite.

3. Antennae with 14 segment.
4. Pair of spiracles are present in all but first and last somites.

Examples. *Geophilus*, *Strigamia*, *Mecistocephalus*.

Order 2. Scolopendromorpha

1. They contain 21 to 23 pairs of legs.
2. Tergal plates correspond to sternal plates and are equally large.
3. Antennae with 17 to 31 segments.
4. Eyes are absent or composed of four ocelli.
5. They contain 9 to 11 pair of spiracles.

Examples. *Scolopendra*, *Theatops*.

Subclass 2. Anamorpha

1. Development is **anamorphic**, *i.e.*, the youngs contain segments less than that of adults.
2. No brooding occurs.
3. Adults have 15 pairs of legs.
4. A pair of **organ of Tomosvary** is present on the head at the base of antennae. It is used in **vibration detection** and **monitoring of humidity**.

This subclass is divided into **two** orders.

Order 1. Lithobiomorpha

1. Antennae contain 19 to 70 segments.
2. Trunk contains nine large and six small somites.
3. Large and small tergal plates alternate.
4. Spiracles are paired and lateral.
5. Eyes with many ocelli.
6. Development is anamorphic.

Examples. *Lithobius*, *Bothropolys*.

Order 2. Scutigeromorpha

1. Legs and antennae are very long.
2. Fifteen sternal plates, but with only eight tergal plates.
3. Eyes are large and compound.
4. Spiracles are unpaired and located mid-dorsally on tergal plates.
5. Development is anamorphic.

Example. *Scutigera.*

Class 2. Symphylla

1. It includes 160 species of myriapods which live in soil of green houses and attack plant roots and can be serious pest to vegetable and flower crops.
2. They are between 1 and 8 mm in length and superficially resemble lithobiomorph centipedes.
3. Somites of trunk are covered by 15 to 24 tergal plates.
4. Trunk contains 12 legs bearing segments. Trunk terminates in a tiny oval telson.
5. First maxilla with no palp. Second maxillae fuse to form a **labium.**
6. A single pair of spiracles opens on to the sides of the head, and the tracheae supply only the first three trunk segments.
7. **Gonopore** is present ventrally on fourth trunk segment.

8. The last (14th) segment bears a pair of spinnerets or cerci and a pair of long, sensory hairs (called **trichobothria**).
9. Eyes are absent.
10. Two organs of Tomosvary are present (Fig. 6.9).
11. Development is anamorphic.

Examples. *Scutigerella* (Fig. 6.8); *Hanseniella.*

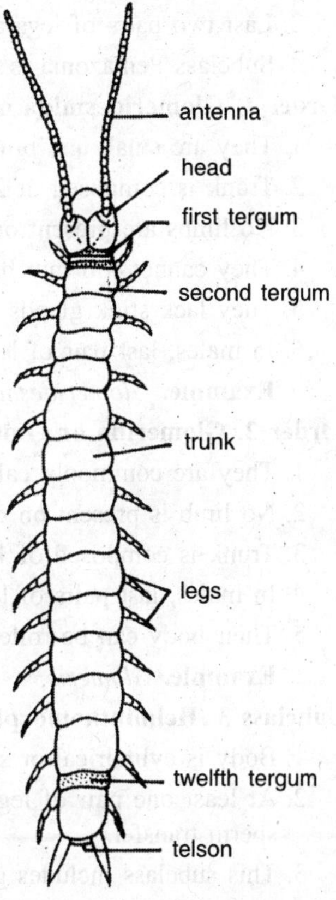

Fig. 6.8. *Scutigerella* (in dorsal view).

Class 3. Diplopoda (*'Thousand legged'* or *millipedes*)

1. It includes about 10,000 species.
2. Millipedes avoid light and live beneath leaves, stones, bark and logs and in soil. They are herbivorous and run much more slowly.
3. Diplopod (double foot) refers to the most important character of millipedes. They contain two pairs of legs in nearly each abdominal segment. This occurs because two embryonic somites fuse to form an adult somite, called **diplosegment**.
4. Body is elongated and cylindrical.
5. Body is divided into **head** (first 5 segments), **thorax** (of 3 segments) and **abdomen** of 20 to 100 segments.
6. Last head segment is called **collum** and forms a large collar behind the head.
7. Thorax contains one pair of legs in each segment.
8. Each of the diplosomites is enclosed in a continuous skeletal ring, with no articular membranes between the **tergal**, **pleural** and **sternal plates**. Each segment is covered with single tergal plate dorsally and two sternal plates ventrally. Tergites are impregnated with calcium salts like the crustacean.
9. **Stink glands** or **repugnatorial glands** are present in many millipedes (*e.g., Julus, Apheloria*) secreting noxious substances which may be **aldehydes, quinones, phenols** and **hydrogen cyanide**.
10. Actual copulatory organs are modified trunk appendages, called **gonopods.**

Class Diplopoda is divided into following **three** subclasses:

Subclass 1. Pencillata (*Pselaphgnatha*)

1. Minute millipedes with broad and soft bodies. Integument bears tufts and rows of serrated scale-like, setae.
2. Eyes are present.
3. Gonopods and stink glands are absent.
4. Trunk bears 13 somites.

Example. *Polyxenus.*

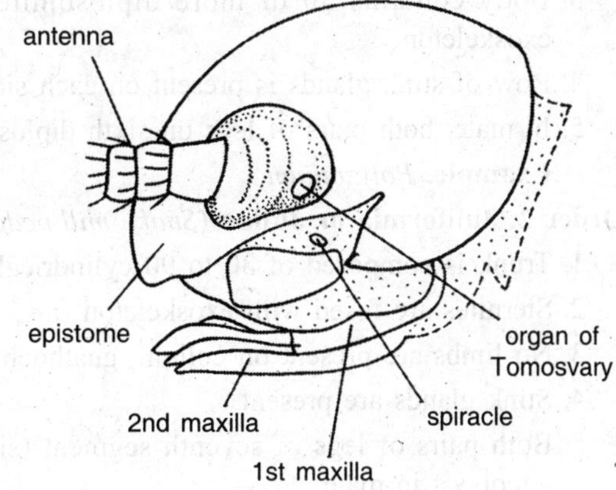

Fig. 6.9. Head of *Hanseniella.*

Subclass 2. Pentazonia

1. Tergal plates are arched.
2. Last two pairs of legs are modified for clasping.

Subclass Pentazonia is divided into **two** orders:

Order 1. Glomeridesmida or Limacomorpha

1. They are small and blind.
2. Trunk is composed of 20 to 22 arched somites.
3. No limbs are present on collum, gnathochilarium.
4. They cannot roll into ball.
5. They lack stink glands.
6. In males, last pair of legs are modified as **gonopods**.

Example. *Glomeridesmus*.

Order 2. Glomerida or Oniscomorpha

1. They are commonly called **pill millipedes.**
2. No limb is present on collum, gnathochilarium.
3. Trunk is composed of 12 to 13 somites which are flattened on the ventral surface.
4. In males, last pairs of legs are modified ; the last pair is used for sperm transfer.
5. Their body can be rolled into ball.

Example. *Glomeris*.

Subclass 3. Helminthomorpha

1. Body is cylindrical or somewhat flattened.
2. At least one pair of legs (gonopods) of the seventh segment in the male is modified for sperm transfer.
3. This subclass includes greatest number of millipedes species.

This subclass is divided into **eleven** orders, out of which following **three** are more important.

Order 1. Polyzoniida or Colobognatha (*Suctorial millipedes*)

1. Largely tropical, suctorial and elongated millipedes.
2. No limbs are present of collum; gnathochilarium.
3. Body contains 30 or more diplosomites. Sternal plates are not fused with rest of exoskeleton.
4. Row of stink glands is present on each side of body.
5. In male, both pairs of legs on sixth diplosomite is modified as gonopod.

Example. *Polyzonium*.

Order 2. Juliformia or Julida (*Snake millipedes*)

1. Trunk is composed of 30 to 90 cylindrical segments.
2. Sternites are fused with exoskeleton (*i.e.*, pleurotergal arch).
3. No limbs are present on collum, gnathochilarium.
4. Stink glands are present.
5. Both pairs of legs of seventh segment (sixth when collum is excluded) are modified as gonopods in male.

Examples. *Julus*, *Nemasoma*, *Blaniulus, Thyroglutus*.

Order 3. Polydesmida (*Flat-backed millipedes*)

1. Relatively short body, with no limbs on collum, gnathochilarium.
2. Trunk contains 19 diplosomites, each covered by two skeletal rings, anterior **prozonite** and posterior **metazonite**.
3. Sternites are fused with exoskeleton (*i.e.*, pleurotergal arch).
4. Stink glands are present on some somites.
5. Eyes are absent.
6. In males, first pair of legs on seventh diplosomite are modified as gonopods.

   Examples. *Polydesmus*, *Oxidus*, *Apheloria*.

Class 4. Pauropoda

1. It includes about 500 species.
2. They are minute, soft-bodied, grub-like animals which live in leaf mold and soil. They feed on fungi or decomposing plant tissue. Some are predatory.
3. The head bears a pair of discoidal sensory organ for moisture detection and two biramous Y-shaped antennae.
4. The trunk consists of 11 segments and a telson. Each segment, except the first, last and telson bears a pair at short legs.
5. Genital pore is present on the third trunk segment.

   Example. *Pauropus*.

Class 5. Hexapoda or Insecta (L., *insectum* = being cut into)

1. This class includes more than 750,000 species of air breathing arthropods; most of which are terrestrial some are aquatic.
2. Body is divided into **head**, **thorax** and **abdomen**.
3. **Head** is formed of six fused segments. It bears a pair of **antennae**, a pair of **compound eyes** and mouth parts which are adapted for chewing, biting, piercing, sucking, siphoning and sponging type.
4. Heads of most insects are oriented so that the mouth parts are directed either downward (**hypognathus**), anteriorly (**prognathous**) or posteriorly (**opisthognathus**).
5. **Thorax** is formed of three segments and bears **three pairs of legs** and **one** or **two pairs** of **wings**.
6. **Abdomen** is formed of not more than 11 segments.
7. Abdomen generally lacks appendages except modified posterior gonopods.
8. Respiration takes place by the help of branched **trachea**.
9. Exoskeleton is composed of a **tergum**, a **sternum** and two **pleura**. **Sclerites** (thickened areas of cuticle) are prominent features of insect body surface. Sclerites are often separated by **sutures**.
10. Sexes are separate, fertilization is internal.

    Class Insecta is divided into **two** subclasses:

Subclass 1. Apterygota (Gr., *a* = without + *ptera* = wings)

1. Wings are absent and wingless condition is primary.
2. Mouth parts either become sunk into a pouch (**entognatha**) or do not sunk into head pouch (**ectognatha**).
3. Compound eyes and Malpighian tubules are reduced or absent.
4. Metamorphosis does not occur.

Subclass Apterygota is divided into **five** orders:

Order 1. Protura

1. Small insects with a cone-shaped head without compound eyes or ocelli. They contain pseudoculi (*i.e.,* receptors for mechanical vibrations).
2. Antennae are absent and first pair of legs are used as antennae. Entognathous mouth parts.
3. They have a 12-segmented abdomen lacking cerci.
4. They contain simple or no tracheal system.
5. Development is anamorphic.
6. They live in damp humus and soil and feed on decayed organic matter.

Examples. *Eosentomon, Acerentulus.*

Order 2. Aptera (or Diplura)

1. Small, soft, wingless insects with slender, flattened bodies.
2. They contain entognathous mouthparts.
3. They contain long antennae but lack eyes.
4. Abdomen is provided with two cerci.

Example. *Campodea.*

Order 3. Collembola (*Spring tails*)

1. Minute, soft bodied with short antennae.
2. Eyes are either absent or represented by isolated ommatidia.
3. Simple, endognathous chewing mouthparts with sharp mandibles.
4. They contain an **abdominal jumping** or **springing organ**. Jumping organ is composed of spring **(furcula)** on fifth abdominal somite and clasp **(tentaculum)** on third abdominal somite.
5. They lack abdominal cerci as well as Malpighian tubules.
6. They feed decaying plant material and fungi.

Examples. *Entomobrya, Achorutes.*

Order 4. Microcoryphia or Archeognatha (*Jumping bristle tails*)

1. Small to medium sized, soft, wingless insects with elongated cylindrical bodies.
2. They contain long antennae and styliform (*i.e.*, slender and elongated) abdominal cerci.
3. They have ectognathous mouthparts and feed on algae, lichens and moss.

Example. *Machilis.*

Order 5. Thysanura (*Silverfish and bristle tails*)

1. Small insects with long, often scaly bodies.
2. They possess ectognathous biting or chewing type mouth parts.
3. Head is with long **antennae** and **compound eyes. Ocelli** may be present or absent.
4. They are fast-running insects having short cursorial legs.
5. Abdomen is with three styliform appendages—one caudal filament and two long abdominal cerci.
6. Malpighian tubules and tracheae are well developed.
7. Some species of them are found in house where they eat books and clothing.

Example. Silver fish (*Lepisma, Thermobius*).

Subclass 2. Pterygota (Gr., *ptera* = wings)

1. Insects are with wings or they are secondarily wingless.
2. They have ectognathous mouthparts.
3. They undergo metamorphosis.

Subclass Pterygota is divided into **24** orders. Some of the important orders are as follows:

Order 1. Ephemeroptera or Ephemerida (*Mayflies*)

1. Slender body with short hair-like antennae and large compound eyes.
2. **Chewing mouthparts** are present in aquatic nymphs and vestigial mouthparts in short-lived (or ephemeral) adults.
3. Wings are net-veined, of which the first is larger than the second and both pairs are held vertically over back when at rest.
4. Abdomen contains 10 distinct somites and 11th vestigial somite.
5. They contain two very long abdominal cerci and caudal filament.
6. Metamorphosis is gradual or incomplete; with aquatic **nymphs** which have **tracheal gills** for respiration.

Example. *Ephemera.*

Order 2. Odonata (= toothed flies) (*Dragon flies and damselflies*)

1. Slender (damselflies) or stout-bodied (dragon flies), predaceous forms with strong **biting and chewing mouthparts.**
2. Head is with huge compound eyes.
3. Two pairs of similar, equal-sized, narrow, net-veined wings held straight out from body when at rest (dragon flies) or held vertically (damsel flies).
4. Long abdomen is with 10 distinct somites and two rudimentary anal somites.
5. Abdominal cerci are rudimentary.
6. Metamorphosis is gradual or incomplete; nymphs are aquatic with or without caudal gills.

Examples. Dragon flies (*Macromia, Sympetrum*).

Order 3. Orthoptera (= straight wings) (*Grasshoppers, katydids, crickets, roaches, mantids and walking sticks*)

1. Large headed insects with **strong biting and chewing mouthparts.**
2. Many species are with long, jumping (or saltatorial) hindlegs.
3. Wings are present or absent. Most winged forms have membranous hindwings folded fanlike beneath leathery forewings (called **tegmina**).
4. Largely herbivorous, sometimes causing vast crop damage (*e.g.*, Locust).
5. Metamorphosis is gradual or incomplete.

Examples. Grasshoppers (*Melanopus, Poecilocerus*); **locust** (*Schistocerca*); **cockroach** (*Periplaneta, Blatta*); **cricket** (*Gryllus*), **mole cricket** (*Gryllotalpa*); **walking sticks** (*Carausius*); **leaf insect** (*Phyllium*) and **katydids**.

(**Note.** Some classifications places the cockroaches, mantids and walking sticks in separate orders: **1. Blattaria** (cockroaches), 2. **Phasmida** (walking sticks and *Phyllium*) and 3. **Mantogea** (praying mantids). All these three orders have been included in single order **Dictyoptera**, see **Meglitsch** and **Schram** 1991).

Order 4. Isoptera (= equal wings) (*Termites*)

1. Social insects with **caste system**; members of society are differentiated as reproductive members, workers and soldiers.

2. Colony includes both winged and wingless individuals.
3. Soft bodied, pale coloured, with abdomen broadly joined to thorax.
4. Wings are of equal size and held **horizontally** over **abdomen.**
5. Metamorphosis is gradual or incomplete.
6. **Biting mouthparts** are present.

Examples. *Termes*, *Zootermopsis*.

Order 5. Dermaptera (= skin wings) (*Earwig*)

1. Elongated body with tough exoskeleton; **chewing mouthparts,** compound eyes and large forceps-like **anal cerci** on the posterior end.
2. Wingless or with short forewings forming tegmina (elytera) that cover all but tip of much folded, membranous, radially-veined, fan-like hindwings.
3. They are **nocturnal** and have **omnivorous** food habits.
4. Metamorphosis is gradual or incomplete.

Example. Earwig (*Forficula*).

Order 6. Plectoptera (*Stone fly*)

1. Adults have long antennae, **chewing mouthparts** and two pairs of well-developed, membranous wings or vestigial wings.
2. Abdomen bears long cerci.
3. Metamorphosis is gradual or incomplete.
4. Nymphs are aquatic, with or without gills.

Example. *Taeniopteryx* **(stone fly).**

Order 7. Mallophaga (= *wool eater*) (*Chewing lice and bird lice*)

1. Wingless, flattened insects that live as ectoparasites on birds and mammals.
2. Eyes are reduced or absent; legs are short having stout claws and thorax is small. Head is hypognathous.
3. Abdomen with five to eight somites.
4. Metamorphosis is gradual or incomplete.
5. They contain **biting and chewing mouthparts** and feed on scales, feathers, hair, skin and sometimes dried blood around wounds.

Example. *Cuclogaster* **(chicken louse).**

Order 8. Anoplura (= unarmed tail) (*Chewing lice and bird lice*)

1. They are similar to chewing lice but mouthparts are adapted for **sucking.**
2. They are ectoparasites of birds and mammals.
3. A number of species are parasitic on domestic animals and the **head louse** or **crab louse** are parasites of humans. They cause irritation and serve as vectors of disease such as typhus fever.

Examples. *Haematopinus* **(sucking louse)**; *Pediculus humanus* **(human louse).**

Order 9. Thysanoptera (*Thrips*)

1. Small, slender insects with **rasping, sucking mouthparts.**
2. They include both winged and wingless species. When winged, forewings are larger and both pairs are narrow, nearly veinless and with fringe of fine hairs on posterior margin.
3. Abdominal cerci are absent.
4. Most species feed on sap in flowers.

Example. *Taeniothrips* **(pear thrips).**

Order 10. Hemiptera (= half wings) (*True bugs*)

1. They contain **piercing** and **sucking beak** which can be folded under the body.
2. **Forewings** with a thickened basal and distal membranous section. The membranous portions of forewings overlap when at rest over the abdomen.
3. Hindwings are entirely membranous.
4. Gradual or incomplete metamorphosis.
5. They are herbivorous and predaceous.

Examples. *Cimex* **(bed bug)**, *Belostoma* **(giant water bug)**, *Nepa* **(water scorpion)**.

Order 11. Homoptera (=same wings) (*Cicadas, leafhoppers and aphids*)

1. They include herbivorous insects related to the hemipterans, but beak arises from back of head.
2. Mouthparts adapted for **piercing** and **sap-sucking**.
3. Forewings are typically membranous. Wings are commonly held in a tent-like position over the body.
4. Metamorphosis is gradual or incomplete.

Example. Buffalo treehopper; *Cicada*.

Order 12. Neuroptera (*Lacewing, dobson flies and snake flies*)

1. They are carnivorous, with **chewing mouthparts.**
2. Antennae are usually long and slender; compound eyes are large.
3. Legs are slender. Forelegs are sometimes raptorial.
4. They contain two pairs of similar, membranous, net-veined wings which are held roof-like over the back at rest.
5. Metamorphosis is complete.
6. Larvae are predaceous and usually terrestrial.

Examples. *Mantispid*; *Chrysopa* **(lacewing)**; *Sialis* **(alder fly or dobson fly)**.

Order 13. Coleoptera (= shield wings) (*Beetles, weevils, fireflies*)

1. It is largest order of insects containing over 300,000 species.
2. They contain tough or hard exoskeleton, **chewing mouthparts,** antennae variable and large compound eyes.
3. Adults have two pairs of wings, of which **forewings** are modified as heavy protective covers **(elytra)**, meeting in straight line at middle of back.
4. **Hindwings** are membranous and usually completely hidden by elytra when at rest.
5. Abdomen is ten-segmented lacking cerci.
6. They have complete metamorphosis.
7. Most of them are plant feeders but some are predatory too.
8. They are mostly terrestrial, a few are aquatic.

Examples. *Hippodamia*, *Melolontha* **(cockchafer beetle)**.

Order 14. Lepidoptera (= scale wings) (*Butterflies, moths*)

1. Soft-bodied insects with two pairs of wings which are often large and showy. Forewings are often larger than hindwings.
2. Body and appendages (wings, etc.) are covered with overlapping pigmented scales.
3. **Mouthparts** are modified as coiled proboscis and used for **sucking** liquid food such as flower nectar.

4. Antennae are long and often feathery.
5. They contain large compound eyes.
6. Abdomen lack cerci.
7. They undergo complete metamorphosis. Larva are **caterpillars** which are usually plant feeders (herbivorous); adults feed little or not at all.

Example.

1. **Butterflies** which are diurnal with large, often brightly coloured wings which fold vertically over back at rest, have knobed antennae and soft bodies, *e.g.*, *Papilio* (**swallow-tail butterfly**); *Pieris*, *Kalima*.
2. **Moths** which are nocturnal and fold wings horizontally and contain pinnate antennae, stout bodies and usually dull coloured wings, *e.g.*, **royal walnut moth**; **gipsy moth**; **silk moth** (*Bombyx mori*) ; **clothes moths** which feed on textiles, clothing and fur.

Order 15. Diptera (= two wings) (*True flies, mosquitoes*)

1. **Forewings** are membranous and used for flying and **hindwings** are reduced to knoblike balancers, called **halters**.
2. Mouth parts are variable. Thus, they are either **lapping, sucking, piercing** type or vestigial.
3. Compound eyes are large; antennae are short and abdomen is without cerci.
4. They undergo complete metamorphosis.
5. Adults are often vectors of diseases; larvae are frequently damaging to vegetables and domestic animals.

Examples. Mosquitoes (*Anopheles*, *Culex*); **house fly** (*Musca*); **fruitfly** (*Drosophila*); **horseflies, midges** and **gnats**.

Order 16. Hymenoptera (= membrane wings) (*Ants, bees, wasps and sawflies*)

1. A large and varied order which includes both winged and wingless species.
2. They contain strong exoskeleton and **chewing, lapping** or **sucking** mouthparts.
3. Head contains slender **antennae** and large compound eyes.
4. Wings are either absent, reduced or two pairs of stiff, membranous, transparent wings are present.
5. Abdomen contains a slender pedicel.
6. They often live in **nests** in large societies with division of labour.
7. They undergo complete metamorphosis.
8. Larvae are caterpillars or are grub-like with chewing mouthparts.

Examples. Wasps (*Polistes*, *Vespa orientalis*); **black ant** (*Compnotus compressus*); **red ant** (*Oecophyla smaragdyna*); **mud dauber** (*Sceliphoran*), **honey bee** (*Apis indica*).

Order 17. Siphonoptera (*Fleas*)

1. They are ectoparasitic on birds and mammals.
2. Small, wingless insects with laterally flattened bodies.
3. Hind legs are long, stout and adapted for jumping.
4. Abdomen is without cerci.
5. They have **piercing** and **sucking mouthparts.**
6. They are vectors of certain epidemics, *e.g.* Indian rat flea transmits *Bacillus pestis*, a causative agent of **bubonic plague**.

7. They undergo complete metamorphosis.

Examples. Indian rat flea (*Xenopsylla cheopis*); **human flea** (*Pulex irritans*).

Box 6.3

The remaining **seven** lesser important orders of subclass **Pterygota** are the following:

1. **Grylloblatodea**, *e.g.*, *Grylloblatta*.
2. **Embioptera**, *e.g.*, **web spinners** such as *Embia, Oligotoma*.
3. **Psocoptera** or **Corrodentia**, *e.g.*, **Book lice** (*Ectopsocus*), **bark lice**.
4. **Zoraptera**, *e.g.*, *Zorotypus*.
5. **Strepsiptera**, *e.g.*, minute beetle-like twisted wing parasites.
6. **Mecoptera**, *e.g.*, **Scorpion flies** (*Panorpa*).
7. **Trichoptera**, *e.g.*, **Caddish flies** (*Rhyacophila*).

The animal groups which zoologist judge to have attained the **"greatest biological success"** are those which

1. Have the largest number of species and of individuals;
2. Occupying the widest stretches of territory and the greatest variety of habitats;
3. Consuming the largest amount and kinds of food; and
4. Most capable of defending themselves against their enemies.

By these standards the phylum which occupies first place among the animals (vertebrates and invertebrates) is the phylum **Arthropoda**. Horny material (**chitin**) of exoskeleton is to the arthropods what steel is to civilised human being, and it is partly to the possession of this hard cuticle that the arthropods owe their success.

6.4 TYPE STUDIES

1. *LIMULUS* (Horse-shoe crab or King crab)

Classification

Phylum	—	**Arthropoda**
Subphylum	—	**Chelicerata**
Class	—	**Merostomata**
Subclass	—	**Xiphosura**
Order	—	**Limulida**
Genus	—	***Limulus***

Habit and Habitat

Limulus or **king crab** is a marine animal and found along the north western Atlantic Coast and the Gulf of Mexico. It lives on soft bottoms in shallow waters. During breeding season, both male and female come on land (called **nuptial visit**) and dig holes in sand at the upper limit of high tides to lay eggs. *Limulus* is a scavenger and feeds on molluscs, worms and other organisms including bottom-dwelling algae.

Limulus is **not a true crab**. It is known as a **living fossil**, because it has remained unchanged since the Triassic period nearly 200 million years ago. It is believed that *Limulus*, together with trilobites, originated from some common arthropod ancestor but remained universally isolated.

Structure

Limulus (*L. polyphemus*) reaches a length of 60 cm and is dark brown in colour. Its body is divisible into an anterior **prosoma** and a posterior **opisthosoma** (Fig. 6.10). The prosoma is convex above, covered by a broad, horse-shoe shaped **carapace** bearing a pair of median simple eyes and a pair of compound eyes. *Limulus* can detect movement through eyes. On its ventral side, prosoma bears six pairs of appendages surrounding the mouth. The first pair or **chelicerae** are small, trisegmented and chelate. The next five pairs are **walking legs** which serve both the function of locomotion and food procurement. The first four pairs of walking legs are chelate. The coxa of each leg contains spines, called **gnathobases**. These macerate and move the food towards mouth. The fifth pair of legs are not chelate. The coxae of these legs bear a short process, the **flabellum**. Flabellum is used for cleaning the gills. A pair of small, oval and plate-like appendages, called the **chilaria**, are also present a little behind the mouth. Chilaria are directed vertically downwards and represent the degenerated seventh pair.

The opisthosoma is hexagonal and movably articulated with prosoma. It consists of six seg-mented **mesosoma**, unsegmented **metasoma** and a long **telson** or caudal spine. Mesosoma bears six pairs of appendages. The first pair is fused and forms the **genital operculum** and bears two genital pores. Remaining five pairs of appendages are flap-like and membranous. The undersurface of each flap forms leaf-like folds called **gills** or **branchial lamellae** (book gills), for respiration.

A long triangular caudal spine or **telson** articulates to the posterior end of abdomen. Telson is highly mobile and is used for correcting the body when it is accidently turned over.

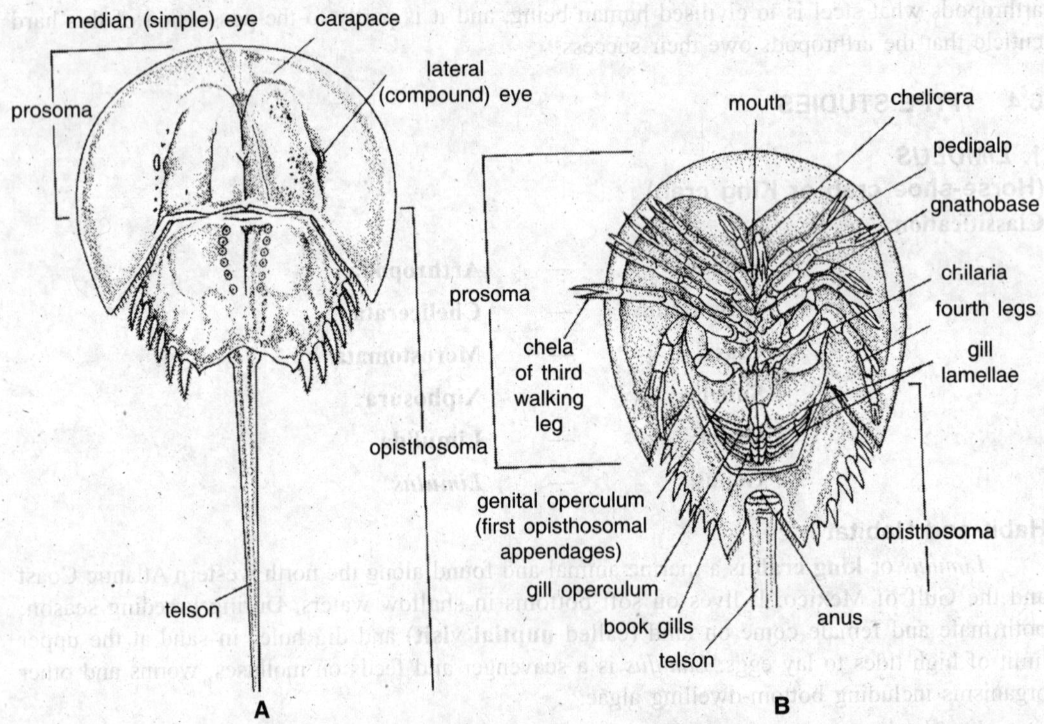

Fig. 6.10. *Limulus* (king crab). A—Dorsal view; B—Ventral view.

In *Limulus*, excretion takes place by **coxal glands** or **brick red glands**. Sexes are separate and sexual dimorphism is distinct. **Fertilization** is **external** and **cleavage** of its centrolecithal eggs is **complete** or **holoblastic**. The larva emerges from the egg as a free swimming **trilobite larva**. Average life span of *Limulus* is 19 years and it attains its sexual maturity within 9 to 12 years. *Limulus* has an economic importance too; it is fed to chickens and pigs and sometimes consumed by human beings.

2. *DAPHNIA* (Water flea)

Classification

Phylum	—	**Arthropoda**
Subphylum	—	**Crustacea**
Class	—	**Branchiopoda**
Subclass	—	**Diplostraca**
Order	—	**Cladocera**
Genus	—	***Daphnia***

Habit and Habitat

Daphnia is commonly called **water flea**. It is found in freshwater ponds, streams and ditches. *Daphnia* is widely distributed in India, Asia, Europe and U.S.A. *Daphnia* forms the food of fishes in aquatic food chain.

Structure

Daphnia is 2 mm in length. Its body is divisible into head, thorax and abdomen (Fig. 6.4). General body segmentation is, however, less distinct. **Head** is formed of six somites. Body except head is enclosed in a bivalved **shell** or carapace which terminates posteriorly as **dorsal** or **caudal spine**. The head is rounded and bent downwards and look like the beak of a bird. Head bears a pair of large biramous **antennae**, a pair of small **antennules**, a large, median, sessile **compound eye** and large **mandibles**. In adult, second pair of maxillae are absent. Head also contains a vestigial median naupliar eye and a pair of **nuchal sense organs**. Antennules contain chemoreceptor bristles. Antennae form the chief organ of locomotion of *Daphnia*. Its movement is **vertical** and **jerky**. The downstroke of the powerful antennae propels the animal upward. Then *Daphnia* uses antennae as a parachute and sinks slowly. Little **setae** present at the end of abdomen act as stabilizers.

Thorax bears five pairs of leaf-like appendages. These appendages bear fine setae which are used in food collection. *Daphnia* is suspension feeder. Each thoracic appendage also bears a thin, vesicular **epipodite** which functions as a **gill**. The **abdomen** is greatly reduced and devoid of appendages. **Postabdomen** is large and bears sensory setae.

Daphnia is **dioecious**, *i.e.*, its sexes are separate. The female carries the eggs and embryos in a large **brood chamber** on her back. It is situated between abdomen and posterior part of carapace. A pair of spine-like processes of postabdomen close the brood chamber and prevent the escape of eggs from it.

During summer when the food is in plenty and water temperature is favourable, *Daphnia* produces only female offsprings from diploid eggs **parthenogenetically**. During winter, change in water temperature decreases the food supply and induces appearance of few males. In this condition, eggs undergo meiosis and are **fertilized**. The fertilized eggs are enclosed in the brood chamber. Fertilized eggs are large and only two are produced in a batch. When female moults, brood chamber along with its fertilized eggs or embryo is cast off. The brood chamber sinks to bottom and can withstand drying and freezing. Under favourable conditions, the eggs start developing and females are produced.

3. *CYPRIS*

Classification

Phylum	—	**Arthropoda**
Subphylum	—	**Crustacea**
Class	—	**Ostracoda**
Subclass	—	**Podocopa**
Order	—	**Podocopida**
Genus	—	***Cypris***

Habit and Habitat

Cypris is commonly called **mussel shrimp**. It is a free-swimming small-sized (1 to 2 mm length) crustacean and occurs in freshwater stagnant ponds. It is **carnivorous** and feeding on small organisms.

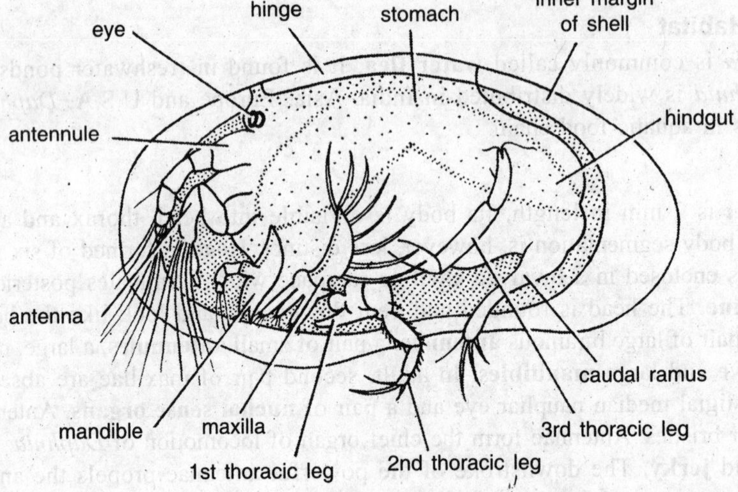

Fig. 6.11. *Cypris*. Mussel shrimp.

Structure

Body of *Cypris* is unsegmented and laterally compressed (Fig. 6.11). It is completely enclosed in a bivalved **carapace**, the two halves of which are articulated in the mid-dorsal line and inter-connected by a strong adductor muscle. A large median eye is present at the anterior end of body. The compound eye and heart is absent.

Head bears four pairs of appendages such as **antennules, antennae, mandibles** and **maxillae**. Antennules and antennae are large biramous and help in swimming. *Thorax* bears three pairs of appendages, *i.e.*, legs. **Abdomen** is devoid of appendages and is terminated by a pair of small **caudal styles**. *Cypris* is **parthenogenetic** and it lacks males.

4. *CYCLOPS*

Classification

Phylum	—	**Arthropoda**
Subphylum	—	**Crustacea**
Class	—	**Copepoda**
Order	—	**Cyclopoida**
Genus	—	***Cyclops***

Habit and Habitat

Cyclops is found throughout the world in fresh and brackish water. Its movement is rapid and jerky.

Structure

Body of *Cyclops* is elongated or pear-shaped and measures 1.5 to 5 mm in length (Fig. 6.12). Its body is divisible into three distinct parts: cephalothorax, thorax and abdomen. The **cephalothorax** is formed by the fusion of head and first thoracic segment. Cephalothorax is broad and has a convex dorsal surface. It is covered with a **carapace**. On the dorsal surface, there is a **median** or **naupliar eye**. *Cyclops*, in fact, derives its name after the great giant of Greek mythology, the **cyclops**, who had a single eye in his forehead.

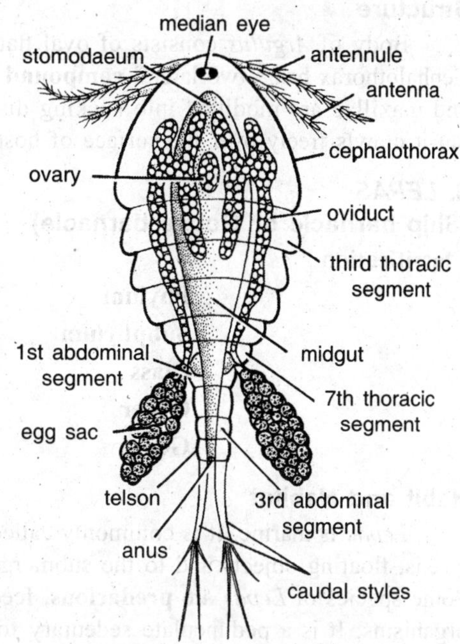

Fig. 6.12. *Cyclops*. Female.

Cephalothorax contains four pairs of appendages : **Antennules** are long, uniramous and horn-like. These are chief organs of locomotion. In the male, they are modified as **clasping organs**; these are used to hold the female during copulation. The **antennae** are short and uniramous. **Mandibles** and **maxillae** are other cephalo-thoracic appendages.

Thorax is formed of five free thoracic segments. First four thoracic segments bear biramous appendages. The fifth thoracic segment bears a pair of vestigial limbs. The last thoracic segment bears the **genital aperture. Abdomen** is formed of four segments and bear no appendages. The last abdominal segment bears anus on its dorsal surface and a pair of branched **caudal styles** having **setae.** In the female, the first abdominal segment is fused with last thoracic segment. The fused segment carries a pair of **ovisacs** or **egg sacs** in mature females.

Cyclops is **dioecious**, *i.e.,* its sexes are separate. Fertilization is internal and ovisacs (=egg sacs) function as brood chambers. Each sac contains up to 50 or more eggs. The eggs hatch as **nauplius larvae.**

Cyclops have great economic importance. They form an important component of aquatic food chain. They are eaten as plankton by fishes. *Cyclops* serve as an intermediate host for guinea worm.

5. *ARGULUS*

Classification

Phylum	—	**Arthropoda**
Subphylum	—	**Crustacea**
Class	—	**Branchiura**
Genus	—	***Argulus***

Habit and Habitat

Argulus is commonly known as **carp lice**. It is an **ectoparasite** of the branchial chamber or skin of freshwater fishes.

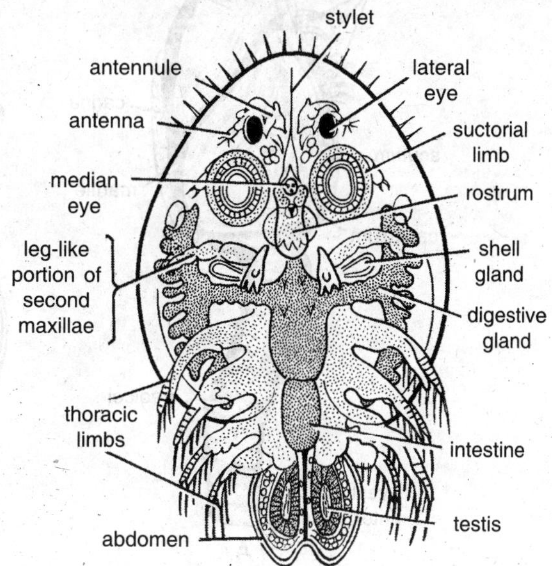

Fig. 6.13. *Argulus.*

Structure

Body of *Argulus* consists of oval flattened cephalothorax and a small bilobed abdomen. Cephalothorax bears two lateral **compound eyes** and one median simple eye. Mouth is suctorial and maxillae are modified into sucking discs. The animal is not permanently attached to host but it crawls freely over the surface of host.

6. *LEPAS*
(Ship barnacle or Goose barnacle)

Classification

Phylum	—	**Arthropoda**
Subphylum	—	**Crustacea**
Class	—	**Cirripedia**
Order	—	**Thoracica**
Genus	—	***Lepas***

Habit and Habitat

Lepas is marine. It is commonly called **ship** or **goose barnacle**. It is found attached to sea weeds, floating objects and to the submerged bottom of the ship. *Lepas* is **suspension feeder**. Some species of *Lepas* are **predacious**, feeding on copepods, isopods and other relatively large organisms. It is a pedunculate sedentary form. Too many *Lepas* remain attached at a place. A particular protein in the exoskeleton of older attached animals has been found to attract settling **cypris larvae** of *Lepas*.

Structure

Body of *Lepas* (Fig. 6.14) is divisible into a muscular, flexible **stalk** or **peduncle** and the **capitulum** or **body proper**. Peduncle is attached to substratum and bears capitulum at its distal end. It is covered with wrinkled skin and represents the preoral end of the animal. Peduncle contains

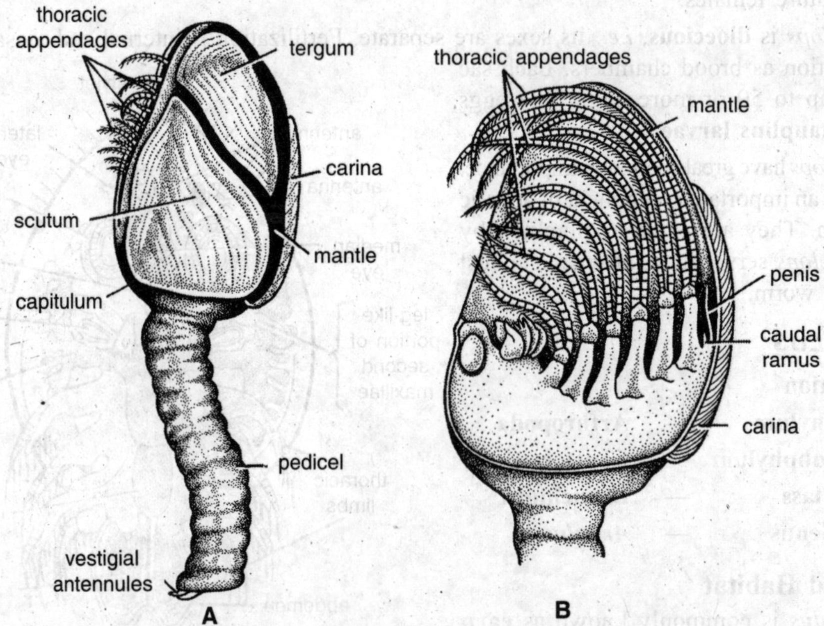

Fig. 6.14. *Lepas*. A–Entire animal; B–Carapace removed.

the vestiges of the larval **antennules** and the adhesive **cement glands** which open through the antennules.

The **body proper** or **capitulum** is enclosed in a bivalved **mantle** or **carapace** (Fig. 6.15). The carapace is covered with five calcareous plates. Of these one is median and dorsal (called **carina**), two are lateral and proximal (**scuta**) and two are lateral and distal (**terga**). A large **adductor muscle** runs transversely between the two scuta. The body proper is divisible into **head** and **thorax**. The head contains vestiges of first antennae (antennules) and mouth which is surrounded by paired mandibles, maxillae and maxillulae. Second pair of antennae are absent.

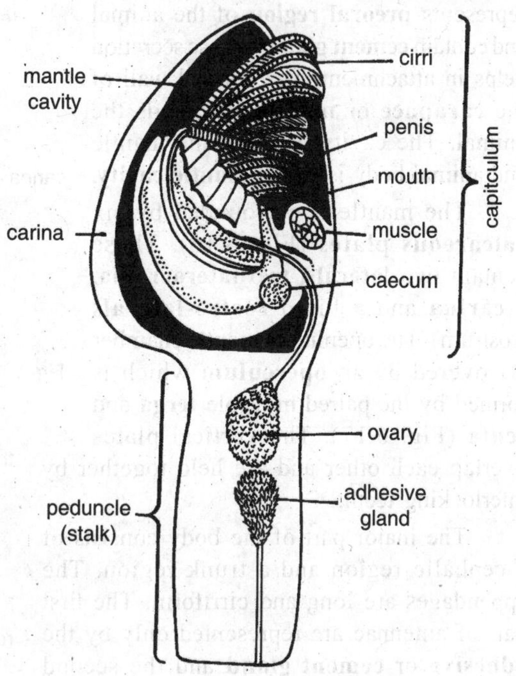

Fig. 6.15. *Lepas*. Internal structure.

Thorax bears six pairs of long, biramous appendages. These are filament-like and bear many long setae. These are called **cirri** and used in **suspension feeding.**

Lepas lacks heart, arteries, gills but contains **maxillary glands** for excretion. It is **hermaphrodite** (monoecious) but **cross-fertilization** is the rule. Fertilization is internal occurring in ovisac and sperm transfer occurs through the **penis** which is inserted into the mantle cavity of another individual. Development of *Lepas* is indirect involving **nauplius larva** and **cypris larva**.

Attached *Lepas* to ship bottom tend to hamper ship's speed, so much efforts and money have been expanded to develop anti-barnacle paints and other measures such as scrapping of barnacles at dry docks.

7. *BALANUS*
(Acorn-shell or rock-barnacle)

Classification

Phylum	—	**Arthropoda**
Subphylum	—	**Crustacea**
Class	—	**Cirripedia**
Order	—	**Thoracica**
Genus	—	***Balanus***

Habit and Habitat

Balanus is commonly known as **rock barnacle** or **acorn barnacle**. It is marine and found attached to rocks in great numbers between tide marks. It is also found attached to molluscan shell, floating objects (ship), whales and other animals. *Balanus* is **suspension feeder**.

Structure

Balanus is sedentary and sessile, *i.e.*, it lacks peduncle so its shell is directly attached to the rocks. Body of *Balanus* is surrounded by a calcareous shell. The broad attached under- surface of the *Balanus* is called **basis** or **disc**. It represents **preoral region** of the animal and contain **cement glands** whose secretion helps in attachment. The vertical wall of the **carapace** or **mantle** surrounds the animal. The cavity between the mantle and animal body is called **mantle cavity.**

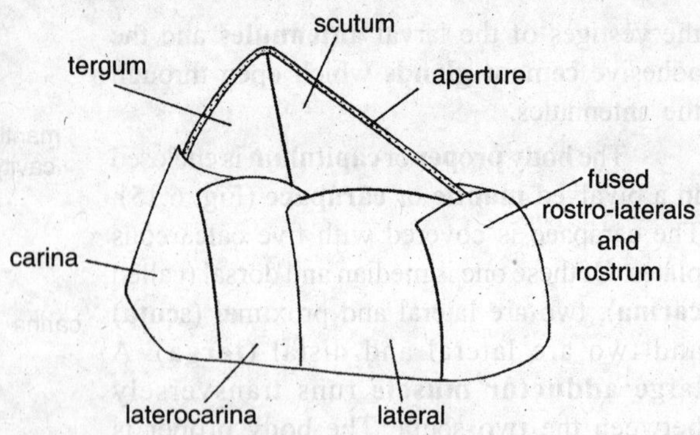

Fig. 6.16. *Balanus*. Number and position of shell plates (diagrammatic lateral view).

The mantle is surrounded by six **calcareous plates** (Fig. 6.16). These contain two **laterals,** two **laterocrania,** a **carina** and a fused **rostro-laterals** (rostrum). The opening of mantle chamber is covered by an **operculum** which is formed by the paired movable **terga** and **scuta** (Fig. 6.17). The vertical plates overlap each other and are held together by interlocking teeth.

The major part of the body consists of a cephalic region and a trunk region. The appendages are long and cirriform. The first pair of antennae are represented only by the **adhesive** or **cement gland** and the second pair is present in the larva. The second pair of maxillae are united. Thoracic appendages are six pairs of delicate cirriform legs that sway in water and propel food organisms towards mouth.

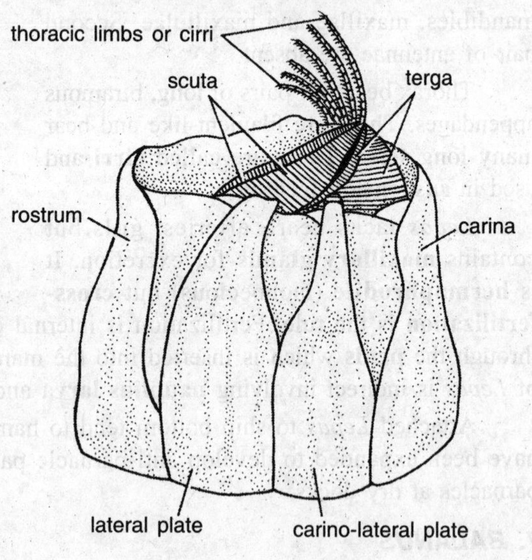

Fig. 6.17. *Balanus*.

Balanus is **hermaphrodite** but **cross-fertilization** is the rule. Fertilization is internal and development is indirect involving **nauplius larva** and **cypris larva**.

8. *SACCULINA*

Classification

Phylum	—	**Arthropoda**
Subphylum	—	**Crustacea**
Class	—	**Cirripedia**
Order	—	**Rhizocephala**
Genus	—	***Sacculina***

Habit and Habitat

Sacculina is commonly known as **root-headed barnacle**. It is marine and found throughout the world. It is **parasitic** on decapod crustaceans such as crab.

Structure

Adult *Sacculina* (called **sacculina externa**) has the appearance of a **fleshy tumour** attached by a peduncle to the abdomen of the crab on its ventral side (Fig. 6.18). From the peduncle numerous branched filaments are ramified in all parts of the body of the host except the heart and the gills.

These roots absorb nourishment from the host. *Sacculina* shows great *degeneration due to parasitic mode of life*. It lacks mouth, anus, carapace, segmentation and appendages. It contains only reproductive organs such as a pair of elongated **testes** and a pair of **ovaries** with accessory **genital glands, genital atrium** and **collateral glands**. There is a single nerve ganglion. The entire body of *Sacculina* is covered with a mantle. Opening of the mantle cavity of *Sacculina* is placed at the hind end of the parasite as genital aperture or cloacal aperture.

The adult *Sacculina* is difficult to recognize as an arthropod. The study of its developmental history justifies its inclusion as a crustacean.

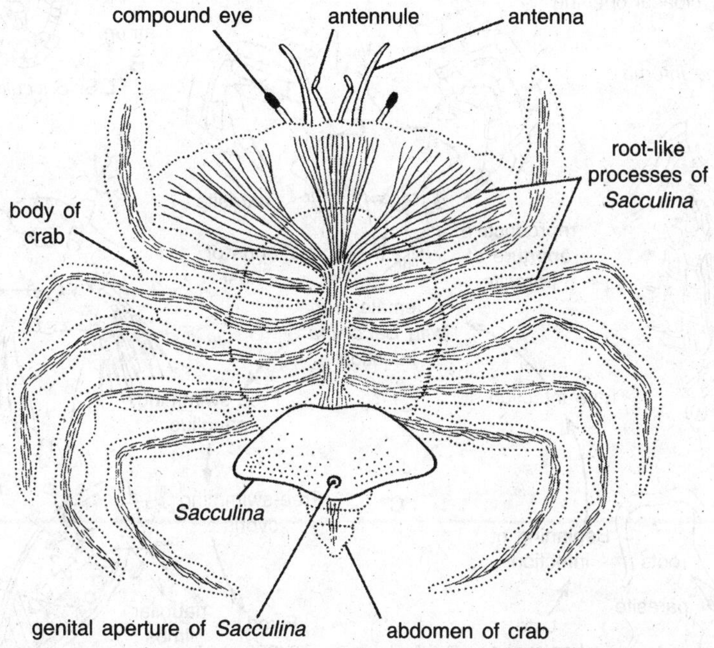

Fig. 6.18. *Sacculina*. A parasite attached to ventral surface of crab.

Life History

Sacculina is **hermaphrodite** and fertilization is internal. The youngs are hatched from the eggs as free swimming **nauplius larvae**. A nauplius larva moults into free-swimming **cypris larva**. A cypris larva on finding a crab (the host), attaches itself to the body of the crab by the help of its hook-like antennule. Now the cypris larva loses its trunk and appendages and becomes a rounded mass of cells, called **kentrogen larva** (Fig. 6.19). Kantrogen larva contains a chitinous rod known as **dart** which convey undifferentiated cells of larva into the body cavity of host. These undifferentiated cells enter the host's body and finally settles on the midgut of the host. This stage of *Sacculina* is called **sacculina interna**. It sends root-like absorptive processes throughout the body of the host except heart and gills. Such a massive parasitic attack sterilizes the host and prevents further moulting of the crab.

Finally the main body of the parasite pushes out as a swelling in the abdomen of the crab. This phase is called **sacculina externa**. *Sacculina* survives for three years on the host and then dies.

Effects of Sacculina on the Host (crab). Parasitic mode of life of *Sacculina* affects both the host and the parasite:

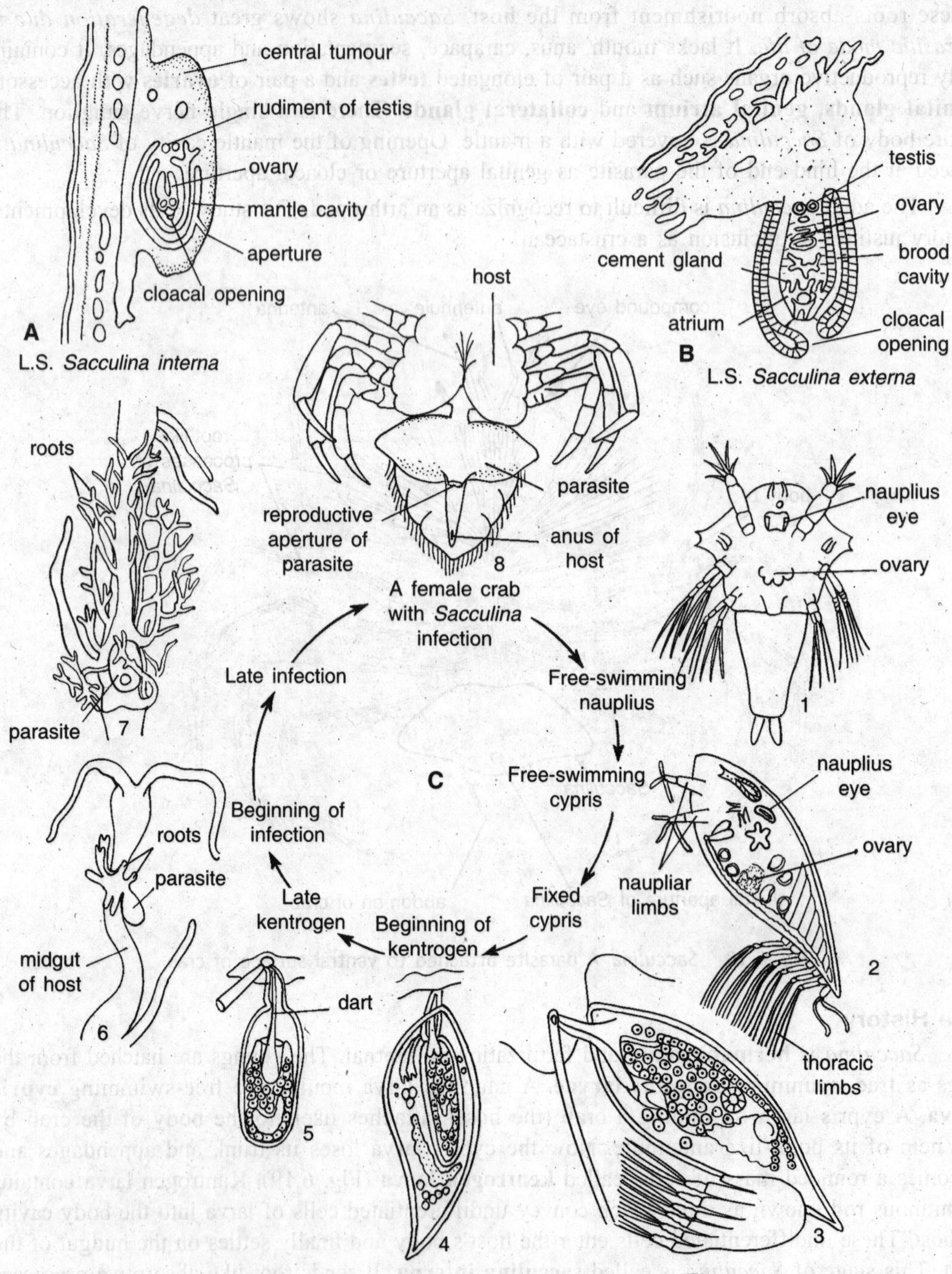

Fig. 6.19. *Sacculina*. Life cycle.

1. The crab does not moult and so does not grow in size.
2. *Sacculina* causes complete degeneration of the gonads of the host.
3. *Sacculina* changes the **secondary sexual characters** of the host. This phenomenon is called **parasitic castration**. In the female crab ovaries degenerate and ovigerous endopodites of

pleopods reduce in size. In the male crab the size of the chela reduces and they become similar to chela of female. Male's copulatory **styles** too reduce in size.

Parasite (*Sacculina*) itself undergoes several significant degenerative changes such as (*i*) loss of segmentation; (*ii*) loss of appendages; (*iii*) loss of different organ systems such as digestive, circulatory, respiratory and excretory systems; (*iv*) loss of typical crustacean features, and (*v*) reduction of metabolic rate.

9. *EUPAGURUS* (Hermit crab)

Classification

Phylum	—	**Arthropoda**
Subphylum	—	**Crustacea**
Class	—	**Malacostraca**
Subclass	—	**Eumalacostraca**
Superorder	—	**Eucarida**
Order	—	**Decapoda**
Genus	—	***Eupagurus***

Habit and Habitat

Eupagurus is commonly called **hermit crab**, because it leads a secluded life like a hermit (*i.e.*, hermit is a person who abandons society and lives alone, especially for religious contemplation). It is cosmopolitan in distribution and is found at Bombay and Goa beaches, from Alaska to lower California and Florida. Hermit crab lives in empty shell of a marine gastropod and often leads a commensal life with sea anemone.

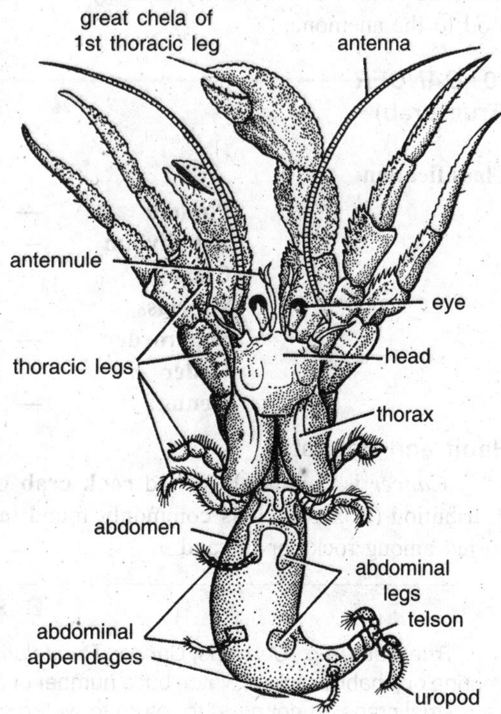

Fig. 6.20. *Eupagurus*.

Structure

Eupagurus is a peculiar crustacean having extreme modification in order to adjust in the coils of molluscan shell. Its body is soft and is divided into **cephalothorax** and **abdomen** (Fig. 6.20). **Cephalothorax** is broad and flattened. It is covered with a hard carapace which is bifid posteriorly. The cepha-lothoracic appendages are well developed and help in locomotion and food capture. **Head** bears a pair of stalked elongated eyes, a pair of short **antennules** and a pair of large **antennae**. **Thorax** bears five pairs of legs. The first pair of chelate legs having unequal chela (*i.e.*, right chela is much larger than the left), the last two pairs are usually reduced. The larger first leg of left side acts as an **operculum**. It closes the mouth of the shell when the animal withdraws itself inside the shell. Normally the cephalothorax protrudes out of the shell and larger chelate leg is used to capture and crush the prey.

The abdomen is cylindrical, soft and without external segmentation. Abdomen contains liver and gonads. It is spirally twisted to suit the shape of the shell. Abdominal appendages of the right side are absent, while those of left side are reduced (in female they are used to carry eggs). **Uropods** (last abdominal appendages) are hook-like and are adapted to cling to the shell. Surface of fourth and fifth pairs of legs and of uropods is covered with tiny tubercles that help in gripping the shell.

Sexes are separate. Spermatophores and ova are released simultaneously and fertilization is external. When the growth of the body demands, *Eupagurus* selects a new shell and deserts the old one.

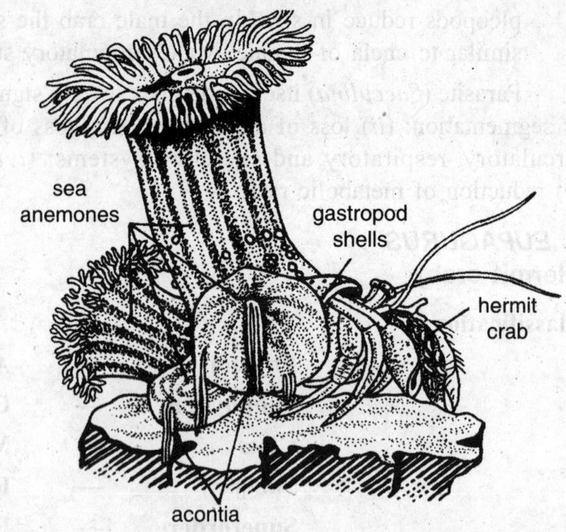

Fig. 6.21. Hermit crab and sea anemone in commensalism.

Commensalism

Hermit crab and sea anemone form a close but non-parasitic association, called **commensalism** (Fig. 6.21). Sea anemone lives upon the mollusc shell while *Eupagurus* lives inside the shell. The shell and the anemone provide protection to hermit crab. Enemies of hermit crab avoid sea anemones due to their stinging cells and bad taste. In return, the hermit crab carries the sea anemone (which is a sedentry animal) from place to place and thus provides fresh variety of food to the anemone.

10. *CANCER*
(True crab)

Classification

Phylum	—	**Arthropoda**
Subphylum	—	**Crustacea**
Class	—	**Malacostraca**
Subclass	—	**Eumalacostraca**
Superorder	—	**Eucarida**
Order	—	**Decapoda**
Genus	—	***Cancer***

Habit and Habitat

Cancer is commonly called **rock crab** or **true crab**. It is marine having cosmopolitan distribution (Box 6.4). It is commonly found in India, Europe and U.S.A. Cancer is found buried among rocks or in sand.

Box 6.4

True crabs, *e.g.*, *Carcinus*, *Cancer*, *Paratelphusa*, etc., are found in all types of habitats. Most are marine or inhabit brackish water, but a number of them are freshwater, amphibious or terrestrial. Only terrestrial crabs do not need to return to water, since they have direct development (*i.e.*, they lack any larval stage in their development). *Paratelphusa masoniana* is freshwater crab of Punjab.

Structure

Body of *Cancer* is oval and flattened and is divided into **cephalothorax** and **abdomen** (Fig. 6.22). Cephalothorax is quite large and is usually much broader than long. Carapace is broad, thick and hard. Eye stalks and antennae are short. **Mandibles**, **maxillae** and **maxillulae** are well developed. Thorax bears three pairs of **maxillipedes** and five pairs of well developed **walking legs**. Third **maxillipedes** are broad, flat and valve-like and cover the other mouth part.

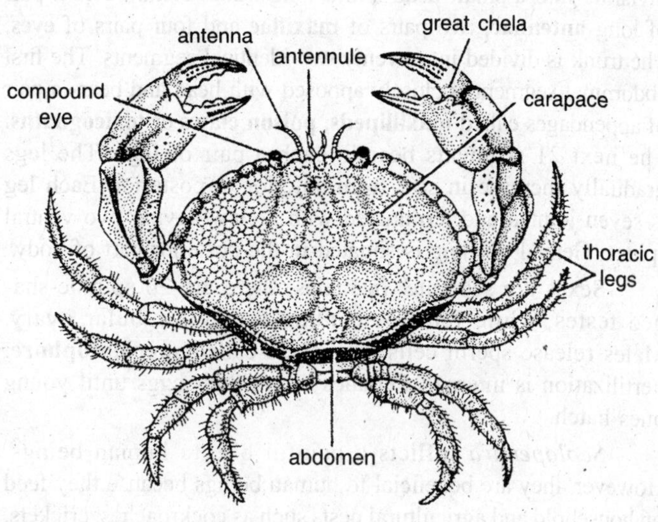

Fig. 6.22. *Cancer.*

The first pair of walking legs are **chelate** forming pinching claws. The abdomen is greatly reduced and lies permanently in a groove in the ventral or sternal surface of the thorax. Male's abdomen is narrow and V-shaped, while abdomen of female is broad and U-shaped and is adapted for carrying eggs. Abdominal appendages (**pleopods**) are much reduced. In males there are only two pairs of pleopods and it uses them as copulatory organs. Females have four pairs of pleopods which are used for carrying eggs. Uropods are absent.

Sexes are separate. Fertilization is internal. Newly laid eggs form a bright orange mass, called a **sponge**. In *Cancer*, development is indirect. Young ones hatches as **zoea** larva, which passes through **megalopa** stage before reaching maturity. Many species of crab are caught and eaten by human beings throughout the world. **Blue crabs** and **cancer crabs** are popular food item.

11. *SCOLOPENDRA* (Centipede)

Classification

Phylum	—	**Arthropoda**
Subphylum	—	**Uniramia**
Class	—	**Chilopoda**
Subclass	—	**Epimorpha**
Order	—	**Scolopendromorpha**
Genus	—	***Scolopendra***

Habit and Habitat

Scolopendra is commonly called **centipede**. It is commonly occurs under stones, in rotten logs and in houses in damp places such as cellars and bathrooms. It is a nocturnal, active and fast moving creature. *Scolopendra* is carnivorous and voracious feeder. It feeds insects, spiders, worms and slugs. It kills its prey with the poison of the poison claws or maxillipedes which are modified first pair of trunk appendages.

Structure

The body of *Scolopendra* is brightly coloured and dorsoventrally flattened and segmented (Fig. 6.23). Body is divisible into a small head and a long trunk. Head bears a pair of long **antennae**, two pairs of **maxillae** and four pairs of **eyes.** The trunk is divided into twenty two identical segments. The first abdominal segment is closely apposed with head and bears a pair of appendages called **maxillipeds, poison claws** or **toxicognaths**. The next 21 segments bear in each a pair of legs. The legs gradually increase in size from anterior to posterior. Each leg is seven jointed and terminated with a claw having two ventral spines. Genital opening is situated in the posterior part of body.

Sexes are separate. In males, there are 20 spindle-shaped testes, while females contain a single tubular ovary. Males release sperm cells in packets called **spermatophore**. Fertilization is internal. Females brood their eggs until young ones hatch.

Scolopendra inflicts a painful bite to human beings. However, they are beneficial to human beings because they feed on household and agricultural pests such as cockroaches, crickets, ants, beetles, moths, caterpillars, flies, bedbugs, silver fish, etc.

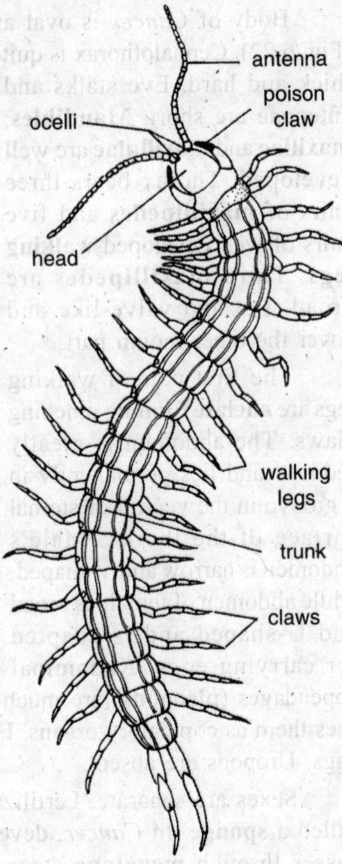

Fig. 6.23. *Scolopendra.*

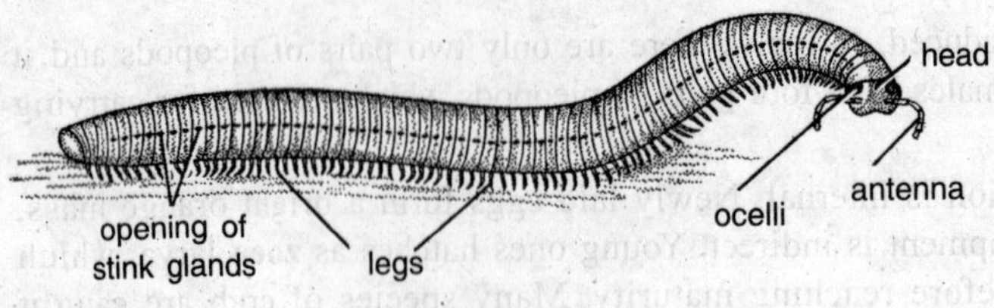

Fig. 6.24. *Julus.*

12. *JULUS* (Millipede)

Classification

Phylum	—	**Arthropoda**
Subphylum	—	**Uniramia**
Class	—	**Diplopoda**
Subclass	—	**Helminthomorpha**
Order	—	**Julida**
Genus	—	***Julus***

Habit and Habitat

Julus or **millipede** is a sluggish and timid creature found hidden usually in dark and damp places under stones or wood or in decaying leaves. The animal is found rolled up under stones. They are **herbivorous.**

cture

ody of *Julus* is elongated and cylindrical, consisting of large number of segments (Fig. 6.24). colour may be yellowish brown or reddish chestnut. The body is divisible into **head, thorax** and **abdomen**. Head bears a pair of sessile eyes, a pair of short **antennae**, a pair of **mandibles** and a pair of **maxillae** forming **gnathochilarium**, forming floor of pre-oral chamber. Poison claws are absent. Thorax contains four segments and each segment bears a pair of appendages. Abdomen has 20 to 100 segments, each abdominal segment bears two pairs of clawed legs. Body is terminated into **telson.** A pair of stink or **repugnatorial glands** occurs on side of each body segment. The secretion of these glands is toxic and repellant to other small animals.

Sexes are separate. Gonopores are located mid-ventrally on third abdominal segment. Fertilization is internal.

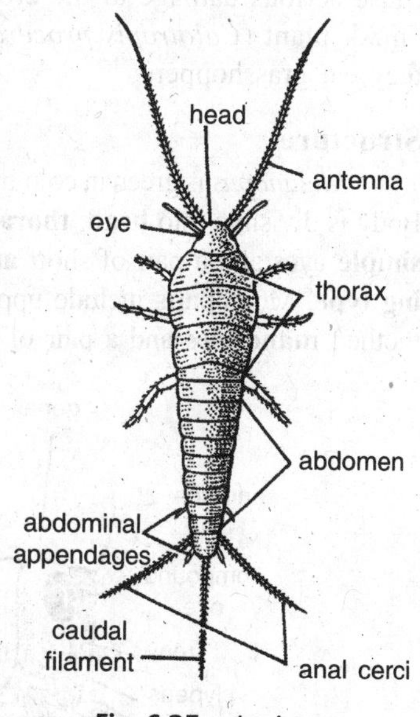

Fig. 6.25. *Lepisma.*

13. *LEPISMA* (Silver fish)

Classification

Phylum	—	**Arthropoda**
Subphylum	—	**Uniramia**
Class	—	**Insecta**
Subclass	—	**Apterygota**
Order	—	**Thysanura**
Genus	—	***Lepisma***

Habit and Habitat

Lepisma saccharina is commonly known as **silver fish**. It is. commonly found in damp, cool places and in cup-boards and book-racks where it devours starchy products.

Structure

The body of *Lepisma* is small (1.25 cm in length), flat and silver-coloured. Its body is divided into head, thorax and abdomen (Fig. 6.25). **Head** bears a pair of many segmented antennae and a pair of ocelli. Mouthparts are adapted for **biting** and **chewing**. The **thorax** is distinct and the **prothorax** forms the largest segment of body. Thorax bears three pairs of legs but lacks wings. Abdomen is eleven segmented. The last abdominal segment carries at the posterior end a pair of long, many jointed **cerci** and a long **tail** in between. The entire body of *Lepisma* is covered by shining scales. Its excrement products are in the form of dry pellets. The adults can moult and possess regenerating power. *Lepisma* cause considerable damage to the books, clothing by chewing off starch, sizing or any other articles containing glue.

14. *POICILOCERUS* (Grasshopper)

Classification

Phylum	—	**Arthropoda**
Subphylum	—	**Uniramia**
Class	—	**Insecta**
Subclass	—	**Pterygota**
Order	—	**Orthoptera**
Genus	—	***Poicilocerus***

Habit and Habitat

Poicilocerus (also *Melanopus*) is commonly known as **grasshopper**. Grasshoppers are found throughout the world in open grasslands. They feed on grass and all other leafy vegetation. They cause serious damage to the crops. They are solitary and residential. *Poicilocerous pictus* lives on Ak plant (*Calotropis procera*) in India. Spiders, frogs, lizards and birds are their enemies, they eat grasshoppers.

Structure

Melanopus is green in colour and *Poecilocerus* is yellowish or brownish in colour containing spots. Body is divisible into **head**, **thorax** and **abdomen**. Head bears a pair of sessile **compound eyes**, three **simple eyes** and a pair of short **antennae**. The mouthparts of a grasshopper are of **biting** and **chewing type**. Mouthparts include upper lip or **labrum**, a tongue-like **hypopharynx**, a pair of triangular, toothed **mandibles** and a pair of **maxillae**. The mandibles are used to bite and crush the food.

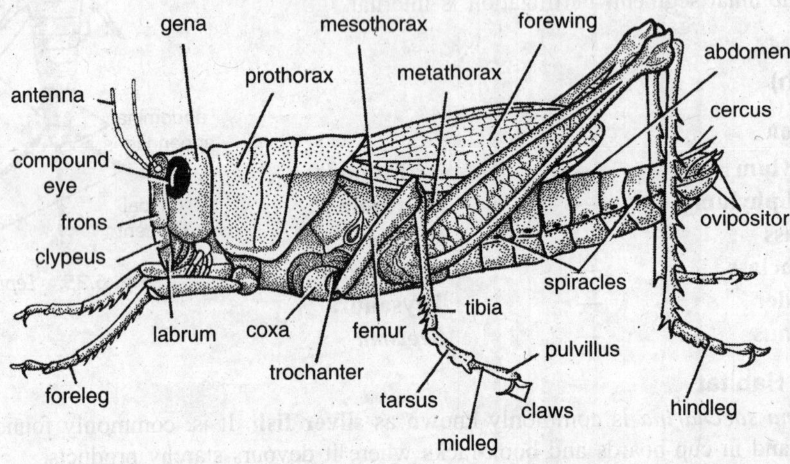

Fig. 6.26. Grasshopper (*Poicilocerus or Melanopus*).

Thorax is formed of three segments: an anterior larger **prothorax**, a middle **mesothorax** and a posterior **metathorax**. Each thoracic segment bears a pair of jointed and clawed legs. All the legs are used in walking and climbing. However hind legs are more powerful and adapted for jumping (**saltatorial legs**).

Mesothorax and metathorax bear a pair of wings each. Forewings are elongated, thick leathery and straight, called **tegmina**, while hindwings are membranous. Tegmina simply covers the hind wings and latter are used in flight. Abdomen is composed of ten segments.

Sexes are separate, sexual dimorphism is well marked. Female is provided with a long **ovipositor**. Fertilization is internal. Development is direct; nymphs are devoid of wings. Both nymph and adult grasshoppers damage crops. They are serious pests. They are eaten by human beings in several parts of world.

15. *SCHISTOCERCA* (Locust)

Classification

Phylum	—	**Arthropoda**
Subphylum	—	**Uniramia**
Class	—	**Insecta**
Subclass	—	**Pterygota**
Order	—	**Orthoptera**
Genus	—	***Schistocerca***

Habit and Habitat

Schistocerca gregaria is commonly known as **locust**. The term *locust* includes gregarious and migratory grasshopper-like forms. *Schistocerca gregaria* is found in the deserts of tropical region of world. It is most widely distributed species and commonly found in Africa, Arabia, India, Afghanistan, Philippines, Japan, Mexico and U.S.A.

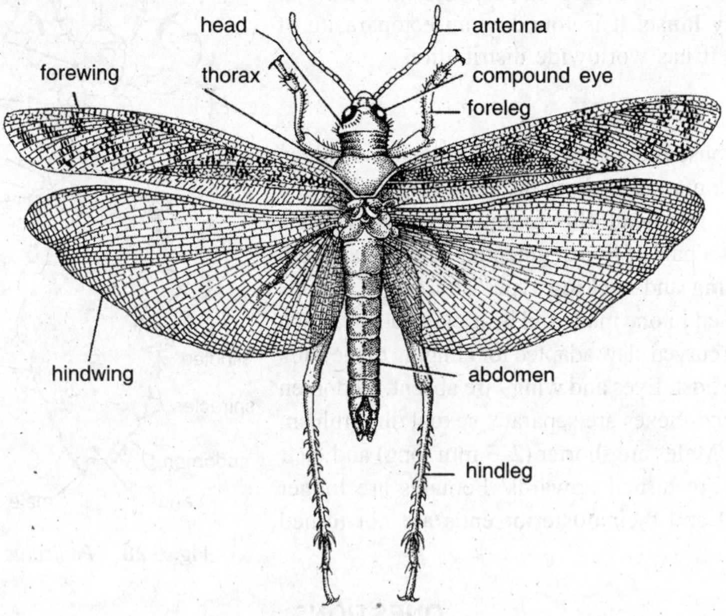

Fig. 6.27. Locust (*Schistocerca*).

Structure

Body of locust is divisible into **head, thorax** and **abdomen** (Fig. 6.27). Head bears short **antennae** and well-developed compound eyes. Its mouth parts are **biting** and **chewing** type. Thorax is well developed and bears three pairs of legs and two pairs of wings. Forewing are elongated, thick, leathery and straight. Hindwings are large and membranous. **Abdomen** comprises ten segments with eleventh vestigial segment. The lateral sides of first abdominal segment carries a **tympanum**. Sexes are separate. Male possesses **anal styles** and female possesses **ovipositor**.

Locusts have two phases in their life: **solitary** or **non-migratory** and **gregarious** or **migratory**. In solitary phase, they have a yellow body colour and in gregarious phase, they have pinkish colour. The migratory phase appears due to some unknown reason and just before migration their population increases rapidly. The swarms migrate toward green fields. The swarms of locusts move like a cloud and the direction is guided by the flow of wind. They damage crops and vegetation severely in course of their journey.

16. *PEDICULUS* (Human head and body louse)

Classification

Phylum	—	**Arthropoda**
Subphylum	—	**Uniramia**
Class	—	**Insecta**

Subclass	—	**Pterygota**
Order	—	**Anopleura**
Genus	—	***Pediculus***

Habit and Habitat

Pediculus humanus is commonly known as **human head and body louse**. It is found as an ectoparasite of human beings. It has worldwide distribution.

Structure

Body of louse is dorsa-ventrally flattened and pale in colour with dark markings along the sides (Fig. 6.28). The body is divisible into **head**, **thorax** and **abdomen**. Head is small and bears a pair of five-segmented antennae. Mouth parts are **piercing** and **sucking** type. Thoracic segments appear to be fused in one mass and bear clawed legs. Each leg bears a large curved claw adapted for clinging to the hairs or tissues of the host. Eyes and wings are absent. Abdomen is nine segmented. Sexes are separate; sexual dimorphism is well marked. Males are shorter (2–3 mm long) and their posterior ends are turned upwards. Females are longer (3–4 mm long) and their posterior ends are not turned upwards.

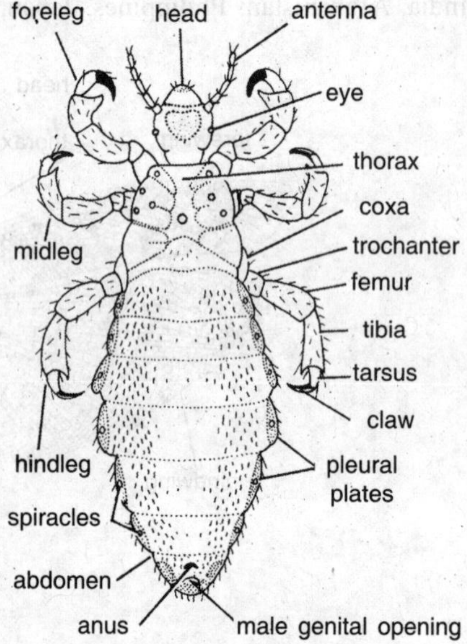

Fig. 6.28. *Pediculus humanus.*

QUESTIONS

Long Answer Questions

1. Give an outline classification of phylum Arthropoda.
2. Give chief characteristics of phylum Arthropoda mentioning the distinctive features of different classes into which the phylum is divided. Give examples of each class.
3. Classify Arthropoda giving characters and examples of each class.
4. Give the distinguishing features of main groups of Arthropoda with two examples of each.
5. What are distinguishing characters of class Crustacea and Arachnida?
6. State the points that indicate that the arthropods are biologically successful groups of animals.
7. Classify Crustacea upto orders. Give important features and examples of each.

Short Answer Questions

1. Give classification, habits, habitat, characters and economic importance of following animals:

(*i*) *Limulus* (King crab)	(*ii*) *Daphnia*
(*iii*) *Cypris*	(*iv*) *Cyclops*
(*v*) *Argulus*	(*vi*) *Lepas*
(*vii*) *Balanus*	(*viii*) *Sacculina*
(*ix*) Hermit crab (*Eupagurus*)	(*x*) *Cancer* (Crab)
(*xi*) *Scolopendra* (Centipede)	(*xii*) Millipede (*Julus*)
(*xiii*) *Lepisma* (Silver fish)	(*xiv*) *Poiceloceros* (Grasshopper)
(*xv*) Desert locust (*Schistocerca*)	(*xvi*) Head louse (*Pediculus*)

Very Short Answer Questions

1. Which animal of subclass Xiphosura is called a living fossil?
2. Name the two main tagmata of the arachnid body.
3. With which appendages does a centipede bite?
4. Name a crustacean parasite of crabs.
5. Cite two examples of order Thoracica.
6. Give the zoological name and class of king crab.
7. What is the common name of *Lepisma*?
8. Name two insects having complete metamorphosis.
9. Which is poisonous, millipede or centipede?
10. From where do locust swarms visit India periodically?
11. Name a crustacean in which female carries eggs in ovisacs.
12. Which *Anopheles* (male or female) causes malaria in human beings? (*C.C.S.U. Meerut 2001*)
13. What is the food of *Anopheles*?
14. Which part of the body of bees carry eggs?

Fill in the Blanks

1. The living arthropods are divided into three subphyla: .... , Crustacea and ....
2. There are two pairs of antennae in ....
3. The appendages of arthropods are ....
4. .... are ingrowths of cuticle to bear muscles.
5. Periodical replacement of exoskeleton is called .... or ....
6. Ecdysis is the phenomenon of ....
7. .... are social and polymorphic insects.
8. The eggs are .... in arthropods.
9. Stomodaeum in arthropods is lined by ....
10. Excretory organs of arthropods are .... glands and .... tubules.
11. Arthropods have .... circulatory system.
12. .... are the only invertebrates to have the power of flight.

Multiple Choice Questions

1. Members of phylum Arthropoda show following characters except one which is not correct. Mark it
   (*a*) appendages are jointed (*b*) cilia are uniformly present
   (*c*) exoskeleton in chitinous (d) body cavity is haemocoel.
2. Trilobites are
   (*a*) extinct arthropods (*b*) connecting link between annelids and molluscs
   (*c*) living fossils (*d*) insects
3. Which is the main characteristic of class Crustacea?
   (*a*) cephalothorax and two pairs of antennae
   (*b*) gills and green glands
   (*c*) both (*a*) and (*b*)
   (*d*) cephalothorax and one pair of antennae

4. Class Arachnida is characterized by
(*a*) one pair each of pedipalpi and chelicertes (*b*) four pairs of legs
(*c*) both (*a*) and (*b*) (*d*) three pairs of legs

5. Abdomen is greatly reduced in?
(*a*) *Palaemon* (*b*) *Palamnaeus*
(*c*) *Cancer* (*d*) *Buthus*

6. The arthropods having three pairs of legs are members of the class
(*a*) Crustacea (*b*) Chilopoda
(*c*) Diplopoda (*d*) Insecta

7. Body cavity of arthropods is
(*a*) coelom (*b*) haemocoel
(*c*) pseudocoel (*d*) coelenteron

8. Excretory organs of crustaceans are
(*a*) nephridia (*b*) malpighian tubules
(*c*) green glands (*d*) flame cells

9. Respiratory organs in Arthropoda is
(*a*) tracheae (*b*) gills
(*c*) book lungs (*d*) all

10. Which of following arthropod respires by book lungs?
(*a*) spider (*b*) *Scolopendra*
(*c*) *Palaemon* (*d*) *Anopheles*

11. Excretory product in an insect is
(*a*) uric acid (*b*) urea
(*c*) ammonia (*d*) guanine

12. The arthropods with a sting at the hind end is
(*a*) *Scolopendra* (*b*) *Galeodes*
(*c*) *Palamnaeus* (*d*) *Lepas*

13. Which of the following has raptorial prolegs?
(*a*) locust (*b*) praying mantis
(*c*) wasp (*d*) bedbug

14. *Julus* belongs to class
(*a*) Arachnids (*b*) Diplopoda
(*c*) Chilopoda (*d*) Crustacea

15. *Galeodus* belongs to the class
(*a*) Arachnida (*b*) Crustacea
(*c*) Chilopoda (*d*) Diplopoda

16. Which of the following insect is not social?
(*a*) honeybee (*b*) ant
(*c*) termite (*d*) locust

17. Drones in honeybee are

(*a*) fertile males (*b*) fertile females

(*c*) sterile males (*d*) sterile females

18. Basic unit of compound eye is

(*a*) retina (*b*) rhabdome

(*c*) corneal facet (*d*) ommatidium

19. Which is a living fossil?

(*a*) *Limulus* (*b*) *Peripatus*

(*c*) *Lepisma* (*d*) *Pediculus*

20. The study of insects is called

(*a*) entomology (*b*) ethology

(*c*) epidemiology (*d*) none of the above

21. Termitarium is

(*a*) bunch of termites (*b*) nest of termites

(*c*) colony of termite (*d*) larvae of termites

22. In phylum Arthropoda, fertilization

(*a*) is external (*b*) is internal

(*c*) takes place in water (*d*) takes place on land

23. Complete metamorphosis is found in

(*a*) silver fish (*b*) bedbug

(*c*) grasshopper (*d*) gypsy moth

24. The moulting hormone or ecdysone is secreted from

(*a*) corpora allata (*b*) corpora cardiaca

(*c*) prothoracic glands (*d*) none of these

25. The juvenile hormone is secreted from

(*a*) corpora allata (*b*) corpora cardiaca

(*c*) corpora adiposa (*d*) none of these

26. Histolysis and histogenesis occur during

(*a*) larval stage (*b*) pupal stage

(*c*) adult stage (*d*) none of these

27. Complete metamorphosis occurs in

(*a*) locust (*b*) silver fish

(*c*) bedbug (*d*) mosquito

ANSWERS

Very Short Answer Questions

1. King crab or *Limulus* of subclass Xiphosura is known as living fossil because it is living on earth since very ancient time, the Triassic period (200 million years) as primitive chelicerate, unchanged.

2. Prosoma and opisthosoma.

3. First pairs of legs modified into poison claws.

4. *Sacculina*.

5. *Lepas* and *Balanus*.

6. *Limulus*, Merostomata.

7. Silverfish.

8. Butterfly and moth.

9. Centipede.

10. Arabia, Persia, Afghanistan and Baluchistan.

11. *Cyclops*.

12. It is the female *Anopheles* mosquito which causes malaria in human beings. The sexual cycle of malarial parasite *Plasmodium vivax* is completed in female mosquito where as asexual cycle is passed in two phases in human beings.

13. Male *Anopheles* mosquito feeds on flower juices where as female *Anopheles* sucks human blood. It is because their mouth parts are modified into elongated piercing and sucking styles lying within proboscis.

14. The abdomen of queen bee (*Apis indica*) carries a large number of eggs. It lives for several years and lays about 1000 eggs per day.

Fill in the Blanks

1. Chelicerata, Urinamia;
2. Crustacea;
3. Jointed;
4. Apodemes:
5. Moulting, ecdysis;
6. Casting of old exoskeleton;
7. Termites;
8. Centrolecithal;
9. Cuticle;
10. Green, Malpighian;
11. Open;
12. Insects.

Multiple Choice Questions

1. (*b*) **2.** (*a*) **3.** (*c*) **4.** (*c*) **5.** (*c*) **6.** (*d*)
7. (*b*) **8.** (*c*) **9.** (*d*) **10.** (*a*) **11.** (*a*) **12.** (*c*)
13. (*b*) **14.** (*b*) **15.** (*a*) **16.** (*d*) **17.** (*a*) **18.** (*d*)
19. (*a*) **20.** (*a*) **21.** (*b*) **22.** (*b*) **23.** (*d*) **24.** (*c*)
25. (*a*) **26.** (*b*) **27.** (*d*).

7

Palaemon (*Macrobrachium*) (Prawn) (Phylum Arthropoda)

The shrimp *Palaemon* (new name *Macrobrachium*) is commonly known as **prawn.** It is a freshwater crustacean arthropod and is commonly found in India. The common Indian species of *Palaemon* are *P. malcolmsonii, P. carcinus, P. idae, P. rudis* and *P. lamarrei*. *Palaemon malcolmsonii* is studied in most of the Indian Universities as a typical representative of the Crustacea.

7.1 CLASSIFICATION

Phylum	–	**Arthropoda**
Subphylum	–	**Crustacea**
Class	–	**Malacostraca**
Subclass	–	**Eumalacostraca**
Superorder	–	**Eucarida**
Order	–	**Decapoda**
Suborder	–	**Pleocyemata**
Infraorder	–	**Carida**
Genus	–	***Palaemon* (*Macrobrachium*)**

Habit and Habitat

Palaemon is found in freshwater streams, rivers, lakes and ponds. It is a nocturnal, bottom-dweller and lives within underwater crevices and aquatic vegetation. It comes out from its hides in the night in the search of food. *Palaemon* is **omnivorous** and feeds on algae, moss, other weeds, small insects and debris (*e.g.*, decaying leaves). It has dual mode of locomotion: slow walking (*i.e.*, crawling) on the bottom with the help of its walking legs and active swimming in water with the help of pleopods. Sexes are separate and sexual dimorphism is distinct. *Palaemon malcolmsonii* breeds in May, June and July and the females carry eggs attached to their pleopods. Development of *Palaemon* is direct.

7.2 EXTERNAL FEATURES

1. Shape and size. The body of *Palaemon* is elongated, laterally compressed, spindle-like and bilaterally symmetrical. Such a shape offers minimum resistance to water during swimming. The size of adult *Palaemon* varies greatly in different species. For example, *P. carcinus* from Travancore measures 90 cm and *P. malcolmsonii* from Madhya Pradesh and Madras is generally 25–38 cm long.

2. Colouration. Fresh specimen is pale-blue or greenish in colour with brown or orange red patches and bands. On exposure to air, it becomes clay-brown or yellow.

3. Division of body. The body of adult *Palaemon* consists of nineteen appendage bearing segments. It has two distinct regions: an anterior rigid (immovable) cephalothorax and a posterior flexible (movable) **abdomen** (Fig. 7.1). The entire outer surface of the body is covered by a hard exoskeleton.

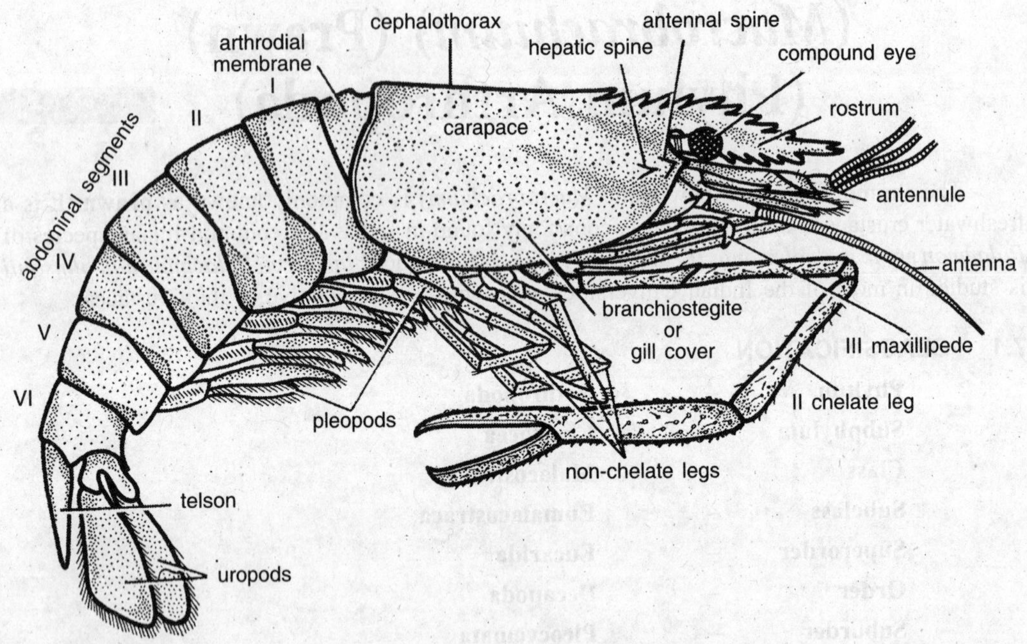

Fig. 7.1. *Palaemon*. External features.

***(i)* Cephalothorax.** It is large, cylindrical, rigid and unjointed. It is formed by the union of 13 segments—5 of the **head** or **cephalic region** and 8 of the **thorax** or **thoracic region.** Each of these segments bears on its ventral surface a pair of appendages; thus, there are in all 13 pairs of cephalothoracic appendages. A pair of movable and stalked **compound eyes** are present in the presegmental region.

***(ii)* Abdomen.** The abdomen is rounded dorsally but laterally compressed. It is jointed having six movable segments and a terminal conical piece, **telson.** Each abdominal segment carries a pair of jointed appendages called **pleopods** or **swimmerets.**

4. Exoskeleton. The entire body and the appendages are covered by a hard and chitinous cuticle which forms **exoskeleton.** Exoskeleton is formed by the underlying cells and becomes hard due to the deposition of calcium salts (calcium carbonate and calcium phosphate) and **scleroprotein.** The exoskeleton besides affording protection against injuries, provides space for muscle insertion. It is not continuous throughout the body but is localised to form thick **plates,** called **sclerites** which are joined by thin softer cuticle forming **arthrodial membranes.** Arthrodial membranes facilitate the movement of sclerites. Each sclerite consists of dorsal **tergal plate,** ventral **sternal plate** and two **lateral plates.**

In the cephalothorax, the tergal and the pleural plates of all sclerites are fused to form a continuous shield-like covering called **carapace.** On both ventro-lateral sides, the carapace hangs freely to form gill-chamber, gill-cover or branchiostegite. On the mid-dorsal surface the carapace is drawn into a long serrated projection called, **rostrum,** towards the anterior end. This is defensive in function.

At the base of rostrum, on either side is an **orbital notch** which accommodates the stalk of compound eye.

5. External Apertures

(1) Mouth. It is a slit-like aperture, situated mid-ventrally below the anterior end of head.

(2) Renal apertures. There are a pair of minute openings, one on a papilla on the inner surface of coxa of antennae.

(3) Female genital apertures. These are also paired crescentic apertures and lie on the inner surface of the coxae of the third pair of walking legs.

(4) Male genital apertures. These are slit-like and lie on the inner side of the arthrodial membrane con-necting the fifth pair of walking legs with the thorax.

(5) Openings of statocysts. These are paired, each situated in a depression at the dorsal surface on the precoxa of antennules.

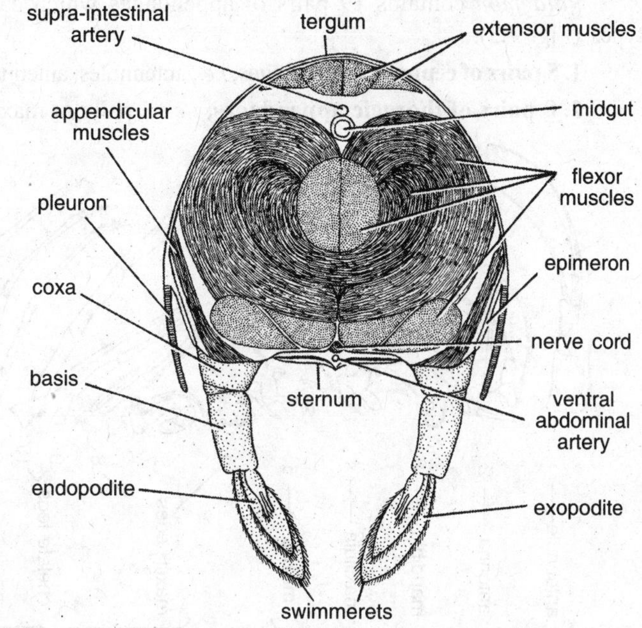

Fig. 7.2. *Palaemon*. T.S. abdominal region showing arrangement of skeleton.

(6) Anus. It is a longitudinal aperture situated mid-ventrally at the base of telson.

6. Appendages. In all there are nineteen pairs of appendages; one pair of appendages in each segment. These are of following types:

(*i*) **Cephalic appendages.** These are five pairs and include **antennules, antenna, mandibles, maxilulla** and **maxillae.**

(*ii*) **Thoracic appendages.** These are eight pairs and include first three pairs of **maxillipedes** and five pairs of **walking legs.**

(*iii*) **Abdominal appendages.** These are six pairs and include first five pairs of **pleopods** or **swimmerets** and last one pair of **uropods.**

The cephalic appendages are modified for searching the food and feeding; walking legs help in locomotion; swimmerets help in swimming and uropods help in change of direction while swimming.

7.3 APPENDAGES

Palaemon possess paired segmentally arranged jointed appendages. Each pair of appendages arises from the ventro-lateral margins of each segment. Except antennules, all appendages of *Palaemon* show a typical **biramous plan** of construction; antennules are **uniramous.** Each appendage has a basal stalk, called **base** or **protopodite** which is made up of two joints or **podomeres,** a lower coxapodite or **coxa** and an upper basopodite or **basis.** From the protopodite arise two rami or branches, an inner **endopodite** and an outer **expodite,** each of which may be composed of one to several podomeres. Appendages of *Palaemon* exhibit marked variations in form and size in response to different functions.

Types of Appendages

Palaemon contains 19 pairs of appendages which can be classified into following three types (Fig. 7.3):

1. 5 pairs of cephalic appendages, *i.e.*, antennules, antennae, mandibles, maxillulae and maxillae.

2. 8 pairs of thoracic appendages, *i.e.*, 3 pairs of maxillipedes and 5 pairs of walking legs.

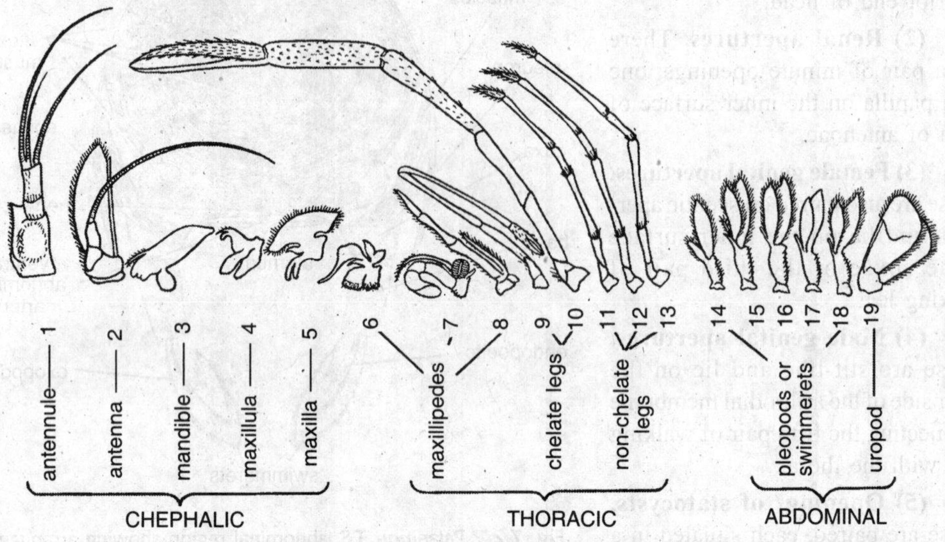

Fig. 7.3. Appendages of *Palaemon*.

3. 6 pairs of abdominal appendages, *i.e.*, 5 pairs of pleopods or swimmerets and 1 pair of uropods.

Structure and Function of Appendages

A. Cephalic Appendages

These are 5 pairs and are of following two types:

***(i)* Pre-oral appendages,** *e.g.*, antennules and antennae;

***(ii)* Post-oral appendages,** *e.g.*, mandibles, maxillulae and maxillae.

1. Antennules. These are uniramous pre-oral cephalic appendages on the first body segment one on either side below the eye-stalks. The **protopodite** is formed of **precoxa, coxa** and **basis.** The precoxa is large and dorso-ventrally flattened. It bears a wide depression to accommodate compound eyes and a spiny lobe, the **stylocerite,** on the outer side of the proximal end. In the depression is situated the **opening of statocyst** protected by a fold of integument. The **coxa** is short and cylindrical. **Basis** is longer than coxa and carries two whip like-elongated and many jointed **feelers.** The outer feeler is divided into an inner smaller branch and outer larger branch. The feelers are not homologous to exo- and endopodite. It is, thus, uniramous (or unbranched).

Coxa and feelers of antennule bear sensory setae. Precoxa contains a statocyst in its cavity. Antennules are **tactile** and **balancing organs.**

2. Antennae. These are also preoral cephalic appendages which are situated one on either side in 2nd segment just below and behind the antennules. The protopodite is small and is formed of **coxa** and **basis.** These two are greatly swollen due to the presence of excretory organs inside them. The excretory organ (green gland) opens to outside by a minute renal aperture on the inner side of the coxa. **Basis** bears a short spine and two rami (branches): **exo-** and **endopodite.** The endopodite is a long, many jointed **feeler.** The exopodite is broad and leaf-like plate called **squama.**

It bears setae along its inner and distal margins. Squama serves the function of a balancer during swimming. The antennae are **tactile, excretory** and **balancing** organ.

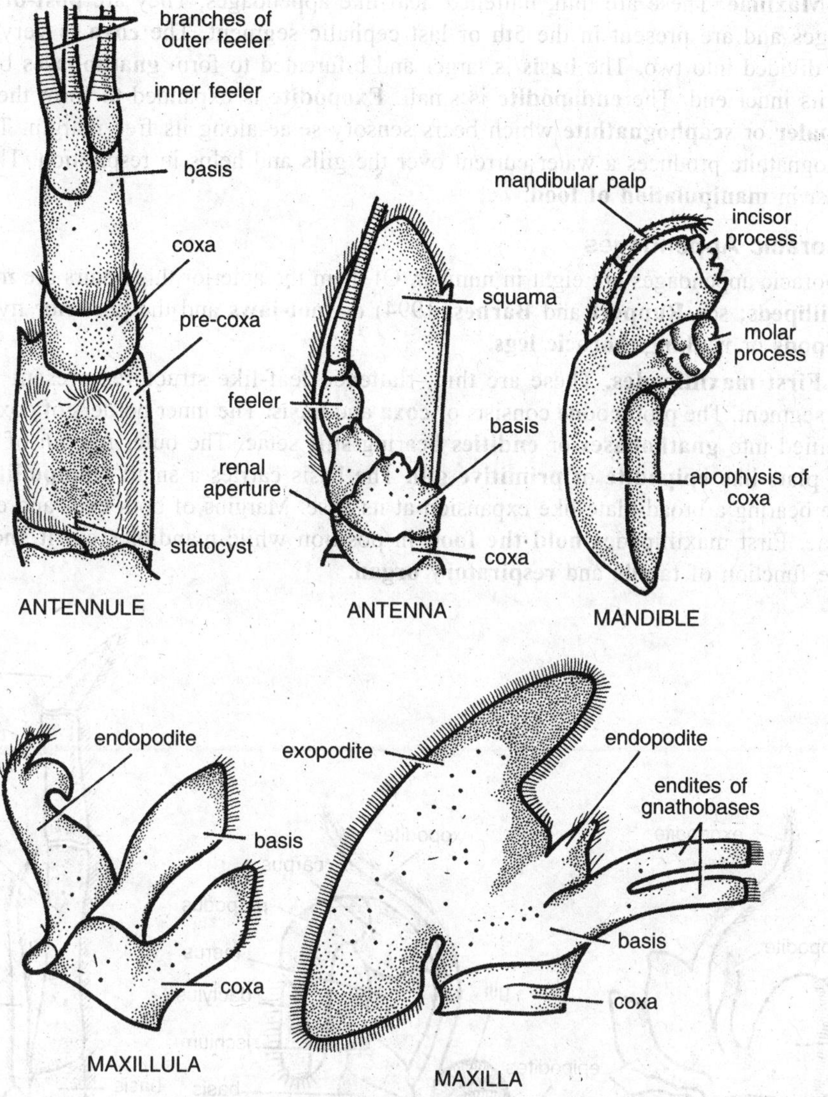

Fig. 7.4. *Palaemon.* Cephalic appendages.

3. Mandibles. These are post-oral cephalic appendages situated in third segment one on either side of mouth. Mandibles are short but stout and highly calcified structures. The greater part of each mandible consists of coxa which is differentiated into a proximal spoon-shaped **apophysis** and a solid distal **head.** The head consists of two parts: a **molar process** with five or six dental plates and a flat plate-like **incisor process** with three teeth. On the outer side, the head carries a three-segmented **mandibular palp.** It consists of a basal segment representing **basis** and the distal two segments representing **endopodite.** The expodite is absent. Mandibles are **masticatory** in function.

4. Maxillulae. These are small delicate (thin) and leaf-like post-oral appendages borne by 4th segment of the body. Each consists of **coxa, basis** and **endopodite.** The **exopodite** is absent. Coxa, and basis are broad and project inwards as jaws or **gnathobases** fringed with bristles.

Endopodite is curved and bifurcated terminally. It has small hooks. The maxillulae are used for passing food to mouth (or manipulation of food).

5. Maxillae. These are thin, flattened, leaf-like appendages. They are **post-oral** cephalic appendages and are present in the 5th or last cephalic segment. The **coxa** is very small and partially divided into two. The **basis** is larger and bifurcated to form **gnathobases** bearing stiff setae at its inner end. The **endopodite** is small. **Exopodite** is expanded to form the large fan-shaped **baler** or **scaphognathite** which bears sensory setae along its free margin. The beating of scaphognathite produces a water current over the gills and helps in respiration. The maxillae also assist in **manipulation of food.**

B. Thoracic Appendages

Thoracic appendages are eight in number. Of them the anterior three pairs are maxillipedes (or **maxillipeds;** see **Ruppert** and **Barnes,** 1994) or **foot-jaws** and the posterior five pairs are the **perepods** or **walking/thoracic legs.**

1. First maxillipedes. These are thin, flattened leaf-like structures present in the first thoracic segment. The protopodite consists of **coxa** and **basis.** The inner borders of coxa and basis are modified into **gnathobases** or **endities** bearing stiff setae. The outer margin of coxa bears bilobed, plate-like **epipodite** or **primitive gill.** The basis carries a small **endopodite** and long **expodite** bearing a broad plate-like expansion at its base. Margins of both exo- and endopodites bear setae. First maxillipedes **hold the food** in position while mandibles cut it and they also serve the function of **tactile** and **respiratory organ.**

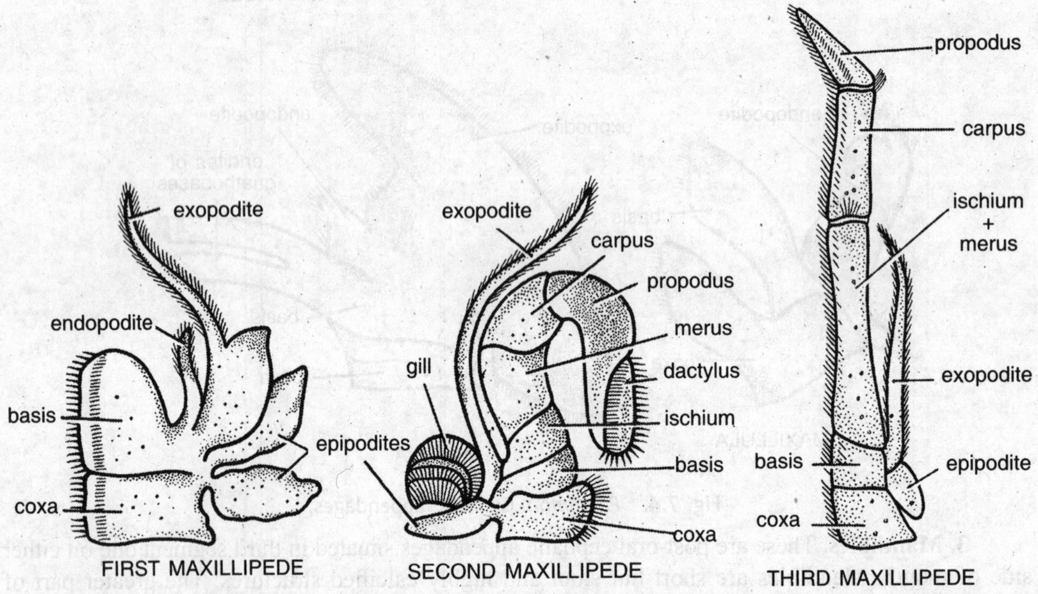

Fig. 7.5. *Palaemon*. Maxillipedes.

2. Second maxillipedes. These are not flattened. The **protopodite** consists of **coxa** and **basis.** The coxa is short and carries an **epipodite** and a **gill** (podobranch) on its outer border. The **exopodite** is long, slender, unjointed and covered with setae. The **endopodite** arises from the basis and consists of five segments or podomeres namely **ischium, merus, carpus, propodus** and **dactylus.** The distal two segments *i.e.*, propodus and dactylus are curved inwards and lie parallel to form a cutting plate with bristles. Second maxillipedes are **tactile, hold food** and **respiratory** in function.

3. Third maxillipedes. These are largest and look-like a thoracic or walking legs. The **coxa** bears a small **epipodite** on its outer side. It lacks a gill but contains setae along its inner border. The basis is small and carries an outer slender unjointed **exopodite** and an inner three-jointed **endopodite.** In the endopodite the first podomere represents fused **ischium** and **merus,** second is the **carpus** and third represents fused **propodus** and **dactylus.** The expodite is thickly covered with setae all along its length. The basis and endopodite both contain setae only along their inner margin. Third maxillipedes are **tactile, hold food** and are **respiratory** in function.

4. Thoracic or walking legs. The five pairs of walking legs of *Palaemon* differ from maxillipedes in the absence of the **exopodites** and **epipodites.** A typical walking leg such as **fourth,** consists of a two-jointed **protopodite** and five-jointed **endopodite.** The protopodite consists of a short ring-shaped **coxa** and a triangular **basis.** The endopodite consists of five podomeres, viz., **ischium, merus, carpus, propodus** and **dactylus.** All the seven podomeres are arranged in a linear series and are movably hinged together.

Walking legs are of two general types:

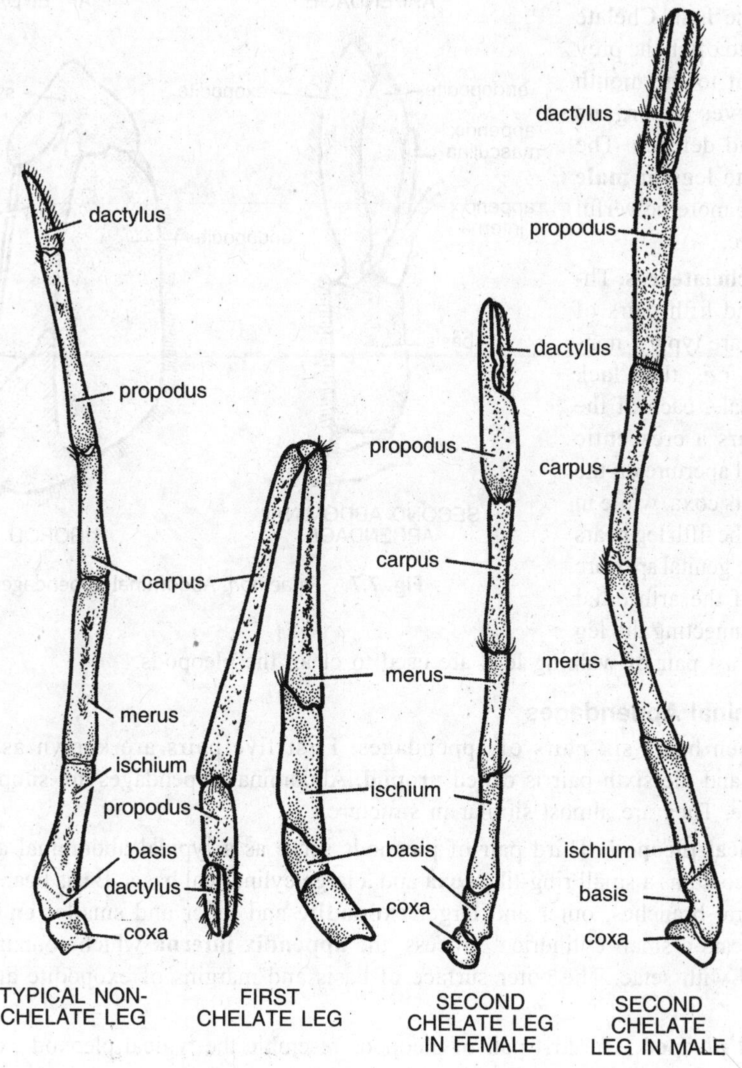

Fig. 7.6. *Palaemon*. Thoracic or walking legs.

1. Chelate legs or cephalopods. First and second pairs of walking legs bear at their distal ends a pincer-like or forceps-like structure called **chela.** In chelate legs, the propodus is very large and it extends beyond its articulation with the dactylus. Dactylus is freely movable and works against the propodus to form the chela. The biting edges of chela bear teeth and their entire surface is covered with setae. The second chelate legs are larger than the first. Chelate legs are used to catch the prey and transfer it to the mouth and also serves as organs of offense and defence. The second chelate legs in **male** are larger and more powerful than in female.

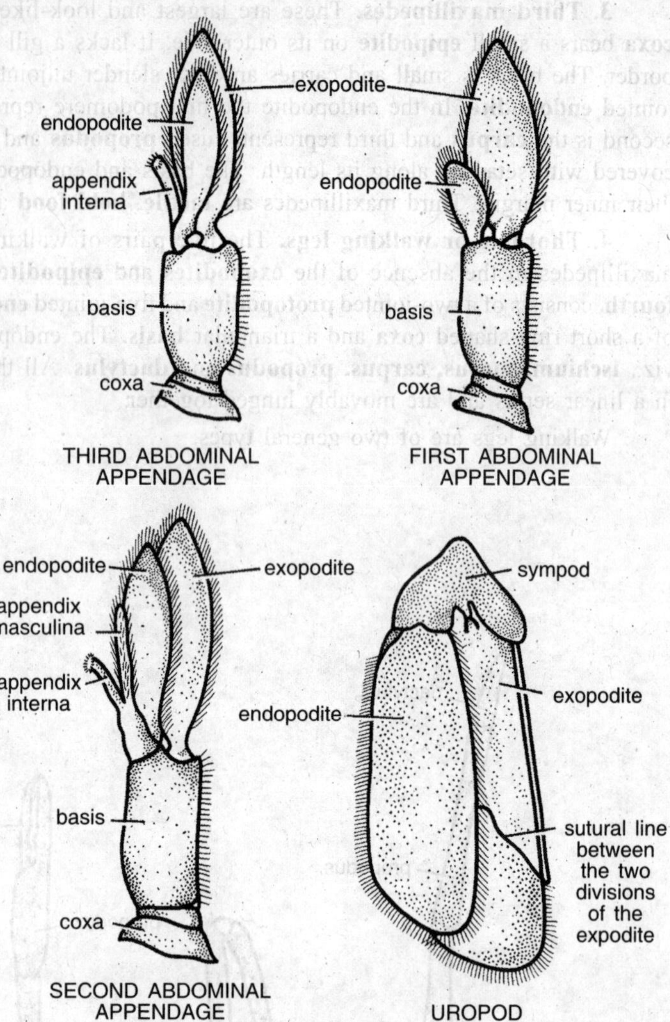

Fig. 7.7. *Palaemon.* Abdominal appendages.

2. Nonchelate legs. Third, fourth and fifth pairs of walking legs are typical nonchelate legs, *i.e.*, they lack chela. In female, each of the third leg bears a crescentic female genital aperture on the inner side of its coxa, while in male each of the fifth leg bears a slit-like male genital aperture on its side of the arthrodial membrane connecting the leg and thorax. Last pair of walking legs are used to clean the pleopods.

C. Abdominal Appendages

Abdomen bears six pairs of appendages. First five pairs are known as **pleopods** or **swimmerets** and the sixth pair is called **uropod.** Abdominal appendages are simple and typical biramous type. They are almost similar in structure.

1. Typical pleopod. Third pair of pleopods serve as a typical abdominal appendage. Its **protopodite** contains a small ring-like **coxa** and a large cylindrical basis. Basis bears two flattened leaf-like lateral branches, outer and larger **exopodite** and inner and smaller **endopodite.** The endopodite bears a small cylindrical process, the **appendix interna** which contains a knob-like head covered with setae. The outer surface of basis and margins of exopodite and endopodite bear setae.

2. First pleopod. The first pair of pleopods resemble the typical pleopod except that they lack appendix interna and the endopodite is very small.

3. Second pleopod. The second pair of pleopods of male contain an additional rod-like process, the **appendix masculina,** in between the appendix interna and endopodite. Appendix masculina bears long spine-like setae. Female prawns do not contain this process.

Functions of pleopods include the following:

(1) These are used as paddles in swimming.

(2) During breeding season, in female, the appendix interna of opposite pleopods become interlocked to form a series of bridges on the ventral side of abdomen. The bridges support the eggs. The female carries the fertilized and developing eggs in these bridges.

(3) The second pair of pleopods of male are used in copulation.

4. Uropods. The last or the sixth pair of abdominal appendages are called **uropods** or **tail feet.** These are large, broad and flattened, stoutly-built structures. They lie on either side of telson. In the uropod, the coxa and basis are fused together to form a triangular **sympod.** The **exopodite** and **endopodite** are oval and oar-shaped. The exopodite is broad and large and is divided into two unequal parts by a oblique transverse **suture.** The endopodite is small sized and undivided. The margins of exopodite and endopodite, except the outer of the exopodite, are fringed with long sensory setae.

The uropods along with telson form a broad fan-shaped **tail fin** which acts as a balancing organ during locomotion. When endangered or disturbed, *Palaemon* suddenly jumps backwards and upwards with the help of uropods.

7.4 BODY WALL AND ECDYSIS

The body wall of *Palaemon* consists of following three parts:

1. An outer thick layer of **cuticle;**
2. A middle single layer of **epidermis** or **hypodermis;**
3. An inner **dermis.**

1. Cuticle. It is composed of two layers: an outer **epicuticle** and an inner **procuticle** (Fig. 7.8).

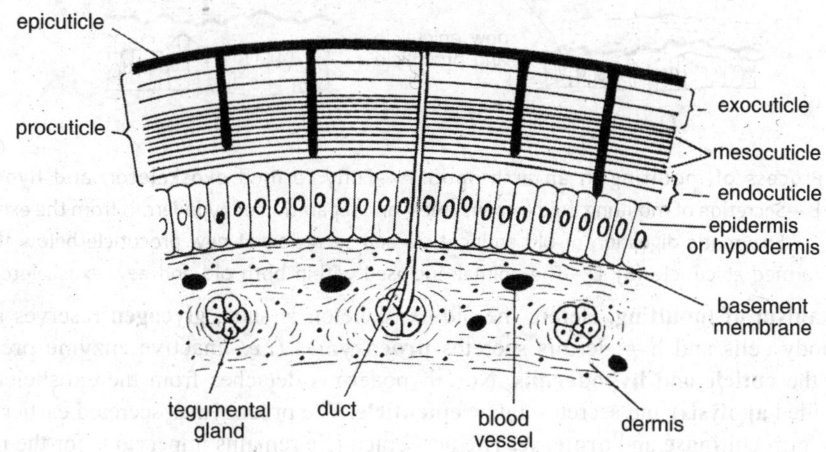

Fig. 7.8. *Palaemon*. T.S. of body wall.

***(i)* Epicuticle.** It is thin non-chitinous layer and composed of lipids, proteins, salts of calcium and pigments.

***(ii)* Procuticle.** It is a thick, laminated, chitinous but non-cellular layer. It is composed of (*i*) an **outer pigmented layer,** called **exocuticle,** of pigments, proteins, chitin and calcium salts; (*ii*) a **middle layer,** called **mesocuticle,** which is most heavily calcified and contains more chitin and less protein; and (*iii*) an **innermost thin layer,** called **endocuticle,** of protein and chitin.

Exocuticle is absent at joints and along lines where the exoskeleton tends to rupture during moulting. Since chitin is permeable to water, it is coated with a thin waxy epicuticle.

2. Epidermis or hypodermis. It lies below the cuticle and is composed of a single layer of columnar and glandular cells. This layer secretes **chitin** and the **cuticle** for the formation of **procuticle.** Hypodermis rests upon a basement membrane. It also contains trichogen cells which form the setae.

3. Dermis. Below the hypodermis is dermis. The dermis is made of connective tissue containing the blood lacunae. Dermis also contains **tegumental glands** which secrete **epicuticle.** These glands are formed by a group of cells which open into a long narrow duct opening outside on the epicuticle. Chromatophores are also present in the connective tissue.

Moulting or Ecdysis

The crustaceans (*Palaemon*) and other arthropods remain covered by a rigid, non-living exoskeleton which does not allow them to grow in size. In order to grow, the arthropods periodically shed their skeleton and secrete a new one. Periodic shedding of the exoskeleton in arthropods is called **moulting** or **ecdysis.**

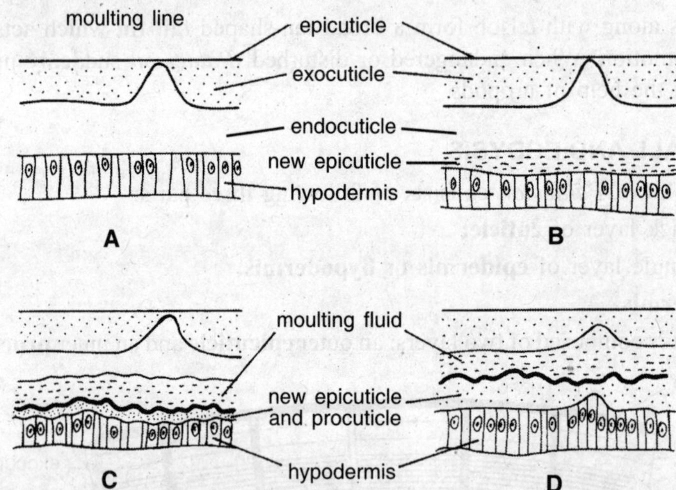

Fig. 7.9. Process of moulting in an arthropod. A—Fully formed exoskeleton and hypodermis; B—Secretion of moulting fluid (proenzymes) and separation of hypodermis from the exoskeleton; C—Enzymatic digestion of old endocuticle and secretion of new procuticle below the newly formed epicuticle (B); D—The animal enclosed within both old and new exoskeletons.

Mechanism of moulting. Before the old exoskeleton is shed, **glycogen** reserves are built up in the body cells and hypodermis secretes **proenzymes** (*i.e.*, inactive enzyme precursors) in between the **cuticle** and **hypodermis.** Now hypodermis detaches from the exoskeleton (this process is called **apolysis**) and secretes a new **epicuticle.** The proenzymes secreted earlier become activated to form **chitinase** and **protease.** The new epicuticle remains impervious for the moulting enzymes. These enzymes, however, digest the old endocuticle. The digestion products are absorbed through the new epicuticle. With the erosion of the old endocuticle, the hypodermis secretes new procuticle beneath the new (Fig. 7.9C). At this point the animal is enclosed with both an old and a new exoskeleton. The old skeleton now ruptures (splits) along certain predetermined lines and animal pulls out of old exoskeleton.

The new exoskeleton is soft and wrinkled (*i.e.*, extensible). It is stretched to accommodate the increased size of the animal. Various adaptations are used to keep the body at a maximum volume while new skeleton hardens: 1. Blood pressure is increased; 2 Insects inflate the crop with air, or 3. Crustaceans imbibe water. The tissues grow rapidly, and the new cuticle hardens during the short **postmolt period.** At this time, the stored mineral resources of crustaceans are rapidly exhausted, for they are deposited in the hardening of cuticle. Hardening of epicuticle is also helped by **tanning** of proteins which occurs by addition of phenols (Box 7.1).

Box 7.1 Moulting

1. Moulting (or ecdysis) is far more than a simple splitting of the cuticle and secretion of new one. Arthropod life consists of alternating periods of **premolt, molt, post-molt** and **intermolt.** Some arthropods (insects) escape into a permanent intermolt when they are adult; this period is known as **anecdysis.** However, others (crustaceans) never achieve stability, moulting occurs at intervals throughout adult life. The period between moults are known as **molts.**
2. Molts are recurrent crisis in the lives of arthropods. Until the new skeleton hardens, the protective case is weak, *and without hardened skeletal plates; so, skeleton is unable to support the weight of body and movement is often restricted.* To avoid danger to their life, arthropods may seek seclusion for moulting, build a protective silken retreat, retire into burrows, or have special mechanism to speed skeletal hardening. In any event, moulting remains an especially dangerous time in arthropod life cycle.
3. In *Palaemon*, the physiological processes involved in moulting are controlled by hormones. A **moult-inhibiting neurohormone** is secreted by a group of brain cells called **X-organs.** When secretion of X-organs are inhibited by **photoperiod** and **temperature,** etc., a moulting hormone or the **ecdysone** is secreted by **Y-organ** which is located beneath the muscles connected with mandibles. Ecdysone begins the process of moulting.

7.5 DIGESTIVE SYSTEM

The digestive system of *Palaemon* consists of an **alimentary canal** and a digestive gland, called **hepatopancreas.**

A. Alimentary Canal

Alimentary canal is a straight tube and is distinguished into following three parts:

1. **Stomodaeum** or **foregut** which comprises mouth, buccal cavity, oesophagus and stomach.
2. **Mesenteron** or **midgut** which comprises the intestine.
3. **Proctodaeum** or **hindgut** which comprises the rectum and the anus.

The foregut and hindgut have an internal lining of **cuticle** (intima) but the midgut has a soft lining of **endoderm.**

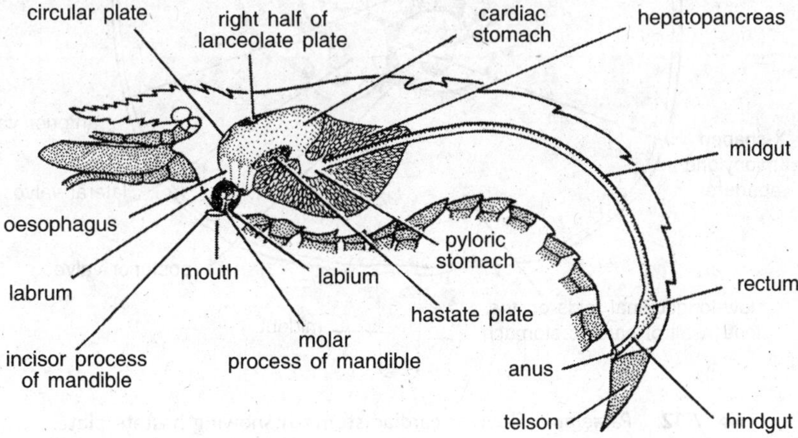

Fig. 7.10. *Palaemon*. Digestive system (alimentary canal and hepatopancreas) in lateral view.

1. Foregut

1. Mouth. It is a large longitudinal slit situated mid-ventrally in third cephalic segment. Mouth is bounded by a shield-like **labrum** in front, bilobed **labium** behind and the **incisor processes** of mandible laterally. It leads into a short buccal cavity.

2. Buccal cavity. It is an antero-posteriorly compressed small chamber. It has a thick chitinous lining which is thrown into irregular folds. The **molar processes** of **mandibles** project into the buccal cavity and lie opposite to each other. These are used to crush the food between them.

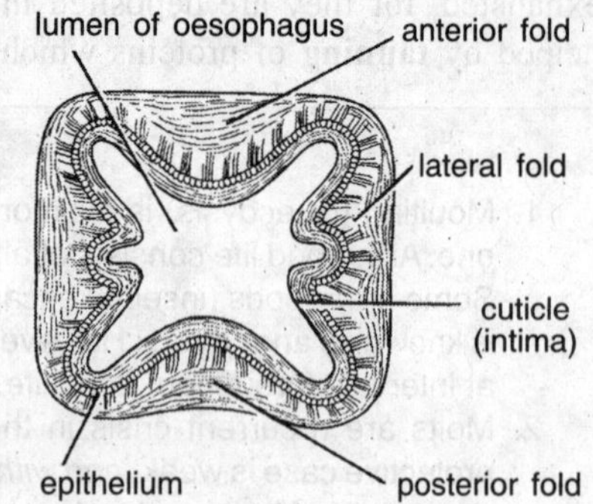

Fig. 7.11. *Palaemon.* T.S. of oesophagus.

3. Oesophagus. It is a short but broad tube which runs vertically upwards from the buccal cavity to the cardiac stomach. Internally the thick muscular walls of the oesophagus is thrown into four prominent longitudinal folds which project into its lumen. Due to these folds, the lumen of oesophagus assumes a star-shaped appearance. Of these folds, the **anterior** one is short but the **posterior** and **lateral folds** are larger and more prominent. Each of the lateral folds becomes further divided into two smaller unequal folds. The chitinous lining of oesophagus bear small bristles (Fig. 7.11).

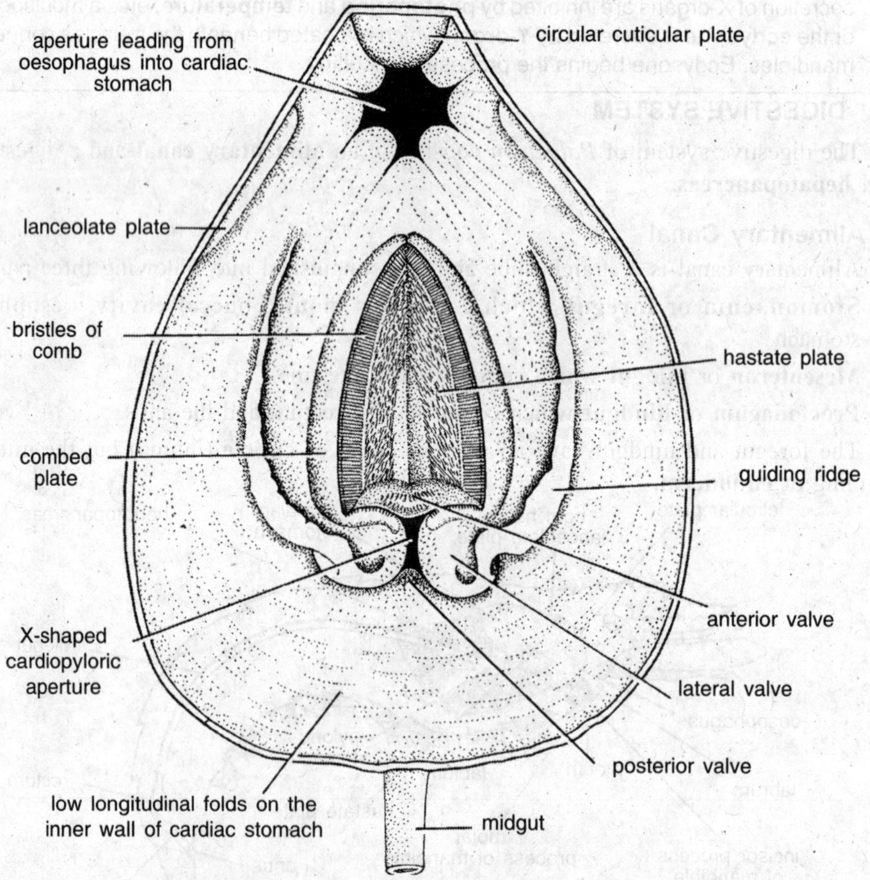

Fig. 7.12. *Palaemon.* Floor of cardiac stomach showing hastate plate.

4. Stomach. It is the largest part of alimentary canal. It occupies most of the cephalothoracic cavity. Stomach lies burried in the hepatopancreas, *i.e.*, it is covered over by the hepatopancreas laterally, posteriorly and ventrally. The stomach is divided into two unequal chambers: cardiac stomach and pyloric stomach.

(1) Cardiac stomach. It is bag-like anterior chamber of stomach. Internal chitinous lining (*i.e.*, intima) of cardiac stomach is thin and produced into numerous inconspicuous longitudinal folds covered by minute bristles (Fig. 7.12). The wall of stomach is supported by a number of calcified cuticular plates such as follows:

(*i*) Circular plate. It is a thin plate present infront of oesophageal opening.

(*ii*) Lanceolet plate. It is present in the roof of cardiac stomach just behind the circular plate.

(*iii*) Hastate plate. It is a large triangular plate which is embedded in the floor of cardiac stomach. It is called **hastate plate** because it looks like the **head of a spear.** Hastate plate contains a distinct median ridge with gradually sloping sides. Its upper surface is covered with thick growth of delicate setae. Its anterior end is somewhat conical. The posterior broad part of hastate plate is depressed, fringed with setae and forms the *anterior valve* of the **cardiopyloric aperture.** Both lateral sides of hastate plate are supported by two longitudinal cuticular **supporting rods.**

(*iv*) Lateral grooves and groove plates. A narrow lateral groove runs along the outer border of each supporting rod (*i.e.*, to each lateral border of the hastate plate). The floor of each lateral groove is covered by a cuticular plate which is like an open drain-pipe and is called the **groove plate** (Fig. 7.13).

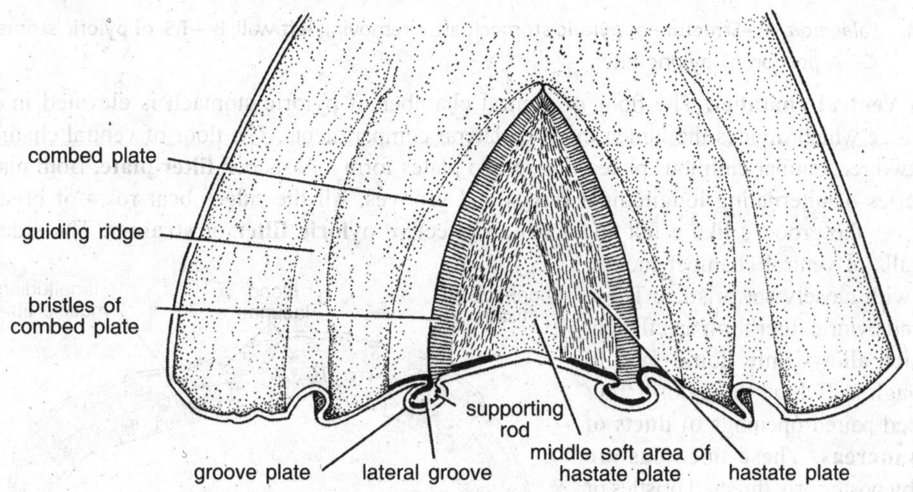

Fig. 7.13. *Palaemon*. Floor of cardiac stomach cut across.

(*v*) Combed plate. Each lateral groove on its outer border is supported by a long **ridge plate** or **combed plate.** The two combed plates are united anteriorly but remain separated posteriorly by **cardio-pyloric aperture.** The inner border of each combed plate contains equally spaced large delicate bristles which bridge over the lateral groove and partially overlap the lateral margins of the hastate plate. The bristles of combed plate move constantly to and fro.

(*vi*) Guiding ridges. Just outside the combed plate, the wall of cardiac stomach is folded on each lateral side to form two chitinous bluish black **lateral longitudinal folds** (Fig. 7.12). These folds are low in front but gradually increase in height behind. Posteriorly, however, these folds bend inwards to form the lateral valve of the cardiopyloric aperture. Since these lateral longitudinal folds guide the food towards the cardio-pyloric aperture, they are called **guiding ridges.**

***(vii)* Cardio-pyloric aperture.** It is a narrow X-shaped aperture and is guarded by four **valves.** The **anterior valve** is formed by the triangular posterior depressed end of hastate plate; the **lateral valves** by the large flap-like posterior ends of guiding ridges and **posterior valve** by semilunar folds of posterior wall of cardiac stomach. All these valves bear numerous setae which form a fine sieve through which only liquid food or very fine food particles can pass and enter pyloric stomach. The cardio-pyloric aperture leads into pyloric stomach.

(2) Pyloric stomach. It is very small narrow posterior chamber of the stomach. It lies below the posterior end of the cardiac stomach. Its lateral walls are thick and muscular. They divide the lumen of the pyloric stomach into a small dorsal chamber and a large ventral chamber. Both chambers are connected by a short narrow vertical slit-like aperture (Fig. 7.14).

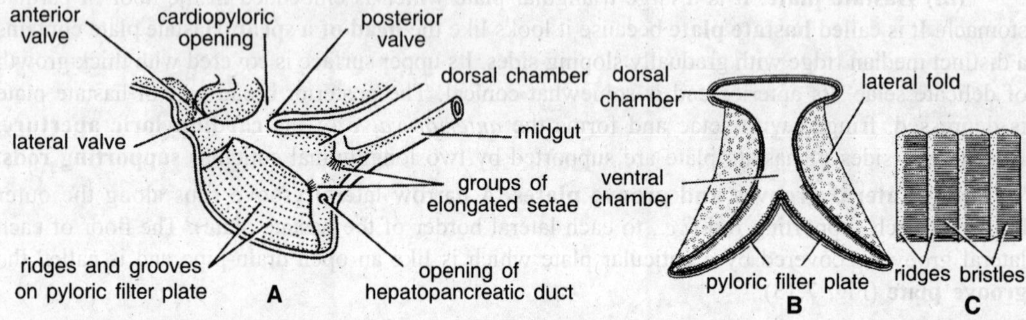

Fig. 7.14. *Palaemon.* A—Structure of pyloric stomach after removing left wall; B—T.S. of pyloric stomach; C—A portion of pyloric filter.

***(a)* Ventral chamber.** The floor of ventral chamber of pyloric stomach is elevated into a median ridge which divides the lumen into two lateral compartments. The floor of ventral chamber contains two rectangular chitinous plates. These two plates form a A-shaped **filter plate.** Both plates bear a series of alternating longitudinal ridges and grooves. All the ridges bear rows of bristles which cover the grooves like a felt and form an effective **pyloric filter** or **strainer.** The side or lateral walls of ventral chamber too are covered with closely set bristles. These bristles and pyloric filter form an efficient filter which allows only liquid food to pass through it. Behind the pyloric filter are situated paired openings of **ducts of hepatopancreas.** These openings are guarded by posteriorly directed bristles or setae which prevent regurgitation of food.

***(b)* Dorsal chamber.** The dorsal chamber bears a small, blind diverticulum or **caecum** dorsally. It leads into the midgut distally.

2. Midgut

It is a long, narrow, straight and slender tube which runs from the posterior part of cephalothorax to the sixth abdominal segment. Midgut lies above the mass of the ventral abdominal muscles.

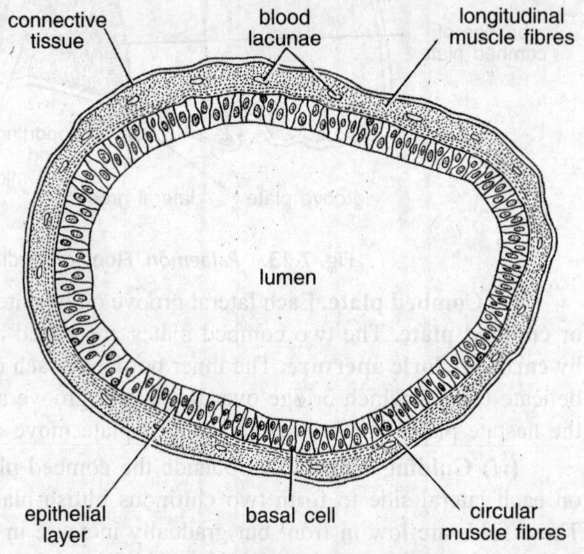

Fig. 7.15. *Palaemon.* T.S. of midgut.

Histology. The outermost wall of the midgut is formed by a thin layer of **connective tissues.** It is followed by a thick layer of **longitudinal muscle fibres** which contains a network of blood lacunae. It is followed by a thin layer of **circular muscle fibres** and an **epithelial layer** of tall columnar cells which rest on a **basement membrane** (Fig. 7.15). Midgut is not lined by a chitinous cuticle.

The epithelial lining of posterior part of midgut is thrown into many longitudinal folds. The folds greatly reduce the cavity of the intestine (or midgut).

3. Hindgut

It extends from the posterior end of the midgut to the anus. It forms shortest part of the alimentary canal. Anteriorly, hindgut is swollen into a thick muscular sac, called the **rectum,** while the posterior part is narrow and tubular which opens to exterior through the anus. The rectum bears several thick longitudinal folds which project into the lumen of the hind gut (Fig. 7.16). Rectum is lined internally by **chitinous cuticle.**

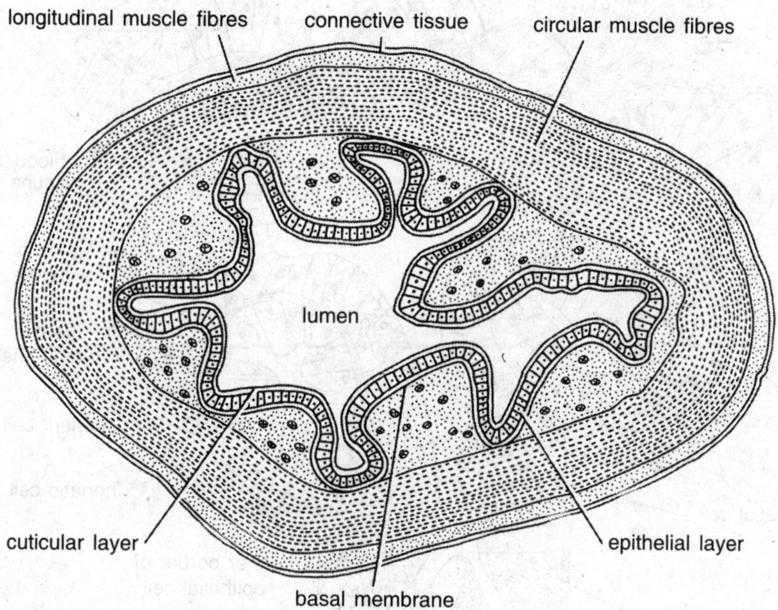

Fig. 7.16. *Palaemon.* T.S. of rectum.

4. Hepatopancreas

The **digestive gland, liver** or **hepatopancreas** is a large bilobed, compact, orange-coloured gland which fills the cephalothoracic cavity. It surrounds the stomach and anterior part of midgut. Developmentally it is formed from one pair of lateral tubular outgrowths, the hepatic caeca, of the midgut. The hepatopancreas of adult consists of numerous **glandular tubules** which are branched in a racemose manner. These branches are held together by the connective tissue as to form a compact mass. The wall of the tubules is formed of a single layer of columnar epithelium. This epithelium contains the following four types of cells: 1. **Granular cells;** 2. **Ferment cells;** 3. **Hepatic cells** and 4. **Basal** or **replacing cells.** All the epithelial cells rest on a basement membrane (Fig. 7.17).

The branching tubules join each other to form larger and larger canals which ultimately fuse to form two large **hepatopancreatic ducts.** Both ducts open into the ventral chamber of pyloric stomach. Via these ducts, digestive secretions which are produced by the hepatopancreas reach both cardiac as well as pyloric stomach.

Functions. The hepatopancreas functions like the liver, pancreas and small intestine of higher animals. It performs the following functions:

1. Like the pancreas, it secretes digestive enzymes for the digestion of carbohydrates, fats and proteins.
2. Like the small intestine it absorbs digested food.
3. Like the liver, it serves as an important storage organ. It stores glycogen, fat and calcium.
4. Some intracellular digestion also occurs in its cells.

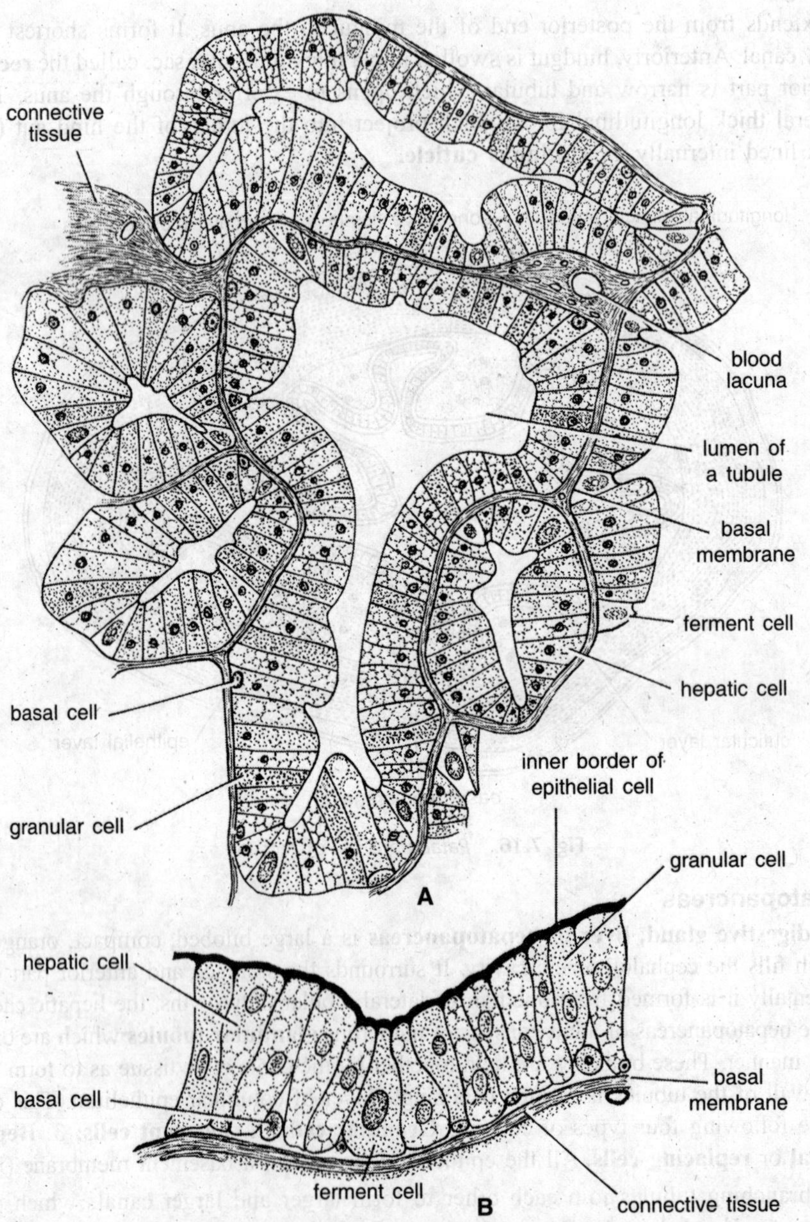

Fig. 7.17. *Palaemon*. A—T.S. through a portion of the hepatopancreas; B—T.S. of a portion of the wall of an hepatopancreatic tubule.

Food and Feeding

Palaemon is nocturnal and **omnivorous** animal. It feeds on algae, mosses and other aquatic weeds. Occasionally, it feeds on small insects, snails, tadpoles, fishes, etc., and debris of bottom. Food is captured by the chelate legs and is passed on to the third maxillipedes which send it to other mouth parts. Coxae of second maxillipedes hold the food, while incisor processes of mandibles cut it into small pieces. These pieces are further torn by maxillae and maxillulae. The food is now swallowed into the buccal cavity with the help of 1st pair of maxillipedes, maxillae and maxillulae. Inside the buccal cavity food is masticated by molar processes of mandibles. Ultimately the food is passed on to the cardiac stomach through the oesophagus. The wall of cardiac stomach work like grinding mills.

Digestion, Absorption and Egestion

In *Palaemon*, the digestion of food is started in the cardiac stomach. The enzymatic secretions of hepatopancreas are poured into the pyloric stomach from where they reach the cardiac stomach and mix with the food. By alternate contraction and expansion of cardiac stomach the food is churned and also digested by the digestive enzymes. Guiding ridges guide the semi-digested food to pass over the hastate plate. When this food passes over the hastate plate, the moving bristles of the combed plates cut it into fine particles. The food in the semi-liquid and semi-digested form is filtered through the bristles of the combed plates into the lateral grooves below; then it is carried into the ventral chamber of pyloric stomach through the cardio-pyloric aperture. The paste-like food is again filtered through the hepatopancreatic ducts and then absorbed. As already described, the hepatopancreas serves for intracellular digestion and as well as for absorption.

The undigested food (having some unabsorbed food) ascend the dorsal chamber and passes into midgut for absorption. The remaining digested food is absorbed in the midgut and residual faecal matter is passes to the hindgut and is finally expelled out through the anus.

7.6 RESPIRATORY SYSTEM

Respiratory Organs

In *Palaemon*, the respiratory organs are well developed and they are meant for its aquatic mode of life. Its respiratory organs are

1. A pair of **branchiostegites;**
2. Three pairs of **epipodites;**
3. Eight pairs of **gills.**

These organs are contained in a pair of **gill chambers** (Fig. 7.19) which are situated one on either side of the thorax and are protected externally by the downwardly grown sides of carapace called **gill covers** or **branchiostegites.** The gill chambers open on all sides except dorsally.

1. Branchiostegites

The inner linings of branchiostegites are thin, membranous and highly vascular containing blood lacunae. These are constantly bathed in freshwater, thus, forming large respiratory surfaces. These absorbs oxygen dissolved in water and gives out carbon dioxide.

2. Epipodites

Epipodites are primitive gills. These are three pairs, simple, thin, leaf-like, highly vascular outgrowths of integument. They are attached on the outer side of coxa of the maxillipedes which are present in first three thoracic segments. Epipodites lie in the anterior part of gill chamber below the scaphognathite of maxillae. The epipodites of first pair of maxillipedes are bilobed and largest

(Fig. 7.19). Respiratory exchange of oxygen and carbon dioxide occurs through these respiratory organs.

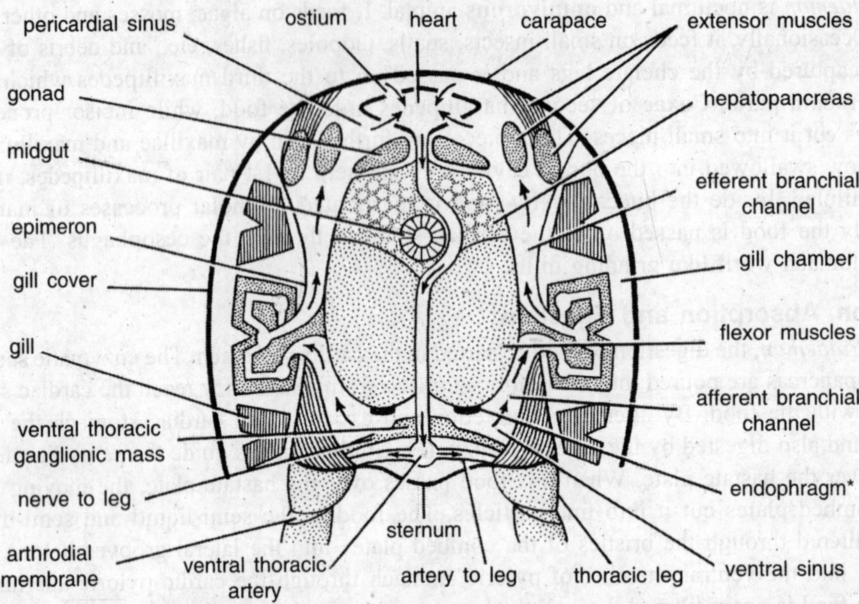

Fig. 7.18. *Palaemon*. Diagrammatic T.S. through thorax showing gill chambers and blood supply in the gills. *Endophragm is a septum formed by union of cephalic and thoracic apodemes. Apodeme is an internal ridge to which organs and muscles attach.

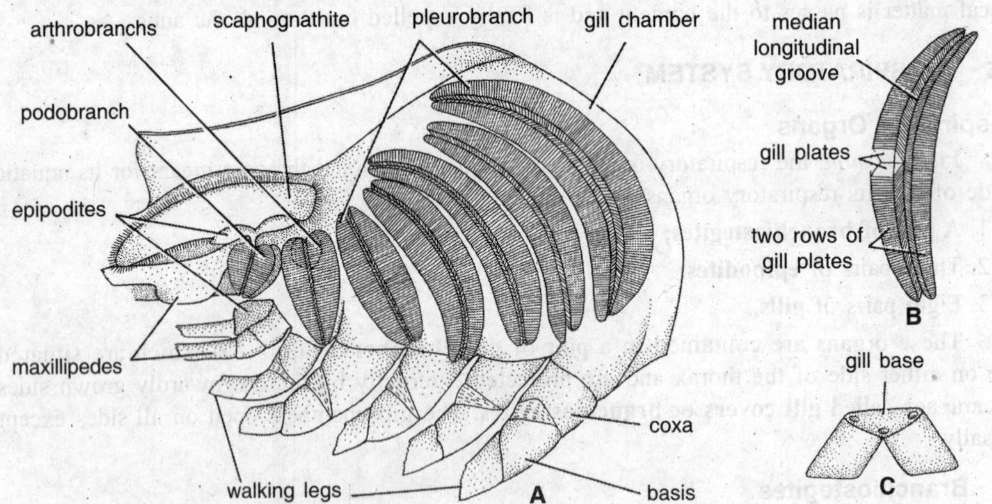

Fig. 7.19. *Palaemon*. A—Left gill-chamber is exposed to show position of epipodites and gills; B—Complete gill (phyllobranch); C—Cross-section of gill.

3. Gills

There are eight **branchiae** or **gills** in each narrow gill chamber. But only seven gills are seen on removing the gill cover as the third gill lies below the second (Fig. 7.19).

(*a*) Types of gills. Gills are named according to their place of origin and location. There are following three types of gills in each of the gill cover:

(*i*) **Podobranchs or foot gills.** The gills attached to the coxae of appendages are called **podobranchs**. There is **one** podobranch attached to the coxa of each second maxillipede.

(*ii*) **Arthrobranchs or joint-gills.** These gills are attached to the arthrodial membranes (which connect the appendages with the body). There are **two** arthrobranchs attached to the arthrodial membrane of each third maxillipede.

(*iii*) **Pleurobranchs or side-gills.** These gills are attached to the lateral walls of the throax above the attachment of the appendages. There are **five** pleurobranchs attached to the lateral walls of the thoracic segments bearing five walking legs.

(*b*) **Branchial formula.** Branchial formula shows the number and location of gills and epipodites (primitive gills) in a single gill chamber. The branchial formula of *Palaemon* is shown in the form of a table (Table 7.1) as follows:

Table 7.1

	Appendages	*Epipodite*	*Podobranch*	*Arthrobranch*	*Pleurobranch*	*Total*
1.	1st Maxillipede	1	–	–	–	1
2.	2nd Maxillipede	1	1	–	–	2
3.	3rd Maxillipede	1	–	2	–	3
4.	1st walking leg	–	–	–	1	1
5.	2nd walking leg	–	–	–	1	1
6.	3rd walking leg	–	–	–	1	1
7.	4th walking leg	–	–	–	1	1
8.	5th walking leg	–	–	–	1	1
	Total	3	1	2	5	11

(*c*) **Structure of gill.** All the eight gills are more or less crescentic or semilunar in shape. The size of gills gradually increases antero-posteriorly, *i.e.*, each gill is larger than in front of it. Each gill is attached to the thorax at its middle by a small connection known as **gill-root.** The branchial nerve and branchial channels enter and leave the gill through the gill root.

In *Palaemon*, all the gill are phyllobranchs, *i.e.*, each gill consists of two rows of leaf-like rhomboidal **gill plates** arranged parallel to one another like the leaves of a book and lie at right angles to the long axis or the base of gill (Fig. 7.19B). The two rows of gill plates are separated by a median longitudinal groove which extends lengthwise in the gill-axis. The gill plates are largest in the middle and gradually reduce at their two ends.

(*d*) **Histology.** A transverse section of a gill (Fig. 7.20) reveals that the gill base or axis is roughly triangular in outline and consists of central core of **connective tissue** enclosed by a layer of **epidermis** or **hypodermis.** Hypodermis is externally protected by a thin chitinous **cuticle.** A gill plate is made up of single layer of cells with cuticle on both sides. These cells are of two types, *viz.*, **pigmented** and **transparent,** alternating with each other.

(*e*) **Blood supply of the gills.** A **median** and two **lateral longitudinal blood channels** run in the central axis (or base) of the gill (Fig. 7.21). The two lateral channels are

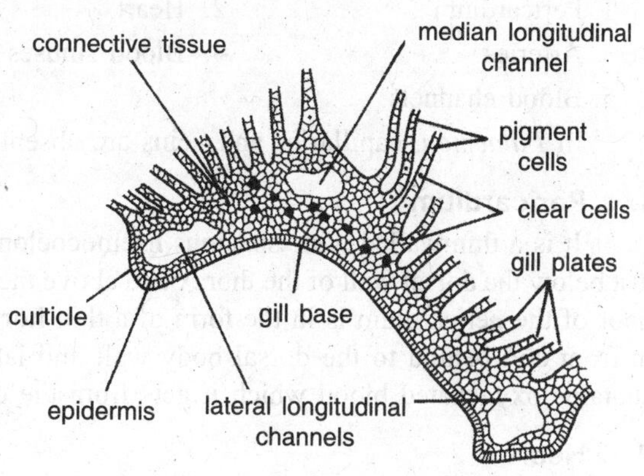

Fig. 7.20. *Palaemon.* An oblique T.S. of a gill.

connected by a series of **transverse channels (or transverse connectives).** These form a ladder-like structure. In each gill plate the lateral longitudinal channel gives off from its outer side a slender **marginal channel** which runs all along the free margins of the gill-plate and then joins the median longitudinal channel.

Course of blood in a gill. Deoxygenated blood to each gill is brought by the **afferent branchial channel** which empties into the tran-sverse connective present just in front of the gill root. The blood then flows into both the **lateral longitudinal channels** and passes through the **marginal channels** and reaches the **median longitudinal channel.** During this course the blood gets oxygenated, *i.e.*, blood leaves carbon dioxide and takes up oxygen. From the median longitudinal channel of each gill all the oxygenated blood is carried by an **efferent branchial channel** to the pericardial sinus.

Fig. 7.21. *Palaemon.* A—Blood supply in gill-plates; B—Blood circulation in principal blood channels in a gill.

Mechanism of Respiration

The oxygen dissolved in water is utilized for respiration. A continuous current of fresh-water is maintained into the gill chamber by the constant vibrations of the scaphognathite of maxilla which bales out the used current from the anterior end of gill chamber. While the water current passes from behind forward into the gill chamber, it directly bathes the gill, epipodites and branchiostegites. The oxygen dissolved in water is taken in by blood and carbon dioxide diffuses out in the water.

7.7 BLOOD VASCULAR SYSTEM

The circulatory or blood vascular system of *Palaemon* is of ***open type, i.e.***, arteries open into the cavities or **lacunae** or spaces in the body. These lacunae lacks walls. The blood vascular system of *Palaemon* consists of following parts:

1. Pericardium
2. Heart
3. Arteries
4. Blood sinuses
5. Blood channels

In *Palaemon*, capillaries and veins are absent.

1. Pericardium

It is a thin-walled wide spacious haemocoelomic chamber or **sinus** around the heart. It lies just below the dorsal wall of the thorax and above the reproductive organs and hepatopancreas. The floor of the pericardium is in the form of a thin **horizontal pericadial septum** which is attached in front and behind to the dorsal body wall and laterally to the thoracic wall. The pericardium contains oxygenated blood which it gets from the channels and passes the blood into the heart.

2. Heart

It is a triangular muscular organ with its apex in front and broad base behind. Heart is enclosed in the pericardium in the median dorsal part of thorax dorsally to the alimentary canal. It is kept in position by means of a number of muscle strands formed of muscle fibres. A median longitudinal **cardiopyloric strand** extends from the apex of the heart to the pyloric stomach and two **lateral strands,** each extending from the lateral angle of the heart to the body wall.

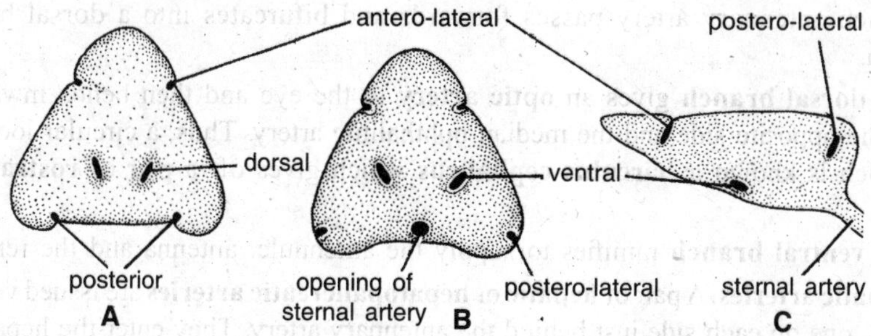

Fig. 7.22. *Palaemon*. Heart in A—Dorsal view; B—Ventral view and C—Lateral view.

The thick and muscular wall (having spongy meshwork of muscle fibres) of the heart is perforated by **five pairs** of apertures or **ostia** (*L. ostium* = a door). These ostia are located as 1. **Mid-dorsal pair,** 2. **Mid-ventral pair;** 3. **Antero-lateral pair;** 4. **Postero-lateral pair;** 5. **Posterior pair** (Fig. 7.22). Each ostium is a small, slit-like opening and its lips act as valves allowing blood to flow into the heart and not vice versa. Heart does not contain a continuous cavity.

3. Arteries

Arteries arise as thick, strong, muscular and narrow tubes from the heart to supply blood to various organs. The main arteries are as follows:

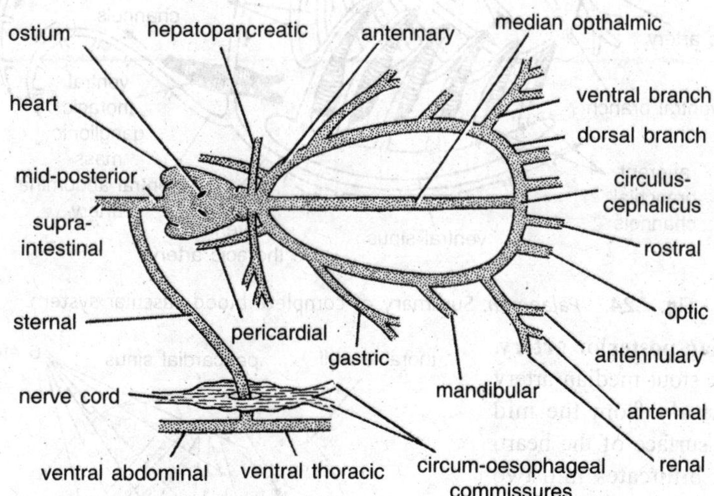

Fig. 7.23. *Palaemon*. Heart and main arteries.

1. Median ophthalmic artery. It is a single slender artery which arises from the apex of the heart. It runs forward to the head region in the mid-dorsal line above the renal sac and below the carapace. Anteriorly median ophthalmic artery reaches the roof of the oesophagus where it joins the two antennary arteries. It supplies blood to the oesophagus, cardiac stomach and the head.

***(ii)* Antennary arteries.** A pair of them originate from the apex of the heart one from either side of the median ophthalmic artery. Each runs obliquely forward along the outer border of the mandibular muscle. During its course it gives out the following three branches:

(*a*) a **pericardial branch** to the pericardium;

(*b*) a **gastric branch** to the cardiac stomach, and

(*c*) a **mandibular branch** to the mandibular muscles.

Then each antennary artery passes forwards and bifurcates into a dorsal branch and a ventral branch.

(*a*) The **dorsal branch** gives an **optic artery** to the eye and then bends inwards to meet its fellow of the opposite side and the median ophthalmic artery. Thus, a circular loop-like artery is formed which is known as **circulus cephalicus** which gives off a pair of **rostral arteries** to the rostrum.

(*b*) The **ventral branch** ramifies to supply the antennule, antenna and the renal organ.

(*iii*) Hepatic arteries. A pair of **hepatic** or **hepatopancreatic arteries** are issued ventrolaterally from the heart, one on each side just behind the antennary artery. They enter the hepatopancreatic mass and ramifies within it.

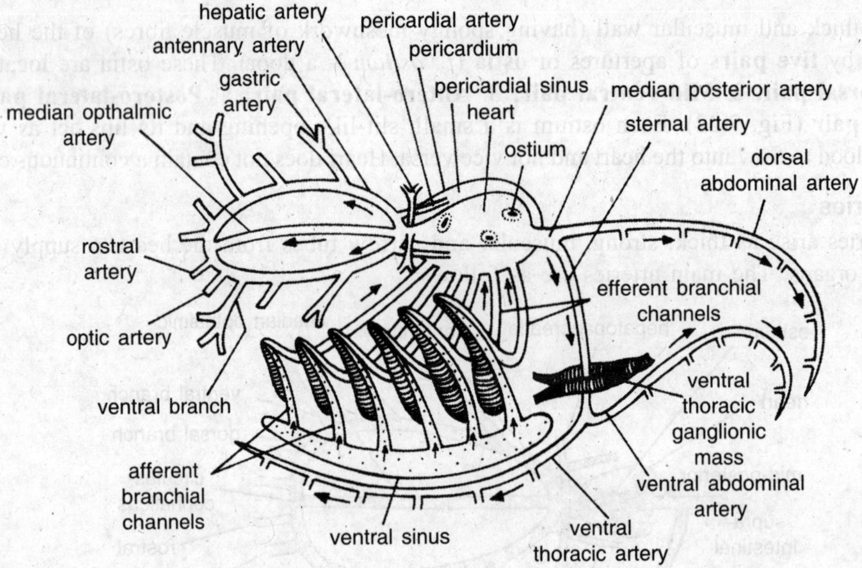

Fig. 7.24. *Palaemon*. Summary of complete blood vascular system.

(*iv*) Median posterior artery. It is a short but stout median artery running posteriorly from the mid postero-ventral surface of the heart. Immediately it bifurcates into two branches: a slender **supra-intestinal artery** and a stout **sternal artery.**

(*a*) Supra-intestinal artery. It is also called **dorsal abdominal artery.** It runs mid-dorsally above the intestine extending upto the hindgut. It supplies blood to the midgut (intestine) and the dorsal muscles of the abdomen.

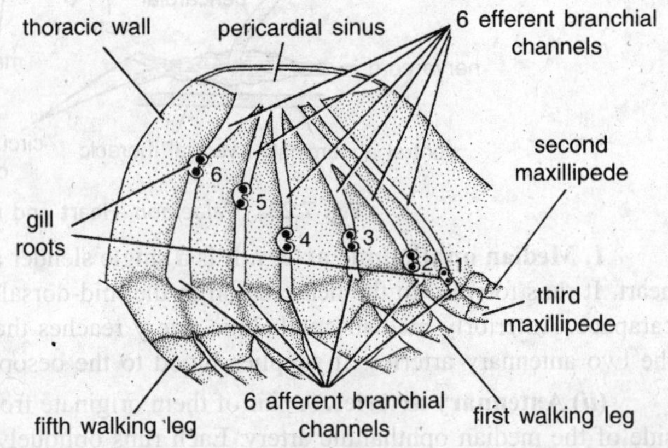

Fig. 7.25. *Palaemon*. Branchial channels and pericardial sinus.

(b) Sternal artery. It is stoutest artery in the blood vascular system of *Palaemon*. It runs downward, traverses the hepatopancreatic mass, pierces the ventral thoracic mass and finally comes to lie above the sternum (*i.e.*, below the ventral nerve cord). Here it bifurcates into the thoracic and abdominal branches. The **ventral thoracic branch** runs forward upto the mouth and supplies the sternal region, the maxillulae, maxillae and the first three pairs of walking legs. The **ventral abdominal branch** runs backwards and supplies blood to the ventral region of abdomen, the last two pairs of walking legs, the abdominal appendages and the hindgut.

4. Blood Sinuses

Various arteries branch repeatedly in vrious organs of body and finally open into **blood sinuses** or **lacunae** of the haemocoel. These lacunae lack walls and exist as large cavities among the muscles and visceral organs. All the sinuses of the body eventually meet into a pair of elongated and ill-defined **ventral sinuses** which are situated below the hepatopancreas and above the floor of the thorax. The ventral sinuses communicate with each other at various places.

5. Blood Channels

Blood channels are the lacunar tubes lacking in definite walls. From each ventral sinus blood is carried to the gill of that side by means of six **afferent branchial channels** (Fig. 7.25). As there are two ventral sinuses, the blood channels of both sides are six pairs.

The channels run upwards from the ventral sinus along the inner side of the lateral thoracic wall and enter the gill through gill roots. The first afferent branchial channel supplies the podobranch and arthrobranchs, while the remaining five channels supply blood to the pleurobranchs. Inside the gill, blood flows in following direction — transverse connective → two lateral longitudinal channels → marginal channels → median logitudinal channels (see Fig 7.26). Oxygenated blood is conveyed from the gills to the pericardium by six pairs of **efferent branchial channels.** These channels leave the gills through their gill-roots.

Blood

The blood of *Palaemon* is a clear fluid. It contains colourless amoeboid corpuscles, the **leucocytes** which are phagocytic in nature. The fluid or the **plasma** contains the respiratory pigment, the **haemocyanin.** Haemocyanin is a compound of copper and protein. It becomes bright blue when combined with oxygen, and colourless when it combines with carbon dioxide. The blood has a remarkable power of coagulation or clotting. If an appendage is injured or cut, the blood clots at once. Oxygen, carbon dioxide, digested food and waste products are transported through the blood.

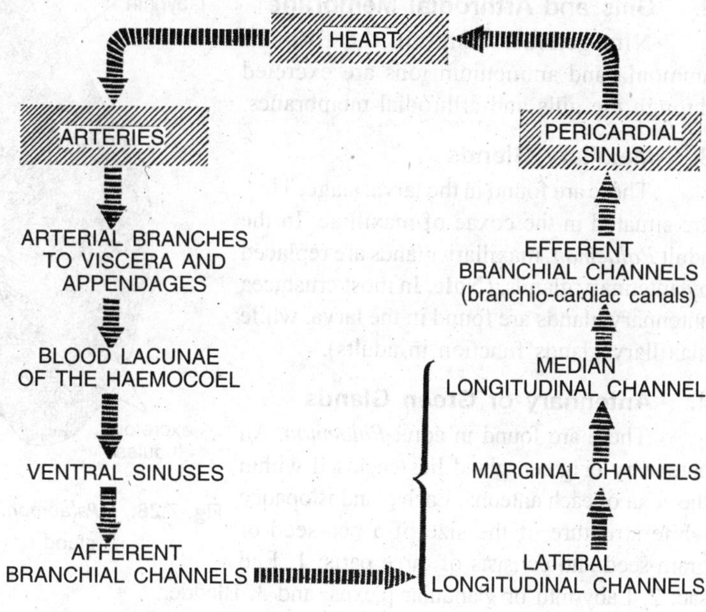

Fig. 7.26. *Palaemon.* A diagrammatic representation of the course of circulation of blood.

Course of Circulation of Blood

Due to rhythmic contraction of the heart, the blood is pumped out of the heart into the arteries which carry the oxygenated blood to all parts of the body. The course of circulation of blood in the body of *Palaemon* has been diagrammatically represented in Fig. 7.26.

7.8 EXCRETORY SYSTEM

Palaemon is an aquatic animal. Its nitrogenous wastes includes ammonia, uric acid and other nitrogen-containing compounds. Excretory system of *Palaemon* comprises the following parts:

1. Exoskeleton
2. Gills and arthrodial membrane
3. Maxillary glands
4. Antennary or green glands
5. Lateral ducts
6. Renal sac

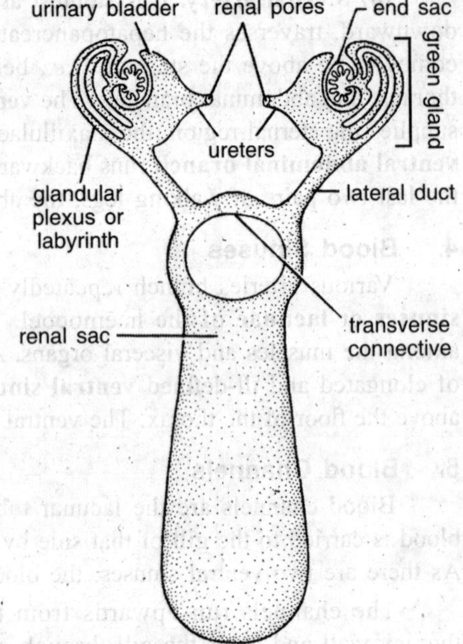

Fig. 7.27. *Palaemon*. Excretory organs: green glands, lateral ducts and renal sac.

1. Exoskeleton

The non-living exoskeleton of thick chitinous cuti-cle is a nitrogenous product secreted by the hypodermis. It is shed during moulting. In this way animal get rid of accumulated nitrogenous wastes in the integument.

2. Gills and Arthrodial Membrane

Nitrogenous wastes in the form of ammonia and ammonium ions are excreted through the gills and arthrodial membranes.

3. Maxillary Glands

These are found in the larval stage. They are situated in the coxae of maxilliae. In the adult *Palaemon*, maxillary glands are replaced by antennary glands. (**Note.** In most crustacea antennary glands are found in the larva, while maxillary glands function in adults).

4. Antennary or Green Glands

These are found in adult *Palaemon*. An antennary or green gland lies enclosed within the coxa of each antenna. Each gland is opaque white structure of the size of a pea-seed or gram-seed and consists of three parts: 1. End sac; 2. Labyrinth or glandular plexus, and 3. Bladder.

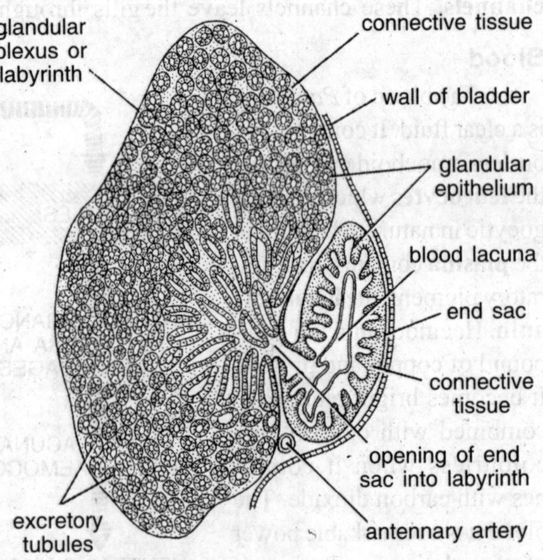

Fig. 7.28. *Palaemon*.T.S. through an antennary or green gland.

(1) End sac. It is a small, yellowish, bean-shaped structure situated between the bladder and labyrinth. Its wall consists of two layers, the **outer thick layer** of connective tissue and **inner thin layer** of excretory epithelium. The outer layer contains many blood lacunae. The excretory epithelium is produced into a number of **radial septa** which project into the central cavity of the end sac. The cavity of end sac contains a large **blood lacuna** which communicates with the labyrinth by a single aperture. The aperture is provided with a sphincter. The end sac serves as a filtering aparatus.

(2) Labyrinth or glandular plexus. It is relatively larger in size and green in colour. It is situated on outer side of end sac. Glandular plexus contains a mass of highly convoluted and branched excretory tubules. These tubules are embedded in the connective tissue which is richly supplied with blood lacunae.The tubules are narrow and are lined by excretory epithelium and collectively opens into the end sac by a single aperture and into bladder by many openings. Labyrinth serves as a site of reabsorption.

(3) Bladder. It is largest part of green gland. It is a thin-walled, spacious sac, lying on the inner side of the end sac. Inner wall of bladder is produced into a short **ureter** which opens to the exterior by a **renal pore.** The renal pore is located on a papilla on the inner side of coxa of the antennae. Bladder is lined by a single layer of excretory epithelium. Bladder serves as urinary bladder for temporary storage of urine.

Physiology of Antennary or Green Glands

The green glands are complex nephridia-like and perform two important functions like the vertebrate kidneys:

1. Elimination of nitrogenous waste products. The excretory epithelium of end sac is in direct contact with the blood contained in the lacuna. It collects the waste nitrogenous products from the blood. The chief excretory products are **ammonia** and **amines.** All the extracted waste is conveyed to the excretory ducts of labyrinth where selected reabsorption occurs and useful substances seprated from the filterate are reabsorbed by the blood lacunae present in the connective tissue. The urine is formed and passed on to the bladder and is expelled out of body from time to time.

2. Osmoregulation. Large quantities of water are separated from the blood by the antennary or green glands to maintain an osmotic equilibrium of body.

5. Lateral Ducts

From the bladder of each antennary gland runs posteriorly a narrow lateral duct. The lateral ducts of both sides are connected by a **transverse connective** in front of brain. These lateral ducts open into renal sac.

6. Renal Sac

Renal sac or **nephriodioperitoneal sac** is a large median, elongated, thin-walled sac situated above the cardiac stomach and just below the carapace in cephalothorax. Renal sac is lined by single cell thick excretory epithelium of flattened cells. Renal sac collects excretory matter from the posterior part of body.

7.9 NERVOUS SYSTEM

Nervous system of *Palaemon* consists of (1) Central nervous system; (2) Peripheral nervous system; and (3) Sympathetic nervous system.

A. Central Nervous System

It consists of (1) **Brain** or a pair of **supra-oesophageal ganglia;** (2) a pair of **circum-oesophageal commissures;** (3) **ventral thoracic ganglionic mass;** and (4) a ventral **nerve cord.**

1. Brain or supra-oesophageal ganglia. The brain is bilobed structure which lies at the base of rostrum anterior to the oesophagus. It is surrounded by a thick mass of fat. It is formed by the fusion of several pairs of ganglia. On each side brain gives off following nerves:

(*i*) **Antennulary nerve.** It arises from below the origin of optic nerve and enters the antennule of its side. Inside the antennule, it sends a **statocystic branch** to the statocyst.

(*ii*) **Optic nerve.** It is a stout nerve which arises from the dorsal surface of the brain and supplies to the compound eye.

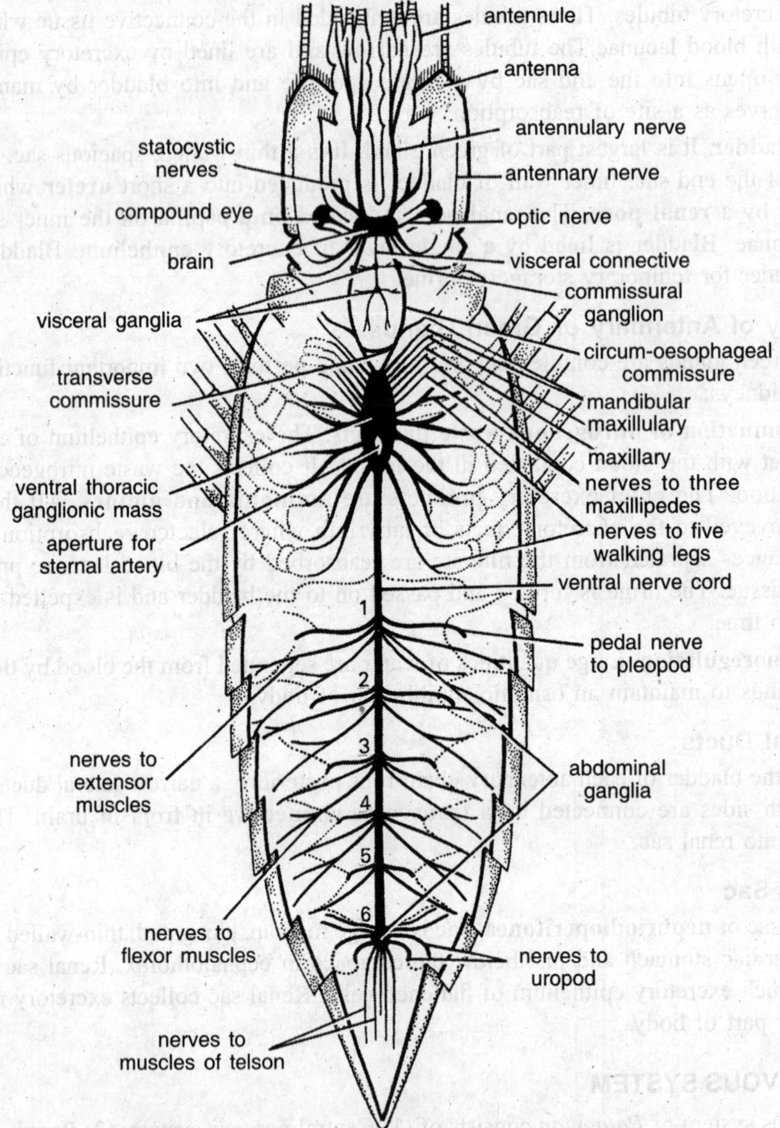

Fig. 7.29. *Palaemon*. Nervous system in dorsal view.

(*iii*) **Ophthalmic nerve.** It arises from the brain and supplies to ocular muscles in the eye-stalk.

(*iv*) **Antennary nerve.** It is a stout nerve which arises from the ventral surface of the brain. It supplies to the antenna.

(*v*) **Tegumental nerve.** It is a slender nerve which arises just behind the antennary nerve. It supplies to the labrum.

2. Circum-oesophageal commissures. These are two stout nerves, which arise from the posterior end of brain. These run backwards and downwards around the oesophagus and unite ventrally with **sub-oesophageal ganglia.** These ganglia form the indistinguishable anterior part of the **ventral thoracic ganglionic mass.** Each commissure bears a small **commissural ganglion** near its anterior end and gives off a small nerve to the mandible of its size.

The two circum-oesophageal commissures are connected with each other by a slender **transverse commissure** near the posterior end.

3. Ventral thoracic ganglionic mass. It is an elongated oval structure situated immediately above the thoracic sternal plates in the mid-ventral line. It is a composite mass which is formed as a result of fusion of **eleven pairs** of segmental nerve ganglia of the cephalothorax. It gives off eleven pairs of nerves on the lateral sides. The first three pairs are **cephalic nerves,** supplying the mandibles, maxillulae and maxillae. The last eight pairs are **thoracic nerves,** of which the first three pairs supply the three pairs of maxillipedes and the remaining five pairs supply the five pairs of walking legs. Each nerve to a leg becomes bifurcated before entering the leg.

4. Ventral nerve cord. The ventral thoracic ganglionic mass extends posteriorly to form a stout **ventral** or **abdominal nerve cord.** It runs along the mid-ventral line of the abdomen and contains an **abdominal ganglion** in each abdominal segment. The ventral nerve and abdominal ganglia are double.

Each of the first five abdominal ganglia gives off three pairs of nerves in its segment— (*i*) a pair of **pedal nerves** to the pleopods; (*ii*) a pair of nerves to the extensor muscle, and (*iii*) one pair of nerves to the flexor muscles of the succeeding segment. The last or **sixth abdominal ganglion** is comparatively large and is formed due to fusion of several ganglia. It supplies two pairs of nerves to the flexor muscles of the sixth segment, two pairs of nerves to uropods, two pairs of nerves to the telson and a single nerve to the rectum and hindgut.

B. Peripheral Nervous System

The various nerves orginating from the central nervous system, as discussed above, to innervate the different parts of the body form the peripheral nervous system.

C. Sympathetic Nervous System

The **sympathetic, visceral** or **autonomic** nervous system comprises of a few small ganglia and nerves. A small nerve arises from the mid-posterior part of the brain and bears two **visceral oesophageal ganglia,** one anterior and another posterior. The anterior viceral ganglion is joined with two **commissural ganglia** by a pair of connectives. The posterior visceral ganglion is free and gives off two pairs of nerves to the muscles of the wall of oesophagus and the cardiac stomach.

7.10 SENSE ORGANS

Palaemon has following two main sense organs.

1. Compound Eyes

Palaemon consists of a pair of black and hemispherical compound eyes. Each eye is borne on a short two-jointed movable **eye stalk** lying in an orbital notch. Eye stalk belongs to a presegmental region of cephalothorax.

Structure. Each eye of *Palaemon* is a composite structure. It is composed of a great number of independent **visual** or **optical units,** called **ommatidia;** each of which is able to form a separate image of the part of the object present in front of it. The ommatidia are all alike in structure and are arranged radially along the long axis of eye (Fig. 7.30).

The eye is protected externally by a transparent covering of cuticle known as **cornea.** It is divided into a large number of square **facets** like those of graph paper. Beneath each facet lies an ommatidium. All ommatidia are simple and are arranged radially lying side by side and separated by dark pigment cells.

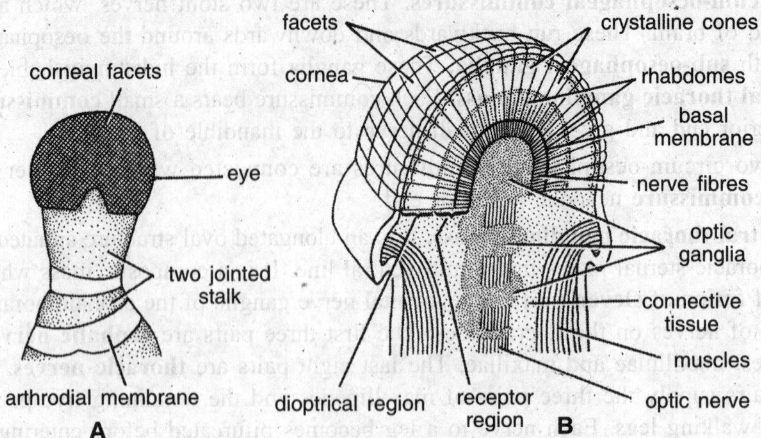

Fig. 7.30. *Palaemon.* A—Stalked compound eye; B—A diagrammatic L.S. of compound eye.

Structure of ommatidium. Each ommati-dium consists of a number of cells arranged in a linear order and comprises the following parts:

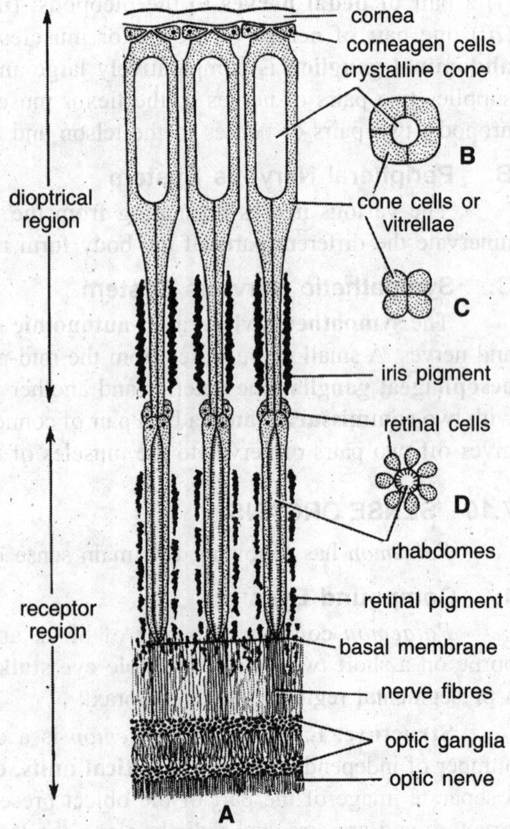

Fig. 7.31. *Palaemon.* A—Three ommatidia in L.S.; B—T.S. of an ommatidium through cone cells; C—T.S. through basal end of cone cells; D—T.S. through retinal cells and rhabdome.

1. Cornea. It is outermost part of the ommatidium. It is the corneal facet which is thickened in the centre to form a translucent **biconvex lens.** Cornea is made of chitinous cuticle of body wall; it sheds of at the time of moulting and is again secreted by the underlying cells.

2. Corneagen cells. Below each cornea occurs a pair of **corneagen cells.** These are modified epidermal cells and secrete new cornea when the old one is cast off during moulting.

3. Cone cells or vitrelle. These are a group of four much elongated cells situated beneath the corneagen cells. Cone cells secrete and enclose a transparent or translucent **crystalline cone** which works like a **second lens.** Distal end of the cone cells are long and tapering. The **crystalline cone** serves as an optic fibre through which light is channeled to the rhabdome.

The cornea, corneagen cells and cone cells together constitute the **dioptrical region.** This region of eye is concerned with the focussing of light rays falling from the object on the inner sensitive region of eye.

4. Retinal or retinular cells. The cone cells are followed by group of seven elongated photoreceptor or **retinular cells** forming the proximal part of the axis of an ommatidium. Each retinal cell acts as a single photoreceptor unit.

5. Rhabdome. It is an elongated, spindle-shaped and transversely striated, translucent cylinder or rod which is secreted and fully enclosed by the retinal or retinular cell (Box 7.3).

6. Basement membrane and optic nerve. The retinal cells and the rhabdome rest on a **basement membrane.** A nerve fibre arises from the base of each retinal cell. Thus, a bundle of seven nerve fibres penetrates the basement membrane and makes connection with the optic ganglia within the eye stalk. A stout **optic nerve** connects the optic ganglia with the brain.

The retinal cells and rhabdome upto the basement membrane constitute the **receptor region.** Its function is to receive the light rays focussed by the dioptrical region and to form the image.

Pigment sheath. Each ommatidium remains separated from its neighbours by a sheath of movable, amoeboid, pigment cells, called **chromatophores.** The pigment cells are arranged in two series. The outer or distal series is called the **iris pigment sheath.** The inner, basal or proximal series surrounds the retinal (or retinular) cells. It is called **retinal** pigment sheath. (**Note.** According to **Ruppert** and **Barnes** (1994) the retinular cells contain black or brown pigment granules, which constitute the proximal retinal pigment). Both proximal and distal pigment cells can migrate according to variation in intensity of light.

Function of compound eye: mosaic vision. Formation of image in a compound eye is called **mosaic vision.** Each ommatidium of the compound eye produces a separate image of a small partof the object present just infront of it. The final image of the object is formed by putting several small pieces together like a mosaic painting in Roman bath or mosaic floor in which separate pieces of same size but of different or same colours are put together. In fact, the image formed by the compound eye is similar to the image produced on a television screen, on which the picture is formed by small dots of light. The apposition image formed by a compound eye is often called a **mosaic image.**

In diurnal crustaceans (and other arthropods such as butterflies, houseflies, bees and wasps), the compound eyes are adapted for bright light and it produces an **apposition** or **mosaic image.** But in nocturnal forms such as *Palaemon* (and moth, fireflies, crabs, lobsters, etc.) it is adapted for seeing in weak light and a **superposition image** is formed.

Box 7.2 Transverse striations of rhabdome

The inner surface of the retinal or retinular cells are light-senstitive and are called **rhabdomeres.** In fact, the light-sensitive surfaces are thrown into **tubular** folds or **microvilli** which are oriented at right angles to the axis of ommatidium (Fig. 7.32). The rhabdomeres fuse to form a cylindrical, translucent rhabdome. The tubular folds provide transverse striated appearance to the rhabdome.

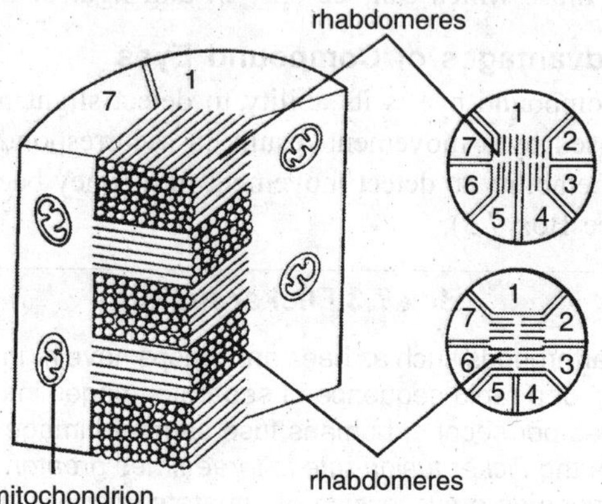

Fig. 7.32. Rhabdome of compound eye of a crayfish. Stereodiagram showing perpendicular orientation of microvilli projecting from two retinular cells. T.S. of two levels showing organisation of the rhabdomeres of seven retinular cells of one ommatidium of the compound eye.

1. Formation of apposition or mosaic image. In the **bright light,** during the day time, the pigment cells spread and completely isolate an ommatidium. As a result, light from an ommatidium cannot pass to another ommatidium. In this condition, the light rays which reach the cornea at an angle (or obliquely) are absorbed by the pigment sheath. These light rays fail to form image. The light rays which fall perpendicularly upon the cornea, travel through the crystalline cone and reach the rhabdome and the image of point of the object is formed (Fig. 7.33A).

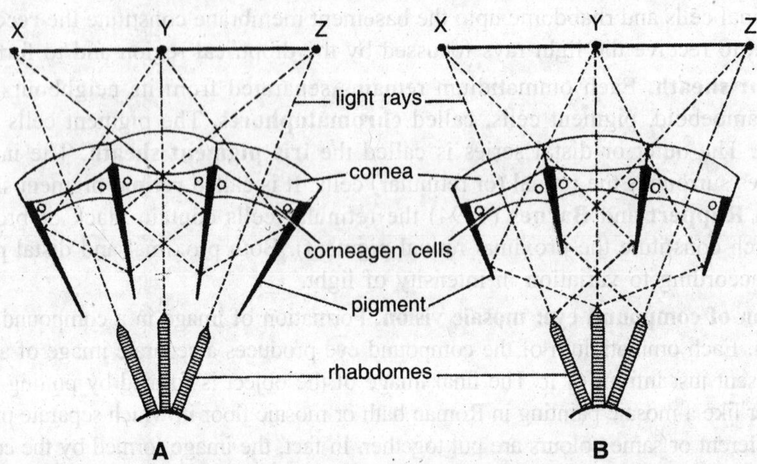

Fig. 7.33. *Palaemon.* Diagrammatic representation of image formation by a compound eye. A—Apposition or mosaic image; B—Superposition image.

2. Formation of superposition image. In the dimlight the pigment cells migrate (retract) towards the distal and basal parts of the ommatidia. In this condition, the neighbouring ommatidia work in unison. The oblique light rays can also form a point of image after passing through a number of ommatidia in their way. As a result, an overlapping of points of images takes place and a continuous image is obtained. This is called **superposition image.** It is not sharp and the animal gets only a sort of idea of the moving object in the surrounding.

Night-blind and day-blind insects. In some insects the eyes are set permanently in former or latter conditions described above. The individuals which can see only in bright light of day time are **night-blind** and those which can see only in dim light of the night are **day-blind.**

Advantages and Disadvantages of Compound Eyes

1. The advantage of compound eye is its ability to detect slight movements. A little shift in a point of light caused by a movement results in a corresponding shift in the ommatidia being stimulated. The ability to detect movement in a honey bee is three times greater than that of a human (see Box 7.3).

Box 7.3 Flicker fusion

Many active, diurnal arthropods such as bees and wasps have high flicker fusion rates. **Flicker fusion** refers to the fusing of a rapid sequence of separate images into one, as in motion picture. At approximately 16 frames per second, humans fuse separate images into a continuous moving image, but for honey bees the *flicker fusion rate is three times greater*. Thus, what we see as one continuous image a bee sees as a succession of separate images. An animal with a high flicker fusion rate can detect short increments of motion more readily than for which the fusion rate is low. A short motion in space or time might be imperceptible to us, but to a bee it might be captured in several separate images and thereby detected.

2. Another advantage of the compound eye is that the corneal surface is greatly convex, resulting in a wide visual field. This is particularly true for the stalked, compound eyes of crustacean (*Palaemon*), in which the cornea may cover an arc of 180 degrees or more.
3. The main disadvantage of compound eye is that they are not very good for distance vision compared with the large eyes of birds and mammals. So most arthropods are **short sighted.** But what matters to most arthropods is the world at close range — within 20 cm around them.
4. Images formed by compound eyes are not sharp, they are crude at best, for the small size of the eye limits the number of photoreceptor units (ommatidia).
5. The smallest arc of the visual field subtended (*i.e.*, the side of a triangle opposite to an angle) by an ommatidium in the honey bee is 1 degree compared with 1 minute of arc (1 degree = 60 minutes) in much larger human eye. Thus, a row of closely spaced vertical bars would appear as a continuous horizontal bar to a bee.

2. Statocyst

The statocysts are a pair of small white, cuticular, bead-like or sac-like structures. Each statocyst lies inside the **precoxa** of each **antennule,** attached to its dorsal wall. The base of each statocyst is slightly depressed and the neck opens to the outside by a minute aperture situated on the concave roof of precoxa of antennule. The opening remains covered by a minute fold of body wall. The statocyst is innervated by a small **statocystic branch** of the antennulary nerve. The statocyst originates as an invagination of body wall (*i.e.*, it is ectodermal in origin).

Structure. The cavity of statocyst contains minute sand particles. On the floor of the statocyst there is an oval ring of elongated delicate **receptor setae.** Each receptor seta is attached to the inner wall of statocyst and is innervated by minute fibres of statocystic nerve. The seta consists of a swollen base and a long tapering **shaft** directed toward the sand grains. The shaft is bent in middle and its inclined part bears fine bristles (Fig. 7.34B).

Since the statocyst is formed by invagination of body wall, its lining is formed of chitinous and calcified cuticle. During moulting, the lining of statocyst and sand grains are also shed. The sand grains are replaced when the head is buried in sand or the animal actually inserts sand grains into the statocyst.

Functions. The statocysts are organs of balance. They detect the change in animal's position during swimming. In fact, they detect the force of gravity and functions as the organs of **orientation**

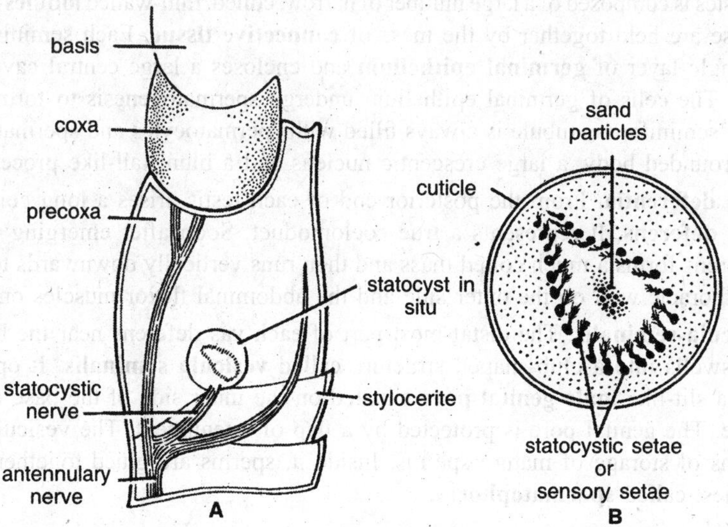

Fig. 7.34. *Palaemon.* A—Statocyst in situ; B—T.S. of statocyst.

and **equilibrium** like the semilunar canals of internal ear of vertebrates. The sand grains function as statolith.

Any change in position of swimming animal causes shift in the position of sand grains. The sand grains press some of the sensory setae and they stimulate them. This stimulus is conveyed to the brain through nerves and the animal corrects its position.

7.11 REPRODUCTIVE SYSTEM

Palaemon is **dioecious**, *i.e.*, sexes are separate and **sexual dimorphism** is well marked. Male and female individuals show the following differences:

1. The male is bigger than the female of same age.
2. Abdomen of male is narrower than that of a female.
3. The second chelate legs of male are larger, stouter and more spiny than that of a female.
4. The second pleopod of male contains an additional process, the **appendix masculina,** between the endopodite and appendix interna. It is not found in the female.
5. The epimera of abdominal segments in the female are larger than those of male. Large epimera are adapted for carrying eggs.
6. Paired genital openings in the male lie on the coxae of 5th pair of thoracic legs. In females the oviducts open on coxae of 3rd pair of walking legs.

Male Reproductive Organs

The male reproductive organs consist of —

1. A pair of **testes**
2. A pair of **vasa deferentia**
3. A pair of **vesiculare seminales**

1. Testes. The two testes are soft, white, elongated structures lying above the posterior half of the dorsal surface of the hepatopancreas and below the pericardial sinus and the heart (Fig. 7.35A). The two testes remain fused at their anterior ends. There is a narrow gap in between the middle and posterior parts of the two testes. The **cardio-pyloric strand** passes through this gap to connect the heart with cardiac stomach.

Each testes is composed of a large number of narrow, coiled, thin-walled **lobules** or **seminiferous tubules.** These are held together by the mass of **connective tissue.** Each seminiferous tubule is lined by a single layer of **germinal epithelium** and encloses a large central cavity representing **true coelom.** The cells of germinal epithelium undergo spermatogenesis to form **spermatozoa.** The lumen of seminiferous tubule is always filled with spermatocytes and spermatozoa. A mature sperm has a rounded body, a large crescentic nucleus and a blunt tail-like process.

2. Vasa deferentia. From the posterior end of each testis arises a long coiled and narrow tube, the **vas deferens.** It represents a true coelomoduct. Soon after emerging from the testis, each vas deferens forms a much coiled mass and then runs vertically downwards towards fifth leg between the thoracic wall on the outer side and the abdominal flexor muscles on the inner side.

3. Vesicula seminalis. The distal most part of each vas deferens near the base of the fifth walking leg swells into a club-shaped structure called **vesicula seminalis.** It opens to exterior by means of a slit-like **male genital pore** situated on the inner side of the base of fifth thoracic leg of its side. The genital pore is protected by a flap of integument. The vesicula seminalis are used as organs of storage of mature sperms. Inside it, sperms are glued together to form white compact bodies, called **spermatophores.**

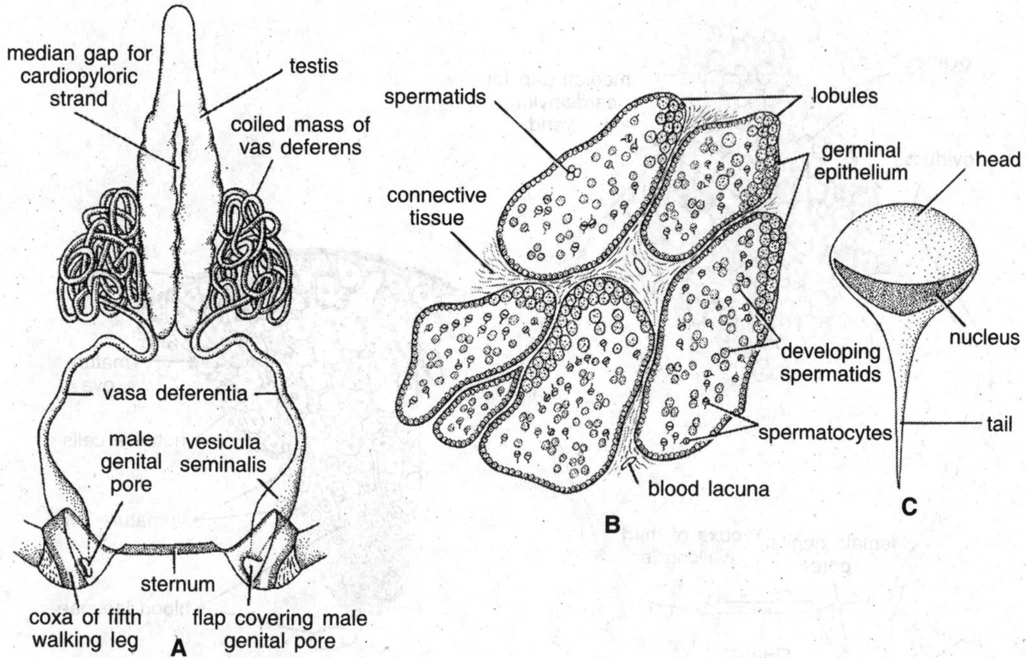

Fig. 7.35. *Palaemon* A—Male reproductive organs in dorsal view; B—T.S. of a portion of testis; C—A mature spermatozoa.

Female Reproductive Organs

The female reproductive organs consist of—

1. A pair of **ovaries**
2. A pair of **oviducts**

1. Ovaries. The ovaries in female occupy the same position as testes in male (Fig. 7.36A). The two ovaries are white, compact and sickle-shaped structures. They are situated above the hepatopancreas and below the pericardial sinus and heart. Both ovaries touch each other at the two ends but are widely separated in the middle by a gap for the passage of cardiopyloric muscle strand. The shape and size of ovaries vary with the age and season of the year. In the breeding season, the ovaries grow in size and extend up to the first abdominal segment.

Each ovary consists of numerous rows of ova arranged radially and enclosed in a membranous capsule. The ova are arranged in a definite sequence under different stages of development. The immature ova lie towards the centre, while the mature ova occur towards the periphery. Each developing ovum is being surrounded by a number of small nutritive cells. The mature ova are large nucleated cell with cytoplasm full of yolk granules; yolk forming a central core (centrolecithal eggs). In the centre of ovary, there is a cavity (a true coelom) which contains blood lacunae.

2. Oviducts. From the middle of the outer surface of each ovary arises a short, thin-walled tube called **oviduct** (Fig. 7.36A). Its anterior end is broad and forms the **oviducal funnel** which lies opposed to the surface of ovary. The oviduct runs vertically downwards to open to the exterior by **female genital pore.** It is situated on the inner side of the coxa of the third thoracic walking leg of its side. Oviduct is a true coelomoduct.

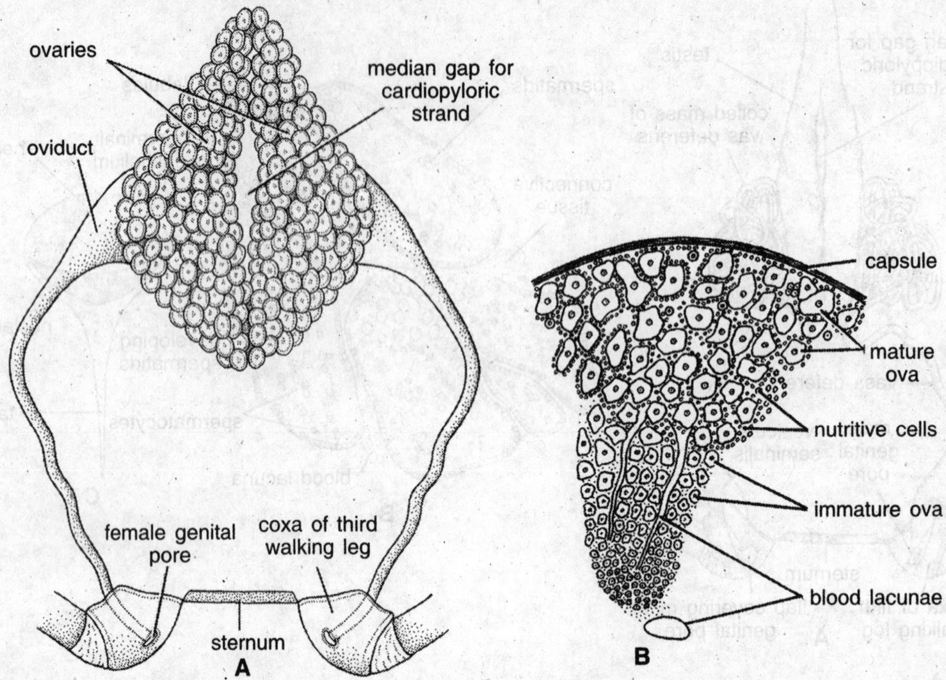

Fig. 7.36. *Palaemon* A—Female reproductive organs in dorsal view; B—T.S. of a portion of ovary.

Fertilization and Development

Fertilization. The breeding season in *Palaemon malcolmsonii* is during the months of May, June and July. Fertilization is **external.** During copulation, a male puts the female down on her back and deposits sperms (spermatophores) on the ventral surface near the female genital pores. As the eggs come out of female genital pore, they are fertilized by the sperms already present there. The female lays eggs in large numbers. The fertilized eggs are attached to the pleopods by a sticky secretion of the tegumental glands.

Development. In *Palaemon*, development is **direct** and the embryos remain attached to the pleopods until they hatch. The newly hatched young offspring looks like the adult. The adult form is reached after a series of moults.

7.12 LARVAL FORMS OF CRUSTACEA

The development of crustacea is either **direct** or **indirect.** The development is direct in case of large eggs with enough yolk and indirect in case of small eggs with a little amount of yolk. For example, some crustaceans (*e.g.*, *Palaemon*, *Astacus*, etc., develop directly from the egg and offsprings

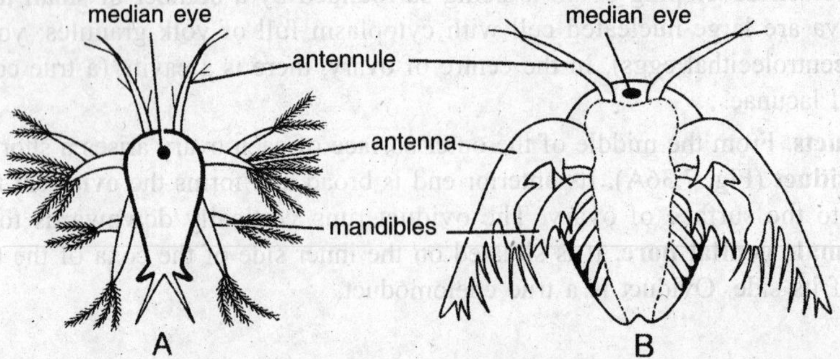

Fig. 7.37. Nauplius larva. A—Nauplius of *Penaeus*; B—Nauplius of *Apus*.

are very similar to the adult. Therefore, there is no larval stage and no metamorphosis. However, in some other crustaceans (*e.g.*, *Nebalia*, *Mysis*, etc.) development is direct but metamorphosis occurs inside the egg case. In majority of crustaceans, development is indirect being accompanied by little or more metamorphosis. In such a case, the individual hatching out of the egg is drastically different from the adult. It undergoes a series of developmental changes to acquire the adult organisation. These different young or immatue stages are known as **larvae.** In crustacea, the following larval stages have been studied:

1. Nauplius larva. Nauplius larva (Fig. 7.37) is the first (*i.e.*, it hatches out of the egg) and most common free-swimming larval stage in many crustacea (*i.e.*, Ostracoda, Copepoda, Branchi-ura, *Cirripedia*, etc). It is microscopic, oval or pear-shaped and unsegmented. Its body has a broad anterior **head** region, middle **trunk** region and bilobed **anal** region. Head bears a single sessile median eye called **nauplius eye** and a pair of **antennules** which are uniramous and unbranched. Nauplius eye is made of 3 or 4 ocelli which are pigmented cups with no lens. Antennules bear two groups of sensory cells forming the **frontal organs.** Trunk has two pairs of biramous appendages (called **feet**), *i.e.*, **antennae** and **mandibles.** Mouth opens at the anterior end, while **anus** lies on the posterior extremity. Alimentary canal is a straight tube.

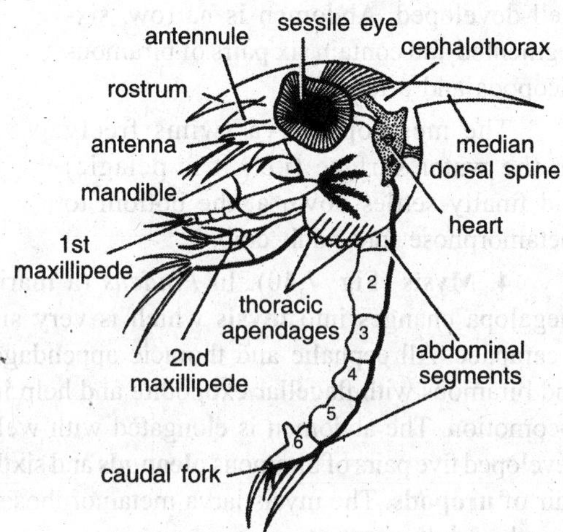

Fig. 7.38. Zoea larva of crab.

In some crustaceans (*e.g.*, most Malacostraca), nauplius exists within egg membranes and develops into the adult, but in some it passes through many interme-diate larval forms (such as **metanuplius, protozoea, zoea, cypris, mysis, megalopa, phyllosoma, alima,** etc.).

2. Zoea larva. It is fourth larval stage of the crusta-cea. Nauplius changes into metanauplius, metanauplius change into porotozoea and protozoea changes into zoea larva. In crabs, however, the young is hatched at the zoea stage.

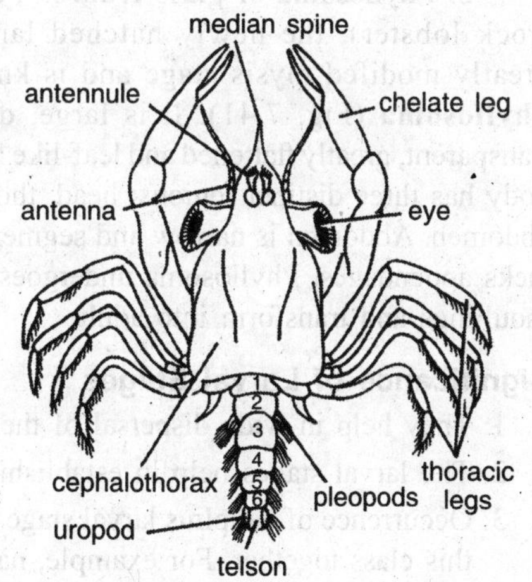

Fig. 7.39. Megalopa larva of crab.

Zoea is a free-swimming stage. Body of zoea comprises a large unsegmented **cephalothorax** and **long segmented abdomen** (Fig. 7.38). The cephalothorax is covered with a helmet-shaped **carapace** which contains two long spines—an anterior **rostral** and a median **dorsal spine.** Two lateral spines are also present. Cephalic region contains a pair of large, **sessile compound eyes.** There are **five** pairs of well-developed head appendages (*i.e.*, antennules, antennae, mandibles, maxillulae and maxillae) and two pairs of thoracic appendages (*i.e.*, two pairs of maxillipedes). Abdomen is six-segmented and devoid of appendages. Last abdominal segment, however, bears caudal fork.

In *Squilla*, zoea is slightly modified and is known as **alima larva.** The zoea after passing through successive molts becomes megalopa larva.

3. Megalopa larva (Fig. 7.39). It is the larva of crab and it develops from zoea larva through successive moults. It has a broad crab-like cephalothorax covered by a dorsal shield of carapace. Carapace contains an anterior median **rostral spine.** The paired compound eyes are large and stalked. The cephalic and thoracic appendages are well-developed. Abdomen is narrow, six-segmented and contain six pairs of biramous pleopods and a telson.

The megalopa larva swims freely on the water surface (*i.e.*, it is pelagic) and finally settles down at the bottom to metamorphose into adult crab.

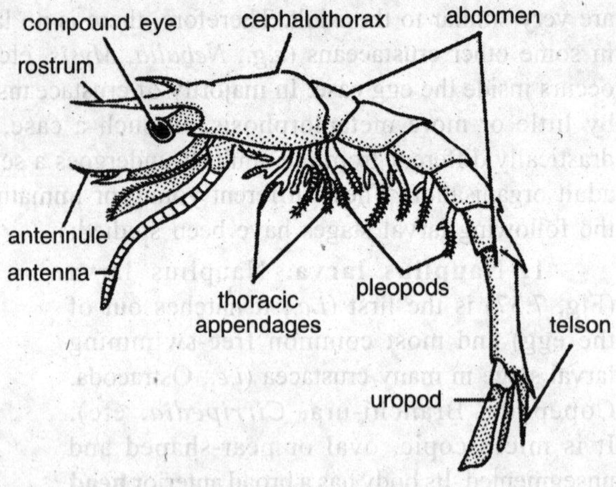

Fig. 7.40. Mysis larva of *Penaeus.*

4. Mysis (Fig. 7.40). In *Pinaeus* (a marine prawn), zoea larva instead of changing into megalopa changes into **mysis** which is very similar to the adult mysis. Head and thorax have a carapace. All cephalic and thoracic appendages are present. All thoracic appendages are alike and biramous with flagellar exopodite and help in locomotion. The abdomen is elongated with well developed five pairs of biramous **pleopods** and sixth pair of **uropods.** The mysis larva metamorphoses into the adult prawn.

In lobster (*Homarus*) both nauplius and zoea are passed within the egg, it hatches as a mysis which changes into an adult.

5. Phyllosoma or glass crab. In *Palinurus* (rock lobster), the newly hatched larva is a greatly modifed mysis stage and is known as **phyllosoma** (Fig. 7.41). It is large, delicate, transparent, greatly flattened and leaf-like larva. Its body has three distinct regions: head, thorax and abdomen. Abdomen is narrow and segmented but lacks appendages. Phyllosoma undergoes several moultings and transform into adult.

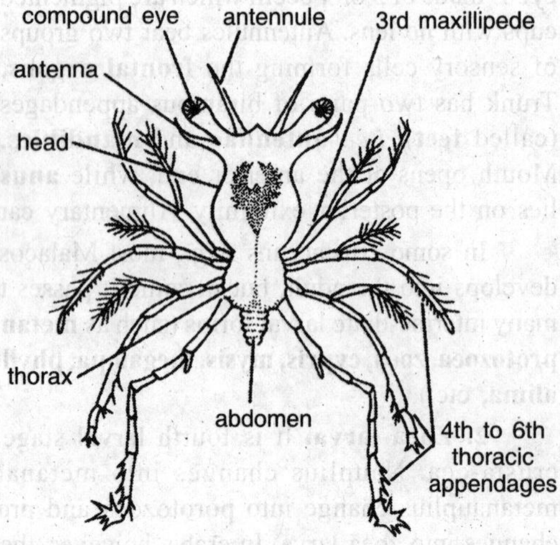

Fig. 7.41. Phyllosoma larva of *Palinurus.*

Significance of Larval Stages

1. They help in wide dispersal of the species.
2. The larval stages help in establishing relationship between various groups.
3. Occurrence of nauplius larval stage in all crustaceans connect the different representatives of this class together. For example, nauplius establishes relationship of some obscure animals such as *Sacculina* where adult has lost the characters of the class and even the phylum.

QUESTIONS

Long Answer Questions

1. Describe the habit and habitat and external features of *Palaemon*.
2. Give an illustrated account of appendages of *Palaemon* and show how these structures have modified to perform different functions.

3. Describe a typical biramous appendage and its modifications in *Palaemon*.
4. Enumerate the appendages of prawn and give well labelled diagrams to explain their structure (No description is required).
5. What is a biramous appendage? How many pairs of total appendages occur in prawn? Describe the cephalic appendages of prawn with their functions.
6. What do you understand by exopodite and endopodite? Discuss the modifications of endopodite in maxillipedes and walking legs of *Palaemon*.
7. Describe the cephalic and thoracic appendages of *Palaemon*.
8. Describe in detail the structure and function of the cephalothoracic appendages of *Palaemon*.
9. Give an account of cephalic appendages of prawn.
10. Describe in detail the structure and function of the abdominal appendages of *Palaemon*.
11. Give an account of the digestive organs of prawn and the mechanism of feeding. What is the role of the hepatopancreas in prawn?
12. Describe the alimentary canal of *Palaemon*.
13. Give an account of the alimentary canal in *Palaemon* and describe the physiology of digestion in this animal.
14. Give an account of the digestive organs of prawn.
15. Give a detailed account of the digestive system of *Palaemon* and explain the structural peculiarities of its different parts.
16. Describe the digestive system of *Palaemon* and point out the importance of hastate plate.
17. Give an account of respiratory organs of prawn (*Palaemon*).
18. Describe the organs and mechanism of respiration in prawn (*Palaemon*).
19. Describe in detail the respiratory system in prawn.
20. What is epipodite? In what respects an epipodite differs from a true gill? Describe the mechanism of respiration in *Palaemon*.
21. Describe the blood vascular system of *Palaemon*.
22. Describe in detail the circulatory system and the course of circulation in *Palaemon*.
23. Describe the blood vascular system and course of blood circulation of prawn with proper diagram.
24. Describe the main arteries of prawn.
25. Give a brief illustrated account of the excretory organs in *Palaemon*.
26. Describe the anatomy and physiology of antennary and maxillary gland of *Palaemon*. How do they resemble or differ from the coxal gland of scorpion.
27. Describe the nervous system of *Palaemon*.
28. Give an account of the structure and the working of compound eye in prawn (*Palaemon*).
29. What is mosaic vision? Give an account of structure and function of compound eye in *Palaemon*.
30. Describe the structure of the compound eye in prawn. Explain the mechanism of image formation by such an eye.
31. Describe the structure of compound eye and mosaic vision.
32. Describe the structure and function of ommatidium.
33. Give an account of the structure of the compound eyes in prawn. Explain apposition image and superposition image.
34. Describe the male and female reproductive organs of prawn. Add a note on its fertilization.
35. Describe the male reproductive organs of *Palaemon*.
36. Describe briefly the different larval forms of Crustacea.

Short Answer Questions

1. Give functions of the following:

(*i*) Antenna

(*ii*) Maxilla

(*iii*) Maxillipede; and

(*iv*) Uropod.

2. Discuss role of maxillipedes.

3. What is scaphognathite and what is its function?

4. Describe structure of a typical biramous appendage of prawn.

5. What is the difference between maxilla and a maxillipede?

6. What is chelate leg?

7. Differentiate between a chelate and nonchelate walking leg.

8. Define moulting or ecdysis.

9. Describe structure of stomach of prawn.

10. Why are lateral folds in the wall of cardiac stomach called 'guiding ridges'?

11. Describe the structure of filter apparatus of *Palaemon* or write a note on pyloric filter.

12. Name the digestive gland found in *Palaemon*. Discuss its structure and function.

13. Differentiate between epipodite and gill.

14. What is mosaic vision?

15. What is the difference between superposition and apposition image?

16. Write a note on antennary gland.

17. What is the function of antennary gland?

18. What is sexual dimorphism?

19. How can you identify male and female prawn?

20. Describe male reproductive organs of *Palaemon*.

21. Distinguish between.

(*i*) Second pleopod of male and female prawn.

(*ii*) Cardiac stomach and pyloric stomach.

(*iii*) Podobranch and arthrobranch.

(*iv*) Antenna and antennules.

22. Describe the structure of body wall of *Palaemon*.

23. Describe the process of moulting or ecdysis.

24. Write short notes on the following:

(*i*) Biramous appendage

(*ii*) Antennule appendage

(*iii*) Mandible of prawn

(*iv*) Maxillula

(*v*) Maxillipede of prawn

(*vi*) Second maxillipede of prawn

(*vii*) Pleopod

(*viii*) Second pleopod of prawn

(*ix*) Uropod

(*x*) Moulting or ecdysis

(*xi*) Stomach of prawn

(*xii*) Hastate plate

(*xiii*) Pyloric filter

(*xiv*) Hepatopancreas

(*xv*) Respiratory gills of *Palaemon*

(*xvi*) Antennary glands

(*xvii*) Green gland

(*xviii*) Compound eyes

(*xix*) Image formation of compound eye

(*xx*) Ommatidium

25. Describe structure and function of statocyst.

26. Write a short note on statocyst of prawn.

27. Write short notes on the following.

(*i*) Eye of *Palaemon* or compound eye.
(*ii*) Ommatidium
(*iii*) Mosaic vision
(*iv*) Statocyst
(*v*) Nauplius larva
(*vi*) Zoea larva
(*vii*) Megalopa larva
(*viii*) Mysis
(*ix*) Spermatophore

28. Draw well labelled diagrams of the following:

(*i*) External features of prawn
(*ii*) Cephalic appendages of prawn
(*iii*) First cephalic appendage of prawn
(*iv*) Antennule of *Palaemon*
(*v*) Mandible of prawn
(*vi*) Thoracic appendages of prawn
(*vii*) Abdominal appendages of prawn
(*viii*) Second Maxillipede
(*ix*) Alimentary system of prawn
(*x*) Inner view of cardiac stomach of prawn
(*xi*) Hastate plate
(*xii*) Branchial (gill) chamber of prawn with gill
(*xiii*) Excretory system of prawn
(*xiv*) Nervous system of prawn
(*xv*) Microscopic structure of prawn's eye
(*xvi*) Structure of ommatidium
(*xvii*) V.S. compound eye
(*xviii*) T.S. of body of *Palaemon* through gills
(*xix*) T.S. *Palaemon* through abdominal region
(*xx*) Male reproductive organs of prawn
(*xxi*) Female reproductive system of prawn
(*xxii*) T.S. statocyst
(*xxiii*) T.S. green gland
(*xxiv*) T.S. of gill of *Palaemon*

Very Short Answer Questions

1. How many cephalic appendages prawn has?

2. Give the names of cephalic appendages?

3. What is the chemical nature of chitin?

4. What is ecdysis?

5. What is the lining of gill-cover of *Palaemon* called?

6. Mention the function of scaphognathite in *Palaemon.*

7. What are the functions of pleopods of *Palaemon*?

8. Mention the functions of hepatopancreas.

9. What types of gills prawn has?

10. Where is the hastate plate found in prawn?

11. Where is statocyst located in prawn?

12. What is the shape of abdomen of *Palaemon*?

13. What is the location of male genital aperture in prawn?

14. Describe the position of female genital apertures in prawn.

15. What is appendix masculina? Where it is located in prawn?

Fill in the Blanks

1. The zoological name of freshwater prawn is ....

2. The ability to break off a part of body in prawn is called ....

3. The exoskeleton of prawn is formed of hard plates known as ....

4. The sclerites are hardened by the deposition of ....

5. The exoskeletal sclerites articulate with one another by uncalcified .... membrane.

6. The body of *Palaemon* consists of two regions .... and abdomen.

7. The abdomen of prawn consists of .... segment and a postsegmental region, the ....

8. In prawn, the andominal appendages are called .... or ....

9. In prawn, ... is organ of orientation and balance.

10. Antennary glands are found in ....

11. The last pair of abdominal appendages are called .... These help in ....

12. First pairs of gills in *Palaemon* are called ....

13. Epipodites are present on the coxa of ....

14. Ecdysis is the phenomenon of ....

15. The last pleopod is very large and named ....

16. The cephalothorax of *Palaemon* is composed of .... segments.

17. The thoracic appendages include .... pairs of maxillipedes and .... pairs peraeopods.

18. The peraeopods of prawn are also called ....

19. In prawn, stomodaeum includes buccal cavity .... and ....

20. Stomodaeum and proctodaeum are lined by ....

21. *Palaemon* has three types of respiratory organs ...., epipodites and ....

22. *Palaemon* has .... pairs of epipodites.

23. There are .... pairs of gills in prawn.

24. Body cavity of *Palaemon* is ....

25. The excretory organs of prawn are .... and ....

Multiple Choice Questions

1. *Palaemon* belongs to the class

(*a*) Insecta (*b*) Chilopoda
(*c*) Diplopoda (*d*) Crustacea

2. *Palaemon* is

(*a*) omnivorous (*b*) carnivorous
(*c*) herbivorous (*d*) insectivorous

3. The crustacean biramous appendages have a basal part known as

(*a*) exopodite (*b*) epipodite
(*c*) protopodite (*d*) endopodite

4. The total number of appendages present in the prawn is

(*a*) sixteen pairs (*b*) nineteen pairs
(*c*) seven pairs (*d*) thirteen pairs

5. How many appendages are present in cephalothorax of prawn?

(*a*) 8 pairs (*b*) 5 pairs
(*c*) 13 pairs (*d*) 19 pairs

6. Which of the following walking leg are chelate?

(*a*) first pair (*b*) third pair
(*c*) fourth pair (*d*) fifth pair

7. In the second maxillipede of prawn which part is not present?

(*a*) protopodite (*b*) scaphognathite
(*c*) gnathobases (*d*) endopodite

8. Variation in appendages is based on

(*a*) structure (*b*) function

(*c*) adaptation (*d*) modification

9. The gill is attached on outer side of

(*a*) I maxillipede (*b*) II maxillipede

(*c*) III maxillipede (*d*) All maxillipedes

10. Epipodites are attached to the

(*a*) maxillulae (*b*) maxillae

(*c*) maxillipedes (*d*) peraeopods

11. The respiratory pigment of *Palaemon* is

(*a*) haemoglobin (*b*) haemocyanin

(*c*) chlorocruorin (*d*) haemerythrin

12. The scientific name of "tiger prawn" is

(*a*) *Penaeus indicus* (*b*) *Penaeus vanunamec*

(*c*) *Penaeus monoden* (*d*) *Macrobrachium rosenbergi*

13. The monograph on *Palaemon malcolmonii* was written by

(*a*) S.N. Dass (*b*) Ronald Ross

(*c*) S.P. Patwardhan (*d*) P. Maheswari

14. The excretory matter of *Palaemon* consists of

(*a*) ammonia (*b*) urea

(*c*) amino acid (*d*) all of these

15. The number of ganglia present in the ventral thoracic ganglionic mass is

(*a*) 10 (*b*) 12

(*c*) 11 (*d*) 16

16. The functional unit in the compound eye of prawn is known as

(*a*) stigma (*b*) ocellus

(*c*) ommatidium (*d*) eye spot

17. The image formed in the compound eye in bright light is

(*a*) superposition image (*b*) mosaic image

(*c*) inverted image (*d*) distorted image

18. In prawn, statocysts are present in the

(*a*) antennules (*b*) antennae

(*c*) mandibles (*d*) rostrum

19. Blood plasma of *Daphnia* (water flea) contains

(*a*) haemoglobin (*b*) haemocyanin

(*c*) haemoerythrin (*d*) chlorocruorin

ANSWERS

Very Short Answer Questions

1. Five.

2. Antennules, atennae, mandibles, maxillulae, and maxilla.

3. Polyglucosamine.

4. Casting off old exoskeleton.

5. Branchiostegite.
6. Turning out water from gill chamber.
7. Swimming, back leaping and egg carrying.
8. Enzyme secretion, storage of glycogen, fat and calcium.
9. Phyllobranchs.
10. Floor of cardiac stomach of *Palaemon*.
11. Statocyst is located in the cavity of precoxa of antennule on either side. It opens to outside on its dorsal surface.
12. The abdomen of *Palaemon* is rounded dorsally and little compressed laterally. It remains bent a little under the cephalothorax. Due to this bent position the animal looks like comma (,) in shape.
13. Male prawn has paired genital apertures present on the inner surface of coxae of 5th pair of walking legs.
14. Female prawn has paired genital apertures which are situated on the inner surface of coxae of 3rd pair of walking legs.
15. The second pair of pleopods of male prawn contains an additional rod-like process called **appendix masculina**. It lies between the appendix interna and endopodite. It bears long spine-like setae. This apparatus is used in copulation. It is lacking in the female prawn.

Fill in the Blanks

1. *Palaemon malcolmsonii*
2. Autotomy
3. Sclerite
4. Calcium salts
5. Arthrodial
6. Cephalothorax
7. Six, telson
8. Pleopods, swimmerets
9. Statocyst
10. Coxa of antenae
11. Uropod, steering
12. Podobranch
13. Maxillipedes
14. Casting off old exoskeleton
15. Uropod
16. Thirteen
17. Three, five
18. Walking legs
19. Oesophagus, stomach
20. Chitin
21. Lining of branchiostegite, gills
22. Three
23. Eight
24. Haemocoel
25. Antennary (green) glands, Renal sac

Multiple Choice Questions

1. (*d*) 2. (*a*) 3. (*c*) 4. (*b*) 5. (*c*) 6. (*a*)
7. (*c*) 8. (*b*) 9. (*d*) 10. (*c*) 11. (*b*) 12. (*d*)
13. (*c*) 14. (*d*) 15. (*c*) 16. (*c*) 17. (*b*) 18. (*a*)
19. (*b*).

8

Scorpion (Phylum Arthropoda)

Scorpions are the arachnids and are the oldest known terrestrial arthropods; their fossils are found from the Silurian period. Today scorpions are restricted to tropical and subtropical regions. Some common genera of scorpions are: *Palamneus, Buthus, Parabuthus, Microbuthus, Scorpio, Isometrus, etc.*

8.1 CLASSIFICATION

Classification

Phylum	—	**Arthropoda**
Subphylum	—	**Chelicerata**
Class	—	**Arachnida**
Order	—	**Scorpiones**
Genus	—	***Palamneus*** **(Scorpion)**

8.2 EXTERNAL FEATURES

Habit and Habitat

Scorpion is **solitary**, **nocturnal**, **carnivorous** and **predaceous** animal and is found in desert. It feeds on insects, spiders and other small animals. It can withstand prolonged fast and hunger but show **cannibalism,** *i.e.*, it devours its young ones. Scorpion is **viviparous** and female carries the newly-hatched young on its back. Predation by vertebrates such as birds, snakes and amphibians is a major cause of scorpion mortality.

Scorpion is fairly common in hilly regions (*i.e.*, slopes of hills) of India. *Palamneus* is an Indian scorpion found all over India. *Palamneus bengalensis* is common in North India.

Shape and Size

The body of scorpion is elongated, narrow and dorso-ventrally flattened. The size varies from 1.3 cm (*e.g., Microbuthus pusillus*) to 15 cm (*e.g., Palamneus swammerdami*).

Colouration

The colouration of body depends upon the surroundings and varies from shining black to pale yellow. The dorsal surface is much darker in colour than that of the ventral surface.

External Structure

The elongated segmented body is divisible into two major regions: **1. Prosoma** (*i.e., cephalothorax*) and **2. Opisthosoma** (*i.e., abdomen*) (Fig. 8.1).

***(i)* Prosoma.** It is anterior, short, broad and flat region of body. It is formed by fusion of **head** and **thorax.** Prosoma of adult is made up of six segments, each of which bears a pair of appendages on the ventral side. Dorsally, the prosoma is covered by a large thick, shield-like, square-shaped exoskeletal plate, called **cephalothoracic shield** or **dorsal carapace**. The dorsal carapace is formed by fusion of terga of this region. The anterior margin of the dorsal carapace has a deep notch forming **right** and **left frontal lobes.** The carapace bears a pair of **median eyes** in the middle and two to five pairs of smaller **lateral eyes** on the antero-lateral margins. These are simple eyes. On the ventral side of prosoma, there is a single narrow, median triangular plate, the **sternum** which is situated between the coxae of third and fourth pairs of legs. A pair of **chelicerae**, a pair of **pedipalpi** and four pairs of **walking legs** can be seen originating from the ventral side.

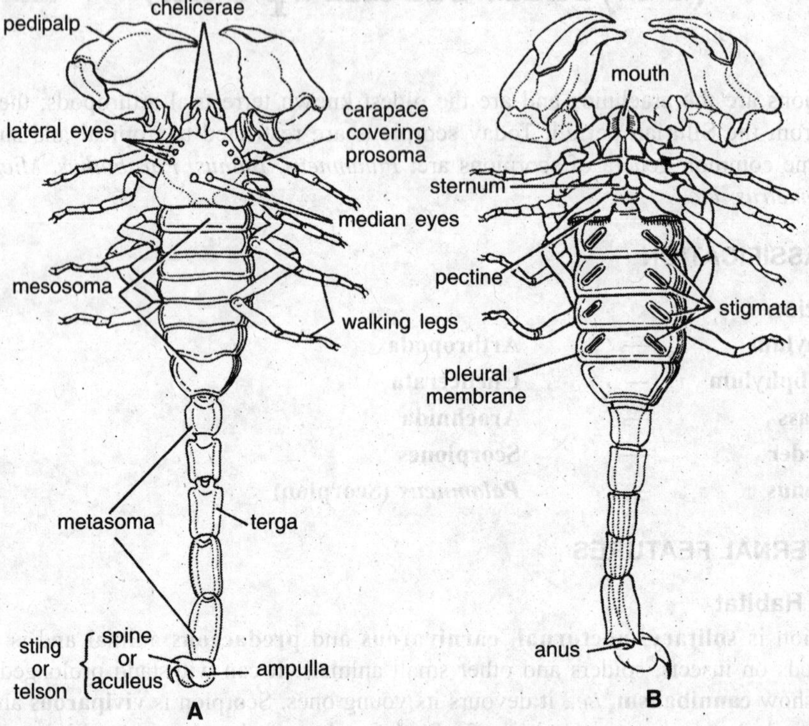

Fig. 8.1. Scorpion. A—Dorsal view; B—Ventral view.

(*ii*) Opisthosoma. The prosoma is followed by a long, distinctly segmented posterior part of body called **opisthosoma**. It is differentiated into two parts: (*a*) an anterior broad **mesosoma** or **preabdomen** and (*b*) a posterior narrow **metasoma** or **postabdomen**.

(*a*) Mesosoma. It consists of **seven** segments. It is as broad as the prosoma anteriorly but slightly narrow posteriorly. Mesosoma lacks appendages but every segment is surrounded by a ring of exoskeleton known as **sclerite**. Each sclerite consists of a dorsal plate, the **tergum**, and a ventral plate, the **sternum**. Both of these plates are joined together by **arthrodial membrane**. On the ventral surface of mesosoma, following structures are visible:

1. Genital operculum. The sternum of first mesosomatic segment is small and bears the median genital aperture which is covered by plate-like, rounded, bifid and movable lid, called **genital operculum**.

2. Pectines (Pectenes). A pair of pectenes are present in the second mesosomatic segment. Each pectene is a comb-like structure consisting of a three-segmented **shaft** or **handle** which is attached to the median line. At its posterior margin, pectene bears a row of 4 to 36 narrow-teeth of a comb. The pectenes are said to be tactile or olfactory organs. These are comparatively larger in males and are curiously modified in females.

3. Stigmata. The sterna of third, fourth, fifth and sixth mesosomatic segments bear a pair of lateral, oblique, slit-like openings, called **stigmata.** Each stigma leads into a pulmonary sac or book-lung.

***(b)* Metasoma.** Mesosoma is followed by metasoma. Metasoma is a cylindrical, narrow (slender), five-segmented structure. It is usually held upwards arching over the back and is erroneously called **tail**. Each segment of metasoma has complete tergal and sternal plates of octagonal shape formed by the complete fusion of terga, sterna and pleurons. These segments are flexibly jointed one behind the other.

The last metastomal segment bears a **sting** or **telson** which consists of two parts—the proximal rounded **ampulla** and the distal pointed recurved spine, the **aculeus**. Inside the ampulla is a pair of **poison glands** enclosed in smooth muscles. The ducts of poison glands open by a pair of minute apertures at the tip of the spine.

External Apertures

1. **Mouth.** It is a small transverse aperture situated antero-ventrally in the prosoma between the cephalic appendages. A small chitinous lobe, called **labrum** hangs above the mouth.
2. **Anus.** It is a small median aperture situated ventrally at the base of the telson.
3. **Genital aperture.** It is situated mid-ventrally in the first segment of mesosoma.
4. **Stigmata.** These are four pairs of slit-like oblique exits of book-lungs, one pair in each mesosomatic segment.
5. **Openings of coxal glands.** These are paired minute openings, one in each coxa of the third pair of legs.
6. **Openings of poison glands.** These are paired openings which, are located on the tips of the spine.

Endoskeleton

Arachnids possess an internal skeleton or endoskeleton in the form of cartilaginous plate, which is called **endosternite**. It is situated inside the body between the prosoma and mesosoma dorsal to nerve cord but ventral to the alimentary canal. In scorpion, it is a triangular structure and produced into several paired processes, the anterior, posterior and horizontal which provide surface for the attachment of muscles. Posteriorly, the body of endoskeleton is fused with a chitinous plate called **diaphragm**, which is perforated by the nerve cord, alimentary canal, muscles and dorsal aorta (Fig. 8.2).

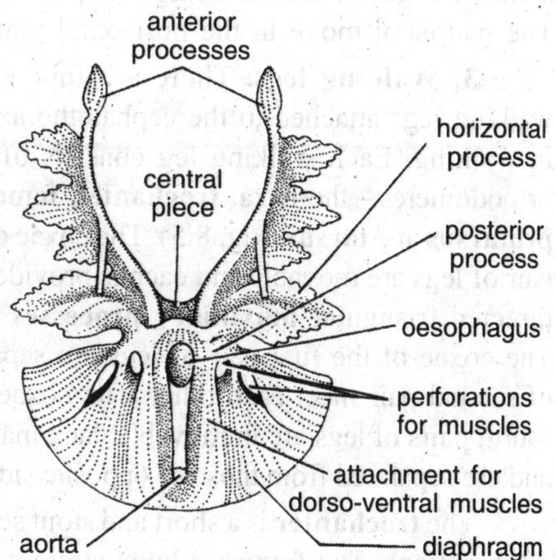

Fig. 8.2. Scorpion. Endoskeleton.

8.3 APPENDAGES

In scorpion, there are six pairs of appendages in the cephalothorax. There are a pair of **chelicerae,** a pair of **pedipalpi** and four pairs of **walking legs.**

1. Chelicerae. These are the anterior-most and pre-oral appendages, lying close to each other on either side of mouth. Chelicerae are homologous to antennae of crustaceans. Each chelicera is a small, three-segmented and chelate structure (Fig. 8.3). The **first**

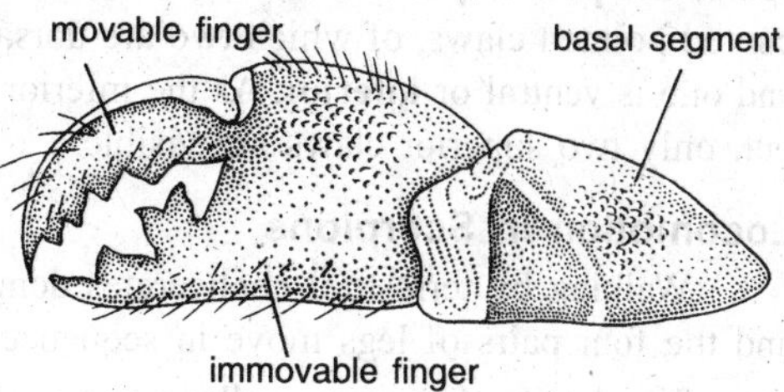

Fig. 8.3. Scorpion. Chelicera.

segment is small, triangular and ring-like and lying concealed beneath the carapace. The **second segment** is large, swollen and bears at its anterior end on the inner side a projection which is pointed and toothed. It forms the **immovable finger** of the chela. The **third segment** arises from the outside of the second segment. It is also curved and toothed and forms the **movable finger** of the chela. The chelicerae are the prehensile appendages of scorpion and are used for holding and tearing the prey.

2. Pedipalpi. The pedipalpi or large pincer claws are a second pair of large powerful and clawed appendages which are post-oral in position. Each pedipalpi is made of six segments, *viz*., **coxa, trochanter, humerus, brachium, manus** and **movable finger** (Fig. 8.4).

(*i*) Coxa. It is small basal segment and is situated on the side of the pre-oral cavity beneath the carapace. Coxa bears towards the mouth a blunt blade-like process, the **gnathobase**. The gnathobases of both sides serve as jaws, protruding in the pre-oral cavity and help in squeezing and crushing the body of the prey.

(*ii*) Trochanter and humerus. Trochanter is also a small segment situated next to coxa. It bears at right angles to the longitudinal axis, a stout and powerful **humerus.**

Fig. 8.4. Scorpion. Pedipalp.

(*iii*) Brachium. It lies above the humerus and is directed forward.

(*iv*) Manus and movable finger. The manus or hand is the longest segment. It forms a powerful chela with the movable finger which is movably articulated on its outer side. Teeth-like serrated edges of movable finger and the immovable finger of padipalpi help in catching the prey. The padipalpi move in the horizontal plane.

3. Walking legs. There are four pairs of identical walking legs attached to the cephalothorax. These are used in walking. Each walking leg consists of seven segments or podomeres—the **coxa, trochanter, femur, patella, tibia, protarsus** and **tarsus** (Fig. 8.5). The coxae of first and second pair of legs are movable and each is provided with forwardly directed triangular **maxilliary processes** or **gnathobases**. The coxae of the first pair of legs are separate while those of second pair meet each other. The coxae of the third and fourth pairs of legs are immovable, lack maxilliary processes and are separated from those of opposite side by the sternum.

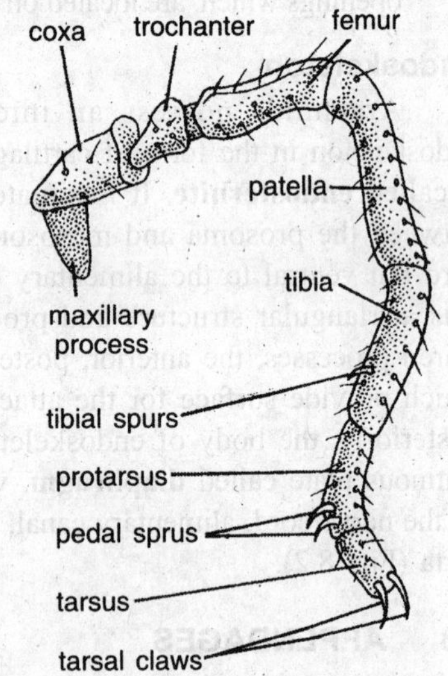

Fig. 8.5. Scorpion. First walking leg.

The **trochanter** is a short and stout segment moving in all directions. The **femur** is long, straight segment moving upward and outward. The **patella** is also long and moves downwards. **Tibia** is elongated and bears a **tibial spur** distally. Protarsus bears a pair of dark tipped claw-like spurs, the **pedal spurs**. The **tarsus** bears three movable curved pointed claws, of which two are dorsal or **superior** and one is ventral or **inferior.** As the inferior claw is worn out, only two superior claws are visible.

Locomotion in Scorpions

Walking by terrestrial chelicerates demands that the body be supported off the substratum and the four pairs of legs move in sequences that maintain the animal's balance.

The details of scorpian walking patterns have been examined by **Root** and **Bowerman** (1978).

During simple forward walking, each of the eight limbs move through the usual power and recovery strokes. In scorpions the joints between the coxae and the body are virtually immovable and do not contribute to the overall motion of limbs. Not all of the walking legs move in the same pattern. The tips of the anterior legs are brought quite high off the ground during their recovery strokes; they may be used to feel ahead as the animal moves (Fig. 8.6A). Also, the tips of each pair of legs extend different distances from the body, allowing for stride overlap without contact between legs (Fig. 8.6B). Leg movements in scorpions do not follow the usual metachronal model. Rather, the typical sequence of movement along one side of the body is leg 4, then 2, then 3, and finally with the legs on the opposite side being generally out of phase. The somewhat staggered overlap of the movement sequences produces a smooth gait movement sequences produces a smooth gait

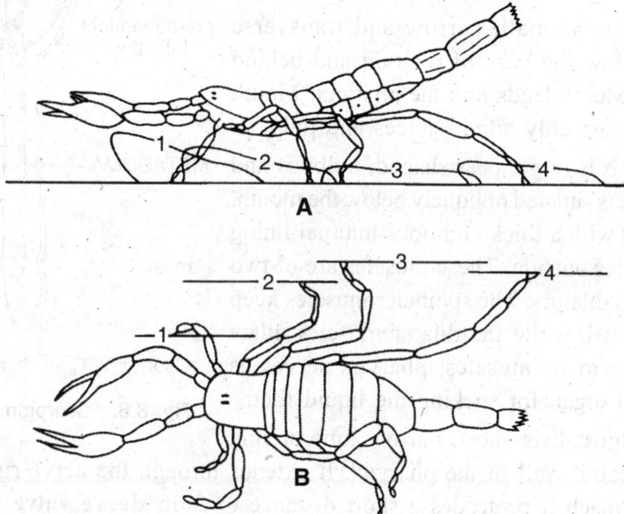

Fig. 8.6. Walking in scorpions. Numbers denote legs 1–4. A—The paths of the leg tips during their recovery strokes. Notice that the anterior legs are lifted higher off the ground than are the posterior legs. B—Distance of each leg tip from the body (after Brusca and Brusca, 2003).

during forward walking, unlike jerking motion of an insect moving by alternating tripod pattern. Like many arthropods, scorpions are also able to change speeds, turn abruptly, walk back-ward and burrow in loose sand.

8.4 DIGESTIVE SYSTEM

1. Alimentary Canal

The alimentary canal (Fig. 8.8) of scorpion is a fairly uniform and straight tube and extends from anterior mouth to posterior anus. It comprises four distinct regions: The **pre-oral cavity, foregut** (stomodaeum), **midgut** (mesenteron) and **hindgut** (proctodaeum).

1. Pre-oral cavity. It is a small cavity formed in front of mouth by the coxae of the first four pairs of appendages of prosoma. It is bounded dorsally by two **chelicerae** and a median **rostrum** which is laterally compressed cushion-like structure. Laterally, the pre-oral cavity is bounded by the **coxae** of pedipalpi and

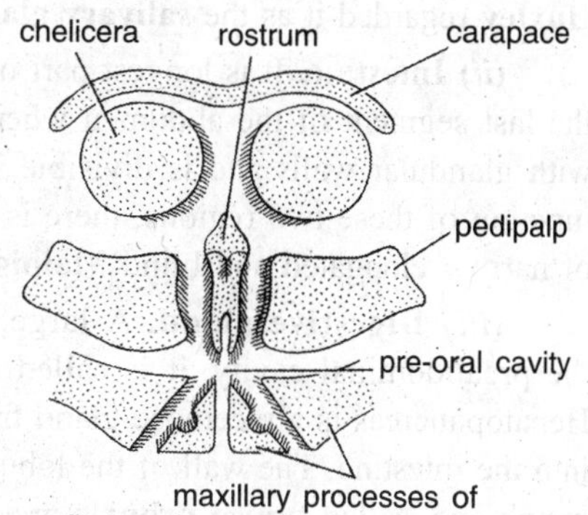

Fig. 8.7. Scorpion. T.S. through the pre-oral cavity.

ventrally its floor is formed by two pairs of **maxillary processes** of first two pairs of walking legs. All these structures bear bristles or hairs which protrude into the pre-oral cavity.

Below the base of the rostrum is located a narrow transverse mouth. Rostrum is also provided with a set of muscles which can roll it into a tube for sucking the fluids into the pharynx.

2. Foregut. It includes **mouth, pharynx** and **oesophagus**. The foregut is internally lined by cuticle.

***(i)* Mouth.** It is a small, narrow and transverse aperture situated below the base of rostrum and behind the pre-oral cavity. Mouth leads into the pharynx. Mouth being a narrow passage only admits juices and pulp.

***(ii)* Pharynx.** It is large, pear-shaped, bulbous and muscular structure. It is situated obliquely below the mouth. Pharynx is provided with a thick chitinous internal lining and numerous muscles outside. These muscles are of two types: sphincters and dilators. The sphincter muscles keep the pharynx contracted, while the dilator muscles dilate it considerably. Due to its muscles, pharynx serves as a powerful suctorial organ for sucking the liquid food.

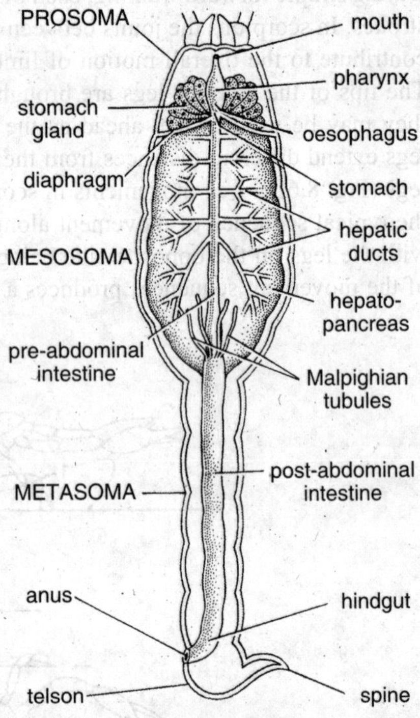

Fig. 8.8. Scorpion. Alimentary canal.

***(iii)* Oesophagus.** It is short, narrow tube arising from the postero-ventral wall of the pharynx. It extends through the nerve ring to open into the stomach. In the stomach it protrudes a short distance to form **sleeve valve** which prevents the food from coming back to the pharynx.

3. Midgut. It includes **stomach, intestine** and **digestive glands**. Midgut is internally lined by epithelium and lacks a lining of chitin.

***(i)* Stomach.** It is slightly dilated thin-walled part of alimentary canal situated in the cephalothorax and extending between the oesophagus and diaphragm. Stomach is covered above and on sides by an irregular brownish gland called **stomach gland**. This gland comprises three lobes—two lateral and one mediodorsal and opens into the stomach by a pair of short lateral ducts. **Huxley** regarded it as the **salivary gland** but **Blanchard** and **Pavlosky** called it stomach gland.

***(ii)* Intestine.** It is longest part of the alimentary canal extending from the diaphragm upto the last segment of the abdomen where it opens into the **hindgut**. The intestine is a wide tube with glandular walls and is divisible into **pre-abdominal** and **post-abdominal regions**. At the junction of these two regions, there is a narrow constriction from where are given out two pairs of narrow elongated and blind **Malpighian tubules**. These tubules are excretory in function.

***(iii)* Digestive gland.** A large, brownish and lobulated gland occupies the whole of the preabdominal cavity, it is called **digestive gland, liver, chylenteron** or **hepatopancreas**. Hepatopancreas is a racemose gland from which five pairs of lateral, narrow **hepatic ducts** open into the intestine. The wall of the lobule is composed of a layer of epithelium arranged on a thin membrane, called **tunica propria** and a peritoneal covering. The cells of the epithelial layer are of two types: 1. **absorptive cells** which are large-sized and numerous; 2. **ferment cells** which are small-sized and less in number.

Box 8.1

According to **Brusca** and **Brusca** (2003), midgut of scorpions bear six pairs of **digestive ceca.** The first pair, called **salivary glands,** lies within the prosoma and produces much of the digestive juice used in preliminary external digestion. The remaining five pairs of digestive ceca are highly convoluted and lie in opisthosoma (*i.e.*, they are so-called hephatopancreas). These ceca produce the enzymes for final digestion and are site of absorption of the digestive products.

4. Hindgut. The hindgut is the smallest and bulbous part of the alimentary canal. It exists only within the last metasomatic segment. It is lined internally by chitin and opens to the exterior by the **anus**. Anus is located at the base of the telson.

Food and Feeding

Scorpions are **carnivorous** and **predaceous** animals. Scorpions feed mostly on insects, although some large species occasionally eat snakes and lizards. Most are nocturnal and detect prey mainly by highly sensitive mechanoreceptors. A fascinating method of **prey location** by the Mojave Desert (of South California) sand scorpion, *Paruroctonus mesaensis,* has been described by **Phillip Brownell** (1977, 1979, 1984). Brownell noticed that *Paruroctonus* ignores both airborne vibrations (*e.g.*, wing beats) and visual input, but responds immediately to nearby prey that are in contact with the sand. This scorpion is even able to detect buried prey, which it immediately uncovers and attacks. Apparently the scorpion senses slight mechanical waves set up in the loose sand by movements of the prey. Special mechanoreceptors in the walking legs of scorpion are stimulated as the waves pass beneath them. The information is processed to determine the direction and approximate distance to the prey source.

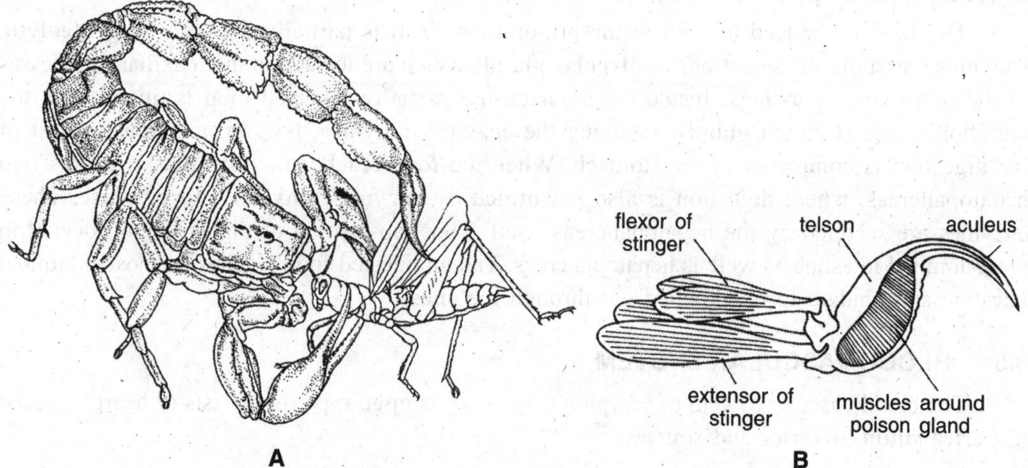

Fig. 8.9. A—The scorpion *Androctonus australis*, applying its stinging apparatus to a grasshopper while holding the prey with its pedipalps. B—The scorpion's telson and stinging apparatus. The telson is normally held flexed; extension drives the aculeus into the prey (after Brusca and Brusca, 2003).

Once a scorpion locates a victim, the prey is grasped by the chelate pedipalps. The opisthosoma is arched over the prosoma, bringing the telson and stinging apparatus into position to inject venom (Fig. 8.9). Muscular contractions force the venom through a pore in the **aculeus** and into the prey. The venom is a **neurotoxin** that can quickly paralyse and kill most prey; in fact, some scorpion venoms can kill large animals, including humans (Box 8.2).

Box 8.2

Perhaps the most familiar of dangerous scorpions are two species that occur in the American Southwest: the bark scorpion (*Centruroides exilicauda*) and the stripe tailed scorpion (*Vaejovis spinigerus*). A north African species, *Androctonus australis,* produces a venom considered to be as potent as that of cobras. Other genera of potentially dangerous scorpions include *Buthus* and *Parabuthus* (Africa and middle east) and *Tityus* (South America) (**H.L. Keegan, 1980**).

Buthus is frequently found in our Indian houses. *Palamnaeus* is the black scorpion that grows to 10–13 cm., in length. *Scorpio, Euscorpio* and *Isometrus* are other common genera. The largest scorpion is the African *Hadogenes troglodytes* which grows to 21 cm.

The neurotoxic venom of scorpions may cause convulsions, paralysis of the respiratory muscles, or cardiac failure. Globally, 5000 people a year are estimated to die from scorpion stings. However, scorpions are immune to their own venom (see **Ruppert** *et al.*, 2004).

The sequence of events in the ingestion of food by scorpions is typical of many arachnids. Once it has been captured and stung, the prey is passed to the chelicerae, which tear it into small bits. The gnathobases grind and mash the food as digestive juices are released through the mouth; this process reduces the food to a semiliquid form. As this organic "soup" is ingested, the hard parts are discarded, and more bits of food are moved between the gnathobases for processing. One chelicera hold the food and other rips it open. The coxae of the pedipalpi and first two pairs of legs now press upon the prey so that the entire body fluid is squeezed out into pre-oral cavity. Meanwhile, the rostrum forms a groove which is applied to the prey and the liquid is sucked into pharynx and then it is driven into oesophagus and stomach.

Digestion, Absorption and Egestion

The food is reduced to pulp in the pre-oral cavity. It is partially digested by **proteolytic enzymes** present in the secretions of **alveolar glands** which are located in the maxillary processes of the first two pair of legs. Inside the stomach the partially digested food is mixed with the secretion of the **stomach gland** containing the enzymes **amylase, trypsin** and **lipase**. Most of the digestion is completed in the stomach. When the food reaches the intestine, it passes into hepatopancreas, where digestion is also performed by enzymes **amylase** and **catalase**. These enzymes are secreted by the hepatopancreas itself. Thus, absorption of digested food occurs in preabdominal intestine as well as hepatopancreas. The undigested food reaches the postabdominal intestine and hindgut to be expelled out through the anus.

8.5 BLOOD VASCULAR SYSTEM

The blood vascular system of scorpion (Fig. 8.10) is **open type**. It consists of heart enclosed in **pericardium**, **arteries** and **sinuses**.

1. Heart and Pericardium

The heart is an elongated and muscular tube lying in groove of the hepatopancreas. It extends back from behind the diaphragm. Heart is incompletely divided into seven distinct chambers by shallow constrictions. Each chamber bears a pair of valvular openings the **ostia** on its dorso-lateral side through which it communicates with the pericardial cavity. Valves of ostia allow flow of blood from the pericardium into the heart but not in the opposite direction. The heart is able to perform slow peristaltic movements due to its muscular walls.

Heart is enclosed in a thin-walled membranous sac, the **pericardium**. Heart is suspended into the pericardium by a pair of connective tissue bands or **ligaments** in each segment, called **ali cordi**.

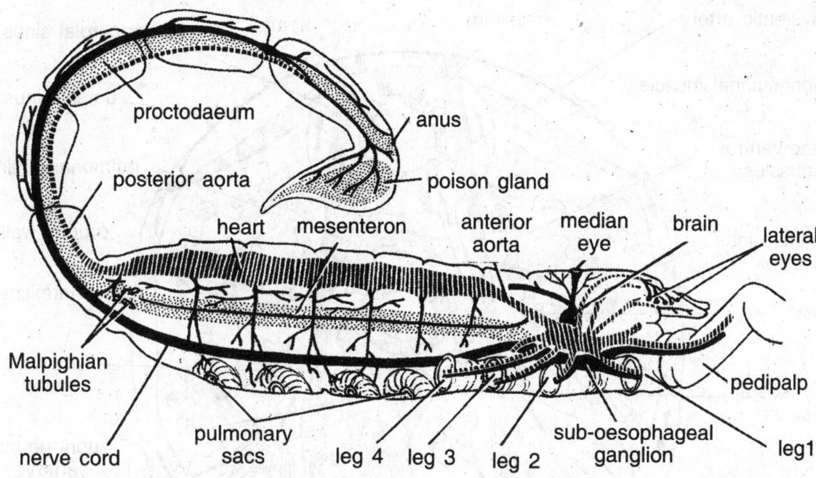

Fig. 8.10. Scorpion. Diagrammatic lateral view of the blood vascular system and other internal organs.

2. Arteries

The heart is continued anteriorly as the **anterior aorta** and posteriorly as the **posterior aorta**. A pair of lateral **systemic arteries** are also given off from each chamber of the heart.

***(i)* Anterior aorta.** It is a short and wide vessel and arises from the anterior end of the first chamber of the heart. It runs along the dorsal surface of the alimentary canal. It sends off the following branches to various organs of prosoma:

(*a*) A pair of **small visceral arteries** to the intestine and hepatopancreas.

(*b*) After piercing the diaphragm, the dorsal aorta gives off a pair of **ventro-lateral arteries** which encircle the oesophagus and unite ventrally to form a **supraneural artery**. Supraneural artery runs medially above the ventral nerve cord supplying the nerve cord and ventral muscles.

(*c*) The anterior aorta also gives of six pairs of arteries which supply the appendages of prosoma, brain and eyes.

***(ii)* Posterior aorta (or caudal artery).** It arises from the posterior end of the heart and runs backward on the dorsal side of intestine upto the last segment. It gives off a pair of arteries in each segment of the postabdomen (metasoma) and then divides into three branches: 1. A **median branch** extends ventrally to join the supraneural artery. 2. Two **lateral branches** run into the posterior lobes of the hepatopancreas and supply the muscles of the fourteenth segment. In telson, posterior aorta terminates into the poison gland.

***(iii)* Systemic arteries.** From each chamber of heart on either side arises a systemic artery. These paired arteries form a network and supply the various parts of preabdomen (mesosoma).

3. Sinuses

All the arteries break up into numerous small branches and sub-branches and pour their blood into a narrow sinus, the **lacunae**. From the lacunae, the blood is collected into larger **sinuses**. There are five sinuses in all–the **dorsal sinus** which lies just above the pericardial sinus; a large **ventral sinus** on the ventral side of the body, two **lateral sinuses** on either lateral side of the body and a **perivisceral** sinus surrounding the alimentary canal. Dorsal sinus is connected to ventral sinus through the lateral sinuses. The perivisceral sinus also communicates with ventral sinus (Fig. 8.10).

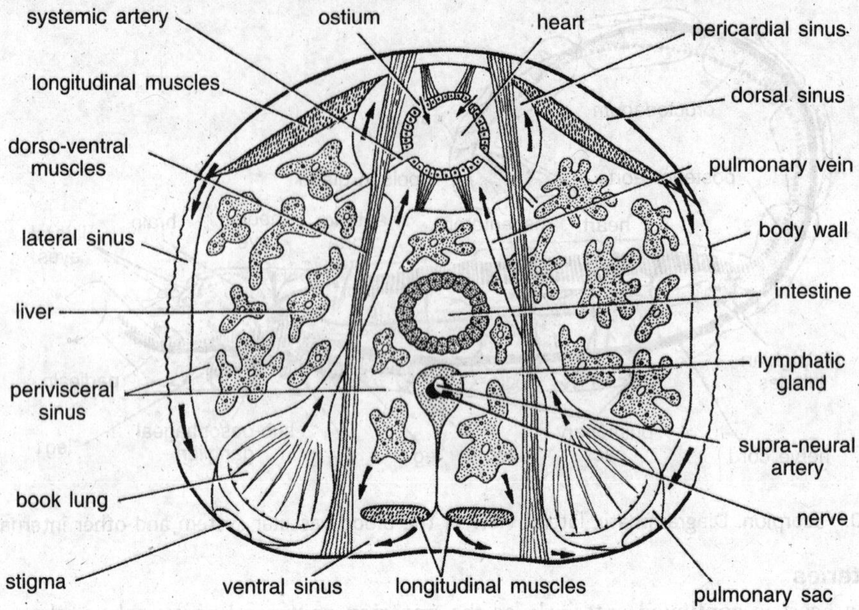

Fig. 8.11. Scorpion. T.S. passing through the middle region of body showing pulmonary and main sinuses of body.

4. Pulmonary Veins

From each pair of book-lungs the oxygenated blood is sent back to the pericardial sinus by a pair of thin-walled **pulmonary veins**. These veins are continued over the dorsal side of the book-lungs as lacunae.

5. Blood

The blood of scorpion is colourless with a bluish tinge. It contains numerous **leucocytes**. The respiratory pigment in blood is **haemocyanin** which is dissolved in the plasma. The blood becomes bluish in colour when oxygenated.

Course of Circulation

With the contraction of ligaments the cavity of the heart enlarges and the blood from the pericardial sinus rushes into the heart through ostia. With the contraction of the heart, ostia are closed and the blood is pumped to various organs of the body through the anterior and posterior aortae and the systemic arteries. The impure (deoxygenated) blood of the body is collected in the ventral

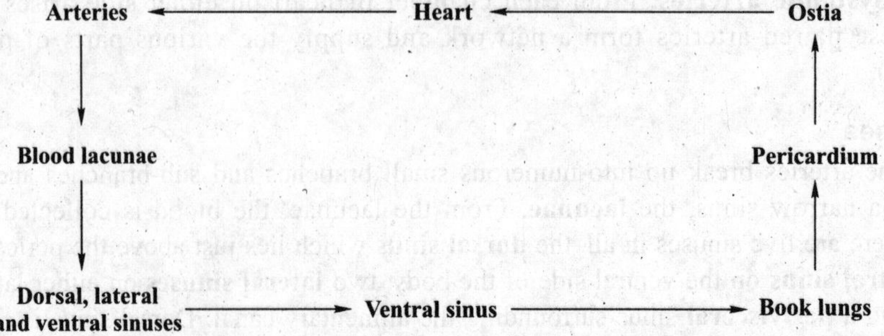

Fig. 8.12. Scorpion. Course of circulation of blood.

sinus. From the ventral sinus blood is sent to the book-lungs and is oxygenated. From the book-lungs pure (oxygenated) blood is carried by the pulmonary veins to pericardial sinus from where it enters the heart through ostia.

8.6 RESPIRATORY SYSTEM

The respiratory system of scorpion is highly characteristic and consists of four pairs of **pulmonary** sacs and **book-lungs.**

Structure of Book-Lungs

The book-lungs are found situated on the ventro-lateral sides in the third, fourth, fifth and sixth mesostomal segments. Each book-lung consists of a compressed sac-like cavity (the pulmonary sac) formed by the invagination of the body wall. This cavity is lined with a thin-layer of cuticle and is divided into two parts: the atrial chamber and the pulmonary chamber (Fig. 8.13).

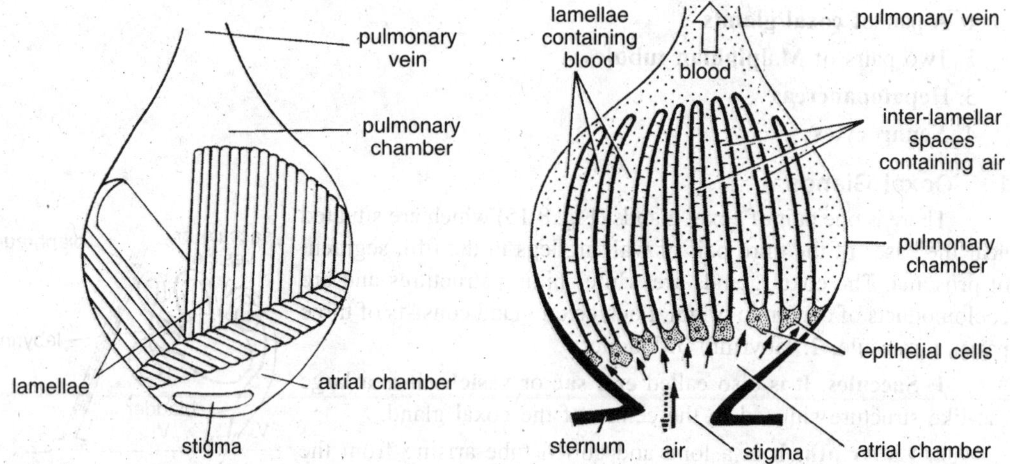

Fig. 8.13. Scorpion. A book-lung.

Fig. 8.14. Scorpion. Vertical section of book-lung to show the mechanism of respiration.

1. Atrial chamber. It is the proximal ventral part of the book-lung. It is in the form of small and dorsoventrally compressed air-chamber. Proximally atrial chamber opens to exterior by a narrow, oblique, slit-like opening, called **stigma**, which is placed ventrolaterally on the sternum. Distally the roof of atrial chamber is perforated by a series of minute slit-like openings or **ostia**. These ostia are placed parallel to one another. Through ostia, the atrial chamber communicates with the **interlamellar air spaces** of the pulmonary chamber.

2. Pulmonary chamber. It is distal part of the cavity of book-lung which is called **pulmonary chamber**. It is dorsal spacious oval chamber filled with about 150 vertical folds or leaf-like **lamellae**, attached to the posterior side of the chamber. The lamellae are arranged parallel to each other and one over the other like the leaves of a book. Each lamella is a hollow structure made of thin layer of cuticle united at their edges and thus enclosing **intralamellar spaces** filled with blood. The outer side of the lamellae bear ridges and bristles which keep the adjacent lamellae apart and little air space, called **interlamellar space,** is left between them for the flow of air. These spaces communicate with the atrial chamber through ostia (Fig. 8.14).

Blood supply. The deoxygenated (impure) blood is sent to each book-lung from the ventral sinus by diverticulum; from where the blood gets filled in the **intralamellar space** of the lamellae. The blood is oxygenated or purified in the lamellae of each book-lung and is collected into a **pulmonary vein** which travels dorsally and opens into the pericardium.

Mechanism of Respiration

The inspiration and expiration of air in the book-lungs is controlled by the action of the, dorso-ventral and atrial muscles. When the **dorso-ventral muscles** are contracted, the pulmonary chamber is compressed and the air filled in interlamellar air spaces is forced out into the atrial chamber. From atrial chamber this air is forced to outside through the stigma due to contraction of the **atrial muscles**. When the atrial muscles relax, the book-lungs resume their normal shape and the fresh air enters through stigma first into the atrial chamber and then into the interlamellar spaces. At this time dorso-ventral muscles relax and pulmonary chamber resumes its normal shape. The **gaseous exchange** takes place through the thin membranous walls of the lamellae and the blood becomes oxygenated and carbon dioxide is expelled out in the air.

8.7 EXCRETORY SYSTEM

Excretory system of scorpion includes the following types of excretory organs:

1. A pair of **coxal glands**
2. Two pairs of **Malpighian tubules**
3. **Hepatopancreas**
4. **Nephrocytes**

1. Coxal Glands

There is one pair of coxal glands (Fig. 8.15) which are situated near the bases of the third pair of walking legs in the fifth segment of prosoma. The coxal glands are white shining structures and are coelomoducts of coelomic origin. Each coxal gland consists of three parts: 1. saccule, 2. labyrinth 3. bladder.

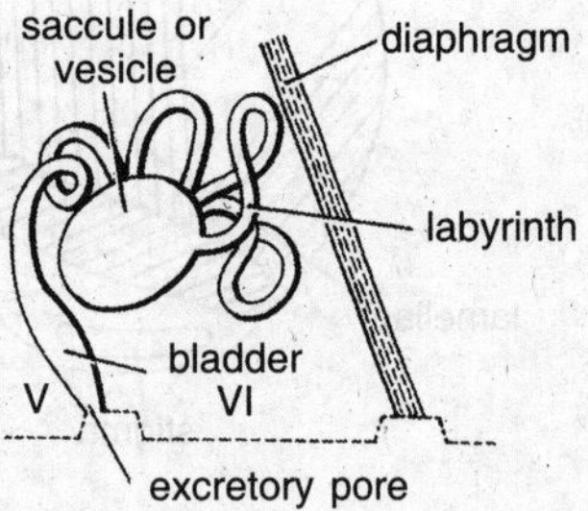

Fig. 8.15. Scorpion. Coxal glands.

1. Saccules. It is also called **end sac** or **vesicle.** It is a large sac-like structure situated in the centre of the coxal gland.

2. Labyrinth. It is a long and coiled tube arising from the saccule and opening in the bladder.

3. Bladder. It represents the swollen terminal part of labyrinth which has enlarged to take the form of a reservoir. It opens to exterior by a small opening situated on the posterior face of the coxa of fifth walking leg.

The saccule and labyrinth of each coxal gland collect excretory nitrogenous wastes from the blood and pass them outside through the excretory pore. **Urate crystals** have been found in the saccule. The coxal glands of scorpion are homologous to green glands or antennary glands of crustacea.

2. Malpighian Tubules

Two pairs of Malpighian tubules are present at the junction of preabdominal and postabdominal parts of intestine. These are endodermal in origin and lie freely in the blood from which they isolated the nitrogenous products (*i.e.*, **guanine crystals**) and discharge them into the intestine. Malpighian tubules possess thin syncytial walls which remove the waste from the blood.

3. Hepatopancreas

Hepatopancreas is regarded as an excretory organ by **Pavlvosky**. It has been found that particles of ammonical carmine injected into the body cavity of scorpion are accumulated in the form of red granules inside cells of hepatopancreas. **Pearson** believes that ferment cells of hepatopancreas pick up excretory products.

4. Nephrocytes

Large nephrocytes and lymphatic organs are specialized excretory structures. Both structures are found under the body wall in the mesosoma. These are believed to be excretory as well as phagocytic in function.

8.8 NERVOUS SYSTEM

Nervous system consists of **central nervous system, peripheral nervous system** and **visceral nervous system**.

1. Central Nervous System

The central nervous system comprises of the (*i*) brain or cerebral ganglion; (*ii*) circumoesophageal connectives; (*iii*) sub-oesophageal ganglion and (*iv*) ventral nerve cord (Fig. 8.16).

(*i*) Brain. The **brain** or **cerebral ganglion** is a small bilobed mass situated in the prosoma just beneath the median eyes in the prosoma. Embryological studies have shown that brain develops from the fusion of eight pairs of ganglia.

(*ii*) Circum-oesophageal connectives. From the brain are given off a pair of thick, short, stout **circumoesophageal connectives** which form a collar around the oesophagus.

(*iii*) Sub-oesophageal ganglion. The circum-oesophageal connectives after forming a collar around the oesophagus unite ventrally in a **sub-oesophageal ganglion**.

(*iv*) Ventral nerve cord. The sub-oesophageal ganglion is continued backwards into the abdomen as a double **ventral nerve cord**. It runs upto the fourth segment of the post-abdomen. The nerve cord bears three ganglia in the pre-abdomen and four in the post-abdomen. Ventral nerve cord is slender and rounded in the pre-abdomen and flattened and ribbon-shaped in the post-abdomen.

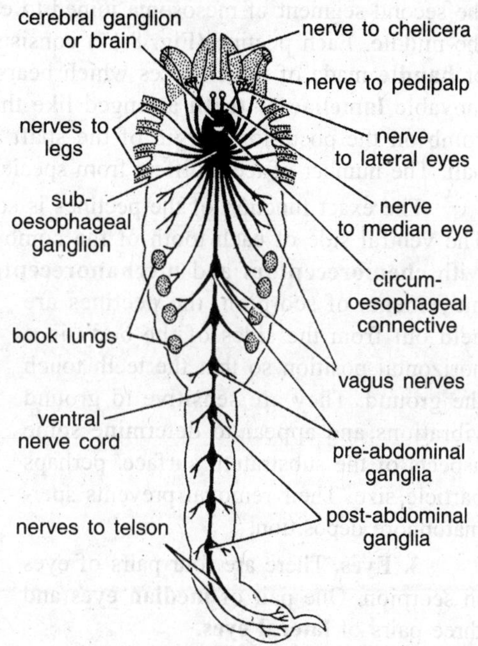

Fig. 8.16. Scorpion. Nervous system.

2. Peripheral and Visceral Nervous System

Several nerves are given out from the brain, circumoesophageal connectives and ganglia of the ventral nerve cord, which constitute these nervous systems.

Brain gives off a pair of **optic nerves** to the *median* eyes, a pair of nerves which branch and supply the *chelicerae* and the *lateral eyes*, a pair of nerves to the *rostrum*, a pair of nerve trunks to the *pedipalpi* and four pairs of nerves to the four pairs of *walking legs*. Besides these, the brain also supplies nerves to the *pharynx, oesophagus* and *heart*. The suboesophageal ganglion may send two to four pairs of **vagus nerves** which run posteriorly into the preabdomen to supply the *genital operculum*, the *pectines* and first two pairs of *book-lungs*. The ganglia of ventral nerve cord of preabdomen send out nerves to the last two pairs of *book-lungs*, *heart*, and *muscles* of abdomen, while those from the post-abdomen supply nerves to the *segmental muscles*. The last ganglion also supplies the sting.

8.9 SENSE ORGANS

The sense organs of scorpion include **sensory setae** or **sensillae, pectines** and **eyes**.

1. Sensory sensillae. These are fine hair-like cuticular structures which are sparsed all over the body. These are much longer in legs and tail. The sensillae are **tactile organs** of scorpion and they are extremely sensitive to touch. Each sensory sensilla is connected to a sensory cell which also remains connected at its base with a nerve fibre. Sensillae are long and bristle-like on the podomeres of pedipalpi.

2. Pectines. The pectines are a pair of modified comb-like appendages lying attached to the sternum of the second segment of mesosoma joined to each other in the middle. Each pectine (Fig. 8.17) consists of a **shaft** or **handle** made of three pieces which bears a series of movable **lamellae** or **teeth** arranged like the teeth of a comb on the posterior margin of the shaft in the outer half. The number of teeth differs from species to species.

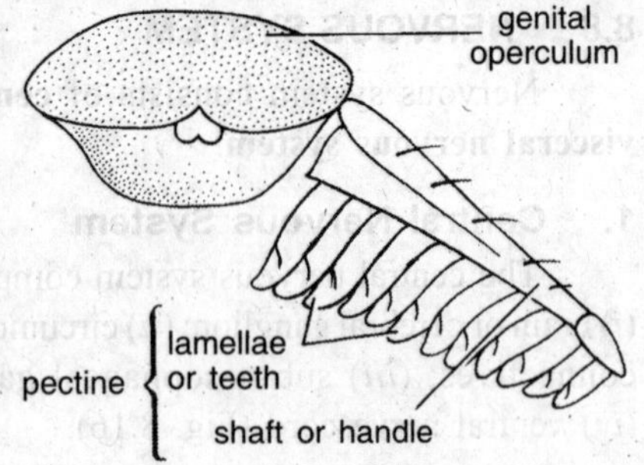

Fig. 8.17. Scorpion. Pectine.

The exact function of the pectines is still not clear. The ventral side of each tooth of the comb is provided with **chemoreceptors** and **mechanoreceptors**. During movements of scorpion, the pectines are held out from the sides of the body in a horizontal position so that the teeth touch the ground. They are sensitive to ground vibrations and appear to determine some aspects of the substratum surface, perhaps particle size. Their removal prevents spermatophore deposition.

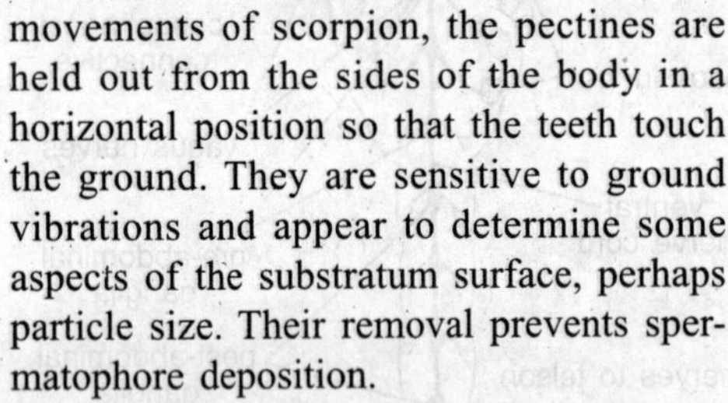

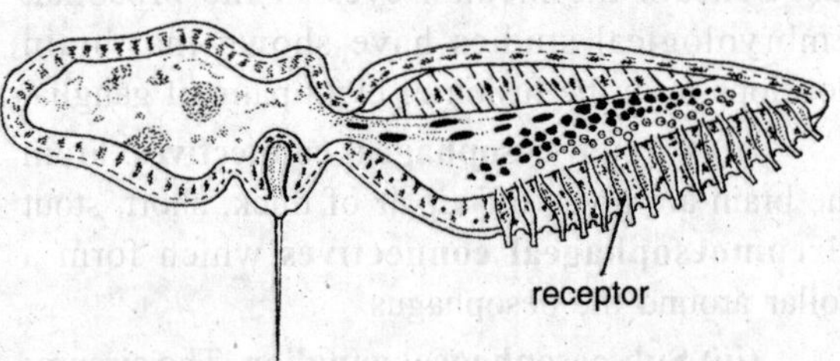

Fig. 8.18. Scorpion. T.S. through pectine.

3. Eyes. There are four pairs of eyes in scorpion. One pair of **median eyes** and three pairs of **lateral eyes**.

(*i*) Median eyes. These are cup-like in shape covered externally by a cuticular **lens** or **cornea**. Inside the pigmented cup are **rhabdomes**. Each rhabdome encloses several **retinal cells** supplied by the nerve fibres of optic nerve (Fig. 8.19).

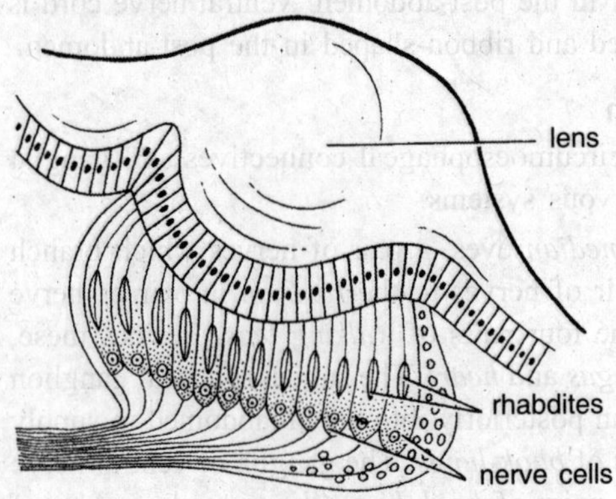

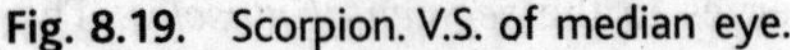

Fig. 8.19. Scorpion. V.S. of median eye.

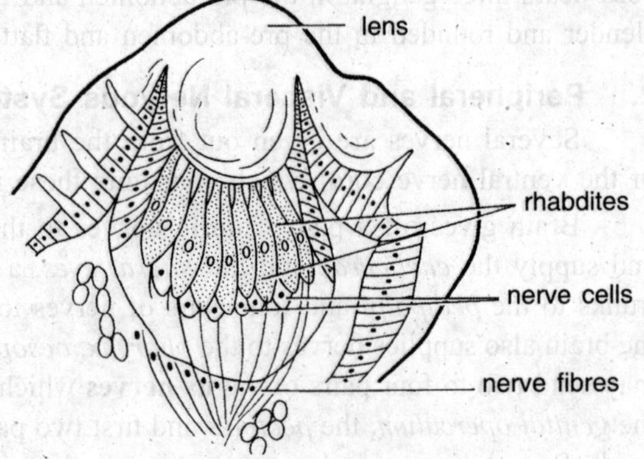

Fig. 8.20. Scorpion. V.S. of lateral eye.

The median eyes of scorpion are said to be intermediate between compound eyes and simple eyes of insects. These resemble the compound eye in having their retinal cells arranged in groups around each rhabdome as in ommatidia. But unlike the compound eyes of insects and crustaceans, the sensory retinal cells serve the function both of receiving the stimuli and transmitting the impulses.

***(ii)* Lateral eyes.** These are similar to simple eyes of insects. These are like a pigmented cup covered externally by a **biconvex lens** formed of transparent cuticle. Inside the cup are several longitudinal optic rods called **rhabdomes** associated with **retinal cells** or **retinulae**. The retinal cells receive nerve fibres (Fig. 8.20).

8.10 REPRODUCTIVE SYSTEM

Sexes are **separate** and scorpions exhibit some **sexual dimorphism**. Females can be identified from males on the following bases:

1. Females are comparatively larger in size than males.
2. The abdomen in females is broader.
3. The padipalpi are smaller in females and longer in males.
4. The pectines are smaller in females having fewer teeth. They are longer in males.
5. Post-abdominal (tail) is short and curved in female.
6. Males can be recognized easily by the hooks which are present on the opercular plates.

Male Reproductive System

Male reproductive system of scorpion includes the following organs (Fig. 8.21):

1. A pair of testes
2. A pair of vasa deferentia
3. A pair of seminal vesicles
4. Genital chambers
5. Accessory glands
6. Paraxial organs

1. Testes. These are elongated ladder-like structures lying embedded in the hepatopanereatic mass in the mesosoma from third to sixth segment. Each testis consists of two slender, thread-like whitish **longitudinal tubules** lying parallel and joined to each other by four **transverse tubules** (the **cross branches**) to form three squares. Each tubule is lined internally by a single layer of germinal epithelium, the cells of which give rise to spermatozoa. The sperms are arranged in bundles, the **spermatophores**. A mature sperm of scorpion is filiform and motile having an oval body and a long tail.

2. Vas deferens. From, the outer angle of each testis arises a narrow duct, the **vas deferens**, which runs forward and opens into **genital chamber** of its side. Prior to its opening, it widens to form **terminal ampulla**. The terminal ampulla receives opening of accessory glands and seminal vesicles.

3. Seminal vesicles (or Vesicula seminalis). It is small club-shaped organ of yellow colour. It stores the mature sperms.

4. Genital chambers. The paired genital chambers are confined to the first and second mesosomatic segments. These run obliquely towards the middle and open into a small median **common genital chamber**. The latter opens to the exterior through **male genital aperture** situated on the ventral surface of first mesosomatic segment and covered over by the bifid genital operculum.

5. Accessory glands. These are four pairs of glands which developed in relation with male reproductive organs. These are **dorsal annex glands, ventral annex glands, oval glands** and **cylindrical glands**. All of them discharge their contents into the terminal ampulla of vas deferens. The fluid secreted by these glands is helpful in reproduction.

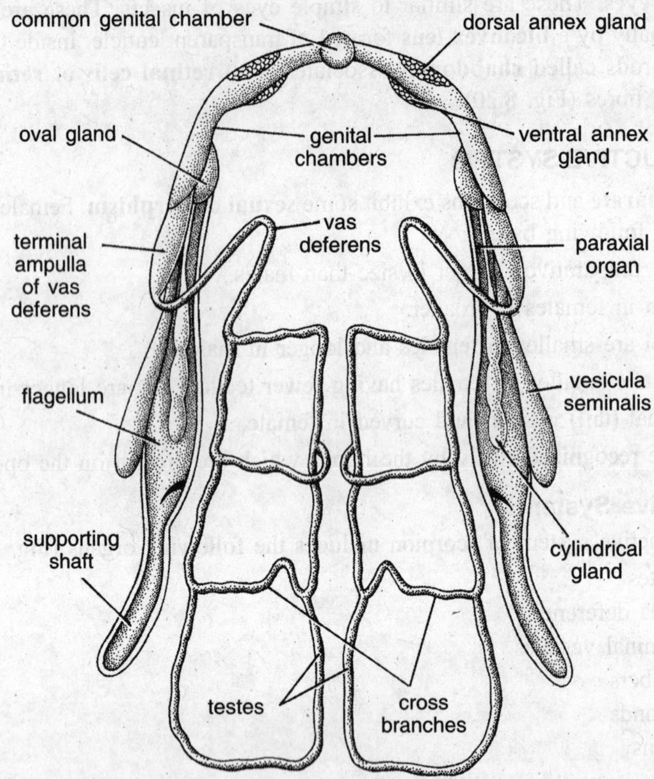

Fig. 8.21. Scorpion. Male reproductive organs.

6. Paraxial organs. Each genital chamber extends backwards on the outer side of testis into a tubular sac, the **paraxial organ**. It consists of an elongated chitinous rod, the **flagellum** which closely fits into its cavity. Flagellum consists of a supporting **shaft** with a longitudinal groove inside it. Few spines are present on the one side of the flagellum. The flagella of both sides can protrude out of the genital operculum and constitute the **double penis** of scorpion. These act as **claspers** during copulation.

Female Reproductive System

The female reproductive system of scorpion consists of following organs:

1. Ovary
2. A pair of oviducts
3. Common genital chamber

1. Ovary. The ovary is a single median structure situated in mesosoma from third to sixth segments embedded in the hepatopancreas. It consists of three elongated narrow **longitudinal tubules** known as **ovarioles**. Of them the smaller one is the median and two large ones are the lateral ovarioles. All ovarioles are connected by four hollow cross branches or the transverse **connectives**, so that the ovary assumes a ladder-like appearance. The ovarioles and their cross branches are lined by germinal epithelium, the cells of which form ova. The mature ova project into hollow bud-like processes on the surface called the **diverticula** or the **follicles** (Fig. 8.22).

In scorpions, as development of a fertilized egg takes place inside the follicle, the latter forms an elongated structure now called **uterus** (Fig. 8.23). The uterus is differentiated into a proximal swollen region or **embryonic chamber** and a distal narrow **appendix** terminating into an enlarged tip, called the **head of appendix**. The follicles are fusiform in *Palamneus* and spherical in *Buthus*.

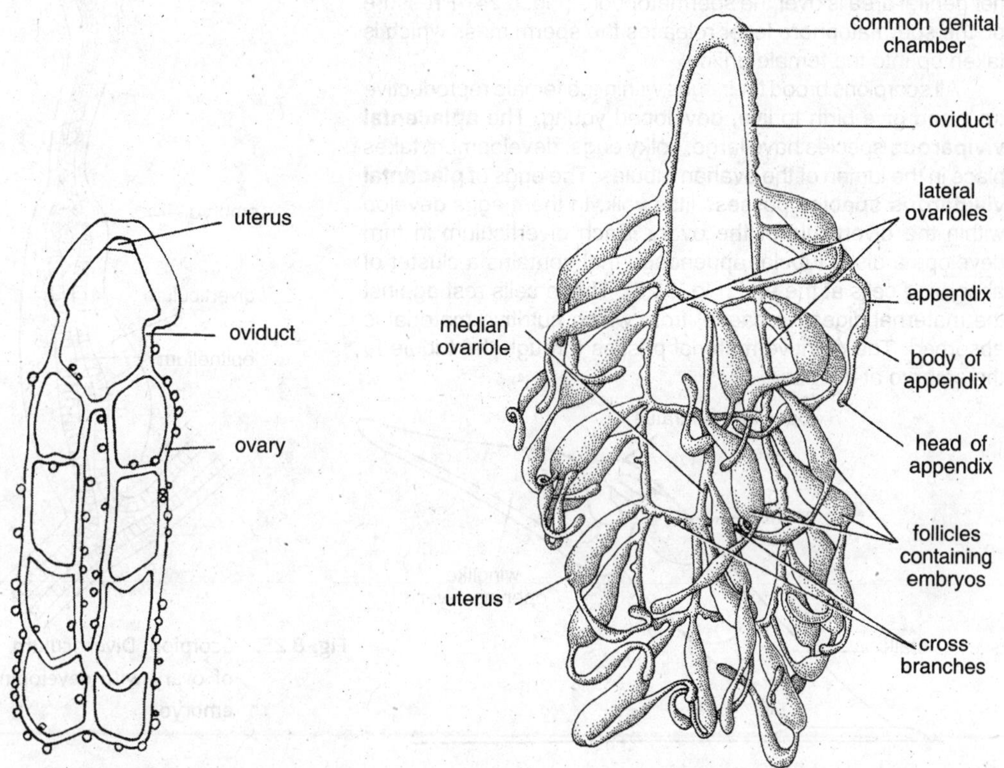

Fig. 8.22. Scorpion. Female reproductive system.

Fig. 8.23. Scorpion. Ovary with mature follicles.

2. Oviducts. The two ovarioles continue anteriorly as the **oviducts** which converge towards the mid-ventral line to open into a small median **common genital chamber**. The common genital chamber opens to outside by the **female genital aperture** situated beneath the bilobed operculum on the sternum of the first preabdominal segment.

Courtship and Development

Scorpions carry on a long courtship dance. Eventually the male persuades the female to retire with him into a burrow (see Box 8.3) In India, scorpions breed during monsoon.

In scorpions, the development is **internal** and very slow. Scorpions are viviparous. The female scorpions give birth to 2 to 95 youngs. The young ones resemble the adults. For some time the mother carries the newly hatched young ones on its back and takes care of them. The young one undergo several moults to grow into the adults.

Box 8.3 Courtship Dance of Scorpions

During the breeding season, the male scorpion roams about and on contacting a female initiates an extended courtship. In some species the male and female face each other; each extends its abdomen high into the air and moves about in circles; in others, the male rocks. The male then seizes the female with his pedipalps, and together they walk backward and forward (called **promenade**). This behaviour may last from 10 min to hours, depending on how long it takes to locate a suitable site for **spermatophore deposition**. Eventually, the male deposits a spermatophore, which he attaches to the ground. A wing-like lever extends from the spermatophore. The male then manoeuvres the female so that her genital area is over the spermatophore (Fig. 8.24). Pressure on the spermatophore lever releases the sperm mass which is taken up into the female orifice.

All scorpions brood their eggs within the female reproductive tract and give birth to live, developed young. The **aplacental viviparous** species have large, yolky eggs; development takes place in the lumen of the ovarian tubules. The eggs of **placental viviparous species** possess little yolk. In them eggs develop within the diverticula of the ovary. Each diverticulum in turn develops a distal tubular appendage that contains a cluster of absorbing cells at the end (Fig. 8.25). These cells rest against the maternal digestive caeca, from which nutritive material is absorbed. The nutritive material passes through the tubule to the embryo at the base.

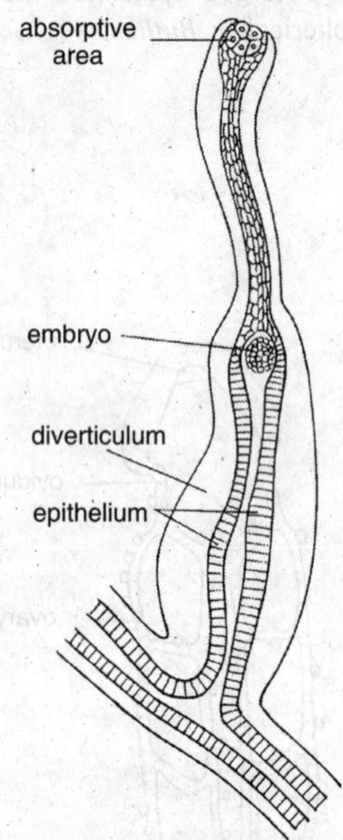

Fig. 8.25. Scorpion. Diverticulum of ovary with developing embryo.

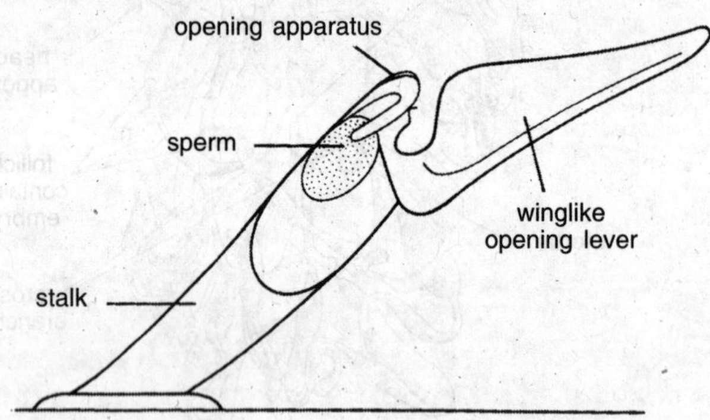

Fig. 8.24. Scorpion. Spermatophore.

QUESTIONS

Long Answer Questions

1. Describe the external features of scorpion.
2. Describe appendages of scorpion.
3. Describe the alimentary canal, mechanism of feeding and digestion of scorpion.
4. Give an account of blood vascular system of scorpion.
5. Give an account of the parts that subserve respiration in scorpion.
6. Give an account of the respiratory organs and mechanism of respiration of scorpion.
7. Describe the excretory organs of scorpion.
8. Write an account of coxal glands of scorpion.
9. Describe the nervous system and sense organs of scorpion.
10. Describe sense organs of scorpion.
11. Describe the male and female reproductive systems of scorpion.
12. Give an illustrated account of the reproductive system of scorpion.
13. Give an illustrated account of female reproductive organs of scorpion.

Short Answer Questions

1. Write functions of chelicerae.
2. Describe excretory organs of scorpion.
3. Describe the mechanism of respiration in *Palamneus*.
4. What is pectine in scorpion? Give its importance to the animal.
5. Define the chelicera.
6. Write short notes on the following:
   (*i*) Pedipalpi
   (*ii*) Endoskeleton of scorpion
   (*iii*) Book-lungs
   (*iv*) Coxal glands of scorpion
   (*v*) Pecten of scorpion
   (*vi*) Chelicerae
   (*vii*) Classification of scorpion
7. Draw well labelled diagrams of the following:
   (*i*) External features of *Palamneus*
   (*ii*) Dorsal view of *Palamneus*
   (*iii*) Ventral view of *Palamneus*
   (*iv*) Appendages of *Palamneus*
   (*v*) Coxal gland of scorpion
   (*vi*) Male reproductive organs of scorpion
   (*vii*) Female reproductive organs of scorpion
   (*viii*) Pectine

Very Short Answer Questions

1. How many divisions of body occur in a scorpion?
2. How many appendages does a scorpion have?
3. What is the function of coxal glands?
4. What kind of vascular system is present in scorpion?
5. How many pairs of eyes are present in scorpion?
6. What are pectines of scorpion?
7. What is cannibalism?

Multiple Choice Questions

1. *Palamnaeus* (scorpion) belongs to class Arachnida because of
   (*a*) four pairs of legs (*b*) chelate legs
   (*c*) sting (*d*) none of these
2. Scorpions are
   (*a*) sanguivorous (*b*) carnivorous
   (*c*) herbivorous (*d*) omnivorous
3. Respiratory organs of scorpion are
   (*a*) lungs (*b*) gills
   (*c*) book-lungs (*d*) ctenidia

4. In how many podomeres are the walking legs of scorpions divided?
(*a*) 5 (*b*) 7
(*c*) 9 (*d*) 11

5. Solid processes of chitin in scorpion are:
(*a*) dermatida (*b*) coelodermatidia
(*c*) epiostracum (*d*) telostracum

6. Hollow processes of chitin in scorpions are
(*a*) hypostracum (*b*) coelodermatida
(*c*) coelostomata (*d*) tracheinida

7. The heart chambers in *Palamnaeus* are
(*a*) 1 (*b*) 3
(*c*) 5 (*d*) 7

8. Respiratory pigment in blood of scorpion is
(*a*) haemocyanin (*b*) erythrocyanin
(*c*) haemoglobin (*d*) none

9. The excretory system of scorpions includes
(*a*) malpighian tubules (*b*) coxal glands
(*c*) hepatopancreas (*d*) nephrocytes
(*e*) all the above

10. A characteristic feature of scorpions is
(*a*) four pairs of walking legs (*b*) vivipary
(*c*) chelicerate (*d*) pedipalpi

ANSWERS

Very Short Answer Questions

1. Two: prosoma or cephalothorax and opisthosoma or abdomen. Opisthosoma includes broad mesosoma and narrow metasoma.

2. In scorpion, there are six pairs of appendages in the cephalothorax (prosoma). Appendages of scorpion include a pair of **chelicerae**, a pair of **pedipalpi** and four pairs of **walking legs**.

3. Excretion of nitrogenous wastes.

4. Open type.

5. Four pairs of eyes (a pair of median eyes and three pairs of lateral eyes).

6. Scorpions have a pair of large, unique, comb-like structures, the pectines, on the ventral surface of the mesosoma. After detailed studies of their innervation, **Foelix** and **Schabronath** (1983) suggested that the pectines act as both mechanoreceptors and chemoreceptors. Other works has shown them to be capable of detecting slight differences in sand grain size (see **Brusca** and **Brusca**, 2003).

7. An animal that devours members of its own species, is called **cannibal.** The phenomenon is called **cannibalism** and is found in scorpions.

Multiple Choice Questions

1. (*a*) **2.** (*b*) **3.** (*c*) **4.** (*b*) **5.** (*a*) **6.** (*b*)
7. (*d*) **8.** (*a*) **9.** (*e*) **10.** (*b*).

9

Zoological Importance of *Limulus*

Limulus polyphemus, the American Horse-shoe crab or king crab belongs to the subclass Xiphosura and class Merostomata of subphylum Chelicerata, Phylum Arthropoda. It is a common marine animal of the Atlantic and Gulf Coasts of North America and a favorite laboratory animal of physiologists in North America. *Limulus* inhabits shallow marine waters, generally on clean sandy bottoms. It is a burrowing, nocturnal and omnivorous animal. It can walk with legs, swim with mesosomatic appendages and leap with telson. *Limulus* often crawls about or burrows just beneath the surface, preying on other animals or scavenging. It feeds upon molluscs, worms and bottom algae. Adults attain sexual maturity in 9 to 12 years and has a life-span of 19 years.

Limulus is a large-sized arthropod, measuring upto 60 cm long (Fig. 9.10). Dorsoventrally flattened body is composed of an anterior **prosoma** (6 fused segments) and a posterior **opisthosoma** (9 segments). Prosoma is covered by a large horse shoe-shaped **carapace** marked by two lateral longitudinal grooves and bearing the median simple eyes (ocelli) and two lateral compound eyes. Ventrally, prosoma bears a pair of 3-jointed preoral **chelicera** followed by five pairs of 6-jointed chelate legs. Each walking leg is formed of seven articles (coxa, trochanter, femur, patella, tibia, tarsus, pretarsus) the last two of which form the chelae. The coxal endites of the pedipalps and first three walking legs are modified as **gnathobases**. Arising from the coxae of the fourth walking legs are tiny appendage-like process called **flabella**, which function as gill cleaners. In addition, just posterior and medial to the last walking legs is a pair of reduced appendages called **chilaria**. Their function is unknown, and there is some controversy about their evolutionary significance–some specialists believe that they may reflect an additional opisthosomal segment (see **Brusca** and **Brusca**, 2003). In male *Limulus*, first pair of walking legs form clasping organs.

The broad hexagonal opisthosoma is divisible into an anterior **mesosoma** (6-segments) covered by a dorsal shield, and a posterior **metasoma** (3-segments) ending in a long pointed **telson**. Mesosoma bears 6 pairs of thin, broad appendages fused in median line. Last five pairs carry 5 pairs of **book gills**. Excretion takes place through a pair of **coxal glands**. Sexes are separate.

Mating and Development

On the onset of the breeding season, horseshoe crabs migrate into the shallow water of protected bays and estuaries. On the east coast of North America, this migration takes place in the spring and summer, and huge numbers of *Limulus* can be seen gathering near the shore in preparation for mating. **Mating** is initiated when the male climbs onto the back of female and grasps her with his modified first walking legs. The clasped pair moves to shallow water, usually at a high spring tide, and the female excavates one or more shallow depressions in the sand and deposits her eggs (2000 to 30,000 eggs per mating). The male releases sperms directly on to the eggs as they are deposited. Then the mates separate, and the females covers the fertilized eggs with sand.

Limulus has indirect development. Its early development takes place in the sand or mud "nest". Cleavage is holoblastic, producing a **stereoblastula** with most of the yolk contained in the inner cells. As development continues, the surface cells at the anterior and posterior ends of the embryo divide rapidly, forming two germinal centers. Some of these rapidly proliferating cells migrate inward as the presumptive endoderm and mesoderm. The anterior germinal center gives rise to the first four segments of the prosoma and the posterior center to the rest of the body. All of the prosomal segments fuse and are eventually covered by the developing carapace-like dorsal-shield.

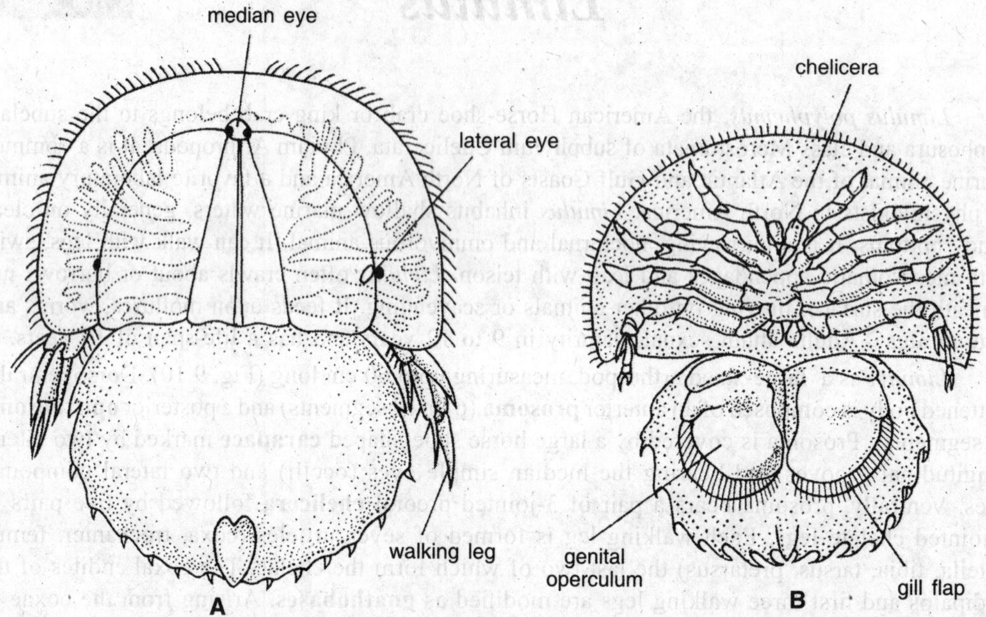

Fig. 9.1. *Limulus*. Euproops larva. A—Dorsal view; B—Ventral view (after Brusca and Brusca, 2003).

As the yolk reserves become depleted, the embryo develops into a **euproops larva** (or **"trilobite larva"**), so named because of its resemblance to the fossil carboniferous xiphosusran *Euproops* (which superficially resembled trilobites; Fig. 9.1). The larvae swim about and periodically burrow in the sand. Segments are formed and appendages added through a series ofmolts until the adult form is reached. Zoologists view this developmental pattern as a mixed life history. The early developmental stages are supplied with an investment of yolk by the female and are protected by the nest she constructs, but the youngs emerge as independent feeding larvae prior to maturation.

9.1 EVOLUTIONARY SIGNIFICANCE

Limulus or king crab is not a crab at all but a primitive marine arthropod resembling the scorpions and spiders. It is an evolutionary relic which has changed very little during last 200 million years. It is a **"living fossil"** whose close relatives became extinct many million years ago. *Limulus* is found to be product of **phyletic evolution** or **anagenesis** (Fig. 9.2)

Box 9.1 Anagenesis

According to Darwin's natural selection, a single species can change through time, such successional changes which occur in a single lineage are called **phyletic evolution** or **anagenesis**. It explains evolution of sub-species level (microevolution) and is contrasted with **cladogenesis** or **phylogenetic branching** (*e.g.*, evolution of cat family).

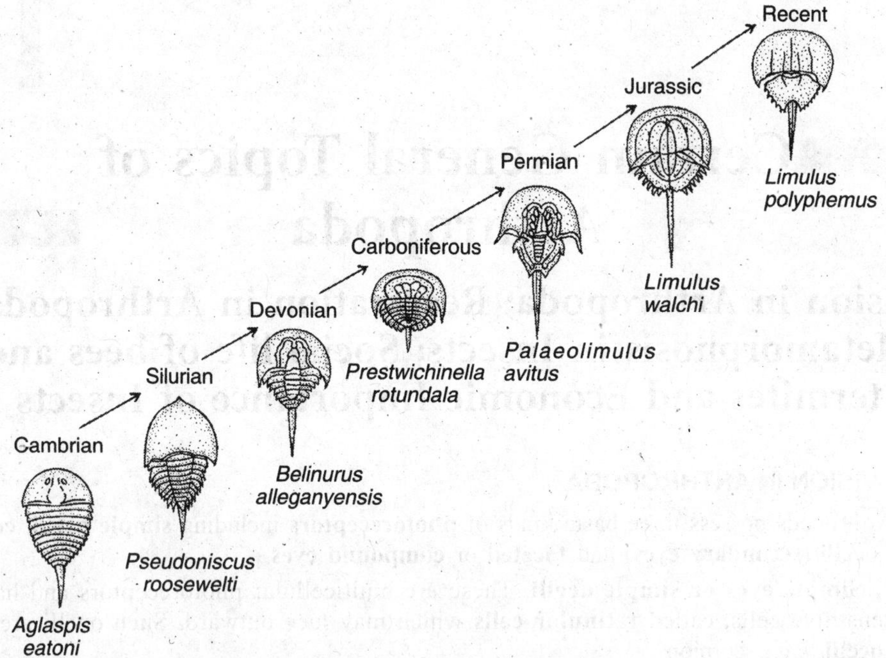

Fig. 9.2. Microevolution or phyletic evolution (anagenesis) in merostom arthropods (horseshoe-crabs), indicating relatively small phenotypic changes over long period of time.

QUESTIONS

Long Answer Question

1. Explain the main features of *Limulus* and write about its evolutionary significance.

Short Answer Questions

1. Why *Limulus* is regarded as a living fossil?
2. Give an account of migration, mating and development of *Limulus*.
3. Write a brief ecological note on king crab.

Very Short Answer Questions

1. What is the name of larva of *Limulus*?
2. *Limulus* is product of which type of evolution?

Fill in the Blanks

1. *Limulus* provides a good example of .... fossil.
2. King crab has .... nutrition.
3. *Limulus* is an aquatic animal. It is .... in habitat.
4. Larva of *Limulus* is called ....

ANSWERS

Very Short Answer Questions

1. Euproops larva (or "trilobite larva").
2. Anagenesis (*i.e.*, microevolution).

Fill in the Blanks

1. Living **2.** Omnivorous **3.** Marine **4.** Euproops or trilobite

10

Certain General Topics of Arthropoda

Vision in Arthropoda; Respiration in Arthropoda; Metamorphosis in Insects; Social life of bees and termites and Economic Importance of Insects

10.1 VISION IN ARTHROPODA

Arthropods possess three basic kinds of **photoreceptors** including **simple ocelli**, **complex lensed ocelli (secondary eyes)** and **faceted** or **compound eyes.**

1. Simple eyes or simple ocelli. These are multicellular photoreceptors and have the light–senstitive cells, called **retinular cells** which may face outward. Such ocelli are called **direct ocelli**, *e.g.*, scorpion.

2. Secondary eyes. In other animals the light–sensitive retinular cells are **inverted**. The inverted type ocelli are common among flatworms (*e.g.*, *Dugesia*, a planarian), nemerteans and spiders and made up of a cup of reflective pigment and retinular cells. Because light can enter only through the cup opening, this sort of ocellus gives the animal a good deal of information about a light direction as well as variations in intensity.

3. Compound eyes. Compound eyes, though found in all four arthropod subphyla, have been lost or modified in various groups throughout the phylum. As their name indicates, compound eyes comprise from a few to many distinct photoreceptive units, called **ommatidia**. Each ommatidium is supplied with its own nerve tracts leading to the major optic nerve and each has its own *field of vision* through square or haxagonal cuticular **facets** on the eye surface. The visual fields of neighboring ommatidia overlap to some extent, such that a shift in the position of an object within the visual field generates impulses from several ommatidia, hence *compound eyes are especially suitable for detecting movement*. However, visual acuity is affected by the degree of overlap among *the* fields of vision of neighboring ommatidia–*the greater the overlap, the poorer the visual acuity*. In general, compound eyes with many small facets probably produce higher–resoultion images than eyes with fewer, larger facets.

1. Vision in Arachnida. The importance of vision varies greatly among the arachnids, although most species possess some sort of **photoreceptors**. At least some species in certain groups are **blind** (*e.g.*, some members of the order Schizomida, Palpigradi, Ricinulei, Pseurdoscorpionida and Acari).Certain spiders depend on photoreception for prey and mate location, particularly the wandering hunters (*Heliophanus* and *Portia*). Vision is relatively unimportant to many sedentary species, such as many web builders, which depend more on tactile cues, vibrations and chemosensitivity. Web builders are not blind, however, and many respond behaviourally to variations in light intensity, while some exhibit distinct escape responses when they visually detect potential predators. Nevertheless, these spiders are generally able to build their webs, capture prey, and mate with little or no visual input. Some spiders are capable of perceiving **polarized light**, presumably as a means of spatial orientation.

Spiders possess **single–lens rhabdomeric eye**, but the sensory units of each are **simple ocelli** in a cluster. Thus, they are quite different from the compound eyes of xiphosaurans and most other arthropods. Each eye includes a thickened cuticular **lens** over a **vitreous body**, which is a layer of cells derived from the epidermis and covering the **retina** (Fig. 10.1). The retina is composed of the **sensory (receptor) cells** and the **pigment cells**. The membranes of the sensory cells bear interdigitating **microvilli**, confirming the rhabdomeric nature of the eyes.

Two forms of this basic eye structure occur among spiders. The anterior median eyes, the **main eyes** which have the light–sensitive portion of the sensory cells directed toward the lens (direct ocelli). Whereas other eyes, called **secondary eyes**, are inverted, with the light receptor

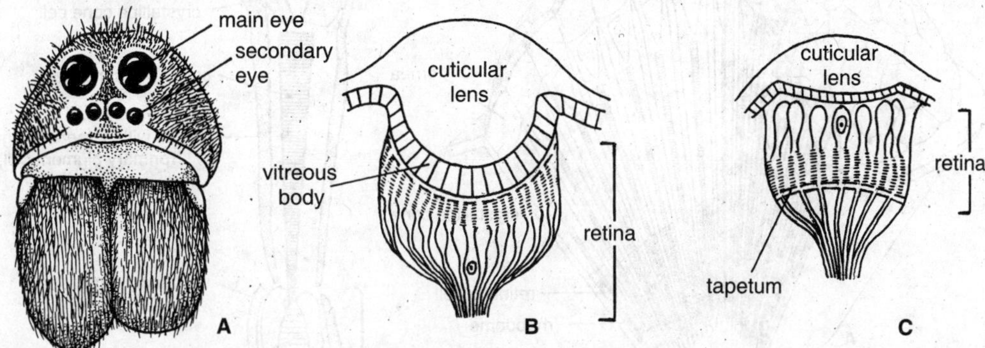

Fig. 10.1. Spider eyes. A–Head–on view of a wolf spider. Note the eye size and position. B–Section through a main eye. C–Section through a secondary eye, showing inverted arrangement of the retinular cells and tapetum.

elements directed away from the lens (Fig. 10.1B,C). Most of the secondary eyes of spiders contain a crystalline reflective layer, called **tapetum**, which may serve to collect and concentrate light in poorly lit conditions (*i.e.*, during night hunting). The reflective nature of the tapetum produces the effect in some spiders of eyes that "shine in the dark".

Scorpions have four pairs of eyes. One pair of median eyes (Fig. 8.19) and three pairs of lateral eyes (Fig. 8.20). These eyes are of direct type and differ from spiders eyes in having the retinal layer external to the epidermis. Most work suggests that scorpions depend much more on mechanoreception and chemoreception than on visual input.

2. Vision of Crustacea. In compound eyes of crustaceans, each ommatidium is covered by a modified portion of the cuticle called the **cornea** (*i.e.*, corneal lens); the special epidermal cells that produce the corneal elements are called **corneagen cells**. The corneagen cells may later withdraw to the sides of the ommatidium to form (usually two) **primary pigment cells**. When viewed externally, the facets on the surface of each cornea produce the characteristic mosaic pattern so frequently photographed by microscopists. The core of each ommatidium comprises a group of **crystalline cone cells** and the **crystalline cone** that they produce, sometime a **crystalline cone stalk**, and a basal **retinula** (= **retinular element**). There are typically four (rarely three or five) crystalline cone cells; an ommatidium with a four–part crystalline cone is highly diagnostic of crustacean–hexapod eye and is termed a **tetrapartite ommatidium** (*Note*. The crystalline cone cells are often called **samper cells** by the entomologists). The crystalline cone is a hard, clear structure bordered laterally by the primary pigment cells, or *iris cells*. The **retinula** is a complex structure formed from several **retinular cells**, which are the actual **photosensitive units** that give rise to the sensory nerve tracts. These retinular cells, usually numbering 8 but ranging from 5 to 13 in various derived conditions, are arranged in a cylinder along the long axis of the ommatidium. The retinular cells are surrounded by **secondary pigment cells**, which isolate each ommatidium from its neighbors. The core of the cylinder is the **rhabdome**, which is made up of **rhodopsin containing microtubular folds** (*i.e.*, microvilli) of the cell membranes of the retinular cells. Each retinular cell's contribution to the rhabdome is called a **rhabdomere**. The microvilli of the rhabdomeres extend toward the central axis of the ommatidium at right angles to the long axis of the retinular cell.

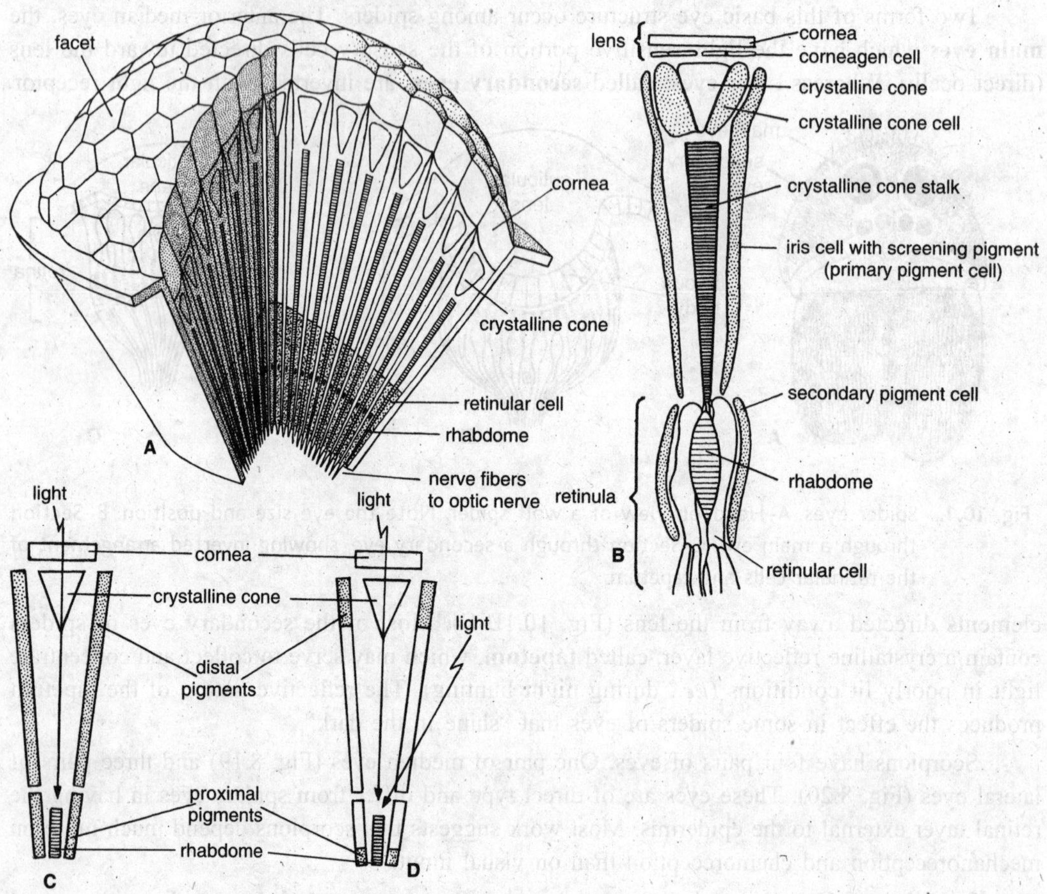

Fig. 10.2. Compound eyes of crustacean arthropods. A–Cutaway view of compound eye; B–A single ommatidial element; C, D–Major ommatidial elements in (c) and oppositional or light–adapted eye and (D) a superpositional or dark–adapted eye.

Working of Compound Eyes

The initiation of impulse depends upon light striking the rhabdome portion of the retinular element. Light that enters through the facet of a particular ommatidium is directed to its rhabdome by the lenslike qualities of cornea and the crystalline cone. The lens has a **fixed focal length**, so accommodation to objects at different distances is not possible. Light is shared among all the rhabdomeres of a given rhabdome, although not necessarily equally.

The versatility of compound eyes is to a large degree a result of the distal and proximal **screening pigments** located in cells that wholly or partially surround the core of the ommatidium (Fig. 10.2B to D). **Distal screening pigments** are located in the iris cells, and **proximal screening pigments** are often located in the retinular cells and secondary pigment cells. In many cases, these screening pigments are capable of migrating in response to varying light conditions and thus changing their positions somewhat along the length of the ommatidium. In **bright light**, the screening pigments may disperse so that nearly all of the light that strikes a particular rhabdomere must have entered through the facet of its ommatidium. In other words, the screening pigment prevents light that strikes the facets at an angle from passing through one ommatidium and into another. Many crustacean eyes are fixed in this condition. Such **appositional eyes** (= light–adapted eyes, Fig.10.2C) are thought to maximize resolution, in that the image from the visual field of each ommatidium is maintained as a discrete unit. Conversely, under conditions of **dim light** screening pigments may concentrate, usually distally, there by allowing light to pass through more than one ommatidium is superimposed on the images formed by neighboring ommatidia. This design has the advantage of producing enhanced irradiances on the retinula, but at the cost of reduced resolution. Many crustacean eyes are fixed in this condition also. Such **superpositional eyes** (dark–adapted eyes; Fig.10.2D) function as efficient light–gathering structures while sacrificing visual acuity and image formation capabilities.

Some arthropod groups possess compound eyes that are always either appositional or superpositional; thus they lack the ability to switch back and forth with varying light conditions. For example, maxillopodans and branchiopods apparently all possess appositional eyes. In Eucarida and Paracardia (of Malacostraca) both types of eyes occur (*e.g.*, isopods and amphipods have appositional eyes, but mysids have superpositional eyes). The crustacean larvae that have compound eyes almost always have the appositional type eyes, which metamorphose into superpositional eyes in those groups that possess them in adulthood.

Significance of stalk of stalked eyes. Among the arthropods, compound eyes elevated on stalks occur only in certain crustaceans (and perhaps a few palaeozoic trilobites). Biologists have long argued over the derivation of such **stalked eyes**, and the matter is still far from settled – are they the primitive condition in arthropods or in crustaceans, or have they been derived multiple times from sessile–eyed ancestors (the latter seems for more likely, see **Brusca** and **Brusca** 2002).

The **eyestalk** is much more than a device to support and move the eye. Eyestalk movement are produced by up to a dozen or more muscles with complex motor innervation. In most malacostracans the eyestalks contain several **optic ganglia** separated by chiasmata, as well as important **endocrine organs**, including the sinus gland and X–organ. Thus, neither loss of eyestalks nor convergent recreation of these structures would have been a simple evolutionary feat. The **loss of functional eyes** is a common evolutionary pathway among the crustacea (and other arthropods), especially in species that inhibit subterranean, deep–sea or interstitial habitats. But among those clades with stalked eyes, the eyestalk remains even when the eye itself degenerates. This indicates significance of the eye stalk.

Compound eyes of insects. We have already describe the fundamental anatomy of hexapod/ crustacean compound eyes. Here we elaborate on some unique aspects of the hexapod eyes. IN some insects the outer surface of the cornea (lens) is covered with minute conical tubercules about 0.2 μm high and arranged in a hexagonal pattern. It is thought that these projections decrease reflection from the surface of the lens, thus increasing the proportion of light transmitted through the facet. Insect eyes in which the crystalline cone is present are called **eucone eyes** (Fig. 10.3). In a eucone eye, immediately behind the crystalline cone are the elongate sensory neurons (= nerve

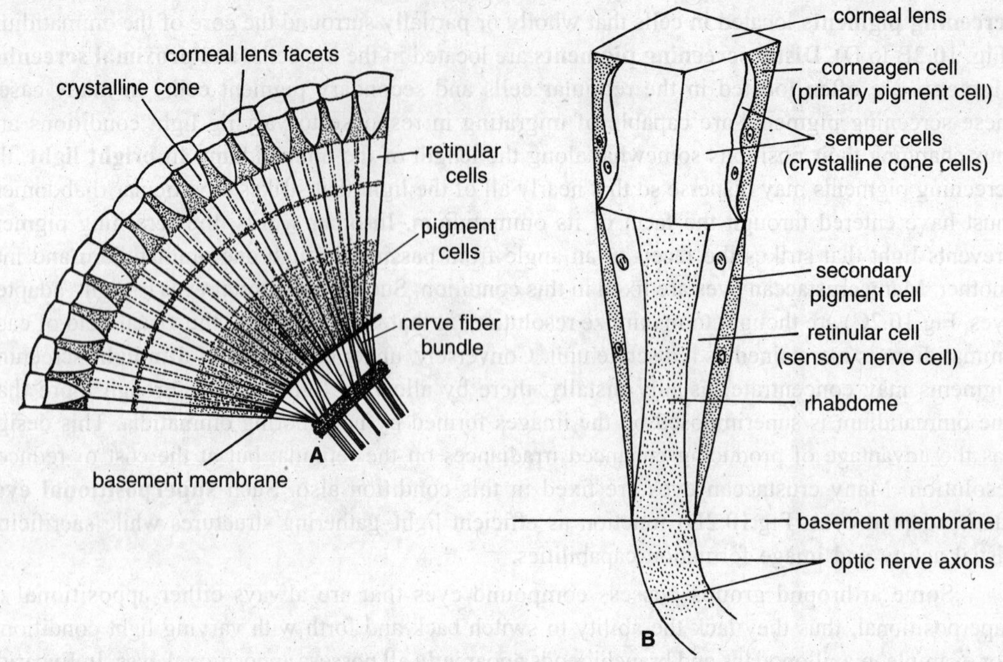

Fig. 10.3. Compound eyes of insects. A–A generalized insect compound eye (cross section). B–A single ommatidium from a eucone compound eyes.

cells) or retinular cells. Primitively, each ommatidium probably contained eight retinular cells arising from three successive divisions of a single cell. This number is found in some insect today, but in most it is reduced to six or seven, with the other one or two persisting as short basal cells in the proximal region of each ommatidium. Arising from each retinular cell is a neuronal axon that passes out through the basement membrane at the base of the eye into the optic lobe. *There is no true optic nerve in insects*; the eyes connect directly with the optic lobe of the brain. In rhadbomeres occur tightly packed microvilli that are about 50 nm in diameter and hexagonal in cross section. The retinular cells are surrounded by 12 to 18 secondary pigment cells, which isolate each ommatidium from its neighbors.

Eyes of Myriapods. Myriapods typically possess eyes at some stage in their life cycle, but zoologists are still debating whether true compound eyes occur in this group. Centipedes possess a few to many eyes that appear to be simple **ocelli** (Fig. 10.4). In the scutigeromorph centipedes, upto 200 of these eyes may cluster to form a sort of **pseudo–compound eye**. However, the eyes of living centipedes appear to function only in the detection of light and dark, and not in image formation. Many parasitic troglobitic myriapods have lost their eyes completely. Symphylans and pauropodans also lack eyes. Many burrowing diplopods are also eyeless, whereas others have 2 to 80 ocelli arranged variously on the head. Some diplopods also possess integumental **photoreceptors** and many eyeless species exhibit negative phototaxis.

The only chelicerates with typical compound eyes, the xiphosurans, have ommatidia that differ in most details from the insect–crustacean design. Silurian scorpions evidently had huge bulging compound eyes (**Brusca** and **Brusca** 2002).

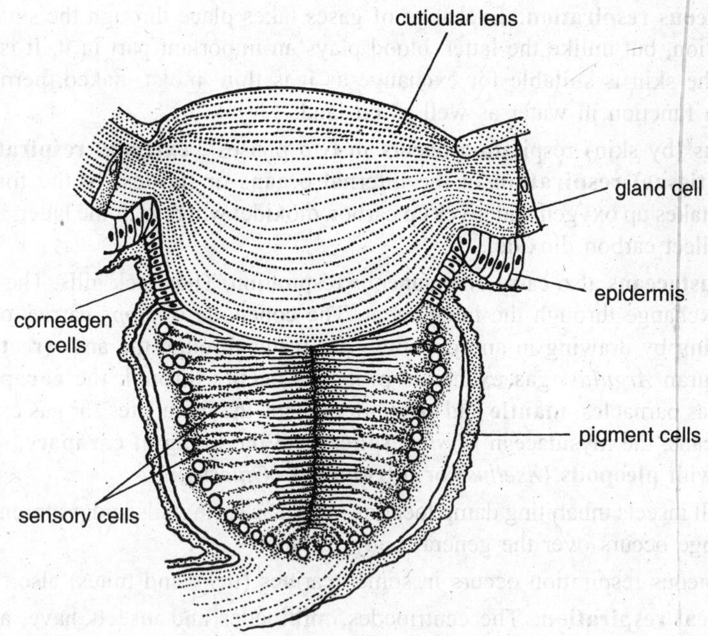

Fig. 10.4. The ocellus of a centipede in section (after Brusca and Brusca 2002).

10.2 RESPIRATION IN ARTHROPODA

Respiration is exchange of environmental oxygen with the carbon dioxide produced in the cells during oxidation. For the nonstop release of energy (during respiration), the cells need uninterrupted supply of oxygen and constant removal of carbon dioxide (Box 10.1)

Box 10.1

The process that involves (*i*) intake of oxygen from the environment (air, water); (*ii*) oxidation of food (*i.e.*, glucose, etc.) with incoming oxygen; (*iii*) elimination of carbon dioxide produced in oxidation, and (*iv*) conservation of energy released during oxidation in biologically useful forms, such as ATP, is known as **respiration**.

In most invertebrates and all vertebrates some organs of the body are specialized for the exchange of gases. These are called **respiratory organs**. Body surface, skin, trachea, branchiae (gills) and pulmonarium (lung) are the main respiratory organs of invertebrates. The surface of these organs at which the exchange of gases takes place is termed **respiratory surface**. For efficient gas exchange, the respiratory surface has large area and are thin, moist, highly vascular and permeable. Water or air is circulated past the respiratory surfaces. This keeps the PO_2 (partial pressure/concentration of oxygen) high and Pco_2 (partial pressure/concentration of carbon dioxide) low in the water or air in contact with respiratory surface to maximize the pressure gradient. Development of respiratory pigment (haemoglobin, haemoerythrin and haemocyanin) in the blood is also an adaptation for gas exchange).

Arthropods do respiration in following ways:

1. Body surface respiration. Body surface respiration occurs in small crustaceans, centipedes, millipedes and insects. Exchange of gases takes place through the general surface of the body and blood plays no role in this respiration. It is also called **direct respiration**.

2. Cutaneous respiration. Exchange of gases takes place through the skin as in the body surface respiration, but unlike the latter, blood plays an important part in it. It is called **indirect respiration**. The skin is suitable for exchange as it is thin, moist, naked, permeable and well vascular. It can function in water as well as on land.

Cutaneous (by skin) respiration occurs in two phases : **external respiration** in the skin and **internal (tissue) respiration** in the internal organs or tissues. In the former, the blood (haemolymph) takes up oxygen and gives up carbon dioxide; whereas in the latter, blood distributes oxygen and collect carbon dioxide.

Many **crustaceans** also carry on cutaneous respiration. They lack gills. The ostracod *Cypris* achieves gas exchange through the integument. The copepode *Cyclops* carries on gas exchange with **rectal lining** by drawing in and expelling out water through the anus (**rectal respiration**). In the branchiuran *Argulus,* gas exchange probably occurs through the **carapace** surface. In cirripeds, such as barnacles, **mantle** and **cirri** are perhaps the main sites for gas exchange. Among the malacostracans, the mysidacean *Mysis* respires with the lining of **carapace**, isopods carry on gas exchange with **pleopods** (*Asellus*) or **integument** (*Oniscus*).

Very small insects inhabiting damp locations, such as springtails and proturans, lack tracheae, and gas exchange occurs over the general body surface.

The cutaneous respiration occurs in some acarines (ticks and mites) also.

3. Tracheal respiration. The centripedes, millipedes and insects have, as an adaptation to land life, thick impermeable integument to minimise the loss of water from the body surface by evaporation. Therefore, they cannot carry out exchange of gases through the skin. Instead, they have evolved a complex system of whitish, shining, intercommunicating air tubes called **tracheae** (Fig. 10.5). The exchange of gases with the help of tracheae open out by paired lateral

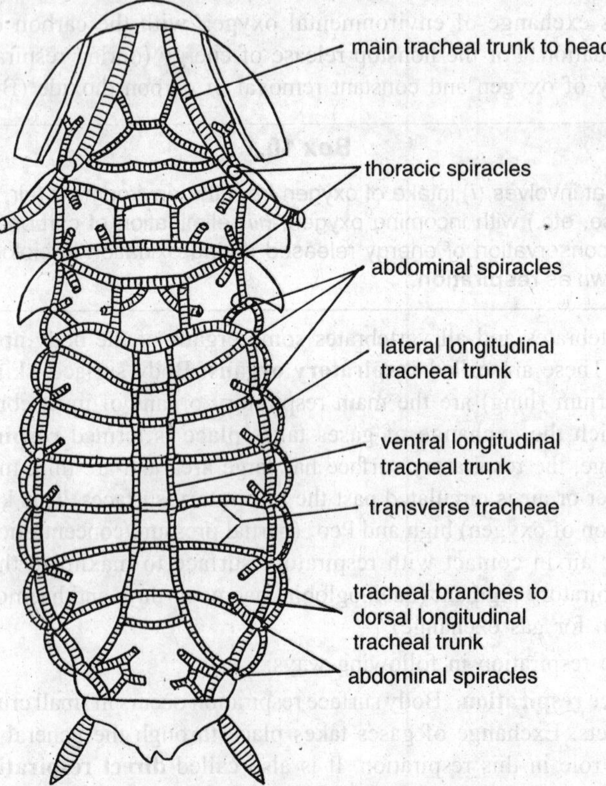

Fig. 10.5. Cockroach. Tracheal system in ventral view.

apertures, the **stigmata** or **spiracles**, in the body wall. The spiracles have a closing device **(valve)** and **filtering apparatus (bristly plates)** to keep out dust particles, parasites and water (Fig. 10.6). The main tracheae divide repeatedly, ultimately forming very fine tubules, the **tracheoles**, which end blindly over the tissue cells (Fig. 10.7). The tracheoles are permeable to fluids. During slow respiratory activity, some tissue fluid diffuses into their distal parts, but is withdrawn during high respiratory activity.

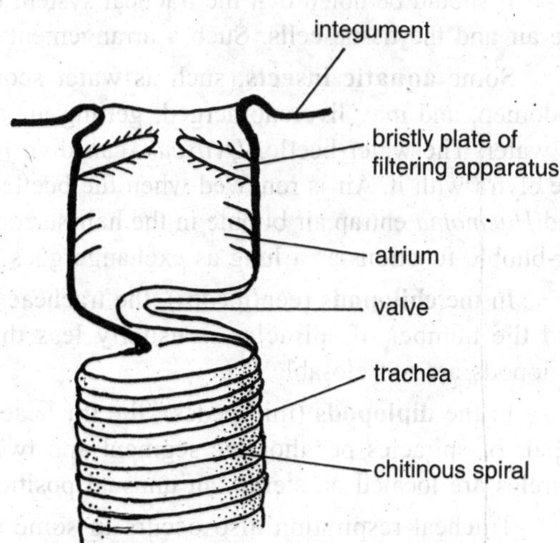

Fig. 10.6. Cockroach. A spiracle with filtering approach and valve.

When an insect is at rest, its oxygen requirement is low and is met without any effort. The oxygen of the air present in the tracheloes dissolves in the fluid in the tracheoles and thence diffuses into the cells. In the cell, oxygen is consumed in oxidation. this set up oxygen gradient between the cells and the tracheoles. The latter draws oxygen from the environmental air through the spiracles and the tracheae.

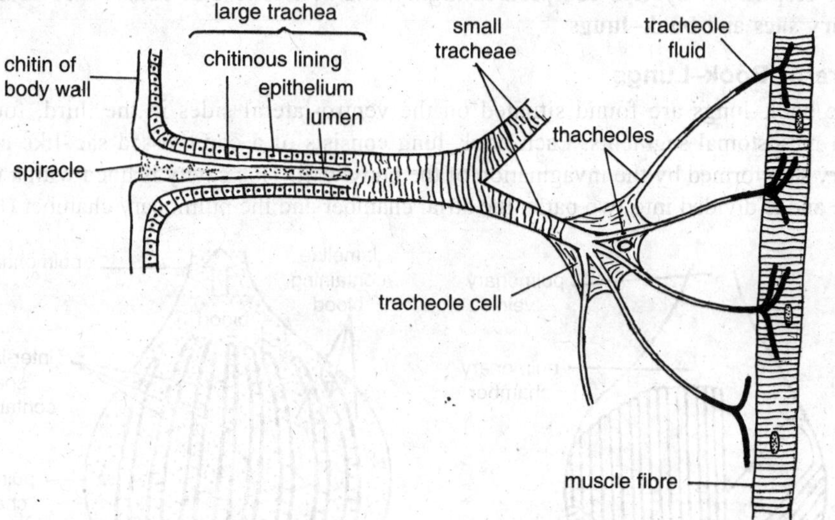

Fig. 10.7. Cockroach. Trachea and its tracheole supply to a muscle fiber.

During locomotion, oxygen requirement of the insect increases and special **respiratory movements** start. These movements consists of alternate **expansion** and **contraction** of the abdomen. Expansion of the abdomen draws fresh air into the tracheal system through the stigmata **(inspiration)** and its contraction expels foul air **(expiration)**. During active respiration, fluid is withdrawn from the tracheoles into the cells. This enables the air to go deep into the tracheoles and reach the cells directly, thereby increasing the efficiency of respiration.

Carbon dioxide (CO_2) produced in oxidation in cells diffuses into the surrounding blood from where it diffuses out through the cuticle which is permeable to it. Some CO_2 is lost through the tracheae during expiration.

It should be noted that the tracheal system brings about exchange of gases directly between the air and the tissue cells. Such a arrangement quickly provides energy for movements.

Some **aquatic insects**, such as water scorpions, have a **respiratory tube** at the end of abdomen, and may live submerged, getting air through this tube by extending it to the surface of water. The water beetle, *Dytiscus*, can dive for a long time by carrying air entrapped under the elytra with it. Air is renewed when the beetle comes to the surface. the aquatic beetles, *Elmis* and *Haemonia* entrap air bubble in the hair surrounding their spiracles to respire under water. The air-bubble functions as a lung as exchange goes on between the air of the bubble and the water.

In the **chilopods** (centipedes), the tracheae anastomose to form a network as in the insects, and the number of spiracles is usually less than the number of segments. The spiracles of chilopods are not closable.

In the **diplopods** (millipedes), the tracheae occur in tufts and do not anastomose. There is a pair of spiracles per thoracic segment and two pairs per abdominal segment. The abdominal spircles are located on sterna, an unusual position among arthropods.

Tracheal respiration also occurs in some **spiders** and certain **acarines** (ticks and mites) also. In spiders, the trachae exist along with the **book lungs**, and usually open out by a single spiracle located just in front of the spinnerets. The mites have one to four pairs of spiracles located on the anterior half of the body. Each spiracle leads into a trachea and not to a tuft of tracheae.

Book Lungs

The respiratory system scorpion is highly characteristic and consists of four pairs of **pulmonary sacs** and **book–lungs**.

Structure of Book–Lungs

The book–lungs are found situated on the ventro-lateral sides in the third, fourth, fifth and sixth mesostomal segments. Each book–lung consists of a compressed sac–like cavity (the pulmonary sac) formed by the invagination of the body wall. This cavity is lined with a thin–layer of cuticle and is divided into two parts: the atrial chamber and the pulmonary chamber (Fig. 10.8).

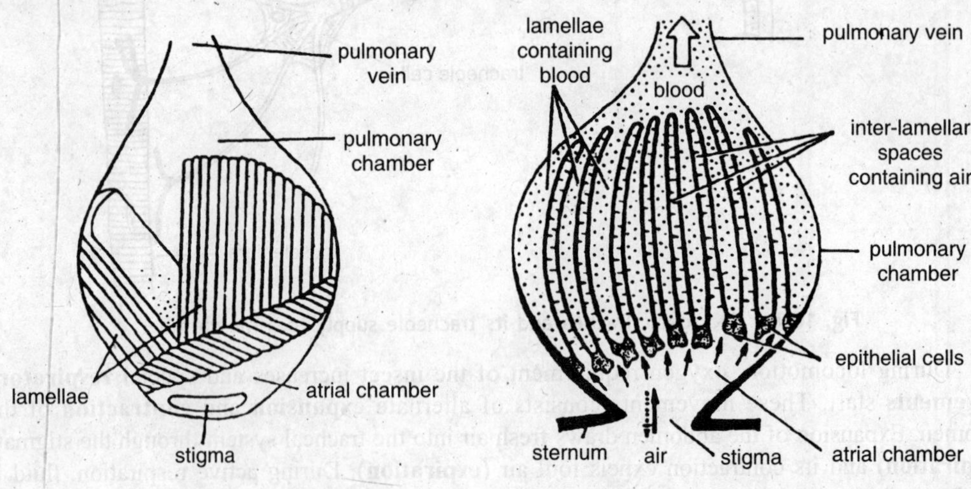

Fig. 10.8. Scorpion. A book-lung.

Fig. 10.9. Scorpion. Vertical section of book-lung to show the mechanism of respiration.

1. Atrial chamber. It is the proximal ventral part of the book–lung. It is in the form of small and dorsoventrally compressed air–chamber. Proximally atrial chamber opens to exterior by a narrow, oblique, slit–like opening, called **stigma**, which is placed ventrolaterally on the sternum.

Distally the roof of atrial chamber is perforated by a series of minute slit–like openings or **ostia**. These ostia are placed parallel to one another. Through ostia, the atrial chamber communicates with the **interlamellar air spaces** of the pulmonary chamber.

2. Pulmonary chamber. It is distal part of the cavity of book–lung which is called **pulmonary chamber**. It is dorsal spacious oval chamber filled with about 150 vertical folds or leaf–like **lamellae**, attached to the posterior side of the chamber. The lamellae are arranged parallel to each other and one over the other like the leaves of a book. Each lamella is a hollow structure made of thin layer of cuticle united at their edges and thus enclosing **intralamellar spaces** filled with blood. The outer side of the lamellae bear ridges and bristles which keep the adjacent lamellae apart and little air space, called **interlamellar space**, is left between them for the flow of air. These spaces communicate with the atrial chamber through ostia (Fig. 10.9).

Blood supply. The deoxygenated (impure) blood is sent to each book–lung from the ventral sinus by diverticulum; from where the blood gets filled in the **intralamellar space** of the lamellae. The blood is oxygenated or purified in the lamellae of each book–lung and is collected into a **pulmonary vein** which travels dorsally and opens into the pericardium.

Mechanism of Respiration. The inspiration and expiration of air in the book–lungs is controlled by the action of the, dorsoventral and atrial muscles. When the **dorso–ventral muscles** are contracted, the pulmonary chamber is compressed and the air filled in interlamellar air spaces is forced out into the atrial chamber. From atrial chamber this air is forced to outside through the stigma due to contraction of the **atrial muscles**. When the atrial muscles relax, the book–lungs resume their normal shape and the fresh air enters through stigma first into the atrial chamber and then into the interlamellar spaces. At this time dorsoventral muscles relax and pulmonary chamber resumes its normal shape. The **gaseous exchange** takes place through the thin membranous walls of the lamellae and the blood becomes oxygenated and carbon dioxide is expelled out in the air.

Branchial Respiration

A large number of aquatic animals have special organs for obtaining oxygen from water. These organs are called **gills** or **branchiae**. The exchange of gases in the gills is known as **gill** or **branchiae respiration**. Among arthropods, branchial respiration is found in many crustaceans, some insects and xiphosurans. The gills are of two main types: external and internal. **External gills** occur in larvae of certain insects. **Internal gills** occur in crustaceans. A gill consists of vascular filaments or plates having a thin epithelium overlying blood capillaries. This arrangement allows a quick gas exchange when water flows over the gill. Gills function only in water.

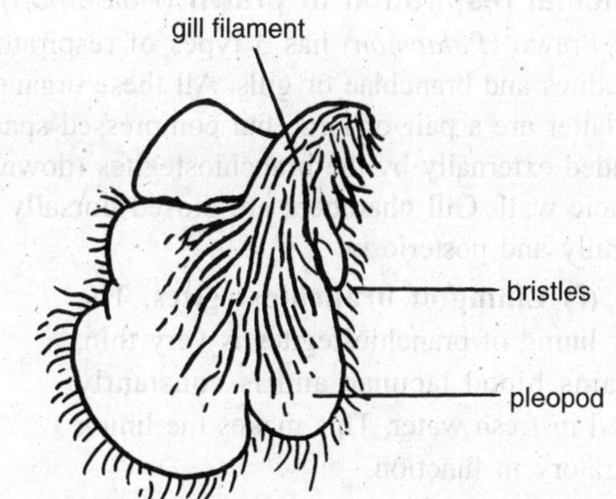

Fig. 10.10. A pleopod in *Squilla empusa* showing filamentous gills.

Branchial respiration in crustaceans. In crustaceans, the gills are typically associated with the appendages. they vary in form, location and derivation. The water current for ventilation is usually set up by the beating of some appendages. Oxygen is transported by the blood (or haemolymph) dissolved in plasma or bound to plasma haemoglobin or haemocyanin.

The stomatopod *Squilla* has 5 pairs of filamentous gills on the anterior 5 pairs of abdominal appendages (pleopods) (Fig. 10.10).

In decapods, the gills are confined to the thorax. They arise from the body wall on or near the point of attachment of the appendage, and are named according to their location. Their number varies in different species. A gill consists of a central axis bearing lateral branches. The latter may be (*a*) **unbranched**, forming a **filamentous** or **trichobranchiate gill** as in crayfish and lobsters (Fig. 10.9); or (*b*) **branched**, forming a **dendritic** or **dendrobranchiate gill**; or plate-like, forming a **lamellar** or **phylobranchiate gill** as in prawn and crabs (Fig. 10.11). The gill axis contains afferent and efferent branchial blood channels. Blood flows from the afferent channel into each filament or lamella of the gill and returns to the efferent branchial channel. Blood of decapods contains haemocyanin dissolved in the plasma.

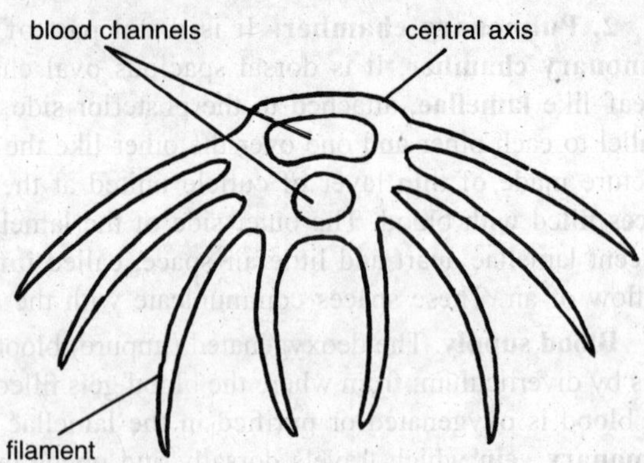

Fig. 10.11. Crustacea. T.S. of filamentous gill.

A paddle–like **scaphognathite** or **gill bailer**, which is an enlarged exopodite of the second maxilla, produces a ventilating current by its beating. Water is drawn forward and the exhalant current flows out anteriorly in front of the head.

The inhalant current enters the branchial chamber containing gills along the ventral and posterior margins of the carapace.

Gas exchange also occurs over the surface of vascular lining of **branchiostegites** and thin, leaf–like, vascular outgrowth, called **epipodites**, of the maxillipedes. The epipodites are considered simple **primitive gills**.

Branchial respiration in prawn (*Palaemon*)

Prawn (*Palaemon*) has 3 types of respiratory organs, namely, lining of branchiostegites, epipodites and branchiae or gills. All these organs are situated in the branchial or gill chambers. The latter are a pair of large but compressed spaces, one on either side of the thorax. They are bounded externally by the branchiostegites (downward extensions of carapace) and internally by thoracic wall. Gill chambers are closed dorsally but communicate with the exterior anteriorly, ventrally and posteriorly.

(*i*) Lining of branchiostegites. The inner lining of branchiostegites is very thin, contains blood lacunae and is constantly bathed in fresh water. This makes the lining respiratory in function.

(*ii*) Epipodite. There are three pairs of epipodites. They lie in the anterior part of gill chamber below the scaphognathites (Fig. 10.12). Epipodites are thin, leaf–like, vascular outgrowths of integument from the outer borders of the coxae of the maxillipedes. The first pair of epipodites are larger than others and are bilobed.

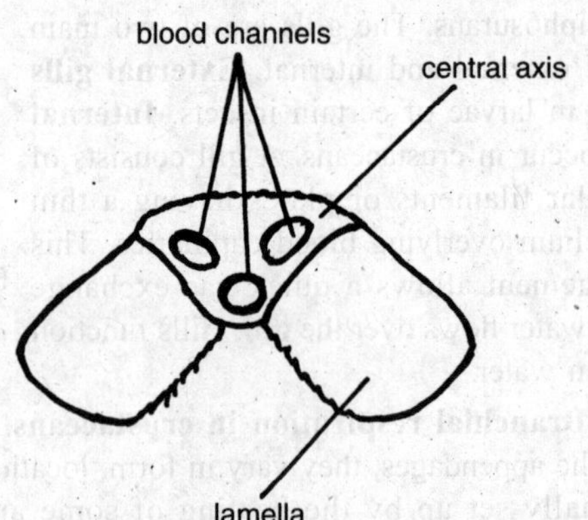

Fig. 10.12. Crustacea. T.S. of lamellar gill.

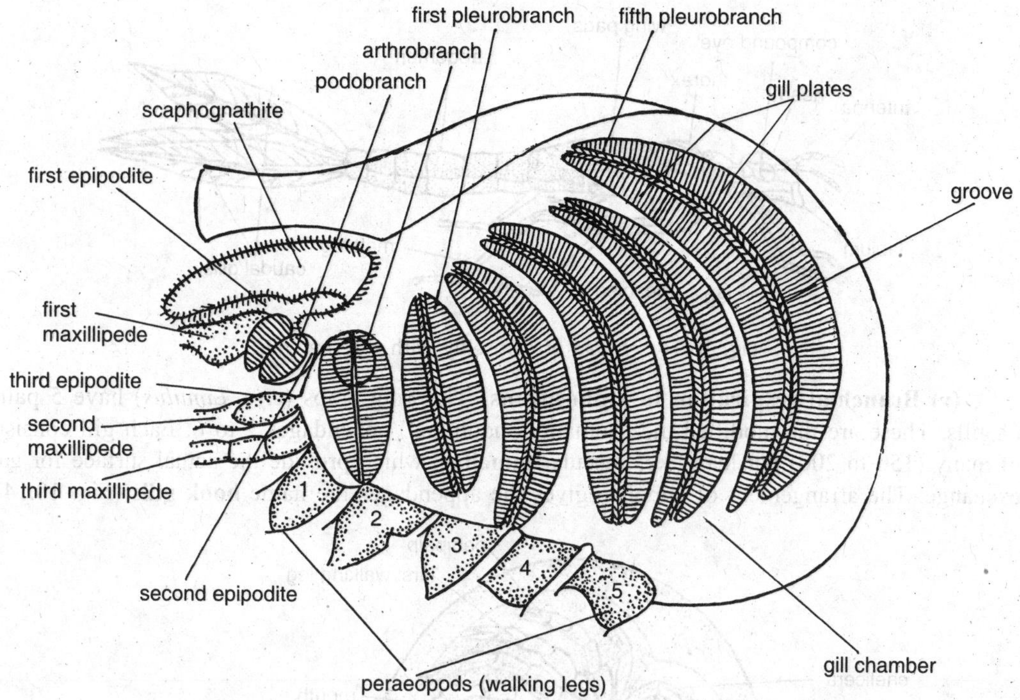

Fig. 10.13. *Palaemon*. Left branchial chamber showing epipodites and gills.

***(iii)* Branchiae or gills**. There are eight pairs of gills. They are somewhat crescentic in form, lie vertically one behind the other in a row and increase in size from front to backwards. The third pair of gills are the smallest and are hidden beneath the upper parts of the gills of second pair. The gills are attached by their middle part, being free in the upper and lower regions. The attached part of the gill is called the **root**. The blood lacunae and nerves enter and leave the gills through its root. According to mode of their attachment, the gills of Palaemon fall into three categories. The first pair of gills are attached to the coxae of the second maxillipedes and are called the **podobranchs**. The second and third pairs of gills are attached to the arthrodial membranes that connect the third maxillipedes to the thorax. They are described as the **arthrobranchs**. The remaining gills are joined to the lateral walls of the thorax, one above each walking leg. These 5 pairs of gills are called **pleurobranchs**.

The gas exchange occurring in the branchiostegite lining, epipodites and gills is external respiration.

The amphibious crabs, such as *Uca* and *Sesarma*, retain water in the gill chambers when on land, and aerate this water to keep the gill functioning.

***(iv)* Branchial respiration in insects**. Naiads of certain aquatic insects, such as damselflies and mayflies, have external **tracheal gills** at the end of abdomen (Fig.10.13). These gills have closed trachae, and gas exchange occurs across the gill surface between water and tracheae. The tracheal gills are finger–like in mayfly naiads and leaf–like in the damselfly naiads. Dragon fly naiads has tracheae containing gills in the form of soft–lamellae on the inner surface of the rectum, and pumps water in and out through the anus to vantilate the gills for gas exchange.

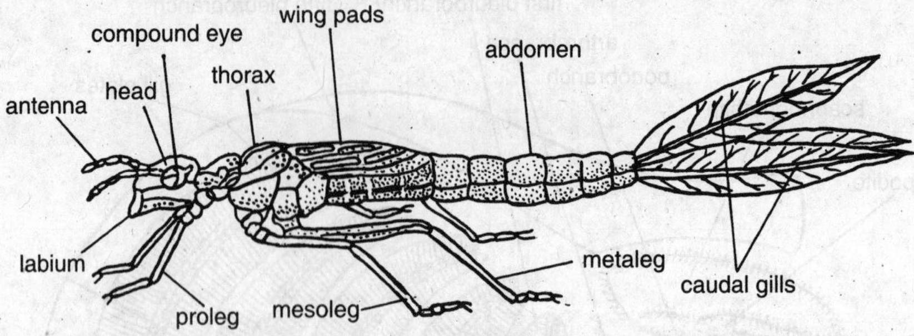

Fig. 10.14. Naiad of damselfly with external gills.

(v) Branchial respiration in xiphosurans. The king crabs (*e.g.*, *Limulus*) have 5 pairs of gills. These are the modified mesosomatic (abdominal) appendages 2 to 6. Each gill consists of many (150 to 200) flat leaf–like, vascular **lamellae**, which provide the actual surface for gas exchange. The arrangement of lamellae gives the appendages the name **book gills** (Fig. 10.14).

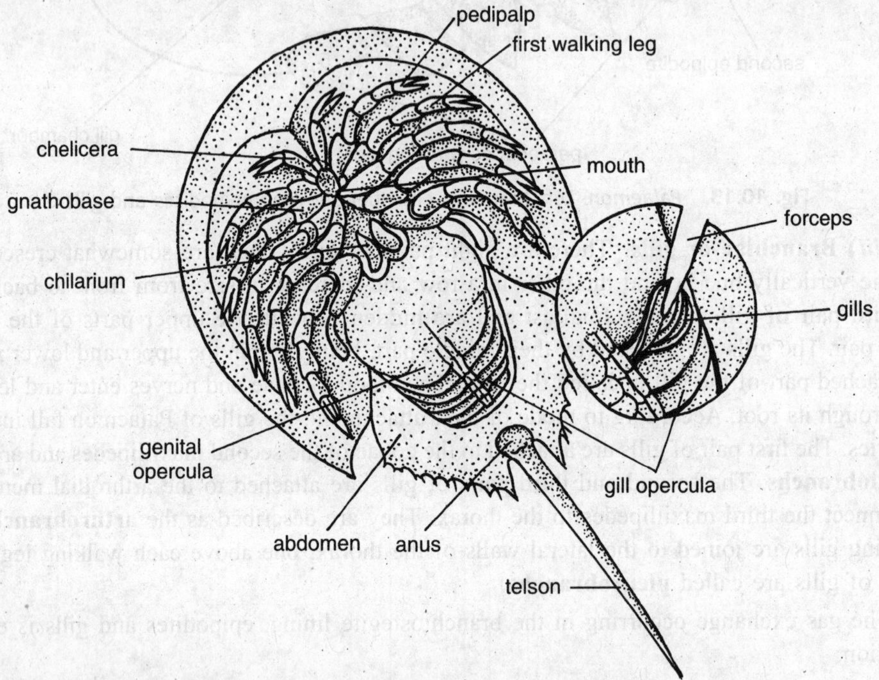

Fig. 10.15. Ventral view of horseshoe crab (*Limulus*) showing book–gills.

The movement of gills maintain a ventilatory current over the lamellae. Water current enters the gap between the cephalothorax and abdomen, and is then directed backward by the flabellum (a spathulate process of the coxa of last or fifth leg).

10.3 METAMORPHOSIS IN INSECTS

Metamorphosis involves dramatic changes in habitat, habit, morphology, physiology and behaviour of the larva to change into adult. A **larva** is a developmental stage of distinctive from that persists for a period of time, lives independently of the adult, differs in form from the adult, and occupies a separate ecological niche. Animals with a life cycle that includes a larval stage have **indirect development**. The larva is anatomically different from the juvenile and adult, reflects its adaptation to the larval, not the adult, ecological niche. Thus, at the close of the larval period, the larva must **locate** and **settle** in a habitat suitable for the adult and then changes its form to that of the **juvenile**, an immature form of the adult. This transformation is the **metamorphosis** (= change of form). Metamorphosis ranges in extent from the gradual addition of new structures and slight rearrangements of tissues to a complete remaking of the juvenile body from undifferentiated **embryonic reserve cell** (or set–aside cells) housed in the larval body. The process by which a juvenile differentiates extent of metamorphosis, whether gradual, cataclysmic (*i.e.*, sudden and violent change), or intermediate, depends on how far the larval and adult body designs have diversed from each other (**Barnes** *et al.*, 2004).

The basic life–history stages of Insecta (Hexapoda) are **egg**, **juvenile** and **adult** (**Imago**), although other stages may be present. The stages are divided into substages, or **instars**, separated by molts (or moults). Usually there is only one adult instar, called the **imago**. Only the imago is sexually mature and has functional wings. Upon hatching from the egg, the first instar juvenile begins a series of alternating instars and molts that eventually culminate in adulthood.

In many taxa, the responsibility for feeding and growth lies with juvenile instars, whereas adults are responsible for reproduction and dispersal. Some insects spend the greater part of the life span, sometimes several years, as a juvenile and the adult stage may be brief, sometimes only a few hours, although these are extremes. In such cases the juvenile accumulates the food reserves that support the adult that follows. In some cases, adults have neither mouth parts nor gut and do not feed.

Types of metamorphosis in Insects

Insects have **progresive metamorphosis**, *i.e.*, in insects simple larval organization changes into more complex organization of adult. It is of following four types:

1. Ametaboly. In the members of subclass Apterygota (wingless insects or hexapods), (*e.g.*, *Lepisma* (silver fish, Order Thysanura) and *Orchesella* (spring tail, Order Collembola), the younger, juvenile, emerging from the egg is exactly similar to adult except smaller in size with immature reproductive organs and external genitalia (*e.g.*, anal styles, etc.). It undergoes several moults (upto 50) and gradually increase in size. These insects do not undergo metamorphosis, so called **ametabolous insects**. Each instar is slightly larger than its predecessor. Adult instars are larger and have functional reproductive system. The life cycle of ametabolous insects includes in it following three stages (Fig. 10.16):

Egg → Juvenile → Adult

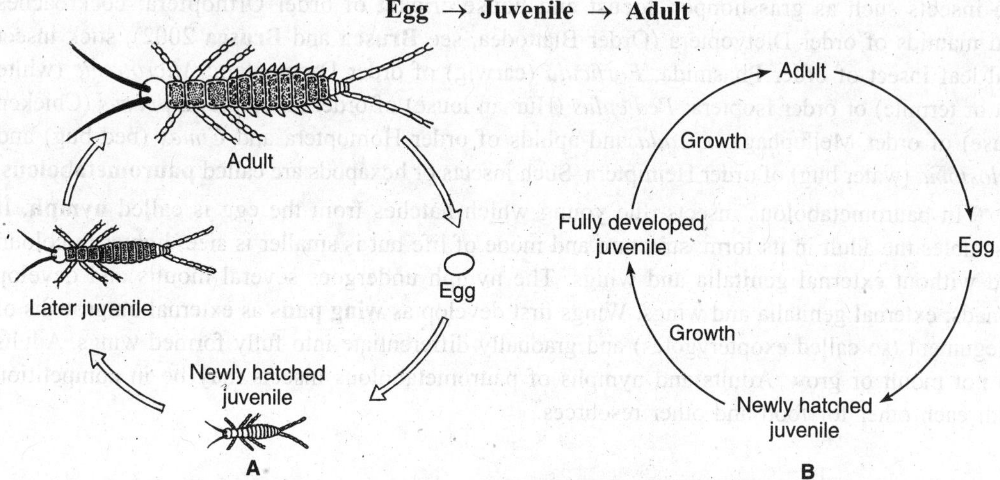

Fig. 10.16. Ametaboly in silverfish. A–Diagrammatic; B–Graphic.

Unlike other hexapods, ametabolous species continue moulting and growing after reaching sexual maturity through several adult instars (Fig. 10.15B). Adult and juveniles have similar niche characteristics and compete for resources.

2. Hemimetaboly or Incomplete Metamorphosis Such type of metamorphosis is found in insects (hexapods) belonging to order Plecoptera (*e.g.*, *Isoperia*–stonefly); Order Odonata (*e.g.*, 1. *Sympetrum*–dragonfly and 2. damselfly); and order Epheroptera (*e.g.*, *Ephemera*–mayfly). Such insects are called **hemimetabolous insects**.

In hemimetabolous insects, the young, called **nymph** or **naiad**, which hatches from the egg is entirely different from the adult in its habitat, habits and morphology. It is aquatic, has tracheal gills for respiration and mandibulate mouth parts. During metamorphosis, naiads shed gills, develop the wings and modify their mouth parts. Adult and nymphs occupy different ecological niches and do not compete for the resources. Thus, life cycle of a hemimetabolous insects include following three stages.

Egg → Naiad → Adult

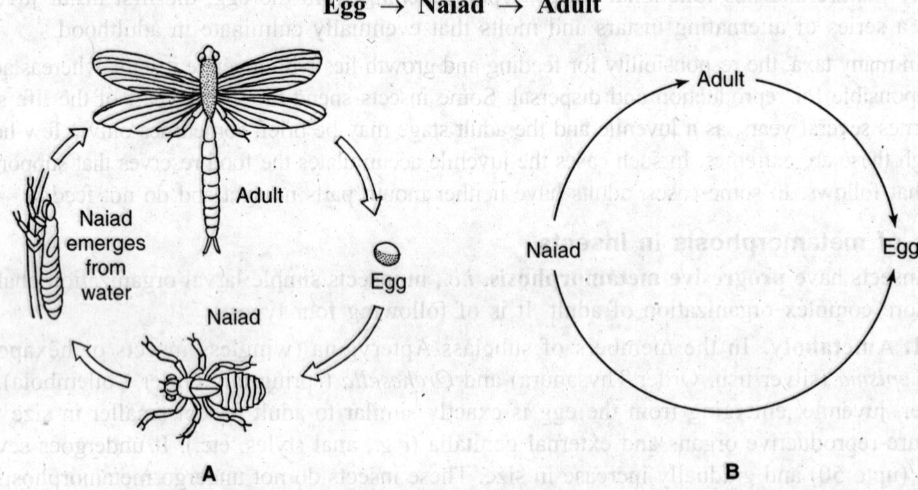

Fig.10.17. Hemimetaboly in dragonfly. A–Diagrammatic; B–Graphic.

3. Paurometaboly or gradual metamorphosis. Such type of metamorphosis is found in the insects such as grasshopper, locust and house cricket of order Orthoptera; cockroaches and mantids of order Dictyoptera (Order Blattodea, see **Brusca** and **Brusca** 2002); stick insect and leaf insect of order Phasmida; *Forficula* (earwig) of order Dermoptera; *Microtreme* (white ant or termite) or order Isoptera; *Pediculus* (Human louse) of order Anoplura; *Lipeurus* (Chicken louse) of order Mellophaga; *Cicada* and aphids of order Homoptera and *Cimex* (bed bug) and *Belostoma* (water bug) of order Hemiptera. Such insects or hexapods are called **paurometabolous**.

In paurometabolous insects, the young which hatches from the egg is called **nymph**. It resembles the adult in its form, structure and mode of life but is smaller is size, lighter in colour, and without external genitalia and wings. The nymph undergoes several moults and develop gonads, external genitalia and wings. Wings first develop as **wing pads** as external outgrowths of integument (so called exopterygotes) and gradually differentiate into fully formed wings. Adults do not moult or grow. Adults and nymphs of paurometabolous insects may be in competition with each other for food and other resources.

Life cycle of a paurometabolous insect includes following three stages:

Egg → Nymph → Adult

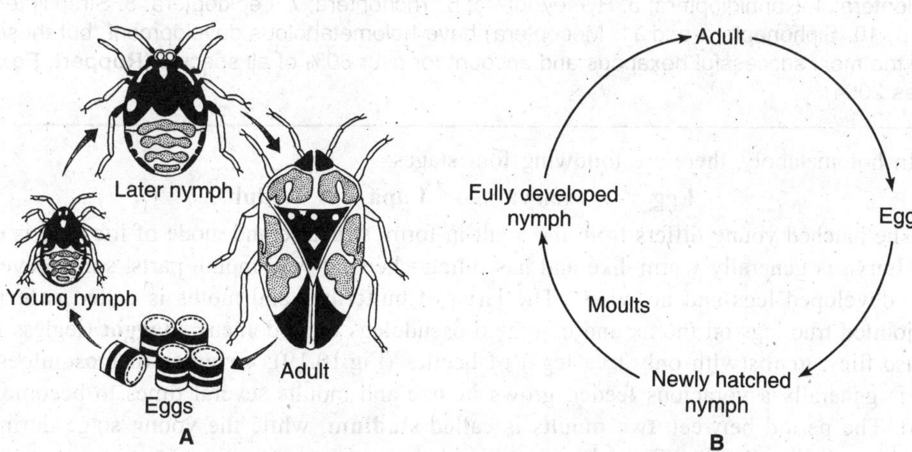

Fig. 10.18. Paurometaboly in stink bug. A–Diagrammatics; B–Graphic.

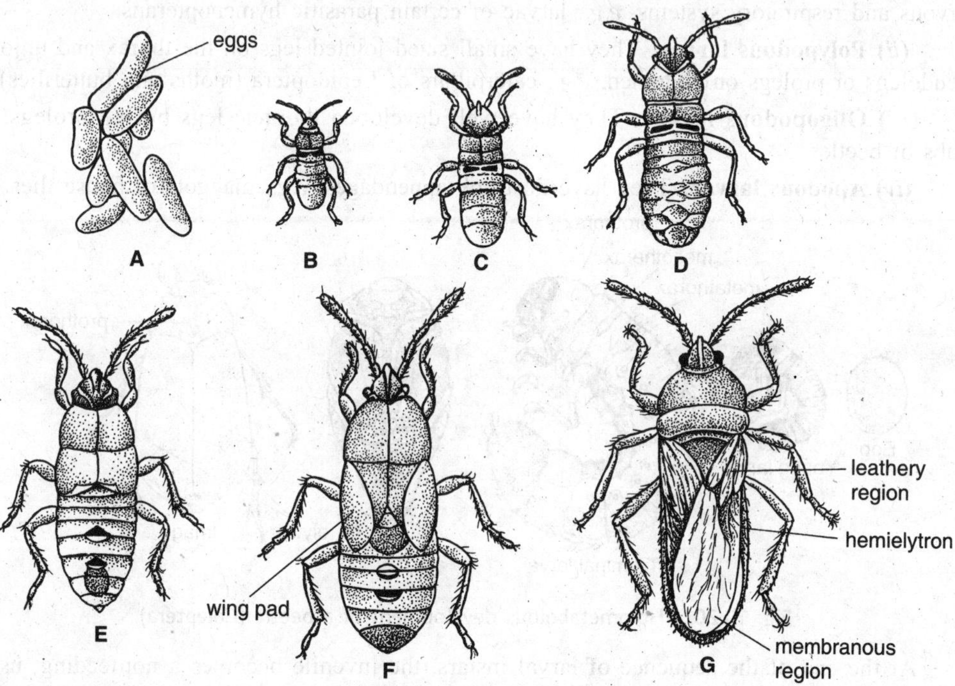

Fig. 10.19. Stages in the paurometabolous development of the chinch bug, *Blissus lencopterus* (Hemiptera).

4. Holometaboly or Complete metamorphosis. It occurs in indirect development. This type of metamorphosis is found in insects belonging to order Coleoptera (*e.g.*, *Coccinella*–lady bird beetle); Order Mecoptera (*e.g.*, *Panorpa*–Scorpion fly); Order Trichoptera (*e.g.*, *Philopotamus*–Caddishfly); Order Lepidoptera (*e.g.*, butterflies and moths; Fig. 10.29); Order Diptera (*e.g.*, mosquitoes, fruit fly and house fly); Order Hymenoptera *e.g.*, *Apis indica*–honey bee; *Polister*–wasp); and order Siphonoptera (*e.g.*, *Xenopsylla*–rat flea) (Box 10.2).

Box 10.2 Holometabolous insects

Of the 33 taxa of recent hexapods (insects), only 11 (*e.g.*, 1. Coleoptera, 2. Neuroptera; 3. Megaloptera; 4. Raphidioptera; 5. Hymenoptera; 6. Trichoptera; 7. Lepidoptera; 8. Strepsiptera; 9. Diptera; 10. Siphonaptera and 11. Mecoptera) have holometabolous development, but these are by far the most successful hexapods and account for over 80% of all species (**Ruppert**, **Fox** and **Barnes** 2004).

In holometaboly, there are following four stages;

Egg → Larva → Pupa → Adult

The hatched young differs from the adult in form, strucutre and mode of life, and is called **larva**. Larva is generally worm–like and has biting–chewing type mouth parts, simple eyes and poorly–developed legs and no wings. The larva of butterflies and moths is called **caterpillar** (with jointed true legs on thorax and unjointed pseudolegs on abdomen), **maggot** (legless larva) of house flies; **grubs**(with only true legs) of beetles (Fig.10.19); **wriggler** of mosquitoes, etc. Larva is generally a voracious feeder, grows in size and moults several times to become fully formed. The period between two moults is called **stadium**, while the young stage during the stadium is called an **instar**. These larva are divided into four categories:

(*i*) Protopodus larvae. They emerge at an early embryonic stage and have undeveloped nervous and respiratory systems, *e.g.*, larvae of certain parasitic hymenopterans.

(*ii*) Polypodous larvae. They have small sized jointed legs on the thorax and unjointed pseudolegs or prolegs on abdomen, *e.g.*, caterpillars of Lepidoptera (moths and butterflies).

(*iii*) Oligopodous larvae. They have well developed thoracic legs but no prolegs, *e.g.*, grubs of beetles.

(*iv*) Apodous larvae. They have no trunk appendages, *e.g.*, maggots of house flies.

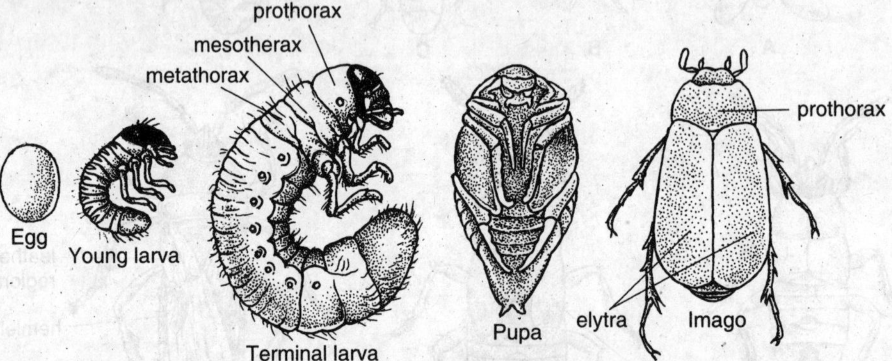

Fig. 10.20. Holometabolous development of a beetle (Coleptera).

At the end of the sequence of larval instars, the juvenile becomes a nonfeeding, usually quiescent **pupa** in a protected location, such as a cocoon or cavity in the ground. The larva secretes a cocoon or pupal case, called **puparium**, from the secretion of its labial galnds and changes into a nonlocomotory and nonfeeding stage, called **pupa**. **Hinton** (1946) described the following two types of pupae:

(*a*) Decticous pupae. They have powerful, sclerotized and articulated mandibles. These functional mandible–containing pupae are found in members of Neuropotera, Mecoptera and Trichoptera.

(*b*) Adecticous pupae. They lack functional pupae. These pupal have non–articulated mandibles which are not used in escaping from the cocoon. These are found in most members of Siphonoptera, Coleoptera, Hymenoptera, Lepidoptera and Diptera.

Inside the pupa the larva undergoes metamorphosis, during which larval organs are destroyed (hydrolyzed) and adult structures develop anew, phoenixlike, from clusters of embryonic reserve cells, called **imaginal discs**. Wing rudiments, for example develop internally from imaginal discs and wings, which have no larval precursors, appear *de novo* with the emergence of the **imago** (young insect) from the pupa (Box 10.3).

Box 10.3

During metamorphosis, inside pupa two types of changes occur:

1. **Histolysis**. It involves degeneration of all those larval organs except the nervous system which are of not much importance in the adult. The organs are degenerated by the action of **phagocytes** or by the activity of lysosomal enzymes.
2. **Histogenesis**. It involves the development of rudiments of adult organs such as wings, legs, antennae, mouth parts and internal organs from the specialized groups of cells called **imaginal buds** or **imaginal discs** or **histoblasts**. These buds are developed directly from the fertilized eggs and remain suppressed through the larval stage. During the pupal stage, these buds grow in size and differentiate into adult structures.

The development of holometaboly was of great adaptive significance in the evolution of insects partly because it allows larvae to utilize different food sources, habitats and life styles than those of adults. It tends to reduce or eliminate **competition** between life history.

Endocrine Control of Insect Metamorphosis

Moulting and metamorphosis of insects are under endocrine control. **Axons** from neurosecretory centers in the protocerebrum deliver the tropic hormone **ecdysiotropin** to the **corpora cardiaca** (Fig. 10.21). Corpora cardiaca are neurohaemal organs associated with aorta that store ecdysiotropin and then release it into the blood. Ecdysiotropin in the blood has two

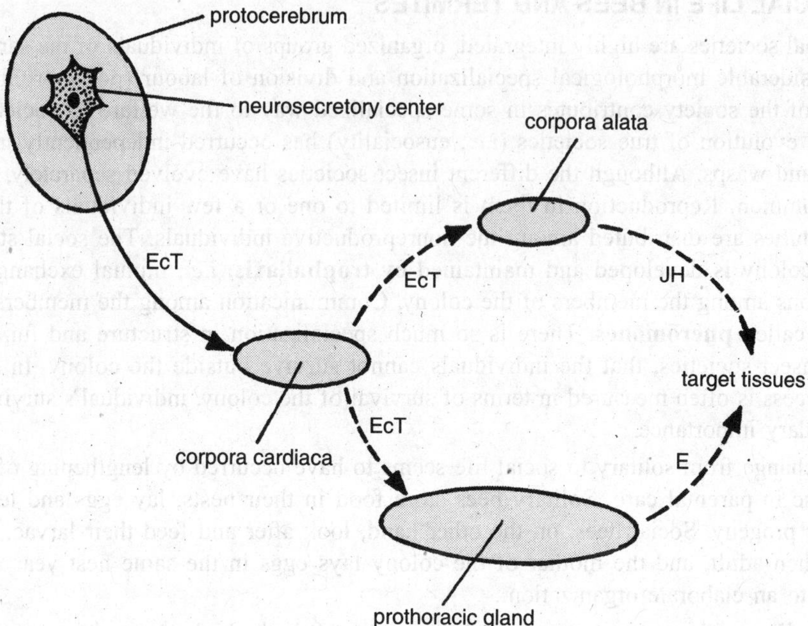

Fig. 10.21. Generalized endocrine system of insects (hexapods). Axonal transport is indicated by the solid arrow and blood transport by dotted arrows. EcT = ecdysiotropin; E = ecdysone; JH = juvenile hormone (after Ruppert, *et al.*, 2004).

effects. It directs **corpora allata** to synthesize and release **juvenile hormone (JH)** and the **prothoracic glands** to release **ecdysone**, both into the blood. Ecodysone initiates a new moult, during which juvenile hormone maintains the juvenile condition.

Under the influence of ecdysone (hormone), the epidermis initiates the moulting cycle. The corpora allata secrete juvenile hormone (Fig.10.21) into the blood during the moult. Juvenile hormone can exert its effect only after the moulting process has been initiated and thus acts only in conjunction with ecdysone. Once the moult cycle begins, its outcome is determined by the level of juvenile hormone in the blood. Juvenile hormone (JH) suppress the expression of genes controlling the development of adult characteristics. If the blood juvenile–hormone level is high, then the moult will be larva → larva; if it is intermediate, larva → pupa; and if low, pupa→adult. In the absence of juvenile hormone, adult characteristics are free to develop and metamorphosis can occur. In addition, without juvenile hormone, the prothoracic gland degenerate and without this source of ecdysone (hormone), moulting ceases. Once moulting ceases, the effects of juvenile hormone are not expressed because it acts only during a moult.

Environmental and physiological events may initiate the onset of moult cycle by stimulating or inhibiting the release of ecdysiotropin from the brain. Stretching of the gut wall due to feeding or changing temperatures or photoperiod (day length), for example, may stimulate the release of ecdysiotropin and initiate a moult.

Diapause. The life cycle of many hexapods, includes diapause, a period of physical and metabolic inactivity that enables the organism to survive such adverse environmental condition as long dry or cold periods. Diapause is under hormonal control and may be passed in any stage of development: egg, juvenile, pupa or imago, depending on the species or environmental condition. It typically is passed in the ground, a cocoon or some other protected place. Many insects cease moulting and enter diapause in response to declining photoperiod in the fall and resume moulting in the spring, with increasing photoperiod.

10.4 SOCIAL LIFE IN BEES AND TERMITES

Animal societies are highly integrated, organized groups of individuals of the same species with a considerable morphological specialization and division of labour (polymorphism). Each individual of the society contributes in some specialized way to the welfare of society. Among insects, the evolution of true societies (*i.e.*, eusociality) has occurred independently in termites, ants, bees and wasps. Although the different insect societies have evolved separately, they have much in common. Reproduction in them is limited to one or a few individuals of the colony, and other duties are distributed among the nonreproductive individuals. The social structure of the insect colony is developed and maintained by **trophallaxis**, *i.e.*, mutual exchange of food and secretions among the members of the colony. Communication among the members occur by chemicals, called **pheromones**. There is so much specialization in structure and functions and habits in insect societies, that the individuals cannot survive outside the colony. In an animal society, success is often measured in terms of survival of the colony, individual's survival having only secondary importance.

The change from solitary to social life seems to have occurred by lengthening of adult life and increase in parental care. Solitary bees store food in their nests, lay eggs and leave never to see their progeny. Social bees, on the other hand, look after and feed their larvae. They live together when adult, and the mother of the colony lays eggs in the same nest year after year. This leads to an elaborate organization.

Animal's social organization is called **eusociality**. It is the highest development of colonial organization and has evolved in only two taxa of insecta, the Isoptera (termites) and the Hymenoptera (bees, ants and wasps). Eusociality has been responsible for the great success of both insectan taxa. All termites and all ants are eusocial, as are many bees and wasps. Eusociality evolved independently in termites and hymenopterans (bees), but it has culminated in similar states of development in both.

Three characteristics of Eusocial Societies

1. Members of the society cooperate in caring for the juveniles. There is a reproductive division of labour, with **sterile** individuals performing the physical labour of the society while reproduction is left to **fertile** individuals. There is an overlap of at least two generations so parents can be assisted by their progeny.

2. No individual can exist independent of the colony nor it can be a member of any other colony but the one in which it developed.

3. All eusocial insects exhibit some degree of polymorphism (Box 10.4) and the different types of individuals in a colony are termed **castes**. The principal castes are **reproductive male** (or "**king**"),

Box 10.4

Polymorphism is the existence of two or more types or individuals within the same species of animal, other than male and female. An example is seen in some social insects such as ants, termites and bees, in which many different types of worker are structurally adapted for different tasks within the colony and in human blood groups.

reproductive female (or "**queen**"), **sterile worker** and **sterile soldier**. Male's function is the **insemination** of the queen, whose role is to produce **new individuals** for the colony. The workers are responsible for the support and maintenance of the colony and soldiers for its defense. Caste is determined by the **ploidy** of the egg (*e.g.*, in bees diploidy for queen and workers and haploidy for the king or drones) and on the **quality** and **quantity** of diet. The queen bee secretes a pheromone called **queen substance** which both suppresses the ovaries of the workers and prevent them form rearing new queens (see **Agarwal, V.K.**, 2009).

A. Social Life in Honeybee (*Apis indica*)

Honeybees are **diurnal**, *i.e.*, they remain active during the daytime. In India three species of *Apis* are commonly found, *viz.*, *Apis dorsata* (Rock bee), *Apis florea* (Little bee) and *Apis indica*(Indian bee). *Apis mellifica* (European bee) occurs in the wild state in Europe. *Apis adamsoni*(African bee) is found in North Africa.

Honeybee society or colony may continue year after year in the same nest, called **hive** or **comb.** The individuals forming a colony are polymorphic having three castes: queen, drones and workers (Fig. 10.21). An average sized colony of honeybee has 10,000 to 16,000 individuals. Of these, only one is queen, 500 to 1000 are drones and the rest are all workers (Box 10.5).

Box 10.5

In honeybee society polymorphism is less highly developed than ants. There is no soldier caste and workers are winged.

1. Queen. The queen of honeybee is diploid fertile female. Her sole duty is to lay eggs. She is thus, the mother of the colony. Queen is unable to form comb or honey. She is fed on a special diet and is perpetually waited upon by the workers.

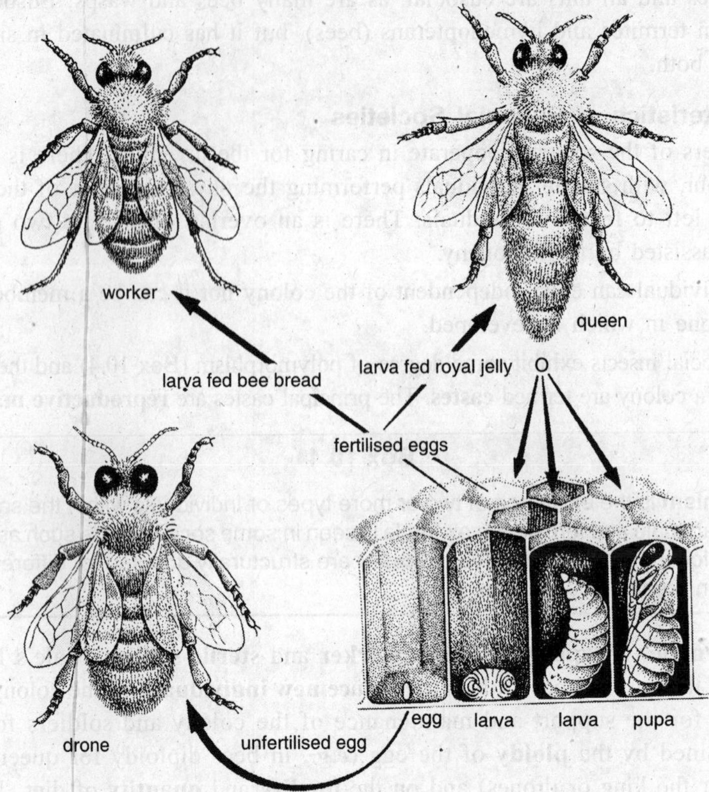

Fig. 10.22. Honeybee. Castes and life cycle.

2. Drones. The drones are the haploid fertile males. They originate from the unfertilized eggs. A new queen takes several **nuptial flights** during which copulation with males (drones) occurs and stores enough sperm to last her life–time (**Ruppert** *et al.*, 2004). The male dies following copulation, for his reproductive organs literally explode into the female. Drones do not work in the hive. They are fed by workers.

3. Workers. The workers are the diploid but sterile females. They perform a variety of tasks for the bee hive. The old workers visit fields, gardens and orchards. They gather and bring nectar, pollen, resin (gum) and water to the hive. Young workers stay in the hive and are called the **house bees**. Among the house bees are the **storekeeper bees** that receive the materials brought by the outdoor workers to store these properly in the cells of hive; the **builder bees** that construct wax cells of the hive; the **repairer bees** that plug cervices between the cells, the **cleaner bees** that remove dead and waste materials; the **fanner bees** that ventilate the hive by beating their wings; the **guard bees** that do fend the colony, the **nurse bees** that feed the larvae (grubs), drones and queen; and the **waiter bees (retinue)** that attend the queen (Box 10.6). The workers also control the production of new queens and disperse the species by founding new colonies. At times of emergency, the workers may lay unfertilized eggs that produce drones.

Box 10.6

Queen bee is looked after with great care by a group of about 50 workers, called **retinue**. These bees wash the queen, comb her, remove her excreta from the hive and periodically feed her with highly nutritious royal jelly.

Box 10.7

Martin Lindauer(1961, 1971) kept a record of the activities of one individual worker throughout her life. He reported two new activities of worker bees besides several already known activities of workers; these are resting and partrolling (see **Agarwal, V.K., 2009**).

Social Activities of Honey Bees

1. Food collection. The fieldworker bees fly from flower to flower and collect two products: nectar and pollen, for using as food.

2. Nectar. The nectar is a sugary secretion in the flowers. It is drawn by worker bees with the help of a **sucking tube** formed by their mouth parts. It is swallowed into a large sac, the **crop** or **honey sac**. At the hind end of this sac is a valve to check the escape of honey, except when needed as food. In the crop, the nectar's sucrose (cane sugar) is acted upon by the enzyme present in the saliva secreted by the honeybee, and changed into **glucose**, **levulose** and **fructose**, which are readily assimilable by human beings. On returning to hive, the worker bees regurgitate the hydrolysed nectar into the storage cells of the hive. Here it is further changed chemically in the mouths of young worker bees that stay home (*i.e.*, store–keepers). The fanner bees then evaporate excess of water from the sweet fluid by vibrating their wings. The final concentrated product is **honey**. The comb cells containing honey are now sealed with wax lids. They are opened at the time of need. Honey is the principal food of adult honey bees and the larvae. Colour and flavour of the honey depends on the source of nectar. It requires 40,000 to 80,000 trips of a worker, and visits to many flowers to collect nectar enough to make 450 grams of honey. An average trip is about 2.4 kilometers. Hence, for a single bee to collect a kilogram of honey would mean travelling four and a half times the distance round the world.

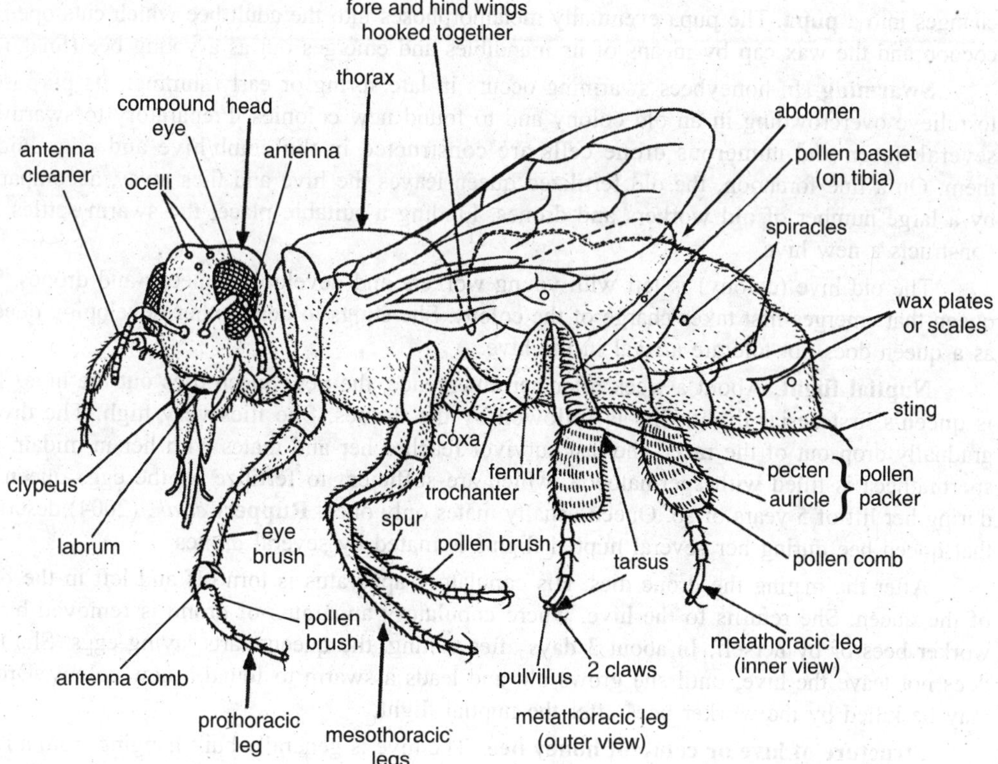

Fig. 10.23. Honeybee. Structure of a worker showing pollen basket and pollen comb.

3. Pollen. As the worker bees suck nectar (by the help of their chewing and lapping mouth parts) from the flowers, the hair covering their entire body get smeared with pollen from anthers of the flowers. The body of the worker is cleaned of the pollen with pollen brushes, present on all the three pairs of legs (Fig. 10.23). From these brushes, the pollen is finally transferred into **pollen baskets** formed on the hindlegs. On return to the hive, the workers push the loads of pollen into storage cells with the assistance of young workers.

(*i*) Royal jelly. A pair of coiled glands, called **maxillary glands** is present in the head of worker bee (which is 7 days old); this gland produces a nitrogenous food, called **royal jelly**. Royal jelly is a white, foamy, yoghurt–like larval food that is necessary for normal queen size and sexual development (**Agarwal, V.K.**, 2009).

Life cycle of Honey Bee

Honey bees undergo **complete metamorphosis**. It includes four stages in its life cycle: egg, larva (grub), pupa and adult (imago). A queen bee is the mother of the whole colony of a beehive. She lays two types of eggs: *fertilized diploid* (32 chromosomes) eggs and *unfertilised haploid* (16 chromosomes) eggs. Diploid eggs develop into *queens* or *workers* and haploid eggs into *drones* (males). Eggs destined to develop into queens, workers and drones are laid in different types of cells of the hive–small cells for workers, medium sized cells for drones, and largest rather conical than hexagonal cells for queens. From respective eggs, the development of queens, drones, and workers respectively takes about 16, 14 and 21 days. A small worm–like **larva**, called **grub**, hatches out from egg in about 3 days. For the first two days, nurse bees feed each larva with royal jelly. Thereafter, the larvae of drones and workers receive honey (for energy) and pollen bread (for protein and for growth), but larvae of future queens continue receiving the royal jelly. Each larva undergoes rapid growth accompanied by five moults. Then, the nurse worker bee close its cell by a wax cap. Within its closed cells, each larva spins a silken **cocoon** around itself and changes into a **pupa**. The pupa eventually metamorphoses into the adult bee which cuts open the cocoon and the wax cap by means of its mandibles and emerges out as a young bee (imago).

Swarming. In honeybees swarming occurs in late spring or early summer. Its purpose is to relieve overcrowding in an old colony and to found new colonies. Preparatory to swarming, several queen and numerous drone cells are constructed in the comb/hive and eggs laid in them. On a fine forenoon, the old fertilized queen leaves the hive and flies away, accompanied by a large number of old workers and drones. Finding a suitable place, the swarm settles and constructs a new hive.

The old hive (colony) is left with young workers and developing queens and drones. The queen that emerges first takes charge of the colony. She stings to death other developing queens, as a queen does not tolerate a rival in the hive.

Nuptial flight. About a week after her emergence, the new queen flies out the hive. This is queen's first aerial flight. She is followed by the drones. She flies very high. The drones gradually drop out of the race. The last survivor reaches her and mates with her in midair. Her spermatheca is filled with spermatozoa, which are sufficient to fertilize all the eggs queen lay during her lift of 5 years or so. Queen usually mates only once. **Ruppert** *et al.*, (2004) described that queen bee during her several nuptial flights is mated by several drones.

After the mating the drone dies. His copulatory apparatus is torn off and left in the body of the queen. She returns to the hive, where copulatory apparatus of drone is removed by the worker bees or by herself. In about 3 days after mating, the queen starts laying eggs. She then does not leave the hive, until she grows old and leads a swarm to found a new colony. Drones may be killed by the worker bees after the nuptial flight.

Structure of hive or comb of honey bee. The hive is generally built hanging from a rock, an arch of a building, hollow of trees, or a branch of a tree. It varies in size from just 15 cm in

Apis florea to about 180 cm in *Apis dorsata*. It consists of thousands of hexagonal cells arranged in two sets opposite to each other on a common base. The cells are thin–walled and so arranged that each side–wall serves for two adjacent cells. The cells of hive are of the two main types:

(*i*) Storage cells. These are about 5 mm wide and commonly lie in the upper portions of hive. They are used for storing honey and pollen.

(*ii*) Brood cells. These cells occupy the central and lower portions of hive. They are used for rearing the brood. They are alike in *Apis dorsata*, but are of three types in other species:

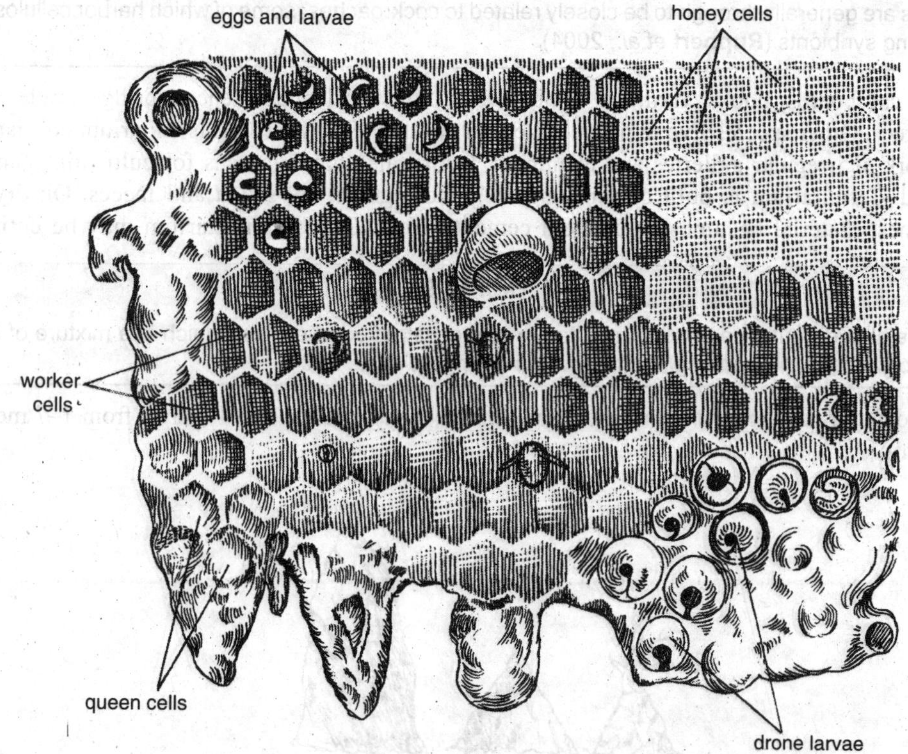

Fig. 10.24. A part of comb of honeybee.

(*a*) Worker cells. These are of the size of storage cells and are used for rearing the workers. Worker cells lie horizontally in the hive.

(*b*) Drone cells. These are about 6mm wide and are used for bringing up the drones. They are also horizontal in position.

(*c*) Queen cells. These cells are still larger and are used to rear the queens. They may be cylindrical or vase–like in form and vertical in position.

There are no cells in the hive for the adults, which keep clustering or roaming about on the surface of the hive.

The hive is built by workers out of **wax** which is secreted by special glands present in their abdomen. The wax collects in the form of small flakes on the undersurface of the abdomen. From here, the flakes are removed by the hindlegs and passed on to the jaws. The jaw convert the wax flakes into a plastic material by masticating and mixing them with secretions of cephalic glands and mould the pastic material into cells. Crevices between the cells are cemented with **propolis**, a sticky resin, collected by the workers from the plant buds. The chambers are polished with **balm**, a special type of propolis, before use. Once built, the hive is used for many years. The various chambers/cells of the hive periodically cleaned, repaired and polished for reuse.

B. Social Life in Termites

Termites (or white ants) are especially fond of cellulose. A symbiotic intestinal protozoan, *Trichonympha*, of termites helps them in digestion of cellulose (Box 10.8). Termites are **nocturnal**, they are adapted for dark places. Some species live in **tunnels** excavated in wooden structures,

Box 10.8

Harvard entomologist, **E.O.Wilson** has referred to termites as **"Social Cockroaches"**, since termites are generally through to be closely related to cockroaches, some of which harbor cellulose–degesting synbionts (**Ruppert** *et al.*, 2004).

such as cardboards, books, furniture, timber, logs, railway sleepers, fences, bridges, poles and living trees. Other species live in nests, called **termitaria** (Fig. 10.25). A termitarium consists of a network of irregular galleries and chambers. It may also have gardens for cultivating fungus for food. Termitarium is built of sand grains glued together by saliva and faeces. On drying up the nest material becomes hard like a cement. (Box 10.9). A termitarium may be entirely

Box 10.9

The aboveground portions of termitarium are constructed of **carton**, which is a mixture of soil and secretions of the frontal gland of workers (**Ruppert** *et al.*, 2004).

underground or may project above the ground like a many–peaked hill. It varies from 1–7 meters in height.

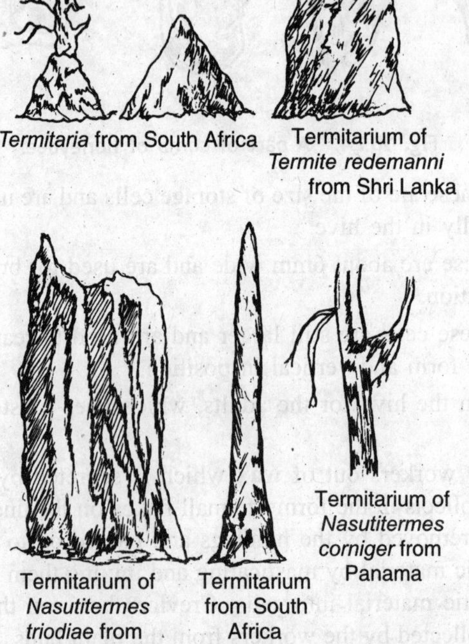

Fig. 10.25. Various types of termite nests.

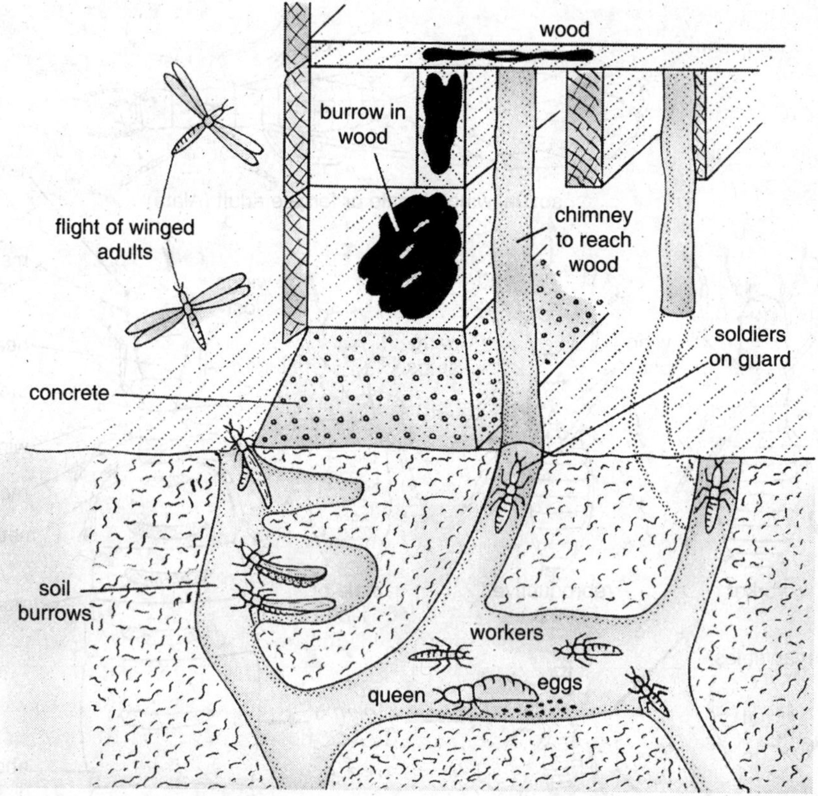

Fig. 10.26. Inner view of a termite colony.

Polymorphism and Social Life. A society or colony of termite has several castes (Fig. 10.26), which are of two main types: fertile and sterile.

1. Fertile castes. The reproductive, sexual or fertile castes (Fig. 10.26) are of following three types:

(*a*) Large–winged (alate) forms or Macropterus castes. These castes are the **king** (male) and **queen** (female), commonly referred to as the **royal couple**. They have a chitinized body of yellow, brown or black colour. The wings extend far beyond the hind end of the body. Wings are used by royal couple only for a brief nuptial flight during which pairing and dispersal occurs. The compound eyes are quite large. The ocelli are usually present. The sole social duty of the queen is to lay eggs. Being mostly filled with eggs in its abdomen, the queen is very large, thick and cylindrical. Being heavy in weight, it lies, almost motionless, in a special chamber of the nest and continue laying for years at a rate of 70 to 80 thousand eggs per year. The king always remains with queen time to time to fertilize the eggs.

Normally there is a single royal couple in a colony, but many couples are formed in the swarming period.

(*b*) Short winged forms or Brachypterous termites. These are **supplementary kings** and **queens**. They have a less chitinized body of pale–yellow colour. The wings are short, pad–like and vestigial. The compound eyes are small. The ocelli may be present or not. The supplementary sexual forms replace the true sexual forms in case the later die.

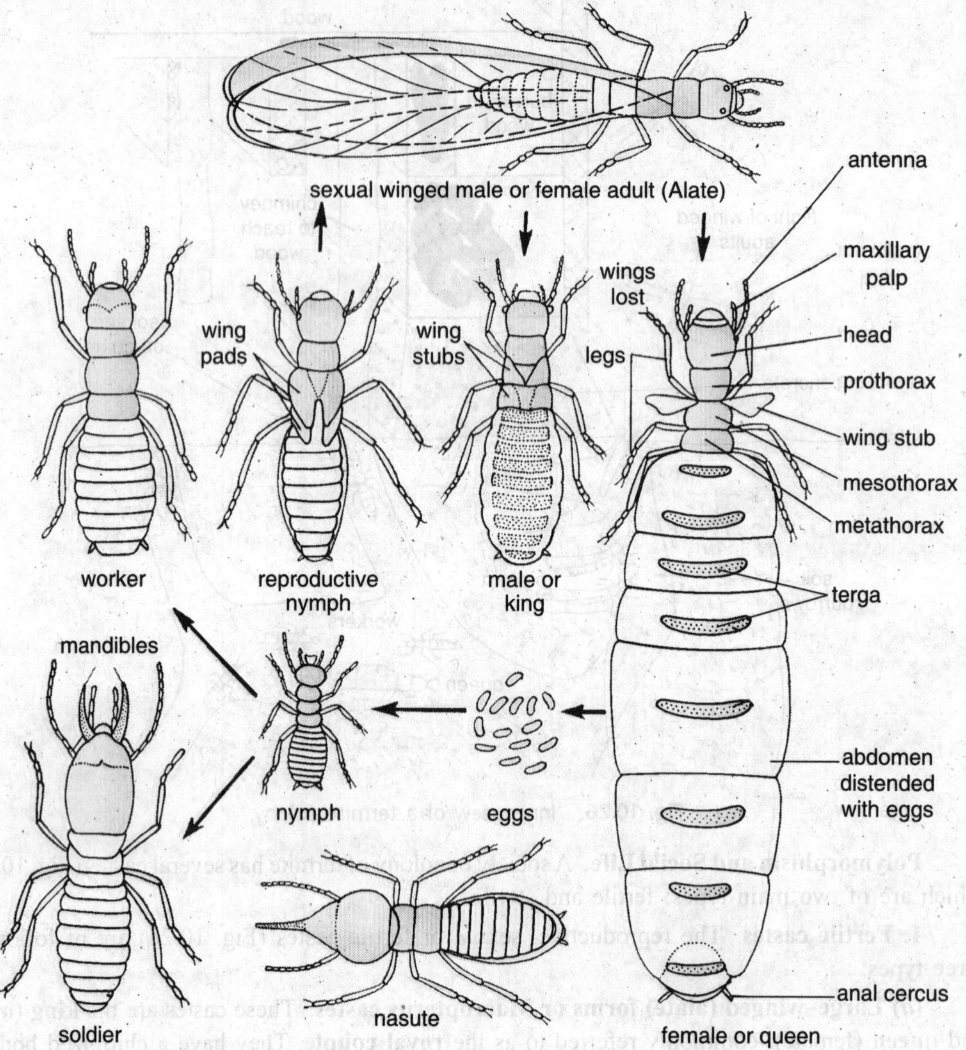

Fig. 10.27. Termite. Polymorphic individuals (castes) and life cycle.

(*c*) Wingless forms or apterous termites. They resemble the workers in appearance. These are rare and occur only in primitive species. Apterous forms have a colourless body. Wings are absent. The compound eyes are vestigial. The ocelli are absent. The sex organs are poorly developed.

2. Sterile castes. The sterile castes are wingless, eyeless (blind) and colourless. The sex organs are rudimentary. Hence, these castes are incapable of reproduction. These are of following three types:

(*a*) Workers. The workers are the most numerous of all castes. Many workers remain engage in queen's service. Workers are small, wingless and sightless (blind). They are generally **dimorphic**, *i.e.*, occur in two forms: small and large. In some cases, they are **trimorphic**, being small, intermediate and large. The workers construct and repair the nest and tunnels of the colony; search for, collect and distribute the food; look after the eggs; and feed the nymphs, soldiers and the royal couples (true king and queen).

(*b*) Soldiers. The soldiers are like the workers, but are somewhat larger and possess stronger mouth parts. Soldiers have a large, sclerotized head with huge, serrated mandibles. Soldiers also may be dimorphic or trimorphic having small, intermediate and large forms. The soldiers protect the nest and colony from the enemies (they kill the intruders with their mandibles) and keep a control over the workers.

(*c*) Nasute. The nasute are also called **proboscideans**. They have vestigial mandibles. They possess a large head prolonged into a **rostrum** which bears at its tip the opening of a **frontal gland**. The nasutes carry on chemical warfare. They throw a jet of toxic, repugnatorial and sticky secretion of the frontal gland on the invaders through their rostrum. This secretion kills or immobilizes the enemy. This secretion is also used to dissolve hard substances, such as stones, coming in the way of the workers during nest building. The nasutes occur in place of the soldiers in higher termites, and may be of 2 or 3 forms: small, intermediate and large.

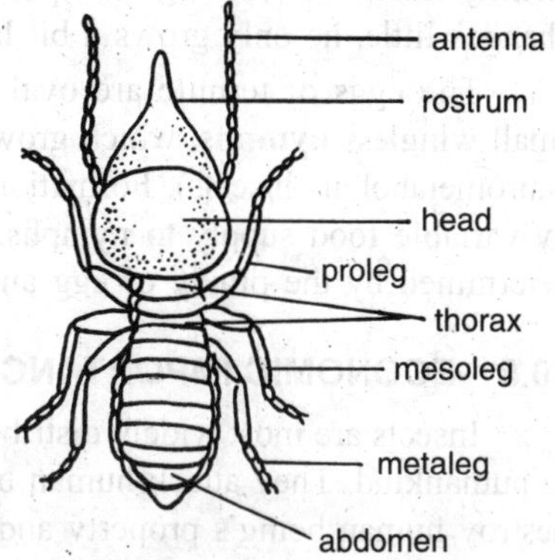

Fig. 10.28. Termite. A nasute.

The soldiers and nasutes are not very successful. This is shown by the fact over 100 species of other insects, arachnids, centipedes and millipedes live regularly as guests in the termite nests.

Feeding. The termites commonly feed on living as well as dead plant material (Box 10.10). They are able to digest cellulose with the help of symbiotic protozoa, such as **trichonympha**, that live in their intestine. The termites may also take fungi and dead bodies of their follows.

Box 10.10 Cellulose utilisation and proctodeal feeding

Termites have specialized in the utilization of **cellulose** and **lignin** of cell walls of wood, grass, leaves, humus, or herbivore dung as food sources. To utilize cellulose, termites apparently are capable of some **cellulase** enzyme synthesis themselves, but they also enlist the aid of flagellates (protozoans), bacteria and fungi. Most termites rely on **endosymbiotic flagellates** in the hindgut as the primary source of cellulose needed for the hydrolysis of wood. Anaerobic respiration of cellulose by the flagellates produces acetic acid, which is then absorbed by the hindgut epithelium and transferred to the blood for distribution to cells, where it will respired aerobically.

Some termites have a gut flora of anaerobic bacteria instead of flagellates. Fungus-growing termites culture a basidiomycete fungus on "combs" of decaying plant material and termite faeces. The fungi break down lignin prior to indigestion of the comb by the termites. Following ingestion, the products of lignin metabolism are metabolized further by gut bacteria into compounds that can be absorbed and utilized by the termite. Young termites must acquire their bacteria or protozoan symbionts via **proctodeal** feeding from the anus of another termite, a process that require **social interaction**. The termite symbiont association thus may have been important in the evolution of social behaviour (**Ruppert** *et al*., 2004).

Swarming. In the rainy season, true kings and queens are produced in large numbers. They leave the termitarium through holes made by the workers, and fly away to new sites. this is called **dispersal flight** or **swarming**.

Mortality is very high during and after swarming. A vast majority of these winged forms is eaten by the dogs, birds, lizards, toads and ants. Only a few lucky ones survive to found a new colony.

Founding new colonies. At the new site, one king and one queen mate, shed wings at the preformed breaking point (dealation), dig out a burrow, and start a new colony. The queen lays eggs. These eggs are tended by the royal couple, and hatch into workers, which soon take up their duties. The stored fat and reabsorbed flight muscles enable the founding couple to go without feeding until the first workers are ready to gather food. However, some eggs or nymphs are eaten up by the parents. The king lives with the queen for life, and they mate at regular intervals. The queen grows her abdomen to an enormous size, 50–70 mm. She now cannot move about and lies passive in the royal chamber. She is tended by the workers and defended by the soldiers. Her abdomen get packed with eggs. The queen may live for 6–15 years, and lay about a million eggs during her life. She is thought to be the most prolific of all insects. When her fertility lessen, her feeding is suspended, and she dies of starvation. her body is eaten up. The king changes little. he only grows a bit larger due to good diet and passive life spent near his queen.

The eggs of termite are oval and pale colour and have smooth surface. They hatch into small wingless **nymphs**, which grow into various castes after several months (Note. Termites are paurometabolous insects). Formation of different castes from similar eggs is perhaps determined by variable food supply to nymphs. According to **Ruppert** *et al.*, (2004) castes of termites are determined by the ploidy of egg and on the quality and quantity of the diet.

10.5 ECONOMIC IMPORTANCE OF INSECTS

Insects are most widely distributed and most powerful. They are useful as well as dangerous to humankind. They attack human beings and their domestic animals by causing disease. Insects destroy human being's property and crops. Hence, insects are of great economic importance for humans. But some insects are beneficial to human beings. Economic importance of insects to humankind can be discussed under the following headings.

A. Injurious or Harmful Insects

1. Injurious to crops. Many insects damage forest trees, growing farm crops, flowers, fruits and stored grain. They mostly belong to orders Lepidoptera, Coleoptera, Diptera and Hemiptera.

(*i*) The **brown tail moth** (*Euproctis*) and **gypsy moth** (*Lymantria*) are serious pests of shade and foliage trees; their larvae are a menace and destroy many types of forest trees.

(*ii*) The **Hessian fly** (*Myetiola*) is a small-sized midge; its larvae bore into and feed upon the pith of wheat plant.

(*iii*) The larvae of moths called **corn borer** (*Chilo* and *Pyrausta*) kill the corn and other monocotyledonous cereal plants by boring into their stems.

(*iv*) The **sugarcane leaf hopper** (*Pyrilla*) sucks the juice of sugarcane both as adult and nymph, causing great loss of sugar.

(*v*) Larvae of moth called **sugarcane stem borer** (*Argyria*), bore through the stem of sugarcane, some cereal plants and fruit trees.

(*vi*) The **Indian rice leaf-hopper** (*Nephotettix*) and **rice bug** (*Leptocorisa*) attack rice in great numbers eating the leaves and ears.

(*vii*) The nymphs and adults of *Hieroglyphus*, an orthopteran eat up the growing shoots of rice plants, thus, preventing formation of grain.

(*viii*) The **Indian cotton bug** or **Red cotton stainer** (*Dysdercus*) sucks the juice of leaves, buds and tender seeds of cotton plants. The cotton produced becomes spotted.

(*ix*) *Aphis*, a hemipteran, is a serious cotton pest of India.

(*x*) **Cabbage aphid** sucks the juice of stem and flowers of the plant of mustard, cauliflower and damage them.

(*xi*) Larvae of two Lepidoptera, *Agrotis* and *Gnorimoschema* are potato cut worms of India; the former feeds on potato leaves and cuts off stems, while the larvae of latter eat the potatoes in the field and stores. Larvae of *Agrotis* are also destructive to pea, cabbage, tobacco, groundnut, wheat and cauliflower.

(*xii*) **Red pumpkin beetle** (*Rhaphidopalpa* or *Aulacophora*) chews and destroys the young and tender plants of following vegetable plants of family Cucurbitaceae such as *Cucurbita* (Kaddu), *Laginaria* (Lauki), *Momordica* (Karela), *Citrullus* (Tindà), *Luffa* (Torai), *Cucumis* (Kakri), etc.

(*xiii*) Larvae of *Bruchus*, a beetle, bores into pods of beans and peas kiling the seed.

(*xiv*) **Grasshoppers** (*Melanopus*, *Poecilocerus*) and locusts (*Schistocerca*) chew the tender leafy plants such as rice, wheat, maize, etc., and cause great damage. Swarms of migratory locusts cause great damage to crops and nightmares to farmers.

(*xv*) Many insects attack fruit trees, they damage roots, trunks, stems, leaves, inflorescence and fruits of these plants. *Drosicha* (**mealy bug**) causes destruction of mangoes and plums. Adults of *Ideocerus* (**mango leaf hopper**) attack the inflorescence of mango plant and suck the sap, thus, they cause considerable damage by preventing formation of mango fruit. *Psylla* (**apple bug**) lays eggs on apple and pear trees, the nymphs on hatching damage the blossoms and shoots. The larvae of *Anthonomus*, a beetle also destroys apple blossom and prevent formation of fruit.

(*xvi*) *Bagrada* (**painted bug**) destroys cruciferous plants by sucking the juice of their tender stems.

(*xvii*) Caterpillar larvae of moths such as *Earias* (**spotted bollworm**) and *Pectinophora* (**pink bollworm**), bore into flowers, young bolls and seeds of cotton, rendering these hollow and useless.

(*xviii*) **Gram caterpillar** and **corn earworm** (*Heliothis*), bore into stems of gram, corn, tomato, tobacco and some other plants.

(*xix*) Many moths (caterpillars) and beetle cause a great deal of damage to stored grains, two beetles, *Tenebrio* and *Tribolium* have similar habits and are commonly found in stores and granaries, the former is found in all stages in meal, flour and stored goods, its larvae are called **meal worms**. *Tribolium* eats stored wheat and grain. *Calandra*, a weevil bores through grain of rice and other stored grain in India.

(*xx*) **Gall insects** produce swellings on the plants to get shelter and food.

2. Injurious to Human Beings and Domestic Animals

1. Parasites. Some insects live upon (**ectoparasites**) or inside (**endoparasites**) the bodies of human beings and domestic animals. The ectoparasites suck the blood of their hosts at short intervals. Examples of such insects are the following:

(*i*) **Lice** (*Pediculus*). These insects occur between hairs of human beings and domestic animals and feathers of birds. **Human lice** are of two subspecies—head louse, *Pediculus humanus capititis* and the body louse, *Pediculus humanus corporis* (it mostly harbours the hairs of arm pits and pelvic region). Besides sucking blood, the lice cause annoyance and itching by their crawling in between hairs or upon host body.

(*ii*) **Fleas.** These are also blood-sucking ectoparasites found amongst the hairs of mammals and feathers of birds. **Indian rat flea** (*Xenopsylla*) transmits **bubonic** and other types of **plague**.

(*iii*) **Bedbugs** (*Cimex*). These are found concealed in cracks and crevices of our mats, furniture, cots, wells, etc. During night these suck human blood and cause intense itching, burning and restlessness, disturb us in our sleep. *Cimex* also transmits typhus fever.

(*iv*) The **forest fly** of cattle and horses (*Hippobosca*) sucks the blood of its hosts and often causes haemorrhage.

(*v*) The **chicken louse** (*Menopon*) sucks blood and causes destruction of fowls. Endoparasitic insects usually occur in cattle. **Botflies** (*Gasterophilus*) are most important amongst these. Larvae of botflies excavate tunnels in host skin, causing great discomfort and pain. These also render the skin useless for leather. **Sheep and horse botflies** lay eggs near the nose of their hosts. Hatched larvae may invade their brain. Larvae of **warble flies** (*Hypoderma*) bore below the skin of oxen and make holes for breathing; they thus, injure the hide but also reduce the meat and milk supply.

3. Disease transmitting insects. The insects not only cause disease by leading parasitic life, but also act as **vector** and play a dreadful role by transmitting disease causing (*i.e.*, pathogenic) viruses, bacteria, protozoans, etc., from host to host either by direct or indirect mechanical way. The various species of genus *Anopheles* (mosquito) transmit parasitic protozoa (*e.g., Plasmodium*) which causes **human malaria**. *Culex* mosquito carries *Filaria bancrofti* (Namatode) which cause **filariasis** in human beings. Deadly disease of **sleeping sickness** is transmitted by **tse-tse fly** (*Glosina*). Other dreadful diseases such as **cholera, typhoid, diarrhea, amoebic dysentery, gangrene, gonorrhea, leprosy** and **tuberculosis** are also spread by **housefly**; **plague** by fleas, **relapsing fever** by **bedbug** and **Kala-azar** by **sandflies** (*Phelbotomous*).

Certain diseases of the domestic animals are also transmitted by certain insects. For example, **nagana** disease in horses and cattle and **surra** disease in domestic animals are conveyed by the flies *Glossina* and *Tabanus*, respectively.

4. Household pests. Human food is spoiled by cockroaches, ants, flies and weevils. Various species of moths (*Tinea, Teniola* and *Trichophaga*) are clothes moths, they lay eggs on warm clothes; the larvae on hatching eat and destroy clothes. They also feed on furs, carpets and dry fruits. *Lepisma*, the silver fish and *Liposcelis*, the book louse live in and destroy books and old manuscripts. *Termites*, the white ants cause great damage to books, carpets, wood furniture and wood-work of the buildings.

B. Beneficial Insects

1. Productive insects. Some insects are of commercial value. They manufacture various products which fetch good market value.

(*i*) **Honeybee** (*Apis*) supply to humankind millions of kilograms of honey and beeswax. **Beeswax** is obtained from its combs and is used in making candles, cream, polish, carbon ribbons, papers, cosmetics and certain ointments. **Honey** is a natural antiseptic and valuable tonic for human body. A drug prepared from the bodies of honeybees too has been used in the treatment of arthritis and snake bite. Rearing and culturing of *Apis* to obtain honey and bees wax is called **apiculture**.

(*ii*) The **lac insects**, *Tacchardia lacca*, secrete lac in the form of protective covering by the integumentary glands of female. The lac is used in the manufacture of **shellac**, varnish, polish, buttons, bangles, pots, gramophone records, sealing wax, toys and some electrical items. A dye is also obtained from dead and dried bodies of females of lac insect which is used by Indian womens for making "alta" or mahavar.

(*iii*) The dark red **cochineal dyes** in Mexico, Spain, etc., are manufactured from the dried bodies of female of the genus *Dactylopius coccus*. This dye is used in cosmetics, colouring the wine and in medicines for whooping cough.

(*iv*) **Cantharidine**, a drug used in healthy growth of hair, is obtained from **blister beetle** (*Lytta* and *Mylabris*).

(*v*) The silk moth, *Bombyx mori* (Fig. 10.29) and other genera, are reared the world over and especially in India, China, Japan and Europe to obtain the best quality of **silk**. The raw silk is obtained from the cocoons of these insects and the thread is reeled off to prepare the silk. The pure silk fetches a good market price and is demanded world over.

(*vi*) Gall insects are helpful in the manufacture of **tannins** which are used variously.

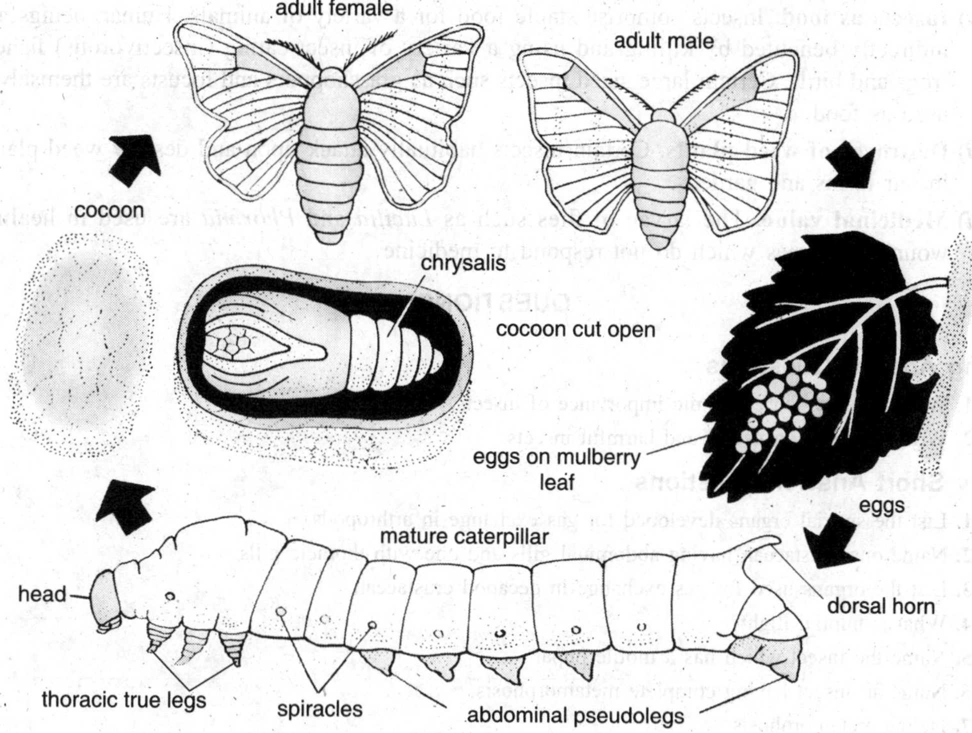

Fig. 10.29. *Bombyx mori* (silkmoth). Life history shows holometaboly.

2. Helpful insects. Many insects are beneficial to human kind in the following ways.

(*i*) **Insect pollinators.** Insects of various groups are of great importance to agriculture, as they are the pollinating agents of various plants. Various fruit, vegetable and garden plants are mostly pollinated by insects such as **honeybee, wasps, butterflies, moths,** etc. These insects sit upon flower to flower for sucking their nectar and pollinate them in this process. Flowers of many plants are structurally adapted for insect-pollination.

(*ii*) **Scavengers.** Some insects are scavengers; they eat up dead animals and vegetable matter, thus, they prevent decay. Some ants, larvae of flies, beetles, crickets, cockroaches, etc., form common example of scavengers.

(*iii*) **Predators.** Some insects, called **entomophagous insects**, feed upon harmful insects and, thus, are useful insects. *Stagomantis*, a mantis is a voracious feeder devouring flies, grasshoppers and caterpillars which are injurious to crops. The larvae and adult of lady bird beetle (*Chilomenes*) feed upon aphids which infect the cotton plants. *Novius* (a **ladybird beetle**) destroys scale worms which are pests of orange and lemon trees. The larvae of **blister beetle** (*Epicauta*) bore through egg-capsules of locusts and destroy them. *Calasoma*, a **ground beetle**, preys upon many kinds of lepidopterous larvae which destroy cereals and cotton. Certain hymenopterous larvae and carnivorous wasps feed upon aphids in large numbers. Large dragon flies feed upon mosquitoes.

(*iv*) **Parasitoid insects.** Certain **parasitic** insects are useful for human beings in the sense that they either inhabit insects which are harmful or feed upon them. *Chaloids* and *Icheumon* flies are parasitic forms which lay eggs in cocoons and larvae of phytophagous lepidoptera. Similarly larvae of *Tachina* fly are parasites of the lepidopterous larvae.

(*v*) **Insects as food.** Insects comprise staple food for a variety of animals. Human beings are indirectly benefited by killing and using a variety of insect-eating (insectivorous) fishes, frogs and birds. Certain large sized insects such as grasshoppers and locusts are themselves used as food.

(*vi*) **Destroyer of weed plants.** Certain insects habitually attack upon and destroy weed plants in our farms and gardens.

(*vii*) **Medicinal value.** The larvae of flies such as *Lucilia* and *Phormia* are used in healing wounds of bones which do not respond to medicine

QUESTIONS

Long Answer Questions

1. Write an essay on economic importance of insects.
2. Write an essay on useful and harmful insects.

Very Short Answer Questions

1. List the special organs developed for gas exchange in arthropods.
2. Name one crustacean having abdominal gills and one with thoracic gills.
3. List the organs used for gas exchange in decapod crustacean.
4. What is nuptial flight?
5. Name the insect which has a motile pupa.
6. Name an insect having complete metamorphosis.
7. Define metamorphosis.
8. What are paurometabolous insects?
9. What is role of corpora allata in metamorphosis of insects?
10. Which social insect has proctodeal feeding?
11. Which insect spreads cholera?
12. From which insect we get the silk?

Fill in the Blanks

1. Simple ocelli have light-sensitive cells, called ....
2. Scorpion has .... ocelli.
3. Spiders have .... eyes in which light-sensitive cells are ....
4. .... eyes of Insecta and Crustacea have many photoreceptive units, called ....
5. Dark-adapted compound eyes of crustaceans are called .... eyes.
6. Insects lack true .... nerve.
7. The .... of .... bring about exchange of gases directly between the air and the tissue cells.
8. The respiratory system of scorpion has .... for respiration.
9. Prawn and crabs have .... or .... gills.
10. Hymenoptera have .... pupae.
11. Polymorphism occurs in social insects such as yellow wasps (*Polistes*), ants, ... and ....
12. *Lepisma* is an .... insect which does not undergo metamorphosis.
13. Examples of .... insects are stonefly and dragonfly.
14. *Wuchereria bancrofti* is transmitted by ....
15. Rat flea spreads ....

Multiple Choice Questions

1. An insect that produces shellac is

(*a*) *Apis indica* (*b*) *Bombyx mori*

(*c*) *Laccifera lacca* (*d*) *Dictyolopius coccus*

2. Insects responsible for transmitting diseases are called

(*a*) pests (*b*) social insects

(*c*) pathogens (*d*) vectors

3. Insects are greatest competitors of human beings because most of them

(*a*) eat and destroy plants and fruits (*b*) are responsible for spreading of serious diseases

(*c*) destroy household goods and properties (*d*) are responsible for all the above

4. Insects responsible for destruction of buildings and wooden structures are

(*a*) moth (*b*) locusts

(*c*) weevils and grubs (*d*) termites

5. The insect vector for kala-azar is

(*a*) tse-tse fly (*b*) sandfly

(*c*) housefly (*d*) fleas

6. Most ticks and mites are

(*a*) free living (*b*) parasites

(*c*) predators (*d*) all of the above

7. Lac is produced as

(*a*) secretion from body (*b*) excretion from body

(*c*) excess fluid oozing out of body (*d*) faeces of lac insects

8. Cantharidin is obtained from

(*a*) ball weevils (*b*) honey bees

(*c*) beetles (*d*) white ants

9. Which of the following is a pest of paddy?

(*a*) *Dacus* (*b*) *Calandra*

(*c*) *Dysdercus* (*d*) *Leptocorisa*

10. One cocoon of silk moth gives the fiber of

(*a*) 10 feet (*b*) 100 feet

(*c*) 1000 feet (*d*) 10,000 feet

11. Bee-venom is used in the treatment of

(*a*) arthritis (*b*) malaria

(*c*) myopia (*d*) cholera

12. Bedbug is

(*a*) endoparasite (*b*) ectoparasite

(*c*) symbiotic (*d*) none of these

13. The feeding habit of bedbug is

(*a*) herbivorous (*b*) carnivorous

(*c*) omnivorous (*d*) sanguivorous

14. The pest called sugarcane leaf hopper is

(*a*) *Dysdercus* (*b*) *Pyrilla*

(*c*) *Hieroglyphus* (*d*) *Chilo*

15. The tracheae are not branched in

(*a*) insects (*b*) centipedes

(*c*) millipedes (*d*) all the above

16. *Palaemon* has for gas exchange
(*a*) podobranchs (*b*) arthrobranchs
(*c*) pleurobranchs (*d*) all the above

17. Wriggler is the larva of
(*a*) housefly (*b*) butterfly
(*c*) moth (*d*) mosquito

18. Which of the following is a social insect?
(*a*) butterfly (*b*) moth
(*c*) honey bee (*d*) locust

ANSWERS

Very Short Answer Questions

1. Branchiae (gills), tracheae, book lungs and book gills.
2. *Squilla*, *Palaemon*.
3. Lining of branchiostegites, epipodites and branchiae.
4. Flight for mating in the air (*e.g.*, honey bee).
5. Mosquito.
6. Butterfly.
7. **Metamorphosis** is changing from one form to another, such as transformation of insect larva to a sexually mature adult or from a nymph to imago (adult).
8. Insects having gradual metamorphosis such as cockroach, termite, grasshopper, etc.
9. It secretes juvenile hormone (JH).
10. Termite.
11. Housefly.
12. Cocoon formed around it by caterpillar of the silk moth (*Bombyx mori*). Cocoon is the source of silk fiber.

Fill in the Blanks

1. Retinular cells;	**2.** Direct;	**3.** Secondary, inverted;
4. Compound, ommatidia;	**5.** Superpositional;	**6.** Optic;
7. Tracheal system, insects	**8.** Book lungs;	**9.** Lamellar, phylobranchiate;
10. Adecticous;	**11.** Termites, bees;	**12.** Ametabolous;
13. Hemimetabolous;	**14.** *Culex* mosquito	**15.** Bubonic plague.

Multiple Choice Questions

1. (*c*)	**2.** (*d*)	**3.** (*d*)	**4.** (*d*)	**5.** (*b*)	**6.** (*b*)
7. (*a*)	**8.** (*c*)	**9.** (*d*)	**10.** (*c*)	**11.** (*a*)	**12.** (*b*)
13. (*d*)	**14.** (*b*)	**15.** (*c*)	**16.** (*d*)	**17.** (*d*)	**18.** (*c*).

11

Invertebrate Hormones

Animals produce and distribute within their bodies chemicals that regulate and coordinate biological activities. These chemicals are termed **chemical coordinators** and they include almost any substance that has some effect on bodily functions. One special category of chemical coordinators is the **hormones**. This term refers to any chemicals that are produced and secreted by some organ or tissue, and are then carried by the blood or other body fluid (*e.g.*, haemolymph) to exert their influence elsewhere in the body. In vertebrates this type of phenomenon is associated with the **endocrine system**, which includes well known glands as production sites. Hormones are divided into two main types: **1. Endocrine hormones** which are produced by more or less isolated glands and released into the circulatory fluid. **2. Neurohormones** which are produced by special neurons called **neurosecretory cells.**

Much remains unknown concerning hormones in invertebrates. Most of our information comes from studies on insects and crustaceans, although hormonal activity has been demonstrated in a few other taxa and is suspected in many others (**Brusca** and **Brusca** 2003). Among the arthropods, hormones are involved in the control or growth, molting, reproduction, eye pigment migration, and probably other phenomena; in at least some other taxa (*e.g.*, annelid worms such as earthworms), hormones influence growth, regeneration and sexual maturation.

Hormones do not belong to any particular class of chemical compounds, nor do they all produce the same effects at their sites of action: some are **excitatory**, some are **inhibitory**. Because endocrine hormones are carried in the circulatory fluid, they reach all parts of an animal's body. The site of action, or **target site**, must be able to recognize the appropriate hormone(s) among the myriad other chemicals in its surroundings. This recognition usually involves an interaction between the hormones and the cell surface at the target site. Thus, under normal circumstances, even though a particular hormone is contacting many parts of the body, it will elicit activity only from the appropriate target organ or tissue that recognize it.

In general sense, **pheromones** (also called **ectohormones**, **Novak**, 1975, see **Agarwal**, 2009) are substances that act as "**interorganismal hormones**". These chemicals are produced by organisms and released into the environment, where they have an effect on other organisms. Most pheromone research has been on *intraspecific actions* such as mate attraction in insects which are frequently related to the airborne chemicals (*e.g.*, bombykol of female silk moth, *Bombyx mori*). Thus, intraspecific pheromones can be viewed as coordinating the activities of populations, just as hormones. There is also great deal of evidence for the existence of **interspecific pheromones**. For example, some predatory species (*e.g.*, some sea stars) release chemicals in the water that elicit extraordinary behavioural responses on the part of potential prey species, generally in the form of escape behaviour (see **Brusca** and **Brusca** 2003).

11.1 HORMONES IN CRUSTACEA

In both crustaceans and insects (and probably all arthropods), the initiation of molting, beginning with the events of premolt or proecdysis, is brought about by the action of a molting hormone, called **ecdysone**. Apparently, however, the pathways controlling the secretion of ecdysone

Box 11.1 Ecdysis

The process of shedding the exoskeleton is called **molting** (or moulting) and it is a phenomenon characteristics of arthropods and a few other invertebrates with thick cuticles (*e.g.*, onychophorans, tardigrades, kinorhynchs, nematodes, nematomorphans, loriciferans and priapulans). In all arthropods (and onychophorans and tardigrades) molting is regulated by a hormone, called **ecdysone**, thus, molting in these groups is referred to as **ecdysis** (**Brusca** and **Brusca**, 2003).

are different in insects and crustaceans (Fig. 11.2 and 11.3). In crustaceans ecdysone is secreted by an endocrine gland called the **Y–organ** located at the base of the antennae or near the mouthparts. The action of the Y–organ is controlled by a complex neurosecretion apparently located near the eyes or eyestalks. During the intermolt period, a **molt–inhibiting hormone (MIH)** is produced by neurosecretory cells of the **X–organ**, located in a region of the eyestalk nerve (or ganglion) called the **medulla terminalis** (Fig. 11.1). MIH is carried by axonal transport to a storage area called the **sinus gland** which appears to control MIH release into the blood (Fig. 11.2). As long as sufficient levels of MIH are present in the blood, the production of ecdysone by the Y–organ is inhibited.

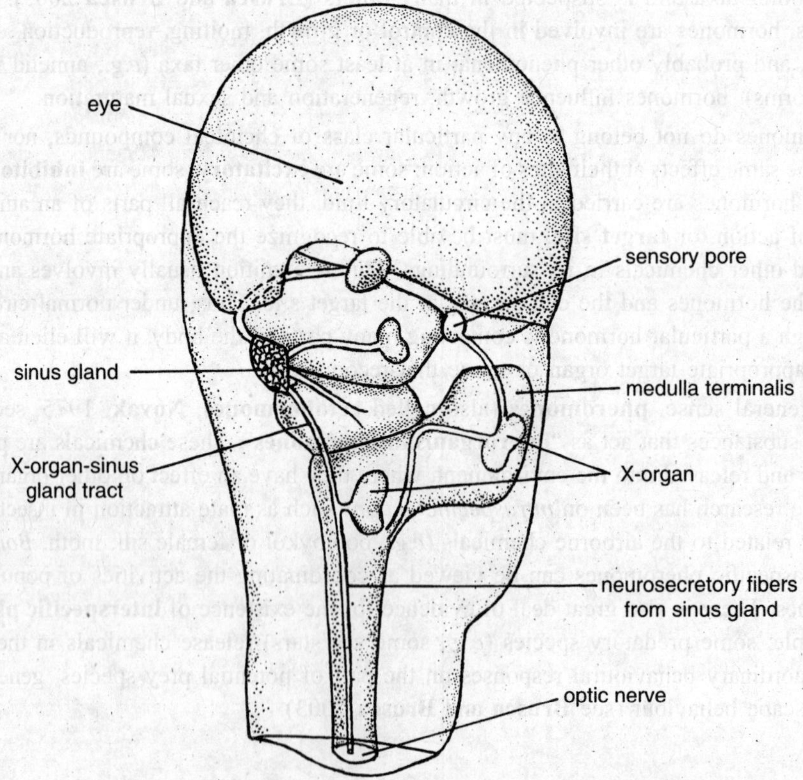

Fig. 11.1. The neurosecretory apparatus in a crustacean eyestalk.

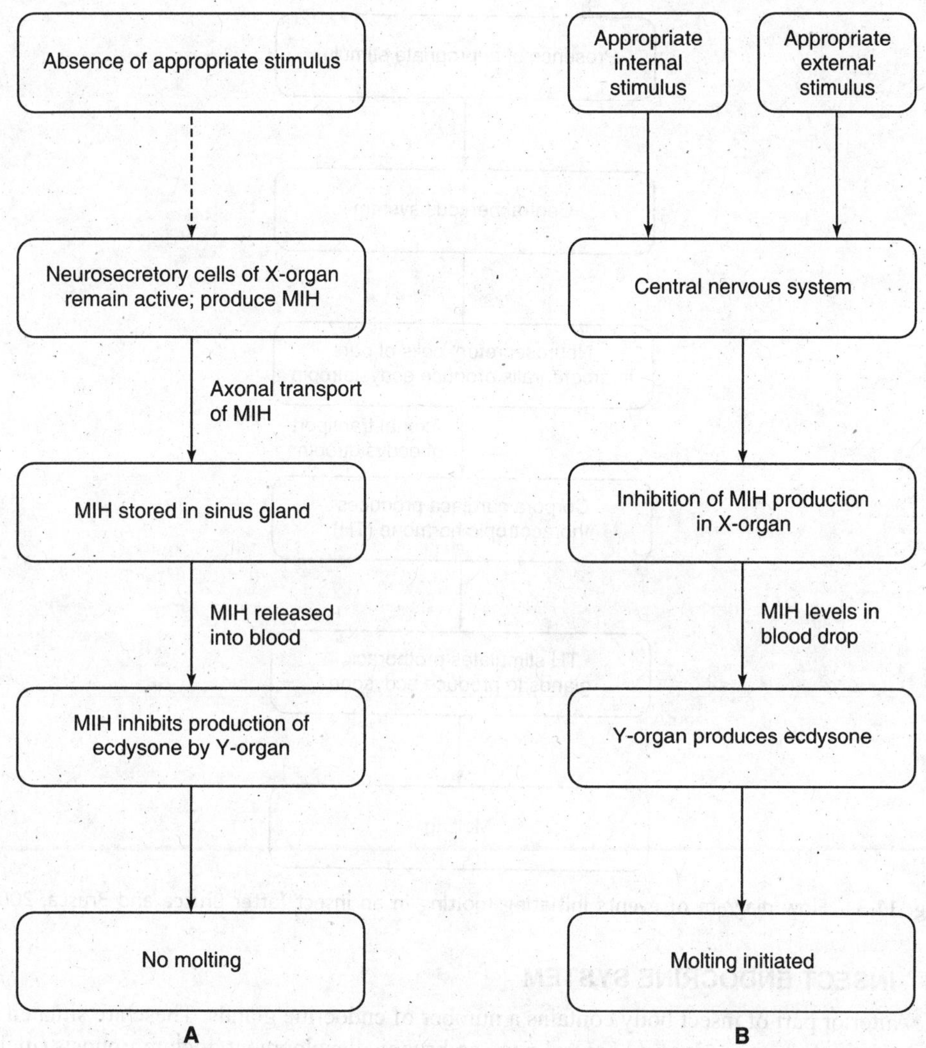

Fig. 11.2. Flow diagram of events inhibiting (A) and initiating (B) molting in crustaceans.

The active premolt and subsequent molt phases are initiated by sensory input to the central nervous system. The stimulus is external for some crustaceans (*e.g.*, day length or photoperiod for certain crayfishes) and internal for others (*e.g.*, growth of soft tissues in certain crabs). External stimuli are transmitted via the central nervous system to the medulla terminalis and X–organ (Fig. 11.2). Appropriate stimuli inhibit the secretion of MIH, ultimately resulting in the production of ecdysone and the initiation of a new molt cycle.

The sequence of events in insects is somewhat different from that in crustaceans in that a molt inhibitor is apparently not involved. When an appropriate stimulus is introduced to the central nervous system, certain neurosecretory cells in the cerebral ganglia are activated. These cells, which are located in the **pars intercerebralis**, secrete **ecdysiotropin**. This hormone is carried by axonal transport to the **corpora cardiaca**, paired neural masses associated with the cerebral ganglia. Here, **thoracotropic hormone** is produced and carried to the **prothoracic glands**, stimulating them to produce and release ecdysone (Fig. 11.3).

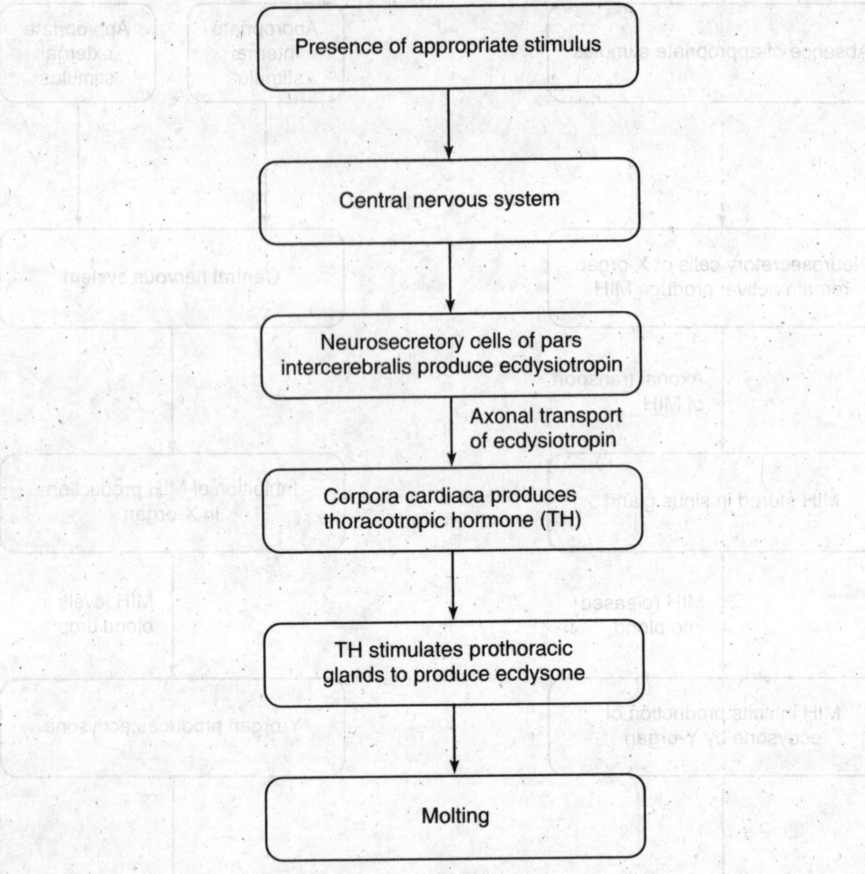

Fig. 11.3. Flow diagram of events initiating molting in an insect (after Brusca and Brusca, 2002).

11.2 INSECT ENDOCRINE SYSTEM

Anterior part of insect body contains a number of endocrine glands. These are situated in the head or in prothoracic region and control post–embryonic development, metamorphosis (including molting) and metabolic regualtion. The endocrine glands of the insects are of following four types:

1. Neurosecretory cells (NSC). These are specialized neurons which are situated in the brain. There are two groups of these cells in the **protocerebral lobes**, a pair of median and a pair of lateral groups present in pars medialis and pars lateralis respectively. These neurons manufacture brain hormone, called **prothoracicotropic hormone (PTTH)**. NSCs release PTTH in response to neural, hormonal or environmental signals.

PTTH is a polypeptide molecule with molecular weight about 40,000 (**Gilbert** 2010). It is transported along the axons (by microtubules) of neurosecretory cells. The axons of these NSCs pass through corpus caridacum and terminate in corpus allatum. PTTH controls the functioning of prothoracic glands and indirectly controls moulting or metamorphosis in insects.

2. Corpora cardiaca (Singular **Corpus cardiacum** or **CC**). These are a pair of small bodies situated just behind the brain in close association with aorta. Corpora cardiaca are **neurohaemal organs** of insects which store ecdysiotropin (EcT) or prothoracicotropic hormone (PTTH) of NSCs of brain. EcT is released into blood (haemolymph) by corpora cardiaca and it has following two effects; 1. It directs corpora allata to synthesize and release JH(= juvenile hormone) in blood; and 2. It directs prothoracic glands to secrete α-ecdysone in blood.

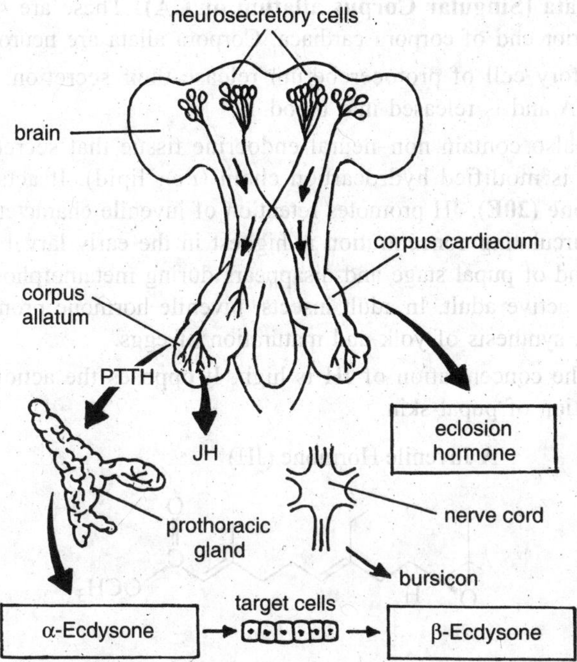

Fig. 11.4. Insect endocrine system and its role in moulting or metamorphosis.

Ecdysone is a peptide and it induces emergence of adult from puparium. Ecdysone also regulates the secretory activity of the prothoracic glands. The corpora cardiaca are said to promote egg development and to increase the frequency of contraction of heart and gut muscles in insects.

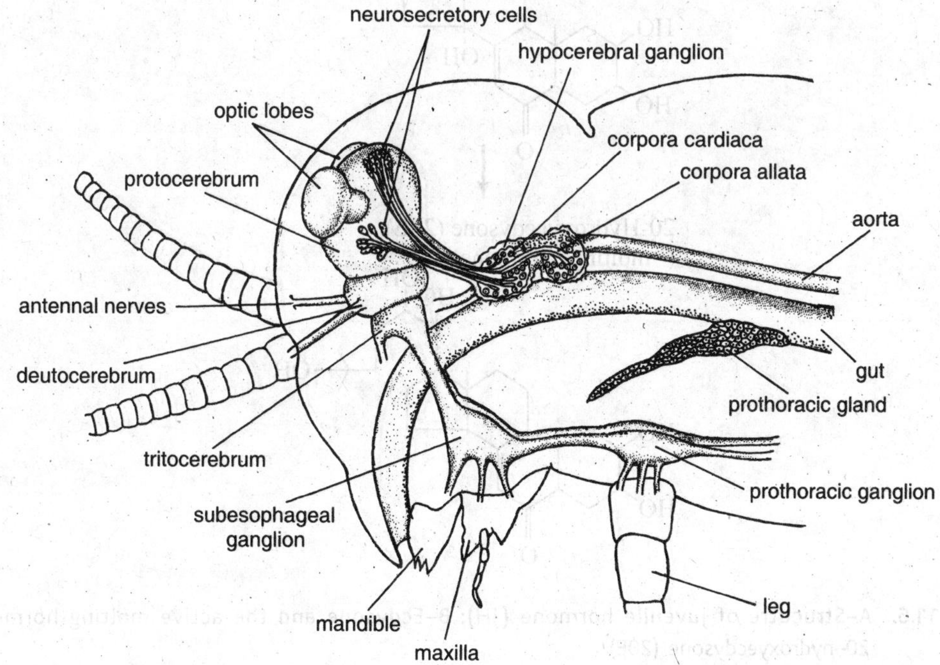

Fig. 11.5. Position of endocrine organs in the head and thorax of a generalized insect (after Brusca and Brusca, 2002).

3. Corpora allata (Singular Corpus allatum or CA). These are small rounded bodies, attached to the posterior end of corpora cardiaca. Corpora allata are neurohaemal organs where axons of neurosecretory cell of protocereberum release their secretion, the PTTH. PTTH is remained stored in CA and is released into blood.

Corpora allata also contain non–neural endocrine tissue that secrete **juvenile hormone (JH)**. Chemically it is modified hydrocarbon chain (*i.e.*, lipid). It acts in association with **20–hydro–oxyecdysone (20E)**. JH promotes retention of juvenile characters of larva postponing metamorphosis. Its circulating concentration is highest in the early larval period. It goes down to minimum at the end of pupal stage and disappears during metamorphosis. It again increases in the reproductively active adult. In adult insects, juvenile hormone promotes development of accessory sex organs, synthesis of yolk and maturation of eggs.

In the larvae, the concentration of JH is high. It opposes the action of ecdysone (20E), preventing the formation of pupal skin.

A Juvenile Hormone (JH)

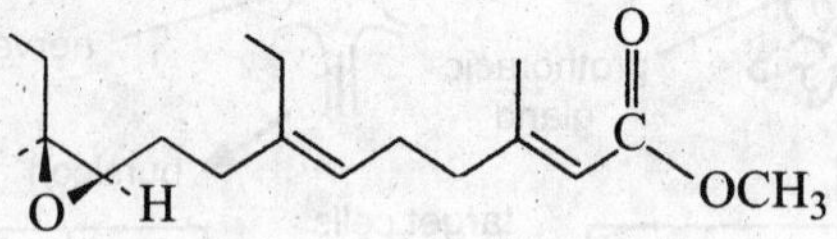

B Ecdysone

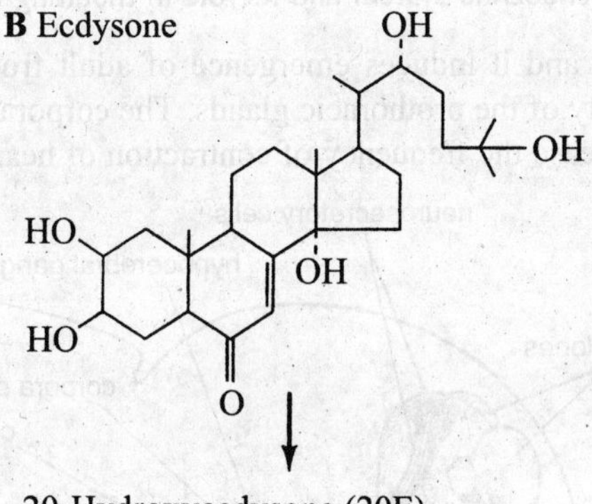

20-Hydroxyecdysone (20E)
("molting" hormone)

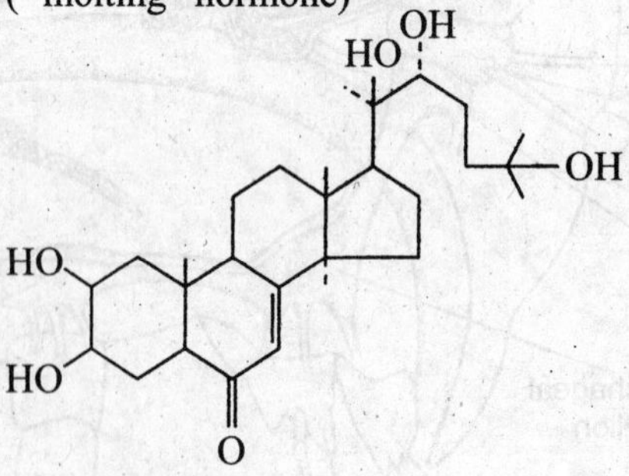

Fig. 11.6. A–Strucutre of juvenile hormone (JH); B–Ecdysone and the active molting hormone 20–hydroxyecdysone (20E).

4. Prothoracic gland (Fig. 11.5). These are paired irregular structures in the prothoracic region.These are present in immature stages in insect. Their cells exhibit cyclical secretory activity, which is maximum between two molts. The prothoracic glands are atrophied in the adult.

Prothoracic glands secrete α–ecdysone hormone which is modified in peripheral tissues to become the active molting hormone 20–hydro–oxyecdysone (20E). The 20E acts on epidermis, initiates the formation of new cuticle and detachement of old cuticle (Box 11.2). PTTH from the brains's neurosecretory cells control the secretory activity of prothoracic gland.

Box 11.2

Each molt of insects is initiated by one or more pulses of 20E. For a larval molt, the first pulse produces a small rise in the 20E concentration in the larval haemolymph (blood) and elicits a change in **cellular commitment** in the epidermis. A second, larger pulse of 20E initiates the differentiation events associated with molting. These pulses of 20E commit and stimulate the epidermal cells to synthesize enzymes that digest the old cuticle and synthesize a new one.

Hormonal Control of Insect Molting: A glimpse of Events. The process of molting in insects involves following steps:

1. Neurohormone PTTH is secreted by NSCs (neurosecretory cells) of brain and is carried to corpora allata (CA). CA releases PTTH in blood.

2. PTTH induces prothoracic gland to release α–ecdysone which peripherally changes into active molting hormone 20E.

3. Corpora allata produces juvenile hormone (JH).

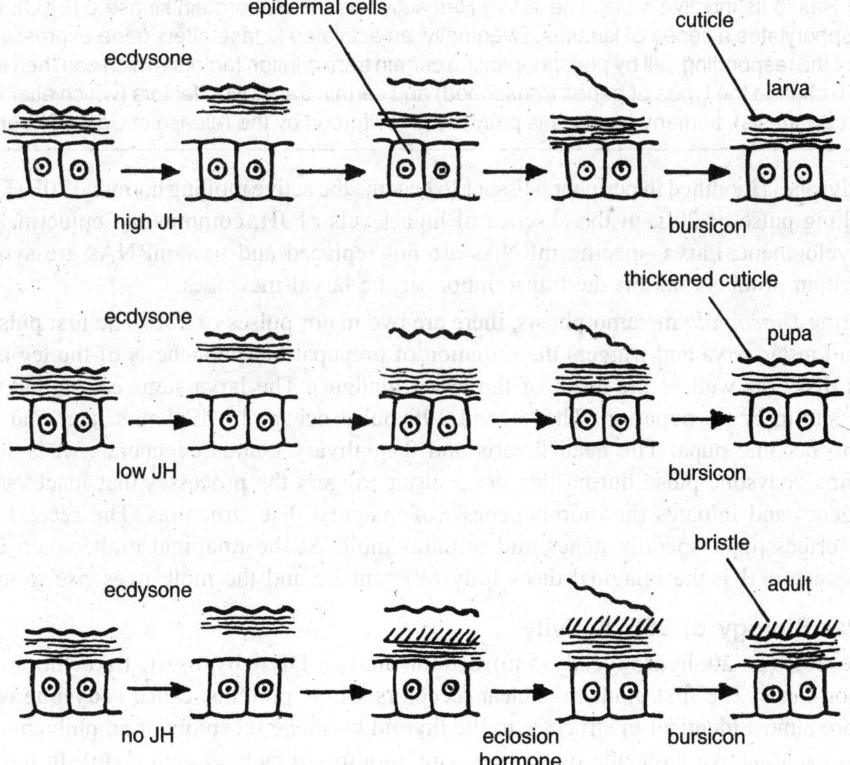

Fig. 11.7. Diagram showing role of various hormones during molting.

4. 20–E (20–hydroxyecdysone) initiates production of new cuticle and detachment of old cuticle, by the epidermis. This is called **apolysis**.

5. Shedding of old cuticle or **ecdysis** is triggered by **eclosion hormone**, secreted by brain.

6. Hormone **bursicon** regulates hardening and darkening (tanning) of new cuticle. It is secreted by the neurosecretory cells of central nervous system (*i.e.*, the brain and the nerve cord).

Mechanism of Hormones in Insect Metamorphosis

In insects, juvenile hormone (JH) is secreted by the corpora allata. The secretory cells of the corpora allata are active during larval molts but inactive during the metamorphic molt and imaginal molt. As long as JH is present in the haemolymph, the 20E–stimulated molt results in a new larval instar. In the last larval instar, however, the medial nerve from the brain to the corpora allata inhibit these glands from producing JH, and there is simultaneous increase in the body's ability to degrade existing JH (**Safranek** and **Williams** 1989). Both these mechanisms cause JH level to drop below a critical threshold value, triggering the release of **PTTH** from the brain. PTTH, in turn, stimulates the prothoracic gland to secrete a small amount of **ecdysone.**

PTTH tends to activate the RTK pathway in the cells of prothoracic gland (**Rewitz** *et al.*, 2009) (Box 11.3).

Box 11.3 RTK pathway

The **receprot tyrosine kinase** (**RTK**) is dimerised by ligand, which causes autophosphorylation of the receptor. An adaptor protein recognises the phosphorylated tyrosines on the RTK and activates an intermediate protein, **GNRP**, which activates the Ras G protein by allowing the phosphorylation of the GDP–bound Ras. At the same time, the GAP protein stimulates the hydrolysis of this phosphate bond, returning Ras to its inactive state. The active Ras activates the Raf protein kinase c (PKC), which in turn phosphorylates a series of kinases. Eventually, an activated kinase alters gene expression in the nucleus of the responding cell by phosphorylating certain transcription factors (which can then enter the nucleus to change the types of genes transcribed) and certain translation factors (which alter the level of protein synthesis). In many cases, this pathway is reinforced by the release of Ca^{2+} (**Gilbert** 2010).

Ecdysone is modified in peripheral tissues to become the active molting hormone 20E (Fig. 11.8). The resulting pulse of 20E, in the absence of high levels of JH, commits the epidermal cells to pupal development. Larva–specific mRNAs are not replaced and new mRNAs are synthesized whose protein products inhibit the transcription of the larval messages.

During *Drosophila* metamorphosis, there are two major pulses of 20E. The first pulse occurs in the thrid instar larva and triggers the initiation of prepupal morphogenesis of the leg and wing **imaginal discs** (as well as the death of the larval hindgut). The larva stops eating and migrates to find a site to begin pupation. The second 20E pulse occurs 10–12 hours later and tells the prepupa to become pupa. The head inverts and the salivary glands degenerate. It is suggested that the first ecdysone pulse during the larval instar triggers the processes that inactivate larva–specific genes and initiates the morphogenesis of imaginal disc structures. The second pulse of 20E transcribes pupa–specific genes and initiates molt. At the imaginal molt, when 20E acts in the absence of JH, the imaginal discs fully differentiate and the molt gives rise to an adult.

Molecular Biology of 20E Activity

The 20E (or 20–hydroxyecdysone) cannot bind to DNA by itself. Like the amphibian thyroid hormones, 20E first binds to nuclear receptors. These proteins, called **ecdysone receptors (EcRs)**, are almost identical in strucutre to the thyroid hormone receptors of amphibians. An EcR protein forms an active molecule by pairing with an ultra–spiracle protein (Usp). In the absence

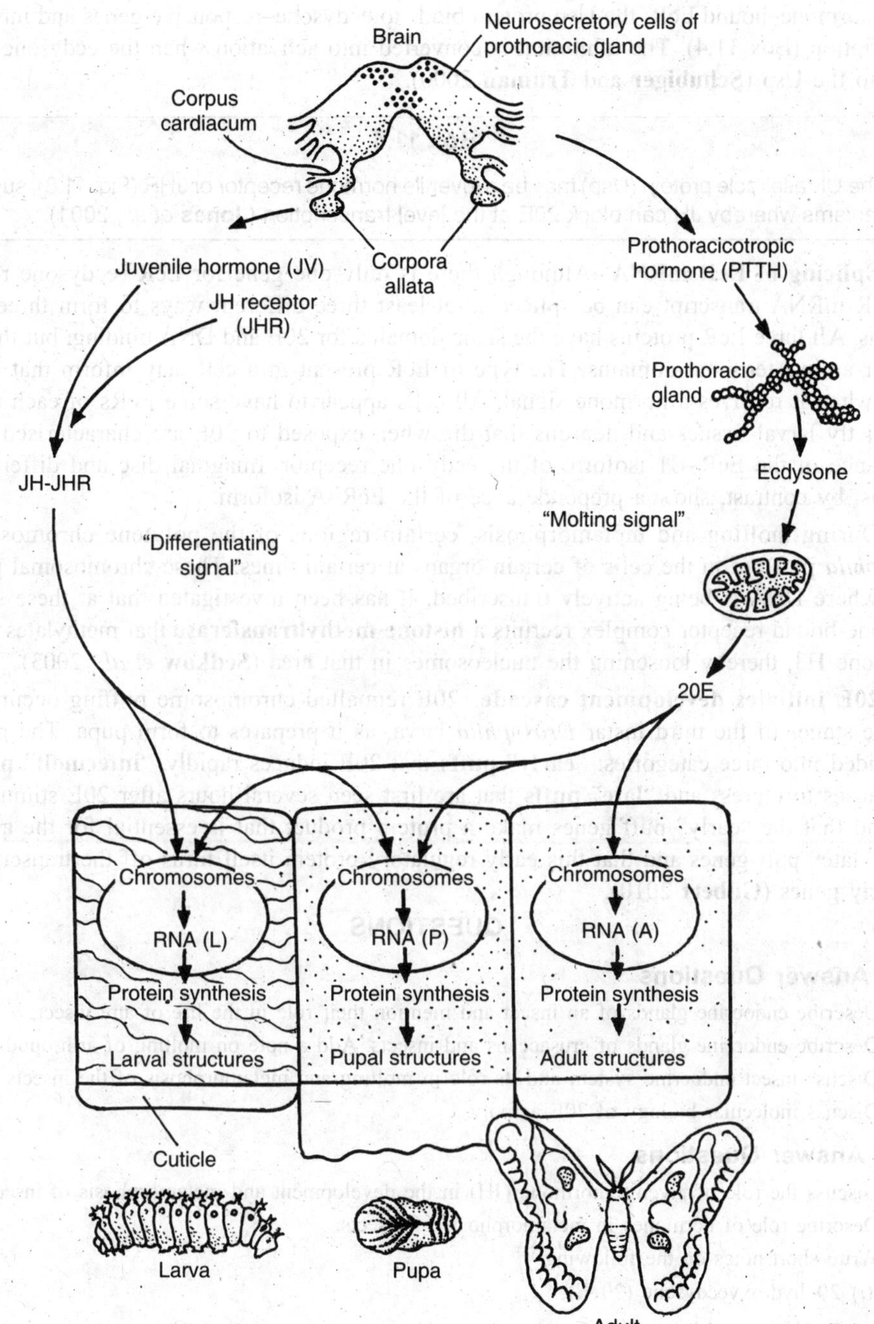

Fig. 11.8. Regulation of insect metamorphosis. General pathway of insect metamorphosis. 20E and JH together cause molts that form the next larval instar. When the concentration of JH becomes low enough, the 20E– induced molt produces a pupa instead of an instar. When ecdysone acts in the absence of JH, the imaginal discs differentiate and the molt gives rise to an adult (imago)(after Gilbert 2010).

of the hormone–bound EcR, the Usp protein binds to ecdysone–responsive genes and inhibit their transcription (Box 11.4). The inhibition is converted into activation when the ecdysone receptor binds to the Usp (**Schubiger** and **Truman** 2000).

Box 11.4

The Ultraspiracle protein (Usp) may be a juvenile hormone receptor or JHR (Fig. 11.8), suggesting mechanisms whereby JH can block 20E at the level transcription (**Jones** *et al.*, 2001).

Splicing of EcR mRNA. Although there is only one gene for EcR (ecdysone receptor), the EcR mRNA transcript can be spliced in at least three different ways to form three distinct proteins. All three EcR proteins have the same domains for 20E and DNA binding, but they differ in their amino–terminal domains. The type of EcR present in a cell may inform that cell how to act when it receives a hormone signal. All cells appear to have some EcRs of each type, but the strictly larval tissues and neurons that die when exposed to 20E are characterised by their abundance of the EcR–B1 isoform of the ecdysone receptor. Imaginal disc and differentiating neurons, by contrast, show a preponderance of the EcR–A isoform.

During molting and metamorphosis, certain regions of the polytene chromosomes of *Drosophila* puff out in the cells of certain organs at certain times. These chromosomal puffs are areas where DNA is being actively transcribed. It has been investigated that at these sites, the ecdysone-bound receptor complex recruits a **histone methyltransferase** that methylates lysine–4 of histone H3, thereby loosening the nucleosomes in that area (**Sedkow** *et al.*, 2003).

20E initiates development cascade. 20E regualted chromosome puffing occurs during the late stages of the third instar *Drosophila* larva, as it prepares to form pupa. The puffs can be divided into three categories: "**early**" **puffs** that 20E induces rapidly, "**intermolt**" **puffs** that 20E causes to regress and "**late**" **puffs** that are first seen several hours after 20E stimulation. It is found that the "early" puff genes make a protein product that is essential for the activation of the "late" puff genes and that this early regulatory protein itself turns off the transcription of the early genes (**Gilbert** 2010).

QUESTIONS

Long Answer Questions

1. Describe endocrine glands of an insect and mention their role in the life of any insect.
2. Describe endocrine glands of crusaceans and insects. Add a note on molting of arthropods.
3. Discuss insect endocrine system and its role in molting and metamorphosis of the insects.
4. Discuss molecular biology of 20E activity.

Short Answer Questions

1. Discuss the role of juvenile hormone (JH) in the development and metamorphosis of insects.
2. Describe role of hormones in metamorphosis of insects.
3. Write short notes on the following:
   (*i*) 20–hydroxyecdysone (20–E);
   (*ii*) Prothoracic glands;
   (*iii*) Newurosecretory cells;
   (*iv*) Endocrine glands of nonchordates.

Very Short Answer Questions

1. Which tissue/organ recretes molt-inhibiting hormone in crustacea.
2. What is the name of brain hormone of insects.
3. Ecdysone hormone in insects in released by which endocrine gland?

4. How many pulses of active molting hormone (*i.e.*, 20E) occur during metamorphosis of insects?
5. What is function of corpora cardiaca?
6. What is function of ecdysiotropin in insects?
7. What is corpora allata?

Fill in the Blanks

1. In crustaceans, ecdysone is secreted by an endocrine gland, called ....
2. Y-organ of crustacea is located at the base of ....
3. MIH of crustacea stands for ....
4. MIH is carried by .... transport to sinus gland.
5. PTTH induces .... to release α-ecdysone which peripherally changes into active molting hormone called ....
6. Corpora allata in insects produce ....
7. Ecdysone initiates a new .... in insects.
8. JH perpetuates .... condition in insects.
9. To act on DNA, 20E first bind to nuclear receptors, called ....

Multiple Choice Questions

1. Juvenile hormone (JH) is secreted by
   (*a*) corpora allata (*b*) intercerebral gland
   (*c*) prothoracic gland (*d*) none of the above
2. In cockroach, the secretion of corpora allata is responsible for
   (*a*) heart beat (*b*) growth
   (*c*) reproduction and metamorphosis (*d*) none of the above
3. The moulting in a nymph of cockroach is controlled by a hormone secreted by
   (*a*) corpora allata (*b*) juvenile hormone
   (*c*) nymph gland (*d*) prothoracic gland
4. Ecdysone regulates
   (*a*) periodic shedding of epidermis of reptiles
   (*b*) ecdysis of cuticle in adult animals
   (*c*) metamorphosis in insects
   (*d*) germination of sponges by gemmules
5. Brain hormone in insects is secreted from
   (*a*) pituitary gland (*b*) neurosecretory cells in the brain
   (*c*) thoracic gland (*d*) nephridia

ANSWERS

Very Short Answer Questions

1. Neurosecretory cells of the X-organ.
2. Ecdysiotropin (EcT) or Protothoracicotropic hormone (PTTH).
3. Prothoracic gland.
4. Two.
5. Corpora cardiaca are neurohaemal organs of insects which store ecdysiotropin (EcT) or PTTH (protothoracicotropic hormone) of NSCs of brain.
6. Ecdysiotropin is released into blood (haemolymph) by corpora cardiaca and it has two effects: 1. It directs corpora allata to synthesize and release juvenile hormone (JH) in blood and 2. It directs prothoracic glands to secrete ecdysone in blood.
7. Insect endocrine glands that secrete juvehile hormone (JH) during larval molts.

Fill in the Blanks

1. Y-organ.

2. Antennae.

3. Moult-inhibiting hormone.

4. Axonal transport, sinus gland.

5. Prothoracic gland, 20E or 20-hydroxyecdysone.

6. Juvenile hormone (JH).

7. Molt.

8. Juvenile.

9. Ecdysone molt receptors.

Multiple Choice Questions

1. (*a*) **2.** (*c*) **3.** (*d*) **4.** (*c*) **5.** (*b*).

12

Insect Pest Management (Insect Control)

In order to prevent damage to crops, horticulture products and to protect ourself and our domestic animals from diseases it becomes essential to control insect pests. Harmful insect pests can be controlled by following three methods:

1. Mechanical control;
2. Chemical control;
3. Biological control.

12.1 MECHANICAL CONTROL

Large-sized harmful insects can be manually picked up, or collected by nets and then destroyed. Besides this, we can also keep such insects, away from us, our cattle and our useful things by using protective meshes on doors and windows of our residences, hospitals, animal houses, storehouses, etc.

Ancient people used to collect, for killing, large numbers of such insects by using gur, sweets, etc., as baits. Modern scientists have synthesized certain sex attractant scents, called **pheromones**, for attracting insects at one place to be killed *en masses*.

Harmful insects can also be controlled through environmental changes such as follows:

1. Localized insects and bettle grubs can be destroyed by the rotation of crops.
2. Houseflies can be controlled by having proper sanitation.
3. Insects that lay eggs in stagnant water (*e.g.*, mosquitoes) can be destroyed by removing stagnant and dirty water in ponds, pools and ditches and drying them.
4. Insects can be destroyed and may be discouraged from breeding by removing heaps of garbage or by drying and burning it. This will destroy the breeding grounds of harmful insects.
5. Insects can also be killed by raising temperature of store houses, godowns or production centre. Insect pest can also be rendered inactive by lowering the temperature such as in the cold storage.

12.2 CHEMICAL CONTROL

Insect control by using various chemicals is simplest and most popular method. These chemicals are called **insecticides,** since they act as insect killers. Insecticides are toxicants and are used in various forms such as powder, sprays, solutions, emulsions or fumes. Insecticides can be divided into the following three categories:

***(i)* Stomach poisons.** The insecticides are sprayed upon food of insects. When ingested, the poisoned food kills the insects. **Lead arsenate, cryolite, sodium arsenate, sodium fluorides, sodium fluorosilicate** and **Bordeaux mixture** are common stomach poisons used for killing insects.

***(ii)* Contact poisons.** These are strong insecticides which kill the insects instantaneously upon mere contact. Some of most popular contact insecticides are **DDT (dichloro-diphenyl-trichloro-ethane), gammexane, pyrethrum, aldrin** and **toxophene**.

***(iii)* Fumigants.** Fumigation is a process of creating lethal concentration of a toxic gas (fumigant) for the time necessary to destroy insect infestation in a closed space. Fumigants consists of volatile chemicals which quickly vapourise and kill the insects without affecting the grain. **Ethylene dibromide, methyl bromide, aluminium phosphide, sulphur, cyanide** are some common fumigants which are used for insect pests.

Disadvantages of chemical control. Human beings have been successful in controlling insects populations by the use of insecticides. But their indiscriminate use has resulted into following problems:

1. When an insect pest is exposed to same insecticide repeatedly it tends to develop resistance to that insecticide. So that after sometime, the use of that particular insecticide becomes ineffective.
2. Ecologically DDT and other insecticides become non-biodegradable pollutant and have polluted our land, water and air. They have produced various health problems for human beings and domestic animals (they tend to result in toxicity due to their biomagnification by various food chains).

12.3 BIOLOGICAL CONTROL

This is self-perpetuating method and by it insects of a specific species can be destroyed. It involves the following methods:

***(i)* Predation.** Biological control of insects pests can be done by introducing insectivorous animals such as lizards, snakes, frogs, toads, spiders or insectivorous birds or mammals. If the population of these insectivorous predators increases, they will devour more and more insects and will keep a check on the increase of insect population.

***(ii)* Parasitism.** Insect pests can also be controlled by introducing those viruses, bacteria, fungus, etc., in insects which produce disease in them. These pathogens ultimately kill the insects and help in reducing their population.

***(iii)* Sterilization.** Male insects are collected and sterilized by exposing them to X-rays or gamma rays. These are, then, released into their populations. These copulate with the females normally, but the ova remain unfertilized. Hence, there are no progenies.

QUESTIONS

Short Answer Question

1. Write a short note on insect control.

Very Short Answer Questions

1. Name two normal agricultural operations that help control insect pests.

2. How does post-harvest stubble removal can control pests?

3. What is a trap crop?

4. What do you mean by biocontrol?

5. Name the agency in India that manages movements of useful insects from one country to another.

6. Cite an example of insect pest control with a virus.

7. What is parasitoid?

8. Give an alternative term for pheromone.

9. Name the synthetic compound that resembles the natural pheromone dispalure of gypsymoth.

10. Mention three types of allelochemicals.

Fill in the Blanks

1. The two main types of insect control are .... and applied.

2. A systemic poison is taken up by .... insects from the plant sap.

3. DDT is a .... poison.

4. Calcium arsenate is a .... poison.

5. Carbon bisulphide and .... are fumigants.

6. Antifeedants protect the crops .... killing, repelling or trapping the insect pests.

7. Mimics of moulting hormone produced by plants are called ....

8. Crop rotation is a .... method of insect control.

9. .... is a predatory insect.

10. Deep and frequent tillage can kill the root-infesting .... of maize.

Multiple Choice Questions

1. Which of the following is not a stomach poison?

(*a*) lead arsenate (*b*) paris green
(*c*) sulphur (*d*) borax

2. Which of the following is both contact poison as well as fumigant?

(*a*) nicotene (*b*) DDT
(*c*) BHC (*d*) naphthalene

3. The chemicals which prevent the insect pests to feed on their host plant is called

(*a*) repellents (*b*) trappers
(*c*) antifeedants (*d*) fumigants

4. Ecdysterones are produced by

(*a*) insect pests (*b*) host plants
(*c*) all plants (*d*) all insects

5. Chlorfluazuron is an

(*a*) insect repellent (*b*) antifeedant
(*c*) chitin synthesis inhibitor (*d*) hormone mimic

6. One of the major difficulties in the biological control of insect pest is that

(*a*) the method is less effective as compared with the use of insecticides
(*b*) the practical difficulty of introducing the predator to specific areas
(*c*) the predator develops a preference to other diets and may itself become a pest
(*d*) the predator does not always survive when transferred to a new environment

7. Biological control of agricultural pests unlike chemical control is

(*a*) self perpetuating (*b*) toxic
(*c*) polluting (*d*) very expensive

8. An insecticide extracted from the leaves of neem is

(*a*) aldicarb (*b*) malathion

(*c*) rotenone (*d*) azardirachtin

9. Insecticides usually act upon

(*a*) digestive system (*b*) nervous system

(*c*) circulatory system (*d*) muscular system

ANSWERS

Very Short Answer Questions

1. Ploughing and hoeing. [Hoeing means to use a hoe to turn (earth) or cut through (weeds)].

2. By eliminating hiding places of pests.

3. An early crop sown around the main crop to attract insect pests common to both.

4. Control of pests by using natural enemies.

5. AICRPBC (= All India Coordinated Research Project on Biological Control of Crop Pests and Weeds).

6. Nuclear polyhedrosis virus is used to control forest defoliating insect pest, the pine sawfly.

7. An insect (*e.g.,* a wasp) that completes its larval development within the body of another insect which it eventually kills and that is free living as a adult.

8. Ectohormone.

9. Gyplure.

10. Allomones, kairomones and synomones.

Fill in the Blanks

1. Natural **2.** Sucking **3.** Contact **4.** Stomach

5. Naphthalene **6.** Without **7.** Phytoecdysone **8.** Cultural

9. Preying mantis **10.** Aphids.

Multiple Choice Questions

1. (*c*) **2.** (*a*) **3.** (*c*) **4.** (*b*) **5.** (*c*) **6.** (*c*)

7. (*a*) **8.** (*d*) **9.** (*b*).

13

Phylum Onychophora and *Peripatus*

Onychophora (Gr., *onychus* = claws; *phorus* = bearing) is an ancient phylum and has not changed greatly since its origin in Cambrian period. It includes those invertebrates which are included into a single genus *Peripatus* having about 110 species. The genus *Peripatus* was established in 1826 by **Guilding.** He regarded it as a mollusc. In 1874, **Moseley** classified it as an primitive arthropoda discovering the tracheae.

13.1 GENERAL CHARACTERS OF ONYCHOPHORA

1. Onychophora are specialized for life in the leaf litter and plant debris and are ideally adapted for squeezing through narrow spaces and cavities. They are found in crevices of rocks, under bark, in stones and in moist places in tropical region.
2. They have a worm-like and indistinctly segmented body.
3. Their **head** is composed of three limb-bearing somites. Thus, head bears three pairs of appendages, *i.e.*, a pair of **antennae,** a pair of **peri-buccal lobes** hiding **jaws** and a pair of **oral** or **slime papillae.** At the base of each antenna is found a simple eye.
4. Each **trunk** segment bears a pair of clawed lobopodial (= lobe-like, unjointed and fleshy) limbs.
5. Body is covered by a thin, flexible and chitinous **cuticle.**
6. Paired **slime** or **adhesive** glands are present, slime is used in food (prey) capturing and self defence.
7. Respiration occurs through unbranched **tracheae.**
8. **Coelom** is reduced to gonadal cavities and nephridial sacs. Principal body cavity is haemocoel.
9. Circulatory system includes heart and haemocoelomic sinus like the artropods.
10. Excretion occurs by segmentally arranged paired **nephridia.** Excretory product most probably is **uric acid.**
11. Sexes are separate. Male sex produces **spermatophores.** Most are **viviparous.** Development is direct and resembles with arthropods.
12. Onychophora forms a connecting link between **Annelida** and **Arthropoda.**

13.2 CLASSIFICATION

Phylum Onychophora contains two families: **Peripatidae** and **Peripatopsidae;** both of which have peculiar geographical distribution. Each family has a wide but discontinuous distribution around the world. The members of two families are never found in the same area. The family **Peripatidae** are equatorial in distribution (*i.e.*, circum-tropical-subtropical distribution), while the family **Peripatopsidae** are limited to the Southern Hemisphere (*i.e.*, circum-south-temperate distribution).

There are several genera (such as *Peripatus*, *Peripatopsis*, *Eoperipatus*, *Opisthoperipatus*, *Paraperipatus*, etc.) but the term *Peripatus* is commonly used as the generic name for all species.

13.3 *PERIPATUS*

1. Distribution

Peripatus exhibits discontinuous geographical distribution and is found in warmer parts of the world such as tropical regions (East Indies, West Indies, Himalayas, Congo and Northern part of South America). It is also found in the temperate regions of Southern Hemisphere (Australia, New Zealand, South Africa and Andes). No species of *Peripatus* occur in the north of the Tropic of Cancer. Common name of *Peripatus* is **velvet worm.**

2. Habit and Habitat

Peripatus is **terrestrial, nocturnal** and **predaceous** animal. It is confined to dark, warm, and humid habitats, *i.e.*, it lives in crevices of rocks, under bark, stones and beneath fallen leaves (= leaf litter). It remains active during night. During cold and dry periods, *Peripatus* becomes inactive and remains hidden in burrows. *Peripatus* is predaceous and feeds on small invertebrates such as snails, insects and worm.

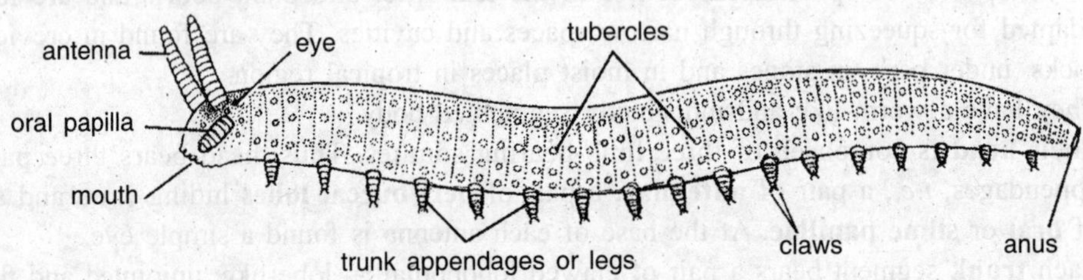

Fig. 13.1. *Peripatus*. External features.

3. Anatomical Peculiarities: External Morphology

1. *Peripatus* is elongated, cylindrical and catterpillar-like measuring 1.5 to 15.0 cm in length.
2. Body colour is blue, green, orange and black. Skin (= outer body covering) is velvety and iridescent in appearance. Skin is thrown into transverse wrinkles bearing numerous small papillae (or tubercles) armed with chitinous spines.

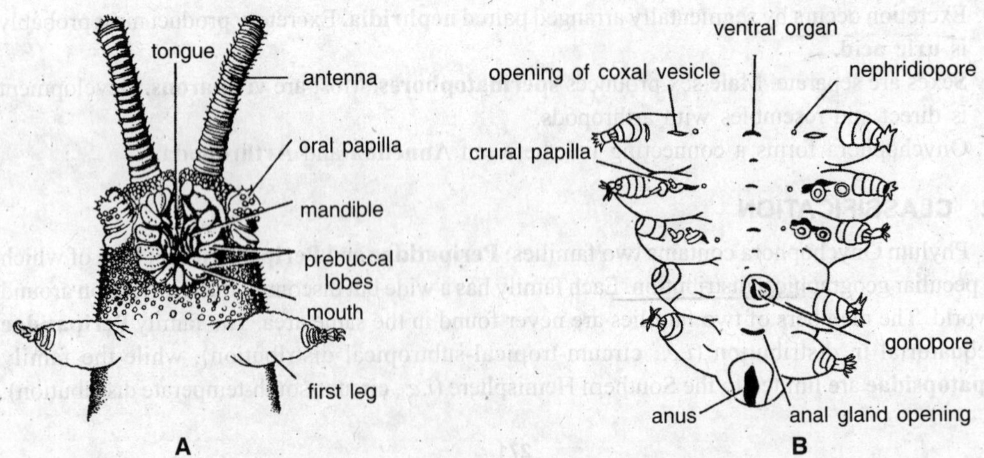

Fig. 13.2. *Peripatus*. A—Anterior end in ventral view; B—Posterior end in ventral view.

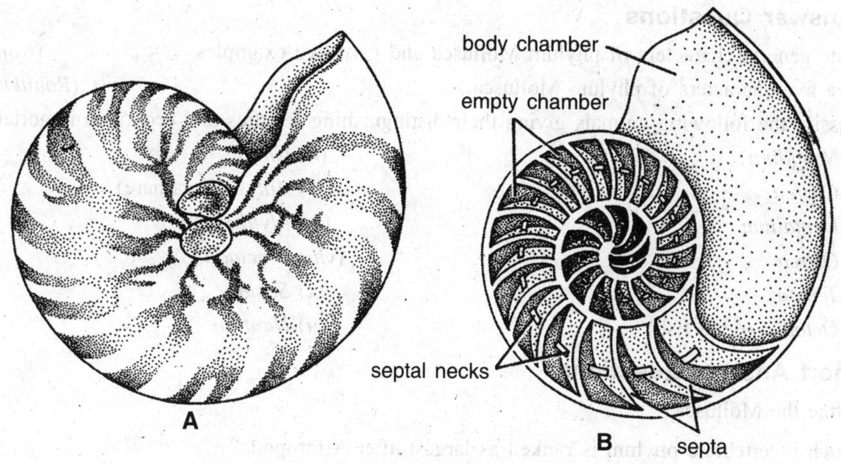

Fig. 14.13. *Nautilus*. A—Complete shell; B—Shell in section.

Head is well developed and bears a pair of simple eyes and 38 to 90 **tentacles** around it. The tentacles are suckerless but bears transverse ridges. An unusual feature of the head of *Nautilus* is the large muscular leathery and protective **dorsal hood** which serves as an operculum of gastropod, *i.e.*, it covers the opening of shell when the animal withdraws into the shell (Fig. 14.14).

In contrast to other cephalopods, in *Nautilus*, food absorption occurs mainly in the digestive gland. *Nautilus* has four **auricles** in the heart, while other cephalopods have only two auricles. Likewise, *Nautilus* has four **nephridia** while the other living cephalopods contain only two nephridia. A pair of **ospharidia** occurs near the gills only in *Nautilus*. **Statocysts** of *Nautilus* lie near the pedal ganglia. Ink glands are absent.

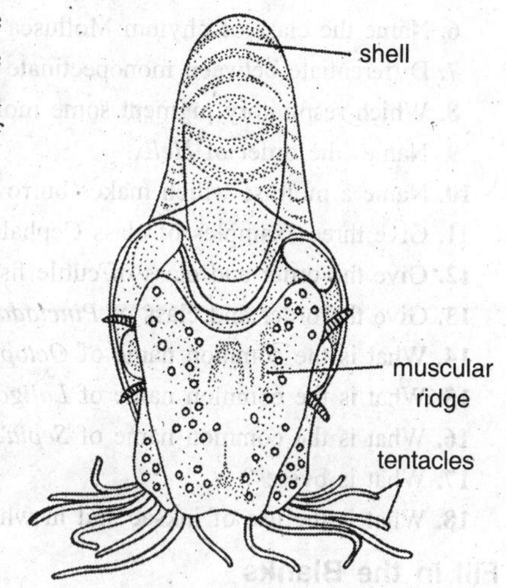

Fig. 14.14. *Nautilus*. Living animal.

Nautilus is dioecious. In male, four arms are perma-nently modified for copulation. Female *Nautilus* has a vestigial left and functional right **oviduct**. In male the left **sperm duct** is rudimentary, while in other cephalopods the right sperm duct is missing.

Economic Importance

Nautilus is considered as an excellent food by the pacific Islanders. Its brilliantly coloured shell is extensively used for decoration.

QUESTIONS

Long Answer Questions

1. Classify Mollusca giving characters and examples of each group. (*C.C.S.U. Meerut 2004*)
2. What are the distinguishing features of class Gastropoda?
3. What is the basis of the classification of phylum Mollusca? Give salient features and classification of phylum Mollusca with suitable examples.
4. Give an outline classification of phylum Mollusca.

Short Answer Questions

1. Write general characters of phylum Mollusca and give five examples. (*Kanpur 2001*)

2. Give five characters of phylum Mollusca. (*Rohilkhand 2001*)

3. Classify the following animals giving their distinguishing features and economic importance.

(*i*) *Neopilina*
(*ii*) *Chiton*
(*iii*) *Cyprea*
(*iv*) *Aplysia* (sea hare)
(*v*) *Dentalium*
(*vi*) *Mytilus*
(*vii*) *Ostrea*
(*viii*) *Pinctada*
(*ix*) *Teredo*
(*x*) *Sepia*
(*xi*) *Octopus* (devil fish)
(*xii*) *Nautilus*

Very Short Answer Questions

1. Define the Mollusca.

2. Which invertebrate phylum is ranked as largest after Arthropoda?

3. Mention a mollusc having a shell formed of eight plates.

4. Name a mollusc with external, coiled shell divided internally into many chambers by septa.

5. Mention the three main body parts of a mollusc.

6. Name the class of phylum Mollusca which include asymmetrical animals and define torsion.

7. Differentiate between monopectinate and bipectinate gills.

8. Which respiratory pigment some molluscs have in their blood?

9. Name the order of *Helix*.

10. Name a mollusc which makes burrows in the wood used in boats and ships.

11. Give three examples of class Cephalopoda.

12. Give the order and class of cuttle fish.

13. Give the order and class of *Pinctada*.

14. What is the common name of *Octopus*?

15. What is the common name of *Loligo*?

16. What is the common name of *Sepia*?

17. What is bysses?

18. What is the use of inksac and in which animal it is found?

Fill in the Blanks

1. .... is a segmented mollusc.

2. *Neopilina* belongs to the class.................. .

3. A mollusc shell consists of 1, 2 or .... pieces.

4. *Chiton* is an example of the class ....

5. Shell in molluscs is secreted by ....

6. Foot in molluscs is mainly a .... organ.

7. Asymmetrical molluscs belong to the class ....

8. Number of arms in *Sepia* is ....

9. Some molluscs having a rasping organ, the .... in the buccal cavity.

10. Circulatory system of molluscs is of .... type.

11. Body cavity of molluscs is ....

12. Excretory organs of molluscs are ....

13. Organs of Bojanus are .... organs.

14. The eggs of molluscs undergo .... and .... cleavage.

15. .... is a living fossil among molluscs.

Multiple Choice Questions

1. Which one is a connecting link between Annelida and Mollusca?
(*a*) *Nautilus* (*b*) *Chaetoderma*
(*c*) *Neopilina* (*d*) *Lymnaea*

2. Pila belongs to the class.............. of phylum Mollusca.
(*a*) Monoplacophora (*b*) Pelecypoda
(*c*) Gastropoda (*d*) Cephalopoda

3. The harmful mollusc is
(*a*) *Chiton* (*b*) *Teredo*
(*c*) *Pinctada* (*d*) *Unio*

4. Superior quality pearls are formed by
(*a*) *Pinctada* (*b*) *Unio*
(*c*) *Ostrea* (*d*) *Pecten*

5. Arms are webbed at the base in
(*a*) *Loligo* (*b*) *Sepia*
(*c*) *Octopus* (*d*) *all of above*

6. Shell is internal and reduced in
(*a*) *Pila* (*b*) *Helix*
(*c*) *Teredo* (*d*) *Limax*

7. One of the following is a living fossil
(*a*) *Unio* (*b*) *Dentalium*
(*c*) *Nautilus* (*d*) *Pecten*

8. Pearl oyster (*Pinctada*) belongs to the class
(*a*) Cephalopoda (*b*) Scaphopoda
(*c*) Gastropoda (*d*) Pelecypoda (or Bivalvia)

9. *Mytilus* belongs to molluscan class
(*a*) Pelecypoda (*b*) Gastropoda
(*c*) Amphineura (*d*) Cephalopoda

10. Sea hare is
(*a*) *Teredo* (*b*) *Pila*
(*c*) *Sepia* (*d*) *Aplysia*

11. The elephant tusk shell is
(*a*) *Cypraea* (*b*) *Limax*
(*c*) *Dentalium* (*d*) *Nautilus*

12. The foot of molluscs is used for
(*a*) burrowing (*b*) swimming
(*c*) creeping (*d*) all of these

13. Locomotion by jet propulsion mechanism is found in class
(*a*) Scaphopoda (*b*) Cephalopoda
(*c*) Gastropoda (*d*) Pelecypoda

14. Worm-like and without shell mollusc is included in class
(*a*) Monoplacophora (*b*) Polyplacophora
(*c*) Aplacophora (*d*) Scaphopoda

15. Torsion is the characteristic of
(*a*) Gastropoda (*b*) Scaphopoda
(*c*) Pelecypoda (*d*) Aplacophora

16. Digestive gland in mollusca is
(*a*) salivary gland (*b*) pancreas
(*c*) liver (*d*) hepatopancreas

17. The larvae of mollusca are
(*a*) veliger (*b*) trochophore
(*c*) glochidium (*d*) all of above

18. The study of Mollusca is generally known as
(*a*) malacology (*b*) concology
(*c*) helminthology (*d*) herpetology

19. Filter feeder molluscs are
(*a*) bivalves (*b*) gastropods
(*c*) *Octopus* (*d*) devil fish

20. Visceral mass of molluscan body contains
(*a*) central nervous system (*b*) gill
(*c*) main body organs (*d*) haemocoel

21. Radula is found in all molluscs except
(*a*) pelecypodes (*b*) cephalopods
(*c*) scaphopods (*d*) aplacophorans

22. The zoological name of razor shell is
(*a*) *Chiton* (*b*) *Loligo*
(*c*) *Dentalium* (*d*) *Solen*

23. The study of molluscan shell is called
(*a*) entomology (*b*) biology
(*c*) conchology (*d*) zoology

24. Shell is internal in
(*a*) Monoplacophora (*b*) Cephalopoda
(*c*) Gastropoda (*d*) Pelecypoda

25. Molluscs have
(*a*) one chambered heart (*b*) two chambered heart
(*c*) three chambered heart (*d*) none of these

ANSWERS

Very Short Answer Questions

1. The molluscs are nonchordates or invertebrates. They are soft-bodied, unsegmented, triploblastic, bilaterally symmetrical coelomate animals. They contain a muscular foot, a visceral hump and a mantle. The body of molluscs is usually covered with a hard shell.
2. The invertebrate animal phylum ranked to be largest after Arthropoda is Mollusca. More than 60,000 living species have been studied from this phylum. In addition about 35,000 fossil species provide the phylum to have its long geological history.
3. *Chiton*.
4. *Nautilus*.
5. Head, visceral mass and foot.
6. The class of phylum Mollusca which contain asymmetrical animals is called **Gastropoda**. The body symmetry of these molluscs is lost due to two reasons: coiling and torsion. Coiling is a normal feature of gastropods in which their body is spirally coiled. Also during embryonic development their visceral mass gets twisted or torted suddenly causing the shifting of posterior end (anus) to the anterior end (mouth). Thus torsion is the phenomenon in which animals body (*i.e.*, viscero-pallium) gets rotated at an angle of 180° during larval development.

7. Molluscs have two types of gills, the aquatic breathing organs. When the gill or ctinedium is provided with a single row of lamellae attached to the axis, it is called **monopectinate gill,** *e.g., Pila*. The gill or ctinedium having two rows of gill lamellae (gill filaments) attached to either side of a central axis is called **bipectinate gills,** *e.g., Patella*.
8. Haemocyanin which is dissolved in the plasma of blood.
9. Pulmonata.
10. *Teredo*.
11. *Sepia, Loligo* and *Octopus*.
12. Decapoda (order), Cephalopoda (class) (Phylum Mollusca).
13. Pseudolamellibranchia (order), Pelecypoda (class) (Phylum Mollusca).
14. Devil fish.
15. Squid.
16. Cuttlefish.
17. Byssus is a bunch of strong thread found in some molluscs, such as *Mytilus,* for attachment to the substratum.
18. An **inksac** is present in *Sepia* and serves for defence. To escape from the enemies, it discharges dark clouds of ink behind it so that the enemy loses sight of it. Sepia ink for the artists is also obtained from *Sepia* (Box 14.5).

Box 14.5 Inksac of Mollusca

In most non-nautiloid cephalopodan molluscs (*e.g., Sepia, Loligo* and *Octopus*), a large inksac is located near the intestine. An ink-producing gland lies in the wall of sac, and a duct runs from the sac to pour into the rectum. The gland secretes a brown or black fluid that contains a high concentration of melanin pigment and mucus; the fluid is stored in the inksac. When alarmed, the animal releases the ink through the anus and mantle cavity and out in the surrounding water. The cloud of inky material hangs together in the water, forming a "dummy" image that serve to confuse predators. The alkaloid nature of the ink may also act to deter predators, particularly fishes, and may interfere with their chemoreception (see **Brusca** and **Brusca**, 2002).

Fill in the Blanks

1. *Neopilina*	2. Monoplacophora	3. Eight
4. Polyplacophora	5. Mantle (pallium)	6. Locomotory
7. Gastropoda	8. Ten	9. Radula
10. Open	11. Haemocoel	12. Kidneys
13. Excretory	14. Spiral, determinate	15. *Neopilina galathea*.

Multiple Choice Questions

1. (*c*)	2. (*c*)	3. (*b*)	4. (*a*)	5. (*c*)	6. (*d*)
7. (*c*)	8. (*d*)	9. (*a*)	10. (*d*)	11. (*c*)	12. (*d*)
13. (*b*)	14. (*c*)	15. (*a*)	16. (*d*)	17. (*d*)	18. (*a*)
19. (*a*)	20. (*c*)	21. (*a*)	22. (*d*)	23. (*c*)	24. (*b*)
25. (*c*).					

15

Pila globosa (The Apple Snail) (Phylum Mollusca)

15.1 CLASSIFICATION

Phylum	—	**Mollusca**
Class	—	**Gastropoda**
Subclass	—	**Prosobranchia**
Order	—	**Mesogastropoda**
Genus	—	***Pila***
Species	—	***globosa***

15.2 HABIT AND HABITAT

Pila globosa is one of the largest freshwater snails. It is found in abundance in freshwater ponds, pools, tanks, rice fields, streams and rivers. It is **herbivorous** and prefer clean water containing succulent aquatic plants such as *Vallisneria, Pistia*, etc. It is **amphibious** being adapted for life in water and on land. During rains it leaves ponds and make long tours on land. During drought conditions it becomes dormant and can remain buried in mud for years. This is called **aestivation.**

Pila globosa is mostly found in Northern India. Other species of *Pila* such as *P. virens, P. nevililana* and *P. theobaldi* are very common in South India.

15.3 EXTERNAL FEATURES

1. Shell

Pila lies enclosed in its shell (Fig. 15.1). The shell is single piece (**univalve**) which encloses the whole body. It is thick globose and lemon-yellow, brownish or blackish coloured. It is hollow, spirally coiled cone and contains six and half coils. A complete coil is called a **whorl.**

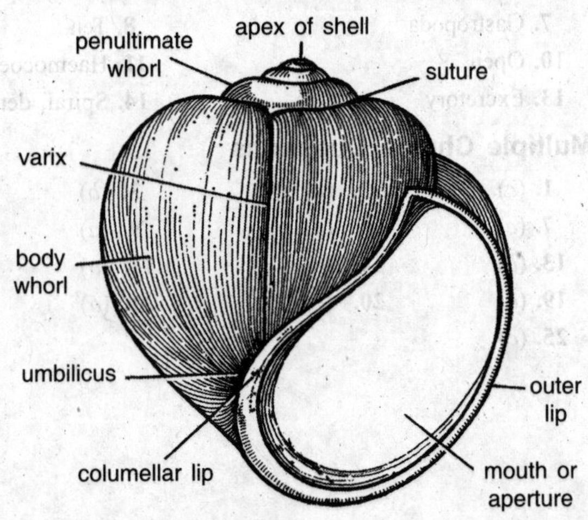

Fig. 15.1. *Pila globosa*. Shell seen from ventral surface.

The shell contains a central axis or

columella. The topmost or the apical whorl (**apex of shell**) is smallest and oldest. Later whorls gradually increase in size and the last whorl is the largest. It is called **body whorl.** Most of the body of *Pila* lies in this whorl. The whorls above the body whorl is called **spire** (last of which is called **penultimate whorl**). There is a **line** or **suture** between every two whorls. The shell surface is marked by numerous **lines of growth** called **varices.** The body whorl opens to outside by a wide **aperture** or **mouth.** Through this mouth, the head and foot protrude. Smooth continuous margin of the mouth of the shell is called **peristome.** It contains outer lip and inner lip or **columellar lip.** Columella opens to outside by a minute opening, the **umbilicus.**

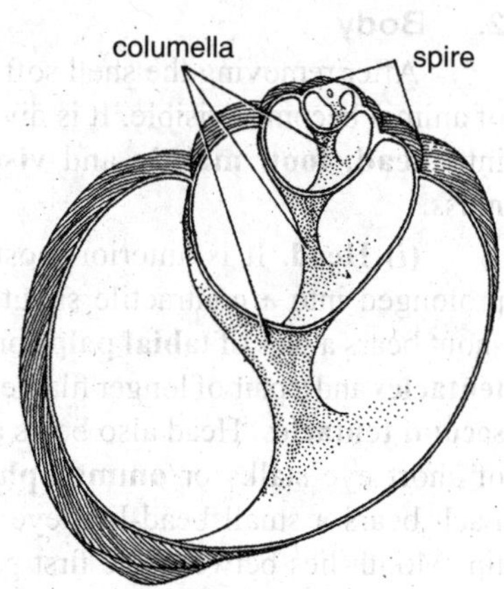

Fig. 15.2. *Pila globosa*. Shell in section showing columella.

The **shell coiling** in *Pila* is right handed, anti-clockwise or **dextral.** When the mouth (of shell) lies on the right-side of columella, the shell is called **dextrally coiled.** If the mouth facing the observer and shell is held with spire up then it is called **sinistrally coiled,** clockwise or left-handed.

Fig. 15.3. *Pila globosa*. Body of animal after the removal of the shell.

Microscopic Structure of Shell

The shell of *Pila* is formed of following three layers:

1. Periostracum. It is thin and pigmented outermost layer. It is not calcified but composed of **conchiolin** (horny protein).

2. Ostracum or **prismatic layer.** It is the middle layer. It is thickest and calcareous having crystalline calcareous plates running vertically.

3. Hypostracum or **nacreous layer.** It is the inner most calcareous layer. In it, the calcareous plates run longitudinally.

2. Body

After removing the shell soft body of animal becomes visible. It is divisible into **head, foot, mantle** and **visceral mass.**

***(i)* Head.** It is anterior most part prolonged into a contractile snout. The snout bears a pair of **labial palps** or **first tentacles** and a pair of longer filamentous **second tentacles.** Head also bears a pair of short eye stalks or **ommatophores.** Each bears a small bead-like eye at its tip. Mouth lies between the first pair of tentacles. Tentacles are highly contractile structures.

***(ii)* Foot.** Foot is large triangular, strongly muscular and ventral in position (Fig. 15.4). The ventral surface of foot is broad, flat and smooth. It is called **sole** which is adapted for creeping movement.

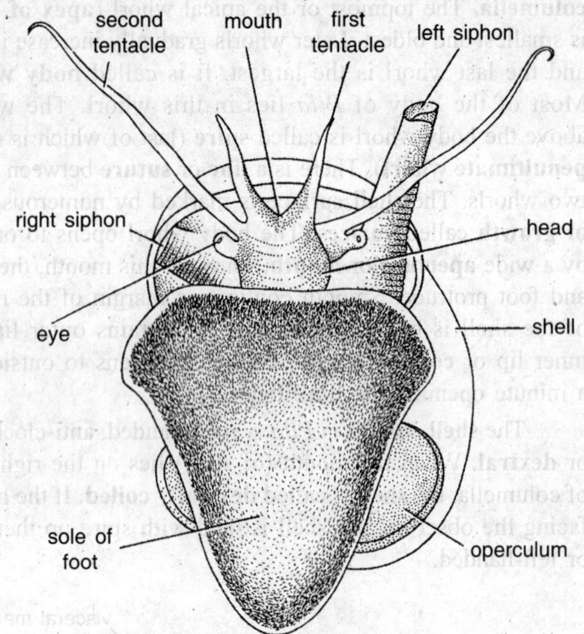

Fig. 15.4. *Pila globosa*. Animal in ventral view showing formation of elongated siphon formed by the left pseudopodium; this is done for aerial respiration.

***(iii)* Operculum.** Foot bears a horny disc, called **operculum,** on the posterior dorsal surface (Fig. 15.5). It is a calcareous plate and formed by the secretions of glandular cells of the foot. Outer surface of the operculum shows many concentric **rings of growth** around a small central axis or **nucleus.** The inner surface of operculum contains an elliptical area the **boss,** to which are attached the opercular muscles. Boss is surrounded by a shallow groove. When the foot is withdrawn into the shell, the operculum fits into the mouth of the shell and closes it.

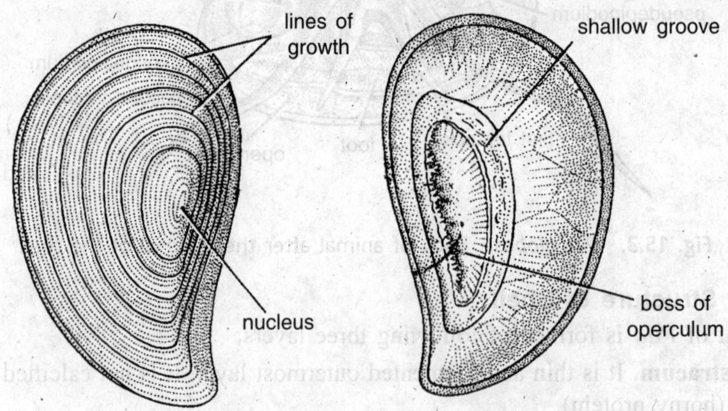

Fig. 15.5. *Pila globosa*. Operculum; A—Outer view; B—Inner view.

***(iv)* Columellar muscles.** A columellar muscle arises from the foot and is inserted in the columella, it attaches the body with the shell and it withdraws the animal inside and closes the operculum. In an expanded animal, the head and foot come out of the shell mouth but viceral mass lies inside the shell.

(*v*) Visceral mass. Between the head and foot the body forms a hump on the dorsal side. The hump is called **visceral mass.** It contains all the organ systems of the body. The visceral mass is spirally coiled and fills up the whorls of the shell. It shows the phenomenon of torsion.

(*vi*) Mantle. Below the shell lies **mantle** or the **pallium.** It is thin and delicate. A long narrow **supra-marginal groove** runs along the thickened free edge of mantle. Behind the groove there is a thick band containing **shell secreting glands** (which secrete the shell). On each side of head the fleshy processes are formed by mantle, called **nuchal lobes** or **pseudepipodia.** Left lobe is longer than right lobe and form **respiratory siphons** (Fig. 15.4).

Pallial Complex (Organs of Mantle Cavity)

Anterior part of body encloses a large space called **mantle cavity.** The organs present inside the mantle cavity are known as **organs of mantle cavity** or **pallial complex.** These are as follows:

1. Epitaenia. A prominent ridge is seen on the floor of the mantle cavity; it is called **epitaenia.** It divides the cavity into two unequal chambers. The chamber of the right side is small and is called **branchial chamber** while the chamber on the left side is large and called **pulmonary chamber.**

2. Ctenidium. A ctenidium or gill lies attached in the branchial chamber to the right side of mantle.

3. Rectum. It lies on the floor of the mantle cavity and opens to the exterior by **anus** which is situated near the right **pseudepipodium** or **right nuchal lobe.**

4. Genital duct. The vas deferens in case of male or the oviduct in female lies in the branchial chamber in between the rectum and epitaenia. Its opening is situated close to the **anus.** A hook-like **penis** or **copulatory organ** in male arises from the edge of mantle in front of male genital aperture. At the base of penis there is found a **hypobranchial gland.**

5. Anterior renal chamber. It projects as a reddish sac at the posterior end of the epitaenia and opens into the mantle cavity by an oblique slit.

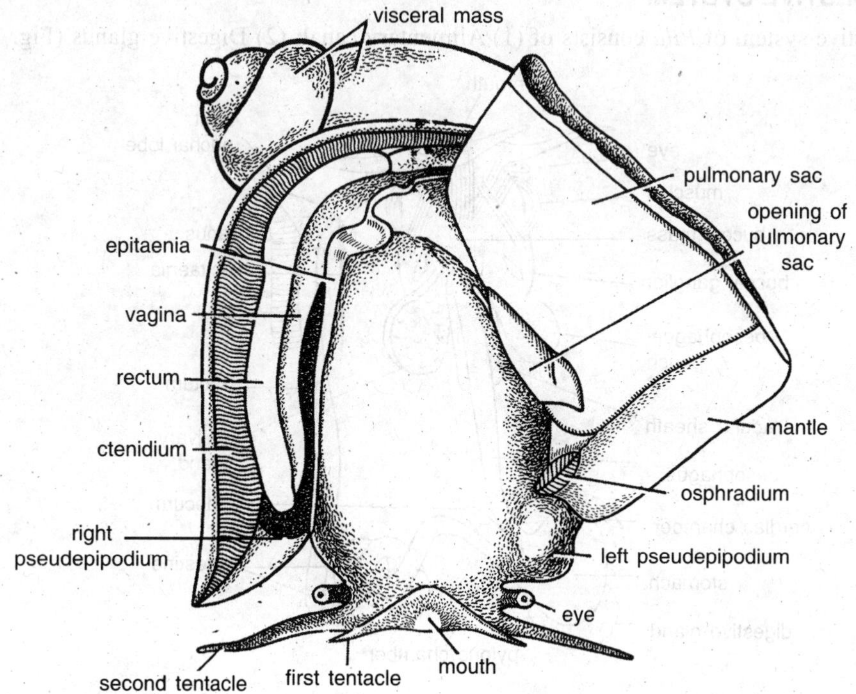

Fig. 15.6. *Pila globosa.* A female individual dissected to show the organs of pallial complex.

6. Pulmonary sac. This is so-called "lung" of *Pila* and found in the pulmonary chamber. It hangs from the roof of mantle cavity and opens into it by a large obliquely placed **pulmonary aperture.**

7. Osphradium. It is a chemosensory structure to test the purity of incoming water currents. It is attached to the roof of mantle cavity just above the left nuchal lobe. Course of water current through mantle cavity has been shown in Fig. 15.7.

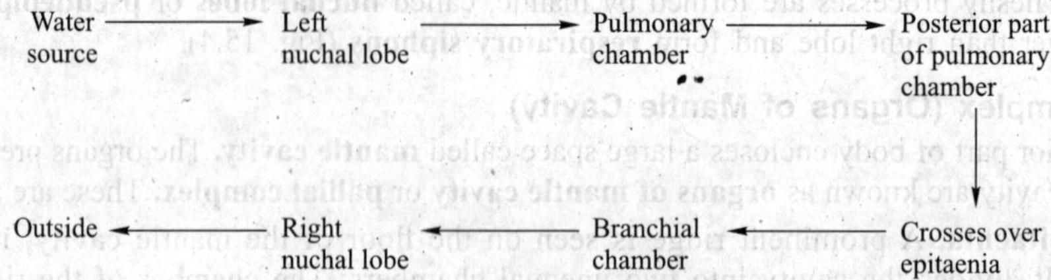

Fig. 15.7. *Pila globosa*. Course of water current through mantle cavity.

Histology of body wall. Body wall of *Pila* is composed of **epidermis** and **dermis.** The epidermis consists of a single layer of epithelial cells. The cells bear cilia in parts not protected by the shell. The epidermis also contains a large number of unicellular glands which secrete mucus, pigments and lime.

The dermis is composed of connective tissues and muscle fibres. It also contains pigment cells.

Coelom

The coelom of *Pila globosa* is represented by the cavities of kidney, gonads and pericardial cavity. The space between the visceral organs is filled with blood. This is called **haemocoel.**

15.4 DIGESTIVE SYSTEM

Digestive system of *Pila* consists of (1) Alimentary canal; (2) Digestive glands (Fig. 15.8).

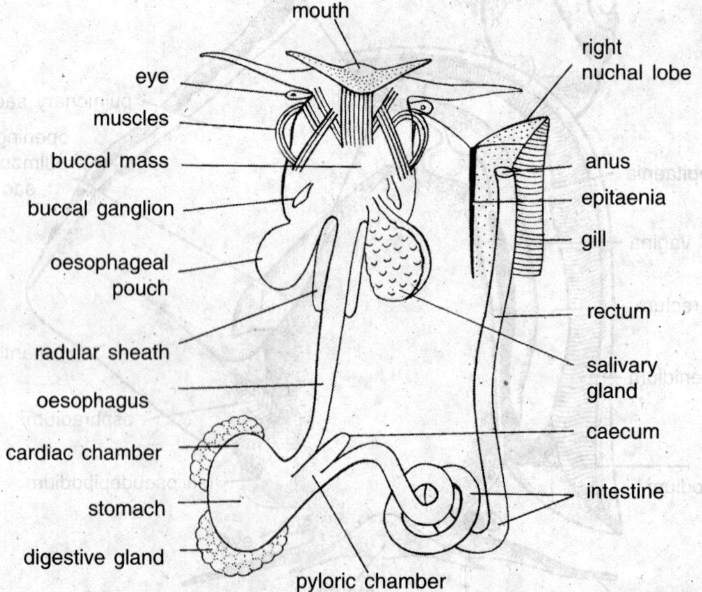

Fig. 15.8. *Pila globosa*. Digestive system: alimentary canal and digestive glands.

A. Alimentary Canal

It is curved and tubular structure and divided into following three parts:

(i) **Fore gut** which includes mouth, buccal cavity and oesophagus.

(ii) **Mid gut** which includes stomach and intestine.

(iii) **Hind gut** which includes rectum and anus.

1. Fore gut or stomodaeum. It is the anterior part of alimentary canal which is lined internally by **ectoderm.**

***(i)* Mouth.** Mouth is a median vertical slit, located at the tip of snout. It is surrounded by a frilled fold of integument. These folds serve as secondary lip.

***(ii)* Buccal cavity.** Mouth opens into **buccal cavity.** It is the cavity of **buccal mass.** The buccal mass is a large muscular pear-shaped organ. Its wall is supported by many cartilages and is provided with seven sets of muscles. The muscles bring about the movement of radula and jaws. The buccal cavity consists of following parts:

***(a)* Jaws.** There are two jaws (Fig. 15.9). Each jaw is a flexible, somewhat elliptical chitinous structure placed dorso-laterally. The jaws form posterior margin of the vestibule. Both jaws are connected together by a thin cuticular membrane. The jaws bear numerous small and two or three large teeth.

cutting edge

ridge of attachment of muscles

Fig. 15.9. *Pila globosa*. Jaws.

***(b)* Odontophore.** The floor of posterior part of buccal cavity is raised up as a fleshy outgrowth called **tongue mass** or **odontophore** (Fig. 15.10). The roof of buccal cavity carries a cartilage and well developed **protractor** and **retractor muscles.** It is supported by a large **'S' shaped lateral cartilage** and a pair of triangular **superior cartilage** (Fig. 15.11). The odontophore bears at its anterior end a rounded process called **sub-radular organ.** Beneath it the floor of buccal cavity has a small pouch-like **sub-lingual cavity.** The odontophore acts as a cartilaginous base of radula.

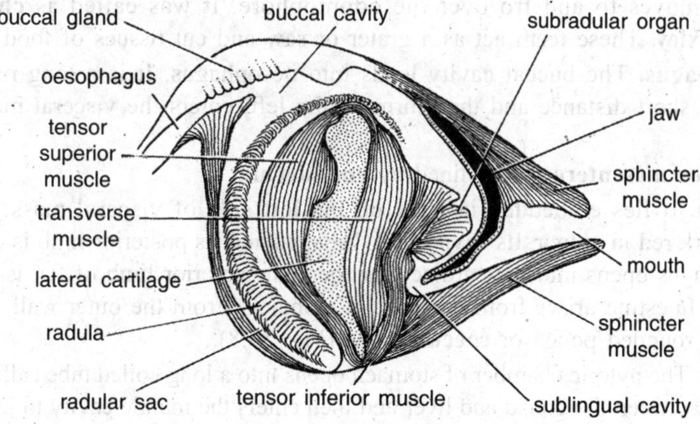

Fig. 15.10. *Pila globosa*. Vertical longitudinal section of the buccal mass about the middle.

***(c)* Radula.** Radula is a narrow, ribbon-like chitinous plate lying in the buccal cavity (Fig. 15.12). It runs longitudinally over the entire length of odontophore (Fig. 15.10). Posterior part of radula is enclosed into a sac called **radular sac.** The radula is covered with numerous horny recurved teeth. All the teeth are arranged in transverse rows. Each row consists of seven teeth in all (Fig. 15.13): the **central** or **median rachidian** (spine-like), followed by a smaller **lateral** and two needle-like **marginal teeth** on each side. The **radular formula** of *Pila* is thus, 2, 1, 1, 1, 2 (*i.e.*, Two marginal teeth + one lateral tooth + one median or central tooth + one lateral tooth + two marginal teeth). The cutting edge of central tooth bears a median and 2-3 lateral cusps on each side. The radular teeth are secreted by the **odontoblast cells** at the bottom of the radular sac.

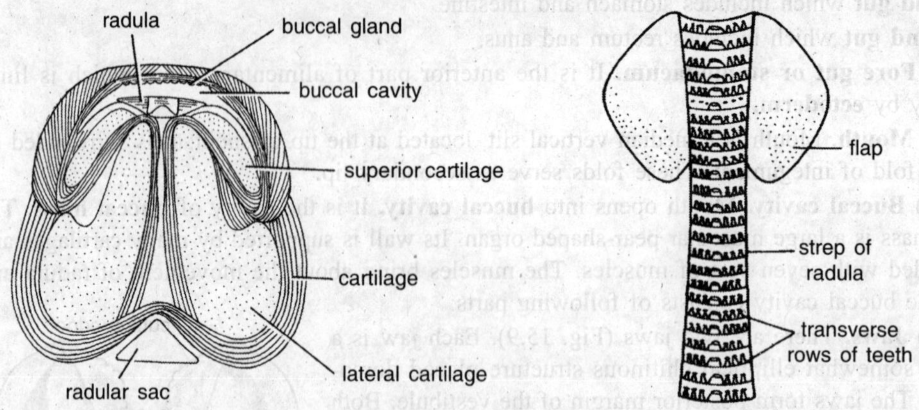

Fig. 15.11. *Pila globosa*. Buccal mass in T.S., showing position of radula on odontophore.

Fig. 15.12. *Pila globosa*. The radula.

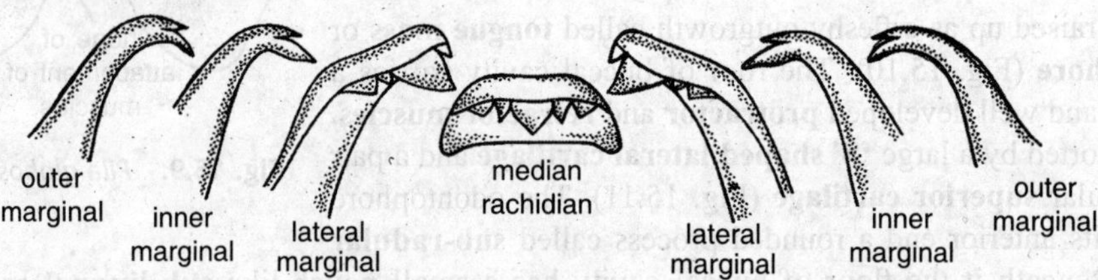

Fig. 15.13. *Pila globosa*. A single row of radular teeth.

The radula moves to and fro over the odontophore. It was called as **chain saw-like movement** by **Huxley.** These teeth act as a grater or saw and cut tissues of food by rasping.

(*iii*) Oesophagus. The buccal cavity leads into oesophagus. It is a long narrow tube. It runs straight for a short distance and then turns to the left, enters the visceral mass and opens into the stomach.

2. Mid gut or mesenteron. It is lined by **endoderm.**

(*i*) Stomach. It lies embedded in liver on the left side of visceral mass. It is roughly rectangular and dark red in colour. Its cavity is 'U' shaped and its posterior limb is called **cardiac chamber.** Oesophagus opens into the cardiac chamber. The anterior limb of 'U' is tubular called **pyloric chamber.** Intestine arises from this part of stomach. From the outer wall of the **pyloric chamber** a small, rounded pouch or **caecum** arise (Fig. 15.8).

(*ii*) Intestine. The pyloric chamber of stomach opens into a long coiled tube called **intestine.** It makes 2 to 3 coils between the gonad and liver and then enters the mantle cavity to join the rectum.

3. Hind gut or proctodaeum. It is lined by ectoderm. It forms the last part of alimentary canal.

Rectum. It lies in the mantle cavity along the left side of the ctenidium. It is thick-walled tube and is lined by cuticle. It opens to outside through **anus** near the edge of right nuchal lobe.

B. Digestive Glands

There are following types of digestive glands:

(1) Buccal glands. It is a pair of glands which lies in the roof of buccal cavity. They secrete accessory digestive juices.

(2) Salivary glands. A pair of white, branched and glandular salivary glands lie on either side of oesophagus just behind the buccal mass. The ducts of salivary glands run anteriorly. The salivary juice contains mucus and digestive fluid. The mucus tends to lubricate the food and radula.

(3) Oesophageal pouches. A pair of simple, rounded, cream-coloured oesophageal pouches lie on each side of oesophagus beneath the salivary glands. They open by small ducts at the junction of buccal cavity and the oesophagus. Their exact function is not known. They probably store the food for some time.

(4) Digestive gland, hepatopancreas or liver. It is a soft; somewhat triangular large gland occupies most of the space of visceral mass. Its colour is brownish and dirty green. It is composed of two lobes (Fig. 15.14). A duct arises from each lobe and unites together before opening into the stomach.

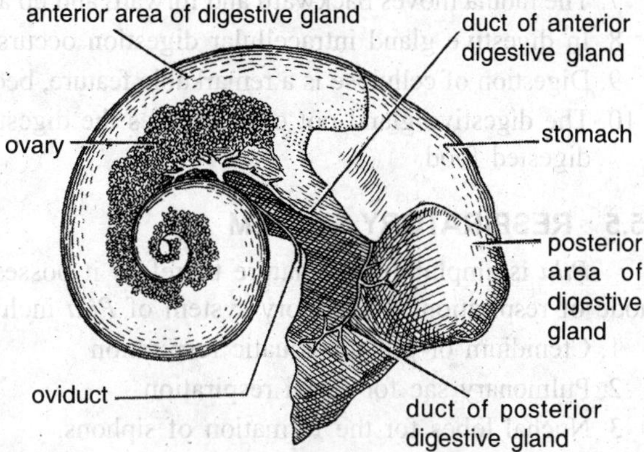

Fig. 15.14. *Pila globosa*. Digestive gland and associated structures seen from inside.

Each lobe is formed of numerous tubules embedded in the connective tissues. The tubes are lined internally by ciliated epithelium. The finer tubules join to form bigger tubules which finally combine into main duct. The terminal part of each tubule is known as **alveolus** and lined by three types of cells:

1. Secretory cells. They produce a ferment (cellulase enzyme) which digests the cellulose of plant cells.

2. Resorptive cells. These produce proteolytic enzymes, engulfs protoplasm of the food and digest it intracellularly.

3. Lime containing cells. These store phosphate of lime.

Food of *Pila*

Pila is mainly herbivorous, rarely it becomes a scavenger. Food of *Pila* includes soft and succulent leaves of aquatic plants such as *Valisneria* and *Pistia*. Sometimes they feed on dead animal tissue also.

Mode of feeding. *Pila* grasps a leaf between the sensory labial palps. By the action of sphincter and protractor muscles of buccal mass, the jaws cut the aquatic plants into pieces. The food is taken into the buccal cavity by the forward and backward movements of radula. In this way food is cut and masticated inside the buccal cavity.

Digestion

Salivary glands pour their secretion containing mucus and starch-digesting enzymes into the buccal cavity where it mixes up with the food. Therefore the digestion of starch begins in the buccal cavity. Mucus lubricates the food for its easy passage to the stomach. In stomach, digestive glands secrete digestive juices containing carbohydrate, protein and fat-digesting enzymes. Thus, the food is digested **extracellularly** in the stomach. Partly digested food particles pass into the resorptive duct of digestive glands, where the digestion of cellulose occur (This is being **intracellular digestion**).

Absorption. Absorption of digested food takes place in digestive glands and intestine.

Egestion. Undigested food passes out of rectum through anus into the branchial chamber from where it is thrown outside along with outgoing water current.

Adaptive features in the digestive system of *Pila*

1. *Pila* is herbivorous having well developed and highly efficient radula and jaws to cut off the plant leaves.

2. Alimentary canal is long with a coiled intestine.
3. For rasping of food from the ground, recurved teeth of radula move to and fro over the odontophore and help in scraping the food.
4. Teeth of radula act like a chain-saw.
5. Stomach is large and here extracellular digestion occurs.
6. Intestine is long and coiled.
7. The radula moves backward and forward and up and down by protractor and retractor muscles.
8. In digestive gland intracellular digestion occurs.
9. Digestion of cellulose is a remarkable feature, because it is a rare process in the animal world.
10. The digestive gland not only secretes the digestive enzymes but helps in the absorption of digested food.

15.5 RESPIRATORY SYSTEM

Pila is amphibious in nature therefore it possesses both **aquatic** and **terrestrial** or aerial mode of respiration. Respiratory system of *Pila* includes the following organs:

1. Ctenidium or gill for aquatic respiration
2. Pulmonary sac for aerial respiration
3. Nuchal lobes for the formation of siphons.

1. Ctenidium

Gill of *Pila* is single and monopectinate (*i.e.*, lamellae are arranged in a single row). It lies on the extreme right side of branchial chamber of mantle cavity.

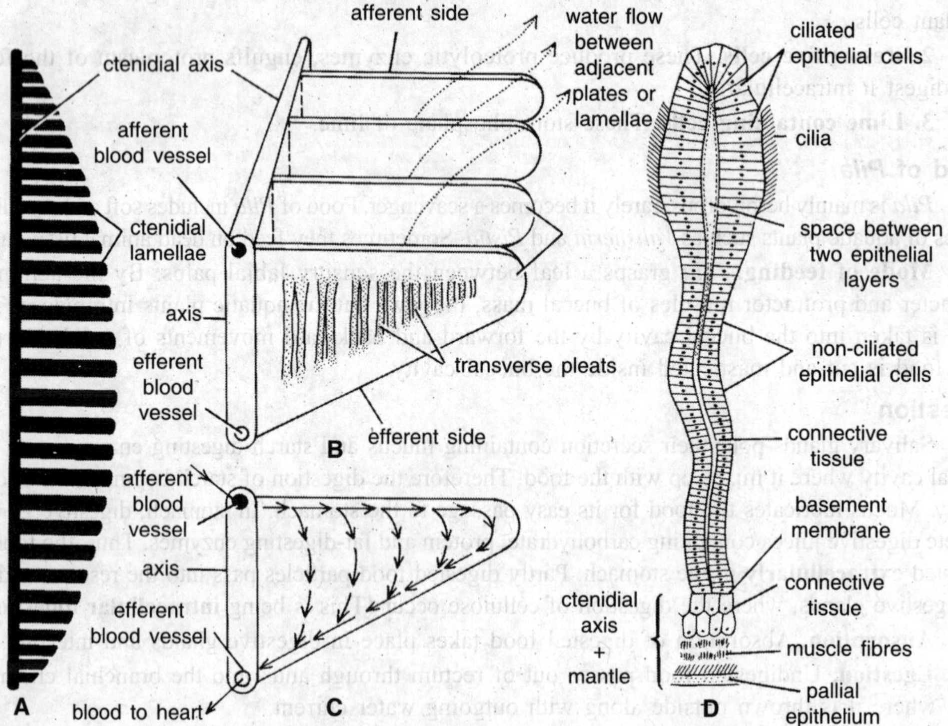

Fig. 15.15. *Pila globosa*. Respiratory organs. A—A monopectinate ctenidium; B—Movement of water currents through gill-lamellae; C—A single lamella showing flow of blood within it; D—A lamella in T.S.

The ctenidium consists of a large number of thin **leaflets** or **lamellae** (Fig. 15.15). Each lamella has a broad base and narrow apex. Lamellae are attached to the mantle through the **ctenidial axis.** The lamellae lie parallel to each other at the right angle to the ctenidial axis. In the middle they are largest and gradually become reduced in size towards the both ends. Each lamella has a smaller right side which receives blood from the afferent vessel and is called the **afferent side** and a longer left side from which blood goes into the efferent vessel called the **efferent side.** Both flat surfaces of each lamella contain transverse **ridges or pleats** which become smaller from the base towards the apex. These are richly supplied with blood vessels.

Blood supply of ctenidium. The afferent vein supplies deoxygenated blood to each gill lamella for oxygenation. The oxygenated blood is collected from all lamellae by the efferent vein.

Histology of ctenidium. Each lamella is formed of two layers of branchial epithelium separated by a narrow space in between. The epithelium consists of three types of cells: 1. Non-ciliated columnar cells; 2. Ciliated columnar cells and 3. Glandular cells. The epithelium rests upon basement membrane followed by a layer of connective tissues and a layer of oblique muscles. The outer most lining is of pallial epithelium or mantle epithelium.

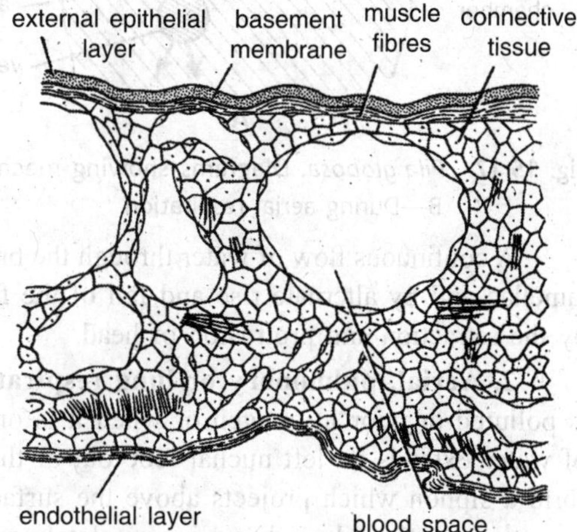

Fig. 15.16. *Pila globosa*. T.S. of a portion of the outer wall of the pulmonary sac.

2. Pulmonary Sac

It is a large bag-like structure which hangs from the roof of mantle cavity in the spacious branchial chamber. The lung or pulmonary sac communicate with pulmonary chamber of the mantle cavity by a large, oval aperture, the **pneumostome,** guarded by two unequal flaps. This sac consists of a densely pigmented dorsal wall and a creamy white ventral wall.

Blood supply of pulmonary sac. The walls of pulmonary sac are richly supplied with blood vessels. A pulmonary sinus lies in the walls of sac and a pulmonary vein collects blood from the walls of the sac and carries it to the heart.

Histology. The outer wall of pulmonary sac consists of an external **epithelial layer** containing fine black pigment granules. The epithelium rests on a thin **basement membrane** below which lies a thick layer of **connective tissue** followed by **endothelium** of the inner surface (Fig. 15.16).

3. Nuchal Lobe

On either side of the head, there is a contractile fleshy projection formed by mantle, called **nuchal lobe** or **pseudopodia**. Left lobe is longer than right. These form elongated siphons during breathing.

Mode of Respiration

Pila is an amphibious animal, therefore, it respires by ctenidium in water and by pulmonary sac when on land or in muddy water. Thus, in *Pila*, the respiration is of two types:

1. Aquatic or ctenidial or branchial respiration. This type of respiration occurs when the animal is under the water. For aquatic respiration, animal relies upon the oxygen dissolved in water. The head and foot are extended out of the shell, while left nuchal lobe increases in size and forms a funnel-like channel. Water enters through this funnel, comes in contact with **ospharidium.** The water reaches the posterior part of pulmonary chamber and enters the branchial chamber after crossing the epithelia and exchange of gases occur (Fig. 15.17A).

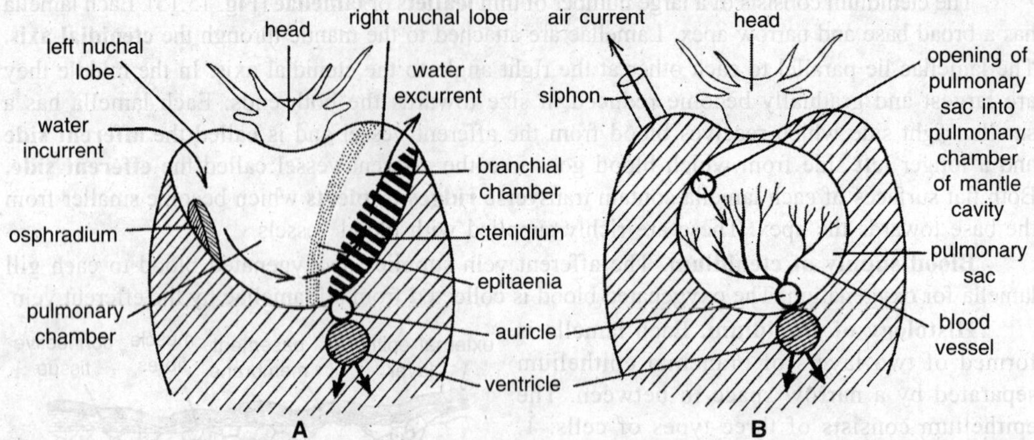

Fig. 15.17. *Pila globosa*. Diagrams showing mechanism of respiration. A—During aquatic respiration; B—During aerial respiration.

A continuous flow of water through the branchial chamber is maintained by cilia of ctenidial lamellae and by alternate rise and fall of the floor of mantle cavity. The latter is brought about by the retraction and protrusion of head.

2. Aerial, pulmonary or lung respiration. *Pila* respires through lung when the water is polluted or poor in oxygen or when it is on land. The animal after coming near the surface of water extends its left nuchal lobe out of the water which becomes elongated and tubular to form a siphon which projects above the surface of water (Fig. 15.4). The pulmonary chamber becomes greatly enlarged and is completely cut off from the branchial chamber. The air enters the pulmonary sac through siphon and the gaseous exchange occurs by the alternate contraction and expansion of the pulmonary sac (Fig. 15.17B).

While *Pila* is on land, the siphon is not formed and air enters pulmonary chamber directly through the expanded nuchal lobe.

During **hibernation** and **aestivation,** the animal lies buried in the mud for long periods. Its shell remains tightly closed by the operculum. Under these conditions, *Pila* respires through the air closed in pulmonary sac.

15.6 BLOOD VASCULAR SYSTEM

The blood vascular system of *Pila* is modified according to its amphibious mode of life. It consists of following parts:

1. Pericardium; 2. Heart; 3. Arteries; 4. Sinuses; 5. Veins.

1. Pericardium

Pericardium is situated between the pulmonary chamber and the posterior renal sac on the left side of the body whorl. It is thin-walled oval chamber enclosing the heart. It is posteriorly surrounded by the stomach and digestive glands, dorsally by the renal sacs (or chambers) and ventrally by the oesophagus. The pericardial cavity represents the true coelom and it communicates with the posterior renal chamber by **reno-pericardial aperture.** The heart, aorta and ampulla lie in the pericardial cavity.

2. Heart

The heart (Fig. 15.18) lies in the pericardial cavity. It is two chambered and consists of (*i*) Auricle; (*ii*) Ventricle.

(*i*) Auricle. It is single, roughly triangular, thin walled and contractile sac. It opens into the ventricle through an **auriculo-ventricular aperture.** The aperture is guarded by two **semi-lunar valves** which allow the blood to flow only from auricle to ventricle but not from ventricle to auricle. The auricle receive blood from three veins:

1. Efferent ctenidial vein from the gill.
2. Efferent renal vein from the posterior renal chamber.
3. Pulmonary vein from the pulmonary sac.

(*ii*) Ventricle. It is the lower part of the heart and lies below the auricle. It is thick walled, ovoid, muscular and spongy. It is provided with muscular strands and has a narrow cavity.

The lower end of the ventricle gives rise a short, muscular tube called **aorta.** The opening between ventricle and aorta is guarded by two **semi-lunar valves** which allow the blood flow only from the ventricle to aorta but not from aorta to ventricle.

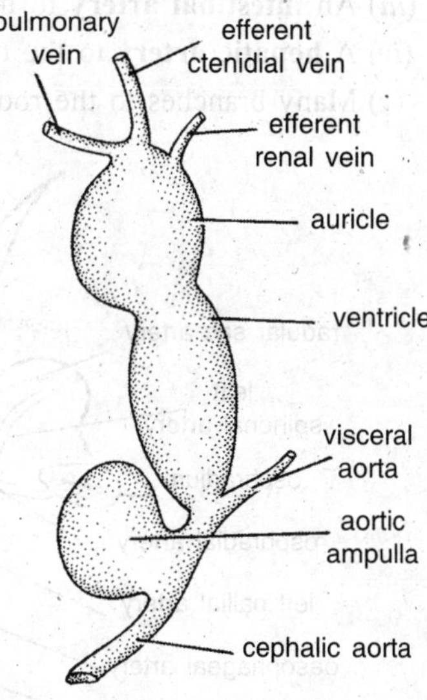

Fig. 15.18. *Pila globosa*. Heart.

3. Arteries

The aorta is divided into two sub-branches (Fig. 15.19):

(*i*) The cephalic aorta; (*ii*) The visceral aorta.

(*i*) The cephalic aorta. It supplies blood to the head region through a number of branches. Its base bears a bulb-like, thick-walled, contractile **aortic ampulla,** which aids in the flow of blood. From the outer side of aorta, three branches are given out: (*i*) a thin **cutaneous artery** which supplies blood to the body wall or skin (*ii*) a thick **oesophageal artery** which supplies blood to oesophagus; (*iii*) a thick **pallial artery** which supplies blood to the left side of the mantle, the left nuchal lobe and the osphradium. From the inner side of aorta only one branch is given out: the **pericardial artery** which supplies blood to the wall of pericardium.

After dividing into above branches cephalic aorta runs along the left side of the oesophagus and cross over to its right side. Meanwhile it sends many fine branches to the wall of oesophagus and the floor of mantle cavity. From the right side of aorta, a large branch arises which supplies the following three branches:

(*i*) A **right pallial artery** which supplies blood to the right part of mantle
(*ii*) A **right siphonal artery** to the right nuchal lobe
(*iii*) A **penial artery** to copulatory organ or penis.

Finally cephalic aorta divides into following four branches:

(*i*) **Radular sac artery** to the radular sac
(*ii*) **Optic arteries** to the eye stalk and eyes
(*iii*) **Tentacular arteries** to the tentacles
(*iv*) **Pedal arteries** to the foot after forming a network.

(2) Visceral aorta. The visceral aorta runs posteriorly into visceral mass and supplies blood through the following branches:

(*i*) A **pericardial artery** to the wall of pericardium
(*ii*) A **gastric artery** to the stomach

(*iii*) An **intestinal artery** to the intestine

(*iv*) A **hepatic artery** to the digestive gland and gonads

(*v*) Many branches to the roof of anterior renal chamber, intestine and rectum, etc.

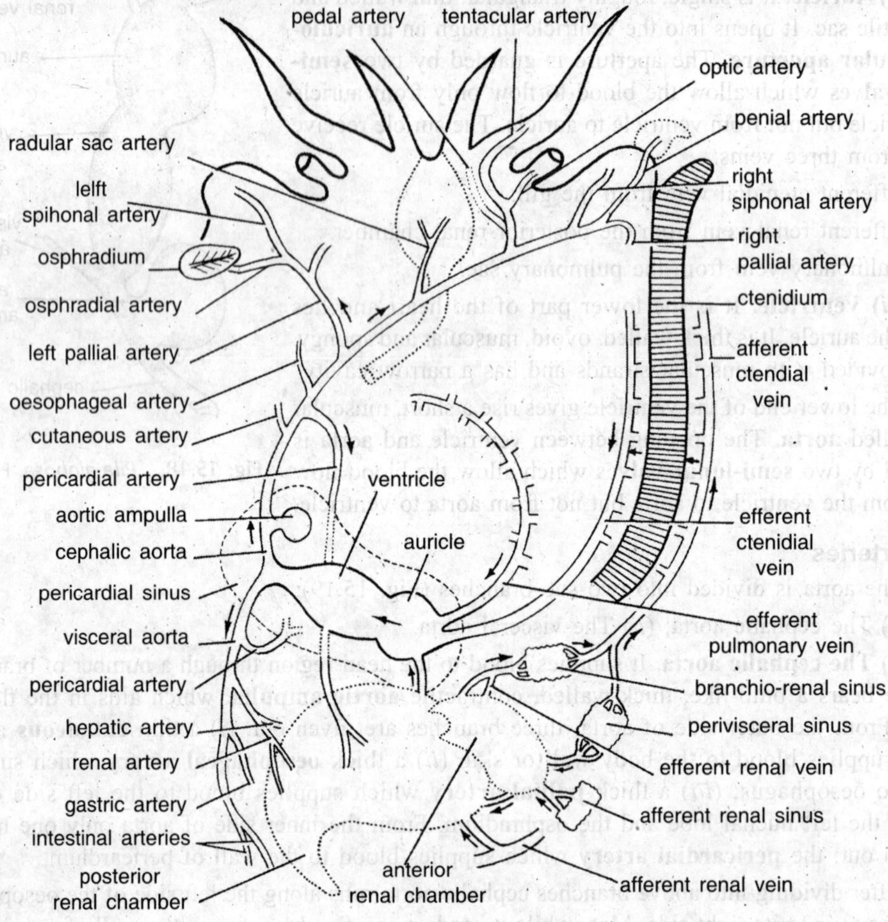

Fig. 15.19. *Pila globosa*. Heart and open type blood vascular system.

4. Sinuses

The ultimate branches of arteries widen inside the tissues or organs into blood spaces which are without definite walls. These spaces are called **blood lacunae.** Generally the blood lacunae connect together to form **sinuses.** They connect arteries to veins and serve as capillaries. The sinuses and lacunae together form the general **haemocoel.** Thus, sinuses are the space between the visceral organs and the body wall. They lack definite walls. There are following four main sinuses:

***(i)* Perivisceral sinus.** It lies in the floor of mantle cavity and occupies the anterior part of the body above the foot and below the floor of mantle cavity. It surrounds the anterior part of alimentary canal. Three channels carry blood from this sinus to the renal chamber and the pulmonary sac. It contains deoxygenated blood.

***(ii)* Peri-intestinal sinus.** It lies within the visceral mass along the coils of intestine. It collects blood from the digestive gland and reproductive organs.

***(iii)* Branchio-renal sinus.** It lies on the right side of anterior renal chamber. It receives blood from the roof of both renal chambers.

(iv) Pulmonary sinus. It lies in the wall of pulmonary sac. It receives blood from the perivisceral sinus.

5. Veins

The following important veins collect blood from the various parts of the body and this blood is carried to the auricles. The main veins are the following:

(i) Afferent ctenidial vein. It collects deoxygenated blood from the perivisceral sinus, branchiorenal sinus, rectum and genital duct through many branches. This blood is carried to ctenidium then to the branchial lamellae for purification.

(ii) Efferent ctenidial vein. It collects oxygenated blood from the lamellae of ctenidium, mantle and copulatory organs and then carries to auricle.

(iii) Afferent renal vein. It collects blood from peri-intestinal sinus and carries to the posterior renal chamber.

(iv) Efferent renal vein. It collects blood from the posterior renal chamber and carries it to the auricle.

(v) Pulmonary vein. It collects blood from the walls of the pulmonary sac and carries it to the auricle.

Blood

Blood is colourless because haemoglobin is absent. However, it contains another respiratory pigment **haemocyanin** which is dissolved in blood plasma. Haemocyanin contain copper, so it becomes blue when combines with oxygen and colourless when combines with carbon dioxide.

Course of Blood Circulation

Blood is pumped to all parts of body through the branches from ventricle (Fig. 15.20). The branches of cephalic and visceral aorta take part in this transportation. From all parts of the body blood reaches the **peri-visceral** and **peri-intestinal sinuses.** From the sinuses blood flows in the following ways:

(1) During aquatic respiration. Afferent-ctenidial vein carries most of the blood from peri-visceral sinus to the ctenidium for aeration. From here oxygenated blood is collected through efferent ctenidial vein and carried to the auricle.

(2) During aerial respiration. The blood from peri-visceral sinus is carried to the pulmonary sac for oxygenation. Pulmonary vein carry this oxygenated blood to the auricle.

(3) Blood from peri-intestinal sinus flows into the renal chamber to get rid of nitrogenous wastes. From here efferent renal vein carry this oxygenated blood to the auricle.

Auricle receives both aerated and non-aerated blood. Mixed blood then enters the ventricle.

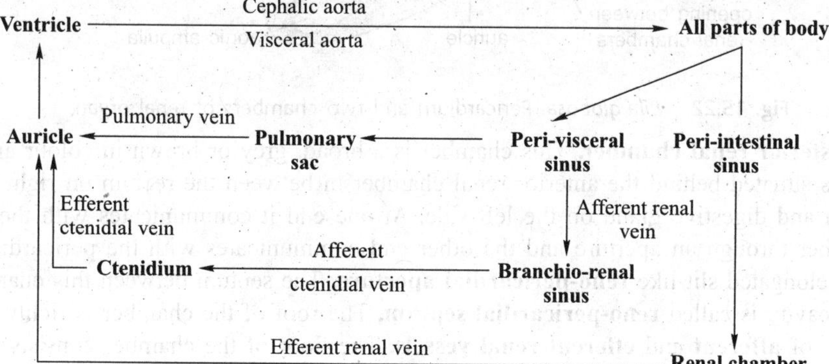

Fig. 15.20. Course of blood circulation in *Pila*.

15.7 EXCRETORY SYSTEM

All the nitrogenous wastes of *Pila* are excreted by a large **kidney, renal organ, organ of Bojanus** or **metanephridia.** It lies behind the pericardium and is a thick-walled sac which is much folded within. The renal organ contains a small anterior chamber and a large posterior chamber (Fig. 15.21 and Fig. 15.22).

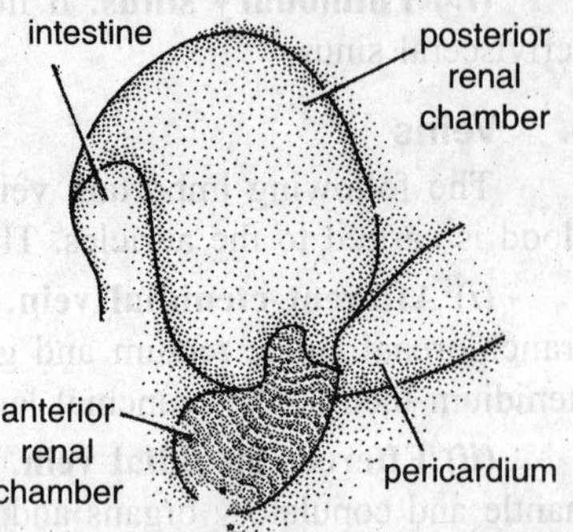

Fig. 15.21. *Pila globosa.* Renal chamber.

1. Anterior renal chamber. It is oval and reddish in colour. Anterior renal chamber is situated anterior to the pericardium. It projects into the mantle cavity and opens into the branchial chamber through a slit like opening near the epitaenia. Its other end opens into the posterior renal chamber through an internal opening. The internal cavity of the anterior chamber is reduced due to presence of triangular leaf-like process or **lamellae.**

The lamellae on the roof are arranged on either side of a median longitudinal axis, the **efferent renal sinus.** On the floor of the chamber, the lamellae are arranged on either side of a similar median axis, the **afferent renal sinus** which is the right branch of the **peri-intestinal sinus.** It branches profusely to supply lamellae on both sides. The roof or floor folds alternate to each other.

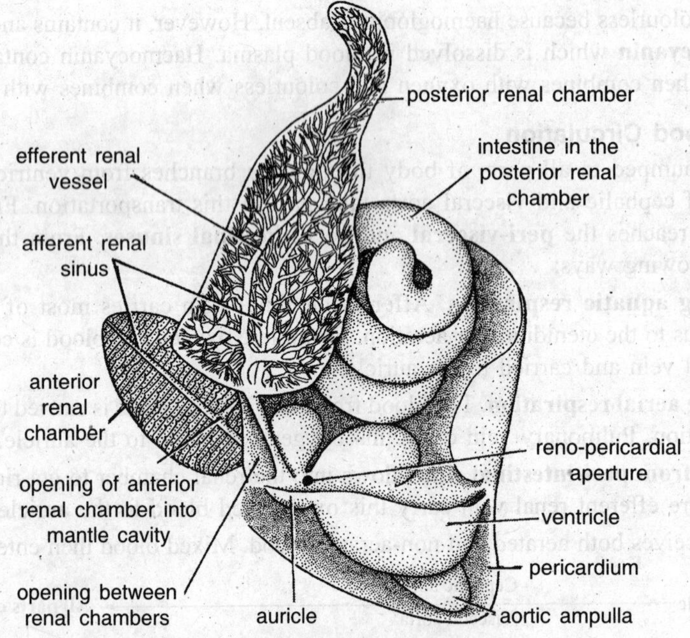

Fig. 15.22. *Pila globosa.* Pericardium and two chambers of renal organ.

2. Posterior renal chamber. This chamber is a broad, grey or brown in colour and hook-shaped. It is situated behind the anterior renal chamber in between the rectum on right side and pericardium and digestive gland on the left side. At one end it communicates with the anterior renal chamber through an aperture and the other end communicates with the pericardial cavity through an elongated slit-like **reno-pericardial aperture.** The septum between this chamber and pericardial cavity is called **reno-pericardial septum.** The roof of the chamber is richly supplied by branches of **afferent** and **efferent renal vessels.** The floor of the chamber consists of a thin lining of epithelium which separates its cavity from the organs projecting in its cavity.

Physiology of excretion. The excretory waste of *Pila* contains **ammonia, ammonium compounds, urea** and **uric acid.** During aquatic life ammonia is excreted in solution while during the terrestrial life *Pila* transforms ammonia into soluble **uric acid;** this is done for the conservation of water. Thus, excretion of *Pila* shows an adaptation for its amphibious mode of life.

These nitrogenous wastes are separated from blood by both chambers of renal organ and discharged into the mantle cavity through the renal aperture. These are ultimately passed out of the body through the right nuchal lobe along with outgoing water current.

Digestive gland also acts as an excretory organ. The cells of digestive gland engulf waste products and later on these are passed out through the stomach and intestine.

15.8 NERVOUS SYSTEM

The nervous system of *Pila* is well developed possessing a concentrated ring of most of the ganglia around the buccal mass and twisting of visceral cords in the figure of '8'. Each ganglion consists of commissures, connectives and nerves. All these structures can be summarised as follows:

1. Ganglia

(*i*) A pair of cerebral or labial ganglia
(*ii*) A pair of buccal ganglia
(*iii*) Paired pleuro-pedal ganglionic mass
(*iv*) Unpaired supra-intestinal ganglion
(*v*) Visceral ganglia.

2. Commissures

(*i*) Cerebral commissure
(*ii*) Buccal commissure
(*iii*) Pedal commissure.

3. Connectives

4. Nerves

A. Ganglia

1. Cerebral ganglia. It is situated on either side of anterior part of buccal mass. A pair of cerebral ganglia are large, flattened, roughly triangular structures connected together by thick, band-like **cerebral commissure** dorsally and by a thin labial **commissure** ventrally. Each cerebral ganglion is further connected with the buccal ganglion of its side through a very slender **cerebro-buccal connective.** Thick band-shaped **cerebro-pleural** and **cerebro-pedal connectives** serve to connect each cerebral ganglion with the corresponding pleural and pedal ganglia.

Nerves. Following nerves are given out in head region:

(*i*) Anteriorly to the skin of snout, tentacles and the buccal mass.
(*ii*) Posteriorly to the tentacles, eyes and the statocyst.

2. Buccal ganglia. A pair of small, triangular buccal ganglia situated dorso-laterally on either side of the junction of buccal mass and oesophagus. The two buccal ganglia are connected together ventrally by **buccal commissure.** A pair of **cerebro-buccal connectives** connect the buccal ganglia with cerebral ganglia.

Nerves. Each buccal ganglia sends nerves to buccal mass, radular sac, salivary glands, oesophagus and oesophageal pouches.

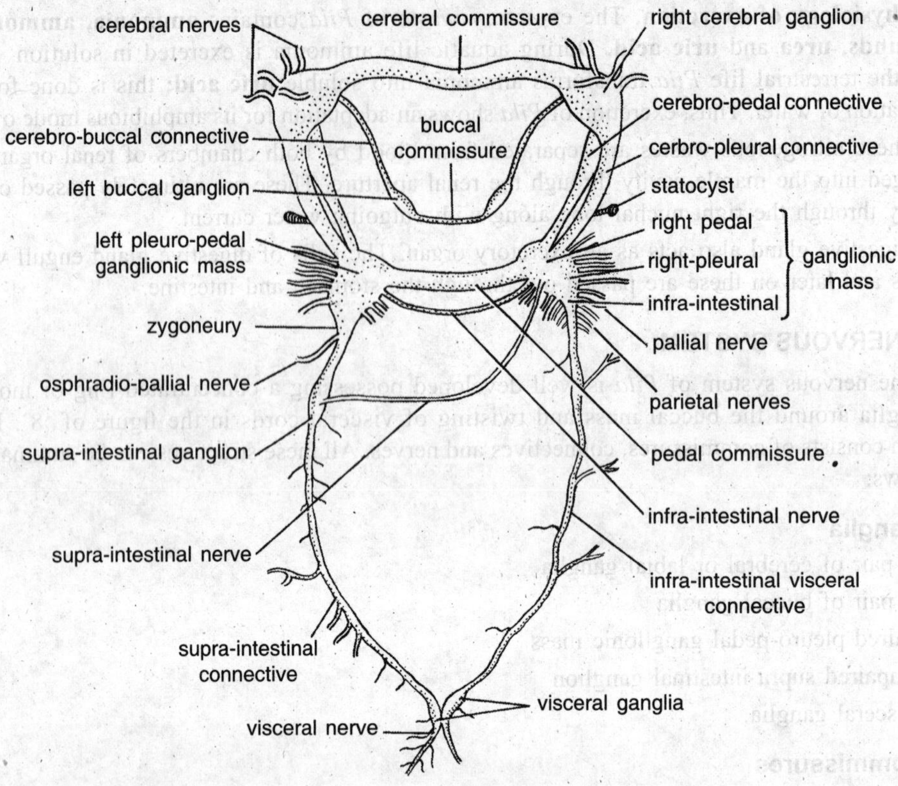

Fig. 15.23. *Pila globosa*. Nervous system.

3. Pleuro-pedal ganglionic mass. It is large, triangular and situated ventro-laterally one on either side of the buccal mass. Each ganglionic mass is formed by the fusion of an outer **pleural** and inner **pedal** ganglion. The line of fusion is shown as slight notch. Pleuro-pedal ganglionic mass of right side remains fused with **sub-intestinal ganglion.** Pedal ganglia are connected by two-thick strap-like **pedal commissures** which run on the ventral side of buccal mass. Two pleural ganglia are connected by a slender loop-like **infra-intestinal nerve.** Paired **cerebro-pleural** and **cerebro-pedal** connectives connect the cerebral ganglia with the pleural and pedal ganglia respectively.

Nerves

(*i*) Left pleural ganglion sends nerves to the body wall, mantle, osphradium, left nuchal lobe, columellar muscles and anterior part of gills.

(*ii*) Right pleural ganglion issues nerves to the body wall, epitaenia, right nuchal lobe, copulatory organs, columellar muscles and rectum.

(*iii*) Pedal ganglia sends nerves to the foot.

4. Supra-intestinal ganglion. It is unpaired, fusiform ganglion which is situated within a haemoc-oelomic sinus. It is connected with the pleuro-pedal ganglionic mass by a stout connection. A thin, fine **supra-intestinal nerve** connects it with the pleural ganglion of right side. It crosses over oesophagus.

Nerves

(*i*) Supra-intestinal ganglion gives off nerves to mantle, anterior part of ctenidium and pulmonary sac.

(*ii*) Supra-intestinal nerve sends a few branches to the body wall.

5. Visceral ganglia. These are spindle-shaped and formed by the fusion of two visceral ganglia. They are situated in the visceral complex near the anterior lobe of the digestive gland. A left **visceral connective** or **supra-intestinal visceral connective** connects the visceral ganglia with the supra-intestinal ganglion. A right visceral connective or **infra-intestinal visceral connective** connects the visceral ganglia with the subintestinal ganglion of the pleuro-pedal ganglionic mass of the right side and runs below the intestine.

Nerves. Visceral ganglia send nerves to the renal organs, genital organs, pericardium, stomach, intestine and the digestive gland and their right side sends few nerves to the body wall.

15.9 SENSE ORGANS

Sense organs of *Pila* consist of sensory cells which are scattered all over the surface and are of following types: 1. Osphradium, 2. Eyes, 3. Statocyst; 4. Tentacles.

1. Osphradium. Osphradium is single, attached to the roof of mantle cavity on its left, hanging over the inhalant siphon. It is small, elongated, somewhat oval in outline with right end bluntly rounded. It is **bipectinate.** It consists of about 28 thick, fleshy, roughly triangular leaflets arranged in two rows on both sides of **central axis.** The middle leaflets are largest. They receive its nerve supply from the left pleural ganglion.

Histology of osphradium. The leaf-lets are covered with a layer of **epithelial cells.** It consists of sensory, ciliated and glandular cells. The epithelium is followed by a thin **basement membrane.** Its interior is filled with connective tissues, nerve fibres and blood spaces.

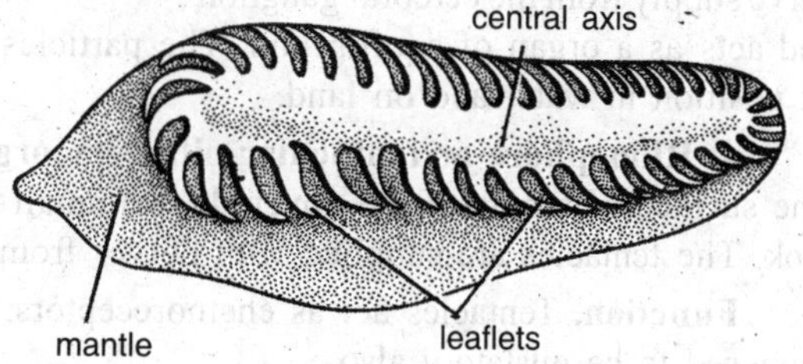

Fig. 15.24. *Pila globosa*. L.S. of osphradium.

Function. The osphradium serves as **olfactory** or **chemoreceptor organ.** It tests the chemical and physical properties of water entering the mantle cavity. If the water is not suitable for breathing, its entrance is checked by the closure of left nuchal lobe. It is also said to exercise selection over the food material.

2. Eyes. *Pila* has a pair of eyes which are short, black and stalked, lies on either side of the head outside the second pair of tentacles. The stalk of eye is called **ommatophore.**

The eye is a cup-like invagination of general epidermis. This invagination is called **optic vesicle** and is surrounded by connective tissues. The eye of *Pila* consists of the following parts:

(1) Inner cornea or pellucida internal. It lies just below the connective tissues.

(2) Lens. The central space of eye is occupied by a hyaline or structureless body called **lens.**

(3) Retina. Retina is made up of large cells. It is a sensitive part of eye. It contains two types of cells: (*i*) **Visual cells.** They are large broad cells having brush-like hair on their outer surface. (*ii*) **Supporting cells** or **packing cells.** They are scattered between the visual cells.

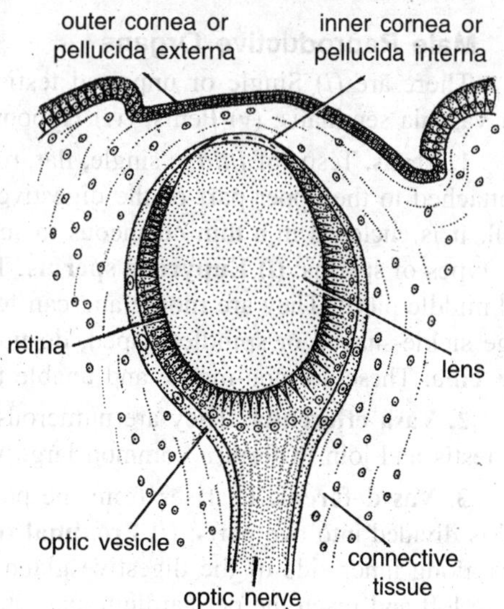

Fig. 15.25. *Pila globosa*. Horizontal longitudinal section of the eye.

4. Optic nerve. Optic nerve enters the eye at its posterior end.

Function. The eyes only respond to change in the intensity of light because sense of light is limited in *Pila*.

3. Statocyst. They are small, paired, pyriform, cream-coloured structures which are located in a depression near the pedal ganglion. Each statocyst is attached to the pedal ganglion of its side by a band of connective tissue. The statocyst is a hollow capsule surrounded by a thick layer of connective tissue. Its cavity is lined by a single layer of ciliated sensory epithelium and filled with a fluid and many minute, oval calcareous particles called as **statoconia.**

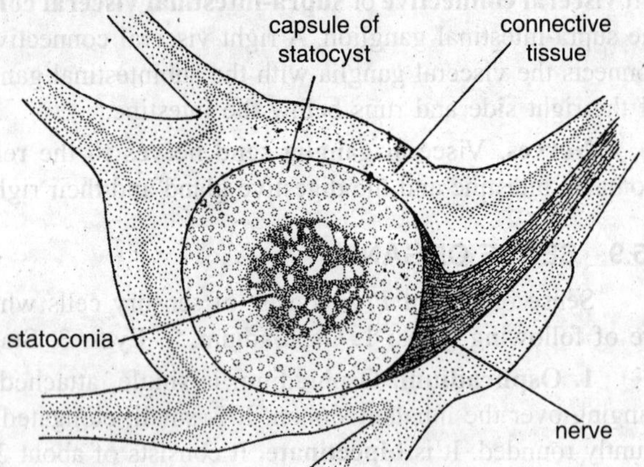

Fig. 15.26. *Pila globosa.* Statocyst.

Function. Statocyst receives its nerve supply from the cerebral ganglion and acts as a organ of equilibrium. The particles of statoconia help the animal in maintaining its position in water and on land.

4. Rhinophore or tentacular olfactory organs. In *Pila*, there are two pairs of tentacles. The surface of tentacles are covered with small ciliated papillae. These provide them a silky look. The tentacles are supplied with nerves from the cerebral ganglia.

Function. Tentacles act as chemoreceptors. They are tactile and chemosensory. They are supposed to be gustatory also.

15.10 REPRODUCTIVE SYSTEM

In *Pila* sexes are separate, *i.e.*, dioecious and **sexual dimorphism** is distinct: the shell of male animal is smaller than the female and male has a copulatory organ or penis.

A. Male Reproductive Organs

There are (*i*) Single or unpaired testis (*ii*) Vasa efferentia; (*iii*) Unpaired vasa deferens (*iv*) Vascula seminalis; (*v*) Penis; (*vi*) Hypobranchial gland (Fig. 15.27).

1. Testis. Testis of *Pila* is single, flat, roughly triangular and whitish-cream coloured. Testis is attached to the upper part of the digestive gland and extending in first or two whorls of the shell. It is enclosed in a thin cutaneous membrane. The **germinal epithelium** of testis produces two types of sperms: **(*i*) Eupyrene sperms.** These are small, thread-like containing head and tail and middle piece. They are motile and can fertilize the eggs. **(*ii*) Oligopyrene sperm.** These are large sickle-shaped or spindle-shaped, do not have head, but have tail. The tail is formed of 4 to 5 cilia. These are non-motile and unable to fertilize the eggs.

2. Vasa efferentia. They are numerous, minute ducts, originating at different levels from the testis and join to form a common large vas deferens.

3. Vas deferens. It arises from the posterior part of the testis and passes under the skin and is divided into two parts: **(*i*) Proximal tubular part.** It is distal thick-walled glandular part, runs along inner side of the digestive gland upto the posterior renal chamber and finally turns to the left and reach the pericardium and after turning to the right opens into the ventral side of vesicula seminalis. **(*ii*) Terminal glandular part.** It is thick-walled glandular part, runs along the left side of the rectum into the branchial chamber and opens into it by male genital pore. The genital pore is situated a little behind the anus to a claw-like projection of **genital papilla.**

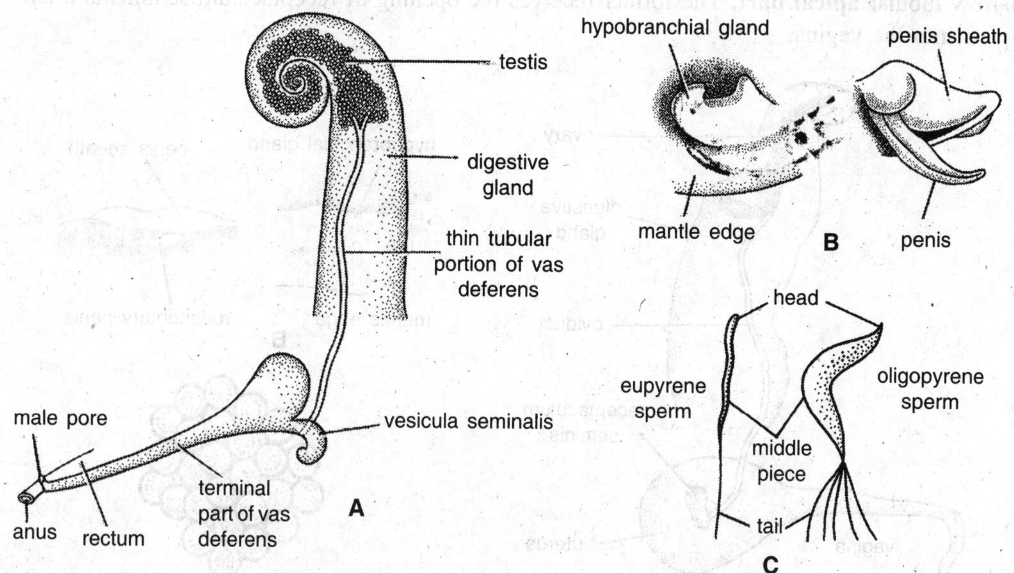

Fig. 15.27. *Pila globosa*. A—Male reproductive organs; B—Male copulatory organ in surface and ventral view; C—Sperms.

4. Vesicula seminalis or Seminal vesicle. It is situated on the right side of pericardium below the junction of anterior and posterior renal chambers. It is swollen flask-shaped sac having a blind rounded prolongation. It serves to store the sperms. The narrow tubular end of seminal vesicle is continuous with the glandular part of vas-deferens.

5. Copulatory organ (penis). Penis arise from the edge of mantle in front of anus. It is long about 1.5 cm, stout, slightly curved, flagellar with a swollen base and narrow tip. It bears a longitudinal groove. The penis is enclosed in a flap-like **penis-sheath** of yellow colour. The sheath is sac-like outgrowth of the mantle at the base of penis.

6. Hypobranchial gland. It is an oval glandular thickening at the base of the penis sheath. It consists of tall cells containing small basal nuclei. The surface of the glandular area is somewhat pleated but there is no duct and the secretions of the gland cells are poured directly on the surface.

B. Female Reproductive Organs

The female reproductive organs consist of following parts: 1. Ovary; 2. Oviduct; 3. Uterus; 4. Vagina; 5. Recepticulum seminis; 6. Copulatory organ; 7. Hypobranchial gland (Fig. 15.28).

1. Ovary. In female, the ovary occupies the same position as the testis in the male, *i.e.*, into first, second and third whorls of the shell. Ovary is a highly branched structure of a light orange colour which becomes darker in fully mature individuals. The branches of ovary, called **acini** are blind, flask-shaped structures having single layer of cells and open into the oviduct.

2. Oviduct. It is semi-transparent, narrow tube arising from the middle of ovary. It runs anteriorly just below the skin along the inner margin of the digestive gland. Near the renal organ it turns downwards.

3. Receptaculum seminis. It is a bean-shaped structure. It is situated at the junction of oviduct with the uterus and is enclosed in the cavity of posterior renal chamber. It stores the sperms received from male during copulation.

4. Uterus. It is a large, deep-yellow coloured and pear shaped structure. It lies inside the body whorl below the intestine. Uterus is differentiated into broad, vesicle-like rounded **basal part** and narrow tubular **apical part.** The former receives the opening of receptaculum seminis and latter opens into the vagina.

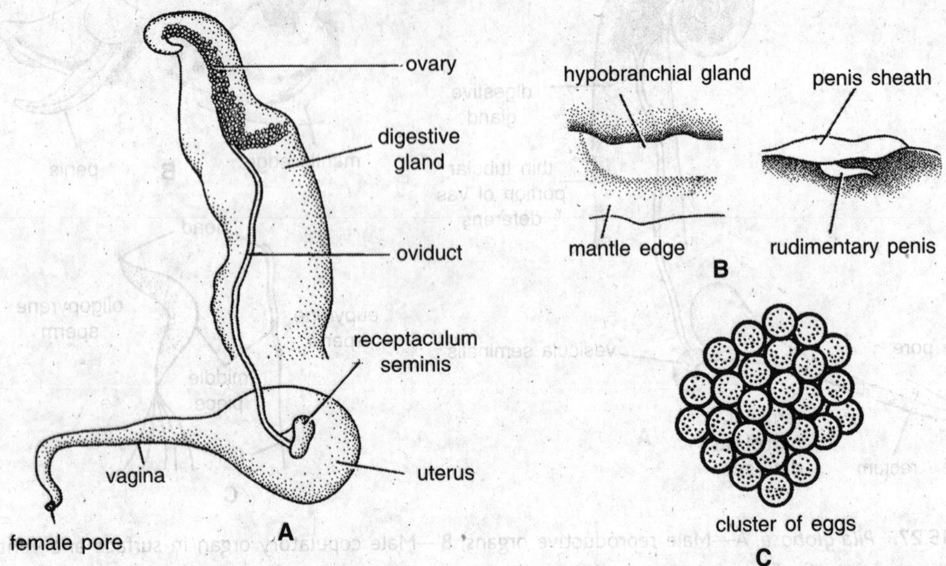

Fig. 15.28. *Pila globosa*. A—Female reproductive organs; B—Female copulatory organs in surface and ventral view; C—Eggs.

5. Vagina. The terminal narrow part of the oviduct is known as **vagina.** It is a cream-coloured tube which runs forward on the right side of rectum. It opens into the branchial chamber by **female genital aperture** which is situated behind the anus on a small papilla.

6. Copulatory organ. The female has a **rudimentary penis.** It is a small, thin flagellar structure of uniform thickness, being enclosed within the penis sheath.

7. Hypobranchial gland. It is a rudimentary glandular thickening.

Copulation

In *Pila globosa*, copulation takes place either in water or on land (Fig. 15.29). It lasts for 3 hours. Male and female snails come together facing each other. The penis of the male is expanded and gets attached to the genital papilla by its base. Then the penis and its sheath are inserted into the mantle cavity of the female. The tip of the penis is put into the female genital aperture and spermatozoa are transferred through the vagina into the receptaculum seminis.

Fertilization. Fertilization is **internal;** eggs are fertilized in the uterus. *Pila* is **oviparous** and oviposition or egg laying is started after a day or two of fertilization. The fertilized eggs are laid in masses of 200 to 800 in moist earth near ponds and lakes by the help of two temporary folds of foot.

Development

Eggs of *Pila* are round and as big as the pea seeds. Development is **direct,** *i.e.*, there is no free swimming larval stage. Young *Pila* hatches out of the egg. During development, due to torsion, the visceral mass and the shell of the embryo become spirally coiled.

Economic Importance

Pila causes great economic loss by damaging the rice crops. However, it is eaten by the poor people.

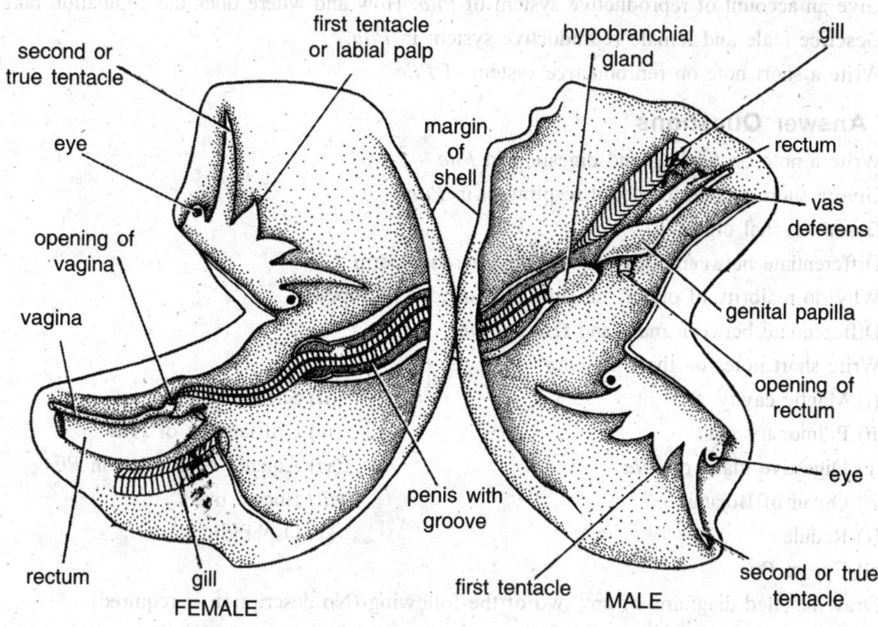

Fig. 15.29. *Pila globosa*. Copulation.

QUESTIONS

Long Answer Questions

1. Describe in detail the shell of *Pila* and give an account of contents of mantle cavity and course of water into it.
2. Describe organs of pallial complex in snail.
3. Describe the digestive system, food, feeding mechanism and digestion in *Pila*. Why is *Pila* called a scavenger?
4. Describe the digestive system of *Pila* and discuss various adaptive features found in its digestive system.
5. What is radula? Give its structure and function.
6. Give an account of the respiratory organs and mode of respiration in *Pila*.
7. Describe the respiratory organs of *Pila* and explain how they are adapted for its mode of life.
8. Give an account of the two types of respiration in *Pila*.
9. Describe the respiratory system of *Pila*.
10. Describe the respiratory organs of *Pila*. Discuss the mechanism of aerial respiration.
11. Give an account of respiratory system and aquatic respiration in *Pila*. (*C.C.S.U. Meerut 2004*)
12. Describe the blood vascular system or circulatory system of *Pila*.
13. Describe the excretory system of *Pila*.
14. Give an account of the organ of Bojanus.
15. Describe the renal organ and physiology of excretion in *Pila*.
16. Describe the nervous system of *Pila*.
17. Describe the nervous system of *Pila* and compare it with that of freshwater mussel.
18. Describe the nervous system of *Pila* related to torsion. (*C.C.S.U., Meerut 2002*)
19. Give an account of sense organs in *Pila*.
20. Give an account of the reproductive system of male and female reproductive organs in *Pila*.

21. Give an account of reproductive system of *Pila*. How and where does the copulation takes place?

22. Describe male and female reproductive system in *Pila*.

23. Write a short note on reproductive system of *Pila*.

Short Answer Questions

1. Write a note on feeding and digestion in *Pila*.

2. Give a short account of aerial respiration in *Pila*.

3. Describe renal organs of *Pila*.

4. Differentiate between eupyrine and oligopyrene sperm.

5. Why do majority of organs on right side of *Pila* disappear?

6. Differentiate between male and female *Pila*.

7. Write short notes on the following:

(*i*) Mantle cavity
(*ii*) Shell of *Pila*
(*iii*) Pulmonary sac
(*iv*) Gill lamella of *Pila*
(*v*) Digestive gland of *Pila*
(*vi*) Aquatic respiration in *Pila*
(*vii*) Organ of Bojanus
(*viii*) Buccal mass
(*ix*) Radula
(*x*) Osphradium
(*xi*) Eye in *Pila*

8. Draw labelled diagrams of any two of the following (No description is required).

(*i*) Organs visible in mantle cavity
(*ii*) Organs of pallial complex of *Pila*
(*iii*) Molluscan shell
(*iv*) V.S. buccal cavity of *Pila*
(*v*) Heart and pericardium of *Pila*
(*vi*) Nervous system of *Pila*
(*vii*) V.S. of eye of *Pila*
(*viii*) Reproductive organs/system of *Pila*

Very Short Answer Questions

1. What is the common name of *Pila*?

2. Name the two parts of foot of *Pila*.

3. Name the structure that divides the mantle cavity of *Pila* into right and left parts.

4. What type of digestion occur in the *Pila*?

5. What is radula? Describe its movement in word.

6. Give the radular formula of *Pila*.

7. Name the digestive glands of *Pila*.

8. What type of respiration is found in *Pila*? (*C.C.S.U. Meerut, 2004*)

9. Which nuchal lobe in *Pila* function as the inhalent channel in respiration?

10. What is the role of pneumostome?

11. Where does the nephrostome opens in *Pila*?

12. Why do majority of organs of right side of *Pila* disappear?

13. Define detortion.

14. Development is direct in *Pila*. Is it true?

Fill in the Blanks

1. Zoological name of apple snail is ....

2. *Pila* feeds on .... vegetation.

3. Period of inactivity of *Pila* in the drought period is called ....

4. *Pila* does locomotion by the help of ....

5. .... refers to the shell whorls except the body whorls.

6. The central hollow axis in *Pila* is known as ....

7. Columella opens out by an aperture, the ....

8. Head of *Pila* bears .... pairs of tentacles.

9. Mantle is also known as ....

10. The operculum of *Pila* is secreted by ....

11. Water testing organ in *Pila* is ....

12. There are .... teeth in each row of radula.

13. Cellulase is present in the secretion of ....

14. Blood of *Pila* contains a respiratory pigment name .... in the ....

15. Ctenidium of *Pila* is ....

Multiple Choice Questions

1. *Pila* shell is dextral, if the coiling is

(*a*) right handed (*b*) left handed

(*c*) anticlock wise (*d*) none of these

2. *Pila* is mainly

(*a*) carnivorous (*b*) herbivorous

(*c*) sanguivorous (*d*) omnivorous

3. How many teeth are found in a row of radula?

(*a*) 7 (*b*) 5

(*c*) 9 (*d*) 11

4. Body cavity in the apple snail is

(*a*) coelom (*b*) pseudocoel

(*c*) haemocoel (*d*) absent

5. Torsion is shown in

(*a*) *Unio* (*b*) *Pila*

(*c*) *Loligo* (*d*) *Mytilus*

6. Which of the following is a chemoreceptor?

(*a*) omnatophore (*b*) umbilicus

(*c*) nuchal lobe (*d*) osphradium

7. Mode of respiration in *Pila* is

(*a*) aquatic (*b*) aerial

(*c*) none (*d*) both

8. Aerial respiration in *Pila* occurs from

(*a*) mantle (*b*) ctenidium

(*c*) pulmonary sac (*d*) branchiae

9. Ctenidium of *Pila* performs respiration in

(*a*) air (*b*) water

(*c*) both the places (*d*) terrestrial

10. Pneumostome is an opening of

(*a*) buccal cavity (*b*) pulmonary sac

(*c*) kidney (*d*) shell

11. Blood vascular system in *Pila* is

(*a*) open (*b*) closed
(*c*) lacking (*d*) greatly reduced

12. Blood of *Pila* contains

(*a*) RBC (*b*) WBC
(*c*) ameoebocyte (*d*) monocyte

13. *Pila* has .... for excretion

(*a*) flame cells (*b*) metanephridial system
(*c*) malpighian tubes (*d*) gland cells

14. Statocyst of *Pila* is an organ of

(*a*) equilibrium (*b*) defence
(*c*) excretion (*d*) none

15. *Pila* is

(*a*) monoecious (*b*) dioecious
(*c*) hermaphrodite (*d*) asexual

16. *Pila* is

(*a*) oviparous (*b*) viviparous
(*c*) ovoviviparous (*d*) none of these

ANSWERS

Very Short Answer Questions

1. Apple snail.

2. Propodium and metapodium.

3. Epitaenia.

4. Extracellular as well as intracellular.

5. Radula is a rasping organ of *Pila*. It is file-like ribbon studded with horny or chitinous teeth, found in the mouth of *Pila* and other molluscs except Bivalvia. Radula runs longitudinally over the entire length of odontophore within the buccal mass of *Pila*. It moves to and fro over the odontophore. **Huxley** described such a movement as radula as **chain-saw** like movement.

6. 2, 1, 1, 1, 2.

7. Oesophageal pouches, salivary glands and liver or hepatopancreas.

8. *Pila* is an amphibious molluscan animal, *i.e.*, it can live both on land as well as in water. To survive under both the situation it has both aquatic as well as aerial mode of respiration. Pila's aquatic respiration occurs by the help of single monopectinate gill or ctenidium and its aerial respiration takes place by the help of pulmonary sac.

9. Left.

10. Passage for the air breathing.

11. Pericardial cavity.

12. Besides torsion, the reason of gastropods being asymmetrical is the loss, reduction or atrophy of paired organs from one or other side. Due to this reason, when the pallial/mantle cavity comes to lie on right, the organs present on the left side get degenerated or lost gradually leaving the organs only on the right side of body.

13. It is phenomenon just opposite to torsion. Some gastropods such as *Aplysia* undergo detortion. When the shell is lost or even reduced. In this process ctenidia (gills) occupy the posterior position and visceral hump gets completely untwisted.

14. Yes.

Fill in the Blanks

1. *Pila globosa*
2. Succulent
3. Aestivation
4. Foot
5. Spire
6. Columella
7. Umbilicus
8. Two
9. Pallium
10. Foot glands
11. Osphradium
12. Seven
13. Hepatopancreas (liver)
14. Haemocyanin, Plasma
15. Monopectinate.

Multiple Choice Questions

1. (*a*) **2.** (*b*) **3.** (*a*) **4.** (*c*) **5.** (*b*) **6.** (*d*)
7. (*d*) **8.** (*c*) **9.** (*b*) **10.** (*b*) **11.** (*a*) **12.** (*c*)
13. (*b*) **14.** (*a*) **15.** (*b*) **16.** (*a*).

16

Unio (Lamellidens) (Phylum Mollusca)

Freshwater mussels occur all over the world. Their typical genera include *Anodonta* common in Englant, *Lamellidens* found in India and *Unio* with universal distribution. The description given here will apply almost fully to these three and other freshwater mussels.

16.1 CLASSIFICATION

Phylum	—	**Mollusca**
Class	—	**Pelecypoda**
Subclass	—	**Palaeoheterodonta**
Order	—	**Unionoida**
Genus	—	***Unio***

16.2 HABIT AND HABITAT AND EXTERNAL FEATURES

Habit and Habitat

Unio is found in freshwater ponds, lakes, rivers and streams. It is especially abundant in water containing lime. *Unio* remains buried in the mud or sand at the bottom in such a way that posterior tip of shell valves remain exposed for the ingress (entry) and egress (exit) of water. During night *Unio* lives in shallow water while in the day it migrates in deeper water. The food of *Unio* is microscopic organisms–plants (diatoms) and animals, which are taken by filter-feeding mechanism.

The flow of water through the mantle cavity by beating of cilia is one of most critical factor in the survival of *Unio* and other pelecypods. Water currents enter and leave the mantle cavity through siphons made by left and right mantle folds; they bring food and oxygen and waft away faeces, carbon dioxide, excretory wastes and genital products.

Siphons open and close in response to light, touch and other stimuli. A slight disturbance causes the withdrawal of foot and siphons and closing of shell valves. Starfishes are the natural enemy of *Unio*; they open the shell and feed on their soft viscera.

Geographical Distribution

Unio is found all over the world. *Lamellidens marginalis* is found in Indian rivers and other parts of the world.

External Features

1. Shape and size. *Unio* measures 5 to 10 cm in length about 5 cm in width. The body is soft, unsegmented, bilaterally symmetrical and laterally compressed. The anterior end of body is roughly oval (broad and blunt) and posterior end is somewhat narrower (Fig. 16.1).

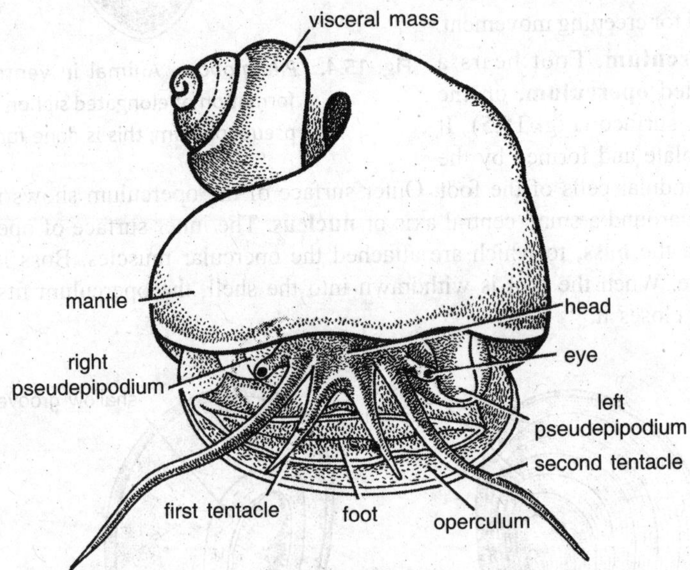

2. Shell. Body of *Unio* is completely enclosed in hard, calcareous, brownish and bivalved shell (hence called Bivalvia). Both of the valves remain connected together dorsally (Fig. 16.1).

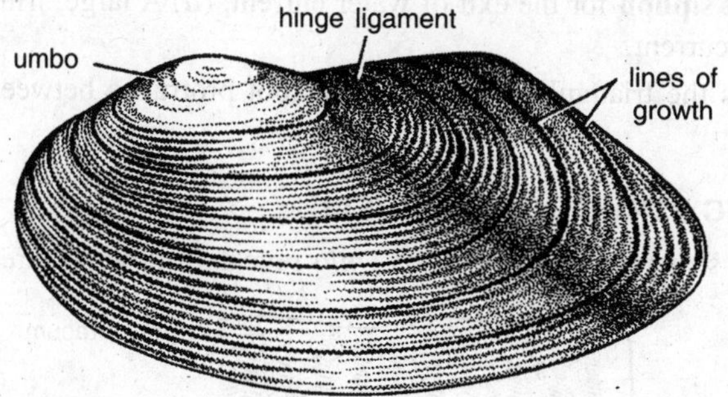

Fig. 16.1. *Unio*. Entire animal with shell (Lateral view).

***(i)* Shell valves.** The shell consists of two similar, equal sized (equivalved) valves which are known as **right** and **left valves**. Each valve has concave inner and convex outer surface.

***(ii)* Umbo.** Each shell has whitish, slightly elevated, dome-shaped structure called **umbo**. It is the oldest and thickest part of the shell.

***(iii)* Hinge ligament.** The two shell valves are joined along dorsal side by hinge ligament. It is an external, brown, tough, elastic and horny structure. It is composed of the protein **conchiolin.** They help in opening and closing of shell valves.

***(iv)* Lines of growth.** There are many crescentic lines around the umbo which run parallel to the margin of the shell. These lines are called **lines of growth.** The shell increases in size by the deposition of lime along outer rim of shell each year.

***(v)* Hinge teeth.** The inner surface of each shell valve along its hinge line at its dorsal margin bears two sharp, blade-like ridges or teeth. Teeth are separated by grooves or sockets. The teeth of one valve fit into corresponding sockets of other valve. These control the gaping of shell valves and hold them from slipping out of position.

***(vi)* Muscle scars.** The inner surface of each shell valve bears scars or impressions of muscles attached to it. There are three scars at the anterior end for **anterior adductor, anterior retractor** and **anterior protractor muscles** (Fig. 16.2). The two scars at the posterior end are those of **posterior adductor** and **posterior retractor muscles**.

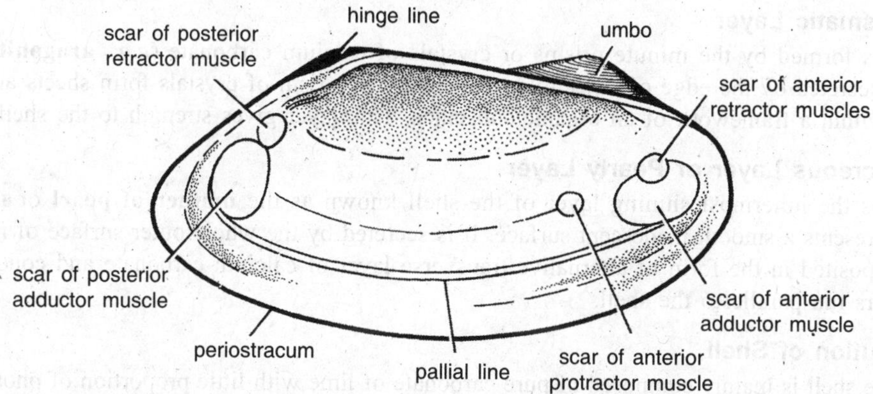

Fig. 16.2. *Unio*. Inner view of shell valve showing muscle scars.

A line joining the scars for adductor muscles of two ends is called **pallial line.** It marks the attachment of mantle to shell valve.

3. Siphons. A delicate **mantle** lies beneath the shell that envelops the whole visceral mass. At the posterior or aboral end, the mantle lobes of both sides form two short siphons: (*i*) A small, smooth edged **exhalent siphon** for the exit of water current. (*ii*) A large, fringed **inhalent siphon** for the entry of water current.

4. Foot. At times the triangular tongue-shaped foot protrudes between the shell valves at the anterior end.

16.3 MICROSCOPIC STRUCTURE OF SHELL

The microscopic structure of the shell of *Unio* reveals following three layers (Fig. 16.3):

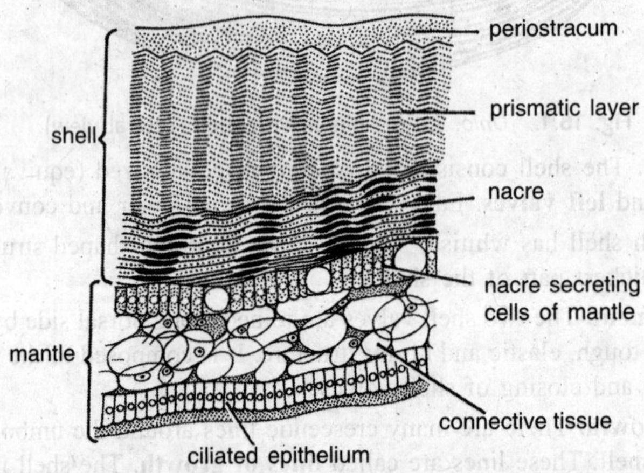

Fig. 16.3. *Unio*. Cross section of shell and mantle.

1. Periostracum

It is outermost, thin, brownish, translucent, horny layer. Periostracum is greenish in colour and formed of a special chitinous substance the **chonchiolin**. It is secreted by the edge of mantle. It is a protective layer for the under-lying calcareous prismatic layer which otherwise can be eroded out by the action of carbolic acid of the surrounding medium. The periostracum is worn away from the older part of the shell such as umbo, leaving the prismatic layer exposed to the exterior.

2. Prismatic Layer

It is formed by the minute prisms or crystals of calcium carbonate (*e.g.*, **aragonite** and **calcite**) secreted by the edge of outer fold of mantle. The prism of crystals form sheets and are always within a framework of an organic substance. This layer gives strength to the shell.

3. Nacreous Layer or Pearly Layer

It is the innermost shining layer of the shell known as the **mother of pearl** or **nacre,** since it presents a smooth irridescent surface. It is secreted by the whole outer surface of mantle and is deposited in the form of alternative transverse layer of calcium carbonate and concholin. The layers run parallel to the shell.

Constitution of Shell

The shell is mainly composed of pure carbonate of lime with little proportion of phosphate of lime and an organic base allied to chitin which is known as **conchiolin**. The proportion of carbonate of lime varies from 88 to 99% and of phosphate of lime 1 to 10%. Conchiolin forms a sort of membranous framework for the shell.

Formation of Shell

Regarding the formation of shell there exist following two views:

1. Bowerbank and Carpenter's view. The shell is an organic formation growing by interstitial deposits in the same manner as teeth and bones in higher animals.

2. Reaumer-Eising view. The shell is of nature of an excretory substance deposited like cuticle outside of the skin. It is formed simply of a number of calcareous particles which are held togetherby a kind of animal glue.

Mantle begins to form shell during the veliger larval stage. The two mantle folds secrete **conchiolin** to form two tiny valves connected by a narrow ligament. New material is added at the margins of the valves, producing a series of concentric growth rings (or lines). As the veliger settles to the bottom, shell growth continues but more rapidly on the ventral side. As a result, the original larval shell is displaced dorsally forming an elevation (**umbo**) directed anteriorly so that the right and left valves can be easily distinguished.

The mantle is pressed against the inner surface of the valves. It ends in three folds: **1. Innermost fold** which contains radial and circular muscles; **2. Middle fold** which is sensory in function; and **3. Outer fold** which secretes the new shell (Fig. 16.4). **Periostracum** forms at the inner margin, calcareous material of the prismatic layer at the outer border and the **nacreous material** on the whole mantle surface, thickening the shell as it ages.

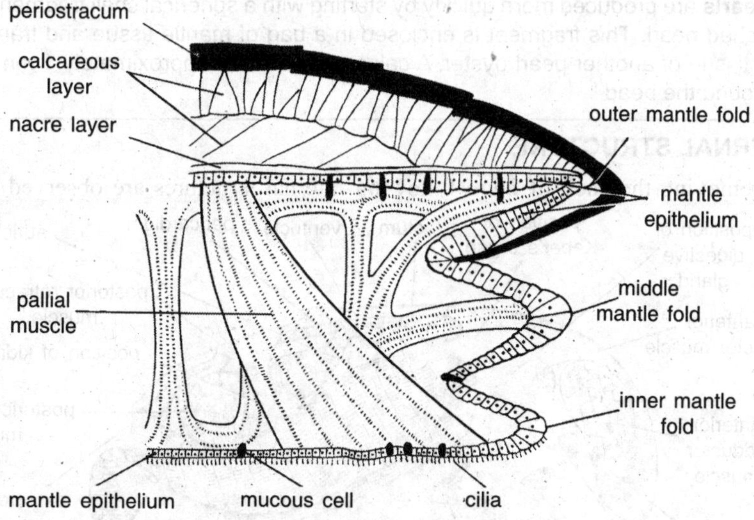

Fig. 16.4. T.S. through margin of shell and mantle of a bivalve showing mantle lobes and places of shell secretion.

There is a minute, **extrapallial space** between mantle and shell, except at points of muscle attachment. The carbonates and phosphates of calcium (lime) are separated from the blood by the cells of mantle and deposited in the extrapallial space. It is in this space that other shell materials (*e.g.*, conchiolin) are first secreted. From the extrapallial fluid of this space both the calcareous elements and the surrounding organic framework are then deposited.

16.4 PEARL FORMATION

Occasionally some foreign object or particle such as sand grain, debris, a parasite, an egg or any microscopic organism comes to lie in between the shell and mantle. The particle causes irritation to the animal. This induces epithelial cells to secrete concentric layers of nacre around it. This results in the formation of a **pearl** (Fig. 16.5). The foreign particle inside the pearl is known as **nucleus** and concentric layers of nacre are known as **mother of pearl**.

If the foreign body is enfolded within the mantle and moved about during secretion, the pearl becomes spherical or ovoid. Commonly, however, the developing pearl adheres to or even becomes completely embedded in the shell. The pearl of *Unio* is not economically precious (see Box 16.1).

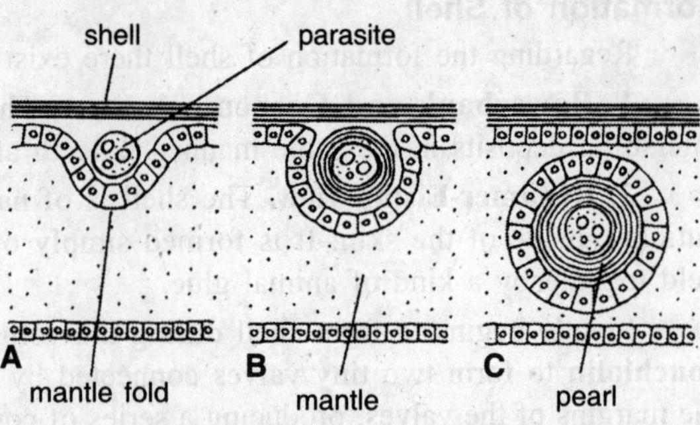

Fig. 16.5. *Unio*. Pearl formation.

Box 16.1 Cultured Pearls

Finest natural pearls are produced by the pearl oyesters, *Pinctada margaritifera*, *Pinctada vulgaris*, *Pinctada mertensi* which inhabit most of the warmer seas. There are two methods of producing **cultured pearls**. Most cultured pearls are started with a microscopic globule of liquid or ground "seeds" of unionid shells placed into a pearl oyster. The resulting year-old seed pearl is then transplanted into another oyster. A pearl of marketable size is obtained three years after transplantation.

Bead pearls are produced more quickly by starting with a spherical shell fragment only smaller than the finished pearl. This fragment is enclosed in a bag of mantle tissue and transplanted into the gonadal tissue of another pearl oyster. A calcareous veneer approximately 1 mm thick is then down laid around the bead.

16.5 INTERNAL STRUCTURE

After removing the shell of *Unio* following internal structures are observed (Fig. 16.6)

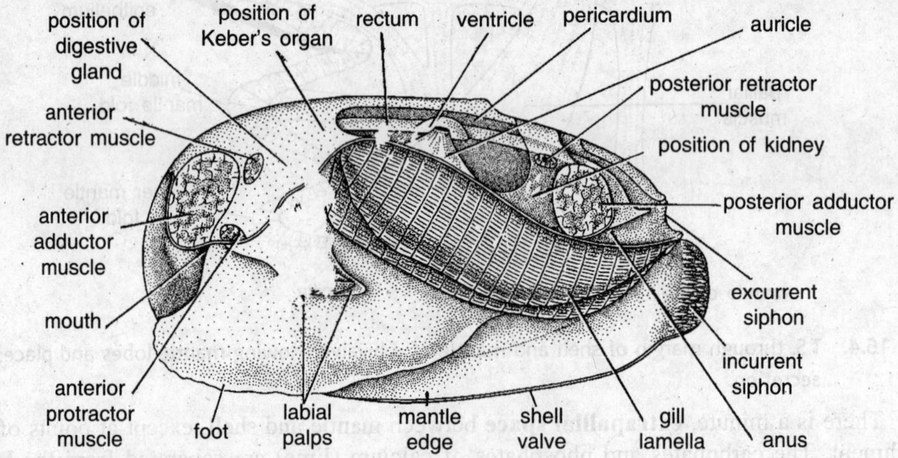

Fig. 16.6. *Unio*. Internal organs seen after the removal of the left shell valve and left mantle lobe.

1. Mantle or Pallium

Just beneath each shell valve lies a thin, creamy white, semi-transparent and soft covering of skin. It encloses the soft body parts and is called **mantle** or **pallium**. Pallium overhangs the body like a skirt. It consists of two lateral halves or **lobes**, one below left and other below right shell valves. The two mantle folds join each other and the body dorsally. The ventral edges of the mantle lobes are thick, folded and muscular. The edge contains three folds (Fig. 16.4) — an **inner**, a **middle** and an **outer fold**. The outer fold secretes the outer (**periostracum**) and middle calcareous

(prismatic) layer of the shell. The inner layer of shell or **nacre** is secreted by the entire surface of mantle.

The space enclosed by mantle folds around the body is called **mantle cavity**. It extends entire length of the body on each side. The mantle cavity is divided into two chambers, a large ventral **infra-branchial chamber** and a small dorsal **supra-branchial chamber**. At the posterior end both mantle lobes are thick and muscular. They meet together and form two short tubes or **siphons**. The upper or dorsal siphon is narrow and is called **exhalant siphon**. Water leaves the body through it. Below the exhalant siphon lies another tube, called **inhalant siphon**. It is wider and has delicate **fimbriae** at its edges. Inner margin of inhalant siphon contains papillae which test the incoming water currents.

Histology. The mantle consists of (*i*) a thin **outer layer** of columnar epithelial cells. It contains many unicellular **nacre secreting gland cells** and branches of blood vessels. It remains in contact with shell; (*ii*) a thick **middle layer** which is formed of connective tissue, muscle fibres (called **pallial muscles**), branches of blood vessels and nerves; and (*iii*) a thin inner layer of **ciliated epithelial cells**. It contains many unicellular **mucous glands** (Fig. 16.4).

Function. The mantle of *Unio* protects the body, secretes shell valves and mucus and forms siphons for the flow of water.

Organs of Mantle Cavity

Mantle cavity of *Unio* contains the following structures:

1. Head. *Unio* lacks a distinct **head**. Eyes and tentacles are absent.

2. Mouth. It is a large and transverse slit. It lies immediately below the **anterior protractor muscle** and is surrounded with two pairs of **labial palps**. Labial palps are large, flattened and fleshy structures forming upper and lower lips of mouth.

3. Visceral mass. It is laterally compressed part of the body and occupies the middorsal part of the mantle cavity. It contains all the system of body such as digestive system, circulatory system, reproductive system, excretory system and nervous system.

4. Muscles. *Unio* contains the following four types of muscles:

***(i)* Adductor muscles.** *Unio* is an **isomyarian** pelecypod, *i.e.*, it has two equal sized adductor muscles (Box 16.2). **Anterior** and **posterior adductor muscles** are large, cylindrical and transverse bands of muscles. They run from one shell valve to another. Their contraction closes the shell valves, and when these muscles relax the hinge ligament opens the shell valves.

Box 16.2 Adductor Muscles in Pelecypoda

1. **Isomyarian.** Two adductor muscles are present. They are more or less equal in size; *e.g., Unio, Mya*.
2. **Anisomyarian.** Reduction of anterior adductor muscle, *e.g., Mytilus*.
3. **Monomyarian.** Loss of anterior adductor muscle, *e.g., Pecten, Pinctada*.

***(ii)* Retractor muscles. Anterior** and **posterior retractor muscles** run from the foot to the shell. Their contraction withdraws the foot into the shell.

***(iii)* Protractor muscles.** It is a small band of muscles. It lies close to the anterior adductor muscle. Its action compresses the visceral mass and causes extension of foot out of the shell.

***(iv)* Pallial muscles.** These extend from the edge of the mantle lobes and becomes attached with the shell valves all along the pallial line.

5. Foot or podium. It is antero-ventral, hatchet-like or wedge-shaped extension of visceral mass. It is a large, muscular, laterally compressed structure. Its thick basal part contains a portion of alimentary canal, digestive gland and gonad. Its free ventral part forms a longest **keel**.

The foot has a large sinus which gets filled with blood when the foot is extended by blood pressure and muscular action of a pair of **pedal protractor muscles.** The foot becomes swollen and turgid; it is then used for crawling and for ploughing through mud in burrowing. Foot is withdrawn into the shell by the action of **anterior** and **posterior retractor muscles.**

6. Gills. A pair of large, plate-like gills or ctenidia hang in the mantle cavity. One gill lie on each side of the foot. Each gill or ctenidium is folded to form two **laminae** and each lamina contains two **lamellae**. The outer gill lamina is a little smaller than the inner one. All lamina bear numerous pores and cilia.

In addition to respiratory gaseous exchange, the gills are also used to collect food and as brooding chambers.

7. Anus. The rectum opens into the exhalant siphon through anus.

16.6 DIGESTIVE SYSTEM

Digestive system of *Unio* is distinguished into: (1) Alimentary canal, and (2) Digestive gland (Fig. 16.7).

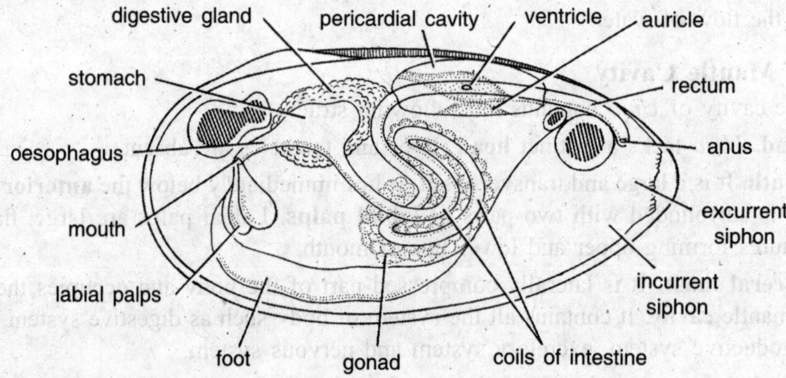

Fig. 16.7. *Unio*. Alimentary canal and digestive gland.

1. Alimentary Canal

It is a long curved tube starting from mouth at one end of the body and ending into the anus situated at the rear end. It is divided into the following parts:

1. Mouth. Mouth is a transverse slit-like aperture, lies behind the anterior adductor muscle or below anterior protractor muscle. The mouth is surrounded by two pairs of triangular ciliated flap-like **labial palps.** The labial palps of anterior-outer pair are joined together over the mouth forming upper lip, while the labial palps of posterior-inner pair are united below the mouth forming the lower lip. The lips form a ciliated oral groove, which leads into the mouth. Jaws and radula are absent. Mouth leads into the oesophagus.

2. Oesophagus. The oesophagus is a short, wide, ciliated tube which leads into the stomach.

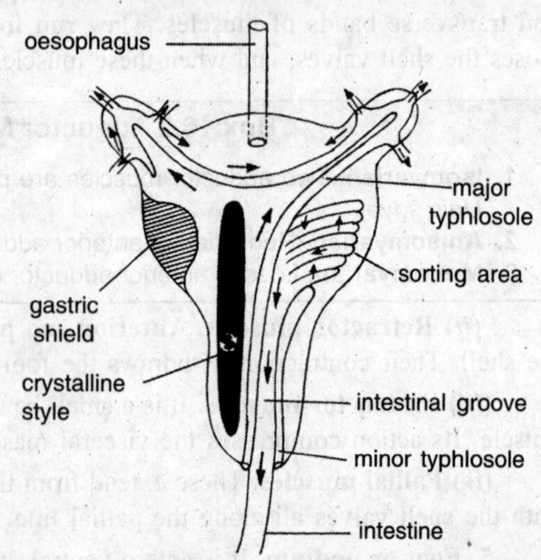

Fig. 16.8. Stomach of a typical unionoid showing typhlosoles and sorting of food particles.

3. Stomach. It is a rounded sac which is wide and thick-walled, situated dorsally in the visceral mass and surrounded by a pair of digestive glands. The stomach is divided into dorsal part (proximal globular region) and ventral part (narrowed distal region). The dorsal part of stomach contains a **sorting area**, prominent **gastric shield** and **pores of digestive gland** (Fig. 16.8). The ventral part of stomach contains a **style sac** (or **pyloric caecum**) with **crystalline style** and **intestinal grooves** with **major** and **minor typhlosoles** (Fig. 16.8).

The internal epithelial lining of the style sac bears cilia. The style sac contains a gelatinous rod called the **crystalline style**. It is secreted by the internal lining of the style sac. It is composed of mucoproteins (= consolidated mucus) and enzymes such as **carbohydrases** (*e.g.*, **amylase, cellulase, glucogenase**) and **lipase**. The cilia of the style sac rotate (clockwise) the style and move it forward (like a screw) due to which its anterior end is rubbed against a chitinous plate (*i.e.*, gastric shield). As the style strikes against the gastric shield, style end is abraded and dissolved, *i.e.*, small fragments containing mucus and enzymes separate from style and mix with food particles. When feeding stops, or when the animal is starved, the crystalline style is gradually dissolved.

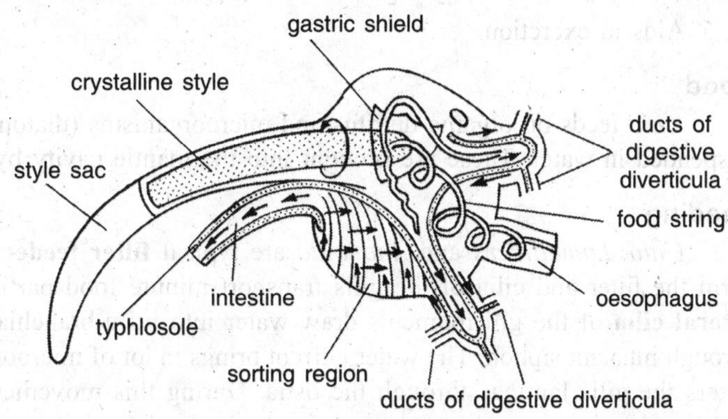

Fig. 16.9. Stomach of a lamellibranch, showing rotation of the crystalline style and winding in of mucous food string. Arrows indicate the ciliary pathways.

4. Intestine. The intestine is long, coiled narrow tube situated in the visceral mass. In its anterior part intestinal lining has two folds known as **typhlosoles**. The intestine is continued behind into the rectum.

5. Rectum. The rectum is a narrow tube running horizontally up to the posterior end in the dorsal part of the body. It passes through the ventricle and pericardium of heart (in fact, ventricle is wrapped around the rectum). Rectum finally opens into the exhalant siphon by **anus**. Rectum also contains a **typhlosole** which runs mid-ventrally throughout the length of rectum. Typhlosole is meant for increasing the surface of food absorption.

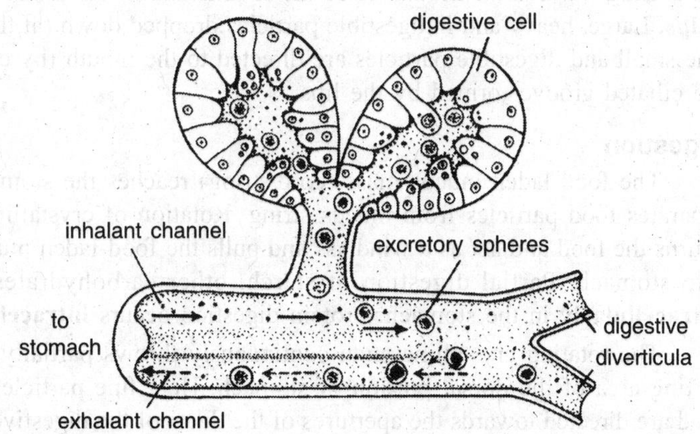

Fig. 16.10. Diagram of a section of digestive diverticulum showing the absorption and intracellular digestion of material passed inward from the stomach (solid arrow) and the outward passage of wastes (dashed arrows).

2. Digestive Gland

It is also called **liver**. It is a large paired gland of dark brown or green colour. Digestive gland surrounds the stomach and is composed of highly branched tubules. It opens into dorsal part of stomach by number of ducts to discharge the digestive enzymes and to receive partially digested food from the stomach. In fact, there is a continuous two way flow in the main duct of digestive gland. Materials for intracellular digestion and absorption enter the ducts and undigested cell fragments and other wastes (excretory waste) are returned to the stomach through the same tube (Fig. 16.10). Digestive gland of *Unio* performs the following functions:

1. Phagocytosis;
2. Intracellular digestion of proteins and lipids;
3. Secretion of enzymes for extracellular digestion in stomach;
4. Absorption of already digested food (*e.g.*, starch, cellulose);
5. Aids in excretion.

Food

Unio feeds on organic detritus and microorganisms (diatoms, protozoa, etc.) which remain suspended in water. These are brought into the mantle cavity by the respiratory water currents.

Feeding

Unio, *Lamellidens* and *Anodonta* are typical **filter feeders** or **ciliary feeders**. Their gills form the filter and cilia of the gills transport minute food particles to mouth. Beating of large **lateral cilia** of the gill filaments draw water into infra-branchial chamber of the mantle cavity through inhalant siphon. The water current brings in lot of microorganisms, detritus and silt. Water enters the gills laminae through the ostia. During this movement the water is filtered. Coarser and heavier sand particles, etc., are simply dropped from the surface of the gills to the edge of mantle (having ventral ciliated tract). These are driven towards the inhalant siphon by the cilia of the mantle edge. These rejected particles form the **pseudofaeces** and leave the mantle cavity through the inhalant siphon.

The lighter food particles are passed on to the outer gill surfaces by **latero-frontal cilia**. These particles stick to the mucus secreted by the gills. The food laden mucus mass in the form of **cord** or **string** is moved ventrally by the beating of frontal cilia of the food groove which is located at the lower edge of 'V' of each gill lamina. Cilia of the food groove move the mucus food string towards the mouth. Here the food is again sorted and conveyed by the **cilia of labial palps.** Large, heavy and indigestible particles dropped down (at the edge of the mantle and foot). The small and digestible particles are directed to the mouth (by cilia of the labial palps) through the ciliated groove formed by the lips.

Digestion

The food laden mucus mass (cord/string) reaches the stomach. Lower pH of the stomach separates food particles from mucus string. Rotation of crystalline style adds enzymes to food, churns the food and act as a windlass and pulls the food-laden mucus string from the oesophagus into stomach. Partial digestion of starch, other carbohydrates (cellulose) and lipids occur **extracellularly** in the stomach. Protein digestion occurs **intracellularly** in the digestive glands.

The rotating crystalline style continuously throws partially digested food into the ciliated sorting area of the dorsal portion of stomach. Here fine particles and fluid containing digested food are directed towards the apertures of the ducts of the digestive glands by the cilia (Fig. 16.9). Coarse and heavy particles are separated from the string and sent to the intestine. Digestion is completed intracellularly in the cells of ductules of digestive glands.

Absorption of digested food occurs chiefly through the walls of the intestine and rectum. The typhlosole increases the absorptive surface. Digestive gland also absorbs digested food.

The phagocytotic amoeboid cells found in the wall of entire alimentary canal also engulf, digest and distribute food particles.

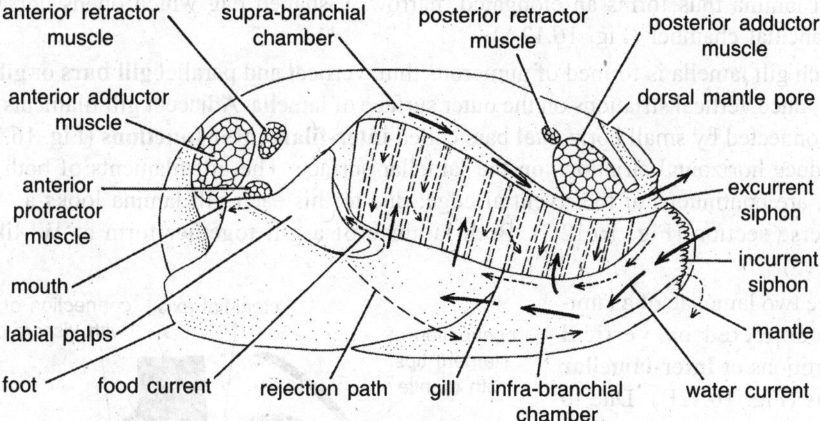

Fig. 16.11. *Unio*. **Feeding mechanism and course of water current. Rejected particles and their path of removal is shown by dotted arrows.**

The undigested waste is ejected through the **anus** in the exhalant siphon from where it goes out along with outgoing water currents.

16.7 RESPIRATORY SYSTEM

Ctenidia or Gills

The ctenidia or gills are highly modified derivative of mantle. A pair of large plate-like gills hang in the mantle cavity. One gill lies on each side of the foot. The gills are attached with the visceral mass and mantle lobes (Fig. 16.6).

Structure of gills. Each gill contains two **demibranchs** or **gill plates** or **laminae** — an outer and inner. The inner lamina which lies towards the visceral mass is larger than the outer lamina which lies towards the mantle lobe. These gills divide the mantle cavity into a smaller dorsal **supra-branchial cham-ber** and a larger ventral **infra-branchial chamber**. The former opens into the exhalant siphon and latter communicates with the inhalant siphon.

Each gill plate or lamina is formed of similar flaps or **gill lamellae**, an outer and an inner. Both lamellae remain connected at their ventral, anterior and posterior ends but they remain separated dorsally. They enclose a narrow space or cavity called **inter-lamellar space**.

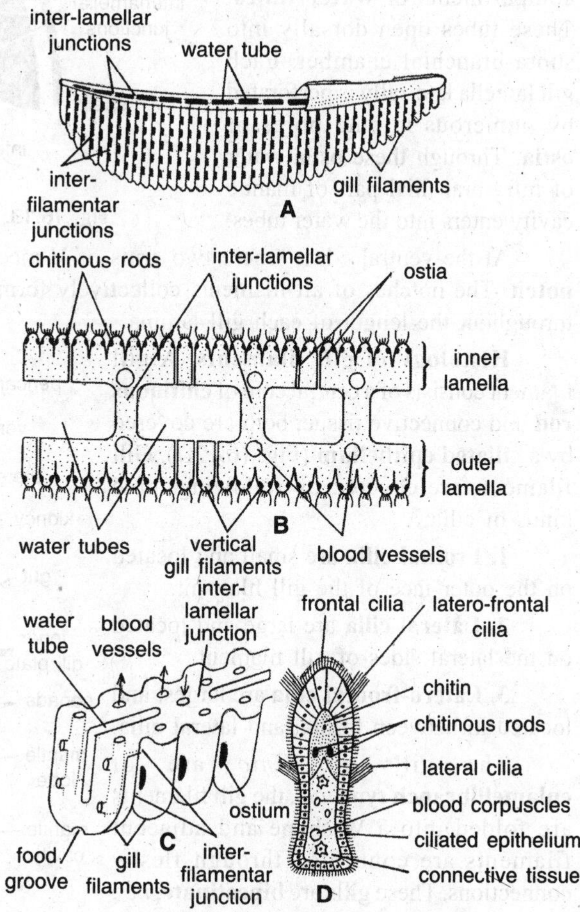

Fig. 16.12. *Unio*. **Structure of gill. A– A demibranch or gill lamina seen from outside; B—A portion of gill lamina in T.S; C—A portion of gill lamina showing water tube; D—Cross section of one gill filament.**

Each gill lamina thus forms an elongated, narrow, V-shaped bag which opens dorsally into the supra-branchial chamber (Fig. 16.12A).

Each gill lamella is formed of numerous thin, vertical and parallel **gill bars** or **gill filaments**. These produce vertical striations on the outer surface of lamella. Adjacent gill filaments of a lamella remain connected by small horizontal bars called **inter-filamentar junctions** (Fig. 16.12A). These bars produce horizontal striations on the lamellar surface. The gill filaments of both lamellae of a lamina are continuous at their ventral edge; due to this each gill lamina looks a 'V' or 'U' in a transverse section (Fig. 16.12C). Both laminae of a gill together form a 'W' like structure (Fig. 16.13).

The two lamellae of a lamina are connected by vertical cross partitions or **inter-lamellar junctions** (Fig. 16.12B). Due to these junctions the inter-lamellar space is divided into a series of compartments or **water tubes**. These tubes open dorsally into supra-branchial chamber. Each gill lamella late-rally is perforated by numerous minute pores or **ostia**. Through these ostia water of infra-branchial part of mantle cavity enters into the water tubes.

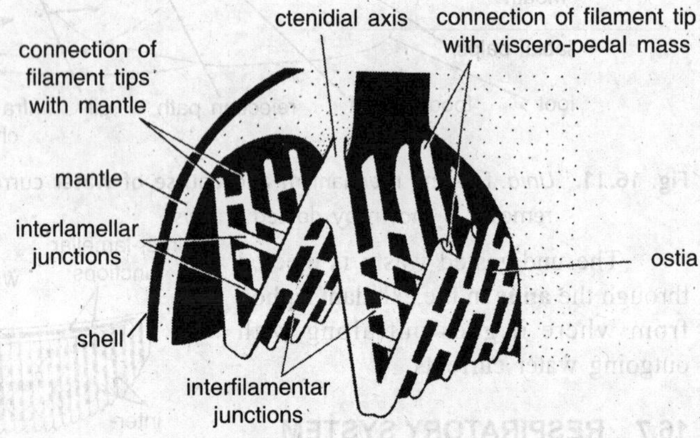

Fig. 16.13. A part of gill showing attachment.

At the ventral edge, where two arms of V meet together, each V-shaped filament bears a **notch**. The notches of all filaments collectively form a continuous **food groove** which extends throughout the length of each gill lamina.

Histology of gill filaments. A gill filament consists of a central core of **chitinous rod** and connective tissue, both are covered by a **ciliated epithelium** (Fig. 16.12D). Gill filaments are covered by following three kinds of cilia;

1. Frontal cilia are small and located on the outer face of the gill filament.

2. Lateral cilia are large and located on the lateral sides of gill filament.

3. Latero-frontal cilia are largest and located in between frontal and lateral cilia.

The gills of *Unio* are of **eulamellibranch type**, *i.e.*, the gill filaments are folded into a W-shape and adjacent filaments are connected through fleshy connections. These gills are **bipectinate**, *i.e.*, there are two rows of gill filaments—one on each side of a central **ctenidial axis**.

Attachment of gills. Each gill has a central **ctenidial axis**. It remains fused to the

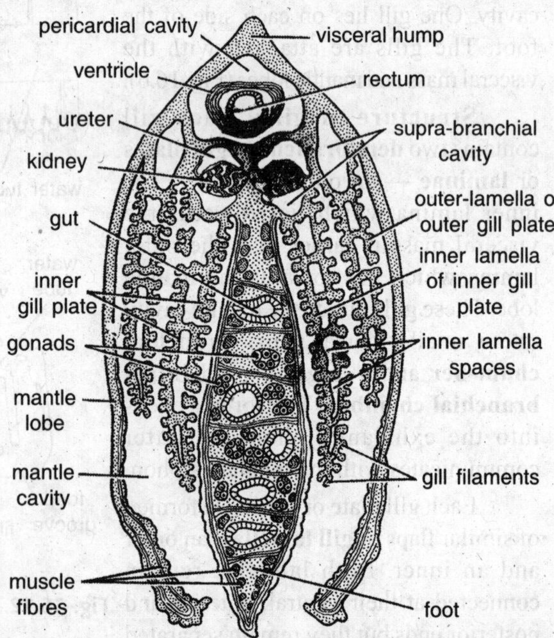

Fig. 16.14. *Unio*. T.S. of body through the middle region of gills and foot.

dorsal wall of the mantle cavity throughout but remains free near the posterior end of body. The arms of 'W' that are attached to the ctenidial axis are called **descending limbs** and the arms close to the mantle or visceral mass are called **ascending limbs**. Tips of ascending lamellae are fused with upper surface of mantle (on the outer side) and with viscro-pedal mass (on the inner side). Behind the foot the tips of ascending lamellae of inner side, fuse with that of other gill.

Blood supply of gills. The ctenidia or gills are supplied by **afferent branchial vessel** carrying deoxygenated blood from the kidneys and divides into branches at the interlamellar junctions. These branches unite to open into the **efferent branchial vessel** which carries away oxygenated blood to the heart.

Course of water current. Constant beating of cilia of the gill filaments causes a continuous current of water into the infra-branchial chamber through the inhalent siphon and enters the gill through ostia (Fig. 16.11). After reaching the gills water fills in the water tubes and flows out into the supra-branchial chamber which opens into cloacal chamber, from here water discharges to exterior through exhalant siphon (Fig. 16.15).

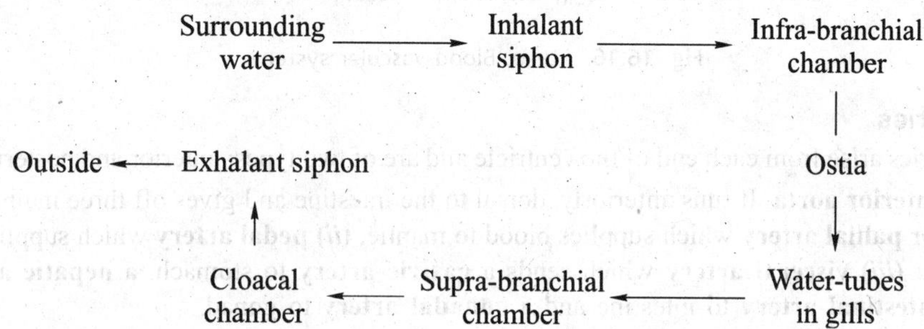

Fig. 16.15. *Unio*. Diagrammatic representation of course of water current.

2. Mantle

Mantle is richly supplied with the blood vessels and it remains in contact with water, hence, gáseous exchange takes place through its thin wall.

Physiology of Respiration

When water passes through the mantle cavity and the water tubes in the gill, gaseous exchange takes place. Thus, carbon dioxide from the blood diffuses out in the water and dissolved oxygen from water diffuses in the blood.

16.8 CIRCULATORY SYSTEM

In *Unio* the blood vascular system is of open type, *i.e.*, the blood vessels open into the spaces present within visceral organs. The circulatory system consists of heart, pericardium, arteries, sinuses, veins and blood. Capillaries are absent.

1. Heart and Pericardium

Heart lies along the mid-dorsal line infront of the posterior adductor muscle, enclosed within a thin walled triangular chamber, the **pericardium**. The cavity of pericardium is called **pericardial cavity**, it contains the **coelomic fluid**.

The heart consists of two auricles or atria and a ventricle. **Auricles** are thin walled, refractile, highly contractile and triangular chambers situated one on either side of the ventricle. The auricles are attached to the pericardial wall by thin broad bases and opens into the ventricle by an **auriculo-ventricular aperture**. The auricle receive blood from the gills, kidney and mantle. The **ventricle** is a large, thick walled and muscular chamber. It is wrapped around the rectum.

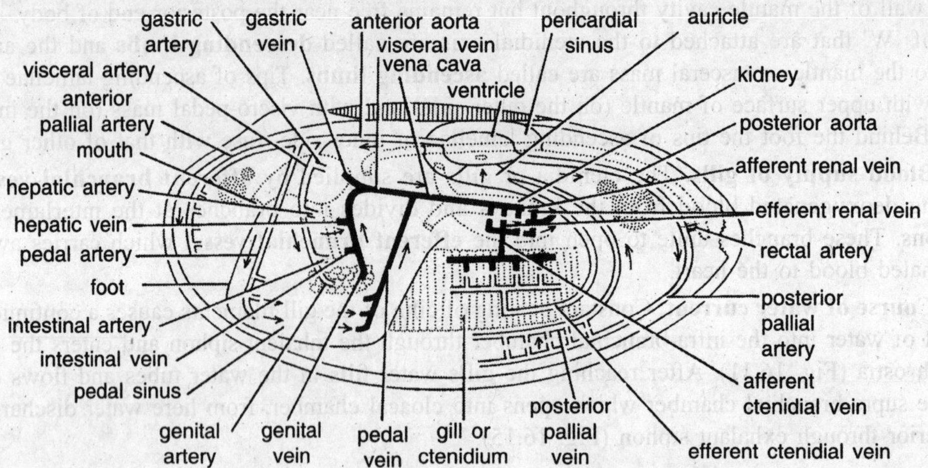

Fig. 16.16. *Unio*. Blood vascular system.

2. Arteries

Arteries arise from each end of the ventricle and are of two types: anterior and posterior aorta.

1. Anterior aorta. It runs anteriorly, dorsal to the intestine and gives off three main arteries: **(*i*) Anterior pallial artery** which supplies blood to mantle; **(*ii*) pedal artery** which supplies blood to the foot; **(*iii*) visceral artery** which sends a **gastric artery** to stomach, a **hepatic artery** to liver, an **intestinal artery** to intestine and a **gonadal artery** to gonad.

2. Posterior aorta. It runs posteriorly, ventral to the intestine. It supplies blood to the pericardium and kidneys through its fine branches. It sends a prominent **rectal artery** to the rectum and a large **posterior pallial** artery to mantle.

3. Sinuses

Sinuses are spaces in the connective tissues of organs. They lack definite walls, *i.e.*, epithelial lining is not present. Arteries open into these sinuses. From them the blood is collected by veins.

4. Veins

Veins collect blood from the different parts of the body such as

(*i*) Visceral vein collects blood from the visceral organs.

(*ii*) Pedal vein collects blood from the pedal sinus.

(*iii*) Efferent renal vein collects blood from the kidney and unite to form two **afferent ctenidial veins.** An afferent ctenidial vein runs through the central axis of the gill and sends a branch to each gill filament. The blood in the gill is oxygenated.

(*iv*) Efferent ctenidial vein. It collects oxygenated blood from the gill filaments and carry the blood to auricles.

(*v*) Vena cava. It directly sends deoxygenated blood to the auricles.

5. Blood

The blood consists of **plasma and leucocytes**. The plasma contains a copper-containing protein (respiratory pigment) called **haemocyanin**. Blood distributes nutrients and oxygen to all parts of body. It collects carbon dioxide and other wastes of metabolism from different parts of the body.

Course of Blood Circulation

When the auricles are filled up with pure blood, the ventricle pumps it to various organs of the body in both forward and backward directions. The impure blood is collected through veins and sinuses from various parts of body and carried to the kidney and gills for purification. The oxygenated blood from gills is carried to the auricles and again pumped through ventricle (Fig. 16.17).

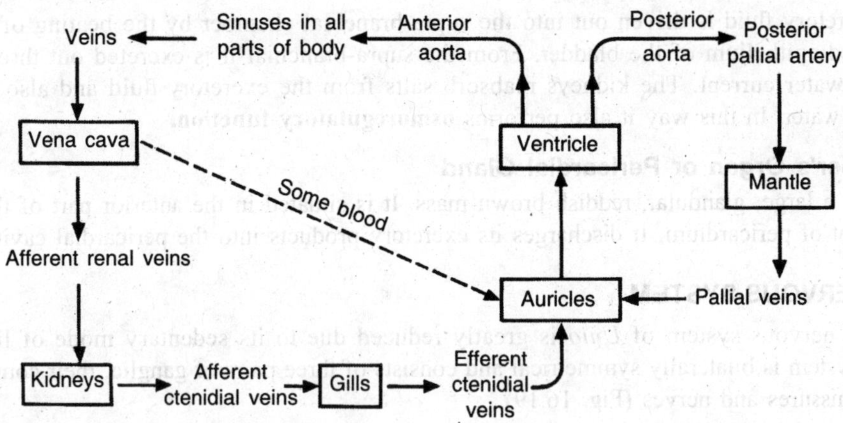

Fig. 16.17. *Unio.* Diagrammatic representation of course of blood circulation.

16.9 EXCRETORY SYSTEM

In *Unio*, excretion takes place through kidney or organs of Bojanus and the Keber's organ or the pericardial glands.

1. Kidneys or Organs of Bojanus

Kidneys or nephridia are also called **organs of Bojanus** (Fig. 16.18) after the name of their discoverer. There are two kidneys situated on both sides of vena cava and between the pericardium and gills. Kidneys are like a hair pin and represent true coelom. Its one arm opens into the pericardial cavity through a ciliated aperture, called the **nephrostome** or **renopericardial aperture**. The other arm opens into the mantle cavity through an **excretory** or **renal pore**. The arm of kidney opening into the pericardial cavity is brown, thick-walled, glandular and spongy. It forms kidney proper. The

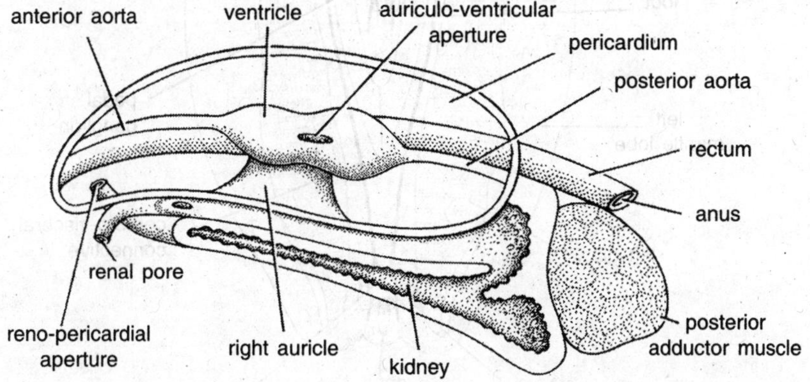

Fig. 16.18. *Unio.* Excretory system.

other arm of kidney is non-glandular, thin-walled having ciliated internal epithelium. It forms the ureter or urinary bladder and opens into the supra-branchial chamber of the mantle cavity. The bladders of both the kidneys communicate with each other through an oval aperture.

Physiology of excretion. Excretory products of *Unio* includes nitrogenous waste (*e.g.*, **ammonia** and traces of **urea** and **uric acid** and **guanine**). These are separated by glandular part of kidney from the coelomic fluid of the pericardial cavity and also from its blood that has been supplied by the vena cava.

Excretory fluid is driven out into the supra-branchial chamber by the beating of cilia of the ciliated epithelium of the bladder. From the supra-branchial it is excreted out through the outgoing water current. The kidneys reabsorb salts from the excretory fluid and also remove excess of water. In this way it also performs **osmoregulatory function.**

2. Keber's Organ or Pericardial Gland

It is a large, glandular, reddish brown mass. It is situated in the anterior part of the body just infront of pericardium. It discharges its excretory products into the pericardial cavity.

16.10 NERVOUS SYSTEM

The nervous system of *Unio* is greatly reduced due to its sedentary mode of life. The nervous system is bilaterally symmetrical and consists of three pairs of ganglia, their connectives and commissures and nerves (Fig. 16.19).

1. Cerebro-pleural ganglia. These are paired small somewhat triangular and yellowish in colour. One of the ganglia is situated on each side of the oesophagus at the base of labial palps. These are connected together by a short **cerebral commissure.** Each ganglion sends nerves to the anterior adductor muscle, labial palp and anterior part of mantle lobes.

2. Pedal ganglia. A pair of pedal ganglia lie in the foot at the junction of foot and visceral mass. The two pedal ganglia are joined to form a bilobed mass which gives off nerves to the foot,

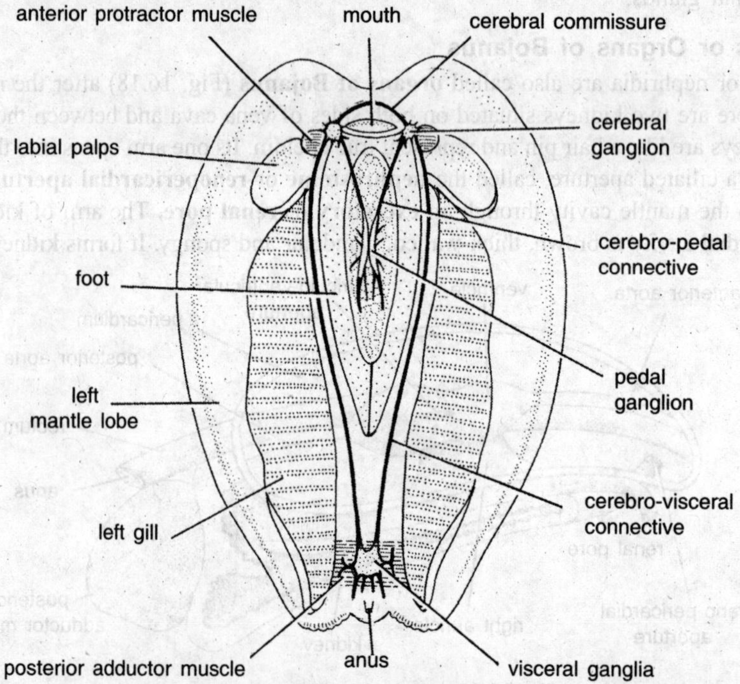

Fig. 16.19. *Unio*. Nervous system in ventral view.

foot muscles and statocysts. Each pedal ganglia is connected with the cerebropleural ganglia through **cerebro-pedal connective.** It passes through the foot.

3. Visceral ganglia. The two visceral ganglia lie mid-ventrally on the posterior adductor muscles. These ganglia are fused together to form a flattened somewhat rectangular mass. On each side it is connected with cerebro-pleural ganglion of each side by long, delicate **cerebro-visceral connectives.**

On each side, the visceral ganglia send (*i*) **dorsal pallial nerve,** (*ii*) a **posterior pallial nerve** to the posterior part of the mantle; (*iii*) a **posterior renal nerve** to the kidney; (*iv*) a **branchial nerve** to the gill; and (*v*) a **posterior adductor nerve.**

Functions. Pedal and cerebropleural ganglia control the foot movement and the anterior adductor muscle. The posterior adductor muscle and siphons are controlled by visceral ganglia. The cerebro pleural ganglia coordinate the movements of foot and shell valves.

16.11 SENSORY ORGANS

In *Unio* sense organs are not well developed. Eyes and tentacles are absent. Its sense organs are of following types:

1. Statocysts. Each statocyst is a minute, hollow vesicle (Fig. 16.20) and lies very close to a pedal ganglia. It is lined by sensory cells and contains a mass of lime called **statolith.** These help the animal in maintaining balance. Statocysts are supplied with nerve from cerebro-pleural connectives.

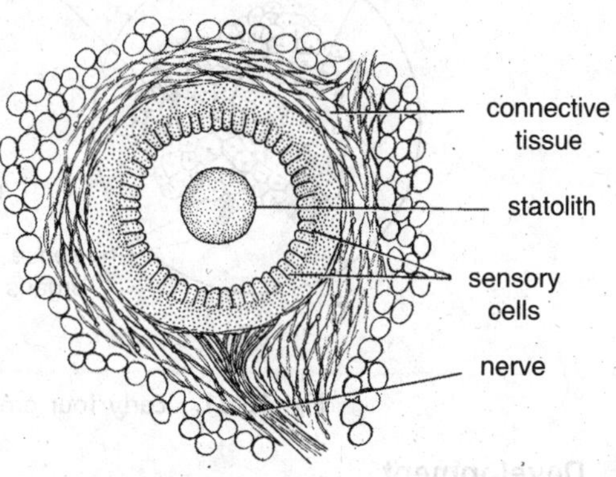

Fig. 16.20. *Unio*. T.S. of statocyst.

2. Osphradium. Beneath the posterior adductor muscles in close contact of visceral ganglia there is a path of sensory epithelium called **osphradium.** It is probably **chemosensory.** It checks the water passing through the mantle cavity.

3. Sensory cells. Sensory cells, probably **tactile** in nature, found on the edges of the mantle and on the **papillae** of the inhalent siphon.

16.12 REPRODUCTIVE SYSTEM

Unio is **dioecious,** *i.e.*, sexes are separate. It does not show sexual dimorphism. The reproductive organs consist of gonads and their ducts.

1. Male reproductive organs. They include a pair of white, branched, glandular mass situated in the visceral mass between the loops of intestine. Each testis gives out a narrow, vas deferens, opens into the supra-branchial chamber by male gonopore situated in front of the renal aperture.

2. Female reproductive organs. There are a pair of red-coloured glandular, branched mass of tubules. Each ovary gives out an oviduct, which opens by female genital aperture, situated in front of renal opening.

Life History

1. Eggs. The eggs are larger, yolkier and each is surrounded by **vitelline envelope** which contains a minute opening, the **micropyle,** through which sperms enters the egg during fertilization.

2. Fertilization. Females of *Unio* brood their ova; the eggs are discharged into the supra-branchial chamber through the genital aperture. These eggs are carried into the water tubes of the outer gill laminae and are held by the mucus secreted by the gill epithelium. The sperm from the male escape the body through the outgoing water current. Some of them goes into the females body through inhalant siphon. On entering the gill laminae, they fuse with eggs and bring about **'internal' fertilization.**

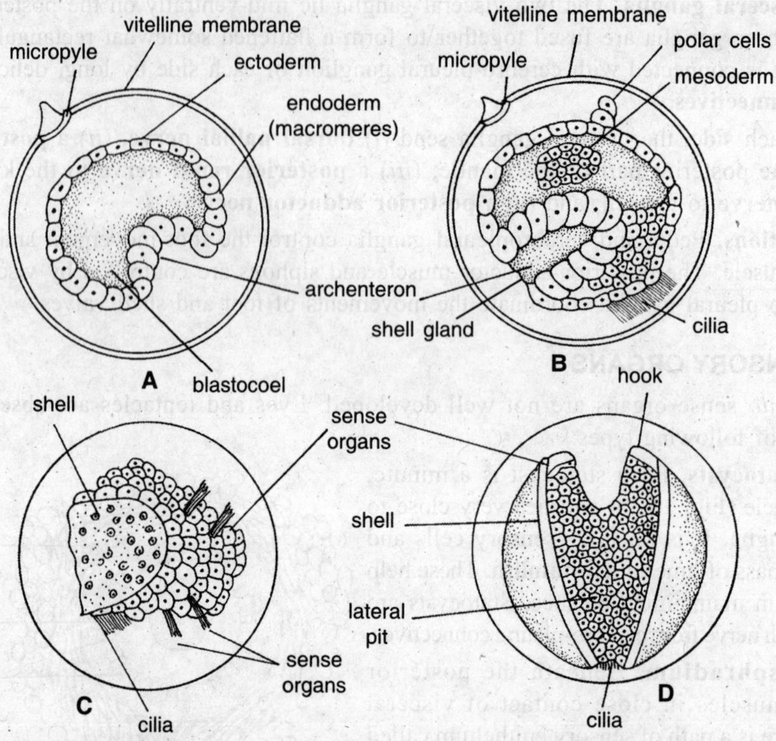

Fig. 16.21. Early four embryonic stages of freshwater mussel.

Development

Development of *Unio* is **indirect** and includes a free-swimming (and ectoparasitic) **glochidium larva** (a modified veliger larva). It occurs inside the water tubes of gills which act as brood pouches. Development of *Unio* resembles greatly with that of *Anodonta*.

1. Cleavage and blastulation. The fertilized egg or zygote undergo **complete (holoblastic)** but **unequal** and **spiral cleavage.** It produces a hollow ball of cells, the **blastula.** It contains small, yolkless and multiplying micromeres on the upper side (animal pole) and large, yolk-filled **macro-meres** on the lower side.

2. Gastrula. Gastrula is formed by invagination of macromeres (Fig. 16.21). It bears a tuft of long cilia posteriorly and a small **archenteron.** Large cells of gastrula are budded off into the blastocoel where they form **mesoderm**. At the same time a deep invagination marks the dorsal surface of the embryo. This invaginated part forms the characteristic **shell glands.**

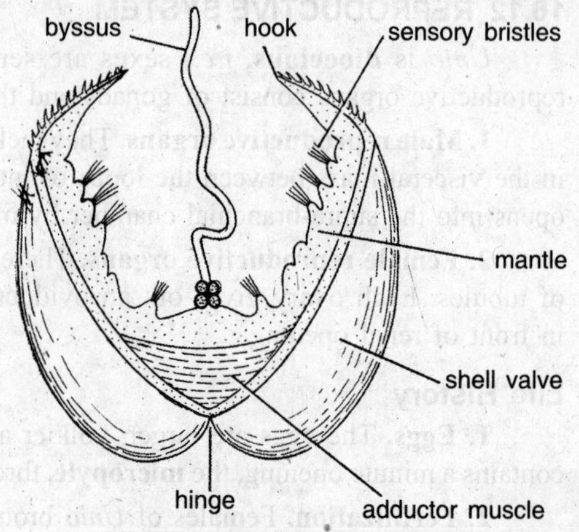

Fig. 16.22. *Unio*. A glochidium.

These shell glands secrete a **univalve embryonic shell** first and later on, replace it by a triangular **bivalved shell.** The lower ends of two shell valves are curved inwards and bear spines and hooks. The mantle develops gradually. A few of the mesoderm cells constitute the **adductor muscle** between the shell valves anteriorly. The mouth of gastrula disappears and a definitive mouth is formed by invagination in the ectoderm. Thus the larva is divided into a **body proper** and right and left **mantle lobes.** Now the larva is called **glochidium larva.**

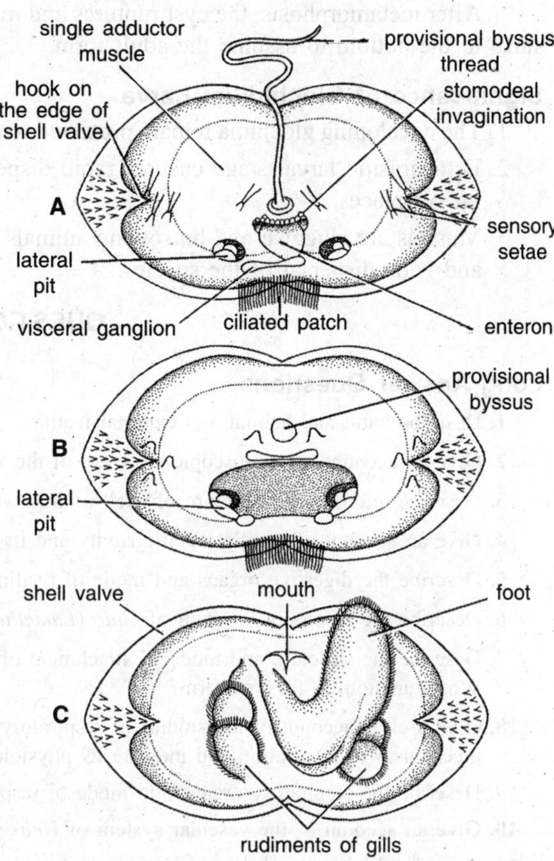

Fig. 16.23. *Unio*. Three stages of metamorphosis.

3. Glochidium larva. The word glochidium means point of an arrow. Glochidium larva ranges from 0.1 mm to 0.4 mm. Its body is enclosed in a triangular, bivalved and porous shell with its valves united dorsally and free ventrally. The free end bears hook and spines. The shell valves are lined by mantle lobes (Fig. 16.22). The edge of mantle lobes bear 3 to 4 tufts of brush like **sensory bristles.** The two valves are connected together by a single large **adductor muscle.** The two valves clap together by the action of adductor muscle. Larva lacks a mouth and an anus. Larva is without foot but contains a **glandular pouch** on the ventral surface to secrete a long, sticky thread called **provisional byssus.**

The glochidia are produced in a large number in the outer laminae and look as stuffed mass. Larvae feed on mucus secreted by gills. After being fully mature the larvae are discharged out of the gill through exhalant siphon along with the outgoing water current. When they find a fresh water fish they lead an ectoparasitic life for about 10 weeks. They become attached to the host's skin, fins or gills, encysted by the host's tissues and feed on nutrients by absorbing from body fluid. The larval mantle contains **phagocytic cells** which feed on the tissues of the host. Growth of larva takes place and metamorphosis occurs.

Metamorphosis

During metamorphosis of glochidium larva the following changes occur (Fig. 16.23):

1. The single adductor muscle is replaced by two adductors.
2. The larval byssus, sensory organs and byssus glands disappear.
3. Foot and gill make their appearance.
4. A stomodaeum is formed as an invagination and opens into archenteron. The tip of the archenteron and overlying ectoderm rupture to form **anus.**
5. The embryonic mantle folds elongate with the increase in size of shell.
6. Rudiments of visceral organs develop.
7. Two papillae or the rudiments of gills appear on each side of the foot.

After metamorphosis, the cyst ruptures and miniature clams are liberated in the water. These sinks to the bottom to assume the adult form.

Significance of Glochidium Larva

1. The developing glochidia remain protected within the cyst. Thus chance of survival increases.
2. Ectoparasitic larval stage ensures rapid dispersal of the species, carried out by the fish to far off places.
3. Mussels are sluggish and burrowing animals. Glochidium is the only way of achieving for and wide dispersal of the species.

QUESTIONS

Long Answer Questions

1. Describe habit and habitat and external features of *Lamellidens*.

2. Give an account of microscopic structure of the shell of *Unio*. How is it formed?

3. Describe microscopic structure of shell of freshwater mussel. How is pearl formed?

4. Give an account of mantle, mantle cavity and its organs of *Unio*.

5. Describe the digestive organs and mode of feeding in the *Lamellidens* or *Unio*.

6. Describe the respiratory system of *Unio* (*Lamellidens*).

7. Describe the structure and mode of attachment of ctenidia of a freshwater mussel you have studied. What function do they perform?

8. Give a clear account of the studies of respiratory organs in freshwater mussel (*Unio*). Describe the mechanism of respiration and indicate its physiological importance to the animal.

9. Describe the respiratory organs and mode of respiration in *Lamellidens* or a bivalve mollusc.

10. Give an account of the vascular system of *Unio*.

11. Describe the heart and the circulatory system of the lamellibranch that you have studied.

12. Give an account of excretory organs of *Unio* (*Lamellidens*).

13. Give an account of nervous system of *Unio* and compare it with that of *Pila*.

14. Describe various sense organs of *Unio*.

15. Give the structure and function of statocyst in that of *Unio*.

16. Describe the reproductive organs and life history of *Lamellidense*.

17. Give an account of the development upto the adult stage of a bivalve mollusc.

18. Give in detail the various development stages, found in the life-history of *Lamellidense*.

19. Describe the life history of fresh-water mussel and show how dispersal is brought about.

Short Answer Questions

1. Decribe the structure and metamorphosis of the glochidium larva and explain how it is beneficial to the animal.

2. How is pearl formed in *Unio*?

3. Write a short account of digestion in *Unio*.

4. Describe structure of gill of *Unio*.

5. Describe the valves of *Unio*.

6. Why *Unio* is called a filter feeder?

7. Describe position of gonads in *Unio*.

8. When and how fertilization occur in *Unio*?

9. Give a short account of pearl formation and pearl industry.

10. Write short notes on the following:

(*i*) Crystalline style.
(*ii*) Gills of *Unio*.
(*iii*) Glochidium larva.
(*iv*) Shell of *Unio*.
(*v*) Nacreous layer.
(*vi*) Keber's organ.
(*vii*) Organ of Bojanus.

11. Draw neat and labelled diagram of the following:

(*i*) V. S. shell of Mollusca.
(*ii*) T. S. of gill lamina of *Unio*.
(*iii*) T. S. of *Unio* through heart.
(*iv*) T. S. *Unio* through middle of body or through gill region.
(*v*) Blood vascular system of *Lamellidens*.
(*vi*) Nervous system of *Unio*.
(*vii*) Glochidium larva.
(*viii*) T. S. *Unio* passing through branchial or gill or middle region.
(*ix*) T. S. of *Unio* through heart and foot.

12. Differentiate between gill-lamella and gill-lamina.

Very Short Answer Questions

1. Give common name of *Unio* or *Lamellidens*.

2. What is chemical composition of shell of *Lamellidens*?

3. How are the two valves of shell of *Unio* united?

4. What is the function of adductor muscles?

5. Name the muscles which draw the foot into the shell.

6. What controls the opening and closing of shell valves of *Lamellidens*?

7. In *Unio*, which is larger, exhalant siphon or inhalant siphon.

8. What is typhlosole?

9. Which kind of ctenidia *Unio* has?

10. Name the respiratory pigment found in the blood of some species of *Unio*.

11. What is the function of Keber's organs in *Unio*?

12. Mention the host of the larva of *Unio*.

13. What are marsupia?

14. Which type of development occurs in *Unio*?

15. Select the correct alternative.

(*i*) Shell valves of *Lamellidens* are lateral/dorsal and ventral.
(*ii*) Water current of *Unio* is not purely/purely respiratory.

Fill in the Blanks

1. Shell of *Unio* consists of .... valves.
2. .... is the outer layer of the shell of *Unio*.

3. Foot of *Unio* is .... shaped.
4. Shell of *Unio* is secreted by ....
5. Mouth of *Unio* is bound on either side by .... and .... labial ....
6. The gills of *Unio* are called ....
7. The gills of *Unio,* besides carrying out gas exchange, also act as ....
8. Heart of *Unio* is three chambered, having .... atria and .... ventricle.
9. Nitrogenous waste material excreted by kidneys in *Unio* is chiefly ....
10. The osphradium of *Unio* functions as ....

Multiple Choice Questions

1. *Unio* is a member of the class

(*a*) Gastropoda (*b*) Cephalopoda
(*c*) Pelecypoda (*d*) Scaphopoda

2. The body of *Unio* is

(*a*) bilaterally symmetrical (*b*) radially symmetrical
(*c*) asymmetrical (*d*) biradially symmetrical

3. The outer layer of shell of *Unio* contains

(*a*) chitin (*b*) conchiolin
(*c*) $CaCO_3$ (*d*) Si_2O_3

4. The middle layer in the shell of *Unio* is

(*a*) prismatic layer (*b*) periostracum
(*c*) nacre (*d*) pallium

5. The muscles which are responsible to open and close the shell valve in *Unio*, are

(*a*) pallial muscles (*b*) protractor muscles
(*c*) adductor muscles (*d*) retractor muscles

6. Body cavity of *Unio* is

(*a*) coelom (*b*) haemocoel
(*c*) none of these (*d*) both of these

7. Pearl producing species of mollusc is

(*a*) *Tridacna maxima* (*b*) *Solen kempi*
(*c*) *Pinctada vulgaris* (*d*) *Mytilus viridus*

8. Typhlosole is present in *Unio* in

(*a*) stomach (*b*) intestine
(*c*) rectum (*d*) oesophagus

9. Gills in *Unio* are

(*a*) a part of visceral mass (*b*) derivatives of mantle
(*c*) derivatives of foot (*d*) modification of tentacles

10. Heart beat of *Unio* per minutes is

(*a*) 10 – 20 (*b*) 20 – 40
(*c*) 40 – 80 (*d*) 20 – 100

11. Keber's organ is the excretory organ of

(*a*) *Pila* (*b*) *Unio*

(*c*) *Neopilina* (*d*) *Doris*

12. "Organs of Bojanus" are derivatives of

(*a*) spleen (*b*) kidney

(*c*) gall bladder (*d*) liver

13. The chemoreceptor osphradium is present in

(*a*) *Pila* (*b*) *Unio*

(*c*) both (*d*) none of them

14. Statocyst in *Unio* is meant for

(*a*) olfaction (*b*) gustation

(*c*) equilibrium (*d*) photoreception

15. Larva of *Unio* is called

(*a*) trochophore (*b*) glochidium

(*c*) veligar (*d*) zoaea

16. Glochidia larva is

(*a*) endoparasite of pig (*b*) ectoparasite of pig

(*c*) ectoparasite of fish (*d*) endoparasite of sheep

ANSWERS

Very Short Answer Questions

1. Freshwater mussel.

2. The chemical composition of shell of *Lamellidens* includes

(*i*) Calcium carbonate – 89 – 99%

(*ii*) Phosphate of lime – 1 – 2%

(*iii*) Conchiolin – Only traces

3. The two valves of shell of *Unio* are united by a dorsal elastic band called a **hinge ligament.** It is continuous with the two shell valves and is made of uncalcified conchiolin.

4. Closing of shell.

5. Retractors.

6. In *Lamellidens*, there are two adductor muscles for opening and closing of shell valves. These muscles are **anterior** and **posterior adductor muscles**. They are large, cylindrical and transverse bands of muscles running from one shell valve to the other. On contraction of these muscles the shell valves are closed and on their relaxation, shell valves are opened.

7. Inhalent siphon is larger than exhalent siphon.

8. Infold of the wall of gut (*i.e.*, rectum) into the lumen.

9. Lamelliform.

10. Haemocyanin.

11. Excretion.

12. Freshwater fish.

13. Water tubes of ctenidia that function as brood pouches.

14. Indirect development since egg of *Unio* develops into a glochidium larva which undergoes metamorphosis to become an adult-like form.

15. (*i*) Lateral.
(*ii*) Not purely.

Fill in the Blanks

1. Two **2.** Periostracum **3.** Hatchet **4.** Mantle (pallium)
5. Outer, inner, palps **6.** Ctenidia **7.** Brood pouches **8.** Two, one
9. Ammonia **10.** Chemoreceptor

Multiple Choice Questions

1. (*c*) **2.** (*a*) **3.** (*b*) **4.** (*a*) **5.** (*c*) **6.** (*d*)
7. (*c*) **8.** (*c*) **9.** (*b*) **10.** (*c*) **11.** (*b*) **12.** (*b*)
13. (*c*) **14.** (*c*) **15.** (*b*) **16.** (*c*).

17

Some General Topics of Mollusca (Respiration and Pearl Formation)

17.1 RESPIRATION IN MOLLUSCA

A. What is Respiration?

All living cells need energy for carrying on their function. The source of energy is the food (carbohydrates, lipids and proteins) brought to the cells by blood from the alimentary canal or liver. The molecules of food hold energy in their chemical bonds. Their bond energy is released by oxidation in the cells. In oxidation, usually oxygen (O_2) is used, carbon dioxide (CO_2) and water (H_2O) are produced, and energy is released. The process of oxidation occurs in many steps (such as glycolysis, oxidative decarboxylation, Krebs cycle, etc.). Carbon dioxide is poisonous and harmful for the cells. It must be removed. Water is useful and retained in the cells. Energy released in oxidation is stored in the high energy bond of ATP molecules. These high energy bonds are later broken by enzymatic hydrolysis and the energy so set free is used by the cells in their activities. The energy obtained from the ATP molecules is called **biologically useful energy** because it drives the life processes. The multistep reaction may be briefly expressed as under:

$$\underset{\text{Glucose}}{C_6H_{12}O_6} + \underset{\text{Oxygen}}{6O_2} \longrightarrow \underset{\text{Carbon dioxide}}{6CO_2} + \underset{\text{Water}}{6H_2O} + \text{Energy}$$

Eliminated ↑ ($6CO_2$); Reused ↑ ($6H_2O$); Energy → Stored in ATP → Hydrolysis → Energy used in life activities

The process that involves (*i*) intake of oxygen (O_2) from the environment, (*ii*) oxidation of food with the incoming oxygen; (*iii*) elimination of carbon dioxide produced in oxidation, and (*iv*) conservation of energy released during oxidation in biologically useful forms, such as ATP (= adenosine triphosphate), is known as **respiration**. Respiration may be briefly defined as the exchange of environmental oxygen with the carbon dioxide produced in the cells during oxidation. For the nonstop release of energy (respiration), the cells need uninterupped supply of oxygen and constant removal of carbon dioxide.

B. Respiratory Organs and gas Exchange

In most invertebrates (or nonchordates) and all vertebrates (or chordates) some organs of the body are specialized for the exchange of gases. These are called **respiratory organs.** Gills, trachea, book-lungs, pulmonary sac (lung) are the main respiratory organs in inverterates.

The surface of the organs at which the exchange of gases takes place is termed **respiratory surface**. For respiratory or gas exchange to occur, water or air must be brought to and removed from the respiratory surface.

The exchange of respiratory gases (Oxygen and Carbon dioxide) in the respiratory organs and tissues takes place according to the law of gases, and is, thus, a physical rather than a biological process. Gas exchange occurs by diffusion, *i.e.*, the net flow of a substance from a region of higher concentration to a region of lower concentration. The diffusion is directly proportional to:

(*a*) Partial pressure gradient (difference) of gases on the two sides of a membrane between them;

(*b*) thinness of the membrane;

(*c*) surface area of the membrane;

(*d*) permeability of the membrane.

Moisture is also necessary because O_2 and CO_2 are more easily exchanged in cells when in a liquid medium.

B. Respiration in Mollusca

In Mollusca, the following types of respiration occurs: 1. Body surface respiration and cutaneous respiration; 2. Branchial respiration; 3. Pulmonary respiration.

(1) Body Surface Respiration and Cutaneous Respiration. In molluscs the naked pteropods, a group of Gastropoda, have no gills. Gas exchange occurs across the general body surface. In an intertidal clam, *Geukensia demissa* (Pterimorpha, Bivalvia), there is a considerable direct uptake of oxygen across the body wall. When out of water at low tide, the clam uses oxygen of air, but the deep tissues respire anaerobically. The scaphopod, *Dentalium*, lacks gills. Gas exchange occurs through the mantle surface. Many Cephalopods with webbed arms have vestigial gills. In them, the gas exchange occurs through the general body surface.

(2) Branchial Respiration in Mollusca Many aquatic animals have special organs for obtaining oxygen from water and releasing carbon dioxide into water. These organs are called **gills** or **branchiae**. The exchange of gases in the gills is known as **gill** or **branchial respiration**. The branchial respiration is found in some annelids, many crustaceans, most molluscs and all echinoderms.

The molluscs have one to numerous gills, called **ctenidia**, for gas exchange. The ctenidia lie in the mantle cavity and function in water only. A ventilatory current of water is produced by gill cilia. Blood of mollusc contain the respiratory pigment called **haemocyanin**.

1. Monoplacophorans. The monoplacophorans, such as *Neopilina*, have 5 to 6 pairs of monopectinate gills (having a single row of lamellae, Fig. 17.1) in the pallial groove present around the foot on the ventral side (Fig. 17.1).

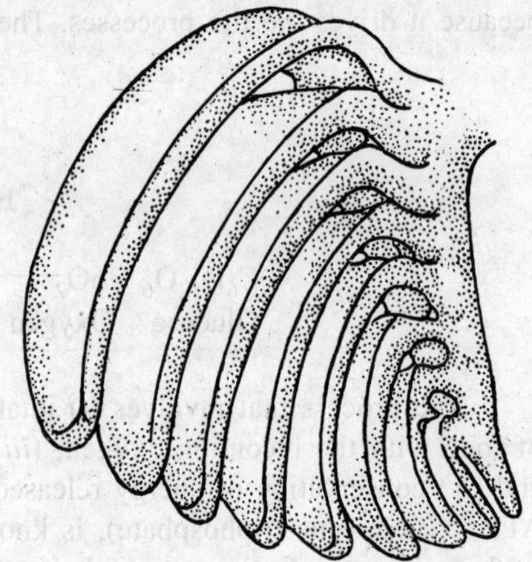

Fig. 17.1. *Neopilina.* One of the gills.

2. Polyplacophorans. The polyplacophorans, such as *Chiton*, have a few to numerous pairs of bipectinate gills (having two rows of lamellae) in the mantle cavity present on the sides of the foot on the ventral side (Fig. 17.2).

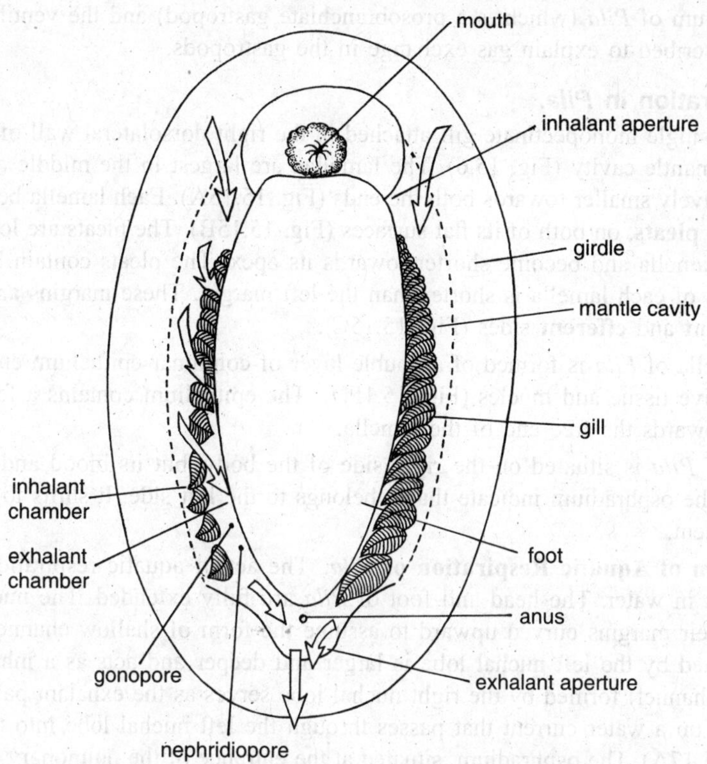

Fig. 17.2. The chiton, *Lepidochitona cinerea*. A ventral view showing the pallial groove and direction of the respiratory water currents. The arrow show water entering through the two anterolateral inhalant apertures and exiting through the single median, posterior exhalant aperture (after Ruppert *et al.*, 2004).

3. Aplacophorans. The aplacophorans, such as *Chaetoderma*, have a pair of bipectinate gills in a small mantle cavity at the hind end (Fig. 17.3).

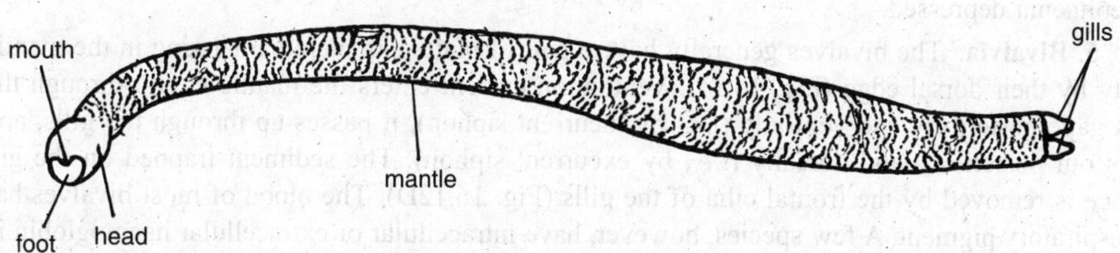

Fig. 17.3. *Chaetoderma*. Gills.

4. Gastropod. Among the prosobranchiate gastropods, the primitive species, such as *Haliotis* (the abalone or ear shell) have two bipectinate gills, where as the others, such as *Pila*, have a single monopectinate gill. In some prosobranchiates, such as *Patella*, the true ctenidia are reduced, and mantle folds form secondary gills which project into the pallial groove on each side of body. The opisthobranchiate gastropods, such as *Aplysia*, have plicate or folded ctenidium and have developed secondary gills for gas exchange.*Doris*, for example, has a ring of branching secondary gills around the anus.

The ctenidium of *Pila* (which is a prosobranchiate gastropod) and the ventilatory water it generates are described to explain gas exchange in the gastropods.

Aquatic Respiration in *Pila*.

Pila has a single monopectinate gill attached in the right dorsolateral wall of the branchial chamber of the mantle cavity (Fig. 15.6). The lamellae are largest in the middle of the gill and become progressively smaller towards both the ends (Fig. 15.15A). Each lamella bears transverse ridges, called the **pleats**, on both of its flat surfaces (Fig. 15.15B). The pleats are longest towards the base of the lamella and become shorter towards its apex. The pleats contain blood vessels. The right margin of each lamella is shorter than the left margin. These margins are respectively called the **affarent** and **efferent** sides (Fig. 15.15C).

A gill lamella of *Pila* is formed of a double layer of columnar epithelium enclosing a thin sheet of connective tissue and mucles (Fig. 15.15D). The epithelium contains a few gland cells and is ciliated towards the free end of the lamella.

The gill of *Pila* is situated on the right side of the body, but its blood and nerve supply and position of the osphradium indicate that it belongs to the left side. It shifts to the right side during development.

Mechanism of Aquatic Respiration of *Pila*. The act of aquatic respiration takes place when the *Pila* is in water. The head and foot of *Pila* are fully extended. The nuchal lobes are expanded and their margins curved upward to assume the form of shallow channels or siphons. The siphon formed by the left nuchal lobe is larger and deeper and acts as a inhalant passage. The siphon (or channel) formed by the right nuchal lobe serves as the exhalant passage. Cilia of gill lamellae set up a water current that passes through the left nuchal lobe into the pulmonary chamber (Fig. 15.17A). The osphradium, situated at the entrance of the pulmonary chamber, tests the nature of the incoming water. A current of dirty water is immediately discontinued. From the pulmonary chamber, the water current crosses the epitaenia near its posterior end and enters the branchial chamber. Here the water flows forward and washes the gill lamellae for exchange of gases. Oxygen of water diffuses through the epithelium of the lamellae into the blood in them. Carbon dioxide of the blood diffuses into the water. The water, thus befouled passes out through the right nuchal lobe.

During aquatic respiration of *Pila*, the opening of the pulmonary sac is kept closed and the epitaenia depressed.

5. Bivalvia. The bivalves generally have a pair of plate–like ctenidia hanging in the mantle cavity by their dorsal edge (Fig. 16.11). Ventilation current enters the mantle cavity through the shell gape posteriorly and ventrally (*i.e.*, by incurrent siphon), it passes up through the gills, and flows out posteriorly and dorsally (*i.e.*, by excurrent siphon). The sediment trapped on the gill surface is removed by the frontal cilia of the gills (Fig. 16.12D). The blood of most bivalves has no respiratory pigment. A few species, however, have intracellular or extracellular haemoglobin in their blood. In some species, haemoglobin is associated with ganglia and is called **neuroglobin**. Haemoglobin and myoglobin impart red colour to the mantle and other tissues. Haemocyanin occurs in some protobranchs. Ctenidium of *Unio* and its ventrilation are explained, before reviewing variation in gill strucutre in the group Bivalvia of Mollusca.

Aquatic Respiration in *Unio*. Gas exchange into *Unio* (and *Anodonta*) occurs in the ctenidia (true gills) and mantle.

***(a)* Ctenidium.** *Unio* has two ctenidia hanging in the mantle cavity, one on either side of the body. A ctendium is a large, flat organ (Fig. 16.12A). It consists of two more or less rectangular plates, the **inner** and **outer laminae** (Fig. 16.12B). Each lamina further comprises two similar plates, the **inner** and **outer lamellae**, united with one another anteriorly, ventrally and posteriorly, but free dorsally (Fig. 16.12C). A lamina, thus, has the form of a long bag with a narrow cavity open above. Each lamella is composed of vertical bars, the **gill filaments**, connected together by thinner horizontal bars, the **interfilamental junctions** (Fig. 17.5; Box 17.1).

Box 17.1 Holobranchs, Demibranchs and Lamellae

In bivalve molluscs, there are two gills, one on the left and one on the right, in the mantle cavity. Each gill is a **holobranch**, or whole gill, composed of two half gills, or **demibranchs**. A demibranch is composed of two lamellae, one ascending and one descending (Fig. 17.4). Each demibranch

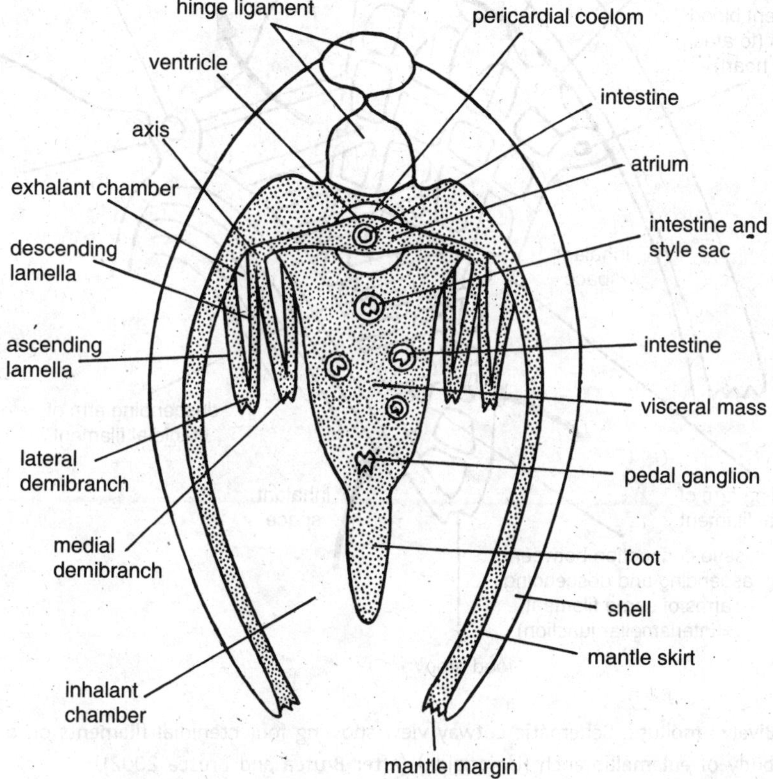

Fig. 17.4. Bivalve mollusc. Cross section of a generalized lamellibranch (after Ruppert, *et al.*, 2004).

gives the impression of being a complete gill but is not, for each is formed of either all the lateral filaments or all the medial filaments, but not both. A holobranch consists of two demibranchs (lateral and medial) and a total of four **lamellae** (two **ascending** and two **descending**). Since there are two gills, one right and one left, a typical clam has two holobranchs, four demibranchs, and eight lamellae.

At the ventral free edge of the lamina, the gill filaments of its two lamellae are continuous. It, thus, follows that the lamina has V–shaped gill filaments whose outer limbs form outer lamella and inner limbs constitute the inner lamella. The lamellae have minute apertures, the **ostia**, bounded in front and behind by the gill filaments and above and below by interfilamentar junctions. The ostia lead from the mantle cavity into the cavity of the lamina. The cavity of lamina is divided into small chambers, called **water tubes**, by vertical **interlamellar junctions** containing branches of afferent and efferent branchial veins. The water tubes open dorsally into a space called **suprabranchial chamber**.

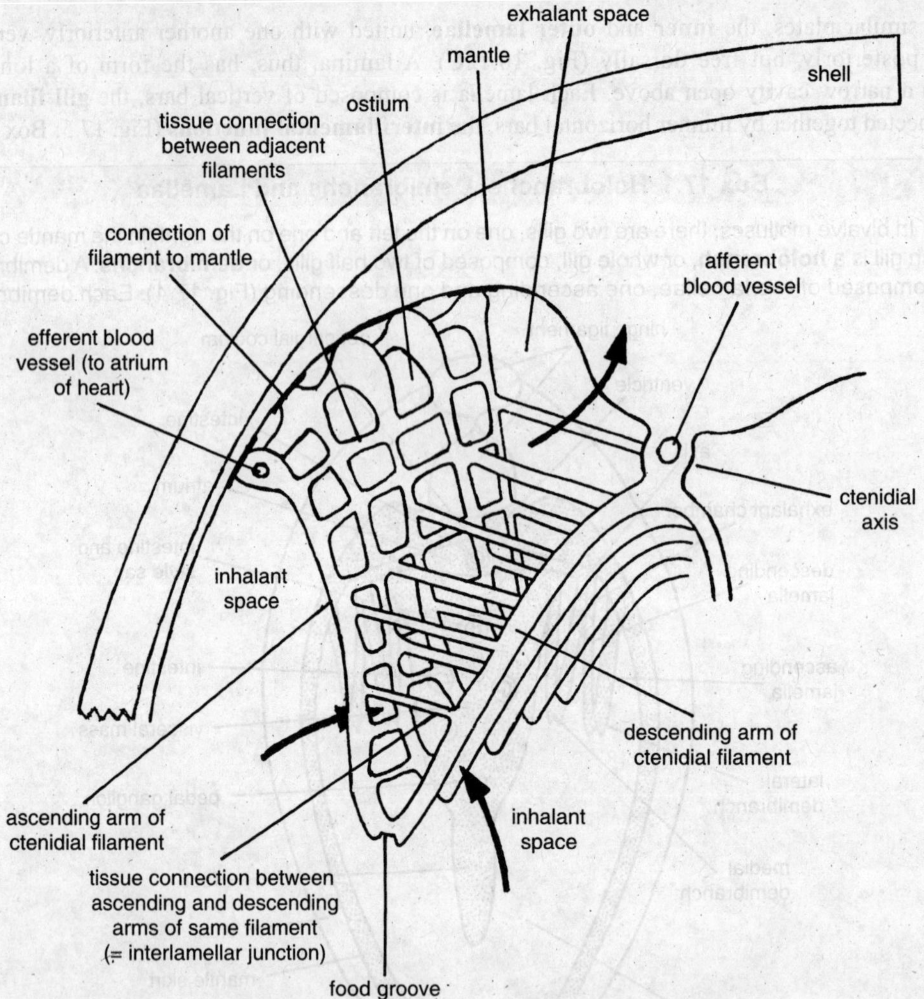

Fig. 17.5. Bivalve mollusc. Schematic cutway view showing four ctenidial filaments on one side of the body of eulamellibranch (Unionoida) (after Brusca and Brusca 2002).

The gill filaments (Fig.16.12D) are composed of connective tissue containing chitinous supporting rods and covered by ciliated epithelium. The cilia are larger on the outer, latero -external and lateral parts of the gill filaments (Fig. 17.6). These larger cilia are referred to as the **frontal, laterofrontal** and **lateral cilia**.

(b)* Attachement of gills or ctenidia in *Unio (Fig. 16.14). The outer lamella of the outer lamina is attached all along the length to the inner surface of the mantle. The inner lamella of the outer lamina and outer lamella of the inner lamina are together fixed to the sides of the visceral mass a little below the origin of the mantle. The inner lamellae of the inner laminae are joined to the visceral mass anteriorly, are free in the middle and are attached to one another posteriorly.

By the above mode of attachment of the gill lamellae to the mantle and to the visceral mass, a long narrow space, the **suprabranchial chamber**, is separated off over each gill lamina from the mantle cavity. The latter communicates with the exterior by the inhalant siphon (Fig. 16.11). The suprabranchial chambers open behind into a common passage, the **cloaca**, which leads out by the exhalant siphon.

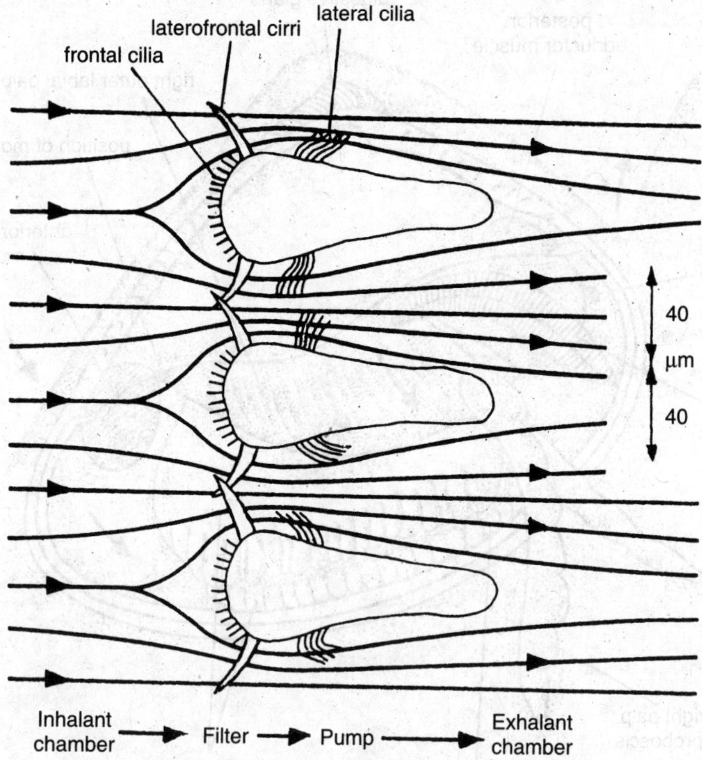

Fig. 17.6. Bivalve mollusc. Gill cilia and their working. The frontal cilia beat perpendicular to the plane of page. The effective stroke of the laterofrontal cilia is to the left and the effective stroke of the lateral cilia is to the right of the image. Arrows indicate the direction of the feeding/ventilating current produced by the lateral cilia (after Ruppert *et al.*, 2004).

(c) Working of ctenedial respiration of *Unio*. The lateral cilia of the gill–filaments set up currents that draw fresh water from outside into the mantle cavity through the inhalant siphon. They are aided in the work by cilia present on the inner surface of mantle and on the palps. From the mantle cavity, the water passes through the ostia into the water tubes. In water tubes, oxygen of the water diffuses into the blood flowing in the branching vessels of the interlamellar junctions and carbon dioxide diffuses in the reverse direction, *i.e.*, from branchial vessels into the water tubes. From the water tubes the foul water passes into the suprabranchial chambers and thence into the cloaca and finally escapes through the exhalant siphon. The frontal cilia of gill–filaments remove the sediment trapped on the gill surface.

Besides being respiratory, the water current brings food and eliminates faeces and nitrogenous wastes.

Aquatic respiration by mantle. The mantle being thin, vascular, devoid of cuticle and in contact with water, also carries on respiration in *Unio*.

Variations in Bivalve Ctenidium

The protobranchs, such as *Nucula*, have small bipectinate gills, each consisting of a central axis bearing two rows of short, flat filaments (Fig. 17.7).

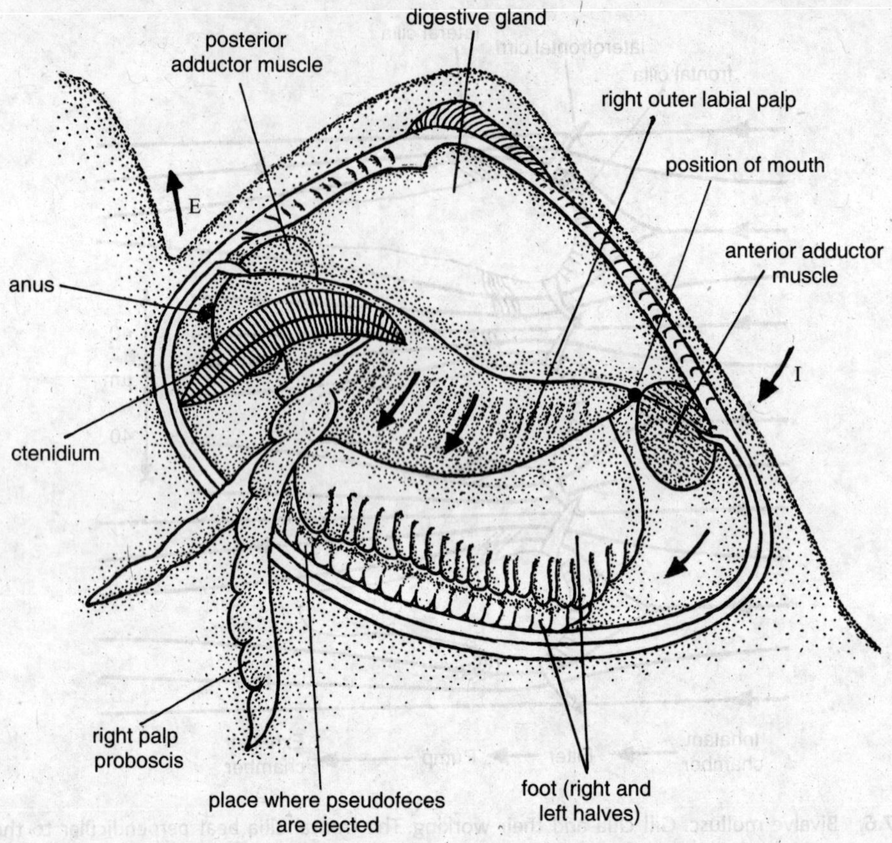

Fig. 17.7. *Nucula* (Nut clam). The bivalve (protobranch) is seen from the right side, it is natural position in the substratum (right shell valve and right mantle lobe removed). Arrows show direction of water currents in the mantle cavity. I = inhalant region; E = exhalant region (after Brusca and Brusca, 2002).

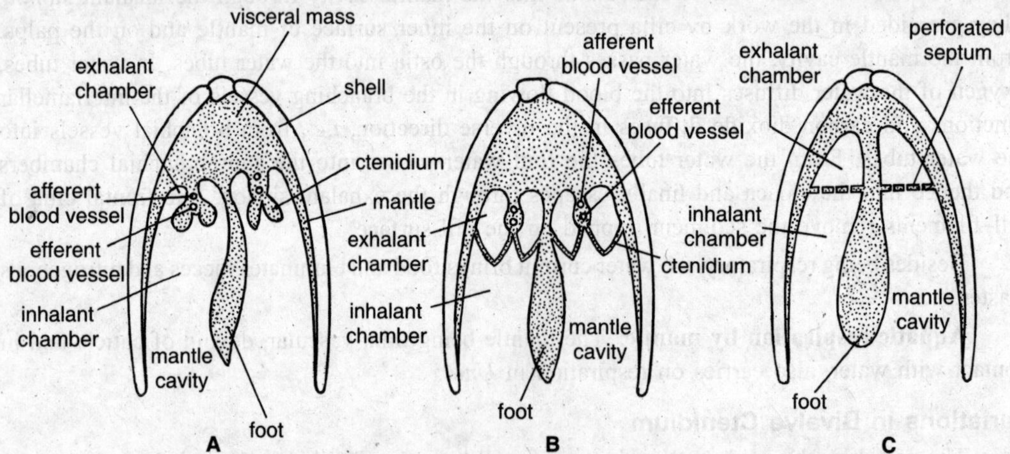

Fig. 17.8. Arrangement of ctenidia in some bivalve molluscs (transverse sections), showing evolution of lamellibranch gills. A–Protobranch; B–Lamellibranchs; C–Septibranchiate anomalodesmatan (Brusca and Brusca, 2002).

The other bivalves, except septibranchs, have W–shaped gills derived from folding of the protobranch gill filaments (Fig. 17.9). In filibranchs, for example, *Mytilus*, the interfilamentar junctions are lacking or formed by groups of interlocking cilia, and the interlamellar junctions are absent or nonvascular.

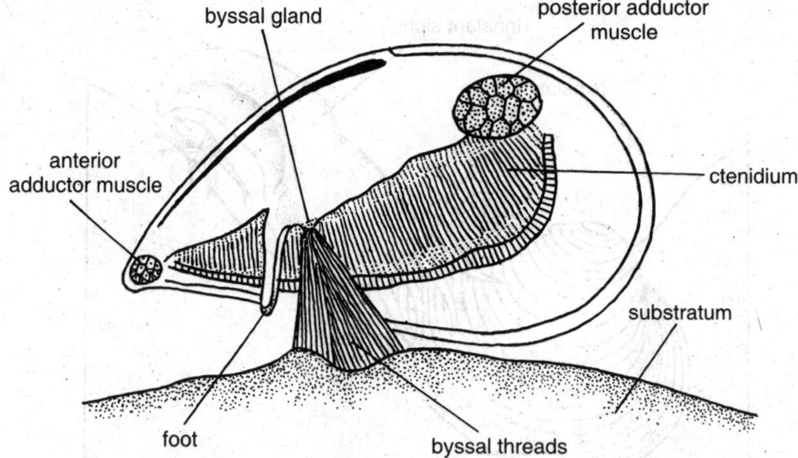

Fig. 17.9. *Mytilus* (mussel) in lateral view with left shell valve removed (Brusca and Brusca, 2002).

In the pseudolamellibranchs, such as *Pecten*, the W–shaped gills are vertically folded to increase the surface area (Fig. 17.10). They are called **plicate gills**. The interfilamentar junctions are vascular or nonvascular. In eulamellibranchs, such as *Unio*, *Anodonta*, and *Cardium*, the

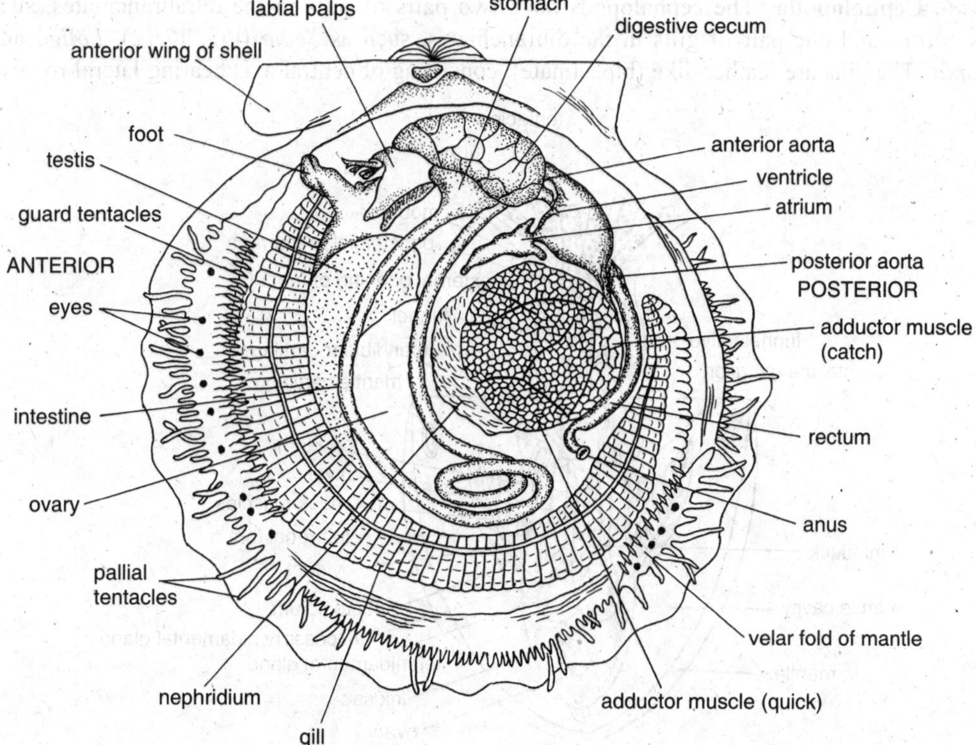

Fig. 17.10. The scallop *Pecten*. Internal structure viewed from the left side, with the left shell valve removed (after Ruppert *et al.*, 2004).

W–shaped gill may be smooth or folded, the interfilamental junctions are vascular, and the interlamellar junctions are vascular and firm. In septibranchs, such as *Poromya*, the gills are reduced to horizontal muscular partitions dividing the mantle cavity into upper and lower chambers. It has a hoodlike inhalant siphon (Fig. 17.11).

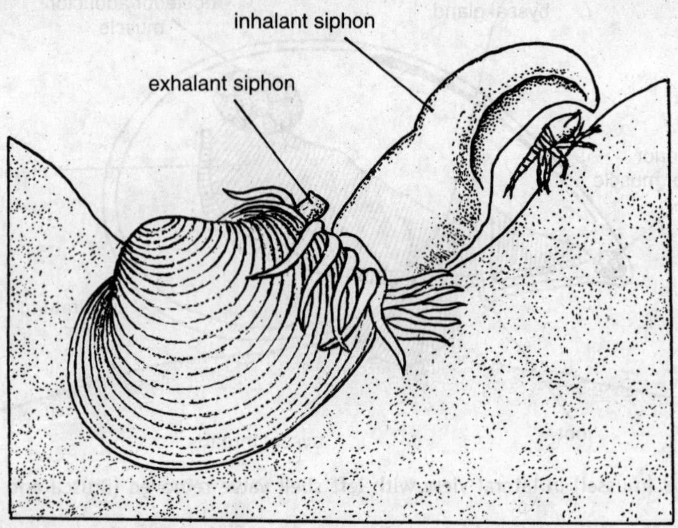

Fig. 17.11. The septibranch *Poromya granulata* capturing a crustacean (a cumacean) with its hoodlike inhalant siphon (after Ruppert *et al.*, 2004).

6. Cephalopoda. The cephalopods have two pairs of gills in the tetrabranchiates, such as *Nautilus*, and one pair of gills in the dibranchiates, such as *Sepia* (Fig. 17.12), *Loligo* and *Octopus*. The gills are feather–like (bipectinate), consisting of central axis bearing lateral rows of

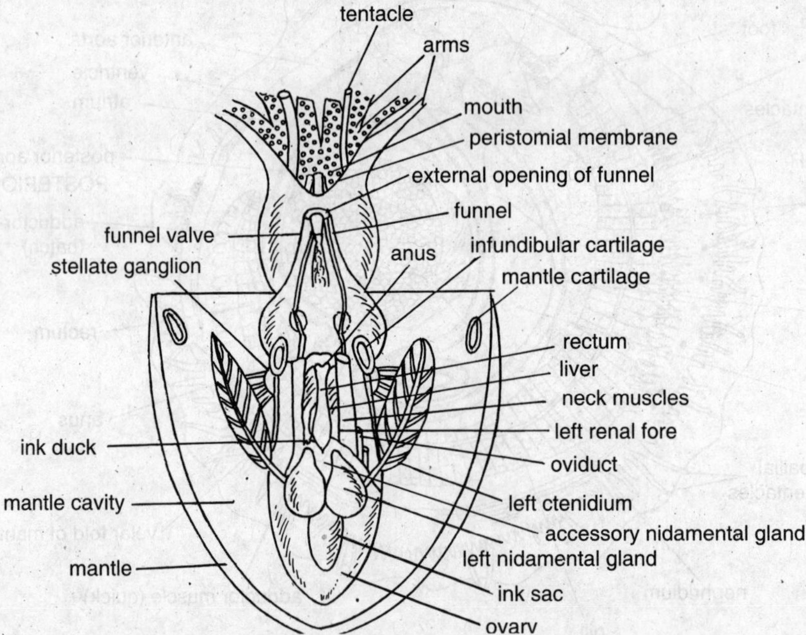

Fig. 17.12. *Sepia.* Mantle cavity of female from postero–ventral side.

lamellae (Fig. 17.12). The lamellae become smaller towards the tip of the gills. The lamellae are folded to increase their respiratory surface. Ventilatory water current is produced by muscular contraction of the mantle rather than by ciliary action, cilia being absent. The water current enters the mantle cavity through the pallial aperture, passes over the gills, and leaves via funnel. The cephalopod blood contains haemocyanin.

Pulmonary Respiration in Mollusca

In mollusca, pulmonary respiration occurs in the pulmonate gastropods and some other gastropods.

***(i)* Pulmonate gastropods**. These include land snails, many aquatic (mostly freshwater, some marine) forms. They have lost the gill. Their mantle cavity is transformed into a **lung** or **pulmonarium** for breathing air. The lung, has a highly vascular roof, and opens out by a small contractile pore, the **pneumostome**, on the right side of the body. Ventilation of the lungs is brought about by arching and flattening of the floor of the mantle cavity. The pneumostome generally remains open all the time, or opens and closes with the ventilating cycle. Most pulmonates have lost operculum, but the marine pulmonate, *Amphibola*, is exceptional in retaining operculum.

Many freshwater species of lower pulmonates (order Basommatophora), such as *Lymnaea* come to the surface of water to get air for gas exchange. *Lymnaea* can extend the edges of the mantle as a long tube for this purpose. The pneumostome is closed when the animal is under water. Pulmonates can remain submerged from 15 minutes to one hour. Some aquatic pumonates have given up air breathing. They either fill their mantle cavity with water or have developed a secondary gill, called **pseudobranch**, as folds of mantle near the pneumostome (*Planorbis*).

The higher pulmonates (order Stylommatophora) include land forms. They have two pairs of tentacles: short anterior and long posterior, carrying eyes at the tips. They may have reduced or no shell. The snail (*Helix*) retain shell, the slug (*Limax*) have lost it. Both have a prominent pneumostome on the right side of the body (Fig. 17.13).

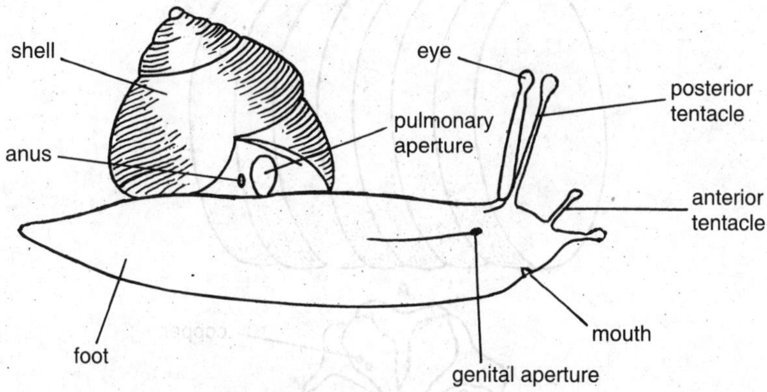

Fig. 17.13. *Helix*. The land snail showing penumostome.

Pulmonary Respiration in *Pila*

All land snails are not pulmonate, many operculate prosobranch snails, such as *Pila*, have adaptation for air breathing. *Pila* has pulmonary sac (or lung) and cetenidium (or gill; Fig. 15.6).

Pulmonary sac or lung of *Pila*. The pulmonary sac or lung is a large sac developed in the roof of the pulmonary chamber. It is a modification of the mantle itself. It has thin, moist, highly vascular and muscular walls. The dorsal wall of the lung is pigmented but the ventral wall is creamy white. The lung communicates with the pulmonary chamber of the mantle cavity by a large oval aperture, the **pneumostome**, guarded by two unequal flaps (Fig. 15.6).

Mechanism of Aerial or Pulmonary Respiration. *Pila* resorts to aerial respiration (*i*) at regular intervals during favourable conditions, (*ii*) when the surrounding water becomes dirty and deficient in oxygen, (*iii*) when it comes on land, and (*iv*) during aestivation.

For aerial respiration in water, *Pila* comes near the surface. The left nuchal lobe is extended and its right edge is rolled over the left edge to form a **tube** or **siphon**. This siphon may be about 2 cm long and is thrust out of water. If the surface of water is covered with vegetation, *Pila* keeps itself submerged and siphon fully extended. But if the water surface is clear of vegetation, it exposes a part of its shell and its siphon changes into a **funnel** by shortening in length and broadening at the end. The epitaenia is tightly pressed against the roof of the mantle cavity so that the pulmonary chamber is cut off from the branchial chamber. The pneumostome becomes almost circular and apposed to the base of the siphon or funnel (left nuchal lobe). This brings the pulmonary chamber in direct communication with the exterior instead of via mantle cavity as before. Now the pulmonary sac expands and contracts alternately and this brings about inspiration and expiration. Exchange of gases between blood of mantle and air takes place through the thin wall of the pulmonary sac.

On land the mechanism of aerial respiration remains the same, except that the nuchal lobe simply expands and does not change into a siphon.

During aestivation, when *Pila* lie doramnt buried in mud for long periods, the air imprisoned in the pulmonary sac or lung in used for respiration.

Respiratory Pigment in Mollusca

Haemocyanin is the respiratory pigment of the molluscs and arthropods and it is found in the blood. Its molecule has many pairs of **copper** atoms bound directly to the protein component. Haemocyanin binds an oxygen molecule between a pair of copper atoms. Haemocyanin has large molecule and occurs extracellularly in the plasma only. Haemocyanin molecules differ in structure

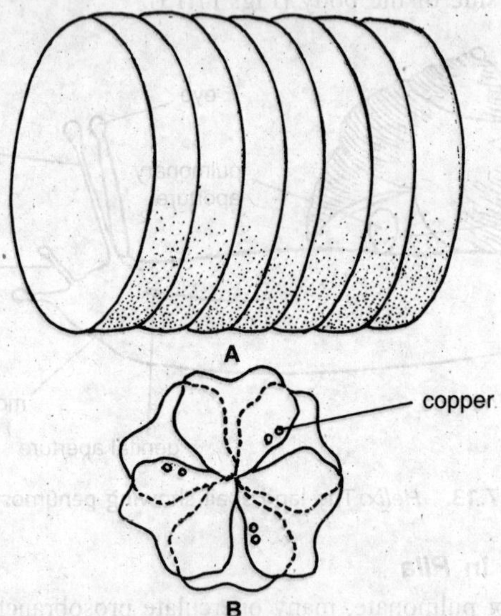

Fig. 17.14. Haemocyanin molecule. A–In molluscs; B–In arthopods.

in the two groups, being larger and cylindrical in the molluscs and smaller and hexamers in the arthropods (Fig. 17.14). Haemocyanin becomes light blue when oxygenated and colourless when deoxygenated.

17.2 PEARL FORMATION IN BIVALVES

Pearl is known to human beings since ancient times. A pearl due to its delicate appearance and shine has retained its position as a costly ornamental object.

Pearl is of **animal origin** and is obtained from **pearl oyster** (Fig. 17.15) and mussels belonging to phylum Mollusca.

Types of Peal Oysters and their Occurrence

The pearl oyster belong to genus *Pinctada*, family Pteriidae, class Bivalvia and phylum Mollusca. The important species producing good quality of pearl are *Pinctada vulgaris* (Schumachar), *Pinctada marigavitifera* (Linn.), *Pinctada chemnitzi* (Phillippi), *Pinctada anamioides* (Reeve) and *Pinctada atropurpurea* (Dunker). All these species are marine and it is the marine forms which produce the pearl of fine quality. Certain freshwater forms of mussels belonging to genera *Unio* and *Anodonta* also produce pearls but they are of inferior quality and hardly of any use. However, the European species belonging to the family Unionidae have been found to produce superior quality pearls.

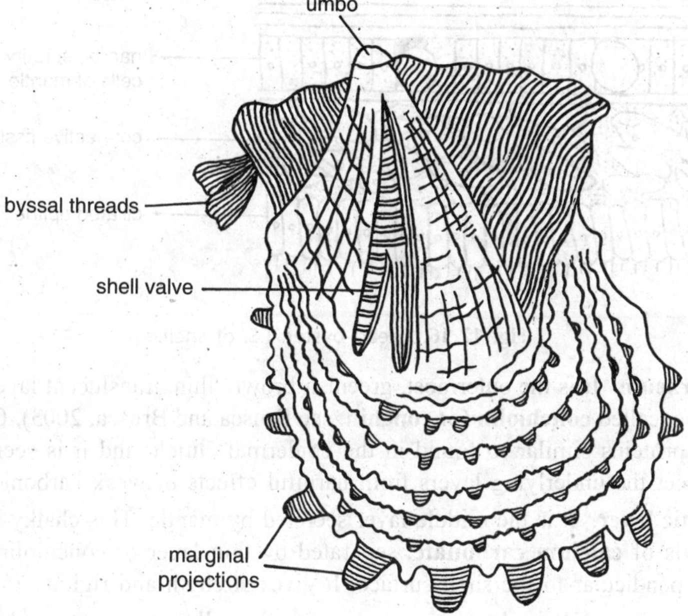

Fig. 17.15. *Pinctada vulgaris*, The Indian pearl oyster.

The main sites of production of pearl are the Persian Gulf, Gulf of Manaar (Shri Lanka), Sulusea (near Philippines). Besides these coast of Australia and shores of Central America (Pennama Bay, Gulf of California) including islands of south Pacific are also important production centers of pearl. Nowadays Japan has surpassed all the countries in the bulk of pearl production by invention of **pearl culture techiniques**.

Pearl oysters are generally found attached to the rocks or dead coral reefs by byssal threads. When present in large number at a particular place they form **pearl oyster banks** or **pearl oyster beds**. Pearl oyster beds in India are present on both eastern and western coasts. However, the east coast is more productive and extensive than the west coast as its extends from Cape Comorin to Kilakarai. Tuticorin is the most productive region of the east coast lies Gulf of Kutch. Hence the major sites of pearl fisheries in India are Gulf of Manar both on Indian as well as ShriLankan sides, Gulf of Kutch, Palk Bay, Baroda and Tuticorin.

Formation and Nature of Pearl

Pearl is secreted by the mantle of the pearl oyster as a protection against foreign objects which may be sand particles, minute forms or other such things. To study the mode of formation of pearl it is essential to know the structure of the shell and mantle.

Structure of shell. Shell of pearl oyster is composed of three distinct layers (Fig. 17.16).

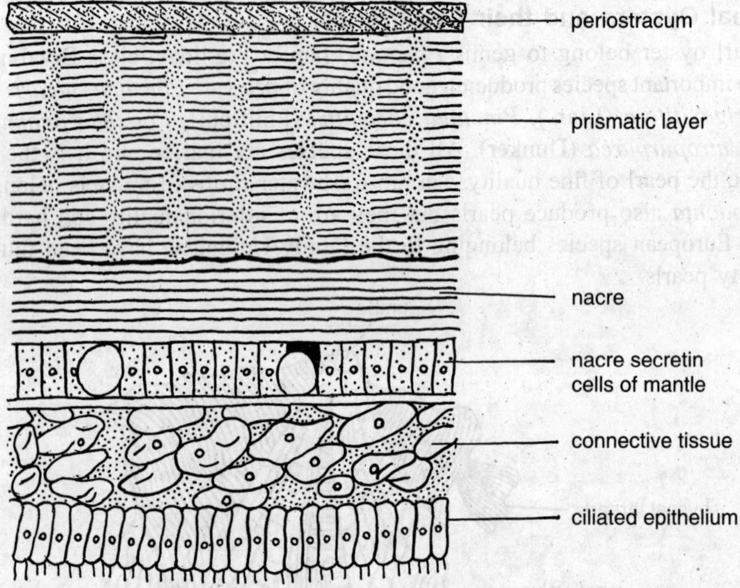

Fig. 17.16. Pearl oyster. T.S. of shell.

1. Periostracum. It is the outermost, greenish-brown, thin, translucent layer made up of an organic substance, called **conchiolin** (or **conchin**, see Brusca and Brusca, 2003). Conchin includes quinone-tanned proteins similar to found in the epidermal cuticle and it is secreted by mantle. It serves to protect the underlying layers from harmful effects of weak carbonic acid in water.

2. Prismatic layer. It is the middle layer secreted by mantle. This chalky layer is made up of minute crystals of **calcium carbonate**, separated by thin layer of conchiolin. These crystals are arranged perpendicular to the shell surface. It gives strength and rigidity to the shell.

3. Nacreous layer. It is the inner most pearly **lamellar** or **nacreous** layer of the shell and is better known as **"mother** of **pearl"** because this layer is responsible for the formation of pearl. It consists of alternate layers of calcium carbonate and conchiolin, which lie parallel to the surface of the shell. This layer is also secreted by mantle and its function is to protect the delicate surface from harmful effects of the foreign particles.

Structure of mantle. Mantle is made up of following three layers:

***(i)* Columnar epithelium**. It contains numerous nacre secreting unicellular glands.

***(ii)* Connective tissue layer**. The middle fibrous layer of mantle is mainly composed of connective tissues.

***(iii)* Ciliated epithelium**. Inner most layer of mantle contains mucous secreting cells.

Structure of the pearl. The pearl contains a central nucleus around which several layers of organic and inorganic matter are laid. A precious, beautiful and a fine quality pearl is that in which the layers are accurately and symmetrically laid. It is also free from any sort of adhesion. The chemical analysis of a pearl shows that it contains about 90% calcium carbonate, 5% organic substances and 5% water and other residues.

Process of Pearl Formation

There are three different views regarding the formation of pearl inside the body of pearl oyster.

1. The common and most accepted view is that pearls are formed due to irritation caused to mantle by the entry of foreign bodies.

2. Internal parasites act as provoking agents in the process of formation of pearls.

3. It is the internal pathological conditions of the host which results in the formation of pearl.

Only the first view is universally accepted now a days. According to this view whenever any sort of foreign body enters and gets between the mantle and shell, it becomes enclosed in a sac of mantle epithelium (Fig. 17.17). The foreign body now acts as an irritant and stimulates the mantle epithelium to secrete concentric layers of nacre around this foreign body. These layers when harden becomes pearl (Fig. 17.17). Size of pearl is directly proportional to the degree of

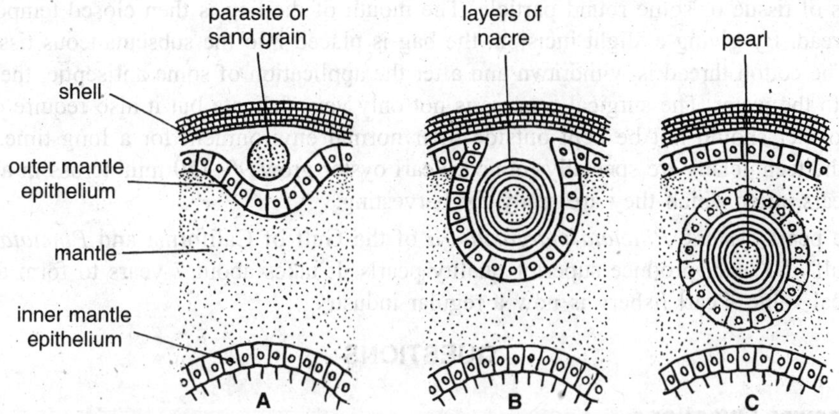

Fig. 17.17. Stages of pearl formation in a bivalvia mollusc.

irritation cause by foreign agents. Time taken for the formation of a pearl of average size is three to five years.

Fishing of pearl oysters for natural pearls in India. Fishing of pearl oysters in order to obtain natural pearl is known to Indians since ancient days. The optimum depth for the occurrence of pearl oyster bed is 22–25 meters, usually 15–20 kilometers away from the shore. In India, pearl oysters are collected by divers. Divers with the help of sinking stones go deep into sea water and collect the oysters in small bags tied around their waist. The amount of collection depends on the skill of the diver and his capacity to hold breath under water. The time of collection is in the morning till mid–day. The best season for pearl fishing is summer during which the sea remains calm and clear.

The days collection are brought to the beach and are left there for few days during which they get partially decomposed. Shell of the decomposing oysters are then removed by hand so that the inner contents get fully exposed for further decomposition. In next 3 to 4 days, the organic mass becomes too soft and well puterified and are then rinsed with several changes of water. The heavier part or debris which may bear pearl settle down in the container in which washing has been done. The sediment is rinsed and washed several times so that the pearls are picked up.

Techniques of pearl culture. It was Japan which has the honour of developing techniques to culture pearl oyster and obtaining pearl out of them by introducing artificial devices. The pearls obtained by this method are of fine quality and are similar to that of natural one.

It was **Mikimoto**, a Japanese, who for the first time discovered the art of cultivating the pearl oysters. In 1913, he produced pearl by culture which were just like that of natural ones.

The first step of pearl culture in Japan needs certain areas of shallow waters to be formed as oyster bed or **oyster farm**. The important oyster beds in Japan are situated in Ago Bay, Katla Bay, Matozoa Bay, Heijo Bay, Mikimoto Bay, Omura Bay, Nanao Bay, etc. Stones, layer's of wood, dead corals, etc., are laid in these shallow coastal areas on which oysters deposit their "spat", New methods for the collection of larvae of oysters have been evolved in which metallic cages are let down in the sea which collect the floating larval from off shores. The collecting spat are transferred to the site of culture where these oysters under constant watch are grown to a considerable size. Only after they attain a moderate size, they undergo the surgical impregnation of foreign bodies (*i.e.*, introduction of some foreign materials between mantle and shell of oyster). The laboratories for this purpose are established just by the side of the cultural ground so that the treated oysters may be quickly placed back to the water. For impregnating the foreign body, a tiny bag made from the mantle parenchyma of fresh water mussel is formed in which are placed fragments of tissue or some round particle. The mouth of the bag is then closed temporarily by cotton thread. By giving a slight incision, the bag is placed into the subscutaneous tissue of the mantle. The cotton thread is withdrawn and after the application of some antiseptic, the oyster is returned to the water. The surgical process is not only very delicate but it also require quickness because oyster should not be kept outside their normal environment for a long time. For this purpose the technicians are specially trained. Pearl oyster grow 90–150 mm across in about four to five years, after which they are ready for harvesting.

The pearl oysters, *Pinctata margaritifera* of the Gulf of California and *Pinctata vulgaris* of the Gulf of Kutch, produce superior quality pearls. It takes about 7 years to form a pearl of commerical value. Pearl fishery is now a regular industry.

QUESTIONS

Long Answer Questions

1. Describe the structure of ctenidum in *Unio* or in *Anodonta.*

2. Explain the process of gas exchange in a bivalve.

3. Give an account of pulmonary respiration in gastropods.

4. Describe the a quatic respiration in molluscs.

5. Explain the pulmonary respiration in *Pila*.

Short Answer Questions

1. What is respiration?

2. Which molluscs have pectinate gills?

3. Write a note on pulmonarium.

4. How do haemocyanin differ in molluscs and arthropods?

5. How are pearls produced?

6. Write a short note on pearl culture technique.

Very Short Answer Questions

1. Name the four major types of respiration found in the invertebrates.

2. List the features which make a surface fit for gas exchange.

3. Name the various types of gills found in bivalve molluscs. Give one example of each.

4. What is the role of respiratory pigment.

5. What is the name of common Indian pearl oyster?

Fill in the Blanks

1. Energy released in oxidation of food is stored in the high–energy bonds of .... molecules.

2. .... respiration occurs without using oxygen.

3. *Unio* has .... ctenidum.

4. Pulmonary respiration is characterisitc of ________ gastropods.

Multiple Choice Questions

1. Branchiae function in
   (*a*) water (*b*) air
   (*c*) both water and air (*d*) neither water nor air
2. True ctenidia has been replaced by secondary gills in
   (*a*) *Neopilina* and *Chiton* (*b*) *Doris* and *Patella*
   (*c*) *Unio* and *Pila* (*d*) *Nucula* and *Mytilus*
3. Ventilation current over ctenidia is produced by muscular action in
   (*a*) gastropods (*b*) pelecypods
   (*c*) cephalopods (*d*) none of the above
4. Pulmonarium found in some molluscs is a modification of
   (*a*) foot (*b*) mantle cavity
   (*c*) ctenidium (*d*) stomach
5. The pearl oyster secretes pearl to
   (*a*) regenerate injured parts (*b*) protect itself against invading parasites
   (*c*) harden its mantle cavity (*d*) isolate damage tissues of pearl
6. Whose secretion forms the pearl ?
   (*a*) prismatic layer (*b*) columnar epithelial cells of mantle
   (*c*) ciliated epithelial cells of mantle (*d*) connective tissue of mantle
7. The best quality of pearl is known as
   (*a*) real moti (*b*) sweta moti
   (*c*) lingha moti (*d*) none

ANSWERS

Very Short Answer Questions

1. Body surface, cutaneous, branchial and pulmonary.
2. Thinness, fluid film, rich vascularization and permeability to gases (oxygen, carbon dioxide).
3. Bipectinate in *Nucula*, filibranch in *Mytilus*, pseudolamellibranch in *Pecten*; lamellibranch in *Unio* and *Anodonta* and septibranch in *Poromya*.
4. To increase greatly the oxygen–carrying capacity of blood.
5. *Pinctada vulgaris*.

Fill in the Blanks

1. ATP (Adenosine triphosphate). 2. Anaerobic.
3. Lamellibranch. 4. Pulmonate.

Multiple Choice Questions

1. (*a*) 2. (*b*) 3. (*c*) 4. (*b*) 5. (*b*) 6. (*b*)
7. (*c*).

18

Torsion in Gastropods

Molluscs are typically bilaterally symmetrical animals but this symmetry is lost in Gastropoda due to two processes: **coiling** and **torsion.** Most gastropods live with the body coiled up in a spiral. In addition, while they are embryonic, the visceral mass is suddenly **torted** or **twisted** about, bringing the back end to front over the head. Most gastropods remain permanently torted while few partially untort during later growth.

18.1 DEFINITION OF TORSION

Torsion is the anticlockwise rotation of viscero-pallium at 180° from its initial position during embryonic/larval development of Gastropoda. As a result of this rotation, the posterior part of mantle cavity with its pallial complex, anus and rectum, etc., are brought to the front of body and anus comes to lie close to mouth. Entire process of torsion generally takes only a few minutes.

18.2 EFFECTS OF TORSION ON BODY STRUCTURE

The torsion introduces the following changes in the organization of *Pila* and other gastropods:

1. Displacement of mantle cavity. Before torsion the mantle cavity opens posteriorly. After torsion the mantle cavity opens in front just behind the head (Fig. 18.1).

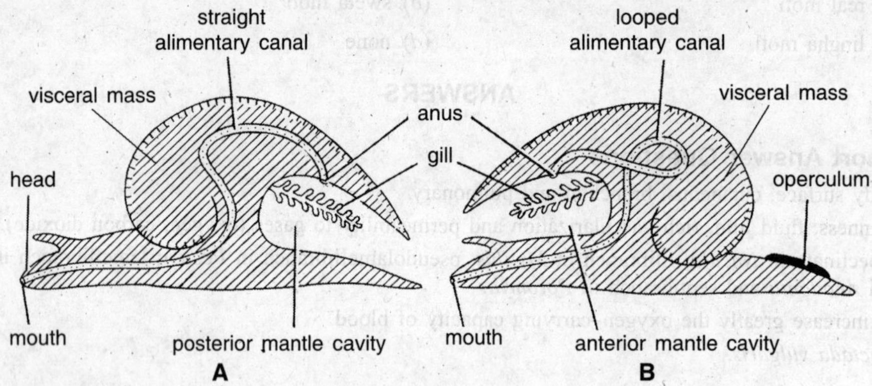

Fig. 18.1. Gastropod showing the phenomenon of torsion. A—Before torsion; B—After torsion.

2. Changes in relative position of organs. Before torsion, ctenidia point backwards, the auricles are behind the ventricle and mouth and the anus are at the opposite ends. After torsion, ctenidia come to lie in front and point anteriorly, the ctenidium of right side comes to lie on the left and that of left on the right; the auricles become anterior to the ventricle, the auricle of right side comes to lie on the left and vice versa.

3. Coiling of shell and visceral mass. The shell and visceral sac which are originally dorsal or **exogastric** become ventral or **endogastric** as a result of torsion.

4. Looping of alimentary canal. Primarily straight alimentary canal of larva becomes looped or 'U' shaped and posteriorly placed larval anus comes to occupy an anterior position.

5. Loss of symmetry. As a result of torsion original bilateral symmetry is lost due to displacement of pallial cavity and visceral sac.

6. Loss of organs or atrophy. Asymmetry of gastropods is also produced due to loss, reduction or atrophy of paired parts from one side or other. If pallial cavity comes on the right, the organs on the left side are degenerated leaving only the organ of right side.

7. Twisting of nervous system (Chiastoneury). Torsion introduces major changes in nervous system. As a result of torsion, except the visceral ganglia, all the ganglia and nerves come to occupy anterior position in body. Because of this pleural, pedal and buccal ganglia lie at the anterior end to form a nerve ring.

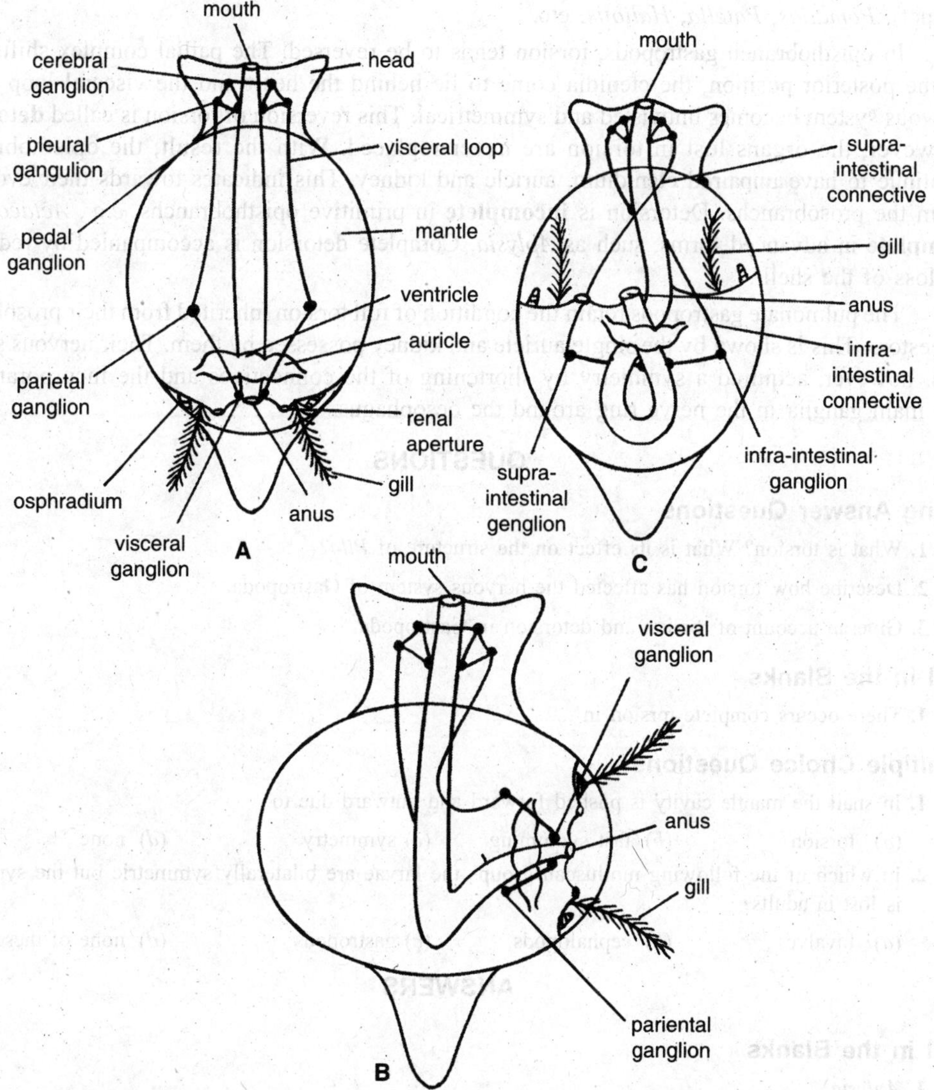

Fig. 18.2. Diagram to show position of different organs in body (A) before and after torsion(C).

The long pleuro-visceral connectives become twisted into figure of '8'. **Right connective** forms the **supra-intestinal nerve** and left connective forms **infra-intestinal nerve.** These nerves are present above and below (dorsal and ventral) to the oesophagus and join. Finally these are removed.

In some forms after the coiling of the visceral hump, torsion occurs by rotation only through 90°, so that the ctenidia and anus point laterally (Fig. 18.1B).

Significance. The anteriorly placed mantle cavity houses the head with sense organs, respiratory structures, etc., in adults due to occurrence of torsion; there is advantage. Thus, it causes an intimate contact for gaseous exchange with incoming water and also helps to test this water (**Morton** 1958).

18.3 DETORSION

Torsion occurs and persists as such only in prosobranch gastropods, *e.g.*, *Acmaea* (marine limpet), *Pomatias*, *Patella*, *Haliotis*, etc.

In opisthobranch gastropods, torsion tends to be reversed. The pallial complex shifts back to the posterior position, the ctenidia come to lie behind the heart and the visceral loop of the nervous system becomes untwisted and symmetrical. This reversion of torsion is called **detorsion**. However, the organs lost in torsion are never replaced. With the result, the opisthobranchs continue to have unpaired ctenidium, auricle and kidney. This indicates towards their evolution from the prosobranchs. Detorsion is **incomplete** in primitive opisthobranchs, *e.g.*, *Actaeon*, but **complete** in advanced forms, such as *Aplysia.* Complete detorsion is accompanied by reduction or loss of the shell.

The pulmonate gastropods retain the condition of full torsion inherited from their prosobranch ancestors. This is shown by the single auricle and kidney possessed by them. Their nervous system has, however, acquired a symmetry by shortening of the connectives and the incorporation of the main ganglia in the nerve ring around the oesophagus.

QUESTIONS

Long Answer Questions

1. What is torsion? What is its effect on the structure of *Pila*?

2. Describe how torsion has affected the nervous system of Gastropoda.

3. Give an account of torsion and detorsion in Gastropoda.

Fill in the Blanks

1. There occurs complete torsion in ....

Multiple Choice Questions

1. In snail the mantle cavity is pushed forward and outward due to

(*a*) torsion (*b*) fast swimming (*c*) symmetry (*d*) none

2. In which of the following molluscan groups the larvae are bilaterally symmetric but the symmetry is lost in adults?

(*a*) bivalves (*b*) cephalopods (*c*) gastropods (*d*) none of these

ANSWERS

Fill in the Blanks

1. *Aplysia.*

Multiple Choice Questions

1. (*a*) **2.** (*c*).

19

Phylum Echinodermata (Characters, Classification and Types)

Phylum **Echinodermata** (Gr., *echinos* = hedgehog; *derma* = skin) includes about 6000 species of marine **spiny skinned animals** such as starfishes, brittle stars, sea urchins, sea cucumbers, sand dollars and sea lilies. They distinctly have a unique **water vascular system** equipped with **podia** or **tube feet** which are used in locomotion and feeding. Echinodermata forms the only major group of **deuterostome** (Box 19.1) Invertebrata.

Box 19.1 Protostomia *vs.* Deuterostomia

1. **Protostomes** (Gr., *proto* = first; *stoma* = mouth). Animals in which **blastopore** (which is a groove of early gastrula through which cells move inside to form endoderm and mesoderm) develops into the mouth, *e.g.*, all metazoan invertebrates except echinoderms.
2. **Deuterostomes** (Gr., *deutero* = second; stoma = mouth). Animals in which blastopore becomes anus and mouth formation is secondary, *e.g.*, Echinodermata and Chordata.

19.1 DEFINITION

The echinoderms are triploblastic, unsegmented, enterocoelous coelomate animals having pentamerous radial symmetry (*i.e.*, biradial symmetry with an oral-aboral axis). Their larvae are bilaterally symmetrical. They contain an endoskeleton of calcareous plates or ossicles and a peculiar hydraulic ambulacral or water vascular system of coelomic origin. They have a complete alimentary canal, diffuse nervous system and no excretory system.

19.2 GENERAL CHARACTERS

1. The echinoderms are exclusively **marine,** free-living and mostly bottom-dwellers.
2. They are **triploblastic, coelomate** and **unsegmented** animals.
3. **Head** is absent but body is differentiated into **oral** and **aboral surfaces.**
4. Adults show **radial symmetry** (usually **pentamerous,** *i.e.*, body parts occurring in fives) but their larvae are bilaterally symmetrical.
5. They are of moderate to considerable size but none are microscopic.
6. The body wall consists of an outer layer of **epidermis,** a middle **dermis** and an inner lining of peritoneum.
7. **Endoskeleton** consists of closely fitted calcareous plates of mesodermal origin forming a shell, many called **theca** or **test** or may be composed of separate small ossicles. Skeleton bears projecting **spines** or tubercles that give the body surface a warty or spiny appearance. Some echinoderms such as sea urchins have long movable spines.

8. A part of embryonic coelom is modified into a unique or distinctive **water vascular system** which consists of an array of radiating canals (ambulacral grooves) and tube-like appendages called **tube feet.**
9. Due to presence of water vascular system, the oral surface remains divided into **ambulacra** and **inter-ambulacra (inter-radii).**
10. Coelom is spacious.
11. Digestive system is well developed. The coiled alimentary canal extends from the **mouth** on oral surface to the **anus** on the aboral surface.
12. Blood vascular system (also called **haemal** or **blood lacunar system**) is present but poorly developed.
13. Respiration occurs through **body wall, podia, respiratory tree** or **gills** (or **papulae**).
14. Excretory organs are absent.
15. Nervous system is simple and consists of networks concentrated into **circum-oral ring** and **radial ganglionated nerve cords.**
16. Sense organs are poorly developed.
17. Sexes are usually separate; sexual dimorphism is not well marked. Fertilization is external.
18. Development is indeterminate and indirect including characteristic larvae which undergo metamorphosis into the radially symmetrical adults.

19.3 OUTLINE CLASSIFICATION

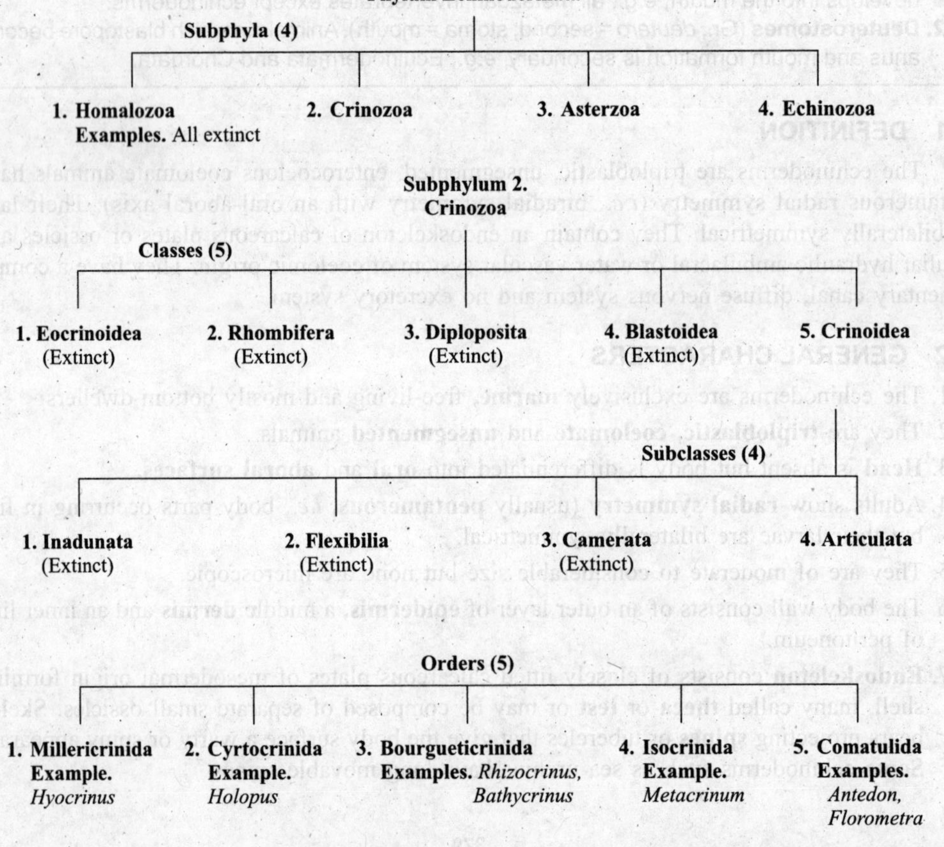

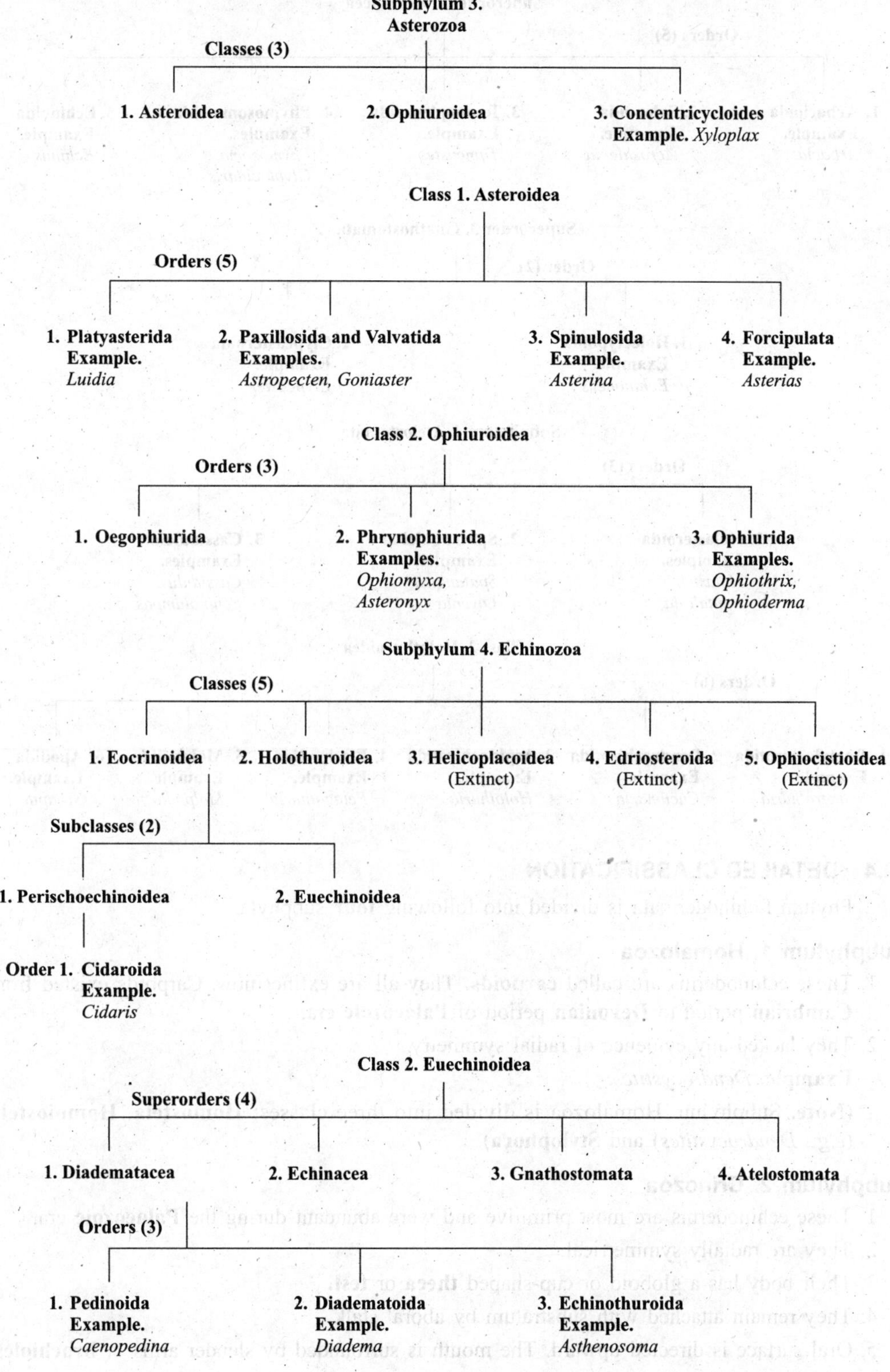
Subphylum 3.
Asterozoa
Classes (3)
1. Asteroidea
2. Ophiuroidea
3. Concentricycloides
Example. Xyloplax
Class 1. Asteroidea
Orders (5)
1. Platyasterida
Example.
Luidia
2. Paxillosida and Valvatida
Examples.
Astropecten, Goniaster
3. Spinulosida
Example.
Asterina
4. Forcipulata
Example.
Asterias
Class 2. Ophiuroidea
Orders (3)
1. Oegophiurida
2. Phrynophiurida
Examples.
Ophiomyxa,
Asteronyx
3. Ophiurida
Examples.
Ophiothrix,
Ophioderma
Subphylum 4. Echinozoa
Classes (5)
1. Eocrinoidea
2. Holothuroidea
3. Helicoplacoidea
(Extinct)
4. Edriosteroida
(Extinct)
5. Ophiocistioidea
(Extinct)
Subclasses (2)
1. Perischoechinoidea
2. Euechinoidea
Order 1. Cidaroida
Example.
Cidaris
Class 2. Euechinoidea
Superorders (4)
1. Diadematacea
2. Echinacea
3. Gnathostomata
4. Atelostomata
Orders (3)
1. Pedinoida
Example.
Caenopedina
2. Diadematoida
Example.
Diadema
3. Echinothuroida
Example.
Asthenosoma

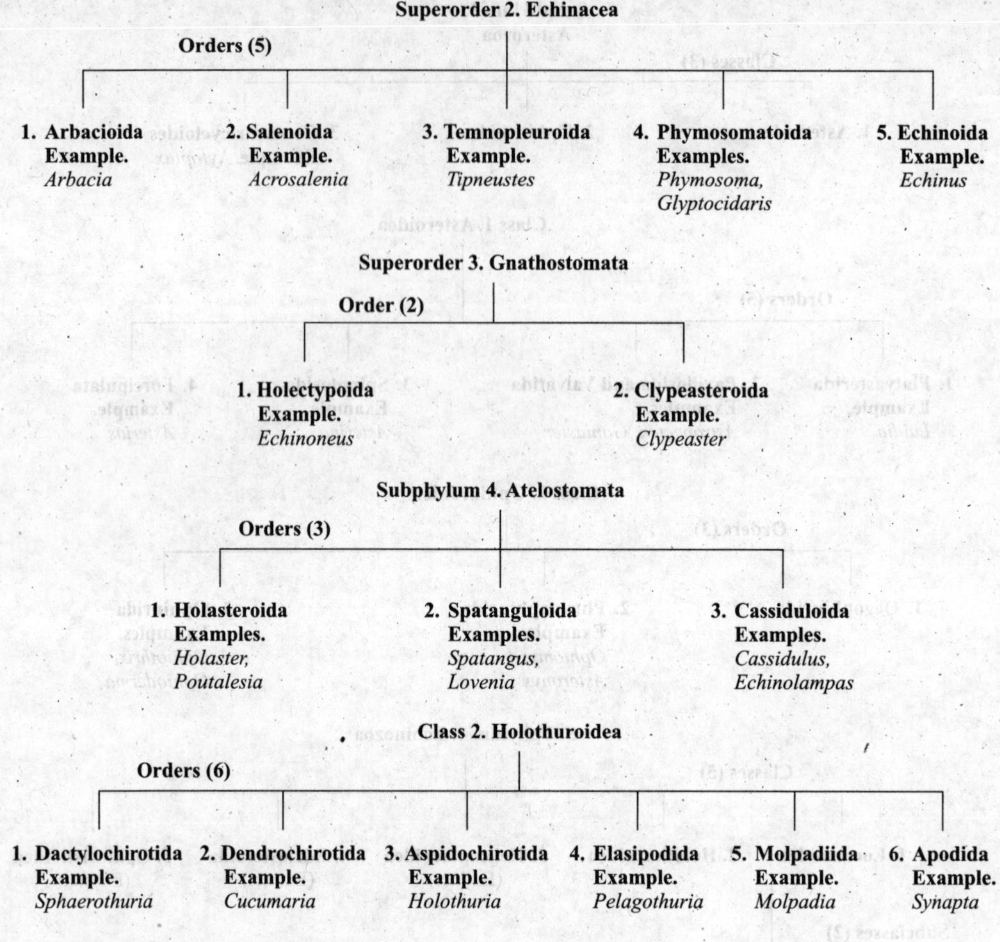

19.4 DETAILED CLASSIFICATION

Phylum Echinodermata is divided into following **four** subphyla.

Subphylum 1. Homalozoa

1. These echinoderms are called **carpoids.** They all are extinct now. Carpoids existed from **Cambrian** period to **Devonian** period of **Palaeozoic** era.
2. They lacked any evidence of radial symmetry.

   Example. *Dendrocystites*.

   (**Note.** Subphylum Homalozoa is divided into three classes: **Homostela, Homoiostela** (*e.g., Dendrocystites*) and **Stylophora**).

Subphylum 2. Crinozoa

1. These echinoderms are most primitive and were abundant during the **Palaeozoic** era.
2. They are radially symmetrical.
3. Their body has a globoid or cup-shaped **theca** or **test.**
4. They remain attached with substratum by aboral stalk.
5. Oral surface is directed upward. The mouth is surrounded by slender arms or **brachioles.**
6. This subphylum includes mostly extinct and some living echinoderms.

Subphylum 3. Asterozoa

1. Unattached (free moving) radially symmetrical, star-shaped echinoderms.
2. Body is composed of a flattened central disc and radially arranged arms or rays.

Subphylum 4. Echinozoa

1. Radially symmetrical globoid or discoid echinoderms.
2. Arms or brachioles are absent.
3. Mostly unattached (*i.e.*, free moving).

SUBPHYLUM 2. CRINOZOA

Subphylum Crinozoa contains the fossil eocrinoids (class **Eocrinoidea**), cystoids (class **Cystoidea**) and the fossil and living crinoids (class **Crinoidea**). Only the classification of the class Crinoidea is covered here.

Class Crinoidea

1. It includes both extinct and living forms.
2. Living members are without stalk and free-moving but extinct forms were attached by a stalk.
3. They are commonly called **sea lilies** and **feather stars.**
4. Sea lilies are attached animals while feather stars are unattached.
5. Pentamerous body of crinoids consists of an aboral cup, the **calyx** and oral cover or roof, the **tegmen.** Oral side or surface of body is directed upward.
6. Oral surface contains a central **mouth** and eccentric **anus.**
7. They contain five movable arms. Each arm is divided into two branches at the base so that there appears to be ten arms.
8. Arms are with or without **pinnules.** Pinnate arms bear a row of pinnules on each side of the ambulacral groove.
9. They contain five ambulacral grooves which are open and extend along each of the five arm and pinnules up to tip.
10. Spines and pedicellariae are absent.
11. Water vascular system is peculiar. There are about 1500 minute ciliated funnels which perforate the membrane that covers the oral surface. **Stone canal** is present but **madreporite** is absent.
12. Sexes are separate. Gonads are indefinite masses of gametes located in the arms or pinnules. **Genital pinnules** generally lack an ambulacral groove. Males shed sperm first; this then stimulates the female to shed the eggs.
13. Larva of crinoids is non-feeding barrel-shaped **vitellaria** or **doliolaria.**

    The class Crinoidea is divided into **four** subclasses.

Subclasses 1 to 3. Inadunata, Flexibilia and Camerata

1. These subclasses include extinct crinoids.
2. All were stalked Palaeozoic echinoderms.
3. They were with or without cirri.
4. Arms were branched, with or without pinnules.
5. Calyx was either (*i*) rigid, monocyclic or dicyclic and with lower arms free from it (Inadunata); (*ii*) flexible, dicyclic and with arms flexibly incorporated in it (Flexibilia); or (*iii*) rigid, mono—or dicyclic, lower arms incorporated into it (Camerata).

   Examples. *Hybocystites* (Inadunata); *Forbesiocrinus* (Flexibilia), *Xenocrinus* (Camerata).

Subclass 4. Articulata

1. Extinct as well as all the living crinoids belong to this subclass.
2. Calyx flexible, pentamerous and dicyclic. Lower arms incorporated into it.
3. Tegmen leathery containing calcareous particles (ossicles) or small plates.
4. Mouth and ambulacral grooves exposed (*i.e.*, they are not covered with tegminal plates).

Living members of subclass Articulata are included in the following **five** orders:

Order 1. Millericrinida

1. This order includes **sea lilies** which lack cirri.
2. They are attached crinoids. Stalk extends upto 1 metre in length.

Examples. *Hyocrinus*, *Calamocrinus*.

Order 2. Cyrtocrinida

1. Includes **sea lilies** without stalk.
2. Aboral end of crown is attached directly to the substratum.
3. There are only two living species found in Caribbean and mid-Atlantic seas.

Example. *Holopus*.

Order 3. Bourgueticrinida

1. Includes small **sea lilies.**
2. Stalk is slender and without cirri.

Examples. *Rhizocrinus*, *Cenocrinus*, *Bathycrinus*.

Order 4. Isocrinida

1. Includes **sea lilies** with cirri.
2. Cirri are short, tendril-like and present aborally.

Examples. *Metacrinum*, *Cenocrinum*.

Order 5. Comatulida

1. Includes **feather stars.**
2. Unattached stalkless crinoids.

Examples. *Antedon*, *Florometra*, *Neometra*, *Capillaster*.

SUBPHYLUM 3. ASTEROZOA

Subphylum Asterozoa is divided into following **three** classes.

Class 1. Asteroidea

1. Body flattened, pentagonal or star-shaped.
2. They are commonly called **starfish** or **sea stars;** there are 1500 species of these.
3. Oral and aboral surface are distinct. The oral surface is directed downward and aboral surface upward.
4. Arms are wide and not sharply set-off from the central disc.
5. Ambulacral grooves are open and have podia or tube feet.
6. A large coelomic cavity is present in the selectively wide arms.
7. Endoskeleton is flexible, made of separate ossicles.
8. Development is indirect and includes **bipinnaria** or **brachiolaria larva.**

Class Asteroidea includes following **five** orders.

Order 1. Platyasterida

1. They are primitive and have tube feet without suckers.
2. Mostly extinct. Only two living genera.

Examples. *Platyasterias*, *Luidia*.

Order 2 and 3. Paxillosida and Valvatida (= Order Phanerozonia)

These two orders were formerly united within the order **Phanerozonia.**

1. Tube feet either with suckers (Paxillosida) or without suckers (Valvatida).
2. Both orders contain arms with two rows of marginal plates.
3. Pedicellariae of the sessile type (without stalk).
4. Podia or tube feet are arranged in two rows.

Examples. *Astropecten*, *Pentaceros*, *Culcita*, *Ctenodiscus*, *Goniaster* (cushion star).

Order 4. Spinulosida

1. Arms with small marginal plates.
2. Pedicellariae are absent.
3. Aboral surface is covered with short spines.
4. Tube feet are in two rows and contain suckers.

Examples. *Asterina*, *Echinaster*, *Henricia*, *Acanthaster*.

Order 5. Forcipulata

1. Marginal plates inconspicuous or absent.
2. Pedicellariae pedunculate type having a short stalk and three ossicles.
3. Tube feet are arranged in four rows and provided with suckers.

Examples. *Asterias*, *Heliaster*, *Zoroaster*, *Pycnopodia* (many-rayed star), *Leptasterias*.

Class 2. Ophiuroidea

1. This class includes **brittle stars,** of which there are about 2000 species.
2. They contain extremely long five or more arms which are sharply set off from the central axis.
3. Arms largely filled with vertebral ossicles.
4. They lack ambulacral grooves, intestine and anus.
5. Madreporite is present on oral surface.
6. Tube feet play no role in locomotion except in *Ophionereis annulata*.
7. Podia help in food collection and transport.
8. Development is indirect and includes a freely swimming pluteus larva.

Class Ophiuroidea includes following **three** orders.

Order 1. Oegophiurida

1. Mostly extinct; only one living species is present.
2. Dorsal and ventral shields or **bursae** are absent.
3. Madreporite exists at the edge of the central disc.

Example. An Indonesian species.

Order 2. Phrynophiurida

1. It includes **brittle stars** and **basket stars.**
2. Dorsal arm shields are absent.

3. In primitive brittle stars (suborder **Ophiomyxina**), disc and arm plates are covered by a thick and soft skin.
4. In basket stars, arms are simple, branched and capable of coiling vertically (suborder **Euryalina**).

Examples. Brittle stars such as *Ophiocanops*, *Ophiomyxa*; and **Basket stars** such as *Asteronyx*, *Gorgonocephalus*.

Order 3. Ophiurida

1. It includes **brittle stars** and **serpent stars.**
2. Mostly small ophiuroids, usually with five simple arms.
3. Arms are capable of transverse movement only.
4. Dorsal arm shields are present.

Examples. *Ophioderma*, *Ophiothrix*, *Ophioscolex*, *Ophiomaza*, *Ophionereis*.

Class 3. Concentricycloides

1. Minute deep-water echinoderms (which were collected in 1983 and 1984 from coast of New Zealand and described in 1986 by **Baker** *et al.*).
2. Body is disc-shaped and covered aborally with plate-like ossicles.
3. Gut is present or absent.
4. Water vascular system have two ring canal with podia arising from the outer one.

Example. *Xyloplax* (Fig. 19.1).

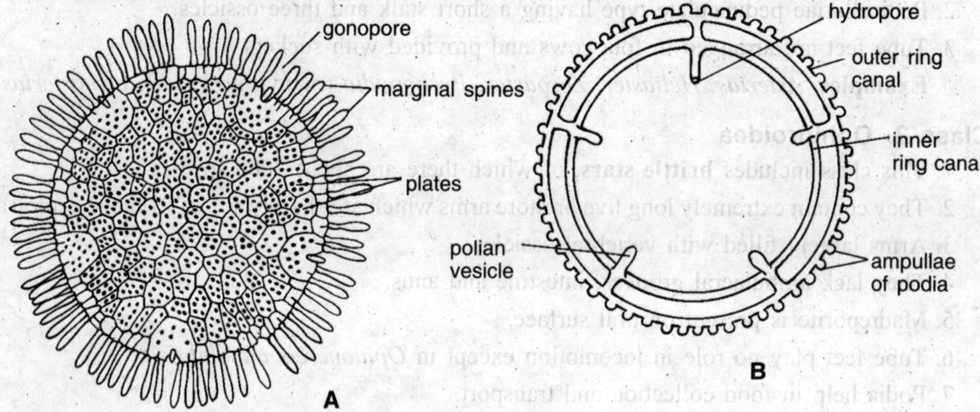

Fig. 19.1. *Xyloplax* (Class Concentricycloides). A—Dorsal view; B—Water vascular system.

SUBPHYLUM 4. ECHINOZOA

Subphylum Echinozoa is divided into **five** classes; three of which namely class **Helicoplacoidea** (*e.g.*, *Helicoplacus*), Class **Edrioasteroidea** (*e.g.*, *Lepidodiscus*) and class **Ophiocistoidea** (*e.g.*, *Volchovia*) include all extinct Echinozoa. The rest two classes namely **Echinoidea** and **Holothuroidea** include living as well as extinct members.

Class 1. Echinoidea

1. It includes about 950 species of **sea urchins, heart urchins** and **sand dollars.**
2. The name Echinoidea means "like a hedgehog". It is called so because it contains movable spines (like the hedgehog) that covers the body.

3. Body is spherical, disc-shaped, oval or heart-shaped and lacks arms.
4. Body is enclosed in an endoskeletal shell or test of closely fitted calcareous plates covered with movable spines.
5. Podia or tube feet come out from the pores of ambulacral plates and are locomotory in function.
6. **Mouth** is centrally placed on oral surface. **Anus** is located at the aboral pole.
7. Mouth is generally provided with an elaborate jaw apparatus (Aristotle's lantern; Box 19.2).

Box 19.2 Aristotle's Lantern

Aristotle's lantern is a unique and highly developed scraping apparatus of sea urchins. It is associated with the digestive system (*i.e.*, mouth). It has been named so after its discoverer and its resemblance to the ancient Greek ship-lantern. The majority of sea urchins are grazers, scraping the substratum surface on which they live with their teeth which are operated by Aristotle's lantern.

Lantern is pentamerous inverted pyramid-like structure lying close to the peristome (Fig. 19.2). It is composed of five large, calcareous plates called **alveoli** or **pyramids,** each of which is shaped somewhat like an arrowhead with the point projecting toward mouth. The pyramids are arranged radially, with each side connected to that of adjacent pyramid by means of transverse muscle fibres (called **comminator muscles**). Passing down the midline along the inner side of each pyramid is a long, calcareous band. The oral end of the band projects beyond the tip of the pyramid as an extremely hard, pointed **tooth.** Because there is one such tooth band for each pyramid, there are five teeth projecting from the oral end of the lantern. The curled upper end of the band is enclosed within a dental sac and is the area of new tooth formation.

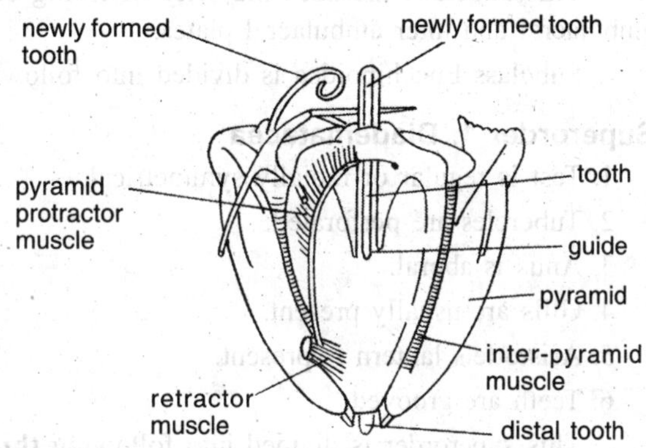

Fig. 19.2. Aristotle's lantern of echinus in lateral view.

In addition to the teeth and pyramids, Aristotle's lantern is composed of a number of smaller, rod-like pieces at the aboral end. The upper end of each pyramid bears a bow-shaped bar, the **epiphysis.** It is formed by the fusion of two ossicles. Five slender bars of radii, called **compasses**, extend radially inwards from the point of union of adjacent epiphysis. Five more pieces, the **rotules,** are also present beneath the compasses. By mean of special muscles (*e.g.*, protractor muscles and retractor muscles) the lantern can be partially protruded and retracted through the mouth. Other muscles control the teeth, which can be opened and closed. Aristotle's lantern is used in scraping and also in pulling and tearing a wide range of plant and animal materials.

8. They contain peripharyngeal coelom.
9. Ambulacral grooves are absent.
10 Pedicellariae (pinching organs) are stalked and three jawed.
11. They contain five pairs of gills for respiration.
12. Sexes are separate. Gonads are pentamerous.
13. Development is indirect and includes free swimming **echinopluteus larva.**

Class Echinoidea is divided into **two** subclasses.

Subclass 1. Perischoechinoidea

Subclass Perischoechinoidea includes mainly primitive fossil urchins of the Palaeozoic seas. It is divided into four orders. Members of three orders are completely extinct now. **Cidaroida** is the only one (order) that survived the Palaeozoic era and became the ancestor of the remaining echinoids, most of which belong to the next subclass.

Order Cidaroida

1. Test is rigid having two rows of narrow and simple plates for each ambulacrum and interambulacrum.
2. Spines are widely separated and each is encircled by small secondary spines.
3. Gills and balancing organs are absent.
4. Tridentate and globiferous pedicellariae are present.

Examples. *Eucidaris*, *Cidaris*, *Notocidaris*.

Subclass 2. Euechinoidea

This subclass includes majority of living echinods. Its members have two columns of ambulacral and inter ambulacral plates.

Subclass Euechinoidea is divided into following **four** superorders.

Superorder 1. Diadematacea

1. Test is regular or radially symmetrical.
2. Tubercles are perforated.
3. Anus is aboral.
4. Gills are usually present.
5. Aristotle's lantern is present.
6. Teeth are grooved.

This superorder is divided into following **three** orders.

Order 1. Pedinoida

1. Test is rigid and with spines.
2. Ambulacral plates ending at peristome margin.
3. Gill notches are slight.
4. Ten buccal plates exist on the peristomial membrane.

Examples. *Caenopedina* (only living genus of order Pedinoida).

Order 2. Diadematoida

1. Rigid or flexible test is globose having two rows of intrambulacral plates and two rows of compound ambulacral plates, both ending at peristome margin.
2. Spines are hollow and bear fine teeth.
3. All types of pedicellariae such as dentate, foliate and ophiocephalous (except globiferous pedicellariae) are present.
4. Ten buccal plates are present on peristomial membrane.

Examples. *Diadema*, *Plesiodiadema*.

Order 3. Echinothuroida

1. Flexible test of separated plates; ambulacral plates with peristome extending to mouth rim.
2. Simple plates are present on peristomial membrane.
3. Poisonous secondary spines are present.
4. Gills are absent.

Examples. *Asthenosoma*, *Echinothuria*.

Superorder 2. Echinacea

1. It includes sea urchins having radial symmetry, rigid test and solid spines.
2. Gills are present.
3. Peristomimal membrane with ten buccal plates.
4. Aristotle's lantern is well developed.

This superorder is divided into following **five** orders.

Order 1. Arbacioidia

1. Preriproct is covered by four or five anal plates.
2. Primary tubercles are without pores.
3. Compound ambulacral plates are composed of a median primary plate and a short plate on either side.

Example. *Arbacia.*

Order 2. Salenoida

1. Stirodont with epiphyses separate and teeth crested on inner surface.
2. Interambulacral plates with single spine.
3. Anus is not centrally located. It is situated within periproct due to the presence of a large plate (called **suranal plate**).

Example. *Acrosalenia.*

Order 3. Temnopleuroida

1. Lantern is camarodont (*i.e.*, large epiphyses are joined across pyramid top).
2. Test is sculptured in some.
3. Ambulacral plates are compound (*i.e.*, each plate is composed of two primary plates and one narrow plate between their narrow ends.)

Examples. *Tripneustes*, *Toxopneustes*, *Temnopleurus*.

Order 4. Phymosomatoida

1. Compound ambulacral plates are three; interambulacral plates with more than one tubercle.
2. Primary tubercles are without pores.
3. Lantern is camarodent.

Examples. *Phymosoma*, *Glyptocidaris*.

Order 5. Echinoida

1. Test is not sculptured.
2. Lantern is camarodent.
3. Compound ambulacral plates are composed of two primary plates and one narrow plate between their outer ends.
4. Three to sixteen pores per plate occur.
5. Gill notches are shallow.

Examples. *Echinus*, *Echinometra*, *Paracentrotus*, *Strongylocentrotus*.

Superorder 3. Gnathostomata

1. It includes irregular (bilaterally symmetrical) urchins.
2. Lantern is present.

3. Ambulacral plates are simple or weakly compound.
4. Teeth are keeled.
5. Mouth is in the centre of oral surface but anus has shifted out of centre.

This superorder is divided into following **two** orders.

Order 1. Holectypoida

1. Test is regular with apical plates centred at the aboral pole.
2. Peristome central.
3. Anus is somewhat separate from apical plates.
4. Ambulacra are not petaloid; they are narrow on oral surface.
5. Lantern teeth with lateral flanges.
6. Mostly extinct.

Examples. *Echinoneus* and *Micropetalon* are only living members.

Order 2. Clypeasteroida

1. It includes **cake urchins** or **sand dollars.**
2. Test is greatly flattened and oval or circular. Test is usually densely covered by small spines.
3. Peristome and apical system are usually central.
4. Petaloid ambulacra, *i.e.*, ambulacral areas are shaped like a petal radiating from the centre, are present.
5. Teeth are without lateral flanges.
6. Phyllodes (= oral ambulacral areas) and gills are absent. Respiration is performed by podia of petaloids.
7. Some sand dollars (*e.g.*, keyhole sand dollars or *Mellita*) contain large elongated notches or openings in testa, called **lunules** (Fig. 19.3).

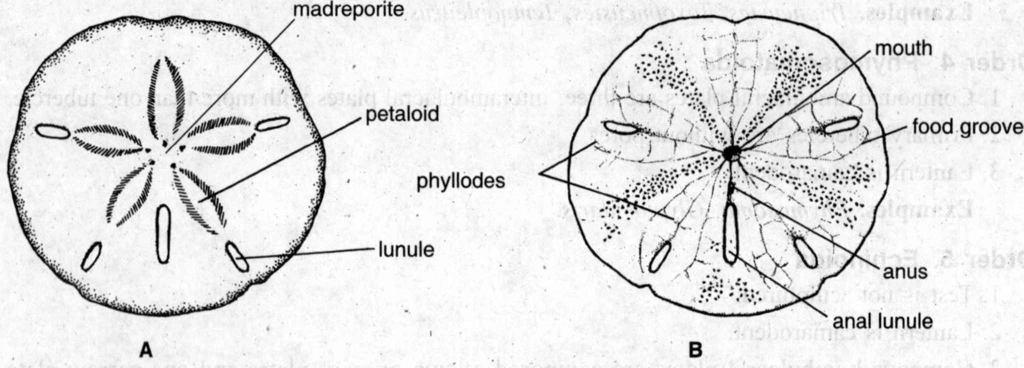

Fig. 19.3. *Mellita quinqniesperforata*, the give slotted sand dollar of the Atlantic coast of USA. A– Dorsal view of test ; B– Ventral (oral) view of test (after Ruppert *et. al.* 2004).

Examples. *Clypeaster* (sea biscuits), *Fibularia*, *Rotula*, *Mellita*.

Superorder 4. Atelostomata

1. It includes regular (bilaterally symmetrical) urchins.
2. Lantern is absent.

3. Ambulacral plates are simple.
4. Test is rigid.

This suborder is divided into following **three** orders.

Order 1. Holasteroida

1. Oval or bottle-shaped urchins with thin delicate test.
2. Petaloids and phyllodes are not developed.
3. Ambulacra of aboral surface are petal-shaped; petals are not similar.

Examples. *Holaster*, *Pourtalesia* (a deep-water species).

Order 2. Spatanguloida

1. It includes **heart urchins.**
2. Test is oval or cordiform, bilaterally symmetrical about long axis. It contains short spines.
3. Ambulacra of aboral surface is petal-shaped; petals are not similar. Petaloids are shunken.
4. Oral centre is shifted anteriorly and anus is shifted out of aboral apical centre.
5. Phylodes are present.
6. Gills are absent.

Examples. *Spatangus*, *Echinocardium*, *Moira*, *Meoma*, *Lovenia.*

Order 3. Cassiduloida

1. Somewhat similar to heart urchins.
2. Body is round or oval.
3. Ambulacra of aboral surface are petal-shaped; all petals are similar.

Examples. *Cassidulus*, *Echinolampas.*

Class 2. Holothuroidea

1. It includes 900 species of **sea cucumbers.**
2. Warty body is elongated along the oral-aboral axis. As a result animal lies with the side of body on the substratum.
3. Endoskeleton is very much reduced. It is represented by microscopic **ossicles.**
4. Oral or buccal podia form a circle of **tentacles** (10 to 30) around the mouth (See Box 19.3).

Box 19.3 Tentacles in Holothuroidea

Oral tentacles of Holothuroidea may be of following four types:

1. **Digitate** or finger-like, *e.g.*, *Molpadia*, *Deima*,
2. **Dendritic** or arborally branched, *e.g.*, *Thyone*.
3. **Pinnate** or with side branches on a central axis, *e.g.*, *Leptosynapta*.
4. **Peltate** or flattened, with a short stalk and terminal disc, often with horizontal branches, *e.g.*, *Actinopyga*, *Peniagone*.

5. Mouth and anus are situated at the opposite ends. Arms are absent. Cloaca contains respiratory tree.
6. Ambulacra are reduced.
7. No ossicle form the jaw apparatus.
8. Tiedman's bodies are absent.
9. Some move on the bottom; others live beneath stones or in crevices.
10. They are deposit–and suspension feeders.

11. Madreporite is located in the coelom.
12. Development is indirect including **doliolaria** larva.

Class Holothuroidea is divided into following **six** orders.

Order 1. Dactylochirotida

1. It includes primitive sea cucumbers.
2. Eight to thirty digitiform (finger-like) tentacles are present.
3. Body is U-shaped and enclosed within a flexible test of imbricating plates.

Examples. *Sphaeothuria*, *Echinocucumis*.

Order 2. Dendrochirotida

1. Buccal podia or tentacles are 10 to 30 in number. They are branched like a tree (dendritic) and not provided with ampullae.
2. Podia (tube feet) occur on the sole, on all the ambulacra, or over the entire surface.

Examples. *Cucumaria*, *Thyone*, *Psolus*.

Order 3. Aspidochirotida

1. Oral tentacles are peltate or shield-like.
2. Many podia are present. They sometimes form a sole.
3. Respiratory trees are present.

Examples. *Holothuria*, *Actinopyga*, *Stichopus*.

Order 4. Elasipodida

1. It includes deep-sea cucumbers which contain large conical papillae and podia.
2. Tentacles are peltate.
3. They lack oral retractor muscles and respiratory trees.

Examples. *Pelagothuria*, *Peniagone*.

Order 5. Molpadiida

1. Posterior end of body tapers to form a tail.
2. Fifteen digitiform tentacles are present.
3. Podia are absent or reduced to anal papillae.
4. Respiratory tree is absent.

Examples. *Molpadia*, *Caudina*.

Order 6. Apodida

1. Worm-like sea cucumbers.
2. They contain only buccal podia or tentacles. Tentacles are digitiform or pinnate.
3. Water vascular system is reduced, lacking tube feet or radial canals.
4. Respiratory tree is absent.

Examples. *Synapta*, *Leptosynapta*, *Euapta*.

19.5 TYPE STUDIES

1. *ANTEDON*

Classification

Phylum	—	**Echinodermata**
Subphylum	—	**Crinozoa**

Class	—	**Crinoidea**
Subclass	—	**Articulata**
Order	—	**Comatulida**
Genus	—	***Antedon***

Habit and Habitat

Antedon is commonly called **feather star.** It is marine and free living crinoid that lives in littoral zones of the sea. Its seven species are found at moderate depths in all seas such as seas of Western Europe, Western Africa, Mediterranean, and West of Tropical America.

Antedon is a gregarious form and clings to the sea bottom for long periods by its cirri. However, it can creep slowly through its cirri and swim by undulating movements (*i.e.*, bending and straightening) of the arms. *Antedon* swims and crawls only for short distances. Swimming is mainly an escape response of *Antedon*.

Antedon is **nocturnal.** It hides in the crevices in coral reefs during day and keeps its arms tightly rolled. At sunset, it crawls upwards and extends it arms to feed. It is a **suspension feeder,** feeding on microscopic living organisms such as diatoms, algae, protozoa and small plankton (see Box 19.4). *Antedon bifida* feeds on the detritus.

Box 19.4 Mode of Feeding in *Antedon*

A hungry feather star (*Antedon*) clings to a rock with its arms spread wide and its pinnules at attention, forming an interlocking net to filter out small food particles carried in passing food currents.

The epidermis is thin over most of body surface but its cells are tall and ciliated in the ambulacral grooves. Food is trapped in mucus and the mucous strings move along the ambulacral groove like conveyer belts (Fig. 19.4). Interambulacral cilia on the tegmen surface cause cleansing currents that sweep towards the grooves, turn and run peripherally just outside the grooves.

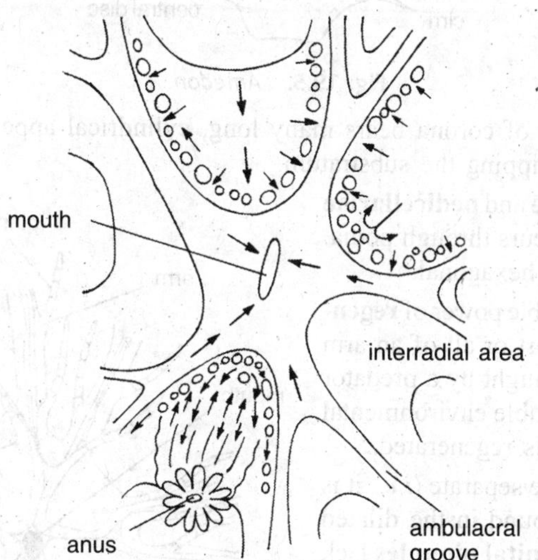

Fig. 19.4. *Antedon*. Ciliary currents on oral surface. Note the cleansing currents and food-collecting currents in food grooves.

Structure

Antedon shows a pentamerous **radial symmetry.** It consists of a **central disc** and a series of five radiating **arms.** The main part of body is called **corona** which is differentiated into an upper **oral surface** and lower **aboral surface.** It contains a cup-shaped aboral **calyx** which is formed of loose-fitting skeletal ossicles. On the oral side of disc occurs a membrane containing small skeletal ossicles, called **tegmen.** Tegmen bears the central **mouth** and the **anus** on a papilla (called **anal cone**) in an inter radius. Five long and slender flexible **arms** arise at the junction of calyx and tegmen. Each arm immediately bifurcate to form in total ten arms. Diminutive branches called **pinnules** are attached to the arm margins to give them a feathery appearance.

The skeleton of arms and pinnules is jointed; **brachial ossicles** support the arms, and **pinnulars** support the pinnules. Five ciliated **ambulacral grooves** radiate from the mouth towards the arms (Fig. 19.5) where each divides into two and the branches extend along the oral surface of the arms and pinnules. The margins of the grooves are bordered by movable flaps, called **lappets** which can expose or cover the groove. Two rows of tentacle-like **tube-feet** or **podia** lacking suckers (= ampulla) are present along the margins of ambulacral grooves. Podia are used to secrete mucus and to push food into the ambulacral groove.

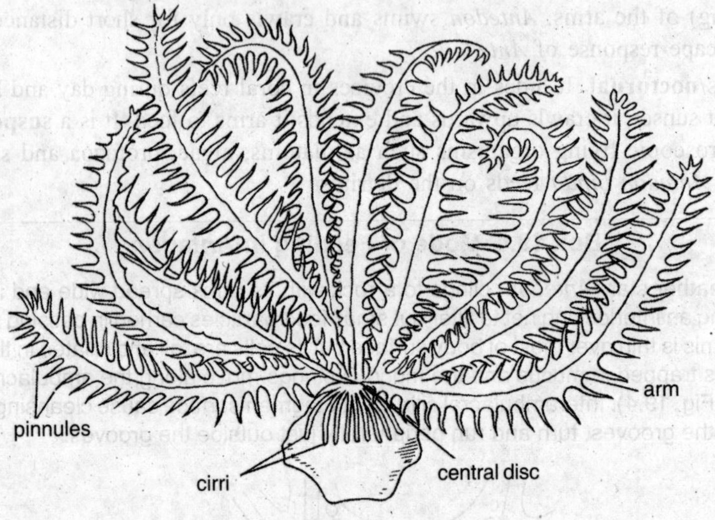

Fig. 19.5. *Antedon.*

The **aboral surface** of corona bears many long, cylindrical appendages or the cirri. Cirri are used by animal for gripping the substratum.

Spines, madreporite and **pedicellariae** are absent. Respiration occurs through podia. Ring canal of *Antedon* is hexagonal.

Antedon has remarkable power of **regeneration** and **autotomy.** Part or all of an arm can be shed if animal is caught by a predator or if subjected to unfavourable environmental conditions. The lost arm is regenerated.

In *Antedon*, sexes are separate (*i.e.*, it is **dioecious**). Gonads are found in the dilated bases of pinnules. Such **genital pinnules** lack an ambulacral groove. **Fertilization** is external and development is **indirect.** In *Antedon*, the eggs are cemented to the outer surface of the pinnules by the secretion of epidermal **cement gland** cells. The larva of *Antedon* is a non-feeding barrel-shaped **vitellaria** larva which develops into a stalked and sessile **pentacrinoid.**

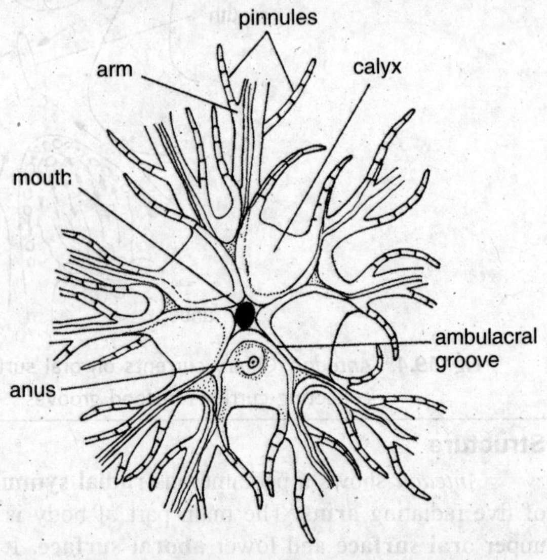

Fig. 19.6. *Antedon.* Oral view of central disc.

2. *ECHINUS* (Sea Urchin)

Classification

Phylum	—	**Echinodermata**
Subphylum	—	**Echinozoa**
Class	—	**Echinoidea**
Subclass	—	**Euechinoidea**
Superorder	—	**Echinacea**
Order	—	**Echinoda**
Genus	—	***Echinus***

Habit and Habitat

Echinus is commonly called **sea urchin** (other common sea urchins are *Arbacia, Strongylocentrotus*, etc.) It is marine and found in the sea in the rocky places. *Echinus* occurs from intertidal zone to a depth of 5000 metres. It is widely distributed in the Atlantic, Mediterranean and Pacific Oceans. In India, sea urchin (*Salmacis bicolor*) is found along Madras Coast.

Echinus moves slowly by means of highly movable spines and tube feet. It sometimes excavates holes in the rocks for hiding. It cleans its body by movement of spines and pedicellariae. *Echinus* feeds on sea weeds, algae, dead animal matter and small organisms. It itself is food for starfishes, crabs, fishes, birds, mammals (including human beings).

Structure

Body of *Echinus* is globular, somewhat flattened at oral pole. It exhibits **radial symmetry.** Body is divided into an **aboral** and an **oral hemisphere.** The oral end bears a centrally located **mouth** and is directed against the substratum. Through the mouth project five teeth of Aristotle's lantern (see Box 19.5). Mouth is surrounded by a soft membrane known as **peristome.** Peristomal membrane bears five pairs of short, heavy, modified **tube feet** or **podia,** called **oral tentacles.** It also bears five pairs of bushy **gills** or **branchiae. Anus** lies in the centre of **aboral surface.** The membrane surrounding the anus is called **periproct.** It is a small, circular membrane having a number of plates.

Box 19.5 Aristotle's lantern of *Echinus*

Sea urchins feed with a highly developed scraping apparatus called Aristotle's lantern (Fig. 19.7B).

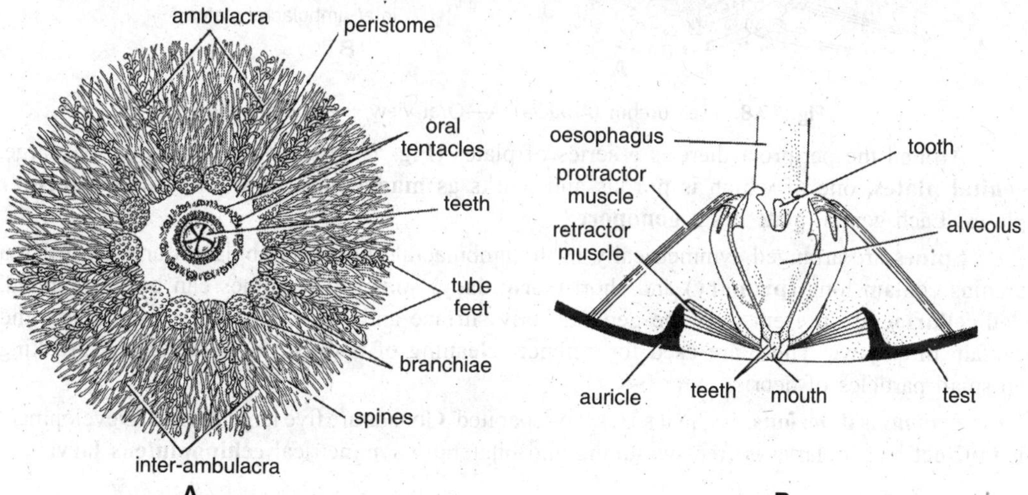

Fig. 19.7. *Echinus*. A—Oral view; B—Aristotle's lantern in situ.

It consists of (*i*) five arrow head-like calcareous plates called **pyramids;** (*ii*) five radial **teeth;** (*iii*) five **epiphyses;** (*iv*) five **rotulae;** and (*v*) five **compasses.** These 25 plates surround the oesophagus and enclose a **lantern coelom.** Various plates are moved by intricate system of muscles. Muscles such as **retractor muscle** and **protractor muscle** also extend between the lantern and five arches arising from the corona, called **auricles.**

The body of *Echinus* is enclosed in a rigid globular shell or **corona.** Corona is formed of closely fitting endoskeletal calcareous plates bearing round tubercles. The surface of corona is divided into five **ambulacral regions** alternating with five **interambulacral regions.** In each ambulacral region, there are two rows of tube feet.

In corona, **skeletal plates** or **ossicles** are arranged in rows. These rows run from oral pole to aboral pole. Each ambulacral area is composed of two rows of **ambulacral plates** and each interambulacral area is composed of two rows of **interambulacral plates.** There are thus 20 rows of plates. The edges of plates fit accurately and firmly together and form a rigid **shell, test** or **corona** around the body. The surface of plates contain **tubercles** for the articulation of movable **spines.** Each ambulacral plate bears pores through which tube feet are extended.

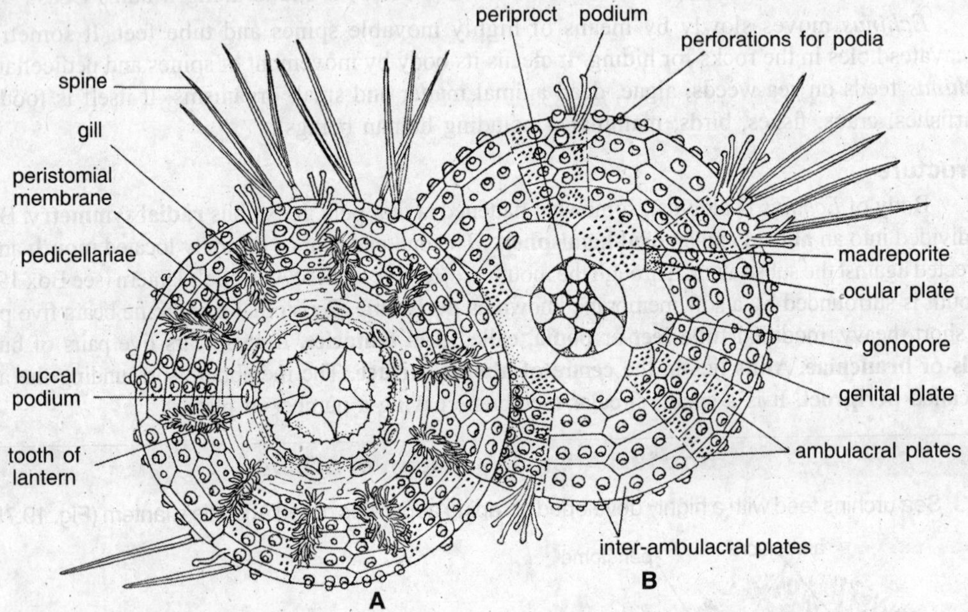

Fig. 19.8. Sea urchin (*Arbacia*). A—Oral view; B—Aboral view.

Around the periproct there is a series of plates (Fig. 19.8B). These consist of five large, **genital plates,** one of which is porous and works as **madreporite** and five smaller **occular plates.** Each genital plate has a **gonopore.**

Spines are arranged symmetrically in the ambulacral and interambulacral areas. Most sea urchins contain long **(primary)** and short **(secondary)** spines. The spines can be regenerated. **Pedicellariae** are present over the general body surface and peristome. They are stalked and contain three jaws. These are used for defence, cleaning of body surface, biting and breaking up small particles of debris.

Echinus is **dioecious,** *i.e.*, in it sexes are separated. Gonads are five large masses. Development is **indirect** and its larva is free-swimming and bilaterally symmetrical **echinopluteus larva.**

3. *CLYPEASTER*
(Cake urchin)

Classification

Phylum	—	**Echinodermata**
Subphylum	—	**Echinozoa**
Class	—	**Echinoidea**
Subclass	—	**Euechinoidea**
Superorder	—	**Gnathostomata**
Order	—	**Cypeasteroida**
Genus	—	***Clypeaster***

Habit and Habitat

Clypeaster is commonly known as **cake urchin.** It is a marine form, found in tropical and subtropical waters—Red sea and North Carolina to Brazil. *Clypeaster* is found buried in sand at the bottom in the sea. The sea biscuit, *Clypeaster rosaceus* of Florida and the West Indies, does not burrow but sits on the sea bottom. It is a selective **deposit feeder** feeding on organic debris.

Structure

Body of *Clypeaster* is flattened and oval or disc-shaped. Test is densely covered with small dark brown **spines** which help in locomotion. Body is differentiated into **oral** and **aboral surfaces.**

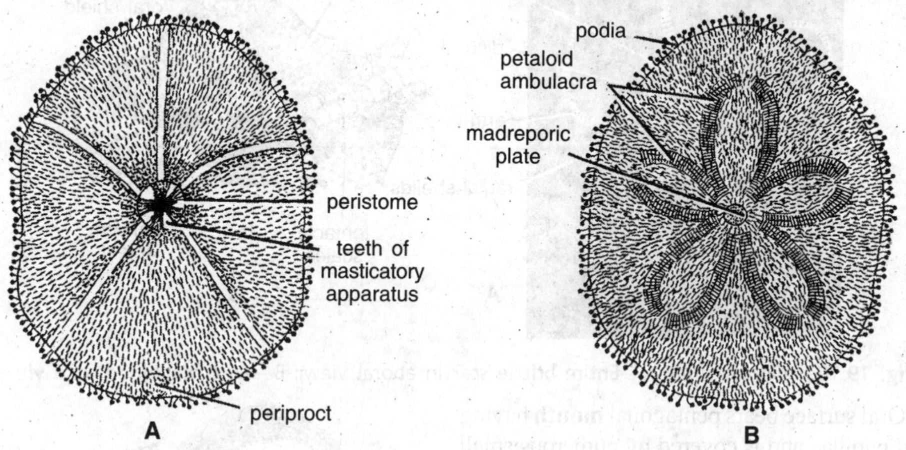

Fig. 19.9. *Clypeaster*. A—Oral view; B—Aboral view.

Mouth and anus both occur on oral surface. Mouth is central in position and anus is eccentric. Tube feet or podia are found on both oral and aboral surface.

The madreporite is central on aboral surface. From madreporite arise five wide and well marked **petaloid ambulacral areas** separated by **interambulacral areas.** In each ambulacral region, there are two rows of **tube-feet.** The podia of petaloids are modified for gas exchange.

The **oral ambulacral areas** or **phyllodes** contain modified podia for obtaining food particles. **Pedicellariae** and **sphaeridia** (ovoid sensory bodies) are present. Aristotle's lantern is present. Peristomial gills are absent.

Clypeaster is **dioeceous.** Development is indirect and includes **echinopleuteus larva.**

4. *OPHIODERMA*
(Brittle star)

Classification

Phylum	—	**Echinodermata**

Subphylum	—	**Asterozoa**
Class	—	**Stelleroidea (Ophiuroidea)**
Order	—	**Ophiurida**
Genus	—	**1. *Ophioderma;***
		2. *Ophiothrix.*

I. Ophioderma

Ophioderma is commonly called **brittle star.** It is marine and found in shallow water and at a depth of 280 metres. *Ophioderma* is littoral, tropical and mainly found in West Indies. It is free swimming, nocturnal and carnivorous echinoderm. It contains great power of **autotomy** and **regeneration,** *i.e.,* it can shed one or more arms when it is disturbed or caught by a predator. The lost arm is regenerated.

Structure

Body of *Ophioderma* is olive green in colour. It comprises a **pentagonal disc** and five slender, solid, cylindrical, muscular and flexible **arms** bearing short spines. The pentagonal disc measures 10 to 20 mm in diameter and arms are 40 to 60 mm in length.

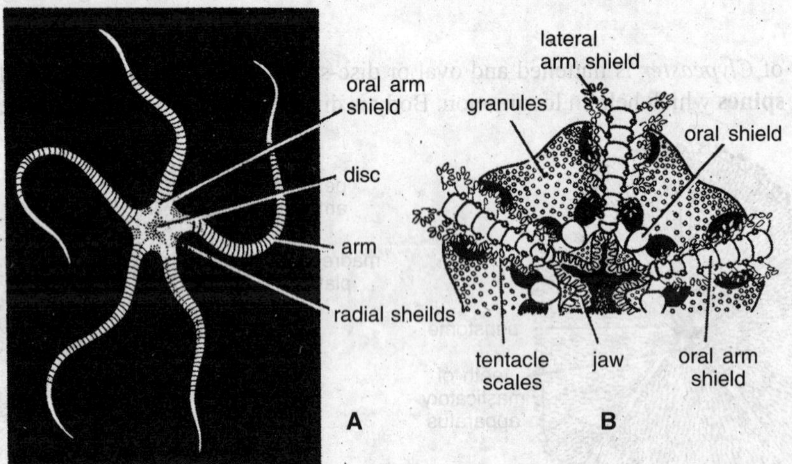

Fig. 19.10. *Ophioderma*. A—Entire brittle star in aboral view;; B—Disc magnified (oral view).

Oral surface bears pentagonal **mouth** having several papillae and is covered by numerous small closely set granules. Teeth are arranged in a single row. Four **genital bursae** occur in each inter-radius. Ambulacral grooves are covered by ossicles.

Aboral surface bears five pairs of **radial shields** and a **madreporite** but lacks **anus.** Tube feet are present on the ventrolateral side of arms in two rows.

The sexes are separate.

II. Ophiothrix

Habit and Habitat

Ophiothrix is commonly called **spiny brittle star.** It is marine and commonly found along Atlantic Coast. It lives in shallow to deep waters, hiding beneath stones and weeds during day time.

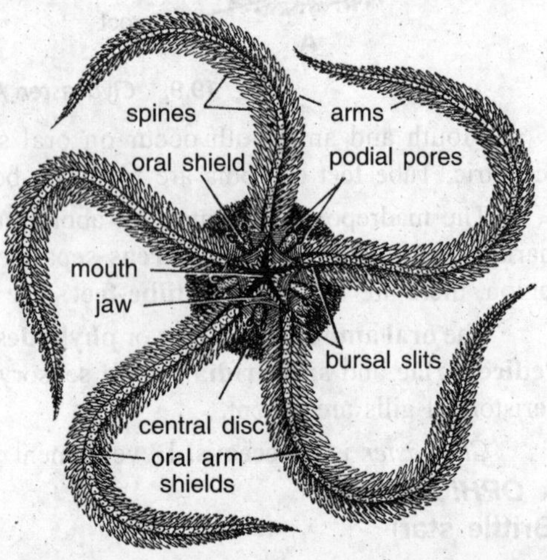

Fig. 19.11. *Ophiothrix*. Oral view.

It is most active echinoderm. It swims by jerking the arms quickly. It creeps and swims by snake-like movements of the arm (*i.e.*, due to this reason *Ophiothrix* is also called **serpent star**). *Ophiothrix* quickly breaks off an arm when seized or injured. Lost arm is soon regenerated. It feeds on bottom debris containing microscopic life and on small crustaceans and molluses. It is eaten by fishes.

Structure

Ophiothrix has a small rounded central **disc** and five slender **arms** which arise from the lower surface of the disc. Central disc on its **oral side** bears a central, large star-shaped **mouth**. In one of the inter-radii is located the **madreporite** which is a modified oral shield/plate. Near the base of each arm are paired ciliated grooves called **genital bursae** serving as gonopore and also helping in respiration and excretion. In the inter-radial positions, there are five **buccal plates** or **oral shields**. Internal and lateral to them are ten **angular shields**, bearing spines projecting into the mouth. Aboral surface of the disc has a series of **radial shields** around the bases of arms and a large number of tiny calcareous plates covered by the vestigial epidermis and cuticle.

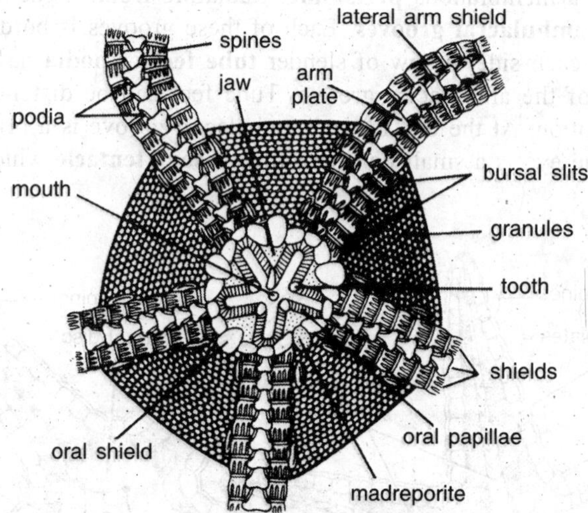

Fig. 19.12. *Ophiothrix*. Aboral view of central disc.

Each arm appears jointed since it is bounded by four longitudinal rows of calcareous plates/ shields. There are two rows of **lateral shields**, one row of **aboral shields** and one row of **oral shields** (oral shields are also called **vertebrae**). Lateral shields bear 2 to 15 large spines in a vertical row. There is no **ambulacral groove** on the oral surface of arm.

The **tube feet** or **podia** are small, lack suckers and arise from tiny pores left in the sutures between ventral (aboral) and lateral plates. The pair of tube feet in each arm is directed towards the mouth and serves to taste and push the food into the mouth. There is a sensory tube foot at the tip of each arm. Pedicellariae are absent.

Ophiothrix is **dioecious**. Gonads are small. Fertilization is external. Development is **indirect**. The larva is called **ophiopluteus.**

5. *PENTACEROS*

Classification

Phylum	—	**Echinodermata**
Subphylum	—	**Asterozoa**
Class	—	**Stelleroidea (Asteroidea)**
Order	—	**Phanerozonia (Valvatida)**
Genus	—	***Pentaceros***

Habit and Habitat

Pentaceros is commonly called **sea pentagon**. It is now included in genus *Oreaster.* It is marine and found commonly in the Indo–Pacific Ocean and around the West Indies, in the Bay of Bengal and Arabian Sea. *Pentaceros* feeds on oysters, so is very harmful to the pearl industry.

Structure

The body of *Pentaceros* is very thick, regular, star-shaped and has a diameter of 25 cm. It has a large **central disc** and five short tapering **arms**. Arms are short with broad base, called **stellate arms**. They are not sharply demarcated from the central disc. The axes of arms are called **radii** and the spaces between them are **inter-radii**. The body is covered by many calcareous endoskeletal plates or **ossicles**.

Body of *Pentaceros* has flat lower **oral surface** and upper convex **aboral surface**. In the centre of the oral surface is a five sided aperture called **actinostome** which encloses a mouth surrounded by a membranous **peristome**. Radiating from the actinosome into the five arms are five narrow **ambulacral grooves**. Each of these grooves is bordered by two rows of ambulacral **spines** on each side. A row of slender **tube feet** or **podia** having sucker-like ends occurs on each side of the ambulacral groove. Tube feet can be distended and fixed to the substratum for locomotion. At the end of each ambulacral groove is a bright red **eye** made of many **ocelli**. Above the eye is a small non-retractile **sensory tentacle** which is **olfactory** and a modified tube foot.

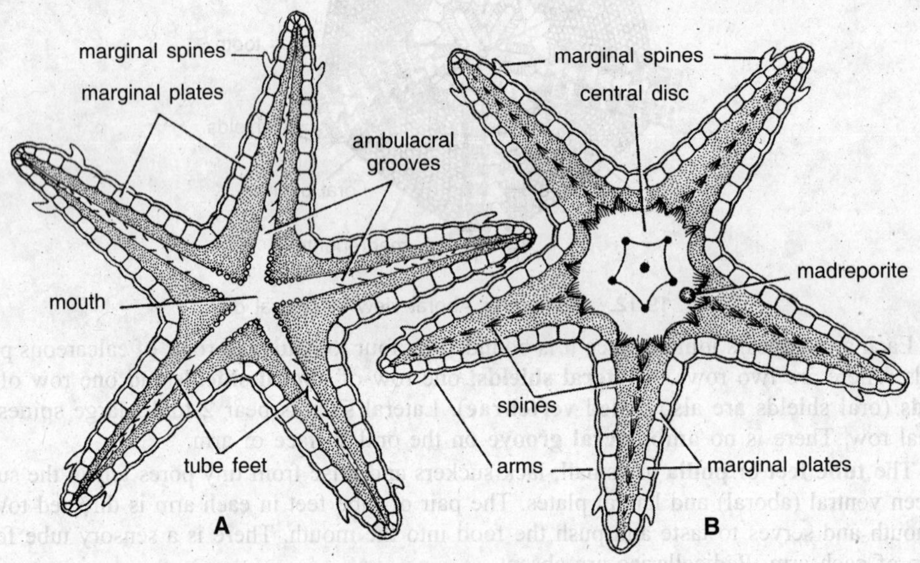

Fig. 19.13. *Pentaceros*. A—Oral view; B—Aboral view.

Aboral surface of *Pentaceros* is yellow reddish brown in colour and has many stout spines in irregular rows; all spines rest on ossicles. Between the spines, **papullae** (dermal branchiae or gills) project singly. Papullae are retractile and coelom-containing thin-walled extension of body wall meant for respiration and excretion. In the centre of aboral surface of disc is located a minute **anus**. Near the anus between two arms is a flat light coloured plate called **madreporite**. Microscopic stalked or sessile **pedicellariae** are located on both oral and aboral sides around the spines.

6. *HOLOTHURIA*

(Sea cucumber)

Classification

Phylum	—	**Echinodermata**
Subphylum	—	**Echinozoa**
Class	—	**Holothuroidea**
Order	—	**Aspidochirotida**
Genus	—	***Holothuria***

Habit and Habitat

Holothuria is commonly called **sea cucumber**. It is found in thousands on the sandy bottom of shallow tropical and subtropical waters of Indo-Pacific area. *Holothuria* is found distributed in India, West Indies and Florida. It is sluggish and sedentary. However, *Holothuria* creeps slowly engulfing microorganisms alongwith mud, *i.e.*, it is **suspension feeder** and **detritivore**.

Structure

The body of *Holothuria* is elongated, bilaterally symmetrical animal. It lacks arms and bears mouth and anus on the opposite ends. The body wall is leathery having an endoskeleton of minute calcareous **ossicles**. Spines and pedicellariae are absent.

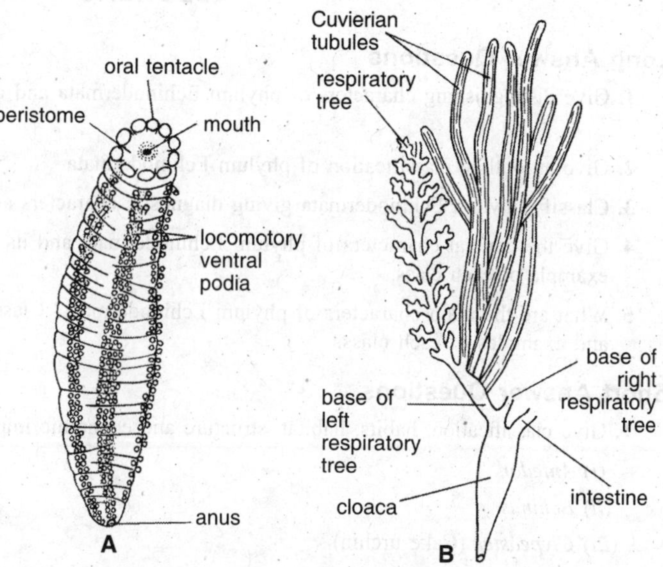

Fig. 19.14. *Holothuria*. A—Entire animal; B—Relation of respiratory tree and Cuvierian tubules with cloaca.

Five double rows of **tube feet** extend from mouth to anus. These rows represent five **ambulacral areas** and they are separated by five **interambulacral areas.** Ventral surface contains three ambulacral regions forming the **sole**. Dorsal surface contains two ambulacral area. The tube feet of ventral surface contains sucker and are used for creeping and attachment.

The mouth is surrounded by a circle of about 20–30 shield-like **tentacles** which are sensitive and contractile. The tentacles are modified tube feet and remain covered with mucus. Tentacles are extended to trap small organisms. The food-ladden tentacles are inserted into the pharynx one by one. The tentacles and mouth (introvert) can be completely withdrawn within the anterior part of body when the animal is disturbed.

The cloaca contains a pair of large, much branched **respiratory trees.** These lie in coelom. Cloaca pumps water in the tubules of respiratory trees. The cloaca dilates due to which sea water is rushed into it through the anus. When the anal sphincter closes, the cloaca contracts and water is forced into the respiratory tree. Pumping is slow. *Holothuria* requires six to ten cloacal dilations and contractions to fill the trees. All the water is expelled in one action. During expulsion whole body of *Holothuria* contracts. Gaseous exchange occurs between the water and coelomic fluid through the respiratory tree. **Haemal system** and **water-vascular system** are well developed. **Madreporite** is internal.

Holothuria exhibits a peculiar way of self-defence. It has a large mass of blind tubules (**tubules of Cuvier**) attached to the base of left tree. When the animal is attacked, the anus is directed towards the attacker, the body wall contracts, cloaca ruptures and Cuvierian tubules are shot out of the anus. During their expulsion, the tubules are greatly elongated and they break free from the respiratory tree. The tubules are sticky. They trap the attacker in a net of adhesive threads and *Holothuria* escapes. After the discharge, the Cuvierian tubules are regenerated.

Sexes are separate, gonad is single and attached to left side of the dorsal mesentery. Development is **indirect** and includes **auricularia larva.**

Economic Importance

The body wall of *Holothuria* is eaten in China.

QUESTIONS

Long Answer Questions

1. Give distinguishing characters of phylum Echinodermata and classify upto orders. *(Awadh 2001; Kanpur 2001)*
2. Give an outline classification of phylum Echinodermata.
3. Classify phylum Echinodermata giving diagnostic characters and examples of each order.
4. Give the general characters of phylum Echinodermata and its various classes. Mention at least one example of each class. *(Allahabad 2001)*
5. What are the main characters of phylum Echinodermata? Classify it by giving diagnostic characters and examples of each class.

Short Answer Questions

1. Give classification, habits, habitat, structure and economic importance of the following:
   (*i*) *Antedon*
   (*ii*) *Echinus*
   (*iii*) *Clypeaster* (Cake urchin)
   (*iv*) *Ophioderma* (Brittle star)
   (*v*) *Pentaceros*
   (*vi*) *Holothuria*
2. Write short note on
   (*i*) Aristotle's lantern (*ii*) Lunules

Very Short Answer Questions

1. Which system is found in Echinodermata only? What is its utility.
2. Name the body symmetry in which body parts occur in fives.
3. Body of echinoderms can be distinguished into how many surfaces? Name them.
4. Give an degenerate character of echinoderms.
5. Mention one character in which echinoderms resemble the chordates.
6. What is the function of pedicellariae?
7. Name the cells found in the coelomic fluid.
8. Radial symmetry of adult echinoderms is a secondarily developed trait. Is it so?
9. Name three classes of subphylum Asterozoa.
10. Name classes of living species of subphylum Echinozoa.
11. Give two examples of class Asteroidea.

12. Mention the order and class of *Clypeaster*.
13. Name one cylindrical echinoderm.
14. Name one disc-like echinoderm.
15. Mention a fixed echinoderm.
16. To which class *Echinus* belongs?

Fill in the Blanks

1. Subphylum Crinozoa has a single living class called ....
2. Adult echinoderms have .... symmetry.
3. Tube feet are mainly useful in ....
4. Sea water enters into ambulacral system through a calcareous plate called ....
5. Tube feet protrude through .... areas.
6. Dermis contains skeletal plates called ....
7. Ambulacral system is also called ....
8. Respiratory organs of echinoderms are ....
9. Circulatory system of echinoderms is called ....
10. Phylum Echinodermata is divided into four sybphyla .... , ... , ... and ....
11. Fertilization in echinoderms is ....
12. Echinoderm larvae have .... symmetry.
13. Common name of .... is cake urchin.

Multiple Choice Questions

1. Water vascular system is peculiarity of
   (*a*) Porifera (*b*) Coelenterata
   (*c*) Mollusca (*d*) Echinodermata
2. Which phylum is exclusively marine
   (*a*) Echinodermata (*b*) Porifera
   (*c*) Protozoa (*d*) Arthropoda
3. *Ophiothrix* belongs to class
   (*a*) Asteroidea (*b*) Echinoidea
   (*c*) Ophiuroidea (*d*) Holothuroidea
4. An echinoderm with melon-shaped body is
   (*a*) sea-star (*b*) brittle star
   (*c*) sea cucumber (*d*) sea urchin
5. Common name of *Antedon* is
   (*a*) sea star (*b*) feather star
   (*c*) star fish (*d*) cake urchin
6. Sea cucumber belongs to the class
   (*a*) Holothuroidea (*b*) Asteroidea
   (*c*) Echinoidea (*d*) Ophiuroidea
7. Which of the following belongs to the class Echinoidea?
   (*a*) *Asterias* (*b*) *Ophiothrix*
   (*c*) *Echinus* (*d*) *Cucumaria*

8. Cloacal respiration occurs in

(*a*) *Clypeaster* (*b*) *Antedon*

(*c*) *Echinus* (*d*) *Cucumaria*

9. Arms, spines and pedicellaria are absent in

(*a*) sea urchin (*b*) sea star

(*c*) sea lily (*d*) sea cucumber

10. The term ambulacral has to do with

(*a*) inter filament bridges (*b*) marine animals

(*c*) myosin filaments (*d*) tube feet

11. The regeneration abilities of echinoderms are

(*a*) slight (*b*) moderate

(*c*) phenomenal (*d*) occasional

12. Autotomy is common amongst

(*a*) sea lilies (*b*) sea stars

(*c*) brittle star (*d*) all of these

13. In echinoderms, skeleton is

(*a*) calcareous (*b*) chitinous

(*c*) siliceous (*d*) tunicin made

14. The animal which is commonly known as 'sand dollar' is

(*a*) *Antedon* (*b*) *Clypeaster*

(*c*) *Echinarachinus* (*d*) *Ophiothrix*

15. *Thyone* is commonly known as

(*a*) brittle star (*b*) sea cucumber

(*c*) sea lily (*d*) cake urchin

16. Sea pentagon is common name of

(*a*) star fish (*b*) sea urchin

(*c*) *Pentaceros* (*d*) *Cucumaria*

17. Aristotle's lantern is found in

(*a*) *Cucumaria* (*b*) *Antedon*

(*c*) *Echinus* (*d*) *Ophiothrix*

18. Brachiolaria larva is characteristic of the class

(*a*) Crinoidea (*b*) Asteroidea

(*c*) Ophiuroidea (*d*) Echinoidea

19. The larva that occurs in the development of holothurians

(*a*) auricularia (*b*) brachiolaria

(*c*) doliolaria (*d*) bipinnaria

20. In which class of phylum Echinodermata, pluteus larva is found?

(*a*) Asteroidea (*b*) Holothuroidea

(*c*) Ophiuroidea (*d*) Echinoidea

21. Aristotle's lantern is useful in

(*a*) photo reception (*b*) chewing

(*c*) excretion (*d*) respiration

22. Excretory products of echinoderms are

(*a*) urea (*b*) urea and creatine

(*c*) ammonia (*d*) uric acid

23. Respiratory organs in Holothuroidea are

(*a*) papulae (*b*) dermal branchiae

(*c*) respiratory tree (*d*) bursae

24. Ophiopluteus is

(*a*) an echinoderm (*b*) a poriferan

(*c*) an annelid (*d*) a larva of echinodermata

ANSWERS

Very Short Answer Questions

1. The water-vascular system (WVS) is the system represented in echinoderms only. It is a modified part of coelom consisting of a system of canals containing sea water and amoeboid corpuscles. It serves to bring about locomotion, feeding and aherence to the substratum.

2. This type of body symmetry is called **pentamerous symmetry**. Adult echinoderms show radial (usually pentamerous) symmetry but their larvae are bilaterally symmetrical.

3. The body of echinoderms can be distinguished into **oral** and **aboral** surfaces. Oral surface bears mouth and water-vascular system where as aboral surface has madreporite and anus on it.

4. Lack of head.

5. Enterocoelous formation of coelom and mesoderm.

6. To rid the body surface of debris and microorganisms.

7. Coelomocytes (amoebocytes).

8. Yes.

9. Asteroidea, Ophiuroidea and Concentricycloides.

10. Two class — Echinoidea and Holothuroidea.

11. *Asterias* and *Oreaster*.

12. Clypeastroida (order), Echinoidea (class).

13. *Cucumaria*.

14. *Clypeaster*.

15. *Antedon* or feather star.

16. *Echinus* belongs to the class Echinoidea of the phylum Echinodermata. This animal is commonly known as **sea urchin.**

Fill in the Blanks

1. Crinoidea **2.** Radial **3.** Locomotion **4.** Madreporite

5. Ambulacral **6.** Ossicles **7.** Water vascular system **8.** Tube feet

9. Haemal system **10.** Homalozoa, Crinozoa, Asterozoa and Echinozoa

11. External **12.** Bilateral **13.** *Clypeaster*.

Multiple Choice Questions

1. (*d*) **2.** (*a*) **3.** (*c*) **4.** (*d*) **5.** (*b*) **6.** (*a*)

7. (*c*) **8.** (*d*) **9.** (*d*) **10.** (*d*) **11.** (*c*) **12.** (*c*)

13. (*a*) **14.** (*c*) **15.** (*b*) **16.** (*c*) **17.** (*c*) **18.** (*b*)

19. (*a*) **20.** (*d*) **21.** (*b*) **22.** (*b*) **23.** (*c*) **24.** (*d*).

20

Asterias (Starfish or Sea Star) (Phylum Echinodermata)

20.1 CLASSIFICATION

Phylum	—	**Echinodermata**
Subphylum	—	**Asterozoa**
Class	—	**Asteroidea**
Order	—	**Forcipulata**
Genus	—	***Asterias***

Asterias is commonly known as **starfish**. But this is not fish. Hence its name is misleading and now its more suitable common name is **sea star**. *Asterias* contains about 150 species, all of which have different geographical distribution. Some other common sea stars are *Pentaceros* (or *Oreaster;* it is an Indian species which is found in the Bay of Bengal and Arabian sea), *Astropecten*, *Heliaster*, *Solaster*, *Luidia*, etc.

20.2 HABIT, HABITAT AND EXTERNAL FEATURES

Habit and Habitat

Asterias is a free-living exclusively marine, bottom dwelling (or **benthonic**) animal. *Asterias* is found in shallow water in North Temperate seas and found abundantly on North-Atlantic Coast. It is found in abundance in India and U.S.A. *Asterias forbesi* occurs on the eastern sea shore from the Maine to the Gulf of Mexico and is found equally abundant on hard, rocky and sandy or soft bottom.

Most species of *Asterias* are **solitary** but under certain ecological constraints (*e.g.*, to avoid direct sunlight) many individuals may gather at some place for the purpose of protection. Most of them are **nocturnal**, remain quiet in day time but become active during night. They remain attached to rocks, shells, piles and piers, etc. Their oral surface faces the substratum. They slowly move by crawling with the help of arms on the rocky bottom. All sea stars are **carnivorous** and feed voraciously on almost any slow moving or sessile animals, chiefly on polychaetes, crustaceans, molluscs (snails, bivalves) and other echinoderms. They also feed on dead animals, so act as scavangers of sea. *Asterias* shows remarkable power of **autotomy** and **regeneration**.

External Features

1. Shape, size and colour. *Asterias* has a **radially symmetrical** star shaped body (Box 20.1). The diameter of body reaches upto 24 cm. Colour of *Asterias* varies from orange to purplish, the upper (aboral) surface being darker than the lower (oral) surface.

2. External structure. The body of *Asterias* consists of a central, pentagonal (5 sided) **central disc** from which radiate out five elongated, tapering, symmetrically spaced projections, the **rays** or **arms.** Each arm has the shape of an isosceles triangle (*i.e.*, triangle having two sides of equal length). The body lacks a well defined head and has two distinct surfaces. The lower surface, the one nor-mally kept towards the substra-tum, is flat and is known as the **oral** or **actinal surface** because it bears the mouth (Fig. 20.1). The upper surface is convex and is called the **aboral** or **abactinal surface** (Fig. 20.6). The oral and aboral surfaces are not the ventral and dorsal surfaces, but corres-pond to the left and right sides of the bilaterally symmetrical larva. The axes occupied by the arms are known as the **radii** and the regions of the central disc between the arms is the **inter-radii**.

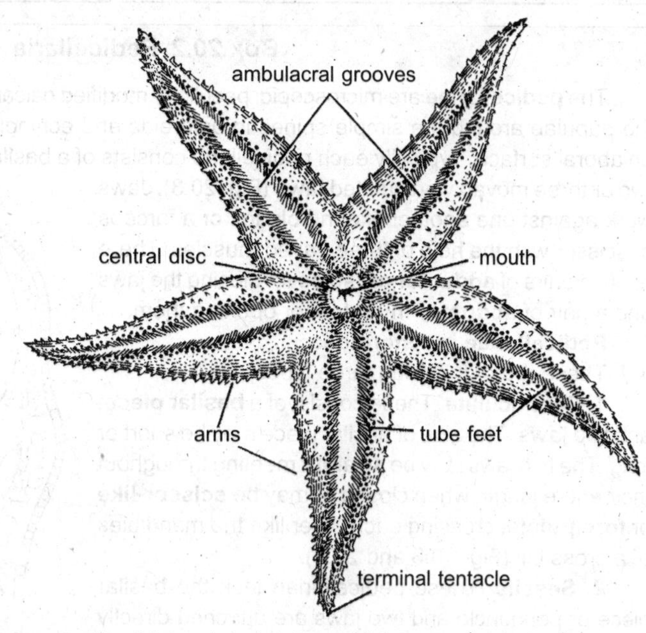

Fig. 20.1. *Asterias*. External features (oral view).

A. Oral Surface. The oral surface of *Asterias* contains following structures.

1. Mouth. Mouth is a circular aperture at the top of 5-rayed (pentagonal) depression at the centre of the oral surface of the central disc. The depression is known as **actinostome**. The mouth is surrounded by a soft membrane, the **peristomial membrane** or **peristome.** Peristome is provided with a sphincter and is guarded by five groups of **oral spines** or **mouth papillae.**

Box 20.1 Symmetry of Echinodermata

Echinoderms are quite distinct in their morphology and differ from other coelomate animals in having a characteristic **pentamerous radial symmetry**. The radial symmetry is seen in majority of echinoderms except for some holothurians. The radial symmetry is not restricted to only the external form of an echinoderm, it is shown even by the arrangements, of internal organs.

In all echinoderms, the external surface is differentiated into five radial and five interradial areas. This gives pentamerous radial symmetry. The radial regions bear tube feet. These are called **ambulacral regions**. The areas between ambulacral regions are called **interambulacral regions** and they contain spines (Fig. 20.2). But in some holothurians, the tube feet are scattered all over the body surface. In Asteroidea and Ophiuroidea, the body is prolonged into arms in the direction of radii.

The radical symmetry of echinoderms is exhibited only in the organisation of the adults and is acquired at the time of metamorphosis. The larval stages of different classes of Echinodermata exhibit a **bilateral symmetry**. In fact, the radial symmetry is not total or perfect, since, most radially symmetrical forms present indication, of bilateral symmetry, *e.g.*, by the presence of a single madreporite, presence of single gonad and gonoduct in Holothruoids. Because of these reasons, it is generally believed that the radial symmetry

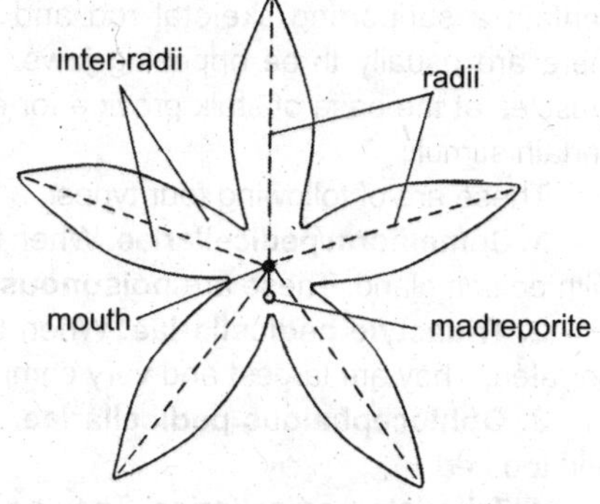

Fig. 20.2. Sea star showing pentamerous radial symmetry.

of echinoderms has been secondarily acquired. It is also believed that the radial symmetry in ancestral echinoderms has evolved due to sedentary habit and that echinoderms have evolved from free-swimming bilaterally symmetrical, triploastic, coelomate ancestor.

Box 20.2 Pedicellaria

The **pedicellariae** are microscopic, peculiarly modified calcareous spines. They are found among the papulae around the simple spines in asteroids and echinoids. These occur on oral as well as on aboral surface. Typically each pedicellaria consists of a basilar piece, the **stalk** or **peduncle** and two or three movably articulated **jaws** (Fig. 20.3). Jaws work against one another like the blades of a forceps or scissor with the help of three paired muscles. There are two pairs of **adductor muscles** for closing the jaws and a pair of **abductor muscles** for opening them.

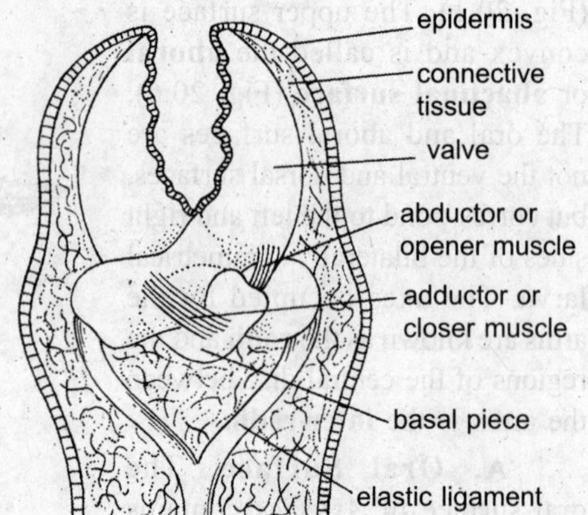

Fig. 20.3. Distal end of a scissor-type pedicellaria from *Asterias*.

Pedicellariae in Asteroidea

These are of following three types:

1. Pedunculate. These consist of a **basilar piece**, and two **jaws**. The size of basilar piece may be short or long. The two jaws may be **straight** meeting throughout their entire length when closed or may be **scissor-like** or **forcipulate,** crossing each other like the mandibles of a cross bill (Fig. 20.3 and 20.4).

2. Sessile. These pedicellariae lack the basilar piece or penduncle and two jaws are attached directly to the ossicles by the muscles. Sessile pedicellariae may be **spiniform**, when present in clusters on the adjacent ossicles or **fascicu-late**, when the cluster of spines is present on the same ossicles.

3. Alveolar. These are similar to sessile pedicellariae but differ from them in being partly embedded in the endoskeletal depression, called **alveolus.**

Pedicellariae in Echinoidea

Pedicellariae are characteristic of all echinoids. They are located over the general body surface as well as on the peristome. The echinoid pedicellaria is composed of a long stalk surmounted by jaws. The stalk may contain a supporting skeletal rod and there are usually three opposing jaws. Muscles at the base of stalk provide for elevation and direction of the pedicellariae in response to certain stimuli.

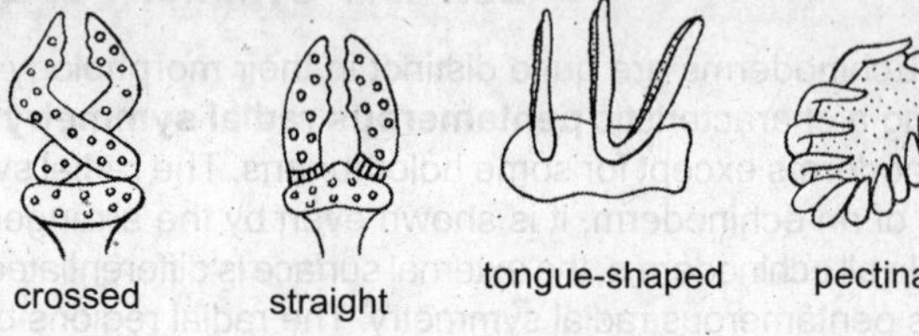

Fig. 20.4. Different type of pedicellariae of Asteroidea.

These are of following four types:

1. Gemmiform pedicellariae. When the stalk is very stiff and head is round. Each jaw is provided with poison gland. These are **poisonous pedicellariae.**

2. Tridactyle pedicellariae. When the stalk is very flexible and jaws are long tappering and serrated. They are largest and very common.

3. Ophiocephalous pedicellariae. When the stalk is very flexible and jaws are short, broad and toothed.

4. Trifoliate pedicellariae. They are smallest. Each consists of a very flexible stalk and short broad toothed jaws. Jaws do not meet at their tips.

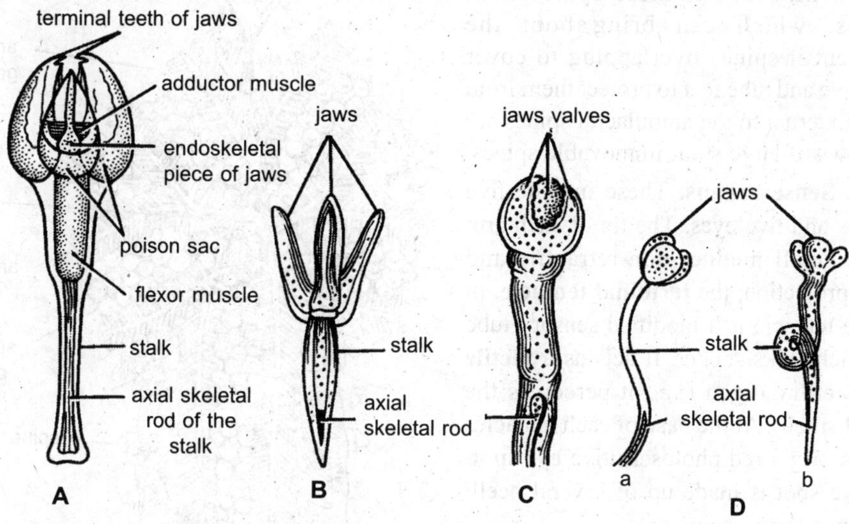

Fig. 20.5. Different types of pedicellariae in sea urchin A—Gemmiform; B—Tridactyl; C—Ophiocephalous; D—Trifoliate (*a*) of *Tripneustes* and (*b*) of *Echinus*.

Function. Pedicellariae are used for defence or for cleaning the body surface, biting and breaking up small particles of debris, which are then removed by the surface cilia. When the pedicellariae are touched on outside, they snap open; when touched on the inside, the snap shut. They also respond to chemical stimuli.

2. Ambulacral groove. There are five narrow grooves that radiate out from the angles of acti-onostome, one extending along the middle of the oral surface of each arm upto the tip.

3. Tube feet or podia. These are soft, thin-walled, tubular, retractile structures which are arranged in four rows in each ambulacral groove. Each podium is provided with terminal sucker. Sucker functions as a suction pump to afford firm attachment on the surface to which it is applied. The tube feet are multipurpose organs. They mainly act as organs of locomotion and capturing food. They also help in respiration and adherence to substratum. Tube feet also act as sensory organs.

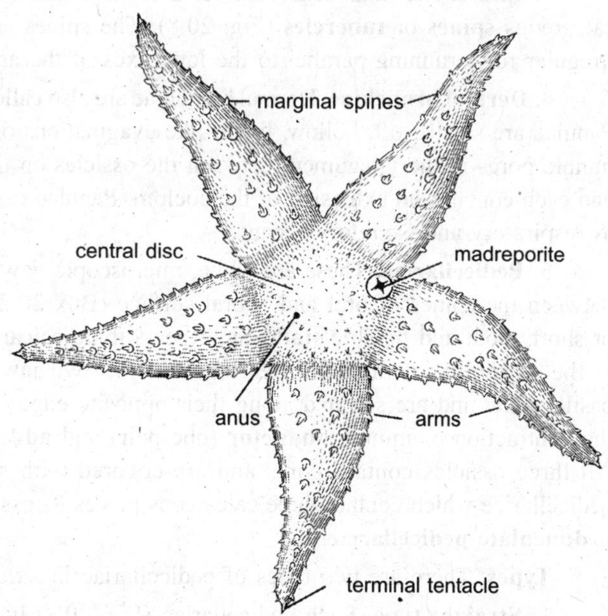

Fig. 20.6. *Asterias*. External features (aboral view).

4. Ambulacral spines. The spines are short, stout outgrowths from the calcareous plates, the **ossicles,** embedded in the body wall and covered by epidermis. On the oral surface, there are two rows of spines along either border of each ambulacral groove. These are called the **ambulacral**

spines. The bases of these spines have muscles, which can bring about the movement of spines overlapping to cover the groove and tube feet to protect them from injury. External to the ambulacral spines are three rows of large stout immovable spines.

5. Sense organs. These include five tentacles and five eyes. The tip of each arm bears a small median non-retractile and hollow projection, the **terminal tentacle**. In fact, the tentacle is a modified sensory tube feet which lacks sucker. It acts as a **tactile** and **olfactory** organ (*i.e.*, it perceives the sense of smell). At the base of each tentacles occurs a bright red photosensitive **eye spot**. Each eye spot is made up of several ocelli but cannot form an image.

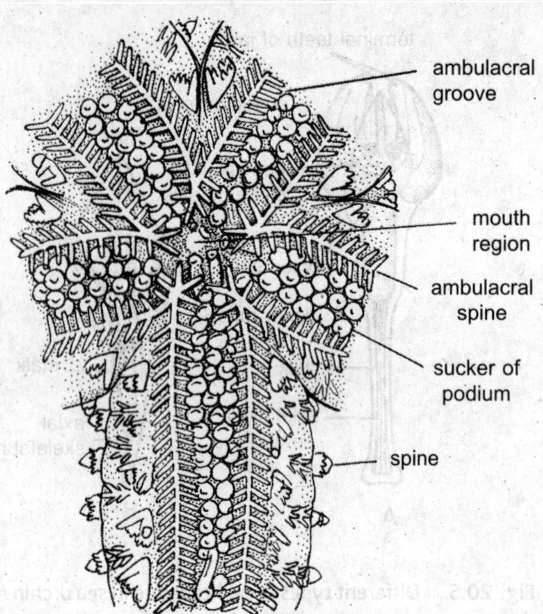

Fig. 20.7. *Asterias*. Oral surface of the disc and arm.

B. Aboral surface. The aboral surface of *Asterias* contains the following structures:

1. Anus. It is a small circular aperture. It is situated close to the centre of the central disc of aboral surface.

2. Madreporite. It is a large circular button-like porous plate. It is situated towards one side of the disc between two arms. These two arms are called **bivium** and remaining three arms—the **trivium**. Thus, madreporite introduces a bilateral symmetry in the radial symmetry of *Asterias*.

3. Spines. The entire aboral surface is covered with numerous short, strong, blunt, immovable, calcareous **spines** or **tubercles** (Fig. 20.8). The spines are variable in size and are arranged in irregular rows running parallel to the long axes of the arms.

4. Dermal branchiae. Dermal branchiae are also called **dermal gills** or **papulae** (Fig. 20.8 A). Papulae are small, soft, hollow, finger-like evaginations of the body wall. They protrude through minute pores in the integument between the ossicles on the aboral surface. They are thin-walled and each contains an extension of the coelom. Papulae can be extended and withdrawn and serve as respiratory and excretory organs.

5. Pedicellariae. These are white, microscopic, jaw-like or pincer-like structures occurring between the spines of oral and aboral surface (Box 20.2). Each pedicellaria consists of a long or short, stout and flexible **stalk**. Stalk bears three calcareous ossicles or plates: a **basilar plate** at the extremity of the stalk bearing at it top two jaws. Jaws are movably articulated with basilar plate and are serrated along their opposite edges. The jaws can be opened or closed by the contraction of minute **abductor** (one pair) and **adductor** (two pairs) muscles respectively. All three ossicles contain pores and are covered with sensory and glandular epithelium. The pedicellariae which contain three calcareous pieces (= ossicles) and a stalk are called **forcipulate pedunculate pedicellariae**.

Types. There are two types of pedicellariae in *Asterias*.

1. Straight type. Such pedicellariae (Fig. 20.8) lie among the dermal branchiae. In them two jaws are straight. When closed they remain parallel and meet throughout their length.

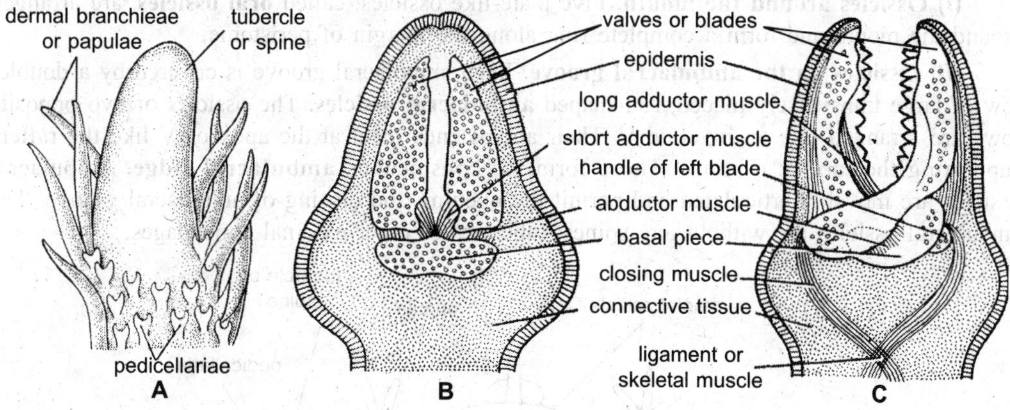

Fig. 20.8. *Asterias*. A—A cluster of pedicellariae, papulae and spine; B—Straight type pedicellaria; C—Crossed type pedicellaria.

The two jaws work against each other like the blades of forceps.

2. Crossed type. These pedicellariae are small and are arranged in rings round the white spines on the aboral surface. In them two jaws cross each other like a pair of scissors.

Function. Pedicellariae are sensitive to contact. They serve as defensive and offensive organs. They help in keeping the body surface free from debris and minute organisms (*e.g.*, larvae of sponges and coelenterates) which may settle on the body surface. In this way, they provide protection to gills. In some starfishes, they may also help in the capture of small prey.

Lastly, entire surface of the body, including that of its outgrowths such as tube feet, spines, pedicellariae and dermal branchiae is ciliated. The body of *Asterias* is enclosed in a hard but flexible integument. The hardness of the integument is due to the presence in it of calcareous ossicles.

20.3 ENDOSKELETON

Skeleton of echinoderms is derived from **mesoderm** or from **dermis** of body wall. Therefore it is called **endoskeleton**. It is formed of crystals of calcium carbonate ($CaCO_3$), *i.e.*, it is calcareous. Endoskeleton occurs in two forms:

1. In the form of **calcareous plates** or **ossicles** found in the soft dermis of bodywall.
2. In the form of **warts** or **tubercles**, attached to the calcareous plates of dermis. These projects out of the skin.

Skeleton of *Asterias* and other echinoderms differ from the skeleton of other invertebrates mainly in one character. The skeleton of invertebrates is developed from epidermis, but in echinoderms it develops from dermis or mesodermal layer. The body of starfish, thus, is enclosed in a tough, flexible integument containing numerous calcareous ossicles.

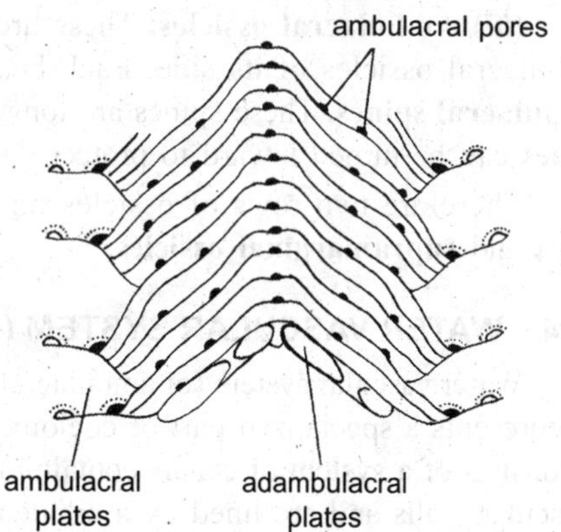

Fig. 20.9. Arrangement of ambulacral ossicles.

Arrangement of ossicles. Dermal ossicles have different but definite shapes and patterns. These are interconnected by connective tissue. In starfish, skeletal ossicles are arranged in a specific orders as follows:

***(i)* Ossicles around the mouth.** Five plate-like ossicles, called **oral ossicles,** are arranged around the mouth and form a complete ring along the margin of peristome.

***(ii)* Ossicles on the ambulacral groove.** Each ambulacral groove is covered by a double-row of large transversely placed, rod-shaped **ambulacral ossicles.** The ossicles of two opposite rows are arranged like an inverted V. Their aboral ends meet at the apex of V like the rafters supporting the roof of a house. These form the conspicuous **ambulacral ridge.** Ambulacral ossicles are movably articulated and permit the opening and closing of ambulacral groove. The ambulacral ossicles are without any spines, tubercles or other external appendages.

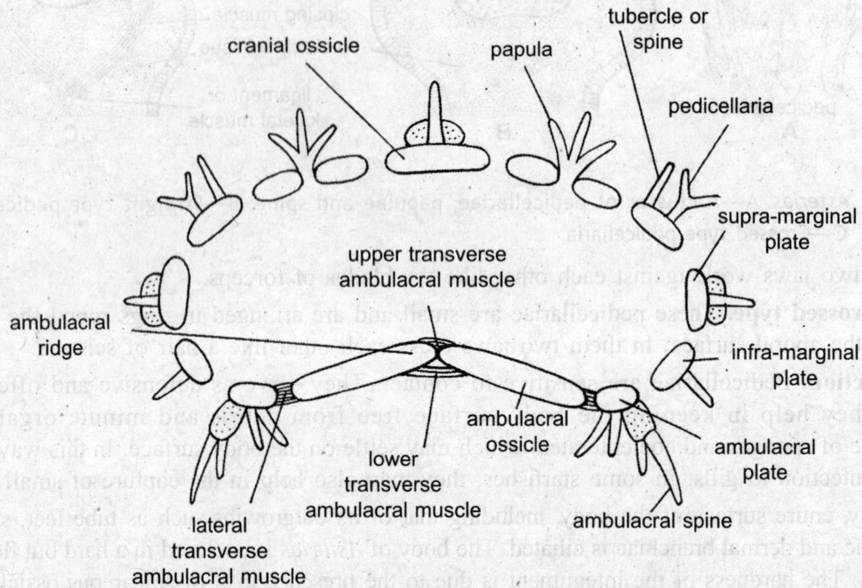

Fig. 20.10. Starfish. C.S. through the arm to show endoskeletal ossicles.

Each ambulacral ossicle has a notch on its outer as well as inner margin. The two notches of the adjacent ossicles together form an oval aperture, the **ambulacral pore** for the passage of tube-feet. The ambulacral pores are so arranged that they form two rows on each side of the ambulacral groove.

***(iii)* Ambulacral ossicles.** These are present along the edges and are articulated with the ambulacral ossicles of its side. Each bears 2 to 3 movable spines on small tubercles, called **ambulacral spines.** These spines are long and bear groups of large straight **pedicellariae.** These spines can be turned inward to protect the groove and podia.

There are two rows of ossicles on the outside of ambulacral ossicles. These are called **infra** and **supramarginal ossicles.**

20.4 WATER VASCULAR SYSTEM (OR AMBULACRAL SYSTEM)

Water vascular system (or ambulacral system) is a typical and unique feature of echinoderms. It represents a specialized part of coelom, being developed from the left hydrocoel of the larva. It consists of a system of canals containing sea water and amoeboid corpuscles. The canals have muscular walls and are lined by a ciliated epithelium. It helps in locomotion.

Structure

Water vascular system (WVS) of *Asterias* consists of the following parts (Fig. 20.11):

1. Madreporite
2. Stone canal

3. Ring canal
4. Tiedmann's bodies
5. Radial canals
6. Lateral or transverse canals
7. Tube feet

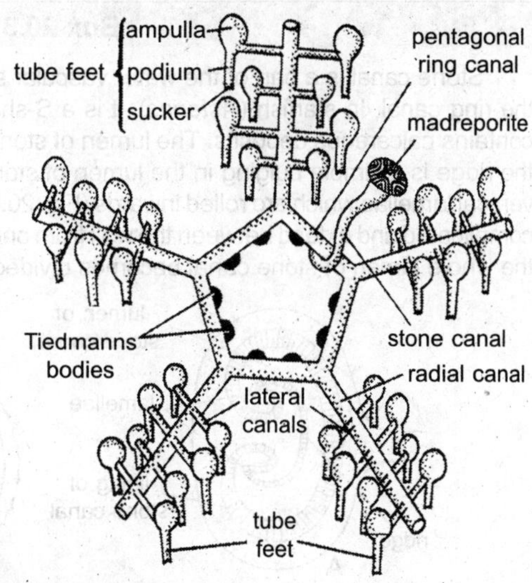

Fig. 20.11. *Asterias*. Water vascular system.

1. Madreporite. It is a light-coloured, sieve-like, button-shaped or circular calcareous skeletal plate. It is located on the aboral surface of the central disc in inter-radial position. The surface of madreporite is marked by a number of radiating narrow, straight or wavy groo-ves or **furrows** (Fig. 20.12). The furrows are covered with ciliated epithelium and at their bottom they contain about 250 minute pores. Each pore opens into **pore canals.** The pore canals unite to form larger **collecting canals** within the substance of madreporite. Collecting canals lead into a sac-like **ampulla** below the madreporite. Ampulla is continued as a canal inside the stone canal.

2. Stone canal. It is S-shaped verti-cal canal which extends downwards (ora-lly) and opens into a ring canal. Its wall is supported by a series of calcified rings and is lined internally with tall ciliated cells. The beating of cilia draw water into the canal. The stone canal is a simple tube in a young starfish, but in adult, its wall produces a prominent **ridge** in the lumen of stone canal. The ridge is bifurcated to form two spirally rolled **lamellae** (Fig. 20.12). They ensure water circulation.

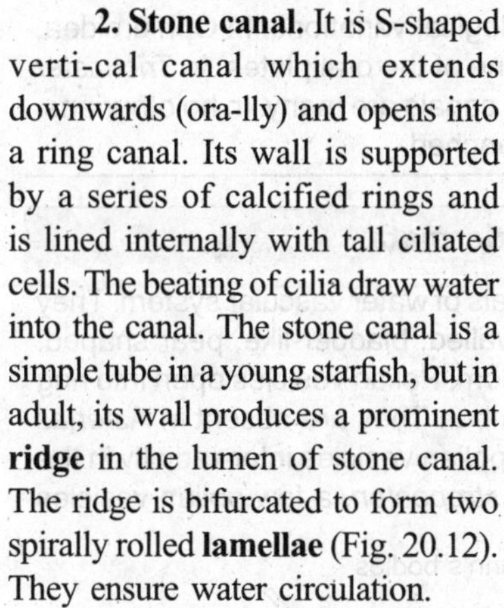

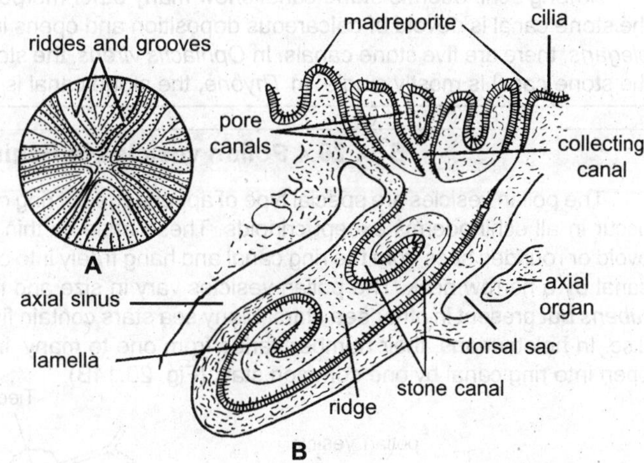

Fig. 20.12. *Asterias*. Madreporite. A—Surface view; B—Vertical section.

During its course, the stone canal along with an **axial gland** ensheathed by a wide, thin-walled tubular coelomic-sac, called **axial sinus.** This entire structure is called **axial complex.**

3. Ring canal. It is also called **ring vessel** or **water ring.** It is a wide five-sided (pentagonal) canal in the oral region around the oesophagus.

In most starfishes, but not in *Asterias*, the ring canal gives off from the inner side in each inter-radius a large, thin-walled, contractile, pear-shaped sac, the **polian vesicle.** The polian vesicles store water and help in regulating pressure in the water vascular system. They also manufacture amoeboid cells of water vascular system (Box 20.4).

4. Tiedmann's bodies. From its inner side, the ring canal produces five pairs of greatly folded pouches, called **racemose vesicles** or **Tiedmann's bodies.** Each pair of these bodies has an inter-radial position. The cavity of each Tiedmann's bodies is divided into a number of chambers which communicate with the ring canal. In *Asterias*, there are only **nine** Tiedmann's bodies. The 10th body is absent and its place is occupied by the stone canal (Fig. 20.11).

Box 20.3 Stone canal

Stone canal is a part of the water vascular system of echinoderms. It connects madreporite with the ring canal. In starfish (*Asterias*), it is a S-shaped tube. It is called **stone canal** because its wall contains calcareous deposits. The lumen of stone canal is divided by a longitudinal ridge. In *Henricia*, the ridge is a simple bulging in the lumen of stone canal. In *Asterias*, this ridge is bifurcated into two vertical lamellae which are rolled inwards (Fig. 20.13A). In *Astropecten*, the coiled lamellae become very complicated and extend between the wall from one side to another of the lumen (Fig. 20.13B). In *Culcita*, the whole lumen of stone canal becomes divided into a number of irregular chambers (Fig. 20.13C).

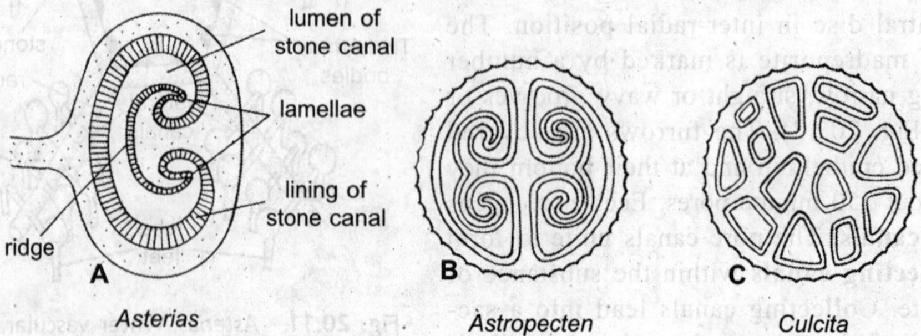

Fig. 20.13. Sectional view of stone canal in various Asteroids. A—*Asterias*; B—*Astropecten*; C—*Culcita*.

Among echinoderms stone canal show many other morpohological variations. In Ophiuroidea, the stone canal is devoid of calcareous deposition and opens in one of the oral plates. In *Trichaster elegans*, there are five stone canals. In *Ophiactis virens*, the stone canals are many. In holothurians, the stone canal is mostly single. In *Thyone*, the stone canal is branched.

Box 20.4 Polian vesicles in Echinodermata

The polian vesicles are special type of appendages of ring canals of water vascular system. They occur in all echinoderms except crinoids. These arise as thin - walled, bladder-like, pear shaped, ovoid or rounded sacs from the ring canal and hang freely into coelom. Polian vesicles open into ring canal by a narrow neck. The polian vesicles vary in size and number. They are absent in *Asterias rubens* but present in other sea stars. Many sea stars contain five polian vesicles inter-radially in the disc. In holothurians, their number varies from one to many. In *Astropecten*, a few polian vesicles open into ring canal by one common stalk (Fig. 20.14B).

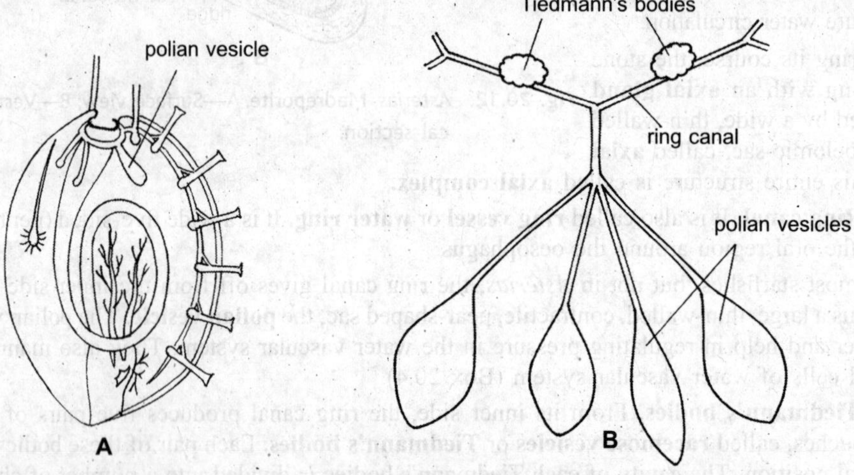

Fig. 20.14. A—Polian vesicle in *Holothuria*; B—Polian vesicles in *Astropecten*.

The actual function of Tiedmann's bodies is still unknown. They are supposed to be lymphatic glands to manufacture the **amoebocytes** of water vascular system. **Furguson** (1984) demonstrated that some sea water passes through these bodies into the perivisceral coelom. According to him these bodies participate in the production of coelomic fluid.

5. Radial canals. From the outer surface of the ring canal are given out five long, ciliated **radial canals.** One radial canal extends into each arm upto its extremity through the ambulacral groove of the arm. Radial canals run on the oral side of the ossicles and terminate as the lumen of terminal median tentacles.

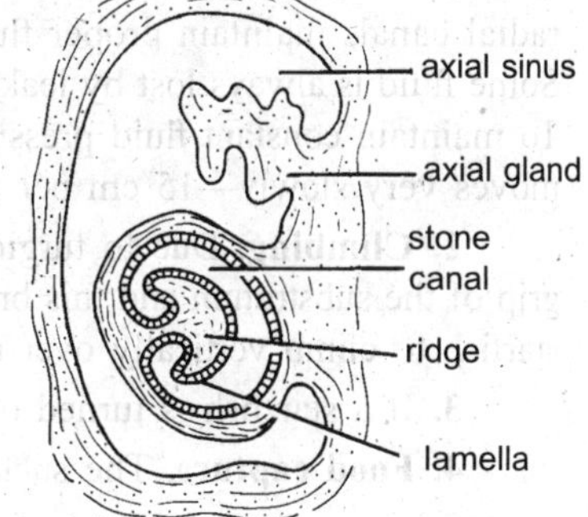

Fig. 20.15. *Asterias.* T.S. of axial complex.

6. Lateral or transverse canals. In each arm, the radial canal gives out two series of short, narrow, transverse branches called **lateral, transverse** or **podial canals.** Each lateral canal contains a **valve** to prevent backward flow of fluid into the radial canal and it opens into ampulla at the base of a tube foot.

7. Tube feet. A **tube foot** or **podium** is a hollow, elastic, thin-walled, closed sac-like structure. It contains an upper sac-like **ampulla,** a middle tubular **podium** and a lower disc-like flattened **sucker.** The ampulla lies within the arm, projecting into the coelom above the **ambulacral pore** which is a gap between the adjacent ambulacral ossicles for the passage of podium (Fig. 20.16). Tube feet, in fact, are external projections of the body wall; each of them has a lining of **ciliated epithelium** and a covering of **coelomic epithelium** or **peritoneum.** Between these two epithelia lie **connective tissue** and **longitudinal muscle fibres.** Contraction of the muscles of a side causes bending of a podium.

The podia are arranged in two double rows along the length of the ambulacral groove. They are chief locomotory and respiratory organs of *Asterias*.

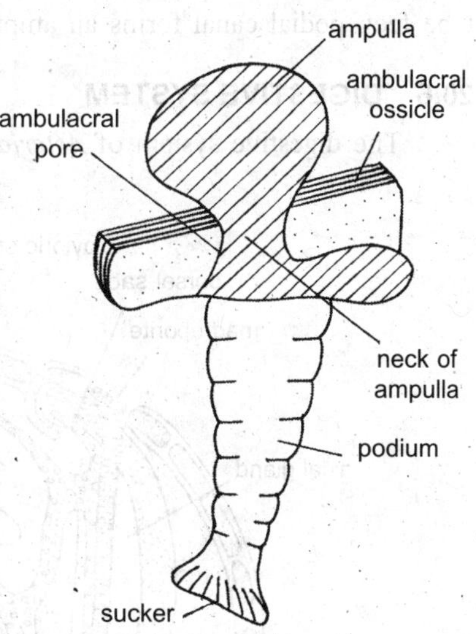

Fig. 20.16. *Asterias.* A tube foot.

Function

The water vascular system is lined with cilia and is filled with a fluid very similar to sea water. This fluid, however, contains **coelomocytes,** a little **protein** and a high concentration of **potassium ions.** The system operates as **hydraulic system** and brings about locomotion, adhesion to the substratum and the capture and handling of food. All these functions are performed by the help of tube feet.

(1) Locomotion. The beating of cilia causes water to enter through madreporite. It passes into the tube feet (= podia) and their ampullae. When the ampulla contracts, the valve in the lateral canal closes, and water (fluid) is pressed into tube feet which get distended (upto 5 cm) due to their muscular walls. The podium comes in contact with the substratum and the sucker adheres to it. Adhesion is also has a chemical base. The podium secretes a substance that makes bonds with the surface of substratum. Another secretion breaks the bonds and sucker becomes free (**Thomas** and **Hermans** 1985).

After their fixation with substratum, the tube feet contract in unison and push the body forward. Due to contraction of muscles, the fluid from each tube foot, returned to the ampulla and tube foot is shortened. This lessens the pressure at the tip so that the suckers detach. The same process is repeated again and again to bring about locomotion. In a section of arm most podia perform the same step and the animal moves forward. The action of podia is highly coordinated. During locomotion one or two arms act as leading arms and podia in all the arms move in same direction.

Other parts of the water vascular system—the madreporite, stone canal, ring canal and radial canals maintain proper fluid pressure necessary for the working of ampulae and podia. Some fluid is always lost by leakage through podial walls when fluid pressure increases in podia. To maintain constant fluid pressure some sea water is drawn through the madreporite. Star fish moves very slowly—15 cm per minute.

2. Climbing. Due to turgidity and terminal suckers, the tube feet are able to take a firm grip of the substratum and this brings about adhesion. Combined action of podial suckers enables starfish to climb vertically over rocks.

3. If a star fish is turned over, it can right itself by folding.

4. Food capture. The suckers of the tube feet are also used to capture food.

20.5 LOCOMOTION

Starfish is a slow moving animal. Its locomotory organs are **tube feet** or **podia** which are arranged in four rows in the ambulacral groove of each arm. Each podium is hollow external projection form the body wall with a terminal sucker (Fig. 20.16). Each tube foot receives a lateral, transverse or podial canal from the radial canal of water vascular system. At the base of tube feet, podial canal forms an ampulla. It is continued as a canal inside the podium.

20.6 DIGESTIVE SYSTEM

The digestive system of *Asterias* comprises the **alimentary canal** and **digestive glands.**

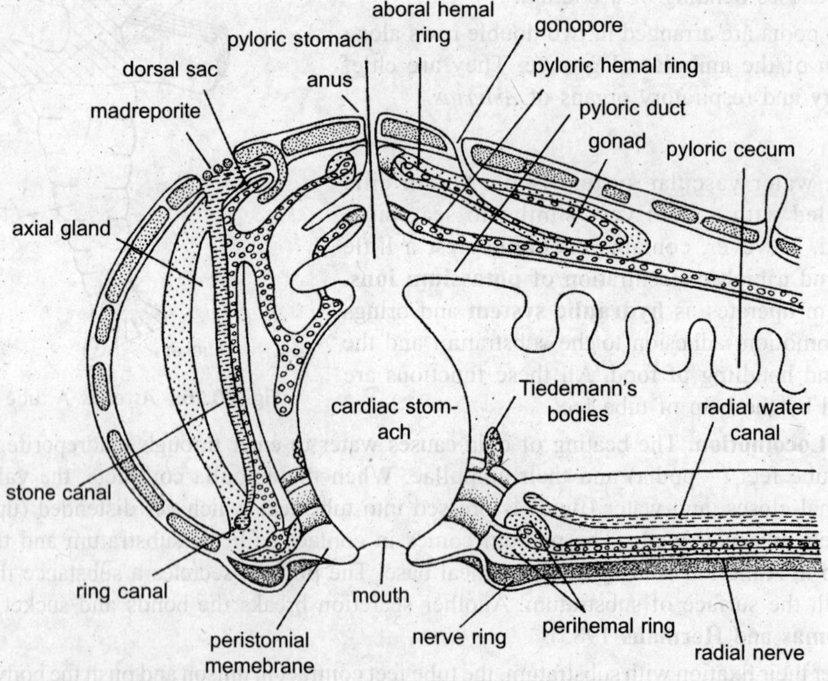

Fig. 20.17. Sea star. The central disc and base of one arm (vertical section).

Alimentary Canal

In *Asterias*, the alimentary canal is tubular, short but spacious canal which extends vertically along the oral-aboral axis in the central disc. It comprises of the following parts:

1. Mouth. The pentagonal actinostome leads into mouth which is situated on the oral surface in the centre of peristomial membrane. The mouth is surrounded by a sphincter muscle. The mouth leads upwards into the oesophagus.

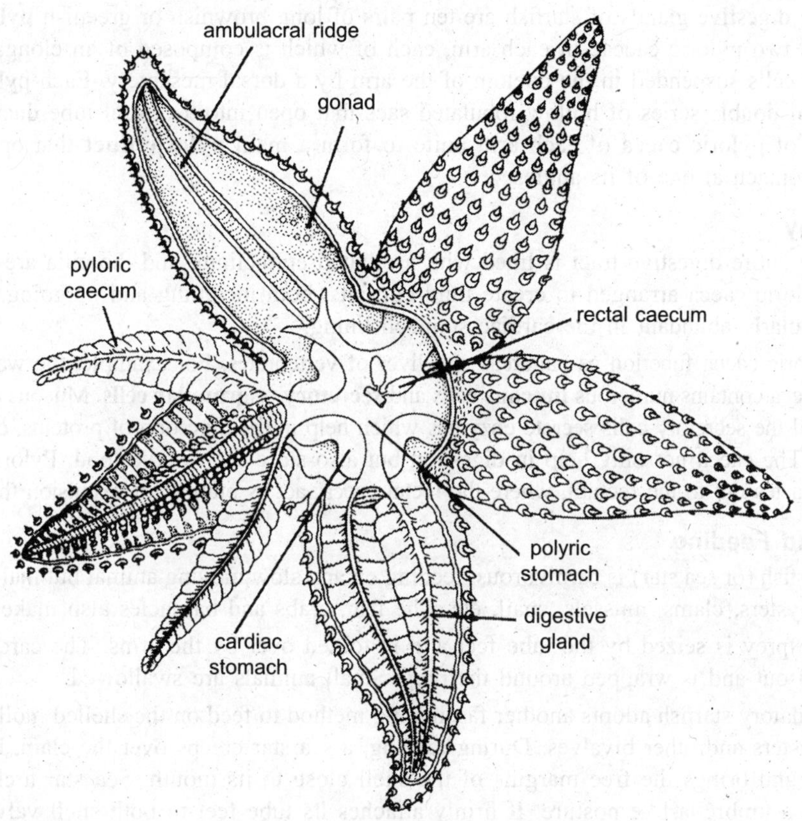

Fig. 20.18. *Asterias*. Digestive system.

2. Oesophagus. It is a very short, wide and vertical tube. It opens aborally in the stomach.

3. Stomach. The stomach is a large, thin-walled spacious five-lobed sac which fill the interior of the disc. It is typically divided by a horizontal constriction into a voluminous oral part, the **cardiac stomach** and a flattened aboral part, the **pyloric stomach.**

***(i)* Cardiac stomach.** The cardiac stomach is connected with oesophagus. It has a muscular, highly folded wall which bulges out to form five lobes, one opposite, each arm. The cardiac stomach is held in position by five pairs of triangular **mesenteries** or **gastric ligaments** which are formed of connective tissue and muscles. A pair of gastric ligaments connect the cardiac stomach to the ambulacral ridge in each arm. During the feeding process, the cardiac stomach can be everted through the mouth by the contraction of muscles of body wall. The retraction of cardiac stomach is brought about by five pairs of **retractor muscles** which arise from the lateral sides of the ambulacral ridge.

***(ii)* Pyloric stomach.** Cardiac stomach leads upwards into a small aboral pyloric stomach which is thin-walled flattened star-shaped sac. It is simply the reception chamber of the ducts from the pyloric caeca. It communicates with the intestine.

4. Intestine. It is a short, tubular, five-sided tube extending vertically from aboral side of pyloric stomach to open through a minute anus in the middle of aboral surface of disc. It gives out two or three small branched and brownish outpocketings called **rectal caecae** which are placed inter-radially. Rectal caecae are excretory glands; they secrete a brownish fluid.

5. Anus. The intestine open on the aboral surface of the central disc by a small rounded aperture, the **anus.**

Digestive Glands

The digestive glands of starfish are ten pairs of long brownish or greenish **pyloric caecae.** There are two pyloric caecae in each arm, each of which is composed of an elongated mass of glandular cells suspended in the coelom of the arm by a dorsal mesentery. Each pyloric caecum consists of double series of hollow lobulated sacs that open into a central tube duct. Two ducts of a pair of pyloric caeca of each arm unite to form a main **pyloric duct** that opens into the pyloric stomach at one of its angles.

Histology

The entire digestive tract is lined with a ciliated epithelium, and the cilia are in the ducts of the pyloric caeca arranged to create fluid currents, both incoming and outgoing. Gland cells are particularly abundant in the cardiac stomach lining.

Pyloric caeca function as pancreas and liver of vertebrates. For example, the wall of sacs of pyloric caeca contains numerous **mucous cells** and **secretory** or **granular cells.** Mucous cells produce mucus and the secretory cells secrete enzymes which help in the digestion of proteins, carbohydrates and fats. The caeca not only help in digestion but also store the reserve food. Pyloric ducts also convey wastes out to the rectum, where the rectal caeca, act as pumps for expulsion through anus.

Food and Feeding

Starfish (or sea star) is carnivorous, feeding on any slow moving animal but mainly molluscs such as oysters, clams, mussels, snail, etc. The fish, crabs and barnacles also make its food.

The prey is seized by the tube feet and is folded over by the arms. The cardiac stomach is everted out and is wrapped around the prey. Small animals are swallowed.

Predatory starfish adopts another fascinating method to feed on the shelled molluscs such as clams, oysters and other bivalves. During feeding, a sea star creeps over the clam, holds it with tube feet and brings the free margins of the shell close to its mouth. Sea star arches its body assuming a umbrella-like posture. It firmly attaches its tube feet to both shell valves and tries to pull apart them (these two shell valves are held tightly together by the powerful adductor muscles). The tube feet gripping the shell valves exert a steady pull till the adductor muscles of the clam are exhausted and give way. As the valves gape, the sea star inserts its everted cardiac stomach into the mantle cavity of clam to devour it. The everted cardiac stomach of some sea stars can squeeze through a space as slight as 0.1 mm. The gape increases as digestion ensues and the clams adductor muscles are attacked.

Digestion, Absorption and Egestion

Digestion in sea star is primarily extracellular. After everting the cardiac stomach over the captured prey, sea star pours out digestive enzymes secreted by the cardiac stomach and pyloric caeca over the prey. The enzymes digest the visceral organs of clam; they thus, converts **proteins** into **peptones** and **amino acids; polysaccharides** (carbohydrates) into **glucose** and **fats** (lipids) into **fatty acids** and **glycerol.** When the digestion is partially completed, the sea star withdraws its stomach along with the digested food by means of its retractor muscles and moves on leaving behind the empty shell of the prey. Remaining extracellular digestion of food occurs in cardiac stomach.

Products of stomach digestion are carried by ciliary tracts up the to stomach wall and through pyloric ducts into the pyloric caeca, where digestion is both **intracellular** and **extracellular** and **absorption** occurs. Products of digestion may be stored in the cells of pyloric caeca or passed through the caeca into the coelom for distribution. The undigested food is egested out through the anus.

20.7 RESPIRATION

Respiration in *Asterias* is brought about by the **dermal branchiae** or **papulae** (Fig. 20.19) and **tube feet.** Papulae are scattered over the aboral surface and tube feet project in the ambulacral grooves. Both these structures are thin-walled and enclose a part of coelom. They bring the coelomic fluid very close to the surrounding sea water for exchange of respiratory gases.

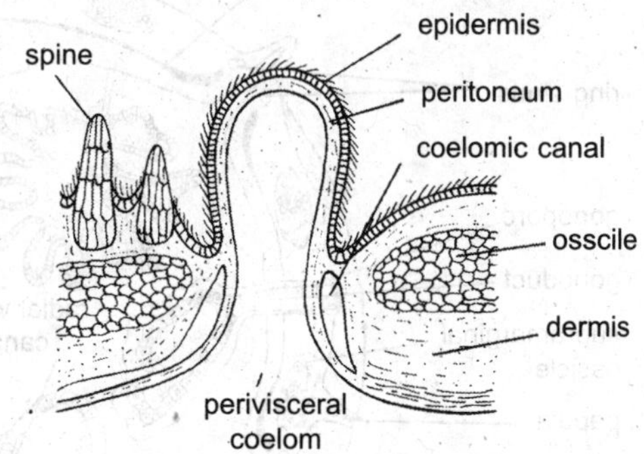

Fig. 20.19. Starfish. Section through papula.

20.8 CIRCULATORY SYSTEM

Asterias lacks a true-blood vascular system. The system which is responsible for circulation of digested food to various body organs is often known as **circulatory system.** The so-called circulatory system of *Asterias* includes following two systems (*i*) **perihaemal system;** (*ii*) **haemal system** (Fig. 20.20).

1. Perihaemal System

It is derived from the coelom and consists of a series of tubular sinuses. These enclose the sinuses of the haemal system except the gastric haemal tufts. The perihaemal system comprises of the following perihaemal sinuses:

(*i*) Axial sinus. It is thin walled, vertical, wide tubular coelomic cavity enclosing the axial gland and stone canal. The axial sinus along with axial gland and stone canal form the **axial complex.**

(*ii*) Aboral perihaemal ring sinus. It is a tubular pentagonal sinus around the intestine, lying just inside the aboral wall of the central disc. It communicates with the axial sinus.

(*iii*) Genital sinuses. Aboral perihaemal ring sinus gives off five pairs of **genital branches,** one pair in each arm. These enclose the gonads and the genital haemal strands.

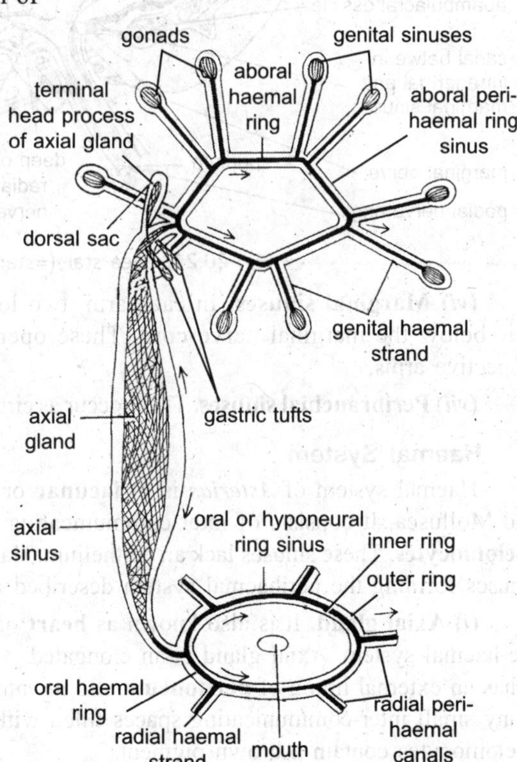

Fig. 20.20. *Asterias*. Haemal and perihaemal system.

(*iv*) Oral hyponeural ring sinuses. At its oral end, the axial sinus opens into the inner division of a circular channel, called the **oral hyponeural ring sinus** which runs around the mouth and above the nerve-ring. It is a large tubular sinus and is divisible into an inner narrow and an outer wide ring by an oblique circular septum called the **haemal strand.**

***(v)* Radial perihaemal sinuses.** From the outer part of the oral ring sinus arise five **radial hyponeural** or **perihaemal sinuses,** each extending through an arm enclosing the radial haemal strands. These sinuses also give out fine channels into the tube feet.

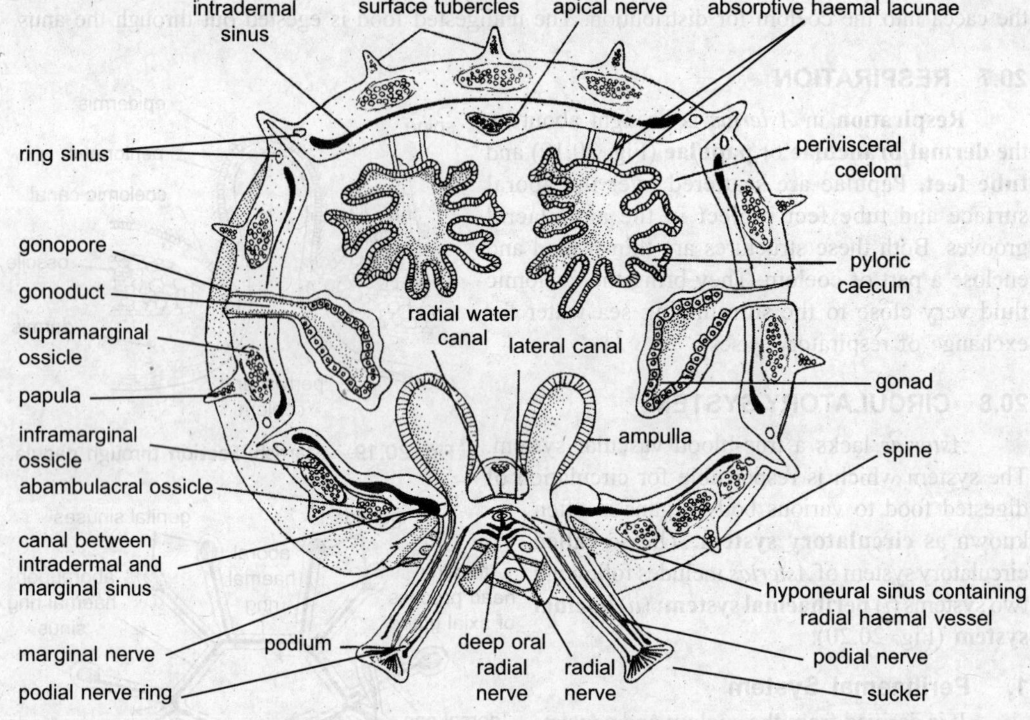

Fig. 20.21. Sea star (=starfish; *Asterias*). T.S. Arm.

***(vi)* Marginal sinuses.** In each arm, two longitudinal marginal sinuses run one on each side below the marginal nerve cord. These open into the radial perihaemal sinuses of their respective arms.

***(vii)* Peribranchial sinuses.** These occur as circular sinuses around the basal part of the papulae.

2. Haemal System

Haemal system of *Asterias* is of **lacunar** or **open type** like the haemocoel of Arthropoda and Mollusca. It consists of inter-communicating sinuses filled with **coelomic fluid** containing **coelomocytes.** These sinuses lack an epithelium. Haemal system is largely enclosed in the coelomic sinuses forming the perihaemal system described above. It comprises the following parts.

***(i)* Axial gland.** It is also known as **heart** or brown **gland** and forms the principal part of the haemal system. Axial gland is an elongated, vertical, fusiform, brownish and spongy gland. It has an external lining of **peritoneum** and its interior is filled with connective tissue containing many small inter-communicating spaces filled with a fluid having amoeboid **coelomocytes.** The coelomocytes contain a brown pigment.

The axial gland is connected with the oral haemal sinus at its oral end and with the aboral haemal sinus at its aboral end. A small terminal **head process** arises from the aboral end of the axial gland. Head process lodged in a separate, closed contractile coelomic sac called **dorsal sac.** The dorsal sac is situated below the madreporite, close to the ampulla of the stone canal. A pair of **gastric tufts** arise from the haemal sinuses in the wall of cardiac stomach and open into the axial gland near its aboral end. Digested food from the stomach passes into the haemal circulation through the gastric tufts.

***(ii)* Aboral haemal ring.** It is a pentagonal ring canal situated beneath the aboral surface of the central disc. It gives off five pairs of **genital haemal strands** to the gonads.

***(iii)* Oral haemal ring.** It is the circular haemal sinus, located around the mouth just below the ring canal of the water vascular system.

***(iv)* Radial haemal sinuses.** These arise radially from the oral haemal ring. In each arm extends one radial haemal sinus along the floor of the ambulacral groove. Each radial haemal sinus gives off branches to the tube feet or podia.

Function. The haemal system acts as a pathway for the distribution of food substances carried by the coelomocytes. The flow of fluid (*i.e.* colourless blood) within it is maintained by the contractile activity of the dorsal sac (or **heart;** see **Rupert** and **Barnes,** 1994). For example, in *Asterias forbesi* dorsal sac beats rhythmically—6 beats per minute. However, the pattern of circulation of fluid is unknown.

The axial gland acts as a genital stolon, producing sex cells which reaches the gonads through the aboral haemal ring and its branches.

20.9 EXCRETION AND OSMOREGULATION

In most echinoderms dissolved nitrogenous wastes (ammonia) diffuse across body surfaces to the outside. This type of excretion occurs across the podia and papulae in sea star and other asteroids. Precipitated material and other particulate wastes are phagocytosed by certain coelomocytes in the body fluids and then discharged by various methods. In sea star, waste-laden coelomocytes accumulate in the papulae, which then pinch off their distal ends, expelling the cells and waste material. Some studies indicate that the rectal glands or rectal caecae may also be involved in excretion (**Brusca** and **Brusca**, 2003).

Echinoderms are generally considered to be strictly marine, **stenohaline** creatures (Box 20.5). Consequently, they do not have problem of osmotic and ionic regulation. The evidence to date suggests that echinoderms are **osmo-conformers.** Both water and ions pass relatively freely across thin body surfaces and the tonicity of the body fluids varies with environmental fluctuations. There appear to be some ionic regulation through active transport, but it is minimal.

Box 20.5 Animal response to osmotic conditions of the medium

There are two extreme patterns of response to osmotic conditions of the environment. Animals may be osmotically labile (dependent) and their body fluid concentration may change with the medium. These are **osmoconformers** or **poikilosmotic animals.** Other animals are osmotically stable (independent) and when the medium changes, the internal concentrations of the body fluid remains constant or unchanged. These are **osmoregulators** or **homoisomotic animals.**

Poikilosmotic animals which can tolerate wide ranges in salinity are known as **euryhaline**, *e.g.*, *Mytilus, Aplysia*. On the other hand poikilosmotic animals which cannot tolerate wide ranges in salinity are known as **stenohaline animals**, *e.g.*, most marine invertebrates (**Singh** and **Kumar** 2005).

20.10 REPRODUCTIVE SYSTEM

Autotomy and Regeneration

Echinoderms possess great power of **regeneration.** *Asterias* is capable to regenerate its any lost part of body at any time. Moreover, if an arm is injured or held up, *Asterias* usually cast it off near the base at the fourth or fifth ambulacral ossicle. This is called **autotomy.** The opening left in the central disc by broken off arm is immediately closed by the contraction of the adjacent body wall musculature for the protection of internal body organs and regeneration of new arm starts at that place. Autonomy is seen in most **ophiuroids,** some **asteroids** and some **holothuroids** but does not occur in **echinoids.** For example, many brittle stars, when are taken out of water, break off portions of their arms into pieces till the central disc completely devoid of arms is left.

A disc deprived of all of its arms regenerates. In *Asternia vulgaris*, a single arm with a portion of disc regenerates an entire animal. But in *Linckia*, an arm totally devoid of disc can regenerate complete animal (Fig. 20.22). Specimens with regenerating arms at the base of the large original arm are called **comets.**

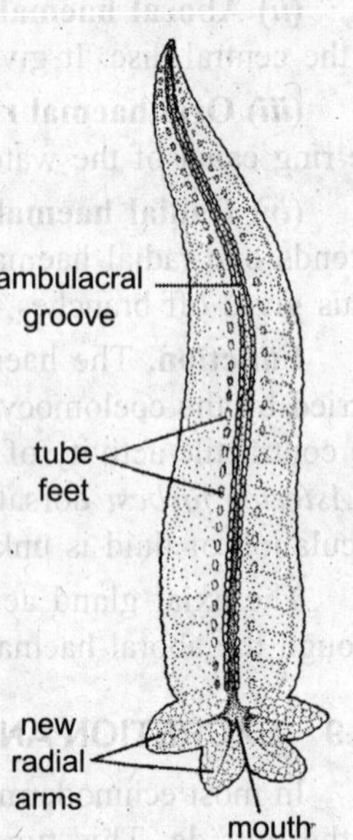

Fig. 20.22. Comet stage of *Linckia*.

In **crinoids,** the arms as well as the visceral sac of the central disc are regenerated when these become accidentally removed. In **holothuroids,** the oesophagus or cloaca with Cuvierian organs is ejected on slightest danger (autotomy) and is regenerated again.

Reproduction in *Asterias* is mainly sexual. Most species of *Asterias* are **unisexual** or **dioecious,** *i.e.*, sexes are separate. Sexual dimorphism is absent. Reproductive organs of *Asterias* are of primitive type. They lack copulatory organs, accessory glands, receptacles for storing ova and reservoirs for storing mature spermatozoa.

Gonads

The **testes** (male gonads) and **ovaries** (female gonads) are morphologically similar but they show colour variation. The testes are pale grey and the ovaries vary from pink to orange. There are five pairs of gonads in either sex, a pair in each arm. The proximal end of gonads are attached to the aboral body wall and they lie laterally freely between pyloric caeca and ampullae of the tube feet.

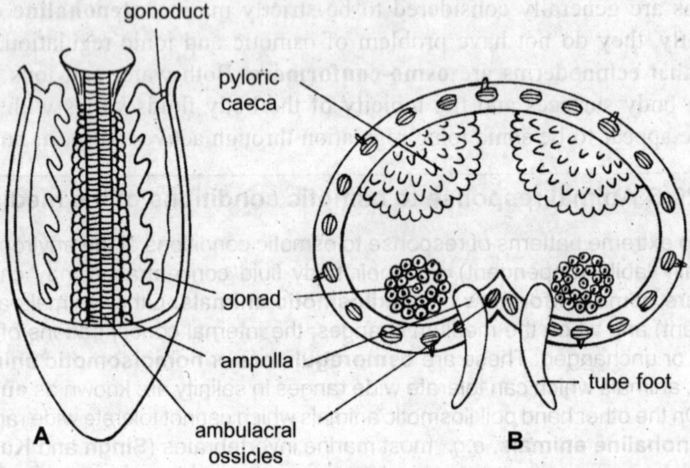

Fig. 20.23. *Asterias*. Gonads. A—An arm is cut open to show gonads. B—T.S. arm showing position of gonads.

The size of gonads varies with season of the year, being largest during the breeding period (late spring). At maturity gonads occupy a considerable portion of the perivisceral space. Each feathery tuft or bunch of grapes consisting of membranous and rounded follicles. It is enclosed in a genital sinus of the perihaemal system. Microscopic examination reveals that the gonad is lined by a germinal epithelium with a connective tissue matrix containing germ cells. From the proximal end of each gonad arises a short ciliated **gonoduct** which opens out laterally on the aboral surface by a minute **gonopore.**

Spermatozoa and ova are discharged by the male and female sea stars respectively into the sea water. Release of gametes is brought about by substances produced in the radial nerve.

In *Asterias*, there is a single breeding season in a year. It is the late spring and appears to be induced by the rising spring temperature. A single female lays about 200 million eggs in a season.

Fertilization

In *Asterias*, fertilization is external; it takes place in sea water.

Development

The embryological development of *Asterias* is **indirect** and includes various larval stages. The fertilized egg or zygote is spherical, 0.5 mm in diameter and contains little amount of yolk. The **cleavage** is rapid, **holoblastic,** equal, indeterminate and radial type (see Box 20.6). As a result, by the second day, a single layered, hollow, spherical, ciliated embryo, called **blastula** or **coeloblastula** is formed. Blastula contains a fluid-filled central cavity, the **blastocoel** and swims about in water freely. The blastula undergoes embolic **invagination** at its vegetal pole and becomes a two-layered, ciliated, cup-like **gastrula.** Its outer layer is formed of **ectoderm** and inner layer of **endoderm** (Fig. 20.24). The cavity of the gastrula which is lined by endoderm is known as **gastrocoel** or **archenteron.** It opens to the exterior by a wide aperture, called **blastopore.** On the ventral side of the embryo, a tubular ingrowth of ectoderm forms the **mouth.**

Box 20.6

1. **Holoblastic cleavage.** When entire egg is divided by each cleavage furrow. When it produces equal-sized blastomeres, it is called **equal holoblastic cleavage.**
2. **Indeterminate cleavage.** When the fate of blastomere is not fixed.
3. **Radial cleavage.** When axes of the cleavage furrows are parallel or at right angles to the axis extending from animal to vegetal pole.

During gastrulation, the advancing tip of archenteron buds off **mesenchyme cells** into blastocoel, which form the **mesenchyme** or **mesoderm.** The advancing blind end of archenteron evaginates as two pouches, which finally get pinched off as the right and left sacs. These sacs enclose anterior part of archenteron, the **enterocoel,** and produce **coelom,** its mesodermal lining and water vascular system. Thus, in *Asterias*, mesoderm has a dual origin and coelom is **enterocoelous.** The embryo at this stage becomes a free-swimming larva.

Larval Development

The development of sea star includes the following larval stages:

1. Dipleurula larva. This is first larval stage and is commonly found in all echinoderms. It is egg - shaped having a bilaterally symmetrical body. The cilia which earlier cover the surface of embryo uniformly, become restricted to two ciliary wavy bands — a **peri-oral band** surrounding a concave area around the mouth and an **adoral band** lying inside the mouth. **Stomodaeum** develops as an ectodermal invagination on the ventral side of embryo and becomes continuous with the archenteron. The opening, thus, developed forms the **mouth** of larva. The archenteron gets differentiated into oesophagus, stomach and intestine. The blastopore becomes the **anus.** With these changes embryo develops into **dipleurula larva** which is capable of independent existence. This larva actively feeds on diatoms, etc. The adoral band of cilia helps in collecting the food particles. Dipleurula larva swims near the surface and rotates clock-wise with the help of cilia of peri-oral band.

2. Bipinnaria larva. In *Asterias*, the free swimming larva which is hatches out of the egg is called **bipinnaria larva.** It develops from zygote in about one week. It is a bilaterally symmetrical larva and possesses two ciliated bands: 1. **Preoral ciliated band** which surrounds the preoral lobe of larva; 2. **Postoral ciliated band** which appears to be longitudinally placed and form a complete

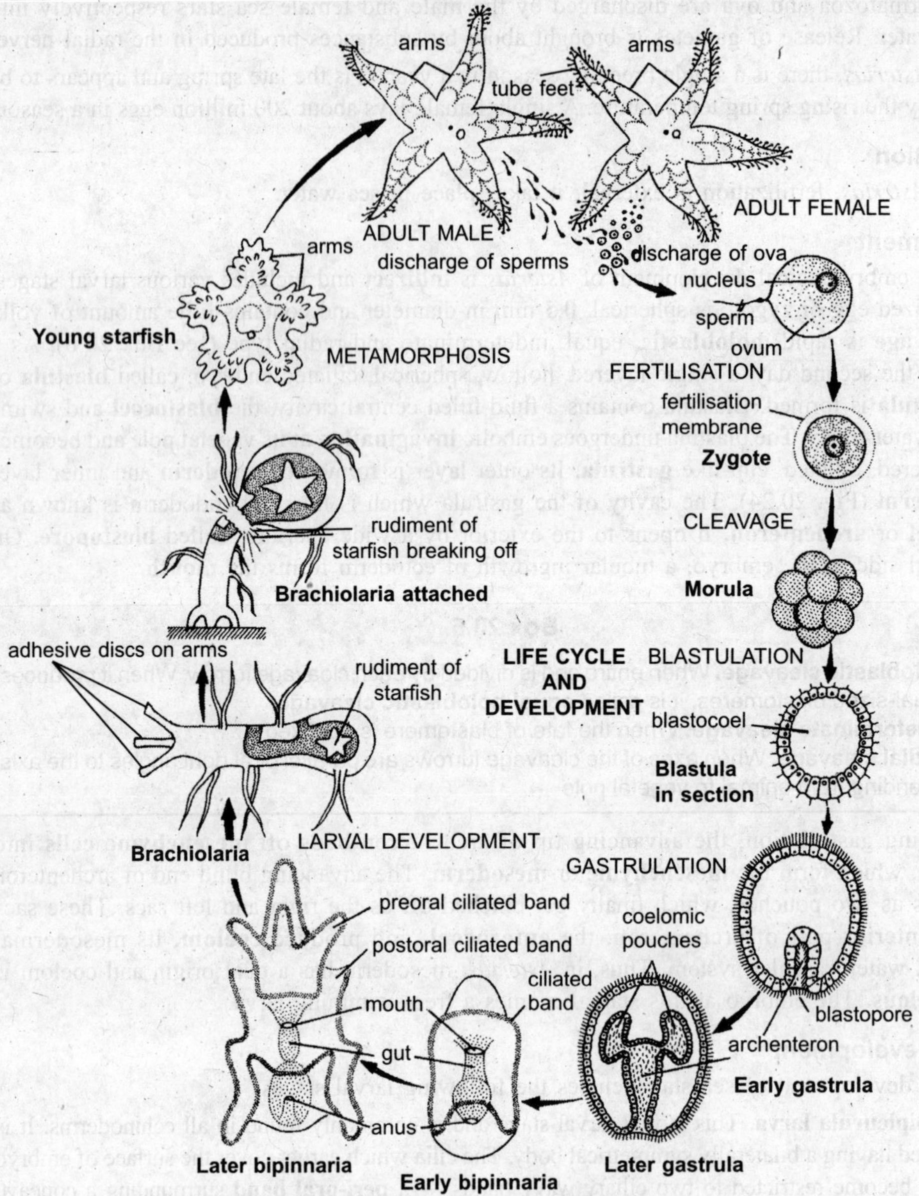

Fig. 20.24. *Asterias*. Development and life history.

ring between the mouth and anus. Both of these ciliated bands are continued over a series of prolongations or projections of the body, called **lobes** or **larval arms** (Fig. 20.25). The name and number of arms developing from preoral and postoral ciliated bands are as follows:

Name of lobe or arm	Number
1. Dorso-median	One
2. Ventro-median	One
3. Pre-oral	Two

4. Antero-dorsal	Two
5. Postero-dorsal	Two
6. Post-oral	Two
7. Postero-lateral	Two

The pre-oral and ventro-median arms develop from the pre-oral ciliated band and rest of the arms develop from the post-oral ciliated band. The arms are provided with muscles and are contractile in a nature. Inside the body occur the coelomic apparatus and the alimentary canal. The bipinnaria larva feeds, on diatoms, etc., by creating food-bearing currents by ciliary tracts in the stomodeal wall. It swims freely by forwarding its anterior end, with a clockwise rotation. After sometime bipinnaria larva develops into the next larval stage called **brachiolaria larva.**

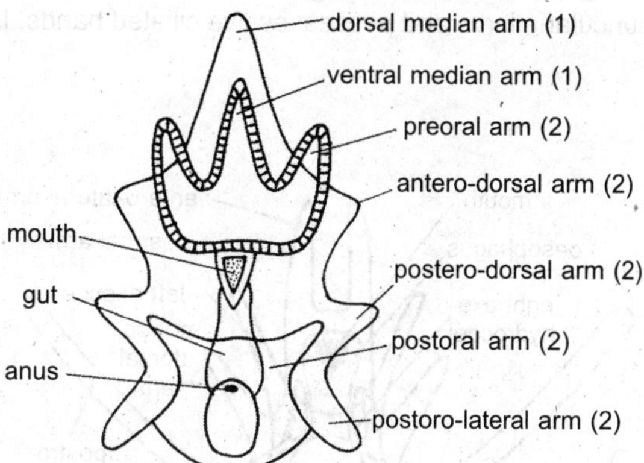

Fig. 20.25. *Asterias*. Bipinnaria larva.

3. Brachiolaria larva. The lobes of bipinnaria larva become modified into long, slender and ciliated **larval arms.** From the preoral lobes arise three short and non-ciliated processes ending into adhesive **fixation discs** or **suckers.** These are called **brachiolar arms.** These arms are devoid of calcareous rods and have prolongations from the coelomic cavity. Its alimentary canal consists of mouth, stomach, intestine and anus. Brachiolaria larva swims and feeds like bipinnaria larva. The appearance of sucker marks the beginning of meta-morphosis.

Other echinoderms produce a variety of larval forms such as **auricularia, doliolaria, ophiopluteus, echino-pluteus,** etc. (see Box. 20.7).

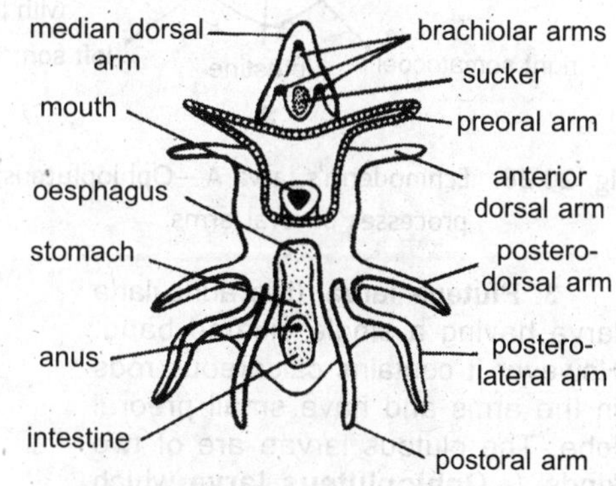

Fig. 20.26. Brachiolaria larva.

Box 20.7 Other larval forms of echinoderms

1. Auricularia larva. In most holothuroids, the eggs are rich in yolk and deve-lopment is direct, but in some cases the egg hatches into a free - swimming bila-terally symmetrical larva. It is known as **auricularia larva** (Fig. 20.27A and Fig. 20.29A). It is transparent pelagic orga-nism of about 1 mm size. The body is provided with single longitudinal ciliated band which forms a **pre-oral loop** around the mouth and **anal loop** encircling anus. These help in swimming. The **pre-oral lobe** is very well - formed. There are no calcareous rods, being replaced by spheroids, star - shaped or wheel - like bodies.

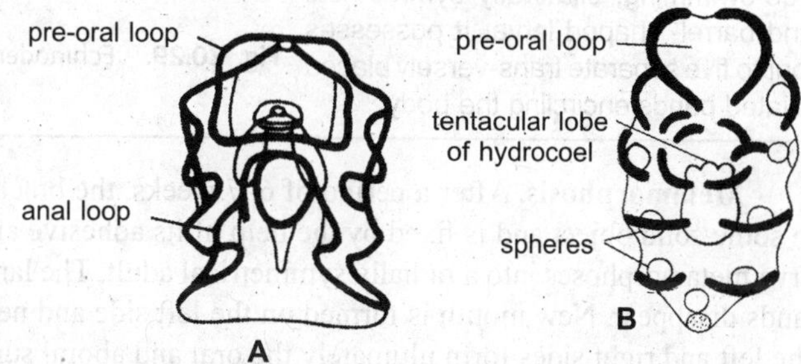

Fig. 20.27. Enhinoderm's larva A—Auricularia; B—Doliolaria.

The interior of larva is occupied by the curved gut with saccular stomach, hydrocoel and right and left stomocoels. The hydrocoel is produced into primary tentacles and communicates with the hydropore by canal.

The auricularia larva metamorphoses into second larval stage known as doliolaria larva (Fig. 20.27B).

2. Doliolaria larva (Fig. 20.27B). It is a barrel-like larva of Holothuroidea which develops from auricularia larva and possesses five ciliated bands. It is also called **pupa stage.**

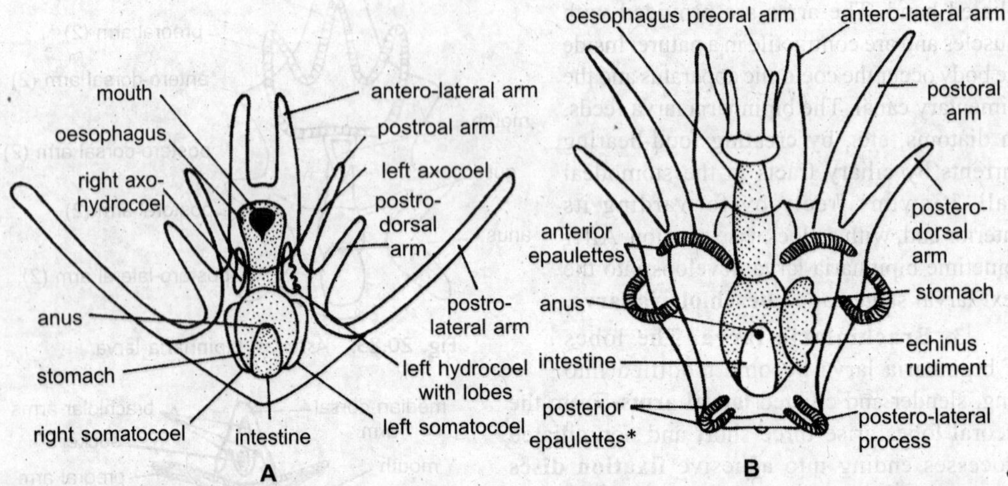

Fig. 20.28. Echinoderm's larva A—Ophiopluteus; B—Echinopluteus. *Epaulette— Any of the knobbed processes on oral arms.

3. Pluteus larva. It is auricularia larva having a single ciliated band. However it contains calcareous rods in the arms and have small preoral lobe. The pluteus larvae are of two kinds 1. **Ophiopluteus larva** which occurs in Oph-iuroidea (Fig. 20.28A) 2. **Echinopluteus larva** which occurs in Echinoidea (Fig. 20.28B).

4. Vitellaria or Yolk Larva. This larval stage is present in *Antedon*. It is free-swimming, bilaterally symmetrical and barrel- shaped larva. It possesses four to five separate trans-versely placed ciliated bands encircling the body.

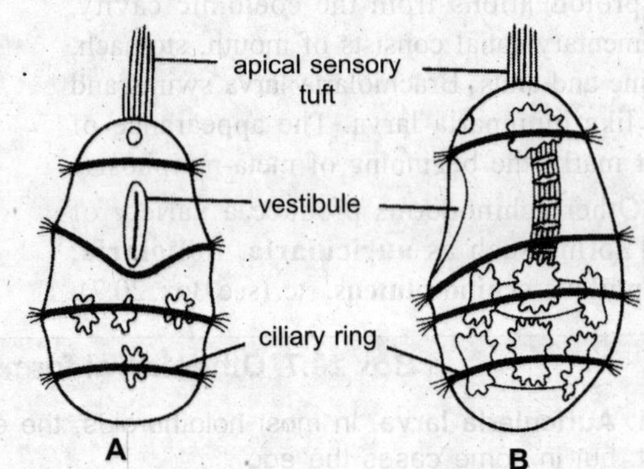

Fig. 20.29. Echinoderm's larva A—Auricularia; B—Doliolaria.

Metamorphosis. After a period of 6–7 weeks, the brachiolaria larva settles on the bottom or on some solid object and is fixed by the help of its adhesive arms. Now the bilaterally symmetrical larva metamorphoses into a radially symmetrical adult. The larval mouth and anus close. The ciliary bands disappear. New mouth is formed on the left side and new anus on the right side of the larva. The left and right sides form ultimately the oral and aboral surfaces respectively. Five lobes, called **arm rudiments,** develop around oral-aboral axis. Skeletal elements develop on the arm rudiments and the radial canals grow into the arms. The coelom of the adult develops from the right and left coelomic pouches of the larva. In each arm two pairs of outgrowths from the coelom form the first **tube feet** and serve for attachment. Further complex reorganisational changes result in the formation of adult *Asterias*. The newly detached rudiment of the body of sea star is about 1 mm in diameter and contains short stubby arms.

QUESTIONS

Long Answer Questions

1. Give an account of habit and habitat of starfish. Describe its external features.
2. Describe structures present on the oral and aboral surfaces of starfish.
3. Describe the external features of *Asterias*. Why the oral-aboral axis of its body is not the antero-posterior axis.
4. Give an account of skeletal system of starfish.
5. Describe in detail the water vascular system or ambulacral system of starfish. (*CCSU Meerut 2009*)
6. Give detailed account of the structure and function of water vascular system of *Pentaceros*.
7. Give an account of ambulacral system of *Asterias*. How does it help in the locomotion of the animal.
8. Describe the locomotory organs and mode of locomotion in starfish.
9. Write the structure of tube feet and how are they extended and retracted.
10. Give an account of the digestive system and mechanism of feeding in a starfish (= sea star).
11. Write a short note on physiology of digestion.
12. Describe the circulatory system of *Asterias*.
13. Describe the reproductive system of *Asterias*. Where does fertilization occur in this animal? Add a note on its development. Describe various larval forms of echinoderms.
14. Describe life-history of *Asterias*.
15. Why is larva of starfish of particular significance?
16. Write an essay about the larval forms of Echinodermata.
17. Describe locomotion, feeding, respiration and habitat in starfish.
18. Describe symmetry of Echinodermata.
19. Describe water vascular system in Echinodermata.

Short Answer Questions

1. Write short notes on the following:
   (*i*) Symmetry of Echinodermata.
   (*ii*) Endoskeleton of *Echinus*.
   (*iii*) Pedicellaria
   (*iv*) Water vascular system
   (*v*) Madreporite
   (*vi*) Stone canal
   (*vii*) Polian vesicle
   (*viii*) Autotomy and Regeneration
   (*ix*) Ambulacral groove
   (*x*) Tube feet
   (*xi*) Tiedmannn's bodies
   (*xii*) Locomotion in *Asterias*
   (*xiii*) Axial organ
   (*xiv*) Axial gland
   (*xv*) Bipinnaria larva
   (*xvi*) Auricularia larva
2. Draw well labelled diagrams of the following:
   (*i*) External features of starfish.
   (*ii*) Aboral view of starfish.

(*iii*) Oral view of starfish.

(*iv*) Position of pyloric caeca in the arm of *Asterias*.

(*v*) T.S. arm of starfish.

(*vi*) Water vascular system or ambulacral system of starfish.

(*vii*) Pedicellaria of starfish.

(*viii*) Haemal and perihaemal system of *Pentaceros*.

(*ix*) Bipinnaria larva.

(*x*) Brachiolaria larva.

(*xi*) Echinopluteus larva.

3. Differentiate between the following:

(*i*) Polian vesicles and podium

(*ii*) Sea lily and brittle star

(*iii*) Sea urchin and sea cucumber

Very Short Answer Questions

1. Name two types of pedicellariae present in *Asterias*.

2. Where is madreporite located in *Asterias*?

3. What is bivium?

4. What is actinostome?

5. Name the type of digestion in *Asterias*.

6. Mention the three parts of a tube foot.

7. What is the function of Tiedmann's body?

8. What is chief excretory matter of *Asterias*?

9. What type of cleavage does an egg of sea star (*Asterias*) undergo?

10. How many Teidmann's bodies are found in sea star (or star fish)? (*C.C.S.U. Meerut 2002*)

11. What do you call the property of regrowing of lost parts?

12. What is autotomy?

13. What is a comet?

14. Where are eyes of *Asterias* located?

15. Describe Tiedmann's bodies. (*C.C.S.U., Meerut 2009*)

Fill in the Blanks

1. *Asterias* is commonly called ....

2. Sea star belongs to the class .... and phylum ....

3. *Asterias* has .... radial symmetry.

4. Tube feet help the sea star in .... , feeding and ....

5. The respiratory organs of starfish are .... or ....

6. Pedicellariae help the sea star in ....

7. In *Asterias*, tube feet are located in the .... grooves.

8. The coelomic fluid of *Asterias* contains phagocytic cells termed ....

9. The pyloric caeca function as the ....

10. In sea star, fertilization takes place in ....

11. Sea star (*Asterias*) moves with the help of ....

12. The coelom of sea star is ....

13. Dermal papulae of sea star are useful in .... and in respiration.

Multiple Choice Questions

1. Locomotory organs of sea star are

(*a*) podia (*b*) polian vesicles

(*c*) vacuole (*d*) none

2. Epidermis of *Asterias* contains

(*a*) epithelial cells (*b*) gland cell

(*c*) neurosensory cells (*d*) all the above

3. The skeletal ossicles of sea star arise from

(*a*) ectoderm (*b*) mesoderm

(*c*) endoderm (*d*) none of these

4. Which is enterocoelic animal

(*a*) *Taenia* (*b*) *Pheretima*

(*c*) *Palaemon* (*d*) *Asterias*

5. An animal which eats its prey after digesting it is

(*a*) *Physalia* (*b*) Leech

(*c*) sea star (*d*) *Sepia*

6. Ambulacral system of *Asterias* helps in

(*a*) locomotion (*b*) attachment

(*c*) respiration (*d*) all of these

7. Ambulacral system of sea star contains

(*a*) blood (*b*) coelomic fluid

(*c*) water (*d*) mucus

8. In sea star stone canal is associated with

(*a*) blood vascular system (*b*) canal system

(*c*) digestive system (*d*) water vascular system

9. Madreporite and Tiedmann's body is associated with

(*a*) ambulacral system (*b*) nervous system

(*c*) digestion (*d*) all of these

10. Stone canal connects the

(*a*) ring canal and polian vesicle (*b*) ring canal and radial canal

(*c*) madreporite and ring canal (*d*) madreporite and polian vesicle

11. Respiration in *Asterias* occurs in the

(*a*) pedicellariae (*b*) ossicles

(*c*) papulae (*d*) madreporite

12. *Asterias* is

(*a*) ammonotelic (*b*) uricotelic

(*c*) ureotelic (*d*) none of these

13. The eggs of *Asterias* are

(*a*) microlecithal (*b*) macrolecithal

(*c*) megalecithal (*d*) oligolecithal

14. Larval stage in life history of sea star is

(*a*) bipinnaria (*b*) brachiolaria

(*c*) both (*d*) none

ANSWERS

Very Short Answer Questions

1. Pedicellariae with parallel jaws and with cross jaws.
2. On the aboral surface of the central disc in one of the inter-radii.
3. Two arms of *Asterias* having madreporite between them.
4. Depression in the middle of the central disc in one of the inter radii.
5. Extracellular as well as intracellular.
6. Ampulla, podium and sucker.
7. Formation of phagocytic coelomocytes present in the water vascular system.
8. Ammonium compounds.
9. Holoblastic, radial and indeterminate (or regulative).
10. There are five pairs of greatly folded pouches called **recemose vesicles** or **Teidmann's bodies** produced from the inner side of the ring canal of water vascular system of sea star (*Asterias*).
11. The property of regrowth of lost parts in animals is called **regeneration**. Sea star (*Asterias*) can regrow any lost part of body any time.
12. Automatic breaking of any body part by an animal is called **autotomy.** Sea stars are capable of casting off its any arm, if injured or held up, from near its base.
13. A disc of *Asterias* which has lost all of its arms may regenerate all of them. A specimen which has regenerating arms at the base of the large original arm is called a **comet.**
14. The tip of each arm bears a small red pigment spot, the eye. Each eye contains several ocelli. They lie below the sensory tentacles and are sensitive to light.
15. These are four or five pairs of pouch-like bodies attached to the ring canal of sea stars functioning in production of coelomocytes.

Fill in the Blanks

1. Sea star (starfish) **2.** Asteroidea, echinodermata
3. Pentamerous **4.** Locomotion, respiration
5. Dermal branchiae, papulae **6.** Clearing the body surface of debris and organisms
7. Ambulacral **8.** Coelomocytes (amoebocytes)
9. Digestive gland **10.** Sea water
11. Tube feet **12.** Enterocoelous
13. Excretion

Multiple Choice Questions

1. (*a*) **2.** (*d*) **3.** (*b*) **4.** (*d*) **5.** (*c*) **6.** (*d*)
7. (*c*) **8.** (*d*) **9.** (*a*) **10.** (*c*) **11.** (*c*) **12.** (*a*)
13. (*a*) **14.** (*c*).

21

Some General Topics of Echinodermata (Water Vascular System of Asteroids, Larval Forms of Echinodermata and Phylogenetic Position of Echinodermata)

21.1 WATER VASCULAR SYSTEM (WVS) OF ASTEROIDS

The water vascular system (WVS) is unique to the echinoderms. It is derived from the coelom, from which it separates during development of the animal. WVS consists of stone canal, ring canal, radial canal, lateral canal and tube feet. It helps in locomotion and adherence to the substratum. It operates during locomotion as a hydraulic system(**Hydraulic** means relating to or operated by a liquid moving in a confined space under pressure). In this chapter water vascular system of Asteriods only will be described.

The asteriod water vascular system (WVS), like that of all echnioderms, consists of the hydraulic locomotory tube feet (Box 21.1) and a pentamerous arrangement of internal coelomic canals (Fig. 21.1A). The internal canals include a circumoral **ring canal**, from which a **radial canal** extends into each arm. Perpendicular to the radial canal is a bilateral sequence of **lateral**

Box 21.1

Like the members of their sister taxon, Enteropneusta, echinoderms are **tricoelomate**, and some of the coelomic cavities develop diverticula that adopt unique functions. The most conspicuous of these is the **water vascular system** (**WVS**), which includes tubular outgrowths of the body wall (the tube feet) that are used for locomotion, feeding, gas exchange and other tasks (**Ruppert** *et al.*, 2004).

canals, each of which terminates in a bulbous **ampulla, tube foot** and sometimes, a **sucker** (Fig. 21.1). A **stone canal** ascends aborally from the ring canal and opens into a small chamber,the **madreporic ampulla**, which joins the **madreporite**, a specialized porous ossicle on the surface of the disc. The pores of the madreporite open to the surrounding sea water. The other colourful and conspicuous madreporite is irregularly furrowed and resembles a solitary stony (*i.e.*, madreporarian) coral. The entire WVS is lined by myoepithelium and thus is ciliated and muscular. The muscles are best developed in the ampullae and tube feet, and the ciliary flows in the internal canals function in transport. The WVS in well developed in asteroids and used for locomotion, prey manipulation and gas exchange.

Each lateral canal has a valve and terminates in an ampulla and a tube foot (Fig. 21.1 B,C). The ampulla is a small, muscular sac that bulges into the perivisceral coelom (Fig. 21.1A, B). The ampulla opens directly into a canal that passes downward between the ambulacral ossicles into the tube foot. The **ambulacral ossicles** form the floor of the ambulacral ridge; a radial canal lies on the oral side of these ossicles (Fig. 21.1 B,C).

The tube feet project into ambulacral groove (Fig. 21.2B) and are arranged in two rows when the lateral canals are all of the equal length or in four rows when they are alternately long and short. A well–developed longitudinal musculature lines the ampulla and tube feet. Commonly, the tube–foot tip is expanded to form a sucker (Fig. 21.1 A,B).

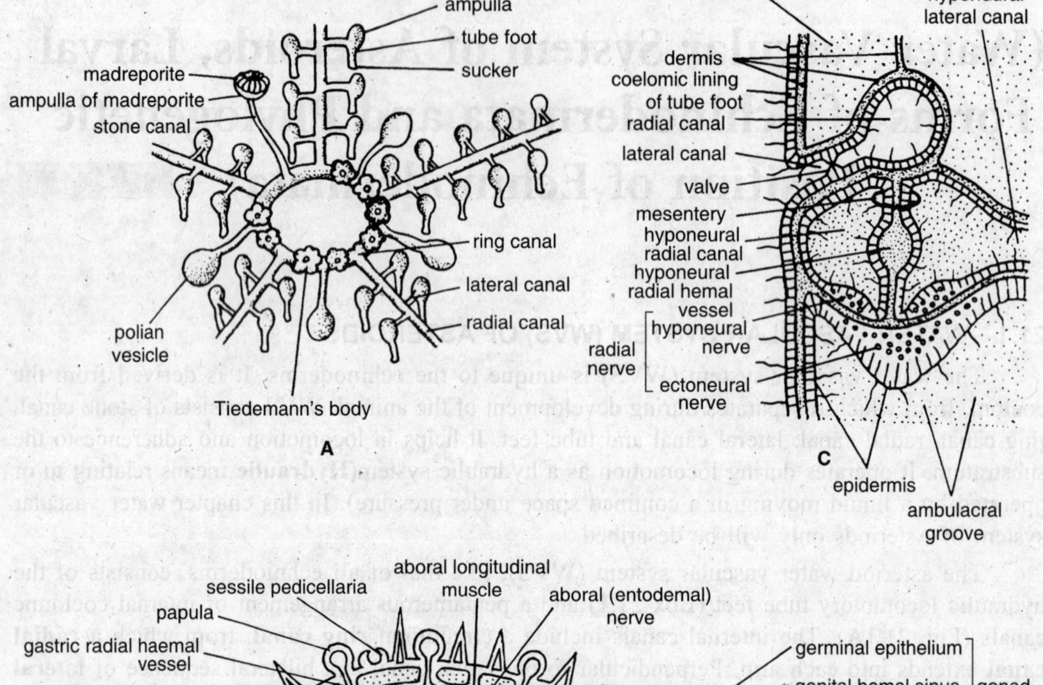

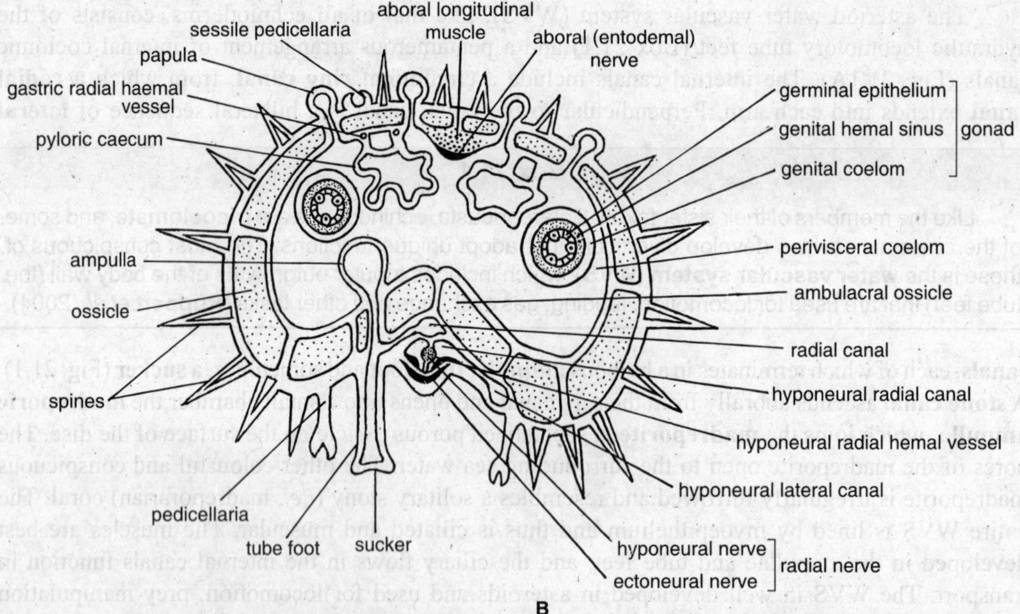

Fig. 21.1. Asteroidea. Water–vascular system and radial organs. A–Diagram of the water vascular system (WVS). B–Diagrammatic cross section of a sea star arm. C–Enlargement of the ambulacral area of B (Ruppert *et. al.,* 2004).

Operation of Tube Feet. The cyclic movement of each tube feet consists of extension–attachment, force generation (pulling), and detachment–retraction. Extension occurs as the longitudinal musculature of the ampulla contracts, the valve in the lateral canal closes, and water is forced into the tube foot, which then elongates. When the tube foot contacts the substratum, the sucker adheres. Adhesion is largely chemical with the tube foot secreting a substance that bonds with the surface. Another secretion later breaks the bond and releases the tube foot. (it is just like the duo–gland adhesive systems of flatworms and gastrotrichs). When attachement is prolonged or a large force is generated, adhesion by suction may also occur. Once attached by its sucker, longitudinal muscles contract and exert a pulling force for *locomotion* or other purposes. The sucker then detaches and the contraction of longitudinal muscles *retracts* the tube feet as fluid is displaced into the ampulla. Contraction of longitudinal muslces on only one side of tube foot *bends* the appendage.

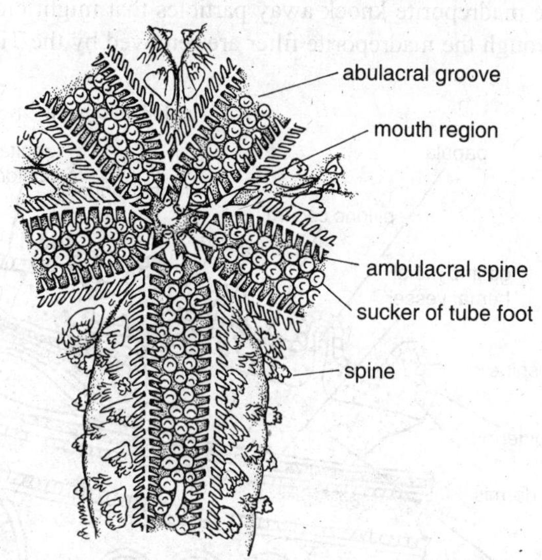

Fig. 21.2. *Asterias forbesi*. Ambulacral oral surface of the disc.

The internal canals of WVS have two primary functions: **circulation** and **fluid–volume maintenance** for hydraulic function. All ciliary flows in WVS, including the tube feet, are bidirection, so each canal or tube foot is simultaneously analogous to both artery and vein. This circulation probably transports oxygen from the tube feet to the internal tissues of the disc, and perhaps nutrients from the gut to the musculature of the tube feet.

Functions of Tiedmann's bodies and polian vesicles. The ring canal is the hub of the system, where all the radial canals ineract. Two organs that play a role in system maintenance are also situated here. On the inner circumference of the ring canal are five pairs of spongy pouches, called **Tiedmann's bodies** (Fig. 21.1A and Fig. 21.3). They are located interradially, between the arms. Tiedmann's bodies function as lymph nodes, removing and destroying bacteria and other unwanted particulates in the WVS fluid. Also attached interradially to the inner side of the ring canal of many asteroids, but not *Asterias*, are one to five elongate, muscular sacs konwn as **polian vesicles** (Fig. 21.1A). As their function is uncertain, but they may be a fluid reservoir for the WVS as a whole or perhaps only for the oralmost tube feet, which often are much larger than the tube feet on the arms but have normal size ampullae.

Functions of stone canal and madreporite. The stone canal and madreporite help maintain the fluid volume of WVS. The stone canal originates on the ring canal, ascends vertically to the aboral body surface, and joins the madreporic ampulla (shared by the axial canal) which opens to the exterior via the madreporite (Fig. 21.1A and Fig. 21.3). The stone canal is named for the calcareous ossicles in its wall. They support the canal wall as do the cartilagenous rings in human trachea. The madreporite is perforated by pores that lie in ciliated surface furrows. The ciliary beat in the stone canal is bidirectional, but the inward (oralward) beating cilia are strongly developed and overwhelm the opposing cilia. This results in an inflow of sea water across the madreporite (Box 21.2), down the stone canal, and into the remainder of the WVS. The stone canal is thus a "**ciliary pump**" that maintains the fluid volume of WVS. Under normal conditions only a trickle of water enters the WVS via the madreporite and stone canal because the system ends blindly at the tube feet and back–pressure oppose inflow. But the pressurised, thin–walled tube feet are leaky and some fluid recovery is required. The cilia on the surface of the madreporite knock away particles that might clog the pore or enter the WVS. Those that slip through the madreporite filter are removed by the Tiedmann's bodies or phagocytic coelomocytes.

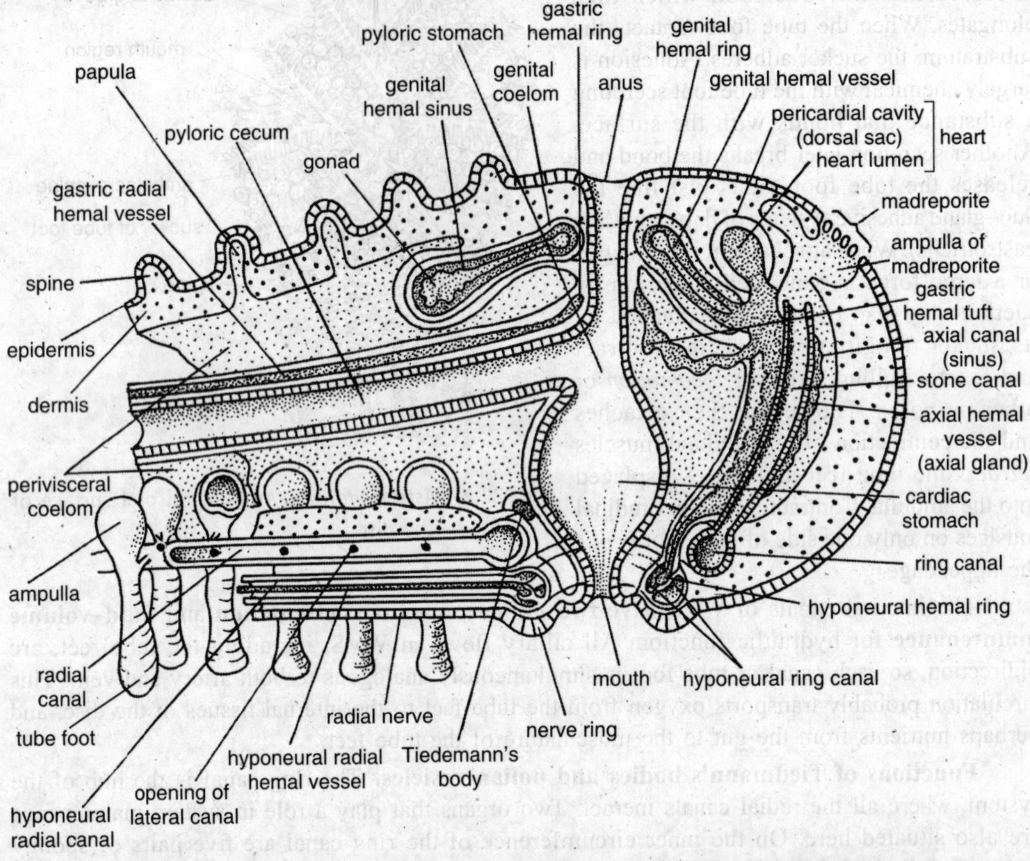

Fig. 21.3. Asteroidea. Diagrammatic internal anatomy of gut, coelom and haemal system (after Ruppert *et al.*, 2004).

Box 21.2

1. Using radioactive tracers, **Ferguson** (1984) demonstrated that water enters through the madreportie.
2. **Ferguson** and **Walker** (1991) described the stone canal in some sea stars as a "ciliary pump" that draws fluid into the WVS from both madreporite and the axial sinus of the haemal system. Thus, it appears that the liquid in the water vascular system is a combination of environmental sea water and body fluid.
3. The operation of the tube feet depends on hydraulic pressure regulation and on muscle action of the individual ampullae and podia (**Brusca** and **Bursca** 2003).

The WVS fluid is similar to seawater except that it contains coelomocytes, some proteins and higher concentration of potassium ions (K^+). The dissolved proteins and elevated K^+ concentration creates an osmotic influx of water across the tube feet. Thus, osmotic recovery of water by the tube feet and physical pumping by the stone canal combine the offset water losses from muscular pressurisation of the tube feet.

WVS of sea daisies. The water vascular system of the sea daisies, *Xyloplax* (asteroids; Box 21.3) is unique among the echinoderms (**Baker** *et al.*, 1986). A madreporite homologue, the hydropore opens on the aboral surface on the CD interambulacrum (Fig. 19.1A and Fig. 21.4) and connects internally to a pair of concentric water canals (Fig. 19.1B). Polian vesicles lie on the other four interambulacra. The outer, marginal water canal gives rise to peripherally located suckerless podia. Each podium bears an internal ampulla. This is the only echinoderm WVS in which the podia are not arranged along the ambulacra (see **Brusca** and **Brusca** 2002).

Box 21.3 "Sea Daisies"

Body discoidal (< 1 cm diameter); with ring of marginal spines, but without arms or rays; skeletal plates arranged concentrically; suckerless podia in a ring near body margin (Fig. 19.1B); two ring canals with hydropore on CD interambulacrum; five large ossicles on aboral surface mark ambulacra (Fig. 19.1A); gut absent or incomplete. Classification of the enigamatic sea daisies (previously class Concentricycloidea) is problematic, but many autorities assign them to the order Sipnulosida, class Asteriodea (**Brusca** and **Brusca** 2002).

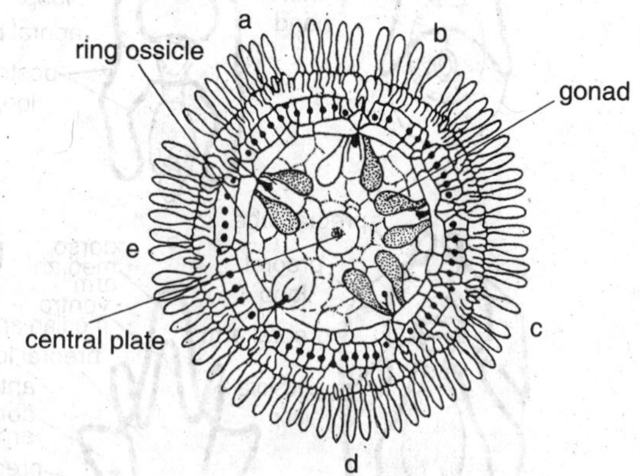

Fig. 21.4. *Xyloplax* (the sea daisy). External features of oral side of the body.

21.2 LARVAL FORMS OF ECHINODERMATA

Echinoderms have following two types of embryonic development.

1. Direct Development. Some echinoderms have **direct development** without a larval stage. These forms generally inhabit Arctic and Antarctic waters, where conditions are not favourable for the larval life. Eggs of these echinoderms are rich in yolk and develop in some part of the body in yolk and develop in some part of the body of the mother.

2. Indirect Development. Majority of echinoderms have indirect development with a larval stage. The eggs of these echinoderms have very little yolk and develop in the sea water (outside mother's body). The zygote undergoes holoblastic, nearly equal, radial and indeterminate cleavage, producing a ciliated blastula (Fig. 21.5). The balstula by invagination changes into a gastrula. The cilia that cover the surface of the gastrula uniformly, become restricted to (*i*) a large **perioral ciliated band** which surrounds a concave area around the mouth on the ventral side and (*ii*) a small **adoral ciliated band** that lines the mouth (stomodaeum). With all these changes, the embryo becomes a larva capable of independent existence. It has a bilateral symmetry and is called the **dipleurula larva**. It is a fundamental larva from which airse the various types of larvae of Echinodermata. Dipleurula larva changes into other larvae by (*a*) the formaton of variously arranged short or long processes, the **arms** and (*b*) modification of the ciliated bands.

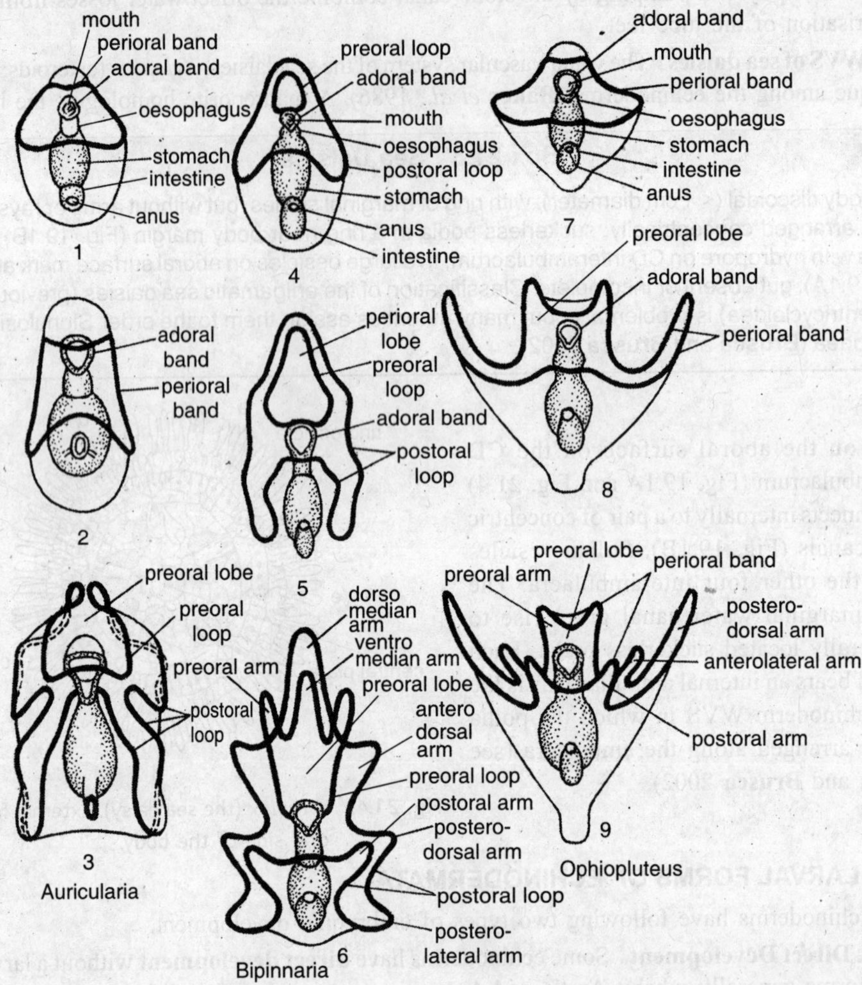

Fig. 21.5. Development of echinoderm larva. A – Dipleurula; B and C– Auricularia; D to F – Bipinnaria and G to L – ophiopluteus.

The most striking feature of the echioderm larvae is their bilateral symmetry, which is in marked contrast to the radial symmetry of the adults. The larvae of echinoderms bear on the surface a varying number of ciliated bands with which they swim about. They have a complete functional digestive tract that consists of ciliated stomodaeum, oesophagus, stomach, intestine and anus. The stomodeal cilia set up a water current that draws food (*e.g.,* diatoms) into the gut. The digestive tract buds off two sacs, which give rise to the coelom, water vascular system and other coelomic cavities of the adult (Fig. 21.6).

After a free–swimming planktonic life of about two weeks to two months, the bilateral larva undergoes **metamorphosis** into an adult with radial symmetry. The larval arms disappear in metamorphosis and are not equivalent to the arm of certain adult echinoderms such as sea–stars and brittle stars. As the metamorphic transformations take place, most of the larval strucutres are lost and the juvenile assumes **benthic life**. Many echinoderm larvae appear to settle preferentially near conspecific adults. In the echinoid, sand dollar (*Dendraster excentricus*) successful metamorphosis is triggered by **pheromones** that are released by adults and act on the larval nervous system (**Brusca** and **Brusca** 2002).

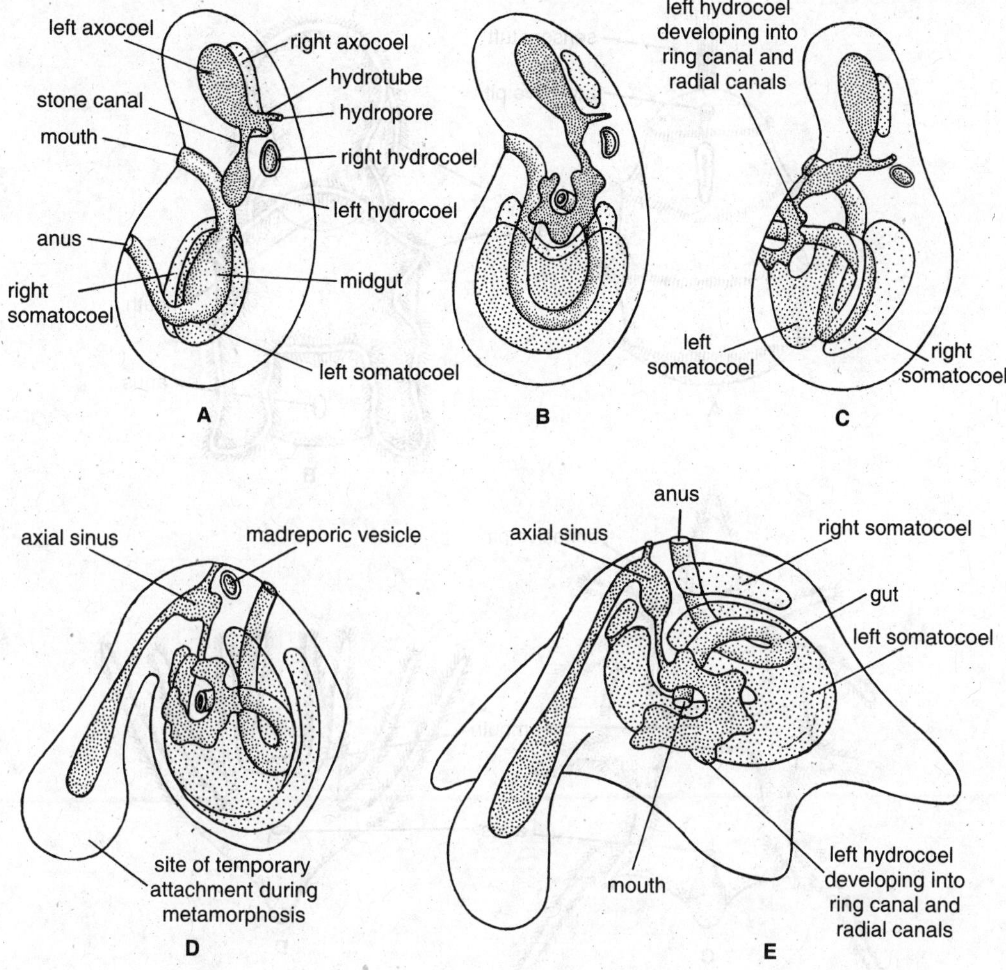

Fig. 21.6. A–D The development of the coelom and its derivatives in an asteriod. E–Metamorphosis in the same animal.

Larvae of Different Classes of Echinodermata

The larvae of the various classes differ in shape of body, size, number and arrangement of the arms and number of ciliated bands. The larva is called **bipinnaria** in the Asteroidea, **auricularia** in the Holothuroidea, **echinopluteus** in the Echinoidea, **ophiopluteus** in the Ophiuroidea, and **crinoid** in the Crinoidea (Fig. 21.5).

1. Bipinnaria larva and brachiolaria larva or Asteroida. (Fig. 21.5 and Fig. 21.7). This larva has a large preoral lobe which gives off three arms: a **ventro–median** and a pair of **preoral** all directed forward. The sides of larva bear four pairs of arms : **anterodorsal, posterodorsal, postoral** and **posterolateral**. These arms give the larva its name. Besides these is a prominent dorsomedian arm extending forwards and continuous with the anterodorsal arm. The perioral ciliated band is split into two : the **preoral loop** that borders the two : the **preoral loop** that

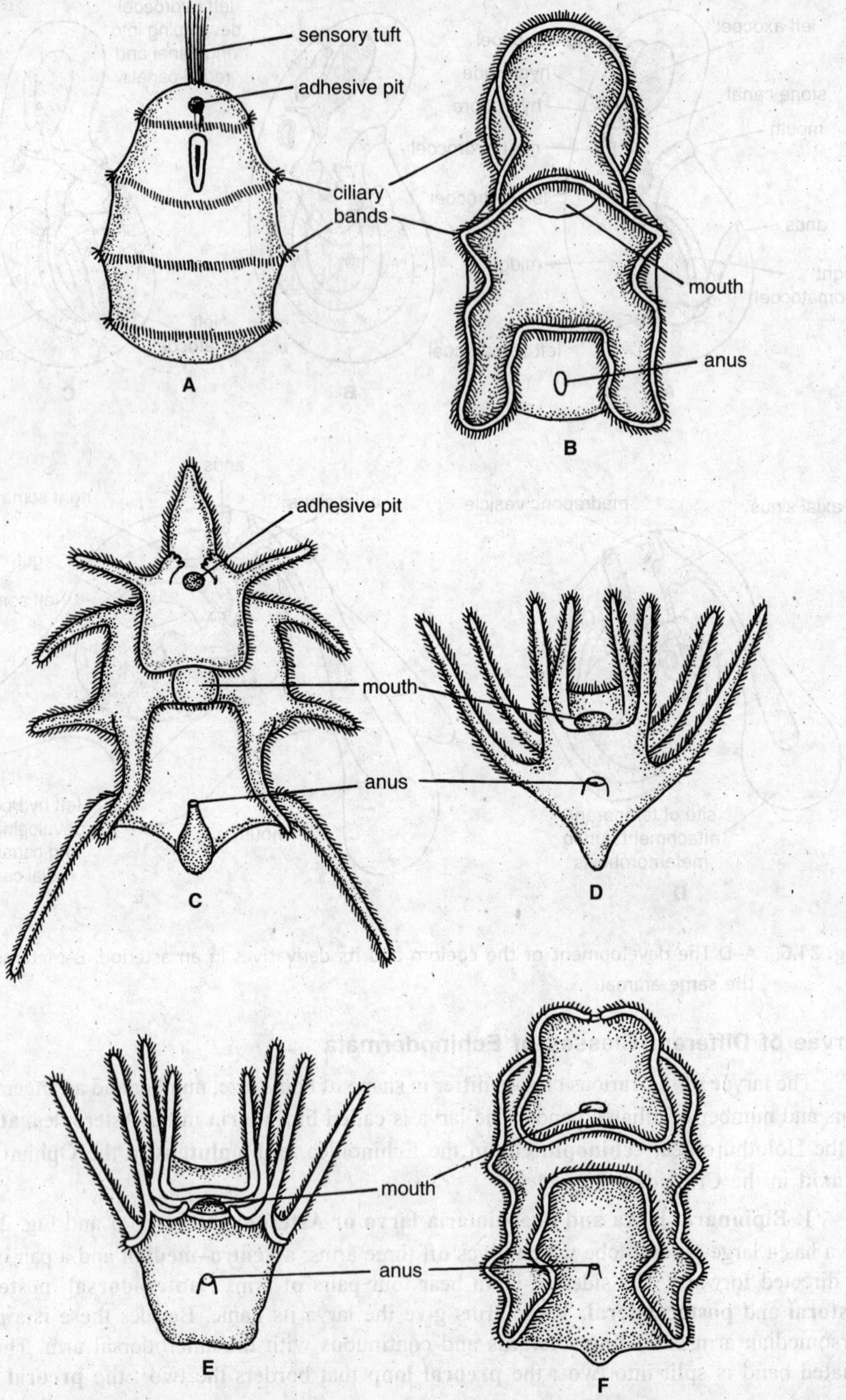

Fig. 21.7. Echinoderm larval types. A–Vitellaria larva of Crinoidea; B, C–Bipinnaria and later Brachiolaria larvae of sea star, D–Ophiopluteus larva of a brittle star; E–Echinopluteus larva of a sea urchin; F–Auricularia larva of a sea cucumber (after Brusca and Brusca 2002).

borders the preoral lobe and its arms and the **postoral loop** which borders all the remaining arms. The two looops may be separate from the beginning or may remain continuous. The arms may be very long and slender.

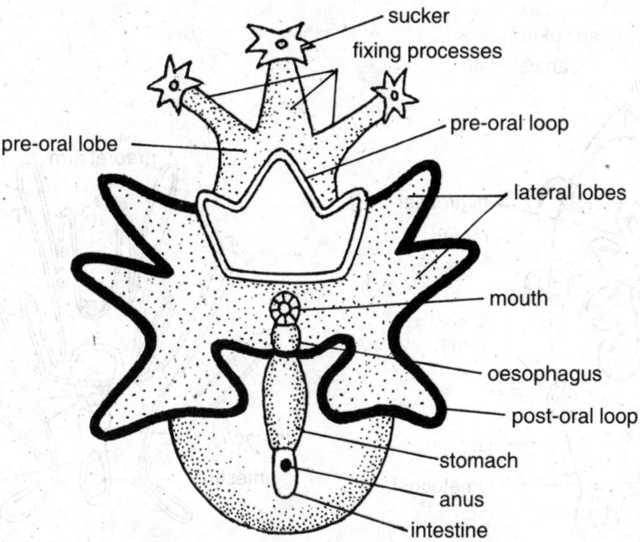

Fig. 21.8. Brachiolaria larva of sea star.

The bipinnaria larva later on develops three **fixing arms** on the ventral side in front of the preoral loop. These arms are relatively short, without ciliated band, have adhesive cells at the tips, and carry an adhesive **sucker** between their bases. The bipinnaria larva with the brachiolar arms is called the **brachiolaria larva** (Fig. 21.8). The brachiolaria larva settles to the bottom and adheres with some object. The tips of the arms provide a temporary attachment till the sucker is fixed. Metamorphosis occurs in the fixed condition.

In some cases (*Astropecten*), brachiolaria stage is omitted and bipinnaria do metamorphosis without attachement. In certain forms (*Asterina*) bipinnnaria larval stage is missing.

2. Auricularia larva and Doliolaria larva of Holothuroidea. The auricularia larva (Fig. 21.9) has a large preoral lobe like the bipinnaria. Its sides bear short paired processes, hence its name. The processes are supported by calcareous particles (rings, spehres and stars). The ciliated band is single. It follows a wavy course along the sides of the larva, passes backward on the ventral side of the anterior region to form the preoral loop, and extend forward on the ventral side of the posterior region to form the postoral or anal loop. A sensory flagellated apical plate is present in some cases.

The auricularia larva later becomes barrel–shaped and its ciliated band breaks up into separate segments. The latter join up by ends, forming 3–5 ciliated rings around the larval body. The auricularia larva with this modification is called the **doliolaria larva** (Fig. 21.9). Metamorphosis takes palce gradually during free swimming life. Tentacles appear earlier than the functional tube feet. At this stage, the metamorphosing larva is sometimes named the **pentactula larva**. The young sea–cucumber, when ready, sinks to the bottom to assume the adult mode of life.

The embryo of holothuroidea may bypass the auricularia stage and develop directly into a doliolaria. It may omit both the larval stages and swim about as a simple, oval, fully ciliated larva.

3. Echinopluteus larva of Echinoidea. The echinopluteus larva (Fig. 21.9B) is conical is shape. The preoral lobe is reduced. The arms are slender, directed forward and supported by a skeleton of calcareous rods. There may be six pairs of arms: **preoral**, **anterolateral**, **anterodorsal**,

postoral, **postero-dorsal** and **postero-lateral**. Usually the echinopluteus larva has five pairs of arms, the posterolateral arms being absent; and may have only four pairs, the anterodorsal being absent. The ciliated band is undivided and borders all the arms.

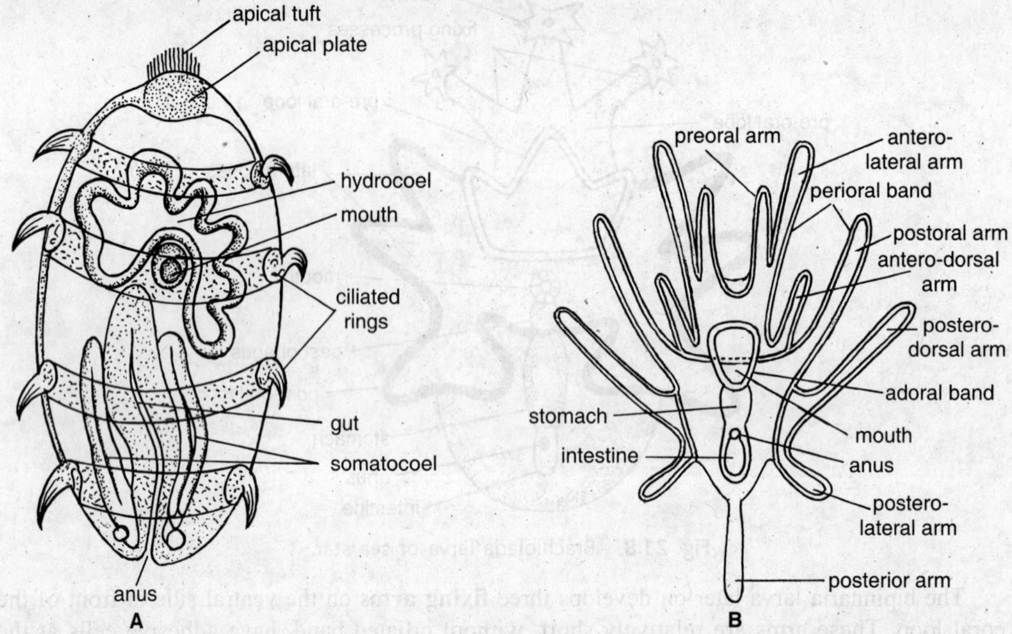

Fig. 21.9. Echinoderm larvae. A–Doliolaria larva of holothuroida; B–Echinopluteus larva of *Spatangus* (heart urchin).

The echinopluteus larvae vary a good deal in their size and angle of the arms with the body. Spatangoid pluteus has a medain posterior arm without the ciliated band. The echinopluteus, During later larval life, gradually sinks to the bottom and undergoes very rapid metamorphosis that takes just an hour. There is no attachment as in asteroids.

4. Ophiopluteus larva of Ophiuroidea. The ophiopluteus larva (Fig. 21.10) largely resembles the echinopluteus. It is conical in shape. The preoral lobe is small. There are four pairs of arms: **anterolateral, postoral, posterodorsal** and **preoral**. The posterodorsal arms are often the longest. They are absent in some cases. All the arms are slender, directed forward and supported by a skeleton of calcareous rods. The ciliated band is single and passes over all the arms. Ophipluteus larva continues to swim during metamorphosis and there is no fixed stage. The tiny brittle star sinks to the bottom to take up the adult mode of life.

5. Crinoid larva of Crinoidea. The crinoid larva of Crinoidea is also called the **doliolaria larva** and it resembles the doliolaria of the holothuroids. It is barrel–shaped, bears at the anterior end an apical sensory plate with a tuft of cilia and is surrounded by 4 or 5 tansverse ciliated rings. It has a midventral **adhesive pit** near the apical plate. It also has a midventral **stomodaeum** or **vestibule** between the first and second or second or third ciliated rings.

After a short free–swimming life, the doliolaria settles to the bottom and attaches itself to some object by the adhesive pit. Prolonged metamorphosis now starts and produces a fixed stalked stage called the **pentacrinoid larva**. This larva resembles a **sea lily** and consists of a stalk and a crown. The crown finally develops cirri and breaks free from the stalk and the young feather star takes up the adult mode of life.

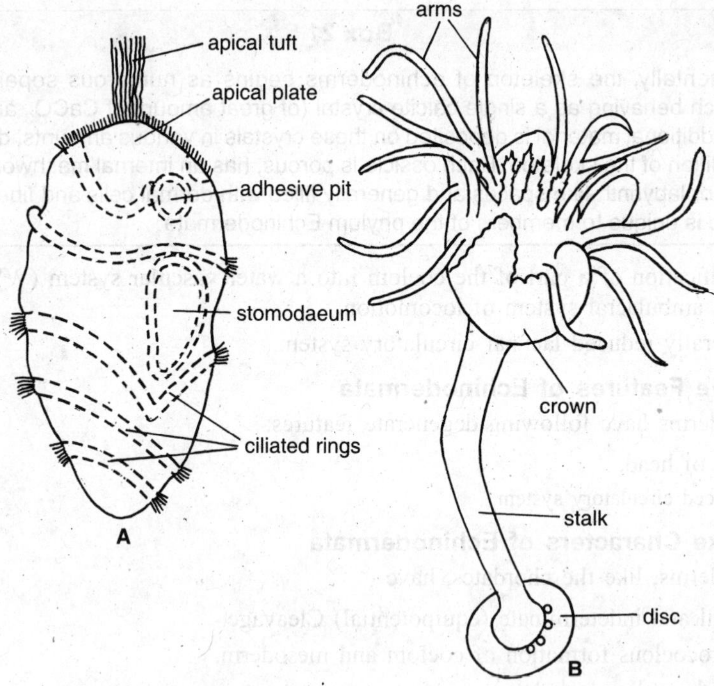

Fig. 21.10. *Antedon* larvae. A–Crinoid larva; B–Pentacrinoid larva.

Significance of Echinoderm's Bilateral Larvae. The formation of active bilateral larvae in the life history shows that the ancestors of the echinoderms were either active bilaterally symmetrical or that the larvae have become adapted to an active life in the water.

21.3 PHYLOGENETIC POSITION OF ECHINODERMATA

The echinoderms are among the most familiar marine invertebrates. They form the only major group of deuterostome invertebrates. They have many peculiar features, not found in other phyla. **Hyman** remarked "echinoderms have been created to puzzle the zoologists."

Unique Features of Echinodermata

Echinoderms have the following unique features:

1. Bilateral symmetry in the larva and pentamerous (pentaradial) symmetry in the adult.
2. Demarcation of the body surface into ambulacral and interambulacral areas.
3. Characteristic tube feet or podia for locomotion.
4. Mutable connective tissue forming dermis (Box 21.4).

Box 21.4

The dermis in Echinodermata is composed of collagenous connective tissue. The latter is **mutable** as the animal can readily alter its stiffness according to its requirement. The mutable connective tissue is made rigid or flexible by the formation or breakdown of cross bridges of calcium ions between the macromolecules in the matrix of the tissue.

5. Presence of mesodermal endoskeleton of calcareous plates (having **stereom structure** Box 21.5), usually with spines.

Box 21.5

Developmentally, the skeleton of echinoderms begins as numerous separate spicule-like elements, each behaving as a single calcite crystal (of great amount of $CaCO_3$ and small amount of $MgCO_3$). Additional material is deposited on these crystals in various amounts, depending on the ultimate condition of the skeleton. Each ossicle is porous, has an internal meshwork (the **stereom**) of lattice-like or labyrinth-like spaces and generally filled with dermal cells and fibers (the **stroma**). This structure is unique to members of the phylum Echinodermata.

6. Modification of a part of the coelom into a water vascular system (WVS), *i.e.*, water-filled ambulacral system of locomotion.
7. Generally reduced lacunar circulatory system.

Degenerative Features of Echinodermata

Echinoderms have following degenerate features:

1. Lack of head.
2. Reduced circulatory system.

Chordate-like Characters of Echinodermata

Echinoderms, like the chordates, have-

1. Radial and indeterminate (equipotential) Cleavage.
2. Enterocoelous formation of coelom and mesoderm.
3. Mesodermal endoskeleton.
4. Deuterostomous mouth.
5. Phosphocreatine for use in energy cycle of muscle contraction.

Evolution of Echinoderm's Symmetry

It is believed that the echinoderms arose from some free-moving, bilateral, coelomate ancestor. The echinoderm larva recapitulates the bilateral condition (Box 21.6). Some individuals

Box 21.6 Role of Hox Gene in Origin of Radiality in Echinoderms

Some echinoderm larvae retain their ancestral bilateral condition, the expression of genes that dictate radiality must occur relatively late in development. According to **Lowe** and **Wray** (1997), this change involved major modifications of the roles of Hox genes during the early evolution of the echinoderm bauplan (see **Brusca** and **Brusca** 2002).

of such an ancestral form may have become attached to the bottom and assumed a sessile mode of life. This new way of life resulted in a shift to the radial symmetry because this symmetry better suites the sessile existence.

Pattern of Echinoderm Phylogeny

Inspite of a rich fossil record and many decades of work, the origin and subsequent evolution of echinoderms remain highly controversial issues.

The echinoderm lineage probably originated with the Precambrian invasion of epibenthic habitats (Box 21.7) by an ancestral burrowing deuterostome. The line diversified rapidly, and most of the fundamental body plans within the phylum were probably established in the early

Box 21.7 Animals of Benthic Habitat

Benthic animals may live on the surface of the substratum of ocean (**epifauna,** or **epibenthic forms**, such as most sea anemones, sponges, many snails and barnacles) or burrow within soft substrata (**infauna**). **Infaunal forms** include many relatively large invertebrates, such as clams and various worms, as well as some specialized very tiny forms that inhabit the spaces between sand grains, termed **interstitial organisms** (the smallest of which are **meiofauna**, animals smaller than 0.5 mm).

Benthic animals may also be categorized by their locomotor capabilities. Animals that generally quite motile and, active, are described as being **errant** (*e.g.*, carbs, many worms), whereas those are that are **sessile** (*e.g.*, sponges, corals, barnacles). Others are unattached or weakly attached but generally do not move around much (*e.g.*, crinoids, solitary anemones, most clams); these animals are said to be **sedentary**.

Cambrian. Echinoderm diversity reached its zenith during the early to mid-Palaeozoic, but by the beginning of the Mesozoic it has declined greatly in terms of higher taxa, leaving only five major groups, namely Crinoidea, Holothuroidea, Echinoidea, Ophiuroidea and Asteroidea, persisting to recent times.

The origin of echinoderms probably involved the evolution of the endoskeleton composed of plates with the unique stereom structure (Box 21.5). Evidence suggest that the skeleton originated prior to the adoption of radial symmetry, the latter marking the "true" echinoderms. The first

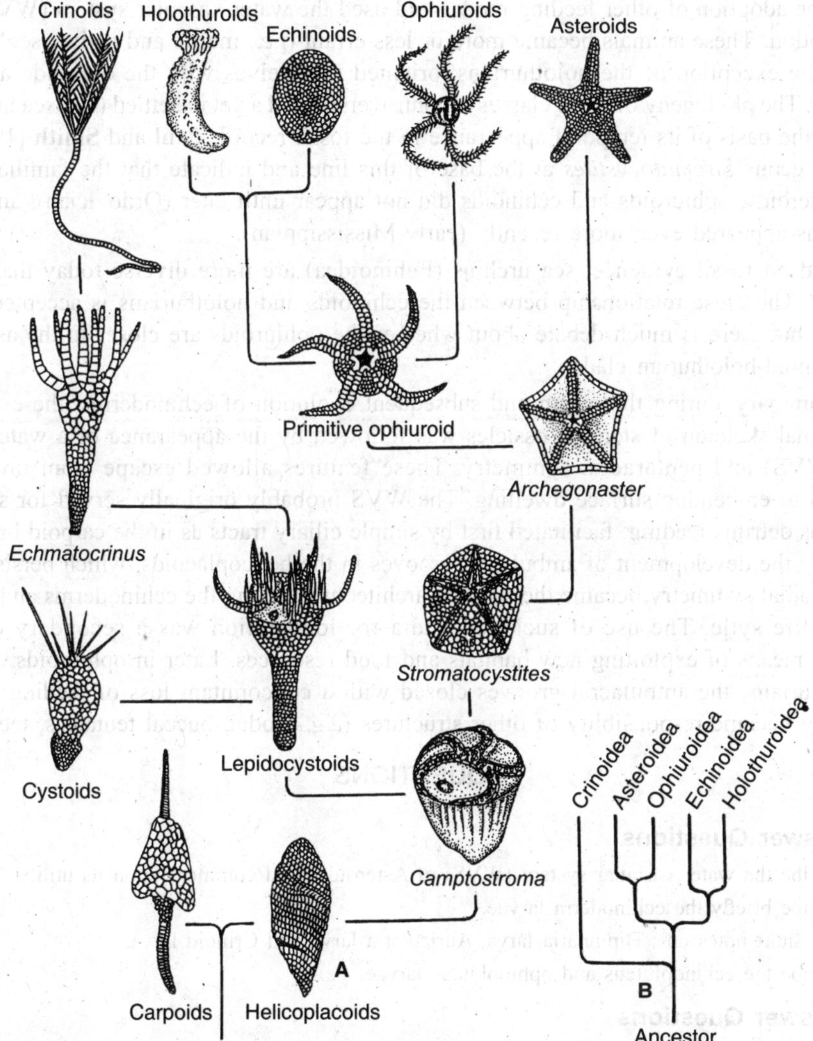

Fig. 21.11. Phylogeny of echinoderms. Two orthodox evolutionary trees of the echinoderms A-Modern tree (includes both extinct and extant classes); B-Traditional tree of extant classes. (after Brusca and Brusca 2002).

dichotomy separates the main echinoderm clade from a now extinct group called the **Carpoids** (Fig. 21.11). The first true echinoderms may have been the **helicoplacoids**. These odd creatures appeared in the early Cambrain and died out soon there after.

The next major dichotomy separates the crinoids and extinct cystoids from other echinoderms. These animals suspension fed by orienting with the oral side upward and using the open ambulacral grooves in the arms or brachioles for transporting food to the mouth. **Paul** and **Smith** (1988) suggest that the extinct genus *Camptostroma* is a similar to what may have been the common ancestor of these two monophyletic sister clades. It is the earliest known pentaradial echinoderm, with the five ambulacra developed, perhaps derived from the triradial pattern of the helicoplacoids. *Camptostroma* is usually assigned to a wholly extinct group called the Edrioasteroidea. The lineage that persists today as the class Crinoidea include extinct forms that bore attachment stalks arising from the aboral surface (*e.g.*, extinct crinoids and cystoids).

The origin of the lineage that includes the asteroids, ophiuroids, echinoids and holothurians involved the adoption of other feeding modes and used the water vascular system (WVS) largely for locomotion. These animals became more or less errant (*i.e.*, motile and active; see Box 21.7) and with the exception of the holothurians, oriented themselves with the oral side against the substratum. The phylogeny of these classes is controversial and a yet unsettled (**Brusca** and **Brusca** 2002). On the basis of its temporal appearance in the fossil record, **Paul** and **Smith** (1988) place the extinct genus *Stromatocystites* at the base of this line and indicate that the familiar benthic-feeding asteroids, ophiuroids and echinoids did not appear until later (Ordovician), and that the holothurians appeared even more recently (early Mississippian).

Based on fossil evidence, sea urchins (Echinoidea) are more diverse today than anytime in the past. The close relationship between the echinoids and holothurians is accepted by most specialists, but there is much debate about whether the ophiuroids are closer to the asteroids or to the echinoid-holothurian clade.

In summary, during the origin and subsequent evolution of echinoderms, the evolution of a mesodermal skeleton of stereom ossicles was followed by the appearance of a water vascular system (WVS) and pentaradial symmetry. These features allowed escape from infaunal life (Box 21.7) to epibenthic surface dwelling. The WVS probably originally served for suspension and perhaps detritus feeding, facilitated first by simple ciliary tracts as in the carpoid branchioles, and later by the development of ambulacral grooves in the helicoplacoids, which persisted in the crinoids. Radial symmetry became the popular architecture among the echinoderms and enhanced their new life sytle. The use of suckered podia for locomotion was a secondary event that provided a means of exploiting new habitats and food resources. Later in ophiuroids, echinoids, and holothurians, the ambulacral grooves closed with a concomitant loss of feeding functions, and feeding became responsiblity of other structures (*e.g.*, podia, buccal tentacles, teeth).

QUESTIONS

Long Answer Questions

1. Describe the water vascular system (WVS) of Asteroides and comment upon its utility.
2. Describe briefly the echinoderm larvae.
3. Write short notes on : Bipinnaria larva, Auricularia larva and Crinoid larva.
4. Describe the echinopluteus and ophiopluteus larvae.

Short Answer Questions

1. Write short notes on the following :
   (*a*) Tube feet; (*b*) Madreporite; (*c*) Stone canal.
2. Describe unique features of WVS of *Xyloplax* (sea daisy).

Very Short Answer Questions

1. Who demonstrated that water enters in WVS through madreporite?
2. Give the role of Tiedmann's body.
3. Give an alternative name for the water vascular system.
4. What type of cleavage does the egg of the echinoderm undergo?
5. Is there any echinoderm which has a direct development?
6. Name the fundamental larva of echinoderms.
7. What is the symmetry of the echinoderm larvae?
8. Mention the kind of larva found in Asteroidea.
9. Name the larva of Holothuroidea.
10. In which group of echinoderms you find ophiopluteus larva.
11. Which echinoderm class has a pentacrinoid larva.

Fill in the Blanks

1. In WVS of .... podia are not arranged along the ambulacra.
2. Stone canal of sea stars is also called ....
3. Tube feet of Asteroidea are located in the .... grooves.
4. There is no madreporite in ....
5. There are .... rows of tube feet in each ambulacral groove or area.

ANSWERS

Very Short Answer Questions

1. Ferguson.
2. Formation of phagoctyic coelmocytes present in the WVS.
3. Ambulacral system.
4. Holoblastic, radial and indeterminate.
5. Yes.
6. Dipleurala (Echinopaedium).
7. Bilateral.
8. Bipinnaria.
9. Auricularia.
10. Ophiuroidea.
11. Crinoidea.

Fill in the Blanks

1. *Xyloplax*.
2. Ciliary pump.
3. Ambulacral.
4. Crinoidea.
5. Two.

22

Phylum Hemichordata (Characters, Classification and Types)

The **hemichordata** (Gr., *hemi* = half; *chorde* = cord) is a small group of worm-like marine invertebrates. They are called hemichordata since notochord in them occurs in the anterior end of the body. Recently, notochord of hemichordates is considered as **buccal diverticulum.** Hemichordata is a group of **deuterostomes** like the Echinodermata and Chordata. It was once considered a subphylum of the chordates (as non-vertebrate chordates). Hemichordates form an evolutionary link between the echinoderms and chordates.

22.1 DEFINITION

Hemichordates are marine, soft-bodied, bilaterally symmetrical, coelomate, worm-like animals that contain diffused nervous system in the epidermis, a buccal diverticulum and paired gill-slits and no definite nephridia. Their body is divided into proboscis (protosome), collar (mesosome) and trunk (metasome).

22.2 OUTLINE CLASSIFICATION

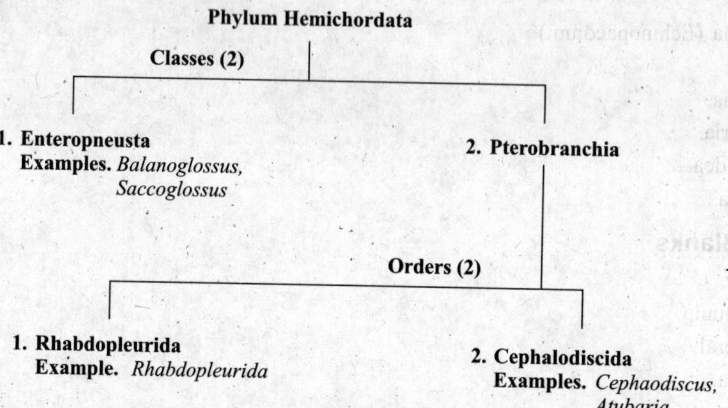

22.3 GENERAL CHARACTERS

1. Simple, solitary or colonial sedentary or free-living and marine animals, found exclusively in sea-water.
2. They are deposit or suspension feeders.

3. They are soft-bodied, bilaterally symmetrical, coelomate and worm-like.
4. Most are burrowers and some live beneath the stones.
5. The body is divided into an anterior muscular **proboscis (protosome),** a short **collar (mesosome)** and a long **trunk (metasome).** Proboscis is separated by a narrow stalk from the collar.
6. Mouth opens infront of collar.
7. Epidermis is ciliated and richly glandular. Cuticle is absent.
8. Coelom is **enterocoelous** and divisible into protocoel, mesocoel and metacoel.
9. A buccal diverticulum, formerly regarded as notochord, present in the proboscis.
10. Alimentary canal is a straight tube having a terminal anus.
11. Gill slits are numerous, paired, present in the anterior part of the trunk.
12. Blood vascular system is simple; blood flows forward in dorsal blood vessel.
13. Excretion through a single **glomerulus** situated in the proboscis.
14. Nervous system consists of an inter-epidermal nerve plexus and a dorsal hollow neural tube.
15. Movement occurs by peristaltic contraction of the proboscis.
16. Sexes are separate, fertilization is external and development may be **indirect** (having **tornaria larva)** or direct.

22.4 DETAILED CLASSIFICATION

The phylum Hemichordata is divided into **two** classes:

Class 1. Enteropneusta

1. Enteropneusta, which means **gut-breather,** includes about 70 species which are known as **acorn worm,** because of the typical shape of their proboscis.
2. They live in shallow sea-water. Most are burrowers (Fig. 22.1) Acorn worms are very **sluggish,** most live in tubes and eat sand or mud; others are free living and ciliary mucus-feeders.

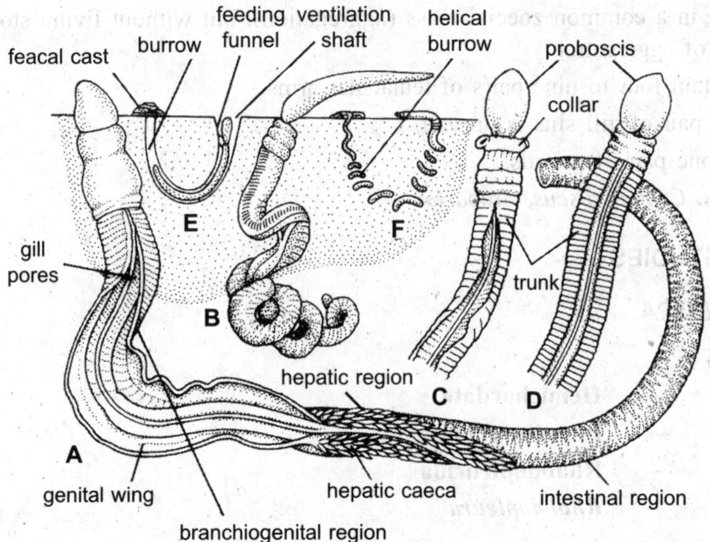

Fig. 22.1. Enteropneust diversity and burrow structure. A—*Balanoglossus*; B—*Saccoglossus*; C—*Ptychodera*; D—*Schizocardium*; E—Burrow arrangement and feeding position of *Balanoglossus*; F—Burrow and feeding position of *Saccoglossus*. Its burrow is helical and trunk is thrown into a permanent twist.

3. Most acorn worms are moderate in size, upto 30 cm or so long, but *Balanoglossus gigas* may reach a length of 2.5 m.
4. They are slimy and very soft animals, *i.e.*, they are fragile. Many smell like iodoform.
5. Many pharyngeal gill slits are present in row on trunk.
6. Alimentary canal is straight.
7. Dioecious. Gonads are located in the trunk coelom just behind the gill slits.

   Examples. *Balanoglossus*, *Saccoglossus*, *Protoglossus*, *Planetosphaera*, *Ptychodera*.

Class 2. Pterobranchia

1. Pterobranchia, which means **wing gills,** includes three genera and fifteen species.
2. They are colonial, sedentary and tubicolous (*i.e.*, living in tubular **zoecium**) hemichordates.
3. Collar bears two curved hollow **arms** which bear ciliated and hollow **tentacles.**
4. Proboscis is flattened like a shield and secretes the tube.
5. Alimentary canal is U-shaped, anus is dorsal, close to the mouth.
6. They are with or without pharyngeal slits.
7. Dioecious. Development includes a ciliated larva which is different from tornaria.

   Class Pterobranchia is divided into **two** orders:

Order 1. Rhabdopleurida

1. Colonial hemichordata, each living in tubular zoecium secreted by the epidermis and connected to other members by living stolon.
2. They contain two tentacular arms.
3. No gill slits are present.
4. Single gonad is present.

   Example. *Rhabdopleura* (It is the only genus).

Order 2. Cephalodiscida (or Cephalodisca)

1. They live in a common zoecial mass (aggregation), but without living stolon connecting members of aggregation.
2. They contain four to nine pairs of tentacular arms.
3. Only one pair of gill slits is present.
4. There is one pair of gonads.

   Examples. *Cephalodiscus*, *Atubaria*.

22.5 TYPE STUDIES

1. *RABDOPLEURA*

Classification

Phylum	—	**Hemichordata**
Class	—	**Pterobranchia**
Order	—	**Rhabdopleurida**
Genus	—	***Rhabdopleura***

Habit and Habitat

Rhabdopleura is sedentary and marine, found at great depths in the seas of Southern Hemisphere, Burmuda and European Coasts. It forms colonies at bottom. The individual is attached by a stolon.

The colony of *Rhabodopleura* measures about 10 cm and is enclosed in a membranous branched tubes, called **coenoecium** which gives off upright tubes for the zooids (Fig. 22.2). The colonies are attached to stones, corals and shells. The creeping tube is somewhat irregular (Fig. 22.2A) and contains the black **stolon** (Fig. 22.2B), formed of a firm tube containing a cellular column from which new buds arise. The upright tubes are built of regular, successive rings (Fig. 22.2D).

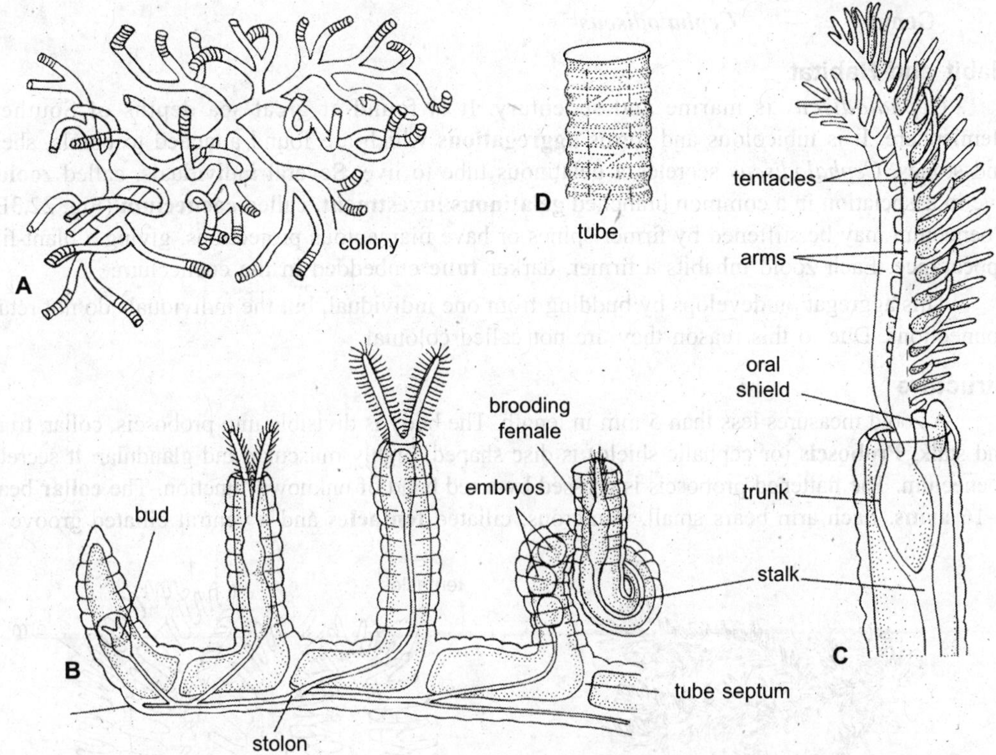

Fig. 22.2. *Rhabdopleura*. A, B—Colonial organization; C—Feeding position of a zooid with cephalic shield folded over the rim of tube; D—Various sections of a tube.

Structure

A zooid measures 1 mm in length. Body colour is brown. The body is divisible into **proboscis (oral shield** or **cephalic shield), collar, trunk** and **stalk. Proboscis** is shield-like, contains a red pigment band and secrete tubes in which animal lives. Peristaltic movement of proboscis helps in locomotion. The **collar** bears two lophophore arms, *i.e.*, they bear ciliated **tentacles.** Both the arms and tentacles are hollow. They help in capturing the food. Collar also bears **mouth** on the ventral side. *Rhabdopleura* lacks gill slits, so it deflects excess water by means of the fold-like oral lamella near the mouth.

The **trunk** is elongated and oval. From its lower surface arises a **stalk** which remains attached to the stolon.

Alimentary canal is U-shaped and anus is located on the dorsal side. **Nervous system** consists of collar-cord with ganglionic cells and nerve fibres.

Sexes are separate but **sexual dimorphism** is not well marked. The saccate gonad lies infront of the stomach. Short gonoducts open behind the collar-trunk boundary. **Fertilization** of large yolky eggs is external. **Development** is **indirect.** The larva is ciliated and becomes adult-like by the metamorphosis. A colony is formed by budding of the stalk (or common stolon).

2. *CEPHALODISCUS*

Classification

Phylum	—	**Hemichordata**
Class	—	**Pterobranchia**
Order	—	**Cephalodiscida**
Genus	–	***Cephalodiscus***

Habit and Habitat

Cephalodiscus is marine and sedentary. It is found at great sea depths of Southern Hemisphere. It is tubicolous and forms **aggregations** which are found attached to corals, shells and stones. *Cephalodiscus* secretes a gelatinous tube to live. Several individuals, called **zooids,** live in association in a common branched **gelatinous investment,** called **coenecium** (Fig. 22.3B). Coenecium may be stiffened by firmer spines or have filamentous projections, giving a plant-like appearance. Each zooid inhabits a firmer, darker **tube** embedded in the coenecium.

This aggregation develops by budding from one individual, but the individuals do not retain connections. Due to this reason they are not called colonial.

Structure

A zooid measures less than 5 mm in length. The body is divisible into proboscis, collar, trunk and stalk. **Proboscis** (or cephalic shield) is disc shaped, highly muscular and glandular. It secretes coenecium. The flattened proboscis is crossed by a red band of unknown function. The **collar** bears 8–16 **arms.** Each arm bears small, numerous, ciliated **tentacles** and a ventral ciliated groove or

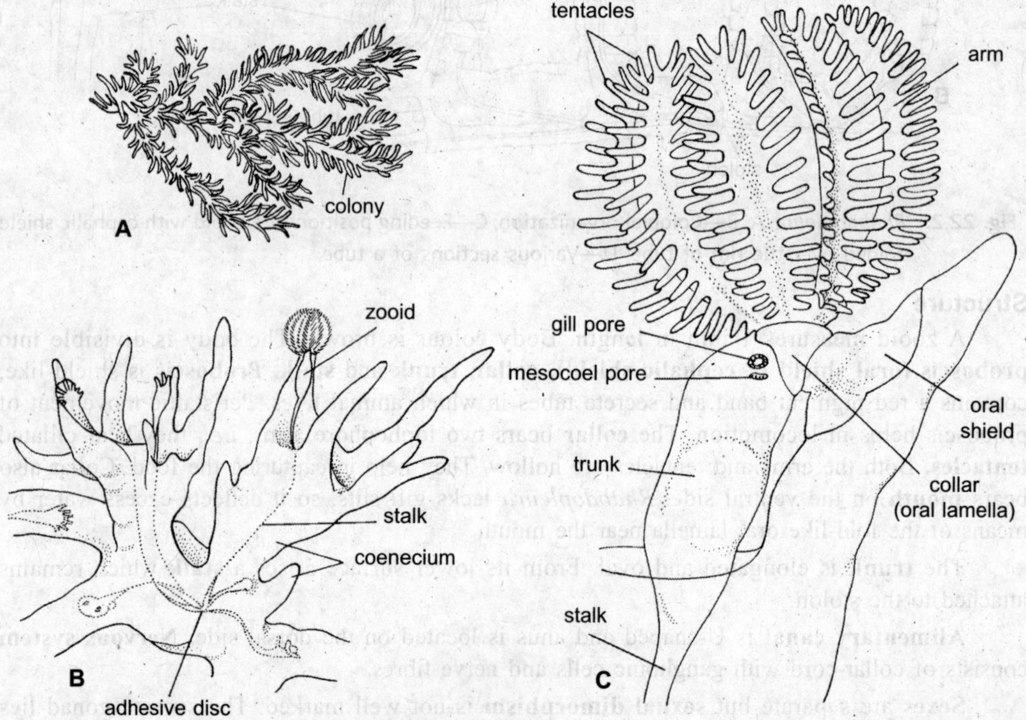

Fig. 22.3. *Cephalodiscus.* A, B—Colonial organization; zooids emerge from openings in the coenecium and perch on branch tips while feeding (B); C—Feeding posture.

gutter. Mucus glands on the dorsal side of the arms help to trap food. Collar also bears **mouth** ventrally and **anus** on dorsal side. Mouth can be closed by proboscis. Flanges of body wall (**oral lamellae**) guide food into the mouth. The **trunk** is oval and elongated, contains a pair of **gill slits.** Posterior part of the trunk continues as a long, contractile **stalk.** Stalk terminates in an adhesive region where the zooid is attached to its tube. Zooids are freely movable. They leave the tubes and cling to coenecium surface while feeding.

Nervous system is epidermal. This consists of a solid collar **ganglion** that join mid-dorsal and latero-dorsal thickenings of the nerve plexus in the cephalic shield.

Gonads are found in the trunk cavities and the development is direct. It also reproduces asexually by budding.

QUESTIONS

Long Answer Questions

1. Give the characters and classification of the phylum Hemichordata.

2. Give classification, habits and habitat and structures of following:
1. *Rabdopleura*. 2. *Cephalodiscus*.

Short Answer Questions

1. Write short note on hemichordata.

Very Short Answer Questions

1. Name the excretory organ of hemichordates.

2. Cite two examples of hemichordates bearing tentacles on the collar.

3. Name the three coeloms found in hemichordates.

4. Which are primitive, enteropneusts or pterobranchs?

5. In which groups of hemichordates, the epidermal cells are monociliated?

6. Name a hemichordate that lacks tentacles on the collar.

7. Which hemichordates have a dorsal hollow nerve cord?

8. Name the coeloms that do not open out in *Balanoglossus*.

Fill in the Blanks

1. Phylum Hemichordata is divided into two classes: .... and ....

2. Acorn worms belongs to the class ....

3. .... are colonial hemichordates.

4. .... is a pterobranch.

5. .... are tricoelomate animals.

6. There is no peritoneum in ....

7. U-shaped digestive tract is found in ....

8. A hemichordate structure considered earlier homologous to the notochord of chordates is named ....

9. Nervous system is largely intraepidermal in ....

10. Collar with tentaculate arms is characteristic of ....

Multiple Choice Questions

1. The acorn worms belong to phylum

(*a*) Chordata (*b*) Hemichordata

(*c*) Urochordata (*d*) Cephalochordata

2. A non-contractile heart occurs in

(*a*) earth worm (*b*) sea star

(*c*) acorn worm (*d*) apple snail

3. Gastrulation in hemichordates takes place by

(*a*) invagination (*b*) ingression

(*c*) delamination (*d*) all these means

4. Stomochord is present in

(*a*) *Amphioxus* (*b*) *Herdmania*

(*c*) *Balanoglossus* (*d*) none of the above

5. *Atubaria* belongs to

(*a*) Rhabdopleurida (*b*) Grapholita

(*c*) Cephalodiscida (*d*) none of these

6. Larva of Hemichordata is

(*a*) tornaria (*b*) glochidium

(*c*) bipinnaria (*d*) nauplius

ANSWERS

Very Short Answer Questions

1. Glomerulus.

2. *Rhabdopleura* and *Cephalodiscus*.

3. Protocoel, a pair of mesocoels, a pair of metacoels.

4. Pterobranchs.

5. Pterobranchia.

6. *Balanoglossus*.

7. Enteropneusts.

8. Metacoels.

Fill in the Blanks

1. Enteropneusta, Pterobranchia **2.** Enteropneusta **3.** Pterobranch

4. *Rhabdopleura* **5.** Hemichordates **6.** Hemichordates

7. Pterobranchia **8.** Stomochord **9.** Hemichordata

10. Pterobranchs

Multiple Choice Questions

1. (*b*) **2.** (*c*) **3.** (*d*) **4.** (*c*) **5.** (*c*) **6.** (*a*).

23

Balanoglossus (Phylum Hemichordata)

23.1 HABIT, HABITAT AND EXTERNAL FEATURES

Classification

Phylum	—	**Hemichordata**
Class	—	**Enteropneusta**
Genus	—	***Balanoglossus***

Habit and Habitat

Balanoglossus (Gr., *balanus* = acorn, fruit of oak + *glossa* = tongue) is a marine, soft bodied, sluggish, burrowing worm-like animal. It is commonly called **acorn** or **tongue worm.** It lives in U-shaped burrow in shallow water or sand (Fig. 23.1). Burrow is 50–75 cm deep, with a funnel-shaped opening at the anterior end and a round posterior opening. Anterior opening of burrow includes a **feeding funnel** and **ventilation shaft.** The burrow is plastered by mucus. The faecal matter (piles of coiled rope-like castings) lies close to round opening of burrow.

Balanoglossus produces an offensive smell resembling to that of decomposing faecal matter. It is a suspension feeder. It exhibits bioluminescence, emitting bright greenish light.

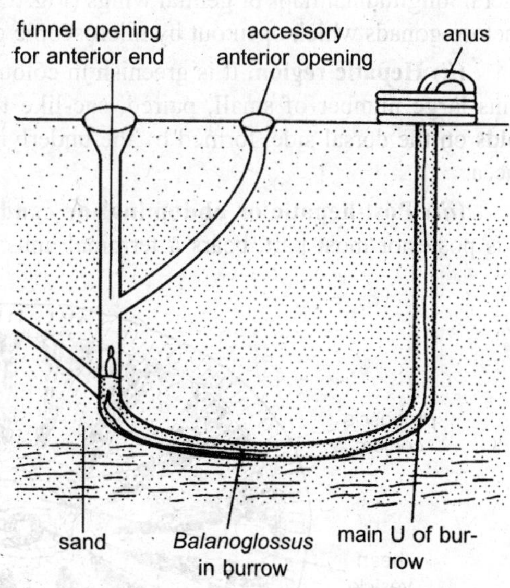

Fig. 23.1. *Balanoglossus*. U-shaped burrow and faecal cast.

External Features

Size, Shape and Colour

Most of species of *Balanoglossus* measures 2–3 cm to 2 metres or more. Body is cylindrical and worm-like. Entire body is covered with cilia. Colour of body is bright with reddish or orange tint. The various body parts are differently coloured.

Division of the Body

The body is unsegmented, bilaterally symmetrical and divided by two constrictions into three parts—(1) proboscis (**protosome**), (2) collar (**mesosome**) and (3) trunk (**metasome**) (Fig. 23.2).

1. Proboscis. It is an anterior club-shaped, highly muscular, tongue-like part. Inside the proboscis, the space is called proboscis-coelom or **proctocoel** which opens by **proboscis pore,** situated near its base (Fig. 23.3). The posterior narrow part of proboscis is called **stalk** which connect it to the collar. Below the stalk is a U-shaped ciliated **preoral ciliary organ** as a depression, to test the quality of food.

2. Collar. It is short, cylindrical middle part of the body. Its anterior part is flap-like or funnel-like called **collarette.** It bears a wide mouth on the ventral surface. The posterior end projects backward and surrounds a part of trunk. The thick collar wall encloses buccal cavity and paired cavities of **collar coelom** (**mesocoel;** Fig. 23.4). Mesocoel opens into the first pair of gill pouch through **collar pores.**

3. Trunk. It is long, cylindrical with flattened posterior part and forms major part of the body. It bears mid-dorsal and mid-ventral longitudinal ridge. These ridges extend upto posterior end and the nerves and blood vessels run through these ridges. Trunk contains the **trunk coelom** or **metacoel.** Trunk is divided into following three parts:

(*i*) Branchiogenital region. It is the anterior part of trunk. It bears a row of **gill pores** on the dorsal ridge. The number of the gill pores range from a few to 100 or more pairs. Its side wall is produced into a pair of thin, flat, lateral longitudinal flaps or genital wings (Fig. 23.5). These enclose gonads which open out by microscopic gonopores.

(*ii*) Hepatic region. It is greenish in colour and contains large number of small, paired, sac-like **transverse folds** on the dorsal side, formed by the underlying hepatic caeca.

(*iii*) Post-hepatic or abdominal or caudal region. It is posterior most part, bears a terminal anus.

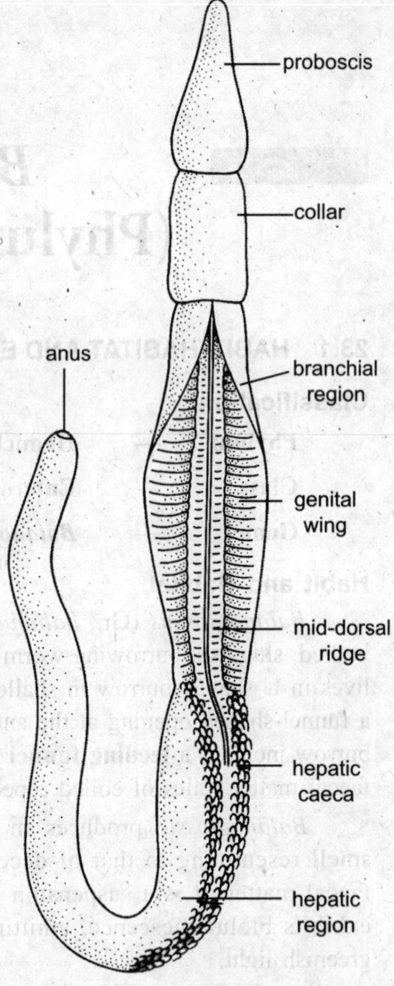

Fig. 23.2. *Balanoglossus.* External features in dorsal view.

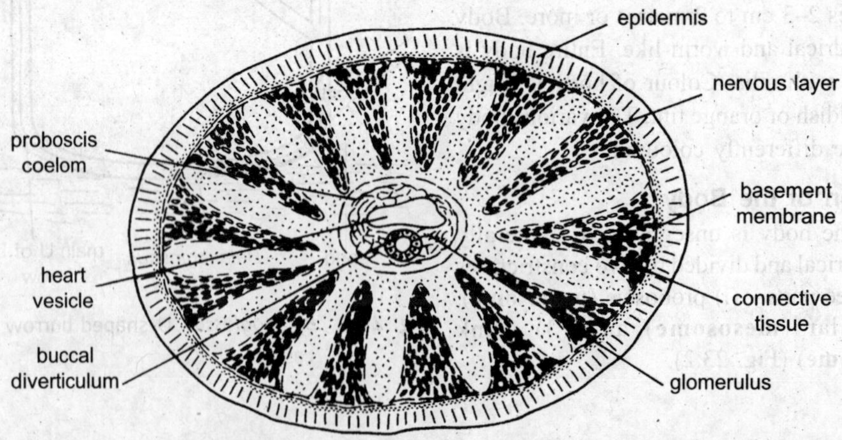

Fig. 23.3. *Balanoglossus.* T.S. of proboscis.

23.2 INTERNAL STRUCTURE

Body Wall

Body wall of *Balanoglossus* is soft, slippery and formed of (*i*) epidermis; (*ii*) muscular layer; (*iii*) peritoneum (Fig. 23.6). It lacks a cuticle.

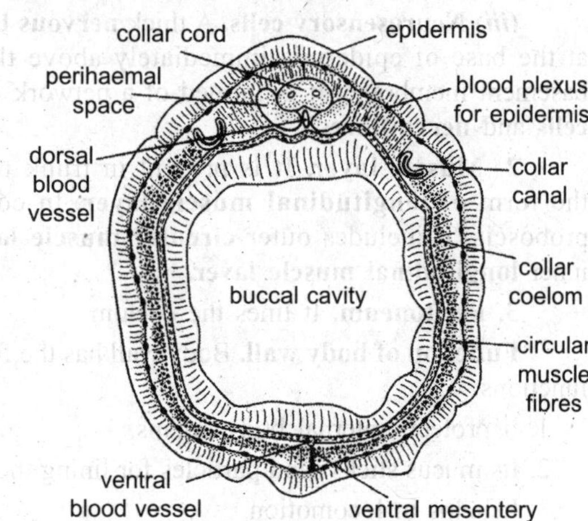

Fig. 23.4. *Balanoglossus.* T.S. of collar.

1. Epidermis. Epidermis is ciliated and richly glandular. It consists of a single layer of tall, slender, columnar cells. They are of three different types:

(*i*) Ciliated columnar cells. They are multiciliated and form bulk of epidermis.

(*ii*) Gland cells. They are most abundant in collar region. They secrete mucus and are of three types:

(*a*) Globlet cells. They are flask shaped and have granular cytoplasm.

(*b*) Coarsely granular gland cells.

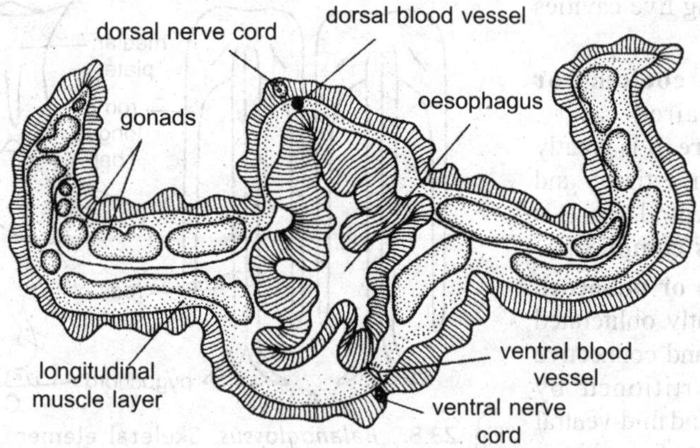

Fig. 23.5. *Balanoglossus.* C.S. through genital wings

(*c*) Reticulated mulberry-like gland cells. Some gland cells produce bromine-containing compounds that impart strong medicinal odour to acorn worms and may protect them from bacterial infection or animal predation.

A **basal lamina** and a connective tissue **dermis** lie beneath the epidermis, but the dermal fibres are well developed. This fact perhaps accounts for the fragility of body.

The collar is marked by five transverse bands of alternating glandular and non-glandular epithelium.

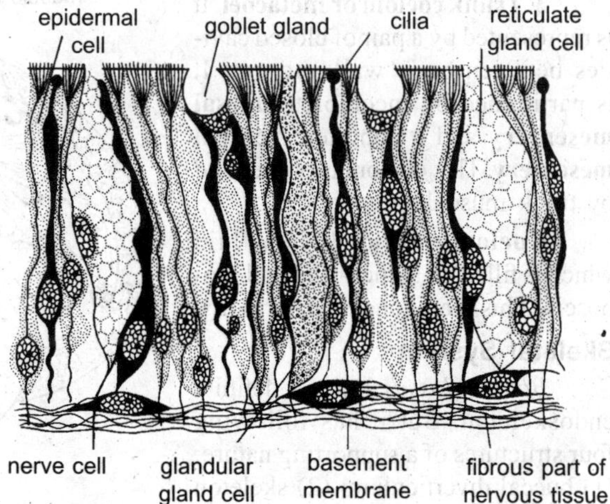

Fig. 23.6. *Balanoglossus.* V.S. of body wall.

(*iii*) Neurosensory cells. A thick **nervous layer** lies at the base of epidermis immediately above the tough basement membrane. It is formed of a network of **nerve cells** and **nerve fibres.**

2. Muscle layer. It is present in trunk region in the form of **longitudinal muscle layer.** In collar and proboscis, it includes outer **circular muscle layer** and inner **longitudinal muscle layer.**

3. Peritoneum. It lines the coelom.

Function of body wall. Body wall has the following functions:

1. It protects internal body organs.
2. Its mucus sticks sand particles for lining the burrow.
3. It helps in locomotion.
4. It receives stimuli.

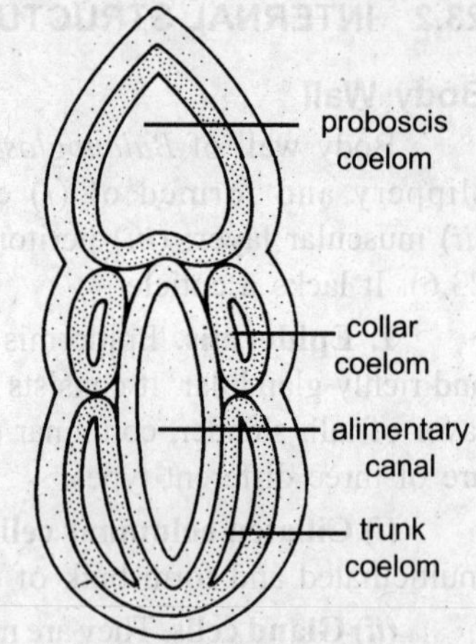

Fig. 23.7. *Balanoglossus*. Diagram of tripartite embryo showing coelomic cavities.

Coelom

Balanoglossus is **tricoelomate** like most lophophorate and echinoderms. Coelom is enterocoelous, having five cavities (Fig. 23.7):

1. Proboscis coelom or protocoel. It is unpaired cavity, opens by **proboscis pore.** It is greatly obliterated by connective tissue and muscle strands. The central space is occupied by **proboscis complex.**

2. Collar coelom or mesocoel. It is paired and is greatly obliterated by collar musculature and connective tissue. It is also partitioned by incomplete mid-dorsal and mid-ventral mesenteries. It opens into first gill sac by a collar canal and collar pore.

3. Trunk coelom or metacoel. It is represented by a pair of closed cavities between body wall and gut. It is partitioned by incomplete **dorsal mesentery** and a complete **ventral mesentery.** This coelom is obliterated by trunk musculature.

Coelomic fluid. It is a fluid which is filled up in the collar and proboscis. This maintains their turgidity.

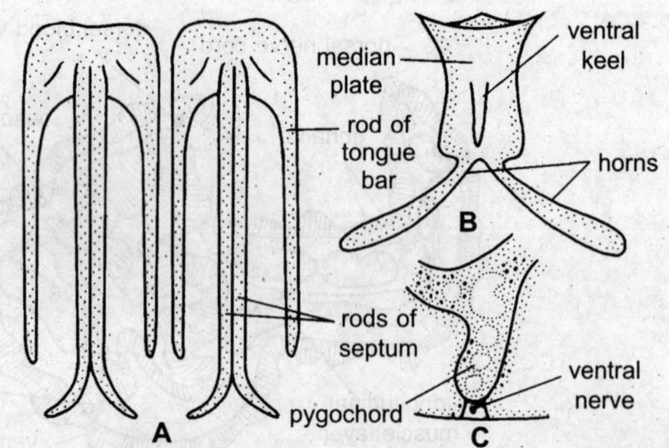

Fig. 23.8. *Balanoglossus*. Skeletal elements. A—Branchial skeleton; B—Proboscis skeleton; C—Pygochord.

Skeletal System

Balanoglossus lacks a definite endoskeleton. But it has following four structures of a supporting nature: (1) buccal diverticulum; (2) skeleton of proboscis; (3) branchial skeleton; (4) pygochord (Fig. 23.8).

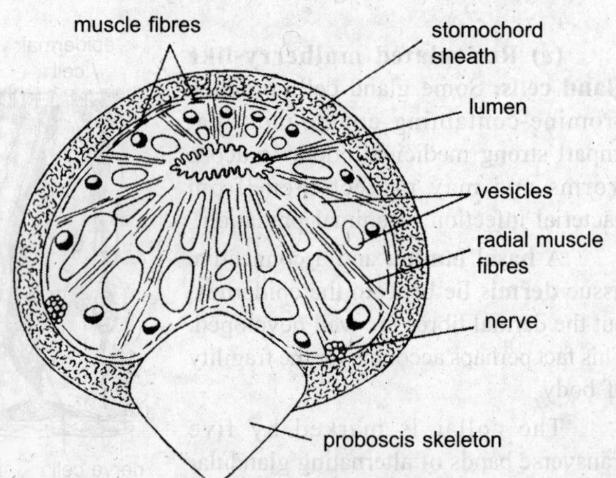

Fig. 23.9. *Balanoglossus*. T.S. stomochord.

1. Buccal diverticulum or stomochord. It is a hollow structure which arises from the roof of buccal cavity and extends upto proboscis coelom. Due to its endodermal origin and being restricted to the anterior part of the body, it is named as **stomochord** (Fig. 23.9).

2. Proboscis skeleton or nuchal skeleton. It lies ventrally close to the buccal diverticulum. It is a Y-shaped chitinous structure. Its median plate has a median-ventral keel and lies beneath the buccal diverticulum and two arms extend along the margin of the buccal cavity in the collar region.

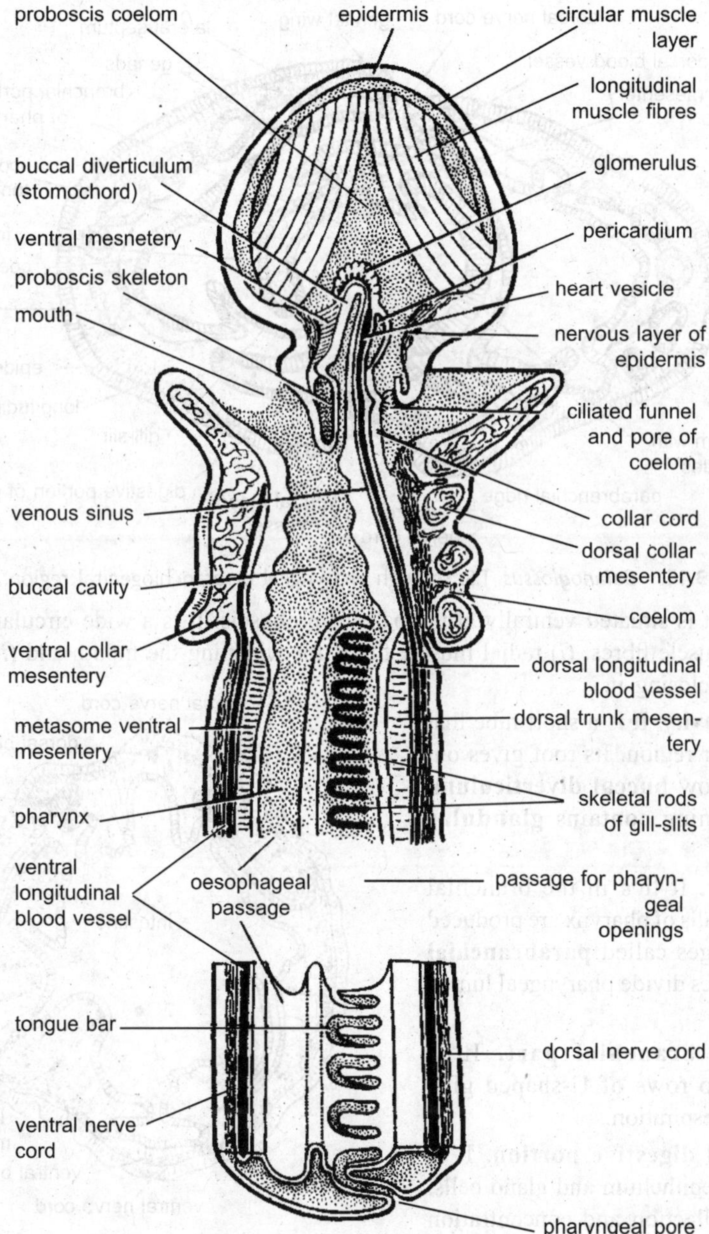

Fig. 23.10. *Balanoglossus*. M.L.S. of anterior region to show the alimentary canal.

3. Gill bars. They are supported by two types of gill-bars—primary and secondary gill bars. They are formed due to thickening of the basement membrane.

4. Pygochord. It is hard longitudinal rod-like structure which extends from the ventral side of the intestine to the body wall. Its cells are vacuolated. Pygochord may support the soft abdominal region.

23.3 DIGESTIVE SYSTEM

Alimentary canal is a straight tube supported by dorsal and ventral mesenteries and consists of the following structures: mouth, buccal cavity, pharynx, oesophagus, intestine and anus.

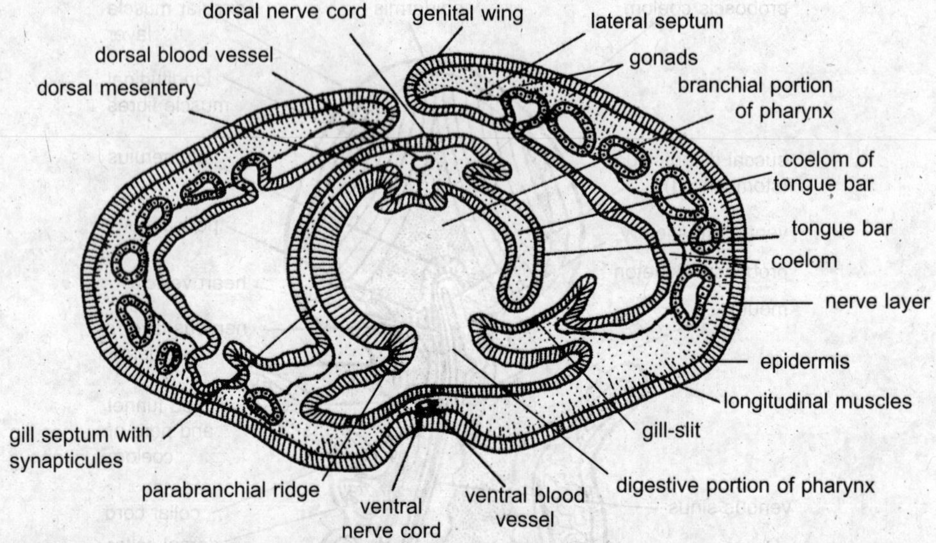

Fig. 23.11. *Balanoglossus*. T.S. through pharyngeal or branchiogenital region.

1. Mouth. It is situated ventrally at the base of proboscis. It is a wide circular aperture. It has two sets of muscle fibres: (*i*) radial muscle fibres for widening the mouth and (*ii*) concentric muscle fibres for closing it.

2. Buccal cavity. It is a short tube that occupies the collar region. Its roof gives out a short, stiff hollow **buccal diverticulum.** Its epithelium lining contains **glandular globlet cells.**

3. Pharynx. It lies in the branchial region. The side walls of pharynx are produced into a pair of ridges called **parabranchial ridges.** These ridges divide pharyngeal lumen into two parts:

(*i*) Dorsal branchial part. It is perforated by two rows of U-shaped gill-slits. It helps in respiration.

(*ii*) Ventral digestive portion. It is lined by cilia-ted epithelium and gland cells. It helps in the collection and concentration of food particles.

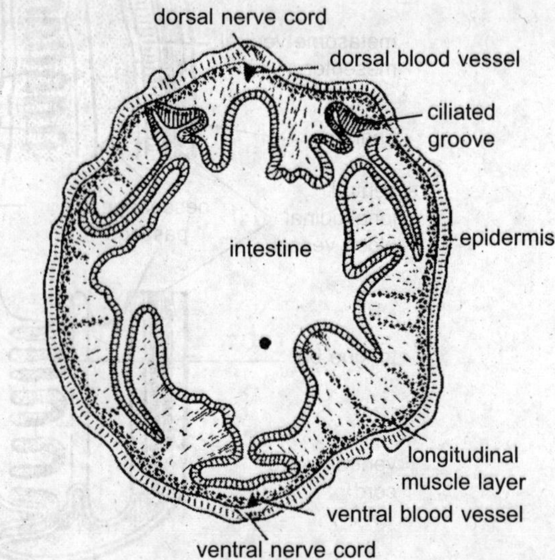

Fig. 23.12. *Balanoglossus*. T.S. through post-hepatic intestine.

4. Oesophagus. This short tube connects pharynx with intestine. Dorsal region of pharynx forms the **post-branchial canal.** It has thick, folded and glandular epithelium. The posterior part of oesophagus is narrow and its epithelium has deep furrows.

5. Intestine. It is situated in the trunk region and has two parts:

***(i)* Hepatic region.** It is highly vascular. Numerous hepatic caeca are produced from its dorsal wall. Their cells are dark green or dark brown.

***(ii)* Post hepatic region.** It is straight, thin walled tube and bears a pair of dorsolateral grooves. It is lined with ciliated epithelial cells.

6. Anus. It is situated at the tip of trunk. It is a circular aperture, surrounded by sphincter muscles.

Food and feeding. *Balanoglossus* is a **ciliary feeder.** It feeds on microscopic, dead and decaying organic matter present in water. Due to movement of lateral cilia of gill slits food laden water currents are directed backward which enters mouth, buccal tube, pharynx, gill slits and branchial sacs and leaves through gill-slits (Fig. 23.13). Food particles are entangled in mucus, forming mucous string. Mucus is secreted by gland cells of proboscis and contains **amylase enzyme.**

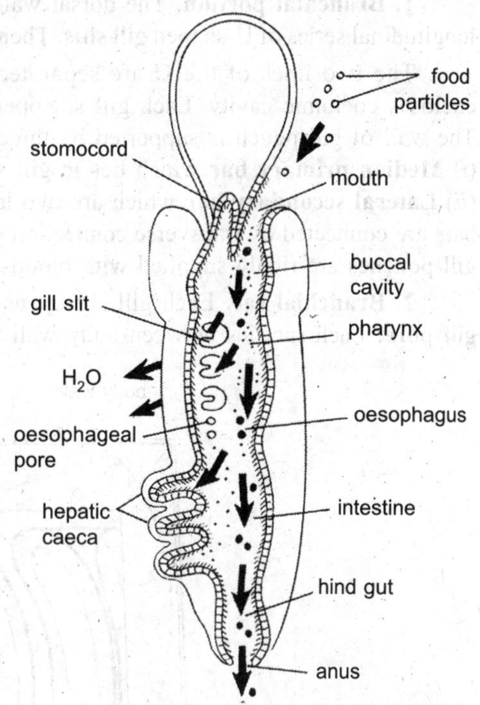

Fig. 23.13. *Balanoglossus*. Diagrammatic path of food and removal of water via gill-slits, in the pharynx.

Digestion. Digestion of food is brought about in the alimentary canal by the help of enzymes such as **amylase, maltase, lipase** and **protease.** They are secreted by pharynx, oesophagus and hepatic caeca of intestine.

23.4 RESPIRATORY SYSTEM

In *Balanoglossus* respiratory organs include:

1. Branchial portion of pharynx and gill pouches.
2. Branchial sacs.

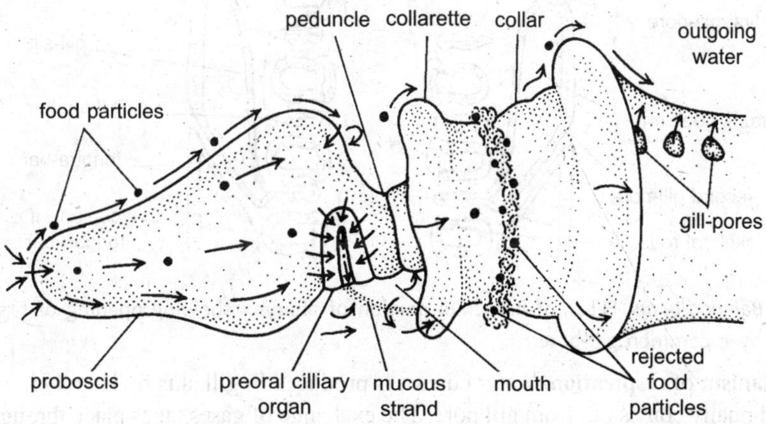

Fig. 23.14. *Balanoglossus*. Feeding current and mode of food testing by preoral ciliary organ.

1. Branchial portion. The dorsal wall of branchial portion of pharynx is perforated by a longitudinal series of U-shaped **gill slits.** Their number increases with increase in the size of animal.

The two lines of the U are separated by a **tongue bar.** The tongue bar is **hollow** and encloses coelomic cavity. Each gill slit opens into a **gill pouch** or **branchial sac** (Fig. 23.15). The wall of gill pouch is supported by three chitinous rods and form M-shaped rods. These are (*i*) **Median primary bar** which lies in gill septa. It forms middle arms of M and its lower end. (*ii*) **Lateral secondary bar** which are two lateral arms of M and lie in tongue bars. Both of the bars are connected by transverse connection called **synapticulae,** supported by chitinous rod. The gill pouches are richly supplied with blood and gill slits are lined with cilia.

2. Branchial sac. Each gill slit opens into a **branchial sac** which opens to exterior by a **gill pore.** Each sac lies between body wall and pharynx.

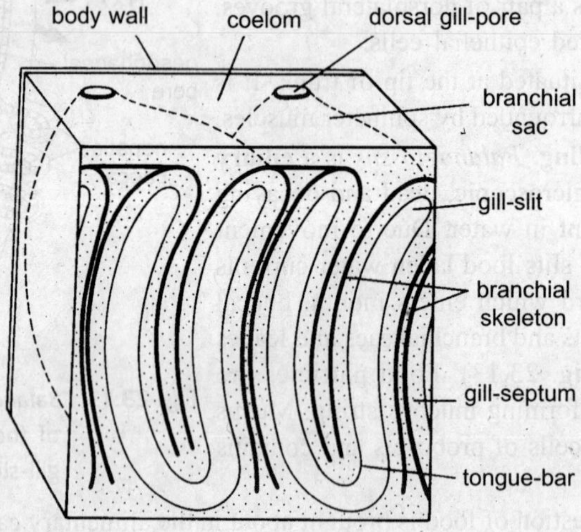

Fig. 23.15. *Balanoglossus.* Diagrammatic three-dimensional view of two gill-slits and two branchial sacs.

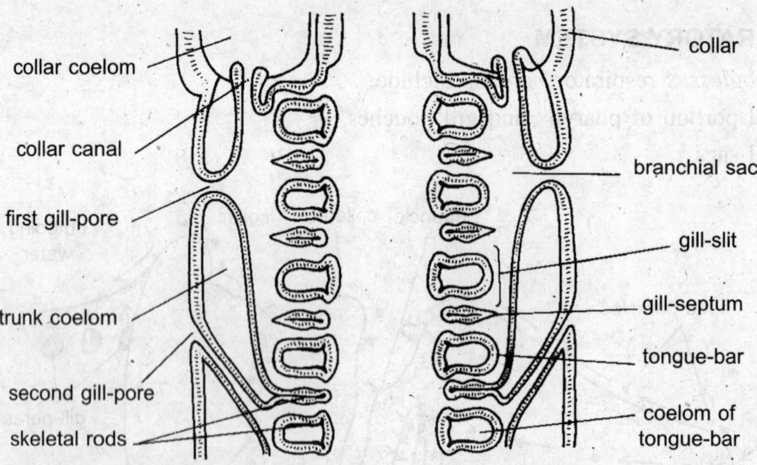

Fig. 23.16. *Balanoglossus.* H.L.S. of branchial portion of pharynx showing opening of first four gill-slits by a common gill-pore.

Mechanism of respiration. Water current is produced in gill slits by lateral cilia and enters the pharynx and finally comes out from gill pore. The exchange of gases takes place through tongue bars.

23.5 BLOOD VASCULAR SYSTEM AND EXCRETORY SYSTEM

Blood vascular system of *Balanoglossus* is of **open type** and consists of—(*i*) blood; (*ii*) dorsal and ventral vessels; (*iii*) central sinus; (*iv*) cardiac vesicle or heart and (*v*) blood vessels (Fig. 23.17).

(1) Blood. Blood is colourless. It contains a few white corpuscles. Respiratory pigment is absent.

(2) Dorsal and ventral vessels. There are two main longitudinal and highly contractile vessels, the **dorsal** and **ventral vessels,** running along the length of the body. The dorsal vessel is situated just below the dorsal nerve cord and above the alimentary canal and runs through the dorsal mesentery. The blood flows anteriorly through the dorsal vessel. The **ventral vessel** is located in the ventral mesentery and blood flows posteriorly in it.

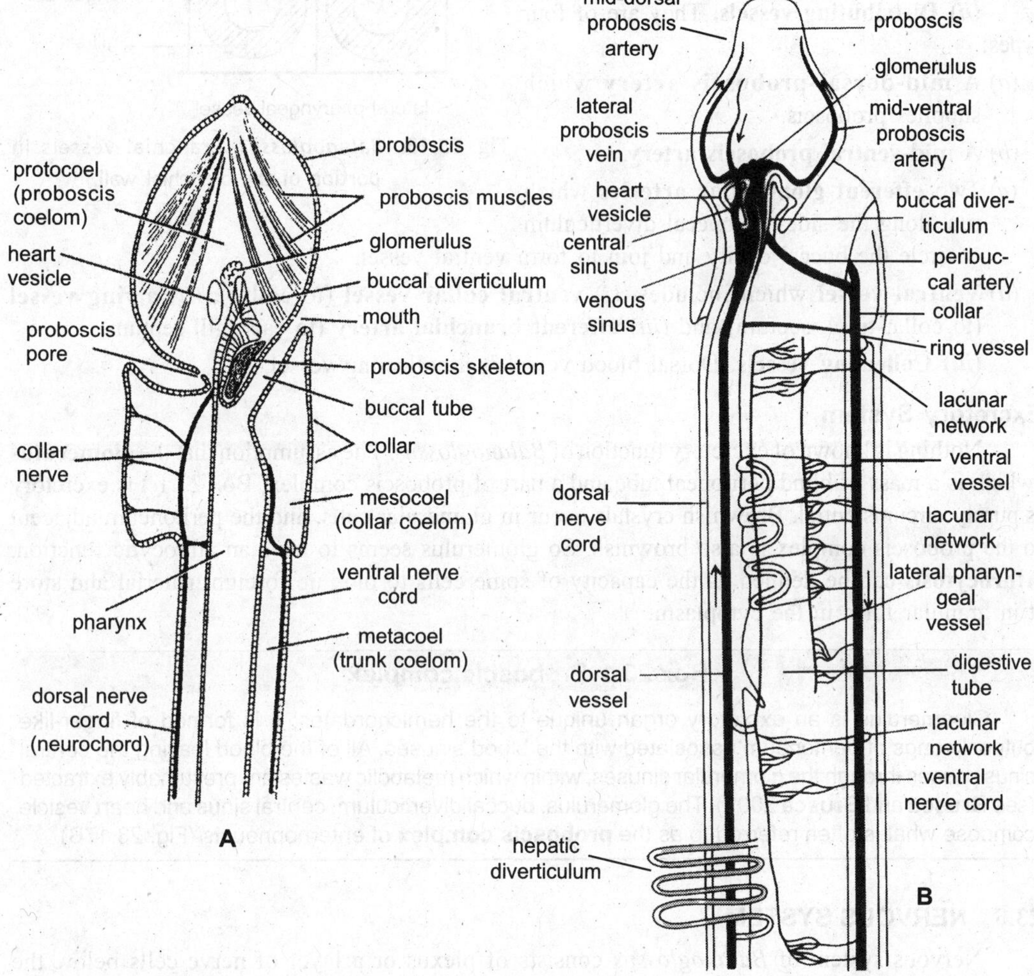

Fig. 23.17. *Balanoglossus*. A. Front end in sagittal section; B. Blood vascular system in anterior end in lateral view.

3. Central sinus. The dorsal vessel is dilated just infront of collar forming a **venous sinus** which passes anteriorly into a **central sinus.** Central sinus receives blood from the collecting vessels at its posterior end. Anteriorly it pumps blood into afferent vessels.

4. Cardiac vesicle or heart. It is situated above the central sinus. It is a closed triangular sac with muscular wall. It produces contraction in central sinus to pump the blood.

5. Blood vessels. These include (*i*) afferent vessels; (*ii*) distributing vessels and (*iii*) collecting vessels.

(*i*) Afferent vessels. They arise from the anterior end of heart (= central glomerulus sinus). These form plexus in **glomerulus.** Here the nitrogenous wastes are removed from the blood.

(*ii*) Distributing vessels. They are of four types:

(*a*) A **mid-dorsal proboscis artery** which supplies proboscis.

(*b*) A **mid-ventral proboscis artery.**

(*c*) Two **efferent glomerular arteries** which run along the sides of buccal diverticulum encircle the buccal cavity and join to form ventral vessel.

(*d*) **Ventral vessel** which includes, (*i*) **ventral collar vessel** (to collar), (*ii*) a **ring vessel** (to collar-trunk septum) and (*iii*) **afferent branchial artery** (to each gill septum).

(*iii*) Collecting vessels. Dorsal blood vessel is a collecting vessel.

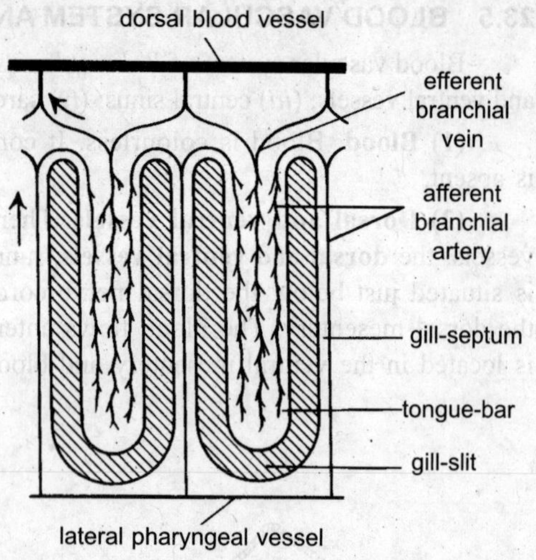

Fig. 23.18. *Balanoglossus.* Branchial vessels in portion of the branchial wall.

Excretory System

Nothing is known of excretory function of *Balanoglossus.* The assumption that the **glomerulus** (which is a mass of blind peritoneal tube and a part of proboscis complex; Box 23.1) is excretory is purely circumstantial. Brownish crystals occur in glomerular cells, and the peritoneum adjacent to the proboscis complex is also brownish. So glomerulus seems to have an athrocytic function. **Athrocytosis** can be defined as the capacity of some cells to pick up foreign material and store it in granular form in the cytoplasm.

Box 23.1 Proboscis complex

Gloomerulus is an excretory organ unique to the hemichordates; it is formed of finger-like outpocketings of peritoneum associated with the blood sinuses. All of the blood leaving the central sinus passes through the glomerular sinuses, within which metabolic wastes are presumably extracted (see **Brusca** and **Brusca** 2003). The glomerulus, buccal diverticulum, central sinus and heart vesicle compose what is often referred to as the **proboscis complex** of enternopneusts (Fig. 23.17B).

23.6 NERVOUS SYSTEM

Nervous system of *Balanoglossus* consists of plexus or a layer of nerve cells below the basement membrane.

Nervous layer is thickened strands forming a **mid-dorsal** and a **mid-ventral nerve cord.** They run the entire length of trunk. In collar region they are connected by a circular **nerve ring.** The dorsal nerve cord is connected by another circular strand called **anterior nerve ring.**

The dorsal cord traverse collar coelom and forms nervous centre. They transport impulses over the body.

Sense Organs

Balanoglossus has simple sense organs which are of following two types:

1. Neurosensory cells. These are present in the epidermis of proboscis and in anterior part of collar.

2. Pre-oral ciliary organ. At the base of the proboscis, on the ventral side, there is a U-shaped depression bordered by long epidermal cells bearing long cilia. This structure is called the **pre-oral ciliary organ.** It is a **chemoreceptor** and it tests the quality of food and water entering the mouth.

23.7 REPRODUCTIVE SYSTEM

Reproduction in *Balanoglossus* is of following three methods:

1. Asexual reproduction. In *Balanoglossus capensis*, during summer young ones if cut into small pieces from tail, each part regenerates into a complete adult.

2. Regeneration. Proboscis, collar and trunk pieces can regenerate into new part.

3. Sexual reproduction. Sexes are separate; sexual dimorphism is not well marked. Gonads may or may not be branched, found in longitudinal rows in the genital wings. The testes and ovaries are similar.

Fertilization. Fertilization is external. Mature ova and sperm are released in surrounding sea water to fertilize each other.

23.8 DEVELOPMENT

In *Balanoglossus*, development is **indirect,** *i.e.*, embryonic development of fertilized egg forms a free swimming, ciliated and microphagous **tornaria larva** which undergoes **metamorphosis** to form an adult-like individual (Fig. 23.19).

1. Cleavage and gastrulation. Cleavage of zygote is **holoblastic** (equal and radial) and results in a sphere of blastomere called **morula.** By developing a central fluid-filled cavity, called **blastocoel,** morula develops into a single-layered **coeloblastula.**

Soon blastula undergoes an **invagination** of cells of vegetal pole to form an **archenteron** which opens outside through the **blastopore.** When ecdoderm and endoderm separates from each other and blastopore is closed, the embryo is called **gastrula.**

2. Formation of larva and hatching. The gastrula elongates along the antero-posterior axis. The anterior tip of archenteron is differentiated as a coelomic vesicle, called **protocoel.** Thus, origin of coelom is **enterocoelic.** The remaining posterior part of the archenteron develops into alimentary canal. The protocoel becomes triangular in shape and opens outside on dorsal side by **hydropore.** Later on, protocoel develops into **proboscis coelom** and hydropore becomes **proboscis pore.** Collar and trunk coeloms (*i.e.*, mesocoel and metacoel) develop much later as solid invaginations of the midgut.

After the formation of protocoel, the inner end of the early gut moves towards the ventral surface and opens to outside through a **mouth.** The gut becomes differentiated into oesophagus, stomach, intestine and **anus.** Anus is formed at the place of closed blastopore. After a day or so, the entire embryo is covered uniformly by small cilia. The ciliated embryo escapes from the egg's fertilization membrane; this is called **hatching.** Free-swimming, ciliated larva is called **tornaria larva.**

3. Structure of tornaria larva. Tornaria larva was first described by **J. Muller** in 1850. It is somewhat conical, transparent, glossy and varying in size from 1–4 mm. Its body is having a number of **ciliary bands.** The **preoral ciliary band** encircling the preoral lobe. A **postoral ciliary band** is present around the mouth. It is also called **circumoral ciliary band.** It collects food from water.

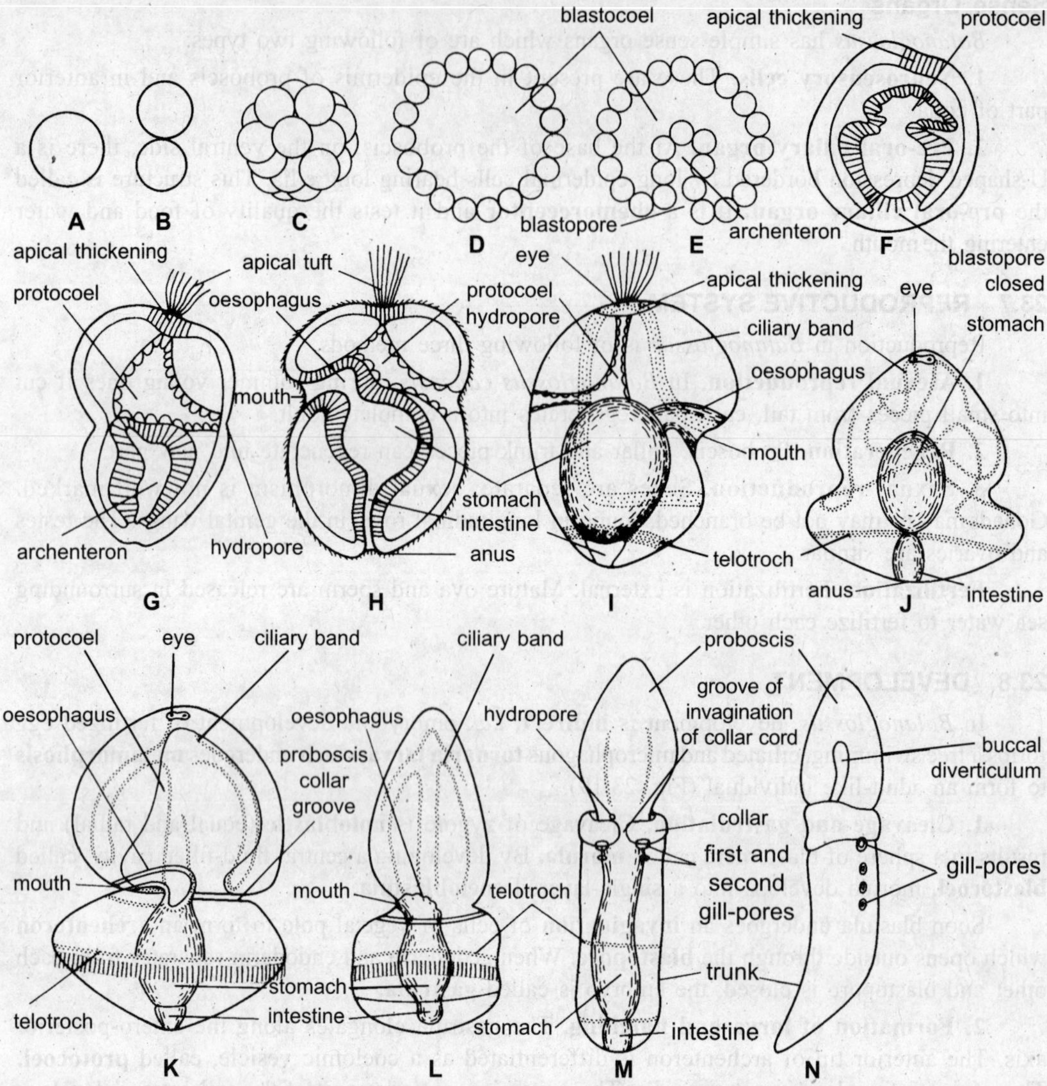

Fig. 23.19. *Balanoglossus*. Stages of development. A—Zygote; B—Two-cell stage; C—Morula; D—Coeloblastula; E—Early gastrula in L.S.; F—Gastrula; G—Early tornaria; H—Young tornaria larva in section; I—Young tornaria; J—Fully formed tornaria; K to N—Metamorphosis of tornaria into a young *Balanoglossus*.

There is another ciliary band, called **telotroch** or circum-anal ring, near the posterior end. It acts as a locomotory organ.

Anterior end of the body bears and ectodermal thickening, the **apical plate,** consisting of the nerve centre and carries nerve cells, eye spot and a band of apical cilia.

Alimentary canal is short but wide. Mouth is ventral but anus is terminal. Internally body is having **protocoel** or **proboscis cavity** which opens outside through the **proboscis pore.**

4. Metamorphosis. Tornaria larva sinks to the bottom and undergoes the following metamorphic changes:

(*i*) It reduces its size due to loss of water.

(*ii*) The ciliary bands, sensory cilia and eye spot are lost.

(*iii*) The preoral parts become uniformly ciliated to form proboscis.
(*iv*) A constriction develops at the base of proboscis to separate it from the collar.
(*v*) The trunk elongates.
(*vi*) The hydropore persists as the proboscis pore.
(*vii*) Buccal diverticulum and gill slits appear as outgrowths of the alimentary canal.

23.9 AFFINITIES

Gegenbaur established group of **Enteropneusta** for *Balanoglossus*. **Sedgwick** and **Huxley** (1814) suggested its chordate nature because of its peculiar anatomy and embryology. **Bateson** (1885) considered hemichordata as a subphylum of the phylum **Chordata.** However, **Dawydoff** (1948), **Marcus** (1958) and **Hyman** (1959) made Hemichordata an independent invertebrate phylum. *Balanoglossus* and other hemichordates show affinities with annelids, nemertines, echinoderms and chordates.

1. Affinities with Annelida

Similarities

1. Both have cylindrical worm-like body.
2. Both are of burrowing habit and tubicolous nature.
3. Both have similar mode of feeding and casting.
4. Circulatory system is similar. Heart is dorsal to gut as in Annelida.
5. Blood vessels are arranged in same manner in both.
6. In hemichordates nervous system is epidermal (partially) but in annelids, it is totally epidermal.
7. Toranaria larva resembles trochophore larva of annelids.
8. The pharyngeal pockets of hemichordates are comparable to the oesophageal pockets present in *Polygordius*.

Differences

1. Annelids do not have pharyngeal gill slits.
2. Presence of dorsal nerve in *Balanoglossus*, but it is absent in annelids.
3. Nervous system is primitive and intra-epidermal in Hemichordata.
4. The nephridium is absent in tornaria larva but it is present in **trochopore larva** of annelids.
5. Preoral proboscis coelom is present in tornaria but absent in trochophore.
6. The **blastopore** becomes **mouth** in trochophore while it forms **anus** in tornaria.
7. Process of embryonic development is entirely different in both groups.

2. Affinities with Echinodermata

Similarities in adults

1. In both coelom has enterocoelic origin.
2. Heart vesicle and glomerulus of hemichordates are homologous to dorsal sac and axial gland of Echinodermata.
3. In both nervous system is poorly developed and form epidermal nerve plexus.
4. Both are similar in habitat and ecological niches.
5. In both similar types of proteins and phosphogens are present.

Similarities in larva

1. Both are pelagic, small, transparent and oval in shape.
2. Longitudinal ciliary band is present in tornaria and in bipinnaria larva.
3. Both contain tripartite division of body.
4. In both development of coelom is similar (*i.e.*, enterocoelic origin).
5. Opening of proboscis by proboscis pore is similar to the hydrocoel of echinoderm dipleurula and shows the beginning of water vascular system.

Differences

1. Apical plate, eye-spot and sensory hair are present in tornaria larva but absent in echinoderms.
2. Transverse ciliary band is present in tornaria while absent in echinoderms.
3. Protocoel is single in tornaria while paired in echinoderm larva.

3. Affinities with Nemertines (Rhynchocephalia)

Similarities

1. Feeding and burrowing habit of both are similar.
2. Both are elongated, vermiform without external segmentation.
3. Anus is terminal in both.
4. In both, the skin is smooth containing unicellular glands.
5. Metamerically arranged simple gonads.

Differences

1. Proboscis is protrusible in nemertines and non-protrusible in *Balanoglossus*.
2. The dorsal nerve is absent in nemertines and present in *Balanoglossus*.
3. The lateral nerve is present in nemertines and absent in *Balanoglossus*.

4. Affinities with Phoronida

Similarities

1. Tornaria larva resembles the actinotrocha larva of Phoronida.
2. In *Phoronis*, the paired gastric diverticula are present but in *Balanoglossus*, single oesophageal diverticulum is present.
3. The nervous system is same in both.
4. Both possess power of regeneration.

Differences

1. The gill-slits and stomochord (= buccal diverticulum) is present in *Balanoglossus* absent in *Phoronis*.

5. Affinities with Pogonophora

Similarities

1. Body and coelom is divided into three regions.
2. Enterocoelous mode of origin of coelom.
3. In both, nervous system intraepidermal.
4. Mesosoma and metasoma are divided by a septum.
5. Pericardial sac is present in some pogonophores.
6. Gonads found in trunk.

Differences

1. Protocoelic nephridial coelomoducts present in pogonophores but not in Hemichordata.
2. Alimentary canal is absent in pogonophores but well developed in hemichordates.
3. Type of nervous system is protosoma in pogonophores but mesosoma in hemichordates.

6. Affinities with Chordates

Similarities

1. Presence of a notochord (*i.e.*, stomochord).
2. Presence of pharyngeal gill slits.
3. Presence of a dorsal tubular nerve cord.
4. Structure and function of pharynx and branchial apparatus are similar to those of Urochordata and Cephalochordata.
5. In both, origin of coelom is enterocoelous.

Differences

(*i*) Notochord

1. Notochord extends upto proboscis, while in chordates notochord extends upto full length.
2. In *Balanoglossus* notochord lies below dorsal blood vessels.
3. In *Balanoglossus* notochord is without any sheath.
4. Notochord is hollow and lined with epithelial cells in Hemichordata but solid rod-like in chordates.
5. Notochord is supporting in chordates and not in *Balanoglossus*.

(*ii*) Dorsal nerve cord

1. Nervous system of hemichordates is distinctly of invertebrate type.
2. It is intra-epidermal and consists of dorsal and a ventral nerve cord and a circumenteric nerve ring.

(*iii*) Gill slits

1. Gill slits are dorsal in *Balanoglossus* but lateral in *Amphioxus* and other chordates.
2. Differences exist in the arrangement of skeletal gill-bars.
3. Differences exist in the mode of blood supply.

Conclusion

By considering the affinities, it can be concluded that echinoderms, hemichordates and chordates have arisen from a common ancestral stock, the dipleura larva. Further, the echinoderms deviated greatly from the ancestral stock and formed blind branch in main line of evolution. The main line of evolution continued to give rise to hemichordates and chordates.

Thus, it is most logical to place hemichordates in the invertebrates as an independent phylum. Hemichordata has arisen from an ancestral stock that has given rise, on the one hand, to echinoderms and on the other hand, to hemichordates and chordates.

QUESTIONS

Long Answer Questions

1. Describe habit, habitat and external features of *Balanoglossus*.

2. Describe the morphology of *Balanoglossus*.

3. Describe the habitat and habit of *Balanoglossus*. Draw only a neat and labelled diagram of the blood vascular system of this animal and show the course of blood circulation only by a graphic representation.

4. Draw a neat and labelled diagram of median longitudinal section (M.L.S.) of anterior region of *Balanoglossus*. Describe its internal structure.

5. Describe the development of *Balanoglossus*.

6. Describe the development of *Balanoglossus* upto the formation of tornaria larva.

7. Give an account of development of *Balanoglossus*.

8. Give an account of structure of tornaria larva of *Balanoglossus* and describe its metamorphosis.

9. Discuss the affinities of *Balanoglossus* or Hemichordata. What is its latest position in the scheme of classification?

10. Give general characters of Hemichordata. Discuss affinities of *Balanoglossus*.

11. Give an account of external morphology of *Balanoglossus* in brief. Discuss its affinities in detail. (*C.C.S.U., Meerut 2009*)

Short Answer Questions

1. Discuss coelom in *Balanoglossus*.

2. Write a note on tornaria larva.

3. Write short notes on the following:

(*i*) *Balanoglossus*

(*ii*) Pharynx of *Balanoglossus*

(*iii*) Tongue bar

(*iv*) Tornaria larva

(*v*) Chordate affinities of *Balanoglossus*

(*vi*) Stomochord

4. Draw neat and labelled diagram of following (No description is needed).

(*i*) L.S. anterior part of *Balanoglossus* or V.S. or M.L.S. *Balanoglossus* through proboscis, collar and trunk part of body

(*ii*) T.S. *Balanoglossus* through proboscis

(*iii*) T.S. *Balanoglossus* through collar

(*iv*) T.S. *Balanoglossus* through branchial region

(*v*) T.S. *Balanoglossus* passing through branchio-genital region

(*vi*) V.S. body wall of *Balanoglossus*

(*vii*) Tornaria larva of *Balanoglossus*

(*viii*) *Balanoglossus*

Very Short Answer Questions

1. Where is *Balanoglossus* found?

2. What is a collarette?

3. Name the five coeloms of acorn worm.

4. Which coelom of *Balanoglossus* do not open out?

5. Give the role of oesophagus of acorn worm.

6. Which part of the gut of *Balanoglossus* has gill slits?

7. Name the site of digestion in acorn worm.
8. What is the function of preoral ciliary organ?
9. What are synapticula?
10. Heart of *Balanoglossus* is non-contractile. Is it so?
11. What type of cleavage does the egg of acorn worm undergo?
12. Name the larva of *Balanoglossus*.

Fill in the Blanks

1. *Balanoglossus* is the familiar example of the phylum ....
2. *Balanoglossus* lives in .... shaped, burrows.
3. Three regions of body of *Balanoglossus* are: .... , mesosome and ....
4. .... is tricoelomate.
5. *Balanoglossus* has a pair of mesocoels in the .... region.
6. Stomochord is a forward diverticulum from the ....
7. Gill slits of acorn worm are .... shaped.
8. Blood flows .... in the dorsal vessel in *Balanoglossus*.
9. Glomerulus of *Balanoglossus* functions as the .... organ in the acorn worms.
10. Sexes are .... in *Balanoglossus*.

Multiple Choice Questions

1. *Balanoglossus* belongs to class ....
   (*a*) Enteropneusta (*b*) Plerobranchia
   (*c*) Graplita (*d*) Planctosphaeroidea
2. The body of *Balanoglossus* is
   (*a*) radially symmetrical (*b*) biradially symmetrical
   (*c*) bilaterally symmetrical (*d*) laterally symmetrical
3. The coelom of *Balanoglossus* is
   (*a*) enterocoelous (*b*) haemocoelous
   (*c*) schizocoelous (*d*) homocoelous
4. *Balanoglossus* is
   (*a*) ciliary feeder (*b*) active feeder
   (*c*) filter feeder (*d*) all of the above
5. The excretory organ of *Balanoglossus*
   (*a*) malpighian tubules (*b*) flame cells
   (*c*) nephridia (*d*) proboscis gland
6. In *Balanoglossus*, the function of preoral ciliary organ is
   (*a*) photoreceptor (*b*) chemoreceptor
   (*c*) reoreceptor (*d*) tangoreceptor
7. Respiration in *Balanoglossus* is
   (*a*) cutaneous (*b*) branchial
   (*c*) tracheal (*d*) pulmonary

8. Acorn worm has a plexus of blood sinuses in
(*a*) gill septa (*b*) tongue bars
(*c*) both (*a*) and (*b*) (*d*) highly pressurized
9. Circulatory system of *Balanoglossus* is
(*a*) open (*b*) closed
(*c*) lacking (*d*) highly pressurized
10. Hollow nerve cord occurs in acorn worms in the
(*a*) proboscis (*b*) collar
(*c*) trunk (*d*) entire body
11. The larva of *Balanoglossus* is
(*a*) tornaria (*b*) miracidium
(*c*) brachiolaria (*d*) none of these

ANSWERS

Very Short Answer Questions

1. *Balanoglossus* is tubiculous and found in shallow waters between the tide marks along the coast of warm and temperate oceans.
2. Anterior part of collar around the neck.
3. Protocoel, a pair of mesocoels, a pair of metacoels.
4. Metacoels.
5. Moulding of food particles and mucus into a cord.
6. Respiratory chamber of pharynx.
7. Hepatic region of intestine.
8. To test the water and food particles entering the mouth.
9. Connections between gill septa and tongue bars.
10. Yes.
11. Holoblastic, complete and radial.
12. Tornaria.

Fill in the Blanks

1. Hemichordata **2.** U **3.** Protosome, metasome **4.** metasome
5. collar **6.** Buccal cavity **7.** U **8.** Forward
9. Excretory **10.** Separate.

Multiple Choice Questions

1. (*a*) **2.** (*c*) **3.** (*a*) **4.** (*a*) **5.** (*d*) **6.** (*b*)
7. (*b*) **8.** (*c*) **9.** (*a*) **10.** (*b*) **11.** (*a*).

(CC–IV)

CELL BIOLOGY

1

Introduction

All living organisms (except viruses) consist of one or more cells. The cells are structural and functional units of life.

The biological science which deals with the study of structure, function, molecular organization, growth, reproduction and genetics of the cells, is called **cytology** (Gr., *kytos* = hollow vessel or cell; *logous* = to discourse) or **cell biology.** Much of the cell biology is devoted to the study of structures and functions of specialized cells. The results of these studies are used to formulate the generalization applied to almost all cells as well as to provide the basic understanding of how a particular cell type carries out its specific functions. The cell biologist, without losing sight of the cell as a morphologic and functional unit within the organism, has to study biological phenomena at all levels of organization and to use all the methods, techniques and concepts of other sciences.

Cytology Versus Cell Biology. The cell biology has been studied by the following three avenues: **classical cytology** dealt with only light microscopically visible structure of the cell; **cell physiology** studied biochemistry, biophysics, and functions of the cell and **cell biology** interpreted the cell in terms of molecules (macromolecules such as nucleic acids and proteins). In recent years distinction between classical cytology, cell physiology and cell biology has become blurred and outmoded and now two term—cytology and cell biology are used as the synonyms.

Discovery of Cell and its Organelles. Cells were discovered in 1665 by an English scientist, **Robert Hooke.** He saw cells for the first time in a thin slice of cork with his crude microscope. Hooke described the cork as made up of hundreds of little hexagonal boxes, giving a kind of honey-comb appearance. He called these little boxes **cells**. In 1833, a Scottish botanist, **Robert Brown** discovered and named the **nucleus** in a cell. **J.E. Purkinje** in 1839, used the term **protoplasm** for a living substance present inside the cell. For coinage of other terms in cell biology see Box 1.1.

1.1 CELL THEORY

In 1838, a German botanist **Mathias Jacob Schleiden** (1804–1881) put forth the idea that cells were the units of structure in the plants. In 1839, his co-worker, a German zoologist, **Theador Schwann** (1810–1882) applied Schleiden's thesis to the animals. Both of them, thus, postulated that the cell is the basic unit of structure and function in all life. This simple, basic and formal biological generalization is known as **cell theory** or **cell doctrine.** In fact, both **Schleiden** and **Schwann** are incorrectly credited for the formulation of the cell theory; they merely made the generalizations which were based on the works of their predecessors such as **Oken** (1805), **Mirbel** (1807), **Lamarck** (1809), **Dutrochet** (1824), **Turpin** (1826), etc. However, **Schleiden** was the first to describe the nucleoli and to appreciate the fact that each cell leads a double life—one independent, pertaining to its own development, and another, as integral part of a multicellular plant. **Schwann** studied both plant and animal tissues and his work with the connective tissues such as bone and cartilage led him to modify the evolving cell theory to include the idea that living things are composed of both cells and the products or secretions of the cells. **Schwann** also introduced the term **metabolism** to describe the activities of the cells.

Box 1.1 The Naming of Various Cell Components

1.	Cell	*Robert Hooke* (1665)
2.	Cell membrane	*C. Nageli* and *C. Cramer* (1855)
3.	Plastids	*Haeckel* (1866)
4.	Karyokinesis	*Schleicher* (1878)
5.	Chromatin	*W. Flemming* (1879)
6.	Mitosis	*W. Flemming* (1882)
7.	Chloroplast	*Schimper* (1883)
8.	Phagocytosis	*Metchinikoff* (1883)
9.	Cytoplasm and nucleoplasm	*E. Strasburger* (1884)
10.	Centrosome	*T. Boveri* (1888)
11.	Chromosome	*Waldeyer* (1888)
12.	Nucleic acid	*Richard Altmann* (1889)
13.	Mitochondrion	*C. Benda* (1897)
14.	Golgi complex	*Camillo Golgi* (1898)
15.	Plasmodesmata	*E. Strasburger* (1901)
16.	Meiosis	*J.B. Farmer* and *J.B. Moore* (1905)
17.	Plasmalemma	*J.Q. Plowe* (1931)
18.	Endoplasmic reticulum	*K.R. Porter* (1945)
19.	Lysosome	*C.de. Duve* (1948)
20.	Microbody	*J. Rhodin* (1954)
21.	Prokaryote and Eukaryote	*Hans Ris* (1960)
22.	Peroxisomes	*De Duve* (1965)
23.	Glyoxisomes	*R.W. Breidenbach* (1967)
24.	Viroids	*T.O. Diener* (1971)
25.	Prion	*Prusiner* (1982)

In the coming years, the cell theory was to be extended and refined further. **K. Negeli** (1817–1891) showed in 1846 that plant cells arise from the division of pre-existing cells. In 1855, a German pathologist **Rudolf Virchow** (1821–1902) confirmed the Nageli's principle of the cellular basis of life's continuity. He stated in Latin that the cells arise only from the pre-existing cells (*viz.,* his actual aphorism was "*omnis cellula e cellula*"–every cell from a cell). **Virchow,** thus, established the significance of cell division in the reproduction of organisms. In 1858, **Virchow** published his classical textbook *Cellular Pathology* and in it he correctly asserted that as functional units of life, the cells were the primary sites of disease and cancer. Later, in 1865, **Louis Pasteur** (1822–1895) in France gave experimental evidence to support Virchow's extension of the cell theory.

The modem version of cell theory states that (1) All living organisms (animals, plants and microorganisms) are made up of one or more cells and cell products. (2) All metabolic reactions in unicellular and multicellular organisms take place in cells. (3) Cells originate only from other cells, *i.e.,* no cell can originate spontaneously or *de novo*, but comes into being only by division and duplication of already existing cells. (4) The smallest clearly defined unit of life is the cell.

The cell theory had its wide biological applications. With the progress of biochemistry, it was shown that there were fundamental similarities in the chemical composition and metabolie activities of all cells. **Kolliker** applied the cell theory to embryology—after it was demonstrated that the organisms developed from the fusion of two cells—the spermatozoon and the ovum. However, in the recent years, large number of sub-cellular structures such as ribosomes, lysosomes, mitochondria, chloroplasts, etc., have been discovered and studied in detail. Consequently, it may appear that cell is no longer a basic unit of life, because life may exist without cells also. Even then, the cell theory remains a useful concept.

Exception to cell theory. Cell theory does not have universal application, *i.e.,* there are certain living organisms which do not have true cells. All kinds of true cells share the following three basic characteristics: 1. **A set of genes** which constitute the blueprints for regulating cellular activities and making new cells. 2 A limiting **plasma membrane** that permits controlled exchange of matter and energy with the external world. 3. **A metabolic machinery** for sustaining life activities such as growth, reproduction and repair of parts. **Viruses** do not easily fit in these parameters of a true cell. Thus, they lack a plasma membrane and a metabolic machinery for energy production and for the synthesis of proteins. However, like any other cellular organism, viruses have (1) a definite genetically determined macromolecular organization; (2) a genetic or hereditary material in the form of either DNA or RNA; (3) a capacity of auto-reproduction; and (4) a capacity of mutation in their genetic substance. In consequence, viruses can only reproduce inside the host cells which may belong to animals, plants or bacteria. They use their own genetic programme for reproduction but rely on the raw materials (*i.e.,* amino acids, nucleotides) and biosynthetic machinery of the host cells (*i.e.,* ribosomes, tRNA, enzymes) for their multiplication. Thus, a virus may be defined as an infectious, subcellular and ultramicroscopic particle representing an obligate cellular parasite and a potential pathogen whose reproduction (replication) in the host cell and transmission by infection cause characteristic reaction in the host cells. Outside the host cells, viruses are just like non-living inert particles and like the salt or sugar, they can be purified, crystallized and placed into jars on a shelf for years. Due to this fact, viruses have been variously described such as "*naked genes that had somehow acquired the ability to move from one cell to another*" or as "*cellular forms that have degenerated through parasitism*" or as "*primitive organisms that have not reached a cellular state.*"

There are certain other organisms such as the protozoan *Paramecium,* the fungus *Rhizopus* and the alga *Vaucheria* (Fig. 1.1B) which do not fit into the purview of the cell theory. All of these organisms have bodies containing undivided mass of protoplasm which lacks cell-like organization and has more than one nucleus. They tend to raise the question that whether cell is a basic unit of structure in them.

1.2 PROTOPLASM THEORY

Up to middle of the 19th century, greater emphasis was given to the cell wall and less to the cellular content. But soon cell biologists started to recognize the importance of "juicy" or "slimy" contents of the cells. In 1835, **Felix Dujardin** termed the jelly-like material within protozoans as **sarcode.** In 1835, **H.von Mohl** (1805–1875) described cell division. In 1839, the Czech biologist **J.E. Purkinje** (1787–1869) coined the term protoplasm to describe the contents of cells (animal embryos). **Von Mohl,** in 1846, applied the name protoplasm to the contents of embryonic cells of the plants. **Max Schultze,** in 1861, established similarity between sarcode and protoplasm of animal and plant cells and, thus, offering a theory which later on was improved and called protoplasm theory by **O.Hertwig** (1849–1922) in 1892.

Protoplasm theory holds that all living matter, out of which animals and plants are formed, is the protoplasm. The cell is an accumulation of living substance or protoplasm which is limited in space by outer membrane and possesses a nucleus. The protoplasm which is filled in the nucleus is called **nucleoplasm** and that exists between the nucleus and the plasma membrane is called **cytoplasm**.

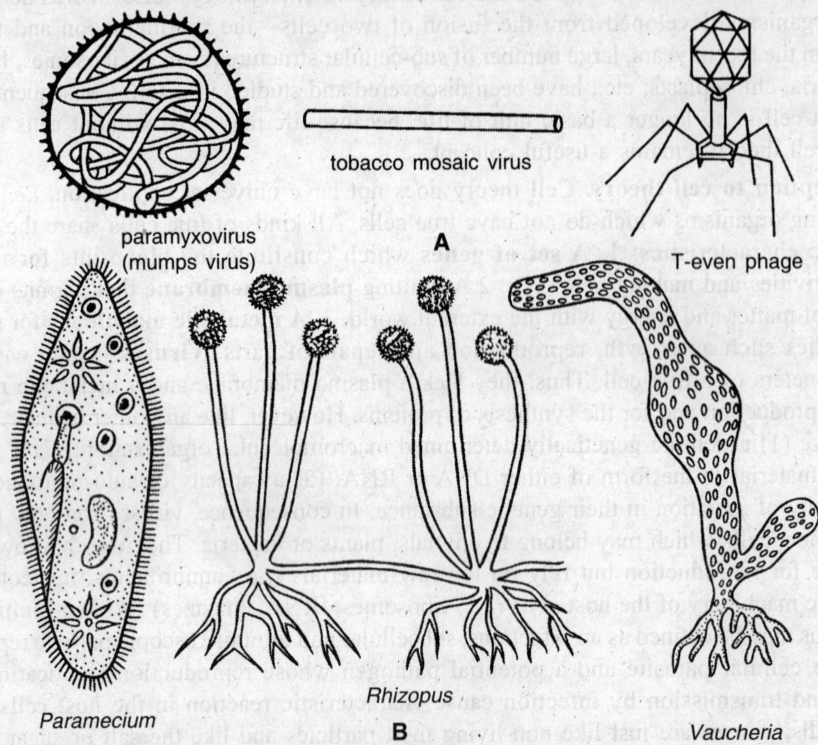

Fig. 1.1. Organisms forming exceptions to the cell theory: A—Three type of viruses; B—Three cases of cellular organization.

1.3 ORGANISMAL THEORY

The organismal theory holds that the body of all multicellular organisms is a continuous mass of protoplasm which remains divided incompletely into small centres, the cells, for the various biological activities. Thus, a multicellular organism is a highly differentiated protoplasmic individual, differing with a unicellular Protozoa only in size and degree of differentiation of the protoplasm. The differentiation involves separation of the protoplasm into subordinate semi-independent compartments, the so-called cells. Even the embryological development of a multicellular individual includes only growth and progressive internal differentiation of a small single protoplasmic individual (egg). Organismal theory too fails to ascertain the position of viruses.

QUESTIONS

Long Answer Questions

1. Give a detailed account of cell theory.

Short Answer Questions

1. What is cell biology? Who had discovered the cell?
2. Write short notes on the following:
   (*i*) Cell theory
   (*ii*) Protoplasm theory
   (*iii*) Organismal theory.

Very Short Answer Question

1. Explain the contribution of the following in the discipline of Cell Biology.

(*i*) Robert Hooke
(*ii*) C. Benda
(*iii*) R. Altman
(*iv*) M. Knoll and Ruska
(*v*) C. Golgi.

Fill in the Blanks

1. .... are an exception to cell theory.

2. Cell theory was first given by ……….. and ……..

Matching of Columns

Column A	Column B
(*i*) Robert Hooke	(*a*) Protoplasm
(*ii*) Robert Brown	(*b*) Cell theory
(*iii*) Schleiden and Schwann	(*c*) The term "cell"
(*iv*) Von Mohl	(*d*) Nucleus

Multiple Choice Questions

1. Which of the following is an exception to cell theory?

(*a*) viruses
(*b*) mycoplasma
(*c*) bacteria
(*d*) RBC

2. The cell was seen for the first time under the microscope by

(*a*) Robert Hooke
(*b*) Malpighi
(*c*) Leeuwenhoek
(*d*) Grew

3. Cell theory that all plants and animals are made up of small cells was proposed by

(*a*) Robert Hooke
(*b*) Knoll and Ruska
(*c*) Schleiden and Schwann
(*d*) Watson and Crick

4. The cell theory states that

(*a*) cells are fundamental structural units of plants and animals
(*b*) cells reproduce by mitosis and meiosis
(*c*) all cells are living
(*d*) all cells have nucleus

ANSWERS

Very Short Answer Questions

(*i*) Coined the term cell.
(*ii*) Coined the term mitochondrion.
(*iii*) Coined the term nucleic acid.
(*iv*) Produced first electron microscope.
(*v*) Discovered Golgi complex.

Fill in the Blanks

1. Viruses

2. Schleiden, Schwann.

Matching of the Columns

(*i*) *c*;
(*ii*) *d*;
(*iii*) *b*;
(*iv*) *a*.

Multiple Choice Questions

1. (*a*) **2.** (*a*) **3.** (*c*) **4.** (*a*).

2

Cell Biological Techniques

2.1 CELL FRACTIONATION AND CENTRIFUGATION

Sometimes it becomes necessary to break up tissues and cells and to isolate various parts of the cell for structural or biochemical analysis. For this purpose, the technique of cell fractionation is employed. Cell fractionation method involves, essentially the homogenisation or destruction of cell boundaries by different mechanical or chemical procedures, followed by the separation of the subcellular fractions according to mass, surface and specific gravity by centrifuges.

The centrifuge is the most important tool available to cell biologists for the separation and analysis of subcellular organelles and their constituent macromolecules. Centrifugal force is generated by placing test tubes containing samples to be fractionated in a holder called a **rotor** and then spinning the rotor at high speeds by the help of a **motor** in a centrifuge. The ultra-centrifuge, developed by the Swedish chemist, **Theodor Svedberg** between 1920 to 1940, is a special high-speed instrument capable of spinning rotors at speeds of 100,000 revolutions per minute or more. The large forces generated by these high speeds make it possible to isolate not just organelles but macromolecules such as proteins and nucleic acids as well.

Stokes formula. The rate of movement of any particle subjected to a centrifugal force is given by **Stokes formula:**

$$\frac{dx}{dt} = \frac{2r^2(\rho_P - \rho_M)}{9\eta} \cdot g$$

where dx/dt is the velocity with which the particle moves towards the bottom of the tube, r is the particle radius, ρ_p and ρ_M are the respective densities of the particle and the suspending medium and g is the centrifugal force exerted on the particle. This formula tells us that when the density of a particle exceeds that of the medium, its velocity will be positive and the particle will migrate in the direction of the force field (towards the bottom of tube). If the particle is less dense than the medium, its velocity will be negative, meaning that it will move towards the top of the tube. Finally, if the density of particle is exactly equal to the density of the suspending medium, the particle will not move because the term, $\rho_p - \rho_M = 0$.

Types of Centrifugation

(*a*) **Velocity centrifugation** (It separates organelles and molecules based on differences in size)

(*i*) **Differential centrifugation**

(*ii*) **Moving zone centrifugation**

(*b*) **Isodensity centrifugation** (It separates organelles and molecules on the basis of differences in density).

For these centrifugations, the following three centrifugation processes are employed:

1. High speed centrifugation. In high speed centrifugation rotor runs at a speed of about 20,000 rpm and generates a gravitational pull of about 50,000 g. At this speed most microorganisms, cell debris and cell organelles form sediments. It is used in **differential centrifugation.**

2. Ultracentrifugation. Ultracentrifuge (Fig. 2.1) of **Svedburg** operates at extremely high speed (about 75,000 rpm) and produces a gravitational pull of about 500,000 g. It is used for **differential centrifugation, zonal centrifugation, density gradient centrifugation** and **isopycnic centrifugation**.

3. Analytical centrifugation. This is the latest and most sophisticated instrument. Its sample container has a quartz window and the rotor has an optical system. It is possible to observe the material sedimenting and measure the sedimentation rate during the run.

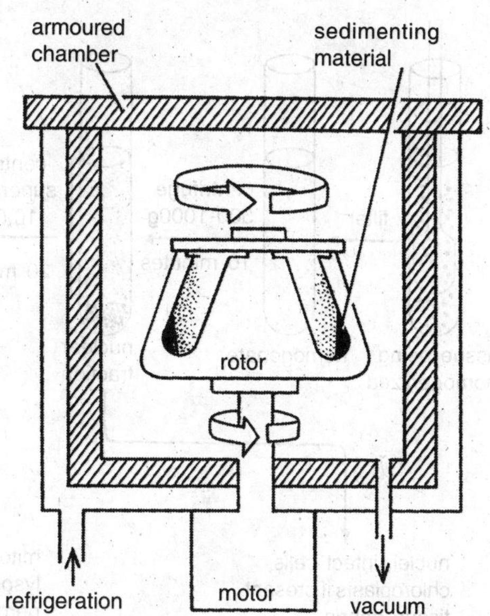

Fig. 2.1. The ultracentrifuge. The sample (*i.e.*, homogenate) is contained in tubes that are inserted into a ring of cylindrical holes in a metal rotor. Rapid rotation of the rotor generates immense centrifugal forces, which cause particles in sample to sediment. The vacuum tends to reduce friction, preventing heating of the rotor and allowing the refrigeration system to maintain the sample at 4°C.

Mechanism of Centrifugation (Cell Fractionation)

1. In the cell fractionation, the cells are gently broken by grinding a small piece of tissue in a homogeniser having a moving close-fitting glass or plastic pestle within a tube that contains a medium such as sucrose solution to preserve the cellular organelles (Fig. 2.3). The solution containing homogenised or disrupted cells, is called **homogenate**. In a homogenate the particles of different sizes are distributed homogeneously. The homogenate is subjected differential centrifugation of increasing velocity.

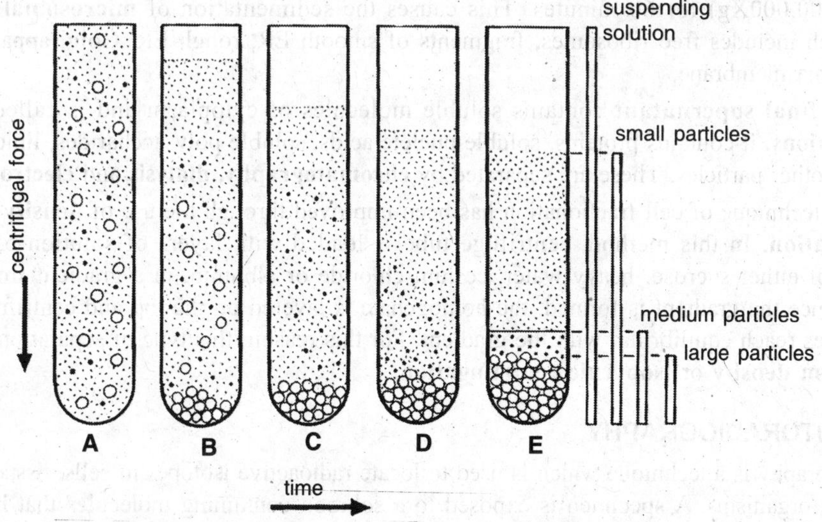

Fig. 2.2. Diagrammatic representation of the process of centrifugation and sedimentation of particles of different sizes.

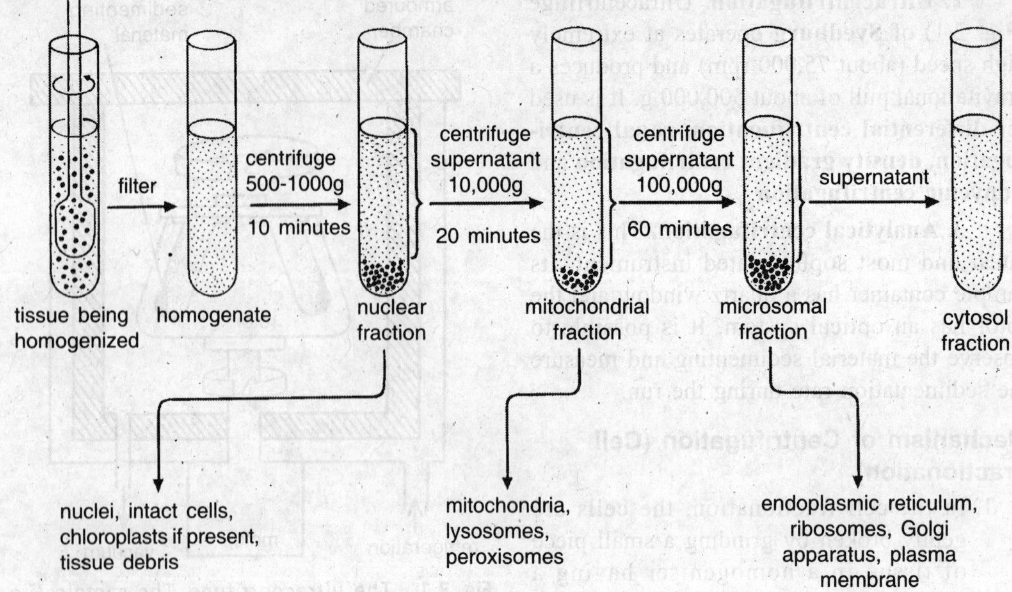

Fig. 2.3. Subcellular fractionation by differential centrifugation.

2. The tissue/liver homgenate is subjected first to low speed centrifugation at 700X g for 10 minutes. This sediments cell nuclei to the bottom of centrifuge tube. This **sediment-I** is celled **nuclear fraction**. Nuclear fraction includes nucleus, intact cells, tissue debris and chloroplasts (in case of plant cells).
3. The overlying **supernatant-I** is transferred to another tube for second centrifugation at 10,000X g for 20 minutes. The **sediment-II** is called **mitochondrial fraction**. It contains mitochondria and also lysosomes and peroxisomes.
4. Again the **supernatant-II** is removed and subjected to third centrifugation at (~ 100,000Xg) for 60 minutes. This causes the sedimentation of **microsomal fraction** which includes free ribosomes, fragments of smooth ER, rough ER, Golgi apparatus and plasma membrane.
5. The **final supernatant** contains soluble molecules of cytoplasm and is called **cytosol fractions.** It contains proteins, soluble nucleic acids, soluble polysaccharides, lipid droplets and other particles. These are separated by **chromatography, dialysis** and **electrophoresis**.

The technique of cell fractionation has been improved through the use of **density gradient centrifugation.** In this method, centrifuge tube is loaded with layers of solution of varying densities of either sucrose, heavy water, cesium chloride or albumin, in a gradient from top to bottom. Once the gradient is formed, the homogenate is layered on the top and centrifuged until the particles reach equilibrium with the gradient. For this reason, this type of separation is called **equilibrium density** or **isopycnic centrifugation.**

2.2 AUTORADIOGRAPHY

Autoradiography is a technique which is used to locate radioactive isotopes in cells, tissues, organs and whole organisms. A specimen is exposed to a solution containing molecules that have been made radioactive by the incorporation of radioactive isotopes, such as tritium (^{3}H), carbon 14 (^{14}C), phosphorus (^{32}P) and sulphur (^{35}S). The tagged molecules are often precursor molecules used by the cell in the synthesis of other needed molecules. At intervals, samples are removed from the solution;

in case of smaller tissues, the samples are sectioned and mounted on glass slides or grids. The sections are then coated with a photographic emulsion and stored in the dark for periods ranging up to several months.

When a radioactive atom emits a beta particle (*i.e.,* electron) the photographic emulsion is affected in a manner similar to the exposure of a photographic emulsion to light. Over a period of time sufficient radioactive emissions occur to affect the silver grains of the emulsion. Black spots will appear at those sites when the emulsion is developed. Such spots will mark sites in the tissues where the radioactive atoms have accumulated. These sites can be identified by examining the stained tissue sections under the light microscope.

In the technique of autoradiography, for the study of DNA metabolism of cell ^{3}H-thymidine is used; for RNA metabolism ^{3}H-uridine is used; for protein synthesis various tritiated (^{3}H-tagged) amino acids are used; and for polysaccharides and glycoproteins tritiated monosaccharides such as ^{3}H-mannose and ^{3}H-fucose are employed.

Pulse-labelling technique. This technique is used for those cases where biological molecules undergo considerable modifications after their synthesis (*e.g.,* ribosomal RNA). Here, actively growing cells are exposed to a radioactive precursor for a short period. The labelled precursor is then removed and replaced by 'cold' (unlabelled) precursor molecules. The unlabelled precursor is then removed and replaced by 'cold' (unlabelled) precursor molecules. The unlabelled precursors are incorporated into the newly synthesized molecules and have the effect of 'chasing' the previously synthesized molecules containing the radioactive precursor through any maturation process. If the molecular species under investigation is sampled shortly after the start of the experiment, only the primary synthetic product will contain radioactivity. After longer time intervals, the original radioactive molecules will have been replaced with non-radioactive molecules. In this way the flow of radioactivity through a maturation process can be followed, together with any movement of molecular species within the cell.

2.3 OPTICAL AND ELECTRON MICROSCOPY

A **microscope** is an optical instrument consisting of a single lens or a combination of lenses for magnifying objects too small to be seen or clearly observed by the naked eye. Thus, it is used to obtain magnified images of small objects such as cells of living organisms. Investigations or studies by means of the microscope is called **microscopy**. In cell biology, various types of microscopes are used to study the structure of cell; some important ones are the following:

1. Optical or light microscopes (simple, monocular, binocular and compound microscopes)
2. Electron microscope
3. Phase contrast microscope
4. X-ray diffraction microscope
5. Interference microscope
6. Polarized light microscope
7. Ultra-violet microscope
8. Ultra-microscope or Dark field microscope

A **simple microscope** (also called dissecting microscope) has a single lens system. It is composed of a single convex lens. **A compound microscope** consists of two or more **lens systems** or **elements.** A **monocular microscope** contains one ocular or eyepiece while a **binocular microscope** has two oculars.

Here, only two types of microscopes, namely light microscope and electron microscope will be described. (For some historical microscopes see Box 2.1.

Box 2.1 Some Historical Microscopes

1. First simplest compound microscope was designed by **Francis** and **Janssen** in 1590.
2. **Robert Hooke** (1665) constructed the compound microscope (Fig. 2.4)
3. **Anton van Leeuwenhoek** (1675) was the first to use microscope for biological studies. He was first to observe living free cells. His microscope (Fig. 2.5) consisted of a single lens with a high power of magnification and was called **simple microscope.** He made 247 microscopes.

Resolving Power of Microscope

In the search for information about the structure and composition of cells, the cell biologists immediately face two limitations: the exceedingly small dimensions of cells and their component parts and the transparent nature of cells. The diameters of the majority of cells fall within a range of 0.2 and 50 μm. The human eyes have limited distinguishing or resolving power. The ability of an observational instrument such as a human eye or a microscope to reveal details of structure is expressed in terms of **limit of resolution** (l) which is defined as *the smallest distance that may separate two points on an object and still permit their observation as distinct separate points*. The unaided human eye under optimal conditions in green light (to which it is most sensitive) cannot distinguish between points less than about 0.1 μm or 100 μm apart. Structural details smaller than this, *e.g.,* cell, is unresolved unless some instrument capable of higher resolution is used. **Magnification,** *the increase in size of optical image over the size of the object being viewed*, is of no use unless the observational system can resolve the various parts of the structure being examined. Increased magnification without improved resolution results only in a large blurred image. The human eye has no power of magnification, so magnifying glasses may be used to magnify images up to about 10 times. A light compound microscope in which many lenses are combined together has a useful magnification of about 1,500 times.

The limit of resolution (l) of any optical instrument (*i.e.,* eye or microscope) is given approximately by the Abbe's relationships:

$$\text{Resolution } (l) = \frac{\text{wavelength } (\lambda)}{\text{numerical aperture } (n \sin \alpha)}$$

where λ (lambda) is the wavelength ("colour") of the illumination or radiation used to form the image, n is refractive index (a function of density) of the material (*i.e.,* mostly air or water) between the specimen and the first lens (or objective lens), and sin α is sine of the semi-angle of aperture of the first lens as viewed from the specimen. The quantity "n sin α" is often called the **numerical** aperture (NA).

Abbe's relationships make it clear that high resolution in a microscope can only be achieved by manipulating a small number of variables: the wavelength of the illuminating radiation, the refractive index and the aperture. The **aperture** is limited to something less than 90° since that would have the lens and specimen in contact with one another. In fact, 85° is about the limit in good optical microscopes. Such angles require an excellent lens. In most cases, the aperture is less because the edges of the lens introduce distortions and so cannot be used. **Refractive index** is easy to alter, but only within narrow limits. It can be increased by using oils to fill the space between the specimen and the objective lens. Transparent immersion oils used in today's microscopes (*i.e.,* **oil immersion lens**) have n up to about 1.6. Still 1.6 is big improvement over air or water (n = 1). In a microscope, the smallest detectable detail is equal to about one-half the **wavelength** of light with which it is observed. The smaller the object, the shorter the wavelength of light required. Hence, the wavelength of light is the area which has great chances of improvement. One can, for example, use ultraviolet light instead of visible light, thus, improving resolution as much as two-fold. In order to do that however, special lenses (*e.g.,* of quartz) must be used since ordinary glass blocks much ultraviolet light. In such a microscope, called ultraviolet microscope, the eyes cannot be used to view the image directly, for they are insensitive to ultraviolet light. Lastly, a specimen cannot absorb light of wavelength below 0.3 µm.

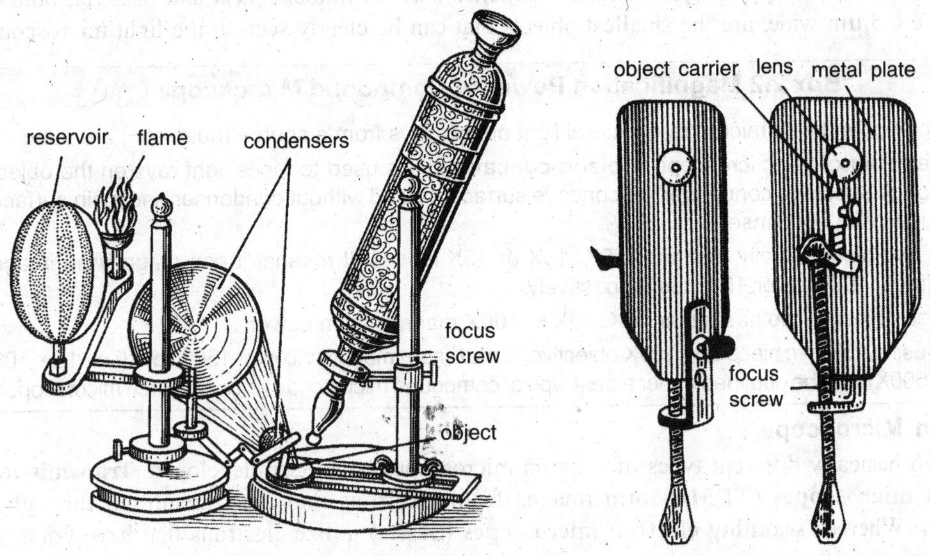

Fig. 2.4. Hooke's compound microscope. **Fig. 2.5.** Leeuwenhoek's microscope.

Thus, a good light microscope, with a numerical aperture of 1.4 and using light of short wavelength (0.4 µm) will resolve two points at about 0.17 µm separations. By such a microscope though, one can see considerable details in most cells, there is also a great deal that cannot be seen. For instance, ribosomes and chromatin threads of nucleus are about 0.02 mm in diameter and quite invisible to the light microscope. For them electron microscope is used. In cell biological studies, the following two types of microscopes are most extensively used:

Light Microscopes (Optical Microscopes)

The compound light microscope is the simplest and most commonly used microscope. It consists of the following three lens system.

***(i)* Condenser lens system** occurs beneath the specimen and its function is the collection and focussing of the light rays on the object or specimen which is placed on the stage of the microscope.

***(ii)* Objective lens system** remains near and above the specimen. It produces and magnifies the image of the specimen.

***(iii)* Eye piece lens system** or ocular lens system remains near the eyes of the observer and it magnifies and forms the image (secondary) of the (primary) image previously produced by the objective.

Lenses. A simple lens can magnify an object only three times (3X). For getting a magnification of more than 3X, a combination of several lenses is used. Such a combination of several lenses (called **elements**) function as a single convex lens and a magnification of about 20X can be obtained. For example, an **eye piece lens** contains 2 to 3 lenses or elements. Likewise, the objective lens is made by placing 8 to 10 lenses or elements. Individual lenses are placed so close to each other that they act as a single lens. Only achromatic lenses are used in a microscope. (For certain other facts see Box 2.2).

Working. Light rays from the light source (sunlight or bulb-light) are focussed on the object by the condenser lens. The objective lens produces the enlarged image, called **primary image,** of the object or material. The eye piece lens further magnifies the primary image into secondary image which is seen by our eye. Cellular organelles such as mitochondria and bacteria, both of which are 0.5 μm wide, are the smallest objects that can be clearly seen in the light microscope.

Box 2.2 Magnification Power of Compound Microscope

1. In compound light microscope natural light or light rays from a source (bulb, etc.) are used.
2. The mirror of the microscope is **plano-concave** and is used to focus light rays on the object through substage condenser. Its concave surface is used without condenser and plain surface is used with condenser.
3. An eye piece usually has either 5X, 10X or 15X lenses. It means, it can magnify an image 5 times, 10 times or 15 times respectively.
4. Each objective too has either 10X, 40X or 100X magnification power.
5. By using 15X eye piece and 100X objective, a maximum magnification of 1500 time (*i.e.*, 15 × 100 = 1500X) may be obtained under a best type of compound microscope (*e.g.*, research microscope).

Electron Microscope

Two basically different types of electron microscopes have been developed. **Transmission electron microscopes (TEMs)** form images from electrons that are transmitted through a specimen. Whereas **scanning electron microscopes (SEMs)** utilize electrons that have bounced off the surface of the specimen. A transmission electron microscope has a very high magnification (500,000 times). The resolution TEM was designed by **Knoll** and **Ruska** of Germany in 1932. It permits direct study of biological ultrastructures (*e.g.*, structure of cell organelles). The practical limit of resolution of electron microscope is about 3 to 5 Å.

In TEM, the illuminating agent is not light but the **electrons** of short wavelength (0.500 Å), The wavelength of electrons is determined by the voltage at which these are generated. For example, at 50,000 volts, the electrons have 0.50 Å wavelength and resolution power of electron microscope can be 0.50 Å/2 = 0.25 Å. Similarly, in an electron microscope with a voltage of 100,000 volts, the wavelength of an electron is 0.004 nm (or .04 Å). Theoretically the resolution of such a microscope should be 0.02 Å. But practical resolving power of most modern electron microscope is 0.5 μm or 5 Å. This is 400 times greater than that of a light microscope.

Structure

The transmission electron microscope comprises the following parts:

1. Electron gun and anode. In TEM, the source of illumination is electrons which are produced and concentrated into an electron beam by **electron gun.** Electron gun consists of an electrically heated **tungsten filament or cathode** that emits electrons. Outside the gun is an anode which attracts electrons. Thus, between the filament and anode the electrons attain a high velocity with wavelength of .005 to .003 nm.

2. Microscope column. Electrons can travel in straight line in vacuum only, because in the air, these collide with oxygen or nitrogen atoms. The electron microscope, therefore, is enclosed in a evacuated metal tube.

3. Condenser lens. It is the **electromagnetic coil** which focusses or condenses the electron beam in the plane of object.

4. Objective lens. It is the **electromagnetic coil** which produces first magnified image of the object. It focusses the electrons which are reflected by the object and forms the first image.

5. Projector lens. It is also the electromagnetic coil which magnifies the first image formed by objective lens. It produces the final image.

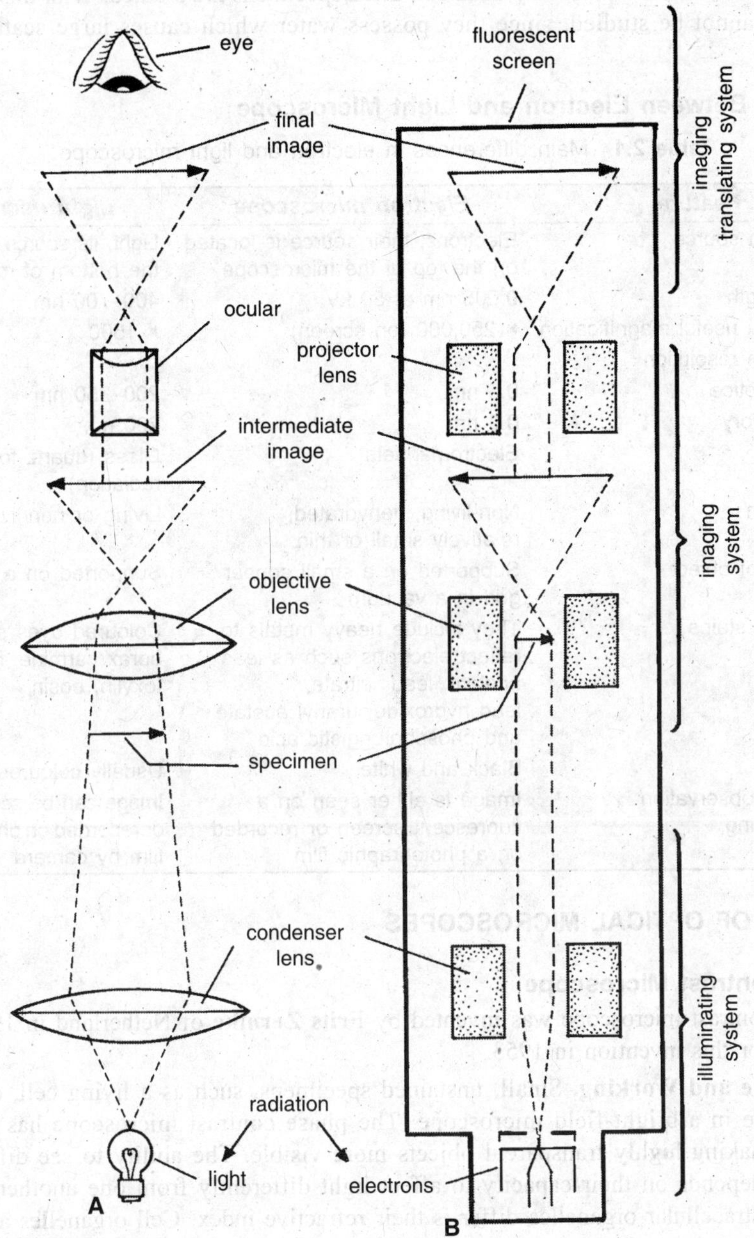

Fig. 2.6. A comparison of the lens systems of light (A) and electron microscope (B).

6. Fluorescent screen or photographic plate. Fluorescent screen is used for observing the magnified image of the object. It remains coated with a chemical (*e.g.,* zinc sulphide) which on being excited form the image as on the screen of television. The final image can also be captured on photographic film; such photographs are known as **electron micrographs**.

In the light microscope, the focussing is done by moving the lens tube up and down. In TEM, focussing is done by changing the flow of current through the coils of the electromagnetic lenses.

Further, in electron microscopy only dead and dried specimens are studied. With this microscope, living cells cannot be studied, since they possess water which causes large scattering of the electrons.

Differences Between Electron and Light Microscope

Table 2.1 Main differences in electron and light microscope.

	Feature	*Electron microscope*	*Light microscope*
1.	Radiation source	Electrons; their source is located on the top of the microscope	Light, its source is located at the bottom of microscope
2.	Wavelength	0.005 nm at 50 kV	400–700 nm
3.	Maximum useful magnification	× 250,000 (on screen)	× 1500
4.	Maximum resolution		
	(*i*) in practice	0.5 nm	200–250 nm
	(*ii*) in theory	0.2 nm	200 nm
5.	Lenses	Electromagnets	Glass (quartz for ultraviolet radiation)
6.	Specimen	Non-living, dehydrated, relatively small or thin.	Living or non-living
7.	Support of object	Supported on a small copper grid in a vacuum	Supported on a glass slide
8.	Common stains	They include heavy metals to reflect electrons such as lead acetate, lead citrate, lead hydroxide, uranyl acetate and phosphotungistic acid	Coloured dyes such as borax carmine, haema-toxylin, eosin
9.	Image	Black and white	Usually coloured
10.	Mode of observation or recording	Image is either seen on a fluorescent screen or recorded on a photographic film	Image can be seen with eye or recorded on photographic film by camera

EXAMPLES OF OPTICAL MICROSCOPES

1. Phase Contrast Microscope

Phase contrast microscope was invented by **Frits Zernike** of Netherland in 1932. He got Nobel Prize for this invention in 1953.

Principle and Working. Small, unstained specimens, such as a living cell, can be very difficult to see in a bright-field microscope. The phase contrast microscope has solved this problem by making highly transparent objects more visible. The ability to see different parts of an object depends on their capacity to affect light differently from one another. One basis upon which intracellular organelles differ is their refractive index. Cell organelles are made up of different proportions of various molecules: DNA, RNA, protein, lipid, carbohydrate, salts and water. Regions of different composition are likely to have different refractive indexes. Normally, however, such differences cannot be detected by our eyes. The phase contrast microscope converts differences in a refractive index into differences in intensity (relative brightness and darkness) which are then visible to the eye. The basis for this conversion centres on the ability to light waves to interact with one another, a property termed **interference**.

In contrast to bright-field microscope, phase contrast microscope performs two functions:

1. It separates the direct light (the background light of the field) from the light diffracted by the object, and,

2. It causes these two types of waves to be approximately one-half wavelength out of phase with one another so that they can destructively interact and cause changes in intensity.

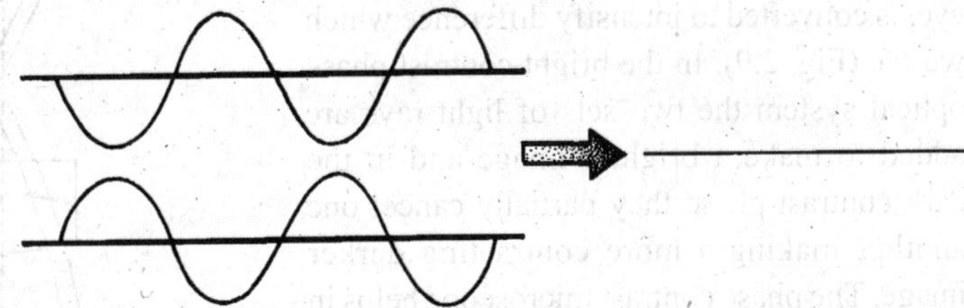

Fig. 2.7. Destructive interference. The two waves of light are one-half wavelength out of phase. When the waves meet, they combine to cancel each other out.

Suppose that all parts of the object being examined have exactly the same refractive index as the medium in which they are suspended. If this were the case, all the light from the object and the direct light would be one-half wave-length out of phase with one another, and **interference** would result in a uniform decrease in intensity (Fig. 2.7). Now consider an object whose various parts have different refractive indexes, all of which are greater than the surrounding medium. Differences in refractive index result in **phase shifts** of light rays passing through different parts of the specimen (Fig. 2.8). The relative brightness **(negative contrast)** or darkness **(positive contrast)** of each part of the image reflects the way in which the light from that part of the specimen interferes with the direct light. Since parts of the specimen have different refractive indexes, the various parts of the specimen appear in the image with varying intensity, thereby providing the necessary contrast to be visualized.

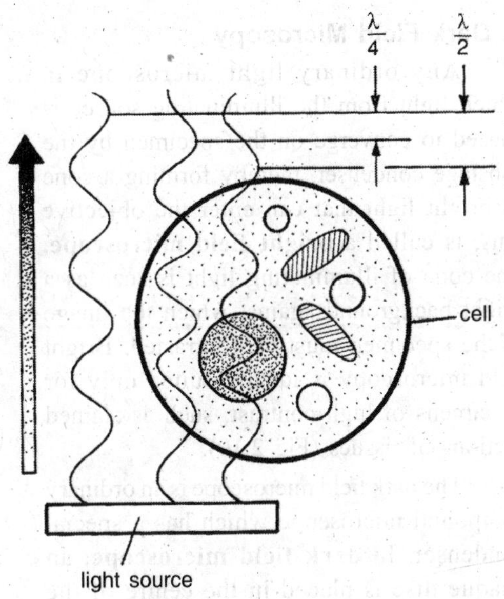

Fig. 2.8. Schematic diagram of retardation of light rays to varying degrees as they pass through different parts of a cell. Because the light rays that emerge from different parts of the cell are in *different phases* of their cycle, they interfere with the direct light to different degrees, producing different degrees of lightness and darkness in the final image.

Structure. The phase contrast microscope contains an annular phase plate in the objective and an annular diaphragm in the condenser. The annular diaphragm of phase contrast microscope permits light to pass through the condenser as a hollow cone, the remaining light being absorbed. This cone is focussed on the object. The phase plate which is placed at the back focal plane of the objective lens, is a transparent disc containing a groove (or elevation) of such a size and shape as to coincide with the direct image of the substage annular diaphragm which is formed at the back focal plane when no object is viewed. If an object is placed between the condenser and the objective, in addition to direct image a number of overlapping diffracted images of the diaphragm then appear at the back focal plane of the objective. The depth of the groove (or the elevation) in the phase plate of the objective is so made that the two sets of rays forming the direct image and the diffracted image differ in optical path by a quarter wavelength of the illuminating beam of light. Under these conditions the phase difference, which is not seen by the eye, is converted to intensify difference which we see (Fig. 2.9). In the bright contrast phase optical system the two sets of light rays are added to make a brighter image and in the dark contrast phase they partially cancel one another making a more contrasting darker image. The phase contrast microscope helps in the observation of behaviour of chromosomes of living cells during mitosis or meiosis and various other cellular components such as mitochondria and vacuoles. Observations of living cells on a television screen using a video camera provide especially dramatic portraits of the dynamic activities of a cell (Table 2.2).

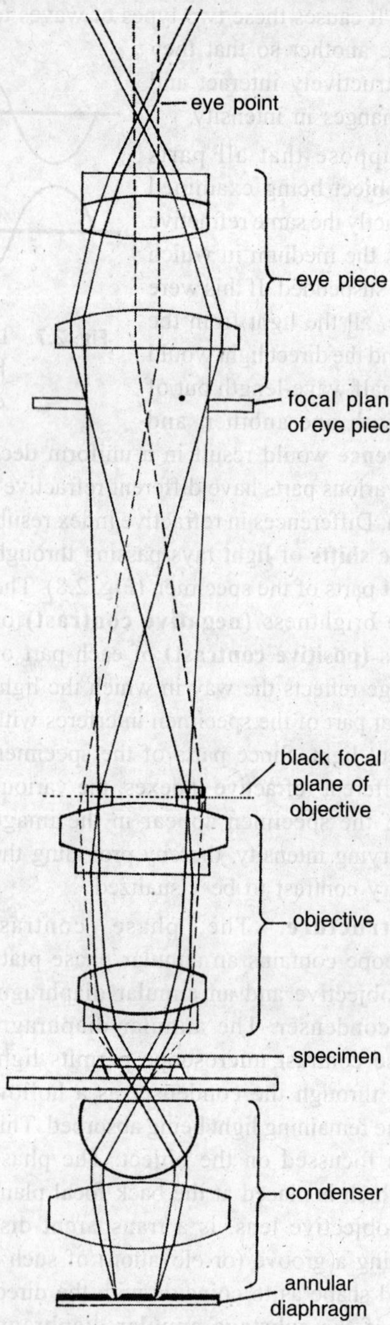

Fig. 2.9. Light path in phase contrast microscope.

2. Dark Field Microsopy

Any ordinary light microscope in which light from the illuminating source is caused to converge on the specimen by the substage condenser, thereby forming a cone of bright light that can enter the objective lens, is called a **bright field microscope.** The cone of illuminating light is seen as a bright background against which the image of the specimen must be contrasted. Bright field microscopy is ideally suited only for specimens of high contrast, such as stained sections of tissues (Fig. 2.10).

The dark field microscope is an ordinary compound microscope which has a special condenser. In **dark field microscope**, an opaque disc is placed in the centre of the condenser. The light rays are thus removed from the centre and no direct light rays enter the objective. Due to opaque disc, light rays are directed from the sides and only scattered (oblique) rays tend to enter the objective lens. As a result, the object appears illuminated (bright) against black background. Dark field microscopy improves visibility and resolution.

3. Fluorescence Microscopy

The fluorescence microscope is based on the phenomenon of fluorescence.

Table 2.2 The differences between light microscope and phase contrast microscope.

Light microscope	*Phase contrast microscope*
1. The refracted light rays are delayed by about one-quarter wavelength (0.25 λ).	1. The refracted light rays are delayed by one-half wavelength (0.50 λ).
2. All points of the image look equally bright.	2. The brightness of different points of image differs. It corresponds to differences in the refractive indexes of various cellular organelles.
3. Difference in the wavelengths of refracted and unrefracted light rays is not changed.	3. This increases the difference to produce maximum contrast (*i.e.*, interference) in the image.
4. Both living and dead (stained) specimens may be seen by this instrument.	4. Living (unstained) structures can be seen by this instrument.
5. There is no phase plate.	5. In this microscope, a glass phase plate is placed above the objective lens to increase the phase difference between the refracted and unrefracted light rays.

Fluorescence. It is found that when certain chemical compounds are illuminated by ultraviolet rays they become excited at molecular level and start to emit light of visible range. This phenomenon is known as **fluorescence**. The fluorescence may be of two kinds: **autofluorescence** and **secondary fluorescence**. Some substance, *e.g.*, chlorophyll becomes readily fluoresces or illuminates and produces red light, is an example of autofluorescence. When the fluorescence is induced in a substance by certain specific dyes or stains, the fluorescence is known as secondary fluorescence, *e.g.*, protein and carbohydrates, etc.

The source of illumination of fluorescence microscope is ultraviolet rays of higher wavelengths (3500 to 4000 Å). The different chemical compounds of the object to be seen when fluoresce by these ultraviolet rays, they emit rays of different wavelength. The rays of fluorescence can be detected and the chemical nature of the given object can be tested.

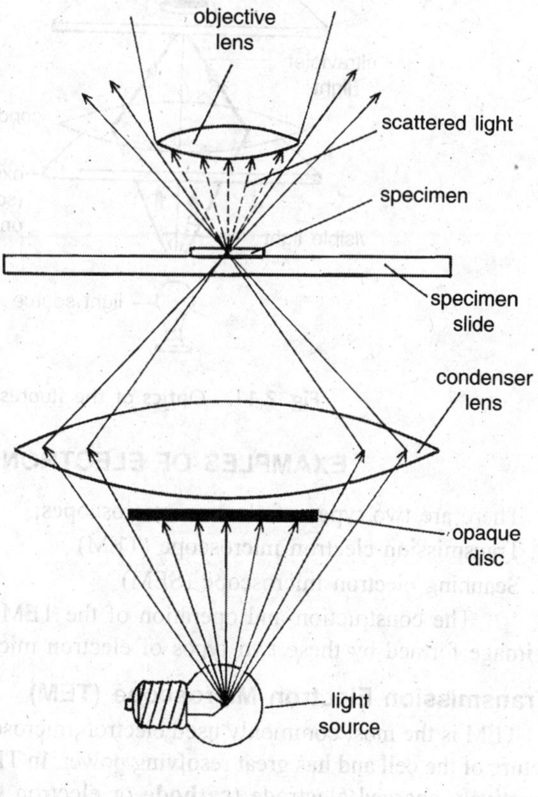

Fig. 2.10. Dark field microscope. Light path from the condenser to the objective lens.

The ordinary compound microscope can be easily converted into the fluorescence microscope by placing a special filter in between the source of illumination and eye piece.

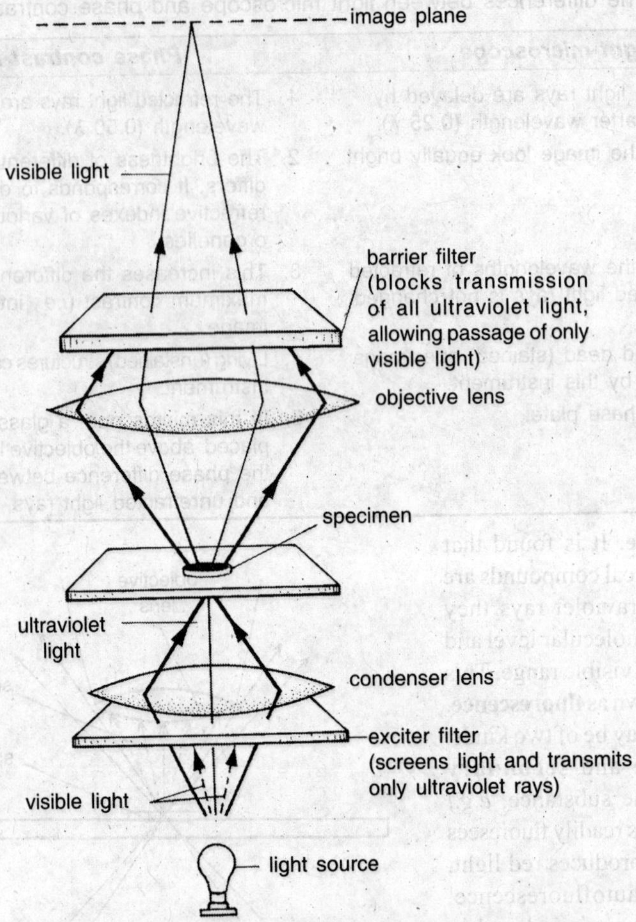

Fig. 2.11. Optics of the fluorescence microscope.

EXAMPLES OF ELECTRON MICROSCOPES

There are two types of electron microscopes:

1. Transmission electron microscope (TEM)
2. Scanning electron microscope (SEM)

The construction and operation of the TEM are very different from that of the SEM. The image formed by these two types of electron microscopes is also quite different.

1. Transmission Electron Microscope (TEM)

TEM is the most commonly used electron microscope. It is widely used to study the internal structure of the cell and has great resolving power. In TEM, an electric field propels electrons from a negatively charged electrode (**cathode** or electron gun). The electrons emitted from the gun are attracted towards the positively charged anode and then are focussed into beam by the electromagnetic **condenser**. This beam of electrons illuminates entire stained specimen. Many of these electrons pass right through the specimen and reach the electromagnetic objective. But some are absorbed or scattered by atoms of metallic stains that have combined with certain atoms of specimen. The **objective** magnifies the image of the specimen. This image is received by the third electromagnetic lens which works as **projection lens**. It further magnifies the image and projects it on a zinc sulphide or fluorescent screen at the bottom of TEM. This screen is similar to screen of a television tube. It is coated with a layer of crystals that respond to electrons by emitting visible light. The final image can also be captured on the photographic film.

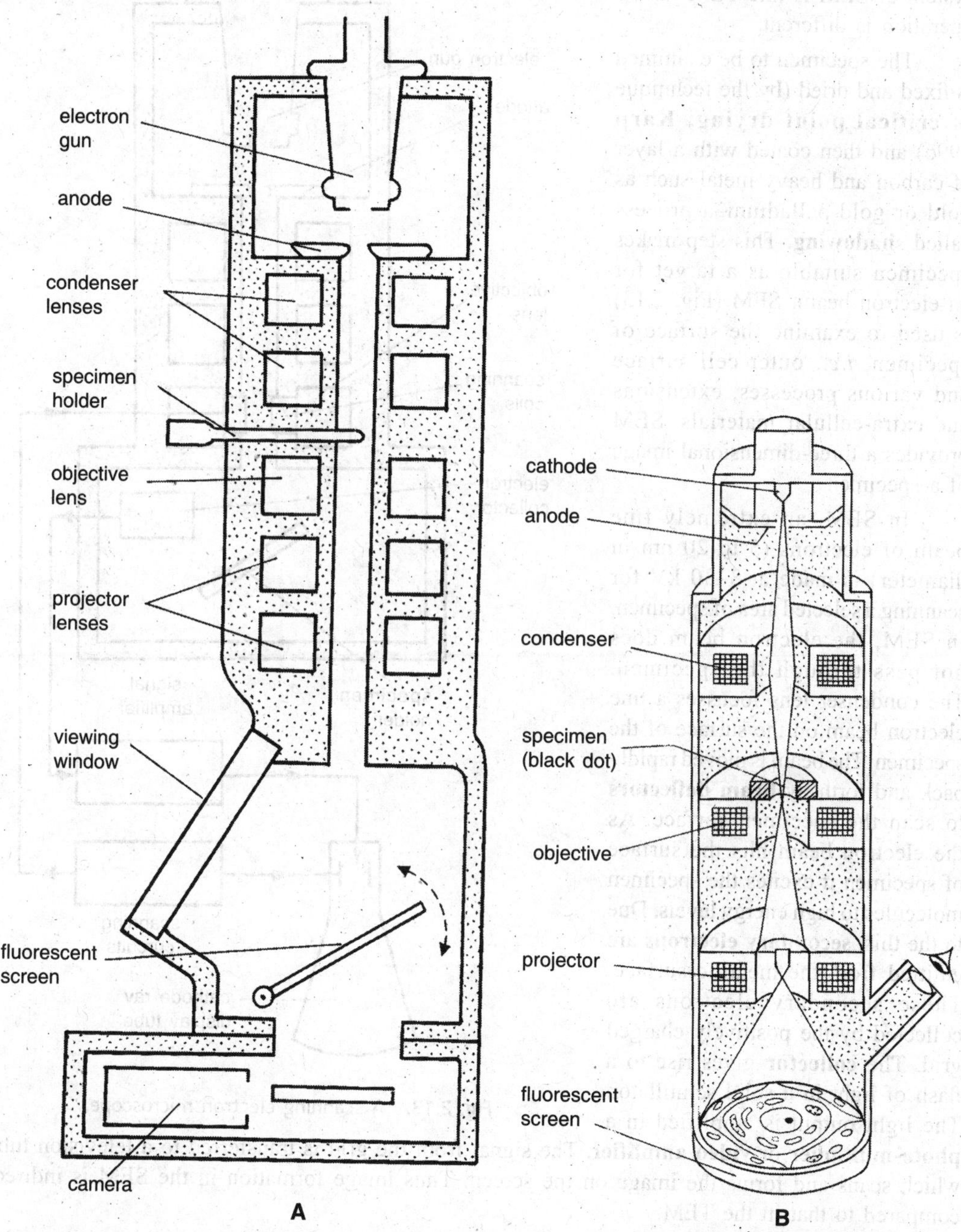

Fig. 2.12. Transmission electron microscope A–Diagrammatic view; B–Pathway of electron beam.

Electrons that are not transmitted by the specimen leave correspondingly dark regions on the viewing plate. Thus, image from a TEM is a pattern of bright and dark areas corresponding to the areas of greater or lesser electron density.

2. Scanning Electron Microscope (SEM)

The electron gun and lens system of SEM is like TEM but its operation is different.

The specimen to be examined is fixed and dried (by the technique of **critical point drying; Karp** 1996) and then coated with a layer of carbon and heavy metal such as gold or gold-palladium–a process called **shadowing.** This step makes specimen suitable as a target for an electron beam. SEM (Fig. 2.13) is used to examine the surface of specimen, *i.e.,* outer cell surface and various processes, extensions and extra-cellular materials. SEM provides a three-dimensional image of a specimen.

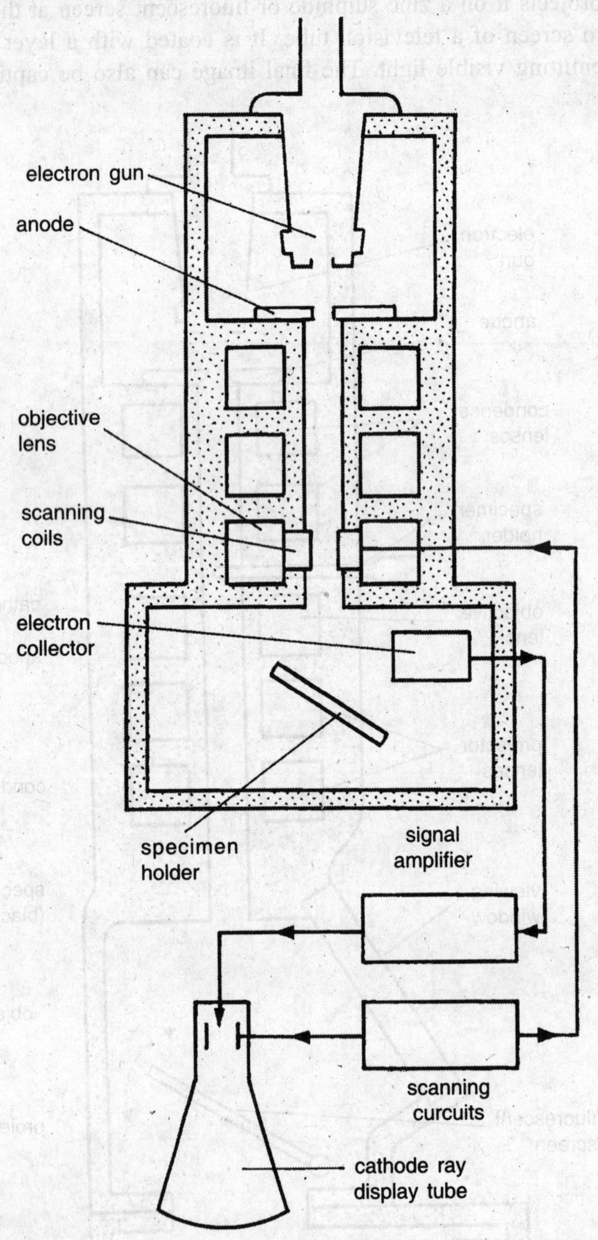

Fig. 2.13. A scanning electron microscope.

In SEM, an extremely fine beam of electrons (5 to 20 nm in diameter) is made at 3–30 kV for scanning a selected area of specimen. In SEM, the electron beam does not pass through the specimen. The condenser lens focusses a fine electron beam on the surface of the specimen. The beam is moved rapidly back and forth by **beam deflectors** to scan the specimen surface. As the electron beam hits the surface of specimen it excites the specimen molecules to high energy levels. Due to the this, **secondary electrons** are emitted from the metallic surface. These secondary electrons are collected by the positively charged grid. The **collector** gives rise to a flash of light in a solid scintillator. The light output is amplified in a **photo-multiplier** or **video amplifier.** The signal from the grid is transferred to a television tube which scans and forms the image on the screen. Thus image formation in the SEM is indirect compared to that in the TEM.

The amount of secondary electrons produced depends on the angle of specimen points with the scanning beam. Surface perpendicular to the beam produces maximum electrons while surfaces at greater angles release less electrons. Thus, the number of electrons produced depends on the three-dimensional shape of the specimen surface (or surface topology, *i.e.,* the crevices, hills and pits of the specimen). Accordingly the image contains bright (which correspond the elevations or ridges in specimen surface) and darker regions (which correspond to the valleys). These shadows give the image a three-dimensional appearance.

The resolving power of SEM is comparatively less than that of transmission electron microscope. It has an effective magnification upto 20,000 times.

2.4 CHROMATOGRAPHY

Chromatography is a technique for separating mixtures of gases, liquids or dissolved substances or the components of **proteins** (*i.e.,* amino acids), of **nucleic acids** (*i.e.,* nucleotides) or of **chlorophyll pigment.** By this technique, a mixture of dissolved components is fractionated as it moves through some type of porous matrix.

Principle of Chromatography

Chromatography is based on the phenomenon of adsorption and differential migration of components of a mixture or a compound. This needs the following two basic things:

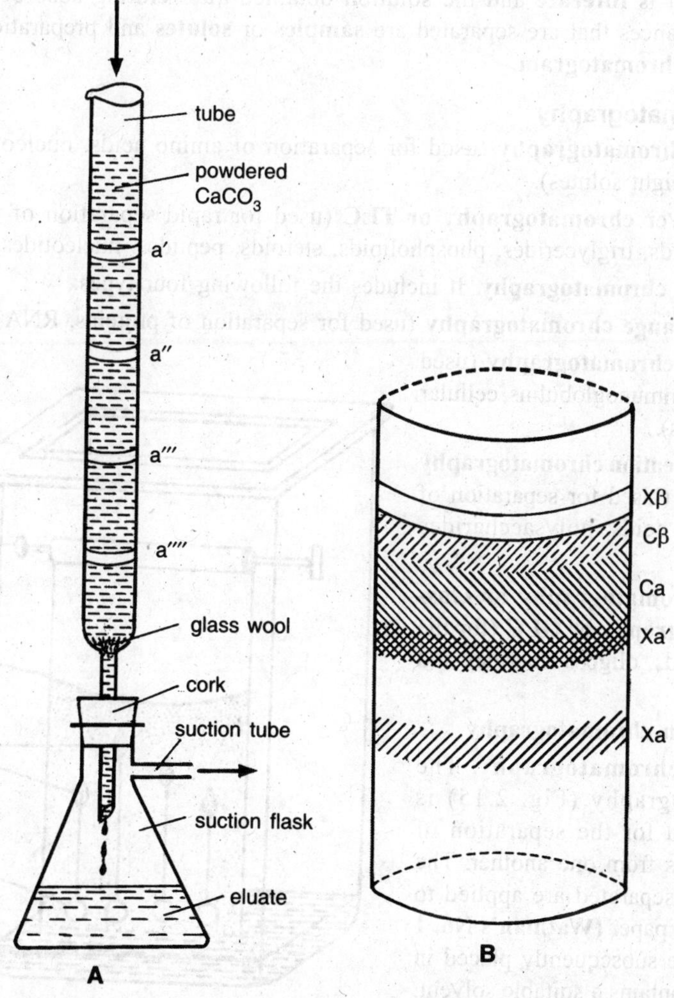

Fig. 2.14. A–An apparatus for column chromatography: B–Chromatogram of chlorophyll.

1. A force for propelling the molecules of any solution, mixture or cytosol. It is called **mobile phase** and it involves moving solvent.
2. A selective impedance causing differential movement of various molecules (*i.e.,* called **impeding medium** or **adsorbent**). It is also **immobile phase** and it involves the matrix (filter paper, $CaCO_3$, etc.) through which the solvent is moving.

During chromatography the mixture or solution is allowed to percolate in an insoluble medium or adsorbent or matrix which has different affinity for molecules of different substances. The molecules migrate through the medium matrix) at different rates and are separated. For example, if a solution of leaf pigment, chlorophyll in petroleum ether is allowed to percolate through a column of powdered calcium carbonate, the pigments are initially absorbed by pure solvent (*i.e.,* petroleum ether), the individual pigments—carotene, chlorophyll and xanthophyll separate and appear as the coloured bands or zones on the calcium carbonate column like the light rays in spectrum. Such a preparation is called **chromatogram**.

The impeding medium (or immobile phase) is called **adsorbent**. The liquid that passes through adsorbent is **filterate** and the solution obtained after eluting adsorbed substances is **eluate**. The substances that are separated are **samples** or **solutes** and preparation having these solutes is called **chromatogram**.

Types of Chromatography

A. Paper chromatography (used for separation of amino acids, nucleotides, and other low molecular weight solutes).

B. Thin layer chromatography or TLC (used for rapid separation of unsaturated and saturated fatty acids, triglycerides, phospholipids, steroids, peptides, nucleotides, etc.).

C. Column chromatography. It includes the following four types:

1. Ion-exchange chromatography (used for separation of proteins, RNA and DNA).

2. Affinity chromatography (used for separation of immunoglobulins, cellular enzymes, mRNAs).

3. Gel permeation chromatography or Gel filtration (Used for separation of proteins, nucleic acids, polysaccharides and lipids).

4. Gas chromatography or Gas-liquid chromatography or **GLC** (for the separation of lipids, oligosaccharides and amino acids).

D. Partition chromatography

1. Paper chromatography. The paper chromatography (Fig. 2.15) is a simple method for the separation of smaller molecules from one another. The molecules to be separated are applied to sheets of suitable paper **(Watman's No. 1 paper)** which are subsequently placed in a vessel which contains a suitable solvent. A distinction is made between ascending and descending paper chromatography

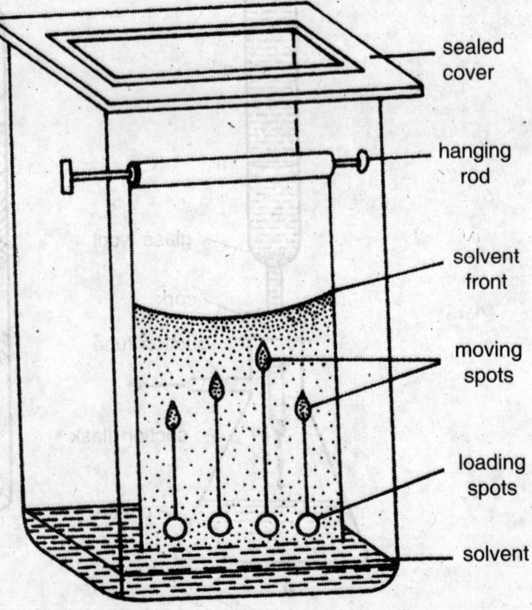

Fig. 2.15. Ascending paper chromatography.

according to whether the solvent migrates on the paper from below or above. Highly soluble components of the sample mixture will migrate at the solvent front; other substances move more slowly according to their solubility. From this it follows that the choice of solvent determines the speed of migration of the individual molecules and is crucial for a successful separation. The paper chromatography method is used for the separation of amino acids, nucleotides and other lower molecular-weight metabolic products.

If the solvent flows towards upper side on the paper, it forms the **ascending paper chromatography** (Fig. 2.15), and if it flows towards lower side then it forms the **descending paper chromatography**.

2. Column chromatography. In column chromatography (Fig. 2.16), an insoluble medium is packed into a glass tube; the length and width of this tube influence the separation of the molecules. The molecules to be separated are applied to the top of the column and their migration is started by adding a solvent. The characteristic separation which results depends on the choice of solvent and carrier material. A positively charged carrier binds negatively charged molecules; other carriers contain pores which are penetrated by the smaller molecules, which are, therefore, slowed down. The solution which flows from the column is collected in small fractions (Fig. 2.16), which contain the separated classes of molecules. Column chromatography is important for the separations of mixtures of proteins, that is, for the isolation of enzymes such as cytochrome C or RNA polymerase.

Adsorbents and solvents. A few of the adsorbents used in column chromatography are carbonates of calcium, magnesium and sodium; various forms of charcoal activated in special ways; Fuller's earth, bentonite, Lloyd's reagent, talcum, clays of different types; alumina, silica gel, cellulose, resin, starch, sucrose, inulin and benzoic acid, etc. The adsorbent column neither dissolves in the developing solvent nor reacts chemically with the solvent or the substance to be separated.

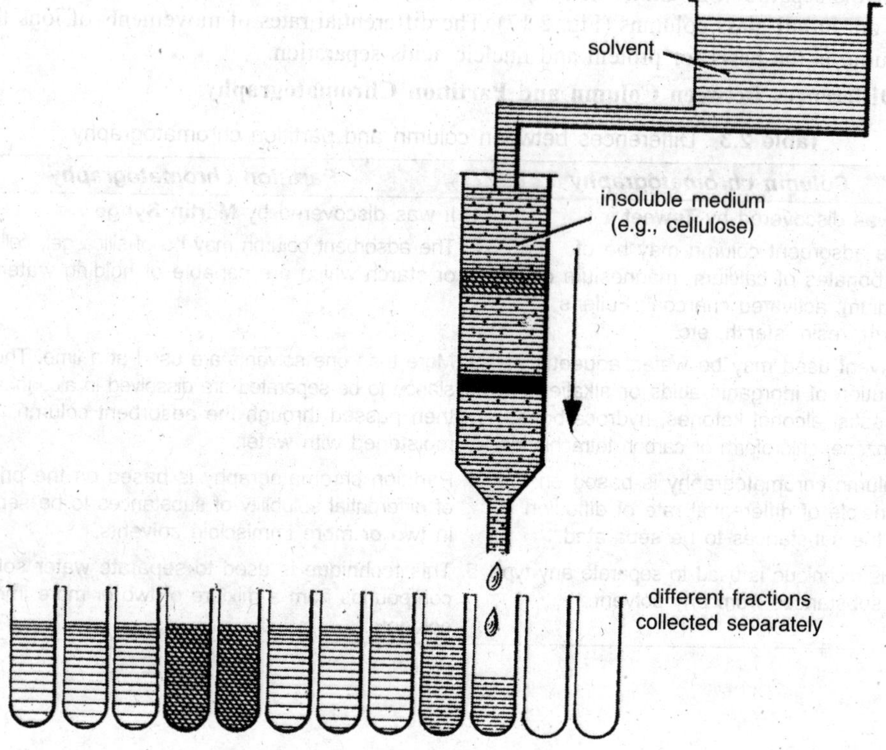

Fig. 2.16. Column chromatography.

The **solvents** commonly used are water, aqueous solutions of various acids, alkalis and salts; alcohols, ketones and cyclohexanone, hydrocarbons such as benzene, chloroform and carbon tetrachloride, etc.

3. Ion exchange chromatography. In ion exchange chromatography molecules are separated by their difference in charge. The ion-exchange resin is used as chromatographic adsorbent consists of porous polystyrene beads (polymers) to which ionizable groups are added chemically. The resins bearing negative charge are called **cation exchanger** (*e.g.,* **DEAE-** or **diethylaminoethyl-cellulose**) and positively charged resins are **anion exchangers** (*e.g.,* **CM- or carboxy methyl-cellulose**).

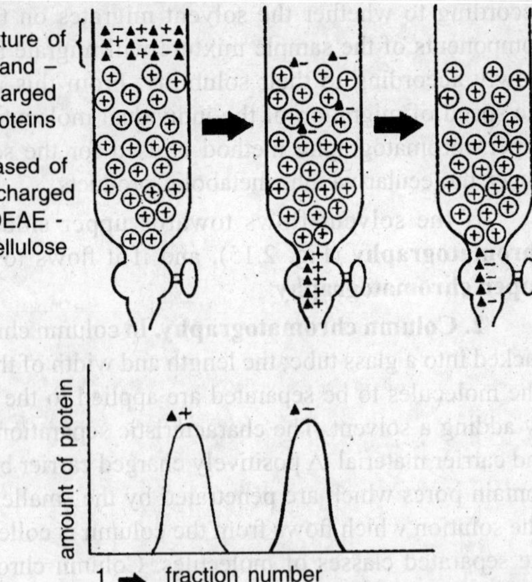

Fig. 2.17. Ion-exchange chromatography. The separation of two species of protein by DEAE-cellulose. In this case, a positively charged ion-exchange resin is used to bind the more negatively charged protein.

When a mixture of molecules possessing different charges passes through a charged column, the molecules possessing opposite charge get associated with the column, while those with same charge or with no charge pass through. The process is used for the separation of amino acids, proteins and other macromolecules and the separation is carried out in tall glass columns (Fig. 2.17). The differential rates of movements of ions through the column is the basis for protein and nucleic acids separation.

1. Differences Between Column and Partition Chromatography

Table 2.3 Differences between column and partition chromatography.

Column chromatography	*Partition chromatography*
1. It was discovered by **Tsweet**.	1. It was discovered by **Martin Synge**.
2. The adsorbent column may be of carbonates of calcium, magnesium or sodium; activated charcoal; Fuller's earth, resin, starch, etc.	2. The adsorbent column may be of silica gel, cellulose or starch which are capable of holding water.
3. Solvent used may be water, aqueous solution of inorganic acids or alkalies or salts, alcohol ketones, hydrocarbons, benzene, chloroform or carbon tetrachloride.	3. More than one solvents are used at a time. The substance to be separated are dissolved in alcohol and are then passed through the adsorbent column, already moistened with water.
4. Column chromatography is based on the principle of differential rate of diffusion of the substances to be separated.	4. Partition chromatography is based on the principle of differential solubility of substances to be separated in two or more immiscible solvents.
5. This technique is used to separate any type of substances from any solvent.	5. This technique is used to separate water soluble compounds from a mixture of two or more immiscible solvents.

2. Differences Between TLC and GLC

Table 2.4 Some differences between TLC and GLC

TLC or Thin-layer chromatography	*GLC or Gas-liquid chromatography*
1. Adsorbent may be cellulose powder, alumina or cellulose.	1. Adsorbent or the immobile phase is represented by liquid.
2. The mobile phase is the solvent (liquid).	2. The mobile phase is inert gas such as nitrogen, carbon dioxide, helium or argon.

2.5 ELECTROPHORESIS

Molecules or macromolecules may be separated in an electric-field if they are charged to different extents. The mixture of compound is applied to supporting films which dip into two containers filled with a salt solution. One of the containers holds a cathode, the other an anode. On passing an electric current, the negatively charged molecules migrate to the anode and the positively charged molecules to the cathode. Paper, agar and starch are examples of substances which may be used as supporting films.

The rate of migration of a molecule in an electric field is determined by its size and the number of charged groups per molecule. The electrophoresis method is used in the separation of proteins, nucleic acids and their building blocks.

Some of the common types of techniques of electrophoresis, which are currently used in cell biology, are the following:

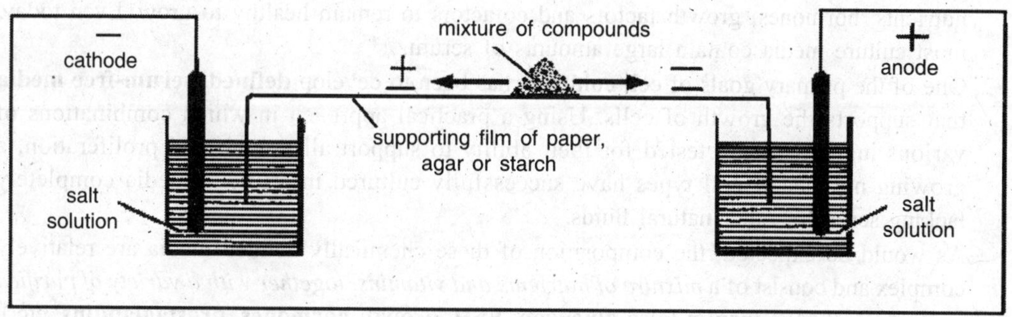

Fig. 2.18. Electrophoresis.

1. **Moving-boundary electrophoresis** (used for proteins).
2. **Gel or zone electrophoresis** (used for proteins).
3. **Discontinuous electrophoresis** (used for isolation of proteins of plasma membrane).
4. **SDS-PAGE or Sodium dodecyl sulphate-polyacrylamide gel electrophoresis** (used for separating and sizing macromolecules such as proteins, *e.g.,* membrane proteins, protein component of cytoskeleton, etc.).
5. **Maxam-Gilbert technique** (used for separation of polynucleotide fragments of RNA and DNA).
6. **Immunoelectrophoresis** (used for antigens and antibodies).

2.6 DIALYSIS

It is a sensitive method for separating lower molecular-weight component from macromolecules. A thin membrane in the form of a tube is filled with the solution containing the molecules to be separated (Fig. 2.19). The pore size of the membrane allows the diffusion of small molecules such as salt or amino acids; larger molecules such as proteins or nucleic acids cannot pass through the pores and so remain inside the dialysis tube.

2.7 CELL CULTURE

The growth of cells or tissues of living organisms outside the body in a suitable sterile culture (or nutrient) medium under controlled environmental conditions is called **cell culture** or **tissue culture**. Depending on the nature of the explanted material, the techniques are called **cell, tissue, organ** or even **whole embryo culture**. The culture medium may be natural or artificial and either in solid form (*e.g.,* agar) or in liquid form (*e.g.,* physiological saline).

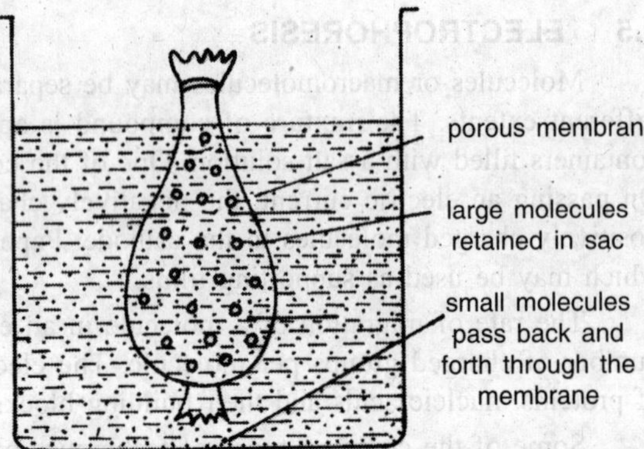

Fig. 2.19. Dialysis.

Certain Basic Requirements of Tissue Culture

1. The early tissue culture studies used media containing a great variety of *unknown* substances. Cell growth was accomplished by adding fluids obtained from living systems, such as **lymph, blood serum** or **embryo homogenate**. It was found that cells required a considerable variety of nutrients, hormones, growth factors and cofactors to remain healthy to grow. Even today, most culture media contain large amounts of serum.
2. One of the primary goals of cell culturists has been to develop **defined, serum-free media** that supports the growth of cells. Using a practical approach in which combinations of various ingredients are tested for their ability to support all growth and proliferation, a growing number of cell types have successfully cultured in artificial media completely lacking serum or other natural fluids.

   As would be expected, the composition of these chemically defined media are relatively complex and consist of a *mixture of nutrients and vitamins, together with a variety of purified proteins, including* **insulin** (also **glucagon, FSH, growth hormones, prostaglandins,** etc.), **epidermal growth factors** and **transferrin** (which provides the cells with usable iron).
3. In addition to a complex culture medium, most cells need an appropriate surface on which to grow. Unlike most cancer cells, which are capable of growing in suspension, normal cells require a solid substratum, such as glass or plastic.

Box 2.3 Historical

1. **Ross Harrison** (1907) of Yale University performed one of the classic experiments in biology. He removed a small piece of tissue from the developing nervous system of a frog embryo and placed the fragment into a tiny drop of lymphatic fluid. Harrison observed that nerve cells remained alive and healthy for many days and some of the cells sprouted processes that grew out into the surrounding medium.
2. **Alexis Carrel** showed that extracts of chicken embryo stimulate cells to grow and divide.
3. **Harry Eagle** spent many years in defining the nutritional needs of mammalian cells and formulating media to optimize their growth.
4. Once the cells settle to the bottom of the culture dish, they secrete extracellular materials such as **collagen, fibronectin** and **laminin,** which help in keeping them attached to their substratum.

Primary and Secondary Culture

The first step in cell culture is to obtain the cells. In most cases, one need only to remove a vial of frozen, previously cultured cells from a tank of liquid nitrogen, thaw the vial and transfer the cells to the waiting medium. A culture of this type is referred to as a **secondary culture** because the cells are derived from a previous culture. A **primary culture,** on the other hand, is one in which the cells are obtained from the organism itself. Most primary cultures of animal cells are obtained from embryos, whose tissues are more readily dissociated into single cells than those of adults.

To accomplish dissociation, the tissue is removed from the embryo and is either mechanically agitated or treated with a proteolytic enzyme, such as **trypsin** or **collagenase,** which is capable of digesting proteinaceous material that help hold the cells together. The tissue is then washed free of the enzyme and usually suspended in a saline solution that lacks Ca^{2+} ions and contains a substance, such as **EDTA (ethylene diamine tetra acetate)** that binds (*i.e.,* chelates) calcium ions. Calcium ions play a key role in cell-cell adhesion and their removal from the tissues greatly facilitates the separation of cells. Cell suspensions obtained by such treatments usually consist of a heterogeneous mixture of cell type.

Once the cells have been dissociated into a single-cell suspension, either they can be cultured directly, or they can be separated according to cell type and then placed in culture. Cells can be separated by a number of techniques, including differential centrifugation or the use of a **fluorescence-activated cell sorter.** In this latter approach, the cell suspension is treated with **fluorescent antibody** that binds specifically to the surface of the cell type to be cultured, and the suspension is passed through an electronic instrument that is able to deflect the fluorescently labeled cells into a separate tube from those cells lacking the label.

Once the cells have been separated, they can be grown either **suspended** in a liquid medium or **attached** to a solid surface. There are basically two types of primary cultures that can be started:

1. In a **mass culture**, a large number of cells are added to a culture dish; they settle and attach to bottom and form a relatively uniform layer of cells. Those cells that survive will grow and divide and after a number of generations, form a **monolayer** of cells that covers the bottom of dish.
2. In the other type of culture, called a **clonal culture**, a relatively small number of cells are added to a dish, each of which, after settling and attaching to surface, is at some distance from its neighbours. In this case, the proliferation of the cells generates individual colonies, or **clones**, of cells, whose members are all derived from the same original cell.

Cell Line

Normal (*i.e.,* nonmalignant) cells are capable of a limited number of cell divisions (*e.g.,* 50 cell divisions in human cells) before they undergo a process of **senescence** and **death.** Because of this, many of the cells that are commonly used in tissue culture studies have undergone **genetic modifications** (due to virus infection, chemical mutagens or radiations) that allow them to be grown indefinitely. Cells of this type (*i.e.,* transformed cells) are referred to as a **cell line** and they are typically capable of growing into malignant tumours when injected into susceptible laboratory animals. Such behaviour is typically shown by mouse cells. Among the best known cell lines are **HeLa cells** (obtained from a human carcinoma), the **L** and **$3T_3$** cells (from mouse embryo), the **BHK cells** (from baby hamster kidney) and the **CHO cells** (from Chinese hamster ovary).

Types of Tissue Culture

1. Culture of bacterial cells. The culture medium of bacteria contains an **energy source** (some simple carbohydrates or some amino acids); **mineral salts** of calcium, potassium, sulphur and phosphorus; **traces of metals** (zinc, copper, magnesium, manganese, etc.) and **growth factors**

(biochemical intermediates). The culture medium is maintained between pH 6 to 8 and at a temperature varying from 15–20°C or 30–37°C according to the species of bacteria.

The culture medium is prepared in *agar* in a petridish. Agar is a carbohydrate material derived from seaweed that is dissolved in a nutritive medium to make a semisolid gel on which bacteria can grow. When it sets, it forms a sheet on the bottom of the petridish. A very dilute suspension of bacteria is cultured in the petridish. As a result of growth and multiplication several colonies of bacteria are formed on the surface of agar. Each colony is derived from a single cell.

2. Culture of mammalian cells. Since mammalian cells are not able to synthesize essential biochemical intermediates from the raw materials, their culture media are very complex. For the culture of mammalian cells, several culture media have been developed; some of them are following:

(*a*) A mixture of serum and chick embryo extract rich in various factors required for growth.

(*b*) **Parker's 199 synthetic medium** contains different metabolites. It is supplemented with some foetal calf serum or horse serum.

Culture techniques. Mammalian cells have been cultured by adopting the following culture techniques:

(*i*) Maximov double coverslip method. A small piece of tissue to be cultured is placed on a coverslip and is covered with a drop of chick plasma and growth medium (Fig. 2.20). The plasma forms a clot over the tissue cells. The coverslip is then mounted in a cavity slide with the clot attached to the undersurface. It lies above the cavity of slide. The cavity of slide provides oxygen and allows exchange of gases. The growing cells form a sheet within the plasma clot.

(*ii*) Monolayer culture method. For the preparation of cell suspension, the chopped tissue is treated with dilute trypsin. The trypsin is removed from the cell suspension by washing in a centrifuge. The cell suspension settles down on the glass and attach to the surface of bottle. The cultured cells may be isolated by the dilute solution of trypsin and the process may be repeated again to get large number of cells.

(*iii*) Trowell method. The small fragments of an organ are placed on a raft or permeable sheet of Trowell chamber and are covered with culture medium. (Trowell is a small hand-held tool with a curved scoop for lifting plants or earth.)

3. Culture of plant cells. Many different types of plant cells such as anther, endosperm, nucellus, embryo, root, shoot tips, leaf, flower buds, etc., are capable of growing in culture. The simple undifferentiated tissues such as **primary meristems** show an organized growth in basic culture medium. But the fully differentiated tissues such as pith, secondary phloem, mesophyll and endosperm, *etc.,* in basic culture medium grow into an unorganized and undifferentiated mass of tissues, called **callus. Gauherst** *et al.,* (1939) succeeded in growing callus in vitro from isolated plant parts. By adding certain hormones, it is now possible to obtain callus culture from most plant tissues.

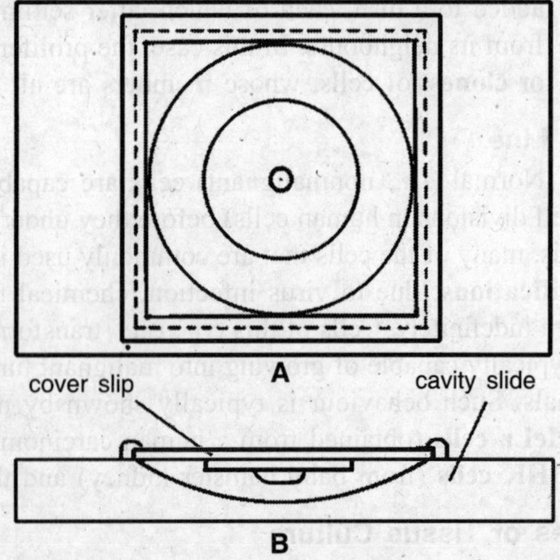

Fig. 2.20. Maximov double coverslip method of cell culture.

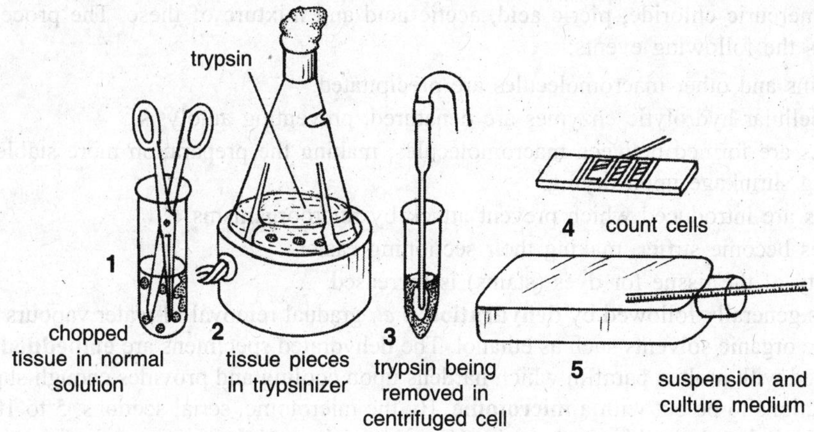

Fig. 2.21. Different steps of monolayer technique of tissue culture.

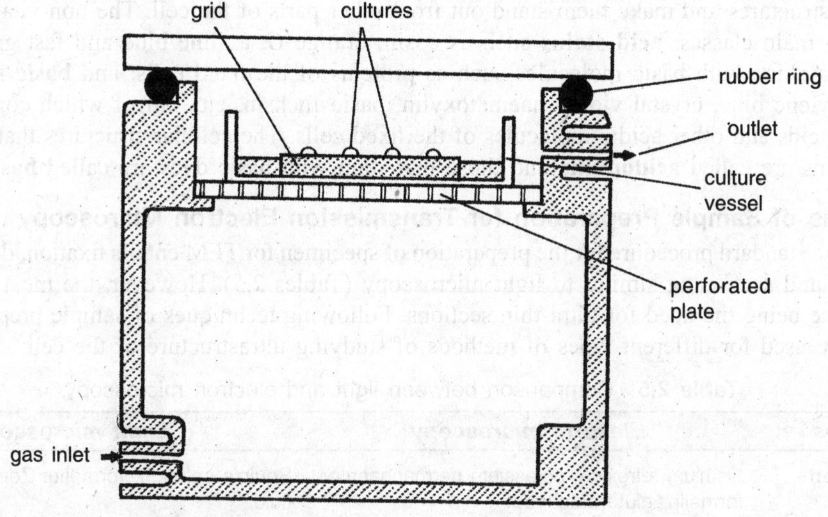

Fig. 2.22. Chamber of organ culture by Trowell method.

Culture technique. The first step in the formation of these cultures is to treat cells with the enzyme **cellulase**, which is capable of digesting away the surrounding cell wall, releasing the naked cell or **protoplast**. The protoplasts can then be grown in a chemically defined medium that promote their growth and division. Under suitable conditions, the cells can be grown into an undifferentiated clump of cells, the **callus**, which can be induced to develop shoots from which a new plant can regenerate.

2.8 SAMPLE PREPARATION FOR MICROSCOPY

Cells are transparent and optically homogeneous: so either they are viewed as such by instruments such as phase contrast microscope or to produce necessary contrast, the cells are passed through various steps of slide preparation such as killing, fixation, dehydration, embedding, sectioning, staining and mounting.

Thus, superior specimens for microscopic examination can be obtained by **killing** the cells and coagulating or **fixing** the protoplasm by preservatives, called **fixatives** such as alcohols, formaldehyde, mercuric chloride, picric acid, acetic acid and mixture of these. The process of fixation involves the following events:

1. The proteins and other macromolecules are precipitated.
2. The intracellular hydrolytic enzymes are denatured, preventing autolysis.
3. Cross links are formed between macromolecules, making the preparation more stable and minimizing shrinkage upon drying.
4. Substances are introduced which prevent attack by microorganisms.
5. The tissues become stiffer, making their sectioning easier.
6. The affinity of the tissue for dyes (stains) is increased.

Fixation is generally followed by **dehydration** (*i.e.,* gradual removal of water vapours from the tissue) by the organic solvents such as ethanol. The dehydrated specimens are **embedded**, *i.e.,* they are infiltrated with molten paraffin which hardens upon cooling and provides enough support to allow thin sections to be cut with a **microtome**. By the microtome, serial sections, 5 to 10 μm thick can be cut and placed on slides in the order of cutting and permitting a sequence of specimens for observation. These sections are stained with a non-vital-stain to increase the contrast.

Stains are the chemicals that can selectively attach to particular molecules of specific cellular structures and make them stand out from other parts of the cell. The non-vital stains fall into two main classes: **acid stains** such as eosin, orange *G*, aniline blue and fast green, all of which combine with basic molecules such as proteins of the fixed cells; and **basic stains** such as methylene blue, crystal violet, haematoxylin, basic fuchsin, etc., all of which combine with nucleic acids and other acidic molecules of the fixed cells. The cellular structures that stain with acid stains are called **acidophilic** and those that stain with basic dyes are called **basophilic**.

Methods of Sample Preparation for Transmission Electron Microscopy

The standard procedure for the preparation of specimen for TEM entails fixation, dehydration, staining and sectioning similar to light microscopy (Tables 2.5). However, the most significant difference being the need for ultra-thin sections. Following techniques of sample preparation are generally used for different types of methods of studying ultrastructure of the cell:

Table 2.5 Comparison between light and electron microscopy.

Steps	*Electron microscopy*	*Light microscopy*
Fixation	Osmium tetroxide, potassium permanganate, formalin, glutaraldehyde.	Bouin's solution; formalin; Zenker's fluid.
Dehydration	Increasing concentration of ethanol (or acetone) followed by propylene oxide.	Increasing concentration of ethanol followed by benzene.
Embedding	Araldite; Vestoplaw, Epan 812; Maraglas; Durcopan.	Paraffin.
Sectioning	Usually 10–100 nm thick sections cut with a glass or diamond knife on an ultramicrotome.	Usually 6 μm thick sections cut with a razor blade on a microtome.
Mounting	On a perforated metal disc (grid) usually covered with formvar or paralodian.	On a glass slide with an egg albumin adhesive. Deparaffinized in xylol for staining.
Staining	With salts of heavy metals such as lead acetate, lead citrate, lead hydroxide, urynyl acetate; phosphotungstic acid. mounted for viewing in Canada balsam	Selective chromatic stains (as haematoxylin and eosin), dehydrated in ethanol series, cleared in xylol and or Permount.
Viewing	Grid is placed between the condenser and objective lenses in vacuum and the image is viewed on phosphorescent screen.	Slide is placed between the condenser and objective lenses and the image viewed in the ocular lens.

1. Monolayer technique. Macromolecules such as DNA and RNA are studied by monolayer technique in which the macromolecules are extended on the air-water interface before being collected on a film.

2. Thin sectioning. This method uses a cutting device known as **ultramicrotome** to remove ultrathin (*i.e.*, 10 nm to 100 nm thick) sections from the specimen. To withstand the passage of ultrafine diamond or glass knife without tearing, the specimen is first embedded in a hard plastic such as, epoxy resin (Table 2.5). The resin is allowed to penetrate the sample before it is polymerized. Sections are floated from the knife of ultramicrotome onto the surface of water and picked up by touching them with a fine wire mesh or small circular copper grid (*i.e.*, small discs perforated with numerous openings). Prior to its use mesh or grid is coated with a thin monolayer film (7.5 to 15 nm thick) of plastic (such as formvar or collodion) or carbon to provide a support to the sections (or sample) (Fig. 2.23). The specimen is visualized through the holes of screen.

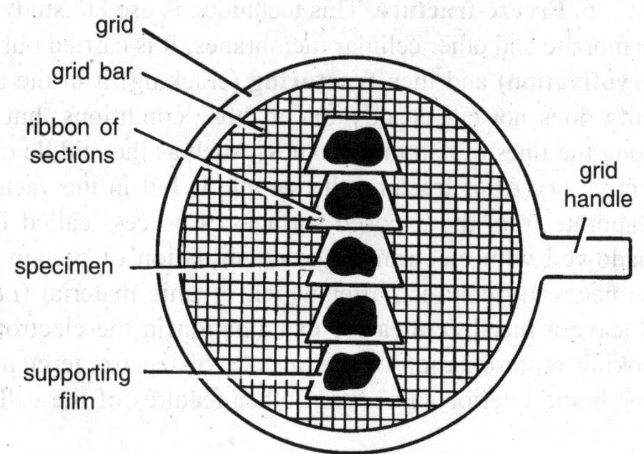

Fig. 2.23. Appearance of a ribbon of ultra-thin sections on the grid.

Sections to be examined with the electron microscope are generally not stained (since no colours are seen with the electron microscope). However, contrast may be improved by "post-staining" with **electron stains** or electron-dense materials such as urynyl acetate, urynyl citrate, lead citrate, osmium tetroxide, etc. The method of thin sectioning is used to study morphology of cell.

3. Negative staining. This technique is used to study small particles such as viruses or macromolecules. Here, the specimen is embedded in a droplet of electron dense material, such as, phosphotungstic acid ($H_3PW_{12}O_{40}$). The electron stain penetrates into all the empty spaces (*i.e.*, openings and crevices) between the macromolecules. The spaces appear well defined in negative contrast. The portions of specimen that exclude stain transmit electrons readily, so their images can be seen.

4. Shadow casting or heavy metal shadowing. This technique is used to study three-dimensional appearance of viruses and certain macromolecules such as DNA molecules and collagen fibres. It involves placing of specimen in an evacuated chamber and evaporating at an angle, a heavy metal such as chromium, palladium, platinum or uranium from a filament of incandescent tungsten. The vapour of heavy metal is deposited on one side of the surface of the elevated particles; on the other side a shadow forms, the length of which permits determination of the height of the particle. In such a specimen, during electron microscopy, the electrons pass readily through the area of light metal content, less readily through the plane on which the particle sits, and are scattered more severely by the side of the particle on which metal has accumulated. Thus, by shadow casting, shape and profile of a particle can be observed.

5. Tracers. Several biological processes such as pinocytosis, phagocytosis and transport of molecules across plasma membrane can be studied by the use of appropriate tracers (*e.g.*, gold, mercuric sulphide, iron oxide, etc.). These tracers are detected by their electron opacity. An ideal tracer should be non-toxic, physiologically inert, composed of small-sized particles of uniform and known size and preserved *in situ* during the processing of the tissue.

6. Freeze-fracture. This technique is used to study the molecular arrangement in the plasma membrane and other cellular membranes. It is carried out by rapidly cooling or freezing the sample **(cryofixation)** and then **fracturing** (cracking) it in the a vacuum while it is still at –100°C. The knife does not cut cleanly under those conditions, but tends to **fracture** (crack) the specimen along the lines of natural weakness, such as the middle of a membrane that runs parallel to the cut (Fig. 2.24). After fracture, the sample is left in the vacuum long enough to allow some water to evaporate from the exposed surfaces, a process called **freeze etching.** The exposed face is then shadowed with electron-dense combination of carbon and metal such as platinum to provide the necessary contrast, after which organic material (*i.e.,* specimen itself) is removed by acids to leave a metal **replica** for examination in the electron microscope. Replica reveals a natural-looking representation of the surface of the freeze-etched object and is the only way of seeing membrane interior and certain other features of the cells.

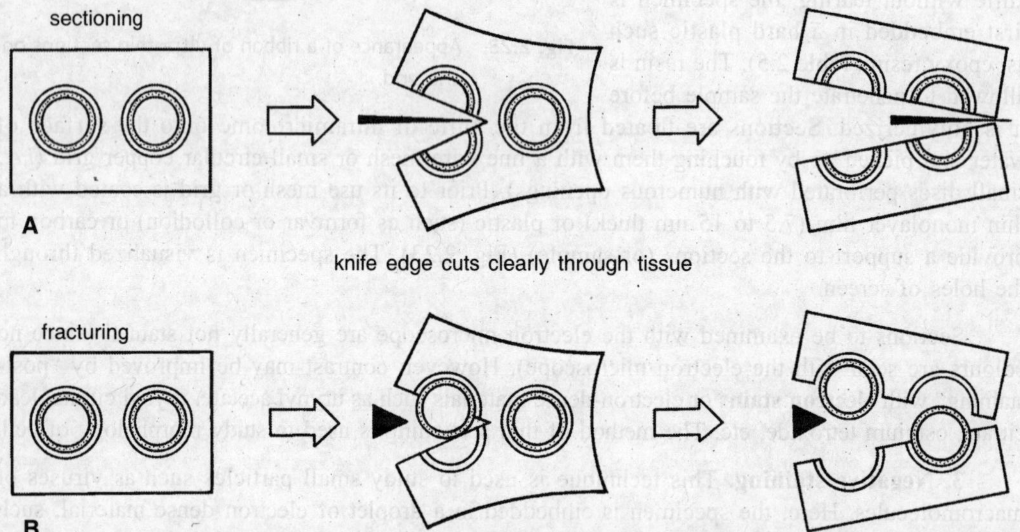

Fig. 2.24. A comparison of sectioning and fracturing tissues.

7. Whole mounts. They are often used to examine chromosomes and other relatively thick objects that can be isolated free of debris. In these methods, the specimen is neither sectioned nor stained. Thick areas will scatter electrons more strongly than thin areas, providing enough contrast to form an image.

Fixation

For microscopical study of dead cells, the process of killing and **fixation** brings about the sudden death of cells or tissues in such a manner that morphological organization and chemical composition of the cell remain unaltered up to greatest extent. The fixative or fixing solution performs following functions: it prevents bacterial decay and autolysis (self digestion) of the cells; renders the components of the cell insoluble, reduces the cell distortion and shrinkage, increases the visibility of different cellular components and prepares the cells for the staining. The fixation of the cells can be achieved either by chemicals or cooling which is as follows:

(*a*) Chemical Fixation. The choice of suitable fixative generally depends on the type of analysis and the chemical nature of different cells. For instance, for the study of the nucleus and chromosomes the **acid fixative** such as **Carnoy's solution** or **Bouin's fluid** are used. The commonly used fixatives are as follows:

***(i)* Acetic acid.** The acetic acid is used from 0.3 to 50 per cent, alone or in mixtures. It does not fix cytoplasmic proteins, but precipitates nucleoproteins, destroys Golgi apparatus and mitochondria of the cells. Moreover, the cells become swell up and soft. The acetic acid is used with basic dyes for the fixation of **nucleus** and **chromosomes**.

***(ii)* Potassium dichromate.** Potassium dichromate is generally used in mixtures and in the percentage of 2–7. It renders proteins insoluble in water and fixes lipids. Potassium dichromate is used for the fixation of chromosomes and some cytoplasmic structures.

***(iii)* Ethanol.** 70 to 100 per cent ethanol is used either alone or in mixture for the fixation of nucleus along with certain basic dyes. It does not precipitate proteins but dissolves lipids. Ethanol makes the cells to become hard but causes shrinkage.

***(iv)* Formaldehyde.** The formaldehyde (4–10 per cent) is used in mixtures for the fixation of Golgi apparatus, mitochondria and enzymes along with some basic stains.

***(v)* Osmium tetroxide.** Osmium tetroxide (0.5–2 per cent fixes cytoplasm, Golgi apparatus, mitochondria and fats of the cell but causes blackening of cell due to its oxidation which renders difficulty in staining. This fixative is widely used for electron microscopy.

***(vi)* Bouin's solution.** The Bouin's solution has 5 part picric acid, 5 parts 40% formaldehyde and one part glacial acetic acid. It is used as the fixative for chromosomes. It precipitates all proteins and causes shrinkage and softening of the cells or tissue.

***(vii)* Carnoy's fluid.** The fixative Carnoy's fluid is most commonly used for the fixation of the nucleoproteins and chromosomes because it has combined properties of ethanol and acetic acid. Carnoy's fluid contains three parts ethanol and one part glacial acetic acid.

***(b)* Fixation by Freezing.** For histochemical and electron microscopic observations sometimes it is required that different chemical substances and organelles remain as such or minimum alteration occur in them. In this method the cells or tissues are killed and fixed by rapid cooling of them. Following methods of freezing fixation are used in recent cytological studies.

***(i)* Fixation by freezing-drying method.** For the electron and X-ray microscopy, the tissues or cells are fixed by rapid freezing and are dehydrated in the vacuum at a low temperature. The rapid freezing of tissues is achieved by placing the living tissue in isopentane or propane cooled upto 160°C or –190°C by liquid nitrogen or helium. After freezing, the material is dried in vacuum at 30°C to –40°C.

By freezing-drying fixation method, the tissue does not shrink, homogeneous fixation of cell takes place, soluble substances are not extracted, the chemical composition of the cell is maintained without any alteration and the fixation takes place very quickly.

***(ii)* Fixation by freezing-substitution method.** In the freezing-substitution fixation method, the living tissue is rapidly frozen and then kept frozen at a low temperature (–20 to –60°C) in a reagent that dissolves the ice-crystals. The reagent may be ethanol, methanol or acetone. This method of fixation is also used in electron microscopy and X-ray microscopy.

***(iii)* Fixation by freezing-etching method.** In the freezing-etching method of fixation the specimen is placed in 20 per cent glycerol and is frozen at –100°C. It is mounted on á chilled holder and splintered with a knife along natural cleavage planes, usually along surfaces of membranes. Occasionally the structure is cross-fractured, giving cross sections of organelles. The splintered preparation is freeze-dried and covered with a platinum and carbon coating in the high, vacuum of a freezing ultramicrotome. When the preparation is dried the vacuum is broken and the preparation is placed in water to float the replica off the carrier. The replica is then washed in basic solution to remove the cellular material and the replica is mounted on a grid and dried for electron microscopy.

Such replica gives the outlines of various cell structures and verify the structures seen in thin fixed and stained sections of various cellular organelles in various types of microscopy.

Staining

The process of colouring the cells, tissues of animal or plant bodies by certain inorganic or organic dyes is known as **staining**. The selection of dye or stain for a particular material usually depends on its chemical nature, the pH value of the fixative used and the chemical reactivity of stain to the material.

Stains. Most cytological stains are solutions of dyes of aromatic organic compounds which have two kinds of active chemical groups such as, **chromophoric** and **auxochromic** groups. The **chromophoric groups** gives the colour to the dye, *e.g.*, carboxyl (–COOH), azo (= N–), nitro ($-NO_2$) quinoid (O =⬡= O) and indamin (–N=) groups. The **auxochromic group** gives to the dye the ability to attach to the tissue or the material and to dissolve and dissociate, in the water, *e.g., Hydroxy* (–OH) group.

The organic stains are classified into three groups:

1. acidic stains,
2. basic stains, and
3. neutral stains.

1. Acidic stains. The acidic stains are usually used for the cytoplasm and proteins. These have great capacity for combining with the tissue at low pH than basic dyes. The most common cytological acidic stains are picric acid, acid fuchsin, Congo red, Janus green *B*, orange *G*, methyl blue, eosin, aniline blue, Bismark brown and fluorescein.

2. Basic stains. The basic stains are used to stain the nucleus, chromosomes and particularly the nucleic acids. The most common cytological basic stains are basic fuchsin, crystal violet, methyl green, safranin, acridine red, azures, methylene blue, thionine and haematoxylin.

3. Neutral stains. The neutral stains have both the properties of acidic and basic stains.

Acidiphilic and basiphilic tissues. The tissue, cell or cellular component taking acidic stains is known as **acidiphilic** tissue, cell or cellular component, *e.g.,* the cytoplasm. The nucleus, chromosomes and DNA have affinity for the basic stains and known as **basiphilic** organelles.

Metachromasia. Some basic dyes stain certain cell components with a different colour than their original colour. This property of stain is known as **metachromasia** and is very useful for **histochemical** and **physiochemical** tests. The property of metachromasia is displayed by the basic stains such as thionine, azure A and toluidine blue which react with mucopolysaccharides, nucleic acids and acidic lipids.

Mordant and lake. Certain dyes stain the proteins and cytoplasm in the presence of some metal or metallic compound which is known as **mordant.** Usually chemically a mordant is a double salt of potassium or ammonium or ferric sulphate. The mordant and the stain are collectively known as the **lake**. The most important mordant is ferric ammonium sulphate (iron alum), which is used along with the stain haematoxylin and carmine.

Staining for Light Microscopy

For light microscopy both acidic and basic stains are used for the staining of cells or tissue. The stains are specific for different types of cells and also for different organelles of cell. The cytoplasmic proteins and caıbohydrates are stained by acidic stains, while the nucleus, chromosomes, etc., are stained by basic stains.

Staining for Electron Microscopy

Electron microscopy usually requires no stain because beside observing visually, the image is photographed and due to the contrast between cellular components a black and white photograph is received. But certain inorganic stains such as lead acetate, lead citrate, lead hydroxide, urynyl acetate, phosphotungstic acid, osmium tetroxide and potassium permanganate are used to increase contrast between cellular components. Sometimes certain organic stains as azure II, leucofuchsin, orcein, etc., are also used to increase the contrast.

Cytochemical Staining

In addition, there are certain specific stains, called **cytochemical stains** that bind selectively to some specific groups of cellular macromolecules such as proteins, nucleic acids, polysaccharides and lipids. For example, Millon reaction, diazonium reaction and Napthol Yellow 5 stain are used for the proteins; alkaline fast green is used for histone (basic protein); Feulgen reaction (using Schiff's reagent) is used for DNA; methyl green-pyronine stain (Unna-Pappenheim stain) is used in distinguishing between DNA and RNA and it stains DNA green and RNA red; acetocarmine and acetoorcein stains are used to stain chromosomes of dividing cells; periodic acid-Schiff (PAS) reaction is used for the demonstration of polysaccharide materials such as starch, cellulose, hemicellulose, and pectin in the plant cells and mucoproteins (glycoproteins), hyaluronic acid and chitin in animal cells; and fat soluble dyes such as Sudan Red and Sudan Black B are used for the lipids. The Sudan Black B is a specific stain for phospholipids and is used to stain Golgi apparatus.

Vital Staining

Vital stains selectively stain the intracellular structures of living cells without serious alteration of cellular metabolism and function. For example, Janus green B selectively stains mitochondria; neutral red stains plant vacuoles and methylene blue stains Golgi apparatus and also nuclear chromatin of dividing cells.

Immunostaining or usage of antibodies in Cell Biology. To use fluorescence microscopy for locating specific molecules or ions within cells, researchers must employ special indicator molecules called **fluorescent probe**. A fluorescent probe is a molecule capable of emitting fluorescent light that can be used to indicate the presence of a specific molecule or ion.

One of the most common application of fluorescent probes is **immunostaining,** a technique based on ability of **antibodies** to recognize and bind to specific molecules (the molecules to which antibodies bind are called **antigens**). Antibodies are proteins produced naturally by the immune system in response to invading microorganisms, but they can also be generated in the laboratory by injecting a foreign protein or macromolecule into an animal such as a rabbit or mouse. In this way, it is possible to produce antibodies that will bind selectively to virtually any protein that a scientist wishes to study. Antibodies are not directly visible using light microscopy, however, if they are linked to a fluorescent dye such as **fluorescein**, they emit a green fluorescence, or to **rhodamine**, they emit a red fluorescence. More recently, antibodies have been linked to **"quantum dots"**, which are tiny light-emitting crystals that are chemically more stable than traditional dyes. To identify the subcellular location of a specific protein, cell are simply stained with a fluorescent antibody directed against that protein and the location of fluorescence is then detected by viewing the cells with light of the appropriate wave-length.

Immunofluorescence microscopy can be performed using antibodies that are **directly labeled with fluorescent dye** (Fig. 2.25A). However, immuno-fluorescence microscopy is more commonly performed using **indirect immunofluorescence** (Fig 2.25B). In indirect immuno-fluorescence, a tissue or cell is treated with an antibody that is not labeled with dye. This antibody is called **primary antibody** and it attaches to specific antigenic sites within the tissue or cell. A second type of antibody, called **secondary antibody**, is then added. The secondary antibody is labelled with a fluorescent dye, and it attaches to the primary antibody. Because more than one primary antibody molecule can attach to an antigen, and more than one secondary antibody molecules can attach to each primary antibody, more fluorescent molecules are concentrated near each molecule that we seek

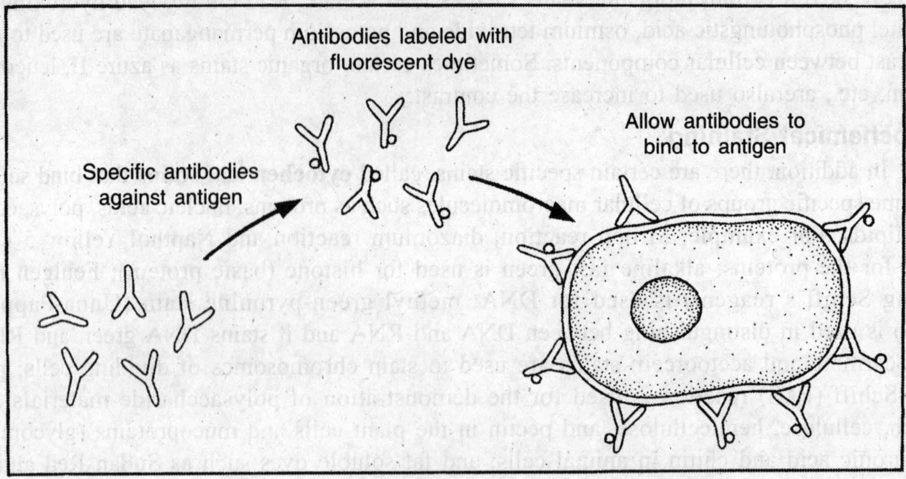

A. Immunofluorescence

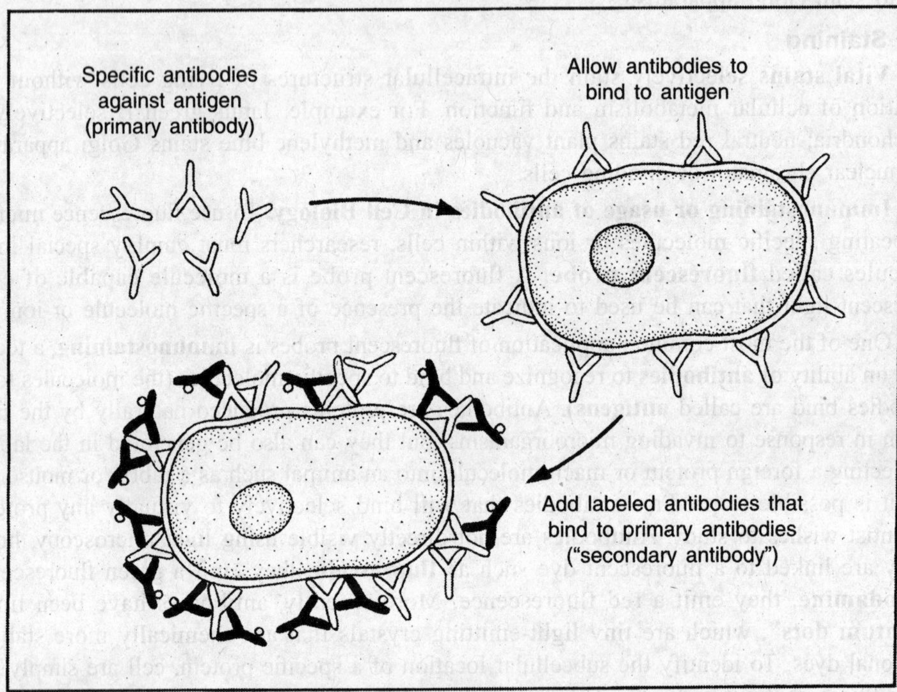

B. Indirect immunofluorescence

Fig. 2.25. Immunostaining using fluorescent antibodies (after Becker *et al.* 2007).

to detect. As a result, indirect immunofluorescence results in signal amplification, and it is much more sensitive than the use of a primary antibody alone.

QUESTIONS

Long Answer Questions

1. Describe the method of cell fractionation.
2. What is centrifugation? Describe in brief the mechanism of centrifugation.

3. Write down the different types of centrifuges.
4. Describe the structure and working of a light microscope. Explain as how does it differ from phase contrast microscope. Enumerate the importance of phase contrast microscope in cell biology.
5. Explain the principle, working and uses of phase contrast microscope.
6. Name different kinds of microscopes used in studying the cell and cell constituents.
7. How does an electron microscope differ from a light microscope?
8. Describe composition, principle of working of an electron microscope. Explain how it differs from light microscope.
9. Briefly explain the principles involved in microscopy. Describe the mechanism of transmission electron microscope.
10. Describe the working of transmission electron microscope and scanning electron microscope. What is the difference between the image formed by two types of electron microscopes.
11. What is resolving power of an optical system? State the working principle of phase contrast and dark field microscopy.
12. Explain the principles and working mechanism of fluorescence microscope.
13. What is a dark field microscope?
14. Name various kinds of chromatographic procedures. Describe the procedure of paper chromatography and column chromatography.
15. What do you mean by fixation? Describe the different types of fixatives used in histological study.
16. What is fixation? Describe the principles of staining with reference to different types of stains used in the study of cell.
17. Write an essay on 'Tissue culture'.

Short Answer Questions

1. Write a short note on the principle of autoradiography.
2. Discuss the principle of autoradiography and its application.
3. What is the limitation of a microscope? Describe the difference between light microscopy and electron microscopy with the help of schematic diagrams.
4. Write down the principle of electron microscope.
5. Describe the principle of working of transmission electron microscope.
6. What is chromatography? Write down the principle. *(H.N.B. Garwal 2008)*
7. Give differences between column chromatography and partition chromatography.
8. What is the difference between TLC and GLC?
9. Write short note on the following:
   (*i*) Electron microscopy
   (*ii*) Resolving power of microscope
   (*iii*) Compound microscope
   (*iv*) Fundamentals of TEM
   (*v*) Chromatography
   (*vi*) Ion exchange chromatography
   (*vii*) Electrophoresis
   (*viii*) Tissue culture
   (*ix*) Dialysis
   (*x*) Phase contrast microscope *(H.N.B., Garwal 2008)*
   (*xi*) Radioisotopes *(H.N.B., Garwal 2008)*

Very Short Answer Questions

1. What is autoradiography?
2. Define chromatography.

3. What is electrophoresis?

4. Define centrifugation.

5. What is cryofixation?

6. What is electron microscope?

7. Define the staining.

Fill in the Blanks

1. Ultrastructure of cell organelles can be studied by ....

Multiple Choice Questions

1. The physical measurement of a molecule in chromatography is

(*a*) Rg value (*b*) Rf value
(*c*) Half life (*d*) Rs value

2. In electrophoresis, the buffer establishes contact between

(*a*) Two electrodes (*b*) The sample and electrodes
(*c*) The sample and supporting medium (*d*) The supporting medium and electrode

3. Separation by electrophoresis in based on *(H.N.B. Garwal 2008)*

(*a*) Charge (*b*) Density
(*c*) Solubility (*d*) Refraction

ANSWERS

Fill in the Blanks

1. Electron microscopes.

Very Short Answer Questions

1. Autoradiography is a procedure for detecting the location of radioactive molecules by overlaying a sample with photographic film, which becomes darkened upon exposure to radioactivity.

2. It is a group of related techniques that utilize the flow of a fluid phase over a non-mobile absorbing phase to separate molecules based on their relative affinities for the two phases, which in turn reflect differences in size, charge, hydrophobicity or affinity for a particular chemical group.

3. It is a group of related techniques that utilize an electrical field to separate electrically charged molecules.

4. It is a process of rapidly spinning a tube containing a fluid to subject its contents to a centrifugal force.

5. It is rapid freezing of small samples so that cellular structures can be immobilized in milliseconds. It is often followed by freeze substitution, in which an organic solvent replaces the frozen water in the sample.

6. It is an instrument that uses a beam of electrons to visualize cellular structures and thereby examine cellular architecture; the resolution is much greater than that of the light microscope, allowing detailed ultrastructural examination.

7. It is incubation of tissue specimens in a solution of dye, heavy metal, or other substance that binds specifically to selected cellular constituents a distinctive colour or electron density.

Multiple Choice Questions

1. (*b*) **2.** (*d*) **3.** (*a*).

3

Prokaryotic Cell and Eukaryotic Cell

The body of all living organisms (bacteria, blue green algae, plants and animals) except viruses has **cellular organization** and may contain one or many cells (Box 3.1). A **cell** is the smallest and complete expression of the fundamental structure and function of all living organisms. It is a unit of structure, function and heredity. The cell is the basic unit of organization or structure, of all living matter. Within a selective and retentive semipermeable membrane, it contains a complete set of different kinds of units necessary to permit its own growth and reproduction from simple nutrients. It has always been quite difficult to define a cell. Different cell biologists have defined the cell differently as follows: **A.G. Loewy** and **P. Siekevitz** (1963) have defined a cell as *a unit of biological activity delimited by a semipermeable membrane and capable of self-reproduction in a medium free of other living systems*. **Wilson** and **Morrison** (1966) have defined the cell as *an integrated and continuously changing system*. **John Paul** (1970) has defined the cell as *the simplest integrated organization in living systems, capable of independent survival*. The organisms with only one cell in their body are called **unicellular organisms** (e.g., bacteria, blue green algae, some algae, Protozoa, etc.). The organisms having many cells in their body are called **multicellular organisms** (*e.g.,* most plants and animals). Any cellular organism may contain only one type of cell from the following types of cells:

A. Prokaryotic cell; B. Eukaryotic cell.

Box 3.1 Why viruses are not true cells?

A virus is neither an organism nor a cell, yet it consists of a core of nucleic acid (DNA or RNA) enclosed in an external mantle of protein. In the free state viruses are quite inert. They become activated only when they infect a living host cell and in the process only the nucleic acid core enter the host's cell. The nucleic acid which is the genetic substance, takes over the metabolic activity of the host cell and utilises the cell machinery for the formation of more viruses, ultimately killing the host cell. In a way, thus, viruses are cellular parasites that cannot reproduce by itself. They are primitive and simpler units of life.

3.1 THREE CELLULAR DOMAINS

The body of all living organisms except viruses has cellular organization. With the advent of electron microscopy, biologists came to recognize two fundamentally different plans of cellular organizations: the simpler one characteristic of bacteria and the more complex one found in all other kinds of cells. Based on structural differences of their cells, organisms were initially divided into two broad groups, the **prokaryotes** (bacteria) and the **eukaryotes** (all other forms of life such as protozoa, plants and animals). The most fundamental distinction between the two groups is that eukaryotic cells have a true, membrane-bounded nucleus (*eu*-is a Greek word for "true" or "genuine", *karyon*-means "nucleus") where as prokaryotic cell do not (*pro*-means "before" suggesting an evolutionarily earlier form of life). The terms prokaryotic and eukaryotic were suggested by

Hans Ris in the 1960's. On the basis of molecular and biochemical criteria, especially sequence analysis of ribosomal RNAs (rRNAs), cell biologists now recognize that the prokaryotes can be further differentiated as either **eubacteria** and **archaea**. The eubacteria ("true bacteria") include most present-day **bacteria** and **cyanobacteria** (also known as blue-green algae)—most of the commonly encountered bacteria, in other words. The archaea (also called **archaebacteria**) are similar to eubacteria in cellular structure, but are as different from eubacteria as they are from eukaryotes in terms of molecular and biochemical distinctions. Archaea are regarded as modern descendants of an evolutionarily ancient form of prokaryote that differed fundamentally from the ancestors of present eubacteria (***Archae***-is a Greek prefix meaning "ancient" or "original"). Present day archaea can be subdivided into theree main groups: the **methanobacteria**, which obtain energy by converting carbon dioxide and hydrogen into methane; the **halobacteria**, which can grow in salty environments (upto 5.5 M NaCl) and the **sulphobacteria**, which obtain energy from sulphur-containing compounds. Some of these archaea are also **thermacidophiles,** which thrive in acidic hot springs with pH as low as 2 and temperatures that can exceed 100°C.

Based on the RNA sequencing work of **Carl Woese**, **C. Fred Fox** and others (1977, 1990), cell biologists now recognize **eukaryotes, eubacteria** and **archaea** as three **fundamentally different groups** or **domains,** of organisms. In addition to the characteristics of their tRNAs, the cells in each group have other distinctive molecular, biochemical and cellular entities and properties in common, including RNA polymerases (the enzymes that synthesize RNA), sensitivity to specific inhibitors of protein and nucleic acid synthesis, and the presence or absence of membrane-bound intracellular compartments (organelles).

Origin of Eukaryotes

The crucial step in the evolution of eukaryotic cells was the acquirement of membrane-enclosed subcellular organelles allowing the development of the **complexity characteristic** of these cells. The organelles of eukaryotes are thought to have arisen by **endosymbiosis**—one cell living inside another. In particular eukaryotic organelles are thought to have evolved from prokaryotic cells living inside the ancestors of eukaryotes.

The hypothesis that eukaryotic cells evolved by endosymbiosis is particularly well supported by studies of mitochondria and chloroplasts, which are thought to have evolved from eubacteria living in larger cells (**Margulis** 1970, 1981). Both mitochondria and chloroplasts are similar to bacteria in size, and like bacteria, they reproduce by dividing in two. Most important, both mitochondria and chloroplasts contain their own DNA, which encodes some of their components. The mitochondrial and chloroplast DNAs are replicated each time the organelle divides, and the genes they encode are transcribed within the organelle and translated on organelle ribosomes. Mitochondria and chloroplasts thus contain their own genetic systems, which are distinct from the nuclear genome of the cells. Furthermore, the ribosomes and ribosomal RNAs of these organelles are more closely related to those of bacteria than to those encoded by the nuclear genomes of eukaryotes.

An endosymbiotic origin of these organelles is now generally accepted with mitochondria thought to have evolved from aerobic eubacteria and chloroplasts from photosynthetic eubacteria such as the cyanobacteria (blue-green algae). The acquisition of aerobic bacteria would have provided an anaerobic cell with the ability to carry out oxidative metabolism. The acquisition of photosynthetic bacteria would have provided the nutritional independence afforded by the ability to perform photosynthesis. Thus, these endosymbiotic associations were highly advantageous to their partners and were selected for in the course of evolution. Through time, most of the genes originally present in these bacteria evidently became incorporated into the nuclear genome of the cell, so only a few components of mitochondria and chloroplasts are still encoded by the organelle genomes (see **Cooper** and **Hausman**, 2007).

A recent hypothesis (**Rivera** and **Lake**, 2004) explained the mosaic nature of eukaryotic genomes by proposing that the genome of eukaryotes arose from a fusion of archaebacterial and eubacterial genomes (Fig. 3.1). According to this model, an endosymbiotic association between an eubacterium and an archaebacterium was followed by fusion of the two prokaryotic genomes, giving rise to an ancestral eukaryotic genome with contributions from both eubacteria and archaebacteria. The simplest version of this hypothesis is that an initial endosymbiotic relationship of an eubacterium living inside an archabacterium gave rise not only to mitochondria but also to the genome of eukaryotic cells, containing genes derived from both prokaryotic ancestors.

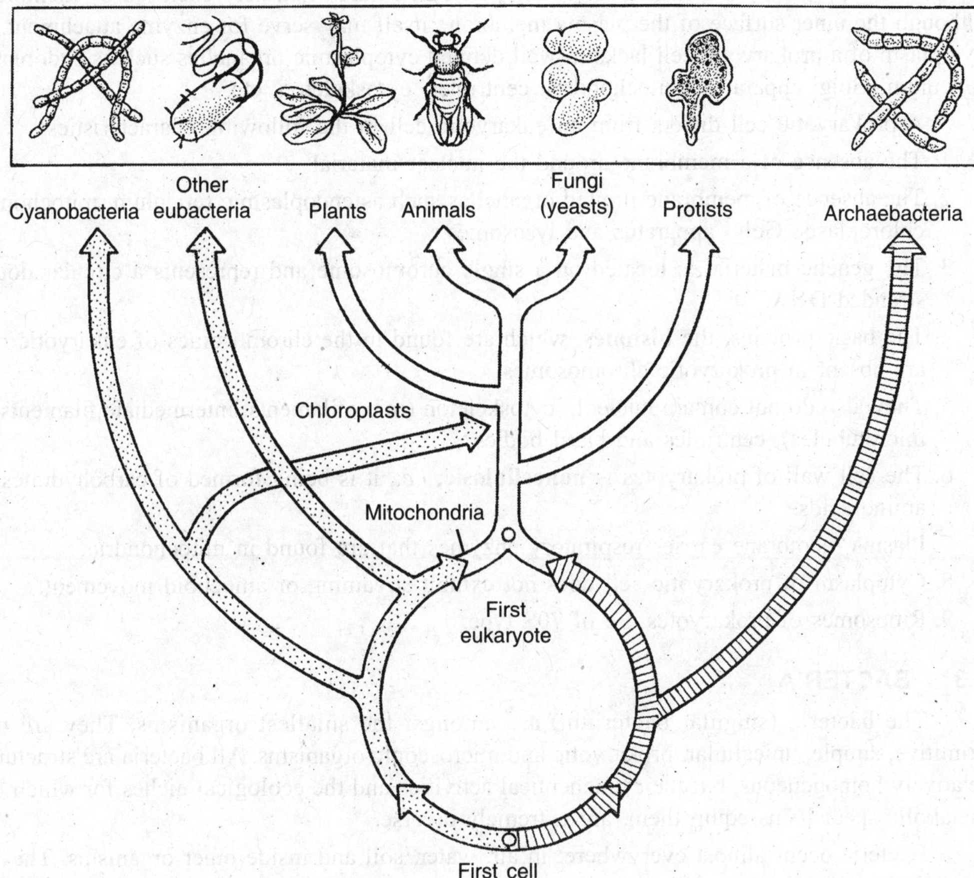

Fig. 3.1. Evolution of cells. Present day cells evolved from a common prokaryotic ancestor which diverged along two lines of descent, giving rise to archaebacteria and eubacteria. Eukaryotic cell may have arisen by endosymbiotic association of an aerobic eubacterium with an archaebacterium, leading to the development of mitochondria as well as formation of a eukaryotic genome with genes derived from both eubacteria and archaebacteria. Chloroplasts subsequently evolved as a result of the endosymbiotic association of a cyanobacterium with the ancestor of plants (after Cooper and Hausman, 2007).

3.2 PROKARYOTIC CELL

The prokaryotic (Gr., *pro* = primitive or before; *karyon* = nucleus) cells are small, simple and most primitive. They are probably the first to come into existence perhaps 3.5 billion years ago. For example, the **stromatolites** (*i.e.*, giant colonies of extinct cyanobacteria or blue green algae) of Western Australia are known to be at least 3.5 billion years old. The **eukaryotic** (Gr.,*eu* = well; *karyon* = nucleus) cells have evolved from the prokaryotic cells and the first eukaryotic (nucleated) cells may have arisen 1.4 billion years ago (**Vidal,** 1983).

The prokaryotic cells are the most primitive cells from the morphological point of view. They occur in the eubacteria (*i.e.*, mycoplasma, bacteria and cyanobacteria or blue-green algae) and archaea. A prokaryotic cell is essentially a **one-envelope system** organized in depth. It consists of central nuclear components (viz., DNA molecules, RNA molecules and nuclear proteins) surrounded by cytoplasmic ground substance, with the whole enveloped by a plasma membrane. Neither the nuclear apparatus nor the respiratory enzyme system are separately enclosed by membranes, although the inner surface of the plasma membrane itself may serve for enzyme attachment. The cytoplasm of a prokaryotic cell lacks in well defined cytoplasmic organelles such as endoplasmic reticulum, Golgi apparatus, mitochondria, centrioles, cytoskeleton, etc.

A prokaryotic cell differs from an eukaryotic cell in the following characteristics:

1. The absence of a membrane around the nuclear material.
2. The absence of membrane limited organelles such as endoplasmic reticulum, mitochondria, chloroplasts, Golgi apparatus and lysosomes.
3. The genetic material is located on a single chromosome and represents a circular double-stranded DNA.
4. The basic proteins, the **histones**, which are found in the chromosomes of eukaryotic cells, are absent in prokaryotic chromosomes.
5. They also do not contain nucleoli, cytoskeleton (microfilaments, intermediate filaments and microtubules), centrioles and basal bodies.
6. The cell wall of prokaryotes is **noncellulosic,** *i.e.*, it is being formed of carbohydrates and amino acids.
7. Plasma membrane carries respiratory enzymes that are found in mitochondria.
8. Cytoplasm of prokaryotic cell does not exhibit streaming or amoeboid movement.
9. Ribosomes of prokaryotes are of 70S type.

3.3 BACTERIA

The bacteria (singular bacterium) are amongst the smallest organisms. They are most primitive, simple, unicellular, prokaryotic and microscopic organisms. All bacteria are structurally relatively homogeneous, but their biochemical activities and the ecological niches for which their metabolic specialisms equip them, are extremely diverse.

Bacteria occur almost everywhere: in air, water, soil and inside other organisms. They are found in stagnant ponds and ditches, running streams and rivers, lakes, sea water, foods, petroleum oils from deeper regions, rubbish and manure heaps, sewage, decaying organic matter of all types, on the body surface, in body cavities and in the internal tracts of humans and animals. Bacteria thrive well in warmth, but some can survive at very cold tops of high mountains such as Alps or even in almost boiling hot springs. They occur in vast numbers. A teaspoonful of soil may contain several hundred million bacteria. They lead either an autotrophic (photoautotrophic or chaemoautotrophic), or heterotrophic (saprotrophic or parasitic) mode of existence. The saprophytic or saprotrophic species of bacteria are of great economic significance for humans. Some parasitic species of bacteria are pathogenic (disease producing) to plants, animals and humans.

Bacteria have a high ratio of surface area of volume because of their small size. They show high metabolic rate because they absorb their nutrients directly through cell membranes. They multiply at a rapid rate. In consequence, due to their high metabolic rate and fast rate of multiplication, bacteria produce marked changes in the environment in a short period.

1. Size of bacteria. Typically bacteria range between 1μm (one micrometre) to 3 μm so they are barely visible under the light microscope. The smallest bacterium is *Dialister pneumosintes* (0.15 to 0.3 μm in length). The largest bacterium is *Spirillum volutans* (13 to 15 μm in length).

2. Forms of bacteria. Bacteria vary in their shapes. Based on their shape, bacteria are classified into the following groups:

(*i*) Cocci (singular coccus). These bacteria are spherical or round in shape. These bacterial cells may occur singly **(micrococci)**; in pairs (**diplococci**, *e.g.*, pneumonia causing bacterium, *Diplococcus pneumoniae*); in groups of four **(tetracocci)**; in a cubical arrangement of eight or more **(sarcinae)**; in irregular clumps resembling bunches of grapes (**staphylococci**,*e.g.*, boil causing bacterium, *Staphylococcus aureus*) or in a bead-like chain (**streptococci**, *e.g.*, sore throat causing bacterium, *Streptococcus pyogenes*; Fig. 3.2).

(*ii*) Bacilli (singular, bacillus). These are rod-like bacteria. They generally occur singly, but may occasionally be found in pairs (**diplobacilli**) or chains (**streptobacilli**). Bacilli cause certain most notorious diseases of humans such as tuberculosis (*Mycobacterium* or *Bacillus tuberculosis*), tetanus (*Clostridium tetani*), typhoid (*Salmonella* or *Bacillus typhosus*), diphtheria (*Corynebacterium diphtheriae*), leprosy (*Mycobacterium leprae*), dysentery and food poisoning (*Clostridium botylinum*). Certain well known diseases of the animals are also caused by bacilli, e.g., anthrax (*Bacillus anthracis*) and black leg (*Clostridium chauvei*).

(*iii*) Spirilla (singular, spirillum). These are also called **spirochaetes**. These are spiral-shaped and motile bacteria (Fig. 3.2). Spirilla cause human disease such as syphilis (*Treponema pallidum*).

(*iv*) Vibrios (singular vibrio). These are comma-shaped or bent-rod like bacteria (Fig. 3.2). Vibrios cause human disease such as cholera (*Vibrio cholerae*).

3. Gram negative and Gram positive bacteria. On the basis of structure of cell wall and its stainability with the Gram stain, the following two types of bacteria have been recognized: Gram positive and Gram negative bacteria. The Gram staining method is named after **Christian Gram** who developed it in Denmark in 1884. In this technique, when heat-fixed bacteria are treated with the basic dye, crystal violet they become blue or purple. Such blue stained cells are treated with a mordant (*i.e.,* agent that fixes stains to tissues) such as iodine (*i.e.,* potassium iodide or KI solution) and ultimately washed with some organic solvent such as alcohol. Some bacteria retain the blue colour, while others lose it and stay colourless. The former are **Gram positive bacteria** (*e.g.,Bacillus subtilis, Staphylococcus,* etc.) and the latter are **Gram negative bacteria** (e.g., *Escherichia coli, Simonsiella*, cyanobacteria, etc.). Colourless Gram negative bacteria may thereafter be stained pink with safranin stain for their better microscopic visibility.

The long search for the chemical basis of this differentiating staining reaction ended in 1950's when it was detected that cell wall of Gram negative bacteria has high lipid content which tends to be dissolved away by alcohol. The alcohol then can enter the cell and leach out the stain, whereas the cell walls of Gram positive bacteria form a barrier (*i.e.*, peptidoglycan layers) that prevent the penetration of the solvent inside the cell.

4. Structure of bacteria. Structural details of a bacterial cell can only be seen with an electron microscope in very thin sections. A typical bacterial cell has the following components:

A. Outer covering: The outer covering of bacterial cell comprises the following three layers:

I. Plasma membrane. The bacterial protoplast is bound by a living, ultrathin (6 to 8 nm thick) and dynamic plasma membrane. The plasma membrane chemically comprises molecules of lipids and proteins which are arranged in a **fluid mosaic pattern**. That is, it is composed of a bilayer sheet of **phospholipid** molecules with their polar heads on the surfaces and their fatty-acyl chains (tails) forming the interior. The **protein** molecules are embedded within this lipid bilayer, some spanning it, some exist on its inner side and some are located on its external or outer side. These membrane proteins serve many important functions of the cell. For example, the transmembrane proteins act as **carriers** or **permeases** to carry on selective transportation of nutrients (molecules and ions) from the environment to the cell or vice versa. Certain proteins of the membrane are involved in oxidative metabolism, *i.e.*, they act as enzymes and carriers for electron flow in respiration and photosynthesis leading to phosphorylation (*i.e.*, conversion of ADP to ATP). The bacterial plasma membrane also provides a specific site at which the single circular chromosome (DNA) remains attached. It is the point from where DNA replication starts. The first stage in nuclear division involves duplication of this attachment, followed by a progressive bidirectional replication of DNA by two replication forks.

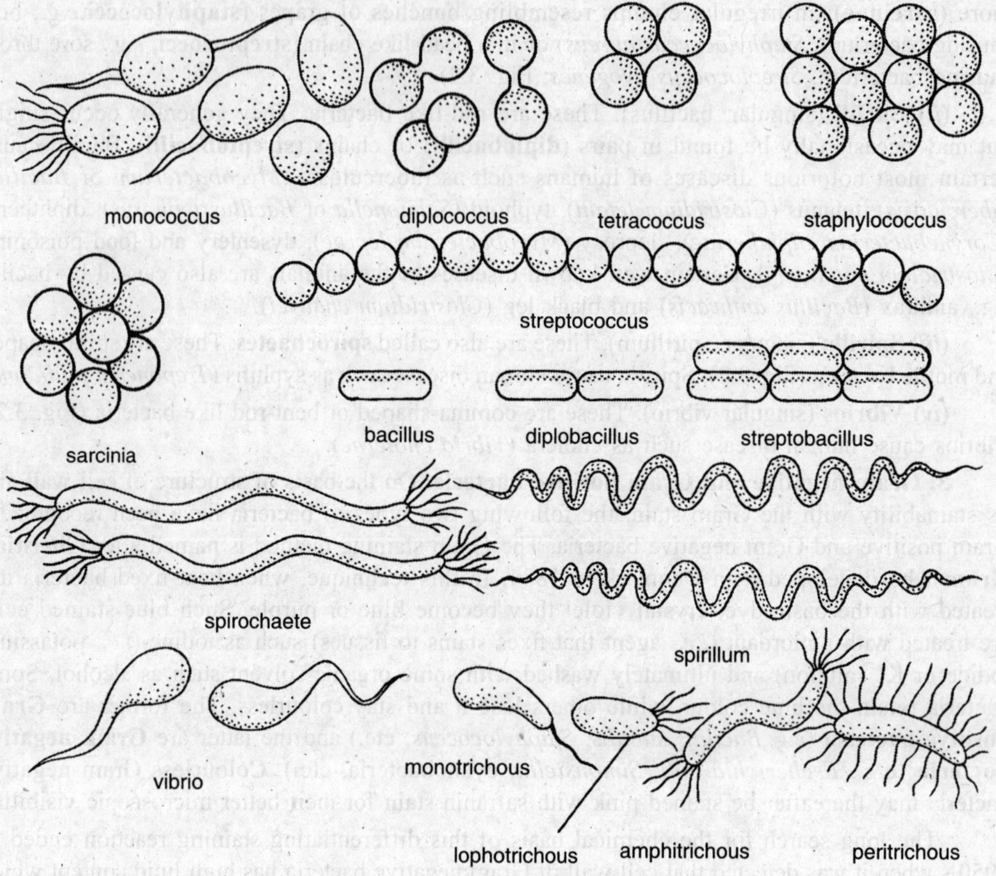

Fig. 3.2. Different forms of bacteria.

Plasma membrane intrusions. Infoldings of the plasma membrane of all Gram-positive bacteria and some Gram-negative bacteria give rise to the following two main types of structures:

1. Mesosomes (or chondrioids). They are extensions of the plasma membrane within the bacterial cell (*i.e.*, cytoplasm) involving complex whorls of convoluted membranes (Fig. 3.3). Mesosomes tend to increase the plasma membrane's surface and in turn also increase their enzymatic contents. They are seen in chemoautotrophic bacteria with high rates of aerobic respiration such as *Nitrosomonas*, and in photosynthetic bacteria such as *Rhodopseudomonas* where they are the site of photosynthetic pigments. Mesosomes are involved in crosswall (septum) formation during the division of cell.

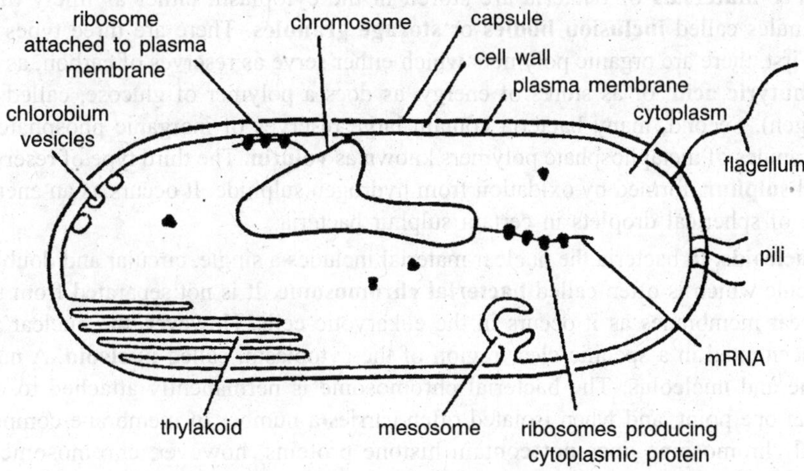

Fig. 3.3. A diagram of structures seen in the prokaryotic (bacterial) cells..

2. Chromatophores. These are photosynthetic pigment-bearing membranous structures of photosynthetic bacteria. Chromatophores vary in form as vesicles, tubes, bundled tubes, stacks, or thylakoids (as in cyanobacteria).

II. Cell wall. The plasma membrane is covered with a strong and rigid cell wall that renders mechanical protection and provides the bacteria their characteristic shapes (the cell wall is absent in *Mycoplasma*). The cell wall of bacteria differs chemically from the cell wall of plants in that it contains proteins, lipids and polysaccharides. It may also contain chitin but rarely any cellulose.

Electron microscopy has revealed the fact that the cell wall of Gram- negative bacteria comprises the following two layers: 1. Gel, proteoglycan or peptidoglycan (*e.g.*, murein or muramic acid) containing **periplasmatic space** around the plasma membrane and 2. The outer **membrane** which consists of a lipid bilayer traversed by channels of **porin** polypeptide. These channels allow diffusion of solutes. The lipids of lipid bilayer are phospholipids and lipopolysaccharides (LPS). LPS have antigenic property and anchor the proteins and polysaccharides of the surrounding capsule. The cell wall of Gram positive bacteria is thicker, amorphous homogeneous and single layered. Chemically it contains many layers of peptidoglycans and proteins, neutral polysaccharides and polyphosphate polymers such as teichoic acids and teichuronic acids.

III. Capsule. In some bacteria, the cell wall is surrounded by an additional slime or gel layer called **capsule**. It is thick, gummy mucilaginous and is secreted by the plasma membrane. The capsule serves mainly as a protective layer against attack by phagocytes and by viruses. It also helps in regulating the concentration, and uptake of essential ions and water.

B. Cytoplasm. The plasma membrane encloses a space consisting of **hyaloplam, matrix** or **cytosol** which is the ground substance and the seat of all metabolic activities. The cytosol consists of water, proteins (including multifunctional enzymes), lipids, carbohydrates, different types of RNA molecules, and various smaller molecules. The cytosol of bacteria is often differentiated into two distinct areas: a less electron dense nuclear area and a very dense area (or dark region). In the dense cytoplasm occur thousands of particles, about 25 nm in diameter, called **ribosomes**. Ribosomes are composed of ribonucleic acid (RNA) and proteins and they are the sites of protein synthesis. Ribosomes of bacteria are 70S type and consist of two subunits (*i.e.*, a larger 50S ribosomal subunit and a small 30S ribosomal subunit). Non-functional ribosomes exist in the form of separated subunits which are suspended freely in the cytoplasm. During protein synthesis many ribosomes read the codes of single mRNA (messenger RNA) molecules and form **polyribosomes** or **polysomes**.

Reserve materials of bacteria are stored in the cytoplasm either as finely dispersed or distinct granules called **inclusion bodies** or **storage granules**. There are three types of reserve materials. First, there are organic polymers which either serve as reserves of carbon, as does **poly-β-hydroxybutyric acid**, or as stores of energy, as does a polymer of glucose, called **granulose** (*i.e.,* glycogen). Second, many bacteria contain large reserves of inorganic phosphate as highly refractile granules of metaphosphate polymers known as **volutin**. The third type of reserve material is elemental **sulphur**, formed by oxidation from hydrogen sulphide. It occurs as an energy reserve in the form of spherical droplets in certain sulphur bacteria.

C. Nucleoids. In bacteria the nuclear material includes a single, circular and double stranded DNA molecule which is often called **bacterial chromosome**. It is not separated from the cytosol by the nuclear membranes as it occurs in the eukaryotic cells. However, the nuclear material is usually concentrated in a specific clear region of the cytoplasm, called **nucleoid**. A nucleoid has no ribosome and nucleolus. The bacterial chromosome is permanently attached to the plasma membrane at one point, and when isolated often carries a number of membrane component with it. Bacterial chromosome does not contain histone proteins, however, chromosomes of some species are found to contain small quantities of a small heat-stable (HU) proteins that may be analogous to eukaryotic histones.

All three classes of RNA (*i.e.*, mRNA, tRNA, and rRNA) are formed (transcribed) by the activity of the single RNA polymerase (RNAP) species in prokaryotes. The messenger RNA formed at the chromosome is directly available for translation without processing, and so ribosomes may attach to the beginning of the mRNA strand and commence translation, while the other end of the mRN A is still being formed by transcription from DNA. Proteins for use within the cell are synthesized at cytoplasmic ribosomes; but ribosomes responsible for the synthesis of membrane proteins or proteins destined for export from the cell to form either the cell wall or secretory products, are attached to the plasma membrane. The resulting exportable polypeptides are ejected directly into or through the membrane as they are formed.

Plasmids. Many species of bacteria may also carry extra chromosomal genetic elements in the form of small, circular and closed DNA molecules, called **plasmids**. Some plasmids are merely **bacteriophage (viral) DNA** which may alternatively be incorporated within the chromosome. Other plasmids may be separated parts of the normal genome from the same or a foreign cell, and may recombine with the main chromosome. One function of some of these plasmids (called **colcinogenic factors**) is the production of antibiotically active proteins or **colicins** which inhibit the growth of other strains of bacteria in their vicinity. Some plasmids may act as **sex** or **fertility factors (F factors)** Which stimulate bacterial conjugation. **R factors** are also plasmids which carry genes for the resistance to one or more drugs such as chloramphenicol, neomycin, penicillin, streptomycin, sulphonamides and tetracyclines.

D. Flagella and other structures. Many bacteria (*e.g., E. coli*) are motile and contain one or more **flagella** for the cellular locomotion (swimming). Bacterial flagella are smaller than the eukaryotic flagella (*i.e.,* they are 15 to 20 nm in diameter and up to 20 µm long) and are also simpler in organization. A bacterial flagellum consists of a helical tube containing a single type of protein subunit, called **flagellin**. The flagellum is attached at its base, by a short flexible **hook** that is rotated, like a propeller of ship, by the flagellar rotatory **motor** *i.e.,* basal body; Fig. 3.4). The flagellar motor comprises four distinct parts: rotor (M ring), stator, bearing (S ring) and rod. The '**rotor**' is a protein disc integrated into the plasma membrane. It is driven by energy stored in the transmembrane proton H^+ gradient (not by ATP breakdown; see **Jones**, 1986) and rotates rapidly (~ 100 revolutions/second) in the lipid bilayer against another protein disc, called the '**stator**'. A **rod** links the 'rotor' to a hook and flagellum, thereby causing them to rotate. The protein **"bearing"** serves to seal the outer membrane of the cell wall as the rotating rod passes through it. The 'stator' and 'bearing' remains stationary.

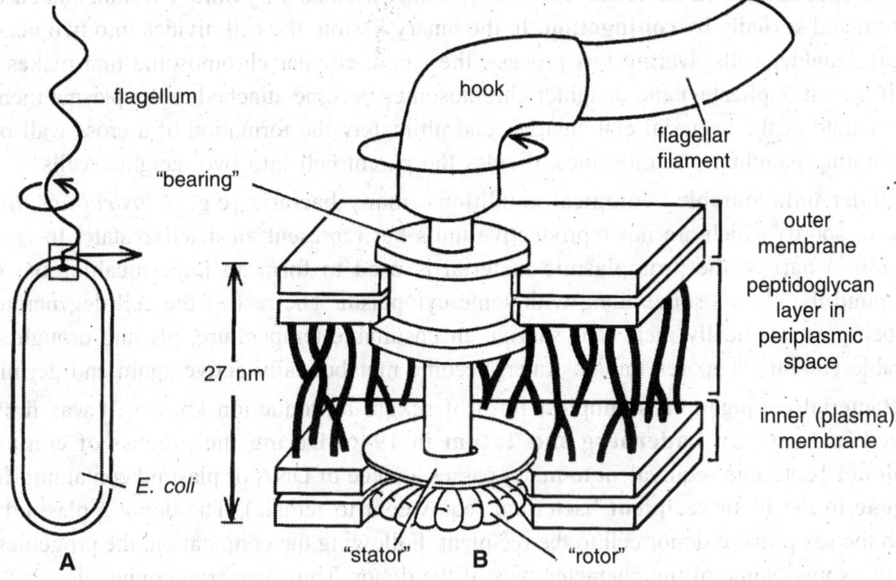

Fig. 3.4. Schematic drawing of the flagellar rotatory 'motor' of *E. coli*.

According to the number and arrangement of the flagella in a bacterial cell, following four types of flagellation patterns have been recognized: **(1) Monotrichous.** There is a single flagellum at one pole of the cell. **(2) Lophotrichous**. There are several flagella at one pole. **(3) Amphitrichous**. The cell bears at least one flagellum at each pole. **(4) Peritrichous**. 'There are flagella all over the surface of cell (Fig. 3.2). Flagella-like **axial filaments** are the characteristics of some spirochetes which move like snakes through the environment. The axial filaments do not project away from the cell but are wrapped around the cell surface.

Fimbriae or pili. Some bacteria (mostly Gram negative bacilli) contain non-flagellar, extremely fine, appendages called **fimbriae** or **pili** (singular **pilus**). Pili are non- motile but adhesive structures. They enable the bacteria to stick firmly to other bacteria, to a surface or to some eukaryote such as mold, plant and animal cells including red blood cells and epithelial cells of alimentary, respiratory and urinary tracts. Pili help in conjugation (*e.g.*, long F-pili or **sex pili** of male bacteria); in the attachment of pathogenic bacteria to their host cells (*e.g.*, attachment of gonorrhea-causing coccus, *Neisseria gonorrhoeae*, to the epithelial cells of the human urinary tract) and in acting as specific sites of attachment for the bacteriophages. Pili are known to be coded by the genes of the plasmid.

Spinae. Same Gram positive bacteria have tubular, pericellular and rigid appendages of single protein moiety, called **spinin**. They are called **spinae** and are known to help the bacterial cells to tolerate same environmental conditions such as salinity, pH, temperature, etc.

5. Nutrition in bacteria. Bacteria show wide diversity in their nutrition. Some are chemosynthetic, some are photosynthetic, but most of them are heterotrophic. Heterotrophic bacteria are mostly either saprophytic or parasitic; Parasitic bacteria live on the body of plants and animals and with few exceptions, most bacteria are pathogenic.

Modes of respiration of bacteria are both aerobic and anaerobic. Some of the end products of bacterial anaerobic respiration are useful to humans, so, they are used in the manufacture of various foods such as butter, cheese and vinegar. *Pseudomonas* is a gram negative heterotrophic aerobic form which can decompose (biodegrade) a wide variety of organic compounds such as hydrocarbons. So it is used in reducing water pollution due to petroleum spillage.

6. Reproduction in bacteria. Bacteria reproduce asexually by **binary fission** and **endospore** formation and sexually by **conjugation.** In the binary fission, the cell divides into two genetically identical daughter cells. During this process, the single circular chromosome first makes a copy of itself (*i.e.*, it duplicates) and daughter chromosomes become attached to the plasma membrane. They separate as the bacterial cell enlarges and ultimately the formation of a cross wall between the separating daughter chromosomes, divides the parent cell into two daughter cells.

Under unfavourable ecological conditions, many bacteria (e.g., *Clostridium, Bacillus,* etc.) form spores which are not reproductive units but represent an inactive state. In endospore formation, a part of the protoplasmic material is used to form an impermeable coat or cyst wall around the chromosome along with some cytoplasm. The rest af the cell degenerates. The spore being metabolically inert can survive an unsuitable temperature, pH and drought. Under favourable conditions spores imbibe water, become metabolically active again and germinate.

Bacterial conjugation is simplest form of sexual reproduction known. It was first of all observed in *E.coli* by **Laderberg** and **Tatum** in 1946. During the process of conjugation, F^+ or **donor** bacterium (equivalent to male) passes a piece of DNA or plasmid containing **fertility** or **F gene** to the F^- or **recipient** bacterium (equivalent to female). The donor's plasmid passes through the sex pilus of donor cell to the recipient. Following the conjugation, the progenies of the recipient express some of the characteristics of the donor. Thus, bacterial conjugation is a means of making new genetic combinations or **recombinations** which are expressed in the progeny.

Examples of Prokaryotic Cells

The following three types of prokaryotic cells are well studied ones:

1. Mycoplasma (PPLO)

Among living organisms that have the smallest mass, are small bacteria called **mycoplasmas** which produce infectious diseases in animals including humans. Mycoplasmas are unicellular, prokaryotic, containing a plasma membrane, DNA, RNA and a metabolic machinery to grow and multiply in the absence of other cells (*i.e.,* they are capable of autonomous growth). They can be cultured *in vitro* like any bacteria, forming **pleomorphic** (Gr., *pleo* = many; *morphe* = forms) **colonies**, *i.e.*, depending on the type of culture medium, mycoplasmas tend to form different shaped colonies such as spheroid (fried-egg-shaped), thin, branching filaments, stellate, asteroid or irregular. They differ from the bacteria in the following respects:

1. Mycoplasmas are filterable through the bacterial filters (this fact was first demonstrated by **Iwanowsky** in 1892).
2. They do not contain cell wall and mesosomes.
3. Like the viruses and animal cells, they are resistant to antibiotics such as penicillin which kills bacteria by interfering with cell wall synthesis.
4. Their growth is inhibited by tetracyclines and similar antibiotics that act on metabolic pathways.

Mycoplasmas were discovered by French scientists, **E. Nocard** and **E. R. Roux** in 1898 while studying pleural fluids of cattle suffering from the disease **pleuropneumonia** (*i.e.*, an infectious disease of warm blooded animals producing pleural and lung inflammation). Similar organisms were later isolated from other animals such as sheep, goats, dogs, rats, mice and human beings and were named as **pleuropneumonia-like organisms (PPLO)**. PPLO were later on included under the genus *Mycoplasma* by **Nowak** (1929) and these organisms are now commonly called **mycoplasmas. W.V. Iterson** (1969) has placed PPLO in the group **Mycoplasmatceae** of bacteria. Currently mycoplasmas are considered as the simplest bacteria. However, some cell biologists still prefer to place PPLO in between the viruses and bacteria.

Mycoplasmas are mostly free-living, saprophytes or parasites. For example, *Mycoplasma laidlawii* (0.1 μm in diameter) is saprophytic and is found in sewage, compost, soil, etc. *Mycoplasma gallisepticum* (0.25 μm in diameter) is parasitic and pathogenic; it is the parasite of cells and cell exudates of respiratory organs of warm-blooded animals causing in them various chronic respiratory diseases.

Mycoplasmas range in size (diameter) from 0.25 to 0.1 μm. They correspond in size to some of the large viruses. The spherical cell of a mycoplasma is bounded at its surface by a 75 Å thick plasma membrane which is composed of molecules of proteins and lipids, but there is no cell wall. Internally the cell's composition is more or less diffuse. The only micro-scopically discernible features within the cell are its genetic component, the DNA and the ribosomes. The DNA molecule is contained in a membraneless and clear-nucleus-like region and it is a double helix which may exit either as the linear strands or a single circular molecule. The nuclear region is surrounded by numerous (50 to 100) 70S type ribosomes existing either freely or in the polysomes (Fig. 3.5). A variety of other cytoplasmic inclusions, such as vacuoles and granules, have also been detected, but their functions are not known. At one side of the cell occurs **bleb** (localised collection of fluid) of ill understood function. As in other prokaryotes, there is no intracellular membranous structure.

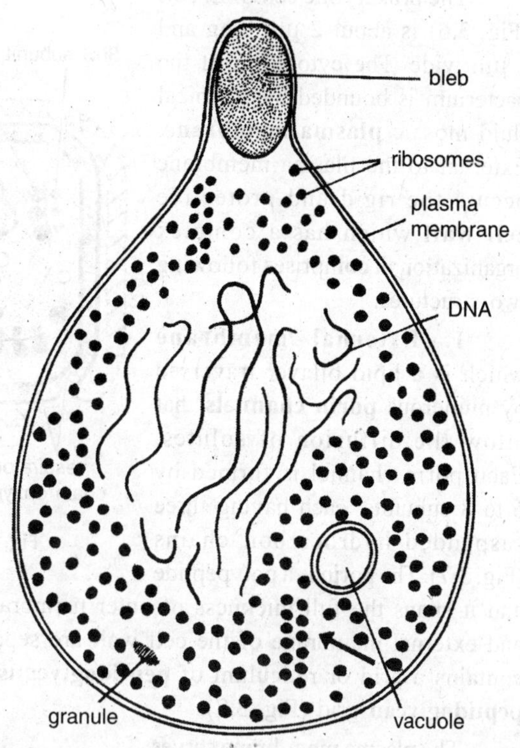

Fig. 3.5. A schematic diagram of typical PPLO cell.

The PPLO cells contain many enzymes which may be required for DNA replication, the transcription of different kinds of RNA molecules and translation involved in protein synthesis, and also in the biosynthesis of adenosine triphosphate (ATP) by anaerobic breakdown of sugars. Unlike viruses, they are free living and do not require host cells for their duplication. PPLO reproduce by binary fission, budding, formation of small spore-like bodies and by growth of large branched filaments that ultimately fragment.

2. *Escherichia Coli* (*E. coli*)

E. coli is a Gram negative, monotrichous, symbiotic bacillus of colon of human beings and other vertebrates. It is heterotrophic and non-pathogenic bacteria producing some vitamins (*e.g.*, vitamin K) for human use. Some strains of *E. coli* are known to recognise and bind specifically to sugar-containing target cells on the surface of gut lining of mammals (*e.g.*, D-mannose residues of epithelial cells of human gut or colon). *E. coli* is one of the best studied bacteria. It has served well in the field of molecular biology, since this bacterium is particularly easy to grow in an artificial medium where it divides every 20 minutes at 37°C under optimal conditions. Thus, a single cell become 10^9 bacteria in about 20 hours.

The prokaryotic cell of *E. coli* (Fig. 3.6) is about 2 μm long and 1 μm wide. The cytoplasm of the bacterium is bounded by a typical fluid mosaic **plasma membrane**. External to the plasma membrane occurs the rigid and protective **cell wall** which has a complex organization; it comprises following two structures:

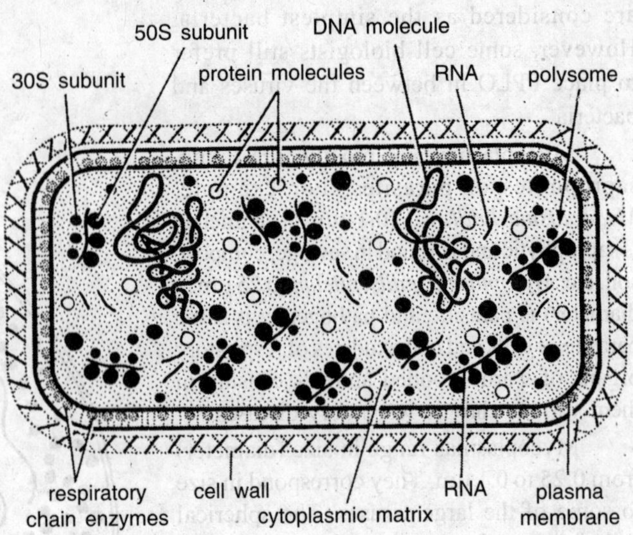

Fig. 3.6. A prokaryotic cell of *Escherichia coli*.

1. External membrane which is a lipid bilayer traversed by numerous **porin channels** that allow the diffusion of solutes. Each porin channel is formed by 6 to 8 subunits, each having three suspended hydrocarbon chains (Fig. 3.7). The porin is a polypeptide and it spans the full thickness of outer membrane. 2. Both membranes-the plasma membrane and external membrane of the cell wall are separated by the **periplasmatic space**. This space contains a grid or reticulum of **peptidoglycans**. Some porin subunits remain attached to the **peptidoglycan** grid (Fig. 3.7).

The plasma mem-brane serves as a molecular barrier with the surrounding medium. It comprises a variety of transport proteins, called **permeases** which control the entrance and exit of small molecules and ions. It contributes to the establishment of bacterial protoplasm. *E. coli* has both oxygen-requiring (aerobic) and non-oxygen-requiring (anaerobic) respiratory machinery for the breakdown of sugar and contains a special group of proteins called the **electron transport chain** for the generation of stored energy in the form of ATP molecules. *E. coli* lacks mitochondria, and respiratory chain enzymes such as cytochromes, enzymes of Krebs cycle, NADH, acid phosphatase, etc., are attached to inner face of the plasma membrane.

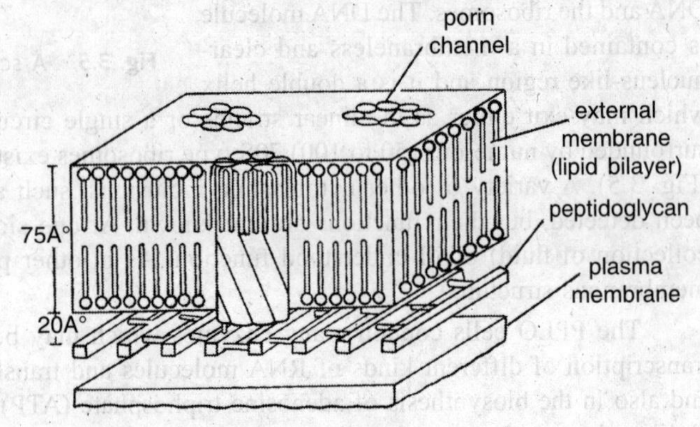

Fig. 3.7. Ultrastructure of the cell wall of a Gram-negative bacterium.

All genes of *E. coli* are contained on a single supercoiled, double-stranded, circular **DNA molecule**, which occurs in a clear zone of cytoplasm, called **nucleoid**, and is attached to the plasma membrane at one point. The total length of the DNA circle is about 1300 μm, comprising about 4.7×10^6 nucleotide pairs; this is enough DNA to code for about 4000 different proteins. The DNA of *E.coli* is naked, lacking histones, but certain polyamines may be bound to some of its phosphates. Electron microscopy of isolated chromosome of *E. coli* has shown that DNA is folded into a series of **looped domins**, *i.e.*, about 45 loops radiate out from a dense proteinaceous **scaffold**

(Fig. 3.8). The DNA of loops is in the so-called supercoiled conformation in which the double helix is itself twisted. The enzyme **DNA gyrase** is responsible for the DNA supercoiling (it is inhibited by the drug called **coumermycin**).

The colloidal cytoplasmic matrix of *E.coli* contains about 5000 distinguishable components, ranging from water to DNA (*i.e.,* three types of RNA, enzymes, glycogen, amino acids, monosaccharides and various other small molecules). Surrounding the DNA is dark dense region of matrix containing 20,000 to 30,000 70S type **ribosomes**, each existing in the form of their two subunits. During protein synthesis numerous complete ribosomes read the codes of mRNA molecules to form the polysomes (Fig. 3.6).

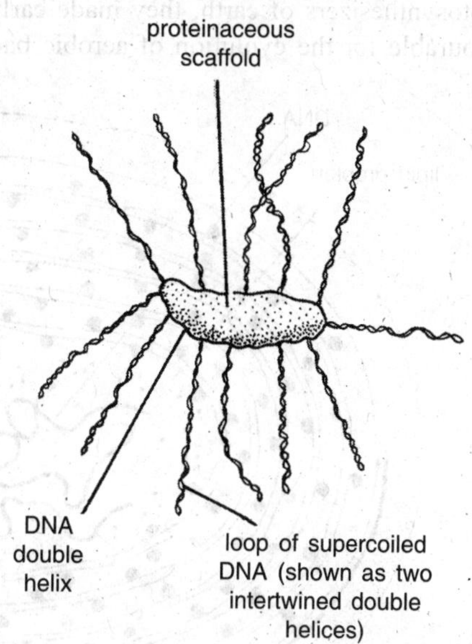

Fig. 3.8 Schematic representation of the chromosome of *E. coli,* showing only 12 of the 45 supercoiled loops.

3. Cyanobacteria (Blue green Algae)

The Gram-negative cyanobacteria or oxyphotobacteria (*i.e.,* oxygen yielding photosynthetic blue green algae) are one of the most successful and primitive (3.5 billion year old) groups of organisms on earth. They even inhabit the steaming hot springs and the undersides of icebergs. Cyanobacteria form another group of prokaryotes which include about 1500 species (85 genera and 750 species are found in India).

Cyanobacteria occur as individual cells, as small clusters or colonies of cells, or as long, filamentous chains. They lack flagella but are able to perform movement by rotatory motion or gliding over a gelatinous layer secreted through the cell surface.

A typical cell of a blue green algae is composed of outer cellular coverings and cytoplasm. The **outer cellular coverings** include an outer most **gelatinous** or **slimy layer**, **the capsule**, a middle **cell wall** and an innermost lipoproteinous **plasma membrane**. The cell wall of blue green algae resembles the cell wall of bacteria and contains an **outer bimolecular membrane** of phospholipids, lipoproteins and lipopolysaccharides, and a grid of peptidoglycans (muramic acid) in the **periplasmatic space** existing in between cell wall and plasma membrane. The **cytoplasm** of cyanobacteria appears more organized than that of other bacteria. The matrix extends throughout the cell. The cytoplasm (or protoplast) is differentiated in two regions:

1. Outer or peripheral pigmented region, the **chromoplasm** having photosynthetic lamellae or **thylakoids**.
2. Inner or central colourless region called **centroplasm** or **DNA plasm** having DNA and crystalline granules (Fig. 3.9).

Because the metabolism of the blue green algae is based on photosynthesis, therefore, the cells of them contain the photosynthetic pigment, viz., the **chlorophyll** and **carotenoid**. In addition to these pigments, these algae contain certain unique pigments collectively called **phycobilin**; one of the phycobilin is blue and called **phycocyanin**, while other type of phycobilin is red and called **phycoerythrin**. The photosynthetic pigments (chlorophylls and carotenoids) occur in flattened sacs called **lamellae** which remain arranged in parallel array. In between the lamellae occur certain granules of 400A° diameter. These granules contain phycobilin pigments and are called **cyanosomes** or **phycobilisomes**. They are attached to the outer lamellar membrane surface. Being earliest oxygenic photosynthesizers of earth, they made early earth's atmosphere aerobic providing the conditions favourable for the evolution of aerobic bacteria and eukaryotes.

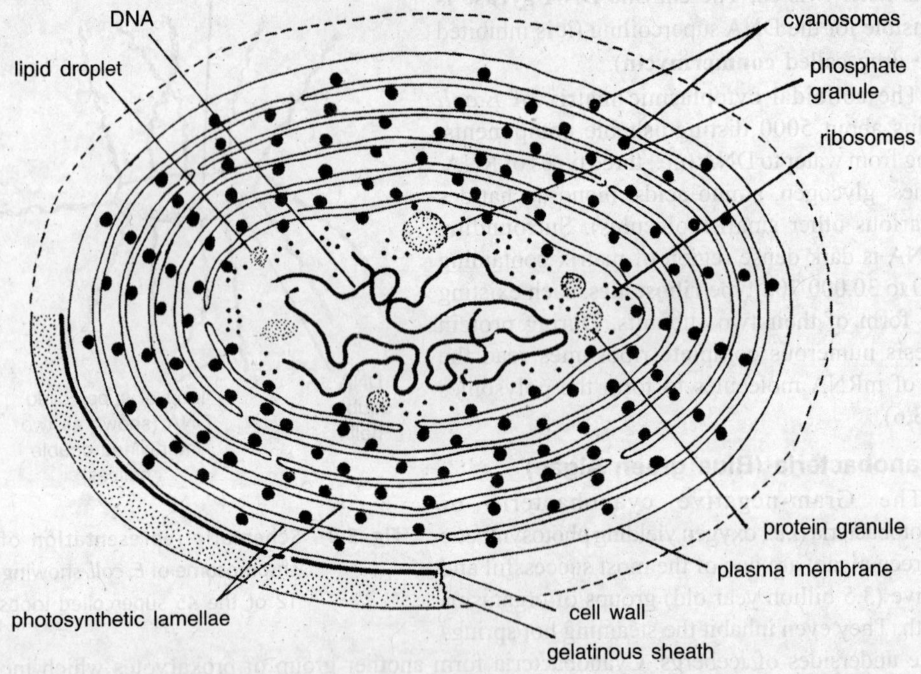

Fig. 3.9. A prokaryotic cell of cyanobacteria (electron microscopic view).

The two subunits of 70S ribosomes of cyanobacteria are freely distributed in the cytoplasm and form polyribosomes during protein synthesis. As in all prokaryotes, the DNA molecule of blue green algae is circular, double-stranded helix and occurs in the centroplasm. This area (nucleoid) is not bound by the nuclear membrane and it does not contain a nucleolus.

Cyanobacteria also contain a variety of **inclusions** in its cytoplasm. Membrane-bound inclusions are the gas vacuoles and the carboxysomes. **Gas vacuoles** are gas-filled cavities which are located in the inner part of chromatoplasm. They occur commonly in planktonic species such as *Nostoc*, *Anabaena*, *Phormidium*, *Calothrix*, *Galaeotrichia*, etc. Gas vacuoles serve the function of flotation or buoyancy. **Carboxysomes** contain enzymes involved in carbon dioxide fixation.

The cytoplasm of blue green algae also contains a variety of membrane-free inclusions such as (1) **cyanophycin granules** which are located in chromatoplasm and are protein storage products, containing large amount of arginine amino acid or copolymers of alanine and aspartic acid; (2) **myxophycean starch** which is the main food storage compound; (3) **polyglucon granules**, polyhedral bodies, lipid droplets, polyphosphate bodies, etc., are some other cytoplasmic inclusions of cyanobacteria.

Lastly, many cyanobacteria (about 20 species) tend to fix atmospheric nitrogen as ammonia, *e.g.*, *Anabaena, Nostoc, Mastigocladus*, etc. Under aerobic condition nitrogen fixation is done principally in special type of cells called **heterocysts**, as in *Nostoc*.

3.4 EUKARYOTIC CELL

The eukaryotic cells (Gr., *eu* = good, *karyotic* = nucleated) are essentially **two envelope systems** and they are very much larger than prokaryotic cells. Secondary membranes envelop the nucleus and other internal organelles and to a great extent they pervade the cytoplasm as the endoplasmic reticulum. The eukaryotic cells are the true cells which occur in the plants (from algae to angiosperms) and the animals (from Protozoa to mammals). Though the eukaryotic cells have different shape, size, and physiology; all the cells are typically composed of plasma membrane, cytoplasm and its organelles, viz., mitochondria, endoplasmic reticulum, ribosomes, Golgi apparatus, etc., and a true nucleus. Here the nuclear contents, such as DNA, RNA, nucleoproteins and nucleolus remain separated from the cytoplasm by the thin, perforated nuclear membranes.

Cell Shape

The basic shape of the eukaryotic cell is **spherical**, however, the shape is ultimately determined by the specific function of the cell. Thus, the shape of the cell may be **variable** (*i.e.,* frequently changing the shape) or **fixed**. Variable or irregular shape occurs in *Amoeba* and white blood cells or leucocytes (In fact, leucocytes are spherical in the circulating blood, but in other conditions they may produce pseudopodia and become irregular in shape). Fixed shape of the cell occurs in almost all protists (*e.g., Euglena, Paramecium*), plants and animals. In unicellular organisms the cell shape is maintained by tough plasma membrane and exoskeleton. In a multicellular organism, the shape of the cell depends mainly on its functional adaptations and partly on the surface tension, viscosity of the protoplasm, cytoskeleton of microtubules, microfilaments and intermediate filaments, the mechanical action exerted by adjoining cells and rigidity of the plasma membrane (*i.e.*, presence of rigid cell wall in plant cells). The shape of the cell may vary from animal to animal and from organ to organ. Even the cells of the same organ may display variations in the shape. Thus, cells may have diverse shapes such as **polyhedral** (with 8, 12 or 14 sides; *e.g.,* squamous epithelium); **flattened** (*e.g.,* squamous epithelium, endothelium and the upper layers of the epidermis); **cuboidal** (*e.g.,* in thyroid gland follicles); **columnar** (*e.g.,* the cells lining the intestine); **discoidal** (*e.g.*, red blood cells or erythrocytes); **spherical** (*e.g.*, eggs of many animals); **spindle shaped** (*e.g.* smooth-muscle fibres); **elongated** (*e.g.*, nerve cells or neurons); or **branched** (*e.g.,* chromatophores or pigment cells of skin). Among plants, the cell shape also depends upon the function of the cell. For example, cells such as glandular hairs on a leaf, the guard cells of stomata and root hair cells have their special shape.

Cell Size

The eukaryotic cells are typically larger (mostly ranging between 10 to 100 μm) than the prokaryotic cells (mostly ranging between 1 to 10 μm). Size of the cells of the unicellular organisms is larger than a typical multicellular organism's cells. For example, *Amoeba proteus* is biggest among the unicellular organisms; its length being 1 μm (1000 μm). One species of *Euglena* is found up to 500 μm (0.5mm) in length. *Euplotes* (a freshwater ciliate) is 120 μm in length. Another ciliate, *Paramecium caudatum* is from 150 to 300 μm (0.15 to 0.3 mm) in length. Diatoms have a length of 200 μm or more. The single-celled alga, *Acetabularia* which consists of a stalk and a cap is exceptionally large-sized and measures up to 10 cm in height.

The size of the cells of multicellular organisms ranges between 20 to 30 μm. Among animals, the smallest cells have a diameter of 4 μm (*e.g.,* polocytes); human erythrocytes being 7 to 8 μm in diameter. Largest animal cell is the egg of ostrich, having a diameter of 18 cm (its yolk or deutoplasm is about 5 cm in diameter); though, some nerve cells of human beings have a meter long "tails" or axons. Among the multicellular plants, the largest cell is the ovule of *Cycas*. The fibre cells (*i.e.,* sclerenchyma cells) of Manila hemp are over 100 cm in length.

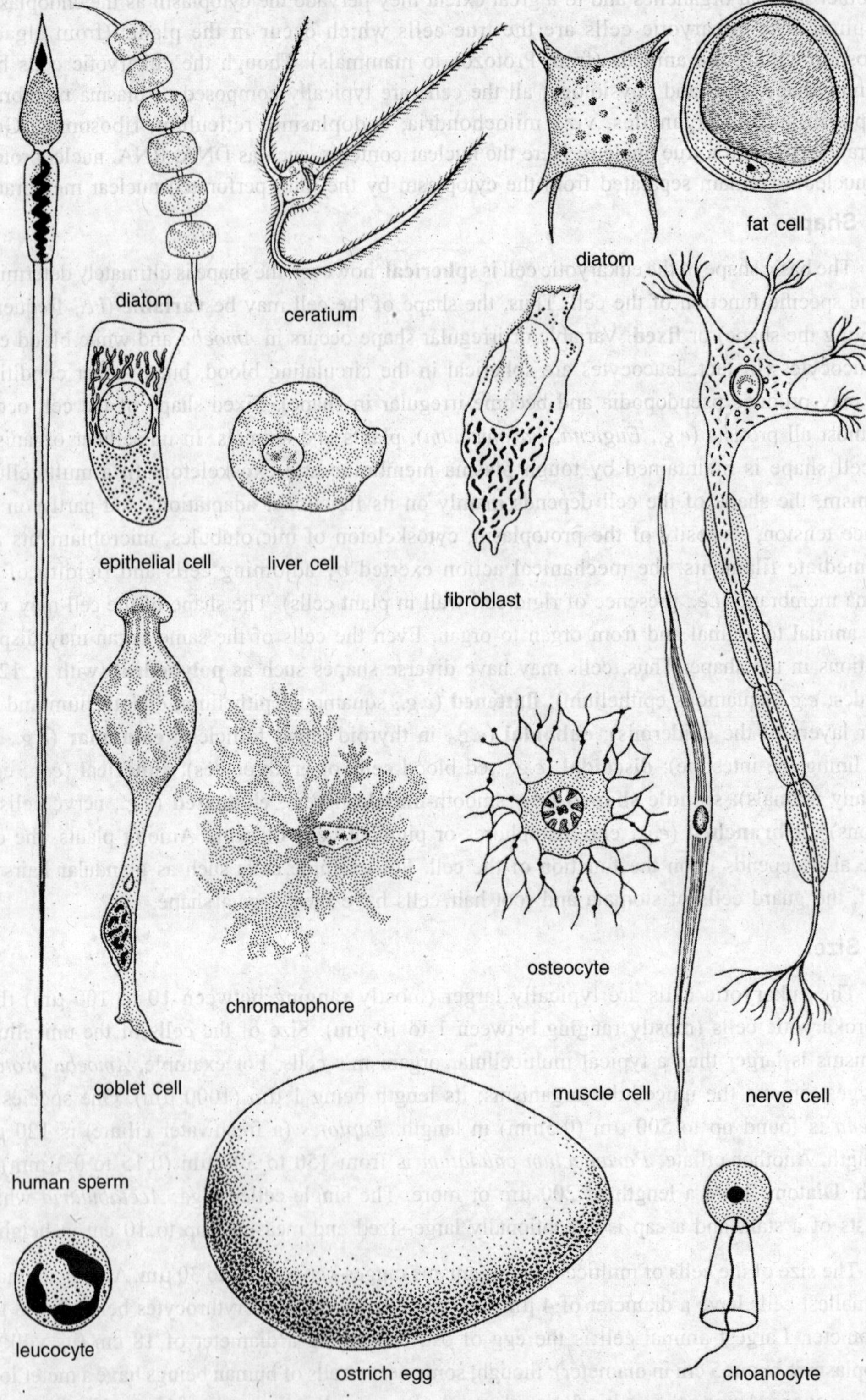

Fig. 3.10. Various types of eukaryotic cells showing different shapes.

Cell Volume

The volume of a cell is fairly constant for a particular cell type and is independent of the size of the organism. (This is called the **law of constant volume**). For example, kidney or liver cells are about the same size in the bull, horse and mouse. The difference in the total mass of the organ or organism depends on the number, not on the volume of the cells. Thus, the cells of an elephant are not necessarily larger than those of other tiny animals or plants. The large size of the elephant is due to the larger number of cells present in its body.

If a cell is to be efficient, the ratio of volume to surface should be within a limited range. An increase in cell volume is accompanied by a much smaller expansion in the surface area of the cell (In fact, volume increases as cube of radius, while surface area increases as square of radius). In other words, a large cell has a proportionately smaller surface and a higher volume: surface ratio than a smaller cell. Further, a large cell volume has to accommodate many organelles simultaneously limiting the exchange of information and materials through the surface. This problem is partially overcome by developing a cylindrical shape or by forming numerous extensions (*e.g.*, microvilli) of the plasma membrane. It is also for this reason that metabolically active cells, tend to be smaller in size.

Cell Number

The number of cells present in an organism varies from a single cell in a **unicellular organism** (Protists such as Protozoa and Protophyta) to many cells in multicellular organisms (most plants, fungi and animals). The number of cells in the multicellular organisms usually remains correlated with size of the organisms and, therefore, small-sized organism has less number of cells in comparison to large-sized organisms. For example, a human being weighing about 80 kg may contain about 60 thousand billion cells in his body. This number would be more in certain other multicellular organisms.

Further, the number of cells in most multicellular organisms is indefinite, but the number of cells may be fixed in some multicellular organisms (*e.g.*, nematodes, rotifers). For example, in rotifers, number of nuclei in the various organs are found to be constant in any given species. This phenomenon of cells or nuclear constancy is called **eutely**. In one species of rotifer, **Martini** (1912) always found 183 nuclei in the brain, 39 in the stomach, 172 in the cornea epithelium, and so on. Among plants, colonial green algae exhibit cell constancy. For example, the green alga, *Pandorina* has a colony consisting of 8, 16 or 32 cells. Likewise, another green alga, *Eudornia*, has 16, 32 or 64 cells in its colony.

Animal Cell

A typical animal cell (Fig. 3.11), which is an eukaryotic cell, consists of following three main parts:

1. Plasma membrane
2. Cytoplasm
3. Nucleus

1. Cell membrane or plasma membrane. Every kind of animal cell is bounded by a living, extremely thin and delicate membrane called **plasmalemma, cell membrane** or **plasma membrane**. In plant cells, plasma membrane occurs just inner to cell wall, bounding the cytoplasm. The plasma membrane exhibits a trilaminar (*i.e.*, three-layered) structure with a translucent layer sandwiched between two dark layers. At molecular level, it consists of a continuous bilayer of lipid molecule (*i.e.*, phospholipids and cholesterol) with protein molecules embedded in it or adherent to its both surfaces. Some carbohydrate molecules may also be attached to the external surface of the plasma membrane, they remain attached either to protein molecules to form **glycoproteins** or to lipids to form **glycolipids**. Such a fluid mosaic plasma membrane is a **selectively permeable membrane;**

its main function is to control selectively the entrance and exit of materials. This allows the cell to maintain a constant internal environment **(homeostasis)**. Transport of small molecules such as water, oxygen, carbon dioxide, ethanol, ions; glucose, etc., across the plasma membrane takes place by various means such as osmosis, diffusion and active transport. The process of active transport is performed by special type of protein molecules of plasma membrane called **transport proteins** or **pumps,** consuming energy in the form of ATP molecules. For bulk transport of large-sized molecules, plasma membrane performs **endocytosis** (*i.e.,* endocytosis, pinocytosis, receptor-mediated endocytosis and phagocytosis) and **exocytosis** both of these processes also utilise energy in the form of ATP molecules.

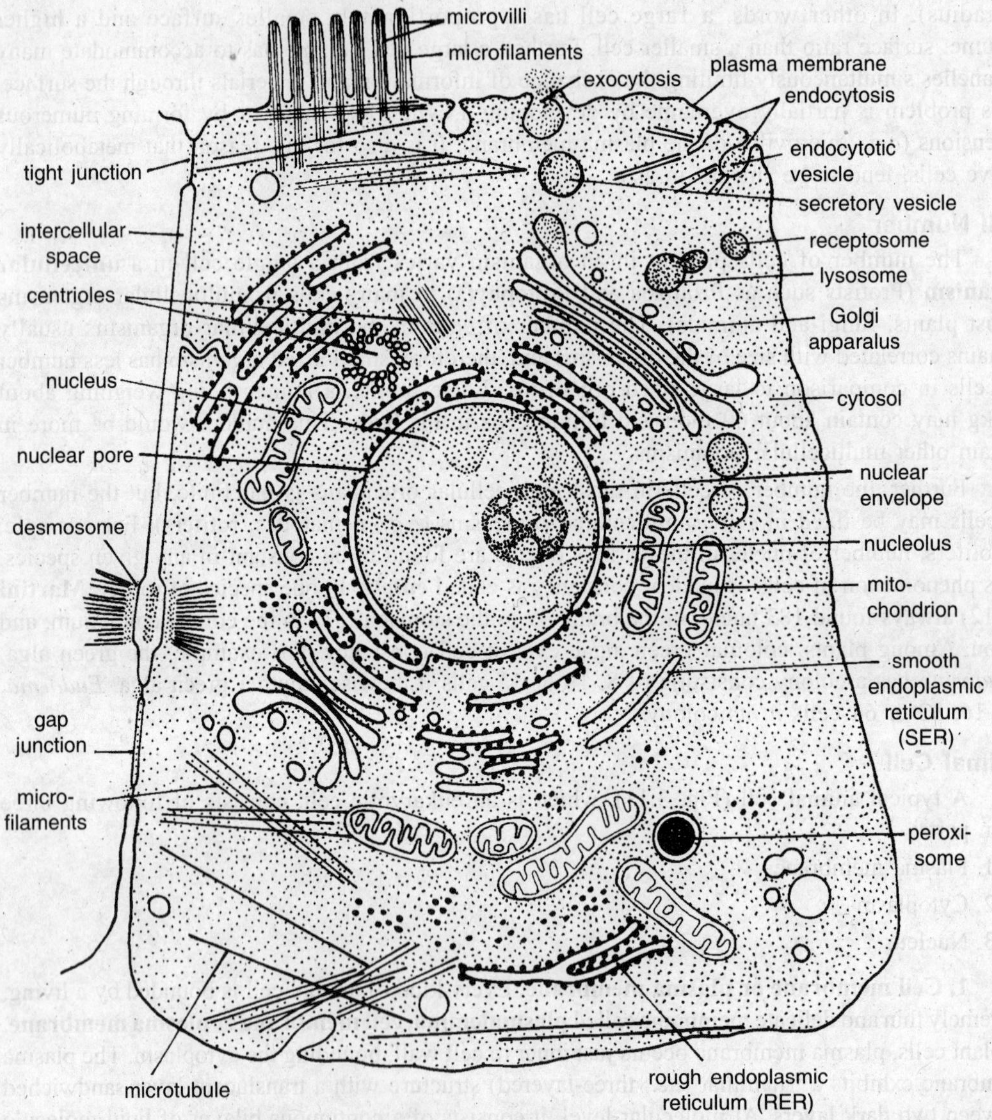

Fig. 3.11. Ultrastructure of a typical animai cell as seen in the electron microscope.

Unit membrane. Various cell organelles such as chloroplasts, mitochondria, endoplasmic reticulum and lysosomes are also bounded by membranes similar to the plasma membrane. All the cellular membranes have a basic trilaminar fluid mosaic **unit membrane** construction. However, their structure and extent of activity are mainly depended on the relative proportion of their constituent protein and lipid molecules. Thus, membranes which are metabolically highly active, *e.g.*, those of mitochondria and chloroplasts have a greater proportion of proteins and more granular appearance than those membranes which are relatively less active, *e.g.*, myelin sheath of certain nerve fibres.

2. Cytoplasm. The plasma membrane is followed by the cytoplasm which is distinguished into following structures:

A. Cytosol. The plasma membrane is followed by the colloidal organic fluid called **matrix** or **cytosol**. The cytosol is the aqueous portion of the **cytoplasm** (the extra nuclear protoplasm) and of the **nucleoplasm** (the nuclear protoplasm). It fills all the spaces of the cell and constitutes its true **internal milieu**. Cytosol is particularly rich in differentiating cells and many fundamental properties of cell are because of this part of the cytoplasm. The cytosol serves to dissolve or suspend the great variety of small molecules concerned with cellular metabolism, *e.g.*, glucose, amino acids, nucleotides, vitamins, minerals, oxygen and ions. In all type of cells, cytosol contains the soluble proteins and enzymes which form 20 to 25 per cent of the total protein content of the cell. Among the important soluble enzymes present in the matrix are those involved in glycolysis and in the activation of amino acids for the protein synthesis. In many types of cells, the cytosol is differentiated into following two parts: (*i*) **Ectoplasm** or **cell cortex** is the peripheral layer of cytosol which is relatively non-granular, viscous, clear and rigid. (*ii*) **Endoplasm** is the inner portion of cytosol which is granular and less viscous.

Cytoskeleton and microtrabecular lattice. The cytosol of cells also contains **fibres** that help to maintain cell shape and mobility and that provide anchoring points for the other cellular structures. Collectively, these fibres are termed as the **cytoskeleton.** At least three general classes of such fibres have been identified. 1. The thickest are the **microtubules** (20 nm in diameter) which consists primarily of the **tubulin** protein. The function of microtubules is the transportation of water, ions or small molecules, cytoplasmic streaming (cyclosis), and the formation of fibres or asters of the mitotic or meitotic spindle during cell division. Moreover, they form the structural units of the centrioles, basal granules, cilia and flagella. 2. The thinnest are the **microfilaments** (7 nm in diameter) which are solid and are principally formed of **actin** protein. They maintain the shape of cell and form contractile component of cells, mainly of the muscle cells. 3. The fibres of middle order are called the **intermediate filaments (IFs)** having a diameter of 10 nm. They have been classified according to their constituent protein such as **desmin filaments, keratin filaments, neurofilaments, vimentin** and **glial filaments**.

Recently, cytoplasm has been found to be filled with a three-dimensional network of interlinked filaments of cytoskeletal fibres, called **microtrabecular lattice.** Various cellular organelles such as ribosomes, lysosomes, etc., are found anchored to this lattice. The microtrabecular lattice being flexible, changes its shape and results in the change of cell shape during cell movement.

B. Cytoplasmic structures. In the cytoplasmic matrix certain non-living and living structures remain suspended. The non-living structures are called **paraplasm** or **inclusions,** while the living structures are membrane bounded and are called **organoids** or **organelles.** Both kinds of cytoplasmic structures can be studied under the following headings:

***(a)* Cytoplasmic inclusions.** The stored food and secretory substances of the cell remain suspended in the cytoplasmic matrix in the form of refractile granules forming the cytoplasmic inclusions. The cytoplasmic inclusions include oil drops, triacylglycerols (*e.g.*, fat cells of adipose tissue), yolk granules (or **deutoplasm,** *e.g.*, egg cells), secretory granules, glycogen granules (*e.g.*, muscle cells and hepatocytes of liver) and starch grains (in plant cells).

***(b)* Cytoplasmic organelles.** Besides the separate fibrous systems, cytoplasm is coursed by a multitude of internal membranous structures, the organelles (literally the word organelle means a tiny organ). Membranes close off at specific regions of the eukaryotic cells performing specialized tasks: oxidative phosphorylation and generation of energy in the form of ATP molecules in mitochondria; formation and storage of carbohydrates in plastids; protein synthesis in rough endoplasmic reticulum; lipid (and hormone) synthesis in smooth endoplasmic reticulum; secretion by Golgi apparatus; degradation of macromolecules in the lysosomes; regulation of all cellular activities by nucleus; organization of spindle apparatus by centrosomes and so forth. Membrane-bound enzymes catalyze reactions that would have occurred with difficulty in an aqueous environment. The structure and function of some important organelles are as follows:

1. Endoplasmic reticulum (ER). Within the cytoplasm of most animal cells is an extensive network (reticulum) of membrane-limited channels, collectively called **endoplasmic reticulum** (or **ER**). Some portion of ER membranes remains continuous with the plasma membrane and the nuclear envelope. The outer surface of **rough ER** has attached ribosomes, whereas **smooth ER** do not have attached ribosomes. Functions of smooth ER include **lipid metabolism** (both catabolism and anabolism; they synthesize a variety of phospholipids, cholesterol and steroids); **glycogenolysis** (degradation of glycogen; glycogen being polymerized in the cytosol) and **drug detoxification** (by the help of the **cytochrome P-450**).

On their membranes, rough ER (RER) contain certain ribosome-specific, transmembrane glycoproteins, called **ribophorins I** and **II**, to which are attached the ribosomes while engaged in polypeptide synthesis. As a growing secretory polypeptide emerges from ribosome, it passes through the RER membrane and gets accumulated in the lumen of RER. Here, these polypeptide chains undergo tailoring, maturation, and molecular folding to form functional secondary or tertiary protein molecules. RER pinches off certain tiny protein-filled vesicles which ultimately get fused to cis Golgi. RER also synthesize membrane proteins and glycoproteins which are cotranslationally inserted into the rough ER membranes. Thus, endoplasmic reticulum is the site of biogenesis of cellular membranes.

2. Golgi apparatus. It is a cup-shaped organelle which is located near the nucleus in many types of cells. Golgi apparatus consists of a set of smooth **cisternae** (*i.e.*, closed fluid-filled flattened membranous sacs or vesicles) which often are stacked together in parallel rows. It is surrounded by spherical membrane bound **vesicles** which transport proteins to and from it.

Golgi apparatus consists of at least three distinct classes of cisternae: **cis Golgi, median Golgi** and **trans Golgi**, each of which has distinct enzymatic activities., Synthesized proteins appear to move in the following direction: rough ER → cis Golgi → median Golgi → trans Golgi → secretory vesicles/cortical granules of egg/lysosomes or peroxisomes. Thus, the size and number of Golgi apparatus in a cell indicate the active metabolic, mainly synthetic, state of that cell. Plant cells contain many freely distributed sub-units of Golgi apparatus, called **dictyosomes**, secreting cellulose and pectin for cell wall formation during the cell division.

Generally, Golgi apparatus performs the following important functions:

1. The packaging of secretory materials (*e.g.,* enzymes, mucin, lactoprotein of milk, melanin pigment, etc.) that are to be discharged from the cell.
2. The **processing** of proteins, *i.e.*, glycosylation, phosphorylation, sulphation and selective proteolysis.
3. The synthesis of certain polysaccharides and glycolipids.
4. The sorting of proteins destined for various locations (*e.g.*, lysosomes, peroxisomes, etc.) in the cell.
5. The proliferation of membranous element for the plasma membrane.
6. Formation of the acrosome of the spermatozoa and cortical granules of an ovum.

3. Lysosomes. The cytoplasm of animal cells contains many tiny, spheroid or irregular-shaped, membrane-bounded vesicles known as **lysosomes**. The lysosomes are originated from Golgi apparatus and contain numerous (about 50) hydrolytic enzymes (*e.g.,* **acid phosphatase** that is cytochemically identified) for intracellular and extracellular digestion. They digest the material taken in by endocytosis (such as phagocytosis, endocytosis and pinocytosis), parts of the cell (by autophagy) and extracellular substances. Lysosomes have a high acidic medium (pH 5) and this acidification depends on ATP-dependent **proton pumps** which are present in the membrane of lysosomes and which accumulate protons (H^+) inside the lysosomes. Lysosomes exhibit great **polymorphism,** *i.e.,* there are following four types of lysosomes: primary lysosomes (storage granules), secondary lysosomes (digestive vacuoles), residual bodies and autophagic vacuoles. The lysosomes of plant cells are membrane-bounded storage granules containing hydrolytic digestive enzymes, *e.g.,* large **vacuoles** of parenchymatous cells of corn seedlings, **protein** or **aleurone bodies** and **starch granules** of cereal and other seeds.

4. Cytoplasmic vacuoles. The cytoplasm of some animal cells (*e.g.,* ciliate protozoans) contains numerous small or large-sized, hollow, liquid-filled structures, the **vacuoles.** These vacuoles are supposed to be greatly expanded endoplasmic reticulum or Golgi apparatus. The vacuoles of animal cells are bounded by a lipoproteinous membrane and their function is the storage, transmission of the materials and the maintenance of internal pressure of the cell.

5. Peroxisomes. These are tiny circular membrane-bound organelles containing a crystal-core of enzymes (such as urate oxidase, peroxidase, D-amino oxidase and catalase) of certain cells (*e.g.,* liver cells and kidney cells). These enzymes are required by peroxisomes in **detoxification** activity, *i.e.,* in the metabolism or production and decomposition, of hydrogen peroxide or H_2O_2 molecules which are produced during neutralization of certain superoxides—the end products of mitochondrial or cytosolic reactions. Peroxisomes are also related with β-oxidation of fatty acids and thermogenesis like the mitochondria and also in degradation of the amino acids. In green leaves of plants, peroxisomes carry out the process of **photorespiration**.

6. Mitochondria. Mitochondria are oxygen-consuming ribbon-shaped cellular organelles of immense importance. Each mitochondrion is bounded by two unit membranes. The outer mitochondrial membrane resembles more with the plasma membrane in structure and chemical composition. It contains **porins,** proteins that render the membrane permeable to molecules having molecular weight as high as 10,000. Inner mitochondrial membrane is rich in many enzymes, coenzymes and other components of electron transport chain. It also contains **proton pumps** and many **permease** proteins for the transport of various molecules such as citrates, ADP, phosphate and ATP. Inner mitochondrial membrane gives out finger-like outgrowths **(cristae)** towards the lumen of mitochondrion and contains tennis-racket shaped $\mathbf{F_1}$ **particles** which contain ATPase enzyme for ATP synthesis.

Mitochondrial matrix which is the liquid (colloidal) area encircled by the inner membrane, contains the soluble enzymes of Krebs cycle which completely oxidize the **acetyl-CoA** (an end product of cytosolic glycolysis and mitochondrial oxidative decarboxylation) to produce CO_2, H_2O and hydrogen ions. Hydrogen ions reduce the molecules of NAD and FAD, both of which pass on hydrogen ions to respiratory or electron transport chain where oxidative phosphorylation takes place to generate energy-rich ATP molecules. Since mitochondria act as the 'power-houses' of cells, they are abundantly found on those sites where energy is earnestly required such as sperm tail, muscle cell, liver cell (up to 1600 mitochondria), microvilli, oocyte (more than 300,000 mitochondria), etc. Mitochondria also contain in their matrix single or double circular and double stranded DNA molecules, called **mtDNA** and also the 55S ribosomes, called **mitoribosomes**. Since mitochondria can synthesize 10 per cent of their proteins in their own protein-synthetic machinery, they are considered as **semi-autonomous organelles.** Mitochondria may also produce heat (brown fat), and accumulate iron-containing pigments (Heme, ferritin), ions of Ca^{2+} and HPO_4^{2-} (or phosphate; *e.g.,* osteoblasts of bones or yolk proteins).

7. Ribosomes. Ribosomes are tiny spheroidal dense particles (of 150 to 200 A° diameter) that contain approximately equal amounts of RNA and proteins. They are primarily found in all cells and serve as a scaffold for the ordered interaction of the numerous molecules involved in protein synthesis. Ribosome granules may exist either in the **free state** in the cytosol (*e.g.,* basal epidermal cells) or **attached** to RER (*e.g.,* pancreatic acinar cells, plasma cells or antibodies-secreting lymphocytes, osteoblasts, etc.). Ribosomes have a sedimentation coefficient of about **80S** and are composed of two subunits namely **40S** and **60S**. The smaller 40S ribosomal subunit is prolate ellipsoid in shape and consists of one molecule of 18S ribosomal RNA (or rRNA) and 30 types of proteins (named as S_1, S_2, S_3, and so on). The larger 60S ribosomal subunit is round in shape and contains a **channel** through which growing polypeptide chain makes its exit. It consists of three types of rRNA molecules, *i.e.,* 28S rRNA, 5.8 rRNA and 5S rRNA, and 40 types of proteins (named as L_1, L_2, L_3 and so on).

8. Microtubules and microtubular organelles. With rare exceptions, such as human erythrocyte, microtubules are found in the cytoplasm of all types of eukaryotic cells. They are long fibres (of indefinite length) about 24 nm in diameter. In cross section each microtubule appears to have a dense wall of 6 nm thickness and a light or hollow centre. In cross section, the wall of a microtubule is made up of 13 globular subunits, called **protofilaments,** about 4 to 5 nm in diameter. Chemically, microtubules are composed of two kinds of protein subunits: **α-tubulin (tubulin A)** and **β-tubulin (tubulin B)**, each of M.W. 55,000 daltons. The wall of a microtubule is made up of a helical array of repeating α and β tubulin subunits. Assembly studies have indicated that the structural unit is an αβ **dimmer** of 8 nm length. Thus, in each microtubule, there are 13 protofilaments, each composed of αβ dimers that run parallel to the long axis of the tubule: The repeating unit is an αβ heterodimer which is arranged 'head to tail' within the microtubule, that is αβ → αβ → αβ. Thus, all microtubules have a defined **polarity:** their two ends are not structurally equivalent.

Microtubules undergo reversible assembly-disassembly (*i.e.,* polymerization and depolymerization), depending on the need of the cell or organelles. Their polymerization is regulated by certain **MAPs** or **microtubule-associated proteins** (*e.g.,* Tau protein). The assembly of microtubules involves preferential addition of subunits (αβ dimers) to one end of tubule, called **A end** (or **net assembly end**); the other end of the tubule is called **D end** (or **net disassembly end**). Such an assembly involves the hydrolysis of GTP to GDP. Thus, assembly of tubulin in the formation of microtubules is a specifically oriented and programmed process. Centrioles, basal bodies and centromeres of chromosomes are the sites of orientation for this assembly. Calcium and **calmodulin** (an acidic protein having four Ca^{2+} binding sites) are some other regulating factors in the *in vivo* polymerization of tubulin. Certain drugs such as **colchicine** and **vinblastin,** are found to block the polymerization of tubulin.

The following cell organelles are derived from special assemblies of microtubules:

1. Cilia and flagella. Ciliary and flagellar cell motility is adapted to liquid media and is executed by minute, specially differentiated appendices, called **cilia** and **flagella.** Both of these organelles have very similar structure; they differ mainly in size and number (*i.e.,* flagella are longer and fewer in number, while cilia are short and numerous). Cilia are used for locomotion in isolated cells, such as certain protozoans (*e.g., Paramecium*) or to move particles in the medium, as in air passages and oviduct. Flagella are generally used for locomotion of cells, such as the spermatozoon and *Euglena* (protozoan). All cilia and flagella are built on a common fundamental plan: a bundle of microtubules called the **axoneme** (1 to 2 nm in length and 0.2 μm in diameter) is surrounded by a membrane that is part of the plasma membrane. The axoneme is connected with the basal body which is an intracellular granule lying in the cell cortex and which originates from the centrioles. Each axoneme is filled with **ciliary matrix,** in which are embedded two central **singlet** microtubules, each with the 13 protofilaments and nine outer pairs of microtubules, called **doublets**. This recurring motif is known as the 9 + 2 array. Each doublet contains one complete microtubule, called the **A subfibre**, containing all the 13 protofilaments. Attached to each A subfibre is a **B subfibre** with 10 protofilaments. Subfibre A has two **dynein arms** which are oriented in a clockwise direction. Doublets are linked together by **nexin links.** Each subfibre A is also connected to the central microtubules by **radial spokes** terminating in fork-like structures, called **spoke knobs** or **heads**.

Propulsion by both cilia and flagella is caused by bending at their base. Cilia move by a whip-like **power stroke** fueled by hydrolysis of ATP, followed by **a recovery stroke.** Flagellar movement is also powered by ATP hydrolysis. In contrast to cilia, flagella generally move by waves that emanate from the base and spread outward toward the tip.

2. Basal bodies and centrioles. Basal bodies and centrioles are similar in structure and function; both act as nucleating centres from which microtubules grow. **Centrioles** are cylinders that measure 0.2 μm × 0.5 μm. This cylinder is open on both ends, unless it carries a cilium or flagellum (then it is called **basal body** or **kinetosome**). The wall of a centriole has nine groups of microtubules arranged in a circle. Each group, called **blade** is a **triplet** formed of three tubules—*A*, *B*, and *C* that are skewed toward the centre. Tubule *A* has 13 protofilaments, while tubules *B* and *C* have only 10 protofilaments. each. There are no central microtubules in the centrioles and no dynein arms like the cilia; however, triplets are linked by connectives. The **procentriole** (or daughter centriole) is formed at right angles to the centriole and is located near the proximal end of the centriole. Both centrioles are found in a specially differentiated region the **centrosome, cell centre** or **centrosphere.** The centrosome is juxtanuclear (L., juxta = near) and firmly attached to the nuclear envelope. At the time of cell division two pairs of centrioles are formed and make up the spindle of microtubules which help in the separation and movement of chromosomes during concluding stages of cell divisions.

3. Nucleus. The nucleus is centrally located and spherical cellular component which controls all the vital activities of the cytoplasm and carries the hereditary material the DNA in it. The nucleus consists of the following three structures:

1. Chromatin. Nucleus being the heart of every type of eukaryotic cell, contains the **genes,** the hereditary units. Genes are located on the **chromosomes** which exist as **chromatin network** in the non-dividing cell, *i.e.,* during interphase. The chromatin has two forms:

***(i)* Euchromatin** is the well-dispersed form of chromatin which takes lighter DNA-stain and is genetically active, *i.e.,* it is involved in gene duplication, gene transcription (DNA-dependent RNA synthesis) and **phenogenesis** or phenotypic expression of a gene through some type of protein synthesis.

***(ii)* Heterochromatin** is the highly condensed form of chromatin which takes dark DNA-stain and is genetically inert. Such type of chromatin exists both in the region of centromere (called **constitutive heterochromatin**) and in the sex chromatin (called **facultative heterochromatin**) and is late-replicating one.

Chemically, the chromatin contains a single DNA molecule, equal amount of histone proteins. (which are of five basic types), some RNA molecules and variable amount of different types of acidic proteins. In fact, the chromatin has its unit structures in the form of **nucleosomes.** The chromatin binds strongly to the inner part of **nuclear lamina,** a 50 to 80 nm thick fibrous lamina lining the inner side of the nuclear envelope. Nuclear lamina is made up of three types of proteins, namely **lamin *A*, *B*** and ***C*.** Lamin proteins are homologous in structure to IF proteins and serve the following functions:

(*i*) They anchor parts of interphase chromatin to the nuclear membrane. They tend to interfere with chromatin condensation during interphase of cell cycle.

(*ii*) Lamins may play a crucial role in the assembly of interphase nuclei after each mitosis.

2. Nuclear envelope and nucleoplasm. Nuclear envelope comprises two nuclear membranes—an **inner nuclear membrane** which is lined by nuclear lamina and an **outer nuclear membrane** which is continuous with rough ER. At certain points the nuclear envelope is interrupted by structures called **pores** or **nucleopores**. Nuclear pores contain octagonal **pore complexes** which regulate exchange between the nucleus and cytoplasm. The number of nucleopores is found to be correlated with the transcriptional activity of the cell. For example, in the frog *Xenopus laevis* oocytes (which are very active in transcription) have 60 pores/μm^2 (and upto 30 million pore complexes per nucleus), whereas frog's mature erythrocytes (inactive in transcription) have only about 3 pores/μm^2 (and a total of only 150 to 300 pores per nucleus).

The nuclear envelope binds the **nucleoplasm** which is rich in those molecules which are needed for DNA replication, transcription, regulation of gene actions and processing of various types of newly transcribed RNA molecules (*i.e.,* tRNA, mRNA and other types of RNA).

3. Nucleolus. Nucleus contains in its nucleoplasm a conspicuous, darkly stained, circular sub-organelle, called **nucleolus.** Nucleolus lacks any limiting membrane and is formed during interphase by the ribosomal DNA (rDNA) of **nucleolar organizer (NO)**. Nucleolus is the site where ribosomes are manufactured. It is here where ribosomal DNA transcribes most of rRNA molecules and these molecules undergo processing before their step-wise addition to 70 types of ribosomal proteins to form the ribosomal subunits.

Plant Cell

All the cytoplasmic and nuclear components found in the animal cells are found in the plant cells (Fig. 3.12). In addition to these cell components, plant cells contain certain structures not found in animal cells. These are **cell wall, plasmodesmata, plastids, large vacuoles** and **glyoxysomes**.

1. Cell wall. The outermost structure of most plant cells is a dead and rigid layer called **cell wall**. It is mainly composed of carbohydrates such as cellulose, pectin, hemicellulose and lignin. and certain fatty substances such as waxes. Ultrastructurally cell wall is found to consist of a microfibrilar network lying in a gel-like matrix. The microfibrils are mostly made up of cellulose. There is a pectin-rich cementing substance between the walls of adjacent cells which is called **middle lamella**. The cell wall which is formed immediately after the division of cell, constitutes the **primary cell wall.** Many kinds of plant cells have only primary cell wall around them. Primary cell wall is composed of pectin, hemicellulose and loose network of cellulose microfibrils. In certain types of cells such as phloem and xylem, an additional layer is added to the inner surface of the primary cell wall at a later stage. This layer is called **secondary cell wall** and it consists mainly of cellulose, hemicellulose and lignin. In many plant cells, there are tunnels running through the cell wall called **plasmodesmata** which allow communication with the other cells in a tissue.

The cell wall constitutes a kind of exoskeleton that provides protection and mechanical support to the plant cell. It determines the shape of plant cell and prevents it from desiccation.

2. Plastids. Plastids occur only in the plant cells. They contain pigments and may synthesize and accumulate various substances. Plastids are of the following types:

***(i)* Leucoplasts** are colourless plastids of embryonic and germ cells lacking thylakoids and ribosomes.

***(ii)* Amyloplasts** produce starch.

***(iii)* Proteinoplasts** accumulate protein.

***(iv)* Oleosomes** or **elaioplasts** store fats and essential oils.

(*v*) **Chromoplasts** contain pigment molecules and are coloured organelles. Chromoplasts impart a variety of colours to plant cells, such as red colour in tomatoes, red chilies and carrots, various colours to petals of flowers and green colour to many plant cells. The green coloured chromoplasts are called **chloroplasts**. They have chlorophyll pigment and are involved in the photosynthesis of food and so act like the kitchens of the cell.

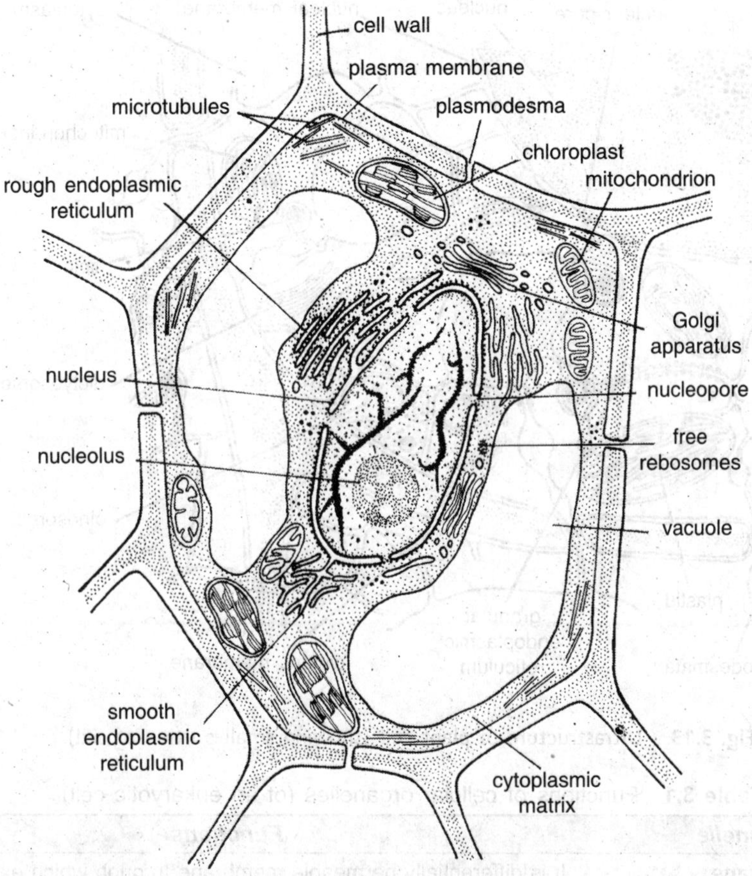

Fig. 3.12. Ultrastructure of a typical eukaryotic cell of a higher plant cell.

Chloroplasts have diverse shapes in green algae but are round, oval or discoid in shape in higher plants. Like mitochondria, each chloroplast is bounded by two membranous envelopes, both of which have no chlorophyll pigment. However, unlike mitochondria there occurs third system of membranes within the boundary of inner membrane, called **grana**. The grana form the main functional units of chloroplast and are bathed in the homogeneous matrix, called the **stroma**. Stroma contains a variety of photosynthetic enzymes and starch grains. Grana are stacks of membrane-bounded, flattened discoid sacs, arranged like neat piles of coins. A chloroplast contains many such interconnected grana on which are located various photosynthetic enzymes and the molecules of green pigment chlorophyll and other photosynthetic pigments to trap the light energy. They contain DNA, ribosomes and complete protein synthetic machinery.

3. Vacuoles. Often most of the volume (up to 90 per cent) of a mature cell is occupied by one or more large vacuoles. The vacuoles of the plant cells are bounded by a single, semipermeable membrane known as **tonoplast**. These vacuoles contain water, phenol, flavonols, anthocyanins (blue and red pigment), alkaloids and storage products, such as sugars and proteins.

4. Glyoxysomes. These organelles develop in a germinating plant seed (*e.g.,* castor bean or *Ricinus*) to utilize stored fat of the seed (*i.e.,* to metabolise the triglycerides). Glyoxysomes consist of an amorphous protein matrix surrounded by a limiting membrane. The membrane of glyoxysomes originates from the ER and their enzymes are synthesized in the free ribosomes in the cytosol. Enzymes of glyoxysomes are used to transform the fat stores of the seed into carbohydrates by way of **glyoxylate cycle**.

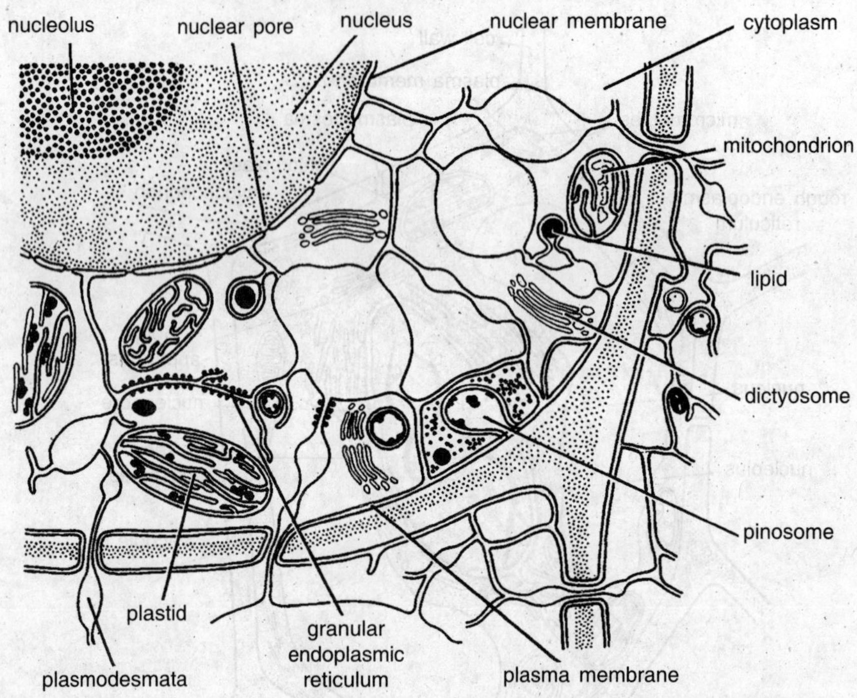

Fig. 3.13. Ultrastructure of plant cell (showing small portion of cell).

Table 3.1 Functions of cellular organelles (of an eukaryotic cell).

Cell organelle	*Functions*
Plasma membrane	It is differentially permeable membrane through which extracellular substances may be selectively sampled and cell products may be liberated.
Cell wall (plant cells)	It forms a thick cellulose wall around the plasma membrane giving strength and rigidity to the cell.
Cytoplasm	It contains machinery for carrying out the instructions sent from the nucleus.
Hyaloplasm or matrix	It contains enzymes for glycolysis and structural materials such as sugars, fatty acids, amino acids, water, vitamins, nucleotides, etc.
Microtubules	They transport water, ions and small molecules of various substances, form structural unit of centrioles, basal granules, cilia, flagella and mitotic spindle and play active role in the cyclosis (amoeboid movement) of cytoplasm and in movement of chromosomes during the cell divisions.
Microfilaments	They maintain the structure of cell and form contractile components of muscle cell.
Centrioles	They form the spindle and have significant role in the movement of chromosomes during cell divisions.
Basal granules	They originate the cilia or flagella and have their importance in ciliary or flagellar movement.

Cell organelle	Functions
Cilia and flagella	These are extracellular projections of the cell which provide movement to the cells and serve various other roles of cells.
Endoplasmic reticulum	It provides greatly expanded surface area for biochemical reactions which normally occur at or across membrane surfaces and provides skeletal support to the colloidal complex of hyaloplasm. It also has transporting business of extracellular and intracellular chemical molecules.
Ribosomes	These are the sites for synthesis of protein molecules.
Golgi apparatus	It stores the synthetic proteins and enzymes of cells and forms secretory vesicles and lysosomes.
Lysosomes (animals only)	It contains intracellular digestive enzymes which aid in disposal of bacteria and other foreign bodies.
Peroxisomes (Microbodies)	They contain enzymes for hydrogen peroxide metabolism, purine metabolism, gluconeogenesis and photorespiration.
Sphaerosomes	They store fats.
Vacuoles	They are storage depots for excess water, waste products, soluble pigments, etc.
Mitochondria	They are the organelles for the oxidation of food by Krebs cycle, production of energy by electron transport chain and beta-oxidation of fatty acids.
Plastids (Plants only)	They are the structures for storage of starch, pigments, and other cellular products. Photosynthesis occurs in chloroplasts.
Nucleus	It regulates growth and reproduction of cell.
Nuclear membrane	It provides selective continuity between nuclear and cytoplasmic materials.
Nucleoplasm	It contains materials for building DNA and messenger molecules (different kinds of RNA molecules, *viz.,* rRNA, mRNA and tRNA) which act as intermediates between nucleus and cytoplasm.
Chromosomes	They are the bearers of hereditary instructions and regulations of cellular processes.
Nucleolus	It disappears during cellular replication and synthesizes the ribosomes.

Differences between Prokaryotic and Eukaryotic Cells

Main differences between prokaryotic cells (of bacteria, blue green algae and mycoplasma) and eukaryotic cells (of protozoa, fungi, plants and animals) have been tabulated in Table 3.2.

Table 3.2 Differences between prokaryotic and eukaryotic cells.

Feature	Prokaryotic cell	Eukaryotic cell
1. Size	Mostly 1–10 mm	Mostly 10–100 μm
2. Multicellular forms	Rare	Common, with extensive tissue formation
3. Cell wall	Present in most but not in all cells	Present in plant and fungal cells only
4. Plasma membrane	Present	Present
5. Nucleus	Absent	Present
6. Nuclear membranes	Absent	Present
7. Chromatin with histone	Absent	Present
8. Genetic material	Circular or linear, double-stranded DNA: genes are not interrupted by intron*	Linear double-stranded DNA: genes frequently interrupted by intron sequences, especially in higher eukaryotes

*Intron is an intervening sequence of nucleotides in DNA, located within a gene that is not included in the mature mRNA.

Feature	Prokaryotic cell	Eukaryotic cell
9. Nucleoli and mitotic apparatus	Absent	Present
10. Plasmids	Commonly present	Rare
11. Cellular organelles:		
(*i*) Mitochondria	Absent	Present
(*ii*) Endoplasmic reticulum	Absent	Present
(*iii*) Vacuoles	Absent	Present
(*iv*) Lysosomes	Absent	Present
(*v*) Chloroplasts	Absent	Present (only in plants)
(*vi*) Centrioles	Absent	Present (absent in higher plants)
(*vii*) Ribosomes	Present (70S)	Present (80S)
(*viii*) Microtubules	Absent	Present
(*ix*) Flagellae	Simple structure composed of the protein flagellin	Complex 9 + 2 structure of tubulin and other proteins
12. Respiration	Many strict anaerobes (oxygen fatal)	All aerobic, but some facultative anaerobes by secondary modifications
13. Metabolic patterns	Great variations	All share cytochrome electron transport chains, Krebs cycle oxidation, Embden-Meyerhof glucose metabolism or glysolysis
14. Photosynthetic enzymes	Bound to plasma membrane as composite chromatophores	Enzymes packaged in plastids bound by membrane
15. Sexual system	Rare: if present one way (and usually partial); transfer of DNA from donor to recipient cell occurs	Both sexes involved in sexual participation and entire genomes transferred; alternation of haploid and diploid generations is also evident

Differences between Animal and Plant Cells

Differences between typical animal and plant cells have been tabulated in Table 3.3.

Table 3.3 Differences between animal and plant cells.

Animal cell	Plant cell
1. Animal cell are generally small in size.	1. Plant cells are larger than animal cells.
2. Cell wall is absent.	2. The plasma membrane of plant cells is surrounded by a rigid cell wall of cellulose.
3. Except the protozoan *Euglena* no animal cell possesses plastids.	3. Plastids are present.
4. Vacuoles in animal cells are many and small.	4. Most mature plant cells have a large central sap vacuole.
5. Animal cells have a single highly complex and prominent Golgi apparatus.	5. Plant cells have many simpler units of Golgi apparatus, called dictyosomes.
6. Animal cells have centrosome and centrioles.	6. Plant cells lack centrosome and centrioles.

QUESTIONS

Long Answer Question

1. Describe the structure or a prokaryotic cell.
2. Describe generalized structure of bacterial cell.
3. Describe the ultrastructure of *E.coli.*
4. What are the chief characteristics of a cyanobacteria?
5. Prokaryotes are considered to be ancestors of eukaryotes. Discuss.
6. Define cell. Give electron microscopic structure of PPLO cell. Give a well labeled diagram of an eukaryotic animal cell.
7. Define the cell. What is an eukaryotic cell? Describe the general shape and size, number of eukaryotic cells in the living organism.
8. Give an account of the structure of an animal cell and state the functions which are attributed to its component parts.
9. Give an illustrated account of the functional anatomy of animal cell.
10. Enumerate the various organelles found in a typical animal cell and briefly describe the functions of Golgi body and ribosomes.
11. Give an illustrated account of ultrastructure of animal cell.
12. Give a detailed labelled diagram of electron microscopic structure of animal cell along with important cell organelles.
13. Give distinguishing characteristics of an eukaryotic plant cell.
14. Describe structure of a typical animal cell and mention the differences between a plant and animal cell.

Short Answer Questions

1. Summarize important diagnostic features of prokaryotes.
2. Write a short note on PPLOs.
3. Write a short note on 'Prokaryotic cell'.
4. Mention the functions of the following cellular components:

   (*a*) Cell wall (*b*) Cell membrane
   (*c*) Golgi apparatus (*d*) Lysosomes
   (*e*) Ribosomes (*f*) Mitochondria
   (*g*) Centrioles (*h*) Plastids
   (*i*) Chromatin
5. Discuss the factors which control the shape of the cells.
6. What are the conditions which set upper limit in cell size?
7. What are the differences between prokaryotic and eukaryotic cells?
8. Draw a neat and labeled diagram of animal cell. Explain differences between prokaryotic and eukaryotic cells.
9. Differentiate between 'Animal cell and Plant cell'.
10. Draw neat and labeled diagram of the following:

    1. Prokaryotic cell
11. Give the structure of Eukaryotic cell. *(H.N.B., Garwal 2008)*

Very Short Answer Questions

1. Name the smallest known cell. Give its size.
2. Enumerate cell organelles that fall under following categories:
   (*i*) Cytomembranes (*ii*) Membrane bound (*iii*) Membraneless
3. What will happen if all the mitochondria of a cell are destroyed?
4. How can you change an animal cell into a plant cell?
5. Name the organelle where photosynthesis occurs in the cell.
6. A cell wall is a part of all plant cells. Is it living or nonliving?
7. Apart from nucleus, which two other cell organelles have their independent DNA?

Fill in the Blanks

1. The longest animal cell is..............
2. Vacuole is generally present in .... cells but absent in .... cells.
3. .... are the fundamental units of life.
4. .... are not surrounded by cell membrane.
5. The nuclear region of prokaryotic cells is called ....

Matching of Columns

Match appropriately the items in column A with those of column B

Column A	Column B
1. Mitochondria	(*a*) Animal cell
2. Centriole	(*b*) Bacteria
3. Nucleus	(*c*) Eukaryotic cell
4. Chromosome without histone	(*d*) Protoplasm
5. The nerve cell of human beings	(*e*) Hyaloplasm
6. The ground substance or cytoplasm	(*f*) Control room of cell
7. DNA	(*g*) 90 cm
8. Colloidal substance	(*h*) Hereditary material

Multiple Choice Questions

1. Which of the following structures is present only in animal cell?
   (*a*) cell membrane (*b*) lysosomes
   (*c*) centrioles (*d*) ribosomes
2. Which of the following structures is associated with the formation of aster during nuclear division?
   (*a*) endoplasmic reticulum (*b*) centrioles
   (*c*) sphaerosome (*d*) ribosome
3. Which of the following statement is not correct?
   (*a*) ribosomes are power house of cell (*b*) lysosomes are called suicide bags
   (*c*) viruses lack their own enzyme system (*d*) bacteria lack a definite cell wall
4. Plant cells differ from animal cells in the absence of
   (*a*) endoplasmic reticulum (*b*) ribosomes
   (*c*) mitochondria (*d*) centrioles
5. A well organized nucleus with a distinct nuclear membrane is absent in
   (*a*) bacterial cells (*b*) eukaryotic cells
   (*c*) protozoan cells (*d*) fungal cells
6. Eukaryotic cells differ from prokaryotic cells in having
   (*a*) endoplasmic reticulum (*b*) ribosomes
   (*c*) true nucleus (*d*) cytoplasm

7. Which of the cell organelle is considered to be rich in catabolic enzymes?
(*a*) lysosomes (*b*) Golgi apparatus
(*c*) mitochondria (*d*) endoplasmic reticulum

8. Lysosomes are called suicide bags because of
(*a*) hydrolytic enzymes (*b*) phagocytic activity
(*c*) proteolytic activity (*d*) respiratory enzymes

9. Cellular organelle containing hydrolytic enzymes are called
(*a*) microsomes (*b*) lysosomes (*c*) oxysomes (*d*) ribosomes

10. Single envelope system is characteristic of
(*a*) prokaryotic cell (*b*) eukaryotic cell (*c*) none (*d*) both

11. Cell boundary of animal cell is
(*a*) cell wall (*b*) plasma membrane
(*c*) both (*d*) none of these

12. The granular structures found attached with endoplasmic reticulum are
(*a*) ribosomes (*b*) lysosome
(*c*) desmosome (*d*) kinetosome

13. The power house of cell is
(*a*) Golgi complex (*b*) histone
(*c*) mitochondria (*d*) mitochondria

14. Prokaryates and eukaryotes have the common
(*a*) mitotic apparatus (*b*) histone
(*c*) genetic code (*d*) mitochondria

ANSWERS

Fill in the Blanks

1. Nerve cell; **2.** Plant, animal; **3.** Cells; **4.** Ribosomes;
5. Nucleoid.

Very Short Answer Questions

1. Smallest cells are bacterium *Dialister pneumogintes* and *Mycoplasma gallisepticum.*
2. (*i*) Endoplasmic reticulum, Golgi and nuclear envelope.
(*ii*) Lysosomes, mitochondria and chloroplasts.
(*iii*) Ribosomes and centrioles.
3. There will be no metabolic energy for metabolic activities and cell will die.
4. By introducing plastids, vacuoles and cell wall.
5. Chloroplast.
6. Nonliving.
7. Mitochondria and chloroplasts.

Matching of Columns

1. (*c*) **2.** (*a*) **3.** (*f*) **4.** (*b*) **5.** (*g*) **6.** (*e*)
7. (*h*) **8.** (*d*).

Multiple Choice Questions

1. (*c*) **2.** (*b*) **3.** (*a*) **4.** (*d*) **5.** (*a*) **6.** (*c*)
7. (*a*) **8.** (*b*) **9.** (*b*) **10.** (*a*) **11.** (*b*) **12.** (*a*)
13. (*c*) **14.** (*c*).

4

Viruses

Viruses (L., venoum or poisonous fluid) are very small submicroscopic biological entities which though lack cellular organization (*viz.,* plasma membrane and metabolic machinery) possess their own genetic material, genetically determined macromolecular organization and characteristic mode of inheritance. For their multiplication, they essentially require the presence of some host cell, *i.e.,* they are obligate cellular parasites of either bacteria, plants or animals.

4.1 STRUCTURE

Viruses are quite a varied group (Fig. 4.l). They range in between 30 to 300 nm or 300 to 3000 Å in size, so they can be observed only by electron microscopy and X-ray crystallography. They have a regular geometrical and macromolecular organization. Basically an infectious virus particle (called **virion**) is composed of a **core** of only one type of nucleic acid (DNA or RNA) which is wrapped in a protective coat of protein, called **capsid**. The capsid consists of numerous **capsomeres**, each having a few **monomers** or **structural units**. Each structural unit is made up of one or more polypeptide chains. The capsomerers are of different shapes such as hollow prism, hexagonal, pentagonal, lobular or any other shape. The specific arrangement of capsomeres in the capsid determines the shape of a virion. Viruses have the following three different types of symmetry:

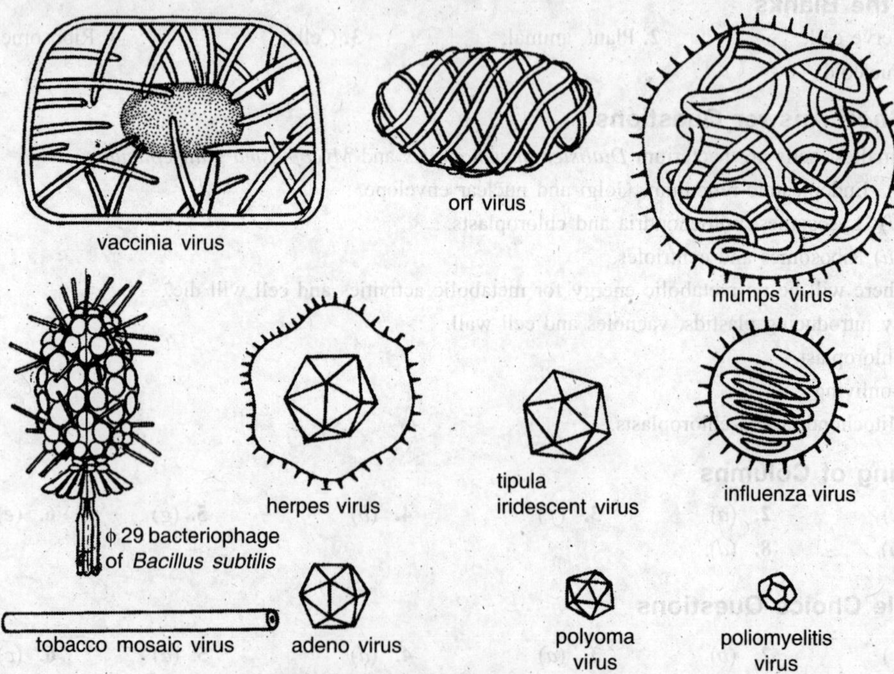

Fig. 4.1. Different kinds of viruses.

1. Icosahedral symmetry. Many viruses have spherical, cubical or polygonal shape which is basically **icosahedral** or 20-sided. Icosahedral symmetry depends on the fact that the assembly of the capsomeres causes the capsid of the virus to be at a state of minimum energy (**Caspar** and **Klug,** 1962). An icosahedral capsid comprises both **pentameres** (*i.e.,* capsomeres containing 5 structural units) and **hexameres** (*i.e.,* capsomeres having 6 structural units). In an icosahedral virus the minimum number of capsomeres is 12 or its multiple such as 32, 42, 72, 92, 162, 252, 362, 492, 642 and 812.

For example, the total number of capsomeres of different icosahedral viruses are:

(*i*) Bacteriophage ϕ (phi) × 174 = 12 pentameres;
(*ii*) Turnip yellow mosaic virus or TYMV = 32 capsomeres;
(*iii*) Poliovirus = 32 capsomeres;
(*iv*) Polyoma virus and papilloma virus = 72 capsomeres;
(*v*) Reovirus = 92 capsomeres;
(*vi*) Herpes virus = 162 capsomeres;
(*vii*) Adenovirus = 252 capsomeres; and
(*viii*) Tipula iridescent virus = 812 capsomeres.

In all of these icosahedral viruses, only 12 capsomeres are pentameres, occupying 12 corners of five-fold symmetry, while the rest are hexameres (Fig. 4.2). Since a polyhedron of 20-sided icosahedron basically has triangular faces, it is also known as **deltahedron.**

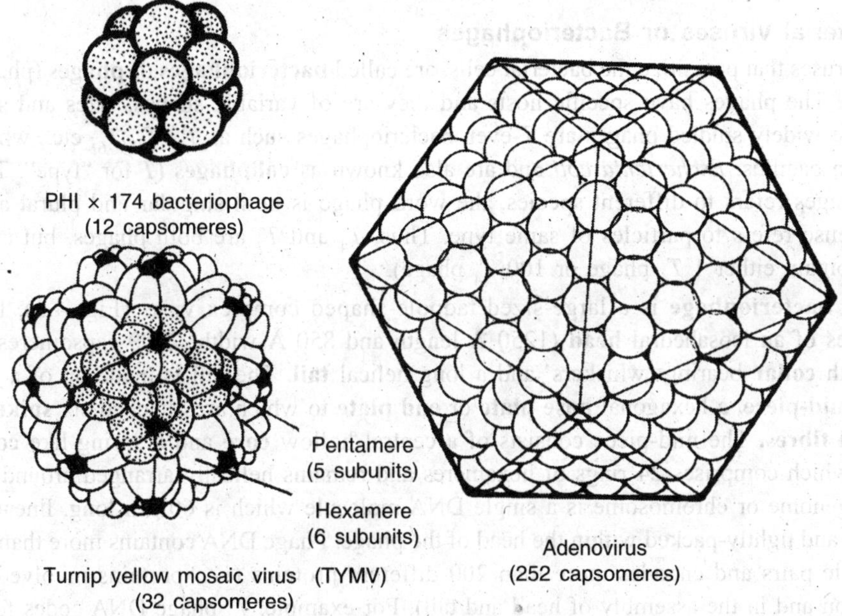

Fig. 4.2. Certain polyhedral (icosahedral) viruses.

2. Helical or cylindrical symmetry. The rod-shaped helical capsid of viruses such as tobacco mosaic virus (TMV), bacteriophage M13 and influenza virus, consists of numerous identical capsomeres arranged into a helix because they are thicker at one end than the other.

3. Complex symmetry. Viruses with complex shaped capsids are of two shapes; those without identifiable capsids (*e.g.,* pox viruses such as vaccinia, cowpox, extromelia and orf viruses) and those with tadpole-shaped structures in which each part has different sort of symmetry (*e.g.,* T-even phages of *E.coli;* T_2 phage has an icosahedral head, helical tail sheath, hexagonal end plate and rod-shaped tail fibres). Some viruses such as rabies virus are bullet-shaped.

Some viruses such as herpes virus, influenza virus, mumps virus and Semliki forest virus are surrounded by a 100–150 Å thick spiked membrane. This membrane contains lipid bilayer of plasma membrane from which projects the virus-specific protein molecules or spikes. It is not made by or specified by the virus itself but is derived from the plasma membrane of the host cell (*i.e.,* animal cell).

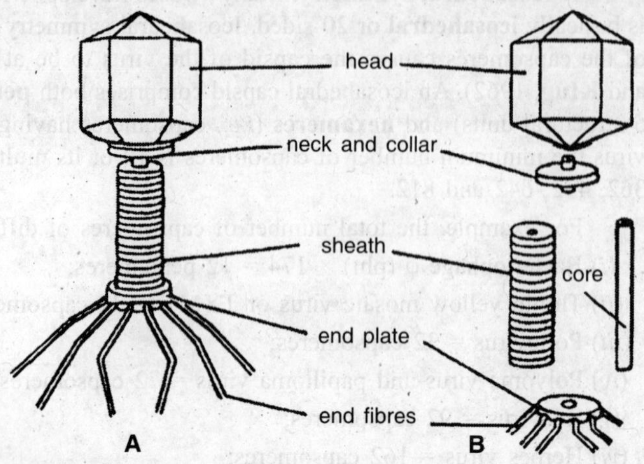

Fig. 4.3. A T_4 bacteriophage. A–External structure; B–Parts.

4.2 TYPES OF VIRUSES

Viruses are of following three types depending on the host cell.

A. Bacterial Viruses or Bacteriophages

Viruses that parasitize the bacterial cells, are called **bacteriophages** or **phages** (phage means 'to eat'). The phages have specific hosts and they are of variable shapes, sizes and structures. The most widely studied phages are *T*-even bacteriophages such as T_2, T_4, T_6, etc., which infect the colon bacillus. *Escherichia coli* and are also known as **coliphages** (*T* for "type". The plural word phages refers to different species; the word phage is both singular and plural and in the plural sense refers to particles of same type. Thus, T_4 and T_7 are both phages, but a test tube might contain either 1 T_4 phage or 100 T_4 phage).

***T_4* bacteriophage** is a large-sized tadpole-shaped complex virus (Fig. 4.3). Its capsid comprises of an icosahedral **head** (1250 Å length and 850 Å width; 2000 capsomeres), a short neck with **collar** bearing 'whiskers' and a long helical **tail.** The tail is made up of a thick and hollow **mid-piece,** a hexagonal **base plate** or **end plate** to which are attached six **spikes** and six long **tail fibres.** The mid-piece consists of a central hollow **core** and a spring-like **contractile sheath** which comprises 24 rings of hexameres and remains helically arranged around the core. The T_4 genome or chromosome is a single DNA molecule which is 60 µm long, linear, double-stranded and tightly-packed within the head of the phage. Phage DNA contains more than 1,66,000 nucleotide pairs and encodes more than 200 different proteins (*i.e.,* proteins involved in DNA replication and in the assembly of head and tail). For example, T_4 phage DNA codes for at least 30 different enzymes (*e.g.,* helicases, topoisomerases, DNA polymerases, DNA ligases, etc.) all of which ensure rapid replication of phage chromosome in preference to DNA of *E.coli* (host cell). Further, during DNA replication, an unusual nitrogen base, called **5-hydroxymethylcytosine** is incorporated in place of cytosine in the phage DNA. This unusual base makes phage DNA recognisable from that of host DNA and selectively protects it from the, nuclease enzymes. The nucleases are encoded in T_4 phage genome to degrade only the DNA of host cell. Some other phage proteins alter host cell's RNA polymerase enzymes, so that they transcribe different sets of T_4 genes at different stages of viral infection according to the phage's needs.

B. Plant Viruses

The plant viruses parasitize the plant cells and disturb their metabolism and cause severe diseases in them. All plant viruses consist of ribonucleoproteins in their organization. The important plant viruses are tobacco rattle virus (TRV), tobacco mosaic virus (TMV), potato virus, beet yellow virus (BYV), southern bean mosaic virus (SBMV) and turnip yellow viruses (TYV). Among plants, few hundred viral diseases are caused, e.g., mosaic diseases of tobacco, cabbage, cauliflower, groundnut and mustard; black-ring spot of cabbage; leaf roll of tomato; leaf curl of papaya, cotton, bean and soyabean; yellow diseases of carrot, peach; little-leaf of brinjal. These diseases are spread mainly by insects such as aphids, leaf hoppers and beetles.

Tabacco mosaid virus (TMV). TMV is the most extensively studied plant virus. It was discovered by **Iwanowski** (1892) and obtained in a pure state (*i.e.,* in paracrystalline form) by **Stanley** (1935). **Bawden** and **Pirie** (1937) extensively purified TMV and showed it to be a nucleoprotein containing RNA. **H. Fraenkel-Conrat** experimentally demonstrated that RNA is the genetic substance of TMV.

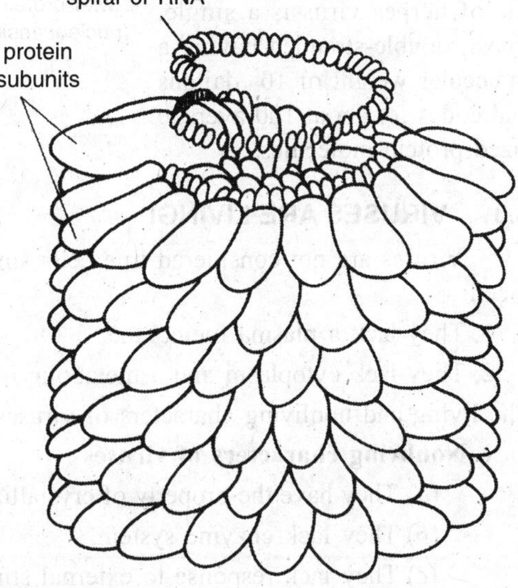

Fig. 4.4. Molecular organization of the tobacco mosaic virus (TMV).

TMV is a rod-shaped, helically symmetrical RNA virus (Fig. 4.4). Each virus particle is elongated, cigarette-like in shape having the length of 3000 Å (300 nm) and diameter of 160 Å (16 nm). In each rod of TMV, there are about 2130 identical elliptical protein subunits or capsomeres. The capsomeres are closely packed and arranged in a helical manner around the RNA helix, forming a hollow cylinder. Thus, there is a hollow core (axial hole) of about 40 Å (4nm) diameter which runs the entire length of the rod and contains the RNA molecule. The RNA molecule does not occupy the hole but is deeply embedded in the capsomeres. RNA of TMV is a single-stranded molecule consisting of 6500 nucleotides and is in the form of a long helix extending the whole length of viral particle. Lastly, there are about 16 capsomeres in each helical turn. Each capsomere contains about 158 amino acids and has a molecular weight of 18000 daltons. The whole TMV capsid has all amino acids found in other plant proteins.

C. Animal viruses. The animal viruses infect the animal cells and cause different fatal diseases in animals including humans. Generally, they have a polyhedron or spherical shape and genetic material in the form of DNA or RNA. The protein coat or capsid of animal viruses is surrounded by an envelope.

Certain common viral infections of human beings are: common cold, influenza, mumps, measles, rubella (German measles), chicken pox, small pox, polio, viral hepatitis, herpes simplex, viral encephalitis, fever blisters, warts and some types of cancer. Among livestock and fowl, viruses cause encephalitis, foot and mouth disease, fowl plague, Newcastle disease, pseudorabis, hog cholera and a variety of warts and other tumors. A virus usually displays some specificity for a particular animal group.

Poliomyelitis is a most extensively studied RNA-containing animal virus. The polio virus has comparatively very simple organization. It consists of a protein shell built up of 60 structurally equivalent asymmetric protein subunits of approximately 60 Å diameter, packed together in such a way that they form a spherical shell of about 300 Å in diameter. The shell or capsid encloses a single stranded RNA molecule of 5,200 nucleotides.

Herpes virus is another most extensively studied animal virus. It possesses a DNA containing core embedded in a regular icosahedral **capsid** (162 capso-meres) and an outer **envelope** of lipids, proteins and carbohy-drates. The DNA molecule of herpes virus is a single, linear, double-stranded having a molecular weight of 10^8 daltons and codes for about 100 average sized protein molecules.

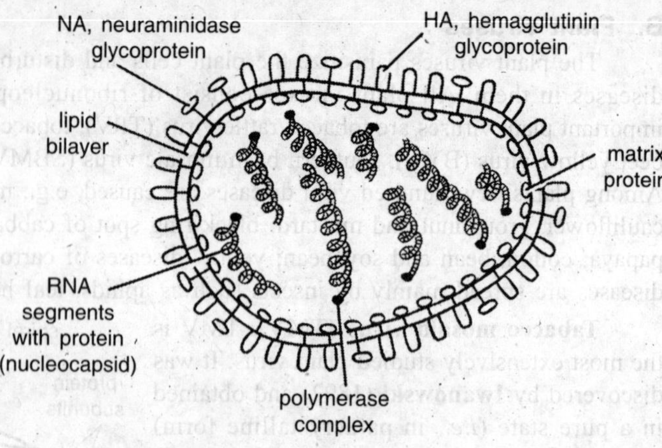

Fig. 4.5. The influenza viron.

4.3 VIRUSES ARE LIVING!

Viruses are not considered true cells since they lack two very important characteristics of a cell:

1. They lack a plasma membrane;
2. They lack cytoplasm and a metabolic machinery to sustain life activities.

The living and nonliving characters of viruses are summarized below:

1. Nonliving characters of viruses

(*a*) They have the property of crystallization like other inorganic and organic compounds.

(*b*) They lack enzyme system.

(*c*) They lack response to external stimuli.

(*d*) They lack growth.

2. Living characters of viruses

(*a*) Viruses have the capacity to multiply (inside a cellular host).

(*b*) They have the property of undergoing mutations.

(*c*) They have the property of recombination and inheritability.

4.4 LIFE CYCLE OF THE BACTERIOPHAGE

Bacteriophages may have the following two types of life cycles:

1. **Lytic cycles**, in which viral infection is followed by **lysis** (bursting and death) of the host cell and release of new infective phages, *e.g.*, virulent phages such as T_4 and all other *T*-even coliphages.
2. **Lysogenic cycles**, in which infection rarely causes lysis, *e.g.*, temperate phages such as P_1 and lambda (λ) phages.

1. Lytic Cycle of a Virulent Phage

Life cycle of a T_4 bacteriophage (Fig. 4.6) involves the following steps:

(*a*) **Attachment** or **adsorption** of phage to bacterial (host) cell.

(*b*) **Injection** or **penetration** of viral genetic material (DNA) into the host cell.

(*c*) **Eclipse period,** during which synthesis of new phage DNA and protein coats takes place.

(*d*) **Assembly** of phage DNA into protein coats.

(*e*) **Lysis** of host cell and release of the infective progeny phages. Such a phage is called **virulent or lytic phage** since it has infectiousness and it causes death of host cell by lysis.

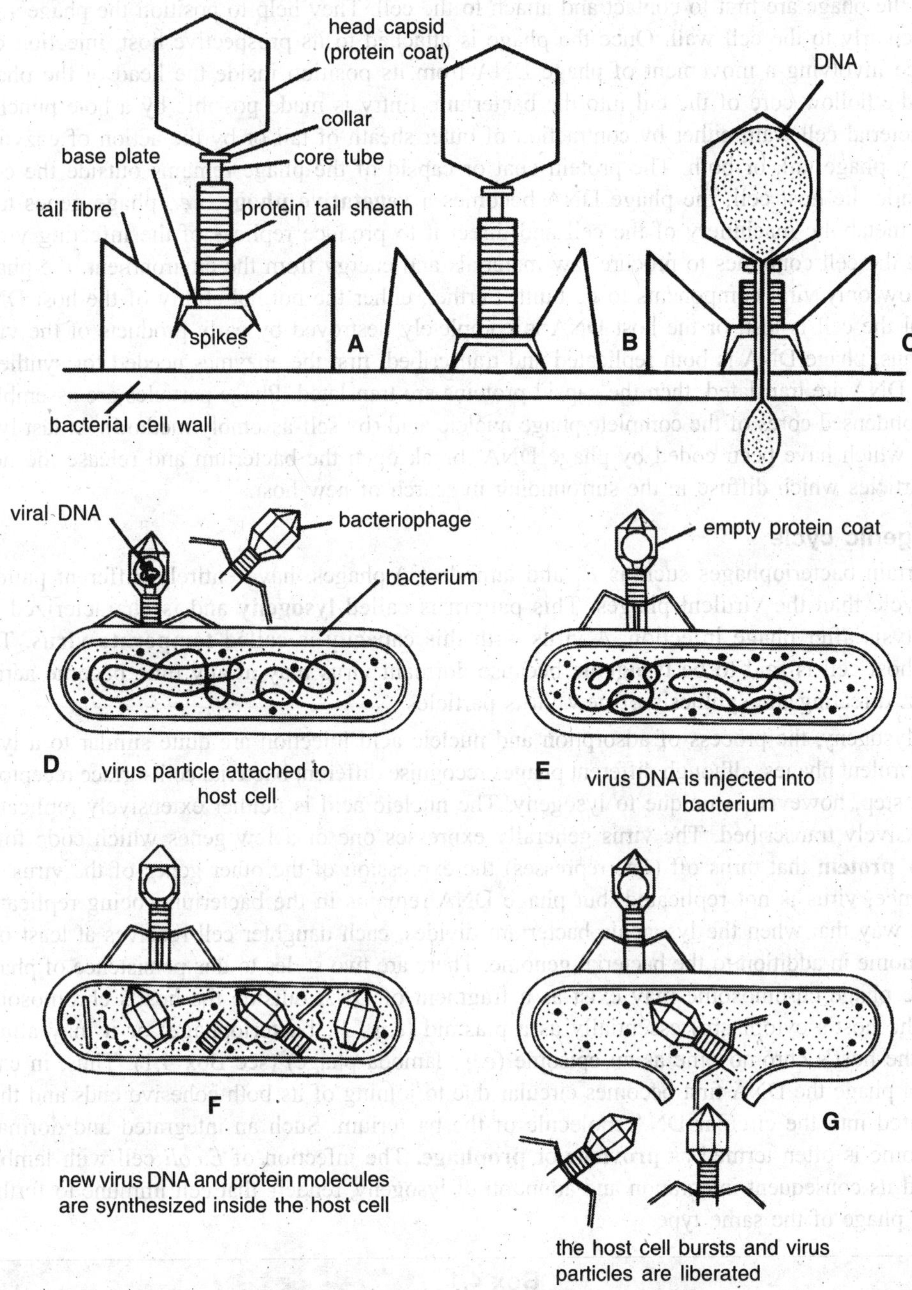

Fig. 4.6. A, B, C—Mode of attachment of a *T*-even phage (bacteriophage) on a bacterial cell wall and injection of DNA into the bacterium (host cell); D, E, F and G—Steps of viral reproduction inside the host cell.

The adsorption of the phage to its host is made possible by a reaction of chemical groups on the two during a random collision. Reactive groups (called **adsorption protein** or **pilot protein; Kornberg**, 1974) at the end of the tail of the phage can join with a complementary set of chemical groups (a **receptor site**) in the cell wall of the bacterium. During adsorption, long tail fibres of the phage are first to contact and attach to the cell. They help to position the phage's tail perpendicularly to the cell wall. Once the phage is attached to its prospective host, injection can take place involving a movement of phage DNA from its position inside the head of the phage through the hollow core of the tail into the bacterium. Entry is made possible by a hole punched in the bacterial cell wall, either by contraction of outer sheath of tail or by the action of enzymes carried by phage tail, or both. The protein coat or capsid of the phage remains outside the cell. Once inside the host cell, the phage DNA becomes a **vegetative phage,** *i.e.,* phage genes take over the metabolic machinery of the cell and direct it to produce replicas of the infecting virus. Although the cell continues to procure raw materials and energy from the environment, the phage genes allow only viral components to be built. Further, either the normal ability of the host DNA to control the cell is lost, or the host DNA is completely destroyed by early products of the viral genes. Thus, phage DNA is both replicated and transcribed; first the enzymes needed for synthesis of phage DNA are translated, then the capsid proteins are translated. Phage particles are assembled around condensed cores of the complete phage nucleic acid (by self-assembly method). At last lytic enzymes which have been coded by phage DNA, break open the bacterium and release the new phage particles which diffuse in the surrounding in search of new host.

2. Lysogenic cycle

Certain bacteriophages such as P_1 and lambda (λ) phages, have entirely different pattern of life cycle than the virulent phages. This pattern is called **lysogeny** and is characterized by delayed lysis after phage infection. A virus with this capacity is called **temperate virus.** The infected host cell is said to be **lysogenic** because dormant virus may at any time become active and begin directing the synthesis of new virus particles.

In lysogeny, the process of adsorption and nucleic acid injection are quite similar to a lytic cycle of virulent phages, although different phages recognise different bacterial cell surface receptors. The next step, however, is unique to lysogeny. The nucleic acid is neither extensively replicated nor extensively transcribed. The virus generally expresses one or a few genes which code for a **repressor protein** that turns off (*i.e.,* represses) the expression of the other genes of the virus. In consequence, virus is not replicated, but phage DNA remains in the bacterium, being replicated in such a way that when the lysogenic bacterium divides, each daughter cell receives at least one phage genome in addition to the bacterial genome. There are two styles to this persistence of phage DNA: the phage chromosome may exist as a fragment of DNA outside the host's chromosome (*i.e.,* in the host's cytoplasm) essentially as a plasmid (*e.g.,* P_1 bacteriophage) or it may attach itself to the host's chromosome as an episome (*e.g.,* lambda phage) (see Box 4.1). Thus, in case of lambda phage the DNA first becomes circular due to joining of its both cohesive ends and then is integrated into the circular DNA molecule of the bacterium. Such an integrated and dormant viral genome is often termed as **provirus** or **prophage.** The infection of *E.coli* cell with lambda phage and its consequent integration and adoption of lysogeny, renders that cell immune to further attack by phage of the same type.

Box 4.1

Episome is an extra-chromosomal, circular, transposable, closed DNA molecule which can exist either integrated into the bacterial chromosome or separately and autonomously in the cytoplasm. **Plasmid** is that bit of autonomous genetic material of bacteria (*i.e.,* circular DNA) that exists only extra-chromosomally and cannot be integrated into the bacterial chromosome (DNA). Now only the term plasmid is used for all kinds of extra-chromosomal autonomous transposable circular DNA fragments.

The calm lysogenic period is ended by some type of shock (*e.g.*, temperature changes, UV irradiation or conjugation of a lysogenic bacterium with a non-lysogenic bacterium) to the lysogenic culture. A shock evidently inactivates the repressor of the phage so that all the phage genes can be expressed. Then the lysogenic bacterium is ruined, for the phage DNA that the host bacterium is harbouring enters the lytic phase. It replicates, transcribes, translates, assembles viron particles and lyses the bacterium (Fig. 4.7).

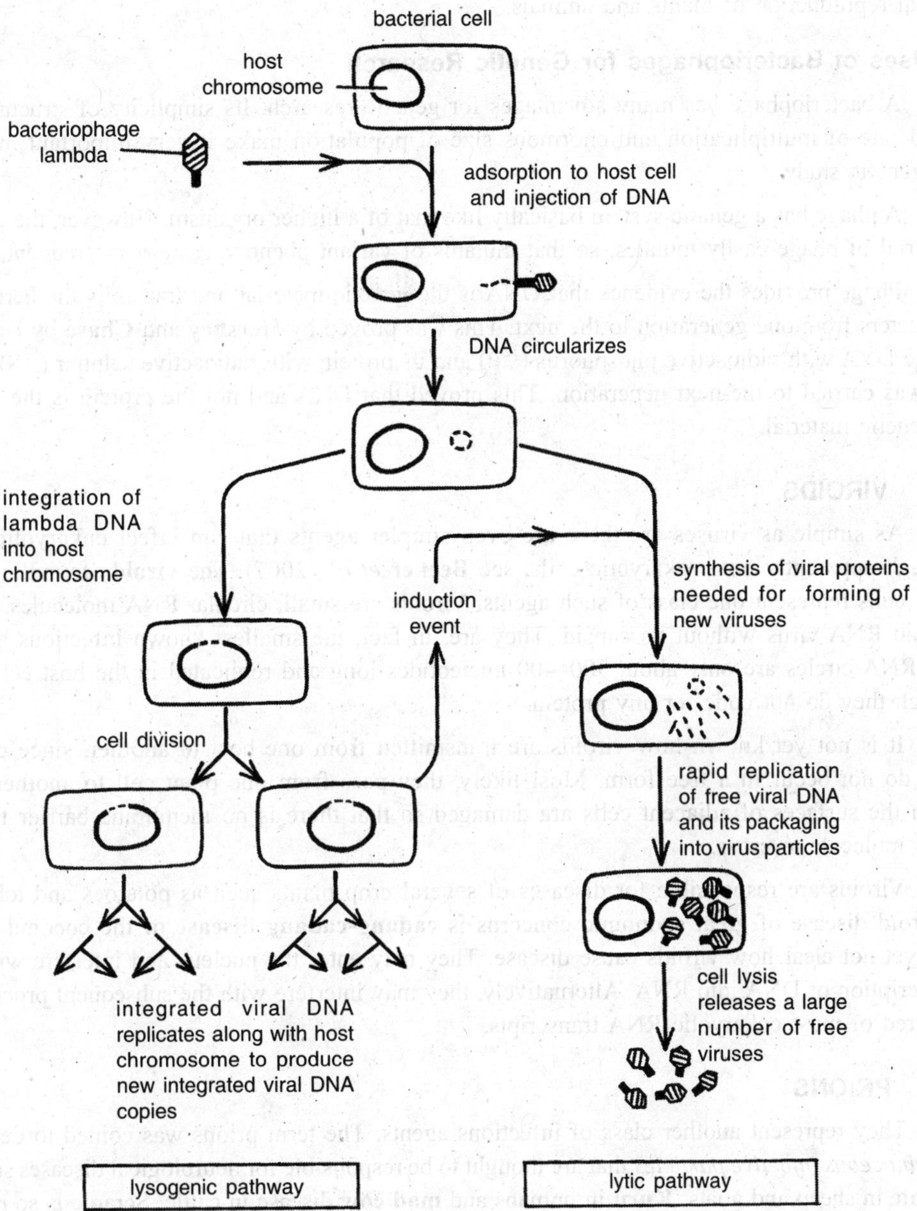

Fig. 4.7. The life cycle of bacteriophage lambda. Its double stranded DNA can exist in both linear and circular forms. The lambda phage can multiply by either a lytic or a lysogenic pathway in the *E.coli*.

3. Recombination in Phages

Recombination is any process generating new combinations of pre-existing genetic material. Recombination in phages was described by **Max** and **Mary Delbruck** (1894). When two different types of phages T_2 and T_4 are placed in a culture of bacteria which is sensitive to both, the new generation of phages consists of the two parental types. But some of them were found to have combination of T_2 and T_4 genes. The recombination in bacteriophages can be equated to the sexual reproduction of plants and animals.

4. Uses of Bacteriophages for Genetic Research

A bacteriophage has many advantages for genetic research. Its simplicity of structure and rapid rate of multiplication and enormous size of population make it very important material for genetic study.

A phage has a genetic system basically like that of a higher organism. However, the genetic material of phage easily mutates, so that mutants of variant phenotype are very frequent.

Phage provides the evidence that DNA is the genetic material and transmits the hereditary characters from one generation to the next. This was proved by **Hershey** and **Chase** by labelling phage DNA with radioactive phosphorus (^{32}P) and its protein with radioactive sulphur (^{53}S). Only ^{32}P was carried to the-next generation. This proved that DNA and not the protein is the carrier of genetic material.

4.5 VIROIDS

As simple as viruses are there are even simpler agents that can infect eukaryotic cells (though apparently not prokaryotic cells; see **Becker** *et al.*, 2007). The **viroids** found in some plant cells represent one class of such agents. Viroids are small, circular RNA molecules, rather like an RNA virus without its capsid. They are, in fact, the smallest known infections agents. The RNA circles are only about 300–400 nucleotides long and replicated in the host cell even though they do not code for any protein.

It is not yet known how viroids are transmitted from one host to another, since clearly they do not occur in a free form. Most likely, they pass from one plant cell to another only when the surfaces of adjacent cells are damaged so that there is no membrane barrier for the RNA molecule to cross.

Viroids are responsible for diseases of several crop plants such as potatoes and tobacco. A viroid disease of great economic concerns is **cadang-cadang disease** of the coconut palm. It is yet not clear how viroids cause disease. They may enter the nucleus and interfere with the transcription of DNA into RNA. Alternatively, they may interfere with the subsequent processing required of most enkaryotic RNA transcripts.

4.6 PRIONS

They represent another class of infections agents. The term prions was coined to describe *proteinaceous infective particles* that are thought to be responsible for neurological diseases such as **scrapie** in sheep and goats, **Kuru** in humans and **mad cow** disease in cattle. Scrapie is so named because infected animals rub incessantly against trees or other objects, scraping off most of their wool in the process. Kuru is a degenerative disease of the central nervous system originally reported among native peoples in New Guinea patients with this or other prion-based disease suffer initially from mild physical weakness and dementia, but the effects slowly become more severe, and the diseases are eventually fatal. Scrapie like brain-wasting disorder occurs in case of mad cow disease. **Stanley Prusiner**, who received a Nobel prize in 1997 for his pioneering work in this field, has proposed that these diseases are transmitted by infectious protein-containing particles called **prions**. Because prions do not appear to contain DNA or RNA, **Prusiner** has formulated a unique theory to explain how prions might transmit disease by causing the infections spread of abnormal **protein folding**. According to this theory, a prion protein (designated Pr Psc) is simply a misfolded version of a normal cellular protein (designated Pr Pc). When the misfolded Pr Psc encounters a normal Pr Pc peptide chain in the process of folding, it causes the normal polypeptide to hold improperly. The resulting, abnormally folded protein triggers extensive nerve cell damage in the brain, leading to uncontrolled muscle movements and eventual death. Both the normal and variant forms of prion protein are found on the surfaces of neurons, suggesting that the protein may somehow affect the receptors that detect nerve signals.

QUESTIONS

Long Answer Questions

1. Describe general organisation of viruses. Why are viruses considered both living as well as non-living?

Short Answer Questions

1. Why viruses are not true cells?
2. Write short notes on the following:
   (*i*) Bacteriophage
   (*ii*) Reproduction in bacteriophage
3. Draw a neat and labelled diagram of the following:
   (*i*) T_4 bacteriophage

Very Short Answer Questions

1. Define the virus.
2. What is viroid?

Fill in the Blanks

1. Viruses are obligate .... of either ...., plants or animals.
2. Capsid of herpes virus has .... capsomeres.
3. TMV has rod-shaped .... symmetry.
4. The capsid of head of T_4 bacteriophage has .... symmetry.
5. TMV is a .... virus and has about .... identical elliptical capsomeres in its capsid.
6. Lambda phage have .... life cycle.

Multiple Choice Questions

1. AIDS causing factors are associated with

(*a*) protozoa (*b*) DNA virus
(*c*) bacteria (*d*) RNA virus

2. Viruses are living because they

(*a*) multiply in the host cells (*b*) carry anaerobic respiration

(*c*) carry metabolic activity (*d*) cause infection

3. Virus parasite in bacteria is

(*a*) bacteriophage (*b*) viroid

(*c*) cyanophage (*d*) mycoplasma

4. Number of capsomeres in TMV is

(*a*) 231 (*b*) 1230

(*c*) 2130 (*d*) 3120

5. Polio virus has

(*a*) ss DNA (*b*) ds DNA

(*c*) ss RNA (*d*) ds RNA

6. T_2 has

(*a*) ds DNA (*b*) ss RNA

(*c*) ds RNA (*d*) ss DNA

7. When a viral DNA integrates with a host DNA, then the process is referred to as

(*a*) lysis (*b*) prophase

(*c*) lysogeny (*d*) none of these

ANSWERS

Very Short Answer Questions

1. Virus is a subcellular parasite composed of a protein coat and DNA or RNA incapable of independent existence; it invades and infects cells and redirects the host cell's synthetic machinery toward the production of more virus.

2. Viroid is a small, circular RNA molecule that can infect and replicate in host cells even though it does not code for any protein (see **Becker** *et al.*, 2007).

Fill in the Blanks

1. Cellular parasites, bacteria. **2.** 162.

3. Helical. **4.** Icosahedral.

5. RNA, 2130. **6.** Lysogenic.

Multiple Choice Questions

1. (*d*) **2.** (*a*) **3.** (*a*) **4.** (*c*) **5.** (*c*) **6.** (*a*)
7. (*c*).

5

Cytosol and Biomolecules

Within the cells of any organism, the living substance, or **protoplasm,** is itself comprised of a multitude of non-living constituents: proteins, nucleic acids, fats (lipids), carbohydrates, vitamins, minerals, waste metabolites, crystalline aggregates, pigments, and many others, all of which are composed of molecules and their constituent atoms. The *protoplasm is alive because of the highly complex organization of these non-living substances and the way they interact with one another*.

5.1 PHYSICAL NATURE OF CYTOSOL

The fluid and soluble portion of the cytoplasm that exists outside the organelles is called **cytoplasmic matrix** or **cytosol.** The cytosol (cytoplasmic matrix) is a colourless or grayish, translucent, viscid, gelatinous or jelly-like **colloidal** substance. It is heavier than water and capable of flowing. In past, there has been a lot of controversy about the physical nature of the matrix. Different workers advanced different theories about the physical characteristics of the matrix. Their theories can be represented as follows:

1. Reticular theory suggests that the matrix is composed of reticulum of fibres or particles in the ground substances (Fig. 5.1A)

2. Alveolar theory was proposed by **Butschili** in 1892 and according to it, the matrix consists of many suspended droplets or alveoli or minute bubbles resembling the foams of emulsion (Fig. 5.1B).

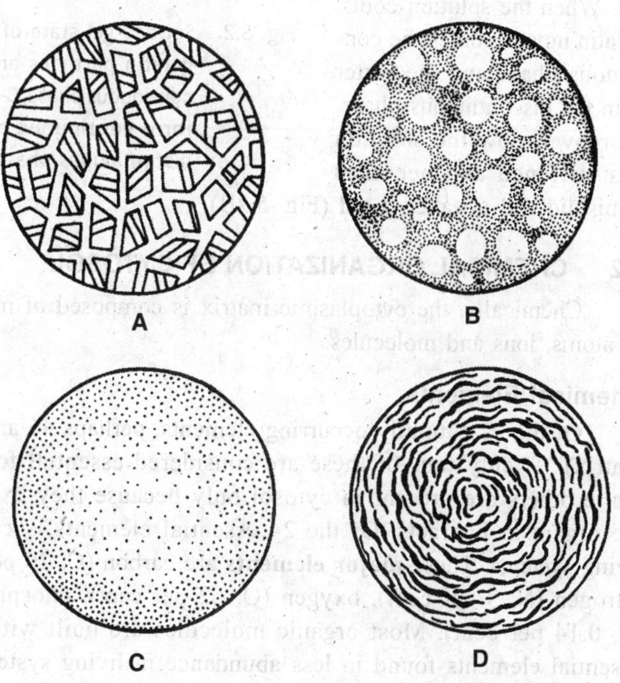

Fig. 5.1. Physical appearance of protoplasm. A—Reticular, B—Alveolar, C—Granular, and D—Fibrillar.

3. Granular theory was propounded by **Altmann** in 1893. This theory supports the view that the matrix contains many granules of smaller and larger size arranged differently. These granules were known as bioplasts (Fig. 5.1C).

4. Fibrillar theory was proposed by **Fleming** and it holds that the matrix is fibrillar in nature (Fig. 5.1D).

5. Colloidal theory has been forwarded recently after the electron microscopical observations of the matrix. According to this concept, the matrix is partly a true **solution**, partly a **colloidal system.**

Phase Reversal

Cytosol (cyto-plasmic matrix) like many colloidal systems, shows the property of **phase reversal**. For example, gelatin particles (discontinuous phase) are dispersed through water (continuous phase) in a thin consistency that is freely shakable (Fig. 5.2A). Such a condition is called a **sol**. When the solution cools, gelatin now becomes the continuous phase and the water is in the discontinuous phase. Moreover, now the solution has stiffened and becomes semisolid and is called a **gel** (Fig. 5.2B).

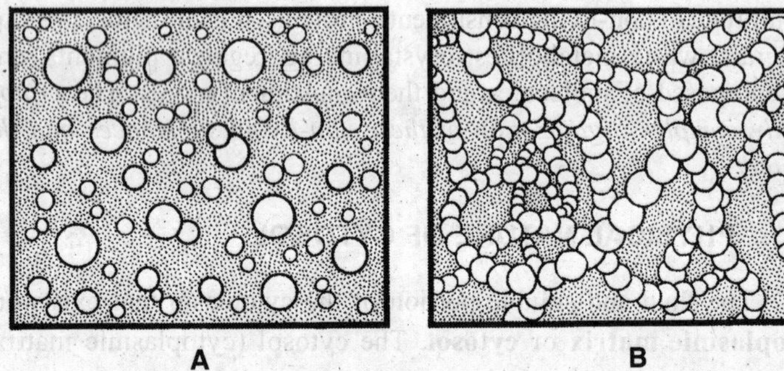

Fig. 5.2. Sol and gel state of the cytosol. A—Sol condition in which gelatin particles are the discontinuous phase, water the continuous phase; B—Gel condition in which gelatin particles form continuous phase (net work), enclosing water as discontinuous phase.

5.2 CHEMICAL ORGANIZATION OF CYTOSOL

Chemically, the cytoplasmic matrix is composed of many chemical elements in the form of atoms, ions and molecules.

Chemical Elements

Of the 92 naturally occurring elements, perhaps 46 are found in the cytosol (cytoplasmic matrix). Twenty four of these are considered essential for life (called **essential elements**), while others are present in cytosol only because they exist in the environment with which the organism interacts. Of the 24 essential elements, six play especially important roles in living systems. These **major elements** are carbon (C, 20 per cent), hydrogen (H, 10 per cent), nitrogen (N, 3 per cent), oxygen (O, 62 per cent), phosphorus (P, 1.14 percent) and sulphur (S, 0.14 per cent). Most organic molecules are built with these six elements. Another five essential elements found in less abundance in living systems are calcium (Ca, 2.5 per cent), potassium (K, 0.11 per cent), sodium (Na, 0.10 per cent), chlorine (C1, 0.16 per cent) and magnesium (Mg, 0.07 per cent). Several other elements, called **trace elements,** are also found in minute amounts in animals and plants, but are nevertheless essential for life. These are iron (Fe, 0.10 per cent), iodine (I, 0.014 per cent), molybdenum (Mo), manganese (Mn), Cobalt (Co), zinc (Zn), selenium (Se), copper (Cu), chromium (Cr), tin (Sn), vanadium (V), silicon (Si), nickel (Ni), fluorine (F) and boron (B).

Ions

The cytoplasmic matrix consists of various kinds of ions. The ions are important in maintaining osmotic pressure and acid-base balance in the cells. Retention of ions in the matrix produces an increase in osmotic pressure and, thus, the entrance of water in the cell. The concentration of various ions in the intracellular fluid (matrix) differs from that in the interstitial fluid. For example, in the cell K^+ and Mg^{++} can be high, and Na^+ and Cl^- high outside the cell. In muscle and nerve cells a high order of difference exists between intracellular K^+ and extracellular Na^+. Free calcium ions (Ca^{++}) may occur in cells or circulating blood. Silicon ions occur in the epithelium cells of grasses. The free ions of phosphate (primary, $H_2PO_4^-$ and secondary, HPO_4^-) occur in the matrix and blood. These ions act as a buffering system and tend to stabilize pH of blood and cellular fluids. The ions of different cells also include sulphate (SO_4^-), carbonate (CO_3^-), bicarbonate (HCO_3^-), magnesium (Mg^{++}) and amino acids. Cellular functions of certain ions have been tabulated in Table 5.1.

Table 5.1 Cellular functions of certain ions.

	Element	*Ionic form present*	*Functions*
1.	Molybdenum	MoO_4^{2-}	Cofactor or activator of certain enzymes (*e.g.*, nitrogen fixation, nucleic acid metabolism, aldehyde oxidation).
2.	Cobalt	Co^{2+}	Constituent of vitamin B_{12}.
3.	Copper	Cu^+, Cu^{2+}	Constituent of plastocyanin and cofactor of respiratory enzymes.
4.	Iodine (Heaviest trace element)	I^-	Constituent of thyroxin, triiodothyronine and other thyroid hormones
5.	Boron	BO_3^{3-}, $B_4O_7^{2-}$	Activates arabinose isomerase.
6.	Zinc	Zn^{2+}	Cofactor of certain enzymes (*e.g.*, carbonic anhydrase, carboxypeptidase).
7.	Manganese	Mn^{2+}	Cofactor of certain enzymes (*e.g.*, several kinases, isocitric decarboxylase).
8.	Iron	Fe^{2+}, Fe^{3+}	Constituent of haemoglobin, myoglobin and cytochromes.
9.	Magnesium	Mg^{2+}	Constituent of chlorophyll; activates ATPase enzyme.
10.	Sulphur	SO_4^{2-}	Constituent of coenzyme A, biotin, thiamine, proteins.
11.	Phosphorus	PO_4^{3-}, $H_2PO^-_4$	Constituent of lipids, proteins, nucleic acids, sugar phosphates, nucleoside phosphates.
12.	Calcium	Ca^{2+}	Constituent of plant cell walls; matrix component of bone tissue; cofactor of coagulation enzymes.
13.	Potassium	K^+	Cofactor for pyruvate kinase and K^+– stimulated ATPase.

Types of Compounds of Cytosol

Chemical compounds are conventionally divided into two groups: **organic** and **inorganic**. Organic compounds form 30 per cent of a typical cell, rest are the inorganic substances such as water and other substances.

Table 5.2 The approximate percentage composition of the human body.

	Substance	*Percentage*
1.	Water	65
2.	Protein	18
3.	Fat	10
4.	Carbohydrate	5
5.	Other organic	1
6.	Inorganic	1

I. Inorganic Compounds

The inorganic compounds are those compounds which normally found in the bulk of the physical, non-living universe, such as elements, metals, non-metals, and their compounds such as water, salts and variety of electrolytes and non-electrolytes. In the previous section, we have discussed a lot about the inorganic substances except the water which will be discussed in the following paragraph.

Water. The most abundant inorganic component of the cytosol is the water (the notable exceptions are seeds, bone and enamel). Water constitutes about 65 to 80 per cent of the matrix. In the matrix the water occurs in two forms, *viz.,* **free water** and **bound water**. The 95 per cent of the total cellular water is used by the matrix as the solvent for various inorganic substances and organic compounds and is known as **free water**. The remaining 5 per cent of the total cellular water remains loosely linked with protein molecules by hydrogen bonds or other forces and is known as **bound water**.

The water contents of the cellular matrix of an organism depend directly on the age, habitat and metabolic activities. For instance, the cells of the embryo have 90 to 95 per cent water which decreases progressively in the cells of the adult organism. The cells of lower aquatic animals contain comparative high percentage of the water than the cells of higher terrestrial animals. Further the percentage of water in the matrix also varies from cell to cell according to the rate of the metabolism.

Unique physical properties of water and their biological utility. There are several extraordinary properties of water that make it especially fit for its essential role in the protoplasmic systems (*i.e.,* cytosol or matrix). Some of the unique properties of water are the following:

1. Water as a solvent. Water is most stable yet versatile of all solvents. Water's properties as a solvent for inorganic substances as mineral ions, solids, etc., and organic compounds such as carbohydrates and proteins, depend on water's dipole nature. Because of this polarity, water can bind electrostatically to both positively and negatively charged groups in the protein. Thus, each amino group in a protein molecule is capable of binding 2.6 molecules of water. The solvency is of great biological importance because all the chemical reactions that take place in the cells do so in aqueous solution. The water also forms the good dispersion medium for the colloidal system of the matrix.

2. Water's thermal properties. Water is the only substance that occurs in nature in the three phases of solid, liquid and vapour within the ordinary range of earth's temperatures. Water has a **high specific heat:** it requires 1 calorie (4.185 joules) to elevate the temperature of 1 gram of water by 1°C (such as from 15 to 16°C). Such a high thermal capacity of water has a great moderating effect on environmental temperature changes and is a great protective agent for all life.

Water also has a **high heat of vaporization.** It requires more than 540 calories (2259 joules) to change 1 gram of liquid water into water vapour. Thus, water tends to have a remarkably high boiling point (100°C) for a substance of such low relative molecular mass. Were it not for this lucky accident, it is likely that liquid water would never have existed on earth and would have been lost to outer space. Further, for terrestrial plants and animals, cooling produced by the evaporation of water is an important means of getting rid of excess heat. Moreover, at the other temperature extreme, large amounts of energy (335 joules or 80 cal per gram) must be lost for water to be converted from the liquid to the solid state. This is called **heat of fusion.** Water's melting point being 0°C.

Another important property of water from a biological standpoint is its unique **density behaviour** during change of temperature. Most liquids become continually more dense with cooling. Water, however, reaches its maximum density at 4°C and then becomes lighter with further cooling. Therefore, ice floats rather than settling on the bottom of lakes and ponds. This protects the aquatic life from freezing.

3. Surface tension. Water has a **high surface tension**. This property, caused by the great cohesiveness of water molecules, is important in the maintenance of protoplasmic form and movement. Despite its high surface tension, water has **low viscosity,** a property that favours the movement of blood through minute capillaries and of cytoplasm inside cellular boundaries.

Molecules dissolved in water, lower its surface tension and tend to collect at the interface between its liquid phase and other phases. This may have been important in the development of the plasma membrane, and certainly plays an important role in the movement of molecules across it.

4. Transparency. The water is transparent to light, enabling the specialized photosynthetic organelles, the chloroplast, inside the plant cell to absorb the sunlight for the process of photosynthesis.

II. Organic Compounds (Micromolecules and Macromolecules)

The chemical substances which contain carbon (C) in combination with one or more other elements as hydrogen (H), nitrogen (N), sulphur (S), etc., are called **organic compounds.** The organic compounds usually contain large molecules which are formed by the similar or dissimilar unit structure known as the **monomers.** A monomer (Gr., *mono* = one, *meros* = part) is the simplest unit of the organic molecule which can exist freely. Some organic compounds such as carbohydrates occur in the matrix as the monomers. The monomers usually link with other monomers to form **oligomers** (Gr., *oligo* = few or little, *meros* = part) and **polymers** (Gr., *poly* = many, *meros* = part). The oligomers contain small number of monomers, while the polymers contain large number of monomers. The oligomers and polymers contain large-sized molecules or macromolecules. When a polymer contains similar kinds of monomers in its macromolecule it is known as **homopolymer** and when the polymer is composed of different kinds of monomers it is known as the **heteropolymer**.

The main organic compounds of the matrix are the carbohydrates, lipids, proteins, vitamins, hormones and nucleotides.

Carbohydrates. The carbohydrates (L., *carbo* = carbon or coal, Gr., *hydro* = water) are the compounds of the carbon, hydrogen and oxygen. They form the main source of the energy of all living beings. Only green part of plants and certain microbes have the power of synthesizing the carbohydrates from the water and CO_2 in the presence of sunlight and chlorophyll by the process of photosynthesis. All the animals, non-green parts of the plants (*viz.,* stem, root, etc.), non-green plants (*e.g.,* fungi), bacteria and viruses depend on green parts of plants for the supply of carbohydrates.

Chemically the carbohydrates are polyhydroxy aldehydes or ketones and they are classified as follows:

A. Monosaccharides (Monomers),

B. Oligosaccharides (Oligomers), and

C. Polysaccharides (Polymers).

A. **Monosaccharides.** The monosaccharides are the simple sugars with the empirical formula $Cn\ (H_2O)n$. They are classified and named according to the number of carbon atoms in their molecules as follows:

(*i*) Trioses contain three carbon atoms in their molecules, *e.g.,* glyceraldehyde and dihydroxy acetone.

***(ii)* Tetroses** contain four carbon atoms in their molecules, *e.g.*, erythrulose, and erythrose.

***(iii)* Pentoses** contain five carbon atoms in their molecules, *e.g.*, ribose, ribulose, deoxyribose, arabinose and xylulose.

***(iv)* Hexoses** contain six carbon atoms in their molecules, *e.g.*, glucose, mannose, fructose and galactose.

***(v)* Heptoses** contain seven carbon atoms in their molecules, *e.g.*, sedoheptulose.

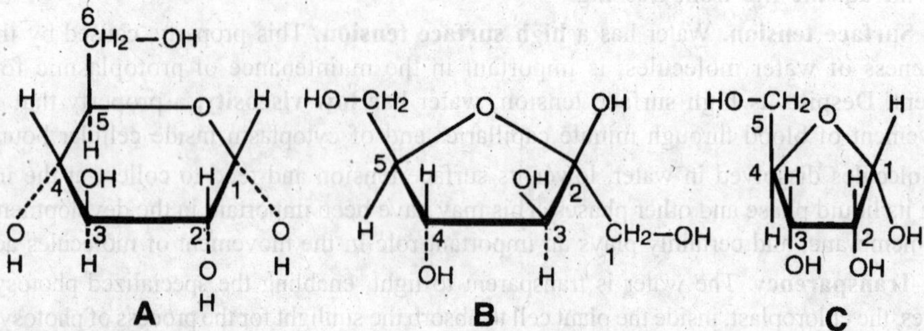

Fig. 5.3. Ring structures of monosaccharides A—Glucose; B—Fructose; C—Ribose.

The monosaccharides are the monomers and cannot split further or hydrolysed into the simpler compounds. The pentoses and hexoses are the most abundantly occurring monosaccharides of the matrix.

The pentose sugar, **ribose** is the important constituent molecule of the ribonucleic acid (RNA) and certain coenzymes as nicotinamide adenine dinucleotide (NAD), NAD phosphate (NADP), adenosine triphosphate (ATP) and coenzyme A (CoA). Another pentose sugar the **deoxyribose** is the important constituent of the deoxyribonucleic acid (DNA). The **ribulose** is a pentose sugar which is necessary for photosynthetic mechanism.

The **glucose**, a hexose sugar, is the primary source of the energy for the cell. The other important hexose sugars of the matrix are the fructose and galactose.

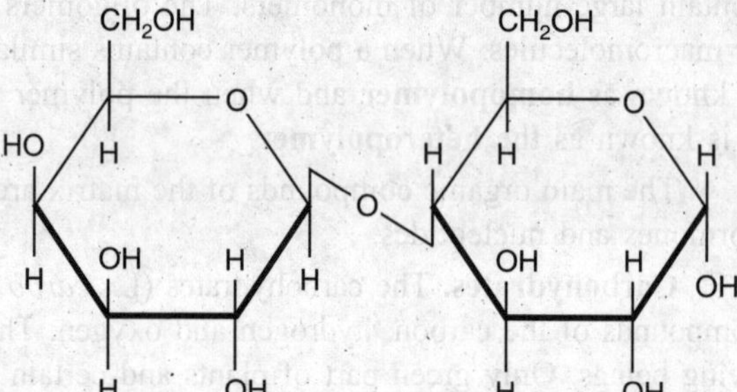

Fig. 5.4. Chemical formula of lactose.

B. Oligosaccharides. The oligosaccharides consist of 2 to 10 monosaccharides (monomers) in their molecules. The monomers remain linked with each other by the **glycosidic bonds** or **linkages**. Certain important oligosaccharides are as follows:

***(i)* Disaccharides** contain two monomers, *e.g.*, sucrose, maltose, lactose, etc.

***(ii)* Trisaccharides** contain three monomers, *e.g.*, raffinose, mannotriose, rabinose, rhaminose, gentianose and melezitose.

***(iii)* Tetrasaccharides** contain four monomers, *e.g.*, stachyose, and scordose.

***(iv)* Pentasaccharides** contain five monomers, *e.g.*, verbascose.

The most abundant oligosaccharides of the animal and plant cells are the disaccharides such as sucrose, maltose and lactose. The sucrose and maltose occur mainly in the matrix of plant cells, while the lactose occurs exclusively in the matrix of animal cells. The molecules of sucrose are composed of D-glucose and D-fructose. The molecules of maltose consist of two molecules of

D-glucose. The molecules of lactose are composed of two monomers, *viz.*, D-glucose and D-galactose. Like monosaccharides all disaccharides are sweet, soluble in water and crystallizable.

C. Polysaccharides. The polysaccharides are composed of ten to many thousands monosaccharides as the monomers in their macromolecules. Their empirical formula is $(C_6H_{10}O_6)$ *n*. The molecules of the polysaccharides are of colloidal size having high molecular weights. The polysaccharides can be hydrolysed into simple sugars.

Polysaccharides can be divided into two main functional groups: the structural polysaccharides and the nutrient polysaccharides. The **structural polysaccharides** serve primarily as extracellular or intracellular supporting elements. Included in this group are **cellulose** (found in plant cell wall), **mannan** (a homopolymer of mannose found in yeast cell walls), **chitin** (in the exoskeleton of arthropods and the cell walls of most fungi and some green algae), **hyaluronic acid, keratin sulphate** and **chondroitin sulphate** (these three are found in cartilage and other connective tissues) and the **peptidoglycans** (in bacterial cell wall). The **nutrient polysaccharides** serve as reserves of monosaccharides and are in continuous metabolic turnover. Included in this group are **starch** (plant cells and bacteria), **glycogen** (animal cells), **inulin** (plants such as artichokes and dandelions) and **paramylum** (an unbranched nutrient and storage homopolymer of glucose found in certain protozoa, *e.g., Euglena*).

Molecules of some polysaccharides are **unbranched** (*i.e.*, linear) chains whose structure may be ribbon-like or helical (usually a left-handed spiral). Other polysaccharides are **branched** and, like many proteins, assume a globular form.

On the chemical basis, the polysaccharides can be divided into two broad classes: the homopolysaccharides and the heteropolysaccharides.

Homopolysaccharides. The homopolysaccharides contain similar kinds of monosaccharides in their molecules. The most important homopolysaccharides of the matrix are the starch, glycogen, paramylum and cellulose.

Heteropolysaccharides. The polysaccharides which are composed of different kinds of the monosaccharides and amino-nitrogen or sulphuric or phosphoric acids in their molecules are known as heteropolysaccharides. The most important heteropolysaccharides are as follows:

(*a*) Hyaluronic acid, keratin sulphate and chondroitin sulphate. Cartilage tissue contains the related acidic heteropolysaccharides such as hyaluronic acid, keratin sulphate and chondroitin sulphate. **Hyaluronic acid** is an unbranched heteropolysaccharide containing repeating disaccharides of **N-acetylglucosamine** (or D-glucosamine) and glucuronic acid. In addition to cartilage, hyaluronic acid is also found in other connective tissues, in the synovial fluid of joints, in the vitreous humor of the eyes, and also in the capsules that enclose bacteria.

Keratin sulphate, like hyaluronic acid, is a repeating disaccharide forming an unbranched chain. Each disaccharide unit of the polysaccharide consists of **D-galactose** and **sulphated N-acetylglucosamine.** It is found in cartilage and cornea.

Chondroitin sulphate is a repeating disaccharide consisting of alternating **glucuronic acid** and **sulphated N-acetyl galactosamine** residues. It is found in cartilage, bone, skin, notochord, aorta and umbilical cord.

(*b*) Heparin. Heparin is a blood anticoagulant and found in the skin, liver, lung, thymus, spleen and blood. Its molecule contains the repeated disaccharide units, each having **D-glucuronic acid** and **D-glucosamine.**

(*c*) Proteoglycans, glycoproteins and glycolipids. Polysaccharides also occur in covalent combination with proteins and lipids, to form the following three types of molecules:

***(i)* Proteoglycans.** The molecules of proteoglycans consist of much longer portion of polysaccharide and a small portion of protein. They are also called **mucoproteins.** The proteoglycans are amorphous and form gels which are able to hold large amounts of water.

The **cartilage proteoglycan** is found extracellularly in cartilage and bone. In its molecule, strands of protein, called **core protein,** extend radially from a long, central hyaluronic acid molecule. In each core protein strand, three carbohydrate bearing regions may be identified. The first region contains numerous oligosaccharides, the second region contains keratin sulphate chains and the third region contains chondroitin sulphate chains. This arrangement gives cartilage its resilience and tensile strength.

***(ii)* Glycoproteins** (or **glycosaminoglycans** or **mucopolysaccharides**). In these molecules, the carbohydrate portion consists of much shorter chains which are often branched. Glycoproteins serve diverse roles in cells and tissues and include certain enzymes, hormones, blood groups, saliva, gastric mucin, ovomucoids, serum, albumins, antibodies or immunoglobins.

***(iii)* Glycolipids.** These molecules are covalent combinations of carbohydrate and lipid. The carbohydrate portion may be a single monosaccharide or a linear of branched chain. Glycolipids form the component of most cell membranes, *e.g.,* cerebrosides and gangliosides.

Lipids (Fats). The lipids (Gr., *lipos* = fats) are the organic compounds which are insoluble in the water but soluble in the nonpolar organic solvents such as acetone, benzene, chloroform and ether. The cause of this general property of lipids is the predominance of long chains of aliphatic hydrocarbons or benzene ring in their molecules. The lipids are non-polar and hydrophobic. The common examples of lipids are cooking oil, butter, ghee, waxes, natural rubber and cholesterol. Like the carbohydrates, lipids serve two major roles in cells and tissues:

1. They occur as constituents of certain structural components of cells such as membranous organelles; plant pigments such as **carotene** found in carrots and **lycopene** that occurs in tomatoes; vitamins such as *A*, *E* and *K*; menthol and eucalyptus oil; and

2. They may be stored within cells as reserve energy sources. Like the starch and glycogen, fat is compact and insoluble and provides a convenient form in which energy-yielding molecules (the fatty acids) can be stored for use when occasion arises.

Lipids are all made of carbon, hydrogen and sometimes oxygen. The number of oxygen atoms in a lipid molecule is always small compared to the number of carbon atoms. Sometimes small amounts of phosphorus, nitrogen and sulphur are also present. Natural fats and oils are compounds of **glycerol** (*i.e.,* glycerine or propane– 1, 2, 3 triol) and **fatty acids**. They are esters which are formed due to reaction of organic acids with alcohols. There is only one kind of glycerol: its molecular configuration shows no variation and it is exactly same in all lipids. The formula of glycerol is $C_3H_8O_3$ and following is its molecular structure:

H
|
H—C—OH
|
H—C—OH
|
H—C—OH
|
H

Glycerol

Fatty acids. A fatty acid molecule is **amphipathic** and has two distinct regions or ends; a long **hydrocarbon chain,** which is **hydrophobic** (water insoluble) and not very reactive chemically, and a **carboxylic acid group** which is ionized in solution (COO^-), extremely **hydrophilic** (water soluble) and readily forms esters and amides. In neutral solutions, salts of the fatty acids form small spherical droplets or **micelles** in which the dissociated carboxyl groups occur at surface and the hydrophobic carbon chains project towards the centre. In cells, the fatty acids only sparingly occur freely; instead, they are esterified to other components and form the saponifiable lipids.

A fatty acid molecule may be either saturated or unsaturated. The **saturated fatty acids** consist of long hydrocarbon chains terminating in a carboxyl group and conform to the general formula:

$$CH_3 - (CH_2)_n - COOH$$

In nearly all naturally occurring fatty acids, n is an even number from 2 to 22. In the **saturated fatty acids,** most commonly found in animal tissues, n is either 12 (*i.e.,* **myristic acid**), 14 (*i.e.,* **palmitic acid**) or 16 (*i.e.,* **stearic acid**). In **unsaturated fatty acids,** at least two but usually no more than six of the carbon atoms of the hydrocarbon chain are linked together by double bonds (–C = C–), *e.g.,* **oleic acid, linoleic, linolenic, arachidonic** and **clupanadonic acids.** Double bonds are important because they increase the flexibility of the hydrocarbon chain, and thereby the fluidity of biological membranes. Unsaturated fatty acids predominate in lipids of higher plants and in animals that live at low temperatures. Lipids in the tissues of animals inhabiting warm climates contain larger quantities of saturated fatty acids.

Essential fatty acids. Some animals, especially mammals, are unable to synthesize certain fatty acids and, therefore, require them in their diet. They are called **essential fatty acids** and include linoleic acid, linolenic acid and arachidonic acid. Such essential fatty acids have to be obtained from plant material by the animal.

Types of lipids. The lipids are classified into three main types:

1. simple lipids,
2. compound lipids and
3. derived lipids.

1. Simple lipids. The simple lipids are alcohol esters of fatty acids:

$$\text{Triglyceride (Simple lipids)} \xrightarrow[H_2O]{\text{Lipase}} \text{Glycerol} + 3\ \text{Fatty acids}$$

Simple lipids are also of following two types:

(*a*) Neutral fats (Glycerides or triglycerides). They are triesters of fatty acids and glycerol. Neutral fats represent the major type of stored lipids and so accumulate in the cytoplasm.

(*b*) Waxes. Waxes have a higher melting point than neutral fats and are the esters of fatty acids of high molecular weight with the alcohol except the glycerol. The most important constituent alcohol of the molecules of waxes is the **cholesterol**, *e.g.,* bees wax.

2. Compound lipids. The compound lipids contain fatty acids, alcohols and other compounds as phosphorus, amino-nitrogen carbohydrates, etc., in their molecules. Some of the compound lipids are important structural components of the cell, in particular of cell membranes. The compound lipids of the cell are of the following types:

(*i*) Phospholipids (or Glycerophosphatides). Such type of lipids form the major constituent of cell membranes. In a molecule of phospholipid two of the —OH or hydroxyl groups in glycerol are linked to fatty acids, while the third —OH group is linked to phosphoric acid. The phosphate is further linked to a hydrophilic compound such as **etanolamine, choline, inositol** or **serine.** Each phospholipid molecule, therefore, has hydrophobic or water-insoluble tail which is composed of two fatty acid chains and a hydrophilic or water-soluble polar head group, where the phosphate is located. Thus, in effect the phospholipid molecules are detergents, *i.e.,* when a small amount of phospholipid is spread over the surface of water, there forms a monolayer film of phospholipid molecules; in this thin film, tail regions pack together very closely facing the air and their head groups are in contact with the water (Fig. 5.5). Two such films can combine tail to tail to make a phospholipid sandwich or self-sealing **lipid bilayer,** which is the structural basis of cell membranes.

Various membranes of cell contain the following four types of phospholipids:

1. **phosphatidyl choline** or **lecithin;**
2. **phosphatidyl ethanolamine** or **cephalin;**

3. phosphatidyl serine; and

4. phosphatidyl inositol.

The other important phospholipids of the matrix are the phosphoinositides (occur mostly in the cells of liver, brain, muscle and soyabean), plasmalogens and isositides.

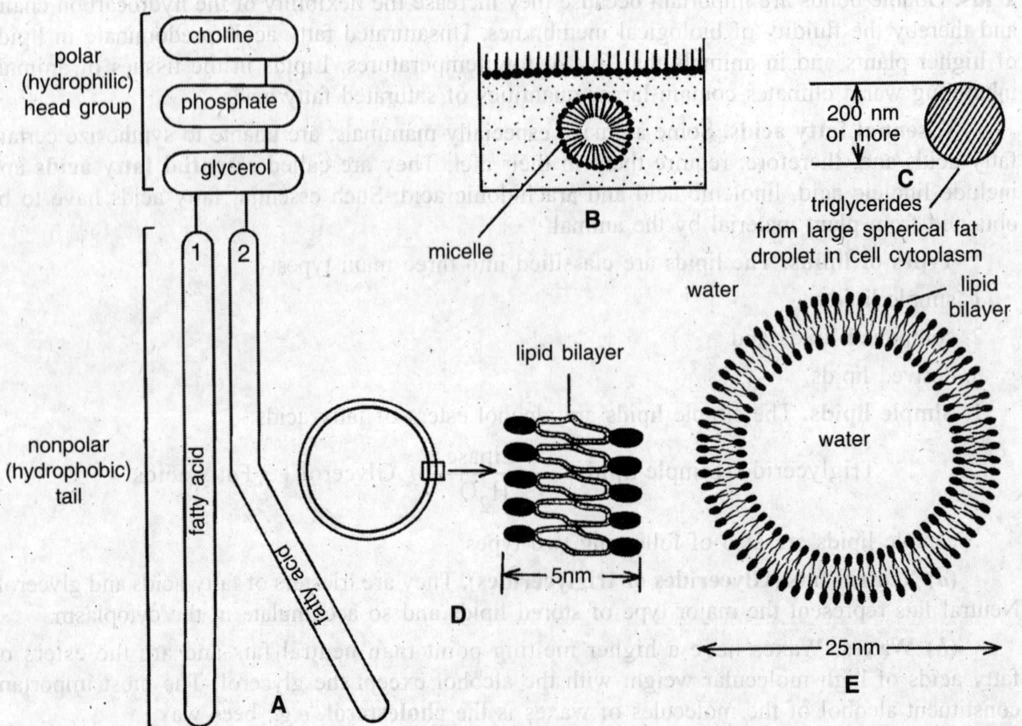

Fig. 5.5. Formation of various types of lipid aggregates. A—Schematic representation of a phospholipid molecule; B—Formation of micelle and monolayer film; C—Formation of a fat droplet by triglycerides; D—Formation of self-sealing lipid bilayer (*e.g.*, liposome); E—Cross-section of a liposome.

Plasmalogens are a special class of phospholipids which are especially abundant in the membranes of nerve and muscle cells and are also characteristic of cancer cells.

Liposomes. When aqueous suspensions of phospholipids are subjected to rapid agitation by using ultrasound (*i.e.*, insonation), the lipid disperses in the water and forms **liposomes** or **lipid vesicles.** Liposomes are small spherical bodies (25 nm to 1 µm in diameter) whose surface is formed by a bilayer of phospholipid molecules enclosing a small volume of the aqueous medium. They exhibit many of the permeability properties of natural membranes, *i.e.*, water soluble small molecules or ions can be enclosed by the liposomes and they can also traverse the lipid bilayer of latter. Recently, liposomes have been found to have great therapeutic promise, since, they can be used as vectors for the transfer of specific drugs, proteins, hormones, nucleic acids, ions or any other molecule into the specific types of animal cells. The contents of the liposomes can enter the target cells by two routes:

1. The liposomes can attach to the surface of target cells and may fuse with the plasma membrane, following which their contents are released into the cytosol or cytoplasmic matrix.
2. The entire liposomes may be endocytosed and degraded intracellularly.

***(ii)* Sphingolipids.** The sphingolipids occur mostly in the cells of the brain. Instead of the glycerol, they contain in their molecules amine alcohol (sphingol or sphingosine). For instance, the myelin sheaths of the nerve fibres contain a lipid known as **sphingomyelin** which contains sphingosine and phospholipids in its molecules.

***(iii)* Glycolipids.** The glycolipids contain in their molecules the carbohydrates and the lipids. The matrix of the animal cells contains two kinds of glycolipids, *viz.*, cerebrosides and gangliosides.

***(a)* Cerebrosides.** The cerebrosides contain in their molecules sphingosine, fatty acids and galactose or glucose. The cerebrosides are the important lipids of the white matter of the cells of brain and the myelin sheath of the nerve. The important cerebrosides are the kerasin, cerebron, nervon and oxynervons.

***(b)* Gangliosides.** The gangliosides have complex molecules which are composed of sphingosine, fatty acids and one or more molecules of glucose, lactose, galactosamine and neuraminic acid. The gangliosides occur in the grey matter of the brain, membrane of erythrocytes and cells of the spleen. Gangliosides act as **antigens**.

One type of ganglioside, called **GM_2**, may accumulate in the lysosomes of the brain cells because of a genetic deficiency that results in the failure of the cells to produce a lysosomal enzyme that degrades this ganglioside. This condition is called **Tay-Sachs disease** and leads to paralysis, blindness and retarded development of human beings.

3. Derived lipids (or Nonsaponifiable lipids). Some types of lipids do not contain fatty acids in their constituents and they are of following three types:

A. Terpenes. The terpenes include certain **fat-soluble vitamins** (*e.g.*, vitamins A, E and K), **carotenoids** (*e.g.*, photosynthetic pigments of plants), and certain **coenzymes** (such as coenzyme Q or ubiquinone). All the terpenes are synthesized from various numbers of a five-carbon building block, called **isoprene unit.** The isoprene units are bonded together in a head-to-tail organization. Two isoprene units form a **monoterpene**, four form a **diterpene**, six a **triterpene**, and so on. The monoterpenes are responsible for the characteristic odours and flavours of plants (*e.g.*, **geraniol** from geraniums, **menthol** from mint and **limoneme** from lemons). **Dolicol phosphate** is a **polyisoprenoid** (*i.e.*, long chain polymer of isoprene) and is used to carry activated sugars in the membrane-associated synthesis of glycoproteins and some polysaccharides.

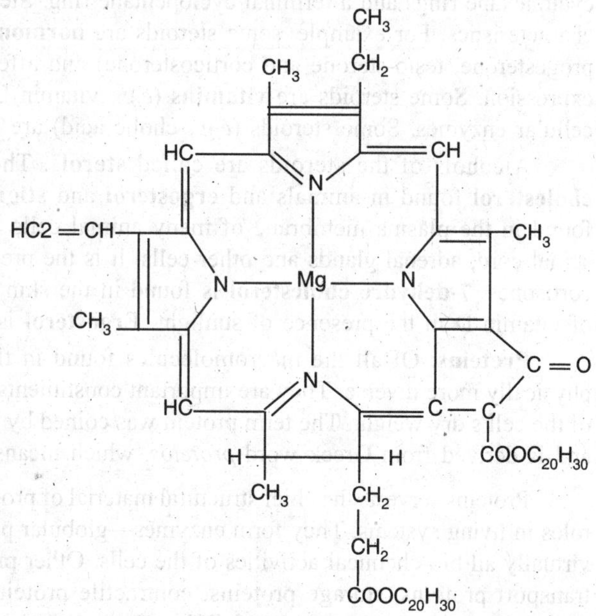

Fig. 5.6. Chemical formula of chlorophyll.

The **carotenoids** are the compound lipids and they form the pigments of the animal and plant cells. There are about 70 carotenoids occurring in both types of cells. The important carotenoids of cells are the α, β and γ carotenes, retinene, xanthophylls, lactoflavin in milk, riboflavin (vitamin B_2), xanthocyanins, coenzyme Q, anthocyanins, flavones, flavonols and flavonones, etc. Chemically all carotenoids are long-chain isoprenoids having an alternating series of double bonds. They are synthesized by plant tissues and are located in the chloroplast lamellae to help in light absorption during photosynthesis. In animal cells, carotenoids serve as precursors of vitamin A.

The **chlorophylls** are essential photosynthetic green pigments of the chloroplasts. A chlorophyll molecule (Fig. 5.6) consists of a **head** and a **tail**. The head consists of a **porphyrin ring** or **tetrapyrrole nucleus** from which extends a hydrophobic tail which is made up of a 20-carbon grouping, called the **phytol**. Phytol ($C_{20}H_{39}$) is a long straight-chain alcohol containing a single double bond. It may be regarded as a hydrogenated carotene (vitamin A). The porphyrins (Gr., *porphyra* = purple) are complex carbon-nitrogen molecules that usually surround a metal, *i.e.*, it is formed from four pyrrol rings linked together by methane bridges and metal atom (Mg or Fe) is linked to pyrrol rings. In chlorophyll molecule, the porphyrin surrounds a magnesium ion, while in haeme of **haemoglobin**, it surrounds an iron ion (Fig. 5.7). Many other pigments of animal cells such as **myoglobin** and **cytochromes** have porphyrin rings in their molecules.

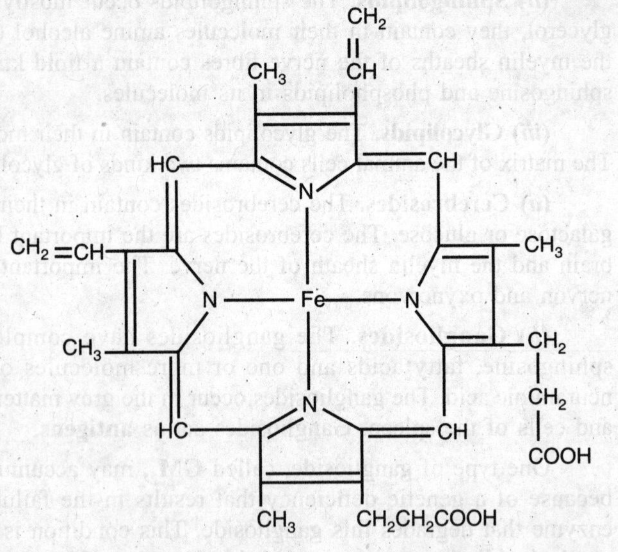

Fig. 5.7. Chemical formula of haeme portion of haemoglobin.

B. Steroids. The steroids consist of a system of fused cyclohexane and cyclopentane rings. All are derivative of **perhydro-cyclopentano-phenanthrene,** which consists of three fused cyclohexane rings and a terminal cyclopentane ring. Steroids have widely different physiological characteristics. For example, some steroids are **hormones** (*e.g.*, sex hormones such as estrogen, progesterone, testo-sterone and corticosterone) and affect cellular activities by influencing gene expression. Some steroids are **vitamins** (*e.g.*, vitamin D_2) and influence the activities of certain cellular enzymes. Some steroids (*e.g.*, cholic acid) are fat emulsifier found in the bile.

Alcohols of the steroids are called **sterols**. The common examples of the sterols are **cholesterol** found in animals and **ergosterol** and **stigmasterol** found in plants. Cholesterol is found in the plasma membrane of many animal cells and also in blood, bile, gallstone, brain, spinal cord, adrenal glands and other cells. It is the precursor of most steroid sex hormones and cortisones. **7-dehydro-cholesterol** is found in the skin where it is responsible for the synthesis of vitamin D in the presence of sunlight. **Ergosterol** is also a precursor of vitamin D.

Proteins. Of all the macromolecules found in the cell, the proteins are chemically and physically more diverse. They are important constituents of the cell forming more than 50 percent of the cell's dry weight. The term protein was coined by Dutch chemist **G.J. Mulder** (1802–1880) and is derived from Greek word *proteios*, which means "of the first rank".

Proteins serve as the chief structural material of protoplasm and play numerous other essential roles in living systems. They form enzymes—globular proteins specialized to serve as catalysts in virtually all bio-chemical activities of the cells. Other proteins are antibodies (immunoglobulins), transport proteins, storage proteins, contractile proteins, and some hormones. In every living organism, there are thousands of different proteins, each fitted to perform a specific functional or structural role. Indeed, a single human cell may contain more than 10,000 different types of protein molecules. Chemically, proteins are polymers of amino acids.

1. Amino acids. Nobel Laureate **Emil Fischer** (1902) discovered that all proteins consist of chains (linear sequence) of smaller units that he named **amino acids.** There are about 20 different amino acids which occur regularly as constituents of naturally occurring proteins (Table 5.3). An organic compound containing one or more amino groups ($-NH_2$) and one or more carboxyl groups (—COOH) is known as amino acid. The amino acids occur freely in the cytoplasmic matrix and constitute the so called **amino acid pool**. Of the 20 commonly occurring amino acids, 19 may be represented by the following general formula (Fig. 5.8).

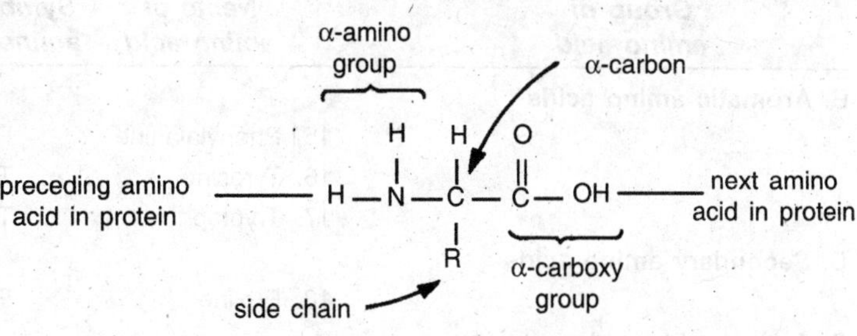

Fig. 5.8. Basic structure of an amino acid.

The sole exception is proline, where the amino group forms part of a ring structure. The central or **alpha** carbon atom of each amino acid is covalently bonded to four groups:

(*i*) A hydrogen atom,

(*ii*) an amino group ($—NH_2$),

(*iii*) an acid (or carboxyl) group, and

(*iv*) a side chain called an **R-group**. It is the particular chemical structure of the R-group that distinguishes one amino acid from another.

In certain amino acids R group is either a hydrogen atom (*e.g.*, glycine, the simplest amino acid) or a hydrophobic aliphatic (*e.g.*, leucine) or aromatic (*e.g.*, phenylalanine) hydrocarbon. In other cases, R group contains either an extra carboxyl group or an extra amino group or its equivalent. Glutamic acid and aspartic acid each have an extra carboxyl (—COOH) group. Lysine and arginine both contain an additional amino group or equivalent structure. Histidine also contains a N group. Other amino acids such as serine and tyrosine have hydroxyl groups in their side chains. Of a particular importance is the amino acid cysteine which possesses a thiol (SH) group.

Table 5.3 The 20 amino acids which occur naturally.

Group of amino acid	*Name of amino acid*	*Symbol of amino acid*	*Single letter code of amino acid*
A. Aliphatic amino acids			
1. Simple amino acids or Mono-amino-monocarboxylic amino acids	1. Glycine	Gly	G
	2. Alanine	Ala	A
	3. Valine	Val	V
	4. Leucine	Leu	L
	5. Isoleucine	Ile	I
2. Acidic amino acids or Monoamino-monocarboxylic acids	6. Aspartic acid	Asp	D
	7. Glutamic acid	Glm	E
3. Basic amino acids or Diamino-monocarboxylic amino acids	8. Lysine	Lys	K
	9. Arginine	Arg	R
	10. Histidine	His	H
4. Hydroxyl containing amino acids	11. Serine	Ser	S
	12. Threonine	Thr	T
5. Sulphur-containing amino acids	13. Cysteine	Cys	C
	14. Methionine	Met	M

Group of amino acid	Name of amino acid	Symbol of amino acid	Single letter code of amino acid
B. Aromatic amino acids			
	15. Phenylalanine	Phe	F
	16. Tyrosine	Tyr	Y
	17. Tryptophan	Try	W
C. Secondary amino acids			
	18. Proline	Pro	P
D. Amino acid amides			
	19. Aspargine	Asn	N
	20. Glutamine	Glu	Q

2. Formation of proteins. Because a molecule of the amino acid contains both basic or amino (—NH_2) and acidic or carboxyl (—COOH) group, it can behave as an acid and base at a time. The molecules of such organic compounds which contain both acidic and basic properties are known as **amphoteric molecules**. Due to amphoteric molecules, the amino acids unite with one another to form complex and large protein molecules. When two molecules of amino acids are combined then the basic group (—NH_2) of one amino acid molecule combines with the carboxylic (—COOH) group of other amino acid and the loss of a water molecule takes place. This sort of condensation of two amino acid molecules by —NH—CO linkage or bond is known as **peptide linkage** or **peptide bond**. A combination of two amino acids by the peptide bond is known as **dipeptide**. When three amino acids are united by two peptide bonds, they form **tripeptide**. Likewise, by condensation of few or many amino acids by the peptide bonds the **oligopeptides** and **polypeptides** are formed respectively. The various molecules of polypeptides unite to form the **peptones**, **proteases** and **proteins**. Thus, protein macromolecules are the polymers of many amino acid monomers. The size (molecular weight), shape, and function of proteins are determined by the number, type and distribution of the amino acids present in the molecule. Proteins occur in a wide spectrum of molecular sizes from small molecules such as the hormone ACTH (or adrenocorticotrophic hormone) which consists of only 39 amino acids and has a molecular weight of 4500, to extremely large proteins such as haemocyanin (an invertebrate blood pigment) which consists of 8200 amino acids and has a molecular weight greater than 900,000.

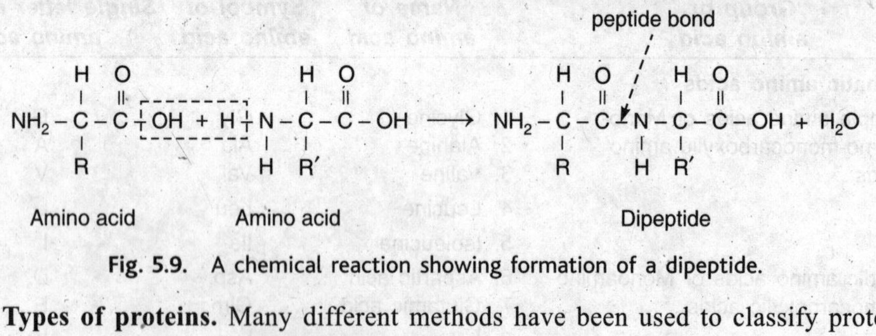

$$\underset{\text{Amino acid}}{NH_2-\underset{R}{\overset{H}{C}}-\overset{O}{\overset{\|}{C}}-OH} + \underset{\text{Amino acid}}{H-\underset{H}{N}-\underset{R'}{\overset{H}{C}}-\overset{O}{\overset{\|}{C}}-OH} \longrightarrow \underset{\text{Dipeptide}}{NH_2-\underset{R}{\overset{H}{C}}-\overset{O}{\overset{\|}{C}}-\underset{H}{N}-\underset{R'}{\overset{H}{C}}-\overset{O}{\overset{\|}{C}}-OH} + H_2O$$

peptide bond

Fig. 5.9. A chemical reaction showing formation of a dipeptide.

3. Types of proteins. Many different methods have been used to classify proteins, no method of their classification being entirely satisfactory:

1. Classification based on biological functions. Based on function proteins fall into nine major classes (Table 5.4). Main proteins are **enzymes,** serving as catalysts that greatly increase the rates of the thousands of chemical reactions on which life depends. **Structural proteins**, on the other hand, provide support and shape to cells and organelles. **Motility proteins** play key roles in the contraction and movement of cells and intracellular structures. **Regulatory proteins** are responsible for control and coordination of cellular functions, ensuring that cellular activities are regulated to meet cellular needs. **Transport proteins** are involved in the movement of other substances into, out of and within the cell. **Hormonal proteins** mediate communication between cells in distant parts of an organism, and **receptor proteins** enable cells to respond to chemical stimuli from their environment. Finally, **defensive proteins** provide protection against disease, and **storage proteins** serve as reservoirs of amino acids.

Thus, with virtually everything that a cell is or does dependent on the proteins it contains (see **Becker** *et al.*, 2007). Most proteins are **monofunctional**, which simply means that a specific protein has a single function, whether catalytic, structural, motile, or regulatory. However, some proteins are **bifunctional**, which means that an individual protein plays two distinctly different roles.

Table 5.4 Protein functions (Source: **Becker** *et al.*, 2007).

Type of protein	*Function*	*Example*
1. Enzymes	Selective catalysis	Digestive enzymes catalyze the hydrolysis of the macromolecules in food.
2. Structural proteins	Support of cellular structures	Collagen and elastin provide fibrous framework in animal connective tissues; keratin is the protein of hair, horns, feathers and other skin appendages.
3. Motility proteins	Movement of cells and cell parts	Actin and myosin are responsible for the movement of muscles; tubulin is the main constituent of cilia and flagella.
4. Regulatory proteins	Regulation of cellular functions	DNA-binding proteins, called transcription factors regulate the read-out of genetic information.
5. Transport proteins	Transport of substances, into, out of, and within cells	Transport of glucose and other small molecules across membranes is mediated by specific transport proteins embeded in the membranes.
6. Hormonal proteins	Communication between distant parts of an organism	Insulin is secreted by the pancreas and regulate the concentration of glucosein the blood of vertebrates.
7. Receptor proteins	Response of cells to chemical stimuli	Receptors in the membrane of a nerve cell detect chemical signals released by other nerve cells.
8. Defensive proteins	Protection against disease	Antibodies in the circulatory system of vertebrates detect and destroy bacteria and viruses.
9. Storage proteins	Storage and release of amino acids	Storage proteins in seeds are broken down during germination to release amino acids.

2. Classification based on shape of proteins. According to the shape or conformation, two major types of proteins have been recognized:

***(a)* Fibrous proteins.** Fibrous proteins are water-insoluble, thread-like proteins having greater length than their diameter. They contain secondary protein structure and occur in those cellular or extracellular structures, where strength, elasticity and rigidity are required, *e.g.,* collagen, elastin, keratin, fibrin (blood-clot proteins) and myosin (muscle contractile proteins).

***(b)* Globular proteins.** Globular protein are water-soluble roughly spheroidal or ovoidal in shape. They readily go into colloidal suspension. They have tertiary protein structure and are usually functional proteins, *e.g.,* enzymes, hormones and immunoglobulins (antibodies). **Actin** of microfilaments and **tubulins** of microtubules are also globular proteins.

3. Classification based on solubility characteristics. According to this criterion proteins can be classified into two main types:

***(a)* Simple proteins.** These proteins contain only amino acids in their molecules and they are of following types:

***(i)* Albumins.** These are water soluble proteins found in all body cells and also in blood stream, *e.g.,* **lactalbumin**, found in milk and **serum albumin** found in blood.

***(ii)* Globulins.** These are insoluble in water but are soluble in dilute salt solutions of strong acids and bases, *e.g.,* **lactoglobulin** found in milk and **ovoglobulin**.

***(iii)* Glutelins.** These plant proteins are soluble in dilute acids and alkalis, *e.g.,* **glutenin** of wheat.

***(iv)* Prolamines.** These plant proteins are soluble in 70 to 80 per cent alcohol, *e.g.,* **gliadin** of wheat and **zein** of corn.

***(v)* Scleroproteins.** They are insoluble in all neutral solvents and in dilute alkalis and acids, *e.g.,* keratin and collagen.

***(vi)* Histones.** These are water soluble proteins which are rich in basic amino acids such as arginine and lysine. In eukaryotes histones are associated with DNA of chromosomes to form nucleoproteins.

***(vii)* Protamines.** These are water soluble, basic, light weight, **arginine** rich polypeptides. They are bound to DNA in spermatozoa of some fishes, *e.g.,* **salmine**, of salmon and **sturine** in sturgeons.

***(b)* Conjugated proteins.** These proteins consist of simple proteins in combination with some non-protein components, called **prosthetic groups**. The prosthetic groups are permanently associated with the molecule, usually through covalent and/or non-covalent linkages with the side chains of certain amino acids. Conjugated proteins are of following types:

***(i)* Chromoproteins.** Chromoproteins are a heterogeneous group of conjugated proteins which are in combination with a prosthetic group that is a pigment, *e.g.,* respiratory pigments such as **haemoglobin, myoglobin** and **haemocyanin; catalase, cytochromes, haemerythrin;** visual purple or **rhodopsin** of rods of retina of eye and yellow enzymes or **flavoproteins.**

***(ii)* Glycoproteins.** Glycoproteins are proteins that contain various amounts (1 to 85 per cent) of carbohydrates. Of the known 100 monosaccharides, only nine are found to occur as regular constituents of glycoproteins (*e.g.,* glucose, galactose, mannose, fucose, acetylglucosamine, acetylgalactosamine, acetylneuraminic acid, arabinose and xylose). Glycoproteins are of two main types: **1. Intracellular glycoproteins** which are present in cell membranes and have an important role in membrane interaction and recognition. They also serve as antigenic determinants and receptor sites. **2. Secretory glycoproteins** are **plasma glycoproteins** secreted by the liver; **thyroglobulin,** secreted by the thyroid gland; **immunoglobulins** secreted by the plasma cells; **ovoalbumins** secreted by the cells of oviduct of hen; **ribonucleases** and **deoxyribonucleases. Mucus** and **synovial fluid** are also glycoproteins with lubricative properties.

***(iii)* Lipoproteins.** Lipid containing proteins are called **lipoproteins.** Their lipid contents are 40 to 90 per cent of their molecular weight and this tends to affect the density of the molecule. There are four types of lipoproteins: 1. **High density lipoproteins (HDL)** or α-lipoproteins; 2. **Low density lipoproteins (LDL)** or β-lipoprotiens; 3. **Very low density lipoproteins (VLDL)** or pre-β-lipoproteins; and 4. **Chylomicrons.** Lipoproteins include some of the blood plasma proteins, various types of membrane proteins, lipovitellin of egg yolk and proteins of brain and nerve tissue.

***(iv)* Nucleoproteins.** Nucleoproteins are proteins in combination with nucleic acids (DNA and RNA). However, these proteins are not true conjugated proteins since the nucleic acid involved cannot be regarded as prosthetic groups. Nucleoproteins are of two types: 1. **Histones** which are quite similar in all plants and animals. Their highly basic nature accounts for the close associations histones form with the nucleic acids. Histones are involved in the tight packing of DNA molecules during the condensation of chromatin into chromosomes for the mitosis. 2. **Nonhistones** have great heterogeneous amino acid composition and are acidic in nature. They have selective combination with certain stretches of nuclear DNA and, thus, are involved in the regulation of gene expression.

(*v*) Metalloproteins. Metalloproteins are proteins conjugated to metal ions which are not part of the prosthetic group, *e.g.*, **carbonic anhydrase** enzyme contains zinc ions and amino acids in its molecule; **caeruloplasmin,** an oxidase enzyme containing copper; and **siderophilin** contains iron.

(*vi*) Phosphoproteins. Phosphoproteins are proteins in combination with a phosphate group, *e.g.*, casein of the milk and ovovitellin of eggs.

4. Structural levels of proteins. The protein as synthesized on the ribosome is a linear sequence of amino acids, polymerized by the elimination of water between successive amino acids to from the peptide bond, and existing as a randomly coiled chain without specific shape and possessing no biological (*i.e.,* catalytic) activity. Within seconds of synthesis being completed, the protein folds into a specific three-dimensional form, which is the same for all molecules of the same type of protein and which now is capable of doing catalysis. According to their mode of folding the following four levels of protein organization have been recognized (Fig. 5.10):

(*a*) **Primary structure of protein,** *e.g.,* insulin.

(*b*) **Secondary structure of protein,** *e.g.,* α-Keratin, collagen, globular proteins.

(*c*) **Tertiary structure of protein,** *e.g.,* myoglobin.

(*d*) **Quaternary structure of protein,** *e.g.,* haemoglobin.

(*a*) Primary protein structure. The primary protein structure is defined as the particular sequence of amino acids found in the protein. It is determined by the covalent peptide bondings

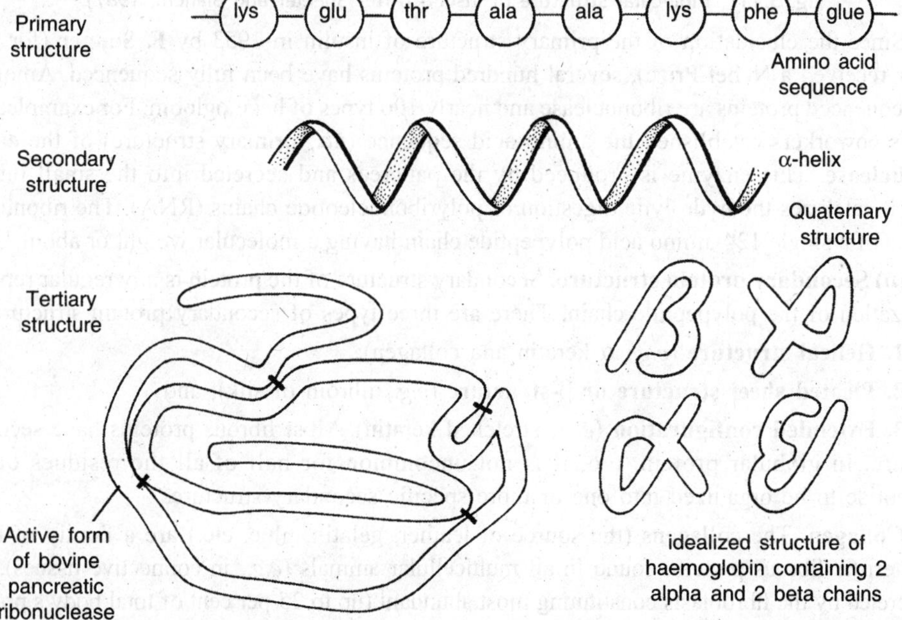

Fig. 5.10. The four levels of structural organization in protein molecules.

between amino acids. Primary structure also includes other covalent linkages in proteins, for example the linkages that may exist between sulphur atoms of cysteine amino acids located in the chain of the protein insulin. The first protein to have its primary structure determined was of insulin, the pancreatic hormone that regulates glucose metabolism in mammals. Insulin has a molecular weight of 5,800 daltons and contains 51 amino acids. Insulin consists of two polypeptide chains of 21 and 30 amino acid residues, called the A and B chains, respectively (Fig. 5.11). (An amino acid residue is that which is left when the elements of water are split out during polymerization).

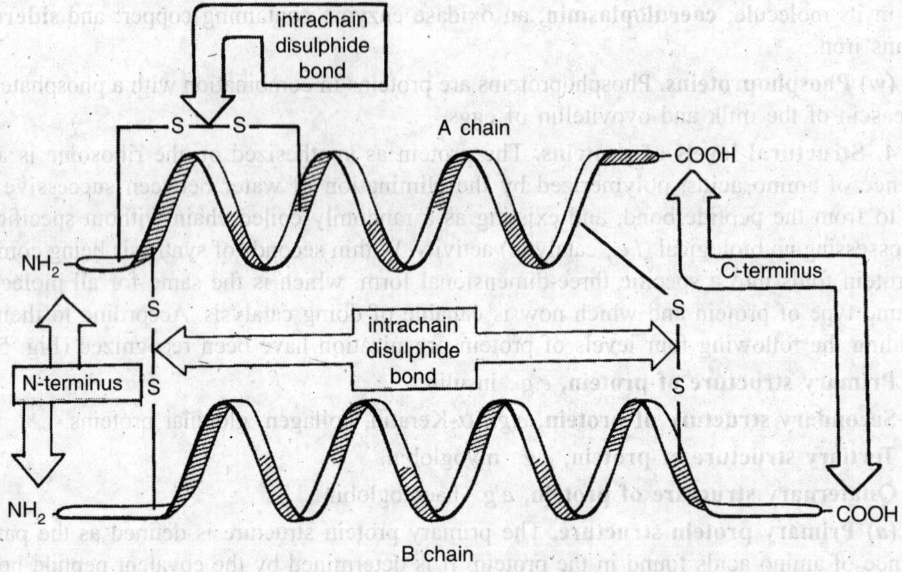

Fig. 5.11. Molecular structure of insulin (After Sheeler and Bianchi, 1987).

Since the elucidation of the primary structure of insulin in 1953 by **F. Sanger** (for which Sanger received a Nobel Prize), several hundred proteins have been fully sequenced. Among the fully sequenced proteins are ribonuclease and nearly 100 types of haemoglobin. For example; **Stein** and his coworkers established the amino acid sequence (*i.e.,* primary structure) of the enzyme **ribonuclease.** This enzyme is produced by the pancreas and secreted into the small intestine where it catalyzes the hydrolytic digestion of polyribonucleotide chains (RNA). The ribonuclease consists of a single 124 amino acid polypeptide chain having a molecular weight or about 12,000.

***(b)* Secondary protein structure.** Secondary structure of the protein is any regular repeating organization of the polypeptide chain. There are three types of secondary protein structure:

1. Helical structure (*e.g.,* α-keratin and collagen);

2. Pleated sheet structure or β-structure (*e.g.,* fibroin of silk); and

3. Extended configuration (*e.g.,* stretched keratin). Most fibrous proteins have secondary structure. In globular protein, too, it is not uncommon for half of all the residues of each polypeptide to be organized into one or more specific secondary structures.

Collagen. The collagens (the source of leather, gelatin, glue, etc.) are a family of highly characteristic fibrous proteins found in all multicellular animals (*e.g.,* in connective tissues). They are secreted by the fibroblasts constituting most abundant (up to 25 per cent of total body's proteins) proteins of mammals. The characteristic feature of collagen (or **tropocollagen**) molecules is their stiff, triple-stranded helical structure (which was discovered by **Rich, Crick** and **Ramachandran**). Three collagen polypeptide chains are left-handed α-helices or alpha chains, each is about 1000 amino acid residues long. These chains are wound around one another in a regular superhelix to generate a rope-like collagen or tropocollagen molecule which is about 300 nm long and 1.5 nm is diameter (Fig. 5.12).

Collagens are exceptionally rich in **proline** (and **hydroxyproline;** both accounting for more than 20 percent of collagen's amino acids) and **glycine**. Other dominant amino acids of collagens are **lysine** and **alanine**.

So far, about 20 distinct collagen-chains have been identified, each encoded by a separate gene. About 10 types of triplet-stranded collagen molecules have been found to assemble from various combinations of 20 types of α-chains. The best defined are types I, II, III and IV. **Type I collagen** is present in the dermis, tendons, ligaments, bone, cornea, dentine of teeth and internal organs and accounts for 90 per cent of body's collagen. **Type II collagen** is present mainly in cartilage, intervertebral disc, embryonic notochord and vitreous humour of eye. **Type III collagen** occurs in skin, cardiovascular system, gastrointestinal tract and uterus. **Type IV collagen** is present in basal laminae or basement membranes of epithelia. Type I, II and III collagens are the fibrillar collagens showing typical striated fibres. Type IV collagen lacks a distinct fibrillar structure (Fig. 5.13).

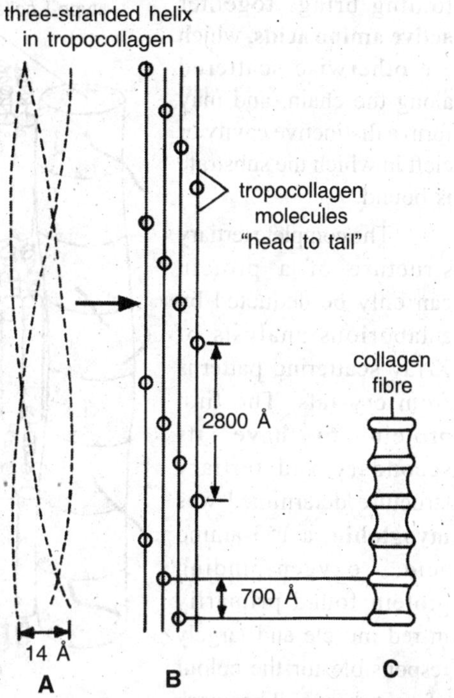

Fig. 5.12. A—A tropocollagen molecule (collagen or superhelix) with three intertwined left-handed α-helices; B—Staggered arrangement of tropocollagen molecules (super helices) in a collagen fibril; C—Striated appearance of a collagen fibril under the electron microscope (after Dyson, 1978).

The individual collagen polypeptide chains (α-chains) are synthesized on membrane bound ribosomes and injected in the lumen of ER as larger precursors, called pro-α-chains. These precursors have distinct polarity, containing terminal propeptides at their *N*-and *C*-terminus. Each pro-α-chain then combines with two others to form a hydrogen-bonded, triple stranded helical molecule, called **procollagen**. Such a procollagen of fibrillar collagens (I, II, and III) is serceted from the fibroblast in the extracellular space. Due to enzyme action its terminal propeptides are removed and the procollagen is converted into a **tropocollagen molecule**. Many tropocollagen molecules spontaneously assemble into the ordered arrays, called **collagen fibrils**. The collagen fibrils are thin (10 to 300 nm in diameter), cable-like structures, many micrometres long, exhibiting cross-striations every 67 nm and are clearly visible in the electron micrographs. The collagen fibrils often aggregate into larger bundles which can be seen in light microscope as **collagen fibres**. Type IV collagen molecules assemble to form a sheet-like **meshwork** that constitutes a major part of all basal laminae (**Martin** *et. al.*, 1985, **Burgeson**, 1988).

(c) Tertiary protein structure. Tertiary protein structure refers to a more compact structure in which the helical and non-helical regions of a polypeptide chain are folded back on themselves. This structure is typical of globular protein structure, in which it is the non-helical region that permits the folding. The folding of a polypeptide chain is not random but occurs in a specific fashion, thereby imparting certain steric (*i.e.,* three-dimensional) properties to the protein. For example, in enzymes folding brings together **active amino acids,** which are otherwise scattered along the chain, and may form a distinctive cavity or cleft in which the substrate is bound.

The complete tertiary structure of a protein can only be deducted by a laborious analysis of X-ray scattering patterns from crystals. The first protein to have its secondary and tertiary structure determined was **myoglobin,** a 153-amino acid, oxygen-binding protein found primarily in red muscle and largely responsible for the colour of that tissue. The work was done at Cambridge under the direction of **J.C. Kendrew** (1961). Although at some points the polypeptide chain does have secondary structure (alpha-helical structure), the chain is mainly characterized by seemingly random loops and folds.

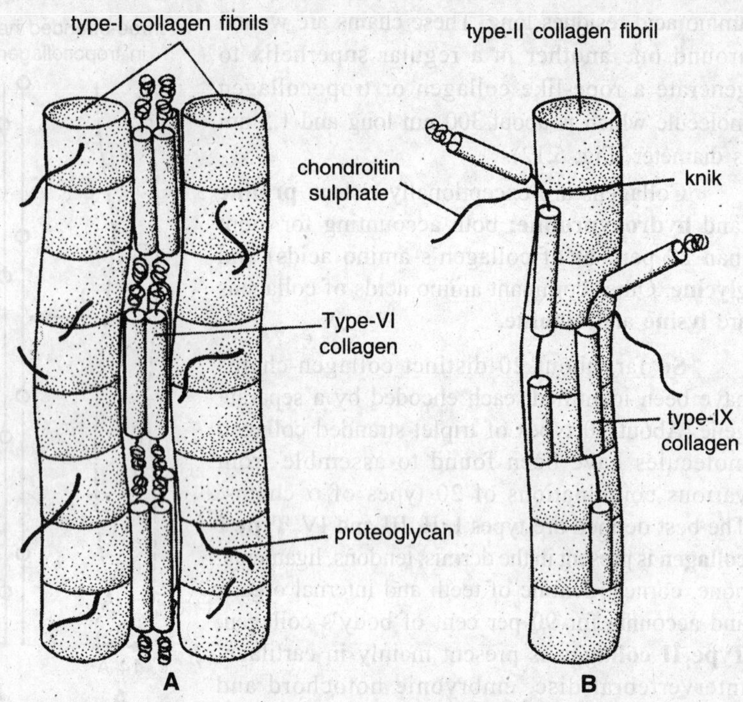

Fig. 5.13. Interactions of fibrous collagens with nonfibrons collagens. A—In tendons, type I collagen fibrils are associated with type-VI collagen. B—In cartilage, type II collagen fibrils are associated with type IX collagen (after Lodish *et. al.,* 2004).

In a tertiary protein the polypeptide chain is held in position by weak secondary bonds which are of different types such as **ionic bonds** (or electrostatic bonds or salt or salt bridges); **hydrogen bonds; hydrophobic bonds** and **disulphide bonds**.

***(d)* Quaternary protein structure.** In proteins that are composed of two or more polypeptide chains, the quaternary structure refers to the specific orientation of these chains with respect to one another and the nature of the interactions that stabilize this orientation. The individual polypeptide chains of the protein are called **sub-units** and the active protein itself is called **multimer**. While multimeric proteins containing up to 32 subunits have been described, the most common multimers are **dimers, trimers, tetramers, pentamers** (*e.g.,* RNA polymerase) and **decamers** (*e.g.,* DNA polymerase III) (Table 5.5). If the protein consists of identical sub-units, it is called **homopolymers** and is said to have **homogeneous quaternary structure,** *e.g.,* the isozymes H_4 and M_4 of lactic dehydrogenase (LDH), enzyme phosphorylase and L-arabinose isomerase. The enzyme β-galactosidase, consists of four identical polypeptide chains. Lastly, when the sub-units of the protein are different, the protein is called **heteropolymer** and is said to have a heterogeneous quaternary structure, *e.g.,* haemoglobin and immunoglobulins. Quaternary proteins are usually joined by hydrophobic forces. Hydrogen bonds, ionic bonds and possibly disulphide bonds may also participate in forming quaternary structures.

Table 5.5 Subunits and molecular weight of some multimer proteins (Source: **Sheeler** and **Bianchi**, 1987).

Protein	Molecular weight	Number	Subunits Designation	Molecular weight
1. Haemoglobin A (human)	64,500	4	Alpha chains (2) Beta chains (2)	15,700 16,500
2. Lactate dehydrogenase	135,000	4	A chain (0 to 4) B chain (4 to 0)	33,600 33,600
3. Immunoglobulin G	150,000	4	Light chains (2) Heavy chains (2)	25,000 50,000
4. Tryptophan synthetase (*E.coli*)	150,000	4	Alpha chains (2) Beta chains (2)	29,500 45,000
5. Aspartate transcarbamylase	306,000	12	C chains (6) R chains (6)	34,000 17,000
6. L-arabinose isomerase (*E.coli*)	360,000	6	(identical)	60,000
7. Apoferritin (iron storage protein)	456,000	24	(identical)	19,000
8. Thyroglobulin	670,000	2	(identical)	3,35,000

Some Examples of Tertiary and Quaternary Proteins

***(i)* Ribonuclease. C.B. Anfinsen** initiated and confirmed the notion that, acting in concert, the specific primary structure of a polypeptide and the innate properties of the side chains of its amino acids cause the polypeptide to spontaneously assume its biologically active tertiary structure. In 1972, he got the Nobel Prize for this definitive work. **Anfinsen** identified four disulphide bridges in the ribonuclease protein, suggesting that the enzyme is highly folded. As is the case with almost all enzymes, the catalytic activity of ribonuclease depends on the maintenance of a particular three-dimensional shape. In concentrated solutions of β-mercaptoethanol and urea, the disulphide bridges of the enzyme are broken and the resultant unfolding of the polypeptide chain is accompanied by a loss of enzyme activity. The enzyme is said to be **denatured**. If the β-mercaptoethanol and urea are removed by dialysis and the denatured ribonuclease reacted with oxygen, the four disulphide bridges reform spontaneously, and essentially all the catalytic activity of the protein is restored. Similar observations have been made with other proteins, that is, they are capable of spontaneously reestablishing their biologically active tertiary (or even quaternary, *e.g.,* haemoglobin) structure after having undergone extensive molecular disorganization.

***(ii)* Haemoglobin.** Haemoglobin is one of the fully sequenced protein. Our present understanding of the structure and function of haemoglobin is the outcome of 50 years of research of **M.F. Perutz**. He got the Nobel Prize in 1962, along with **J.C. Kendrew**, for their studies of haemoglobin and myoglobin.

The haemoglobin is a conjugated globular protein, that is, it contains some non-protein part. In all but the lowest vertebrates, haemoglobin is a tetramer (a heteropolymer). In lampreys, however, haemoglobin is monomeric, that is, it contains a single globin chain like the myoglobin. In humans, most common type of haemoglobin is haemoglobin A (HbA), which consists of 574 amino acid residues and has a molecular weight of 64,500. Its secondary, tertiary and quaternary structure is typical of all higher vertebrate haemoglobins. The protein portion of the haemoglobin molecule, called **globin**, is composed, of four polypeptide chains, each of which is also globular in shape.

The four globin chains consist of, two identical pairs: two **alpha chains** (141 amino acids each) and two **beta chains** (146 amino acids each). The non-protein portion of haemoglobin consists of four iron-containing haem groups, one associated with each of the four globin chains. Nineteen of the twenty biologically important amino acids are included in the globin of haemo-globin.

Haemoglobin molecule is highly symmetric; it can be divided into two identical halves, each consisting of an αβ-dimer. The com-plete tetramer is similar to a mildly flattened sphere having a maximum diameter of about 5.5 nm. The four polypeptide chains are arranged in such a way that unlike chains have numerous stabilizing interactions, whereas, like chains have few. A cavity about 2.5 nm long and varying in width from about 5 to 10 Å passes through the molecule along the axis. Each globin chain envelops its haem group in a deep cleft.

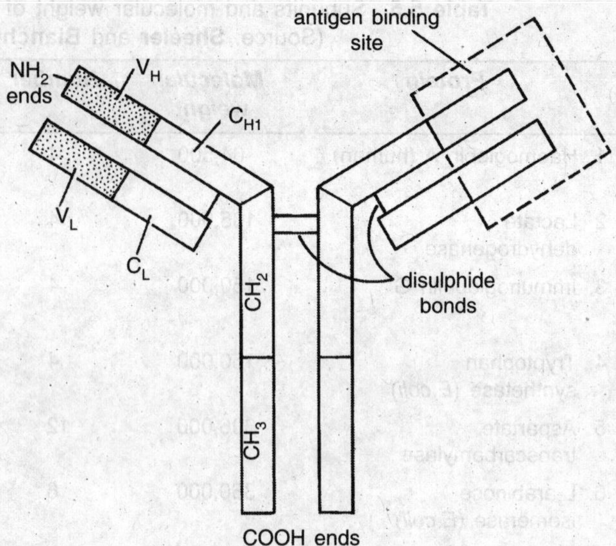

Fig. 5.14. Molecular structure of an immunoglobulin molecule (after Stansfield 1986).

(*iii*) Immunoglobulins. The ability of animals to resist infection by pathogens (viruses, bacteria and other unicellular parasites) and by multicellular endoparasites, is called **immunity**. Specific immune response function by recognizing particular chemical structures, known as **antigens**—on the surface of invading cells. An antigen can be a protein, lipid, carbohydrate or any other molecule. These antigens interact with protein molecules produced by the host, the **immunoglobulins**, which bind the antigen in much the same way as an enzyme binds its substrate. Specific immune responses involve many different types of cells. One type, the **B-lymphocyte** or **B-cell**, is capable of producing **free immunoglobulins**, called **antibodies**.

An immunoglobulin (Ig) molecule is a Y-shaped heteropolymer and is composed of two identical **H (heavy)** polypeptide chains and two smaller identical **L (Light)** polypeptide chains (Fig. 5.14). Heavy chains contain antigenic determinants in the "tail" (carboxyl) segments by which they can be classified as IgG, IgM, IgA, IgD or IgE. Light chains can likewise be typed as kappa or lambda. Within a H chain class or L chain type, these segments exhibit very little variation in primary structure from one individual to another and are called **constant regions (C)**. The amino (—NH_2) ends, however, are extremely diverse in primary structure, even within a class and are called **variable (V) regions.** The $\mathbf{V_H}$ and $\mathbf{V_L}$ regions together form two **antibody-combining sites** (called **antigen-binding sites**) for specific interactions with homologous antigen molecules. The $\mathbf{C_H}$ region consists of three or four similar segments, presumably derived evolutionarily by duplication of an ancestral gene and subsequent modification by mutations; the similar segments are called **domains** and are labelled CH_1, CH_2, CH_3, etc. A mature lymphocyte (plasma cell) produces antibodies with a single class H chain and a single type of L chain, hence, also a single antigen-binding specificity. The first antibodies produced by a developing plasma cells are usually of class IgM.

Enzymes. The cytoplasmic matrix and many cellular organelles contain very important organic compounds known as the **enzymes.** The word enzyme (Greek, "in yeast") had been proposed by **Kuhne** in 1878. The enzymes are the specialized proteins and they have the capacity to act as catalysts in chemical reaction. Like the other catalysts of chemical world, the enzymes are the catalysts of the biological world and they influence the rate of a chemical reaction, while themselves remain quite unchanged at the end of the reaction. The substance on which the enzymes act is known as **substrate**. The enzymes play a vital role in various metabolic and biosynthetic activities of the cell such as synthesis (anabolism) of DNA, RNA and protein molecules and catabolism of carbohydrates, lipids, fats and other chemical substances. The enzymes of the matrix and cellular organelles are classified as follows:

1. Oxireductases. The enzymes catalyzing the oxidation and reduction reaction of the cell are known as oxireductases. These enzymes transfer the electrons and hydrogen ions from the substrates, *e.g.*, hydrogenases or reductases, oxidases, oxygenases and peroxidases.

2. Transferases. The enzymes which transfer following groups from one molecule to other are known as transferases: one carbon, aldehydic or ketonic residues, acyl, glycosyl, alkyl, nitrogenous, phosphorus containing groups and sulphur containing groups.

3. Hydrolases. These enzymes hydrolyse a complex molecule into two compounds by adding the element of the water across the bond which is cleaved. These enzymes act on the following bonds–ester, glycosyl, ether, peptide, other C–N bonds, acid anhydride, C–C, halide and P–N bonds. Certain important hydrolase enzymes are the proteases, esterases, phosphatases, nucleases and phosphorylases.

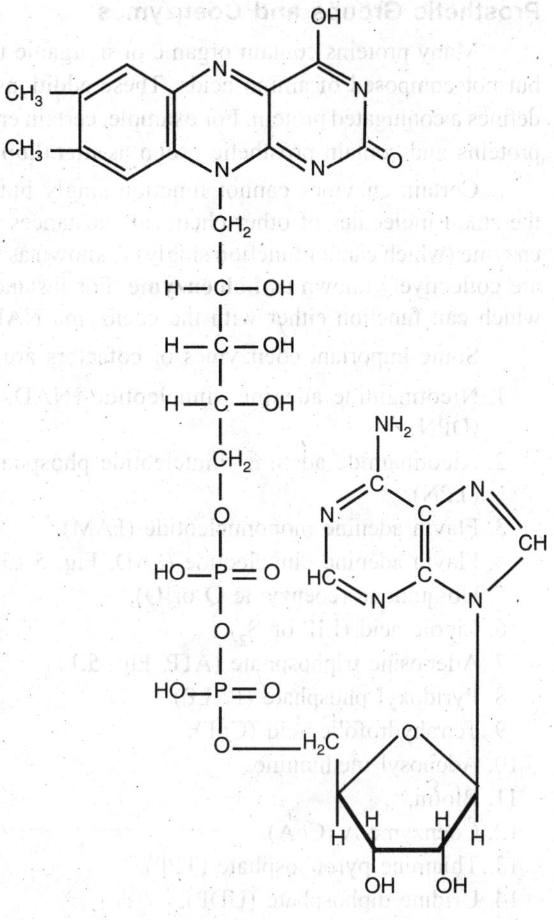

Fig. 5.15. Chemical formula of flavin adenine dinucleotide (FAD).

4. Lyases. The lyase enzymes add or remove group to or from the chemical compounds containing the double bonds. The lyases act on C–C, C–O, C–N, C–S and C–halide bonds.

5. Isomerases. These enzymes catalyse the reaction involving in the isomerization or intramolecular rearrangements in the substrates, *e.g.*, intramolecular oxidoreductases, intramolecular transferases, intramolecular lysases, cis-trans-isomerases, racemases and epimerases.

6. Ligases or synthetases. These enzymes catalyze the linkage of the molecules by splitting a phosphate bond. The synthetase enzymes form C–O, C–S, C–N and C–C bonds.

According to the chemical nature of the substrate the enzymes have also been classified as follows: 1. Carbohydrases, 2. Proteases (endopeptidases and exopeptidases), 3. Amylases, 4. Esterases, 5. Dehydrogenases, 6. Oxidases, 7. Decarboxylases, 8. Hydrases, 9. Transferases, and 10. Isomerases.

The enzymes are **specific** in action and many factors such as pH, temperature and concentration of the substrate affect the rate of the activity of enzymes. Certain enzymes occur in the inactive form (called **proenzymes** or **zymogens**) and these are activated by other enzymes known as **kinases** to perform catalytic activities. Likewise, the enzyme **trypsinogen** of the pancreatic cells is activated in the intestine by the enzyme **enterokinase** and the enzyme **pepsinogen** of the Chief cells of the stomach is activated by the **hydrochloric acid** which is secreted by parietal cells.

Prosthetic Groups and Coenzymes

Many proteins contain organic or inorganic units that are the integral part of the molecule, but not composed of amino acids. These additions are called **prosthetic groups;** their presence defines a conjugated protein. For example, certain enzymes such as **cytochromes** are the conjugated proteins and contain prosthetic group as **metalloporphyrins** complex in their molecules.

Certain enzymes cannot function singly but they can function only by the addition with the small molecules of other chemical substances which are known as **coenzymes**. The inactive enzyme (which cannot function singly) is known as the **apoenzyme**. The apoenzyme and coenzyme are collectively known as **holoenzyme**. For instance, the enzyme **hydrogenase** is an apoenzyme which can function either with the coenzyme NAD^+ or NADP.

Some important coenzymes or cofactors are as follows:

1. Nicotinamide adenine dinucleotide (NAD, Fig. 5.16) or Diphosphopyridine nucleotide (DPN),
2. Nicotinamide adenine dinucleotide phosphate (NADP) or Triphosphopyridine nucleotide (TPN),
3. Flavin adenine mononucleotide (FAM),
4. Flavin adenine dinucleotide (FAD, Fig. 5.15)
5. Ubiquinone (coenzyme Q or Q),
6. Lipoic acid (LIP or S_2),
7. Adenosine triphosphate (ATP, Fig. 5.17),
8. Pyridoxyl phosphate (PALP),
9. Tetrahydrofolic acid (CoF),
10. Adenosyl methionine,
11. Biotin,
12. Coenzyme A (CoA),
13. Thiamine pyrophosphate (TPP),
14. Uridine diphosphate (UDP).

Nucleic Acids

The nucleic acids are the complex macromolecular organic compounds of immense bio-logical importance. They control the important biosynthetic activities of the cell and carry hereditary informations from generation to generation. There occur two types of nucleic acids in living organisms, *viz.,* **ribonucleic acid (RNA)** and **deoxyribonucleic acid (DNA)**. Both types of nucleic acids are the polymers of the nucleotides. A nucleotide is composed of nucleoside and phosphoric acid. Even the nucleoside is composed of the pentose sugars (Ribose or Deoxyribose) and nitrogen bases (Purines or Pyrimidines). The **purines** are **adenine** and **guanine** and the **pyrimidines** are the **cytosine, thymine** and **uracil.** The cytoplasmic matrix contains only RNA, while DNA exclusively remains concentrated in the nucleus.

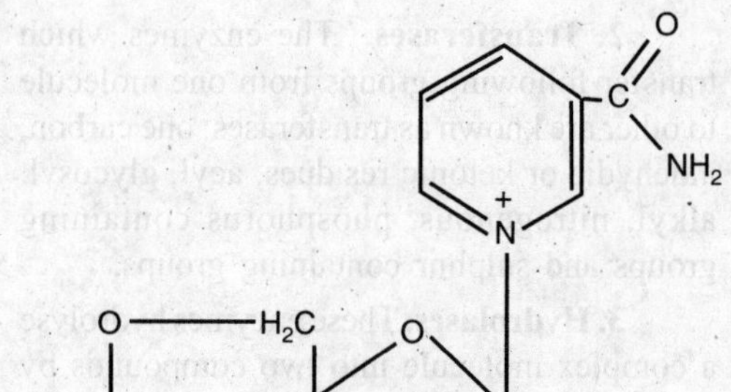

Fig. 5.16. Chemical formula of nicotinamide adenine dinucleotide (NAD^+).

The DNA and RNA have almost similar chemical compositions except a few differences. Both have been compared in Table 5.6.

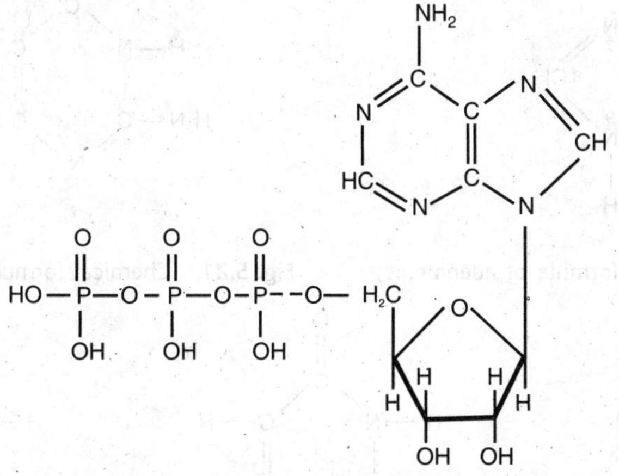

Fig. 5.17. Chemical formula of adenosine triphosphate (ATP).

Table 5.6 Comparison of DNA and RNA.

DNA	RNA
1. It contains pentose sugar known as deoxyribose.	1. It contains pentose sugar called the ribose.
2. The molecule contains the phosphoric acid (phosphate) molecule which connects various sugars with one another.	2. The molecule contains the phosphoric acid (phosphate) molecule which connects various sugars with one another.
3. The nitrogen bases are: in its molecule: (*i*) **Purines**—adenine and guanine. (*ii*) **Pyrimidine**—cytosine and thymine.	3. The molecule contains following nitrogen bases (*i*) **Purines**—adenine and guanine. (*ii*) **Pyrimidines**—cytosine and uracil.
4. Molecules have four nucleotides as deoxyadenosine monophosphate, deoxyguanosine monophosphate, deoxycytidine monophosphate and thymidine monophosphate.	4. Molecules have four nucleotides as adenosine monophosphate, guanosine monophosphate, cytidine monophosphate and uridine mono-phosphate.
5. The molecule contains a double stranded helix structure in which many nucleotides remain arranged in pair.	5. The molecules consist of single chain of polynucleotides.
6. DNA is a genetic material and occurs in chromosomes, nucleoplasm and mitochondria, chloroplasts, etc.	6. RNA is a carrier of genetic informations and it plays very significant role in the mechanism of protein synthesis. It mostly occurs in nucleolus, nucleoplasm and cytoplasm.

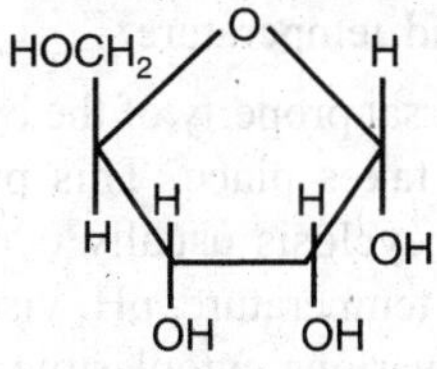

Fig. 5.18. Chemical Formula of ribose sugar.

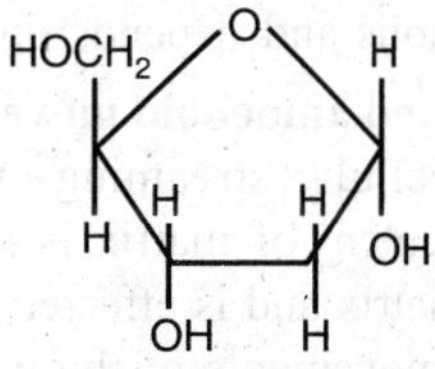

Fig. 5.19. Chemical formula of deoxyribose sugar.

Fig. 5.20. Chemical formula of adenine.

Fig. 5.21. Chemical formula of guanine.

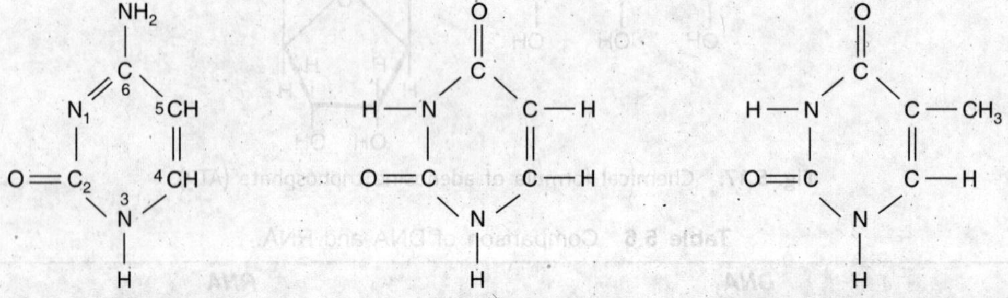

Fig. 5.22. Chemical formula of cytosine.

Fig. 5.23. Chemical formula of uracil.

Fig. 5.24. Chemical of formula of thymine.

5.3 PHYSICAL AND BIOLOGICAL PROPERTIES OF CYTOPLASMIC MATRIX

The matrix is a living substance and it has following physical and biological properties:

Physical Properties

The most of the physical properties of the matrix are due to its colloidal nature and these are as follows:

1. Tyndall's effect. When a beam of strong light is passed through the colloidal system, of the matrix at right angles in the dark room, the small colloidal particles which remain suspended in the colloidal system reflect the light. The path of the light appears like a cone. This light cone is known as **Tyndall's cone** because, this phenomenon has been first of all reported by **Tyndall** (1820–1893) in colloids.

2. Brownian movement. The suspended colloidal particles of the matrix always move in zig-zag fashion. This movement of molecules is caused by moving water molecules which strike with the colloidal molecules to provide motion to them. This type of movement was first of all observed by Scottish botanist **Robert Brown** in 1827 in the colloidal solution. Therefore, such movements are known as **Brownian movement**. The Brownian movement is the peculiarity of all colloidal solutions and depends on the size of the particles and temperature.

3. Cyclosis and amoeboid movement. Due to the phase reversal property of the cytoplasmic matrix, the intracellular streaming or movement of the matrix takes place. This property of intracullular movement of matrix is known as the **cyclosis**. The cyclosis usually occurs in the sol-phase of the matrix and is effected by the hydrostatic pressure, temperature, pH, viscosity, etc. The intracellular movements of the pinosomes, phagosomes and various cytoplasmic organelles such as the lysosomes, mitochondria, chromosomes, centrioles, *etc.,* occur only due to cyclosis of the matrix. The cyclosis has been observed, in most animal and plant cells.

The amoeboid movement depends directly on the cyclosis. The amoeboid movement occurs in the protozoans, leucocytes, epithelia, mesenchymal and other cells. In the amoeboid movement the cell changes its shape actively and gives out cytoplasmic projections known as **pseudopodia**. Due to cyclosis matrix moves these pseudopodia and this causes forward motion of the cell.

4. Surface tension. The molecules in the interior of a homogeneous liquid are free to move and are attracted by surrounding molecules equally in all directions. At the surface of the liquid where it touches air or some other liquid, however, they are attracted downward and sideways or inward, more than upward; consequently they are subjected to unequal stress and are held together to form a membrane. The force by which the molecules are bound is called the **surface tension** of the liquid. The cytoplasmic matrix being a liquid possesses the property of surface tension. The proteins and lipids of matrix have less surface tension, therefore, occur at the surface and form the membrane, while the chemical substances such as NaCl have high surface tension, therefore, occur in deeper part of the matrix.

5. Adsorption. The increase in the concentration of a substance at the surface of a solution is known as adsorption (*L.,ad*=to, *sorbex*=to draw in). The phenomenon of adsorption helps the matrix to form protein boundaries.

6. Other mechanical or physical properties of matrix. Besides surface tension and adsorption, the matrix possesses other mechanical properties, *e.g.,* elasticity, contractility, rigidity and viscosity which provide to the matrix many physiological utilities.

7. Polarity of the egg. The colloidal system due to its stable phase determines the polarity of the cell matrix which cannot be altered by centrifugation of other mechanical means.

8. Buffers and pH. The matrix has a definite pH value and it does not tolerate significant variations in its pH balance. Yet various metabolic activities produce small amount of excess acids or bases. Therefore, to protect itself from such pH variation the matrix contains certain chemical compounds as carbonate-bicarbonate system known as **buffers** which maintain a constant state of pH in the matrix.

Biological Properties

The matrix is a living substance and it has following biological properties:

1. Irritability. The irritability is the fundamental and inherent property of the matrix. It possesses a sensitivity to stimulation, an ability to transmission of excitation and ability to react according to stimuli. The heat, light, chemical substances and other factors stimulate the cytoplasmic matrix to contract.

2. Conductivity. The conductivity is the process of conduction or transmission of excitation from the place of its origin to the region of its reaction. The matrix of nerve cells possesses the property of the conductivity.

3. Movement. The cytoplasmic matrix can perform movement due to cyclosis. The cyclosis depends on the age, water contents, heredity factors and composition of the cells.

4. Metabolism. The matrix is the seat of various chemical activities. These activities may be either constructive or destructive in nature. The constructive processes such as biosynthesis of proteins, lipids, carbohydrates and nucleic acids are known as **anabolic processes,** while the destructive processes such as oxidation of foodstuffs, *etc.,* are known as **catabolic processes.** The anabolic and catabolic processes are collectively known as **metabolic process**.

5. Growth. Due to the secretory or anabolic activities (*Gr., anabolism* = a throwing up) of the cell, new protoplasm continuously increases in its volume. The increase in the volume of the matrix causes into the growth of the cell which ultimately divides into daughter cells by the cell division.

6. Reproduction. The cytoplasm has the property of asexual and sexual reproduction.

QUESTIONS

Long Answer Questions

1. Write in brief the physical properties of the protoplasm.

2. Write in brief the chemical properties of the protoplasm.

3. Describe biological properties of protoplasm.

Short Answer Questions

1. Describe in brief nine types of proteins according to function.

2. Differentiate between DNA and RNA.

Very Short Answer Questions

1. In what forms carbohydrates are stored in plants and animals?

2. Why amino acids are called amphoteric?

3. Name two animal hormones that are peptides (or proteins) and steroids.

4. What is cholesterol?

5. Explain the antibody.

6. Define antigen.

7. What is adenine?

8. Define cytosine.

9. What is guanine?

10. Define thymine.

11. What is uracil?

Fill in the Blanks

1. The monomers link with other monomers to form oligomers and ....

2. When a polymer is composed of different kinds of monomers it is known as ....

3. In a oligosaccharide .... are linked by the .... linkages.

4. Chondroitin sulphate is a .... polysaccharide and is found in .... and other connective tissues.

5. Phospholipids are the .... lipids.

6. In .... molecule of haemoglobin, the porphyrin surrounds an .... ion.

7. M.F. Perutz got Nobel Prize in 1962, along with ...., for their studies on .... and myoglobin.

Multiple Choice Questions

1. The four elements that form the 99% of the protoplasm are

(*a*) N, O, P, S (*b*) C, H, N, O

(*c*) H, N, Ca, Cl (*d*) C, O, Ca, S

2. Which of the following contain iodine?

(*a*) insulin (*b*) thyroxine

(*c*) ocytocin (*d*) adrenalin

3. Sucrose is composed of

(*a*) fructose + galactose (*b*) glucose + fructose

(*c*) glucose + glucose (*d*) glucose + galactose

4. Which of the following lipids are found in largest number in cell membrane?
   (*a*) glycolipids (*b*) phospholipids (*c*) sphingolipids (*d*) mucolipids
5. Name the gland which acts both as an endocrine as well as exocrine gland
   (*a*) pancreas (*b*) pituitary (*c*) thyroid (*d*) adrenal
6. The "repeating unit" of glycogen is
   (*a*) fructose (*b*) mannose (*c*) glucose (*d*) galactose
7. Which of the following is basic amino acid?
   (*a*) threonine (*b*) glycine (*c*) lysine (*d*) cysteine
8. Collagen is
   (*a*) phosphoprotein (*b*) globulin (*c*) derived protein (*d*) scleroprotein

ANSWERS

Very Short Answer Questions

1. In the form of starch in plants and in the form of glycogen in animals.
2. Since amino acids contain at least one amino group and at least one carboxyl group, they are called amphoteric.
3. Peptide (or protein) hormones: Insulin and Growth hormone. Steroid hormone: Estrogen and Testosterone.
4. Cholesterol is a lipid constituent of animal cell plasma membrane; serves as a precursor to the steroid hormones.
5. Antibody is a class of proteins produced by lymphocytes that bind with extraordinary specificity to substances, referred to as antigens, that provoke an immune response.
6. Antigen is a foreign or abnormal substance that can trigger an immune response.
7. Adenine (A) is a nitrogen-containing aromatic base, chemically designated as a purine, that serve as an informational monomeric unit when present in nucleic acids with other bases in a specific sequence; forms a complementary base pair with thymine (T) or uracil (U) by hydrogen bonding.
8. Cytosine (C) is a nitrogen-containing aromatic base, chemically designated as a pyrimidine, that serves as an informational monomeric unit when present in nucleic acids with other bases in a specific sequence; forms a complementary base pair with guanine (G) by hydrogen bonding.
9. Guanine (G) is a nitrogen-containing aromatic base, chemically designated as a purine, which serves as an informational monomeric unit when present in nucleic acids with other bases in a specific sequence; forms a complementary base pair with cytosine (C) by hydrogen bonding.
10. Thymine (T) is a nitrogen-containing aromatic base, chemically designated as a pyrimidine, which serves as an informational monomeric unit when present in DNA with other bases in a specific sequence; forms a complementary base pair with adenine (A) by hydrogen bonding.
11. Uracil (U) is a nitrogen-containing aromatic base, chemically designated as a pyrimidine, that serves as an information monomeric unit when present in RNA with other bases in a specific sequence; forms a complementary base pair with adenine (A) by hydrogen bonding.

Fill in the Blanks

1. Polymers;
2. Heteropolymer.
3. Monosaccharides; glycosidic.
4. Structural; cartilage.
5. Compound.
6. Haeme; iron.
7. J.K. Kendrew; haemoglobin.

Multiple Choice Questions

1. (*b*) 2. (*b*) 3. (*b*) 4. (*b*) 5. (*a*) 6. (*c*)
7. (*c*) 8. (*d*).

6

Plasma Membrane and Cell Wall

6.1 PLASMA MEMBRANE

A plasma membrane encloses every type of cell, both prokaryotic and eukaryotic cells. It physically separates the cytoplasm from the surrounding cellular environment. Plasma membrane is a ultrathin, elastic, living, dynamic and selective transport-barrier. It is a fluid-mosaic assembly of molecules of lipids (phospholipids), proteins and carbohydrates. Plasma membrane controls the entry of nutrients and exit of waste products, and generates differences in ion concentration between the interior and exterior of the cell. It also acts as a sensor of external signals (for example, hormonal, immunological, etc.) and allows the cell to react or change in response to environmental signals. The cells of bacteria and plants have the plasma membrane between the cell wall and the cytoplasm. For cells without cell walls (*e.g.*, mycoplasma and animal cells), plasma membrane forms the cell surface.

The plasma membrane is also called **cytoplasmic membrane, cell membrane,** or **plasmalemma.** The term cell membrane was coined by **C. Nageli** and **C. Cramer** in 1855 and the term plasmalemma has been given by **J.Q. Plowe** in 1931.

Isolation and Analysis

The plasma membrane is so thin that it cannot be observed by the light microscope. Structure of the plasma membrane of various cells has been studied by their isolation from the living systems and also by their artificial synthesis by using their constituent molecules (*e.g.*, liposome, see Chapter 57). The pure and isolated membranes are then studied by biochemical and biophysical methods. The purity of isolated membranes is controlled by **electron microscopy, enzyme analysis** and the **study of surface antigens**. A variety of cells such as mammalian red blood cell (erythrocytes), medullated nerve fibres, Ehrlich mouse ascites tumor cells, liver cells, striated muscle, *Amoeba proteus*, sea urchin eggs and bacteria, have been used in studying the ultrastructure of the plasma membrane. The mammalian erythrocytes and the myelin sheath of the nerve fibre, however, have provided the bulk of information regarding the structure and properties of the plasma membrane. For such experiments, human red blood cells or erythrocytes have been selected by **E. Gorter** and **F. Grendel** (1925) for following advantages: these cells are easy to obtain and are known to be extremely simple. Since these cells contain no intracellular organelles or membrane, so the only membrane structure to be considered is almost entirely that of the cell surface. Lastly, the plasma membrane of erythrocytes is relatively tough and does not readily fragment (See **Lucy**, 1975).

Plasma membranes are more easily isolated from erythrocytes subjected to haemolysis. The cells are treated with hypotonic solutions (to be discussed elsewhere in the chapter) that due to endosmosis produce swelling and then loss of the heamoglobin content (*i.e.*, haemolysis). The resulting membrane is called a **red cell ghost**. If haemolysis is mild, permeability functions of the membrane can be restored by certain treatment, such a ghost is called **resealed ghost**. But if haemolysis is more drastic (*i.e.*, there is complete removal of the haemoglobin) and there is no chance of its resealing, the resulting membrane is called **white ghost**. While the resealed ghosts can be used for the study of physiological as well as biochemical properties, white ghosts can only be used for the study of biochemical properties.

The cell wall of yeast, *Saccharomyces cerevisiae*, can be enzymatically removed by the help of a snail gut enzyme, and the resultant protoplast serves as a source of plasma membrane in a manner similar to that of mammalian erythrocytes.

Chemical Composition

Chemically, plasma membrane and other membranes of different organelles are found to contain proteins, lipids and carbohydrates, but in different ratios (Table 6.1). For example, in the plasma membrane of human red blood cells proteins represent 52 per cent, lipids 40 per cent and carbohydrates 8 per cent.

Table 6.1 Chemical composition of some purified membranes (in percentages).

	Membrane	*Protein*	*Lipid*	*Carbohydrate*
1.	Myelin (Nerve cell)	18	79	3
2.	Plasma membrane:			
	(*i*) Mouse liver	44	52	4
	(*ii*) *Amoeba*	54	42	4
	(*iii*) Human erythrocyte	52	40	8
3.	Spinach chloroplast lamellae	70	30	0
4.	Mitochondrial inner membrane	76	24	0

1. Lipids. Four major classes of lipids are commonly present in the plasma membrane and other membranes: **phospholipids** (most abundant), **sphingolipids, glycolipids** and **sterols** (*e.g.*, **cholesterol**). All of them are amphipathic molecules, possessing both hydrophilic and hydrophobic domains. The relative proportions of these lipids vary in different membranes. Phospholipids may be **acidic phospholipids** (20 per cent) such as **sphingomyelin** or **neutral phospholipids** (80 per cent) such as **phosphatidyl choline, phosphatidyl serine,** etc. Many membranes contain cholesterol. Cholesterol is especially abundant in the plasma membrane of mammalian cells and absent from prokaryotic cells. **Cardiolipin** (diphosphatidyl glycerol) is restricted to the inner mitochondrial membrane.

2. Proteins. The amount and types of proteins in the membranes are highly variable: in the myelin membranes which serve mainly to insulate nerve cell axons, less than 25 per cent of the membrane mass is protein, whereas, in the membranes involved in energy transduction (such as internal membranes of mitochondria and chloroplasts), approximately 75 per cent is protein. Plasma membrane contains about 50 per cent protein.

According to their position in the plasma membrane, the proteins fall into two main types: **integral** or **intrinsic proteins** and **peripheral** or **extrinsic proteins**, both of which may be either **ectoproteins**, lying or exposing to external or extracytoplasmic surface of the plasma membrane or **endoproteins**, lying or sticking out at the inner or cytoplasmic surface of the plasma membrane. The intrinsic proteins tend to associate firmly with the membrane, while the extrinsic proteins have a weaker association and are bound to lipids of membrane by electrostatic interaction. On the basis of their functions, proteins of plasma membrane can also be classified into three main types: structural proteins, enzymes and transport proteins (permeases or carriers). Some of them may act as **antigens**,

receptor molecules (*e.g.*, insulin-binding sites of liver plasma membrane), **regulatory molecules** and so on. **Structural proteins** are extremely lipophilic and form the main bulk (*i.e.*, backbone) of the plasma membrane. **Enzymes** of plasma membrane are either **ectoenzymes** or **endoenzymes** and are of about 30 types (Table 6.2). **Transport proteins** transport specific substances across the plasma membrane and other cellular membranes.

Table 6.2 Some important enzymes present in the plasma membrane.

1.	Acetyl phosphatase	10.	Aminopeptidase
2.	Acetyl cholinesterase (Ectoenzyme of erythrocyte)	11.	Cholesterol esterase
		12.	Guanylase cyclase
3.	Acid phosphatase	13.	Monoglyceride lipase
4.	Adenosine triphosphatase	14.	NAD-ase (Ectoenzyme of erythrocyte)
5.	Mg^{2+} ATPase (Endoenzyme of erythrocyte)	15.	Protein kinase (Endoenzyme of erythrocyte)
6.	Na^+-K^+-ATPase (Ectoenzyme of erythrocyte)	16.	Phospholipase A
		17.	Lactase
7.	Adenylate cyclase	18.	Maltase
8.	RNAase	19.	Sialidase
9.	Alkaline phosphatase	20.	UDP glycosidase

3. Carbohydrates. Carbohydrates are present only in the plasma membrane. They are present as short, unbranched or branched chains of sugars (**oligosaccharides**) attached either to exterior ectoproteins (forming **glycoproteins**) or to the polar ends of phospholipids at the external surface of the plasma membrane (forming **glycolipids**). No carbohydrate is located at the cytoplasmic or inner surface of the plasma membrane. All types of oligosaccharides of the plasma membrane are formed by various combinations of six principal sugars (all of which are glucose-derivatives): **D-galactose, D-mannose, L-fucose, N-acetylneuraminic acid** (also called **sialic acid**), **N-acetyl-D-glucosamine** and **N-acetyl-D-galactosamine**.

Structure of Plasma Membrane

1. Evolution of Fluid Mosaic Model of Membrane

The existence of the plasma membrane of the cell was difficult to prove by direct examination before 1930's (when electron microscopy was invented) because of technological limitations. The membrane is beyond the resolution of the light microscope, rendering a morphological approach of its study quite unfeasible with this instrument. Thus, most of the experimental approaches have been provided by only indirect evidences of the existence of such a membrane around the cells. Let us narrate in brief the saga of evolution of presently well accepted fluid-mosaic model of structure of the plasma membrane:

1. The plasmolysis of plant cells in hypertonic solutions suggests the existence of the plasma membrane in the plants.
2. The very fact that a cell, especially an animal cell which has no cell wall, can exist as a physically defined entity suggests that it must have some sort of boundary around it.
3. The presence of plasma membrane can be inferred because protoplasm leaks out of animal cells when cell surface is punctured.
4. After performing some 10,000 experiments with more than 500 different chemicals, in 1899, **Overton** concluded that the peculiar osmotic properties of living protoplasts are due to a **selective solubility mechanism**. Hydrophobic compounds entered cells more rapidly than hydrophilic ones. **Overton** believed this was because of an outer lipoid layer in which

hydrophobic compounds were more soluble. He correctly speculated that this layer might contain cholesterol, lecithin and fatty oils.

5. **Hober** (1910) and **Fricke** (1925) found that the intact cell had low electrical conductivity, indicating the presence of a lipid layer around it.
6. If a lipid containing **hydrophilic groups** (such as the carboxyl groups of fatty acids or the phosphate groups of phospholipids) is dissolved in a highly volatile solvent (*e.g.*, benzene) and several drops of it are then carefully applied to the surface of the water, the lipid spreads out to form a thin, one-molecule-thick or **monomolecular film.** In this film, it is found that the hydrophilic parts of each molecule project into the water surface and the hydrophobic parts are directed up, away from the water.
7. In 1917, **Langmuir** (Nobel Laureate of 1932 in chemistry) fabricated a **trough** or **film balance** (Fig. 6.1) for measuring the specific minimum surface area occupied by a monomolecular film of lipid and the force necessary to compress all the lipid molecules into this area. Langmuir trough consists of a shallow trough filled with water on which lipid substance can be spread to make a monomolecular film. A barrier can be pushed across the trough to compress the film.

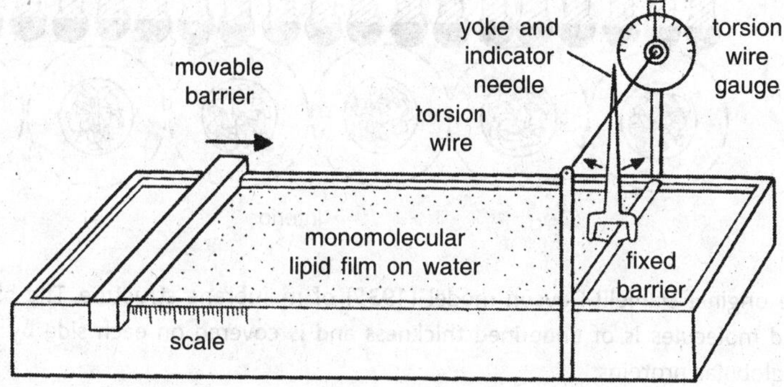

Fig. 6.1. Langmuir trough.

8. In 1925, **Gorter** and **Grendel** extracted the lipids from erythrocyte ghosts of a variety of mammals (such as dogs, sheep, rabbits, guinea pigs, goats and humans) and spread them out on monolayers in the Langmuir trough. These investigators discovered that the area covered by the lipid monomolecular layer film was twice than what was needed to cover the surface of the cells from which the lipid was extracted. Consequently, they safely concluded that *erythrocytes were covered by a layer of lipids two molecules thick* (**lipid bilayer or bimolecular lipid layer**) oriented with polar groups toward the inside and outside of the cell.
9. By studying the surface tension of cells **Harvey** (**Harvey** and **Cole**, 1931, **Danielle** and **Harvey**, 1935) suggested the presence of proteins in the plasma membrane, in addition to the lipids.
10. In 1935, **Danielli** and **Davson**, proposed a model, called **sandwich model**, for membrane structure in which a lipid bilayer was coated on its either side with hydrated proteins (globular proteins). Mutual attraction between the hydrocarbon chains of the lipids and electrostatic forces between the protein and the "head" of the lipid molecules, were thought to maintain the stability of the membrane. From the speed at which various molecules penetrate the membrane, they predicted the lipid bilayer to be about 6.0 nm in thickness, and each of the protein layer of about 1.0 nm thickness, giving a total thickness of about 8.0 nm.

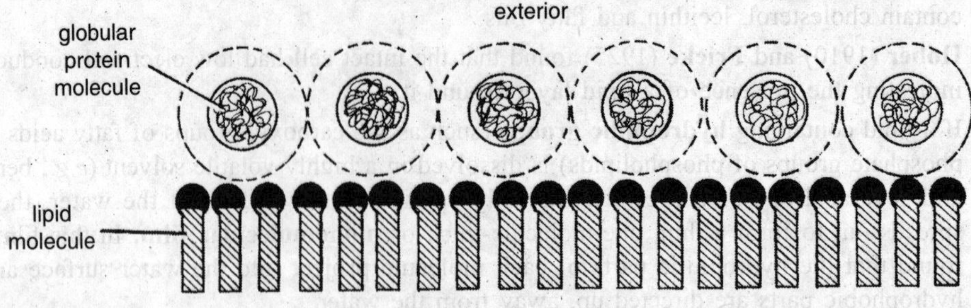

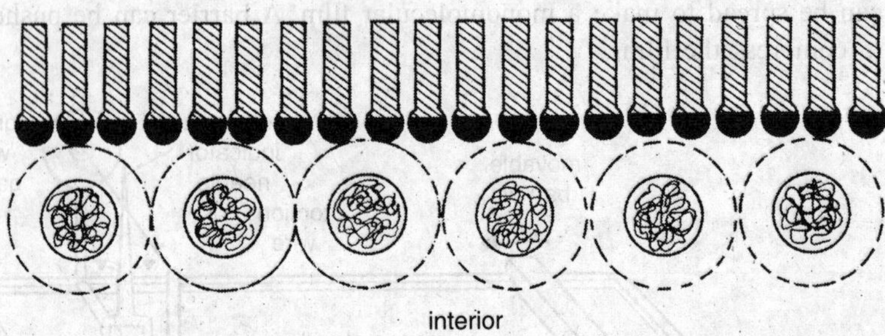

Fig. 6.2. The original Danielli-Davson model (1935) of membrane structure. The bimolecular layer of lipid molecules is of undefined thickness and is covered on each side by a continuous layer of globular proteins.

The Danielli-Davson model got support from electron microscopy. Electron micrographs of the plasma membrane showed that it consists of two dark layers (electron dense granular protein layers), both separated by a lighter area in between (the central clear area of lipid bilayer). The total thickness of the membranes is also turned out to be about 7.5 nm.

11. **Myelin as a model for unit membrane model.** To understand the relationship between the trilaminar appearance of plasma membrane and its underlying molecular structure, **Robertson** chose myelin as the model system for study. **Myelin** surrounds the axons of certain nerve cells and is composed of multiple layers of plasma membrane successively wound around one another. This stack of membrane is derived from the specialised **Schwann cells** that are intimately associated with the axon (Fig. 6.3). During the early stages of myelin formation, axons become surrounded by Schwann cell. The plasma membrane of Schwann cell start wrapping around the axon in a spiral manner. At the later stage of development, the plasma membrane of the Schwann cell become closely stacked upon each other and the cytoplasmic content of the cell becomes disorganised from the space between the membranes. This stack of closely packed membranes around the axon is called **myelin**.

Robertson used myelin, rather than a typical membrane as a model for study because, the repeating layers of membranes present in myelin creates a quasicrystalline structure agreeable to analysis by X-ray diffraction. The X-ray diffraction analyses, carried out by **F.O. Schmitt** and later on **J.B. Fineam** indicated that myelin consists of lipid bilayer alternating with layers of protein. These studies revealed that the general pattern of plasma membrane had one of three possible arrangements of protein and lipid (Fig. 6.4).

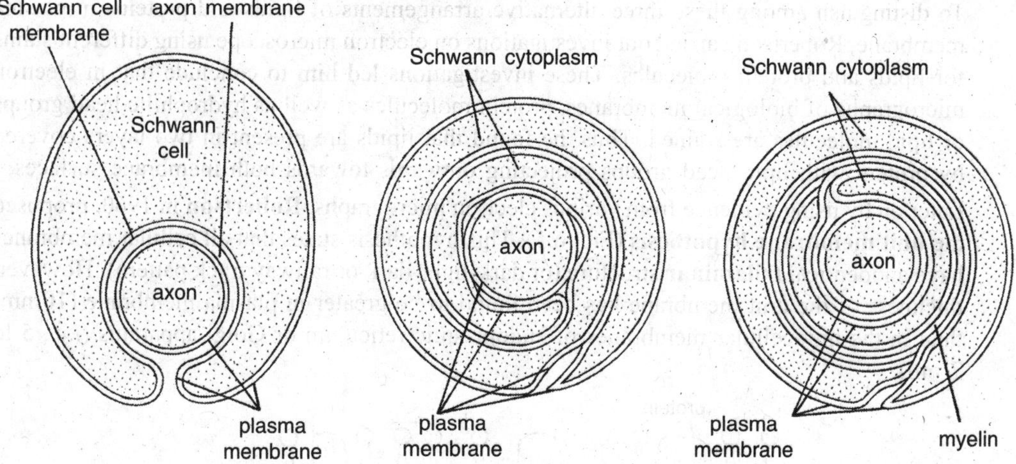

Fig. 6.3. Myelin formation around axon of neuron by Schwann cell.

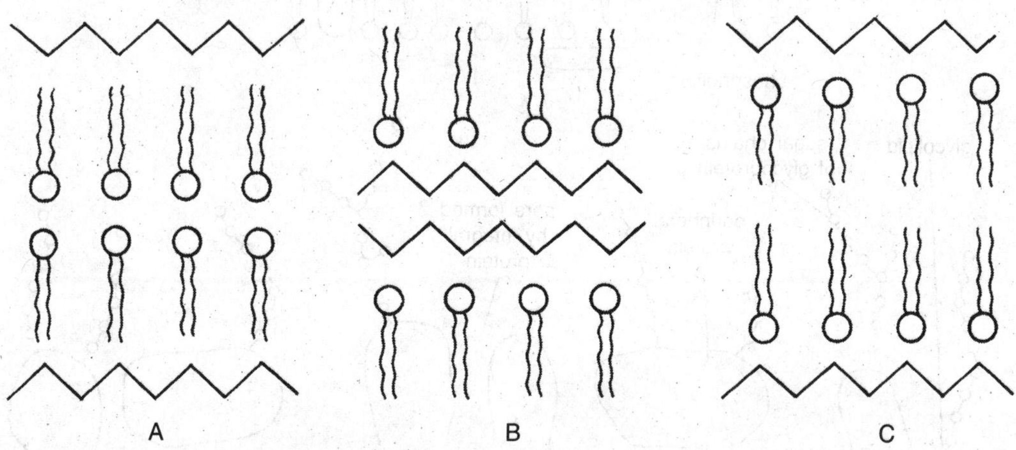

Fig. 6.4. Three possible arrangements of protein and lipid in plasma membrane.

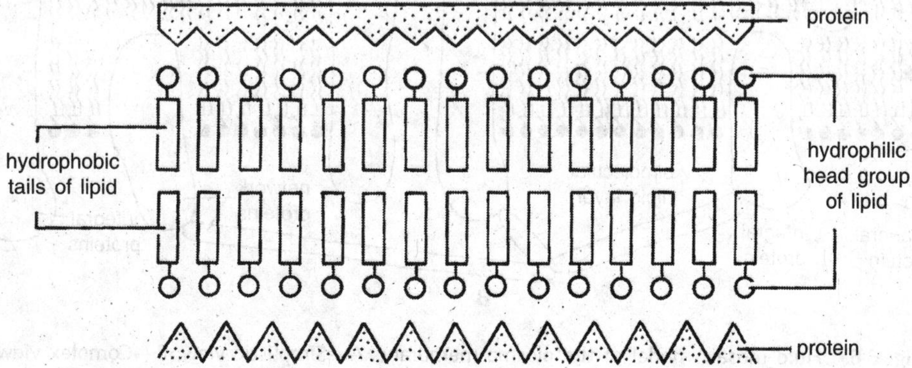

Fig. 6.5. Schematic diagram of the Robertson model of membrane structure. The lipid layer is defined as bimolecular, and the protein is extended but different on the two faces of the membrane.

To distinguish among these three alternative arrangements of lipids and proteins in plasma membrane, Robertson carried out investigations on electron microscope using different stains for lipids and protein molecules. These investigations led him to conclude that in electron micrographs of biological membranes, protein molecules as well as hydrophilic head groups of lipid molecules are stained. Thus, he found that lipids are present in two layers covered by protein with lipid head groups projecting outwards towards both membrane surfaces. Later on by using evidence from various electron micrographs, **Robertson** in 1960, proposed the **unit membrane hypothesis** (Fig. 6.5). This hypothesis states that all cellular membranes have an identical **trilaminar** structure (or dark-light-dark or railway track pattern). However, thickness of the unit membrane has been found to be greater in plasma membrane (10 nm) than in the intracellular membranes of endoplasmic reticulum or Golgi apparatus (*i.e.*, 5 to 7 nm).

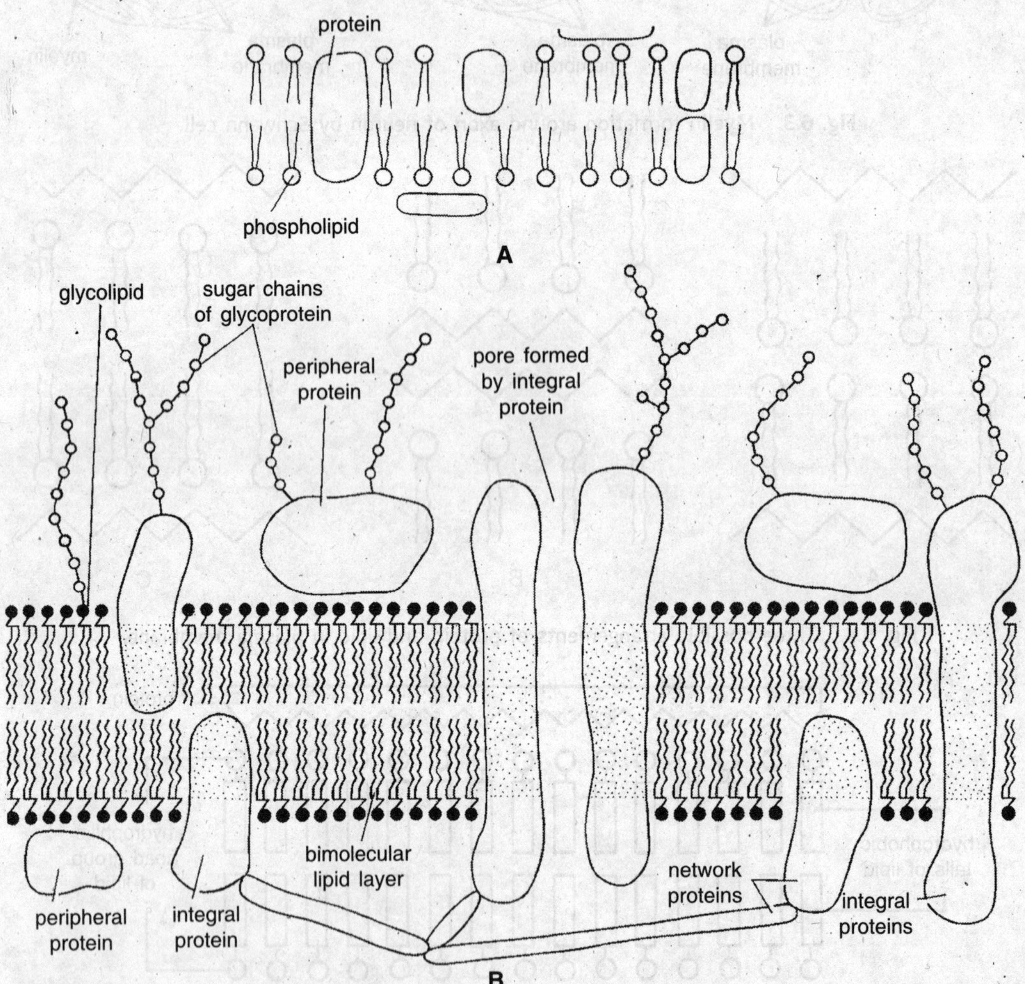

Fig. 6.6. Fluid mosaic model of the plasma membrane. A—Simplistic view; B—Complex view.

12. **S.J. Singer** and **G.L. Nicolson** (1972) suggested the widely accepted **fluid mosaic model** of biological membranes. According to this model (Fig. 6.6), the plasma membrane contains a bimolecular lipid layer, both surfaces of which are interrupted by protein molecules. Proteins occur in the form of globular molecules and they are dotted about here and there in a mosaic

pattern. Some proteins are attached at the polar surface of the lipid (*i.e.*, the extrinsic proteins); while others (*i.e.*, integral proteins) either partially penetrate the bilayer or span the membrane entirely to stick out on both sides (called **transmembrane proteins**). Further, the peripheral proteins and those parts of the integral proteins that stick on the outer surface (*i.e.*, ectoproteins) frequently contain chains of sugar or oligosaccharides (*i.e.*, they are glycoproteins). Likewise, some lipids of outer surface are glycolipids.

The fluid-mosaic membrane is thought to be a far less rigid than was originally supposed. In fact, experiments on its viscosity suggest that it is of a fluid consistency rather like the oil, and that there is a considerable sideways movement of the lipid and protein molecules within it. On account of its fluidity and the mosaic arrangement of protein molecules, this model of membrane structure is known as the "fluid mosaic model" (*i.e.*, it describes both properties and organization of the membrane). The fluid mosaic model is found to be applied to all biological membranes in general, and it is seen as a dynamic, ever-changing structure. The proteins are present not to give it strength, but to serve as enzymes catalysing chemical reactions within the membrane and as pumps moving things across it.

Experimental Evidence in Support of Fluid Mosaic Model of Plasma Membrane

There is a good deal of evidence to support the fluid mosaic model of the plasma membrane:

A. Evidence in support of mosaic arrangement of proteins. Freeze-fracture electron micro-scopy of the plasma membrane by **Branton** (1968) revealed the presence of bumps and depressions (7 to 8 nm in diameter) which are randomly distributed. These were later shown to be transmembrane integral protein particles (Fig. 6.7). (For details of freeze-fracture technique, see Chapter 2).

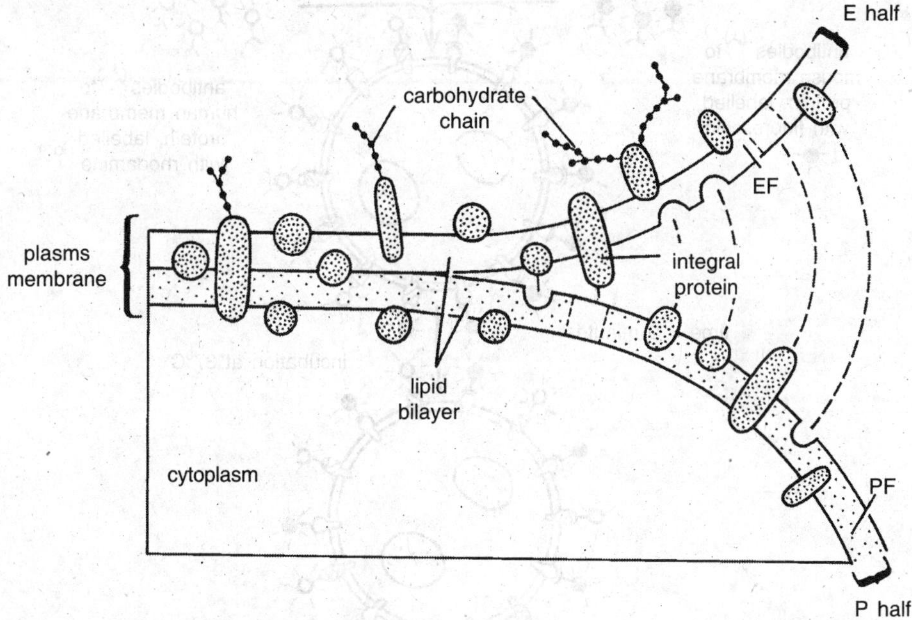

Fig. 6.7. Freeze fracturing of the plasma membrane. The fracture plane occurs at the centre of the lipid bilayer and passes over (or under) the integral membrane proteins. E = exterior, P = protoplasmic (cytosolic) side, F = fracture face.

B. Evidence in support of fluid property of lipid bilayer. Mobility of membrane proteins due to fluid property of lipid bilayer was demonstrated by a classical experiment of **D. Frye** and

M. Edidin (1970). They fused two different types of cultured cells having differe... surface antigens (proteins). The **cell fusion** is achieved by the use of some fusogen such as an inactivated parainfluenza virus, called Sendai virus (named after a city of Japan). A **fusogen** is a membrane fusion promoting factor such as Sendai virus, lysophosphatides, oleic acid or an electric field. Sendai virus facilitates

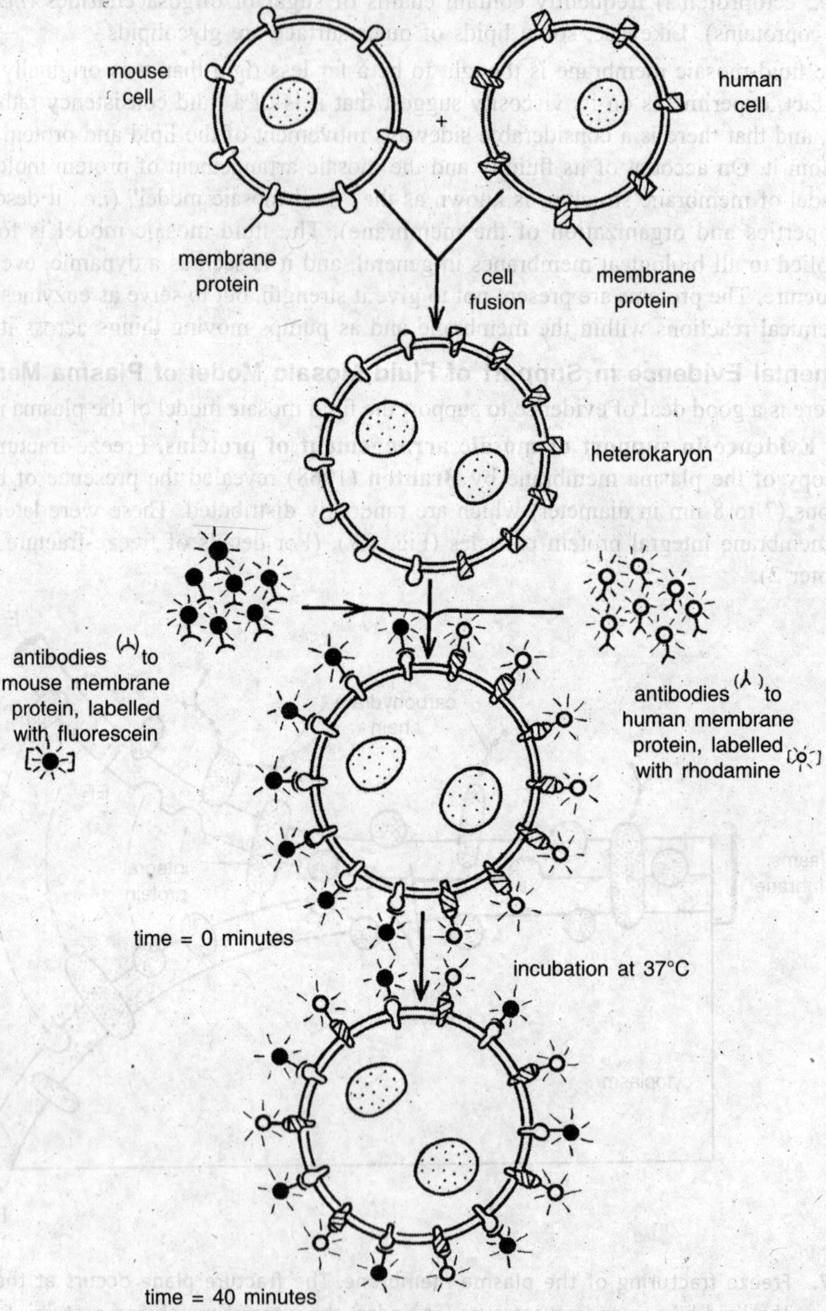

Fig. 6.8. Frye and Edidin's experiment demonstrating the mixing of plasma membrane proteins on mouse-human hybrid cells. In the plasma membrane of newly formed heterokaryon, the mouse and human proteins initially confined to their own halves, but they intermixed with time.

fusion of the plasma membranes and cytoplasms of both cells to produce a **hybrid cell** or **heterokaryon** with two types of nuclei. If the two cells are originally labelled with fluorescent antibodies of different colours, such as fluorescein (green) and rhodamine (red), it is possible at the onset of fusion to recognise the parts of the plasma membrane corresponding to each cell. However intermixing occurs as the antigens are dispersed and the two colours become less and less detectable. After 40 minutes (at 37° C) the intermixing of two colours is complete and the two antigens can no longer be distinguished (Fig. 6.8).

Role of Lipid Molecules in Maintaining Fluid Property of Membrane

(*i*) Types of movements of lipid molecules. Lipid molecules very rarely migrate from one lipid monolayer to other monolayer of lipid bimolecular layer. Such a type of movement is called **flip-flop** or **transbilayer movement** and occurs once a month for any individual lipid molecule. However, in membranes where lipids are actively synthesized, such as smooth ER, there is a rapid flip-flop of specific lipid molecules across the bilayer and there are present certain membrane-bound enzymes, called **phospholipid translocators** (*e.g.*, flippase) to catalyze this activity. On the other hand, lipid molecules readily exchange places with their neighbours within a monolayer (~ 10^7 times a second). This results in their rapid **lateral diffusion**. Individual lipid molecules **rotate** very rapidly about their long axes and their hydrocarbon chains are flexible, the greatest degree of **flexion** occurring near the centre of the bilayer and the smallest adjacent to the polar head groups (Fig. 6.9).

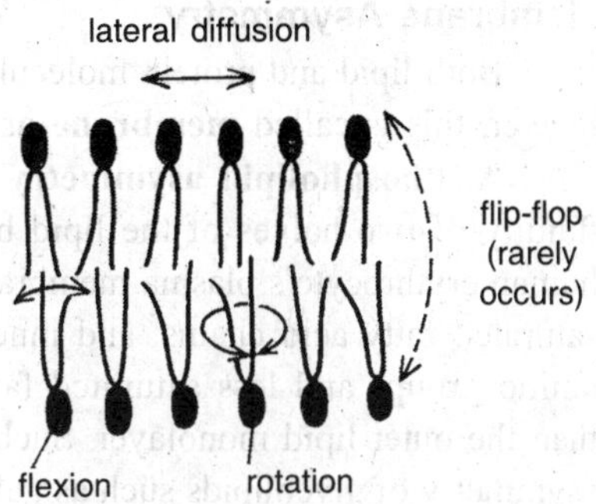

Fig. 6.9. The types of movement possible for phospholipid molecules in a lipid bilayer.

(*ii*) Role of unsaturated fats in increasing membrane fluidity. A synthetic bilayer made from a single type of phospholipid changes from a liquid state to a rigid crystalline or gel (viscous) state at a characteristic freezing point. This change of state is called a **phase transition** and the temperature at which it occurs becomes lower if the hydrocarbon chains are short or have double bonds. Double bonds in unsaturated hydrocarbon chains tend to increase the fluidity of a phospholipid bilayer by making it more difficult to pack the chains together. Thus, to maintain fluidity of the membrane, cells of organisms living at low temperatures have high proportions of unsaturated fatty acids in their membranes, than do cells at higher temperatures.

In fact, certain membrane **transport processes** and **enzyme activities** are found to cease when the lipid bilayer's viscosity increases beyond a threshold level. In contrast, if lipid bilayer's fluidity is increased, the membrane's receptors for the hormone are withdrawn from the cell surface, thereby hampering hormone action.

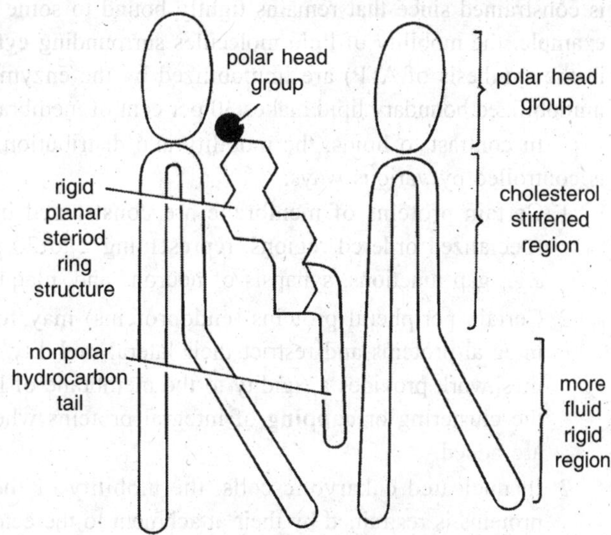

Fig. 6.10. Cholesterol molecule (schematic) interacting with two phospholipid molecules in a monolayer.

***(iii)* Role of cholesterol in maintaining fluidity of membrane.** Eukaryotic plasma membranes (in animals) are found to contain a large amount of cholesterol; up to one molecule for every phospholipid molecule. Cholesterol molecules orient themselves in the lipid bilayer in such a way that their hydroxyl groups remain close to polar head groups of the phospholipids, their rigid plate-like steroid rings interact with and partly immobilise those regions of hydrocarbon chains that are closest to the polar head groups, leaving the rest of the chain flexible (Fig. 6.10). Cholesterol inhibits phase transition by preventing hydrocarbon chains from coming together and crystallizing. Cholesterol also tends to decrease the permeability of lipid bilayers to small water-soluble molecules and is thought to enhance both the flexibility and the mechanical stability of the bilayer.

Membrane Asymmetry

Both lipid and protein molecules have irregular distribution in both monolayers of the lipid bilayer, this is called **membrane asymmetry**.

A. Phospholipid asymmetry in plasma membrane. The lipid composition and state of fluidity of two halves of the lipid bilayer are found to be strikingly different. For example, in human erythrocyte's plasma membrane, outer half contains those phospholipids which have more saturated fatty acid chains, and inner half contains those phospholipids which contain terminal amino groups and less saturated fatty acid chains. As a result, inner monolayer is more fluid than the outer lipid monolayer. Such a phospholipid asymmetry is generated in smooth ER. The asymmetry of glycolipids such as galactocerebroside, ganglioside, etc., in myelin sheath of nerves (*i.e.,* they are found only in the outer half of lipid bilayer) is found to be originated in lumen of Golgi apparatus. The specific role of lipid asymmetry of the membrane is still not clear.

B. Protein asymmetry in plasma membrane. The outer and inner sides of the plasma membrane and other membranes do not contain either the same types or equal amounts of the various peripheral and integral proteins, *e.g.,* erythrocyte's plasma membrane.

Constraints on the Mobility of Membrane Molecules

In the fluid mosaic plasma membrane, there is not complete and independent freedom of movement for its different component molecules. The mobility of some part of lipid molecules is constrained since that remains tightly bound to some of the integral membrane proteins. For example, the mobility of lipid molecules surrounding **cytochrome oxidase** (an enzyme involved in the synthesis of ATP) are immobilized by the enzyme and makes boundary lipid layer. The immobilized boundary lipid makes 30 per cent of membrane lipid in the mitochondrial membrane.

In contrast to lipids, the mobility and distribution of protein molecules in the membrane is controlled by various ways:

1. Certain proteins of membrane are constrained by protein-protein interactions to form specialized ordered regions, representing 2 to 20 per cent of the membrane of a system, *e.g.,* gap junctions, synapsis of neurons and, plaques of halobacteria.
2. Certain peripheral proteins (endoproteins) may form a bridge-like lattice work between integral proteins and restrict their lateral mobility, *e.g.,* spectrin-ankyrin-actin cytoskeletal meshwork provides a rigidity to the membrane of human erythrocytes and does not permit the clustering or **capping** of integral proteins when the appropriate antibodies or lectins are added.
3. In nucleated eukaryotic cells, the mobility of the peripheral endoproteins and integral proteins is restrained by their attachment to the ectoplasmic cytoskeleton. The cytoskeleton is extensive, including **myosin** filaments, **actin** filaments and **microtubules**. Rearrangement of cytoskeletal components just below the cell surface manifests in the distribution of integral membrane proteins and also in the cellular motions, endocytosis and exocytosis.

Functions of Plasma Membrane

Plasma membrane performs the following vital functions for the cell:

1. Regulation of passage of materials. The plasma membrane regulates the passage of materials into and out of the cell. It is **selectively permeable**. It facilitates the entrance of required nutrients into the cells and permits the exit of digestive and nutritive wastes. It checks the exit of useful substances from the cell.

Passage of materials across the plasma membrane occurs by **passive transport** (such as osmosis, simple diffusion), **facilitated diffusion** (*e.g.,* ion transport through charged pores, Ca^{2+}-channels and D-hexose and anion exchange permeases of RBCs), **active transport** (by Na^+ – K^+ ATPase, calcium ATPase, proton pump) and **bulk transport** (by exocytosis, phagocytosis, pinocytosis and receptor mediated endocytosis).

2. Maintenance of differential distribution of ions. Plasma membrane regulates and maintains the differential distribution of ions inside and outside the cell. For example, potassium ions are concentrated inside a living cell, whereas sodium and chloride ions are distributed outside. Such an unequal distribution of ions leads to a potential difference. This electric potential energy in all living cells is known as **membrane potential.**

3. Response to environment. The plasma membrane responds to changes in its environment with the help of **receptor proteins**. These receive chemical messages from other cells and thus help in cell recognition. The receptor proteins respond to **hormones, growth factors** and **neurotransmitters.**

4. Contact with neighbouring cells. The plasma membrane maintains structural and chemical relationships with the neighbouring cells by the help of certain glycoproteins.

5. Protection. Plasma membrane protects the cell and may be involved in cell movement, cell secretion and transmitting impulses.

Origin of Plasma Membrane

There is hardly any cell structure more important to the immediate health of the cell than the plasma membrane. If it is weakened or injured, the cell loses its ability to maintain gradients, to carry out the selective transport of nutrients, and to contain the pool of enzymes and organelles essential for the homeostasis. In consequence, new membranes may be added to existing membranes without altering the functions as a barrier and selective transporter. Also for maintaining the characteristic membrane asymmetry, the membrane must be assembled with precisely the correct molecular topography.

Thus, all cellular membranes grow from pre-existing membranes which act as **templates** for the addition of new precursors. All cells divide, daughter cells receive a full complement of membrane systems which undergo growth until the next division, to be passed on to subsequent progeny. Meanwhile the molecules within the membrane undergo continuous replacement.

The protein molecules of the plasma membrane are synthesized on both attached and free ribosomes. Proteins synthesized by free ribosomes may be inserted into the plasma membrane following their completion and release from the ribosomes. Proteins of plasma membrane synthesized on attached ribosomes of rough ER are **inserted** first into the membrane of RER and then **transferred** to the Golgi apparatus, **processed** there (*e.g.*, glycosylation) and ultimately are dispatched to the plasma membrane via the secretory vesicles. Likewise, the synthesis of phospholipid molecules of the plasma membrane takes place by the smooth ER (SER). Like the proteins, newly synthesized lipids are inserted into SER membranes, then they are passed to Golgi apparatus for the processing and ultimately are dispatched to the plasma membrane via small secretory vesicles. The cytosol also contains a number of **phospholipid transport proteins** that function to transfer phospholipid molecules from one cellular membrane to another (*e.g.*, from ER membranes to plasma membrane).

In fact, the process of glycosylation (or glycosidation, *i.e.*, addition of oligosaccharides containing the sugars such as galactose, fucose and/or sialic acid, to the molecules of proteins and phospholipids of the plasma membrane) is completed at the level of Golgi apparatus. However, some sugars are added to the proteins in the lumen of RER.

6.2 MEMBRANE TRANSPORT

The transport of substances into the cell is performed by the following methods:

A. Transport of Water

1. Osmosis

B. Transport of Ions and Small Molecules

2. Passive transport
   (*i*) Simple diffusion
   (*ii*) Facilitated diffusion
3. Active transport

C. Transport of Large Molecules (Bulk Transport)

4. Phagocytosis
5. Endocytosis
   (*i*) Pinocytosis
   (*ii*) Receptor mediated endocytosis

The plasma membrane acts as a semipermeable barrier between the cell and the extracellular environment. This permeability must be highly **selective** if it is to ensure that essential molecules such as glucose, amino acids and lipids can readily enter the cell, that these molecules and metabolic intermediates remain in the cell, and that waste compounds leave the cell. In short, the **selective permeability** of the plasma membrane allows the cell to maintain a constant internal environment (**homeostasis**). Transport across the membrane may be passive or active. It may occur via the phospholipid bilayer or by the help of specific integral membrane proteins, called **permeases** or **transport proteins.**

A. Passive transport. It is a type of **diffusion** in which an ion or molecule crossing a membrane moves down its electrochemical or concentration gradient. *No metabolic energy is consumed in passive transport*. Passive transport is of following three types:

1. Osmosis

The plasma membrane is permeable to water molecules. The to and fro movement of water molecules through the plasma membrane occurs due to the differences in the concentration of the solute on its either sides. The process by which the water molecules pass through a membrane from a region of higher water concentration to the region of lower water concentration is known as **osmosis** (*Gr.*, *osmos* = pushing). The process in which the water molecules enter into the cell is known as **endosmosis**, while the reverse process which involves the exit of the water molecules from the cell is known as **exosmosis**. In plant cells due to excessive exosmosis the cytoplasm along with the plasma membrane shrinks away from the cell wall. This process is known as **plasmolysis** (*Gr.*, *plasma* = molded, *lysis* = loosing) (Fig. 6.11).

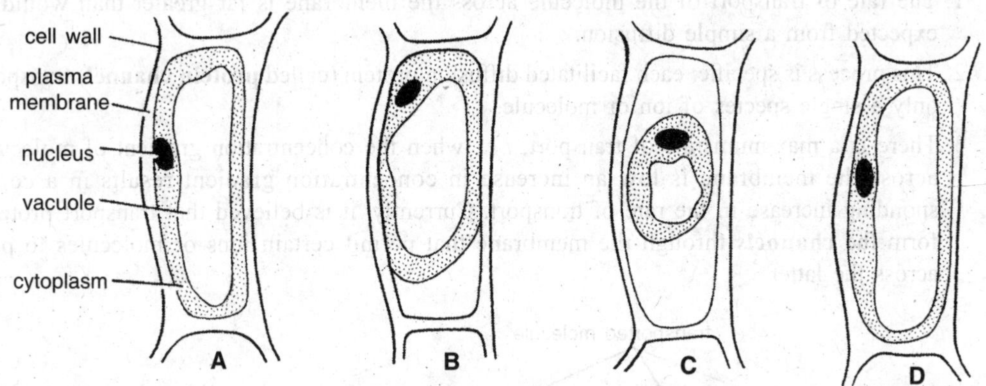

Fig. 6.11 Plasmolysis and deplasmolysis in plant cells. Plasmolysis occurs when a normal plant cell (A) is placed in a hypertonic solution. Water leaves the cell and the plasma membrane shrinks away from the cell wall (B, C). If solutes can penetrate the plasma membrane, the cell will eventually regain water—a process termed deplasmolysis (D).

Box 6.1 Plasma Membrane and Isotonic, Hypotonic and Hypertonic Solutions

A cell contains a variety of solutes in it, for instance, the mammalian erythrocytes contain the ions of potassium (K^+), calcium (Ca^+), phosphate (PO^-_4), dissolved haemoglobin and many other substances. If the erythrocyte is placed in a 0.9% solution of sodium chloride (NaCl), then it neither shrinks nor swells. In such case, because the intra-cellular and extra-celluar fluids contain same concentration, so no osmosis takes place. This type of extra-cellular solution or fluid is known as **isotonic solution** or **fluid**. If the concentration of NaCl solution is increased above 0.9% then the erythrocytes are shrinked due to excessive exosmosis. The solutions which have higher concentrations of solutes than the intracellular fluids are known as **hypertonic solutions**. Further, if the concentration of NaCl solution decreases below 0.9% the erythrocytes will swell up due to endosmosis. The extra-cellular solutions having less concentration of the solutes than the cytoplasm are known as **hypotonic solutions**.

Due to endosmosis or exosmosis the water molecules come in or go out of the cell. The amount of the water inside the cell causes a pressure known as **hydrostatic pressure**. The hydrostatic pressure which is caused by the osmosis is known as **osmotic pressure**. The plasma membrane maintains a balance between the osmotic pressure of the intra-cellular and inter-cellular fluids.

2. Simple Diffusion

In simple diffusion (Fig. 6.12), transport across the membrane takes place unaided, *i.e.*, molecules of gases such as oxygen and carbon dioxide and small molecules (*e.g.*, ethanol) enter the cell by crossing the plasma membrane without the help of any permease. During simple diffusion, a small molecule in aqueous solution dissolves into the phospholipid bilayer, crosses it and then dissolves into the aqueous solution on the opposite side. There is little specificity to the process. The relative rate of diffusion of the molecule across the phospholipid bilayer will be proportional to the concentration gradient across the membrane.

3. Facilitated Diffusion

This is a special type of passive transport, in which ions or molecules cross the membrane rapidly because specific permeases in the membrane facilitate their crossing. Like the simple diffusion, facilitated diffusion does not require the metabolic energy and it occurs only in the direction of a concentration gradient. Facilitated diffusion is characterized by the following special features:

1. The rate of transport of the molecule across the membrane is far greater than would be expected from a simple diffusion.
2. This process is specific; each facilitated diffusion protein (called **protein channel**) transports only a single species of ion or molecule.
3. There is a maximum rate of transport, *i.e.,* when the concentration gradient of molecules across the membrane is low an increase in concentration gradient results in a corresponding increase in the rate of transport. Currently, it is believed that transport proteins form the **channels** through the membrane that permit certain ions or molecules to pass across the latter.

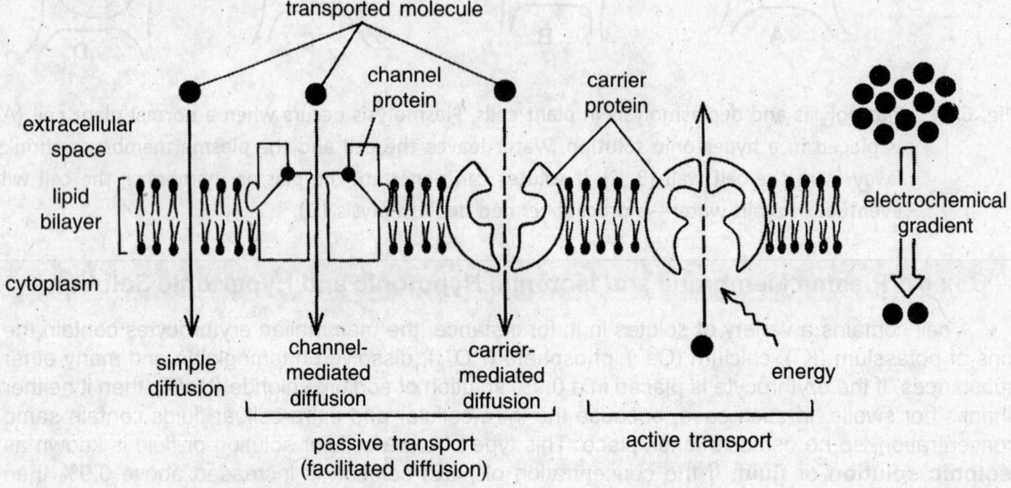

Fig. 6.12. Schematic diagram showing various types of transports across the membrane: simple diffusion, passive transport (down an electrochemical gradient) and active transport (against an electrochemical gradient).

Examples of Facilitated Diffusion

***(i)* Ionic transport through charged pores.** Nerve conduction is propagated along the axonal membrane by **action potential** which regulates opening and closing of two main types of **ion channels** (*i.e.,* channel proteins with water filled pores): **Na^+ channels** (or **voltage-gated Na^+ channels**) and **K^+ channels** (or **K^+ leak channels**). At the point of stimulation there is a sudden and several hundred fold increase in permeability to Na^+, which reaches its peak in 0.1 millisecond (*i.e.,* the membrane potential may depolarise from –90 mV and overshoot to +50 mV). At the end of the period, the membrane again becomes essentially impermeable to Na^+, but the K^+ permeability increases and this ion leaks out of the cell, repolarising the nerve fibre. In other words, during the rising phase of the spike, Na^+ enter through the Na^+ channels, and in descending phase K^+ is extruded through the K^+ channels.

Such ion channels also occur in other types of cells such as muscle, sperm and unfertilized ovum. They are not coupled to an energy source (ATP), so the transport they mediate is always passive ("down hill"), allowing specific ions mainly Na^+, K^+, Ca^{2+} and Cl^- to diffuse down their electrochemical gradient across the lipid bilayer.

Calcium ion channels (Ca^{2+} channels) occur in axonal membranes and other membranes for the entrance of Ca^{2+} ions in the cell. Ca^{2+} ions have a fundamental role in many cellular activities such as exocytosis, endocytosis, secretion, cell motility, cell growth, fertilization and cell division.

In the neuronal membrane, there are a number of Ca^{2+} channels that are driven by the membrane potential and are essential in the release of neurotransmitters (acetylcholine).

***(ii)* D-hexose permease of erythrocyte.** The plasma membrane of mammalian erythrocytes (Fig. 6.13) and other body cells, contain specific channel proteins for the facilitated diffusion of glucose into the cells. They are called **glucose transporter, glucose permease** or **D-hexose permease**. After the glucose is transported into the erythrocyte, it is rapidly phosphorylated (by **hexokinase** enzyme and ATP) to form **glucose-6-phosphate.** Once phosphorylated, the glucose no longer leaves the cell; moreover, the concentration of the simple glucose in the cell is lowered. As a result, the concentration gradient of glucose across the membrane is increased, allowing the facilitated diffusion to continue to import glucose. Since no cellular membrane (except the mitochondrial membranes) contains any permease for facilitated diffusion of phosphorylated compounds, so a cell can retain any type of molecule by phosphorylating them, *e.g.*, ATP and phosphorylated nucleosides are never released from the cells containing a normal intact plasma membrane. However, permeases for ATP and ADP do exist in a mitochondrial membrane to allow these molecules to move across it.

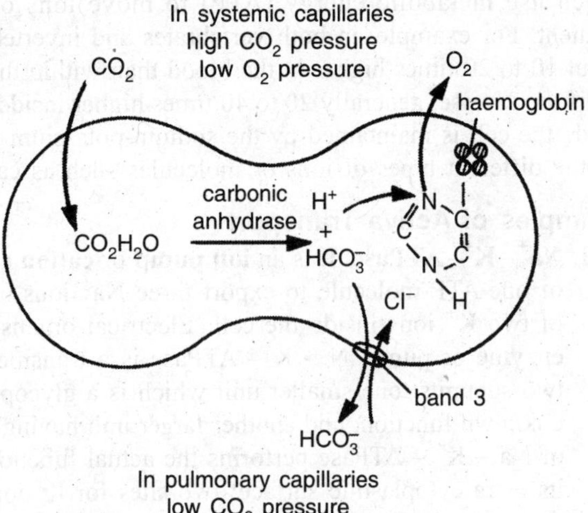

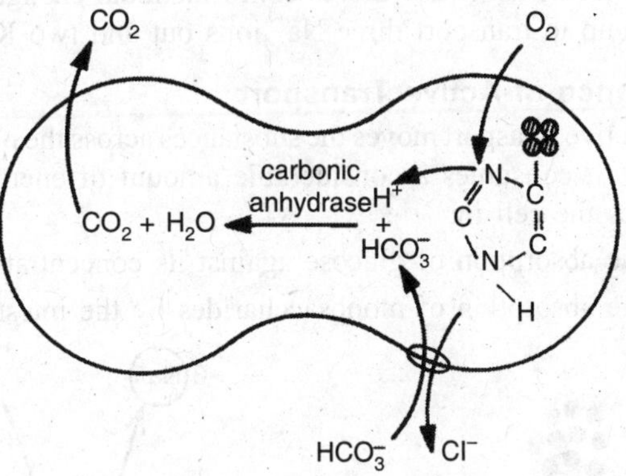

Fig. 6.13. Schematic drawing showing anion transport through the erythrocyte membrane in the capillaries and in the lungs. Band 3 protein (= anion exchange permease) catalyzes the exchange of the anions: Cl^- and HCO_3^- across the erythrocyte membrane.

***(iii)* Anion exchange permease of erythrocyte.** Band 3 polypeptide of plasma membrane of the erythrocytes and other cells is an **ion exchange permease protein** which catalyzes an one-for-one exchange of anions such as chloride (Cl^-) and bicarbonate (HCO_3^-) across the membrane (called **chloride shift;** erythrocyte has 100,000 times more permeability of Cl^- than other cells). The rapid flux of anions in the erythrocyte facilitates the transport in the blood of CO_2 from the tissues to the lungs. Waste CO_2 that is released from cell into the capillary blood, diffuses across the membrane of erythrocyte. In its gaseous form, CO_2 dissolves poorly in aqueous solutions such as blood plasma, but inside the erythrocyte the potent enzyme **carbonic anhydrase** converts it into a bicarbonate anion:

$$CO_2 + H_2O \underset{\text{anhydrase}}{\overset{\text{carbonic}}{\rightleftharpoons}} \underset{\text{proton}}{H^+} + \underset{\text{bicarbonate}}{HCO_3^-}$$

This process occurs while the haemoglobin in the erythrocyte is releasing its oxygen into the blood plasma. The removal of oxygen from haemoglobin induces a change in its conformation that enables a globin histidine (amino acid) side chain to bind to the proton produced by carbonic anhydrase enzyme. The bicarbonate anion formed by carbonic anhydrase is transported out of the erythrocyte in exchange for a chloride (Cl^-) anions:

$$\underset{(in)}{HCO_3^-} + \underset{(out)}{Cl^-} \rightleftharpoons \underset{(out)}{HCO_3^-} + \underset{(in)}{Cl^-}$$

B. Active transport. Active transport uses specific transport proteins, called **pumps,** which use metabolic energy (ATP) to move ions or molecules against their concentration gradient. For example, in both vertebrates and invertebrates, the concentration of sodium ion is about 10 to 20 times higher in the blood than within the cell. The concentration of the potassium ion is the reverse, generally 20 to 40 times higher inside the cell. Such a low sodium concentration inside the cell is maintained by the sodium-potassium pump. There are different types of pumps for the different types of ions or molecules such as calcium pump, proton pump, etc.

Examples of Active Transport

1. **Na^+– K^+– ATPase**. It is an **ion pump** or **cation exchange pump** which is driven by energy of one ATP molecule to export three Na^+ ions outside the cell in exchange of the import of two K^+ ions inside the cell. Electrical organs of eels are found to be very rich in this enzyme or pump. N^+– K^+– ATPase is a transmembrane protein which is a dimer having two subunits: one smaller unit which is a glycoprotein of 50,000 daltons M.W., having an unknown function; and another larger unit having 1,20,000 daltons M.W. The larger subunit of Na^+– K^+ – ATPase performs the actual function of cation transport. It has three sites on its extra cytoplasmic surface: two sites for K^+ ions and one site for the inhibitor **ouabain**. On its cytosolic side, the larger subunit contains three sites for three Na^+ ions and also has one catalytic site for a ATP molecule. It is believed that the hydrolysis of one ATP molecule somehow drives conformational changes in the Na^+– K^+– ATPase that allows the pump to transport three Na^+ ions out and two K^+ ions inside the cell (Fig. 6.14).

Importance of Active Transport

Active transport moves the substances across the plasma membrane against their concentration gradient. It consumes a considerable amount of energy and is very essential for the cell since it enables the cell in:

1. The absorption of glucose against its concentration gradient in the uriniferous tubules.
2. The absorption of monosaccharides by the intestinal cells.

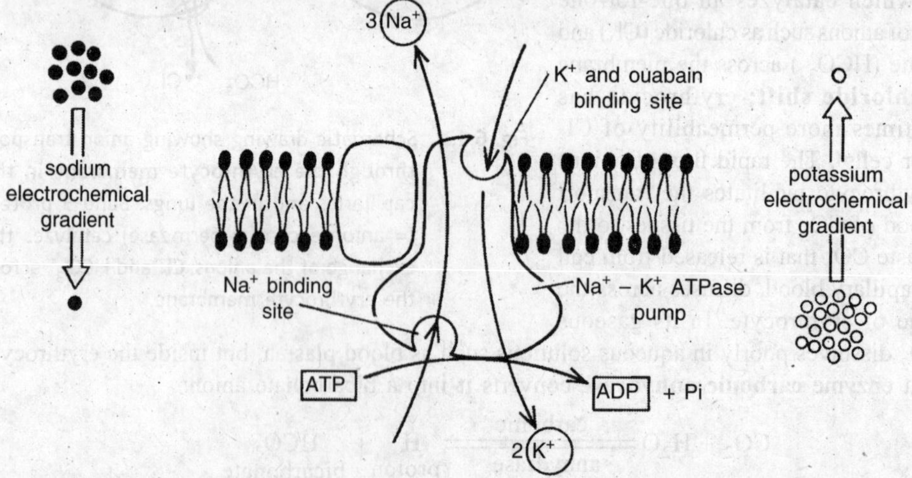

Fig. 6.14. The Na^+ – K^+ ATPase in the plasma membrane actively pumps Na^+ out and K^+ into a cell against their electrochemical gradients. For every molecule of ATP hydrolyzed inside the cell, 3 Na^+ ions are pumped out and $2K^+$ ions are pumped in.

3. Maintenance of $Na^+ - K^+$ pump for establishing the unequal distribution of ions between inside and outside the plasma membrane. It helps in the conduction of nerve impulse in nerve cells and in maintaining a specific level of Na^+ in the blood.

Functions of Proteins of Membrane

Biological membranes (*e.g.*, plasma membrane and unit membrane) contain the following five different classes of proteins that carry out different functions:

1. Structural proteins. These form backbone of plasma membrane. These proteins provide elasticity and mechanical stability to plasma membrane. These have little catalytic activity.

2. Enzymes. These are catalytic proteins. They carry out different biological reactions in different parts of the cell and also in various types of cells.

3. Membrane transport proteins. These serve for the transport of polar molecules of various substances such as ions, monosaccharides, amino acids, nucleotides and certain metabolites across the plasma membrane in and out of the cell. These occur in many forms. Each transport protein is designed to transport a particular class of molecules. These are basically **multipass transmembrane proteins**, projecting on both the sides of lipid bilayer and traversing it many times. Membrane transport proteins are of following types:

(*i*) **Carrier proteins** bind the specific solute to be transported by active transport, *i.e.*, these undergo conformational change in order to transfer the solute across the membrane. These carrier proteins are of following three types:

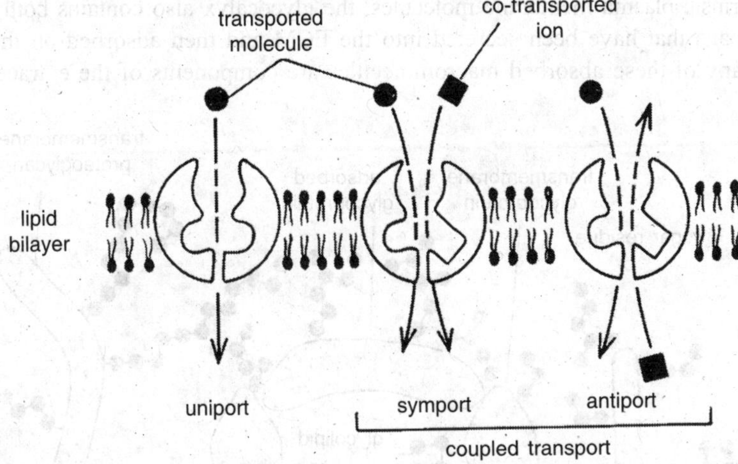

Fig. 6.15. Carrier proteins of membrane functioning as uniports, symports and antiports.

Uniport, symport and antiport. Those carrier proteins which simply transport a single solute from one side of the membrane to the other are called **uniports**. Others function as **coupled transporters**, in which the transfer of one solute depends on the simultaneous transfer of a second solute, either in the same direction (**symport**) or in the opposite direction (**antiport**). Both symport and antiport collectively form the **cotransport**. Most animal cells, for example, must take up glucose from the extracellular fluid, where the concentration of the sugar is relatively high, by passive transport through the glucose carriers (such as **D-hexose permease**) that operate as the uniports. By contrast, intestinal and kidney cells must take up glucose from the lumen of the intestine and kidney tubules, respectively, where the concentration of the sugar is low. These cells actively transport glucose by symport with Na^+ ions whose extracellular concentration is very high. The anion exchange permease of human erythrocytes operates as an antiport to the exchange of Cl^- for HCO_3^-.

(*ii*) **Channel proteins.** These are water filled pores that extend across the lipid bilayer. When these pores are open, they allow specific solutes to pass through them across the membrane.

4. Glycoproteins act as **cell receptors** (*e.g.*, **integrins**) and **cell antigens**.
5. Lipoproteins are **drug receptor proteins**.

Cell Coat or Glycocalyx

Plasma membrane proteins, as a rule, do not protrude naked from the exterior of the cell. They are usually decorated by carbohydrates, which coat the surface of all eukaryotic cells. These carbohydrates occur as **oligosaccharide chains** covalently bound to membrane proteins (glycoproteins) and lipids (glycolipids). They also occur as the polysaccharide chains of integral membrane proteoglycan molecules. Proteoglycans, which consist of long polysaccharide chains linked covalently to a protein core, are found mainly outside the cell as part of the extra-cellular matrix or ECM (discussed in Chapter 7). But for some proteoglycans, the protein core either extends across the lipid bilayer or is attached to the bilayer by a GPI (= glycosyl phosphotidyl inositol) anchor.

The term **cell coat** or **glycocalyx** is often used to describe the carbohydrate-rich zone of the cell surface. This zone can be visualized by a variety of stains, such as **ruthenium red**, as well as by its affinity for carbohydrate-binding proteins called **lectins,** which can be labelled with a fluorescent dye or some other visible marker. The cell coat is a 10 to 20 nm thick layer and is in direct contact with the outer leaflet of the plasma membrane. In *Amoeba*, the cell coat is formed by fine filaments—5 to 8 nm thick and 100 to 200 nm long. Although most of the carbohydrate is attached to intrinsic plasma membrane molecules, the glycocalyx also contains both glycoproteins and proteoglycans that have been secreted into the ECM and then adsorbed on the cell surface (Fig. 6.16). Many of these absorbed macromolecules are components of the extracellular matrix,

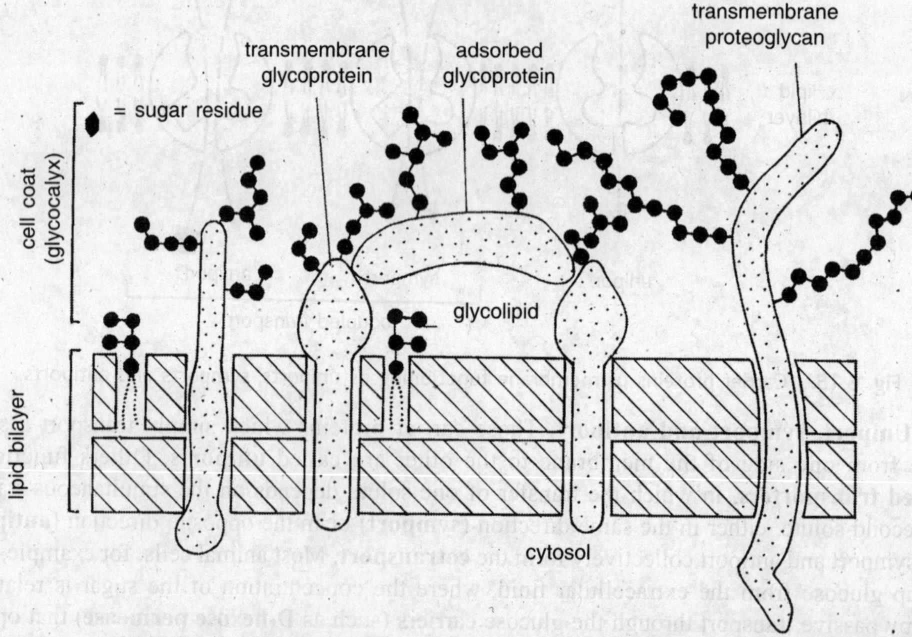

Fig. 6.16. Simplified diagram of the cell coat (glycocalyx). The cell coat is made up of the oligosaccharide side chains of glycolipids and integral membrane glycoproteins and the polysaccharide chains on integral membrane proteoglycans. In addition, absorbed glycoproteins and absorbed proteoglycans (not shown) contribute to the glycocalyx in many cells. Note that all of the carbohydrate is on the noncytosolic surface of the membrane (after Alberts *et al.*, 2002).

so that where the plasma membrane ends and the ECM begins is largely a matter of semantics. One of the possible functions of cell coat is to protect cells against mechanical and chemical damage and to keep foreign objects and other cells at a distance, preventing undesirable protein-protein interaction. The cell coat of the intestinal epithelium is quite strong—it resists vigorous mechanical and chemical attacks; in other cells the coat is labile and may be depleted by washing or enzyme exposure.

Cell Surface Differentiations

In addition to specific molecular components plasma membrane consists of various cell surface dif-ferentiations such as microvilli, invaginations, etc.

1. Invaginations. The bases (inner ends) of certain cells, such as the cells of the kidney, perform active transpor-tation and contain many **invaginations** or **infoldings** of the plasma membrane. At the base of these folds, there deve-lops a septa and, thus, narrow compartments of basal cytoplasm are formed. These infoldings contain many mitochondria. These mitochondria along with the enzymes of plasma membrane possibly provide energy rich compound, *viz.*, ATP to the plasma membrane for the active transportation of solutes.

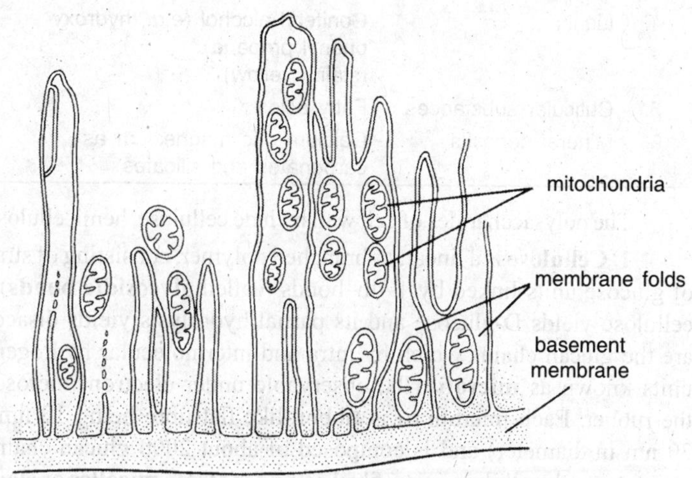

Fig. 6.17. Infoldings of plasma membrane.

2. Microvilli. Microvilli are finger-like, slender projections of plasma membrane which are found in mesothelial cells, hepatic cells, epithelial cells of intestine (**striated border**), uriniferous tubules (**brush border**), gall bladder, uterus, growing oocyte and yolk sac. Microvilli increase the effective surface of absorption. For example, a single epithelial cell of intestine may have as many as 3000 microvilli and in a square millimetre of intestine there may be 200,000,000. These microvilli are 0.6 to 0.8 μm long and 0.1 μm in diameter. The narrow spaces between the microvilli form a kind of sieve through which substances may pass during the process of absorption. Within the cytoplasmic core of a microvillus fine microfilaments are observed which in the underlying cytoplasm form a **terminal web.** The microfilaments contain actin and are attached to the tips of the microvilli by α-actinin; their function is to produce contraction of microvilli.

6.3 CELL WALL

The plant cell is always surrounded by a **cell wall** and this feature distinguishes them from animal cells. The cell wall is a non-living structure which is formed by the living protoplast (A plant cell without its cell wall is called a **protoplast**). In most of the plant cells, the cell wall is made up of cellulose, hemicellulose, pectin and protein. In many fungi, the cell wall is formed of chitin and in bacteria, the cell wall contains protein-lipid-polysaccharide complexes. Thus, the cell wall is a rigid and protective layer around the plasma membrane which provides the mechanical support to the cell. The cell wall also determines the shape of plant cells. Due to the shape of cell walls many types of plant cell as the parenchymatous, collenchymatous, etc., have been recognised.

Chemical Composition of Cell Wall

Chemically speaking, the plant cell wall is composed of a variety of polysaccharides (carbohydrates), lipids, proteins and mineral deposits, all exhibiting distinct staining reactions (Table 6.3).

Table 6.3 Chemical nature and staining reaction of various components of plant cell wall.

	Substance	*Chemical unit*	*Staining reaction*
1.	Cellulose	Glucose	Chlorzinc iodide (stains violet)
2.	Hemicellulose	Arabinose, xylose, mannose, glucose and galactose	No specific stain
3.	Pectin	Glucuronic and galacturonic acid	Ruthenium red
4.	Lignin	Coniferyl alcohol (*e.g.*, hydroxy-phenyl propane) (stains yellow)	Phloroglucinol hydrochloride (stains rose); chlorzinc iodide
5.	Cuticular substances	Fatty acids	Sudan III (stains orange)
6.	Mineral deposits	Calcium and magnesium as carbonates and silicates	

The polysaccharides of cell wall include cellulose, hemicelluloses, pectin compounds and lignins.

1. Cellulose is a linear, unbranched polymer, consisting of straight polysaccharide chains made of glucose units linked by 1–4b–bonds (called **glycosidic bonds**). (**Note.** Complete hydrolysis of cellulose yields D-glucose and its partial hydrolysis yields disaccharide units, **cellobiose**). These are the glucan chains which by intra and intermolecular hydrogen bonding produce the structural units known as **microfibrils,** observable under electron microscopy and having toughness like the rubber. Each microfibril is ribbon-like flat fibre being 10 nm wide and 3 nm thick (or 25 to 30 nm in diameter) and is composed of about 2000 glucan chains in it. According to a classical estimate, each cellulose microfibril comprises three **micelles** or **elementary fibrils:** each elementary fibril contains about 100 cellulose molecules and each cellulose molecule is made up of 40 to 70 glucan chains (*i.e.*, one microfibril = 3 × 100 × 70 = 21000 glucan chains). Often numerous microfibrils get associated to form the **macrofibrils** having up to 0.5 μm diameter and observable under the light microscopy. Cellulose is synthesized by a wide variety of cells that include bacteria (*e.g.*, acetobacter, agrobacter and rhizobium), algae, fungi, cryptogams and seed plants.

2. Hemicelluloses are short but branched heteropolymers of various kinds of monosaccharides such as arabinose, xylose, mannose, galactose, glucose and uronic acid. Some of the common hemicelluloses go under the names xylans, arabinoxylans, glucomannans, galactomannans and xyloglucans.

3. Pectins are water soluble, heterogeneous branched polysaccharides that contain many negatively charged D-galacturonic acid residues along with D-glucuronic acid residues. Because of their negative charge, pectins are highly hydrated and intensely bind cations. When Ca^{2+} is added to a solution of pectin molecules, it cross-links them to produce a semirigid gel. Such Ca^{2+} cross-links are thought to help hold cell-wall component together.

4. Mannan is a homopolysaccharide of mannose and is found in the cell wall of yeast, fungi and bacteria.

5. Agar is a polysaccharide, found in the cell wall of sea weeds and containing D and L-galactose residues.

6. Lignin is a biological plastic and non-fibrous material. It occurs only in mature cell walls and is made of an insoluble hydrophobic aromatic polymer of phenolic alcohols (*e.g.*, hydroxyphenyl propane).

7. The **chitin** is a polymer of glucosamine.

Glycoproteins (present up to 10 per cent in primary cell wall) are hydroxyproline-rich proteins (like the collagen). In them, many short oligosaccharide side chains are attached to hydroxyproline and serine side chains. Thus, more than half the weight of glycoprotein is carbohydrate. These glycoproteins are known to act like the glue to increase the strength of the cell wall.

Cutin is also a biological plastic and is made of fatty acids (waxes). **Suberin** is a water-resistant substance, comprising of fatty acids and found in the cork and cell wall of many plants. **Sporopollenin** is a lipoidal polymer forming tough wall (with species-specific patterns) of pollen grains.

Mineral deposits occur in cuticle in the form of calcium and magnesium carbonates and silicates. Deposits of calcium compounds are found in the cell wall of cruciferous and cucurbitaceous plants. Silicate deposits are common in the cell wall of Graminae family.

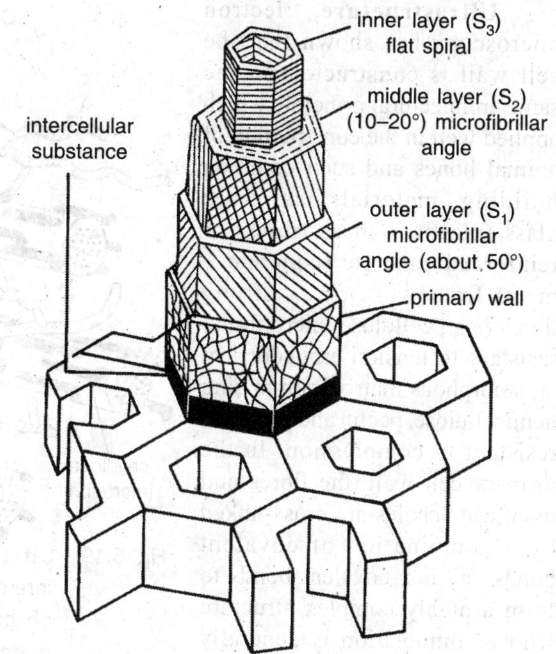

Fig. 6.18. Structure of a cell wall showing middle lamella, primary cell wall and three regions (S_1, S_2 and S_3) of the secondary cell wall.

Structure. The cell wall is complex in nature and is differentiated in the following layers (Fig. 6.18):

(*i*) Primary cell wall;

(*ii*) Secondary cell wall;

(*iii*) Tertiary cell wall.

(*i*) Primary cell wall. The first formed cell wall is known as primary cell wall. It is the outermost layer of the cell and in the immature meristematic and parenchymatous cells it forms the only cell wall. The primary cell wall is comparatively thin and permeable. Certain epidermal cells of the leaf and the stem also possess the cutin and cutin waxes which make the primary cell wall impermeable. The primary cell wall of the yeast and the fungi is composed of the chitin.

(*ii*) Secondary cell wall. The primary cell wall is followed by secondary cell wall. The secondary cell wall is thick, permeable and lies near the plasma membrane of the tertiary cell wall, if the latter occurs. It is composed of three concentric layers (S_1, S_2 and S_3) which occur one after another. Chemically the secondary cell wall is composed of compactly arranged macrofibrils of the cellulose, in between which sometimes occurs lignin as a inter-fibrillar material.

(*iii*) Tertiary cell wall. In certain plant cells, there occurs another cell wall beneath the secondary cell wall which is known as tertiary cell wall. The tertiary cell wall differs from the primary and secondary cell wall in its morphology, chemistry and staining properties. Besides the cellulose, the tertiary cell wall consists of another chemical substance known as the xylan.

Middle lamella. The cells of plant tissues generally remain cemented together by an intercellular matrix known as the **middle lamella.** The middle lamella is mainly composed of the pectin, lignin and some proteins.

Ultrastructure. Electron microscopy has shown that the cell wall is constructed on the same architectural principle which applied well in the construction of animal bones and such common building materials as fibre glass (plastic + glass fibres) or reinforced concrete (concrete + metal framework), *i.e.*, strong fibres (*e.g.*, cellulose microfibrils) resistant to tension embedded in an amorphous matrix (comprising hemicellulose, pectin and proteins) resistant to compression. In the primary cell wall, the fibres and matrix molecules are cross-linked by a combination of covalent bonds and non-covalent bonds to form a highly complex structure whose composition is generally cell-specific (Fig. 6.19). In fact, hemicellulose molecules (*e.g.*, **xyloglucans**) are linked by hydrogen bonds to the surface of the cellulose microfibrils. Some of these hemicellulose molecules are cross-linked in turn to acidic pectin molecules (*e.g.*, rhamno-galacturonans) through short neutral pectin molecules (*e.g.*, arabinogalactans). Cell wall glycoproteins are tightly woven into the texture of the wall to complete the structure of matrix.

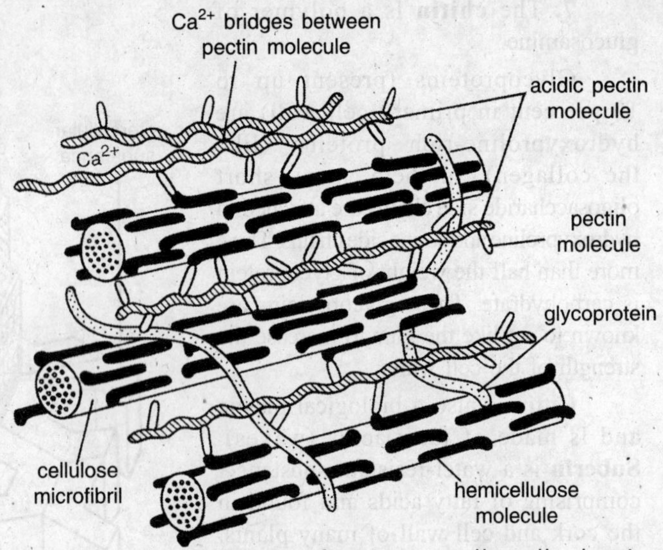

Fig. 6.19. Ultrastructure of primary cell wall showing interconnections between the two major components of the primary cell wall: the cellulose microfibrils and the matrix.

In the multilamellar secondary cell wall, cellulose microfibrils are laid down in layers, the microfibrils of each layer running roughly parallel with each other but at an angle to those in other layers (Fig. 6.18).

Plasmodesmata. Every living cell in a higher plant is connected to its living neighbours by fine cytoplasmic channels, each of which is called a **plasmodesma** (Gr., *desmos* = ribbon; ligament; plural, **plasmodesmata**) which pass through the intervening cell walls. The plasma membrane of one cell is continuous with that of its neighbour at each plasmodesma. A plasmodesma is a roughly cylindrical, membrane-lined channel with a diameter of 20 to 40 nm. Running from cell to cell through the centre of most plasmodesmata is a narrower cylindrical structure, the **desmotubule**, which remains, continuous with elements of the SER membranes of each of the connected cells. Between the outside of the desmotubule and the inner face of the cylindrical plasma membrane is an **annulus of cytosol**, which often appears to be constricted at each end of the plasmodesmata. These constrictions may regulate the flux of molecules through the annulus that joins the two cytosols.

Plasmodesmata are formed around the elements of smooth endoplasmic reticulum that become trapped during cytokinesis (of mitotic cell division) within the new cell wall that will bisect the parental cell. Plasmodesmata function in intercellular communication, *i.e.*, they allow molecules to pass directly from cell to cell. For example, plasmodesmata are especially common and abundant in the walls of columns of cells that lead toward sites of intense secretion, such as in nectar-secreting glands (trichomes of *Abutilon* nectaries). In such cells there may be 15 or more plasmodesmata per square micrometer of wall surface, whereas there is often less than 1 per square micrometer in other cell wall.

In fact, experimental evidence has suggested that the plasmodesmata mediate transport between adjacent plant cells, much as gap junctions of animal cells. They allow the passage of molecules with molecular weights of less then 800 daltons. Transport through the plasmodesmata is also found under complex regulations which may involve Ca^{2+} and protein phosphorylation. Thus, no dye movement is observed between the cells of the root cap and the root apex or between epidermal and cortical cells in either roots and shoots. However, certain plant viruses such as TMV can enlarge plasmodesmata in order to use this route to pass from cell to cell. Tobacco mosaic virus is known to synthesize a protein, called **P30** (30,000 dalton M.W.) that nullifies the normal regulatory mechanisms of plasmodesmata.

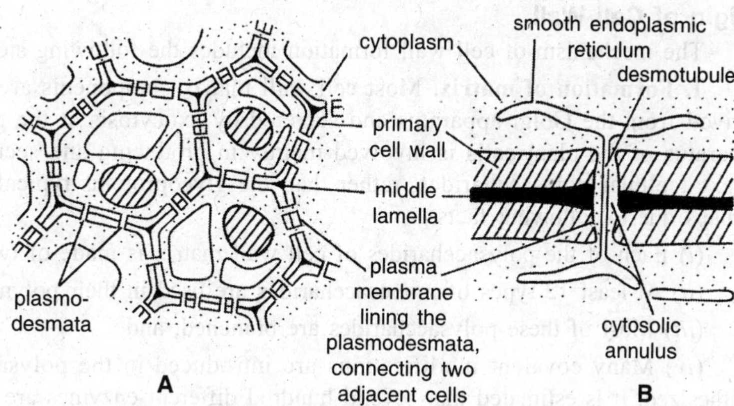

Fig. 6.20. A—Plasmodesmata pierce the cell wall and connect all cells in the plant together to form symplast; B—Details of the structure of a plasmodesma.

Lignification. The structure of cell wall is stabilized by the deposition of lignin in the cell wall matrix. Such a process of lignification was required in connection with the transition from aquatic to the terrestrial plant life during organic evolution of plants. A lignified cell wall is composed of microfibrils of cellulose embedded in the matrix containing large amount of lignin. Usually the primary cell wall becomes more lignified than secondary cell wall.

Functions of Cell Wall

The chief function of cell wall in plant cells is that it provides mechanical strength to the latter. Like the exoskeleton or endoskeleton of animals, cell wall acts like a skeletal framework of plants. Particularly in vascular plants, the cell walls provide the main supporting framework.

Despite its strength, the plant cell wall is fully **permeable** to water and solutes. This is because the matrix of cell wall is perforated by minute water-filled channels through which free diffusion of water and water soluble substances such as gases, salts, sugars, hormones and like can take place. Moreover, the molecules of the matrix are strongly hydrophilic ("water-loving") with the result that in normal circumstances the cell wall is saturated with water like a sponge (*e.g.*, primary cell wall is 60 per cent water by weight): The cross-linked structure of the cell wall is, however, found to slightly block the diffusion of small molecules such as water, sucrose and K^+. The average diameter of the spaces between the cross-linked macromolecules in most cell wall is about 5 nm, this is small enough to make the movement of any globular macromolecules with a M.W. much above 20,000 daltons extremely slow. Therefore, plants must survive on molecules of low molecular weight, and any intercellular signaling molecules that have to pass through the cell wall must also be small and water soluble. In fact, most of the known plant signaling molecules, such as growth regulating substances—auxins, cytokinins and gibberellins—have molecular weights of less than 500 daltons.

During lignification, lignin is deposited in spaces between the cellulose molecules, making the cell wall much more rigid, and making it **impermeable**. Once lignification is complete the protoplasm can no longer absorb materials from outside the cell, which, therefore, dies. Hence, lignified tissue is always dead. Thus, a lignified tissue becomes well adapted for two types of functions:

1. It provides the mechanical strength due to its ligno-cellulose composition.
2. It transports water and salts, since, lignification involves loss of the protoplasm resulting in the formation of a hollow waterproof tube.

Origin of Cell Wall

The mechanism of cell wall formation includes the following steps:

1. Formation of matrix. Most cell wall matrix components are transported via vesicles derived from the Golgi apparatus and secreted by exocytosis at the plasma membrane. Golgi apparatus of the plant cells is involved mainly in producing and secreting a very wide range of extracellular polysaccharides, rather than the glycoproteins typical of animal cells. This is perhaps due to following facts:

(*i*) Each of the polysaccharides of cell wall matrix is made of two or more sugars;

(*ii*) At least 12 types of monosaccharides are used in their polymerization;

(*iii*) Most of these polysaccharides are branched; and

(*iv*) Many covalent modifications are introduced in the polysaccharides after they are synthesized. It is estimated that several hundred different enzymes are engaged in the assembly of the polysaccharide components needed to form a typical primary cell wall. Most of these enzymes are found in the endoplasmic reticulum and Golgi apparatus. Some enzymes, which are concerned with later covalent modifications of the polysaccharides, are found in the cell wall itself.

Since, the cell wall varies in composition and morphology at different locations around the cell, the Golgi-derived vesicles are directed at specific regions of the plasma membrane by the help of cytoskeleton (*i.e.*, microtubules and microfilaments). For example, we can consider the case of formation of new primary cell wall that separates two daughter cells after the karyokinesis of mitosis. At the end of telophase, a barrel-shaped or disc-like region, called **phragmoplast** (Gr., *phragma* = hedge, enclosure; *plasso* = to form) forms in the plane of former spindle equator. The phragmoplast comprises a double-ring of short microtubules on either side of, and terminating at,

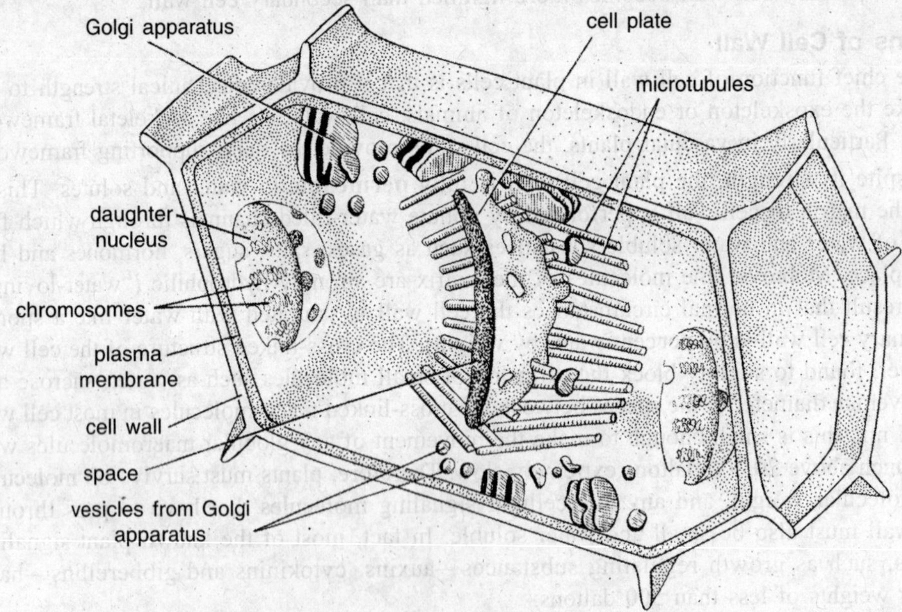

Fig. 6.21. Schematic representation of a cell of a higher plant as seen at telophase in mitosis. In the phragmoplast region, a region of membranes and microtubules, a cell plate forms and grows until it separates the cytoplasm into two daughter cells. The cell plate develops as a membrane-delimited structure enclosing a space in which new cell wall will form. The Golgi apparatus contributes many vesicles to the phragmoplast membrane; the vesicle membranes apparently are incorporated into the membrane of the cell plate and the vesicle contents enter the forming cell wall.

the division plane, and a set of microfilaments are coaligned with the microtubules. Golgi derived vesicles containing cell wall precursors, especially **pectin**, are guided inward along these oriented microtubules until they reach the phragmoplast, where they fuse with one another to form the **cell plate** (Fig. 6.21). Cell plate, at this stage, comprises following structures: (1) central middle lamella; (2) primary cell walls on both sides of middle lamella; and (3) plasma membrane lining cytoplasm of each daughter cell. Ultimately, the microtubular ring moves centrifugally outward as Golgi vesicles continue to add precursors to the growing cell plate. The cell plate fuses with the mother cell wall to create two separate daughter cells. It is not clear which component of the phragmoplast—the microtubules or the actin filaments (or both)—are responsible for the movement and guidance of the Golgi derived vesicles.

2. Synthesis and orientation of cellulose microfibrils. In most plants, cellulose is synthesized on the external surface of the cell by a plasma membrane bound enzyme complex, called **cellulose synthetase** which uses a sugar nucleotide precursor supplied from the cytosol, probably UDP-glucose. As they are being synthesized, the nascent cellulose chains spontaneously assemble into **microfibrils** that form a layer on the surface of the plasma membrane (a lamella) in which all the microfibrils have more or less the same alignment (**Note.** Each cellulose molecule has a polarity, having a 1′ and a 4′ end). Cellulose synthetase complexes are thought to be associated with the ends of growing microfibrils and the sugars present in the extracellular matrix are polymerized into cellulose at these "terminal complexes". Extension of a cellulose microfibril is presumably achieved by lateral movement of the enzyme complex in the fluid phase of plasma membrane, with the microfibril "spun out" on the outer surface of the membrane behind the moving enzyme complex (Fig. 6.22). The direction in which the complex moves and the orientation of the microfibril depend on some interactions between the membrane complex and the underlying cytoplasmic microtubules (*i.e.*, microtubules of cell, cortex). Because the cellulose is synthesized at the plasma membrane, each new wall lamella forms internally to the last formed lamella. The cell wall, therefore, consists of concentrically arranged lamellae, with the oldest on the outside.

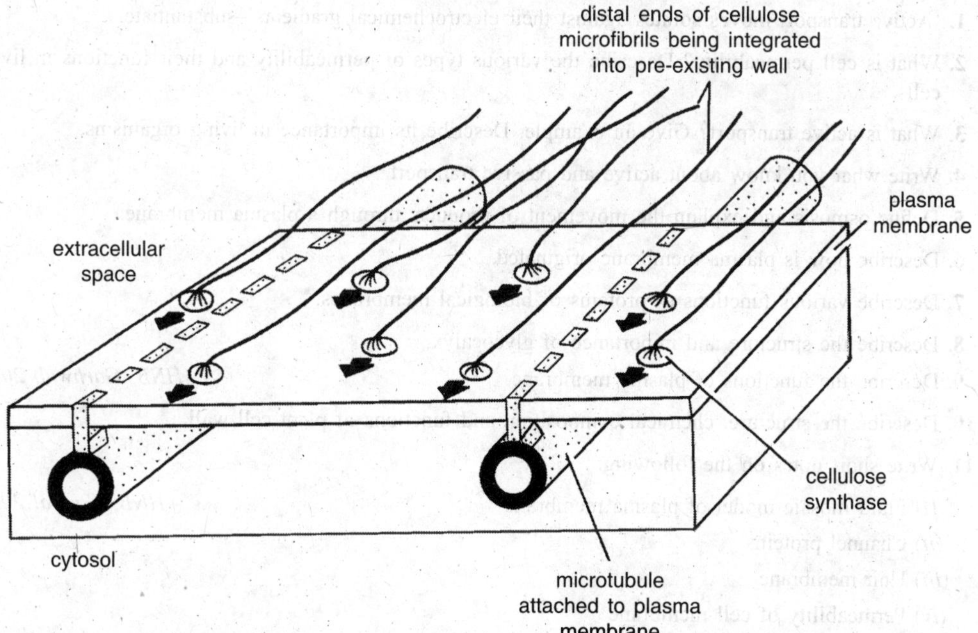

Fig. 6.22. A model explaining the mode of orientation of newly deposited cellulose microfibrils according to orientation of cortical microtubules.

QUESTIONS

Long Answer Questions

1. Describe the structure, nature and functions of plasma membrane.
2. Describe the ultrastructure and functions of plasma membrane.
3. Describe the structure of plasma membrane in the light of recent findings.
4. How many models of plasma membrane do you know? Explain which of the model of it is more dynamic and why? Describe the mechanism of active transport.
5. Describe the accepted model of the structure and chemical composition of cell membrane. State briefly the functional significance of cell membrane.
6. Describe briefly the ultrastructure of cell membrane based upon various theories.
7. Give an account of the fluid mosaic model of plasma membrane. Briefly describe the important functions of this organelle.
8. What are the characteristic features of a cell membrane? Describe the permeability and chemical composition of a cell membrane.
9. Describe fluid mosaic model of structure of plasma membrane. Discuss various experimental evidences in support of this model.
10. Describe the fluidity of plasma membrane.
11. Discuss various types of diffusion that occur across the cell membrane.
12. What is passive transport? Describe the mechanism of transport of small molecules through the cell membrane. (*Guwahati 2007*)
13. Why did Robertson used myelin for his unit membrane model of plasma membrane? What issues were raised against this model?

Short Answer Questions

1. 'Active transport moves solutes against their electrochemical gradients'–substantiate.
2. What is cell permeability? Describe the various types of permeability and their functions in living cells.
3. What is active transport? Give an example. Describe its importance in living organisms.
4. Write what you know about active and passive transport.
5. Define osmosis and explain the movement of products through a plasma membrane.
6. Describe how is plasma membrane originated.
7. Describe various functions of proteins of biological membranes.
8. Describe the structure and importance of glycocalyx.
9. Describe the functions of plasma membrane. (*HNB, Garhwal, 2009*)
10. Describe the structure, chemical composition and functions of plant cell wall.
11. Write short notes on the following:
    (*i*) Fluid mosaic model of plasma membrane (*HNB, Garwal 2008*)
    (*ii*) Channel proteins
    (*iii*) Unit membrane
    (*iv*) Permeability of cell membrane
    (*v*) Active transport

(*vi*) Osmosis

(*vii*) Role of semipermeability of plasma membrane

(*viii*) Passive transport.

12. Differentiate between the following terms:

(*i*) Simple diffusion and facilitated diffusion.

(*ii*) Active transport and passive transport.

(*iii*) Plasma membrane and endoplasmic reticulum.

(*iv*) Cell wall and karyotheca.

(**Hint.** Cell wall is nonliving, porous, cellulose-containing part of cell that occurs outside plasma membrane of the cell of plants. Karyotheca is nuclear envelope which consists of two unit membranes).

13. Differentiate between uniport transport and passive diffusion. (*M.S.U., Tamil Nadu 2006*)

14. Draw labelled diagrams of the following (No description is required):

(*i*) Fluid-mosaic model of plasma membrane.

(*ii*) Cell wall

Very Short Answer Questions

1. Who gave the concept of unit membrane?

2. Name the scientists who proposed bimolecular lipid layer model of plasma membrane.

3. Who proposed fluid mosaic model of plasma membrane?

4. Define passive transport.

5. Define diffusion.

6. What is flip-flop movement?

7. Name three substances passing the tunnels in the integral proteins of plasma membrane.

8. What cause diffusion of ions across the plasma membrane?

9. What is the name given to the proteins which are separated from plasma membranes of RBC?

10. Name any three sugars found in the glycocalyx.

11. Is glycocalyx found in prokaryotic plasma membrane.

12. How do large molecules of sugars and amino acids diffuse through plasma membrane quickly along the concentration gradient?

13. Name the integral components acting as a sodium-potassium pump.

14. Name the processes responsible for bulk transport of substances across the plasma membrane.

15. Name any two substances taken in by the plasma membrane through receptor-mediated endocytosis.

16. Name the organelle present in both prokaryotic and eukaryotic cells.

Mark the Statements True of False

(*i*) Cell membrane is composed of liporotein.

(*ii*) Cellulose of cell wall is a protein.

(*iii*) Plasma membrane is present in all cells.

(*iv*) Cell wall of plant cell is a living structure.

Yes or No Questions

(*i*) Thickness of plasma membrane ranges from 100 Å to 200 Å.

(*ii*) The concept of 'Unit membrane' was put forward by Robertson.

Fill in the Blanks

1. Fluid mosaic model of plasma membrane was proposed by .... and .... in ....
2. Robertson in 1960 proposed the .... hypothesis.
3. In .... of the plant cells due to excessive .... the cytoplasm along with the plasma membrane shrinks away from the cell wall.
4. The Ca^{2+}– .... are the example of .... and they occur in axonal membrane for .... of calcium ions.
5. In active transport, ATP is used to move ions or molecules against their ....
6. D-hexose permease of plasma membrane operates as the ....
7. Na^{+}– K^{+} ATPase is an ion pump or .... with exports three .... outside the cell in exchange of the import of two ....
8 The cell coat or .... is a .... zone of plasma membrane of animal cells.
9. .... of cell wall of the plant cells is stained violet by .... stain.
10. Cellulose of cell wall is synthesized on the external surface of the plant cell by the plasma membrane-bound enzyme, called ....

Multiple Choice Questions

1. Plasma membrane is made up of
   (*a*) glycoproteins and lipids (*b*) phospholipids and glycolipids
   (*c*) phospholipids and proteins (*d*) phosphoproteins and phospholipids
2. Phospholipid molecules are
   (*a*) acidic (*b*) alkaline
   (*c*) amphoteric (*d*) amphipathic
3. In lipid bilayer, the phospholipid molecules are arranged in such a fashion that:
   (*a*) their hydrophobic ends are directed outwards
   (*b*) their hydrophobic ends are directed inwards
   (*c*) the hydrophobic ends of molecules of one layer make contact with the hydrophilic ends of the molecules of other layer
   (*d*) their hydrophobic ends are scattered indiscriminately
4. The hydrophobic ends of phospholipid molecules are
   (*a*) polar (*b*) nonpolar
   (*c*) neutral (*d*) bipolar
5. Spectrin, cytochrome-C and enzymes present on the plasma membrane are examples of
   (*a*) integral or intrinsic proteins (*b*) peripheral or extrinsic proteins
   (*c*) glycoproteins (*d*) lipoproteins
6. The membrane proteins that extend through both sides of lipid bilayer
   (*a*) acidic proteins (*b*) glycoproteins
   (*c*) intrinsic proteins (*d*) glycocolic acid
7. According to fluid mosaic model, the plasma membrane presents
   (*a*) a quasifluid structure (*b*) a micellar structure
   (*c*) a polar structure (*d*) a unit membrane structure

8. Fluidity of plasma membrane at body temperature is due to

(*a*) fluidity of lipid bilayer

(*b*) fluidity of glycolipids

(*c*) presence of numerous unsaturated fatty acids in the lipid bilayer

(*d*) presence of integral proteins in the form of glycophorins

9. Fluid mosaic model to explain structure of plasma membrane was proposed by

(*a*) Hoffman (*b*) Singer and Nicolson

(*c*) Robertson (*d*) Davson and Danielli

10. Unit membrane model of plasma membrane was proposed by

(*a*) Nicolson (*b*) Danielli and Davson

(*c*) Robertson (*d*) Mitchel

11. Egestion of residual substances through plasma membrane of cell is called

(*a*) egestion (*b*) ephagy

(*c*) exocytosis (*d*) emeiocytosis

12. Active transport is

(*a*) movement of molecules against concentration gradient

(*b*) movement of molecules along the concentration gradient

(*c*) movement of molecules to and fro of plasma membrane

(*d*) mone of the above

13. Plasmodesmata allow direct exchange of materials

(*a*) between two distant cells (*b*) between neighbouring plant cells

(*c*) within the cell (*d*) all these.

14. Element necessary for the middle lamella

(*a*) Ca (*b*) Zn

(*c*) K (*d*) Cu

ANSWERS

Very Short Answer Questions

1. Robertson.

2. Gorter and Grendel.

3. Singer and Nicholson.

4. Transport of substances through a membrane along their concentration gradients is called passive transport.

5. Movement of molecules and ions along an electrochemical gradient is diffusion.

6. Movement of lipid molecules from one layer to another of plasma membrane.

7. Water, oxygen, carbon dioxide, urea and glycerol.

8. Electrical difference between the inside and outside of a cell.

9. Tektins.

10. D–galactose, D–mannose, L–fucose, sialic acid, N–acetyl–D–glucosamine. (Note. Students may choose any three of them).

11. No.

12. Transportation of these substances across the plasma membrane takes place by specific carrier protein molecules of the plasma membrane.

13. Na^+, K^+–ATPase.

14. Pinocytosis, phagocytosis, endocytosis, receptor-mediated endocytosis and exocytosis.

15. Cholesterol, steroid hormones, antibodies, some proteins and phagocytosis.

16. Ribosome or Plasma membrane.

True and False Statements

1. True **2.** False **3.** True **4.** False.

Yes or No Questions

1. No **2.** Yes.

Fill in the Blanks

1. S.J. Singer, G.L. Nicolson, 1972.

2. Unit membrane.

3. Plasmolysis, osmosis.

4. Channels, facilitated diffusion, entrance.

5. Concentration gradient.

6. Uniport.

7. Cation exchange, Na^+ ions, K^+ ions.

8. Glycocalyx, carbohydrate-rich.

9. Cellulose, chlorzinc iodide.

10. Cellulose synthetase.

Multiple Choice Questions

1. (*c*) **2.** (*d*) **3.** (*b*) **4.** (*b*) **5.** (*b*) **6.** (*c*)
7. (*a*) **8.** (*c*) **9.** (*b*) **10.** (*c*) **11.** (*d*) **12.** (*a*)
13. (*b*) **14.** (*a*).

7

Cell Junctions, Cell Adhesion and the Extracellular Matrix (ECM)

The building technologies of multicellular animals and plants are different, and each type of organism is formed of many types of **tissues**, in which the cells are assembled and bound together in different ways. In both animals and plants, however, an essential part is played in most tissues by the **extracellular matrix (ECM)**. This complex network of secreted extracellular macromolecules has many functions, but the first and salient is that it forms a supporting framework. ECM helps hold cells and tissues together, and in animals, it provides an organized environment within which migratory cells can move and interact with one another in orderly ways. The extracellular matrix, however, is only half the story. In animals especially, the cells of most tissues are bound directly to one another by **cell-cell junctions**. These junctions too are of many types, serving many purposes in addition to mechanical attachment; but without them, bodies of animals would disintegrate.

In vertebrates, the major tissue types are nerve, muscle, blood, lymphoid, epithelium, and connective tissues. Connective tissues and epithelial tissue represent two extremes of organization. In *connective tissue*, the extracellular matrix is plentiful, and cells are sparsely distributed within it. The matrix is rich in fibrous polymers, especially **collagen**, and it is the matrix-rather than the cells—that bears most of the mechanical stress to which the tissues is subjected. Direct attachments between one cell and another are relatively rare.

In **epithelial tissue**, by contrast, cells are tightly bound together into sheets called **epithelia**. The extracellular matrix is scanty, consisting mainly of a tiny mat called the **basal lamina**, which underlies the epithelium. The cells are attached to each other by cell-cell adhesions, which bear most of the mechanical stresses. For this purpose, strong intracellular protein filaments (component of cytoskeleton) cross the cytoplasm of each epithelial cell and attach to specialized junctions in the plasma membrane. The junctions, in turn, tie the surfaces of adjacent cells either to each other or to the underlying basal lamina.

Epithelial cell sheets line all the cavities and free surfaces of the animal body. The specialized junctions between the cells enable epithelia to form barriers that inhibit the movement of water, solutes and cells from one body compartment to another. As shown in (Fig. 7.1), epithelia almost always rest on a supporting bed of connective tissue. This supporting bed may in turn attach them to other tissues, such as the muscle shown in figure. In this way, tissues join together in various combinations to larger functional units called **organs**.

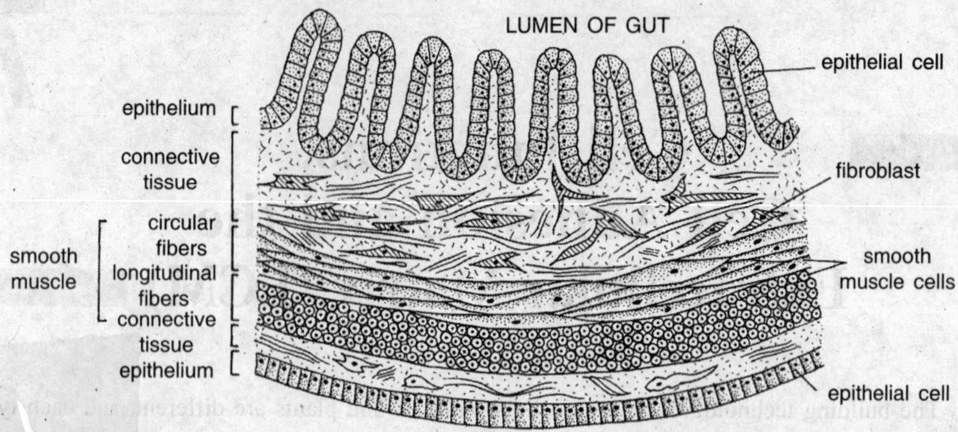

Fig. 7.1. A cross-sectional view of part of the wall of the intestine. This long, tube-like organ is constructed from epithelial tissue, connective tissue and muscle tissue. Each tissue is an organized assembly of cells held together by cell-cell adhesions, extracellular matrix or both (after Alberts *et al.*, 2002).

7.1 CELL JUNCTIONS

Specialized **cell junctions** occur at points of cell-cell and cell-matrix contact in all tissues, and they are particularly abundant in epithelia. Cell junctions are best visualised using either conventional or freeze-fracture electron microscopy, which reveals that the interacting plasma membranes (and often the underlying cytoplasm and the intervening intercellular space as well) are highly specialized in these regions.

Cell junctions (Fig. 7.2) are often divided into following three functional groups:

1. Occluding junctions. They seal cells together in an epithelium in a way that prevents even small molecules from leaking from one side of the sheet to the other. Indeed they form a selectively permeability barrier across epithelial cell sheet.

2. Anchoring junctions. They mechanically attach cells (and their cytoskeleton) to their neighbors or to the extracellular matrix. Thus, they connect the cytoskeleton of a cell to the cytoskeleton of it neighbors or to the ECM.

3. Communicating junctions. They mediate the passage of chemical or electrical signals from one interacting cell to its partner.

The major kinds of intercellular junctions within each group are listed in Table 7.1.

Table 7.1 The functional classification of cell junctions (after **Alberts** *et al.*, 2002).

A. Occluding Junctions
1. Tight junctions (vertebrates only)
2. Septate junctions (invertebrates only)
B. Anchoring Junctions
I. Actin Filament Attachment Sites
1. Cell-cell junctions (adherens junctions)
2. Cell-matrix junctions (focal adhesions)
II. Intermediate Filament Attachment Sites
1. Cell-cell junctions (desmosomes)
2. Cell-matrix junctions (hemidesmosomes)
C. Communicating Junctions
1. Gap junctions
2. Chemical synapses
3. Plasmodesmata (plants only)

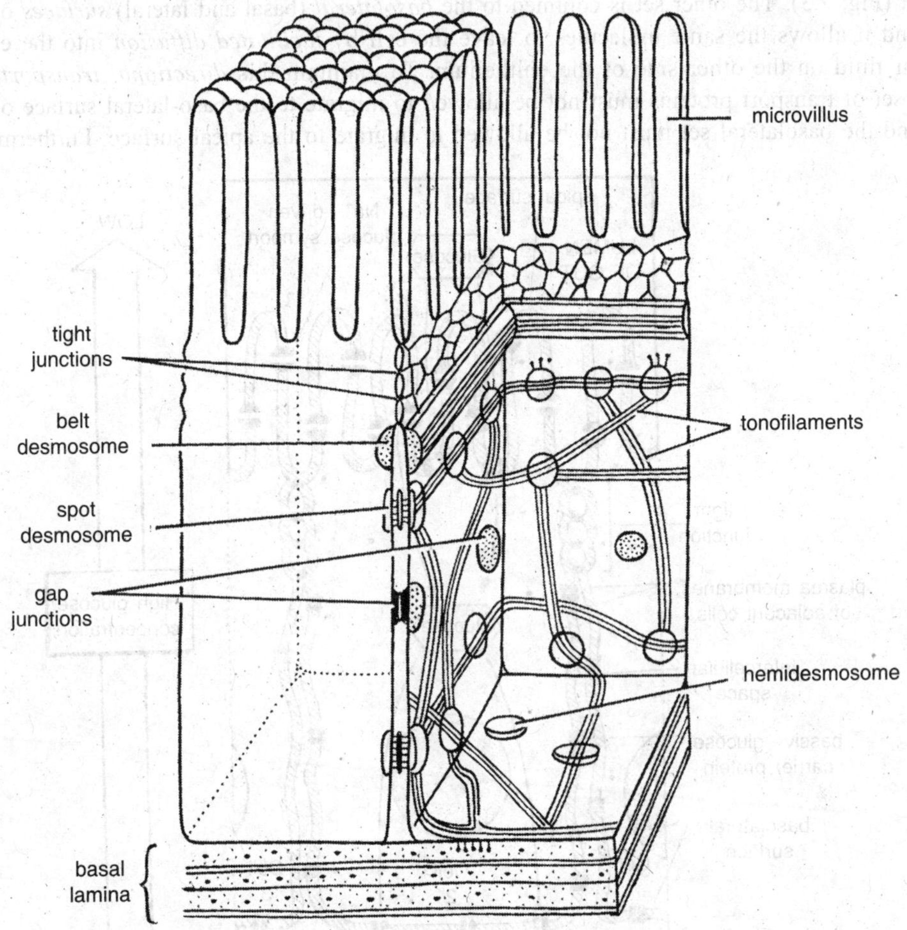

Fig. 7.2. Schematic diagram of the principal types of cell junctions, as found in the intestinal epithelial cell.

A. Occluding Junctions.

All epithelia have at least one important function in common: they serve as selective permeability barriers, separating fluids on either side that have a different chemical composition. This function requires that the adjacent cells be sealed together by occluding junctions.

Occluding junctions are of following two types:

1. Tight junctions. They occur in vertebrates in the organs such as mammalian small intestine, or gut. Tight junctions are also abundant in the ducts and cavities of glands, such as liver and pancreas, that connect with the digestive tract, as well as in the urinary bladder, where they ensure that urine stored in the bladder does not seep out between cells.

The epithelial cells lining the small intestine form a barrier that keeps the gut contents in the gut cavity, the **lumen**. At the same time, however, the cells must transport selected nutrients across the epithelium from the lumen into the extracellular fluid that permeates the connective tissue on the other side (Fig. 7.1). From there these nutrients diffuse into small blood vessels to provide nourishment to the organism. This is called **transcellular transport** and it depends on two sets of membrane-bound **transport proteins**. One set is confined to the apical surface of the epithelial cell (*i.e.*, the surface facing the lumen) and *actively transports* selected molecules into the cell from the gut (Fig. 7.3). The other set is confined to the *basolateral* (basal and lateral) *surfaces* of the cell, and it allows the same molecules to leave the cell by *facilitated diffusion* into the extra-cellular fluid on the other side of the epithelium. To maintain this *directional transport,* the apical set of transport proteins must not be allowed to migrate to the baso-lateral surface of the cell, and the basolateral set must not be allowed to migrate to the apical surface. Furthermore,

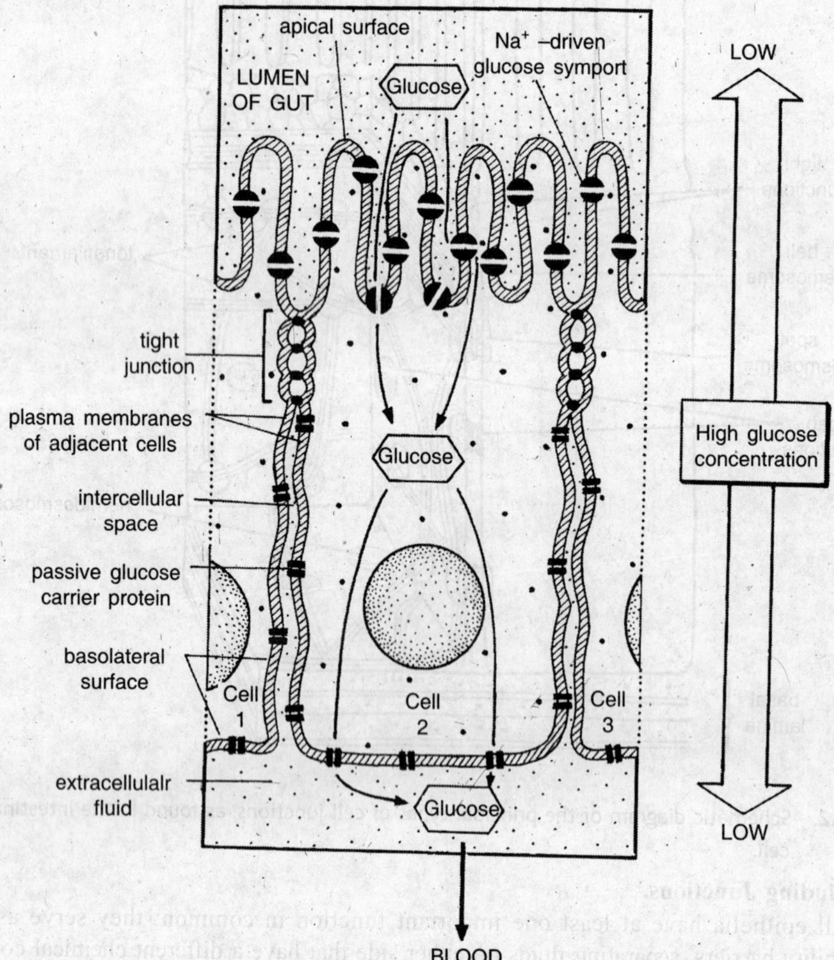

Fig. 7.3. The role of tight junctions in transcellular transport. The transport proteins are confined to different regions of the plasma membrane in epithelial cells of the small intestine. This separation permits a vectorial* transfer of nutrient across the epithelium from the gut lumen to the blood. In the example shown, glucose is actively transported into the cell by Na^+–driven glucose symports at the apical surface, and it diffuses out of the cell by facilitated diffusion mediated by glucose carriers in the basolateral membrane. Tight junctions are thought to confine transport proteins to their appropriate membrane domains by acting as diffusion barriers within the lipid bilayer of the plasma membrane; these junctions also block the backflow of glucose from the basal side of the epithelium into the gut lumen (after Alberts *et al.*, 2002).

***Vectorial.** Vector-like; vector represents a quantity that has magnitude and direction and that is usually represented by part of a straight line with the given direction and with a length representing a magnitude.

the space between the epithelial cell must be tightly sealed, so that the transported molecules cannot diffuse back into the gut lumen through these spaces.

The tight junctions between epithelial cells are thought to have following two types of functions: 1. They function as barriers to the diffusion of some membrane proteins (and lipids) between apical and basolateral domains of the plasma membrane (Fig. 7.3). Mixing of such proteins and lipids occurs if tight junctions are disrupted experimentally, for example, by removing the extracellular Ca^{2+} that is required for tight junction integrity. 2. Tight junctions seal neighboring cell together so that, if a low-molecular-weight tracer is added to one side of an epithelium, it will not generally pass beyond the tight junction (Fig. 7.4). This seal is not absolute, however. Although, all tight junctions are impermeable to macromolecules, their permeability to small molecules varies greatly in different epithelia. For example, tight junctions in the epithelium lining the small intestine are 10,000 times more permeable to inorganic ions, such as Na^+, than the tight junctions in the epithelium lining the urinary bladder. These differences reflect differences in tight junction proteins that form the junctions.

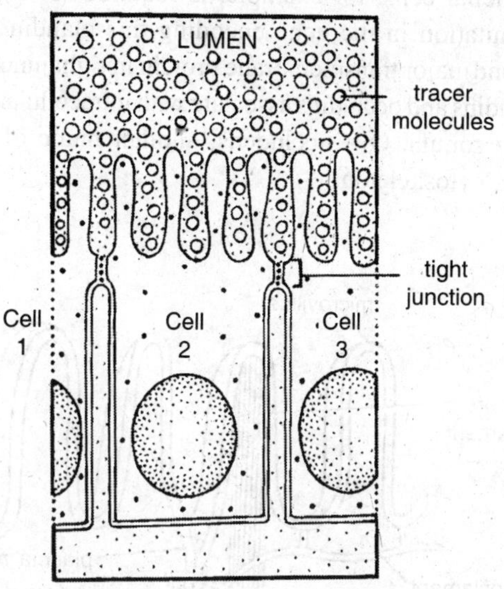

Fig. 7.4. The role of tight junctions in allowing epithelia to serve as barriers to solute diffusion. The drawing shows how a small extra-cellular tracer molecule added on one side of an epithelium cannot traverse the tight junctions that seal adjacent cell together (after Albert *et al.*, 2002).

Epithelial cells can temporarily alter their tight junctions to permit an increased flow of solutes and water through cracks in the junctional barriers. Such **paracellular transport** is especially important in the absorption of amino acids and monosaccharides from the lumen of the intestine, where their concentration can increase enough after a meal to drive passive transport in the desired direction. Thus, tight junctions serve as a regulatory center to help in coordination of multiple cell processes. In pancreatic acinar tissue, they prevent the leakage of pancreatic secretary proteins, including digestive enzymes, into the blood.

Tight junctions are composed of thin bands that completely encircle a cell and are in contact with thin bands of adjacent cells. In a thin section, in the tight junction two adjacent plasma membranes appear to be fused at a series of points. However, in three-dimensional structure, revealed by freeze-fracture technique the tight junctions appear as a network of ridges on the cytoplasmic half of the membrane, with complementary grooves in the outer half. The ridges appear to be composed of two rows of protein particles, as in zipper, each one belonging to the adjacent cells. The lines of these particles produce the sealing and for this reason have been named **sealing strands.** Often, sealing strands form a series of interconnected and anastamosing lines, like a row of stitches in a quilted surface.

Each tight junction sealing strand is composed of long rows of transmembrane adhesion proteins embedded in each of the two interacting plasma membranes. The extracellular domains of these proteins joint directly to one another to occlude the intercellular space (Fig. 7.5). The major transmembrane proteins in a tight junction are the **claudins**, which are essential for tight junction formation and function and differ in different tight junctions. A specific claudin found in kidney epithelial cells, for example, is required for Mg^{2+} to be resorbed from the urine into the blood. A mutation in the gene encoding this claudin results in excessive loss of Mg^{2+} in the urine. A second major transmembrane protein in tight junctions is **occludin**, the function of which is uncertain. Claudins and occludins associate with intracellular peripheral membrane proteins called **ZO proteins** S (Z = zonula, O = occludens, another name of tight junctions) which anchor the strands to the actin cytoskeleton.

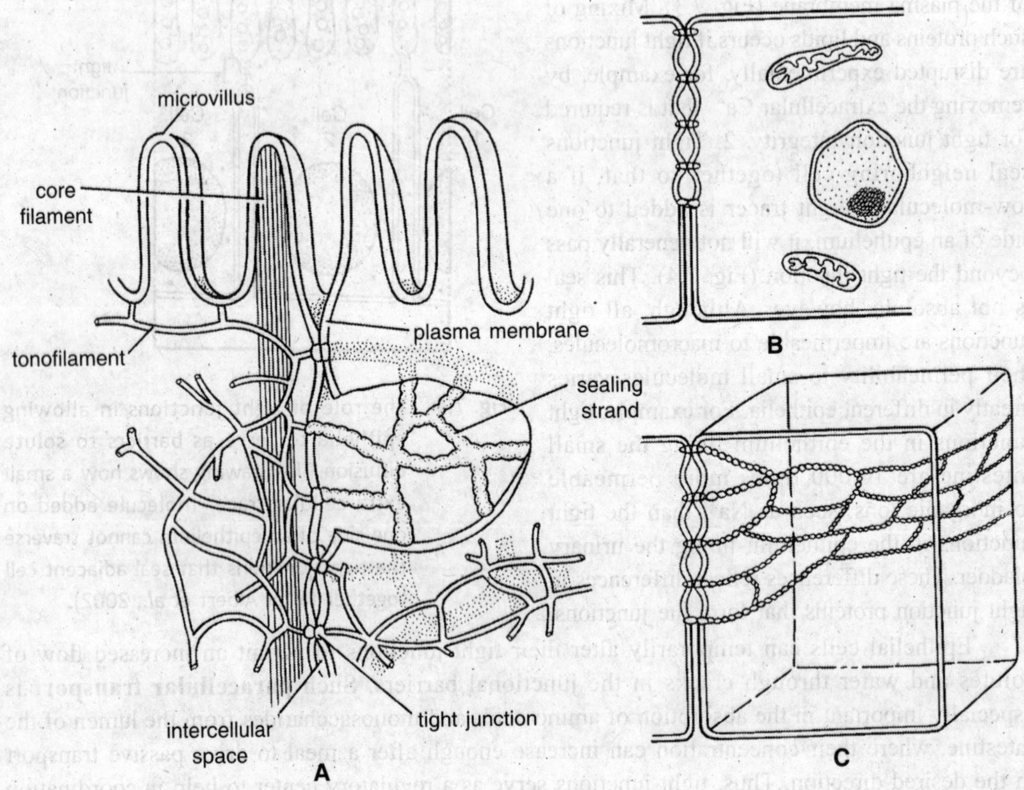

Fig. 7.5. Tight junctions: A—The adjacent plasma membranes are held firmly at the sealing strand which is composed of two rows of particles, as in zipper; B, C—Three-dimensional representations.

2. Septate junctions. These occluding junctions occur in invertebrates. Septate junctions are more regular in structure than the tight junctions and they form a continuous band around each epithelial cell. But their morphology is distinct because the interacting plasma membranes are joined by proteins that are arranged in parallel rows with a regular periodicity. A protein called **Discslarge**, which is required for the formation of septate junction in *Drosophila*, is structurally related to the ZO proteins found in vertebrate tight junction. Mutant flies that are deficient in this protein not only lack septate junctions but also develop epithelial tumors.

B. Anchoring Junctions.

The lipid bilayer is flimsy and cannot by itself transmit large forces from cell to cell or from cell to extracellular matrix. Anchoring junctions solve the problem by forming a strong membrane-spanning structure that is tethered (*i.e.*, fastened) inside the cell to the tension-bearing filaments of the cytoskeleton (Fig. 7.7).

Anchoring junctions are widely distributed in animal tissues and are most abundant in tissues that are subjected to severe mechanical stress, such as heart, muscle and epidermis. They are composed of two main classes of proteins (Fig. 7.8). **Intracellular anchor proteins** form a distinct plaque on the cytoplasmic face of the plasma membrane and connect the junctional complex to either actin filaments or intermediate filaments. **Transmembrane adhesion proteins** have a cytoplasmic tail that binds to one or more intracellular anchor proteins and an extracellular domain that interacts with either the extracellular matrix or the intra-cellular domains of specific transmem-brane adhesion proteins on another cell.

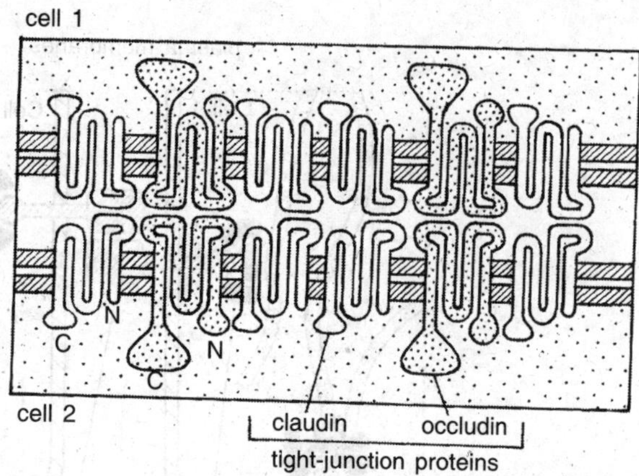

Fig. 7.6. A current model of a tight junction. Drawing showing the transmembrane claudin and occludin proteins in a tight junction. The claudins are the main components of the sealing strands; the function of the occludins is uncertain (after Alberts *et al.*, 2002).

In addition to anchor proteins and adhesion proteins, many anchoring junction contain intracellular signalling proteins that enable the junctions to signal to the cell interior.

Anchoring junctions are of following types:

1. Adherens junctions and **desmosomes** hold cells together and are formed by transmembrane adhesion proteins that belongs to the **cadherin** family.

2. Focal adhesions and **hemidesmosomes** bind cells to the extracellular matrix and are formed by transmembrane adhesion proteins of the **integrin**.

On the intracellular side of the membrane, adherens junctions and focal adhesions serve as connection sites for actin filaments while desmosomes and hemidesmosomes serve as connection sites for intermediate filaments.

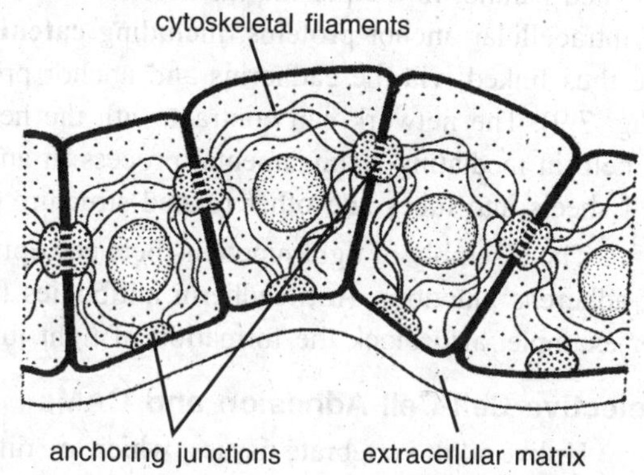

Fig. 7.7. Anchoring junctions in an epithelium. This drawing shows in a general way, how anchoring junctions join cytoskelet filaments from cell to cell and from cell to the extracellular matrix (after Alberts *et al.*, 2002).

I. Actin Filament Attachment Sites

1. Adherens Junctions (Zonula adherens). Adherens junctions commonly occur in epithelia where they often form a continuous **adhesion belt** or **zonula adherens** just below the tight junctions, encircling each of the interacting cell in the sheet. The adhesion belt are directly apposed in adjacent epithelial cells, with the interacting plasma membranes held together by the **cadherins** that serve as transmembrane adhesion proteins.

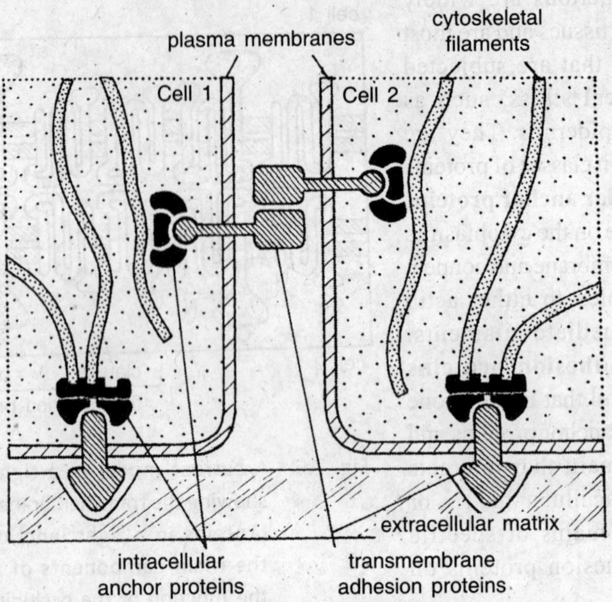

Fig. 7.8. The construction of an anchoring junction from two classes of proteins. This diagram shows how intracellular anchor proteins and transmembrane adhesion proteins form anchoring junctions (after Alberts *et al.*, 2002).

Within each cell, a contractile bundle of actin filaments lies adjacent to the adhesion belt, oriented parallel to the plasma membrane. The actin is attached to this membrane through a set of intracellular anchor proteins, including **catenins, vinculin** and **a-actinin.** The actin bundles are thus linked, via the cadherins and anchor proteins, into an extensive transcellular network (Fig. 7.9). The network can contract with the help of **myosin** motor proteins and it is thought to help in mediating a fundamental process in animal morphogene-sis—the folding of epithelial cell sheets into tubes and other related structure (Fig. 7.10).

The assembly of tight junctions between epithelial cells seems to require the prior formation of adherens junctions. Anti-cadherin antibodies that block the formation of adherens junctions, for example, also block the formation of tight junctions.

Selective Cell-Cell Adhesion and CAMs

Unlike adult vertebrate tissues, which are difficult to dissociate, embryonic vertebrate tissues are easily dissociated. This is usually done by treating the tissue with low concentrations of a proteolytic enzyme such as trypsin. Sometimes this technique is combined with the removal of extracellular Ca^{2+} and Mg^{2+} with a divalent-cation chelator (such as EDTA). These reagents disrupt the protein-protein interactions (many of which are divalent-cation-dependent) that hold cells together. Remarkably, the dissociated cell often reassemble *in vitro* into structures that resemble the original tissue. Findings of this nature reveal that tissue structure is not just a product of history (*i.e.*, origin and development); it is actively maintained and stabilized by the system of affinities that cells have for one another and for the extracellular matrix (ECM).

A salient example of this phenomenon is seen when dissociated cells from two embryonic vertebrate organs, such as liver and the retina, are mixed together and artificially formed into a pellet:

the mixed aggregates gradually sort out according to their origin. More generally, disaggregated cells are found to adhere more readily to aggregates of their own organ than to aggregates of other organs. Evidently there exist **cell-cell recognition systems** that make cells of the same differentiated tissue preferentially adhere to one another; these adhesive preferences are presumably important in stabilizing tissue architecture.

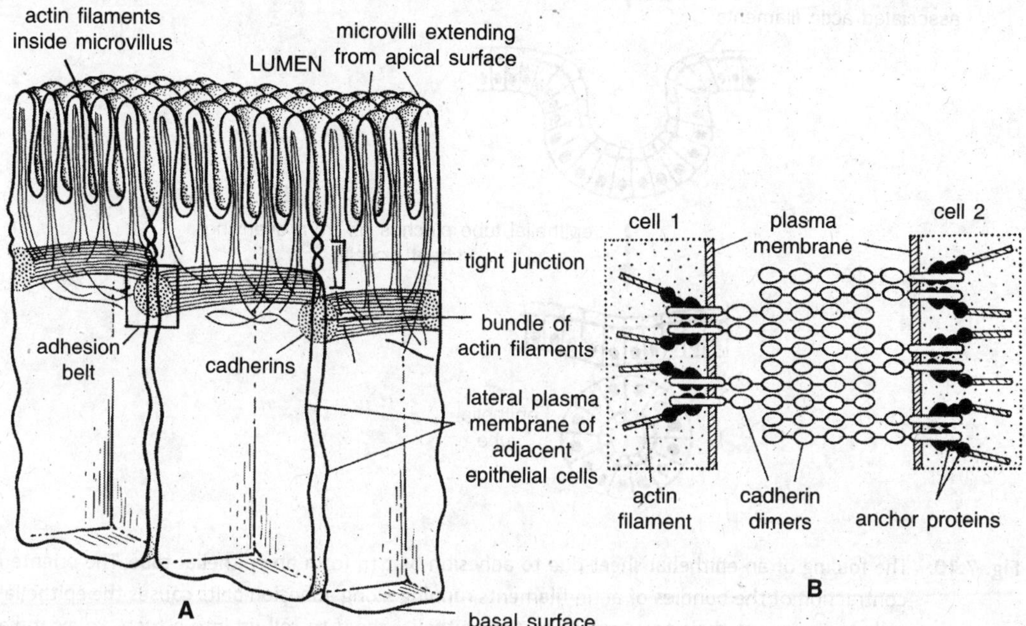

Fig. 7.9 Adherens junctions. A—Adherens junctions in the form of adhesion belts between epithelial cells in the small intestine. The belt-like junction encircles each of the interacting cells. Its most obvious feature is a contractile bundle of actin filaments running along the cytoplasmic surface of the junctional plasma membrane. B—The actin filaments are joined from cell to cell by transmembrane adhesion proteins called **cadherins.** The cadherins form homodimers in the plasma membrane of each interacting cell. The extracellular domain of one cadherin dimer binds to the extracellular domain of an identical cadherin dimer on the adjacent cell. The intracellular tails of the cadherins bind to anchor proteins that tie them to actin filaments. These anchor proteins include α-catenin, β-catenin, γ-catenin (also called plakoglobin), α-actinin and vinculin (after Alberts *et al.*, 2002).

Cells adhere to each other and to the extracellular matrix through cell surface proteins called **cell adhesion molecules (CAMs)**—a category that includes the transmembrane adhesion proteins such as cadherins and integrins. CAMs can be **cell-cell adhesion molecules** or **cell matrix adhesion molecules.** Some CAMs are Ca^{2+}-dependent (*e.g.*, cadherins), whereas others are Ca^{2+}-independent (*e.g.*, immunoglobulins). The Ca^{2+}-dependent CAMs seem to be primarily responsible for the tissue-specific cell-cell adhesion seen in early vertebrate embryos, explaining why these cells can be disintegrated with Ca^{2+}- chelating agents.

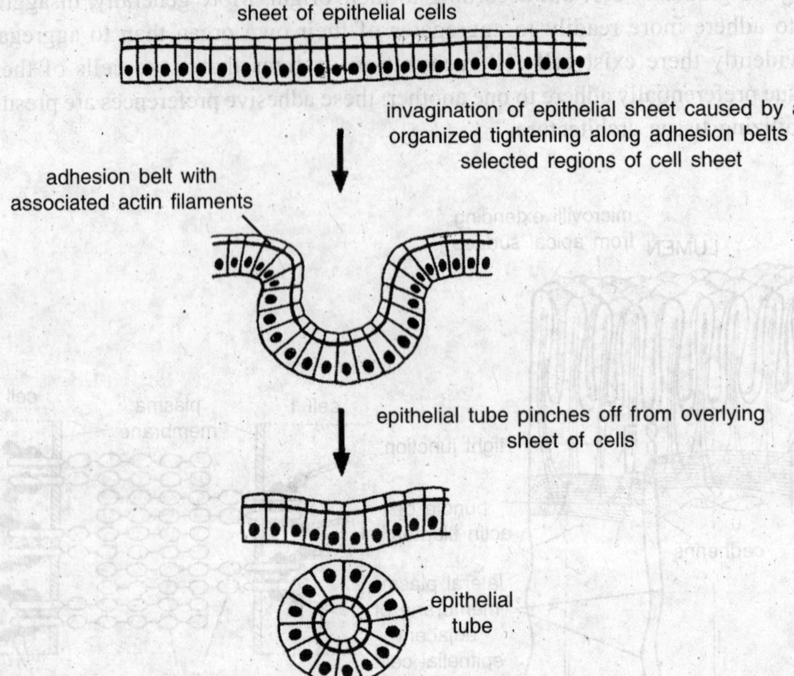

Fig. 7.10. The folding of an epithelial sheet due to adhesion belt to form an epithelial tube. The oriented contraction of the bundles of actin filaments running along adhesion belts causes the epithelial cells to narrow at their apex and helps the epithelial sheet to roll up into a tube. An example is the formation of the neural tube in early vertebrate development (during neurula stage). However, rearrangements of the cells within the epithelial sheet are also thought to have an important role in the process (after Alberts *et al.*, 2002).

Examples of CAMs

1. Cadherins.

The cadherins are the major CAMs which are responsible for Ca^{2+}-dependent cell-cell adhesion in vertebrate tissues. They are of following two main types (Table 7.2).

***(i)* Classical cadherins.** The first three catherins that were discovered were named according to the main tissues in which they were found: **E-cadherin** is present on many types of epithelial cells; **N-cadherin** on nerve, muscle and lens cells; and **P-cadherin** on cells in the placenta and epidermis. All are also found in various other tissues. For example, N-cadherin is expressed in fibroblasts and E-cadherin is expressed in parts of the brain. All these cadherins are **classical cadherins** and they are related in sequence throughout their extracellular and intracellular domains.

***(ii)* Nonclassical cadherins.** They are of many types; more than 50 nonclassical cadherins are expressed in the brain alone. The nonclassical cadherins include proteins with known adhesive function, such as the **desmosomal cadherins** and diverse **proto-caderins** found in the brain. They also include proteins that appear to have nonadhesive functions, such as **T-cadherins**, which lack a transmembrane domain and is attached to the plasma membrane of nerve or muscle cells by a glycosulphosphatidylinositol (GPI) anchor, and the **Fat protein**, which was first identified as the product of a tumor-suppressor gene in *Drosophila*. Together, the classical and nonclassical cadherin proteins form the **cadherin superfamily**.

Table 7.2 Some members of the cadherin superfamily (Source. Alberts *et al.*, 2002).

Name	*Main location*	*Junction association*	*Phenotype when inactivated in mice*
A. Classical cadherins			
1. E-cadherins	Epithelia	Adherens junctions	Die at blastocyst stage; embryos fail to undergo compaction
2. N-cadherins	Neurons, heart, skeletal muscle; lens and fibroblasts	Adherens junctions and chemical synapses	Embryos die from heart defects
3. P-cadherins	Placenta, epidermis, breast epithelium	Adherens junctions	Abnormal mammary gland development
4. VE-cadherins	Endothelial cells	Adherens junctions	Abnormal vascular development (apoptosis of endothelial cells)
B. Nonclassical cadherins			
1. Desmocollin	Skin	Desmosomes	Unknown
2. Desmoglein	Skin	Desmosomes	Blistering skin disease due to loss of keratinocyte cell-cell adhesion
3. Cadherin	Neurons, muscle	None	Unknown
4. Fat (in *Drosophila*)	Epithelial and CNS	None	Enlarged imaginal discs and tumors
5. Protocadherins	Neurons	Chemical synapses	Unknown

Occurrence. Cadherins are found in both invertebrates and vertebrates. Virtually all vertebrates seem to contain one or more cadherins, according to the cell type. They are the main adhesion molecules holding cells together in early embryonic tissues. In culture, the removal of extracellular Ca^{2+} or treatment with anti-cadherin antibodies disrupts embryonic tissues. Mutations that inactivate the function of E-cadherin cause mouse embryos to fall apart and die early in development.

Molecular structure. Most cadherins are single-pass transmembrane glycoproteins about 700–750 amino acids long. Structural studies suggest that they associate in the plasma membrane to form **dimers** or larger **oligomers**. The large extracellular part of the polypeptide chain is usually folded into five or six **cadherin repeats**, which are structurally related to immunoglobulin (Ig) domains (Fig. 7.11A). The crystal structures of E-cadherin and N-cadherin have helped to explain the importance of Ca^{2+} binding for cadherin function. The Ca^{2+} ions positioned between each pair of cadherin repeats, locking the repeats together to form a stiff, rodlike structure: the more Ca^{2+} ions that are bound, the more rigid the structure is. If Ca^{2+} is removed, the extracellular part of the protein becomes floppy and is rapidly degraded by proteolytic enzymes (Fig. 7.11).

Functions of Cadherins

***(i)* Role of cadherins in development.** E-cadherin is the best characterized cadherin. It is usually concentrated in adherens junctions in mature epithelial cells, where it helps connect the cortical actin cytoskeletons of the cell it holds together (Fig. 7.9). E-cadherin is also the first cadherin expressed during mammalian development. It helps cause **compaction**, an important morphological change that occurs at the eight cell stage of mouse embryo development. During compaction, the loosely attached cells, called **blastomeres**, become tightly packed together and joined by intercellular junctions.

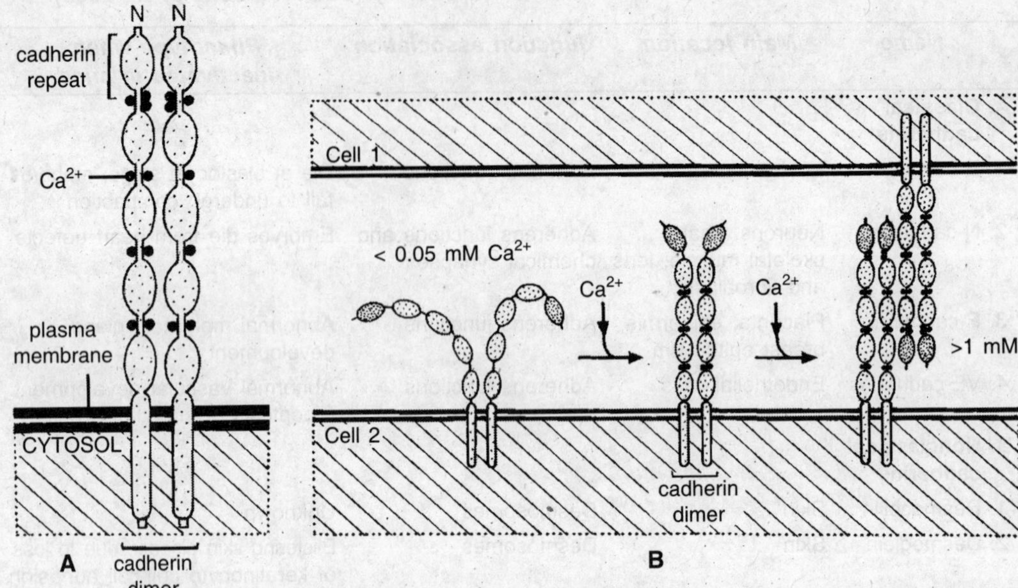

Fig. 7.11. The structure and function of cadherins. A—A classical cadherin molecule. The protein is a homodimer, with the extracellular part of each polypeptide folded into five cadherin repeats. There are Ca^{2+}-binding sites between each pair of repeats. B—The influence of extracellular Ca^{2+}. As the amount of Ca^{2+} increases, the extracellular parts of the cadherin chains become more rigid. When enough Ca^{2+} is bound, the cadherin dimer extends from the surface, where it can bind to a cadherin dimer on a neighboring cell. If Ca^{2+} is removed, the extracellular part of the protein becomes floppy and is degraded by proteolytic enzymes (after Alberts *et al.*, 2002).

Further, as the neural tube forms during the tabulation embryonic stage and pinches off from the overlying ectoderm, neural tube lose E-cadherin and acquire other cadherins, including N-cadherin, while the cells in the overlying ectoderm continue to express E-cadherin.

(*ii*) Role of cadherins in cell-cell adhesion. How do cell-cell adhesion molecules such as the cadherins bind cells together? There can occur following three mechanisms (Fig. 7.13): 1. In **homophilic binding**, molecules on one cell bind to other molecules of the same kind on adjacent cells; 2. In **heterophilic binding**, the molecules on one cell bind to molecules of a different kind on adjacent cells; 3. In **linker-dependent binding**, cell surface receptors on adjacent cells are linked to one another by secreted multivalent linker molecules. Although all three mechanisms have been found to operate in animals, cadherins usually link cells by the homophilic mechanism.

(*iii*) Role of cadherins in their linkage to actin filaments. Most cadherins, including all classical and some nonclassical ones, function as transmembrane adhesion proteins that indirectly link the actin cytoskeletons of the cell they join together. This arrangement occurs in adherens junctions (Fig. 7.9B). The highly conserved cytoplasmic tail of these cadherins interacts indirectly with actin filaments by means of a group of intracellular anchor proteins called **catenins** (Fig. 7.14). This interaction is essential for efficient cell-cell adhesion, as classical cadherins that lack their cytoplasmic domain cannot hold cells strongly together.

Some embryonic cells (*e.g.*, neural tube formation) are known to regulate the adhesive activity of their cadherins. This is done by phosphorylation of anchor protein.

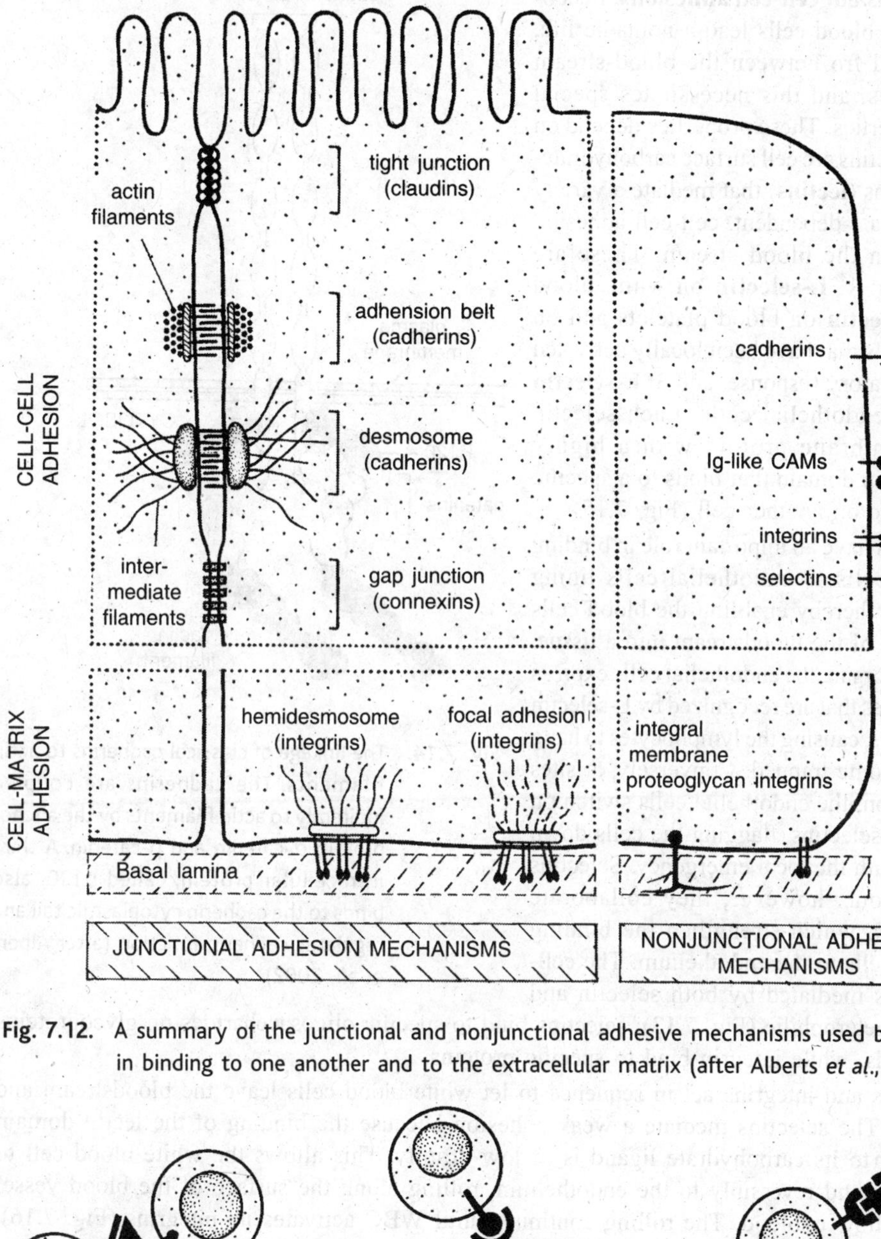

Fig. 7.12. A summary of the junctional and nonjunctional adhesive mechanisms used by animal cells in binding to one another and to the extracellular matrix (after Alberts *et al.*, 2002).

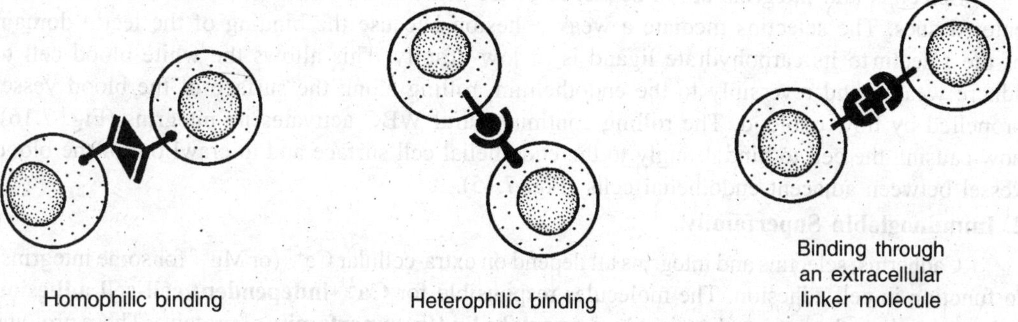

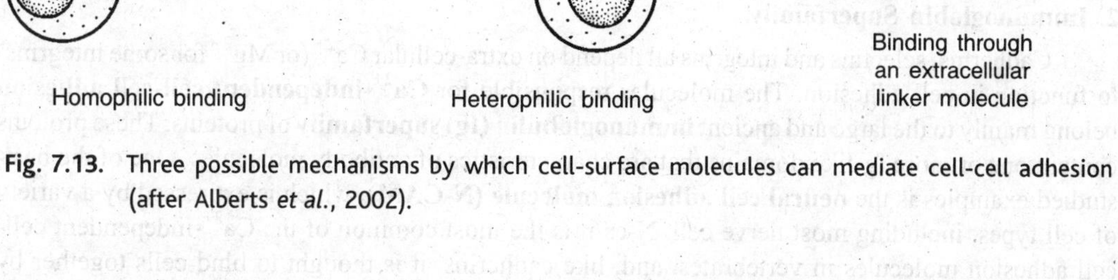

Fig. 7.13. Three possible mechanisms by which cell-surface molecules can mediate cell-cell adhesion (after Alberts *et al.*, 2002).

(*iv*) Transient cell-cell adhesions. Leucocytes or white blood cells lead a nomadic life, moving to and fro between the blood-stream and the tissues, and this necessitates special adhesive properties. These properties depend on **selectins.** Selectins are cell surface carbohydrate-binding proteins (**lectins**) that mediate a variety of transient, Ca^{2+}-dependent, cell-cell adhesion interactions in the blood stream. These are of three types: 1. **L-selectin** on white blood cells, 2. **P-selectin** on blood platelets and on endothelial cells that have been locally activated by an inflammatory response, and 3. **E-selectin** on activated endothelial cells. Each selectin is a transmembrane protein with a highly conserved lectin domain that binds to a specific oligosaccharide on another cell (Fig. 7.15).

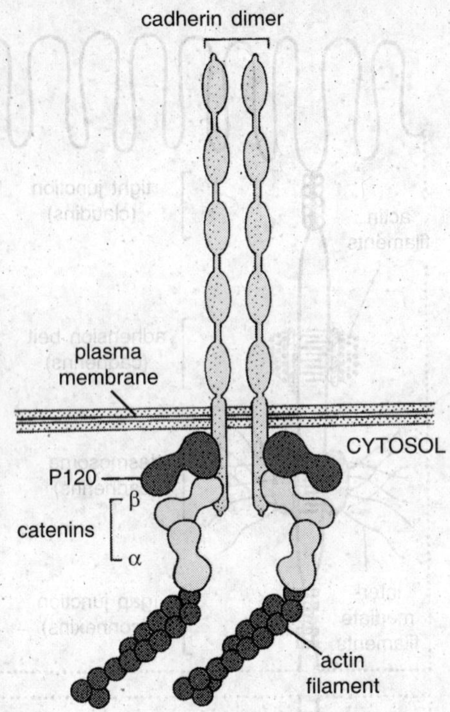

Fig. 7.14. The linkage of classical cadherins to actin filaments. The cadherins are coupled indirectly to actin filaments by the anchor protein α-catenin and β-catenin. A third intracellular protein, called p120, also binds to the cadherin cytoplasmic tail and regulates cadherin function (after Albert *et al.*, 2002).

Selectins have an important role in binding white blood cells to endothelial cells lining blood vessels, thereby enabling the blood cells to migrate out of the bloodstream into a tissue. In a lymphoid organ, the endothelial cells express oligosaccharides that are recognized by L-selectin on lymphocytes, causing the lymphocytes to hang about and become trapped. Conversely, at sites of inflammation, the endothelial cells switch on expression of selectins, flagging the cells down to help deal with the local emergency. Selectins do not act alone, however; they collaborate with *integrins,* which strengthen the binding of the blood cells to the endothelium. The cell-cell adhesions mediated by both selectin and integrins are heterophilic (Fig. 7.13): selectins bind to specific oligosaccharides on glycoproteins and glycolipids, while integrin bind to specific proteins.

Selectins and integrins act in sequence to let white blood cells leave the bloodstream and enter tissues. The selectins mediate a weak adhesion because the binding of the lectin domain of the selectin to its carbohydrate ligand is of low affinity. This allows the white blood cell to adhere weakly and reversibly to the endothelium, rolling along the surface of the blood vessel propelled by flow of blood. The rolling continues until WBC activates its integrins (Fig. 7.16), now causing the cell to bind strongly to the endothelial cell surface and to crawl out of the blood vessel between adjacent endothelial cells (Fig. 7.15).

2. Immunoglobin Superfamily.

Cadherins, selectins and integrins all depend on extra-cellular Ca^{2+} (or Mg^{2+} for some integrins) to function in cell adhesion. The molecules responsible for **Ca^{2+}-independent cell-cell adhesion** belong mainly to the large and ancient **immunoglobulin (Ig) superfamily** of proteins. These proteins contain one or more Ig-like domains that are characteristics of antibody molecules. One of the best-studied examples is the **neural cell adhesion molecule (N-CAM),** which is expressed by a variety of cell types, including most nerve cell. N-cam is the most common of the Ca^{2+}-independent cell-cell adhesion molecules in vertebrates, and, like cadherins, it is thought to bind cells together by a homophilic mechanism (between N-CAM molecules on adjacent cells). Some Ig-like cell-cell adhesion proteins, however, use a heterophilic mechanism. For example, **intercellular adhesion molecules** or **ICAMs** on endothelial cells bind to integrins on white blood cells when WBCs migrate out of the blood stream.

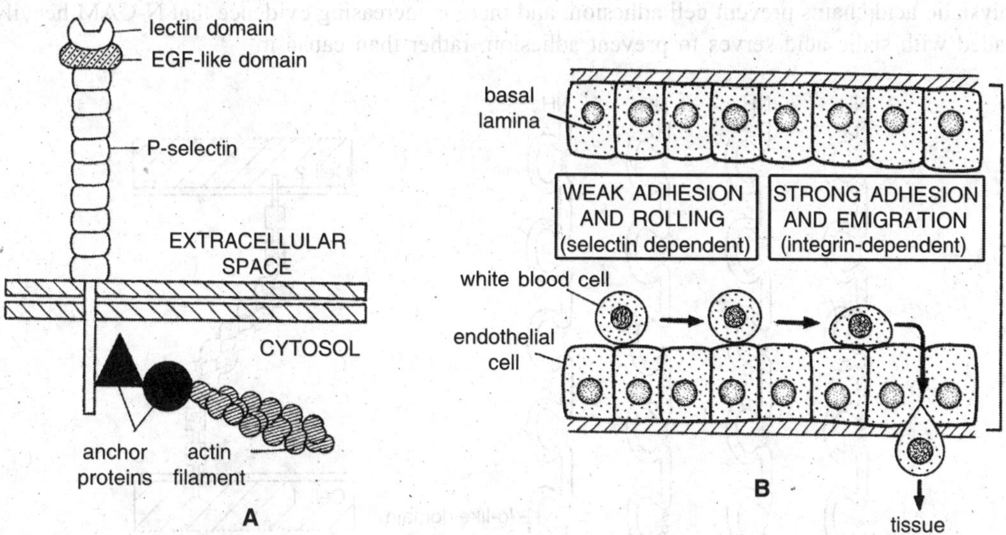

Fig. 7.15. The structure and function of selectins. A—The structure of P-selectin. The selectin attaches to the actin cytoskeleton through anchor proteins that are still poorly characterized. B—How selectins and integrins mediate the cell-cell adhesions required for a white blood cell to migrate out of the blood stream into a tissue (after Alberts *et al.*, 2002).

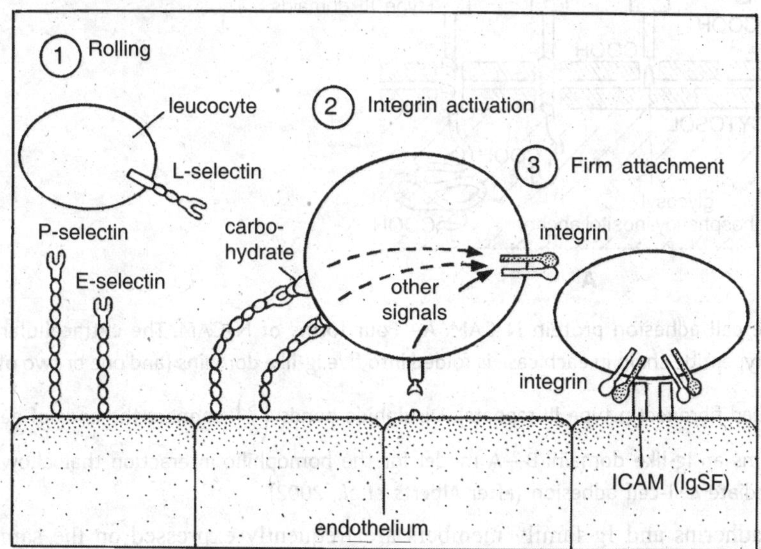

Fig. 7.16. Leucocyte adhesion and selectin. The initial attachment of leucocytes to endothelial cells that line blood vessels is mediated by selectins. (1) Attached leucocytes or white blood cells roll along the blood vessel wall, where they sense activating factors that are deposited on the surface of endothelial cells. (2) This leads to activation of leucocyte integrins, which (3) bind to IgSF proteins (such as.ICAMs), allowing leucocytes to adhere firmly to endothelial cells. Such adhesion allows leucocytes to stop and to pass through blood vessels to sites of inflammation (after Becker *et al.*, 2007).

Types of N-CAM. These are about 20 forms of N-CAM, all generated by alternative splicing of an RNA transcript produced from a single gene. In all forms, the large extracellular part of the polypeptide chain is folded into five Ig-like domains (Fig. 7.17). Some forms of N-CAM carry on unusually large quantity of sialic acid. By virtue of their negative charge, these long polysialic acid chains prevent cell adhesion, and there is increasing evidence that N-CAM heavily loaded with sialic acid serves to prevent adhesion, rather than cause it.

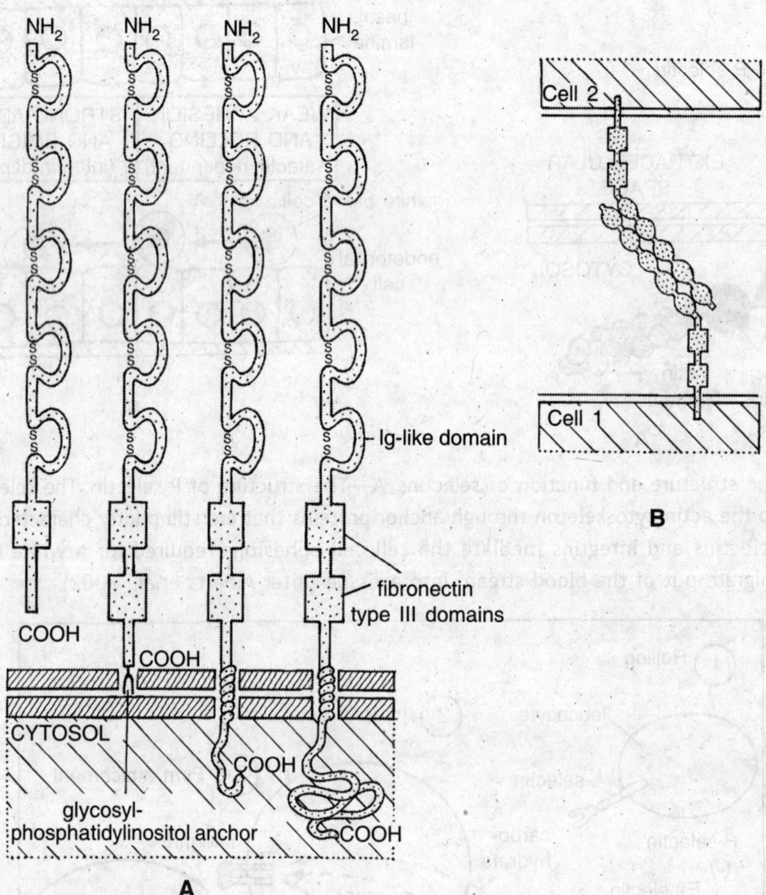

Fig. 7.17. The cell adhesion protein N-CAM. A—Four forms of N-CAM. The extracellular part of the polypeptide chain in each case is folded into five Ig-like domains (and one or two other domains called fibronectin type III repeats). Disulphide bonds $\left(\begin{smallmatrix} S \\ S \end{smallmatrix}\right)$ connect the ends of each loop that forms an Ig-like domain. B—A model for the homophilic interaction that allows N-CAM to mediate cell-cell adhesion (after Alberts *et al.*, 2002).

While cadherins and Ig family members are frequently expressed on the same cells, the adhesions mediated by cadherins are much stronger, and they are largely responsible for holding cell together, segregating cell collectives into discrete tissues, and maintaining tissue integrity. N-CAM and other members of the Ig family seem to contribute more to the **fine-tuning** of these adhesive interactions during development and regeneration. For example, in the developing rodent pancreas, the formation of the **islets of Langerhans** requires cell aggregation, followed by cell sorting. Whereas inhibition of cadherin function prevent cell aggregation and islet formation, loss of N-CAM only impairs the cell sorting process, so that disorganized islets forms.

Likewise, whereas mutant mice that lack N-cadherin die early in development, mutant mice that lack N-CAM develop normally, although they do have some defects in neural development. Mutations in other genes that encode Ig-like cell adhesion proteins, however, can cause more severe neural defects. *L1* gene mutations in human, for example, cause mental retardation and other neurological defects resulting from abnormalities in the migration of nerve cells and their axons.

The significance of Ig-like cell adhesion proteins in connecting the neurons of the developing nervous system has been demonstrated dramatically in *Drosophila*. An N-CAM-like protein called **fasciclin III (FAS3)** is expressed temporarily on some motor neurons, as well as on the muscle cells they normally innervate. If FAS3 is genetically removed from these neurons, they fail to recognize their muscle targets and do not make synapses with them. Conversely, if motor neurons that normally do not express FAS3 are made to express this protein, they now synapse with FAS3-expressing muscle cells to which they normally do not connect. It seems that FAS3 mediates these synaptic connections by a **homophilic "matchmaking" mechanisms** (see **Alberts** *et al.*, 2002).

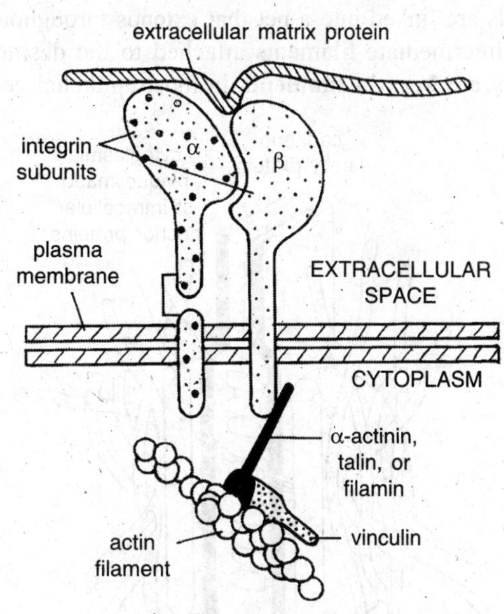

Fig. 7.18. Focal adhesion. Diagram showing some of the proteins that form focal adhesions. The transmembrane adhesion protein is an integrin heterodimer, composed of an α and a β subunit. Its extracellular domains bind to components of the extracellular matrix, while cytoplasmic tail of β subunit binds indirectly to actin filament via several intra-cellular anchor proteins such as vinculin and α-actinin, talin or filamin (after Alberts, *el al.*, 2002).

2. Cell-Matrix Junctions (Focal Adhesions). Some type of anchoring junctions bind cells to the extracellular matrix (ECM) rather than to other cells. The transmembrane adhesion proteins in these cell-matrix junctions are **integrins**. Integrins form a large family of proteins distinct from the cadherins. **Focal adhesion** enable cells to get a hold on the ECM through integrins that intracellularly to actin filaments. In this way muscle cells, for example, attach to their tendons at the **myotendinous junctions**. Likewise, when cultured fibroblasts migrate on an artificial substratum coated with extracellular matrix (ECM) molecules, they also grip the substratum as focal adhesions, where bundles of actin filaments terminate. At all such adhesions, the extra-cellular domains of transmembrane integrin protein bind to a protein component of the extracellular matrix, while their intracellular domains bind indirectly to bundles of actin filaments via the intracellular anchor protein **talin, α-actinin, filamin** and **vinculin** (Fig. 7.18).

II. Intermediate Filament Attachment Sites of Anchoring Junctions

1. Desmosomes (or Maculae adherens). Desmosomes are disc-shaped adhesive junctions approximately 1 μm in diameter having a narrow (30 nm) extracellular gap. They are particularly numerous in tissues that are subjected to mechanical stress, such as cardiac muscle and the epithelial layers of the skin and uterine cervix (see **Karp**, 2002). In a way, desmosomes are buttonlike points of intercellular contact that rivet cells together (Fig. 7.19). Inside the cell, desmosomes serve as anchoring sites for ropelike **intermediate filaments (IFs)**, which form a structural framework of great tensile strength (Fig. 7.19A). Through desmosomes, the intermediate filaments of adjacent cells are linked into a net that extends throughout the many cells of a tissue. The particular type of intermediate filaments attached to the desmosomes depends on the cell type: for example, they are **keratin filaments** in most epithelial cells and **desmin filaments** in heart muscle cells.

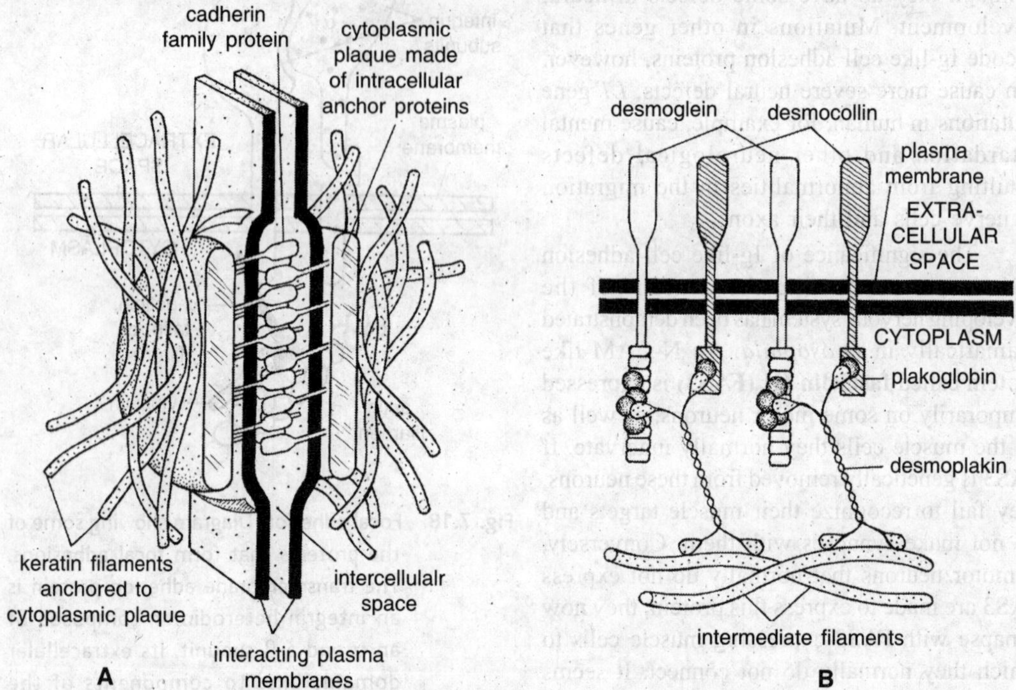

Fig. 7.19. Desmosomes. A—The structural component of a desmosome. On the cytoplasmic surface of each interacting plasma membrane is a dense plaque composed of a mixture of intracellular anchor proteins. A bundle of keratin/desmin intermediate filaments is attached to the surface of each plaque. Transmembrane adhesion proteins of the cadherin family bind to the plaques and interact through their extracellular domains to hold the adjacent membranes together by a Ca^{2+}-dependent mechanism. B—Some of the molecular components of a desmosome. Desmoglein and desmocollin are members of the cadherin family. Their cytoplasmic tails bind plakoglobin, which in turn binds to desmoplakin. Desmoplakin also binds to the sides of intermediate filaments, thereby tying the desmosome to these filaments (after Alberts *et al.*, 2002).

The desmosome has a dense cytoplasmic plaque composed of a complex of intracellular anchor proteins (**plakoglobin** and **desmoplakin**) that are responsible for connecting the cytoskeleton to the transmembrane adhesion proteins (Fig. 7.19B). These adhesion proteins (**desmoglein** and **desmocollin**) like those at an adherens junction belong to the cadherin family. They interact through their extracellular domains to hold the adjacent plasma membrane together.

The importance of desmosome junctions is demonstrated by some forms of the potentially fatal autoimmune skin disease called **pemphigus vulgaris** of human beings. Affected individuals make antibodies against one of their own desmosomal cadherin protein (desmogleins). These antibodies bind to and disrupt the desmosomes that hold their skin epithelial cells (keratinocytes) together. This results in a severe blistering of the skin, with leakage of body fluids into the loosened epithe-lium.

2. Hemidesmosomes. Hemidesmosomes or half-desmosomes resemble desmosomes morphologically and in connecting to intermediate filaments and, like desmosomes, they act as rivets to distribute tensile and shearing forces through an epithelium (Box 7.1). Instead of joining adjacent epithelial cells, however, hemidesmosomes connect the basal surface of an epithelial cell to the underlying basal lamina (Fig. 7.20). The extracellular domains of the $\alpha_6\beta_4$ **integrins** that mediate the adhesion bind to a **laminin** protein in the basal lamina, while an intracellular domain binds via an anchor protein (**plectin**) to keratin intermediate filaments. Whereas the keratin filaments associated with desmosomes make lateral attachments to the desmosomal plaques, many keratin filaments associated with hemidesmosomes have their ends buried in the plaque.

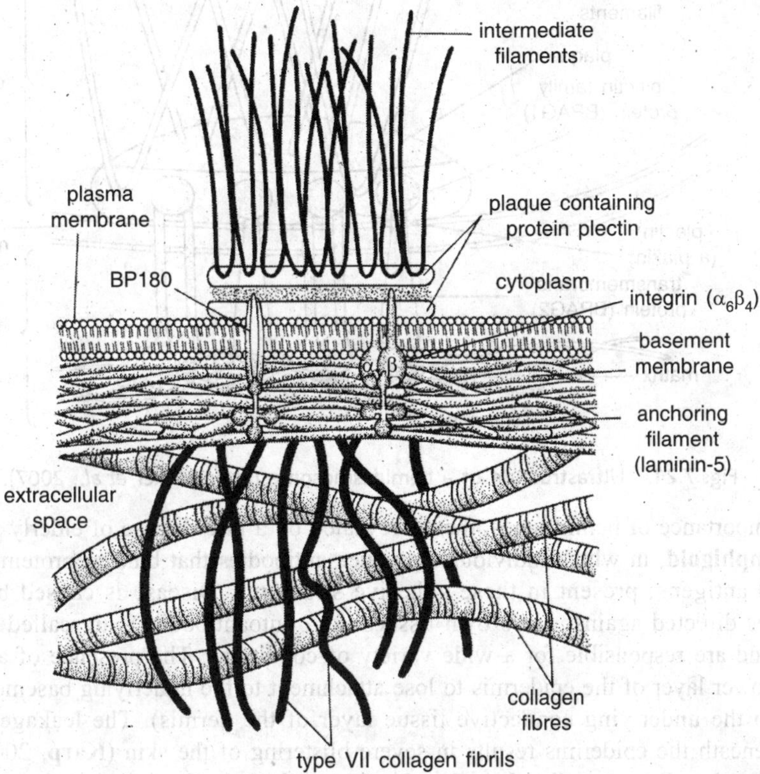

Fig. 7.20. Hemidesmosomes are differentiated sites at the basal surfaces of epithelial cells where the cells are attached to the underlying basement membrane. Schematic diagram showing the major components of a hemidesmosome connecting the epidermis to the underlying dermis. The $\alpha_6\beta_4$ integrin molecules of the epidermal cells are linked to cytoplasmic keratin intermediate filaments by a protein called plectin that is present in plaque and to the basement membrane by anchoring filaments of a particular type of laminin. A second transmembrane protein (BP180) is also present in hemidesmosomes. The collagen fibers are part of the underlying dermis (after Karp, 2003).

Box 7.1

Within body where occur **tightest attachment** between a cell and its extracellular matrix is seen at the basal surface of epithelial cells where the cells are anchored to the underlying basement membrane by a specialised adhesive structure called the **hemidesmosome**.

In addition, another transmembrane protein called BPAG2* and its associated plakin, called BPAG1 can serve as bridge between keratin and laminin (Fig. 7.21). (* BPAG stands for *b*ullous *p*emphigoid *a*ntigen).

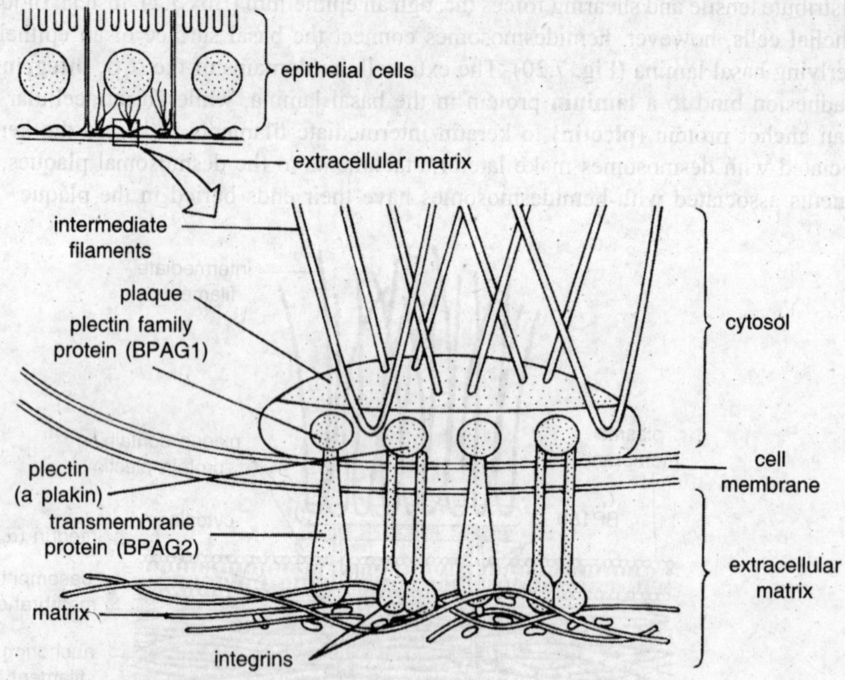

Fig. 7.21. Ultrastructure of a hemidesmosome (after Becker *et al.*, 2007).

The importance of hemidesmosomes is revealed by a rare disease of elderly people, called **bullous pemphigoid**, in which individuals produce antibodies that bind to proteins (the bullous pemphigoid antigens) present in these adhesive structures. Disease is caused by production of antibodies directed against one's own tissues (*i.e.*, autoantibodies), are called **autoimmune disorders** and are responsible for a wide variety of conditions. The presence of autoantibodies causes the lower layer of the epidermis to lose attachment to the underlying basement membrane (and thus to the underlying connective tissue layer of the dermis). The leakage of fluid into the space beneath the epidermis results in severe blistering of the skin (**Karp**, 2002). A similar inherited blistering disease, called **junctional epidermolysis bullosa**, can occur in patients with mutation in β_4 subunit of integrin of the hemidesmosome (see **Becker** *et al.*, 2007).

C. Communicating Junctions

1. Gap Junctions. Many cells of the tissues of higher animals are coupled together by interconnecting **gap junctions, nexus** or **communicating junctions**. The presence of gap junctions explains the ionic or electronic connections between adjacent cells, *i.e.*, there are some cells which are **electrically coupled** and have regions of low resistance in the membrane through which there is a rather free flow of electrical current carried by ions. Such electrical coupling is found extensively in embryonic cells. In adult tissues it is usually found in epithelia, cardiac cells and liver cells. Skeletal muscles and most neurons do not show electrical coupling.

Gap junctions are found to permit molecules such as inorganic ions, sugars, amino acids, nucleotides and vitamins to pass with comparative freedom between one cell and another within a tissue, but they prevent larger molecules, such as proteins, nucleic acids and polysaccharides from being transferred. This observation also explains the phenomenon of **metabolic cooperation** or **metabolic coupling** between cells, *i.e.*, cells can transfer to neighbouring cells, the molecules which cannot be synthesized by the recipient cells. For example, in the tissue-culture experiments, the mutant cells which are deficient in the enzyme **thymidine kinase** can be shown by autoradiography to be capable of DNA synthesis only when grown in a culture vessel together with the wild type cells. This observation shows that required thymidine has been passed from a wild-type cell to a mutant cell, presumably via gap junctions. There are certain other molecules such as IP_3, AMP, ADP, ATP and cAMP that can pass through gap junctions.

A gap junction appears as a plaque-like contact in which the plasma membranes of adjacent cells are in close apposition, separated by a space of only 2 to 4 nm. Structurally, gap junctions consist of hollow channels around which a series of six protein subunits are located; a channel has a diameter of about 1.5 to 2 nm. A single major protein (a macromolecular unit, called **connexon**) of 27000 daltons has been isolated from rat liver preparations consisting of almost pure gap junction material. A connexon appears as an annulus of six **connexin** subunits surrounding the channel. It is believed that the sliding of the subunits caused the channel to open and close. The permeability of channel of gap junction is regulated by Ca^{2+} ions; if the intercellular Ca^{2+} ion level increases, the permeability is reduced or abolished. The gap junctions or connexons of adjacent cells are believed to line up to provide a continuous channel, made up of two connexons opposed end to end (Fig. 7.22).

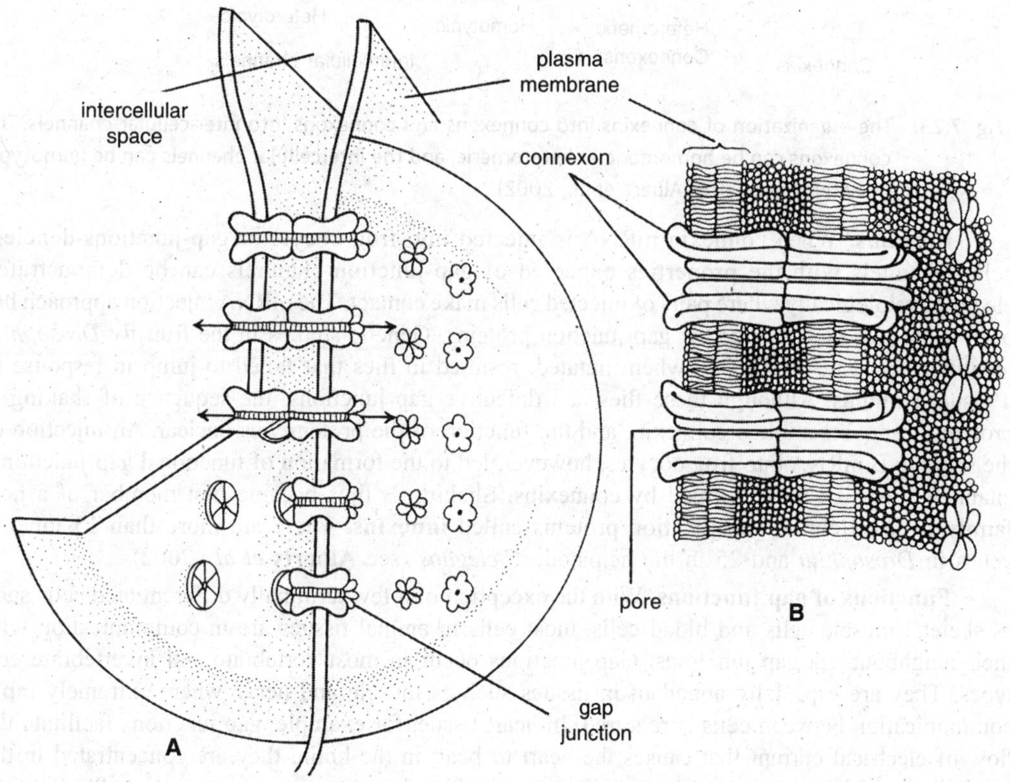

Fig. 7.22. Gap junctions. A—Location of gap junctions between two cells. The channels are made by particles in each membrane that traverse the intercellular space. The flow of fluid between the cells is indicated by arrows. B—Finer structure of unit structure or connexion of gap junctions. The channel has a pore about 2 nm and is formed by two hexamers (six subunits) traversing the lipid bilayers of two plasma membranes.

Gap junctions in different tissues can have different properties. The permeability of their individual channels can vary, reflecting differences in the connexins that form the junctions. In humans, for example, there are 14 distinct connexins, each encoded by a separate gene and each having a distinctive, but sometimes over-lapping, tissue distribution. Most cell types express more than one type of connexin, and two different connexin protein can assemble into a **heteromeric connexon,** the properties of which differ from those of a **homomeric connexon** (Fig. 7.23) constructed from a single type of connexin.

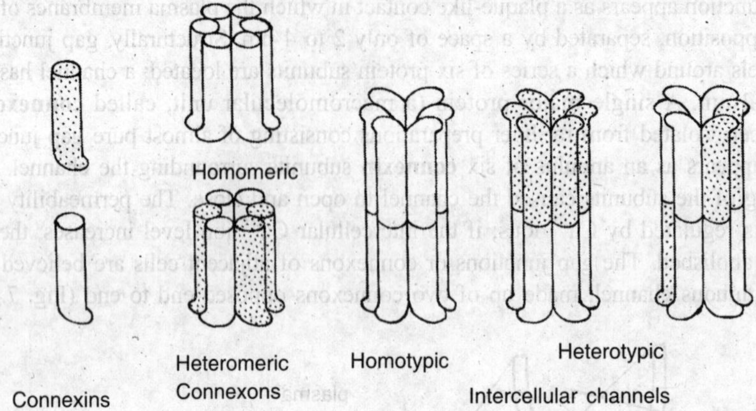

Fig. 7.23. The organization of connexins into connexons and connexons into inter-cellular channels. The connexons can be homomeric or heteromeric, and the intercellular channels can be homotypic or heterotypic (after Albert *et al.*, 2002).

Innexins. When connexin mRNA is injected into frog oocyte or gap-junctions-deficient cells, channels with the properties expected of gap-junction channels can be demonstrated electrophysiologically where pairs of injected cells make contact. The mRNA injection approach has been useful for identifying new gap-junction proteins. Genetic studies in the fruit fly *Drosophila* identified the gene *shaking B*, when mutated, resulted in flies that failed to jump in response to a visual stimulus. Although these flies had defective gap junctions, the sequence of shaking B protein did not resemble a connexin, and the function of the protein was unclear. An injection of the shaking B mRNA into frog oocytes, however, led to the formation of functional gap-junctional channels, just-like those formed by connexins. Shaking B thus became first member of a new family of invertebrate gap-junction proteins called **innexins.** There are more than 15 innexin genes in *Drosophila* and 25 in the nematode *C.elegans* (see **Alberts** *et al.*, 2002).

Functions of gap junctions. With the exception of a few terminally differentiated cells such as skeletal muscle cells and blood cells, most cells in animal tissues are in communication with their neighbour via gap junctions. Gap junctions occur in most vertebrate and invertebrate cell types. They are especially abundant in tissues such as muscle and nerve where extremely rapid communication between cells is required. In heart tissue, for example, gap junctions facilitate the flow of electrical current that causes the heart to beat; in the brain, they are concentrated in the cerebellum, which is involved in coordinating rapid nuclear (*i.e.,* gray matter related) activities. These roles are confirmed by analyzing mutations in connexins and in nexins. For example, mice lacking in one type of connexin have defects in conducting electrical impulses in the heart.

Role of gap junctions in the liver. Gap junctions also occur in many tissues that do not contain electrically excitable cells. In principle, the sharing of small metabolites and ions provides a mechanism for coordinating the activities of individual cells in such tissues and for smoothing out random fluctuations in small molecule concentrations in different cells. For example, in the liver, the release of noradrenaline from sympathetic nerve endings in response to a fall in blood glucose levels stimulates hepatocytes to increase glycogen break-down and release of glucose into blood.

Not all liver cells (hepatocytes) are innervated by sympathetic nerves, however. By means of the gap junctions that connect hepatocytes, the signal is transmitted from the inner-vated hepatocytes to the non-innervated ones. Thus, mice with a mutation in the major connexin gene expressed in the liver fail to mobilize glucose normally when blood glucose levels fall.

2. Role of gap junctions in mammalian ovary. The normal development of ovarian follicles also depends on gap-junction-mediated communication—in this case, between the oocyte and the surrounding granulosa cell. A mutation in the gene that encodes the connexin that normally couples these two cell types causes **infertility** (Fig. 7.24).

3. Role of gap junctions in embryogenesis. Cell coupling via gap junctions also seem to be important in embryogenesis. In early vertebrate embryos, beginning with the late eight call stage in mouse embryos, most cells are electrically coupled to one another. As specific groups of cells in the embryo develop their distinct identities and begin to *differtiate*, they commonly uncouple from surrounding tissue. As the neural plate folds up and pinches off to form the neural tube (Fig. 7.10) for example, its cells uncouple from the overlying epidermal ectoderm. Meanwhile, the cells within each group remain coupled with one another and therefore tend to behave a cooperative assembly all following a similar developmental pathway in a coordinated fashion.

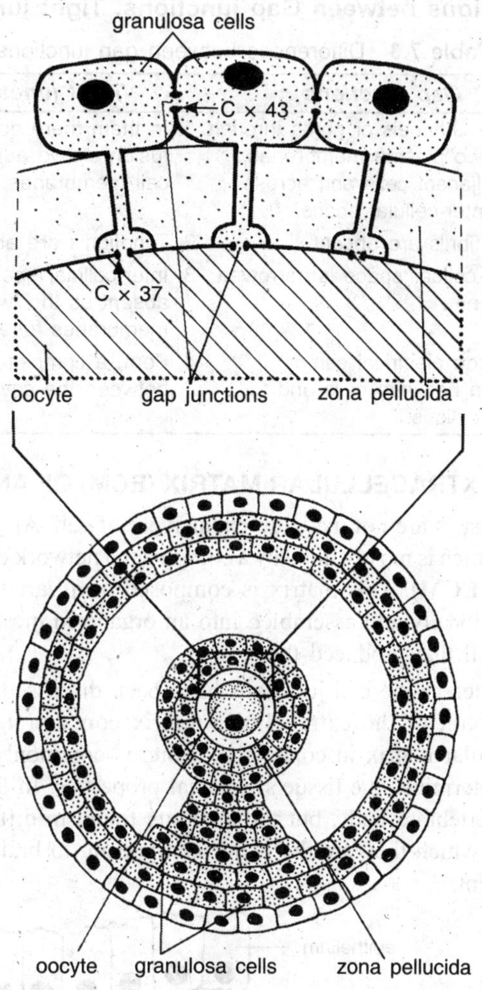

Fig. 7.24. Gap junction coupling in the ovarian follicle. The mammalian ovum is surrounding by a thick layer of extracellular or matrix called the zona pellucida. The surrounding granulose cells are coupled to each other by gap junctions formed by protein connexin 43 (*Cx* 43). In addition, the granulosa cells extend processes through the zona pellucida and make gap junctions with the oocyte. These gap junctions contain a different connexin protein, called Cx 37. Mulation in the gene encoding Cx 37 causes infertility by disrupting the development of both the grannlosa cells and the oocyte (after Alberts *et. al.,* 2002).

Distinctions between Gap junctions, Tight junctions and Desmosomes

Table 7.3 Differences between gap junctions, tight junctions and desmosomes.

Gap junctions	*Tight junctions*	*Desmosomes*
1. Have channels or pores through the two plasma membranes of adjacent cells and across the inter-cellular space.	1. In them there occurs a fusion of two adjacent cell membranes.	1. They contain buttonlike electron-dense regions in each of the opposing cell membranes.
2. Tonofibrils are absent.	2. Tonofibrils are absent.	2. Tonofibrils are present.
3. Inter-celluar space is narrowed to 5 nm.	3. Inter-cellular space is absent as the two adjacent membranes fuse.	3. Inter-cellular space is 30 to 35 nm.
4. Provide electrical communication between cells and for flow of ions.	4. Form a complete barrier between the cavities.	4. Hold cells tightly together.

7.2 EXTRACELLULAR MATRIX (ECM) OF ANIMAL CELLS

Tissues are not made up exclusively of cell. An ample part of their volume is **extracellular space,** which is mostly filled by a complicated network of macromolecules forming the **extracellular matrix (ECM)**. This matrix is composed of a variety of proteins and polysaccharides that are secreted locally and assembled into an organized meshwork in close association with the surface of the cell that produced them.

Whereas the cell junctions have been discussed mainly in the context of epithelial tissues, this account of the extracellular matrix concentrates on connective tissues (Fig. 7.25). The extracellular matrix in connective tissue is commonly more plentiful than the cell it surrounds, and it determines the tissue's physical properties. In fact, connective tissues form the framework of the vertebrate body, but the amounts found in different organs vary greatly-from cartilage to bone, in which they are the major component, to brain and spinal cord, in which they are minor component.

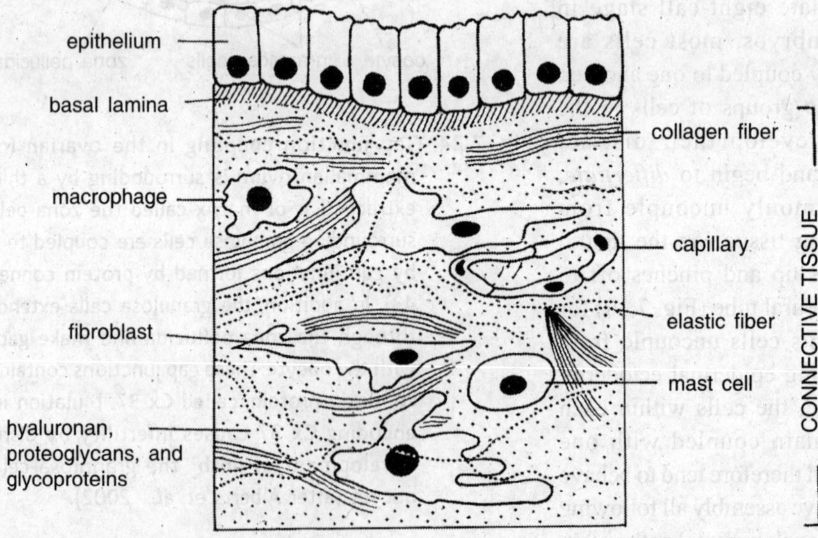

Fig. 7.25. The connective tissue underlying an epithelium. This tissue contains a variety of cells and ECM components. The predominant cell type is the fibroblasts, which secretes abundant extracellular matrix (after Alberts *et al.*, 2002).

Variations in the relative amounts of the different types of matrix macromolecules and the way in which they are organised in the ECM give rise to a surprising diversity of forms, each adapted to the functional requirements of the particular tissue. For example, matrix can become calcified to form the rock-hard structures of bone or teeth, or it can form the transparent matrix of the cornea, or it can adopt the ropelike organization that gives tendons their massive tensile strength. At the interface between an epithelium and connective tissue, the matrix forms a **basal lamina,** which is important in controlling cell behaviour.

Once it was believed that ECM of vertebrates serve mainly as a relatively inert scaffold to stabilize the physical structure of tissue. But now, it is well established that the matrix has a far more active and complex role in regulating the behaviour of the cells that contact it, influencing their survival, development, migration, proliferation, shape and function.

The origins of the ECM are very ancient and virtually all multicellular organisms make it; examples include the cuticles of worms and insects, the shell of mollusks and the cell walls of plants.

1. Cells and Molecules of ECM. The macromolecules that constitute the ECM are mainly produced locally by cells in the matrix. In most connective tissues, for example the matrix macromolecules are secreted largely by cells called **fibroblasts**. In certain specialized types of connective tissues, such as cartilage and bone, however, they are secreted by cells of the fibroblast family that have more specific names: **chondroblasts** form cartilage and **osteoblasts** form bone.

Two main classes of extracellular macromolecules make up the matrix: 1. polysaccharide chains of the class called **glycosaminoglycans (GAGs)**, which are usually found covalently linked to protein in the form of **proteoglycans;** and 2. **fibrous proteins** including **collagen, elastin, fibronectin** and **laminin**, which have both structural and adhesive functions.

The proteoglycan molecules in connective tissue form a highly hydrated, gel-like **"ground substance"** in which fibrous proteins are embedded. The polysaccharide gel resists compressive forces on the matrix while permitting the rapid diffusion of nutrients, metabolites and hormones between the blood and the tissue cells. The collagen fibres both strengthen and help organize the matrix, and rubber-like elastic fibers give it resilience. Finally, many matrix proteins help cells attach in the appropriate locations.

***(i)* GAGs (Glycosaminoglycans).** These are unbranched polysaccharide chains composed of repeating disaccharide units. They are called GAGs because one of two sugars in the repeating disaccharide is always on **amino sugar** (*N*-acetylglucosamine or *N*-acetylgalactosamine), which in most cases is sulphated. The second sugar is usually a **uronic acid** (glucuronic or iduronic). Because there are sulphate or carboxyl groups on most of their sugars, GAGs are highly negative charged. Indeed, they are *the most anionic molecules produced by animal cells.* Four main groups of GAGs are distinguished according to their sugars, the type of linkage between the sugars and the number and location of sulphate groups: 1. **hyaluronan** (or hyaluronic acid); 2. **chondroitin sulphate** and **dermatan sulphate**; 3. **heparan sulphate** and 4. **keratin sulphate.**

GAGs are strongly hydrophilic. They tend to adopt highly extended conformations that occupy a huge volume relative to their mass, and they form **gels** even at very low concentrations. Their high density of negative charges attracts a cloud of cations, most notably Na^+, that are osmotically active causing large amounts of water to be sucked into the matrix. This creates a swelling pressure or **turgor**, that enables the matrix to withstand compressive forces (in contrast to collagen fibrils, which resist stretching forces). The cartilage, matrix that lines the knee joint, for example, can support pressures of hundreds of atmospheres in this way.

The GAGs in connective tissue usually constitute less than 10% of the weight of the fibrous proteins. But, because they form porous hydrated gels, the GAG chains fill most of the extracellular space, providing mechanical support to the tissue. In some rare human genetic disease, there is a severe deficiency in the synthesis of the **dermatan sulphate** disaccharide (Fig. 7.26). The affected individuals have a short stature, permanently aged appearance and generalized defects in their skin, joints, muscles and bones.

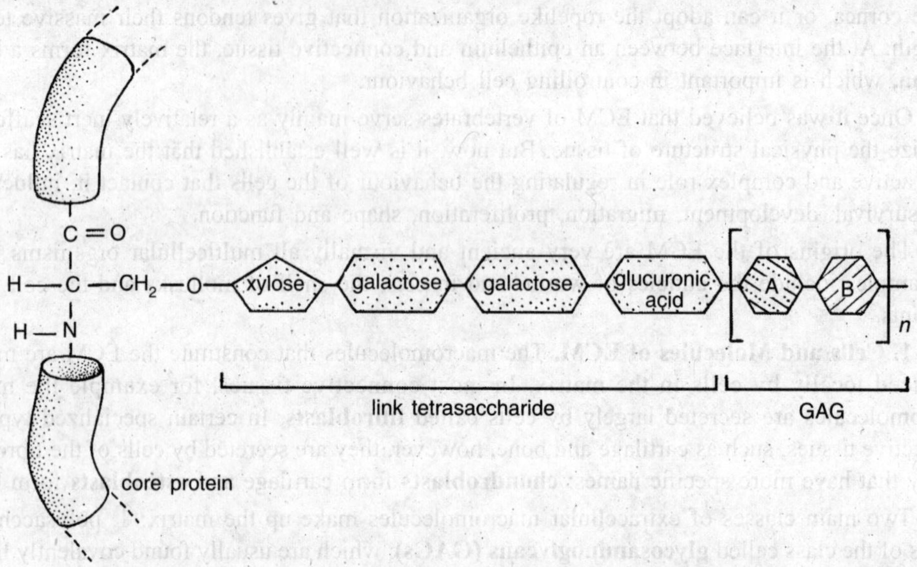

Fig. 7.26. The linkage between a GAG chain and its core protein in a proteoglycan molecule. A specific link tetrasaccharide is first assembled on a serine side chain. The rest of the GAG chain, consisting of mainly of a repeating disaccharide unit is then synthesized, with one sugar being added at a time (after Alberts *et al.*, 2002).

***(ii)* Proteoglycans.** Except for hyaluronan, all GAGs are found covalently attached to proteins in the form of **proteoglycans**, which are made by most animal cells. The polypeptide chain, called **core protein**, of a proteoglycan is made on membrane-bound ribosome and threaded in the lumen of the endoplasmic reticulum. The polysaccharide chains are mainly assembled on this core protein in the Golgi apparatus. First a special link **tetrasaccharide** is attached to a serine side chain on the core protein to serve as a primer for polysaccharide growth; then one sugar at a time is added by specific glycosyl transferase (enzyme) (Fig. 7.27). While still in the Golgi apparatus, many of the polymerized sugars are covalently modified by a sequential and coordinated series of reactions.

The GAG chains of proteoglycans can form gels of varying pore size and charge density; so, they serve as **selective sieves** to regulate the traffic of molecules and cells according to their size, charge, or both. Evidence suggests that a heparan sulphate proteoglycan called **perlecan** has this role in the basal lamina of the kidney glomerulus, which filters molecules passing into the urine from the blood stream.

Proteoglycans are thought to have a major role in chemical signalling between cell. They bind various secreted signal molecules, such as certain protein growth factors and can enhance on inhibit their signaling activity.

Proteoglycans as coreceptors. Not all proteoglycans are secreted components of the ECM. Some (*e.g.*, syndecans, betaglycans, etc.) are integral components of plasma membranes and have their core protein either inserted across the lipid bilayer or attached to the lipid bilayer by a GPI (*i.e.*, glycosylphosphatidylinositol) anchor. Some of these plasma membrane proteoglycans act as

coreceptors that collaborate with conventional cell-surface receptor proteins, in both binding cells to the ECM and initiating the response of cells to some extracellular signal proteins. **Syndecans** are located on the surface of many types of cells, including fibroblasts and epithelial cells, where they serve as receptors for matrix proteins (fibronectin, integrins). Likewise, **betaglycan** binds TGF-β (transforming growth factor-β).

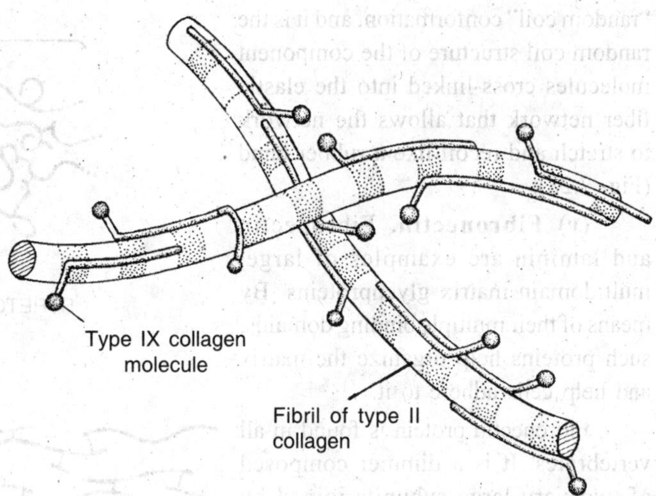

Fig. 7.27. Type IX collagen. Type IX collagen molecules binding in a periodic pattern to the surface of a fibril containing type II collagen (after Alberts *et al.*, 2002).

(*iii*) Collagen. The **collagens** are a family of fibrous protein found in all multicellular animals. They are secreted by connective tissue cells as well as by a variety of other cell types. As a major component of skin and bone, they are the most abundant protein in mammals, constituting 25% of the total protein mass in these animals (see Chapter 7 for more details).

The **fibrillar collagens** (type I, II, III, V and XI collagens) are ropelike, triple-stranded helical molecules that aggregate into long fibrils in the extracellular space. The fibrils in turn can assemble into a variety of highly ordered arrays. **Fibril-associated collagen** molecules, such as type IX and XII, decorate the surface of collagen fibrils and influence the interactions of the fibrils with one another and with other matrix components (Fig. 7.28). In contrast, **type IV collagen** molecule assemble into a sheetlike meshwork that is a crucial component of all mature basal laminae.

(*iv*) Elastin. Many vertebrate tissues, such as skin, blood vessels and lungs, need to be both strong and elastic in order to function. A network of **elastic fibers** in the extra-cellular matrix of these tissues gives them the required resilience so that they can recoil after transient stretch. Elastic fibers are at least five times more extensible than a rubber band of the same cross-sectional area. Long, inelastic collagen fibrils are interwoven with the elastic fibers to limit the extent of stretching and prevent the tissues from tearing.

The main component of elastic fibers is **elastin,** a highly hydrobhobic protein (about 750 amino acid long), which like collagen, is unusually rich in proline and glycine, but unlike collagen, is not glycosylated and contain some hydroxy-proline but no hydroxylysine. The biosynthetic precursor of elastin is soluble **tropoelastin** which is secreted into the extracellular space and assembled into elastic fibers close to the plasma membrane, generally in cell-surface infoldings. After secretion, the tropoelastin molecules, generating an extensive network of elastin fibers and sheets. The cross-links are formed between lysines.

The elastin protein is composed mostly of two types of short segments that alternate along the polypeptide chain: hydrophobic segments, which are responsible for the elastic properties of the molecule; and alanine-and lysine-rich α-helical segments, which form cross-links between adjacent molecules. Each segment is encoded by a separate exon of a split gene. According to one popular model, the elastin polypeptide chain, like the polymer chains in ordinary rubber, adopts a loose "random coil" conformation, and it is the random coil structure of the component molecules cross-linked into the elastic fiber network that allows the network to stretch and recoil like a rubber band (Fig. 7.28).

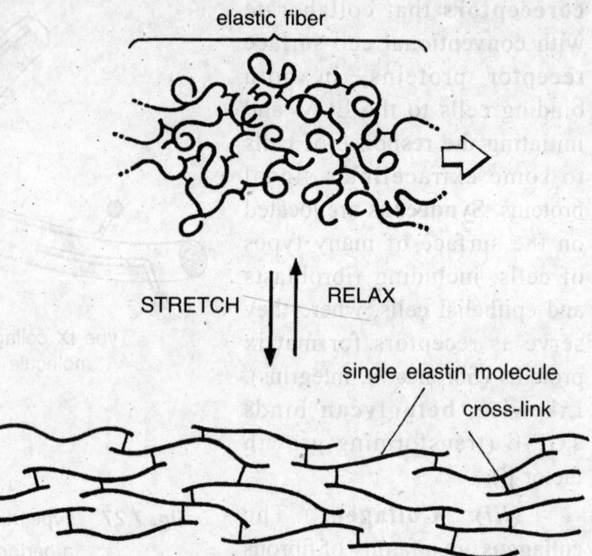

Fig. 7.28. Stretching a network of elastin molecule. The molecules are joined together by covalent bonds to generate a cross-linked network. Each elastin molecule in the network can expand and contract as a random coil.

(v) Fibronectin. Fibronectin and laminin are examples of large, multidomain matrix glycoproteins. By means of their multiple binding domains, such proteins help organize the matrix and help cells adhere to it.

Fibronectin protein is found in all vertebrates. It is a dimmer composed of two very large subunits joined by disulphide bonds at one end. Each subunit is folded into a series of functionally distinct domains separated by regions of flexible polypeptide chain (Fig. 7.29). The domains in turn consist of smaller modules, each of which is serially repeated and usually uncoded by a separate exon, suggesting that the fibronectin gene, like the collagen gene, evolved by multiple exon duplications. All forms of fibronectin are encoded by a single large gene that contains about 50 exons of similar size. Transcription produces a single large RNA molecule that can be alternatively spliced to produce the various isoforms of fibronectin. The main type of module, called the **type III fibronectin repeat,** binds to integrins. It is about 90 amino acid long and occurs at least 15 times in each subunit. The type III fibronectin repeat is the most common of all protein domains.

The RGD sequence. One way to analyze a complex multifunctional protein molecule such as fibronectin is to chop it into pieces and determine the function of its individual domains. When fibronectin is treated with a low concentration of a proteolytic enzyme, the polypeptide chain is cut in the connecting regions between the domains, leaving the domains themselves intact. One can then show that one of its domain binds to collagen, another to heparin, another to specific receptors on the surface of various types of cells and so on (Fig. 7.29) Synthetic peptide corresponding to different segments of the cell-binding domain have been used to identify a specific **tripeptide sequence-Arg-Gly-Asp** or **RGD,** which is found in one of the type III repeats as a central feature of the binding site. Even very short peptides containing this **RGD sequence** can compete with fibronectin for the binding site on cells, thereby inhibiting the attachment of the cells to a fibronectin matrix. If these peptides are coupled to a solid surface, they cause cell to adhere to it.

The RGD sequence is not confined to fibronectin. It is found in a number of extracellular protein, including, for example, the blood-clotting factor *fibrinogen*. Fibrinogen peptides containing this RGD sequence have been useful in the development of anti-clotting drugs that mimic these peptides. Snakes use a similar strategy to cause their victims to bleed: they secrete RGD-containing anti-clotting proteins called **disintegrins** into their venom.

Forms and functions of fibronectin. There are multiple isoforms of fibronectin. One, called **plasma fibronectin,** is soluble and circulates in the blood and other body fluids, where it is thought to enhance blood clotting, wound healing and phagocytosis. All of the other forms assemble on the surface of cells and are deposited in the extracellular matrix as highly insoluble **fibronectin fibrils.** In these cell-surface and matrix forms, fibronectin dimers are cross-linked to one another by additional disulphide bonds.

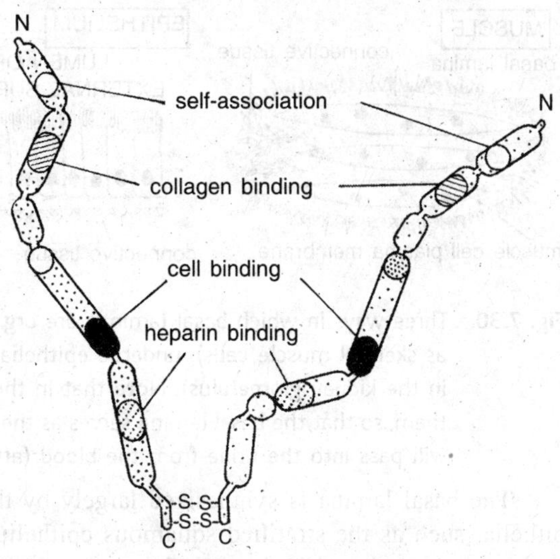

Fig. 7.29. The structure of a fibronectin dimmer. The two polypeptide chains are similar but generally not identical (being made from the same gene but from different spliced mRNAs). They are joined by two disulphide bonds near the C-termini. Each chain is almost 2500 amino acids long and is folded into five or six domains connected by flexible polypeptide segments. Individual domains are specialized for binding to a particular molecule or to a cell as indicated for five of the domains (after Alberts *et al.*, 2002).

Fibronectin molecules assemble into fibrils only on the surface of certain cells. This is because additional proteins are needed for fibril formation, especially fibronectin-binding integrins. In the case of fibroblasts, fibronectin fibrils are associated with integrins at sites called **fibrillar adhesions.** These are distinct from focal adhesions, in that they are more elongated and contain different intracellular anchor proteins. The fibronectin fibrils on the cell surface are highly stretched and under tension. The tension is exerted by the cell and is essential for fibril formation. Some secreated proteins function to prevent fibronectin assembly in inappropriate places. **Uteroglobin,** for example, binds to fibronectin and prevent it form forming fibrils in the kidney. Mice that have a mutation in the uteroglobin gene accumulate insoluble fibronectin fibrils in their kidneys.

Fibronectins are also found to have a role in embryogenesis. Mutant mice that are unable to make fibronectin die early in embryogenesis because their endothelial cells fail to form proper blood vessels.

Fibronectin is important not only for cell adhesion to the matrix but also for guiding cell migrations in vertebrate embryos. Large amounts of fibronectin, for example, are found along the pathway followed by migrating prospective mesodermal cells during amphibian gastrulation (*i.e.*, involution of pharyngeal endoderm and chordamesoderm in frog's gastrulation).

Many matrix proteins are believed to have a role in guiding cell movements during development. The **tenascins** and **thrombospondins,** for example, are composed of several types of short amino acid sequences that are repeated many times and form functionally distinct domains. They can either promote or inhibit cell adhesion, depending on the cell type. Indeed, anti-adhesive interactions are as important as adhesive ones including cell migration (see **Alberts** *et al.*, 2002).

2. Basal Laminae and Laminin. Basal laminae are flexible, thin (40–120 nm thick) mats of specialized extracellular matrix that underlie all epithelial cell sheets and tubes. They also surround individual muscle cells, fat cells and Schwann cells. The basal lamina thus separates these cells and epithelia from the underlying or surrounding connective tissue. In other locations such as the kidney glomerulus, a basal lamina lies between two cell sheets and functions as a highly selective filter (Fig. 7.30). Basal laminae have more than simple structural and filtering functions, however. They are able to determine cell polarity, influence cell metabolism, organize the proteins in adjacent plasma membranes, promote cell survival, proliferation, or differentiation, and serve as specific highways for cell migration.

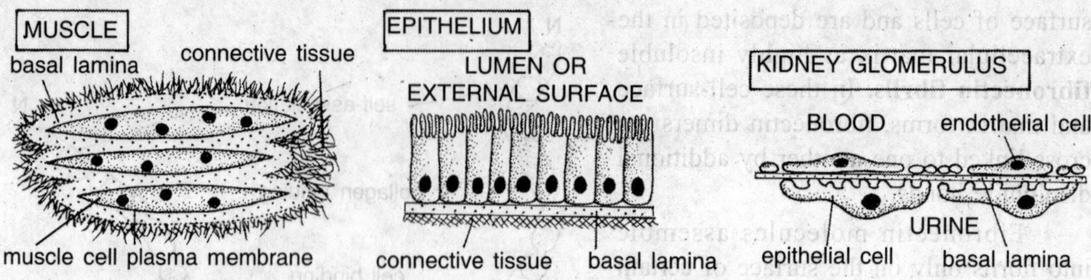

Fig. 7.30. Three ways in which basal laminae are organized. Basal laminae surround certain cells (such as skeletal muscle cells), underlie epithelia and are interposed between two cell sheets (as in the kidney glomerulus). Note that in the kidney glomerulus both cell sheets have gap in them, so that the basal lamina serves as the permeability barrier determining which molecules will pass into the urine from the blood (after Alberts *et al.*, 2002).

The basal lamina is synthesized largely by the cells that rest on it. In some multilayered epithelia, such as the stratified squamous epithelium that forms the epidermis of the skin, the basal lamina is tethered to the underlying connective tissue by specialized anchoring fibrils made of type VII collagen molecules. The term **basement membrane** is often used to describe the composite of the basal lamina and this layer of collagen fibrils (see **Alberts** *et al.*, 2002).

Although precise composition of the basal lamina varies from tissue to tissue and even from region to region, most mature basal laminae contain type IV collagen, the large heparan sulphate proteoglycan perlecan (*e.g.*, agrin) and the glycoproteins laminin and nidogen (or entactin).

Laminin. Early in development, basal laminae contain little or no type IV collagen and consist mainly of laminin molecules. **Laminin-1** (classical laminin) is a large, flexible protein composed of three vary long polypeptide chain (α, β, and γ) arranged in the shape of an asymmetric cross and held together by disulphide bonds (Fig. 7.31). Several isoforms of each type of chain can associate in different combinations to form a large family of laminins. The laminin γ-1 chain is a component of most laminin heterotrimers, and mice lacking it die during embryogenesis because they are unable to make a basal lamina. Like many other proteins in the ECM, the laminin in basement membranes consists of several functional domains: one binds to *perlecan*, one to *nidogen*, and two or more laminin receptor proteins on the surface of cells.

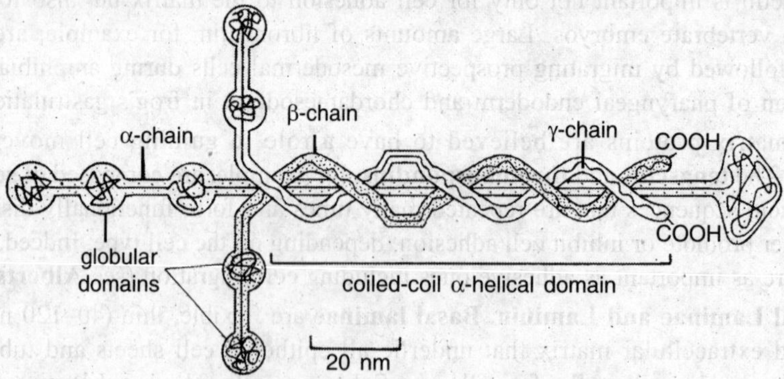

Fig. 7.31. The structure of laminin. The subunits of a laminin-I molecule. This multidomain glycoprotein is composed of three polypeptides (α, β, and γ) that are disulphide bonded into an asymmetric cross-like structure. Each of the polypeptide chains, three types of β chains and three types of γ-chains are known; in principle, they can assemble to form 45 (5 × 3 × 3) laminin isoforms. Several such isoforms have been found each with a characteristic tissue distribution (after Alberts *et al.*, 2002).

Integrins. The linkage of the extracellular matrix to the cell requires transmembrane cell adhesion protein that act as **matrix receptors** and tie the matrix to the cells cytoskeleton. Though some transmembrane proteoglycans function as coreceptors for matrix components, the main receptors on animal cells for binding most extracellular matrix proteins–including collagen, fibronectin and laminins–are the **integrins**. Integrins form a large family of homologous transmembrane, cell-matrix adhesion receptors. In blood cells, integrins also serve as cell-cell adhesion molecules, helping the cells bind to other cells, as well as the ECM.

Integrin are crucially important because they are the main receptor proteins that cells use to both bind to and respond to the extracellular matrix. An integrin molecule is composed of two noncovalently associated transmembrane glycoprotein subunits called α and β (Fig. 7.32). Because the same integrin molecule in different cell types can have different ligand-binding specificities, it seems that additional cell-type-specific factors can interact with integrins to modulate their binding activity.

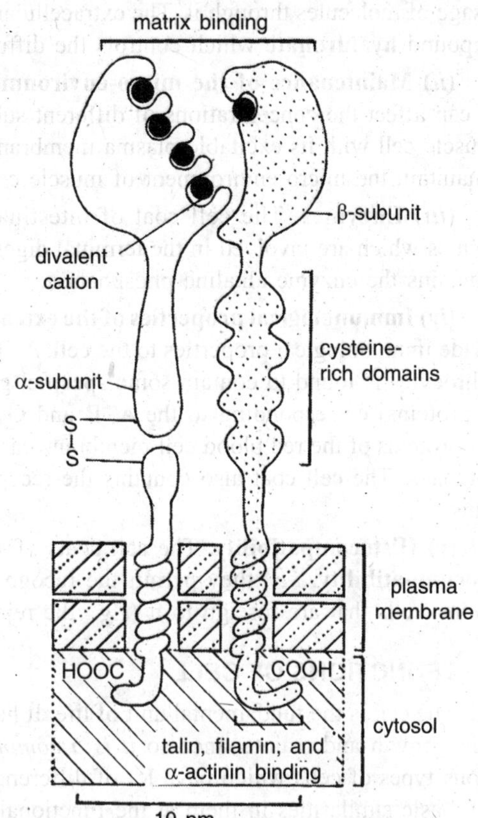

Fig. 7.32. The subunit structure of an integrin cell-surface matrix receptor. Electron micrographs of isolated receptors suggest that the molecule has approximately the shape shown here, with the globular head projecting more than 20 nm from the lipid bilayer. By binding to a matrix protein outside the cell and to the actin cytoskeleton (via the anchor proteins indicated) inside the cell, the protein serves as a transmembrane linker. The α and β subunits are held together by noncovalent bonds. In the fibronectin receptor shown, the α subunit is made initially as a single 140,000 dalton polypeptide chain, which is then cleaved into one small transmembrane domain and one large extracellular domain that contains four divalent-cation-binding sites; the two domains held together by a disulphide (S–S) bond. The extracellular part of the β subunit contains a single divalent-cation-binding site, as well as repeating cysteine-rich region where intrachain disulphide bonding occurs (after Alberts *et al.*, 2002).

The binding of integrins to their ligands depend on extracellular divalent cations (Ca^{2+} or Mg^{2+}), depending on the integrin, reflecting the presence of divalent-cation-binding domains in the extracellular part of the α and β subunits. The type of divalent cation can influence both the affinity and the specificity of the binding of an integrin to its ligands.

Many matrix proteins in vertebrates are recognized by multiple integrins. At least 8 integrins bind fibronectin, for example, and at least 5 bind laminin. A variety of human integrin heterodimers are formed from 9 types of β subunit and 24 types of α subunits.

Function of ECM

In addition to the protection of the plasma membrane, the cell coat performs the following important functions:

(i) **Filtration.** The extraneous coats sometimes act as filters. For instance, the extraneous coats surrounding the blood

capillaries of most vertebrates, especially the kidney glomerulus act as filter and regulate the passage of molecules through it. The extracellular coats of connective tissues contain the chemical compound **hyaluronate** which controls the diffusion.

***(ii)* Maintenance of the micro-environment of the cell.** The extraneous coats of animal cell can affect the concentrations of different substances at the surface of the cell. For example, a muscle cell with its excitable plasma membrane which is surrounded by a glycocalyx is found to maintain the micro-environment of muscle cell by trapping the sodium ions.

***(iii)* Enzymes.** The cell coat of intestinal microvilli are found to contain a variety of enzymes which are involved in the terminal digestion of carbohydrates and proteins. For example it contains the enzyme alkaline phosphatase. .

***(iv)* Immunological properties of the extraneous coats.** Some substances of extraneous coats provide immunological properties to the cell. As for instance the plasma membrane of mammalian erythrocytes is found to contain some specific, genetically determined substances (carbohydrates and proteins) corresponding to the A, B and O blood groups. The major sialic acid containing glycoproteins of the red blood cell membrane carry the M and N antigens that appear infrequently in humans. The cell coat also contains the receptor sites for the influenza virus and for various lectins.

(*v*) Histocompatibility. The cell coats of some cells contain some antigens which provide **histocompatibility**, *i.e.*, they permit the recognition of the cells of one organism and rejection of other cells that are foreign to it (*e.g.*, the rejection of grafts from another organism).

7.3 FUNCTIONS OF CELL

The cell is the fundamental unit of life. It has all the chemical and physical factors necessary for its growth and maintenance so *it is a complete unit of metabolism*. Living organisms have various types of cells which have lot of differences in their shape and size. However, there exist some basic similarities in them at the functional level. Cells are found to contain the following four similar basic functions.

1. Functions as barrier. All cells maintain a barrier that protects the contents of cell from the external environment. Both prokaryotic and enkaryotic cells are enclosed by a thin membrane called **plasma membrane**. This membrane as a barrier maintains the concentration of the solutes in the cell by regulating the transport of materials from in and out of the cell. According to the requirement of the cell sometimes the plasma membrane is folded or invaginated to increase the surface area for more exchange of materials. The plasma membrane has the following roles in the cell:

(*i*) Membranes are selectively permeable allowing certain molecules to pass in and out of the cell.

(*ii*) Membranes can selectively allow certain specific molecules against a concentration gradient (active transport).

(*iii*) The cell receives substances from its environment by the process of **endocytosis.**

(*iv*) Undesired and excess amounts of substances present in the cell are thrown out with the formation of vesicles. These vesicles are pinched off and released from the cell. This type of outward movement is called **exocytosis.**

(*v*) The cell perform certain specific metabolic functions by the help of enzymes and proteins present on plasma membrane.

2. Functions of genetic material. Inheritance and transmission of genetic material from one generation to another are performed in the cell through cell division (mitosis and meiosis). In eukaryotic cells the genetic material is present in the nucleus which is enclosed by nuclear envelope. The nuclear envelope regulates the transport of materials (chemical substances) from the nucleus to the cytoplasm and vice versa. It also serves as an anchoring point of chromosomes/choromatin fibers during interphase. In prokaryotic cells, organised nucleus is absent. The genetic material (DNA) is folded into a compact structure known as **nucleoid**.

The genetic material (DNA) both in prokaryotes and enkaryotes control the synthesis of protein on small cytoplasmic structures called **ribosomes**. Three type of RNA molecules (*i.e.*, rRNA, tRNA and mRNA) are transcribed from DNA to form specific proteins in adequate amounts in correct form. The structure and function of ribosomes, in both prokaryotes and eukaryotes are similar.

3. Functions in cellular metabolism. All cells carry out series of chemical reactions for the synthesis of macromolecules, tapping energy, degradation of unused molecules, converting food substances into sugars, amino acids, etc. Most of the reactions are performed in the cytoplasm of the cell. The initial reactions of the cell start through the degradation of reserve foods (*e.g.,*; glycogen) in the fluid phase of the cytoplasm (cytosol). The main biochemical reactions, which release energy such as ATP occur in the cytoplasmic organelle called **mitochondria**. The desired proteins are synthesized on the ribosomes according to genetic information of genes.

In photosynthetic eukaryotes, the energy from sunlight is used by converting the light energy into chemical energy. This conversion is done by some specialized organelles called **chloroplasts**. In prokaryotes such as photosynthetic bacteria and blue green algae (cyanobacteria) similar trapping of energy from sunlight is done by folding of the plasma membrane.

4. Functions in motility. Cells perform different types of motility, *i.e.*, from locomotion to the movement of some components of the cell. The movements may be either the movement of cell from one place to another (*e.g.*, amoeba, paramecium, white blood cells, sperm, etc.) or the movement of liquid over the surface of a cell by cilia and flagella (*e.g.*, ciliated surfaces of trachea, oviduct, etc.).

In eukaryotic cells, microtubules (Mts) are present which consist of a special protein called **tubulin**. The microtubules play an important role in forming spindle apparatus that helps in the movement of the chromosomes during the cell division. Microtubules also provide a rigid set of tracks for the transport of a variety of membrane-enclosed organelles and vesicles. The mechanical work needed for movement depends on **microtubule-associated motor proteins (motor MAPs)**, which attach to vesicles or organellels and then "walk" along the microtubule, using ATP to provide the needed energy. **Motor protein** or **mechanoenzyme** is a protein that uses energy derived from ATP to change shape in a way that exerts force and causes attached structures to move; includes three families of proteins (myosin, dynein, and kinesin) that interacts with cytoskeletal elements (microtubules and microfilaments) to produce movement (see **Becker** *et al.*, 2007).

In eukaryotic cells, other smaller filaments are found. These are made up of actin proteins and are called **microfilaments**. These microfilaments help in muscle contraction and cytoplasmic streaming (cyclosis).

In this way, the cells are capable of performing all the necessary metabolic activities essential for continuity of life.

QUESTIONS

Long Answer Questions

1. Describe the structure and functions of animal cell junctions. What is extracellular matrix?
2. Describe the main functions of cell.

3. What is cell adhesion? Describe the various components of extracellular matrix which are responsible for this process.
4. Describe various types of cell adhesion molecules which provide differential affinity.

Short Answer Questions

1. What are glycocalyx?
2. What are desmosomes? How many layers of desmosomes are found? Describe their function.
3. Write a short note on extracellular matrix.
4. How does adhesion takes place?
5. Differentiate between gap junctions, tight junctions and desmosomes.
6. Write a short note on tight junctions or zona occludens.

Very Short Answer Questions

1. What is full form of ECM?
2. What are cadherins?
3. Write the names of main cell adhesion proteins.
4. What is fibronectin?
5. What is laminin?
6. Define the basal lamina?
7. What is collagen?
8. Define proteoglycans.
9. Define the plaque.
10. Define integrins.
11. In which tissue the gap junctions are mostly found?
12. What prevents leakage of pancreatic secretary proteins in blood?
13. Name the structure that provide mechanical adhesion between adjacent cells.

Fill in the Blanks

1. The ECM of the epithelium is very thin, it is called ....
2. Septate cell junctions of invertebrates are a type of ....
3. Desmosomes are cell-cell junctions and are a type of............. junctions.
4. Anchoring junctions having .... filament attachment sites between cell and matrix are called ....
5. The particular type of intermediate filaments attached to the cytoplasmic plaque of the desmosome depends on the cell type. For example they are .... filaments in most epithelial cells and .... filament in heart muscles.
6. Tightest attachment between a epithelial cell and its ECM is formed by ....
7. The fibroblasts of ECM of cartilage are called ....
8. The fibroblasts of ECM of a bone are called ....
9. Fibrons proteins of extra-cellular matrix include ...., ..., ... and laminin.
10. Motor proteins of the cells are also called ....

Multiple Choice Questions

1. The cell junctions called tight, adhering and gap junctions are found in
   (*a*) connective tissue (*b*) epithelial tissue
   (*c*) neural tissue (*d*) muscular tissue

2. Desmosomes are concerned with

(*a*) cell division (*b*) cellular excretion

(*c*) cytolysis (*d*) cell adherance

3. Cellular recognition and adhesion are facilitated by components of plasma membranes. These components are generally

(*a*) lipid molecules (*b*) protein molecules

(*c*) both lipid and protein molecules (*d*) glycolipids and glycoproteins

4. Tonofibrils are anchored at this cell junction

(*a*) desmosome (*b*) nexus

(*c*) zonula occludens (*d*) zonula adherens

5. Collagen contains

(*a*) hydroxyproline (*b*) tryptophan

(*c*) arginine (*d*) peptone.

ANSWERS

Very Short Answer Questions

1. The extracellular matrix.
2. Cadherinrs are Ca^{2+} dependent cell adhesion glycoproteins.
3. Adhesion proteins such as fibronectin, laminin and tenacin.
4. It is adhesive glycoprotein which is found in the extracellular matrix and loosely associated with the surface; it binds cells to the ECM and is important in determining the cell shape and guiding cell migration.
5. It is a glycoprotein of the extracellular matrix localized predominantly in the basal lamina of epithelial cells.
6. It is a thin sheet of specialized extra-cellular matrix material that separates epithelial cells from underlying connective tissue.
7. Collagen is a family of closely related proteins that form high-strength fibers found in high concentration in the ECM of animals.
8. They are complex between proteins and glycosaminoglycans found in the extracellular matrix.
9. It is a dense layer of fibrous material located on the cytoplasmic side of adhesive junctions such as desmosomes, hemidesmosomes, and adherens junctions; composed of intracellular attachment proteins that link the junction to the appropriate type of cytoskeletal filament.
10. Integrins are any of several plasma membrane receptors that bind to extracellular matrix components at the outer membrane surface and interact with cytoskeletal components at the inner membrane surface; includes receptors for fibronectin, laminin and collagen.
11. Embryonic tissue.
12. Zona occludens.
13. Desmosomes.

Fill in the Blanks

1. Basal lamina. 2. Occluding junctions. 3. Anchoring.
4. Intermediate, hemidesmosome. 5. Keratin, desmin. 6. Hemidesmosome.
7. Chondroblasts. 8. Osteoblasts. 9. Collagen, elastin, fibronectin.
10. Mechanoenzymes.

Multiple Choice Questions

1. (*b*) 2. (*d*) 3. (*d*) 4. (*a*) 5. (*a*).

8

Endoplasmic Reticulum (ER)

The cytoplasmic matrix is traversed by a complex network of inter-connecting membrane bound vacuoles or cavities. These vacuoles or cavities often remain concentrated in the endoplasmic portion of the cytoplasm; therefore, known as **endoplasmic reticulum,** a name derived from the fact that in the light microscope it looks like a "net in the cytoplasm." (Eighteenth-century European ladies carried purses of netting called **reticules**).

The name "endoplasmic reticulum" was coined in 1953 by **Porter**, who in 1945 had observed it in electron micrographs of liver cells.

8.1 OCCURRENCE

The occurrence of the endoplasmic reticulum varies from cell to cell. The erythrocytes (RBC), egg and embryonic cells lack in endoplasmic reticulum. [**Note**. In the reticulocytes (immature red blood cells) which produce only proteins to be retained in the cytoplasmic matrix (cytosol) (*e.g.*, haemoglobin, the ER is poorly developed or non-existent, although the cell may contain many ribosomes]. The spermatocytes have poorly developed endoplasmic reticulum. The adipose tissues, brown fat cells and adrenocortical cells, interstitial cells of testes and cells of corpus luteum of ovaries, sebaceous cells and retinal pigment cells contain only **smooth endoplasmic reticulum (SER)**. The cells of those organs which are actively engaged in the synthesis of proteins such as acinar cells of pancreas, plasma cells, goblet cells and cells of some endocrine glands are found to contain **rough endoplasmic reticulum (RER)** which is highly developed. The presence of both SER and RER in the hepatocytes (liver cells) is reflective of the variety of the roles played by the liver in metabolism.

8.2 ER AND ENDOMEMBRANE SYSTEM

The endoplasmic reticulum is the main component of the **endomembrane system**, also called the **cytoplasmic vacuolar system** or **cytocavity network**. This system comprises following structures: 1. The **nuclear envelope**, consisting of two non-identical membranes, one opposed to the nuclear chromatin and other separated from the first membrane by a perinuclear space (both forming a cisternae), the two membranes being in contact at the nuclear pores; 2. The **endoplasmic reticulum**; and 3. The **Golgi apparatus**, which is mainly related to some of the terminal processes of cell secretion. **GERL** (or Golgi, ER and lysosome) refers to a special region of endomembrane system, which is more related to the Golgi apparatus and is involved in the formation of lysosomes.

The entire endomembrane system represents a barrier separating cytoplasmic compartments. The membrane of each component of this system has two faces: (*i*) The **cytoplasmic** or **protoplasmic face** and (*ii*) The **luminal face** (Fig. 8.1). The luminal face borders the perinuclear cisternae, the cavities of ER and SER, and the Golgi elements. It also corresponds to the interior of the secretory granules, the lysosomes and peroxisomes and also to faces of mitochondrial membranes confronting to outer mitochondrial chamber.

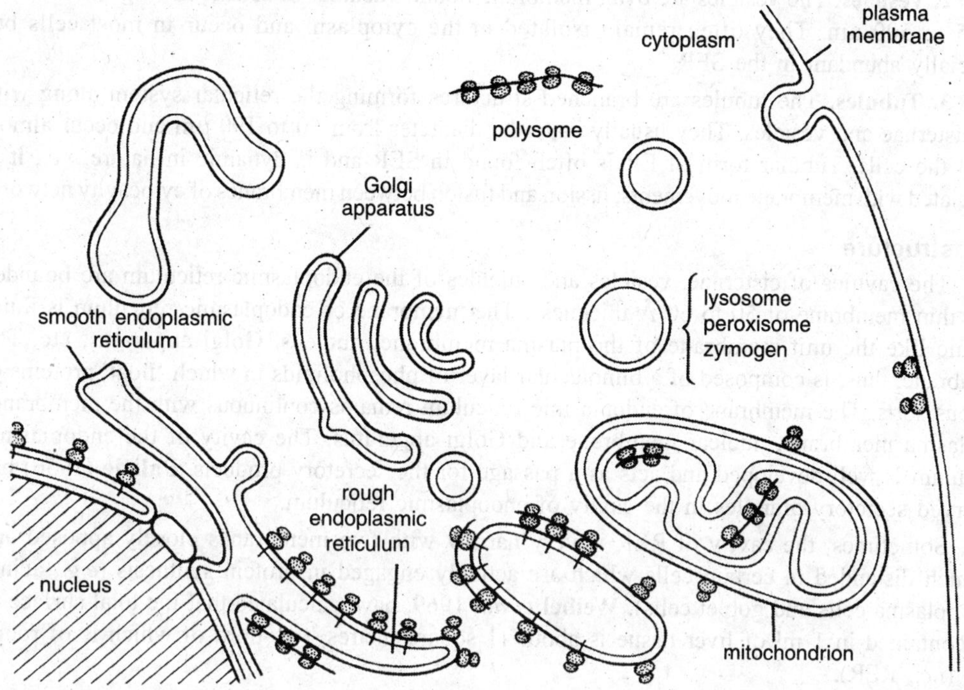

Fig. 8.1. Two faces of membranes of endomembrane system. In each membrane the luminal faces are shown in thick lines, while the cytoplasmic faces are depicted by thin lines. Ribosomes are always located on the cytoplasmic or matrix side.

8.3 MORPHOLOGY (STRUCTURE)

Morphologically, the endoplasmic reticulum may occur in the following three forms: 1. Lamellar form or cisternae (A closed, fluid-filled sac, vesicle or cavity is called **cisternae**); 2. Vesicular form or vesicle and 3. Tubular form or tubules.

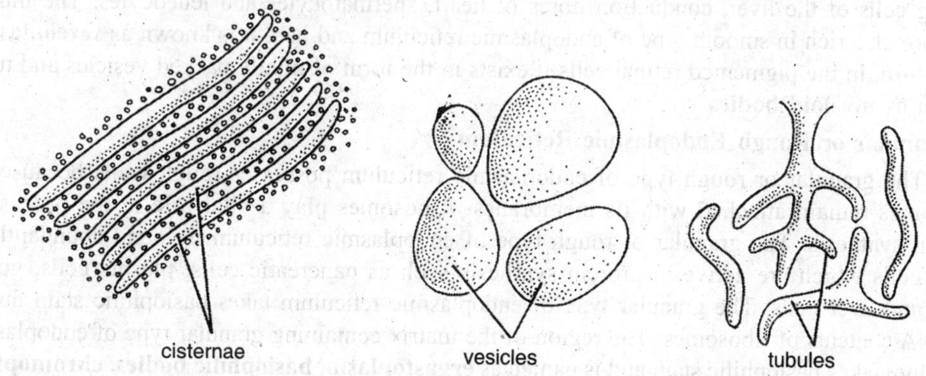

Fig. 8.2. Various components of the endoplasmic reticulum.

1. Cisternae. The cisternae are long, flattened, sac-like, unbranched tubules having the diameter of 40 to 50 μm. They remain arranged parallelly in bundles or stakes. RER usually exists as cisternae which occur in those cells which have synthetic roles as the cells of pancreas, notochord and brain.

2. Vesicles. The vesicles are oval, membrane-bound vacuolar structures having the diameter of 25 to 500 μm. They often remain isolated in the cytoplasm and occur in most cells but especially abundant in the SER.

3. Tubules. The tubules are branched structures forming the reticular system along with the cisternae and vesicles. They usually have the diameter from 50 to 190 μm and occur almost in all the cells. Tubular form of ER is often found in SER and is dynamic in nature, *i.e.*, it is associated with membrane movements, fission and fusion between membranes of cytocavity network.

Ultrastructure

The cavities of cisternae, vesicles and tubules of the endoplasmic reticulum are bounded by a thin membrane of 50 to 60 Å thickness. The membrane of endoplasmic reticulum is fluid-mosaic like the unit membrane of the plasma membrane, nucleus, Golgi apparatus, etc. The membrane, thus, is composed of a bimolecular layer of phospholipids in which 'float' proteins of various sorts. The membrane of endoplasmic reticulum remains continuous with the membranes of plasma membrane, nuclear membrane and Golgi apparatus. The cavity of the endoplasmic reticulum is well developed and acts as a passage for the secretory products. **Palade** (1956) has observed secretory granules in the cavity of endoplasmic reticulum.

Sometimes, the cavity of RER is very narrow with two membranes closely apposed and is much distended in certain cells which are actively engaged in protein synthesis (*e.g.*, acinar cells, plasma cells and goblet cells). **Weibel** *et al.*, 1969, have calculated that the total surface of ER contained in 1 ml of liver tissue is about 11 square metres, two-third of which is of rough type (*i.e.*, RER).

8.4 TYPES OF ENDOPLASMIC RETICULUM

Two types of endoplasmic reticulum have been observed in same or different types of cells which are as follows:

1. Agranular or Smooth Endoplasmic Reticulum

This type of endoplasmic reticulum possesses smooth walls because the ribosomes are not attached with its membranes. The smooth type of endoplasmic reticulum occurs mostly in those cells, which are involved in the metabolism of lipids (including steroids) and glycogen. The smooth endoplasmic reticulum is generally found in adipose cells, interstitial cells, glycogen storing cells of the liver, conduction fibres of heart, spermatocytes and leucocytes. The muscle cells are also rich in smooth type of endoplasmic reticulum and here it is known as **sarcoplasmic reticulum**. In the pigmented retinal cells it exists in the form of tightly packed vesicles and tubes known as **myeloid bodies**.

2. Granular or Rough Endoplasmic Reticulum

The granular or rough type of endoplasmic reticulum possesses rough walls because the ribosomes remain attached with its membranes. Ribosomes play a vital role in the process of protein synthesis. The granular or rough type of endoplasmic reticulum is found abundantly in those cells which are active in protein synthesis such as pancreatic cells, plasma cells, goblet cells, and liver cells. The granular type of endoplasmic reticulum takes basiophilic stain due to its RNA contents of ribosomes. The region of the matrix containing granular type of endoplasmic reticulum takes basiophilic stain and is named as **ergastoplasm, basiophilic bodies, chromophilic substances** or **Nissl bodies** by early cytologists.

In RER, ribosomes are often present as polysomes held together by mRNA and are arranged in typical "rosettes" or spirals. RER contains two transmembrane glycoproteins (called **ribophorins I** and **II** of 65,000 and 64,000 dalton MW, respectively), to which are attached the ribosomes by their 60S subunits.

3. Annulate Lamellae

Usually the endoplasmic reticulum has no pores or annuli in it but in certain cases the pores or annuli have been reported, *e.g.*, ER of invertebrates, oocytes and spermatocytes of the vertebrates. These annuli resemble with the pores or annuli of the nuclear membranes. Like the annuli of nuclear membranes it contains a diaphragm across it and possesses an octagonal symmetry. The annulate lamellae (pores) of the ER arise by the evagination from the nuclear envelope and have their association with the ribosomes.

4. Glycosomes

Although the SER forms a continuous system with RER, it has different morphology. For example, in liver cells it consists of a tubular network that pervades major portion of the cytoplasmic matrix. These fine tubules are present in regions rich in glycogen and can be observed as dense particles, called **glycosomes**, in the matrix. Glycosomes measure 50 to 200 nm in diameter and contain glycogen along with enzymes involved in the synthesis of glycogen. Many glycosomes attached to the membranes of SER have been observed by electron microscopy in the liver and conduction fibre of heart.

8.5 ISOLATION

The membranes of the endoplasmic reticulum can be isolated by subjecting homogenized tissues to differential centrifugation. Electron microscopy of such ER preparations reveals that the membranes disrupt to form closed vesicles (~100 nm diameter) of either a rough or a smooth form. These membranous entities were coined the term "**microsomes**" by **Claude** in 1940, and the relationship between microsomes and the elements of endoplasmic reticulum in the intact cell

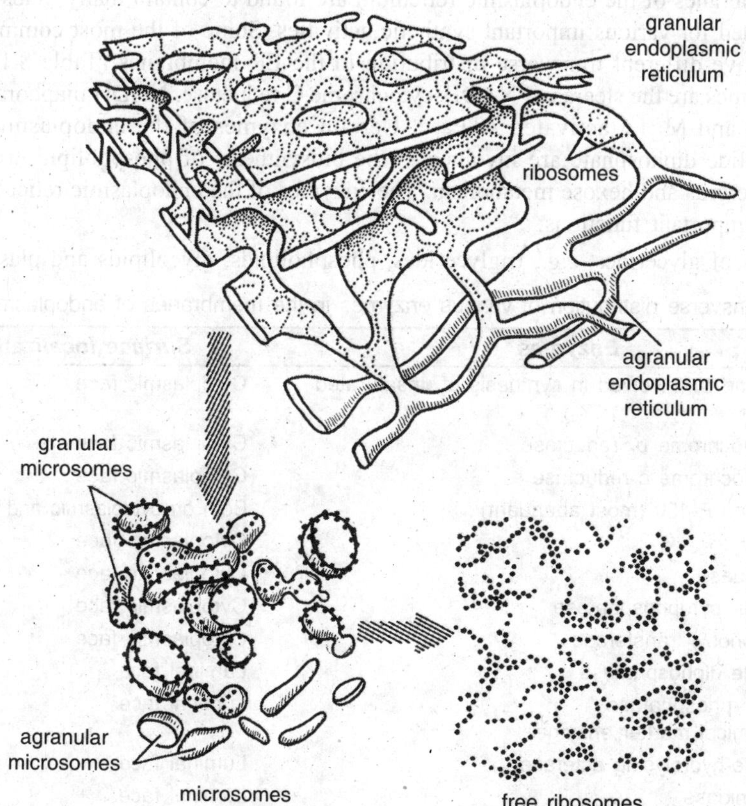

Fig. 8.3. Three-dimensional structure of endoplasmic reticulum showing microsomes and ribosomes.

was established by **Palade** and **Siekevitz** in 1956. Microsomes are not found in intact cells, *i.e.*, they are not natural cell organelles.

Microsomes derived from rough ER are studded with ribosomes and are called **rough** or **granular microsomes**. The ribosomes are always found on the outside surface, the interior being biochemically equivalent to the luminal space of the ER. Homogenate also contains **smooth** or **agranular microsomes** which lack attached ribosomes. They may be derived in part from smooth portion of the ER and in part from fragments of plasma membrane, Golgi apparatus, endosomes and mitochondria. Thus, while rough microsomes can be equated with rough portions of ER, the origin of smooth microsomes cannot be so easily assigned. However, since the hepatocytes of liver contain exceedingly large quantities of smooth ER, therefore, most of the smooth microsomes in liver homogenates are derived from smooth ER.

As rough microsomes can be readily purified in functional form, they are especially useful for studying many biochemical processes carried out by the ER, *e.g.*, protein synthesis, glycosylation and lipid synthesis.

In rat liver, the membranes of microsomes are 60 to 70 per cent protein and 30 to 40 per cent phospholipid by weight.

8.6 CHEMICAL COMPOSITION

ER membranes contain more proteins, both in amount and kind (having about 33 types of polypeptides) than the plasma membrane. They are also richer in phosphotidyl-choline and poorer in sphingomyelin.

The membranes of the endoplasmic reticulum are found to contain many kinds of enzymes which are needed for various important synthetic activities. Some of the most common enzymes are found to have different transverse distribution in the ER membranes (Table 8.1). The most important enzymes are the stearases, NADH-cytochrome C reductase, NADH diaphorase, glucose-6-phosphotase and Mg++ activated ATPase. Certain enzymes of the endoplasmic reticulum such as nucleotide diphosphate are involved in the biosynthesis of phospholipid, ascorbic acid, glucuronide, steroids and hexose metabolism. The enzymes of the endoplasmic reticulum perform the following important functions:

1. Synthesis of glycerides, *e.g.*, triglycerides, phospholipids, glycolipids and plasmalogens.

Table 8.1 Transverse distribution of various enzymes in the membranes of endoplasmic reticulum.

Enzymes	***Surface localization***
1. Cytochrome b_5 (involved in synthesis of unsaturated fatty acids)	Cytoplasmic face
2. NADH-cytochrome b_5 reductase	Cytoplasmic face
3. NADP-cytochrome c reductase	Cytoplasmic face
4. Cytochrome P-450 (most abundant)	Both on cytoplasmic and luminal face
5. ATPase	Cytoplasmic face
6. 5-nucleotidase	Cytoplasmic face
7. Nucleoside pyrophosphatase	Cytoplasmic face
8. GDP-mannosyl transferase	Cytoplasmic face
9. Nucleoside diphosphatase	Luminal face
10. Glucose-6-phosphatase (histochemical marker enzyme)	Luminal face
11. Acetanilide-hydrolysing esterase	Luminal face
12. β-glucuronidase	Luminal face

2. Metabolism of plasmalogens.
3. Synthesis of fatty acids.
4. Biosynthesis of the steroids, *e.g.*, cholesterol biosynthesis, steroid hydrogenation of unsaturated bonds.
5. $NADPH_2$ + O_2—requiring steroid transformations: Aromatization and hydroxylation, side-chain oxidation, deamination, thio-ether oxidations, desulphuration.
6. L-ascorbic acid synthesis.
7. UDP-uronic acid metabolism.
8. UDP-glucose dephosphorylation.
9. Aryl-and steroid sulphatase.

8.7 ORIGIN OF ENDOPLASMIC RETICULUM

The exact process of the origin of endoplasmic reticulum is still unknown. But because membranes of ER resemble with the nuclear membrane and plasma membrane and also at the telophase stage the ER membranes are found to form the nuclear envelope. Therefore, it is normally assumed that the ER has originated by evagination of the nuclear membranes. **Seikevitz** and **Palade** (1960) have reported that the granular type of ER has originated first and later it synthesizes the agranular or smooth type of endoplasmic reticulum. According to **Leskes** (1971) and **Eytan** and **Ohad** (1972), the cell receives ER from its parent cell. Later on, ER grows by expansion.

The synthesis of membranes of ER is found to proceed in the following direction: RER → SER. In fact, membrane biogenesis is a multi-step process involving , first, the synthesis of a basic membrane of lipid and intrinsic proteins and thereafter the addition of other constituents such as enzymes, specific sugars, or lipids. The process by which a membrane is modified chemically and structurally is called **membrane differentiation**. The ER (especially SER) is the organelle containing the main phospholipid synthesizing and translocating enzymes (*i.e.*, there occurs an intense flip-flop of lipid components). The insertion of proteins into ER membranes occurs at the level of RER. Most of these proteins are formed on membrane-bound ribosomes. However, some of these are synthesized by free ribosomes in the cytosol (cytoplasmic matrix) and then are inserted into the membrane. For example, the enzyme **NAD-cytochrome-b_5-reductase** is synthesized in the cytosol (cytoplasmic matrix) and then becomes incorporated in various parts of the endomembrane system (*i.e.*, RER, SER and Golgi apparatus) and in the outer mitochondrial membrane.

8.8 FUNCTIONS OF ENDOPLASMIC RETICULUM

The endoplasmic reticulum acts as secretory, storage, circulatory and nervous system for the cell. It performs following important functions:

A. Common Functions of Granular and Agranular Endoplasmic Reticulum

1. The endoplasmic reticulum provides an ultrastructural skeletal framework to the cell and gives mechanical support to the colloidal cytoplasmic matrix.
2. The exchange of molecules by the process of osmosis, diffusion and active transport occurs through the membranes of endoplasmic reticulum. Like plasma membrane, the ER membrane has permeases and carriers.
3. The endoplasmic membranes contain many enzymes which perform various synthetic and metabolic activities. Further the endoplasmic reticulum provides increased surface for various enzymatic reactions.
4. The endoplasmic reticulum acts as an intracellular circulatory or transporting system. Various secretory products of granular endoplasmic reticulum are transported to various organelles as follows: Granular ER → agranular ER → Golgi membrane → lysosomes, transport vesicles

or secretory granules. Membrane flow may also be an important mechanism for carrying particles, molecules and ions into and out of the cells. Export of RNA and nucleoproteins from nucleus to cytoplasm may also occur by this type of flow.

5. The ER membranes are found to conduct intra-cellular impulses. For example, the sarcoplasmic reticulum transmits impulses from the surface membrane into the deep region of the muscle fibres.
6. The ER membranes form the new nuclear envelope after each nuclear division.
7. The sarcoplasmic reticulum plays a role in releasing calcium when the muscle is stimulated and actively transporting calcium back into the sarcoplasmic reticulum when the stimulation stops and the muscle must be relaxed.

B. Functions of Smooth Endoplasmic Reticulum.

Smooth ER performs the following functions of the cell:

1. Synthesis of lipids. SER performs synthesis of lipids (*e.g.*, phospholipids, cholesterol, etc.) and lipoproteins. Studies with radioactive precursors have indicated that the newly synthesized phospholipids are rapidly transferred to other cellular membranes by the help of specific cytosolic enzymes, called **phospholipid exchange proteins**.

2. Glycogenolysis and blood glucose homeostasis. The process of glycogen synthesis (glycogenesis) occurs in the cytosol (in glycosomes). The enzyme **UDPG-glycogen transferase**, which is directly involved in the synthesis of glycogen by addition of **uridine diphosphate glucose (UDPG)** to primer glycogen is bound to the glycogen particles or glycosomes.

SER is found related to **glycogenolysis** or breakdown of glycogen. An enzyme, called **glucose-6-phosphatase** (a marker enzyme) exists as an integral protein of the membrane of SER (*e.g.*, liver cell). Generally, this enzyme acts as a glucogenic phosphohydrolase that catalyzes the release of free glucose molecule in the lumen of SER from its phosphorylated form in liver (Fig. 8.4). Thus, this process operates to maintain homeostatic levels of glucose in the blood for the maintenance of functions of red blood cells and nerve tissues.

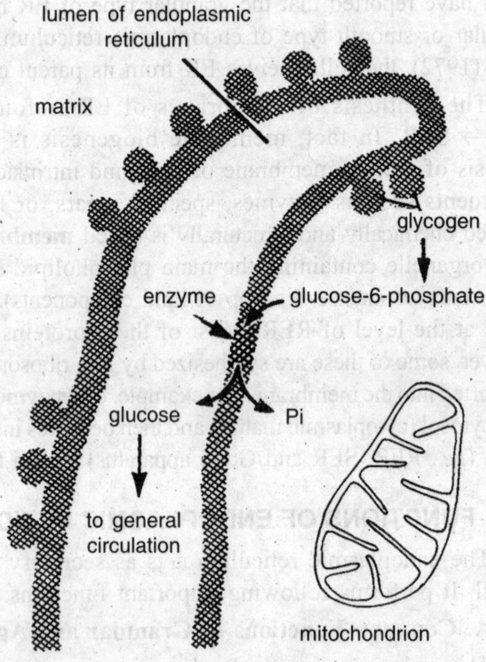

Fig. 8.4. Diagram of the intervention of the smooth endoplasmic reticulum in glycogenolysis with the consequent release of glucose. The enzyme glucose-6-phosphatase, is present in the membrane and has a vectorial deposition by which it receives the glucose-6-phophate from the matrix surface. The product glucose penetrates the lumen of the endoplasmic reticulum.

3. Sterol metabolism. The SER contains several key enzymes that catalyze the synthesis of **cholesterol** which is also a precursor substance for the biosynthesis of two types of compounds—the steroid hormones and bile acids:

(*i*) **Cholesterol biosynthesis.** The cholesterol is synthesized from the acetate and its entire biosynthetic pathway involves about 20 steps, each step catalyzed by an enzyme. Out of these twenty enzymes, eleven enzymes are bounded to SER membranes, rest nine enzymes are the soluble enzymes located in the cytosol and mitochondria. Examples of SER-bound enzyme include **HMG-Co A reductase** and **squalene synthetase**.

(*ii*) **Bile acid synthesis.** The biosynthesis of the bile acids represents a very complex pattern of enzymes and products. Enzymes involved in the biosynthetic pathway of bile acids are hydroxylases, mono-oxygenases, dehydrogenases, isomerases and reductases. For example, by the help of the enzyme **cholesterol 7α-hydroxylase**, the cholesterol is first converted into 7α-hydroxyl cholesterol, which is then converted into bile acids by the help of hydroxylase enzymes. The latter reaction requires NADPH and molecular oxygen and depends on the enzymes of electron transport chains of SER such as **cytochrome P-450** and **NADPH-cytochrome-c-reductase**.

(*iii*) **Steroid hormone biosynthesis.** Steroid hormones are synthesized in the cells of various organs such as the cortex of adrenal gland, the ovaries, the testes and the placenta. For example, cholesterol is the precursor for both types of sex hormones—estrogen and testosterone—made in the reproductive tissues, and the adrenocorticoids (*e.g.*, corticosterone, aldosterone and cortisol) formed in the adrenal glands. Many enzymes (*e.g.*, dehydrogenases, isomerases and hydroxylases) are involved in the biosynthetic pathway of steroid hormones, some of which are located in SER membranes and some occur in the mitochondria. This biosynthetic pathway has the following steps:

Acetate (Cytosol) → Cholesterol (SER) → Pregnenolone (Mitochondria) → Progesterone (SER) → 11-Deoxycortisol (SER) → Aldosterone (Mitochondria)

Progesterone (SER) → 11-Deoxycortisterone (SER) → Cortisol (Mitochondria)

4. Detoxification. Protectively, the ER chemically modifies **xenobiotics** (toxic materials of both endogenous and exogenous origin), making them more hydrophilic, hence, more readily excreted. Among these materials are drugs, aspirin (acetyl-salicylic-acid), insecticides, anaesthetics, petroleum products, pollutants and carcinogens (*i.e.*, inducers of cancer; *e.g.*, **3-4-benzopyrene** and **3-methyl cholanthrene**).

The enzymes involved in the detoxification of aromatic hydrocarbons are **aryl hydroxylases.** It is now known that benzopyrene (found in charcoal-broiled meat) is not carcinogenic, but under the action of aryl hydroxylase enzyme in the liver, it is converted into **5, 6-epoxide**, which is a powerful carcinogen.

A wide variety of drugs (*e.g.*, phenobarbital), when administrated to animals, they bring about the proliferation of the ER membranes (first RER and then SER) and/or enhanced activity of enzymes related to detoxification.

5. Other synthetic functions. SER plays a role in the synthesis of triglycerides in intestinal absorptive cells and of visual pigments from vitamin A by pigmented epithelial cell of retina. In plant cells, SER forms the surface where cellulose cell walls are being formed.

C. Functions of Rough Endoplasmic Reticulum

1. Protein synthesis. The major function of the rough ER is the synthesis of protein. It has long been assumed that proteins destined for secretion (*i.e.*, export) from the cell or proteins to be used in the synthesis of cellular membranes are synthesized on rough ER-bound ribosomes, while cytoplasmic proteins are translated for the most part on free ribosomes. In fact, the array of the rough endoplasmic reticulum provides extensive surface area for the association of metabolically active enzymes, amino acids and ribosomes. There is more efficient functioning of these materials to synthesize proteins when oriented on a membrane surface than when they are simply in solution, mainly because chemical combinations between molecules can be accomplished in specific geometric patterns.

The membrane-bound ribosomes are attached with **specific binding sites** or **receptors** of rough ER membrane by their large 60S subunit, with small or 40S subunit sitting on top like a cap. These receptors are membrane proteins which extend well into and possibly through the lipid bilayer. The receptor proteins with bound ribosomes can float laterally like other membrane proteins and may facilitate formation of the polysome and probably translation which requires that mRNA and ribosome move with respect to each other.

Further, the secretory proteins, instead of passing into the cytoplasm, appear to pass instead into the cisternae of the rough ER and are, thus, protected from protease enzymes of cytoplasm. It is calculated that about 40 amino acid residues long segment at the—COOH end of the nascent protein remains protected inside the tunnel of 'free' or 'bound' ribosomes and rest of the chain, with—NH_2 end, is protected by the lumen of RER. The passage of nascent polypeptide chain into the ER cisterna takes place during translation leaving only a small segment exposed to the cytoplasm at any one time.

How the polypeptide chain gets through the lipid bilayer is not so clear, but it is quite reasonable to propose that the membrane proteins serving as ribosomal receptors also has a very fine channel through its core that opens into the cisterna of the rough ER. The chain may have great flexibility, permitting the amino acids to snake their way single file through the proposed pore. As soon as growing polypeptide chain reaches the cisterna, it folds into its secondary and tertiary structures and thus trapped in the cisterna of the rough ER.

2. Protein glycosylation. The covalent addition of sugars to the secretory proteins (*i.e.*, glycosylation) is one of the major biosynthetic functions of rough ER. Most of the proteins that are isolated in the lumen of RER before being transported to the Golgi apparatus, lysosomes, plasma membrane or extracellular space, become glycoproteins (a notable exception is albumin). In contrast, very few proteins in the cytosol (cytoplasmic matrix) are glycosylated and those that carry them have a different sugar modification.

The process of **protein glycosylation** in RER lumen is one of the most well understood cell biological phenomena. During this process, a single species of **oligosaccharide** (which comprises N-acetyl-glucosamine, mannose and glucose, containing a total of 14 sugar residues) is transferred to proteins in the ER. Because it is always transferred to the NH_2 group on the side chain of an asparagine residue of the protein, this oligosaccharide is said to be **N-linked** or **asparagine-linked** (Fig. 8.5A). The transfer is catalyzed by a membrane-bound enzyme (*i.e.*, **glycosyl transferase**) with its active site exposed on the luminal surface of the ER membrane. The preformed precursor oligosaccharide is transferred *en bloc* to the target asparagine residue in a single enzymatic step almost as soon as that residue emerges in the lumen of ER during protein translocation (Fig. 8.5B). Since most proteins are co-translationally imported into the ER, N-linked oligosaccharides are almost always added during protein synthesis, ensuring maximum access to the target asparagine residues, which are present in the sequences—*Asn-X-Ser* or *Asn-X-Thr* (where *X* is amino acid except proline). *These two sequences, thus, function as signals for N-linked glycosylation.*

The precursor oligosaccharide is held in the ER membrane by a special lipid molecule, **dolicol** (the carrier). The oligosaccharide is linked to the dolicol by a high-energy **pyrophosphate bond** which activates the oligosaccharide for its transfer from the lipid to an asparagine side chain (*i.e.*, it provides activation energy for the glycosylation reaction). The oligosaccharide is built up sugar by sugar on the membrane-bound dolicol (towards the cytosolic side) prior to its transfer to a protein. Sugars are first activated in the cytosol (cytoplasmic matrix) by the formation of **nucleotide-sugar**

intermediates (*e.g.*, UDP-glucose, UDP-N-acetylglucosamine, and GDP-mannose), which then donates their sugar (directly or indirectly) to the lipid in an orderly sequence. At some step of this process, the lipid-linked oligosaccharide is flipped from the cytosolic to the luminal side of the ER membranes. Dolicol is long and very hydrophobic: its 22 five-carbon units can span the thickness of lipid bilayer more than three times, so that the attached oligosaccharide is firmly anchored to the membrane.

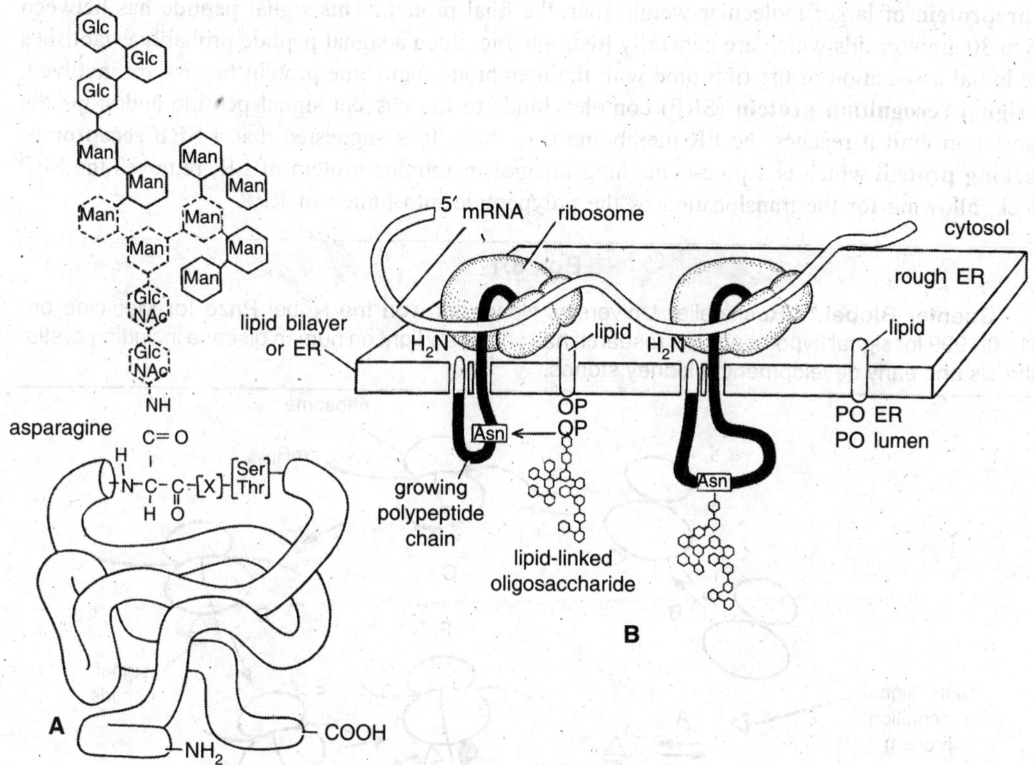

Fig. 8.5. N-linked glycosylation of protein in RER. A—The structure of asparagine-linked oligosaccharide. The sugars shown in shaded-form the 'core-region' of this oligosaccharide. For many glycoproteins, only the core sugars survive the extensive oligosaccharide trimming process in the Golgi apparatus; B—Mode of transfer of the oligosaccharide to the asparagine residues of the nascent protein inside RER lumen.

While still in RER lumen, three glucose residues and one mannose residue are quickly removed from the oligosaccharides of most glycoproteins. Such oligosaccharide "trimming" or "processing" continues in the Golgi apparatus. If a glycoprotein is to contain a terminal glucose, fucose or sialic acid, then those sugars are added in the Golgi apparatus where the appropriate sugar transferase enzymes are localized.

3. Transportation of secretory proteins. From RER secretory proteins pass to Golgi apparatus, secretory vesicles and then to the outside of the cell (exocytosis).

8.9 SIGNAL HYPOTHESIS

The proteins for the secretion, the lysosomes and the membrane formation, are synthesized on the membrane bound ribosomes. The free and bound ribosomes were found to be continuously interchanging and show no differences between them. The **signal hypothesis** was proposed by

Blobel and **Sabatini** (1971; see Box 8.1) to explain how the ribosomes which are meant for the biosynthesis of secretory type proteins get specifically attached to RER membranes. According to this hypothesis, the mRNA is able to recognize free or bound ribosomes. It is postulated that the mRNA for secretory proteins contain a set of **special signal codons** localized after the initial codon AUG. Once the ribosome "recognizes" the signal the ribosome becomes attached to the membrane of ER and the polypeptide penetrates. It is also postulated that at the luminal surface there is a **signal peptidase enzyme** that removes the signal peptide. Thus, the mRNA produces a **preprotein** of larger molecular weight than the final protein. This signal peptide has between 15 to 30 amino acids which are generally hydrophobic. Such a signal peptide probably establishes the initial association of the ribosome with the membrane, but some protein factors are involved. A **signal recognition protein** (SRP) complex binds to the nascent signal peptide and stops the translation until it reaches the ER membrane (Fig. 8.6). It is suggested that a **SRP receptor** or **docking protein** which is a pore-containing integral membrane protein of ER, removes the SRP block, allowing for the translocation of the polypeptide into lumen of RER.

Box 8.1

Guenter Blobel of Rockefeller University, New York won the Nobel Prize for Medicine on 11.10.1999 for signal hypothesis. His research has shed new light on human disease including cystis fibrosis and early development of kidney stones.

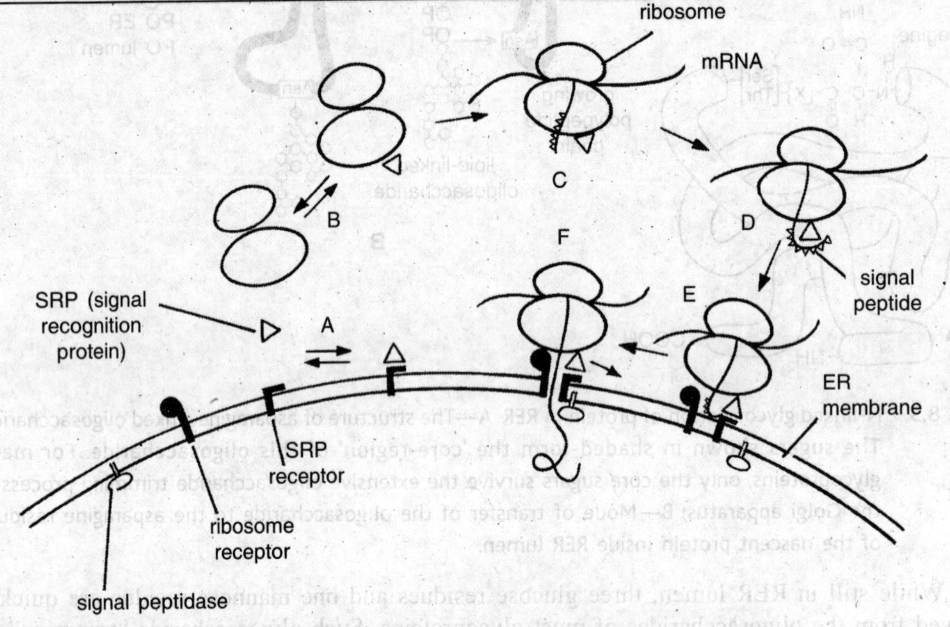

Fig. 8.6. Diagram explaining the signal hypothesis.

8.10 ER EXPORT SIGNALS

Most proteins that enter the transitional ER move through the ER-Golgi intermediate compartment (ERGIC) and on the Golgi. These proteins are marked by sequences that signal either their export from or retention within the ER (Fig. 8.7). Many transmembrane proteins possess di-acidic or di-hydrophobic amino acid sequences in their cytosolic domains that function as ER export signals. Both GPI-anchored proteins and lumenal secretory proteins appear to be recognised and sequestered by these transmembrane receptor proteins (Box 8.2). Very few ER export signals have been detected on lumenal secretory proteins and their recognition may depend on the shape of the correctly folded proteins. It is also possible that there is a default pathway where otherwise unmarked proteins in the ER lumen move to the Golgi and beyond.

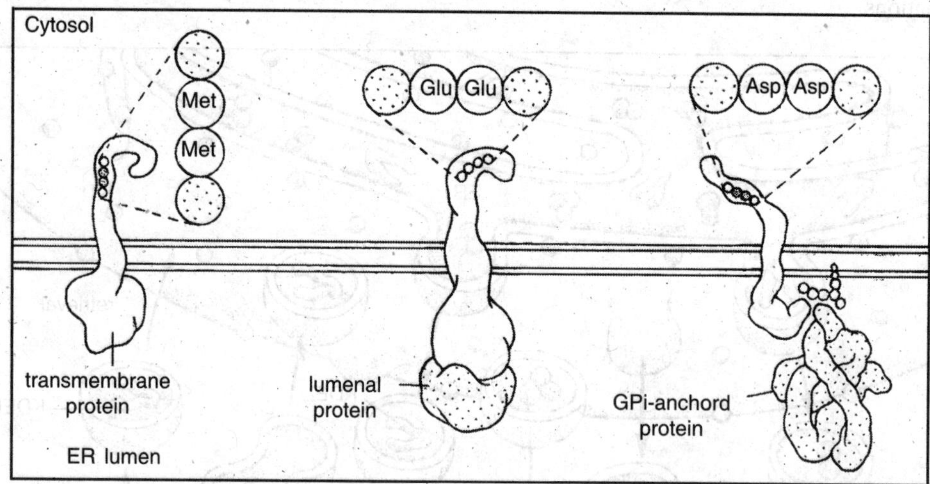

Fig. 8.7. ER export signals. Three types of secretory proteins are recruited into budding vesicles by cytosolic adaptor proteins. Transmembrane proteins are recognised by di-acidic (*e.g.*, Asp-Asp or Glu-Glu) or di-hydrophobic (*e.g.*, Met-Met) signal sequence in their cytosolic segments. Some of these are also receptor for lumenal and GPI-anchored membrane secretory proteins. The transmembrane receptors recognize the GPI-anchors and signal sequences or signal patches characteristic of the folded-lumenal proteins. (Asp=Aspartic acid; Glu=Glutamic acid; Met=Methionine) (after Cooper and Hausman, 2007).

Box 8.2

The GPI-anchored proteins are marked for export. GPI anchor stands for glycosylphosphosphatidylinositol anchor. These are glycolipids containing phosphatidylinositol that anchor proteins to the external face of the plasma membrane (**Cooper** and **Hausman**, 2007).

If proteins that function within the ER (including BiP, signal peptidase, protein disulphide isomerase, and other enzyme discussed earlier) are allowed to proceed along the secretory pathway, they will be lost to the cell. Thus, many such proteins have a targeting sequence Lys-Asp-Glu-Leu (KDEL, in the single letter code) at their carboxy terminus that directs their retrieval back to the ER. If this sequence is detected from a protein that normally function in the ER (*e.g.*, BiP or protein disulphide isomerase), the mutated protein is instead transported to the Golgi and secreted from the cell. Conversely, addition of KDEL sequence to the carboxy terminus of proteins that are normally secreted blocks their secretion. Some ER transmembrane proteins are similarly marked by short C-terminal sequences that contain two lysine residues (KKXX sequences).

Curiously, the KDEL and KKXX signals do not prevent ER proteins from being packaged into vesicles and carried to the Golgi. Instead, these signals cause these ER resident proteins to be selectively retrieved from the ER-Golgi intermediate compartment or the Golgi-apparatus and returned to the ER via a recycling pathway (Fig. 8.8). Proteins bearing the KDEL and KKXX sequences bind to specific recycling receptors in the membranes of these compartments and are then selectively transported back to the ER. The KDEL and KKXX sequences are the best characterised retention/retrieval signals but there may be others. Other proteins are retrieved because they specifically bind to KDEL-bearing proteins such as BiP. Thus continued movement along the secretory pathway or retrieved back from the Golgi to the ER is the first branch point encountered by proteins being sorted to their correct destinations in the secretory pathway. Similar branch points arise at each subsequent stage of transport, such as retention in the Golgi versus export to lysosomes or the plasma membrane. In each case specific localization signals target proteins to their correct intracellular destinations.

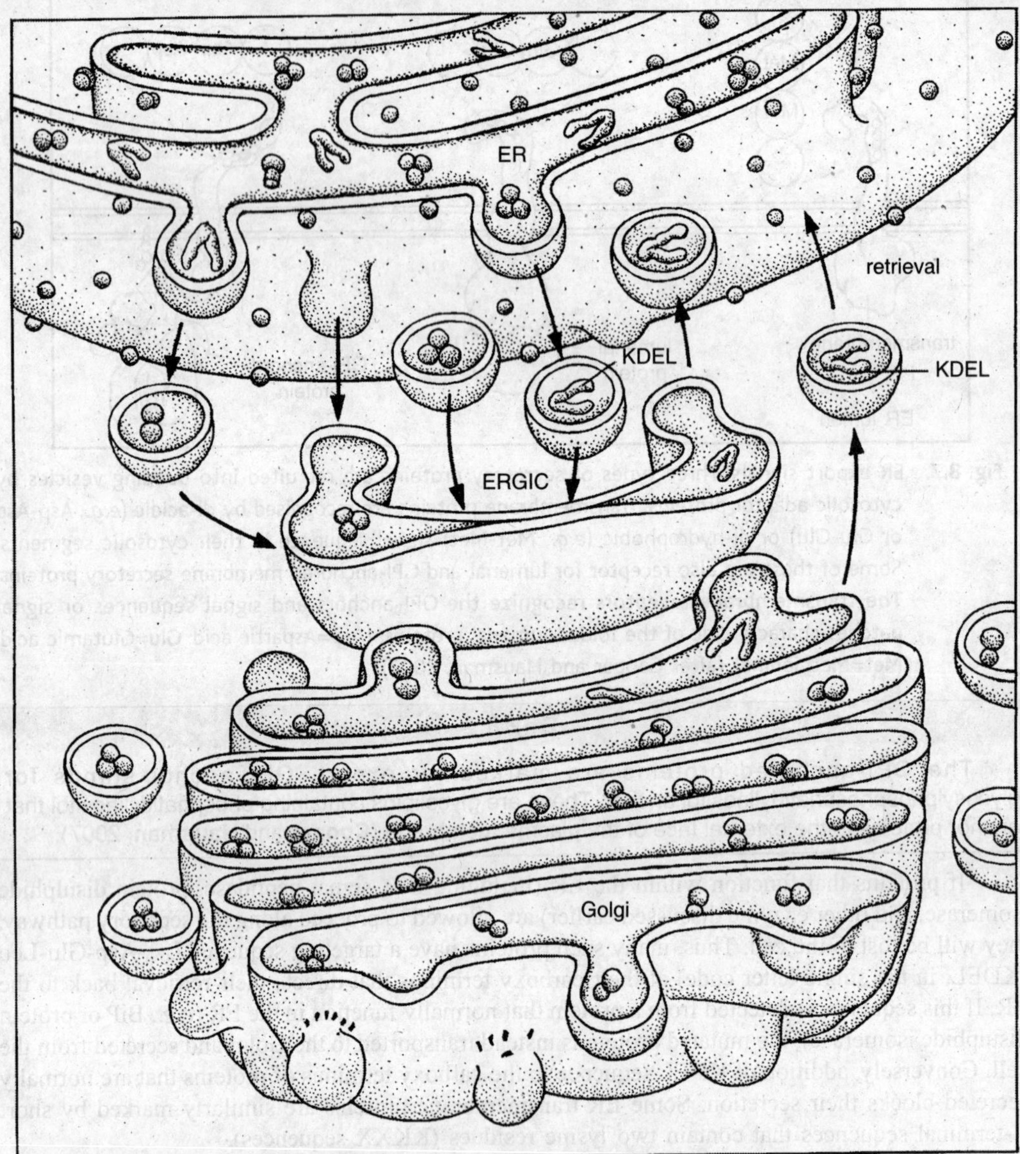

Fig. 8.8. Retrieval of resident ER proteins. Proteins destined to remain in the lumen of the ER are marked by the sequence Lys-Asp-Glu-Leu (KDEL) at their carboxy terminus. These proteins are exported from the ER to the Golgi, but they are recognised by a receptor in the ERGIC (ER Golgi intermediate compartment) or the Golgi apparatus and selectively returned to the ER (after Cooper and Hausman, 2007).

Differences in Endoplasmic Reticulum and Golgi Body/Apparatus

Table 8.2 Differences in endoplasmic reticulum and Golgi apparatus.

Endoplasmic reticulum	*Golgi apparatus*
1. It forms tubular network which traverses endoplasmic region of cytoplasm.	1. It contains stack of flattened tubules and has localized distribution in cytoplasm.
2. It is of two types: (*a*) Smooth ER (*b*) Rough ER	2. It is of one type.
3. It is involved in synthesis and transportation.	3. It is involved in processing and sorting of secretory products which are synthesized by smooth ER and rough ER.
4. Smooth ER is involved in metabolism of glycogen, lipids and steroids.	4. It is involved in carbohydrate synthesis, *e.g.*, cellulose, oligosaccharides.
5. Rough ER is involved in synthesis of protein for membrane, lysosomes and secretory vesicles. Process of glycosylation begins in it.	5. Process of glycosylation and sorting/targetting of proteins is completed in it.
6. It forms new nuclear envelope after each cell division.	6. It forms lysosomes.

QUESTIONS

Long Answer Questions

1. Describe the ultrastructure, types and functions of endoplasmic reticulum. (*H.N.B. Garhwal 2009*)
2. Give the location, structure and function of endoplasmic reticulum.
3. What are the functions of endoplasmic reticulum?
4. What do you mean by membrane system? Describe the same with reference to ER and write about its functions.
5. Describe the endoplasmic reticulum and ribosomes.
6. Give an account of the structure and functions of endoplasmic reticulum. (*Bihar University, 2001*)

Short Answer Questions

1. Give a brief account of microsomes.
2. Mention two important functions of RER.
3. What interrelation exists between SER and RER?
4. A cell has more of SER but just a few RER cisternae. What inference about the cell functions can be drawn?
5. Describe the mechanism of protein glycosylation in the lumen of RER.
6. Differentiate between rough ER and smooth ER.
7. Distinguish between endoplasmic reticulum and Golgi body. (*C.C.S.U. Meerut 2009*)
8. Write short notes on the following:
   (*i*) Endoplasmic reticulum or transporting system
   (*ii*) Functions of endoplasmic reticulum
   (*iii*) Functions of smooth ER.

Very Short Answer Questions

1. Who coined the term endoplasmic reticulum?
2. Which eukaryotic cell is devoid of ER?
3. Name the type of ER associated with glycogen metabolism.
4. What do you understand by sarcoplasmic reticulum?
5. Which cell organelle is responsible for the transport of secretory products to various other cell organelles?

Fill in the Blanks

(*i*) The type of ER is associated with glycogen metabolism is ....

(*ii*) It was possible to observe elements of ER with the help of ....

(*iii*) .... discovered endomembrane system.

(*iv*) .... coined the term endoplasmic reticulum.

(*v*) The eukaryotic cell which is without ER is ....

(*vi*) SER stands for ....

(*vii*) .... membranes develops from ER during cell division.

(*viii*) .... ER is associated with glycoprotein synthesis.

(*ix*) .... ER is developed in cell that secretes mucus.

(*x*) The appearance of ER is dependent on the .... of the cell.

Multiple Choice Questions

1. What part of the cell forms nuclear envelope during telophase?

(*a*) cytoskeleton (*b*) centriole

(*c*) golgi complex (*d*) endoplasmic reticulum

2. Internal membranes or cytomembranes are extensions of cell membrane.

(*a*) the membranes have unit membrane structure

(*b*) all the organelles are bounded by the same membrane

(*c*) have same thickness as that of cell membrane

(*d*) the membrane of ER are connected at certain points with the cell membrane

3. The enzyme glucose-6-phosphatase is present in

(*a*) endoplasmic reticulum (*b*) lysosome

(*c*) mitochondria (*d*) nucleus

4. RER is well developed in cells engaged in synthesis of (*Bihar University, 2004*)

(*a*) nucleotides (*b*) proteins

(*c*) lipids (*d*) secretory products

5. The cells lacking the endoplasmic reticulum are (*Bihar University 2001*)

(*a*) amphibian monocytes

(*b*) mature erythrocytes of mammals

(*c*) mature leucocytes

(*d*) mammalian monocyte

ANSWERS

Very Short Answer Questions

1. Porter.

2. RBC or erythrocyte of mammals.

3. Smooth endoplasmic reticulum.

4. Smooth E.R. of muscle cells is called sarcoplasmic reticulum.

5. Endoplasmic reticulum.

Fill in the Blanks

(*i*) SER (*ii*) Electron microscope

(*iii*) Porter (*iv*) Porter

(*v*) Red blood cells of mammal (*vi*) Smooth endoplasmic reticulum

(*vii*) Nuclear (*viii*) Rough

(*ix*) Rough (*x*) Secretion.

Multiple Choice Questions

1. (*d*) **2.** (*a*) **3.** (*a*) **4.** (*b*) **5.** (*b*).

9

Golgi Apparatus

For the performance of certain important cellular functions such as biosynthesis of polysaccharides, packaging (compartmentalizing) of cellular synthetic products (proteins), production of exocytotic (secretory) vesicles and differentiation of cellular membranes, there occurs a complex organelle called **Golgi complex** or **Golgi apparatus** in the cytoplasm of animal and plant cells. The Golgi apparatus, like the endoplasmic reticulum, is a canalicular system with sacs, but unlike the endoplasmic reticulum it has parallelly arranged, flattened, membrane-bounded vesicles which lack ribosomes and stainable by osmium tetraoxide and silver salts.

9.1 HISTORICAL

An Italian neurologist (*i.e.*, physician) **Camillo Golgi** in 1873 discovered and developed the **silver chromate method** (termed *la reazione nera*) for studying histological details of nerve cells. He, thus, opened a new field of scientific inquiry, called **neuromorphology**. In 1898, Golgi found that Purkinje cells (*i.e.*, nerve cells of cerebral cortex of brain) of barn owl contained an internal reticular network which stains black with the silver stain. He called this structure *apparato reticolare interno* (= internal reticular apparatus).

Due to their presumed high lipid contents, Golgi apparatuses were called **lipochondria** (**Baker**, 1951, 1953). Since originally these were known to be networks, they were also called "**dictyosomes**" (Gr., *dictyes*=net). Currently, the term **Golgi apparatus** is more prevalent one, than many other names such as **Golgi complex, Golgiosome, Golgi bodies, Golgi material, Golgi membrane**, etc. The Golgi apparatus of the cells of plants and lower invertebrates is usually referred to as **Golgi body** or **dictyosome**.

9.2 OCCURRENCE

The Golgi apparatus occurs in all cells except the prokaryotic cells (*viz.*, mycoplasmas, bacteria and blue green algae) and eukaryotic cells of certain fungi, sperm cells of bryophytes and pteridiophytes, cells of mature sieve tubes of plants and mature sperm and red blood cells of animals. Their number per plant cell can vary from several hundred as in tissues of corn root and algal rhizoids (*i.e.*, more than 25,000 in algal rhizoids), to a single organelle in some algae. Certain algal cells such as *Pinularia* and *Microsterias*, contain largest and most complicated Golgi apparatuses. In higher plants, Golgi apparatuses are particularly common in secretory cells and in young rapidly growing cells.

In animal cells, there usually occurs a single Golgi apparatus, however, its number may vary from animal to animal and from cell to cell. Thus, *Paramoeba* species has two Golgi apparatuses and nerve cells, liver cells and chordate oocytes have multiple Golgi apparatuses, there being about 50 of them in the liver cells.

9.3 DISTRIBUTION

In the cells of higher plants, the Golgi bodies or dictyosomes are usually found scattered throughout the cytoplasm and their distribution does not seem to be ordered or localized in any particular manner. However, in animal cells the Golgi apparatus is a localized organelle. For example, in the cells of ectodermal or endodermal origin, the Golgi apparatus remains polar and occurs in between the nucleus and the periphery (*e.g.*, thyroid cells, exocrine pancreatic cells and mucus-producing goblet cells of intestinal epithelium) and in the nerve cells it occupies a circum-nuclear position.

9.4 ULTRASTRUCTURE (ELECTRON MICROSCOPIC STRUCTURE)

The Golgi apparatus is morphologically very similar in both plant and animal cells. However, it is extremely **polymorphic**: in some cell types it appears compact and limited, in others spread out and reticular (net-like). Its shape and form may vary depending on cell type. Typically, however, Golgi apparatus appears as a complex array of interconnecting *tubules*, *vesicles* and *cisternae*. There has been much debate concerning the terminology of the Golgi's parts. The classification given by **D.J. Morre** (1977) is most widely used. In this scheme, the simplest unit of the Golgi apparatus is the **cisterna**. This is a membrane-bound space in which various materials and secretions may accumulate. Numerous cisternae are associated with each other and appear in a stack-like (lamellar) aggregation. A group of these cisternae is called the **dictyosome**, and a group of dictyosomes makes up the cell's Golgi apparatus. All dictyosomes of a cell have a common function.

The detailed structure of three basic components of the Golgi apparatus can be studied as follows (Fig. 9.1 and Fig. 9.2):

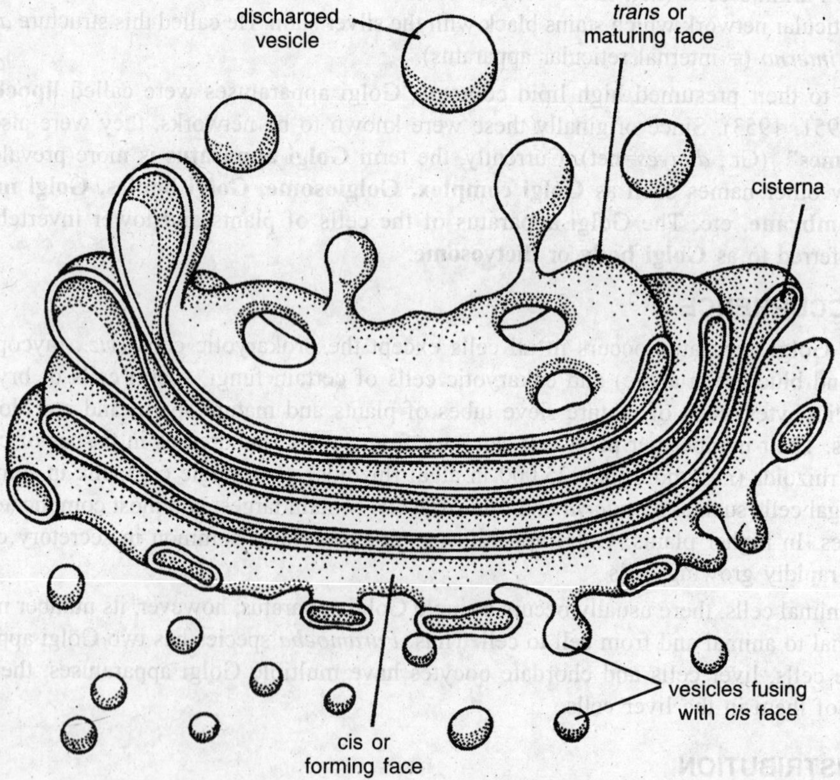

Fig. 9.1. Structure of Golgi apparatus.

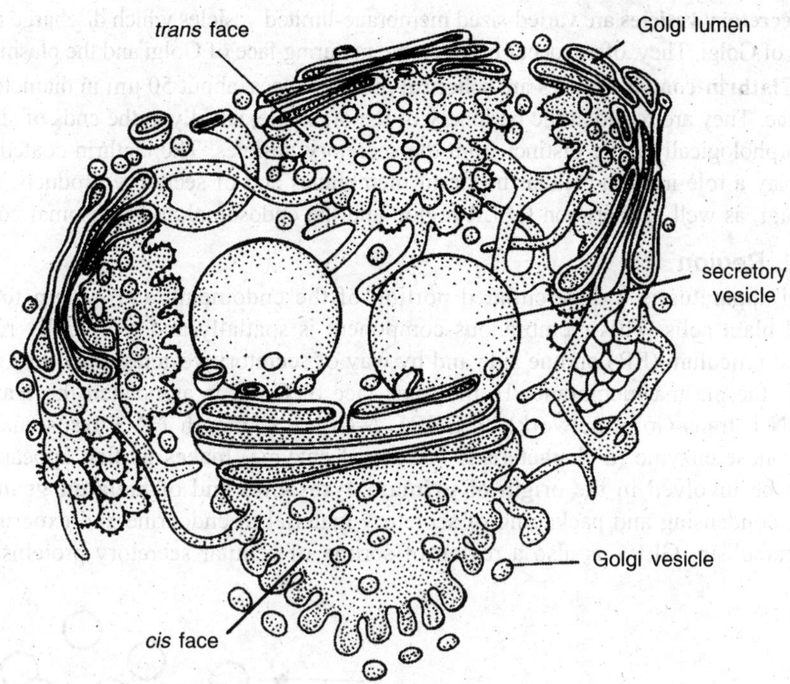

Fig. 9.2. Three-dimensional drawing of a Golgi apparatus; it is drawn from electron micrographs of a secretory animal cell.

1. Flattened Sac or Cisternae

Cisternae (about 1 μm in diameter) are central, flattened, plate-like or saucer-like closed compartments which are held in parallel bundles or stacks one above the other. In each stack, cisternae are separated by a space of 20 to 30 nm which may contain rod-like elements or fibres. Each stack of cisternae forms a dictyosome which may contain 5 to 6 Golgi cisternae in animal cells or 20 or more cisternae in plant cells. Each cisterna is bounded by a smooth unit membrane (7.5 nm thick), having a lumen varying in width from about 500 to 1000 nm.

Polarity. The margins of each cisterna are gently curved so that the entire dictyosome of Golgi apparatus takes on a bow-like appearance. The cisternae at the convex end of the dictyosome comprise **proximal, forming** or **cis-face** and the cisternae at the concave end of the dictyosome comprise the **distal, maturing** or **trans-face**. The forming or cis face of Golgi is located next to either the nucleus or a specialized portion of rough ER that lacks bound ribosomes and is called **"transitional" ER**. Trans face of Golgi is located near the plasma membrane. This polarization is called **cis-trans axis** of the Golgi apparatus.

2. Tubules

A complex array of associated **vesicles** and anastomosing **tubules** (30 to 50 nm diameter) surround the dictyosome and radiate from it. In fact, the peripheral area of dictyosome is fenestrated (lace-like) in structure.

3. Vesicles

The vesicles (60 nm in diameter) are of three types:

***(i)* Transitional vesicles** are small membrane limited vesicles which are thought to form as blebs from the transitional ER to migrate and converge to cis face of Golgi, where they coalasce to form new cisternae.

***(ii)* Secretory vesicles** are varied-sized membrane-limited vesicles which discharge from margins of cisternae of Golgi. They, often, occur between the maturing face of Golgi and the plasma membrane.

***(iii)* Clathrin-coated vesicles** are spherical protuberances, about 50 μm in diameter and with a rough surface. They are found at the periphery of the organelle, usually at the ends of single tubules, and are morphologically quite distinct from the secretory vesicles. The clathrin-coated vesicles are known to play a role in intra-cellular traffic of membranes and of secretory products, *i.e.*, between ER and Golgi, as well as, between GELR region and the endosomal and lysosomal compartments.

The GERL Region

Golgi apparatus is a differentiated portion of the endomembrane system found in both animal and plant cells. This membranous component is spatially and temporally related to the endoplasmic reticulum (ER) on one side and by way of secretory vesicles, may fuse with specific portions of the plasma membrane. To the trans face of Golgi is associated the **trans-reticular Golgi, TGN** (=trans-Golgi-network) or **GERL** (=Golgi + smooth ER + lysosomal), in which acid phosphatase enzyme (a characteristic lysosomal enzyme) makes its first appearance. GERL is found to be involved in the origin of **primary lysosomes** and of **melanin granules**; in the processing, condensing and packaging of secretory material in endocrine and exocrine cells; and in lipid metabolism. GERL is also a region of sorting of cellular secretory proteins.

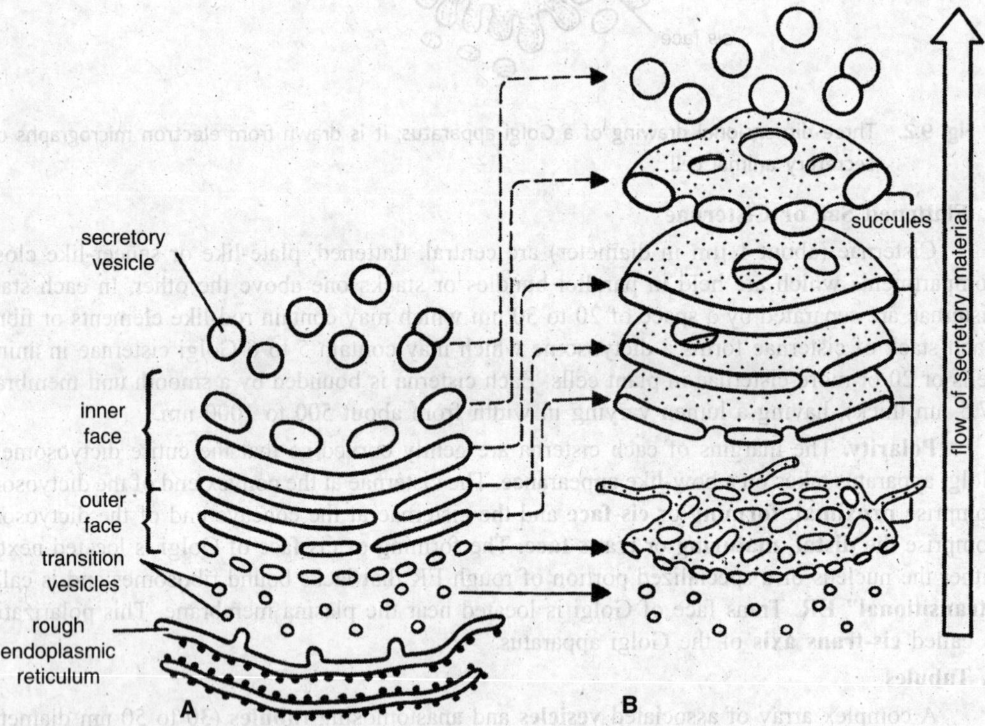

Fig. 9.3. The position and orientation of the Golgi apparatus in the secretory pathway. A—The components as they might be seen by electron microscopy of thin sections; B—A three-dimensional reconstruction of Golgi apparatus.

Zones of Exclusion

A Golgi body or Golgi apparatus is surrounded by a differentiated region of cytoplasm where ribosomes, glycogen, and organelles such as mitochondria and chloroplasts are scarce or absent.

This is called **zone of exclusion** (**Morre** *et al.*, 1971). Endoplasmic reticulum within the zone of exclusion has a smooth surface (lacking ribosomes), and coated vesicles of the Golgi apparatus are restricted to this region. Similar zones of exclusion are associated with microtubules, centrioles, and regions of centriole formation.

9.5 CHEMICAL COMPOSITION

Chemically, Golgi apparatus of rat liver contains about 60 per cent lipid material. The Golgi apparatus of animal cells contains phospholipids in the form of **phosphatidyl choline**, whereas, that of plant cells contains **phosphatidic acid** and **phosphatidyl glycerol**. The Golgi apparatus also contains a variety of enzyme (Table 9.1), some of which have been used as cytochemical markers.

Table 9.1 Some important enzymes of the Golgi apparatus of animal cells.

Enzymes: class and types	*Function*
A. Glycosyl transferases: Glycoprotein biosynthesis	
1. Sialyl transferases	Transfers sialic acid from CMP-sialic acid
2. Galactosyl transferases	Transfer galactose to lipids or proteins
B. Sulpho- and glyco-transferases: Glycolipid biosynthesis	
3. Sulphotransferase	Transfer of sulphate from activated donor
4. Lysolecithin acetyltransferase	Transfer of acyl groups to phospholipid
5. Glycero-phosphate phosphatidyl transferase	Transfer of phosphatidyl group
C. Oxireductases: Oxidation and reduction	
6. NADH-cytochrome c-reductase	Removal or addition of hydrogen
7. NADPH-cytochrome c-reductase	Removal or addition of hydrogen
D. Phosphatases: Hydrolysis of phospholipids	
8. Glucose-6-phosphatase	Removal of phosphate
9. Thiamine pyrophosphatase (Nucleoside diphosphatase)	Hydrolysis of inorganic pyrophosphate
10. ATPase	Removal or addition of phosphate
11. Acid phosphatase	Removal of phosphate
E. Phospholipases: Hydrolysis of lipids	
12. Phospholipase A_1	Removal of non-specific fatty acid chains from phospholipids
13. Phospholipase A_2	Removal of fatty acid chains
F. Kinases: Phosphorylation	
14. Casein phosphokinases	Phosphorylation of casein
G. Mannosidases: Removal of mannose	
15. Mannosidase I and II	Removal of mannose residue from oligosaccharide

Cytochemical Properties of Golgi Apparatus

Different parts of Golgi apparatus have been histochemically identified by specific staining properties.

1. Osmium tetroxide (OsO_4) selectively impregnates the outer face (cis face) of the Golgi apparatus. This stain adheres well to lipids, especially phospholipids and unsaturated fats.

2. Phosphotungstic acid (H_3PO_4. 12 WO_3. 24 H_2O) selectively stain the maturing or trans face of Golgi stack. This stain is an anionic stain having special affinity for polysaccharides and proteins.

3. Glycosyl transferase and **thiamine pyrophosphatase** can be localized cytochemically in the trans cisternae of Golgi apparatus. Transferase enzymes are found to be located in the membranes of Golgi, not in the lumen of cisternae.

4. Acid phosphatase enzyme is cytochemically marked in the GERL region.

9.6 ORIGIN

Origin of Golgi apparatus involves the formation of new cisternae and there is great variation in shape, number and size of cisternae in each stack (dictyosome). The process of formation of new cisternae may be performed by any of the following methods: 1. Individual stacks of cisternae may arise from the pre-existing stacks by division or fragmentation. 2. The alternative method of origin of Golgi is based on *de novo* formation. In fact, various cytological and biochemical evidences have established that the membranes of the Golgi apparatus are originated from the membranes of the smooth ER which in turn have originated from the rough ER. The proximal Golgi saccules are formed by fusion of ER-derived vesicles, while distal saccules "give their all" to vesicle formation and disappear. Thus, Golgi saccules are constantly and rapidly renewed.

9.7 FUNCTIONS

Golgi vesicles are often, referred to as the **"traffic police"** of the cell (**Darnell** *et al.*, 1986). They play a key role in **sorting** many of cell's proteins and membrane constituents, and in **directing** them to their proper destinations. To perform this function, the Golgi vesicles contain different sets of enzymes in different types of vesicles—**cis, middle** and **trans cisternae**—that react with and modify secretory proteins passing through the Golgi lumen or membrane proteins and glycoproteins that are transiently in the Golgi membranes as they are *en route* to their final destinations (Fig. 9.4). For example, a Golgi enzyme may add a "signal" or "tag" such as a carbohydrate or phosphate residues to certain proteins to direct them to their proper sites in the cell. Or, a proteolytic Golgi enzyme may cut a secretory or membrane protein into two or more specific segments (*e.g.*, molecular processing involved in the formation of pancreatic hormone insulin: preproinsulin → proinsulin → insulin).

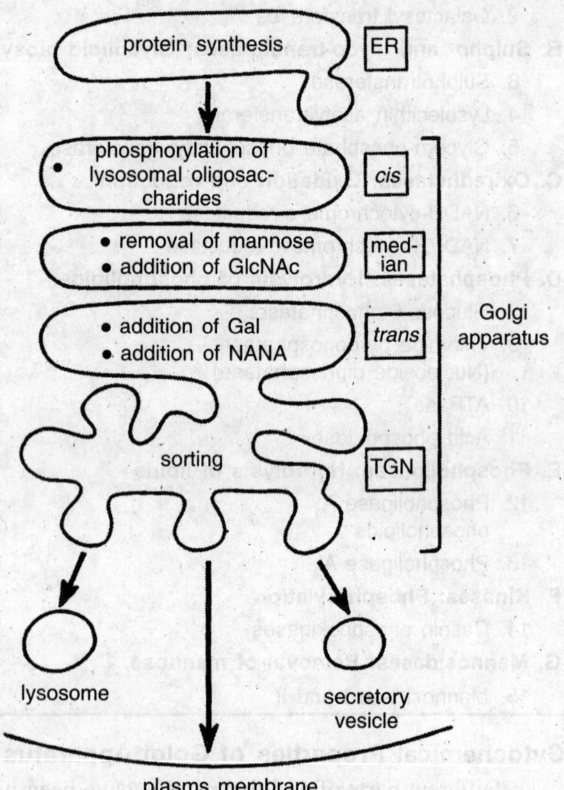

Fig. 9.4. The compartmentalization of the Golgi apparatus. GlcNAc = N-actyl-glucosamine galactose; NANA = N-acetyl neuraminic acid (sialic acid).

Recently, in the function of Golgi apparatus, subcompartmentalization with a division of labour has been proposed between the *cis* region (in which proteins of RER are sorted and some of them are returned back possibly by coated vesicles), and the *trans* region in which the most refined proteins are further separated for their delivery to the various cell compartments (*e.g.*, plasma membrane, secretory granules and lysosomes) (See Box 9.1).

Box 9.1

Following two contrasting models have been forwarded regarding the movement of materials through the Golgi complex.

1. Maturation model. Until the mid-1980s, it was generally accepted that Golgi cisternae were transient structures; it was supposed that they formed at the *cis* face of the stack by fusion of vesicles from the ER and that each cisterna physically moved from the *cis* to *trans* end of the stack, changing in composition as it progressed. This is known as the **maturation model,** since according to it, each cisterna "matures" into a trans cisterna.

2. Rothman and Orci model. According to **J.E. Rothman** and **L. Orci** (1990, 92), the cisternae of a Golgi stack remain in place as stable entities held together by a protein scaffold. In this model, materials moves through the Golgi stack, from the *cis* face to the TGN, by vesicles that bud from one membrane compartment and fuse with a neighbouring compartment farther along the stack.

The Golgi apparatus is primarily a **processing plant** of the cell. It is a centre of *reception, finishing, packaging*, and *dispatch* for a variety of materials in animal and plant cells. Thus, Golgi apparatus performs the following functions in the eukaryotic cells.

1. Glycosylation. Golgi apparatus facilitates linking of carbohydrates and proteins in the formation of **glycoproteins** or **proteoglycans**. Such addition of carbohydrates to proteins is called **glycosylation**. The process of glycosylation is started in the endoplasmic reticulum. Mainly one species of oligosaccharides is added to proteins in the ER. These proteins are transferred to Golgi by coated vesicles. In Golgi, the oligosaccharide chains of proteins are modified by removal (*e.g.*, mannoses) and addition of selected sugars in order to produce the final product. For this purpose the Golgi apparatus contains a variety of **glycosyl transferase enzymes** and sugar nucleotides. More glucose and other sugars such as fucose, galactose and sialic acid are added at the Golgi. Thus, glycosylation process begins in the ER and is completed in the Golgi apparatus.

2. Polysaccharide synthesis. Golgi apparatus is involved in synthesis of complex polysaccharides (**Becker** *et al.*, 2007).

3. Sulphation. A number of secretory products also contain sulphates. The sulphation process requires the activity of **sulphotransferases** enzymes which are present in Golgi apparatus.

4. Concentration and storage of secretory products. The proteins, glycoproteins and lipoproteins are concentrated inside the **secretion vesicles** or in **condensing vacuoles** which bud off from the Golgi cisternae. Once concentrated, the products may be stored in the cytoplasm (inside the vesicle) or may be secreted to the outside of the cell through exocytosis.

5. Secretion. Golgi is mainly associated with the secretory activity of the cell. Various hormones, proenzymes or zymogens, skeletal components, mucus, etc., are secreted through Golgi apparatus. All these products are synthesized in different parts of the cell. Through endoplasmic reticulum these products reach Golgi apparatus where these are packed and concentrated for final secretion. Golgi apparatus of different organs produce following secretion (Table 9.2).

Table 9.2 Summary of secretions which are produced in Golgi apparatus.

Cell type	*Golgi function*
1. Exocrine cells of pancreas	Secretion of zymogens (proenzymes of digestive enzymes such as proteases, lipases, carbohydrases and nucleases)
2. Parotid (salivary) gland cells	Secretion of mucus and zymogen
3. Goblet cells of intestinal epithelium	Secretion of mucus and zymogen
4. Goblet cells of respiratory epithelium	Secretion of mucus (*i.e.*, mucin, a glycoprotein)
5. Paneth cells of intestine (*i.e.*, of crypts of Lieberkuhn)	Secretion of digestive enzymes (*e.g.*, chitinase)

Cell type	Golgi function
6. Brunner's gland cells of intestine	Secretion of enzymes, mucus and hormones
7. Follicle cells of thyroid	Prethyroglobulin
8. Plasma cells of blood	Immunoglobulins
9. Hepatic cells of liver	Transformation and secretion of lipids
10. Cells of alveolar epithelium of mammary gland	Secretion of milk proteins (casein)
11. Nervous tissue (myelocytes, sympathetic ganglia, Schwann cells)	Sulphation reaction
12. Endothelial cells of blood vessels	Sulphation reaction
13. Cornea of birds	Secretion of tropocollagen and collagen
14. Most plant cells	Secretion of pectin and cellulose
15. Melanocytes	Secretion of melanin
16. Primary oocyte	Secretion of yolk and vitelline membrane
17. Root cap cells	Mucus

Types of Secretion

Molecular transport and diffusion across membranes account for the transfer of some materials from cytoplasm to extra cellular space. Cellular secretion may be of the following three types: holocrine, apocrine and merocrine.

(*i*) Holocrine secretion. In this kind of cellular secretion, the entire cell fills with secretory product and then is broken up and sacrificed. An example is the sebaceous gland, which secretes a fatty discharge onto adjacent hair and skin. It is this gland that, when overstimulated, gets plugged to form blackheads and ultimately the pimples of adolescent acne.

(*ii*) Apocrine secretion. In this kind of secretion, the secretion vesicle buds through the plasma membrane carrying with it a portion of that membrane and hence a thin layer of cytoplasm. The secretion vesicle is thus covered by two layers of membrane, its own plus the plasma membrane. Milk fat droplets (MFD) of the cells of the lactating mammary gland are released as apocrine secretion (Fig. 9.5).

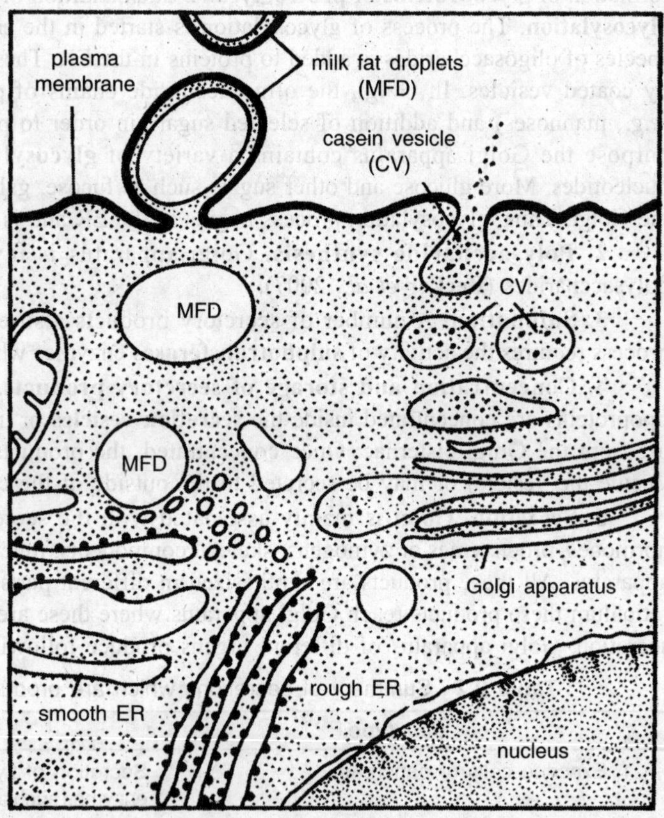

Fig. 9.5. Secretion in lactating mammary gland. Milk fat droplets are released by the apocrine mechanism, whereas casein vesicles normally release their protein by merocrine secretion. The former removes plasma membrane, the latter adds to it.

(*iii*) **Merocrine or eccrine secretion.** It is a most common type of cellular secretion. This process is also called **exocytosis** because it is the function opposite of endocytosis. Thus, it involves the fusion of secretory vesicles with the plasma membrane and their opening to the exterior. It is the usual mode of secretion of most glands and of most other cells, both animal and plant. It is, for example, the mode of secretion of the **true sweat glands**, which are distributed over most of the human skin. In cells of lactating mammary gland, casein vesicles (CV) release their protein by merocrine secretion (Fig. 9.5).

Pancreas which includes exocrine as well as endocrine cells has provided a best example of merocrine secretion.

6. Recycling of plasma membrane components. There occurs a continuous exchange of membrane between plasma membrane and Golgi apparatus due to exocytosis and endocytosis. The membrane material is reprocessed in the Golgi for reuse.

7. Intracellular sorting of proteins. Proteins are exported from the Golgi apparatus to different intracellular and extracellular destinations. The accurate sorting of proteins for selective export is one of the major functions of Golgi. This sorting is done by **coated vesicles** which surround the cisternae. These contain specific receptors which recognize specific markers on the proteins to be exported.

8. Formation of cell plate and cell wall in plants. In plants, Golgi apparatus is mainly involved in the secretion of materials of primary and secondary cell walls (*e.g.*, formation and export of glycoproteins, lipids, pectins and monomers for hemicellulose, cellulose, lignin, etc.). During cytokinesis of mitosis or meiosis, the vesicles originating from the periphery of Golgi apparatus, coalesce in the phragmoplast area to form a semi-solid layer, called **cell plate**. The unit membrane of Golgi vesicles fuses during cell plate formation and becomes part of plasma membrane of daughter cells.

9. Development of acrosome in sperm. Acrosome is a membrane-bounded cap-like structure at the forwarding end of a spermatozoon (in animals). It contains proteolytic enzymes. The acrosome is formed from vesicles derived from the Golgi apparatus during sperm maturation.

10. Formation of primary lysosomes. Hydrolytic lysosomal enzymes produced in rough ER are sorted in TGN and targeted for lysosomes by virtue of phosphorylated mannose residues within their oligosaccharide chains.

11. Lipid transport. When digested lipids are absorbed as fatty acids and glycerol in the small intestine. They are resynthesized to lipids in smooth ER, coated with protein and then transported through the Golgi apparatus to the plasma membrane where they leave the cell (exocytosis), mainly to enter lymph system.

QUESTIONS

Long Answer Questions

1. Give an account of the ultrastructure, chemical composition and functions of Golgi body, Golgi complex or Golgi apparatus. (*Bihar 2004*)
2. Describe ultrastructure of Golgi complex and discuss its role in cell secretion.
3. State the physical and chemical structure and functions of Golgi apparatus.
4. Give an account of discovery, ultrastructure and functions of Golgi complex. (*Bihar 2002*)

Short Answer Questions

1. Describe main functions or role of Golgi complex.
2. What is Golgi? Mention its functions.
3. Write short note on Golgi complex.
4. Write short notes on the following

(*i*) Golgi apparatus (*ii*) Functions of Golgi body

(*iii*) Acrosome formation (*iv*) Zones of exclusion

Very Short Answer Questions

1. Who discovered the Golgi apparatus/Golgi complex?
2. Which face of Golgi cisternae lie towards the plasma membrane?
3. From which face the proteins enter in the Golgi apparatus?
4. Which cell organelle is responsible for glycosylation and sulphation processes?
5. Which cell organelle is responsible for concentration and storage of secretory products?
6. Which part of mammalian sperm is formed by Golgi apparatus?
7. In which organelle synthesis of polysaccharides takes place? *(H.N.B. Garhwal 2009)*

Yes or No Questions

(*i*) Golgi was first described by Bateson. (*ii*) Acrosome is formed by Golgi body.

(*iii*) Golgi apparatus was first described by Morgan.

Fill in the Blanks

1. Golgi apparatus is stainable with .... and ....
2. .... face of Golgi apparatus is located near the plasma membrane.
3. At .... face of Golgi apparatus, ... vesicles coalase to form new cisternae.
4. Acid phosphatase enzyme is a .... marker and it makes its first appearance in .... region.
5. GERL stands for ...., smooth .... and lysosomal.
6. Golgi apparatus is the center of reception, finishing, ... and .... of cellular secretions.
7. Sebaceous glands which are involved in adolscent acne have ....
8. In cells of lactating mammary gland, casein vesicles (CV) release their protein by .... secretion.
9. In cells, the process of glycosylation of proteins is started in .... and finished in ....

Multiple Choice Questions

1. Golgi apparatus is concerned with *(Bihar 2004)*

   (*a*) excretion (*b*) secretion (*c*) ATP synthesis (*d*) RNA synthesis

2. Golgi apparatus is absent in *(H.N.B. Garhwal 2008)*

   (*a*) yeast (*b*) liver cell (*c*) higher plant (*d*) blue green algae

3. Golgi apparatuses present in invertebrates and plant cells are called as

   (*a*) polysomes (*b*) golgisomes (*c*) dictyosomes (*d*) cisternae

ANSWERS

Very Short Answer Questions

1. Camillo Golgi. 2. Trans face. 3. Cis face.
4. Golgi apparatus. 5. Golgi apparatus. 6. Acrosome.
7. Golgi apparatus.

Yes or No Questions

1. No 2. Yes 3. No

Fill in the Blanks

1. Osmium tetroxide, silver salts. 2. Trans.
3. Cis, transitional. 4. Cytochemical, GERL.
5. Golgi, ER. 6. Packaging, dispatch.
7. Holocrine secretion. 8. Merocrine.
9. Endoplasmic reticulum, Golgi apparatus.

Multiple Choice Questions

1. (*b*) 2. (*d*) 3. (*c*).

10

Lysosomes

The lysosomes (*Gr., lyso*=digestive + *soma*=body) are tiny membrane-bound vesicles involved in intracellular digestion. They contain a variety of hydrolytic enzymes that remain active under acidic conditions. The lysosomal lumen is maintained at an acidic pH (around 5) by an ATP-driven proton pump in the membrane. Thus, these remarkable organelles are primarily meant for the digestion of a variety of biological materials and secondarily cause aging and death of animal cells and also a variety of human diseases such as cancer, gout, Pompe's disease, silicosis and I-cell disease.

Box 10.1

Lysosomes were observed and named by **Christian de Duve** in 1955 as cytoplasmic rounded dense bodies. They were termed **"suicide bags"** by **de Duve** in 1959.

Lysosomes were investigated according to following two schools: 1. **C. de Duve** and his co-workers (1963, 1964, 1974) worked in Belgium and their approach was biochemical one. 2. **Alex Novikoff** and his research group (1962, 1964) worked in United States and their approach was morphological and cytochemical. For the discovery of lysosomes and a brilliant series of experiments on them, **de Duve** shared the 1974 Nobel Prize for physiology with **Palade** and **Claude**, both of them were pioneer cell biologists.

10.1 OCCURRENCE

The lysosomes occur in most animal and few plant cells. They are absent in bacteria and mature mammalian erythrocytes. Few lysosomes occur in muscle cells or in acinar cells of the pancreas. Leucocytes, especially granulocytes are a particularly rich source of lysosomes. Their lysosomes are so large-sized that they can be observed under the light microscope. Lysosomes are also numerous in epithelial cells of absorptive, secretory and excretory organs (*e.g.*, intestine, liver, kidney, etc.). They occur in abundance in the epithelial cells of lungs and uterus. Lastly, phagocytic cells and cells of reticuloendothelial system (*e.g.*, bone marrow, spleen and liver) are also rich in lysosomes.

10.2 STRUCTURE

The lysosomes are round vacuolar structures which remain filled with dense material and are bounded by single unit membrane. Their shape and density vary greatly. Lysosomes are 0.2 to 0.5 µm in size. Since, size and shape of lysosomes vary from cell to cell and time to time (*i.e.*, they are polymorphic), their identification becomes difficult. However, on the basis of the following three criteria, a cellular entity can be identified as a lysosome: 1. It should be bound by a limiting membrane; 2. It should contain two or more acid

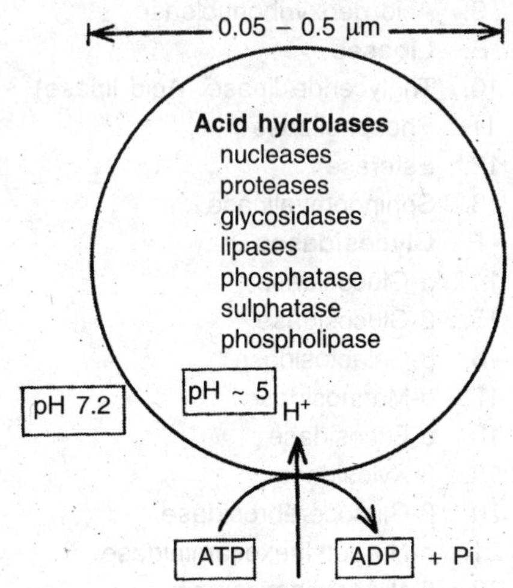

Fig. 10.1. A lysosome showing its various characteristics.

hydrolases; and 3. It should demonstrate the property of enzyme latency when treated in a way that adversely affects organelle's membrane structure.

10.3 CHEMICAL COMPOSITION

The **acid phosphatase** is the principal enzyme which is used as a marker for the lysosomes by the use of **Gomori' staining technique (Gomori,**1952). Specific stains are also used for other lysosomal enzymes such as **B-glucuronidase, aryl sulphatatase, N-acetyl-B-glucosaminidase** and **5-bromo-4-chloroindolacetate esterase**.

Lysosomal Enzymes

According to a recent estimate, a lysosome may contain up to 50 types of hydrolytic enzymes. They include **proteases** (*e.g.*, cathepsin for protein digestion), **nucleases, glycosidases** (for digestion of polysaccharides and glycosides), **lipases, phospholipases, phosphatases** and **sulphatases** (Table 10.1). All lysosomal enzymes are acid hydrolases, optimally active at the pH5 maintained

Table 10.1 Some lysosomal enzymes and their substrates.

	Enzyme	*Substrate*
A.	**Phosphatases**	
1.	Acid phosphatase	Phosphate monoesters
2.	Acid pyrophosphatase	ATP, FAD
3.	Acid phosphodiesterase	Oligonucleotides, phosphodiesters
B.	**Sulphatase**	
4.	Arylsulphatase	Aryl sulphates
C.	**Proteases and peptidases**	
5.	Cathepsin A, B, C, D, and E	Various proteins and peptides
6.	Collagenase	Collagen
7.	Peptidases	Peptides
D.	**Nucleases**	
8.	Acid ribonuclease	RNA
9.	Acid deoxyribonuclease	DNA
E.	**Lipases**	
10.	Triglyceride lipase (Acid lipase)	Triacyglycerols
11.	Phospholipase	Lecithin, phosphatidyl ethanolamine
12.	Esterase	Fatty acid esters
13.	Sphingomyelinase	Sphingomyelin
F.	**Glycosidases**	
14.	α-Glucosidase	Glycogen
15.	β-Glucosidase	Glycoproteins (β-Glucosides)
16.	β-Galactosidase	β-Galactosides
17.	α-Mannosidase	Mannosyl oligosaccharides
18.	α-Fucosidase	Glycoprotein
19.	β-Xylosidase	Glycoprotein
20.	β-Glucocerebrosidase	Glycolipids
21.	α-N-Acetylhexosaminidase	Heparin
22.	β-N-Acetylhexosaminidase	Glycoproteins, glycolipids
23.	Sialidase	Sialic acid derivatives
24.	Lysozyme	Mucopolysaccharides, bacterial cell wall
25.	Hyaluronidase	Hyaluronic acid, chondroitin sulphates
26.	β-Glucuronidase	Polysaccharides, mucopolysaccharides

within lysosomes. The membrane of the lysosome normally keeps the enzymes latent and out of the cytoplasmic matrix or cytosol (whose pH is about ~ 7.2), but *the acid dependency of lysosomal enzymes protects the contents of the cytosol (cytoplasmic matrix) against any damage even if leakage of lysosomal enzymes should occur.*

The so-called **latency** of the lysosomal enzymes is due to the presence of the membrane which is resistant to the enzymes that it encloses. Most probably this is due to the fact that most lysosomal hydrolases are membrane-bound, which may prevent the active centres of enzymes to gain access to susceptible groups in the membrane.

Lysosomal Membrane

The lysosomal membrane is slightly thicker than that of mitochondria. It contains substantial amounts of carbohydrate material, particularly **sialic acid**. In fact, lysosomal membranes contain two groups of acidic highly glycosylated integral proteins, called **lgp-A** and **lgp-B**, whose function may be to protect the membrane from attack of its enclosed enzymes.

The entire process of digestion is carried out within the lysosome. Most lysosomal enzymes act in an acid medium. Acidification of lysosomal contents depends on an ATP-dependent proton pump which is present in the membrane of the lysosome and accumulates H^+ inside the organelle. Lysosomal membrane also contains transport proteins that allow the final products of digestion of macromolecules to escape so that they can be either excreted or reutilized by the cell.

10.4 KIND OF LYSOSOMES (POLYMORPHISM IN LYSOSOMES)

Lysosomes are extremely dynamic organelles, exhibiting polymorphism in their morphology. Following four types of lysosomes have been recognized in different types of cells or at different times in the same cell. Of these, only the first is the **primary lysosome**, the other three have been grouped together as **secondary lysosomes**.

1. Primary Lysosomes

These are also called **storage granules, protolysosomes** or **virgin lysosomes**. Primary lysosomes are newly formed organelles bounded by a single membrane and typically having a diameter of 100 nm. They contain the degradative enzymes which have not participated in any digestive process. Each primary lysosome contains one type of enzyme or another and it is only in the secondary lysosome that the full complement of acid hydrolases is present.

2. Heterophagosomes

They are also called **heterophagic vacuoles**, **heterolysosomes** or **phagolysosomes**. Heterophagosomes are formed by the fusion of primary lysosomes with cytoplasmic vacuoles containing **extracellular substances** brought into the cell by any of a variety of endocytic processes (*e.g.*, pinocytosis, phagocytosis or receptor-mediated endocytosis). The digestion of engulfed substances takes place by the enzymatic activities of the hydrolytic enzymes of the secondary lysosomes. The digested material has low molecular weight and readily passes through the membrane of the lysosomes to become the part of the matrix (Fig. 10.2).

3. Autophagosomes

They are also called **autophagic vacuole**, **cytolysosomes** or **autolysosomes**. Primary lysosomes are able to digest **intracellular structures** including mitochondria, ribosomes, peroxisomes and glycogen granules. Such autodigestion (called **autophagy**) of cellular organelles is a normal event during cell growth and repair and is especially prevalent in differentiating and dedifferentiating tissues (*e.g.*, cells undergoing programmed death during metamorphosis or regeneration) and tissue under stress. Autophagy takes several forms. In some cases the lysosome appears to flow around the cell structure and fuse, enclosing it in a double membrane sac, the lysosomal enzymes being initially confined between the membranes. The inner membrane then breaks down and the enzymes are able to penetrate to the enclosed organelle. In other cases, the organelle to be digested is first encased by smooth ER, forming a vesicle that fuses with a primary lysosome (Fig. 10.4). Lysosomes also regularly engulf bits of cytosol (cytoplasmic matrix) which is degraded by a process, called **microautophagy**.

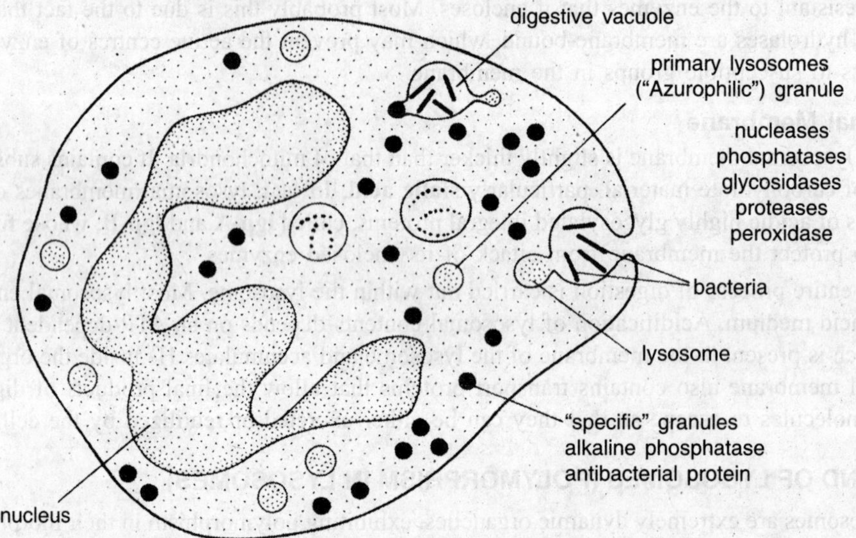

Fig. 10.2. Diagram of a white blood cell (neutrophil) ingesting bacteria. Two types of granules fuse with the phagocytotic (digestive) vacuoles and contribute digestive enzymes and other components.

As digestion proceeds, it becomes increasingly difficult to identify the nature of the original secondary lysosome (*i.e.*, heterophagosome or autophagosome) and the more general term **digestive vacuole** is used to describe the organelle at this stage.

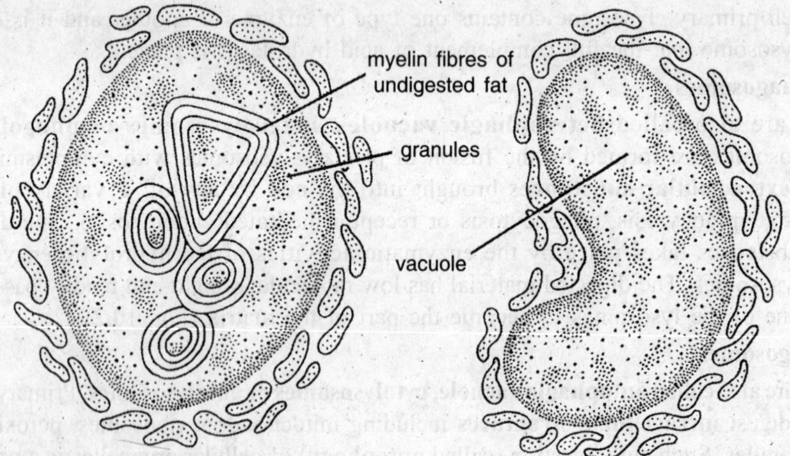

Fig. 10.3. Lysosomes of the kidney cells of rat, showing the presence of residues.

4. Residual Bodies

They are also called **telolysosomes** or **dense bodies**. Residual bodies are formed if the digestion inside the food vacuole is incomplete. Incomplete digestion may be due to absence of some lysosomal enzymes. The undigested food is present in the digestive vacuole as the residues and may take the form of whorls of membranes, grains, amorphous masses, ferritin-like or myelin fibres (Fig. 10.3).

Residual bodies are large, irregular in shape and are usually quite electron-dense. In some cells, such as *Amoeba* and other potozoa, these residual bodies are eliminated by **defecation**. In other cells, residual bodies may remain for a long time and may load the cells to result in their aging. For example, pigment inclusions (age pigment or **lipofuscin granules**) found in nerve cells (also in liver cells, heart cells and muscle cells) of old animals may be due to the accumulation of residual bodies.

10.5 ORIGIN

The biogenesis (origin) of the lysosomes requires the synthesis of specialized lysosomal hydrolases and membrane proteins. Both classes of proteins are synthesized in the ER and transported through the Golgi apparatus, then transported from the trans Golgi network to an intermediate compartment (an **endolysosome**) by means of **transport vesicles** (which are coated by clathrin protein; Fig. 10.4). The lysosomal enzymes are glycoproteins, containing N-linked oligosaccharides that are processed in a unique way in the cis Golgi so that their **mannose** residues are phosphorylated. These **mannose 6-phosphate (M6P)** groups are recognized by **M6P-receptors** (which are transmembrane proteins) in the trans Golgi network that segregates the hydrolases and helps to package them into budding clathrin-coated vesicles which quickly lose their coats. These transport vesicles containing the M6P-receptors act as shuttles that move the receptors back and forth between the trans Golgi network and endolysosomes. The low pH in the endolysosome dissociates the lysosomal hydrolases from this receptor, making the transport of the hydrolases unidirectional.

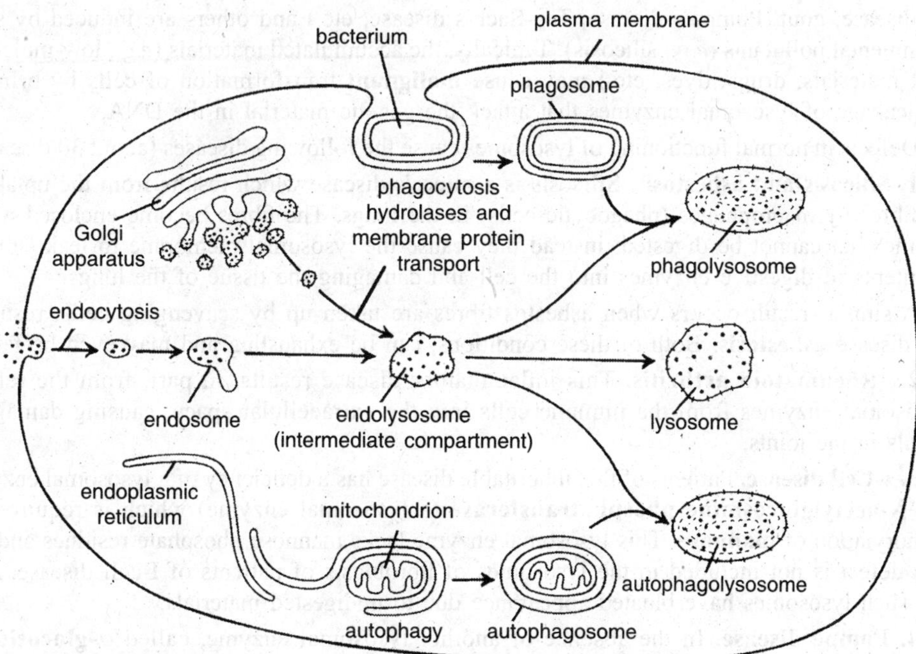

Fig. 10.4. Recently understood mechanism of the origin of three types of lysosomes : phagolysosome, lysosome (the classical secondary lysosome) and autophagolysosome. Transport vesicles (the classical primary lysosomes) originate from trans Golgi network to fuse with endolysosome which contains already endocytosed materials for digestion.

10.6 FUNCTIONS OF LYSOSOMES

The important functions of lysosomes are as follows:

1. Digestion of large extracellular particles. The lysosomes digest the food contents of the phagosomes or pinosomes. The lysosomes of leucocytes enable the latter to devour the foreign proteins, bacteria and viruses.

2. Digestion of intracellular substances. During the starvation, the lysosomes digest the stored food contents, *viz.*, proteins, lipids and carbohydrates (glycogen) of the cytoplasm and supply to the cell necessary amount of energy.

3. Autolysis. In certain pathological conditions the lysosomes start to digest the various organelles of the cells and this process is known as **autolysis** or **cellular autophagy**. When a cell dies, the lysosome membrane ruptures and enzymes are liberated. These enzymes digest the dead cells. In the process of metamorphosis of amphibians and tunicates many embryonic tissues, *e.g.*, gills, fins, tail, etc., are digested by the lysosomes and utilized by the other cells.

4. Extracellular digestion. The lysosomes of certain cells such as sperms discharge their enzymes outside the cell during the process of fertilization. The lysosomal enzymes digest the limiting membranes of the ovum and form penetration path in ovum for the sperms. Acid hydrolases are released from **osteoclasts** and break down bone for the reabsorption; these cells also secrete lactic acid which makes the local pH enough for optimal enzyme activity. Likewise, preceding ossification (bone formation), **fibroblasts** release cathepsin D enzyme to break down the connective tissue.

10.7 LYSOSOMAL STORAGE DISORDERS (LYSOSOMES AND DISEASE)

Malfunctioning of lysosomes often results in various pathological disorders affecting the life of the cell or an individual. Some of these are **inborn diseases**, caused by gene mutation (*e.g.*, I-cell disease, gout, Pompe's disease, Tay-Sach's disease, etc.) and others are induced by some environmental pollutants (*e.g.,* silicosis). Typically, the accumulated materials (*e.g.*, low-molecular weight materials, drugs, dyes, etc.) may cause **malignant** transformation of cells by bringing about leakage of lysosomal enzymes that attack the genetic material in the DNA.

Defects in normal functioning of lysosomes cause the following diseases (about 30 diseases):

1. Silicosis and asbestosis. Silicosis is a miner's disease which results from the uptake of silica fibres by macrophages (phagocytic cells) in the lungs. The fibres become enclosed within lysosomes but cannot be digested; instead they cause the lysosomal membrane to leak, spilling the contents of digestive enzymes into the cell and damaging the tissue of the lungs.

A similar result occurs when asbestos fibres are taken up by scavenging cells, resulting in the disease **asbestosis**. Both of these conditions, can be exhausting and may even be fatal.

2. Rheumatoid arthritis. This inflammatory disease results, in part, from the release of lysosomal enzymes from the immune cells into the extracellular space, causing damage to materials in the joints.

3. I-Cell disease. Patients of this inheritable disease has a deficiency of a lysosomal enzyme, called **N-acetylglucosmine phosphotransferase** (a lysosomal enzyme) which is required for phosphorylation of **mannose**. This lysosomal enzyme lacks mannose phosphate residues and due to this defect is not included in the lysosomes of fibroblasts of patients of I-cell disease. As a result, such lysosomes have bloated appearance due to undigested materials.

4. Pompe disease. In the absence of another lysosomal enzyme, called **α-glucosidase**, undigested glycogen is accumulated in the lysosomes. This results in swelling of these organelles and irreversible damage to the cells and tissues.

5. Tay-Sachs disease. This disease is caused from the deficiency of the enzyme **β-N-hexosaminidase A**, an enzyme required to degrade the ganglioside G_{M2}, a common component of the membranes of brain cells. In its severest form, which strikes during infancy, the disease is characterized by progressive mental and motor retardation, as well as skeletal, cardiac and respiratory abnormalities. Tay-Sachs disease results in mental retardation, blindness and death by the age of 3.

10.8 LYSOSOMES IN PLANTS

Plants contain several hydrolases, but they are not always as neatly compartmentalized as they are in animal cells. Many of these hydrolases are found bound to and functioning within the vicinity of the cell wall and are not necessarily contained in membrane-bound vacuoles at these sites. Many types of vacuoles and storage granules of plants are found to contain certain digestive enzymes and these granules are considered as lysosomes of plant cell. According to **Matile** (1969) the plant lysosomes can be defined as membrane-bound cell compartments containing hydrolytic digestive enzymes. **Matile** (1975) has divided vacuoles of plants into following three types:

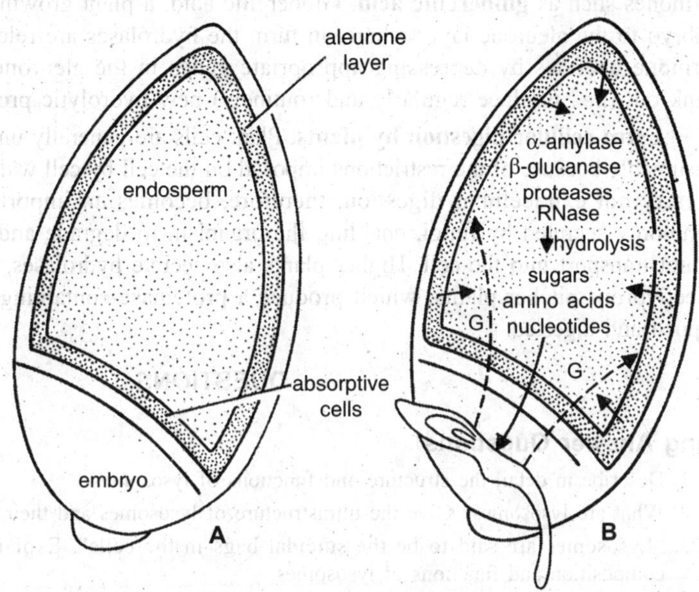

Fig. 10.5. Role of hormone and hydrolytic enzymes in seed germination. A—Ultrastructure of seed before germination; B–During germination, gibberellins (G) pass from the embryo to the aleurone layer, where the *de novo* synthesis of hydrolytic enzymes is induced. These enzymes break down the macromolecular stored reserves and the low molecular weight products are transported to the embryo, where they function as nutrients.

1. Vacuoles. The vacuole of a mature plant cell is formed from the enlargement and fusion of smaller vacuoles present in meristomatic cells; these **provacuoles**, which are believed to be derived from the ER and possibly the Golgi and contain acid hydrolases. These lysosomal enzymes are associated with the tonoplast of large vacuole of differentiating cells. Sometimes, mitochondria and plastids are observed inside the vacuole suggesting autophagy in plants.

2. Spherosomes. The spherosomes are membrane-bounded, spherical particles of 0.5 to 2.5 µm diameter, occurring in most plant cells. They have a fine granular structure internally which is rich in lipids and proteins. They originate from the endoplasmic reticulum (ER). Oil accumulates at the end of a strand of ER and a small vesicle is then cut off by constriction to form particles, called **prospherosomes**. The prospherosomes grow in size to form spherosomes. Basically, the spherosomes are involved in lipid synthesis and storage. But, the spherosomes of maize root tips and spherosomes of tobacco endosperm tissue have been found rich in hydrolytic digestive enzymes and so have been considered as lysosomes. Like lysosomes they are not only responsible for the accumulation and mobilization of reserve lipids, but also for the digestion of other cytoplasmic components incorporated by phagocytosis.

3. Aleurone grain. The aleurone grains or protein bodies are spherical membrane-bounded storage particles occurring in the cells of endosperm and cotyledons of seeds. They are formed during the later stages of seed ripening and disappear in the early stages of germination. They store protein (*e.g.*, globulins) and phosphate in the form of phytin. **Matile** (1968) has demonstrated that aleurone grains from pea seed contain a wide range of hydrolytic enzymes including protease and phosphatase which are required for the mobilization of stored protein and phosphate, although the presence of other enzymes such as β-amylase and RNAase suggest that other cell constituents may also be digested. Thus, like spherosomes, aleurone grains store reserve materials, mobilize them during germination and in addition form a compartment for the digestion of other cell components. The aleurone grains are derived from the strands of the endoplasmic reticulum.

During germination of barley seed, the activity of hydrolases is found to be controlled by hormones such as **gibberellic acid**. Gibberellic acid, a plant growth hormone, is released by the embryo to the aleurone layer where, in turn, the hydrolases are released to the endosperm. This hormone operates by depressing appropriate genes in the aleurone cells, which then begin to crank out (*i.e.,* produce regularly and routinely) new hydrolytic proteins.

Extra-cellular digestion by plants. Plant cells are generally unable to engulf large particles, presumably because of the restrictions imposed on the cell by cell wall. The secretion of hydrolases to carry out extracellular digestion, therefore, becomes an important process. Hydrolases are commonly secreted by fungi, enabling the organism to degrade and grow on macromolecules it cannot transport into the cell. Higher plants also secrete hydrolases, a notable example being the insectivorous pitcher plants, which produce a proteinase-containing liquid in which victims are trapped and digested.

QUESTIONS

Long Answer Questions

1. Describe in detail the structure and functions of lysosomes.

2. What are lysosomes? Give the ultrastructure of lysosomes and their functional significance.

3. "Lysosomes are said to be the suicidal bags in the cells". Explain it. Give structure, chemical composition and functions of lysosomes.

4. Describe lysosomes in plants.

5. Write an essay on lysosome. (*Bihar 2003*)

Short Answer Questions

1. Give an account of lysosomal polymorphism.

2. Explain the involvement of lysosomes in human disease.

3. What are suicide bags of cell? Describe their functions. (*C.C.S.U. Meerut 2007*)

4. Write short notes on the following:

(*i*) Lysosome. (*Bihar 2004; H.N.B. Garhwal, 2008*)

(*ii*) Lysosomes or Suicide bags

(*iii*) Polymorphism in lysosome

(*iv*) Enzymes of lysosomes

5. Lysosomes are suicide bags of the cell. Explain.

6. Write a brief account of polymorphism in lysosomes.

7. Give a brief account of the functions of lysosomes.

8. How and where lysosomes are formed in the cell?

9. Give significance of autophagic vacuole.

10. Lysosomal enzymes are inactive as long as they remain inside the lysosomes but become active as soon as they come in contact with some other substance. Why?

11. Differentiate between the following:

(*i*) Primary and secondary lysosomes (*ii*) Autophagic vacuole and digestive vacuole

(*iii*) Peroxisomes and lysosomes (*iv*) Lysosomes and zymogen granules.

Very Short Answer Questions

1. What is the role of acrosomal granule in sperm penetration?
2. Name the animal cells which do not contain lysosomes.
3. Where does lysosome formed in the cell?
4. Which secondary lysosome stores undigested wastes of cellular digestion?
5. What is autophagy?
6. What is pH of lumen of lysosomes?
7. Give the name of marker enzyme of the lysosome.

Fill in the Blanks

1. Lysosomal function is related to .... of substance.
2. Lysosomes with phagosomes form ....
3. Microbodies containing oxidative enzymes are called ....
4. The microbodies with specific lipid contents in their matrix are known as ....

Yes or No Questions

Write Yes or No as answers of the following:

1. Lysosomes help in protein synthesis.
2. Vacuoles of a cell are known as "suicide bags".

Multiple Choice Questions

1. Lysosomes are absent in *(Bihar 2004)*
   (*a*) erythrocytes (*b*) plasma cells (*c*) nerve cells (*d*) muscle cells
2. Which one of the following is called as the suicidal sac of the cell? *(Bihar 2001, 2003)*
   (*a*) centrosome (*b*) lysosomes (*c*) microsomes (*d*) desmosomes
3. At which pH, enzymes of lysosomes are usually active
   (*a*) pH5 (*b*) pH7 (*c*) pH8 (*d*) at any pH
4. Hydrolytic enzymes are stored in
   (*a*) Golgi bodies (*b*) lysosomes (*c*) endoplasmic reticulum (*d*) mitochondria

ANSWERS

Very Short Answer Questions

1. Lysis.
2. Red blood corpuscles.
3. Golgi apparatus.
4. Residual bodies.
5. Digestion of portions of a cell's own organelles by lysosomes.
6. 5.
7. Acid phosphatase.

Fill in the Blanks

1. Hydrolysis (*i.e.,* intracellular digestion)
2. Heterolysosomes (or food vacuoles)
3. Peroxisomes
4. Sphareosomes.

Yes or No Questions:

1. No 2. No.

Multiple Choice Questions

1. (*a*) 2. (*b*) 3. (*c*) 4. (*b*).

11

Ribosomes

The ribosomes are small, dense, rounded and granular particles of the ribonucleoprotein. They occur either freely in the matrix of mitochondria, chloroplast and cytoplasm (*i.e.*, cytoplasmic matrix) or remain attached with the membranes of the endoplasmic reticulum and nucleus. They occur in most prokaryotic and eukaryotic cells and are known to provide a scaffold for the ordered interaction of all the molecules involved in protein synthesis.

Box 11.1 Some Historical Facts.

1. **George E. Palade**, in 1952, described the ribosomes.
2. **Philip Siekevitz** and **Paul Zamecnik**, in 1952, showed that radioactive amino acids were incorporated into proteins on ribosomes.
3. In 1953, ribosomes were observed in plant cells (roots of bean) by **Robinson** and **Brown.**
4. Presence of ribosomes in both free and membrane attached form was confirmed in the animal cells by **Palade** and **Siekevitz** by electron microscopy.
5. In 1956, **Palade** detected RNA in ribosomes.
6. In 1958, the term **ribosome** was coined by **R.B. Roberts.**
7. **Tissieres** and **Watson** (1958) isolated *E.coli* ribosomes and showed that they consist of two subunits, 50S and 30S.
8. The first definitive evidence linking the nucleolus with eukaryotic ribosome formation was provided in 1960s by **Robert Perry.**
9. **James A. Lake**, in 1981, suggested the well accepted **asymmetrical model** for 70S ribosome.

11.1 OCCURRENCE

The ribosomes occur in cells, both prokaryotic and eukaryotic cells. In prokaryotic cells the ribosomes often occur freely in the cytoplasm. In eukaryotic cells the ribosomes either occur freely in the cytoplasm or remain attached to the outer surface of the membrane of endoplasmic reticulum. The yeast cells, reticulocytes or lymphocytes, meristamatic plant tissues, embryonic nerve cells and cancerous cells contain large number of ribosomes which often occur freely in the cytoplasmic matrix. The cells in which active protein synthesis takes place, the ribosomes remain attached with the membranes of the endoplasmic reticulum. Such cells are the pancreatic cells, plasma cells, hepatic parenchymal cells, nissl bodies, osteoblasts, serous cells, or the submaxillary gland, chief cells of the glandular stomach, thyroid cells and mammary gland cells. The cells which synthesize specific proteins for the intracellular utilization and storage often contain large number of free ribosomes. Such cells are the erythroblasts, developing muscle cells, skin and hair.

11.2 TYPES OF RIBOSOMES

Recently according to the size and the sedimentation coefficient (S) two types of ribosomes have been recognised (Fig. 11.1).

1. 70S ribosomes. The 70S ribosomes are comparatively smaller in size and have sedimentation coefficient 70S and the molecular weight 2.7 × 106 daltons. (**Dalton** is the unit of molecular weight (**MW**); one dalton equals the weight of hydrogen atom. For example, a water molecule weighs 18 daltons). They occur in the prokaryotic cells of the blue green algae and bacteria and also in mitochondria and chloroplasts of eukaryotic cells.

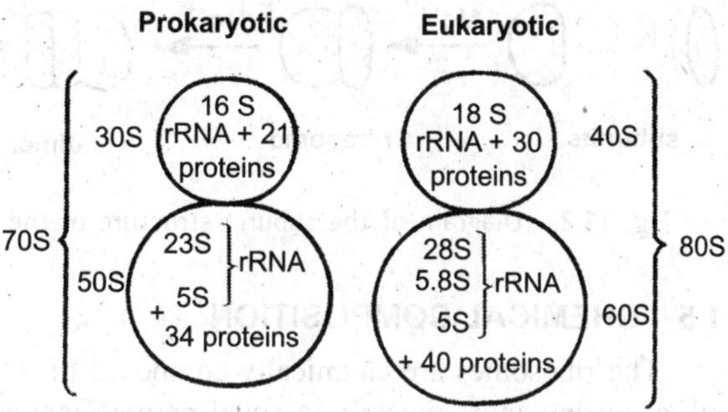

Fig. 11.1. Various components of prokaryotic (70S) and eukaryotic (80S) ribosomal subunits.

2. 80S ribosomes. The 80S ribosomes have the sedimentation coefficient of 80S and the molecular weight 40 × 10^6 daltons. The 80S ribosomes occur in eukaryotic cells of the plants and animals.

The ribosomes of mitochondria and chloroplasts are always smaller than 80S cytoplasmic ribosomes and are comparable to prokaryotic ribosomes in both size and sensitivity to antibiotics, although their sedimentation values vary in different phyla, *e.g.*, 77S in mitochondria of fungi, 77S in mitochondria of mammals and 60S in mitochondria of animals in general. The ribosomes of chloroplasts are 70S type.

11.3 NUMBER OF RIBOSOMES

An *E. coli* cell contains 10,000 ribosomes, forming 25 per cent of the total mass of the bacterial cell. In contrast, mammalian cultured cells contain 10 million ribosomes per cell, each of which is about twice as large as a prokaryotic ribosome.

11.4 STRUCTURE OF RIBOSOMES

The ribosomes are oblate spheroid structures of 150 to 250 Å in diameter. Each ribosome is porous, hydrated and composed of two subunits. One ribosomal subunit is large in size and has a dome-like shape, while the other ribosomal subunit is smaller in size and occurring above the larger subunit and forming a cap-like structure.

The 70S ribosome consists of two subunits, *viz.*, 50S and 30S. The 50S ribosomal subunit is larger in size and has the size of 160 Å to 180 Å. The 30S ribosomal subunit is smaller in size and occurs above the 50S subunit like a cap.

The 80S ribosome also consists of two subunits, *viz.*, 60S and 40S. The 60S ribosomal subunit is dome-shaped and larger in size. In the ribosomes which remain attached with the membranes of endoplasmic reticulum and nucleus, etc., the 60S subunit remains attached with the membranes. The 40S ribosomal subunit is smaller in size and occurs above the 60S subunit forming a cap-like structure. Both the subunits remain separated by a narrow cleft (Fig. 11.1).

The two ribosomal subunits remain united with each other due to high concentration of the Mg^{++} (.001M) ions. When the concentration of Mg^{++} ions reduces in the matrix, both ribosomal subunits get separated. Actually in bacterial cells the two subunits are found to occur freely in the cytoplasm and they unite only during the process of protein synthesis. At high concentration of Mg^{++} ions in the matrix, the two ribosomes (called **monosomes**) become associated with each other and known as the **dimer**. Further, during protein synthesis many ribosomes are aggregated due to common messenger RNA and form the **polyribosomes** or **polysomes** (Fig. 11.2).

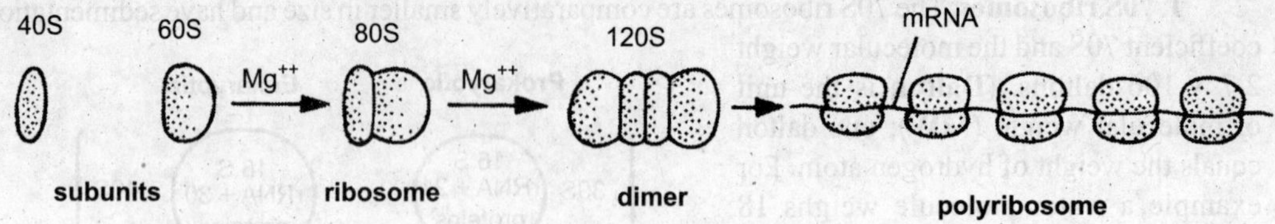

Fig. 11.2. Diagram of the subunit structure of the ribosome and the influence of Mg^{++} ions.

11.5 CHEMICAL COMPOSITION

The ribosomes are chemically composed of RNA and proteins as their major constituents; both occurring approximately in equal proportions in smaller as well as larger subunit. However, the 70S ribosomes contain more RNA (60 to 40%) than the proteins (36 to 37%), *e.g.*, the ribosomes of *E. coli* contain 63% rRNA and 37% protein. While the 80S ribosomes contain less RNA (40 to 44%) than the proteins (60 to 56%), *e.g.*, yeast ribosomes have 40 to 44% RNA and 60 to 56% proteins; ribosomes of pea seedling contain 40% RNA and 60% proteins. There is no lipid content in ribosomes.

1. Ribosomal RNAs

The 70S ribosomes contain three types of rRNA, *viz.*, **23S rRNA, 16S rRNA, 5S rRNA**. The 23S and 5S rRNA occur in the larger 50S ribosomal subunit, while the 16S rRNA occurs in the smaller 30S ribosomal subunit. Assuming an average molecular weight for one nucleotide to be 330 daltons, one can calculate the total number of each type of rRNA. Thus, the 23S rRNA consists of 3300 nucleotides, 16S rRNA contains 1650 nucleotides and 5S rRNA includes 120 nucleotides in it.

The 80S ribosomes contain four types of rRNA, *viz.*, **28S rRNA** (or **25-26 rRNA** in plants, fungi and protozoa), **18S rRNA, 5S rRNA** and **5.8S rRNA**. The 28S, 5S and 5.8S rRNAs occur in the larger 60S ribosomal subunit, while the 18S rRNA occurs in the smaller 40S ribosomal subunit. About 60 per cent of the rRNA is helical (*i.e.*, double stranded) and contains paired bases. These double stranded regions are due to hairpin loops between complimentary regions of the linear molecule.

The 28S rRNA has the molecular weight 1.6×10^6 daltons and its molecule is double stranded and having nitrogen bases in pairs. The 18S rRNA has the molecular weight 0.6×10^6 daltons and consists of 2100 nucleotides. The 18S and 28S ribosomal RNA contain a characteristic number of methyl groups, mostly as 2′-O-methyl ribose. The molecule of 5S rRNA has a clover leaf shape and a length equal to 120 nucleotides. The 5.8S rRNA is intimately associated with the 28S rRNA molecule and has, therefore, been referred to as **28S-associated ribosomal RNA (28S A rRNA)**.

The 55S ribosomes of mammalian mitochondria lack 5S rRNA but contain **21S** and **12S rRNAs**. The 21S rRNA occurs in larger or 35S ribosomal subunits, while 12S rRNA occur in smaller or 25S ribosomal subunit.

It is thought that each ribosomal subunit contains a highly folded ribonucleic acid filament to which the various proteins adhere. But as the ribosomes easily bind the basic dyes so it is concluded that RNA is exposed at the surface of the ribosomal subunits, and the protein is assumed to be in the interior in relation to non-helical part of the RNA.

2. Ribosomal Proteins

Each 70S ribosome of *E. coli* is composed of about 55 **ribosomal proteins**. Out of these 55 proteins, about 21 different molecules have been isolated from the 30S ribosomal subunit, and some 32 to 34 proteins from the 50S ribosomal subunit. The primary structure of several of these proteins has been elucidated. Most of the recent knowledge about the structure of ribosomal proteins has been achieved by dissociation of ribosomal subunits into their component rRNA and protein molecules. When both 50S and 30S ribosomal subunits are dissociated by centrifuging both of them in a gradient of 5M cesium chloride, then there are two inactive core particles (40S and 23S, respectively) which contain the RNA and some proteins called **core proteins (CP)** at the same time several other proteins—the so-called **split proteins (SP)** are released from each particle (Fig. 11.3). There are SP50 and SP30 proteins which may reconstitute the functional ribosomal subunit when added to their corresponding core. Some of the split proteins are apparently specific for each ribosomal subunit. The split proteins have been further fractionated and divided into acidic (A) and basic (B) proteins. **Nomura** *et al.*, (1968) fractionated at least six different groups of proteins in the ribosome (Fig. 11.3).

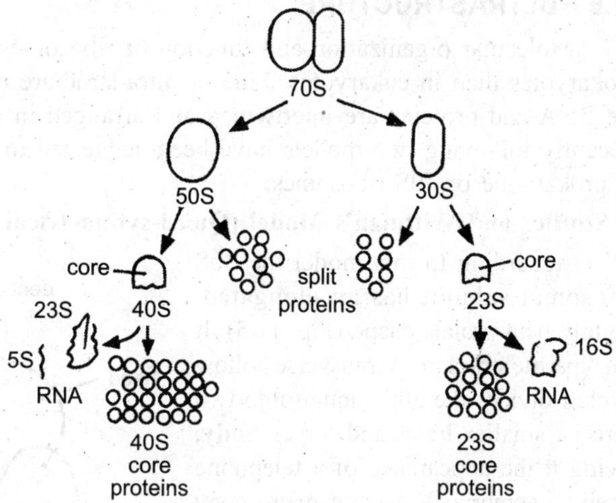

Fig. 11.3. Diagram showing the stepwise dismantling of the two subunits of 70S ribosome. Note that the proteins may be separated into split and core proteins. The 50S subunit contains 23S and 5S RNAs, and the 30S subunit has 16S RNA.

In all, 21 types of proteins have been isolated in smaller subunit (30S) of ribosome of *E. coli*. These are designated as S1 to S21. Similarly, in larger subunit (50S) 34 different proteins designated as L1 to L34, have been isolated. Thus, the 70S ribosome was thought to consist of 55 different proteins. However it was later shown that protein S20 is identical to L6, thus, the correct number of S proteins is 20. Likewise, L8 was shown to be an aggregate of proteins L7, L12 and L10; thus, the correct number of L proteins is 33. Thus, the prokaryotic 70S ribosome consists of 53 different proteins (20S + 33L = 53 proteins). Similar organization of ribosomal proteins and RNA is found in 80S ribosomes (Fig. 11.4). Small 40S subunit comprises 33 proteins and large 60S subunit contains more than 45 proteins (*i.e.*, 50 proteins; **Twyman** 1998).

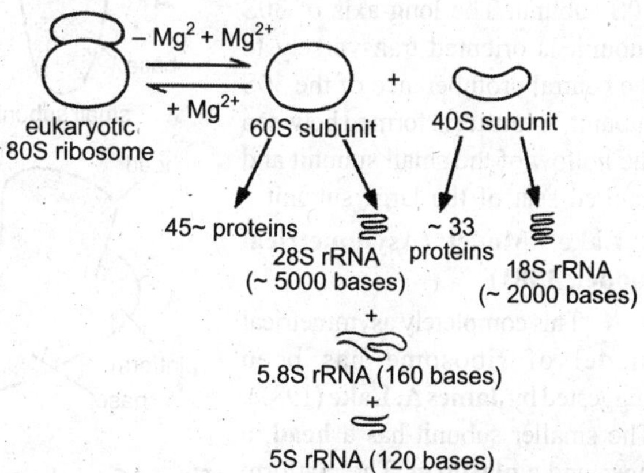

Fig. 11.4. Different RNA and protein components of eukaryotic ribosomes.

Different rRNA molecules evidently play a central role in the catalytic activities of ribosomes in the process of protein synthesis. Various ribosomal proteins have been found to mainly enhance the catalytic function of the rRNA in the ribosomes.

3. Metallic Ions. The most important low molecular weight components of ribosomes are the divalent metallic ions such as Mg^{++}, Ca^{++} and Mn^{++}.

11.6 ULTRASTRUCTURE

Molecular organization and function of ribosomes have been studied more intensively in prokaryotes than in eukaryotes. Fine or ultra-structure of 70S ribosome is very complex. In it, the RNA and proteins are intertwined and arranged in a complex manner in the two subunits. Recently following two models have been suggested to explain the three-dimensional structure of prokaryotic or 70S ribosomes:

1. Stoffler and Wittman's Model (Quasi-symmetrical model, 1977)

According to this model the 30S ribosomal subunit has an elongated, slightly bent prolate shape (Fig. 11.5). It is a bipartite structure. A transverse hollow or cleft divides the 30S subunit into two parts, a smaller **head** and larger **body**, giving it the appearance of a telephone receiver or embryo. In electron microscopy 50S ribosomal subunit showed various shapes depending on structure seen in different views such as frontal-maple leaf, lateral-kidney shaped or rear view-rounded. In a frontal view, the 50S subunit appears bilaterally symmetrical and shows three protuberances arising from a rounded base (maple leaf structure). The **central protuberance** being the most prominent. The 50S subunit is often compared with an armchair, with the rounded base forming a **vaulted seat**, the central protuberance forming the **back** and the lateral protuberances the **arms** of chair.

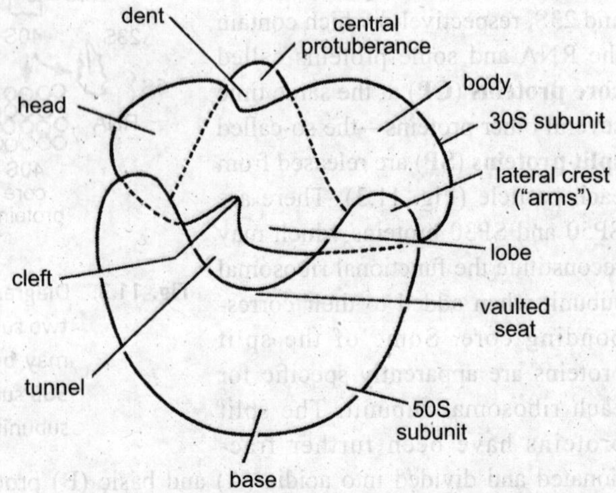

Fig. 11.5. Stoffler and Wittman's model of 70S ribosomes.

When 30S and the 50S subunits become associated to form the 70S ribosome, the frontal face of the 30S subunit with its hollow faces the vaulted seat of the 50S subunit. The long axis of 30S subunit is oriented transversely to the central protuberance of the 50S subunit. A **tunnel** is formed between the hollow of the small subunit and vaulted seat of the large subunit.

2. Lake's Model (Asymmetrical model, 1981)

This completely asymmetrical model of ribosome has been suggested by **James A. Lake** (1981). The smaller subunit has a **head**, a **base** and a **platform**. The platform separates the head from the base by the help of a **cleft**. This cleft is an important functional region; it is

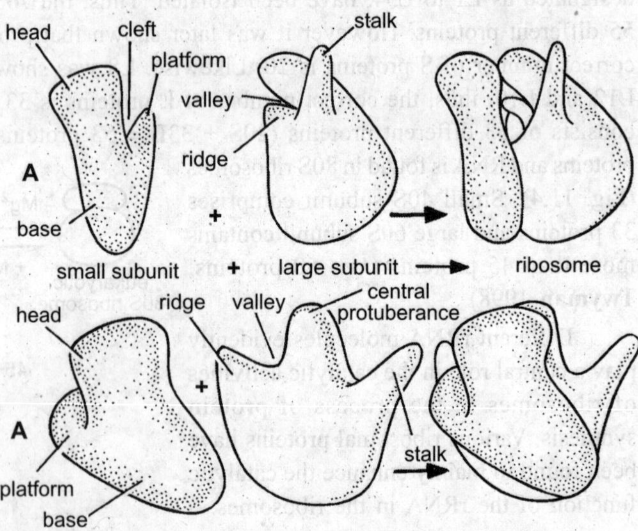

Fig. 11.6. Lake's model of the 70S ribosome. Note the three dimensional structure of the ribosome in two different orientations.

suggested to be the site of codon-anticode interaction and as a part of binding site for initiation factors of protein synthesis.

The large subunit consists of a **ridge**, a **central protuberance** and a **stalk**. The ridge and the central protuberance are separated with the help of a valley (Fig. 11.6).

11.7 THREE DIMENSIONAL MODEL OF 80S RIBOSOME

In spite of the difference in overall sizes (as manifested in the greater molecular weights, sedimentation constants, sizes and numbers of rRNAs and proteins), the cytoplasmic ribosomes of eukaryotes (80S) are remarkably similar in morphology to those of prokaryotes. As in 30S subunits of prokaryote ribosomes, the 40S ribosomal subunit of eukaryotes is divided into **head** and **base** segments by a transverse groove (Fig. 11.7). The 60S ribosomal subunit is generally rounder in shape than the small subunit, although its one side is flattened; this is the side that becomes confluent with the small subunit during the formation of the monomer or monosome (*i.e.*, functional 80S ribosome).

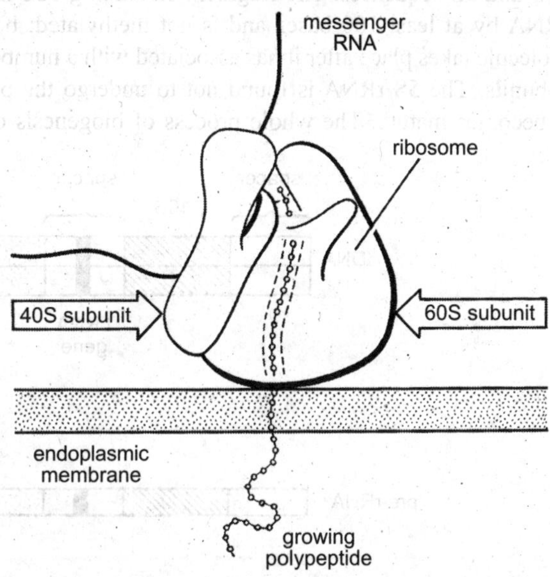

Fig. 11.7. A three dimensional model of eukaryotic ribosome.

Several **domains** of the ribosomes have particularly important functions during protein synthesis. The small subunit contains a binding site for tRNA. The **A-site (amino-acyl-tRNA site)** binds incoming charged tRNAs during elongation, the **P-site (peptidyl-tRNAs site)** binds the tRNA carrying the nascent polypeptide chain. Bacterial 70S ribosomes possess a third **E-site (exit site)** to which spent tRNA are dispatched prior to ejection. The large subunits possess a **peptidyl-transferase domain**, which provides the catalytic activity for peptide bond formation and a GTPase domain whose activities are required for translocation of the ribosome.

11.8 COMPARISON OF 70S AND 80S RIBOSOMES

Eukaryotic 80S ribosomes differ from prokaryotic 70S ribosomes in the following respects: 1. they are considerably larger; 2. they contain a large number of proteins (70–80 types of proteins instead of 53); 3. they have four types of RNA molecules instead of three types; 4. their proteins and nucleic acids are large-sized; 5. the RNA-protein ratio is near to 1 : 1 instead of 2 : 1 and 6. several antibiotics, such as **chloramphenicol**, inhibits bacterial but not eukaryotic ribosomes (this is the basis of the use of many antibiotics in medical treatment). Protein synthesis by eukaryotic ribosomes is inhibited by **cycloheximide**.

However, eukaryotic ribosomes do not differ functionally from those in prokaryotes in a fundamental way; they perform the same functions, by the same set of chemical reactions. The genetic code is the same for all living organisms, and eukaryotic ribosomes are able to translate bacterial mRNAs efficiently, provided that a "cap" is added enzymatically.

11.9 BIOGENESIS OF RIBOSOMES

Ribosomes are not self-replicating particles. Synthesis of various component of ribosomes such as rRNAs and proteins, are under genetic control, *i.e.*, rRNAs and mRNAs (for various ribosomal proteins) are transcribed by genes (DNA). Since the mechanism of biogenesis of 70S and 80S ribosomes differ greatly, so can be studied separately as follows:

1. Biogenesis of 70S Ribosomes

In bacteria a single gene transcript containing the sequences of 16S, 23S and 5S rRNAs, is synthesized by a rRNA operon and this larger molecule is thought to undergo both tailoring and chemical modifications before each rRNA molecule assumes its mature form. During tailoring of larger rRNA molecule, 16S rRNA sequence is first of all cleaved off and is separated from the 23S and 5S sequences. The fragment containing 16S information is still larger than the mature 16S rRNA by at least 100 bases and is not methylated: both the methylation and the tailoring of this molecule takes place after it has associated with a number of proteins to form the precursor ribosomal subunits. The 5S rRNA is found not to undergo the processes of tailoring and methylation before it becomes mature. The whole process of biogenesis of 70S ribosomes takes place in cytoplasm.

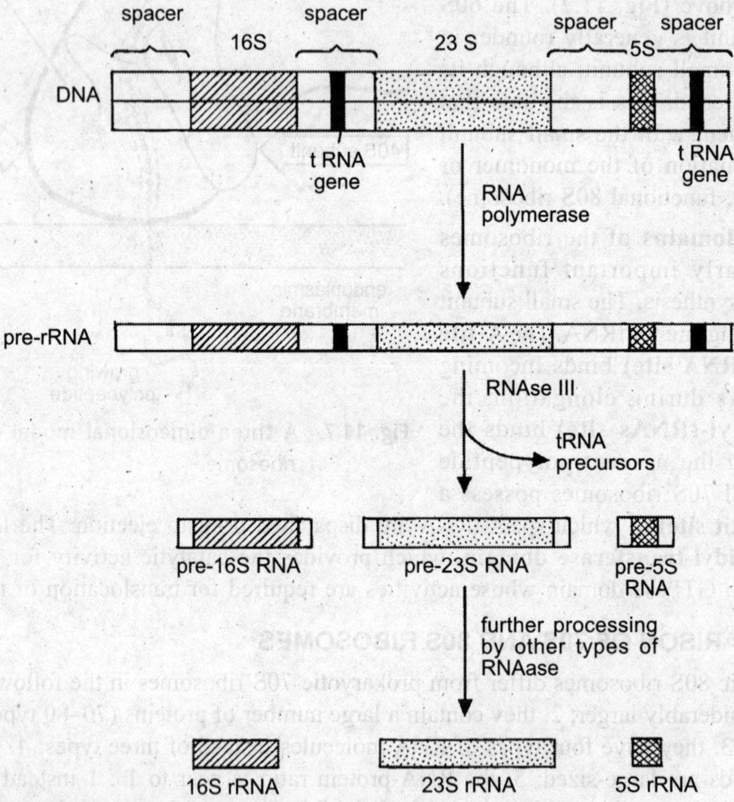

Fig. 11.8. Ribosomal RNA processing in the bacterium *E.coli.*

2. Biogenesis of 80S Ribosomes.

In eukaryotes, the biogenesis of ribosomes is much more complex and involves a long-lasting process in which several regions of cell are involved. The 5.8S, 18S and 28S rRNAs are transcribed as a much larger molecule in the **nucleolar organizer (NO)** which contains many copies 5.8S, 18S and 28S rRNA genes or **ribosomal DNA** (*i.e.*, there is gene redundancy or amplification). The DNA coding for the 5S rRNA is also highly repetitive, but the molecule is synthesized outside the nucleolus. It is in the nucleolus that newly synthesized rRNA accumulates and becomes associated, presumably by a self-assembly process, with 50 or more ribosomal proteins that have been synthesized in the cytoplasm by usual mechanism of protein synthesis and then migrate to the cytoplasm of cell, in the form of ribosomal subunits.

A. Ribosomal RNA synthesis inside nucleolus. All organisms have multiple rRNA genes. In case of *Xenopus*, each nucleolar organizer contains 450 rRNA genes. These genes are **tandemly repeated** or **reiterated** along the DNA molecule (in a head to tail arrangement) and are separated from each other by stretches of **spacer DNA**, which is not transcribed. These rRNA genes are being actively transcribed and the nascent RNA chains are spread perpendicularly to the DNA axis. Each gene is transcribed into a long RNA molecule (which varies in size from 40S to 45S according to species) which will eventually be processed to give rise to 18S, 28S and 5.8S rRNA.

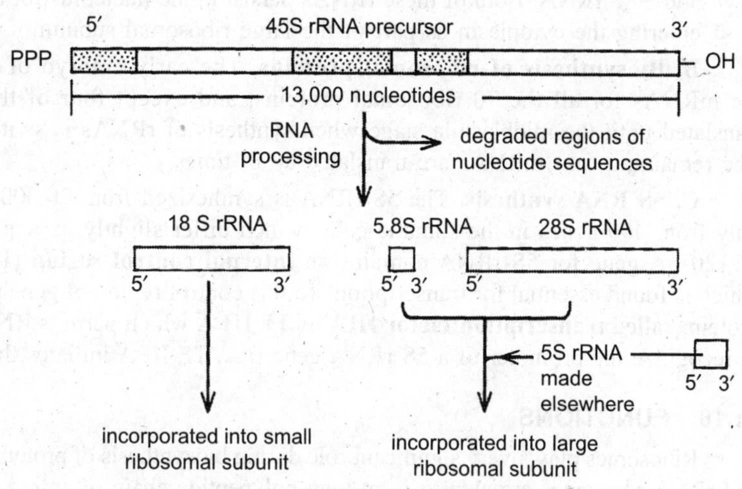

Fig. 11.9. The pattern of processing of 45S rRNA precursor molecule into three separate ribosomal RNAs.

Because each rRNA gene has a fixed initiation site (promoter) and a fixed termination site, the transcripts adopt the characteristic "christmas tree" or "fern leaf" configuration (Fig. 11.8). Nucleolar rRNA genes are transcribed by **RNA polymerase I** (about 100 enzymes per gene). RNA polymerase I molecules are found to remain bound to the nucleolar organisers during mitotic metaphase and anaphase. During this period there is no RNA synthesis and so the enzyme molecules must remain in an inactive state.

Processing of rRNAs inside nucleolus. As already described, rRNA genes are transcribed into a long precursor RNA (which is 40S in *Xenopus* and 45S in HeLa and other human cells); this precursor must be cleaved into 18S, 28S and 5.8 rRNA. In the cleavage process about 50 per cent of the precursor RNA is degraded. In HeLa cell, the processing of rRNA involves the following steps:

(*i*) The first ribosomal RNA in HeLa cells is a large 45S molecule of 14,000 nucleotides. Within this precursor molecule the rRNAs are separated by stretches of spacer RNA and the order of transcription is: 5′ end—18S—5.8S—28S—3′ end. On a fully active gene about 100 RNA polymerase I enzymes (along with transcription factor I or TFI) are transcribing simultaneously on the rRNA gene.

(*ii*) In nucleolus, 45S RNA is rapidly **methylated**, even before transcription is completed. Methylations occur mostly on the ribose moiety (producing 2′-O- methyribose) and occur only in the 18S (46 methylations) and 28S (71 methylations) sequences that have to be conserved. Those segments of 45S which have to be degraded are not methylated.

(*iii*) 45S RNA has a lifetime of about 15 minutes and is then cleaved into smaller components as follows:

$$45S \longrightarrow 41S \begin{cases} \nearrow 32S \longrightarrow 28S + 5.8S \\ \searrow 20S \longrightarrow 18S \end{cases}$$

(*iv*) 20S RNA is rapidly processed in 18S rRNA and probably due to this reason the small ribosomal subunits appear in the cytoplasm earlier than the large ribosomal subunits. The large ribosomal subunits have a slower RNA processing.

(*v*) 32S RNA remains in the nucleolus for about 40 minutes and is then cleaved into 28S rRNA and 5.8S rRNA. Both of these rRNAs persist in the nucleolus for another 30 minutes before entering the cytoplasm as part of the large ribosomal subunit.

B. Biosynthesis of ribosomal proteins. The early embryo of frog is found to contain the mRNAs for all the 70 ribosomal proteins, and except four of these mRNAs, all are not translated until the midblastula stage when synthesis of rRNAs is switched on in the nucleolus. The remaining four mRNAs are translated at all times.

C. 5S RNA synthesis. The 5S rRNA is synthesized from 20,000 genes in the oocytes but only from 400 genes in the somatic cells, which differ slightly in sequence (six nucleotides out of 120). A gene for 5S rRNA contains an **internal control region (ICR)** in its middle region which is found essential for transcription. To this control region of gene remains attached a special protein, called **transcription factor IIIA** or **TF III A** which permits RNA polymerase III enzyme to recognize the promoter of a 5S rRNA gene (*i.e.*, TF III A initiates the synthesis of 5S rRNA).

11.10 FUNCTIONS

Ribosomes play a very significant role during biosynthesis of proteins. With the help of mRNA and rRNA ribosomes are able to form long polypeptide chain of amino acids from the cytoplasm.

11.11 DISTINCTION BETWEEN LYSOSOMES AND RIBOSOMES

Ribosomes are the granules of protein and RNA present in the cytoplasm. They lack any limiting membrane and are often attached to endoplasmic reticulum. They are involved in protein synthesis.

Lysosomes are cytoplasmic organelles (they are limited by a unit membrane) containing about 50 digestive enzymes for intracellular digestion of bacteria and foreign bodies which enter the cells by the process of phagocytosis or pinocytosis. They also remove worn out cellular organelles from the cells. Lysosomes are not found in prokaryotic cells.

QUESTIONS

Long Answer Questions

1. Describe the structure, biogenesis and function of ribosomes.

2. Define ribosome. Give an account of the ultrastructure, chemistry and functioning of ribosomes. Explain their role in protein synthesis.

3. Describe the types and roles of ribosomes.

4. Give an account of structure and functions of eukaryotic ribosomes

5. Describe the ultrastructure of ribosomes and discuss the function of nucleolus as a ribosome factory.

6. Write an essay on ribosomes. (Bihar 2000)

Short Answer Questions

1. Write short notes on the following:

(*i*) Ribosomes.

(*ii*) Polysomes.

(*iii*) Ultrastructure of ribosome.

(*iv*) Ribosome biogenesis.

(*v*) Types of ribosomes.

(*vi*) Different types of RNA.

(*vii*) 80S subunit of ribosome.

2. Distinguish between ribosome and lysosome.

3. Define ribosome. Describe Lake's asymmetrical model of ribosome. (*CCSU Meerut 2002*)

4. Prepare a well-labelled diagram of J.A. Lakes model of 70S ribosome.

Very Short Answer Question

1. Define the ribosome.

2. What is polyribosome?

3. What is rRNA?

4. What is the main function of ribosome?

5. What are the main chemical constituents of ribosome?

6. Who gave the asymmetrical model of the ribosome?

7. Name the three parts of smaller subunit of ribosome.

Yes or No Answer

(*i*) RNA plays an important role in protein synthesis.

Fill in the Blanks

(*i*) Amino acid are linked together to form polypeptide chain by ....

(*ii*) Ribosome is the site of ....

(*iii*) Ribosome were first observed by ....

(*iv*) The term ribosome was used by ....

(*v*) The principal site of ribosomal RNA in the cell is the ....

(*vi*) During protein synthesis mRNA gets attached to .... subunit of ribosome.

(*vii*) Ribosome are formed in .... as .... and .... subunits.

(*viii*) The bacterial ribosomes are formed of .... and .... subunits.

(*ix*) Ribosomal RNA of large subunit is synthesized as .... RNA.

Multiple Choice Questions

1. Ribosomes help in

(*a*) lipogenesis (*b*) cellular digestion

(*c*) protein synthesis (*d*) photosynthesis

2. The sedimentation constant of eukaryotic ribosome is generally 80S. What is the sedimentation constant of its subunits?

(*a*) 40S and 30S (*b*) 50S and 30S

(*c*) 60S and 30S (*d*) 60S and 40S

3. Proteins are synthesized by

(*a*) ribosomes (*b*) mitochondria

(*c*) Golgi apparatus (*d*) mesosomes

4. Ribosomes on hydrolysis produce

(*a*) acidic proteins and rRNA (*b*) basic proteins and rRNA

(*c*) split proteins and rRNA (*d*) proteins, rRNA and DNA

5. If all the ribosomes of a cell are destroyed

(*a*) respiration will not take place (*b*) protein synthesis will not take place

(*c*) photosynthesis will stop (*d*) it will have no effect

6. Cell organelle devoid of DNA is (*Bihar 2004*)

(*a*) ribosomes (*b*) nucleus

(*c*) plastids (*d*) mitochondria

7. Prokaryotic ribosomes are (*Bihar 2004*)

(*a*) 30S (*b*) 50S (*c*) 70S (*d*) 80S

8. Which of the following are non-membranous organelles?

(*a*) microbodies (*b*) vacuoles (*c*) ribosomes (*d*) chloroplasts

9. Ribosomes are attached to ER through

(*a*) ribophorins (*b*) rRNA

(*c*) tRNA (*d*) hydrophobic interaction

ANSWERS

Very Short Answer Question

1. Ribosome is a small particle composed of rRNA and protein that function as the site of protein synthesis in the cytoplasm of prokaryotes and in the cytoplasm, mitochondria and chloroplasts of eukaryotes. It is composed of large and small subunits.

2. Polyribosome is a cluster of two or more ribosomes simultaneously translating a single mRNA molecule.

3. The rRNA is ribosomal RNA which is any of several types of RNA molecules used in the construction of ribosome.

4. Protein synthesis.

5. RNA and proteins.

6. J.A. Lake in 1981.

7. Head, base and platform.

Yes or No Answer

1. Yes.

Fill in the Blanks

(*i*) Peptide bond (*ii*) Protein synthesis

(*iii*) Palad (*iv*) R.B. Roberts

(*v*) Ribosome (*vi*) Small

(*vii*) Nucleolus, ribonucleoproteins, ribosomal (*viii*) 30S, 50S

(*ix*) 45S.

Multiple Choice Questions

1. (*c*) **2.** (*d*) **3.** (*a*) **4.** (*c*) **5.** (*b*) **6.** (*a*)

7. (*c*) **8.** (*c*) **9.** (*a*).

12

Vesicular Traffic in the Secretory and Endocytic Pathway

Transport vesicles play a central role in the **traffic** or **transport** of molecules between different membrane-enclosed compartments of the secretory pathway. Vesicles are similarly involved in the transport of materials taken up at the cell surface (*i.e.,* at the level of plasma membrane through endocytosis). **Vesicular traffic (= transport)** is thus a major cellular activity, responsible for molecular traffic between a variety of specific membrane-enclosed compartments. The selectivity of such transport is therefore key in maintaining the functional organization of the cell. For example, lysosomal enzymes must be transported specifically from the Golgi apparatus to lysosomes not to the plasma membrane or to the ER. There exist some signals that target proteins to specific organelles, such as lysosomes. These proteins are transported within vesicles, so the specificity of transport is based on the selective packaging of the intended cargo into vesicles that recognize and fuse only with the appropriate target membrane. Because of the central importance of vesicular transport to the organization of eukaryotic cells, understanding the molecular mechanisms that control vesicle packaging, budding and fusion is a major area of research in cell biology.

Experimental Approaches for Understanding Vesicular Transport.

Progress toward explaining the molecular mechanisms of vesicular transport has been made by three distinct experimental approaches: (*i*) Isolation of yeast mutants that are defective in protein transport and sorting; (*ii*) Reconstitution of vesicular transport in cell-free systems; and (*iii*) Biochemical analysis of synaptic vesicles, which are responsible for the regulated secretion of neurotransmitters by neurons (see **Cooper** and **Hausman**, 2007).

1. In 1980, **Randy Schekman** and his colleagues have pioneered the isolation of yeast mutants defective in vesicular transport. These include mutants that are defective at various stages of protein secretion (*sec* mutants), mutants that are unable to transport proteins to the vacuole, and mutants that are unable to retain resident ER proteins. The isolation of such mutants in yeasts led directly to the molecular cloning and analysis of corresponding genes, thereby identifying a number of proteins involved in various steps of the secretory pathway. For example, the protein Sec 61 forms a major component of the protein acting as translocation channel in RER.
2. Biochemical studies of vesicular transport using reconstituted systems have complemented these genetic studies and have enabled the direct isolation of proteins from mammalian cells. The first cell-free transport system was developed in 1984 by **James Rothman** and his colleagues, who analysed protein transport between compartments of the Golgi apparatus. Similar reconstituted systems have been developed to analyse transport between other compartments, including transport from ER to Golgi apparatus and transport from the Golgi apparatus to secretory vesicles, vacuoles and the plasma membrane. The development of these *in vitro* systems has enabled biochemical studies of the transport process and functional analysis of proteins identified by mutations in years, as well as direct isolation of some of the proteins involved in vesicles budding and fusion.

3. Insights into the molecular mechanisms of vesicular support have come from studies of synaptic transmission in neurons, which represents a highly specialised form of regulated secretion. A **synapse** is the junction of a neuron with another cell, which may be either another neuron or an effector, such as muscle cell. Information is transmitted across the synapse by chemical neurotransmitters to be released into the synapse and stimulating the post-synaptic neuron of effector cell. Synaptic vesicles are extremely abundant in the brain, allowing them to be purified in large amounts for biochemical analysis. Some of the proteins isolated from synaptic vesicles are closely related to proteins that have been shown to play critical roles in vesicular transport by yeast genetics and reconstitution experiments, so biochemical analysis of these proteins has provided important insights into the molecular mechanisms of vesicle fusion.

Lastly, recent studies using GFP fusion proteins (Box 12.1) have allowed transport vesicles carrying specific proteins to be visualized by immunofluorescence as they move through the secretory pathway. In these experiments, cells are transfected with cDNA constructs encoding secretory proteins tagged with green fluorescent protein (GFP). The progress of the GFP-labelled proteins through the secretory pathway can then be followed in living cells, allowing characterisation of many aspects of the dynamics and molecular interactions involved in vesicular transport.

Box 12.1

GFP or green fluorescent protein is derived from the pacific jellyfish, *Aequoria victoria*.

12.1 SECRETORY PATHWAYS

Integral to the vesicular traffic shown in Figure 12.1 are **secretory pathways** by which proteins move from the ER through the Golgi apparatus to **secretory vesicles** and **secretory granules**, which then discharge their contents to the exterior of the cell. The collective roles of the ER and the Golgi apparatus in secretion were first demonstrated by **Georg Palade** and his colleagues, who used autoradiography to trace the fate of radioactively labelled protein through cells in the salivary gland.

In eukaryotic cells, mode of secretion may be of two types: 1. Constitutive secretion; and 2. Regulated secretion.

1. Constitutive Secretion

After budding from the TGN (trans Golgi network), some secretory vesicles move directly to the cell surface where they immediately fuse with the plasma membrane and release their contents. This process, which is continuous and independent of specific extracellular signals, occurs in most eukaryotic cells and is called **constitutive secretion** (Fig. 12.2). Examples include the continuous release of mucus by cells that line human intestine and the secretion of the glycoproteins of the extracellular matrix.

Default pathway model. Constitutive secretion was once assumed to be a **default pathway** for proteins synthesized by ribosomes attached to the rough ER (see **Becker** *et al.*, 2007). According to this model, all proteins destined for transport from the ER to the Golgi apparatus must have a tag that diverts them from constitutive secretion. Unless an amino acid sequence of oligosaccharide chain identifies a protein for retention in, or transport to, a specific organelle, the protein moves through the endomembrane system and is released outside the cell by default. Support for this model came from studies in which the **retrieval tags** on ER-specific proteins were removed and the fates of the proteins were traced. Removal of the KDEL sequence from resident protein of the ER generally led to secretion of the protein. More recent evidence, however, points to a variety of short amino acid sequence, that identify proteins for constitutive secretion by one of several distinct pathways for the process.

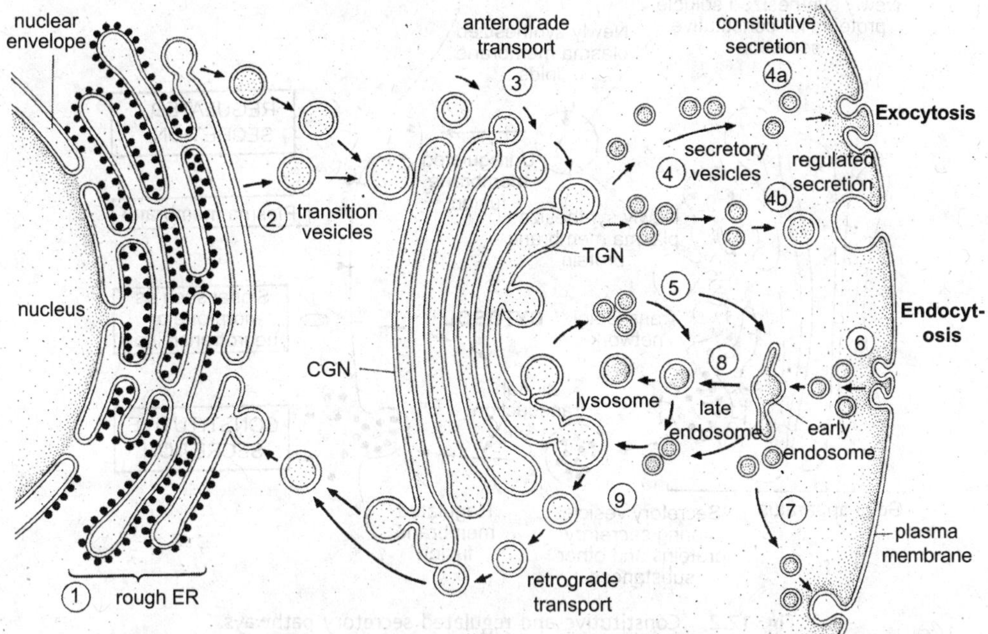

Fig. 12.1. Trafficking through the endomembrane system. Vesicles carry lipids and proteins along several routes from the ER through Golgi apparatus to various destinations, including secretory vesicles, endosomes, and lysosomes. 1. Proteins are synthesized by ribosome attached to the cytosolic surface of the rough ER. Initial glycosylation steps occur within the ER lumen. 2. Transition vesicles carry newly synthesized lipids and glycosylated proteins to the CGN. 3. Lipids and proteins move through the cisternae of the Golgi stack via shuttle-vesicles or as cisternae mature. At the TGN, vesicles bud off to form secretory vesicles (4) or endosomes (5), depending on their protein content. Secretory vesicles move to the plasma membrane, where they release their contents by exocytosis, either constitutively (4a) or in response to an appropriate signal (4b). 6. Protein and other materials are taken into the cell by endocytosis, forming endocytic vesicles that fuse with early endosomes. 7. Cellular components not destined for digestion following endocytosis are recycled to the plasma membrane. 8. Early endosomes containing material for digestion mature to form late endosomes and then lysosomes. 9. Retrograde traffic returns compartment-specific proteins to earlier compartments (after Becker *et al.*, 2007).

2. Regulated Secretion

While vesicles containing constitutively secreted proteins move continuously and directly from the TGN to the plasma membrane, secretory vesicles involved in **regulated secretion** accumulate in the cell and then fuse with the plasma membrane only in response to specific extracellular signals. An important example of regulated secretion is the release of neurotransmitters. Two additional examples of regulated secretion are the release of insulin from the β cells of the pancreatic islets of Langerhans and the release of digestive enzymes from acinar cells of the pancreas.

Regulated secretory vesicles form by budding from the TGN as immature secretory vesicles, which lose their clathrin coats and undergo a maturation process. Maturation involves concentration of the proteins-referred to as **condensation** and frequently also the proteolytic processing of secretory proteins. The mature secretory vesicles then move close to the site of secretion and remain near the plasma membrane, until receiving the signal that triggers release of their contents by exocytosis.

Mature regulated secretory vesicles are usually quite large and contain much more highly concentrated proteins than the constitutive secretory vesicles. Such large dense vesicles are often called **secretory** or **zymogen granules** to distinguish them from other secretory vesicles. For example, in an acinar (secretory) cell of exocrine pancreas of the rat zymogen granules (ZG) are concentrated in the region of the cell between the Golgi stacks from which they arise and the portion of the plasma membrane bordering the lumen into which contents of the granules are eventually discharged.

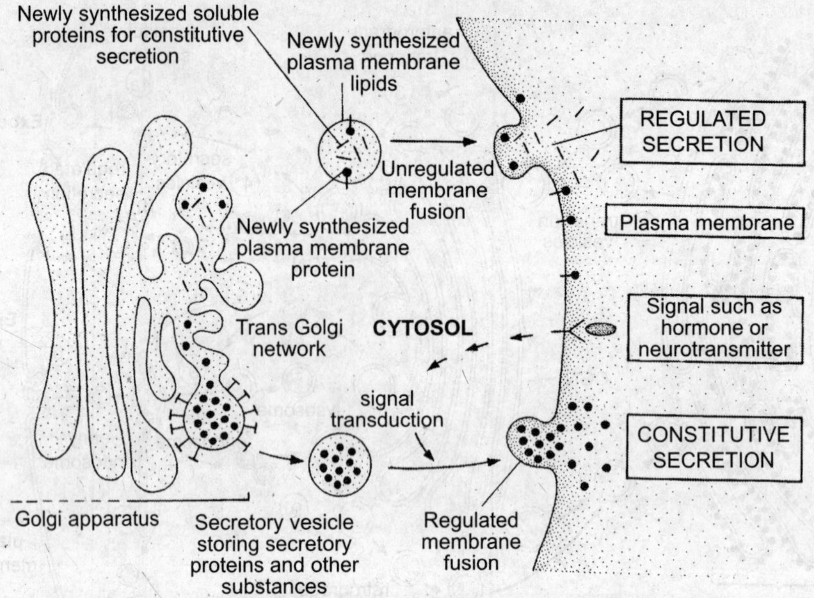

Fig. 12.2. Constitutive and regulated secretory pathways.

The information needed to direct a protein to a regulated secretory vesicle is presumably inherent in the amino acid sequence of the protein, though the precise signals and mechanisms for **sorting proteins** to secretory vesicles are not clear. Current evidence suggests that high concentrations of secretory proteins in secretory granules promote the formation of large **protein aggregates** that exclude non secretory proteins. This could occur in the TGN, where only aggregates would be packaged in vesicles destined for secretory granules, or it could occur in the secretory granules itself. The pH of the TGN and the secretory granule lumens may serve as a trigger favouring aggregation as material leaves the TGN. The soluble proteins that do not become part of an aggregate in the TGN or a secretory granules would be carried by transport vesicle to other locations.

Anterograde and Retrograde Transport

The movement of material from the ER through the Golgi apparatus toward the plasma membrane is called **anterograde transport** (*antero* is derived from a Latin word meaning "front" and *grade* is related to word meaning "step"). Included among the lipids and proteins carried by vesicles are molecules essential only for packaging cargo and directing vesicles to their destinations. For example, every time a secretory granule fuses with the plasma membrane and discharges its contents by exocytosis, a bit of membrane that originated from the ER becomes a part of the plasma membrane. To balance the flow of lipids towards the plasma membrane and to ensure a supply of components for forming new vesicles, the cell recycle lipids and proteins no longer needed during the late stages of anterograde transport (Fig. 21.1). This is accomplished by **retrograde transport** (*retro* is a Latin word meaning "back"), the flow of vesicles from Golgi cisternae back toward the ER (see **Becker** *et al.*, 2007).

12.2 TRANSPORTING MATERIAL ACROSS THE PLASMA MEMBRANE (BULK TRANSPORT BY THE PLASMA MEMBRANE)

Cells routinely import and export large molecules across the plasma membrane. Macromolecules are secreted out from the cell by **exocytosis** and are ingested into the cell from outside through **phagocytosis** and **endocytosis**.

1. Exocytosis.

It is also called **emeiocytosis** and **cell vomiting**. In all eukaryotic cells, **secretory vesicles** are continually carrying new plasma membrane and cellular secretions such as proteins, lipids and carbohydrates (*e.g.*, cellulose) from the Golgi apparatus to the plasma membrane or to cell exterior by the process of exocytosis. The protein to be secreted are synthesized on the rough endoplasmic reticulum (RER). They pass into the lumen of the ER, glycosidated and are transported to the Golgi apparatus by ER-derived **transport vesicles.** In the Golgi apparatus the proteins are modified, concentrated, further glycosidated, sorted and finally packaged into vesicles that pinch off from trans Golgi tubules and migrate to plasma membrane to fuse with it and release the secretion to cell's exterior. In contrast, small molecules to be secreted (*e.g.*, histamine by the mast cells) are actively transported from the cytosol (where they are synthesized on the free ribosomes) into preformed vesicles, where they are complexed to specific macro-molecules (*e.g.*, a network of proteoglycans, in case of histamine), so that, they can be stored at high concentration without generating an excessive osmotic gradient.

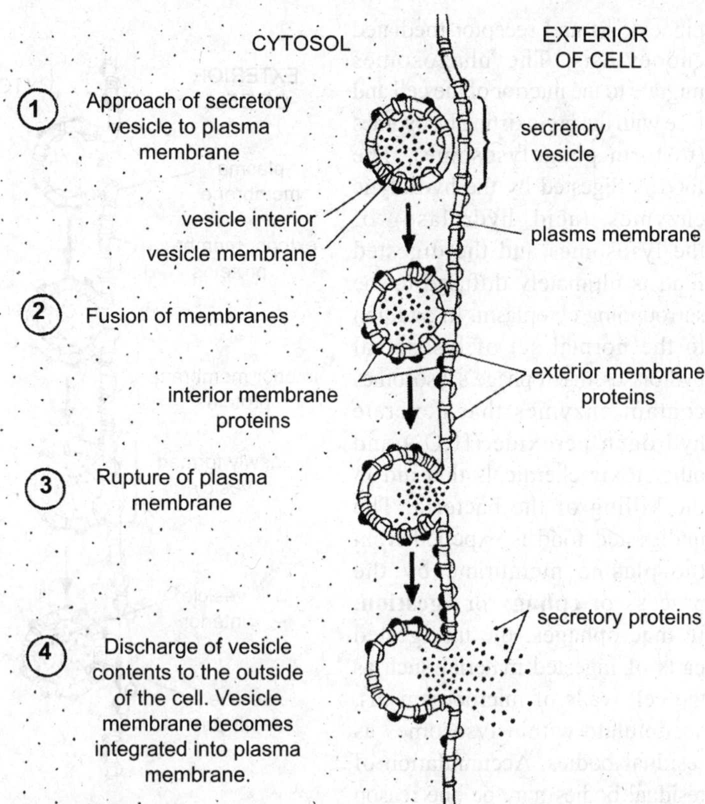

Fig. 12.3. Exocytosis. Release of the contents of a secretory vesicle or granule to the exterior of the cell (after Becker *et al.*, 2007).

During exocytosis the vesicle membrane is incorporated into the plasma membrane. The amount of secretory vesicle membrane that is temporarily added to the plasma membrane can be enormous: in a pancreatic acinar cell discharging digestive enzymes, about 900 μm^2 of vesicle membrane is inserted into the apical plasma membrane (whose area is only 30 μm^3) when the cell is stimulated to secrete.

2. Phagocytosis

Sometimes the large-sized solid food or foreign particles are taken in by the cell through the plasma membrane. The process of ingestion of large-sized solid substances (*e.g.*, bacteria and parts of broken cells) by the cell is known as **phagocytosis** (Gr., *phagein*=to eat, *kytos*=cell or hollow vessel).

Occurrence of phagocytosis. The process of phagocytosis occurs in most protozoans and certain cells of multicellular organisms. In multicellular organisms such as mammals, the phagocytosis occurs very actively in granular leucocytes and in the cells of mesoblastic origin. The cells of the mesoblastic origin are collectively known as the cells of **macrophagic** or **reticuloendothelial system**. They can ingest bacteria, protozoa, cell debris or even colloidal particles by the process of phagocytosis.

Process of phagocytosis. In phagocytosis, first the target particle is bound, to the specific receptors on the cell's surface (process is called **adsorption**), then the plasma membrane expands along the surface of the particle and eventually engulfs it. Vesicle formed by phagocytosis is called **phagosome** and it is typically 1 to 2 μm or larger in diameter, much larger than those formed during pinocytosis and receptor-mediated endocytosis. The phagosomes migrate to the interior of the cell and fuse with the pre-existing lysosomes (to form phagolysosome). The food is digested by the hydrolytic enzymes (acid hydrolase) of the lysosomes and the digested food is ultimately diffused to the surrounding cytoplasm. In addition to the normal set of lysosomal hydrolases, macrophage's lysosomes contain enzymes that generate hydrogen peroxide (H_2O_2) and other toxic chemicals that aid in the killing of the bacteria. The undigested food is expelled from the plasma membrane by the process of **ephagy** or **egestion**. In macrophages, the undigested parts of ingested material such as the cell walls of microorganisms, accumulate within lysosomes as residual bodies. Accumulation of residual bodies may be one reason why macrophages have a very short life time (*i.e.*, less than a few days).

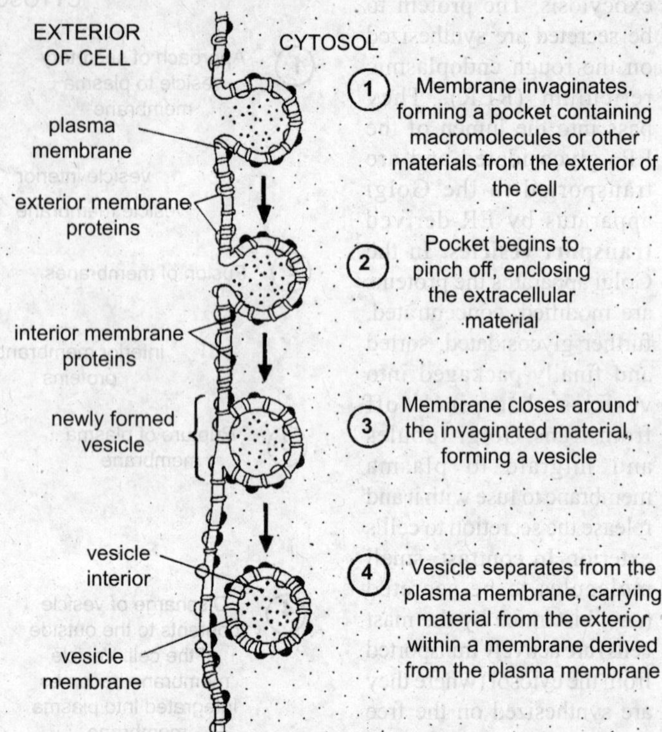

Fig. 12.4. Endocytosis. Uptake of materials from the exterior of the cell (after Becker *et al.*, 2007).

3. Endocytosis

In endocytosis, small regions of the plasma membrane fold inwards or **invaginate** until it has formed new intracellular membrane limited vesicles. In eukaryotes, the following two types of endocytosis can occur: pinocytosis and receptor-mediated endocytosis.

***(i)* Pinocytosis.** Pinocytosis (Gr., *pinein*=to drink; 'cell drinking') is the non-specific uptake of small droplets of extracellular fluid by **endocytic vesicles** or **pinosomes**, having diameter of about 0.1 µm to 0.2 µm. Any material dissolved in the extracellular fluid is internalized in proportion to its concentration in the fluid.

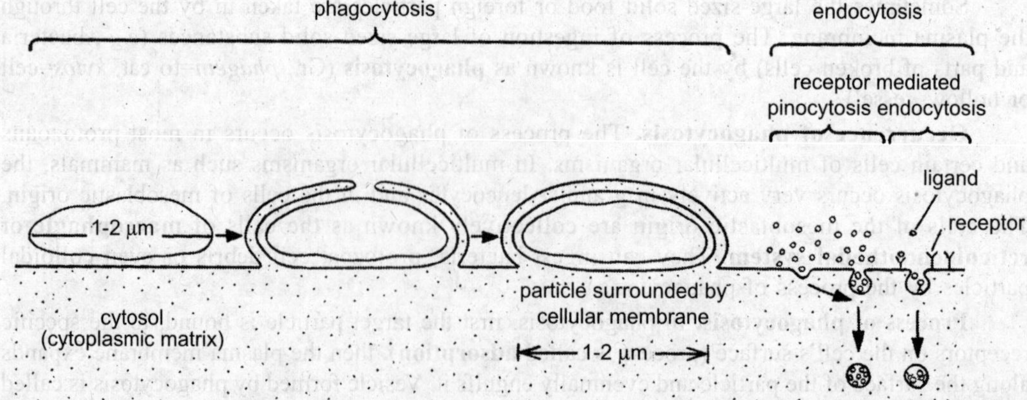

Fig. 12.5. The process of phagocytosis and endocytosis.

The light microscopy has shown that in *Amoeba* tiny **pinocytic channels** are continually being formed at the cell surface by invagination of the plasma membrane. From the inner end of each channel small vacuoles or pinosomes are pinched off, and these move towards the centre of the cell; where they fuse with primary lysosomes, form **food vacuoles**. Ultimately, ingested contents are digested, small breakdown products such as sugars and amino acids diffuse to cytosol.

***(ii)* Receptor-mediated endocytosis.** In this type of endocytosis, a specific receptor on the surface of the plasma membrane "recognizes" an extracellular macromolecule and binds with it. The substance bound with the receptor is called the **ligand.** Examples of ligands may include viruses, small proteins (*e.g.*, insulin, vitellogenin, immunoglobin, transferrin, etc.), vitamin B_{12}, cholesterol containing LDL or low density lipoprotein, oligosaccharide, etc. The region of plasma membrane containing the receptor-ligand complex undergoes endocytosis. The whole process of receptor-mediated endocytosis, includes the following events:

1. Interaction of ligands and cell surface receptors. The macromolecules (ligands) bind to complementary cell-surface receptors. There are more than 25 different types of receptors which are involved in receptor-mediated endocytosis of different types of molecules. Such a receptor is a transmembrane protein which contains two specific binding sites: (1) **ligand-binding site** at the external surface of plasma membrane; and (2) **coated-pit binding site** at the inner or cytosolic face of the plasma membrane.

2. Formation of coated-pits and coated-vesicles. The endocytic cycle begins at specialized regions of the plasma membrane, called **coated-pits**. Coated-pits are depressions of plasma membrane having a coat of bristle-like structure towards their cytosolic side. The ligand-loaded receptors diffuse into these coated-pits. They invaginate into the cell and pinches off to form the **coated-vesicles.**

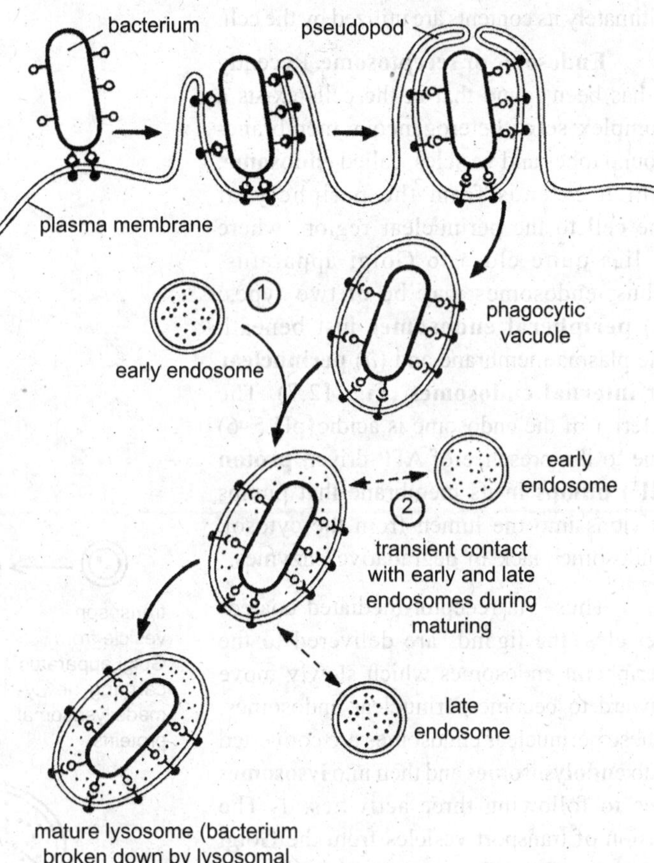

Fig. 12.6. Phagocytosis. Particle or microorganisms bind to receptors on the cell surface, triggering the onset of phagocytosis. In a process involving polymerization of actin, folds of membrane called pseudopods gradually surround the particle. Eventually, the pseudopods meet and engulf the particle, forming a phagocytic vacuole. The vacuole then (1) Fuses with an early endosome or (2) Forms transient connections (indicated by dashed lines) with early and late endosomes and matures into a lysosome, in which digestion of the internalised material occurs. (after Becker *et al.*, 2007).

The coat of coated vesicles is made up of protein, called **clathrin**.

3. Fusion of endocytic vesicle and endosome. Once a coated vesicle is formed, the clathrin and associated proteins dissociate from the vesicle membrane and return to the plasma membrane to form a new coated-pit. The resultant **endocytic vesicle** gets fused with pre-existing endosomes and ultimately its contents are utilized by the cell.

Endosome or receptosome. Recently it has been found that in the cells exists a complex set of heterogeneous membrane-bound tubes and vesicles, called **endosome**, which extends from the periphery of the cell to the perinuclear region, where it lies quite close to Golgi apparatus. Thus, endosomes may be of two types: (*i*) **peripheral endosomes** just beneath the plasma membrane and (*ii*) **perinuclear** or **internal endosomes** (Fig. 12.7). The interior of the endosome is acidic (pH 5–6) due to the presence of ATP-driven **proton (H^+) pumps** in its membrane that pumps H^+ ions into the lumen from the cytosol. Endosomes lack in degradative enzymes.

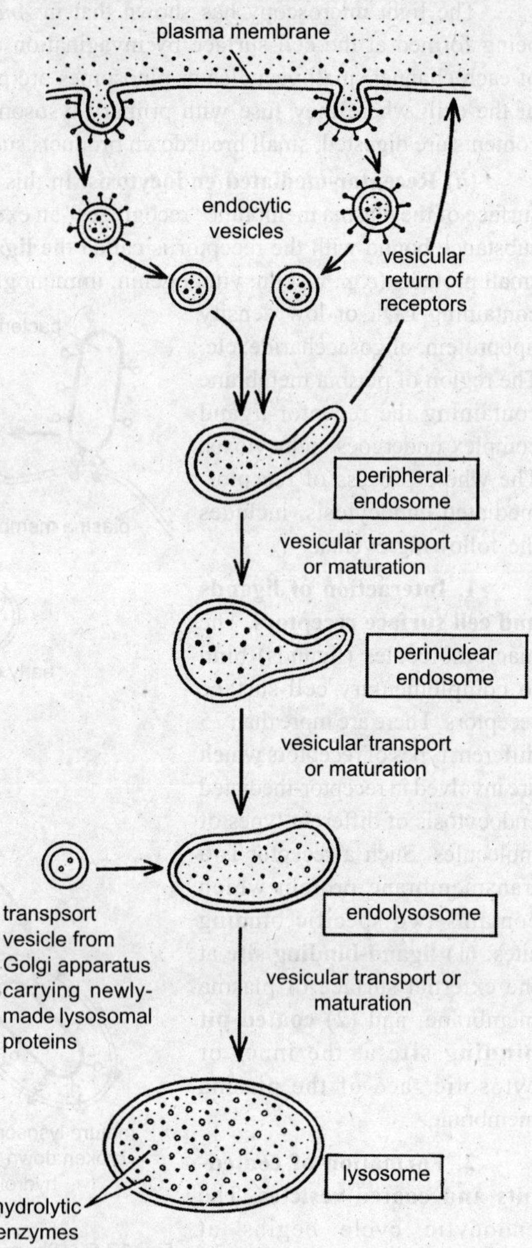

Fig. 12.7. An endocytic pathway: ligands which are endocytosed via coated pits, are delivered to the peripheral endosomes and then move sequentially to perinuclear endosomes, endolysosomes and lysosomes.

Thus, via receptor-mediated coated-vesicles, the ligands are delivered to the peripheral endosomes which slowly move inward to become perinuclear endosomes. These perinuclear endosomes are converted into **endolysosomes** and then into **lysosomes** due to following three activities: 1. The fusion of transport vesicles from the Golgi apparatus. (**Note. Transport vesicles** capture a cargo of molecules. *e.g.*, proteins, from the lumen of one compartment as they pinch off from its membrane and then discharge that cargo into another compartment as they fuse with it. Thus, in such vesicular transport, the transported proteins do not cross any membrane and they are transferred from lumen to lumen). 2. Continuous membrane retrieval, and 3. Increased acidification. The endosomal compartment also acts as the main **sorting station** in the endocytic pathway. The acidic environment of the endosome causes dissociation of ligands from their receptors. Such ligands are destined for destruction in the lysosomes

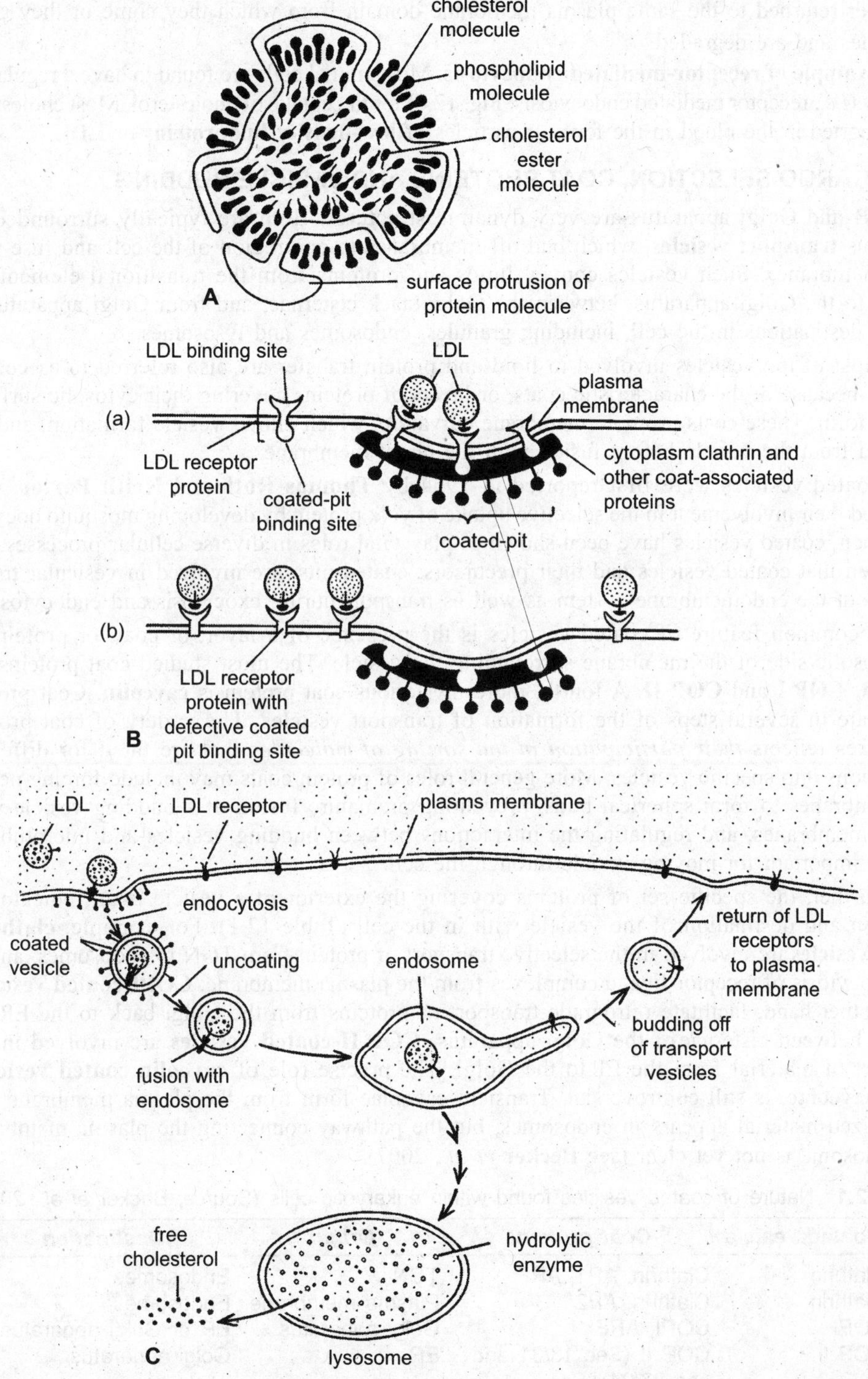

Fig. 12.8. Receptor-mediated endocytosis. A—A LDL particle in cross section; B—LDL receptor proteins binding to a coated-pit in the plasma membrane; C—Mechanism of receptor-mediated endocytosis of LDL particles.

along with the other non membrane-bound contents of the endosome: The receptor-proteins are either returned to the same plasma membrane domain from which they come or they go to lysosomes and are degraded.

Example of receptor-mediated endocytosis. Most animal cells are found to have a regulatory pathway (*i.e.*, receptor mediated endocytosis; Fig. 12.8) for the uptake of cholesterol. Most cholesterol is transported in the blood in the form of particles of **low-density lipoproteins** or **LDL**.

12.3 CARGO SELECTION, COAT PROTEINS AND VESICLE BUDDING

ER and Golgi apparatus are very dynamic structures. Both are typically surrounded by numerous **transport vesicles**, which bud off membranes in one region of the cell and fuse with other membranes. Such vesicles convey lipids and proteins from the transitional elements of the ER to the Golgi apparatus, between the Golgi stack cisternae, and from Golgi apparatus to various destinations in the cell, including granules, endosomes and lysosomes.

Most of the vesicles involved in lipid and protein transfer are also referred to as **coated vesicles** because of the characteristic coats, or layers of proteins covering their cytosolic surfaces as they form. These coats promote membrane curvature, which allows vesicle formation, and are removed from the vesicle before fusion with the target membrane.

Coated vesicles were first reported in 1964 by **Thomas Roth** and **Keith Porter**, who described their involvement in the selective uptake of yolk protein by developing mosquito oocytes. Since then, coated vesicles have been shown to play vital roles in diverse cellular processes. We have seen that coated vesicles and their precursors, coated pits, are involved in vesicular traffic throughout the endomembrane system as well as transport during exocytosis and endocytosis.

A common feature of coated vesicles is the presence of a layer, or coat, or protein on the cytosolic side of the membrane surrounding the vesicle. The most studied coat proteins are **clathrin, COP I** and **COP II**. A fourth, more mysterious coat protein is **caveolin**. Coat protein participate in several steps of the formation of transport vesicles. The variety of coat protein complexes reflects *their participation in the sorting of molecules* that are fated for different destinations into specific vesicles. More general roles of protein coats may include forcing nearly flat membranes to form spherical buds, preventing premature fusion of a budding vesicle with nearby membranes, and regulating the interactions between budding vesicles and microtubules that are important for moving vesicle through the cell.

Further, the specific set of proteins covering the exterior of a vesicle is an indicator of the origin and destination of the vesicle with in the cell (Table 12.1). For example, **clathrin-coated vesicles** are involved in the selective transport of protein from TGN to endosomes, and in the endocytosis of receptor-ligand complexes from the plasma membrane. **COPI-coated vesicles**, on the other hand, facilitate retrograde transport of proteins from the Golgi back to the ER, as well as between cisternae of the Golgi apparatus. **COP II-coated vesicles** are involved in the transport of material from the ER to the Golgi. The precise role of **caveolin-coated vesicles**, called **caveolae**, is still controversial. Transient caveolae form from the plasma membrane and internalized material appears in endosomes, but the pathway connecting the plasma membrane and endosome is not yet clear (see **Becker** *et at.*, 2007).

Table 12.1 Nature of coated vesicles found within eukaryotic cells (Source: Becker *et al.*, 2007).

Coated vesicle	***Coat protein****	***Origin***	***Destination***
1. Clathrin	Clathrin, AP1, ARF	TGN	Endosomes
2. Clathrin	Clathrin, AP2	Plasma membrane	Endosomes
3. COPI	COPI, ARF	Golgi apparatus	ER or Golgi apparatus
4. COP II	COP II (Sec 13/31 and sec 23/24), Sar 1	ER	Golgi apparatus
5. Caveolin	Caveolin	Plasma membrane	?

* *ARF designates ADP ribosylation factor 1; AP1 and AP2 designate different adaptor protein complexes (also called assembly protein complexes).*

1. Clathrin-coated vesicles. Clathrin-coated vesicles are surrounded by coats composed of two multimeric proteins, **clathrin** and **adaptor protein (AP)**. The term *clathrin* comes from *clathratus* the Latin word for a "lattice" as clathrin and AP assemble to form polygonal protein lattices. *Flat clathrin lattices* are composed entirely of hexagons and pentagons.

Clathrin-coated vesicles readily dissociate into soluble clathrin complexes, adaptor protein complexes, and uncoated vesicles. These complexes and vesicles, in turn, can spontaneously reassemble under appropriate conditions. In a slightly acidic solution containing calcium ions, clathrin complexes will even assemble independent of adaptor protein and membrane-bounded vesicles, resulting in empty shells called **clathrin cages.**

***(i)* Components of clathrin lattices.** In 1981 **Ernst Ungewickell** and **Daniel Branton** visualised the basic structural units of clathrin lattices, three-legged structures called **triskelions** (Fig. 12.9A). Each triskelion is a multimeric protein composed of three large polypeptides and three small polypeptides radiating from a central vertex. The large polypeptides, each with a molecular weight of 192,000 are **clathrin heavy chains**; they form the legs of triskelion. Each leg is slightly curved near the middle, or knee, and has a globular domain at its outer tip. The small polypeptides, with a molecular weights in the range of 30,000 – 36,000, are **clathrin light chains**. The light chains are found to be associated with the inner half of each leg.

By combining information collected from electron microscopy and X–ray cyrstallography, researchers have assembled a model for the organization of triskelions into the characteristic hexagons and pentagons of clathrin-coated pits and vesicles (Fig. 12.9B). According to one popular model, one clathrin triskelion is located at each vertex of the polygonal lattice, with the knee of the clathrin heavy chains are located at adjacent vertices, each edge of the clathrin lattice is composed of overlapping legs from three triskelions. This arrangement of triskelions into overlapping networks ensures extensive longitudinal contact between clathrin polypeptides. Such contact may impart the mechanical strength needed when a coated vesicle forms from a membrane. The knees, however, do not interact with other polypeptides. This may provide flexibility for forming both hexagons and pentagons and also may accomodate vesicles of different sizes.

***(ii)* Adaptor protein.** The second major component of clathrin coats-the **adaptor protein (AP)**–was originally identified simply by its ability to promote the assembly of clathrin coats around vesicles; thus, AP is also called **assembly proteins**. Eukaryotic cells contain at least-four types of AP complexes, each composed of four polypeptides—two adaptin subunits, one medium chain, and one small chain. The four polypeptides, which are highly different in each type of AP complex, bind to different transmembrane receptor proteins and *impart specificity during vesicle budding and targeting*. In addition to ensuring that appropriate macromolecules will be concentrated in coated pits, AP complexes also mediate the attachment of clathrin to the plasma membrane. AP complexes are found to be sites for regulation of clathrin assembly and disassembly. For example, the ability of AP complexes to bind to clathrin is affected by pH, phosphorylation and dephosphorylation.

***(iii)* The assembly of clathrin coats drives the formation of vesicles from the plasma membrane and TGN.** The binding of AP complexes to the plasma membrane and the concentration of receptors or receptor-ligand complexes in coated pits require ATP and GTP—though perhaps only for regulation of the process. However, the assembly of clathrin coat around budding vesicles can occur without additional expenditure of ATP and GTP. An exception is the assembly of clathrin coats around vesicles forming from the TGN, which depends on the hydrolysis of nucleotide triphosphates.

The accumulation of clathrin triskelions and the assembly of a clathrin coat on the cytosolic side of a membrane appear to provide part of the **driving force** for formation of a vesicle at the site. In the case of receptor-mediated endocytosis, the assembly of the clathrin coat on the inner side of the plasma membrane causes the membrane to fold inward. Initially, all clathrin units are hexagonal and form a planar, two-dimensional structure. As more clathrin triskelions are incorporated into the growing lattice, a combination of hexagonal and pentagonal units allows the new clathrin coat to curve around the budding vesicle.

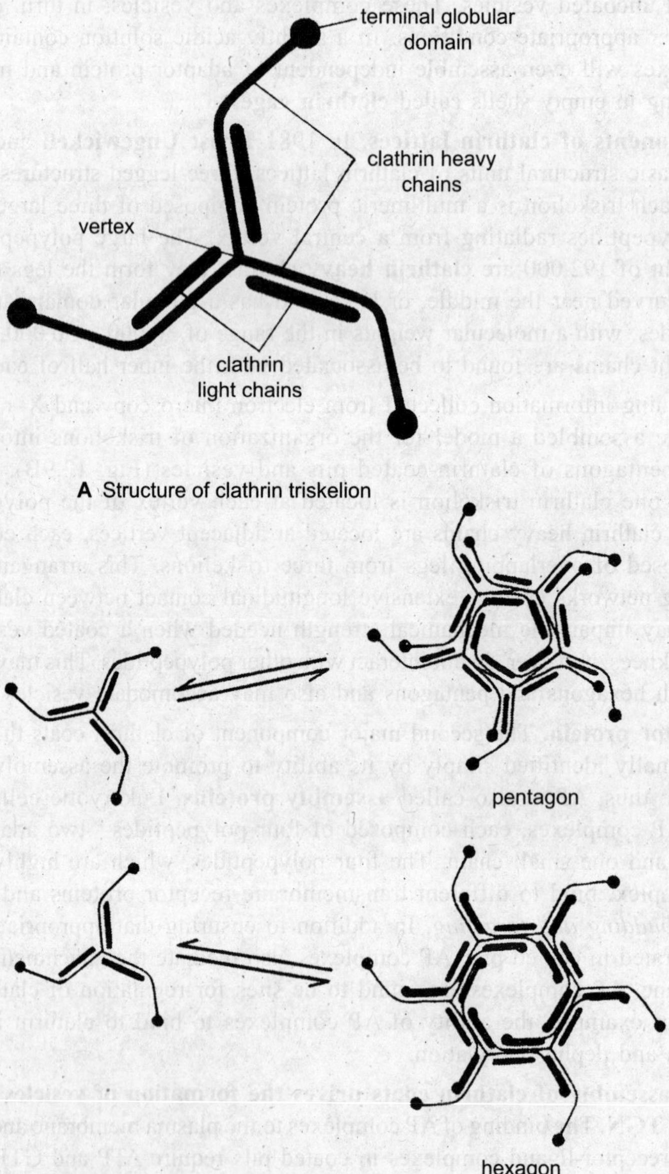

Fig. 12.9. Clathrin triskelion. A—Each triskelion consists of three clathrin heavy chains radiating from a central vertex, with a terminal globular domain at the tip of each triskelion leg and a clathrin light chain bound to the inner half of each leg. B—Under appropriate conditions, clathrin triskelions assemble into the pentagonal and hexagonal structures characteristic of coated pits and vesicles (after Becker *et al.*, 2007).

***(iv)* Sealing of the coated-vesicles.** As clathrin accumulates around the budding vesicle, at least one more protein-**dynamin**–participates in the process. Dynamin is a cytosolic GTPase (enzyme) required for coated pit *constriction and closing of the budding vesicle.* This essential protein was first identified in *Drosophila*. Flies expressing a temperature-sensitive form of dynamin were instantly paralyzed after a temperature shift. Further investigation revealed an accumulation of coated pits in the presynaptic membranes of neuromuscular junctions in affected flies. Binding of GTP probably allows dynamin to form helical rings around the neck of the coated pit. As GTP is hydrolyzed, the dynamin rings tighten and separate the fully sealed endocytic vesicle from the plasma membrane.

***(v)* Uncoating of coated-vesicles.** Uncoating is done in a regulated manner because, in most cases, the clathrin coat remains intact as long as the membrane is part of a coated pit or budding vesicle but dissociate rapidly once the vesicle is fully formed. Like assembly, dissociation of the clathrin coat is an energy-consuming process, accompanied by the hydrolysis of about three ATP molecules per triskelion. At least one protein, an **uncoating ATPase** is essential for this process, though the uncoating ATPase releases only the clathrin triskelion from the APs; the factors responsible for releasing APs from the membrane have not yet been identified.

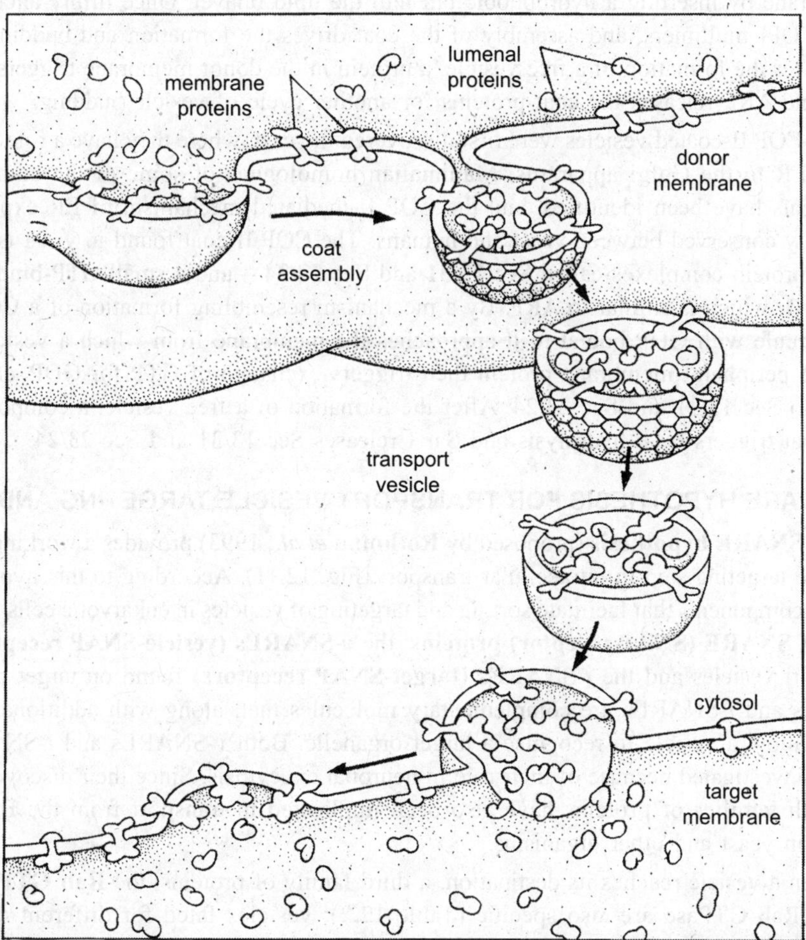

Fig. 12.10. Formation and fusion of a transport vesicle. Membrane and lumenal secretory proteins are collected into selected regions of a donor membrane where the formation of a cytosolic coat results in the budding off of a transport vesicle. During transport the coat is disassembled, and the transport vesicles docks and fuses with the target membrane (after Cooper and Hausman, 2007).

2. Functions of COP I-coated and COP II-coated vesicles. COPI-coated vesicles (COP indicates coat protein) have been found in all eukaryotic cells examined, including mammalian, insect and yeast cells. Such vesicles are surrounded by coats composed of COPI and **ADP ribosylation factor (ARF)**. Viewed by electron microscopy, COPI-coated vesicles do not display polyhedral lattices like those surrounding clathrin-coated vesicles. Instead, they have dense "fuzzy" coats. The major component the coat, COPI, is a protein multimer composed of seven subunits.

COPI-coated vesicles are likely involved in transport from the Golgi apparatus back to the ER as well as between Golgi apparatus cisternae, though their precise role is controversial. Early evidence strongly suggested that COPI-coated vesicle, do not bud from the ER, lending support to the view that COPI-coated vesicles are involved in retrograde transport.

Assembly of a COPI coat is mediated by ARF, which is a small GTP-binding protein. In the cytosol, ARF occurs as part of an ARF-GTP complex. When ARF encounters a specific *guanine nucleotide exchange factor* associated with the membrane from which a new coated vesicle is about to form, however, the GDP is exchanged for GTP. ARF is then able to bind to the membrane by inserting a hydrophobic tail into the lipid bilayer. Once firmly anchored, ARF binds to COPI multimers, and assembly of the coat drives the formation and budding of a new vesicle. After the formation of a free vesicle, a protein in the donor membrane triggers hydrolysis of GTP, and ARF releases the coat proteins for another cycle of vesicle budding.

The COP II-coated vesicles were first discovered in yeast, where they have a role in transport from the ER to the Golgi apparatus. Mammalian homologues of some of the components of COP II coats have been identified, and the COP II-mediated mechanism of ER export appears to be highly conserved between yeast and humans. The COP II coat found in yeast is assembled from two protein complexes-called **Sec 13/31** and **Sec 23/24** – and a small GTP-binding protein called **Sar I**, which is similar to ARF. By a mechanism resembling formation of a COPI coat, a Sar I molecule with GDP bound to it approaches the membrane from which a vesicle is about to form. A peripheral membrane protein then triggers exchange of GTP for GDP, enabling Sar I to bind to Sec 13/31 and Sec 23/24. After the formation of a free vesicle, a component of the COP II coat triggers GTP hydrolysis and Sar I releases Sec 13/31 and Sec 23/24.

12.4 SNARE HYPOTHESIS FOR TRANSPORT VESICLE TARGETING AND FUSION

The **SNARE hypothesis** (proposed by **Rothman** *et al.*, 1993) provides a working model for sorting and targeting step in intracellular transport (Fig. 12.11). According to this hypothesis, the molecular components that facilitate sorting and targeting of vesicles in eukaryotic cells include two families of **SNARE (SNAP receptor) proteins**: the ***v*-SNAREs (vesicle-SNAP receptors)** found on transport vesicles and the ***t*-SNAREs (target-SNAP receptors)** found on target membranes. *V*-SNAREs and *t*-SNAREs are complementary molecules that, along with additional **tethering proteins**, allow a vesicle to recognise a target organelle. Both *v*-SNAREs and *t*-SNAREs were originally investigated because of their role in neuronal exocytosis. Since their discovery in brain tissue, both families of proteins have also been implicated in transport from the ER to Golgi apparatus in yeast and other organisms.

When a vesicle reaches its destination, a third family of proteins, the **Rab GTPases**, come into play. Rab GTPase are also specific (Table 12.2); vesicles fated for different destinations have distinct members of the Rab family associated with them. As shown in Fig. 12.11, the affinity of complementary *v*-SNAREs and *t*-SNAREs for one another forms a four-helix bundle. This ensures that when they collide they will remain in contact long enough for a Rab protein associated with the vesicle to lock the complementary *t*- and *v*-SNARE, together.

Following the vesicle fusion, another set of proteins, including **N-ethylmaleimide-sensitive factor (NSF)** and a group of **SNAPs (soluble NSF attachment proteins)**, mediates release of the *v*- and *t*-SNAREs of the donor and target membranes possibly accompanied by the hydrolysis of

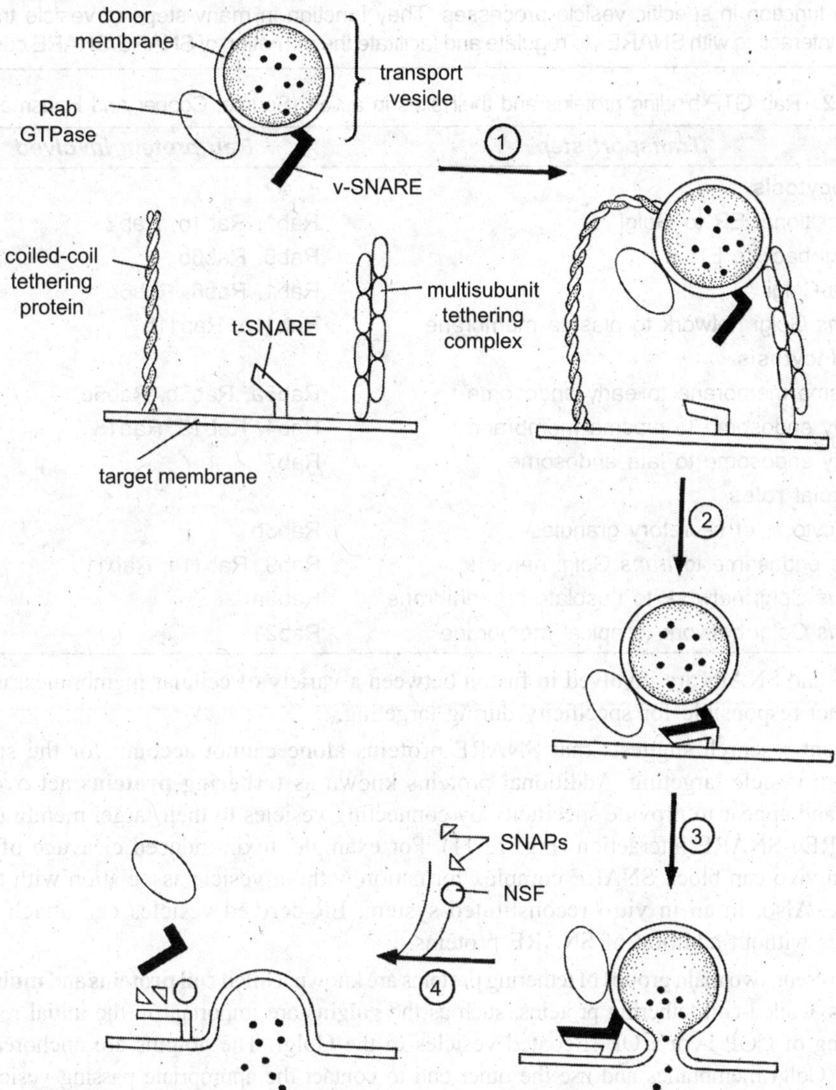

Fig. 12.11. The SNARE hypothesis for transport vesicle targeting and fusion. The basic molecular components that mediate sorting and targeting of vesicles in eukaryotic cells include tethering proteins, *v*-SNAREs on the transport vesicles, t-SNAREs on target membranes, Rab GTPase, NSF, and several SNAPs. 1. The proper vesicle is recognised and bound by particular membrane-anchored tethering proteins: a coiled coil protein and/or a multisubunit tethering complex. 2. A Rab GTPase bound to the incoming vesicle stimulates formation of a stable four-helix bundle composed of one *v*-SNARE helix and three *t*-SNARE helices (the tethering proteins are no longer shown). 3. Fusion of the vesicle membrane with the target membrane is promoted by the *v*-SNARE/*t*-SNARE interaction. 4. Binding of NSF and SNAPs promotes dissociation of SNARE complexes. The exact timing of GTP or ATP hydrolysis is still unclear, but most likely occurs after vesicle fusion (after Becker *et al.*, 2007).

Box 12.2 Rab Proteins

Rab proteins, like the ARF family, participate in many of the vesicle budding and fusion reactions during vesicular transport. More than 60 different mammalian Rab proteins have been identified and shown to function in specific vesicle processes. They function in many steps of vesicle trafficking, including interacting with SNAREs to regulate and facilitate the formation of SNARE/SNARE complexes.

Table 12.2 Rab GTP-binding proteins and their sites in action (Source: Cooper and Hausman, 2007).

Transport step	*Rab protein involved*
I. Exocytosis	
1. Transitional ER to Golgi	Rab1, Rab1b, Rab2
2. Golgi back to ER	Rab6, Rab6b
3. Intra-Golgi	Rab1, Rab6, Rab6b
4. *Trans* Golgi network to plasma membrane	Rab11a, Rab11b
II. Endocytosis	
1. Plasma membrane to early endosome	Rab5a, Rab5b, Rab5c
2. Early endosome to plasma membrane	Rab4, Rab15, Rab18
3. Early endosome to late endosome	Rab7
III. Special roles	
1. Exocytosis of secretory granules	Rab8b
2. Late endosome to *trans* Golgi network	Rab9, Rab11a, Rab11b
3. *Trans* Golgi network to basolateral membrane	Rab8a
4. *Trans* Golgi network to apical membrane	Rab21

ATP. NSF and SNAPs are involved in fusion between a variety of cellular membranes, indicating they are not responsible for specificity during targeting.

Recent research suggests that SNARE proteins alone cannot account for the specificity observed in vesicle targeting. Additional proteins known as **tethering proteins** act over longer distances and appear to provide specificity by connecting vesicles to their target membranes prior to *v*-SNARE/*t*-SNARE interaction (Fig. 12.11). For example, toxin-induced cleavage of SNARE proteins in vivo can block SNARE complex formation without vesicle association with the target membrane. Also, in an in vitro reconstituted system, ER-derived vesicles can attach to Golgi membranes without addition of SNARE proteins.

At present, two main groups of tethering proteins are known: **coiled coil proteins** and **multisubunit complexes**. Coiled-coil tethering proteins, such as the **golgins** are important in the initial recognition and binding of COP I- or COP II-coated vesicles to the Golgi. The golgins are anchored by one end to the Golgi membranes and use the other end to contact the appropriate passing vesicle. These proteins are also important in connecting Golgi cisternae to each other. Antibodies to certain golgian block the action of the golgins and disrupt the structure of the Golgi medial cisternae.

A second class of tethering proteins consists of several families of multisubunit protein complexes containing four to eight or more individual polypeptides. For example, the **exocyst** complex of yeast and mammals is important for protein secretion, binding both to the plasma membrane and to vesicles from the TGN whose contents are destined for export. Other types of multisubunit tethering complexes such as **COG (conserved oligomeric Golgi)** complex, the **GARP (Golgi-associated retrograde protein)** complex and the **TRAPP (transport protein particle)** complex are implicated in the initial recognition and specificity of vesicle-target membrane interaction. Most of these proteins in these complexes are will conserved among organisms, as different as yeast and humans (see **Becker** *et al.*, 2007).

12.5 TRANSPORT OF NEUROTRANSMITTERS ACROSS NERVE SYNAPSES

Nerve cells (neurons) communicate with one another and with glands and muscles at **synapses**, which are of following two types:

1. Electrical synapse. Here, an axon of one neuron, called the **presynaptic neuron**, is connected to the dendrite (s) of another cell, the **postsynaptic neuron**, by gap junctions. As ions move back and forth between the two cells, the depolarisation in one cell spreads passively to the connected cell. Electrical synapses provide for transmission with virtually no delay and occur in places in the nervous system where speed of transmission is critical.

2. Chemical synapse. In this case, the presynaptic and postsynaptic neurons are close to each other but not directly connected (Fig. 12.12). Typically, the presynaptic membrane is separated from the postsynaptic membrane by a gap of about 20–25 nm, known as the **synaptic cleft**. A nerve signal arriving at the terminals of the presynaptic neuron cannot bridge the synaptic cleft as an electrical

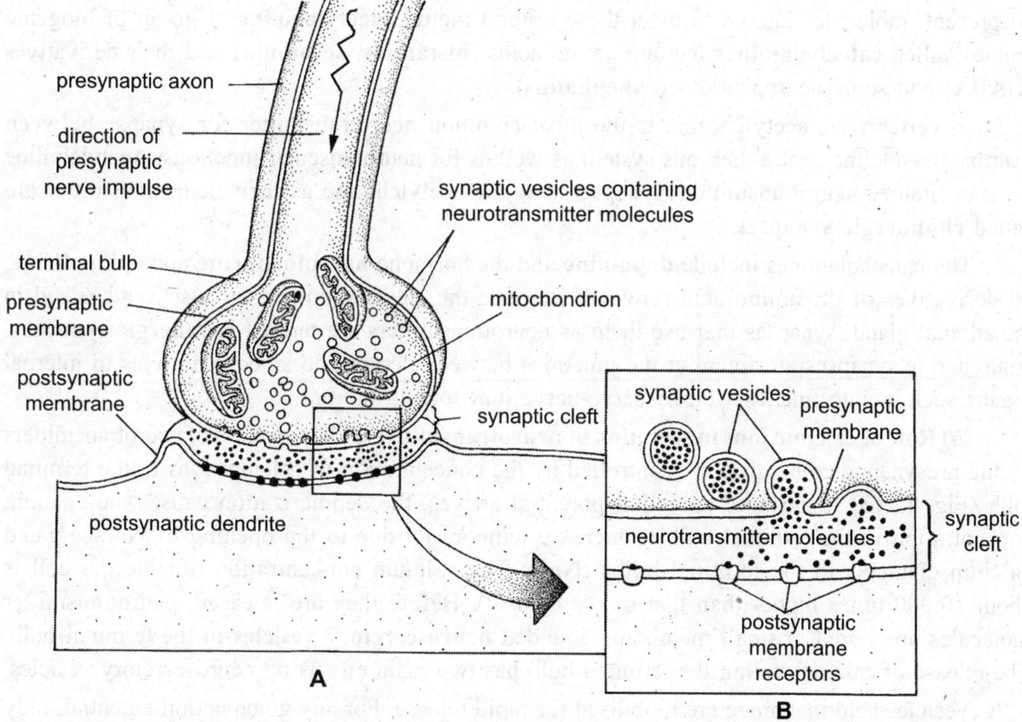

Fig. 12.12. Structure of a chemical synapse. A—When a nerve impulse from the presynaptic axon arrives at the synapse, it causes synaptic vesicle containing neurotransmitters in the terminal bulb to fuse with the presynaptic membrane, releasing their contents into the synaptic cleft. B—Neurotransmitter molecules diffuse across the cleft from the presynaptic (axonal) membrane to the postsynaptic (dendritic) membrane, where they bind to specific membrane receptors and change the polarisation of the membrane, either exciting or inhibiting the postsynaptic cell (after Becker *et al.*, 2007).

impulse. For synaptic transmission to take place, the electrical signal must be converted at the presynaptic neuron to a chemical signal carried by a neurotransmitter. Neurotransmitter molecules are stored in the **terminal bulbs** (also called **synaptic knobs**) of the presynaptic neuron. An action potential arriving at the terminal bulb causes the neurotransmitter to be secreted into and diffuse across the synaptic cleft. The neurotransmitter molecules then bind to specific proteins (called **receptors**) embedded within the plasma membrane of the postsynaptic neuron and are converted back to electrical signals, setting in motion a sequence of events that either stimulates or inhibits the production of an action potential in the postsynaptic neuron, depending on the kind of synapse.

***(i)* Neurotransmitters.** A **neurotransmitter** is a small molecule whose function is to bind to a cell surface protein called a **receptor** within the membrane of a postsynaptic neuron. When a neurotransmitter molecule binds to its specific receptor, the properties of receptor are altered and the postsynaptic neurons responds accordingly. An **excitatory neurotransmitter** causes depolarization of the postsynaptic neuron, whereas an **inhibitory neurotransmitter** typically causes the postsynaptic cell to hyperpolarise.

A neurotransmitter should have the following three properties: 1. It must elicit the appropriate response when introduced into the synaptic cleft; 2. it must occur naturally in the presynaptic neuron; and 3. it must be released at the right time when the presynaptic neuron is stimulated. At present, molecules known to meet these criteria include **acetylcholine**; a group of biogenic amines called **catecholamines**; certain amino acids (**histamine**, **serotonin**) and their derivatives (**GABA**) and some neuropeptides (**eukephalins**).

In vertebrates, acetylcholine is the most common neurotransmitter for synapse between neurons outside the central nervous system as well as for neuromuscular junctions. Acetylcholine is an excitatory neurotransmitter. Synapses that use acetylcholine as their neurotransmitter are called **cholinergic synapses**.

The catecholamines include **dopamine** and the hormone **norepinephrine** and **epinephrine**, all derivatives of the amino acid tyrosine. Because the catecholamines are also synthesized in the adrenal gland, synapses that use them as neurotransmitters are termed **adrenergic synapses**. Andrenergic synapses are found at the junctions between nerves and smooth muscles in internal organs such as intestine, as well as nerve-nerve junctions in the brain.

***(ii)* Role of calcium ions in secretion of neurotransmitters.** The secretion of neurotransmitters by the presynaptic cell is directly controlled by the concentration of calcium ions in the terminal bulb (Fig. 12.13). Each time an action potential arrives, the depolarization causes the calcium concentration in the terminal bulb to increase temporarily due to the opening of voltage-gated calcium channels in the terminal bulbs (**Note.** The calcium concentration outside the cell is about 10,000 times higher than that of the cytosol). Before they are released, neurotransmitter molecules are stored in small membrane-bounded **neurosecretory vesicles** in the terminal bulb. The release of calcium within the terminal bulb has two main effects on neurosecretory vesicles.

1. Vesicles held in storage are mobilised for rapid release. For any given action potential, only a tiny fraction of the total number of vesicles stored in the terminal release their contents. *Neurons hold vesicles in reserve by linking them to the cytoskeleton so that they cannot move close to the synaptic membrane for secretion.*
2. Calcium ions induced sensitive effect on the neurotransmitter release is the rapid docking and fusion of neurosecretory vesicles with the plasma membrane in the terminal bulb region. During this process, the membrane of a vesicle moves into close contact with the plasma membrane of the axon terminal and then fuses with it to release the content of the vesicle.

***(iii)* Mechanism of docking and exocytosis.** For the neurotransmitter to act on a postsynaptic cell, it must be secreted by the process of **exocytosis**. The key event in secretion is the fusion of neurosecretory vesicles with the plasma membrane, which discharges the vesicle contents into the synaptic cleft. The process requires ATP and proceeds through several steps, one of which is calcium dependent.

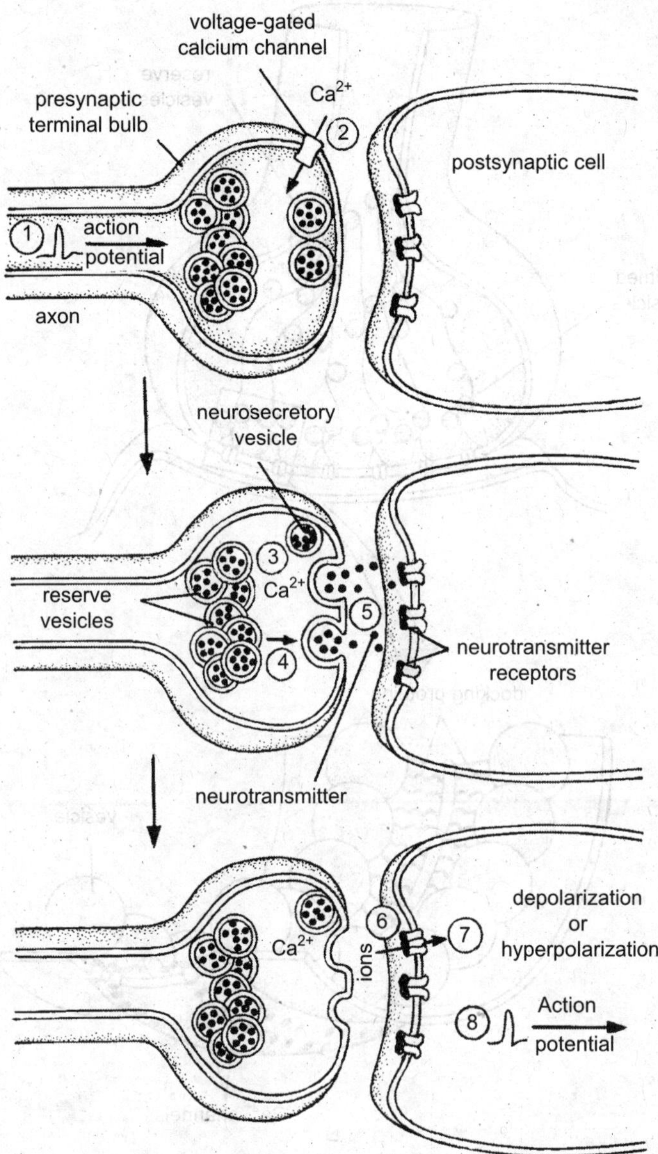

Fig. 12.13. The transmission of a signal across a synapse. 1. An action potential arrives at the presynaptic terminal bulb, resulting in a transient depolarization. 2. Depolarization opens voltage-gated calcium channels, allowing calcium ions to rush into the terminal. 3. This increase in the calcium concentration in the terminal bulb induces the secretion of a fraction of the neurosecretory vesicles. 4. Calcium also causes reserve vesicles to be released from the actin cytoskeleton so that they are ready for secretion. 5. Secreted neurotransmitter molecules diffuse across the synaptic clefts to receptors on the postsynaptic cell. 6. Binding of neurotransmitter to the receptor alters the receptor properties. 7. For receptors that are ligand-gated channels, the channel opening leads to either depolarization or hyperpolarization of the postsynaptic plasma membrane. 8. If depolarization results, a sufficient amount of excitatory neurotransmitter will result in an action potential in the postsynaptic cell (after Becker *et al.*, 2007).

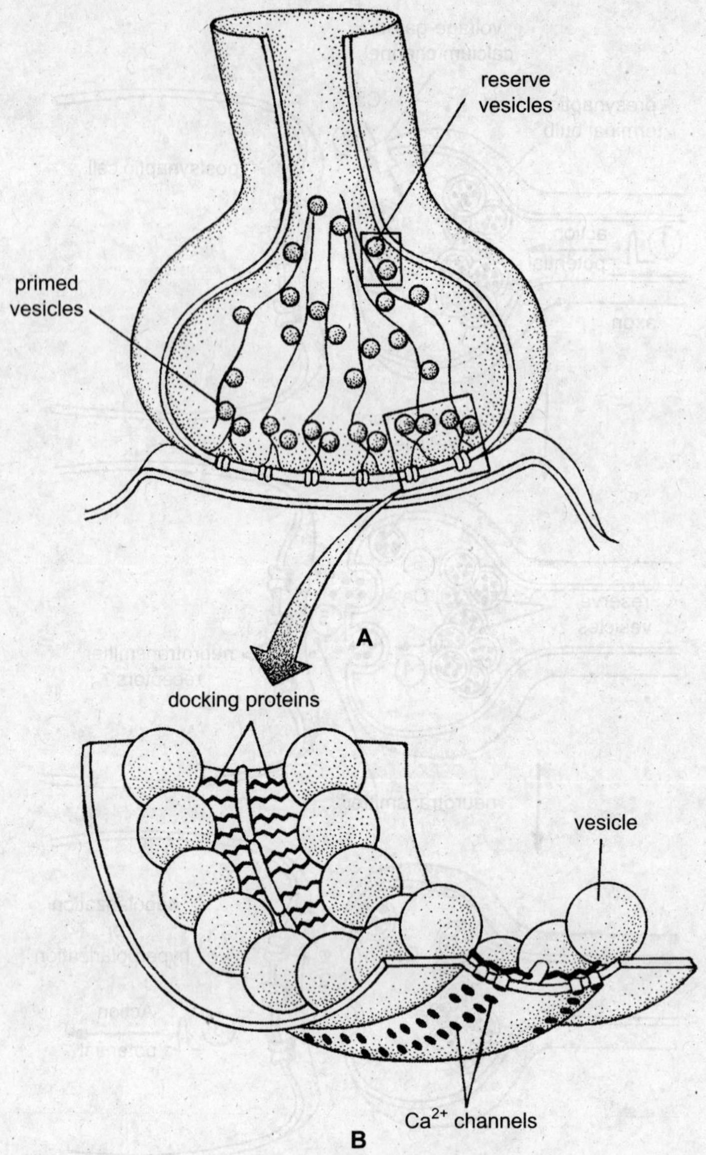

Fig. 12.14. Docking of synaptic vesicles with the plasma membrane of the presynaptic neuron. A—In response to local elevation of calcium in the presynaptic neuron, synaptic vesicles are released from the cytoskeleton, allowing them to fuse with the plasma membrane of the terminal bulb within the active zone. Vesicles bound to the cytoskeleton via synapsin form a reserve pool of vesicles. Once released from the cytoskeleton, vesicles move near the plasma membrane and become "primed" (*i.e.*, charged or prepared). Some vesicles become tightly associated with the plasma membrane, or "docked". When nearby calcium channels open, such docked vesicles fuse with the plasma membrane, releasing their contents. B—A drawing based on an actual TEM reconstruction of the active zone of a motor neuron from a frog. Docked secretory vesicles are arranged in rows and connected by a complex of proteins associated with docking. Calcium channels lie immediately beneath the vesicles (after Becker *et al.*, 2007).

When the action potential arrives at an axon terminal and triggers the opening of voltage-gated calcium channels, calcium enters the terminal bulb. For neurotransmitter vesicles to become available for secretion, they must become disengaged from the cytoskeleton. This process is stimulated by calcium release, and the addition of phosphate groups to **synapsin**, an integral membrane protein found in the membrane of neurosecretory vesicles. When phosphorylated synapsin no longer binds to the cytoskeleton, freeing the vesicles so that they can proceed to the next step in neurotransmitter release.

Once synaptic vesicles are released from the cytoskeleton, they are not quite ready to fuse with the plasma membrane. Such vesicles cluster near, but are not in contact with, the plasma membrane (Fig. 12.14). In order to be ready to fuse with the plasma membrane, such released vesicles must be "primed". Once such priming has occurred, the freed vesicles are now capable of docking and fusing with the plasma membrane of the presynaptic neuron in response to elevated calcium.

Docking takes place at a specialized site, called **active zone**, within the membrane of the presynaptic neuron. The active zone is a highly organised structure. Synaptic vesicle, the proteins to which the vesicles attach and the calcium channels that elicit their release are all clustered in ordered rows poised for secretion (Fig. 12.14B). The close proximity of calcium channels to the vesicles helps to explain the extremely rapid fusion of the primed, releasable population of vesicles with the presynaptic neuron's plasma membrane when that neuron is stimulated.

The processes of docking and fusion of neurosecretory vesicle are mediated by docking proteins within the vesicle and within the plasma membrane of the active zone. Within the vesicle membrane exist **synaptotagmin**, **synaptobrevin** and other proteins which dock with proteins, such as **syntaxin** at the plasma membrane of the active zone (**Koh** and **Bellen**, 2003). Once the initial docking occurs, additional protein from the cytosol stabilise the attachment, forming a multiprotein docking complex. Fusion depends on the successful docking of vesicles at the active zone and the formation of a complete docking complex.

Human diseases due to defects in docking events. Two common and prospective fatal illnesses of humans result from interference with the docking events. Both **tetanus** and **botulism** result from interference by **neurotoxins** with vesicle docking and release. **Tetanus toxin** prevents the release of neurotransmitter from inhibitory neurons in the spinal cord, resulting in uncontrolled muscle contraction (which is why tetanus has been familiarly referred to as "lockjaw"). **Botulinum toxin** prevents release of neurotransmitter from motor neurons, resulting in muscle weakness and paralysis (see **Becker** *et al.*, 2007).

***(iv)* The acetylcholine receptor.** Acetylcholine binds to a ligand-gated sodium channel called **acetylcholine receptor**. When two molecules of acetylcholine bind with receptor, the channel opens and lets sodium ions rush into the postsynaptic neuron causing a depolarization (Fig. 12.15).

The purified acetylcholine receptor has molecular weight of about 300,000 and consists of four kinds of subunits-α, β, γ and δ-each containing about 500 amino acids (Fig. 12.15). Acetylcholine receptors play a key role in the transmission of nerve impulses to muscle. In some cases, human patients develop a autoimmune response to their own acetylcholine receptors (in other words, their immune system attacks their own receptors as if they were foreign invaders). When this happens, the patient can develop a condition known as **myasthenia gravis**, in which degenerative muscle weakness occurs.

Whether excitatory or inhibitory, once the neurotransmitter has been secreted, it must be rapidly removed from the synaptic cleft. If it were not, stimulation or inhibition of a postsynaptic neuron would be abnormally prolonged even in the absence of further signals from presynaptic neurons. In fact, the persistence of an excitatory neurotransmitter such as acetylcholine renders muscles unable to relax, ultimately leading to death.

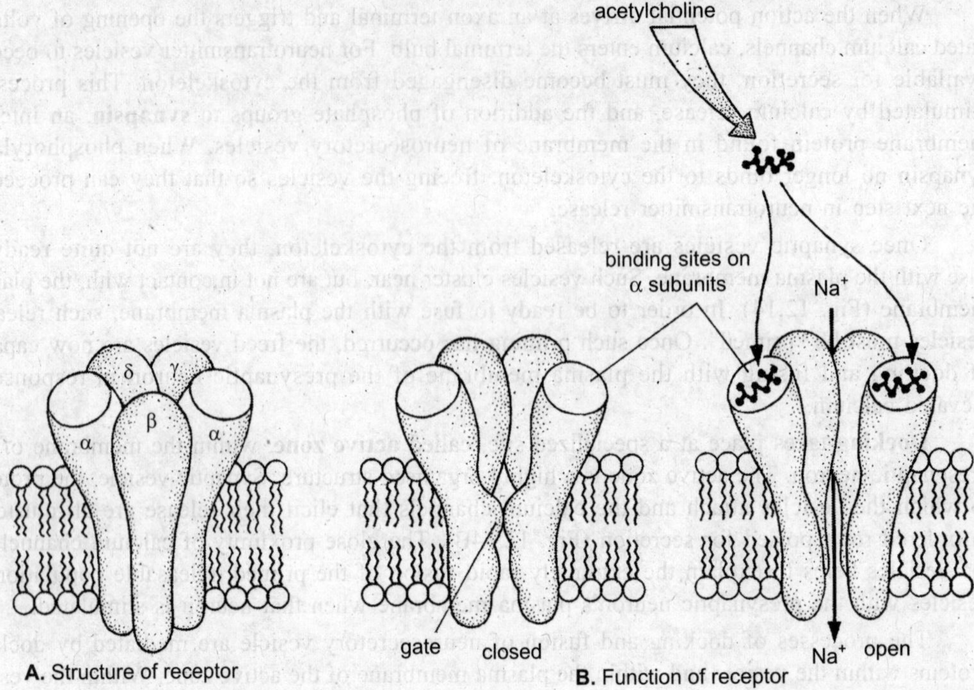

Fig. 12.15 The acetylcholine receptor which is a primary excitatory receptor of central nervous system. A—This receptor contains five subunits, including two α subunits with binding sites for acetylcholine and one each of β, γ and δ. The subunits aggregate in the lipid bilayer in such a way that the transmembrane portion form a channel. B—The channel (shown here with β subunit removed) is normally closed, but when acetylcholine binds to the two sites on the α subunits, the subunits are altered in such a way that the channel opens to allow sodium ions to cross (after Becker *et al.*, 2007).

Neurotransmitters are removed from the synaptic cleft by two specific mechanisms: degradation into inactive molecules or reuptake into the presynaptic terminals (*e.g.*, Serotonin). An example of the first mechanism is found in the case of the acetylcholine into acetic acid (or acetate ions) and choline, neither of which stimulatory to the acetylcholine receptor.

QUESTIONS

Long Answer Questions

1. What is bulk transport? Describe the process of endocytosis and exocytosis in brief.
2. What is vesicular transport? Give an outline of biosynthetic (secretory) pathway.

Short Answer Questions

1. What is the fate of receptor proteins that take part in receptor mediated endocytosis?
2. Write short notes on the following:
   (*i*) Phagocytosis
   (*ii*) Receptor-mediated-endocytosis
   (*iii*) Coated vesicles
   (*iv*) Exocytosis
   (*v*) Pinocytosis

3. Describe the mechanism of transport of neurotransmitters in nerve cells.
4. Differentiate between the following terms:
   (i) Phagocytosis and pinocytosis.

Very Short Answer Questions

1. What is early endosome?
2. What is late endosome?
3. Define secretory vesicle.
4. What is secretory granule?
5. Define secretory pathway.
6. What is exocytosis?
7. Define endocytosis.
8. What is endocytic vesicle?
9. What is SNARE hypothesis?
10. Define transition vesicles.
11. What is transport vesicle?
12. What is coated vesicle?
13. Define phagocytosis.
14. What is receptor-mediated endocytosis?
15. Define synapse.
16. What is neurosecretory vesicle?
17. Write the names of two types of cellular secretions?
18. Name of the cells in mammals that are engaged in phagocytosis.

Fill in the Blanks

1. TGN stands for ...., ...
2. In .... secretion, secretory vesicles accumulate in the cell and then fuse with the plasma membrane only in response to specific extracellular signals.
3. The enzyme mannosidase II, which is involved in trimming the .... residues from the core oligosaccharide is preferentially localized (cytochemically) in the .... cisternae of Golgi apparatus.
4. In .... transport, the flow of secretory vesicles from Golgi cisternae is back toward the ....
5. Both tetanus and botulism in humans result from interference by .... with vesicle docking and release in synapses.
6. Docking of synaptic vesicles takes place at .... within membrane of the .... neuron.

Multiple Choice Questions

1. Unidirectional transmission of a nerve impulse through nerve fibre is due to the fact that
   (*a*) nerve fibre has nodes of Ranvier
   (*b*) nerve fibre is insulated by a medullary sheet
   (*c*) neurotransmitters are released by dendrites and not by axon endings
   (*d*) neurotransmitters are released by the axon endings and not by dendrites
2. Among which one of the following groups of chemicals, all are neurotransmitters?
   (*a*) somatostatin, serotonin, acetylcholine
   (*b*) nor-adrenaline, somatostatin, threonine
   (*c*) glycine, dopamine, melatonin
   (*d*) acetylcholine, nor-adrenaline, dopamine

3. The process by which a cell secretes macromolecules through fusion of secretory vesicles to the plasma membrane is called

(*a*) pinocytosis (*b*) endocytosis

(*c*) phagocytosis (*d*) exocytosis

4. Membrane receptors are

(*a*) carbohydrates (*b*) nucleic acids

(*c*) proteins (*d*) lipids

5. Pinocytosis means endocytosis of

(*a*) droplet of extracellular fluid (*b*) sub-microscopic particles

(*c*) solid particles (*d*) extracellular fluid with particles

6. The neurotransmitter produced at electric synapse is

(*a*) GABA (*b*) acetylcholine (*c*) nor-epinephrine (*d*) none of these

7. A person with Parkinson's disease shows defective action of which neurotransmitter in brain

(*a*) dopamine (*b*) serotonin (*c*) noradrenaline (*d*) eukephalin

ANSWER

Very Short Answer Questions

1. Vesicles budding off the trans-Golgi network that are sites for the sorting and recycling of extracellular material brought into the cell by endocytosis.
2. Late endosome is a vesicle containing newly synthesized acid hydrolases plus material fated for digestion; activated either by lowering the pH of the late endosome or transferring its material to an existing lysosome.
3. Secretory vesicle is a membrane bounded compartment of a eukaryotic cell that carries secretory proteins from the Golgi apparatus to the plasma membrane for exocytosis and that may serve as a storage compartment for such proteins before they are released.
4. It is a large, dense secretory vesicle.
5. Secretory pathway is a pathway in which newly synthesized proteins move from ER through the Golgi apparatus to secretory vesicles and secretory granules, which then discharge their contents to the exterior of the cell.
6. Exocytosis is fusion of vesicle membranes with the plasma membrane so that contents of the vesicle can be expelled or secreted to the extracellular environment.
7. Endocytosis is uptake of extracellular materials by infolding of the plasma membrane, followed by pinching off of a membrane-bound vesicle containing extracellular fluid and materials.
8. It is a membrane vesicle formed by pinching off of a small segment of plasma membrane during the process of endocytosis.
9. It is a model explaining how membrane vesicles fuse with the proper target membrane; based on specific interactions between *v*-SNAREs (vesicle-SNAP receptor) and *t*-SNAREs (target-SNAP receptors) (SNAP=Soluble NSF attachment proteins).
10. It is a membrane vesicle that shuttles lipids and proteins from the endoplasmic reticulum to Golgi apparatus.
11. Vesicle that buds off from a membrane in one region of the cell and fuses with other membranes; includes vesicles that convey lipids and proteins from the ER to the Golgi apparatus, between the Golgi stack cisternae, and from the Golgi apparatus to various destinations in the cell, including secretory vesicles, endosomes and lysosomes.
12. Any of several types of membrane vesicles involved in vesicular traffic within the endomembrane system; surrounded by a coat protein such as clathrin, COP I, COP II or caveolin.
13. It is a type of endocytosis in which particulate matter or even an entire cell is taken up from the environment and incorporated into vesicles for digestion.

14. It is a type of endocytosis initiated at coated pits and resulted in coated vesicles; believed to be a major mechanism for selective uptake of macromolecules and peptide hormones.

15. Synapse is a tiny gap between a neuron and another cell (neuron, muscle fibre, or gland cell), across which nerve impulse is transferred by direct electrical connections or by chemicals called neurotransmitters.

16. It is a small vesicle containing neurotransmitter molecules; located in the terminal bulb of an axon.

17. Constitutive secretion and regulated secretion.

18. Macrophages and neutrophils.

Fill in the Blanks

1. Trans Golgi network.
2. Regulated.
3. Mannose, Medial.
4. Retrograde, ER.
5. Neurotoxins.
6. Active zone, presynaptic.

Multiple Choice Questions

1. (*d*) **2.** (*d*) **3.** (*d*) **4.** (*c*) **5.** (*d*) **6.** (*b*)
7. (*a*).

13

Cellular Energy Transductions (Role of Mitochondria and Chloroplasts)

A living cell bustles with activity. Macromolecules of all types are assembled from raw materials, waste products are produced and excreted, genetic instructions flow from the nucleus to the cytoplasm, vesicles are moved from the Golgi apparatus to the plasma membrane, ions are pumped across cell membranes, and so forth. To maintain such a high level of activity, a cell must acquire and expend energy. The study of various types of energy transformations that occur in living organisms is referred to as **bioenergetics**.

13.1 ENERGY TRANSDUCTIONS

Energy is defined as the capacity to do work, that is, the capacity to change or move some thing. Energy exists in two alternate states: potential and kinetic. A rock perched on the edge of a cliff possesses **potential energy** because it has the potential to perform work. Rock has this potential because it exists in a field of force, in this case, a gravitational field. If the rock is pushed over the edge, then gravity can act on it and cause it to fall, while falling, the rock has **kinetic energy** and can accomplish work, for example, by lifting another object (*i.e.*, fish in Fig. 13.1). Similarly, a resting nerve cell, has potential energy when it maintains a high concentration of sodium ions on the outer side of its plasma membrane and a low concentration of sodium ions on the inner side. Like the flow of water through a dam, the opening of specific channels in the plasma membrane allows sodium ions to flow across the membrane into the cell. The directed movement of sodium ions into the cell is a form of kinetic energy that can be used to accomplish work, as occurs when a nerve impulse passes along the membrane to the nerve cell.

In both of these cases just described–the falling rock and the directed movement of sodium ions–as in all cases of measurement of energy at work, two factors must be considered: a potential factor and a capacity factor (Table 13.1). The **potential factor** is proportional to the intensity of the field of force, whereas the **capacity factor** provides some measure of the "size" of the subject being considered. In the case of falling rock, the **potential factor** is the distance it will fall, and the **capacity factor** is the rock's mass. For the movement of charged ions, the potential factor is the voltage, and the capacity factor is the combined charged of the particles. The work, or energy released during these events, is a multiple of these two factors; as either one increases, so too does the amount of energy.

Rock has potential energy

Falling rock has kinetic energy

Rock has accomplished work

Fig. 13.1. Accomplishing work. The rock perched on top of the cliff has potential energy. The energy available for work is proportional to the mass of the rock and the distance it can fall. Once it leaves the edge of the cliff, the rock has kinetic energy that can be used to accomplish work, in this by lifting both water and fish (after Karp, 2002).

Table 13.1 Potential and capacity factors in measuring energy (Source: Karp 2002).

	Type of energy	*Potential factor*	*Capacity factor*
1.	**Mechanical**		
	Falling of rock	Height	Mass
	Compression	Pressure	Volume
	Stretching	Tension	Length
2.	**Electrical**		
	Movement of ions	Electric potential (voltage)	Charge
3.	**Osmotic**		
	Movement of water into a cell	Solute concentration gradient	Mass
4.	Energy available to do work	Temperature	Entropy

The Laws of Thermodynamics and the Concept of Entropy

Thermodynamics is the study of the changes in energy that accompany events in the universe.

The First Law of Thermodynamics. The first law of thermodynamics is the law of conservation of energy. It states that **energy** can neither be created nor destroyed. Energy can, however, be converted (**transduced**) from one form to another. Electric energy is converted to mechanical energy when we plug in a fan, and chemical energy is converted to thermal energy when fuel is burned in an oil heater. Cells are also capable of energy transduction. The chemical energy present in certain biological molecules, such as ATP, is converted to mechanical energy when organelles are moved from place to place in a cell, to thermal energy when heat is released during muscle cell contraction, or to electrical energy when ions flow across a membrane. *The most important energy transduction in the biological world is the conversion of sunlight into chemical energy–the process of photosynthesis*–which provides the fuel that directly or indirectly powers the activities of nearly all forms of life. A number of animals including fireflies and luminous fish are able to convert chemical energy back into light. Regardless of the transduction process, however, the total amount of energy in the universe remains constant.

To discuss energy transformations involving matter, we need to divide the universe into two parts: the **system** under study and the remainder of the universe, which we will refer to as the

surroundings. A system can be defined in various ways: it may be a certain space in the universe or a certain amount of matter. For example, the system may be a living cell. The changes in a system's energy that occur during an event are manifested in two ways—as a change in heat content of a system and in the performance of work. Even though the system may gain or loss, energy, the first law of thermodynamics indicates that the loss or gain must be balanced by a corresponding gain or loss in the surroundings, so that the amount in the universe as a whole remains constant. The energy of the system is termed as **internal energy (E)**, and its change during a transformation in **ΔE**. One way to describe the first law of thermodynamics is that $\Delta E = Q - W$, where Q is the heat energy and W is the work energy.

Depending on the process, the internal energy of the system at the end can be greater than, equal to, or less than its internal energy at the start, depending on its relationship to its surroundings (Fig. 13.2). In other words ΔE (*i.e.*, change during the transformation) can be positive, zero, or negative.

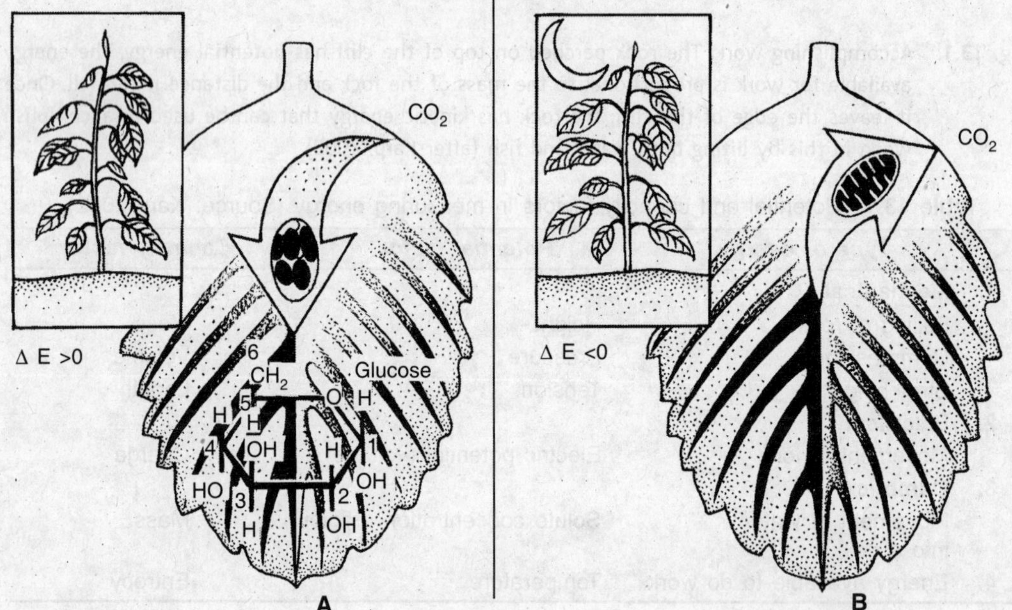

Fig. 13.2. A change in system's internal energy. In this example, the system will be defined as a particular leaf of a plant. A–During the day, sunlight is absorbed by photosynthetic pigments in the leaf's chloroplasts and used to convert CO_2 into carbohydrates, such as glucose molecule shown in the drawing (glucose is subsequently incorporated into sucrose or starch). As the cell continues to absorb light, its internal energy increases; the energy present in the remainder of the universe has to decrease. B–At night, the relationship between the cell and its surroundings is reversed as the carbohydrate produced during the day are oxidized to CO_2 in the mitochondria and the energy is used to run the cell's nocturnal activities (after Karp, 2002).

We can consider a system to be the contents of a reaction vessel. As long as there is no change in pressure or volume of the contents, there is no work being done by the system on its surroundings, or vice versa. In that case, the internal energy at the end of the transformation will be greater than at the beginning if heat is absorbed and less of heat is released. Reactions that lose heat are termed **exothermic**, and ones that gain heat are **endothermic**, and there are many reactions of both types.

The Second Law of Thermodynamics. The second law of thermodynamics expresses the concept that events in the universe have direction; they tend to proceed "downhill" from a state of higher energy to a state of lower energy. Thus, in any energy transformation, there is a decreasing availability of energy for doing additional work. Rocks fall off cliffs to the ground below, and once at the bottom, their ability to do additional work is reduced; it is very unlikely that they will lift themselves back to the top of the cliff. Similarly, opposite charges normally move together, not apart, and heat flows from a warmer to a cooler body not the reverse. Such events are said to be spontaneous, a term that indicates that they are thermodynamically favourable and *can occur without the input of external energy.*

The concept of the second law of thermodynamics was formulated originally for engines, and the law carried with the idea that it is thermodynamically impossible to construct a perpetual-motion machine. In other words, it is impossible for a machine to be 100 percent efficient, which would be required if it were to continue functioning without the input of external energy. Some of the energy is inevitably lost as the machine carries out its activity. A similar relationship holds true for living organisms. For example, when a giraffe browses on the leaves of a tree or a lion preys on the giraffe, most of the chemical energy in the food never becomes available to the animal having the meal. The energy that is unavailable for doing additional work after an event has occurred has a potential factor and a capacity factor (Table 13.1). The potential factor is temperature (in degrees) and the capacity factor is **entropy** (**S**), which has the dimensions of energy per degree (*e.g.*, calories per degree). The unavailable energy is equal to $T\Delta S$, where ΔS is the change in entropy between the initial and final states.

The loss of available energy during a process is a result of a tendency for the randomness, or disorder, of the universe to increase every time there is a transfer of energy. Entropy provides a measure of this disorder; it is also associated with the random movements of the particles of matter, which, because they are random, cannot be made to accomplish a *directed* work process. According to the second law of thermodynamics, every event is accompanied by an increase in the entropy of the universe. When a sugar cube is dropped into a cup of hot water, for example, there is a spontaneous shift of the molecules from an ordered state in the crystal to a much more disordered condition when the sugar molecules are spread throughout the solution (Fig. 13.3A). As the molecules of the sugar cube dissolve into solution, their freedom of movement increases, as does the entropy of the system. The change from the concentrated to the dispersed state results from the random movements of the molecules. The sugar molecules eventually spread themselves equally through the available volume because the state of uniform distribution is the most probable state.

For example, the release of heat from the oxidation of glucose within a cell or from the friction generated as blood flows through a vessel, is another example of an increase in entropy. The release of thermal energy by living organisms increases the rate of random movements of atoms and molecules; it cannot be redirected to accomplish additional work. Since the energy of molecular and atomic movements increases with temperature, so too does the entropy. It is only at absolute zero (0°K), when all movements ceases, that the entropy is zero.

As with other spontaneous events, one must distinguish between the system and its surroundings. The second law of thermodynamics indicate only that the total entropy in the universe must increase; the disorder within one part of the universe (the system) can decrease at the greater expense of its surroundings. The dissolved sugar can decrease in entropy it can be recrystallized by evaporating the water (Fig. 13.3B). The consequence of this process, however, is an increase in the entropy of the surroundings. The increased freedom of movement of the water molecules in the gaseous phase more than balances the decrease in freedom of the molecules of the sugar crystals.

Life operates on in a similar principle. Living organisms are able to decrease their own entropy by increasing the entropy of the environment. Entropy is decreased in an organism when relatively simple molecules, such as amino acids, are ordered into more complex molecules, such as the protein myoglobin in a muscle cell. For this to occur, however, the entropy of the environment must increase, which is accomplished as complex, ordered molecules such as glycogen stored in liver or muscle tissue are converted into heat and smaller less ordered compounds (such as CO_2 and H_2O) are released to the environment. It is this feature of metabolism that allows living organisms to maintain such a highly ordered and improbable state.

Another measure of the energy state of a living organism is provided by the information content of its macromolecules. **Information** is a subject that is difficult to define but easy to recognise. Information can be measured in terms of the ordered arrangement of a structure's subunits. For example, proteins and nucleic acids, in which the specific linear sequence of the subunits is highly ordered, are low in entropy and high in information content. Maintaining a state of high information content (low entropy) requires the input of energy. We can consider here just one molecule of DNA located in one cell of our liver. That cell has dozens of different proteins whose sole job is to patrol the DNA, looking for damage and repairing it. Nucleotide damage in an active cell can be so great that, without this expenditure of energy, the information content of DNA would rapidly deteriorate (See **Karp** 2002).

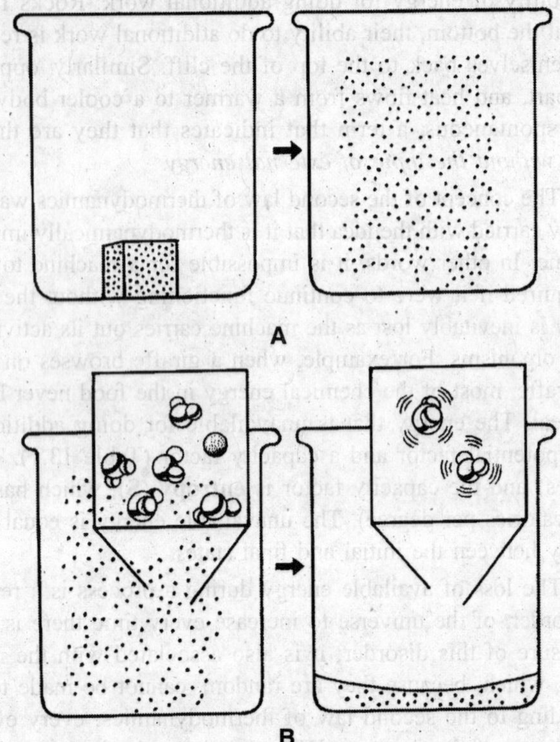

Fig. 13.3. Events are accompanied by an increase in the entropy of the universe. A–A sugar cube contains sucrose molecules in highly ordered arrangement in which the freedom of movement of the sucrose molecules is greatly increased, and their random movement causes them to become equally distributed throughout the available space. Once this occurs, there will be no further tendency for redistribution, and to entropy of the system is at a maximum. B–sugar molecules spread randomly through a solution can be returned to a ordered state, but only if the entropy of the surroundings is increased, as occur when the more ordered water molecules of the liquid phase become disordered following evaporation (after Karp, 2002).

Free Energy. Together, the first and second laws of thermodynamics indicate that the energy of the universe is constant, but the entropy continues to increase toward a maximum. The concepts inherent in first two laws were put together by the American chemist, **Josiah Willard Gibbs** in 1878 into the expression $\Delta H = \Delta G + T\Delta S$, where ΔG is the change in **free energy**, that is, the change during a process in the energy available to do work; ΔH is the change in **enthalpy** or total energy content of the system (equivalent to ΔE), T is the absolute temperature (°K = °C + 273), and ΔS is the change in the entropy of the system. The equation states that the total energy change is equal to the sum of the changes in useful energy (ΔG) and energy that is unavailable to do further work ($T\Delta S$).

Processes that can occur spontaneously, that is, processes that are thermodynamically favoured (have a $-\Delta G$), are described as **exergonic**. In contrast, if the ΔG for a given process is positive, then it cannot occur spontaneously. Such processes are thermodynamically unfavourable and are described as **endergonic**.

13.2 METABOLISM

Metabolism is the collection of biochemical reactions that occur within a cell, which include a tremendous diversity of molecular conversions. Most of these reactions can be grouped into **metabolic pathways** containing a sequence of chemical reactions in which each reaction is catalyzed by a specific enzyme, and the product of one reaction is the substrate of the next. The enzymes constituting a metabolic pathway are usually confined to a specific region of the cell, such as mitochondria or the cytosol. Increasing evidence suggests that the enzymes of a metabolic pathway are often physically linked to one another, a feature that allows the product of one enzyme to be delivered directly as a substrate to the active sites of the next enzyme in the reaction sequence.

The compound formed each step along the pathway are **metabolic intermediate** (or **metabolites**) that lead ultimately to the formation of an **end product**. End products are the molecules with a particular role in the cell, such as an amino acid that can be incorporated into a polypeptide, or a sugar that can be consumed for its energy content. The metabolic pathways of a cell are inter-connected at various points so that a compound generated by one pathway may be shuttled in a number of directions depending on the requirements of the cell at the time.

Metabolic pathways can be divided into two broad types. **Catabolic pathways** lead to the disassembly of complex molecules to form simpler products. Catabolic pathways serve two functions: they make available the raw materials from which other molecules can be synthesized, and they provide chemical energy. The energy released by catabolic pathways is stored temporarily in two forms: a high energy phosphate (primarily ATP) and high-energy electrons (primarily in NADPH). **Anabolic pathways** lead to the synthesis of more complex compounds from simple starting materials. Anabolic pathways are energy requiring and utilize chemical energy released by the exergonic catabolic pathways.

Oxidation and Reduction

Both catabolic and anabolic pathways include key reactions in which electrons are transferred from one reactant to another. Reactions that involve a change in the electronic state of the reactants are called **oxidation-reduction** (or **redox**) **reactions**. Changes of this type involve the gain or loss of electrons, *e.g.*, the conversion of metallic iron (Fe^0) to the ferrous state (Fe^{2+}), by which iron atom loses a pair of electrons, thereby attaining a more positive state. When an atom loses one or more electrons, it is said to be **oxidized**. The reaction is reversible, meaning ferrous ions can be converted to metallic iron, a more negative state, by the acquisition of a pair of electrons. When an atom gains one or more electrons, it is said to be **reduced**.

For metallic iron to be oxidized, there must be some substance to accept the electrons that are released. Conversely, for ferrous ions to be reduced, there must be some substance to donate the necessary electrons. In other words, the oxidation of one reactant must be accompanied by the simultaneous reduction of some other reactant, and vice versa. One possible reaction involving iron might be

$$Fe^0 + Cu^{2+} \rightleftharpoons Fe^{2+} + Cu^0$$

The substance that is oxidized during an oxidation-reduction reaction, that is, the one that loses electrons, is called a **reducing agent**, and the one that is reduced, that is, the one that gains electrons, is called an **oxidizing agent**.

The oxidation or reduction of metals, such as iron or copper, involves the complete loss or gain of electrons. The same cannot occur with most organic compounds for the following reason; the oxidation and reduction of organic substrates during cellular metabolism involve carbon atoms that are covalently bonded to other atoms. When a pair of electrons are attracted more strongly to one of the two atoms of the polarized bond. In a C – H bond, the carbon atom has the strongest pull on the electrons; thus, it can be said that carbon atom is in a reduced state. In contrast, if a carbon atom is bonded to a more electronegative atom, as in a C – O or C – N bond, the electrons are pulled away from the carbon atom, which is thus is an oxidised state, since carbon has four outer-shell electrons it can share with other atoms, it can exist in a variety of oxidation states. This is illustrated by the carbon atom in a series of one carbon molecules (Fig. 13.4) ranging from the fully reduced state in methane (CH_4) to the fully oxidized state in carbon dioxide (CO_2). The relative oxidation state of an organic molecule can be roughly determined by counting the number of hydrogen versus oxygen and nitrogen atoms per carbon atom. The oxidation state of the carbon atoms in an organic molecule provides a measure of the molecules free-energy content.

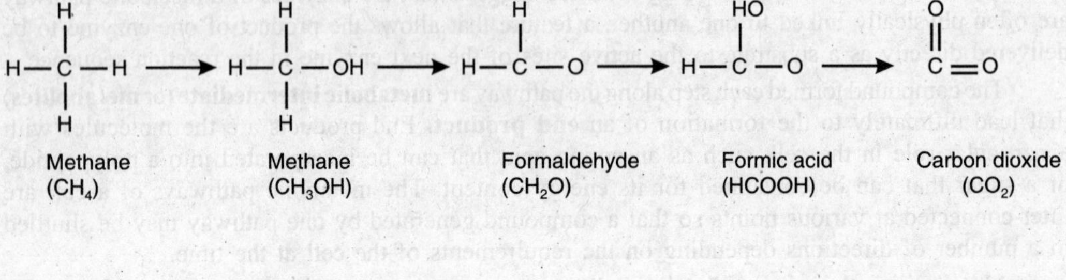

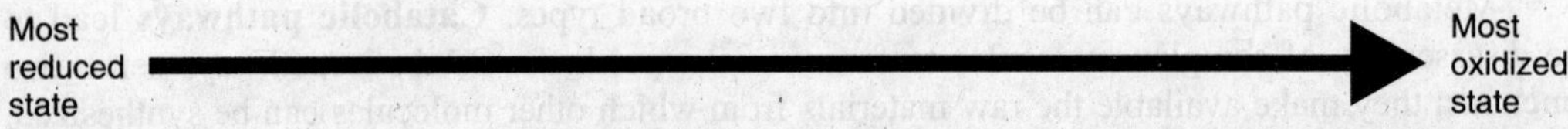

Fig. 13.4. The oxidation state of a carbon atom depends on the other atoms to which it is bonded. Each carbon atom can form a maximum of four bonds with other atoms. This series of simple, one-carbon molecules illustrates the various oxidation states in which the carbon atom can exist. In its most reduced state, the carbon is bonded to four hydrogens (forming methane); in its most oxidized state, the carbon atom is bonded to two oxygen (forming carbon dioxide) (after Karp 2002).

The Capture and Utilization of Energy

The compounds we use as chemical fuels to run automobiles are highly reduced organic compounds, such as natural gas (CH_4) and petroleum derivatives. Energy is released when these molecules are burned in the presence of oxygen, converting the carbon to more oxidized state, as in carbon dioxide and carbon monoxide gases. The degree of reduction of a compound is also a measure of its ability to perform chemical work within the cell. The more hydrogen atoms that can be stripped from a "fuel" molecule, the more ATP that ultimately can be produced. Carbohydrates are rich in chemical energy because they contain strings of (H–C–OH) units. Fats contain even greater energy per unit weight because they contain strings of more reduced (H–C–H) units.

As the sole building block of both starch and glycogen, glucose is a key molecule in the energy metabolism of plants and animals. The free energy released by the complete oxidation of glucose is very large:

$$C_6 H_{12} O_6 + 6O_2 \longrightarrow 6CO_2 + 6H_2O$$

$$\Delta G^{0'} = -686 \text{ kcal/mol}$$

By comparison, the free energy required to convert ADP to ATP is relatively small:

$$ADP + Pi \longrightarrow ATP + H_2O$$

$$\Delta G^{0'} = 7.3 \text{ kcal/mol}$$

It is evident from these numbers that the complete oxidation of a molecule of glucose to CO_2 and H_2O can release enough energy to produce a large number of ATPs. As we will see in this chapter later on, upto 36 molecules of ATP are formed per molecule of glucose oxidized under conditions that exist in most cells. For this many ATP to be produced, the sugar molecule is disassembled in many small steps. Those steps in which the free-energy difference between the reactants and products is relatively large can be coupled to reactions that lead to the formation of ATP.

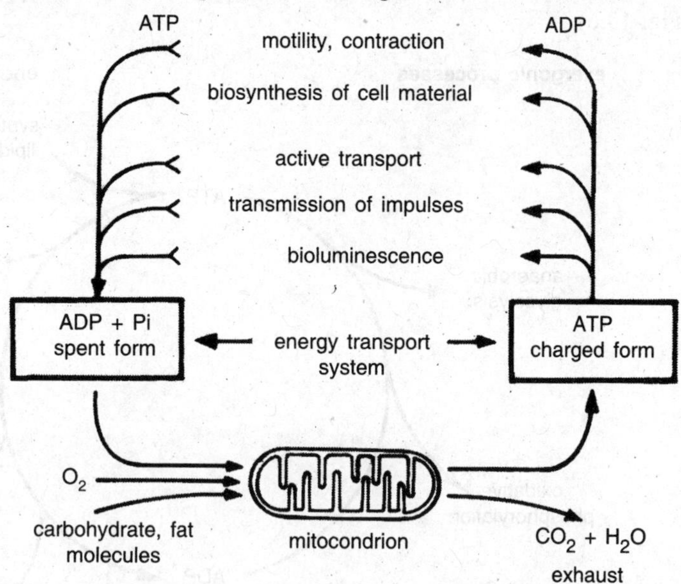

Fig. 13.5. Diagrammatic representation of the energy liberation in mitochondrion and its utilization in various cellular functions.

There are basically two stages in the catabolism of glucose, and they are virtually identical in all aerobic organisms. The first stage, **glycolysis**, occurs in the soluble phase of the cytoplasm (the cytosol) and leads to the formation of pyruvate. The second stage is the **tricarboxylic acid (TCA) cycle** or **Krebs cycle**, which occurs within the mitochondria of eukaryotic cells and the cytosol of prokaryotic cells and leads to the final oxidation of the carbon atoms to carbon dioxide. Most of the chemical energy of glucose is stored in the form of high-energy electrons, which are removed as substrate molecules are oxidized during both glycolysis and the TCA cycle. It is the energy of these electrons that is ultimately used to synthesize ATP.

Adenosine Triphosphate or ATP

The ATP consists of a purine base **adenine,** a pentose sugar **ribose** and three molecules of the **phosphoric acids.** The adenine and ribose sugar collectively constitute the nucleoside **adenosine** which by having one, two or three phosphate groups forms the adenosine monophosphate (AMP), adenosine diphosphate (ADP) and adenosine triphosphate (ATP) respectively. In ATP the last phosphate group is linked with ADP by a special bond known as **"energy rich bond"** because when the last phosphate group of the ATP is released the large amount of energy is released as shown by the following reaction [**Note. F.A. Lipmann**, the 'Father of ATP cycle', introduced the **squiggle notation** (~) to designate the energy-rich bonds of ATP and ADP (see **Jain**, **Jain** and **Jain** 2012)].

A-P~P~P =	A–P~P +	Pi +	7300 calories
ATP	ADP	phosphate group	Energy

In the above reaction, we have seen that by the breaking of the energy rich bond about 7300 calories of energy are released, while the common chemical bond releases only 300 calories of energy. The chemical reactions which synthesize the energy rich bond or ~P bond require great amount of energy which is supplied by the oxidation of the food-stuffs in the mitochondria. The utility of energy rich phosphate bond (~P) of the ATP is that great amount of energy is kept stored in the ready state in a very limited space of the cell. The stored chemical energy is disposed of very quickly at the time of the need in various cellular functions such as respiratory cycle, protein and nucleic acid synthesis, nervous transmission, cell division, transportation and bioluminescence, etc. (Fig. 13.6).

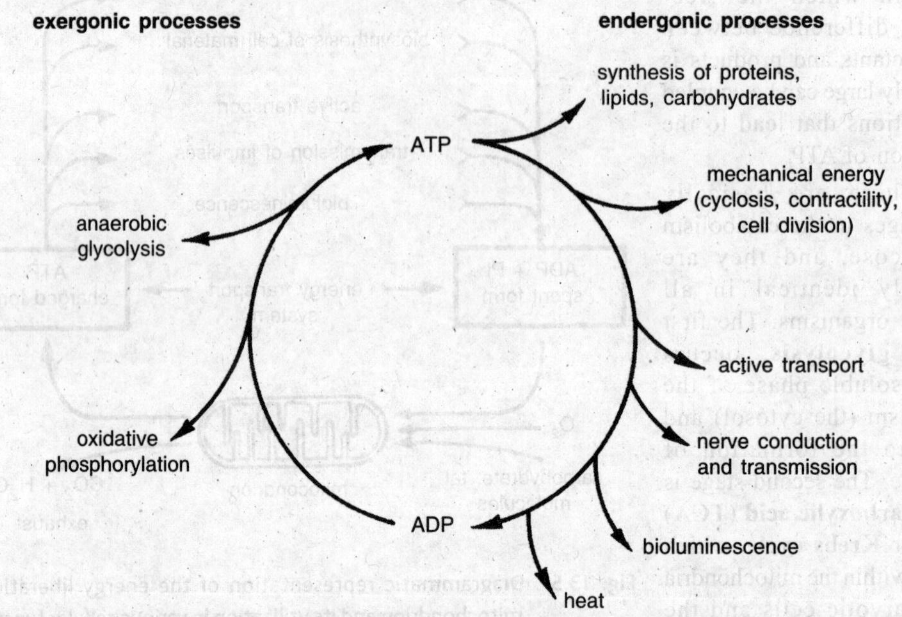

Fig. 13.6. Diagrammatic representation of uses and synthesis of ATP.

Because the terminal phosphate linkage in ATP is easily cleaved with release of free energy, ATP acts as an efficient phosphate donor in a large number of different phosphorylation reactions. In this way, ATP acts as a carrier molecule like the acetyl CoA and as coenzyme like the CoA or NAD.

Later on, besides ATP, certain other energy rich chemical compounds have been found to be active in the cellular metabolism. These are **cytosine triphosphate (CTP), uridine triphosphate (UTP)** and **guanosine triphosphate (GTP)**. These compounds, however, derive the energy from the ATP by nucleoside diphosphokinases. The energy for the production of ATP or other energy rich molecules is produced during the breakdown of food molecules including carbohydrates, fats and proteins **(catabolic** and **exergonic activities).**

Energy transducers. The cell organelles, such as chloroplasts and mitochondria, which converts one form of energy to another are known as **energy transducers**. The plants and animals have not only evolved highly effective energy transducers but have also developed very efficient control systems to regulate the energy transformations and to enable the cells to adjust to variations in environmental conditions.

13.3 MITOCHONDRIA

The mitochondria (Gr, *mito* = thread, *chondrion* = granule) are filamentous or granular cytoplasmic organelles of all aerobic cells of higher animals, plants and also of certain microorganisms including algae, Protozoa and Fungi. These are absent in bacterial cells and red blood cells of mammals. The mitochondria have lipoprotein framework which contains many enzymes and coenzymes required for energy metabolism. They also contain a specific DNA for the cytoplasmic inheritance and ribosomes for the protein synthesis.

Box 13.1 Historical

1. The mitochondria were first observed by **Kolliker** in 1850 as granular structures in the striated muscles. In 1888, he isolated them from insect muscles and named them **sarcosomes**.
2. **Richard Altmann** (1890) developed a specific stain that had useful specificity for the mitochondria. He named this organelle, the **bioblast**. He suggested that bioblasts were independent organelles.
3. The present name mitochondria was assigned by **Benda** (1897–98) to them. He stained mitochondria with alizarin and crystal violet.
4. **Sir Hans Adolph Krebs** (German biochemist in England; born 1900), in 1937 worked out various reactions of the **citric acid cycle** (or tricarboxylic acid or TCA cycle). Krebs received the Nobel Prize in 1953 along with **Lipmann** for his discovery of the citric acid cycle.
5. **Hogeboom** and coworkers (1948) first demonstrated that cellular respiration occurs in mitochondria.
6. **Kennedy** and **Lehninger** (1948–1950) showed that the citric acid cycle (Krebs cycle), oxidative phosphorylation and fatty acid oxidation took place in the mitochondria. In 1951, **Lehninger** proved that oxidative phosphorylation requires electron transport.
7. **Palade** (1954) described the ultra structure of mitochondrial cristae.
8. In 1961, **Mitchell** proposed the highly acclaimed **"chemiosmotic-coupling hypothesis"** for the ATP-production in mitochondria. He got the Nobel Prize in 1978 for the development of this model.
9. In 1963, **Nass** and **Nass** demonstrated the presence of **DNA fibres** in the matrix of mitochondria of embryonic cells.
10. **Attardi, Attardi** and **Aloni** (1971) reported the 70S type ribosomes inside the mitochondria.

Mitochondria are called **"power houses"** or **'atom bombs'** of the cell, since they produce and provide chemical energy (ATP) to the cell.

Distribution or Localization

The mitochondria move autonomously in the cytoplasm, so they generally have uniform distribution in the cytoplasm, but in many cells their distribution is very restricted. The distribution and number of mitochondria (and also of mitochondrial cristae) are often correlated with type of function the cell performs. Typically mitochondria with many cristae are associated with mechanical and osmotic work situations, where there are sustained demands for ATP and where space is at a premium, *e.g.,* between muscle fibres, in the basal infoldings of kidney tubule cells, and in a portion of inner segment of rod and cone cells of retina.

Morphology

Number. The number of mitochondria in a cell depends on the type and functional state of the cell. It varies from cell to cell and from species to species. Certain cells contain exceptionally large number of the mitochondria, *e.g.,* the *Amoeba, Chaos chaos* contain 50,000; eggs of sea urchin contain 140,000 to 150,000 and oocytes of amphibians contain 300,000 mitochondria. Certain cells, *viz.,* liver cells of rat contain only 500 to 1600 mitochondria. The cells of green plants contain less number of mitochondria in comparison to animal cells because in plant cells the function of mitochondria is taken over by the chloroplasts. Some algal cells may contain only one mitochondrion.

Shape. The mitochondria may be filamentous or granular in shape and may change from one form to another depending upon the physiological conditions of the cells. Thus, they may be of club, racket, vesicular, ring or round-shape. Mitochondria are granular in primary spermatocyte of rat, or club-shaped in liver cells (Fig. 13.7).

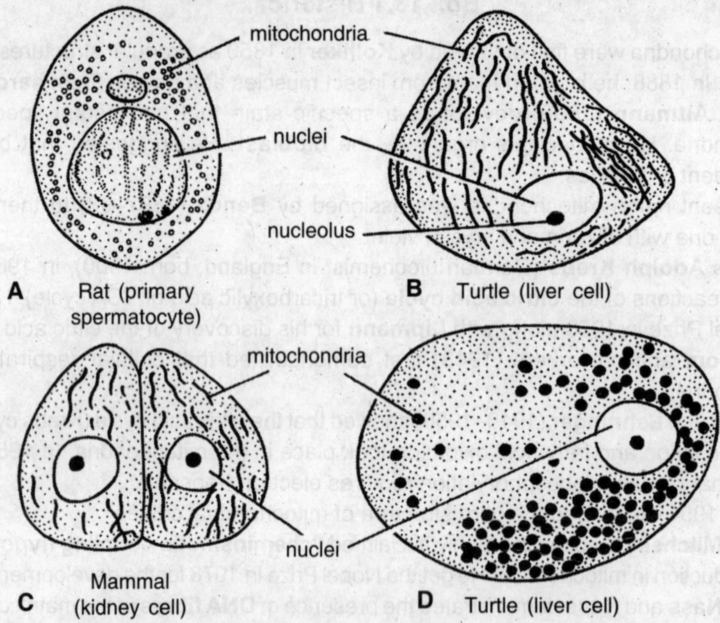

Fig. 13.7. Mitochondria of different type of animal cells.

Size. Normally mitochondria vary in size from 0.5 μm to 2.0 μm and, therefore, are not distinctly visible under the light microscope. Sometimes their length may reach up to 7 μm.

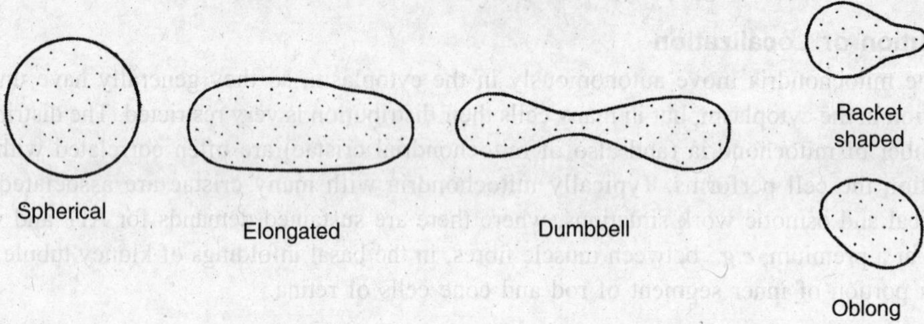

Fig. 13.8. Different shapes of mitochondria.

Structure. Each mitochondrion is bound by two highly specialized membranes that play a crucial part in its activities. Each of the mitochondrial membrane is 6 nm in thickness and fluid-mosaic in ultra-structure. The **outer membrane** is quite smooth and has many copies of a transport protein called **porin** which forms large aqueous channels through the lipid bilayer. This membrane, thus, resembles a sieve that is permeable to all molecules of 10,000 daltons or less, including small proteins. Inside and separated from the outer membrane by a 6–8 nm wide space is present the **inner membrane** (Fig. 13.9). The inner membrane is not smooth but is impermeable and highly convoluted, forming a series of infoldings, known as **cristae**, in the matrix space.

Thus, mitochondria are double membrane envelopes in which the inner membrane divides the mitochondrial space into two distinct chambers:

1. The **outer compartment, peri-mitochondrial space** or the **inter-membrane space** between outer membrane and inner membrane. This space is continuous into the core of the crests or cristae.

2. The **inner compartment, inner chamber** or **matrix space**, which is filled with a dense, homogeneous, gel-like proteinaceous material, called **mitochondrial matrix.** The mitochondrial matrix contains lipids, proteins, circular DNA molecules, 55S ribosomes and certain **granules** which are related to the ability of mitochondria to accumulate ions. Granules are prominent in the mitochondria of cells concerned with the transport of ions and water, including kidney tubule cells, epithelial cells of the small intestine, and the osteoblasts of bone-forming cells. Further, the inner membrane has an outer **cytosol** or **C face** toward the perimitochondrial space and an inner **matrix** or **M face** toward matrix.

In general, the cristae of plant mitochondria are tubular, while those of animal mitochondria are lamellar or plate-like.

Attached to M face of inner mitochondrial membrane are repeated units of stalked particles, called **elementary particles, inner membrane subunits** or **oxysomes** (Fig. 13.10). They are also identified as **F_1 particles** or **F_0 – F_1 particles** and are meant for ATP synthesis (phosphorylation) and also for ATP oxidation (*i.e.,* acting as ATP synthetase and ATPase). $F_0 - F_1$ particles are regularly spaced at intervals of 10nm on the inner surface of inner mitochondrial membrane. According to some estimates, there are 10^4 to 10^5 elementary particles per mitochon-drion.

Mitoplast. When outer mitochondrial membrane is removed with digitonin (a detergent), the so-called **mitoplast** is formed. Mitoplast includes inner mitochondrial membrane with unfolded cristae and matrix. Mitoplast is found to carry out oxidative phosphory-lation. The isolated outer membrane is revealed by negative staining and shows a "folded-bag" appearance (Fig. 13.11). Such isolation of two membranes and compartments has enabled localization of various enzyme systems of mitochondria.

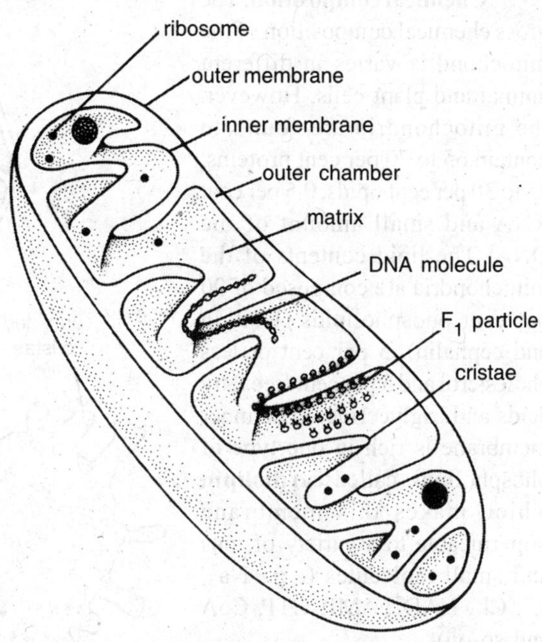

Fig. 13.9. A longitudinally cut mitochondrion showing its internal structure.

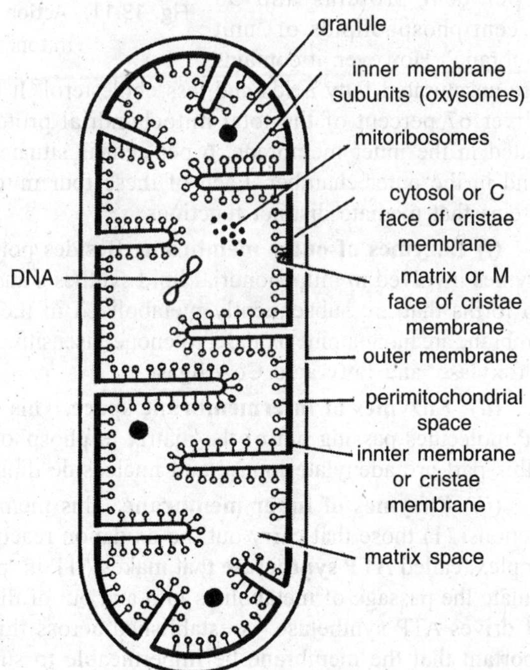

Fig. 13.10 A mitochondrion in sectional view to show its numerous oxysomes.

Chemical composition. The gross chemical composition of the mitochondria varies in different animal and plant cells. However, the mitochondria are found to contain 65 to 70 per cent proteins, 25 to 30 per cent lipids, 0.5 per cent RNA and small amount of the DNA. The lipid contents of the mitochondria are composed of 90 per cent phospholipids (lecithin and cephalin), 5 per cent or less cholesterol and 5 per cent free fatty acids and triglycerides. The inner membrane is rich in one type of phospholipid, called **cardiolipin** which makes this membrane impermeable to a variety of ions and small molecules (*e.g.*, Na^+, K^+, Cl^-, NAD^+, AMP, GTP, CoA and so on).

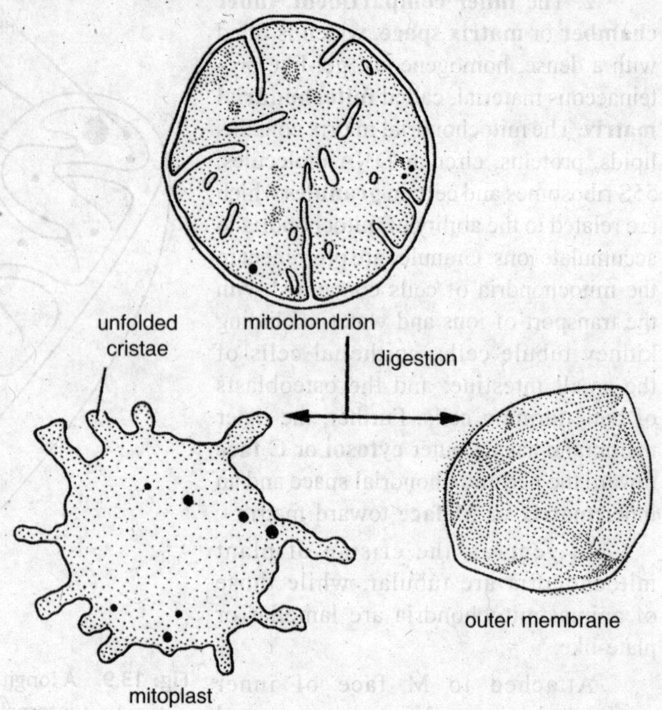

Fig. 13.11. Action of digitonin on liver mitochondrion to produce mitoplast and "folded-bag" of outer membrane.

The outer mitochondrial membrane has typical ratio of 50 per cent proteins and 50 per cent phospholipids of 'unit membrane'. However, it contains more unsaturated fatty acids and less cholesterol. It has been estimated that in the mitochondria of liver 67 percent of the total mitochondrial protein is located in the matrix, 21 per cent is located in the inner membrane, 6 per cent is situated in the outer membrane and 6 per cent is found in the outer chamber. Each of these four mitochondrial regions contains a special set of proteins that mediate distinct functions:

(*i*) Enzymes of outer membrane. Besides porin, other proteins of this membrane include enzymes involved in mitochondrial lipid synthesis and those enzymes that convert lipid substrates into forms that are subsequently metabolized in the matrix. Certain important enzymes of this membrane are monoamine oxidase, rotenone-insensitive NADH-cytochrome-C-reductase, kynurenine hydroxylase, and fatty acid CoA ligase.

(*ii*) Enzymes of intermembrane space. This space contains several enzymes that use the ATP molecules passing out of the matrix to phosphorylate other nucleotides. The main enzymes of this part are adenylate kinase and nucleoside diphosphokinase.

(*iii*) Enzymes of inner membrane. This membrane contains proteins with three types of functions: (1) those that carry out the oxidation reactions of the respiratory chain; (2) an enzyme complex, called **ATP synthetase** that makes ATP in matrix; and (3) specific transport proteins that regulate the passage of metabolites into and out of the matrix. Since an electrochemical gradient, that drives ATP synthetase, is established across this membrane by the respiratory chain, it is important that the membrane be impermeable to small ions. The significant enzymes of inner membrane are enzymes of electron transport pathways, *viz.,* nicotinamide adenine dinucleotide (NAD), flavin adenine dinucleotide (FAD), diphosphopyridine nucleotide (DPN), four cytochromes (Cyt. *b*, Cyt. *c*, Cyt. c_1, Cyt.*a* and Cyt.a_3), ubiquinone or coenzyme Q_{10}, non-heme copper and iron, ATP synthetase, succinate dehydrogenase; β-hydroxy butyrate dehydrogenase; carnitine fatty acid acyl transferase (Fig. 13.12).

***(iv)* Enzymes of mitochondrial matrix.** The mitochondrial matrix contains a highly concentrated mixture of hundreds of enzymes, including those required for the oxidation of pyruvate and fatty acids and for the citric acid cycle or Krebs cycle. The matrix also contains several identical copies of the mitochondrial DNA, special 55S mitochondrial ribosomes, tRNAs and various enzymes required for the expression of mitochondrial genes. Thus, the mitochondrial matrix contains the following enzymes: malate dehydrogenase, isocitrate dehydrogenase, fumarase, aconitase, citrate synthetase, α-keto acid dehydrogenases, β-oxidation enzymes. Moreover, the mitochondrial matrix contains different nucleotides, nucleotide coenzymes and inorganic electrolytes—K^+, HPO_4^-, Mg^{++}, Cl^- and SO_4^-.

Box 13.2 Coenzymes

Some coenzymes have a central role in mitochondrial function. **Coenzyme A** (CoA) is part of a group (Table 13.2) that is derived from a nucleoside (adenine-D-ribose) and contains pantothenic acid (a vitamin of B complex) linked to the ribose by pyrophosphoric acid. CoA can be easily transformed into an ester at the thiol end (–SH) by acetyl group making acetyl-CoA. Acetyl-CoA is a carrier molecule in which acetyl group is linked by reactive bonds so that they can be transferred efficiently to other molecules. The same carrier molecule will often participate in many different biosynthetic reactions in which its group (*i.e.,* acetyl group) is needed, *e.g.,* growing fatty acid.

Other mitochondrial coenzymes are **nicotinamide adenine dinucleotide (NAD^+)** which contains the vitamin nicotinic acid of B complex and **flavin mononucleotide (FMN)** and **flavin adenine dinucleotide (FAD)**, both of which contain riboflavin or vitamin B_2. NAD^+, FMN and FAD are important coenzymes not only in mitochondria but also in chloroplasts.

Table 13.2 Some important coenzymes which act as carrier molecules in transfer of a group.

	Coenzyme	*Group transferred*
1.	ATP	phosphate
2.	NADH, NADPH	hydrogen and electron (hydrogen ions)
3.	Coenzyme A	acetyl
4.	Biotin	carboxyl
5.	5′–Adenosyl-methionine	methyl

Functions

The mitochondria perform most important functions of the cell such as oxidation, dehydrogenation, oxidative phosphorylation and electron transport through the respiratory chain. Their structure and enzymatic system are fully adapted for their different functions. They are the actual respiratory organs of the cells where the foodstuffs *i.e.,* carbohydrates and fats are completely oxidised into CO_2 and H_2O. During the biological oxidation of the carbohydrates and fats large amount of energy is released which is utilized by the mitochondria for synthesis of the energy rich compound known as **adenosine triphosphate** or **ATP**. Because mitochondria synthesize energy rich compound ATP, they are also known as **"power houses"** of the cell. In animal cells mitochondria produce 95 per cent of ATP molecules, remaining 5 per cent is being produced during anaerobic respiration outside the mitochondria. In plant cells, ATP is also produced by the chloroplasts.

Besides the ATP production, mitochondria serve the following important functions in animals:

1. Heat production or thermiogenesis. Only 45 per cent of the energy released during the oxidation of glucose is captured in the form of ATP, the rest 55 per cent is either lost as heat or used to regulate body temperature of warm-blooded animals. In some mammals, especially young animals and hibernating species, there is a specialized tissue called **brown fat**. This tissue, typically located between the shoulder blades, is especially important in temperature regulation; it produces large quantities of body heat necessary for arousal from hibernation. The colour of brown fat comes from its high concentration of mitochondria, which are sparse in ordinary fat cells. The mitochondria of brown fat catalyze electron transport in the usual way but are much less efficient at producing ATP. Hence, a higher than usual fraction of the oxidatively released energy is converted directly to heat (called **non-shivering thermiogenesis**).

2. Biosynthetic or anabolic activities. Mitochondria also perform certain biosynthetic or anabolic functions. Mitochondria contain DNA and the machinery needed for protein synthesis. Therefore, they can make less than a dozen different proteins. The proteins so far identified are subunits of the ATPase, portions of the reductase responsible for transfer of electrons from CoQ to the iron of Cyt c, and three of the seven subunits in cytochrome oxidase. Altogether, no more than 5–10 per cent of mitochondrial components can be attributed to mitochondrial genes.

Some biosynthetic functions of mitochondria are of primary benefit to the rest of the cell. For example, the synthesis of **haeme** (needed for cytochromes, myoglobin and haemoglobin) begins with a mitochondrial reaction catalyzed by the enzyme, delta or δ-aminolevulinic acid synthetase. Likewise, some of the early steps in the conversion of cholesterol to steroid hormones in the adrenal cortex are also catalyzed by mitochondrial enzymes.

3. Accumulation of Ca^{2+} and phosphate. In the mitochondria of **osteoblasts** present in tissues undergoing calcification large amount of Ca^{2+} and phosphate (PO_4^-) tend to accumulate. In them microcrystalline, electrone-dense deposits may become visible. Sometimes, the mitochondria assume storage function, *e.g.*, the mitochondria of amphibian ovum store large amounts of yolk proteins and are transformed into **yolk platelets**.

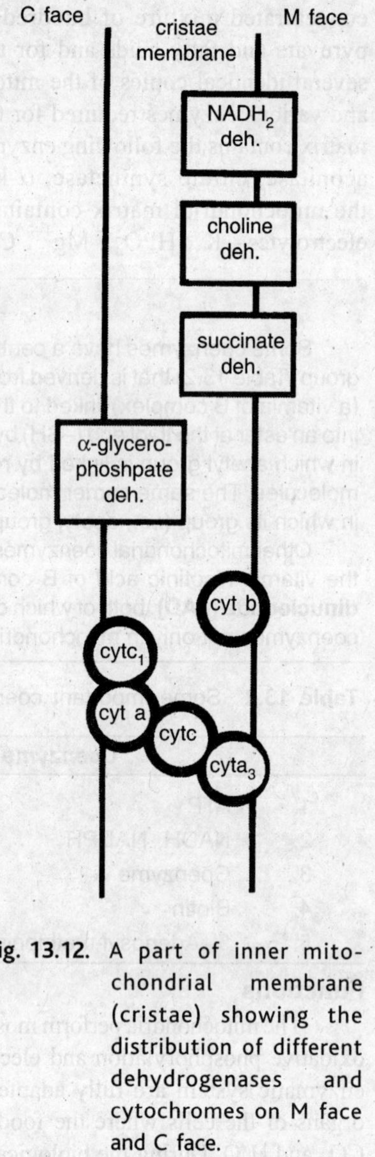

C face
cristae membrane
M face
$NADH_2$ deh.
choline deh.
succinate deh.
α-glycero-phoshpate deh.
cyt b
$cytc_1$
cyt a
cytc
$cyta_3$

Fig. 13.12. A part of inner mitochondrial membrane (cristae) showing the distribution of different dehydrogenases and cytochromes on M face and C face.

13.4 MODE OF ENERGY PRODUCTION

The mitochondrion is the centre of oxidative metabolism in the cell, converting the products from carbohydrate, fat and protein catabolism into chemical energy stored in ATP.

Oxidation of Carbohydrates

The carbohydrates enter in the cell in the form of monosaccharides such as glucose. The hexose sugars are first broken down into 3-carbon compound (pyruvic acid) by a series of chemical reactions by many known enzymes. The pyruvic acid enters in the mitochondria for its complete oxidation into CO_2 and water. The reactions which involve in the oxidation of glucose in to CO_2 and water are known to form the **metabolic pathways** and they can be grouped under the following heads:

1. Glycolysis or Embden-Meyerhof pathways (EMP) or Embden-Meyerhof Parnas pathways (EMPP);
2. Oxidative decarboxylation;
3. Krebs cycle; citric acid cycle or tricarboxylic acid cycle;
4. Respiratory chain and oxidative phosphorylation.

1. Glycolysis. Under anaerobic conditions (*i.e.,* in the absence of oxygen) glucose is degraded into lactic acid or lactate by a process called **glycolysis** (*i.e.,* lysis or splitting of glucose), *e.g.,* it commonly occurs in vertebrate muscles when the energy demand in heavy exercise exceeds the available oxygen.

If glycolysis is carried out under aerobic conditions the final products are pyruvic acid and coenzyme NADH. Glycolysis is achieved by a series of 10 enzymes all of which are located in the cytosol (cytoplasmic matrix). As shown in (Fig. 13.13), in this chain of reactions, the product of one enzyme serves as a substrate for the next reaction. To facilitate its analysis, the sequence of glycolysis can be subdivided into following three main steps: (*i*) activation (stage I); (*ii*) cleavage (stage II); and (*iii*) oxidation (stage III).

***(i)* Activation.** In reactions 1 to 3 the glucose molecules is converted into **fructose-1-6-diphosphate**. This step uses two molecules of ATP and involves the following enzymes: hexokinase, phosphoglucose isomerase (or phosphohexoisomerase) and phosphofructokinase.

***(ii)* Cleavage.** In reactions 4 and 5, fructose-1-6-diphosphate splits into two (3-carbon) end products, **glyceraldehyde-3-phosphate** molecules. During the step of cleavage only two enzymes are used: aldolase (fructoaldolase) and triose isomerase (triosephosphate isomerase).

***(iii)* Oxidation.** In reactions 6 to 10, two molecules of glyceraldehyde-3-phosphate are oxidized and ultimately converted into two molecules of **pyruvic acid**. This step produces four molecules of ATP by **substrate-level-phosphorylation** and involves the following enzymes: phosphoglyceric dehydrogenase (glyceraldehyde phosphate dehydrogenase), phosphoglyceric kinase, phosphoglyceromutase, enolase and pyruvic kinase.

The net energy yield of chain reactions of glycolysis is the production of two ATP molecules from one molecule of glucose. Under aerobic conditions, the end products of glycolysis are pyruvic acid and reduced coenzyme NAD (*i.e.,* NADH). NADH carries two electrons, taken from glyceraldehyde-3-phosphate and contains little energy. However, under anaerobic conditions, pyruvic acid remains in the cytosol (cytoplasmic matrix) and is used as a hydrogen acceptor and converted into lactic acid:

$$\text{Pyruvic acid} \xrightarrow[\text{+NADH}]{\text{Lactic dehydrogenase}} \text{Lactic acid + NAD + Energy}$$

In above case, the following equation represents the overall reaction of glycolysis:

$$\underset{\text{Glucose}}{C_6H_{12}O_6} + Pi + 2ADP \longrightarrow \underset{\text{Lactic acid}}{2C_3H_6O_3} + 2ATP + H_2O$$

2. Oxidative decarboxylation. In aerobic organisms, since pyruvic acid still contains a larger amount of energy, it must undergo further degradation, but this time inside the mitochondria. This is done in three consecutive steps: **oxidative decarboxylation** (removal of carboxyl or—COOH group), **Krebs cycle** and **oxidative phosphorylation**. Pyruvic acid directly enters the mitochondrial matrix and is converted into **acetyl-CoA** by the help of a huge enzyme, called **pyruvic acid dehydro-genase** (Fig. 13.14). The two NADH molecules (which are generated during glycolysis) cannot penetrate directly into the mitochondria, so their electrons are transferred to **dihydroxyacetone phosphate**, which shuttles them into the mitochondria. This process utilizes, one ATP molecule for each NADH; in all two ATP molecules are consumed for two NADH molecules. When both of these NADH pass through ETS, they tend to generate 6 ATP molecules.

Pyruvic acid dehydrogenase and its action. Sometimes two enzymes that catalyze sequential reactions form an **enzyme complex** and the product of the first enzyme does not have to diffuse through the cytoplasm to encounter the second enzyme. The second reaction begins as soon as the

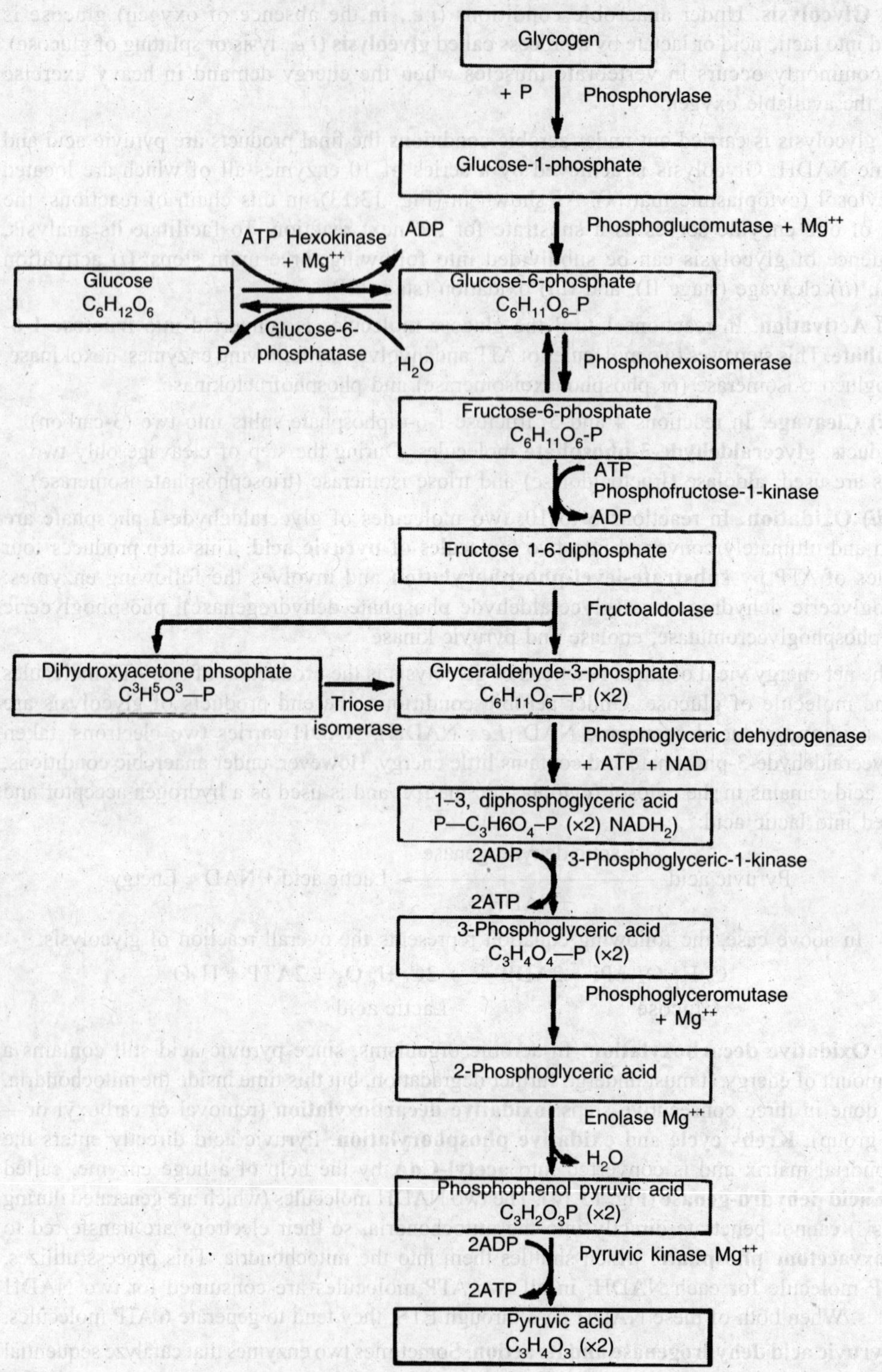

Fig. 13.13. Glycolysis or Embden-Meyerhof pathway showing the step-wise degradation of glucose to pyruvic acid.

first is over. Some large enzyme (multienzyme) aggregates carry out whole series of reactions without losing contact with the substrate. For example, the conversion of pyruvic acid to acetyl CoA proceeds in *three* chemical steps, all of which take place on the same large multienzyme complex (*i.e.,* pyruvate dehydrogenase). Pyruvic acid dehydrogenase occurs in the mitochondrial matrix and is larger than a ribosome in size. It contains multiple copies of **three** enzymes namely **pyruvic acid dehydrogenase, dihydrolipoyl transacetylase** and **dihydrolipoyl dehydrogenase**. It also contains five coenzymes (*e.g.,* NAD, coenzyme A, etc.) and **two regulatory proteins** (*e.g.,* protein kinase and protein phosphatase; both regulating the activity of pyruvic acid dehydrogenase, turning it off whenever ATP levels are high).

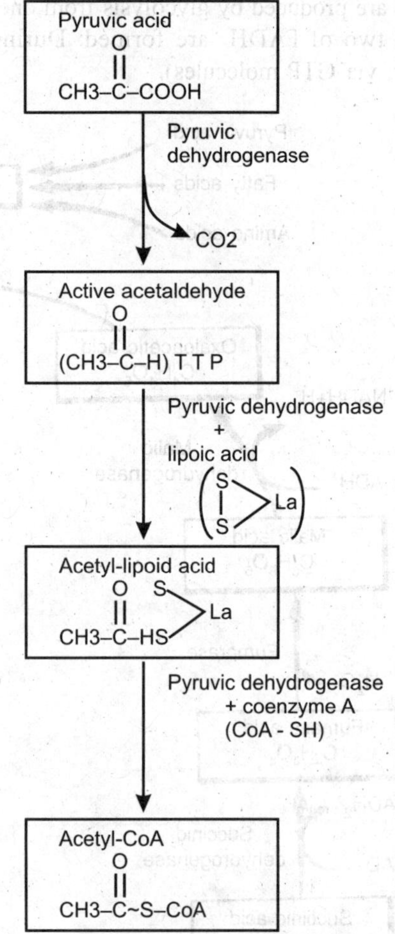

Fig. 13.14. The oxidative decarboxylation of pyrurvic acid to acetyl coenzyme A.

Thus, during oxidative decarboxylation of one molecule of pyruvic acid, one mole of CO_2 is produced and one NAD is reduced to NADH. The end product of this reaction is a 2-carbon compound, the **acetyl group** which is attached to coenzyme A to produce the carrier molecule, called **acetyl CoA**.

3. Krebs cycle. Two acetyl CoA molecules, produced above by oxidation of one molecule of glucose pass through a series of reactions of Krebs cycle to produce CO_2, H_2O and electrons. Enzymes and coenzymes of Krebs cycle are located in the mitochondrial matrix (some of them such as succinic dehydrogenase, are attached to M face of inner mitochondrial membrane, Fig. 13.12). As illustrated in (Fig. 13.15), the Krebs cycle involves the condensation of the acetyl group with oxaloacetic acid to make **citric acid** (6-carbon compound). This step is directed by the enzyme **citrate synthase**. From citric acid, H_2O is released twice by **aconitase** enzyme to produce **isocitric acid**. This is followed by a decarboxylation (loss of CO_2) by **isocitric dehydrogenase**, producing 5 carbon **α-ketoglutaric acid**. CO_2 is released by **α-ketoglutarate dehydrogenase** in the presence of CoA to produce **succinyl CoA** which changes by **succinyl kinase** enzyme (also called succinyl CoA synthetase) into a 4-carbon compound, the **succinic acid** (at this stage one GTP is generated by substrate level phosphorylation). The next enzyme of Krebs cycle, the **succinic dehydrogenase** converts succinic acid into **fumaric acid** and then **fumarase** enzyme produces **malic acid.** The mediation of **malate dehydrogenase** enzyme produces **oxaloacetic acid**, and thereby closes the Krebs cycle.

At each turn of the Krebs cycle, four pairs of hydrogen atoms are removed from the substrate intermediates by enzymatic dehydrogenation and two CO_2 molecules are released. These hydrogen atoms (or equivalent pairs of electrons) enter the respiratory chain, being accepted by either NAD^+ or FAD. Three pairs of hydrogen molecules are accepted by NAD^+, reducing it into NADH, and one pair by FAD, reducing it into $FADH_2$ (this pair of electrons come directly from the succinic dehydrogenase reaction). Since it takes two turns of the cycle to metabolize the two acetyl groups that are produced by glycolysis from one molecule of glucose, a total of six molecules of NADH and two of $FADH_2$ are formed. During Krebs cycle are also produced two ATP molecules (*i.e.*, via GTP molecules).

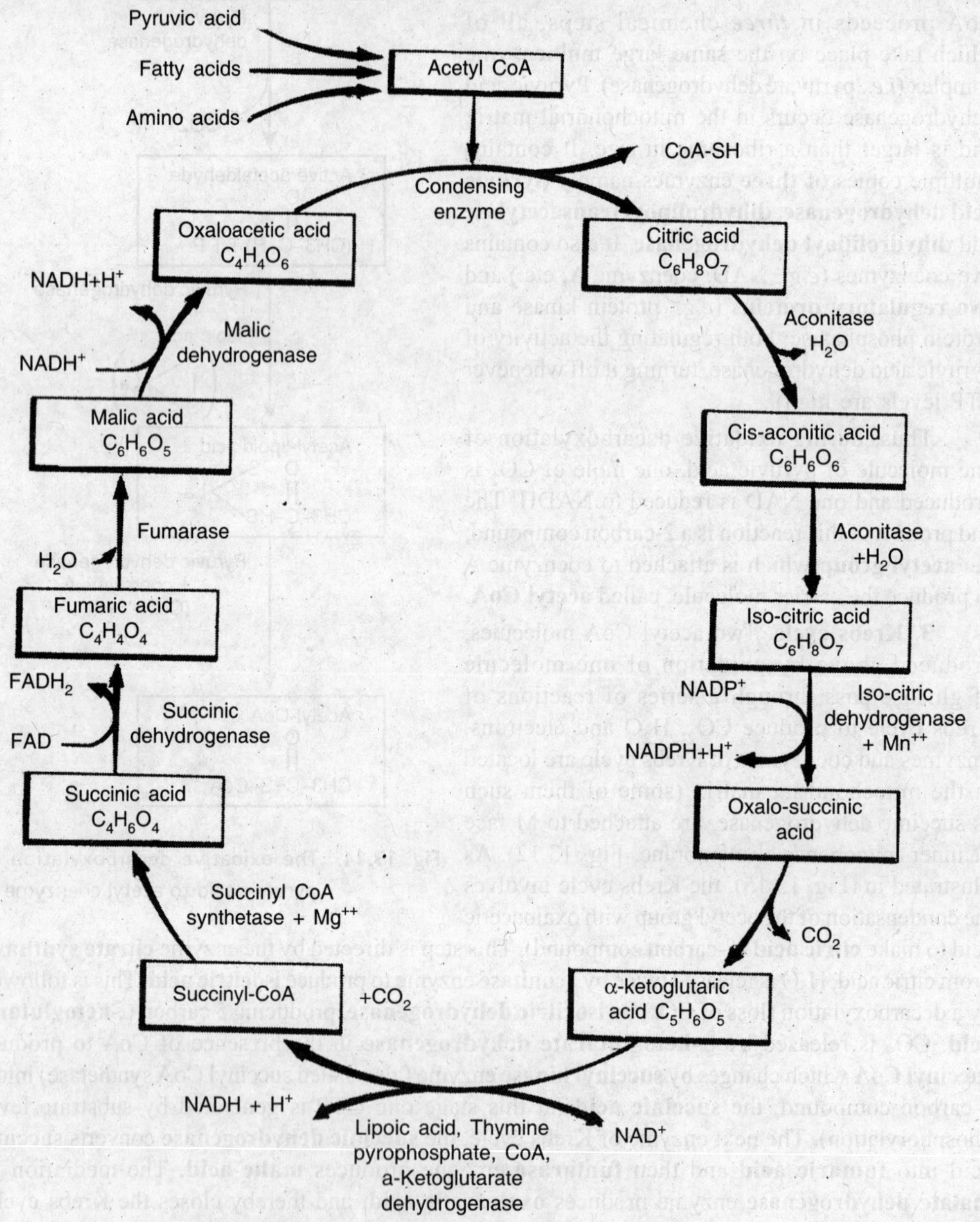

Fig. 13.15. Krebs cycle or tricarboxylic acid cycle.

The Krebs cycle itself releases no energy, but as the glucose breaks up, it also frees the hydrogen atoms attached to the carbons. Each hydrogen atom contains one proton (H^+) and one electron (e^-) The electrons which are released during complete oxidation of glucose (*i.e.*, glycolysis, oxidative decarboxylation and Krebs cycle) carry most of the energy of the glucose. In the final stage of cell respiration, the electron transport system, these electrons will at last release their energy to the cell.

4. Respiratory Chain and Oxidative Phosphorylation. Two molecules of $FADH_2$ and six molecules of NADH produced in Krebs cycle (from two molecules of acetyl-CoA) are oxidized by molecular O_2 in a **respiratory chain** or **electron transport system** or **ETS** involving a series of enzymes and coenzymes.

In the electron transport system, the successive electron acceptors are at lower and lower energy levels. With each transfer to a lower energy level, the electrons release some of their potential energy. That is why this series is called an **electron cascade**, like a cascade of falling water. At each stage, the released energy is used to form ATP. Since electron transport involves oxidation as well as phosphorylation (*i.e.*, ADP + P = ATP) this process by which cell system traps chemical energy is called **oxidative phosphorylation** (Fig. 13.16). The passage of electrons from NAD to oxygen generates 3 ATP molecules, whereas the passage of electrons from FAD to oxygen generates only 2 ATP molecules.

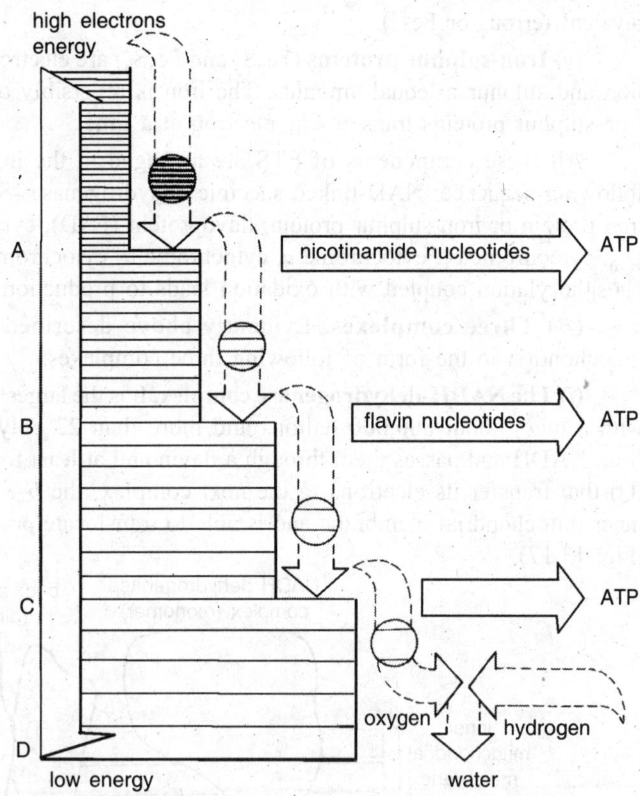

Fig. 13.16. Electron transfer in respiratory cascade or electron transport system.

(*A*) Compounds that occur in ETS. Following five types of compounds are associated with electron transport system of inner mitochondrial membrane:

(*i*) Pyridine-linked dehydrogenases require as their coenzyme either NAD^+ or $NADP^+$ both of which can accept two electrons at a time. There are about 200 dehydrogenases for which NAD^+–linked compounds such as pyruvic acid dehydrogenase, are involved in ETS.

(*ii*) Flavin-linked dehydrogenases (often called **flavoproteins** or **FPs**) require either FAD or FMN. Both are prosthetic groups whose isoalloxazine ring can accept two hydrogen atoms. Flavin-linked enzymes are commonly involved in a number of enzyme systems such as fatty acid oxidation, amino acid oxidation and Krebs cycle activity (*e.g.*, succinic dehydrogenase or SDH).

(*iii*) Ubiquinones were so named because of their occurrence in so many different organisms and their chemical resemblance to quinone. They are found in several different forms including the **plastoquinones** of chloroplasts. The form of ubiquinones present in mitochondria is often called **coenzyme Q_{10}** (CoQ_{10} or Q). CoQ_{10} is a lipid soluble and accepts two hydrogen atoms (or two protons and two electrons) at a time.

(*iv*) Cytochromes are proteins containing iron-porphyrin (haem) groups. There are a large number of cytochrome in cells; most are found in mitochondria, although some also function in the *ER* and in chloroplasts. Mitochondria have five types of cytochromes which are arranged in the following order in an inner membrane: cyt. *b*, cyt. c_1, cyt. *c*, cyt. *a* and cyt. a_3. All of them transfer electrons by reversible valence changes of the iron atom (trivalent ferric or Fe^{3+} ⇌ bivalent ferrous or Fe^{2+}).

(*v*) Iron-sulphur proteins (Fe_2S_2 and Fe_4S_4) are electron carriers of mitochondria containing iron and sulphur in equal amounts. The iron is reversibly oxidized during the electron transfer. Iron-sulphur proteins transfer one electron at a time.

All these components of ETS are arranged in the inner mitochondrial membrane in the following sequence: NAD-linked succinic dehydrogenase (SDH), flavoprotein (FAD), non-haem iron protein or iron-sulphur protein, flavoprotein (FAD), cytochrome *b*, ubiquinone or coenzyme Q_{10}, cytochrome c_1, cytochrome *c*, cytochrome *a*, cytochrome a_3 and three coupling sites, where phosphorylation coupled with oxidation leads to production of ATP.

(*B*) Three complexes. Evidently above described components of ETS occur in the mitochondria in the form of following three complexes:

(*i*) The NADH-dehydrogenase complex. It is the largest of the respiratory enzyme complexes, with a mass about 800,000 daltons and more than 22 polypeptide chains. It accepts electrons from NADH and passes them through a flavin and at least five iron-sulphur centres to ubiquinone (Q) that transfer its electrons to the next complex, the *b*–c_1, complex. This complex spans the inner mitochondrial membrane and is able to translocate protons across it from M side to C side (Fig. 13.17).

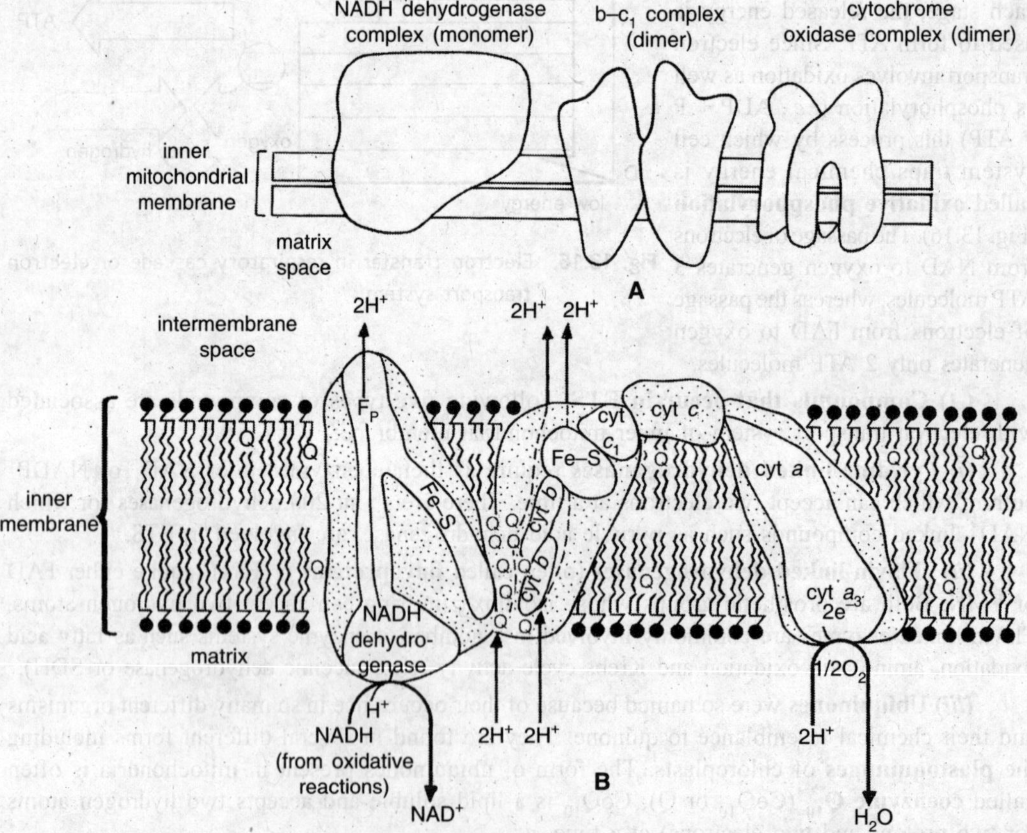

Fig. 13.17. Respiratory enzyme complexes. A—The relative sizes and shapes of the three respiratory enzyme complexes; B—Mode of transfer of H^+ through the three respiratory enzyme complexes of inner mitochondrial membrane.

***(ii)* The *b*–c_1 complex.** It contains at least 8 different polypeptide chains and is thought to function as a dimer of about 500,000 daltons. Each monomer contains three haemes bound to cytochromes and iron-sulphur protein. This complex accepts electrons from ubiquinone (Q) and passes them to cytochrome, a small peripheral membrane protein that carries its electrons to the cytochrome oxidase complex. In the topology of this complex the Q-site may be in the middle of the membrane in the hydrophobic area and the cytochrome c-site on the C side.

***(iii)* The cytochrome oxidase complex.** It comprises at least eight different polypeptide chains and is isolated as a dimer of about 300,000 daltons; each monomer contains two cytochromes (a, a_3) and two copper atoms. This complex accepts electrons from cytochrome *c* and passes them to oxygen and is thought to traverse the mitochondrial membrane, protruding on both surfaces. Such a trans-membrane orientation is associated with the vectorial transport of protons across the membrane.

The cytochrome oxidase reaction is estimated to account for 90 per cent of the total oxygen uptake in most cells. The toxicity of the poisons such as cyanide and azide is due to their ability to bind tightly to this complex and thereby block all electron transport.

***(C)* F_0–F_1 complex or coupling factors.** One of the main proteins in the inner mitochondrial membrane is the multisubunit **coupling factor** (Fig. 13.18), the enzyme that actually synthesizes ATP and simultaneously acting as a **proton pump.** A quite similar enzyme complex is located in the thylakoid membranes of chloroplasts and in the plasma membrane of bacterial cell. The coupling factor has two principal components:

***(a)* F_0-complex.** It is an integral membrane complex, composed of very hydrophobic proteins—3 or 4 distinct polypeptides and one proteolipid—which together span the mitochondrial membrane. F_0–complex possesses the proton translocating mechanism. F_0-complex can be extracted only with strong detergents.

***(b)* F_1-particle.** Attached to the F_0 complex is F_1 particle, a complex of five distinct polypeptides: alpha (α), beta (β), gamma (γ), delta (δ) and epsilon (ε), with the probable composition of $\alpha_3\beta_3\gamma\delta\varepsilon$. F_1 forms the knob or 'tadpole' that protrudes on the matrix side of the inner mitochondrial membrane. F_1 particle can be detached from the membrane by mechanical agitation and is water soluble. When physically separated from the membrane, F_1 particle is capable only of catalysing the hydrolysis of ATP into ADP and phosphate. Hence, it is often called the **F_1-ATPase**. However, its natural function is the synthesis of ATP.

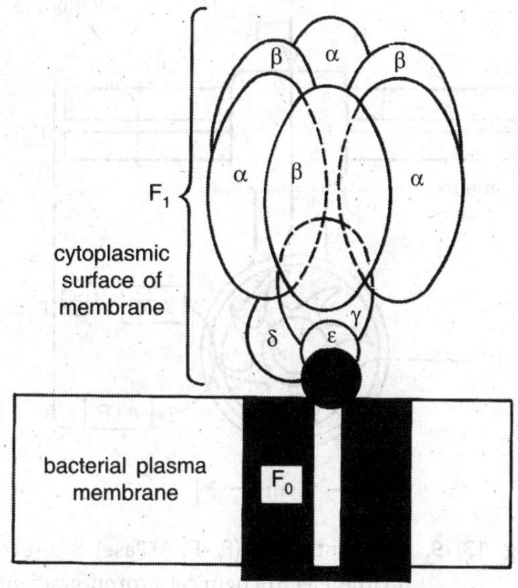

Fig. 13.18. Model of molecular structure of F_0 and F_1 particles in the bacterial plasma membrane. The core of F_1 particle is an $\alpha_3\beta_3$ complex. The δ polypeptide of F_1 and probably the γ are involved in binding F_1 to the membrane-embedded F_0-particle. The role of ε subunit of F_1 complex is not known.

***(D)* Redox reactions and redox couples.** The movements of electrons between cellular reductants and oxidants represent a form of energy transfer in cells. A reductant (or **reducing agent**) is a substance that loses or donates electrons to another substance; the latter substance is the oxidant (or **oxidizing agent**). Conversely, an oxidant is a substance that accepts electrons from another substance, the latter being the reductant. Reactions that involve the movement of electrons between reductants and oxidants are called **redox reactions**.

***(E)* ATP synthesis.** The potentials drop in three large steps, one across each major respiratory enzyme complex. The change in redox potential between any two electron carriers is directly proportional to the free energy released by an electron transfer between them. Each complex acts as an **energy-conversion-device** to harness this free-energy change, pumping H^+ across the inner membrane to create an electrochemical proton gradient as electrons pass through. The energy conversion mechanism underlying oxidative phosphorylation requires that each protein complex be inserted across the inner mitochondrial membrane in a fixed orientation, so that all protons are pumped in the same direction out of the matrix space. Such a **vectorial organization** of membrane proteins has been experimentally proved.

Just as a flow of water from a higher to a lower level can be utilized to turn a water wheel or a hydroelectric turbine, the energy released by the flow of the protons down the gradient is utilized in the synthesis of ATP. Similarly, resultant electrochemical proton gradient is harnessed to make ATP by F_0–F_1 complex (acting as ATP synthetase and proton pump), through which the protons flow back into the matrix (Fig. 13.19). The ATP synthetase is a reversible coupling device that normally converts a back-flow of protons into ATP phosphate-bond energy, but it can also hydrolyze ATP (into ADP and phosphate) to pump protons in the opposite direction, if the electrochemical proton gradient is reduced.

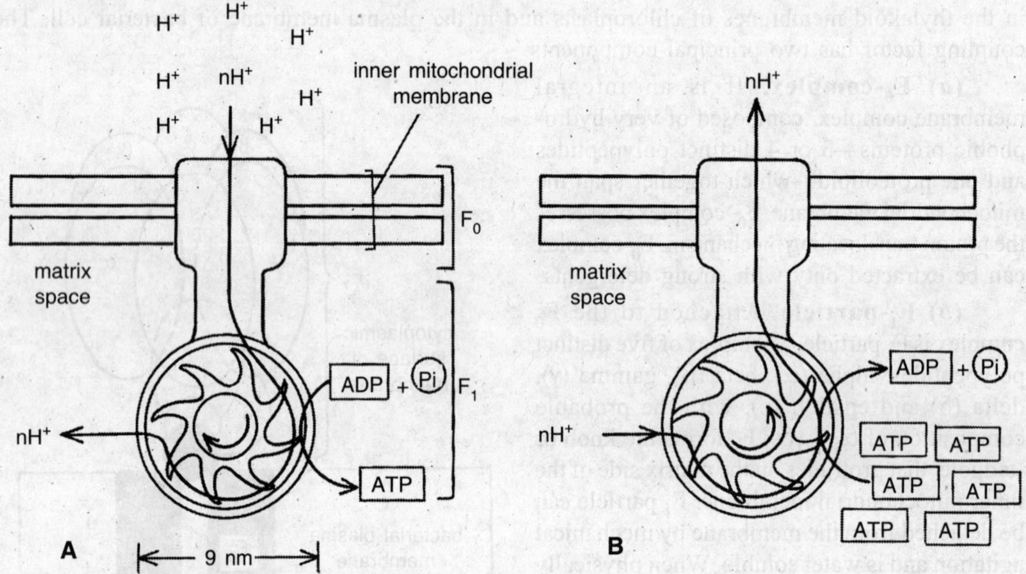

Fig. 13.19. ATP synthetase (F_0–F_1 ATPase) is a reversible coupling device that interconverts the energies of the electrochemical proton gradient and chemical bonds. The ATP synthetase can either synthesize ATP by harnessing the proton motive force (A) or pump protons against their electrochemical gradient by hydrolyzing ATP (B). The direction of operation at any given instant depends on the net free-energy change for the coupled processes of proton translocation across the membrane and the synthesis of ATP from ADP and Pi.

Chemiosmotic mechanism. The controlled movement of protons back across the membrane through ATP-synthesizing enzyme provides the energy required to phosphorylate ADP to ATP. The importance of proton movements on the formation of ATP was first proposed in 1961 by **Peter Mitchell** of the University of Edinburgh. This proposed mechanism is called **chemiosmotic mechanism.** For this contribution, Mitchell was awarded Nobel Prize in Chemistry in 1978.

According to the chemiosmotic model, the main energetic driving force for the synthesis of ATP is an ion gradient across a selectively permeable biological membrane; in chloroplasts and mitochondria, this gradient is known to be a **proton gradient** that is generated as a consequence of electron transfer. This gradient establishes a concentration difference of protons and can also result in difference in electric potential of the membrane. These sources of potential energy can be used for the phosphorylation of ADP by ATP synthase, an enzyme that energetically favourable proton flow to the synthesis of ATP (**Buchanan** *et al.*, 2007).

Paul Boyer in 1964 proposed **rotary motor driving model** for ATP generation in mitochondria. He got Nobel Prize in Chemistry in 1997 for this model. According to this model a conformational change occurs in ATP synthase (*i.e.*, F_0 F_1 complex) which generates ATP molecules (**Karp**, 2010).

***(F)* Energetics of glucose oxidation.** Of the 686,000 calories contained in a mole of glucose, less than 10 per cent (*i.e.*, 58,000 calories) can be released by anaerobic glycolysis. The cell is able to store only 45 per cent of the chemical energy liberated by the combustion of glucose in the form of ATP (*i.e.*, only 36 ATP molecules). The rest of the energy is dissipated as heat or used for other cell functions.

At this stage, let us do stocktaking of ATP generation during aerobic respiration of one mole of glucose. We have seen that glycolysis and Krebs cycle can each generate 2 molecules of ATP per molecule of glucose by substrate level phosphorylation (total 4 ATP molecules). In addition 10 NADH (*i.e.*, 2NADH in glycolysis, 2NADH in oxidative phosphorylation and 6NADH in Krebs cycle) and 2FADH$_2$ are produced which are equivalent to 34 ATP molecules. Thus, a total of 38 ATP molecules are produced per glucose molecule oxidized. However, in most eukaryotic cells 2 molecules of ATP are used in the transportation of 2 mole of NADH produced during glycolysis into the mitochondrion (*via* the malate shuttle) for their further oxidation (*via* ETS). Hence, the net gain of ATP is 38 – 2 = 36 molecules; since one high energy phosphate bond is equal to 36.8 kJ; so 36ATP = 1325 kJ or 360,000 calories :

$$\underset{\text{Glucose}}{C_6H_{12}O_6} + \underset{\substack{\text{Inorganic}\\ \text{phosphate}}}{36Pi} + 36ADP + 6O_2 + 6H_2O \longrightarrow$$

$$6CO2 + 36\ ATP\ (360{,}000\ \text{calories or } 1325\ \text{kJ}) + 12H_2O$$

Mitochondria as Semi-autonomous Organelles

Recently the study of mitochondrial and chloroplast biogenesis became of great interest because it was demonstrated that these organelles contain DNA as well as ribosomes and are able to synthesize proteins. The term **semiautonomous organelles** was applied to the two structures in the recognition of these findings. This term also indicated that the biogenesis was highly dependent on the nuclear genome and the biosynthetic activity of the ground cytoplasm. It is well established now that the mitochondrial mass grows by the integrated activity of both genetic systems, which cooperate in time and space to synthesize the main components. The mitochondrial DNA codes for the mitochondrial, ribosomal and transfer RNA and for a few proteins of the inner membrane. Most of the proteins of the mitochondrion, however, result from the activity of the nuclear genes and are synthesized on ribosomes of the cytosol (cytoplasmic matrix). The cooperation of two genomes has been greatly clarified by studies on the molecular assembly of cytochrome oxidase. This cytochrome, as studied in *Saccharomyces cerevisiae* is made up of seven polypeptide subunits for a combined molecular weight of 139,000 daltons. Three of the polypeptides are coded by mtDNA and assembled on mitochondrial ribosomes. They are very hydrophobic and high in molecular weight (23,000–40,000 daltons). The remaining four subunits are coded by nuclear DNA and made on cytoplasmic ribosomes. These are hydrophilic polypeptides of lower molecular weight (4500–14,000 daltons).

Mitochondrial DNA. Mitochondrial DNA (mtDNA) molecule is relatively small, simple, double-stranded and except for the DNA of some algae and protozoans, it is circular. The size of mitochondrial genome is very much large in plants than in animals. Thus, mtDNA varies in length from about 5 μm in most animal species to 30 μm or so in higher plants. The mtDNA is localized in the matrix and is probably attached to the inner membrane at the point where DNA duplication starts. This duplication is under nuclear control and the enzymes used (*i.e.*, polymerases) are imported from the cytosol.

Mitochondrial ribosomes. Mitochondria contain ribosomes (called **mitoribosomes**) and polyribosomes. In yeast and *Neurospora*, ribosomes have been ascribed to a 70S class similar to that of bacteria; in mammalian cells, however, mitoribosomes are smaller and have a total sedimentation coefficient of 55S, with subunits of 35S, and 25S. In mitochondria, ribosomes appear to be tightly associated with the inner membrane.

Mitochondrial protein synthesis. As already described, mitochondira can synthesize about 12 different proteins, which are incorporated into the inner mitochondrial membrane. These proteins are very hydrophobic (*i.e.*, they are proteolipids). Thus, on the mitoribosomes are made the following proteins: three largest subunits of cytochrome oxidase, one protein subunit of the cytochrome $b–c_1$ complex, four subunits of ATPase and a few hydrophobic proteins. One of the best known differences between the two mechanisms of protein synthesis (*i.e.*, in the cytosol and in the mitochondrial matrix) is in the effect of some inhibitors. The mitochondrial protein synthesis is inhibited by **chloramphenicol**, while synthesis in the cytosol (cytoplasmic matrix) is not affected by this drug. In contrast, **cycloheximide** has the reverse effect.

Import mechanism of mitochondrial proteins. Most mitochondrial proteins are coded by nuclear genes and are synthesized on free ribosomes in the cytosol (cytoplasmic matrix). The import of these polypeptides involves similar mechanism both in mitochondria, and chloroplasts. The transport processes involved have been most extensively studied in mitochondria, especially in yeasts.

Mitochondrial lipid biosynthesis. The biogenesis of new mitochondria and chloroplasts requires lipids in addition to nucleic acids and proteins. Chloroplasts tend to make the lipids they require. For example, in spinach leaves, all cellular fatty acid synthesis takes place in the chloroplast. The major glycolipids of the chloroplast are also synthesized locally.

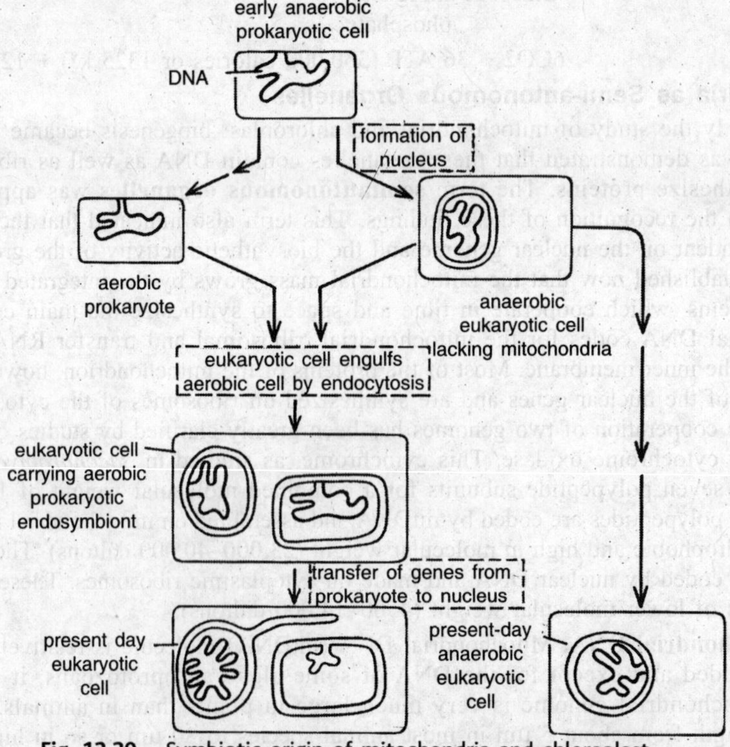

Fig. 13.20. Symbiotic origin of mitochondria and chloroplast.

Mitochondria, on the other hand, import most of their lipids. In animal cells the phospholipids—**phosphatidyl-choline** and **phosphatidyl-serine**—are synthesized in the ER and then transferred to the outer membrane of mitochondria.

Symbiont Hypothesis or Prokaryotic Origin of mitochondria. Early cytologists such as **Altmann** and **Schimper** (1890) have suggested the possibility of origin of the mitochondria from the prokaryotic cells. According to their hypothesis, the mitochondria and chloroplasts may be considered as intra-cellular parasites of the cells which have entered in the cytoplasm of eukaryotic cells in early evolutionary days, and have maintained the symbiotic relations with the eukaryotic cells. The mitochondira are supposed to be derived from the bacterial cells (purple bacteria) while chloroplasts are supposed to be originated from the blue green algae (**Margulis**, 1981). Due to these reasons **Altmann** suggested the name **"bioblasts"** to the mitochondria and he also hinted about their self-duplicating nature.

Recent cytological findings have also suggested many homologies between the mitochondria and the bacterial cells. The similarities between the two can be summarised as follows:

1. Similarity in inner mitochondrial membrane and bacterial plasma membrane.

(*i*) In the mitochondria the enzymes of the respiratory chain are localized on the inner mitochondrial membrane like the bacteria in which they remain localized in the plasma membrane. The bacterial plasma membrane resembles with the inner mitochondrial membrane in certain respects.

(*ii*) The plasma membrane of certain bacterial cells gives out finger-like projections in the cytoplasm known as mesosomes. The mesosomes can be compared with mitochondrial crests. **Salton** (1962) has reported respiratory chain enzymes in the mesosomes.

(*iii*) Because the outer mitochondrial membrane resembles with the plasma membrane, therefore, it may be assumed that the mitochondrial matrix and the inner mitochondrial membrane represent the symbiont which might be enclosed by the membrane of the cellular origin (outer mitochondrial membrane).

2. Similarity in DNA molecule. The DNA molecule of the mitochondria is circular like the DNA molecule of the bacterial cells. Further the replication process of the mitochondrial DNA is also similar to bacterial DNA.

3. Similarity in ribosomes. The mitochondrial ribosomes are small in size and resemble the ribosomes of the bacteria.

4. Similarity in the process of protein synthesis. The process of protein synthesis of both mitochondria and bacteria is fundamentally same because in both, the process of protein synthesis can be inhibited by same inhibitor known as chloramphenicol.

Further, the mitochondria for the process of protein synthesis depend partially on the mitochondrial matrix and DNA and partially on the nucleus and cytoplasm of the eukaryotic cells. This shows the symbiotic nature of the mitochondria.

Due to the above-mentioned similarities between the bacteria and mitochondria, the symbiont hypothesis postulated that the host cell (eukaryotic cell) represented an anaerobic organism which derives the required energy from the oxidations of food by the process of glycolysis. While the mitochondria represent the symbionts which respire **aerobically** and contain the enzymes of **Krebs cycle** and **respiratory chain.** The symbionts seem to be capable to get the energy by **oxidative phosphorylation** from the partially oxidised food (pyruvic acid) of the host cell.

13.5 PLASTIDS

Plastids are present in all living plant cells and in *Euglena* (a protozoan). They are small bodies found free in the cytoplasm. Plastids are often more or less spherical or disc-shaped (1 μm to

1 mm in diameter), but may be elongated or lobed or show amoeboid characteristics. Their other identifying features are their double bounding membranes, the possession of **plastoglobuli** (spherical lipid droplets; store of lipids surplus to current requirements) and an internal membrane fretwork of many discrete internal vesicles. All plastids in a particular plant species contain multiple copies of same relatively small genome (DNA) and 70S-type ribosomes. They are self-replicating organelles containing a protein-synthesizing capacity comparable to that of mitochondria. They perform most important biological activities as the synthesis of food and storage of carbohydrates, lipids and proteins. Plastids are absent in the cells of fungi, bacteria, animals and male sperm cells of certain higher plants.

Types of Plastids. The term 'plastid' is derived from the Greek word "*plastikas*" (= formed or moulded) and was used by **A.F.W. Schimper** in 1885. **Schimper** classified the plastids into following types according to their structure, pigments and the functions:

1. Leucoplasts. The leucoplasts (Gr., *leuco* = white; *plast* = living) are the colourless plastids which are found in embryonic and germ cells. They are also found in meristematic cells and in those regions of the plant which are not receiving light. Plastids located in the cotyledons and the primordium of the stem are colourless (leucoplasts) but eventually become filled with chorophyll and transform into chloroplasts. **True leucoplasts** occur in fully differentiated cells such as epidermal and internal plant tissues. They never become green and photosynthetic. True leucoplasts do not contain thylakoids and even ribosomes. They store the food materials as carbohydrates, lipids and proteins and accordingly are of following types:

***(i)* Amyloplasts.** The amyloplasts (L., *amyl* = starch; Gr., *plast* = living) are those leucoplasts which synthesize and store the starch. The amyloplasts occur in those cells which store the starch. The outer membrane of the amyloplast encloses the stroma and contains one to eight starch granules. In some plant tissues such as potato tuber, amyloplasts can grow to be as large as an average animal cell. In them starch granules may become so large that they rupture the encasing membrane. Starch granules of amyloplasts are typically composed of concentric layers of starch.

***(ii)* Elaioplasts.** The elaioplasts store the lipids (oils) and occur in seeds of monocotyledons and dicotyledons. They also include sterol-rich **sterinochloroplast**.

***(iii)* Proteinoplasts.** The proteinoplasts are the protein storing plastids which mostly occur in seeds and contain few thylakoids.

2. Chromoplasts. The chromoplasts (Gr., *chroma* = colour; *plast* = living) are the coloured plastids containing **carotenoids** and other pigments. They impart colour (*e.g.*, yellow, orange and red) to certain portions of plants such as flower petals (*e.g.*, daffodils, rose), fruits (*e.g.*, tomatoes) and some roots (*e.g.*, carrots). Chromoplast structure is quite diverse; they may be round, ellipsoidal, or even needle-shaped, and the carotenoids that they contain, may be localized in droplets or in crystalline structures. The function of chromoplasts is not clear but in many cases (*e.g.*, flowers and fruits) the colour they produce probably plays a role in attracting insects and other animals for pollination or seed dispersal.

In general, chromoplasts have a reduced chlorophyll content and are, thus, less active photosynthetically. The red colour of ripe tomatoes is the result of chromoplasts that contain the red pigment **lycopene** which is a member of carotenoid family. Chromoplasts of blue-green algae or cyanobacteria contain various pigments such as **phycoerythrin, phycocyanin, chlorophyll a** and **carotenoids**. Chromoplasts are of following two types:

***(i)* Phaeoplast.** The phaeoplast (Gr., *phaeo* = dark or brown; *plast* = living) contains the pigment **fucoxanthin** which absorbs the light. The phaeo-plasts occur in the diatoms, dinoflagellates and brown algae.

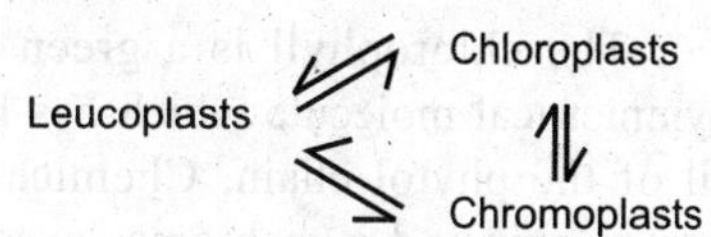

Fig. 13.21. Interconversion of three kinds of plastids.

***(ii)* Rhodoplast.** The rhodoplast (Gr., *rhode* = red; *plast* = living) contains the pigment **phaeoerythrin** which absorbs the light. The rhodoplasts occur in the red algae.

3. Chloroplasts. The chloroplast (Gr., *chlor* = green; *plast* = living) is most widely occurring chromoplast of the plants. It occurs mostly in the green algae and higher plants. The chloroplast contains the pigment chlorophyll *a* and chlorophyll *b* and DNA and RNA.

According to **Schimper**, different kinds of plastids can transform into one another, as shown in Fig. 13.21.

13.6 CHLOROPLASTS

Chloroplasts are **chlorophyll**-containing specialized organelles of plant cells which are involved in the photosynthesis of food (*i.e.*, carbohydrates). So chloroplasts act as **"kitchens of cells"**. They contain their own DNA, ribosomes and protein synthetic machinery to synthesize certain proteins of their use; that is why, they are termed as **semi-autonomous organelles**.

Distribution. The chloroplasts remain distributed homogeneously in the cytoplasm of plant cells. But in certain cells, the chloroplasts become concentrated around the nucleus or just beneath the plasma membrane. The chloroplasts have a definite orientation in the cell cytoplasm. Since chloroplasts are motile organelles, they show passive and active movements.

Morphology

Shape. Higher plant chloroplasts are generally biconvex or plano-convex. However, in different plant cells, chloroplasts may have various shapes, *viz.*, filamentous, saucer-shaped, spheroid, ovoid, discoid or club-shaped. They are vesicular and have a colourless centre.

Size. The size of the chloroplasts varies from species to species. The chloroplasts generally measure 2–3 μm in thickness and 5–10 μm in diameter (*e.g.*, *Chlamydomonas*). The chloroplasts of polyploid plant cells are comparatively larger than the chloroplasts of the diploid plant cells. Generally, chloroplast of plants grown in the shade are larger and contain more chlorophyll than those of plants grown in sunlight.

Number. The number of the chloroplasts varies from cell to cell and from species to species and is related with the physiological state of the cell, but it usually remains constant for a particular plant cell. The algae usually have a single huge chloroplast. The cells of the higher plants have 20 to 40 chloroplasts. According to a calculation, the leaf of *Ricinus communis* contains about 400,000 chloroplasts per square millimeter of surface area. When the number of chloroplasts is inadequate, it is increased by division; when excessive, it is reduced by degeneration.

Chemical Composition

The chloroplasts are composed of the carbohydrates, lipids, proteins, chlorophyll, carotenoids (carotene and xanthophylls), DNA, RNA and certain enzymes and coenzymes. The chloroplasts also contain some metallic atoms as Fe, Cu, Mn and Zn.

The carbohydrates occur in very low percentage in the chloroplasts. The most common carbohydrates of the chloroplasts are the starch and sugar phosphates.

The chloroplasts contain 20–30 per cent lipids of its dry weight. The lipids are composed of 50 per cent fats, 20 per cent sterols, 16 per cent waxes and 7 to 20 per cent phospholipids. The most common alcohols of the lipids are the choline, inositol, glycerol, ethanolamine.

The proteins constitute 35 to 55 per cent of the chloroplast. About 80 per cent proteins are insoluble and forming the unit membranes of the chloroplasts along with the lipids. The 20 per cent proteins are soluble and occur in the form of the enzymes.

The chlorophyll is a green pigment of the chloroplasts. The chlorophyll contains an asymmetrical molecule which has hydrophilic head of four rings of the pyrols and hydrophobic tail of the phytol chain. Chemically the chlorophyll is a porphyrin like the animal pigment haemoglobin and cytochromes except besides the iron (Fe), it contains Mg atom in between the rings of the pyrols which remain connected with each other by the methyl groups. The chlorophyll consists of 75 per cent **chlorophyll *a*** and 25 per cent **chlorophyll *b***.

The carotenoids are carotenes and xanthophylls, both of which are related to vitamin A. The carotenes have hydrophobic chains of unsaturated hydrocarbons in their molecules. The xanthophylls contain many hydroxy groups in their molecules.

DNA of chloroplast of *Chlamydomonas* represents non-chromosomal genetic system and has been found to be related with cytoplasmic heredity. **Ruth Sager**, who is pioneer on nonchromosomal genes, was able to prepare genetic map of *Chlamydomonas* chloroplast. She has shown that the genome of the chloroplast of *Chlamydomonas* is circular like that of bacteria.

Ultrastructure

A chloroplast comprises the following three main components (Fig. 13.22):

1. Envelope. The entire chloroplast is bounded by an **envelope** which is made up of double unit membranes. Across this double membrane envelope occurs exchange of molecules between chloroplast and cytosol (cytoplasmic matrix). Isolated membranes of envelope of chloroplast lack chlorophyll pigment and cytochromes but have a yellow colour due to the presence of small amounts of carotenoids. They contain only 1 to 2 per cent of the total protein of the chloroplast.

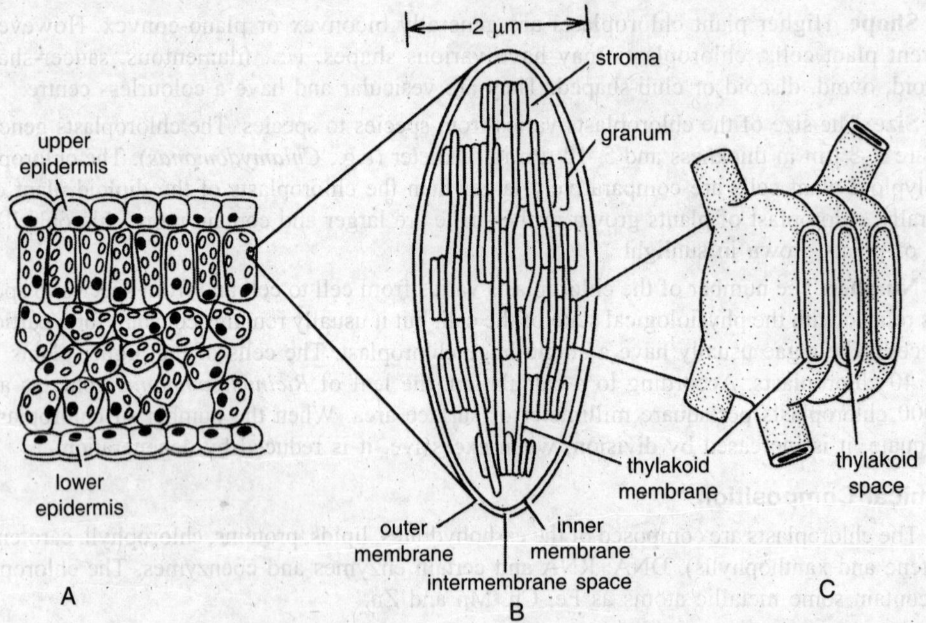

Fig. 13.22. A—Distribution of chloroplasts in mesophyll cells of a leaf; B—Ultrastructure of a chloroplast; C—Details of a granum.

2. Stroma. The matrix or stroma fills most of the volume of the chloroplasts and is a kind of gel-fluid phase that surrounds the thylakoids (grana). It contains about 50 per cent of the proteins of the chloroplast, most of which are soluble type. The stroma also contains ribosomes and DNA molecules (*i.e.*, 80 DNA molecules per chloroplast per cell of *Chlamydomonas*; 20 to 40 DNA molecules per chloroplast per cell of leaf of maize), both of which are involved in the synthesis of some of the structural proteins of the chloroplast. The stroma is the place where CO_2 fixation occurs and where the synthesis of sugars, starch, fatty acids and some proteins takes place.

3. Thylakoids. The thylakoid (thylakoid = sac-like) consists of flattened and closed vesicles arranged as a membranous network. The outer surface of the thylakoid is in contact with the stroma, and its inner surface encloses an **intrathylakoid space** (the third compartment). Thylakoids may be stacked like a neat pile of coins, forming **grana** or they may be unstacked, **intergranal**, or **stromal thylakoids**, forming a system of anastomosing tubules that are joined to the **grana thylakoids**. There may be 40 to 80 grana in the matrix of a chloroplast. The number of thylakoids per granum may vary from 1 to 50 or more. For example, there may be single thylakoid (*e.g.*, red alga), paired thylakoids (*e.g.*, Chrysophyta), triple thylakoids and multiple thylakoids (*e.g.*, green algae and higher plants).

Molecular organization of thylakoids. Molecular organization of the membrane of thylakoids is based on the fluid-mosaic model of the membrane which represents following main characteristics: **fluidity, asymmetry** and **economy** (*i.e.*, lack of movement in the third dimension). Lipids represent about 50 per cent of the thylakoid membrane; these include those directly involved in photosynthesis (called **functional lipids**) such as **chlorophylls, carotenoids** and **plastoquinones. Structural lipids** of thylakoids include glycolipids, sulpholipids and a few phospholipids. Most of these structural lipids are highly unsaturated which confer to the membrane of thylakoids a high degree of fluidity.

The protein components of thylakoid membrane are represented by 30 to 50 polypeptides which are disposed in the following five major supramolecular complexes (Fig. 13.23), which can be isolated with mild detergent:

1. Photosystem I (PS I). This complex contains a reactive centre composed of P700 (Type of pigment which is bleached at the wavelength of 700 nm), several polypeptides, a lower chlorophyll *a*/*b* ratio and β-carotene. It acts as a light trap and is present in unstacked thylakoid membranes. In it light induced reduction of $NADP^+$ takes place.

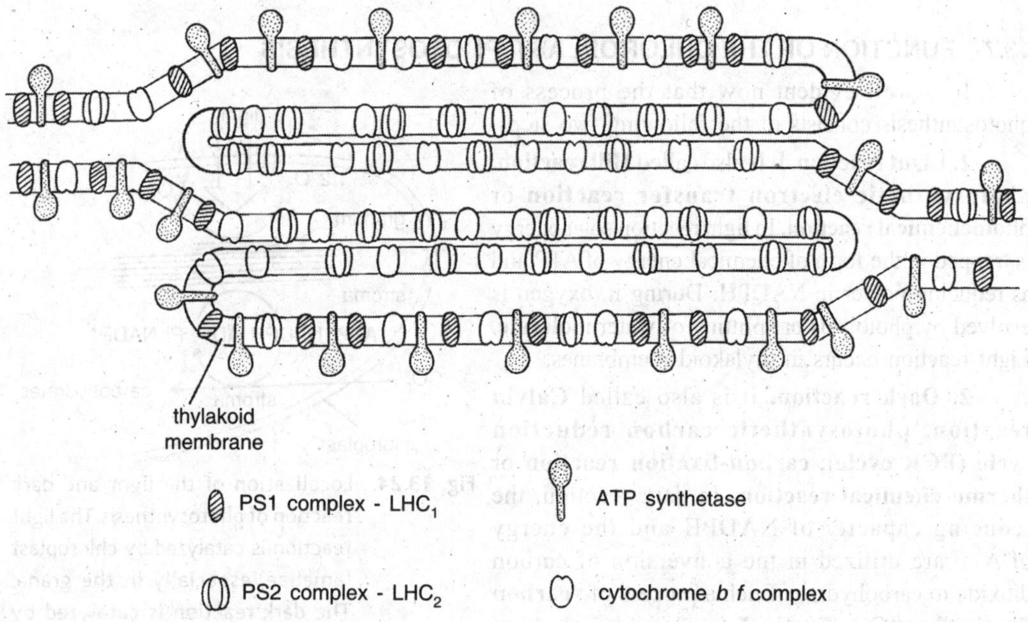

Fig. 13.23. Diagram showing the distribution of the main complexes within the thylakoid membranes both in the granal or stacked and stromal or unstacked regions.

2. Photosystem II (PS II). This complex comprises two intrinsic proteins that bind to the reaction centre of chlorophyll P680 (The pigment that bleaches when absorbing light at 680 nm). It contains a high ratio of chlorophyll *a*/*b* and β-carotene. Frequently, the PS IIs are associated with the light-harvesting complex and are involved in light induced release of O_2 from H_2O (*i.e.*, photolysis of water). PS II works as a light trap in photosynthesis and is mainly present in the stacked thylakoid membranes of grana.

3. Cytochrome b/f. This complex contains one cytochrome *F*, two cytochromes of *b*563, one FeS centre and a polypeptide. It is uniformly distributed in the grana and acts as the electron carrier.

These three complexes are related to the electron transport and are linked by **mobile electron carriers** (*i.e.*, plastoquinone, plastocyanin and ferredoxin). Electron transport through PS II and PS I finally results in the reduction of the coenzyme $NADP^+$. Simultaneously, the transfer of protons from the outside to the inside of the thylakoid membrane occurs.

4. ATP synthetase. As in mitochondria, this complex consists of a $\mathbf{CF_0}$ hydrophobic portion, a proteolipid that makes a proton channel, and a $\mathbf{CF_1}$ (or coupling factor one) that synthesizes ATP from ADP and Pi, using the proton gradient provided by the electron transport. ATP synthetase complexes are located in stacked membrane (grana).

5. Light harvesting complex (LHC). The main function of LH complex is to capture solar energy. It contains two main polypeptides and both cholorophyll *a* and *b*. LH complex is mainly associated with PS II, but may also be associated with PS I. LHC is localized in stacked membranes and lacks photochemical activity.

Mutation and chloroplast structure. The organization of chloroplasts and other plastids is often modified due to mutation. **D.Von Wettstein** (1956) reported that the plastids of normal barley plant have a well organized system of grana and stroma. But the plastids of an albino mutant of barley, fail to develop beyond a particular stage and there occurs no differentiation of grana and stroma. Further, the plastids of a yellow-green mutant of barley develop somewhat further than plastids of an albino plant.

13.7 FUNCTION OF THE CHLOROPLAST: PHOTOSYNTHESIS

It is well evident now that the process of photosynthesis consists of the following two steps:

1. Light reaction. It is also called **Hill reaction, photosynthetic electron transfer reaction** or **photochemical reaction.** In light reaction solar energy is trapped in the form of chemical energy of ATP and as reducing power in NADPH. During it, oxygen is evolved by photolysis or splitting of water molecule. Light reaction occurs in thylakoid membranes.

2. Dark reaction. It is also called **Calvin reaction, photosynthetic carbon reduction cycle (PCR cycle), carbon-fixation reaction** or **thermo-chemical reaction.** In dark reaction, the reducing capacity of NADPH and the energy of ATP are utilized in the conversion of carbon dioxide to carbohydrate. Such a process of **"carbon fixation"** or **"CO_2-fixation"** occurs in the stroma of chloroplast (Fig. 13.24).

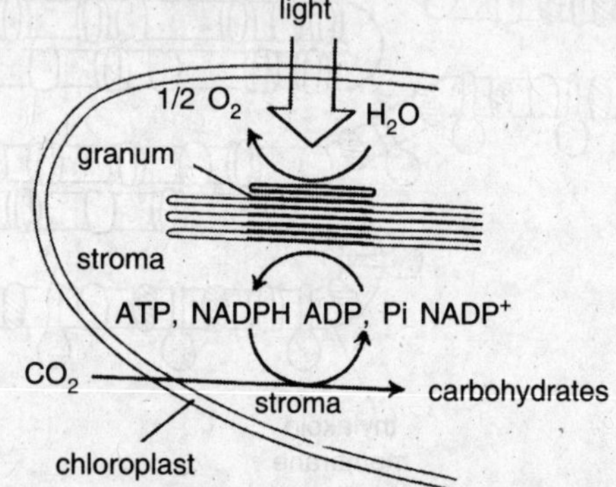

Fig. 13.24. Localization of the light and dark reaction of photosynthesis. The light reaction is catalyzed by chloroplast lamellae, especially in the grana. The dark reaction is catalyzed by enzymes of the stroma.

1. Light Reaction

The most important step of light reaction is harvesting of the maximum amount of solar energy for conversion into chemical energy. When the P680 of photosystem II acquires a sufficient quantum of energy, it emits a pair of electrons. These electrons with high potential energy move down an electron transport chain (of thylakoid membrane) and during this process ATP molecules are formed (Fig. 13.25). Two electrons are passed through an electron acceptor Q (which is a quinone) to an electron transport chain involving four electron carriers (**plastoquinone, cytochrome-559, cytochrome-553** or **cytochrome *f*** and **plastocyanin**), before being passed on to PS I. The electrons are passed through four carriers successively at lower energy levels, so that at each step energy is released, which is harvested in the production of two ATP molecules (from ADP + Pi). The electron lost by P 680 is ultimately accepted by P700 of photosystem I. P700 (PS I) also captures light, and for absorbing each photon, it ejects an electron. This electron is replaced by an electron from PS II and this flow of electron continues as long as the light is available. The electrons liberated from P700 are passed through acceptor X, to an electron transport chain **(ferredoxin, ferredoxin NADP reductase)** at successive lower energy levels. Finally these electrons reach NADP coenzyme, each molecule of which receives an electron, enabling it to pick up a H^+ ion (proton), thus, producing two molecules of NADPH from one molecule of H_2O used in PS II:

$$2e^- + 2H^+ + 2NADP^+ \xrightarrow{2h\nu} 2\,NADPH$$

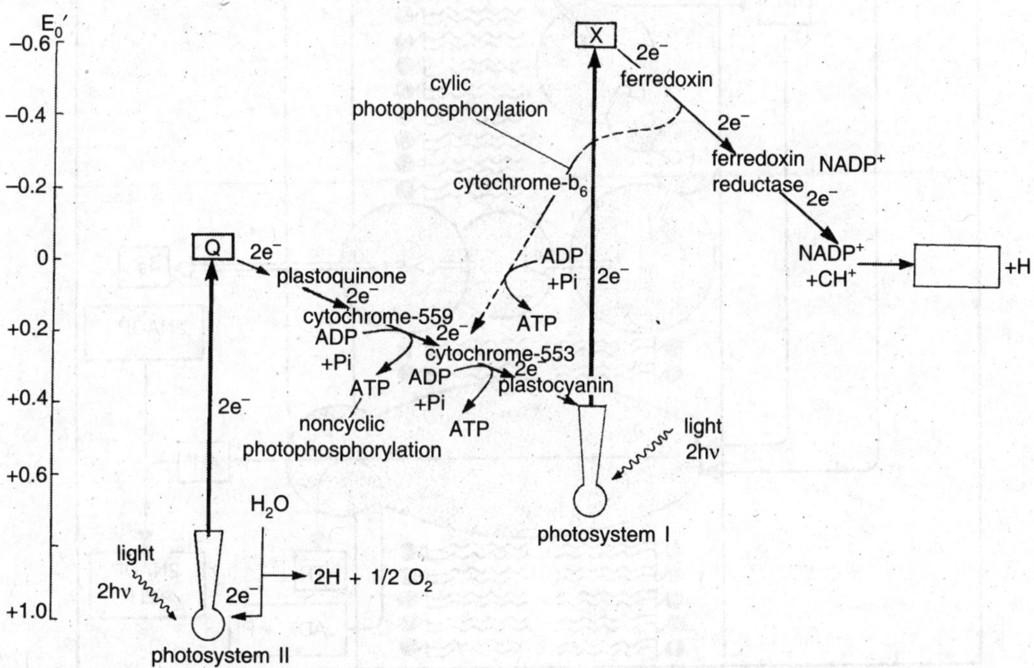

Fig. 13.25. Photosynthetic light reaction showing major events occurring in photosystem I and photosystem II of chloroplasts.

The oxidized P680 regains its electrons by the photolysis of water into $2H^+$, $2e^-$ and oxygen:

$$H_2O \xrightarrow{2\text{ quanta or }2h\nu} \tfrac{1}{2}O_2 + 2H^+ + 2e^-$$

Oxygen is given out by photosynthesizing plants. The protons (H^+) accumulate inside the thylakoid membrane resulting in a **proton gradient.** The energy released by the protons when they diffuse across the thylakoid membrane into the stroma (along H^+ concentration gradient) is used to produce ATP molecules, by **CF_0 – CF_1 ATP synthetase** in the membrane. As synthesis of ATP occurs in light and the process is not cyclic (*i.e.*, it needs a constant supply of water molecules to be oxidized and NADP to be reduced), the process is called **non-cyclic photophosphorylation**.

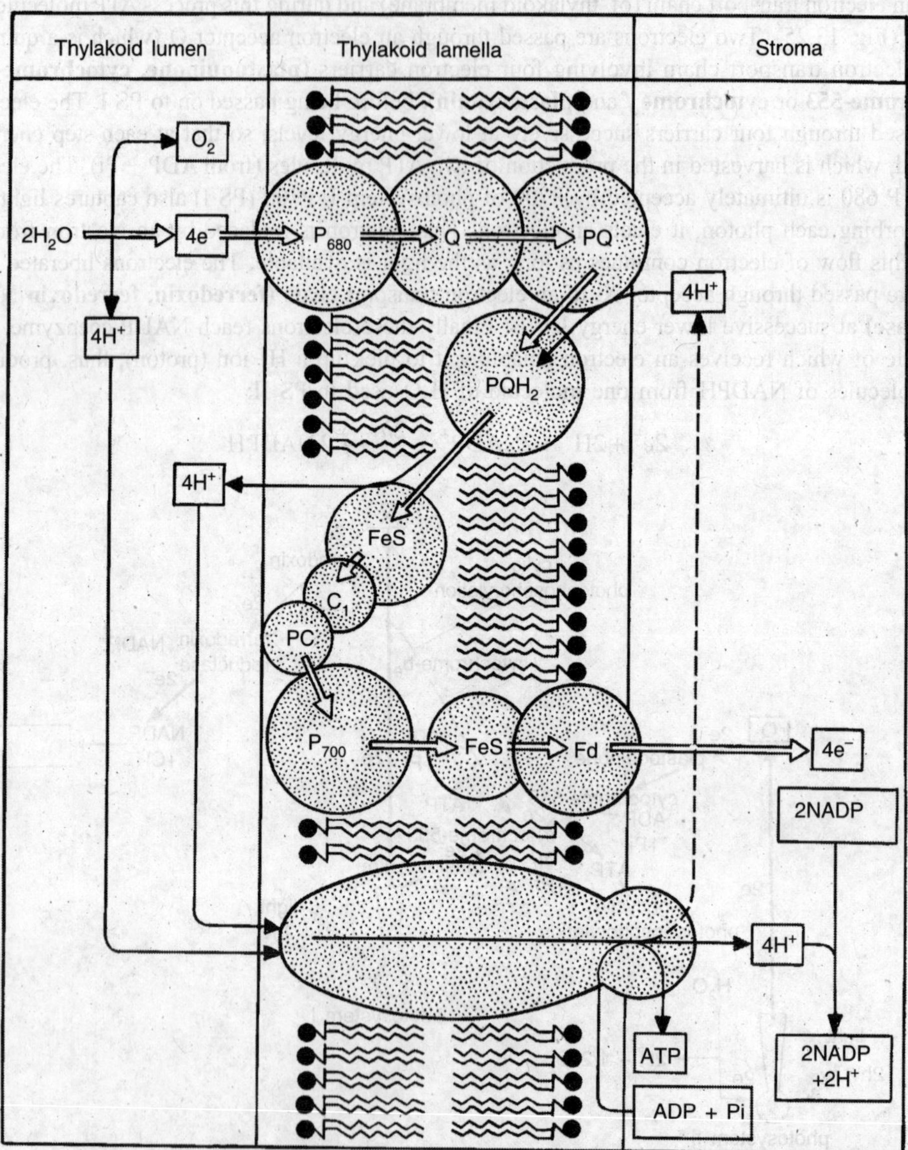

Fig. 13.26. Model for chemiosmotic coupling. Locations of electron transfer intermediates in the thylakoid lamella (membrane) and enzymes involved in ATP synthesis and proton (H^+) transport have been tentatively depicted.

ATP production by **cyclic photophosphorylation** also occurs during the light reaction. This process involves another electron transfer mechanism involving **cytochrome b_6** and starting with P700; the ultimate acceptor of the de-energized electron is also P700 of photosystem I.

All the molecules of ATP and NADP generated in the light reaction of photosynthesis are used by soluble enzymes of stroma of chloroplast during the dark reaction. Total light reaction can be summarized as follows:

$$H_2O + 2\,NADP + 2\,ADP + 2\,Pi \xrightarrow{4h\nu} \tfrac{1}{2}O_2 + 2\,NADPH + 2\,ATP$$

2. Dark Reaction

The dark reaction is completed by passing through following three main phases:

***(i)* Phase 1: Carboxylation.** During this phase of dark reaction, three molecules of carbon dioxide (3C) are attached to three molecules of **ribulose 1, 5, biphosphate (RuBP;** this pentose sugar was previously termed RuDP or ribulose diphosphate; 15C) to produce short-lived six-carbon intermediates. This process is called **carboxylation** and is catalyzed by the enzyme **RuBP carboxylase, carboxydismutase** or **"Rubisco"** (which is widely acclaimed as one of the most abundant proteins present on the planet Earth). Rubisco is a large protein molecule (500,000 dalton MW) comprising 16 per cent of chloroplast protein. The six carbon intermediates are immediately broken down into six molecules of **PGA** or **3-phosphoglyceric acid** (*i.e.*, a *C*–3 or three carbon compound, 6C × 3C = 18C).

***(ii)* Phase 2: Glycolytic reversal.** By utilizing six ATP molecules, these six molecules of PGA are transformed into six molecules of **1, 3, diphosphoglyceric acid** (*i.e.*, 6C × 3C = 18C; 3C from $3CO_2$ and 15C from 3 RuBP). These in turn get converted into six molecules of **glyceraldehyde-3-phosphate, 3-phosphoglyceraldehyde (PGAL)** or **3-phosphoglyceric acid** (triose) by utilizing six NADPH molecules.

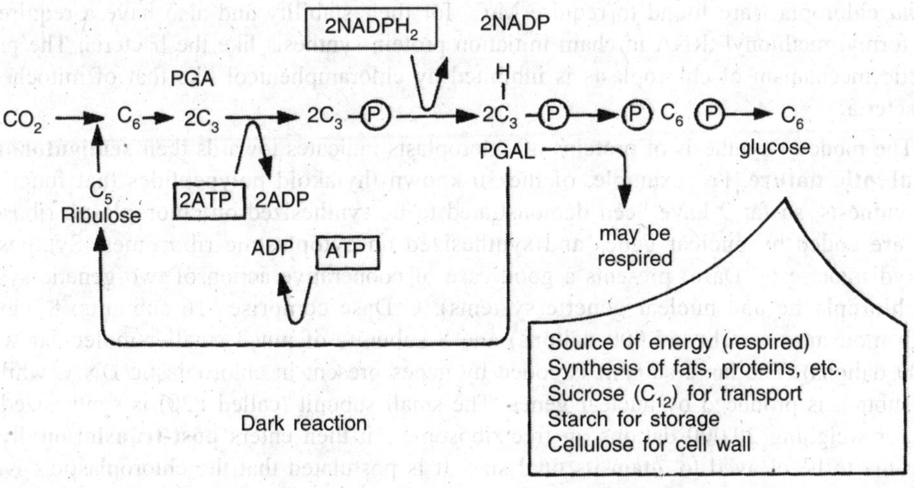

Fig. 13.27. Synthesis of carbohydrate during dark reaction.

***(iii)* Phase 3: Regeneration of RuBP.** For the continuous running of Calvin cycle, there must be a regular supply of ATP and NADPH and also sufficient amount of RuBP. Three molecules of RuBP (3C × 5C = 15C) are regenerated by a complex series of reactions which utilize three ATP molecules and five molecules of 3-phosphoglyceric acid (5C × 3C = 15C). This leaves one molecule of PGAL as the net gain of one Calvin cycle. Two turns of Calvin cycle result in the production of one molecule of **glucose**. This glucose is used by the plant to form a large variety of organic compounds required for its structure and function (*e.g.*, starch, cellulose, lipids, amino acids and proteins, etc.). The dark reaction may be summed up as follows:

$$6\,RuBP + 6CO_2 + 18\,ATP + 12\,NADPH \longrightarrow 6\,RuBP + C_6H_{12}O_6 + 18\,ADP + 18\,Pi + 12\,NADP$$

Chloroplast as Semiautonomous Organelle

Like the mitochondria, the chloroplasts have their own DNA, RNAs and protein synthetic machinery and are semiautonomous in nature.

1. DNA of chloroplast. The chloroplasts of the algae and higher plants are found to contain DNA molecules. First of all **Ris** and **Plant** (1962) have reported DNA molecule in the chloroplast of the *Chlamydomonas*. Later on DNA molecule has been reported from the chloroplasts of other algae and higher plants. In general, chloroplasts have a double helical DNA circle with an average length of 45 µm (about 135,000 base pairs). The replication of chloroplast DNA has been followed with ^3H– thymidine. Maps of the location of genes (genetic maps) have been made in several chloroplast DNAs by the help of restriction enzymes. The gene for the large subunit of carboxydismutase enzyme has been fully sequenced and is found to contain 1425 nucleotides.

2. Ribosomes of chloroplasts. The chloroplasts contain the ribosomes which are smaller than the cytoplasmic ribosomes. The ribosomes of the chloroplast are of 70S type and resemble with the bacterial ribosomes. The ribosomes of the chloroplasts consist of two ribosomal RNAs, 23S rRNA and 16S rRNA. **Lyttleton** (1962) has separated polyribosomes or polysomes from the chloroplast. The chloroplasts also contain aminoacyl-tRNAs, aminoacyl-tRNA synthetases, methionyl- tRNA.

3. Protein synthesis. The DNA of chloroplast codes for chloroplast mRNA, rRNA, tRNA, and ribosomal proteins. It also codes for certain structural proteins of thylakoid membranes. The synthesis of other chloroplast components as chlorophyll, carotenoids, lipids and photosynthetic and starch synthesizing enzymes, is controlled by nuclear genes. The 70S ribosomes of *Euglena* chloroplast are found to require Mg^{++} for their stability and also have a requirement for N-formyl methionyl-tRNA in chain initiation protein synthesis like the bacteria. The protein synthetic mechanism of chloroplasts is inhibited by chloramphenicol like that of mitochondria and bacteria.

The mode of synthesis of proteins of chloroplasts indicates towards their **semiautonomous** or **symbiotic nature**. For example, of the 30 known thylakoid polypeptides that function in photosynthesis, so far 9 have been demonstrated to be synthesized on chloroplastic ribosomes and 9 are coded by nuclear genes and synthesized on cytoplasmic ribosomes. Synthesis of carboxydismutase (C Dase) presents a good case of cooperative action of two genetic systems (*i.e.*, chloroplastic and nuclear genetic systems). C Dase comprises 16 subunits: 8 subunits of high molecular weight (55,000 daltons) and 8 subunits of much smaller molecular weight (14,000 daltons). The large subunit is coded by genes present in chloroplastic DNA, while the small subunit is produced by nuclear genes. The small subunit (called P20) is synthesized as a precursor weighing 20,000 daltons on free ribosomes; it then enters **post-translationally** into the stroma to be cleaved to attain its final size. It is postulated that the chloroplastic envelope has receptor sites that recognize the proteins that are to be incorporated into the organelle. The extra sequence (acting as the **signal**) that is present in P20 is composed of acidic amino acids, in contrast to the hydrophobic ones in the signal sequence of secretory proteins. After entering the chloroplast the signal sequences are removed by a protease enzyme, which is present in the envelope of chloroplast, and the small subunit of C Dase is released into the stroma (**Ellis**, 1981). Thus, chloroplast proteins may be synthesized by three avenues:

1. by an exclusive chloroplastic mechanism,
2. by a mechanism involving nuclear genes and chloroplastic ribosomes, and
3. by nuclear genes and cytoplasmic ribosomes.

Protein transport into chloroplasts resembles transport into mitochondria in many respects: both occur post-translationally, both require energy, and both utilize hydrophilic amino-terminal signal peptides that are removed after use. However, there is at least one important difference that while mitochondria exploit the electrochemical gradient across their inner membrane to help drive the transport, chloroplasts (which have an electrochemical gradient across their thylakoid but not their inner membrane) appear to employ only ATP hydrolysis to import across their double-membrane outer envelope.

Translocation of proteins into the thylakoid space of chloroplasts requires two signal peptides and two translocation events. The precursor polypeptide contains an amino-terminal chloroplast signal peptide followed immediately by a thylakoid signal peptide. The **chloroplast signal peptide** initiates translocation into the stroma through a membrane contact site by a mechanism similar to that used for translocation into mitochondrial matrix. The signal peptide is then cleaved off, unmasking the thylakoid signal peptide, which initiates translocation across the thylakoid membrane (Fig. 13.28).

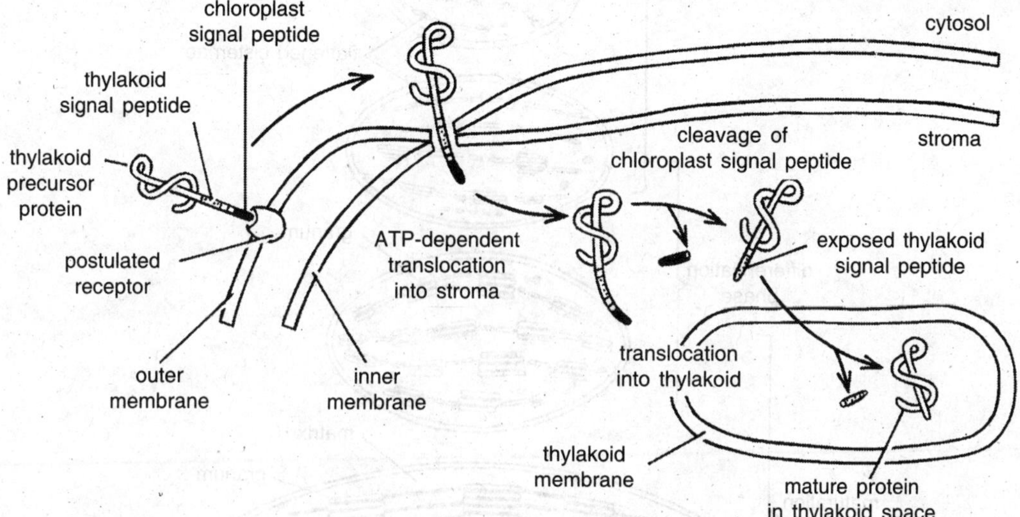

Fig. 13.28. Mode of translocation of proteins into stroma and thylakoid of chloroplast by the help of signal peptides.

Biogenesis of Chloroplast

The chloroplasts never originated *de novo*. Since the classic work of **Schimper** and **Meyer** (1883) it has been accepted that chloroplasts multiply by fission, a process that implies growth of the daughter organelles. This is easily observed in the alga *Nitella,* which contains a single huge chloroplast. In *Nitella* a division cycle of 18 hours has been cinematographically recorded for the chloroplast.

During the development of the chloroplast, the first structure to appear is the so-called **proplastid**, which has a double membrane. Development of proplastid into chloroplast takes place in the following steps:

1. In the presence of light, the inner membrane grows and gives off vesicles into the matrix that are transformed into discs (Fig. 13.29). These intrachloroplastic membranes are the thylakoids which, in certain regions, pile closely to form the grana. In the mature chloroplast the thylakoids are no longer connected to the inner membrane, but the grana remain united by inter-granal thylakoids.
2. In the absence of light, a reverse sequence of changes takes place. This is the process of **etiolation,** in which the leaves lose their green pigment and the chloroplast membranes become disorganized. The chloroplasts are transformed into **etioplasts**, in which there is a paracrystalline

arrangement of tubules forming the so-called **prolamellar body**. Attached to these bodies are young thylakoid membranes that lack photosynthetic activity.

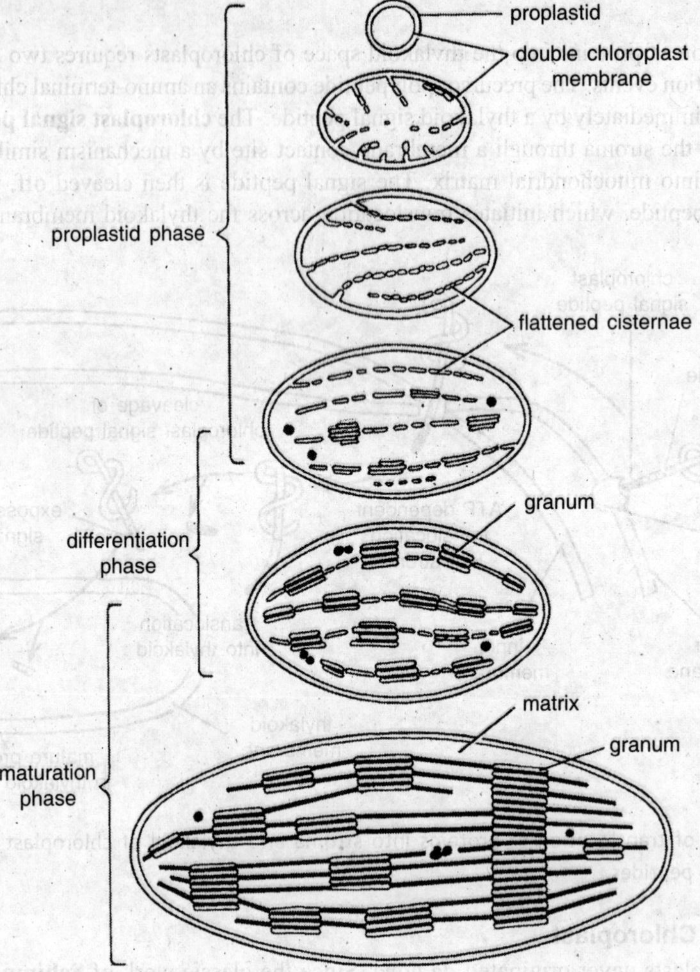

Fig. 13.29. Development of a chloroplast from a submicroscopic proplastid in the presence of light.

The regular crystal lattice of two prolamellar bodies surrounded by young thylakoid membranes is observed by **Osumi** *et al.*, (1984). If etiolated plants are re-exposed to light, thylakoids are reformed and the prolamellar material is used for assembly.

The symbiotic origin of the chloroplast. In certain characteristics, the chloroplasts are comparable with that of a semiautonomous or symbiotic organism living within the plant cells. They divide, grow and differentiate; they contain circular DNA, ribosomal RNA, messenger RNA and are able to conduct protein synthesis. By visualizing these similarities between chloroplast and micro-organism; it has been suggested that chloroplast might have resulted from a symbiotic relationship between an autotrophic microorganism, one which is able to transform radiant energy from sunlight and heterotrophic host cell. The symbiotic origin of the chloroplast appears very justified but **Kirk** (1966) has shown that certain important enzymes which are necessary for the development of the chlorophyll and for the photosynthetic mechanism are synthesized according to the codes of the nuclear DNA.

13.8 COMPARISON OF CHLOROPLASTS AND MITOCHONDRIA

Chloroplasts carry out their energy inter-conversions by chemiosmotic mechanisms in much the same way that mitochondria do and they are organized on the same principles. Thus, each chloroplast contains three distinct membranes which define three separate internal compartments—the intermembrane space, the stroma and the thylakoid space. The thylakoid membrane contains all the energy generating systems of chloroplasts.

Like the mitochondria, chloroplasts have a highly permeable outer membrane; a much less permeable inner membrane, in which special carrier or transport proteins are embedded; and a narrow intermembrane space in between. The inner membrane surrounds a large space called the **stroma**, which is analogous to the mitochondrial matrix and contains various enzymes, ribosomes, RNAs and DNA.

However, there is an important difference between the two: the inner membrane of the chloroplast is not folded into cristae and does not contain an electron-transport chain. Instead, the photosynthetic light-absorbing system, the electorn-transport chain and an ATP synthetase are all contained in a third distinct membrane that forms a set of flattened disc-like sacs, the **thylakoids** (Fig. 13.31).

In a general way, one might view the chloroplast as a greatly enlarged mitochondria in which the cristae are converted into a series of interconnected submitochondrial particles in the matrix space. The knobbed ends of the chloroplast ATP synthetases ($F_0 - F_1$ coupling factors), where ATP is made, protrude from the thylakoid membrane into the stroma, just as they protrude into the matrix from the membrane of each mitochondrial crista.

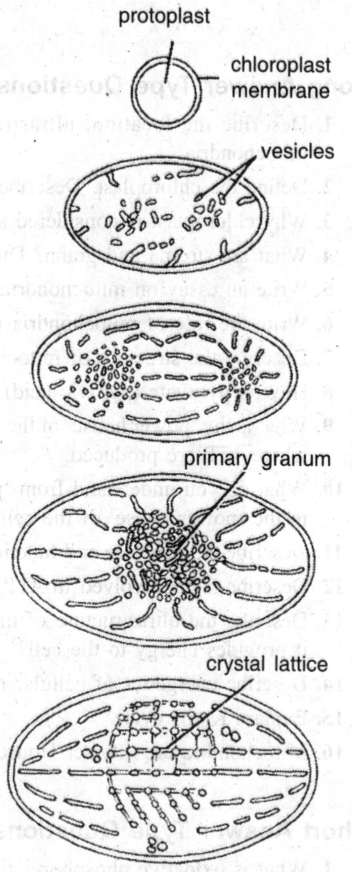

Fig. 13.30. Development of a chloroplast from submicroscopic proplastid in dark.

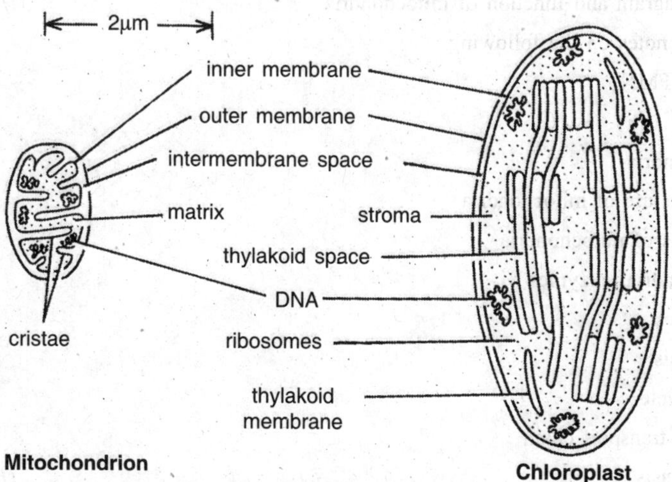

Fig. 13.31. Comparison of a mitochondrion and a chloroplast.

QUESTIONS

Long Answer Type Questions

1. Describe the location, ultrastructure, biochemistry (or chemical composition) and functions of mitochondria.
2. Define the chloroplast. Describe its structure, chemical composition and main functions.
3. Why chloroplast is considered as a semiautonomous organelle. Discuss it.
4. What are stroma and grana? Discuss their role in the process of photosynthesis.
5. Write an essay on mitochondria.
6. Write the role of mitochondria in cellular activities.
7. Describe the structure of mitochondria. Why is it known as power house of the cell?
8. How is pyruvate (pyruvic acid) metabolized via Krebs cycle? Describe the sequence of reactions.
9. What is the 'power house' of the cell? Describe the structure and chemical composition of mitochondria where ATP are produced.
10. What do you understand from 'power house' and 'atom bomb' of the cell? How energy is produced in the 'power house' of the cell?
11. Describe the structure and function of mitochondria with special reference to electron transport system.
12. Describe steps involved in ATP synthesis in mitochondrion.
13. Describe the ultrastructure of mitochondrion with the help of neat and labeled diagrams. How does it provides energy to the cell?
14. Describe energetics of cellular respiration.
15. Explain Krebs cycle.
16. Mitochondria are "Power House" of the cell. Justify this statement with cellular respiration. *(CCSU Meerut 2005)*

Short Answer Type Questions

1. What is oxidative phosphorylation? Where does it occur in the cell?
2. What is the role of elementary particles?
3. Where are the enzyme of respiratory or ETS system located in mitochondria?
4. Give a brief account of the plastids.
5. Write a short note on chloroplast.
6. Give the diagram and function of mitochondria. *(HNB., Garwal 2008)*
7. Write short notes on the following:

   (*i*) FAD or FMN

   (*ii*) ATP

   (*iii*) Mitochondria

   (*iv*) Ultrastructure of mitochondria

   (*v*) Functions of mitochondria

   (*vi*) Mitochondrial enzymes

   (*vii*) Oxysomes

   (*viii*) Glycolysis

   (*ix*) Krebs cycle

   (*x*) Electron transport chain

   (*xi*) Chloroplasts *(HNB Garwal 2005)*

8. Compare the chloroplasts with mitochondria.

9. Differentiate between leucoplasts and chromoplasts.

10. Draw a well labelled diagram of ultrastructure of mitochondria.

11. 'Mitochondrion as a semi-autonomous organelle'.

12. Describe the origin or biogenesis of chloroplasts.

Very Short Answer Questions

1. Who coined the term mitochondria?

2. Name the cell type in which mitochondria are absent.

3. What will happen if all the mitochondria of a cell are destroyed?

4. What will happen if elementary particles of mitochondria are removed?

5. Name the organism which possess only one mitochondria.

6. Which cells possess highest number of mitochondria?

7. For which structure the name bioblast was used by Altman?

8. What is metabolism?

9. Where are enzymes of ETS located in mitochondria?

10. What is the location in enzymes involved in oxidation of pyruvic acid in the mitochondria?

11. Where is the site of oxidative phosphorylation?

12. Who gave the chemiosmotic hypothesis?

13. State the significance of thylakoids of chloroplasts.

14. Name the type of chlorophyll found in all green plant.

15. Name the site of occurrence of photochemical or light reaction.

16. What is the use of sunlight in light reaction?

17. Where does dark or synthetic reaction occur in chloroplasts?

18. What is antenna complex?

19. Define the chlorophyll.

20. What is Calvin cycle?

21. What is the biological significance of carbon atom?

22. Define the chloroplast.

23. What is cellular respiration?

24. What is bioenergetics?

25. Define the metabolic pathway.

26. What is free energy?

27. Define exergonic.

28. What is endergonic?

29. What is ATP?

30. Define coenzyme?

31. What is coenzyme A (CoA)?

32. Define coenzyme Q.

33. What is electron transport system (ETS)?

34. What are cytochromes?

35. Define FAD.

36. Define F_0–F_1 complex.

37. What is chemiosmotic coupling model?

True and False Statements

Indicate true or false statements by a tick mark:

1. ATP synthesis takes place in ribosome

(*a*) True (*b*) False

2. In a biological system an important source of energy is glucose

(*a*) True (*b*) False

3. Krebs cycle takes place in mitochondrial matrix

(*a*) True (*b*) False

Yes or No Questions

Write Yes or No for the following statements:

1. Mitochondrion is made up of double membrane.

2. Cytochrome oxidase enzyme is found in mitochondria.

3. Citric acid cycle was first described by Bateson.

Fill in the Blanks

1. Metabolic reactions in the body are catalysed by ....

2. Citric acid cycle was worked out by ....

3. In dark reaction of photosynthesis carbon fixation occurs in the .... of chloroplast.

4. The light reaction or photochemical reaction of photosynthesis is also called .... reaction and it occurs in .... membranes.

Multiple Choice Questions

Choose the correct answer from the four alternatives given:

1. One of the following is a coenzyme

(*a*) glucose (*b*) protease

(*c*) ATP (*d*) NAD

2. Unit of carbohydrate is

(*a*) monosaccharide (*b*) amino acid

(*c*) fatty acid (*d*) all

3. Cellular respiration, *i.e.*, transformation of chemical energy into usable bioenergy takes place in

(*a*) lysosomes (*b*) ribosomes

(*c*) mitochondria (*d*) endoplasmic reticulum

4. Mitochondria are not found in

(*a*) liver cells (*b*) nerve cells

(*c*) maturing RBCs (*d*) mature RBCs

5. In mitochondria, cristae act as sites for

(*a*) protein synthesis (*b*) oxidation-reduction reactions

(*c*) breakdown of macromolecules (*d*) phosphorylation of flavoproteins

6. Cristae helps in

(*a*) respiration (*b*) transpiration

(*c*) photosynthesis (*d*) photo-oxidation

7. Cytochrome oxidases are found

(*a*) on cristae of mitochondria (*b*) on outer wall of mitochondria

(*c*) in the matrix of mitochondria (*d*) in the lysosomes

8. If the outer membrane is removed from a mitochondrion, the remaining structure is called

(*a*) mitoribosome (*b*) mitoplast

(*c*) microsome (*d*) peroxisome

9. The two types of cellular organelles that transform energy are:

(*a*) chromoplasts and leucoplasts (*b*) mitochondria and chloroplasts

(*b*) mitochondria and chromoplasts (*d*) chloroplasts and leucoplasts

10. Carbon dioxide is fixed in

(*a*) light reaction (*b*) dark reaction

(*c*) aerobic respiration (*d*) anaerobic respiration

11. Calvin cycle is

(*a*) inhibited by light (*b*) dependent upon light

(*c*) independent of light (*d*) supported by light

12. Which of the following occurs in dark reaction of photosynthesis

(*a*) formation of ATP (*b*) release of O_2

(*c*) release of H_2 (*d*) synthesis of PGAL

13. End product of glycolysis is

(*a*) acetyl Co-A (*b*) 1 molecule of pyruvic acid

(*c*) 2 molecules of pyruvic acid (*d*) glucose 1-phosphate

14. At the end of the Krebs cycle, most of the energy removed from the glucose molecule has been transferred to the

(*a*) NADH and $FADH_2$ (*b*) ATP

(*c*) oxaloacetic acid (*d*) citric acid

15. When a molecule is reduced it always

(*a*) loses electrons (*b*) loses both proteins and electrons

(*c*) loses neutrons (*d*) gains electrons

16. Last electron acceptor in ETS in

(*a*) O_2 (*b*) water

(*c*) cytochrome *C* (*d*) cytochrome a_3

17. Enzymes of glycolysis are present in

(*a*) mitochondrial matrix (*b*) cristae

(*c*) cytosol (*d*) all of the above

18. Oxidative phosphorylation means

(*a*) photo-oxidation during photosynthesis (*b*) photolysis of water

(*c*) ATP production in respiration (*d*) anaerobic respiration

ANSWERS

Very Short Answer Questions

1. C. Benda in 1897.

2. Mammalian red blood corpuscles (RBCs).

3. Cell will not get energy for performing various activities.

4. ATP molecules will not be produced.

5. *Micromonas.*

6. Muscle cells.

7. Mitochondria.

8. All chemical reactions occurring within a cell.

9. Inner membrane of mitochondria.

10. Mitochondrial matrix.

11. F_1 particles (of mitochondria).

12. Peter Mitchell.

13. Thylakoids have light absorbing system, electron transport system and ATP synthetase enzyme.

14. Chlorophyll–*a*.

15. Thylakoids of the chloroplasts.

16. Radiant energy present in sunlight is used by chloroplasts to form ATP and to oxidize water molecules.

17. Stroma of chloroplasts.

18. Clusters of hundreds of chlorophyll molecules which trap sunlight energy is called **antenna complex**.

19. Chlorophyll is a light-absorbing molecule that donates photoenergized electrons to organic molecules, initiating photochemical events that lead to the generation of the NADPH and ATP required for the Calvin cycle; because of its absorption properties, chlorophyll gives plants their characteristic green colour.

20. It is a cyclic series of reactions used by photosynthetic organisms for the fixation of carbon dioxide and its reduction to form carbohydrates.

21. Carbon is most important atom in biological molecules, capable of forming upto four covalent bonds.

22. It is a double membrane-enclosed cytoplasmic organelle of plants and algae that contains chlorophyll and enzymes necessary to carry out photosynthesis.

23. It is oxidation-driven flow of electrons from reduced coenzymes to an electron acceptor, usually accompanied by the generation of ATP.

24. It is an area of science that deals with the application of thermodynamic principles to reactions and processes in the biological world.

25. It is a series of cellular enzymatic reaction that convert one molecule to another via a series of intermediates.

26. Free energy (G) represents thermodynamic function that measures the extractable energy content of a molecule under conditions of constant temperature and pressure, the change in free energy is a measure of the ability of the system to do work.

27. It is pertaining to an energy-releasing reaction characterized by a negative free energy change ($\Delta G < O$).

28. It is pertaining to an energy-requiring reaction characterized by a positive free energy change ($\Delta G > O$).

29. ATP or adenosine triphosphate is a adenosine with three phosphates linked to each other by phosphoanhydride bonds and to 5′ carbon of the ribose by a phosphoester bond; it is a principal energy storage compound of most cells, with energy stored in the high-energy phosphoanhydride bonds.

30. It is a small organic molecule that functions along with an enzyme by serving as a carrier of electrons or functional groups.

31. It is a organic molecule that serves as a carrier of acyl groups by forming a high-energy thioester bond with an organic acid.

32. Coenzyme Q (CoQ) is a nonprotein (quinone) component of the mitochondrial electron transport system that serves as a collection point for electrons from both FMN and FAD-linked dehydrogenases; also called **ubiquinone.**

33. It is a group of membrane-bound electron carriers that transfer electrons from the coenzymes NADH and $FADH_2$ to oxygen.

34. These are heme-containing proteins of the ETS involved in the transfer of electrons from coenzyme Q to oxygen by the oxidation and reduction of the central iron atom of the heme group.

35. FAD or flavin adenine dinucleotide is a coenzyme that accepts two electrons and two photons from an oxidizable organic molecule to generate the reduced form, $FADH_2$; it is an important electron carrier in energy metabolism.

36. It is a protein complex in the mitochondrial inner membrane and the bacterial plasma membrane that consists of the F_1 complex bound to F_0 complex; the flow of proton through the F_0 component leads to the synthesis of ATP by the F_1 component.

37. It is a model postulating that electron transport pathways establish proton gradients across membranes and that the energy stored in such gradients can then be used to drive ATP synthesis.

True and False Statements

1. (*b*) **2.** (*a*) **3.** (*a*)

Yes or No Questions

1. Yes **2.** Yes **3.** No

Fill in the Blanks

1. Enzymes **2.** Hans A. Krebs. **3.** Stroma. **4.** Hill, thylakoid.

Multiple Choice Questions

1. (*d*) **2.** (*a*) **3.** (*c*) **4.** (*d*) **5.** (*b*) **6.** (*a*)
7. (*a*) **8.** (*b*) **9.** (*b*) **10.** (*b*) **11.** (*d*) **12.** (*d*)
13. (*c*) **14.** (*a*) **15.** (*d*) **16.** (*a*) **17.** (*c*) **18.** (*c*).

14

Peroxisomes

Peroxisomes are small, simple, single-membrane enclosed cell organelles with a diameter of 0.1 to 1.0 μm. They often contains a dense, crystalline core of oxidative enzymes. Peroxisomes are multifunctional organelles, so contain enzymes involved in a variety of metabolic reactions, including several aspects of energy metabolism. They are found in all eukaryotic cells; most human cells contain about 500 peroxisomes. Peroxisomes do not have their own genomes and ribosomes and all their proteins, called **proxins** (Pex 1, Pex 2, etc.) are synthesized from the nuclear genome. Most proxins are synthesized on free ribosomes and then imported into peroxisomes as completed polypeptide chains.

Comparison of Peroxisomes and Mitochondria.

Peroxisomes share several properties with mitochondria:

1. Both mitochondria and peroxisomes form by splitting from preexisting organelles. (However, unlike mitochondria, peroxisomes can also be regenerated even if entirely lost to the cell).
2. Both types of organelles import performed proteins.
3. Both types of organelles engage in similar types of oxidative metabolism. In fact, at least one enzyme, alanine/glyoxylate aminotransferase, is found in the mitochondria of some mammals (*e.g.*, cats and dogs) and the preoxisomes of other mammals (*e.g.*, rabbit and humans; see **Cooper** and **Hausman**, 2007).
4. Peroxisomes use molecular oxygen like mitochondria, but instead of having cytochromes and capacity of ATP synthesis like them, they contain flavin-linked oxides and catalases for the hydrogen peroxide metabolism and also enzymes for fatty acid metabolism.

   Lastly, while mitochondrial (and plastid) proteins resemble those of prokaryotes, reflecting their endosymbiotic origin, the peroxins resemble typical eukaryotic proteins.

Peroxisomes differ from mitochondria and chloroplasts in many ways. Most notably, these organelles are surrounded only by a single membrane, and they do not contain DNA (genome) or ribosomes. However, they resemble ER in being self-replicating membrane bound organelle.

14.1 HISTORICAL

Since the mid-1950s electron microscopists have observed small structures or bodies in cells that on morphological grounds have been aptly termed **microbodies**. **C.De Duve** and **P. Baudbuin** (1966) coined the term **peroxisome** for the microbodies of mammalian systems and studied their structure and function. **Leaf peroxisomes** were first isolated from spinach leaf homogenate (*i.e.*, from mesophyll cells) by **Tolbert's** group in Michigen in 1968. Glyoxylate cycle containing peroxisomes, called **glyoxysomes**, were discovered in 1969 by **Beevers** in the endosperm cells of germinating castor bean (*Ricinus*). Biogenesis of peroxisomes has been investigated by following cell biologists: **Purdue** and **Lazarow** (2001); **Mullen, Flynn** and **Trelease** (2001); and **Subramani** (2001) and **Lazarow** (2003).

14.2 OCCURRENCE

Peroxisomes is found in all eukaryotic cells but is especially prominent in mammalian kidney and liver cells, in algae and photosynthetic cells of plants, and in germinating seedlings of plant species that store fat in their seeds.

14.3 STRUCTURE

Peroxisomes are variable in size and shape, but usually appear circular in cross section having diameter between 0.2 and 1.5µm (0.15 to 0.25 µm diameter in most mammalian tissues; 0.5 µm in rat liver cells). They have a single limiting unit membrane of lipid and protein molecules, which encloses their granular matrix. In some cases (*e.g.*, in the festuciod grasses) the matrix contains numerous threads or fibrils, while in others they are observed to contain either an amorphous nucleoid or a dense inner core which in many species shows a regular crystalloid structure (*e.g.*, tobacco leaf cell, **Newcomb** and **Frederick**, 1971). Little is known about the function of the core, except that it is the site of the enzyme urate oxidase in rat liver peroxisomes and much of the catalase in some plants (see **Hall** *et.al.*, 1974).

14.4 ISOLATION AND CHEMICAL COMPOSITION

The technique of **isolation** of microbodies from animal and plant tissues includes the following steps: 1. Tissues are ground very carefully to save microbodies from disruption. 2. The homogenate is treated with differential centrifugation to obtain a fraction of the cell homogenate which is rich in microbodies. 3. The enriched fraction is subjected to isopycnic ultra-centrifugation on discontinuous or continuous sucrose density gradient.

Animal peroxisomes often contain a distinct **crystalline core**, which is usually consists of a crystalline form of *urate oxidase.* Crystalline cores are also often present in the peroxisomes of plant leaves, but these usually consist of *catalase* instead. When such cores are present, it is easy to identify microbodies as peroxisomes, since urate oxidase and catalase are two of the enzymes by which peroxisomes are defined. Indeed, these organelles were named "peroxisomes" because they are the site of synthesis and degradation of hydrogen peroxide.

In the absence of crystalline core, it is not always easy to spot peroxisomes ultrastructurally. A useful technique in such cases is a **cytochemical test** for catalase called the **diaminobenzidine (DAB) reaction**. This assay depends on the ability of catalase to oxidize DAB to a polymeric form that causes deposition of electron dense osmium atoms when the tissue is treated with osmium tetroxide (OsO_4). The resulting electron-dense deposits can be readily seen in cells from stained tissue. In animal peroxisomes, the entire internal space often stains intensely with DAB, indicating that catalase exists as a soluble enzyme uniformly distributed throughout the matrix of the organelle. In plant leaf cells, DAB treatment preferentially stains the crystalline cores of the peroxisomes, thereby definitively identifying the cores as crystalline catalase. Because catalase is the single enzyme present in all peroxisomes and does not routinely occur in any other organelle, the DAB reaction is a very reliable and highly specific means of identifying organelles unambiguously as peroxisomes.

Peroxisomes contain at least 50 different enzymes, which are involved in a variety of biochemical pathways in different types of cells. However, peroxisomes are diverse organelles and even in various cell types of a single organism they may contain different sets of enzymes. They can also adapt remarkably to changing conditions. Yeast cells grown on sugar, for example, have small peroxisomes. But when some yeasts are grown on methanol, they develop large peroxisomes that oxidise methanol; and when grown on fatty acids, they develop large peroxisomes that break down fatty acids to acetyl CoA by β–oxidation (see **Alberts** *et al.,* 2002).

14.5 FUNCTIONS

There are at least **five** general categories of peroxisomal functions:

1. Hydrogen peroxide metabolism;
2. Detoxification of harmful substances;
3. Oxidation of fatty acids;
4. Metabolism of nitrogen-containing compounds, and
5. Catabolism of unusual substances.

1. Hydrogen peroxide metabolism. The most obvious role of peroxisomes in eukaryotic cells is the detoxification of H_2O_2, which is accomplished by the *coexistence* in the same organelle of catalase and the oxidases that generate H_2O_2. The oxidases in peroxisomes vary considerably in the specific reactions they catalyze but they all share the property of **transferring** electrons from their respective substrates directly to molecular oxygen (O_2), forming H_2O_2. Using RH_2 to represent an oxidizable organic substrate, the general reaction catalyzed by oxidases can be written as follows:

$$RH_2 + O_2 \longrightarrow R + H_2O_2 \qquad \text{...(Equation 14.1)}$$

The hydrogen peoxide formed in this manner is broken down by catalase in one of two ways. Usually catalase functions in what is called its **catalytic mode**, in which one molecule of H_2O_2 is oxidized to oxygen and a second is reduced to water:

$$2H_2O_2 \longrightarrow O_2 + 2H_2O \qquad \text{...(Equation 14.2)}$$

Dividing reaction 14.2 and adding it to reaction 14.1 yields a summary reaction for the two-step process:

$$RH_2 + \frac{1}{2}O_2 \longrightarrow R + H_2O \qquad \text{...(Equation 14.3)}$$

Alternatively, catalase can function in its **peroxidatic mode**, in which electrons derived from an organic donor are used to reduce H_2O_2 to water:

$$R'H_2 + H_2O_2 \longrightarrow R' + 2H_2O \qquad \text{...(Equation 14.4)}$$

(The prime on R group (*i.e.*, R′) simply indicates that this substrate is likely to be different from the substrate in reaction 14.1) The corresponding summary reaction in this case is

$$RH_2 + R'H_2 + O_2 \longrightarrow R + R' + 2H_2O \qquad \text{...(Equation 14.5)}$$

The result is the same in either case: Hydrogen peroxide is degraded without ever leaving the peroxisome. Given the toxicity of hydrogen peroxide (which is the main active ingredient in a variety of disinfectants), it makes good sense for the enzymes responsible for peroxide generation to be compartmentalized together with the catalase that catalyzes its degradation. Indeed, catalase is the most abundant protein in most peroxisomes, representing upto 15% of the total protein content of this organelle. In this way, virtually every molecule of H_2O_2 generated by oxidases will encounter a molecule of catalase almost immediately be promptly degraded.

H_2O_2 and aging. Most cytosolic H_2O_2 is produced by mitochondria and membranes of ER, although there are also H_2O_2 producing enzymes localized in the cytoplasmic matrix. Catalase acts as a **"safety valve"** for dealing with the large amounts of H_2O_2 generated by peroxisomes; however, other enzymes such as **glutathione peroxidase;** are capable of metabolizing organic hydroperoxides and also H_2O_2, in the cytosol (cytoplasmic matrix) and mitochondria. The production of superoxide anion (O_2–) in mitochondria and cytosol (cytoplasmic matrix) is regulated mainly by the enzyme

superoxide dismutase. All of these protective enzymes are present in high levels in aerobic tissues.

Recently, a possible relationship has been stressed between peroxides and free radicals (such as superoxide anion–O_2^-) with the process of aging. These radicals may act on DNA molecule to produce **mutations** altering the transcription into mRNA and the translation into proteins. In addition, free radicals and peroxides can affect the membranes by causing peroxidation of lipids and proteins. For these reasons reducing compounds such as vitamin E or enzymes such as superoxide dismutase could play a role in keeping the healthy state of a cell.

2. Detoxification of harmful compounds. In its peroxidatic mode (reaction 14.4), catalase can use a variety of substances as electron donors, including methanol, ethanol, formic acid, formaldehyde, nitrites and phenols. Because all of these compounds are harmful to cells, *their oxidative detoxification by catalase* may be a vital peroxisomal function. The prominent peroxisomes of liver and kidney cells are thought to be important in such detoxification reactions.

3. Oxidation of fatty acids. Peroxisomes found in animal, plant and fungal cells contain enzymes necessary for oxidising fatty acids. This process, called **β-oxidation**, also occurs in the mitochondrion as was discussed in chapter 58. About 25 to 50% of fatty acid oxidation in animal tissues occurs in peroxisomes, with the remainder localized in mitochondria. In plant and yeast cells, on the other hand, all β-oxidation is confined to peroxisomes.

In animal cells, peroxisomal β-oxidation appears to be especially important for the catabolism of long-chain (16 to 20 carbons), very long chain (24–26 carbons), and branched fatty acids. The primary product of β-oxidation, acetyl-CoA, is then exported to the cytosol and enters biosynthetic pathways. Once fatty acids are shortened to fewer than 16 carbon, further oxidation usually occurs in the mitochondria. Thus, in animal cells, the peroxisome is important for shortening fatty acids in preparation for subsequent metabolism in the mitochondrion rather than completely breaking them down to acetyl-CoA. In plants and yeast, on the other hand, peroxisomes are essential for the complete catabolism of fatty acids to acetyl-CoA.

4. Metabolism of nitrogen containing compounds. Except for primates, most animals require urate oxidase (also called uricase) to oxidize urate, a purine that is formed during the catabolism of nucleic acids and some proteins. Like other oxidases, urate oxidase catalyzes the direct transfer of electrons from the substrate to molecular oxygen, generating H_2O_2:

$$\text{Urate} + O_2 \longrightarrow \text{Allantoin} + H_2O_2 \qquad \text{...(Equation 14.6)}$$

As noted earlier, the H_2O_2 is immediately degraded in the peroxisome by catalase. The allantoin is further metabolized and excreted by the organism, either as allantoic acid, or in case of crustaceans, fish, and amphibians, as urea.

Additional peroxisomal enzymes involved in nitrogen metabolism include aminotransferase. Members of this collection of enzymes catalyze the transfer of amino acids to α-keto acids.

$$\underset{\text{Amino acid}}{R-\overset{H_3N^+}{\overset{|}{C}}-\overset{O}{\overset{\|}{C}}-O^-} + \underset{\alpha\text{-keto acid}}{R'-\overset{O}{\overset{\|}{C}}-\overset{O}{\overset{\|}{C}}=O^-} \rightleftharpoons \underset{\alpha\text{-keto acid}}{R-\overset{O}{\overset{\|}{C}}-\overset{O}{\overset{\|}{C}}-O^-} + \underset{\text{Amino acid}}{R'-\overset{H_3N^+}{\overset{|}{C}}-\overset{O}{\overset{\|}{C}}=O^-}$$

...(Equation 14.7)

Such enzymes play important roles in the biosynthesis and degradation of amino acids by moving amino groups from one molecule to another.

5. Catabolism of unusual substances (e.g., xenobiotics). Some of the substrates for peroxisomal oxidases are rare compounds for which the cells has no other degradive pathways. Such compounds include D-amino acids, which are not recognized by enzymes capable of degrading the L-amino acids that constitute polypeptides. In some fungi, the peroxisomes also contain enzymes that break down unusual substances called **xenobiotics,** chemical compounds foreign to biological organisms. This category include **alkanes**, short-chain hydrocarbon compounds found in oil and other petroleum products. Fungi containing enzymes capable metabolizing such xenobiotics may turn out to be useful for cleaning up oil spills that would otherwise contaminate the environment (see **Becker** *et.al.*, 2007).

6. Synthetic activities. In addition to providing a compartment for oxidation reactions, peroxisomes are involved in biosynthesis of lipids and the amino acid, lysine. In animal cells, cholesterol and dolicol are synthesized in peroxisomes as well as in ER. In the liver, peroxisomes are also involved in the synthesis of bile acids, which are derived from cholesterol. In addition, peroxisomes contain enzymes required for the synthesis of **plasmalogens** a family of phospholipids in which one of the hydrocarbon chains is joined to glycerol by an **ether bond** rather than by ester bond (Fig. 14.1). Plasmalogens are important membrane components in some tissues. Particularly heart and brain, although they are absent in others. Deficiency of plasmalogens causes profound abnormalities in the **myelination** (formation of myelin-sheath) of nerves, which is one reason why many peroxisomal disorders lead to neurological disease (See **Alberts** *et.al.,* 2002) (Box 14.1).

$$H_2C - O - \overset{H}{C} - \overset{H}{C} - R_1$$
$$HC - O - \overset{O}{\overset{\|}{C}} - R_2$$
$$H_2C - (P) - CH_2 - CH_2 - N^+(CH_3)_3$$

Fig. 14.1. Structure of a plasmalogen.

Box 14.1

The most common peroximal disorder is **X-linked adrenoleukodystrophy.** The defective protein causing this disorder is an integral membrane protein that may be responsible for transporting very long-chain fatty acids into the peroxisomes for β-oxidation. Accumulation of these long-chain fatty acids in human body fluids destroys myelin sheath in nervous system (See **Becker** *et. al.,* 2007).

Roles of Plant Specific Peroxisomes

1. Leaf peroxisomes. In green leaves, there are peroxisomes that carry out a process called **photorespiration** which is a light-stimulated production of CO_2 that is different from the generation of CO_2 by mitochondria in the dark. In photorespiration, **glycolic acid** (**glycolate**), a two-carbon product of photosynthesis is released from chloroplasts and oxidized into **glyoxylate** and $\mathbf{H_2O_2}$ by a peroxisomal enzyme called **glycolic acid oxidase**. Later on, glyoxylate is oxidized into CO_2 and **formate**:

$$CH_2OH.\ COOH + O_2 \longrightarrow CHO - COOH + H_2O_2$$
$$CHO - COOH + H_2O_2 \longrightarrow HCOOH + CO_2 + H_2O$$

Photorespiration is so-called because light induces the synthesis of glycolic acid in chloroplasts. The entire process involves intervention of two basic organelles : chloroplasts and peroxisomes.

Lastly, photorespiration is driven by atmospheric conditions in which the O_2 tension is high and the CO_2 tension low. Apparently O_2 competes with CO_2 for the enzyme **ribulose diphosphate carboxylase** which normally is the key enzyme in CO_2 fixation during photosynthesis. When O_2 is used by the enzyme, an unstable intermediate is formed which breaks down into **3-phosphoglycerate** and **phosphoglycolate**. The latter tends to increase the glycolate concentration by removal of its phosphate group and, therefore, more glycolate is available for additional oxidation and CO_2 release.

Photorespiration is a wasteful process for the plant cell, since, it significantly reduces the efficiency of the process of photosynthesis (*i.e.*, it returns a portion of fixed CO_2 to the atmosphere).

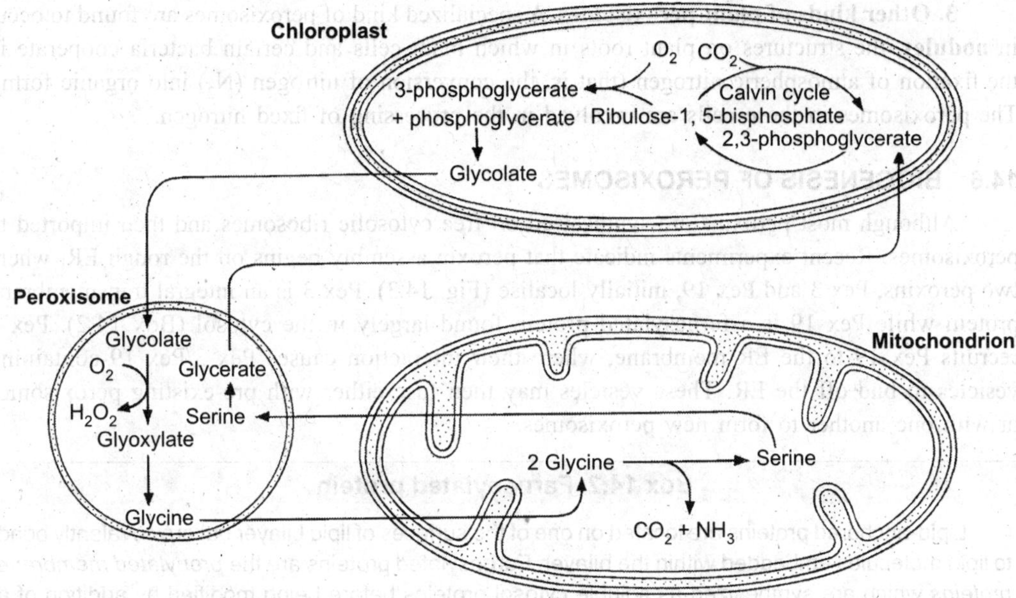

Fig. 14.2. Role of peroxisomes in photorespiration. During photosynthesis CO_2 is converted to carbohydrates by the Calvin cycle, which initiates with the addition of CO_2 to the five-carbon sugar ribulose–1, 5–diphosphate. However, the enzyme involved sometimes catalyzes the addition of O_2 instead, resulting in production of the two-carbon compound phosphoglycolate. Phosphoglycolate is converted to glycolate, which is then transferred to peroxisomes, where it is oxidized and converted to glycine. Glycine is then transferred to mitochondria and converted to serine. The serine is returned to peroxisomes and converted to glycerate, which is transferred back to chloroplasts (after Cooper and Hausman, 2007).

Glycolate cycle. Peroxisomes of plant leaves contain catalaze together with the enzymes of **glycolate pathway**, as glycolate oxidase, glutamate glyoxylate, serine-glyoxylate and asparate-α-ketoglutarate aminotransferases, hydroxy pyruvate reductase and malic dehydrogenase. They also contain FAD, NAD and NADP coenzymes. The glycolate cycle is thought to bring about the formation of the amino acids-glycine and serine-from the non-phosphorylated intermediates of photosynthetic carbon reduction cycle, *i.e.*, glycerate to serine, or glycolate to glycine and serine in a sequence of reactions which involve chloroplasts, peroxisomes, mitochondria and cytosol (**Tolbert,** 1971). The glycolate pathway also generates C_1 compounds and serves as the generator of precursors for nucleic acid biosynthesis.

2. Glyoxysomes. It occurs transiently in seedlings of plant species that store carbon and energy reserves in the seed of fat (primarily triacylglycerols). In such species, stored triglycerols are mobilized and converted to sucrose during early postgerminative development by a sequence of events that includes β-oxidation of fatty acids as well as a pathway known as the glyoxylate (= glycolate) cycle. All of the enzymes needed for these processes are localized to specialized peroxisomes called **glyoxysomes**. Glyoxysomes are found only in the tissues in which the fat is stored (endosperm or cotyledons, depending on the species) and are present only for the relatively short period of time required for the seedling to deplete its supply of stored fat. *Once they fulfill their role in the seedlings, the glyoxysomes are converted to peroxisomes*. Glyoxysomes have been reported to appear again in the senescing (aging) tissues of some plant species, presumably to degrade lipids derived from the membranes of the senescent cells.

3. Other kinds of plant peroxisomes. A specialized kind of peroxisomes are found to occur in **nodules**, the structures on plant roots in which plant cells and certain bacteria cooperate in the fixation of atmospheric nitrogen (that is, the conversion of nitrogen (N_2) into organic form). The peroxisomes in these cells are involved in the processing of fixed nitrogen.

14.6 BIOGENESIS OF PEROXISOMES

Although most peroxins are synthesized on free cytosolic ribosomes and then imported to peroxisomes. Recent experiments indicate that peroxin assembly begins on the rough ER, where two peroxins, Pex 3 and Pex 19, initially localise (Fig. 14.3). Pex 3 is an integral transmembrane protein while Pex 19 is a fernesylated protein found largely in the cytosol (Box 14.2). Pex 3 recruits Pex 19 to the ER membrane, where their interaction causes Pex 3/Pex 19-containing vesicles to bud off the ER. These vesicles may then fuse either with pre-existing peroxisomes or with one another to form new peroxisomes.

Box 14.2 Farnesylated protein

Lipid-anchored proteins are located on one of the surfaces of lipid bilayer but are covalently bond to lipid molecules embedded within the bilayer. Farnesylated proteins are the *prenylated membrane proteins* which are synthesized as soluble cytosol proteins before being modified by addition of a prenyl group, usually a 15 carbon **farnesyl** group. After attachment, the farnesyl group is inserted into the lipid bilayer of the membrane (see **Becker** *et al.*, 2007).

Pex 3, Pex 19, and other peroxisomal membrane proteins then act as receptors for import of the other peroxins, which are translated on free cytosolic ribosomes and then transported into peroxisomes as completed and folded polypeptides. They are targeted to the interior of peroxisomes by at least two pathways, which are conserved from yeasts to human.

Most peroxins are targeted by the simple amino acid sequence Ser-Lys-Leu at their carboxyl terminus (**peroxisome targeting signal 1 or PTS1**). A small number of peroxins are targeted by a sequence of nine amino acids at their amino terminus (**peroxisome targeting signal 2** or **PTS2**). PTS1 and PTS2 are recognized by distinct cytosolic receptors and then passed through a poorly understood channel in the peroxisomal membrane into the matrix. The receptors are subsequently retrieved fróm the peroxisome and recycled. Unlike the translocation of polypeptide chains across the membranes of the endoplasmic reticulum, mitochondria and chloroplasts, the targeting signals are usually not cleaved during the import of proteins into peroxisomes and the mechanism of translocation is not known (see **Cooper** and **Hausman** 2007).

Protein import together with the continuing addition of lipids from the rough ER, results in peroxisome growth, and new peroxisomes can be formed by division of old ones. In addition, peroxisomes undergo a complex **maturation process** that involves the import of different classes of proteins from the cytosol at different times. As a result, the enzyme content, and the metabolic activities of peroxisomes may change as they mature.

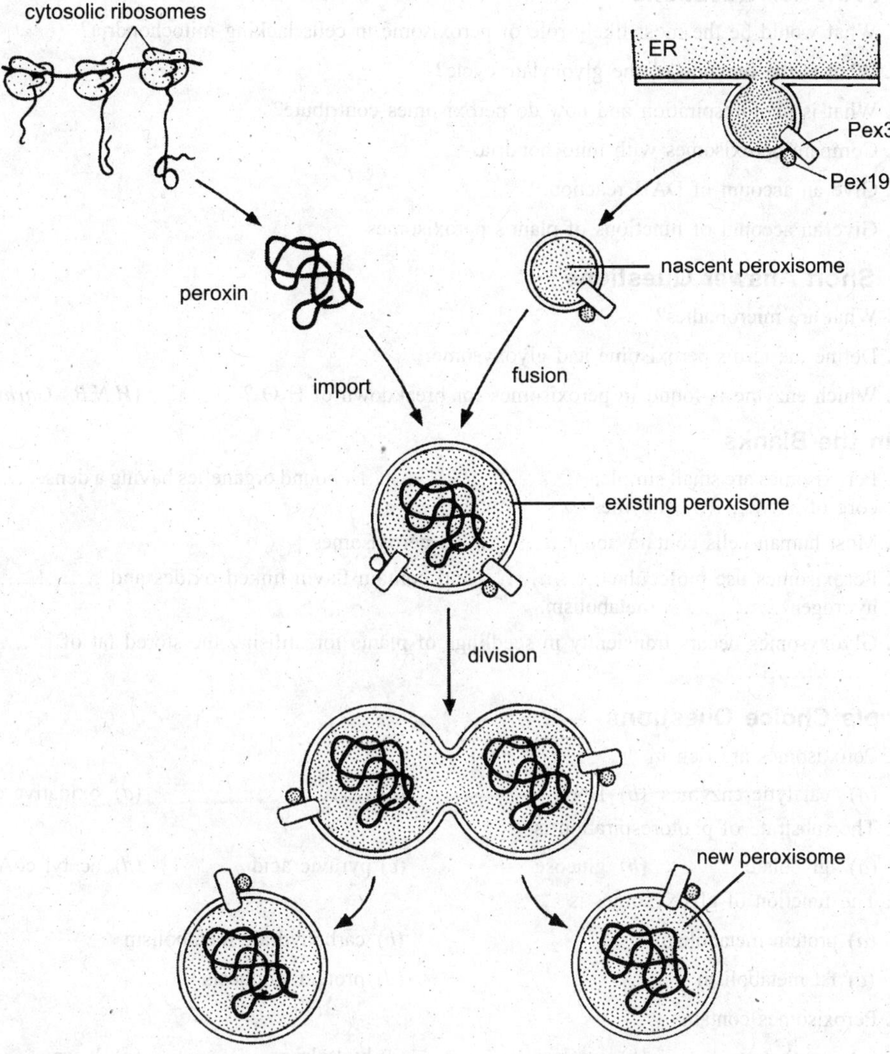

Fig. 14.3. Assembly of peroxisomes. Initiation of peroxisome assembly begins in the rough ER when the transmembrane protein peroxin (Pex 3) recruits the soluble farnesylated protein peroxin (Pex 19) and initiate budding of a nascent peroxisome. The nascent peroxisomes fuse either with each other or with existing peroxisomes. Additional peroxins are synthesized on free cytosolic ribosomes and imported as completed polypeptide chains to form functional peroxisomes, which grow larger and divide (after Cooper and Hausman, 2007).

QUESTIONS

Long Answer Questions

1. Write an account of structure, synthesis, function and histochemical localization of the catalase enzyme in eukaryotic cells.
2. Describe the mode of biogenesis and various functions of peroxisomes in animal cells.

Short Answer Questions

1. What would be the most likely role of peroxisome in cells lacking mitochondria?
2. What is the function of the glyoxylate cycle?
3. What is photorespiration and how do peroxisomes contribute?
4. Compare peroxisomes with mitochondria.
5. Give an account of DAB reaction.
6. Give an account of functions of plant's peroxisomes.

Very Short Answer Questions

1. What are microbodies?
2. Define the terms peroxisome and glyoxysome.
3. Which enzyme is found in peroxisomes for breakdown of H_2O_2? (*H.N.B., Garhwal 2009*)

Fill in the Blanks

1. Peroxisomes are small simple, ... bound organelles having a dense, ... core of .... enzymes.
2. Most human cells contain about .... peroxisomes.
3. Peroxisomes use molecular .... and contain flavin-linked oxides and .... for the hydrogen .... metabolism.
4. Glyoxysomes occurs transiently in seedlings of plants for utilising the stored fat of .... or ....

Multiple Choice Questions

1. Peroxisomes are rich in

   (*a*) catalytic enzymes (*b*) DNA (*c*) RNA (*d*) oxidative enzymes
2. The substrate of photorespiration is

   (*a*) glycolate (*b*) glucose (*c*) pyruvic acid (*d*) acetyl coA
3. The function of glyoxysomes is

   (*a*) protein metabolism (*b*) carbohydrate metabolism

   (*c*) fat metabolism (*d*) protein synthesis
4. Peroxisomes contains

   (*a*) synthetase (*b*) catalase (*c*) hydrolase (*d*) lyase.

ANSWERS

Very Short Answer Questions

1. Microbody is an early term for a peroxisome based on its appearance in electron micrographs.
2. *Peroxisome.* It is a single membrane-bounded cell organelle that contains catalase and one or more hydroxide-generating oxidases and is therefore involved in the metabolism of hydrogen peroxide.

   Glyoxysome. This is a specialized type of plant peroxisome that contains some of the enzymes responsible for conversion of stored fat to carbohydrate in germinating seeds.
3. Catalase.

Fill in the Blanks

1. Single-membrane, crystalline, oxidative.
2. 500.
3. Oxygen, catalases, peroxide.
4. Endosperm, cotyledons.

Multiple Choice Questions

1. (*d*) 2. (*a*) 3. (*c*) 4. (*b*).

15

Cytoskeletal Systems and Cellular Movement

(Microtubules, Microfilaments, Intermediate filaments, Centrioles, Motor MPA, Cilia and Flagella)

Advances in microscopy and other investigative techniques have revealed that the interior of an eukaryotic cell is highly structured. Part of this structure is provided by the **cytoskeleton:** a complex network of interconnected filaments and tubules that extends throughout the cytosol, from the nucleus to the inner surface of the plasma membrane (Box 15.1).

The three major elements of the cytoskeleton are **microtubules (MTs), microfilaments (MFs)** and **intermediate filaments (IFs)**, each of which is unique to eukaryotic cells. However, prokaryotes, such as rod-shaped bacteria, have recently been shown to have proteins that function in a manner very similar to microfilaments (proteins from the *MreB* family), microtubules (the *FtsZ* protein) and intermediate filaments (one type is called *crescentin*) (see **Becker** *et al.*, 2007).

Box 15.1

The existence of an organized fibrous array or cytoskeleton in the structure of the protoplasm was postulated in 1928 by **Koltzoff**. He conceived of a cytoskeleton that determines both the shape of the cell and the changes in its form.

The main proteins that are present in the cytoskeleton are **tubulin** (in the microtubules), **actin**, **myosin**, **tropomyosin** and other (in the microfilaments) and **keratins**, **vimentin**, **desmin**, **lamin** and others (in intermediate filaments). Tubulin and actin are globular proteins, while subunits; of intermediate filaments are fibrous proteins.

Cellular architecture depends on the unique properties of the different cytoskeletal components working together. Microtubules generally thought to resist bending when a cell is compressed, while microfilaments serve as contractile elements that generate tension. Intermediate filaments are elastic and can withstand tensile force. The mechanical integration of IFs, MFs and MTs is made possible by specific linker proteins that connect them. **Plectin**, for example, is a versatile linker protein that is found at sites where IFs are connected to MFs and MTs.

Functions of Cytoskeleton

The cytoskeleton plays important roles in cell movement and cell division, and it positions and actively moves membrane-bounded organelles within the cytosol. It also plays a similar role for messenger RNA and other cellular components. Many enzymes in the cytosol are in fact probably not soluble at all, but are physically clustered and attached to the cytoskeleton in close proximity to other enzymes involved in the same pathways, thereby facilitating the channeling of intermediates within each pathway. The cytoskeleton is also involved in many forms of cell movement and is intimately related to other processes such as cell signalling and cell-cell adhesion. The cytoskeleton is altered by events at the cell surface and, at the same time, appears to participate in and modulate these events.

15.1 MICROTUBULES

Microtubules were first of all observed in the axoplasm of the myelinated nerve fibres by **Robertis** and **Franchi** (1953). They called them **neurotubules**. The exact nature of microtubules was brought into light when **Sabatini**, **Bensch** and **Barnett** (1963) made use of the glutaraldehyde fixative in the electron microscopy. Microtubules of plant cells were first described in detail by **Ledbetter** and **Porter** (1963).

Occurrence

With rare exceptions such as the human erythrocytes, microtubules are found in all eukaryotic cells, either free in the cytoplasm or forming part of centrioles, cilia and flagella. The most abundant source of microtubules for the biochemical studies is vertebrate brain—high densities of microtubules exist in axons and dendrites of nerve cells. In the cytoplasm of animal and plant cells, microtubules occur at following seven sites: 1. Cilia and flagella, 2. Centrioles and basal bodies, 3. Nerve processes, 4. The mitotic apparatus, 5. The cortex of meristematic plant cells, 6. Elongating cells such as during the formation of the lens or during spermatogenesis of certain insects. 7. Selected structures in Protozoa such as the axostyle of parasitic flagellates, the axoneme of *Echinosphaerium*, the fibre systems of *Stentor*, and the cytopharyngeal basket of *Nassula*.

Types of Microtubules

Eukaryotes have following two types of microtubules:

1. Axonemal microtubules. These include the highly organized, stable microtubules found in specific subcellular structures associated with cellular movement, including cilia, flagella, basal bodies and centrioles to which these appendages are attached. The central shaft, or **axoneme**, of a cilium or flagellum consists of a highly ordered bundle of axonemal MTs and associated proteins.

2. Cytoplasmic microtubules. They include loosely organized, dynamic microtubules which pervade the cytosol of most eukaryotic cells. They were detected by the help of fluorescence microscopy around early 1960's.

Structure

Microtubules constitute a class of morphologically and chemically related filamentous rods which are common to both plant and animal cells. A microtubule consists of a long, unbranched, hollow tubules 24–25 nm in diameter, several micrometers long and with 6 nm thick wall having 13 subunits or **protofilaments**. Thus, the wall of the microtubule consists of 13 individual linear or spiralling filamentous structures about 5 nm in diameter, which in turn, are composed of tubulin. These protofilaments have a centre-to-centre spacing of 4.5 nm. Application of negative staining techniques has shown that microtubules have a lumen 14 nm wide and a protofilament or subunit structure in the wall (Fig. 15.1).

Chemical Composition

Biochemically, a protofilament of microtubule is made of a protein called **tubulin**. Tubulin is an acidic protein with a molecular weight of 55,000 and a sedimentation constant of 6S. It occurs in two different forms, called **α-tubulin** and **β-tubulin**, each containing about 450 amino acids. Both of these proteins have a distinct, though closely related, amino acid sequences and are thought to have evolved from a single ancestral protein. The two proteins show very little divergence from the lowest to the highest eukaryotes; for example, the β-tubulins of sea urchin flagella and chick brain cells differ only in one amino acid. Similarities such as this suggest that most mutations disrupt the functions of microtubules and are thus lethal and are eliminated by selection (see **King**, 1986).

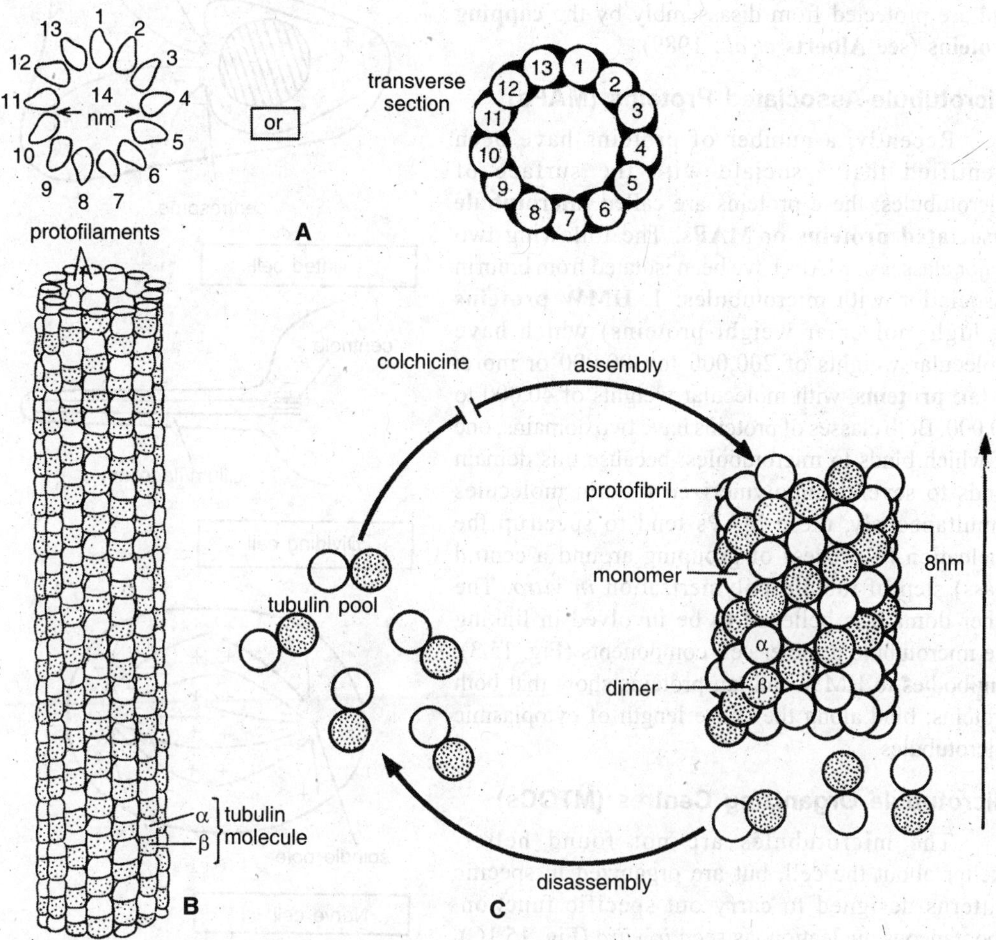

Fig. 15.1. Schematic diagrams of a microtubule, showing how the tubulin molecules pack together to form the cylindrical wall. A—13 tubulin molecules (subunits of protofilaments) in cross section; B—Side view of a short section of a microtubule, with the tubulin molecules aligned into rows, or protofilaments. Each of the 13 protofilaments is composed of a series of tubulin molecules, each with an α/β heterodimer; C—Assembly and disassembly of the microtubule. The microtubule is being disassembled at the bottom while being simultaneously assembled at the top. Colchicine, by blocking the assembly process, produces depolymerization of the microtubules (after De Robertis and De Robertis, Jr., 1987; Alberts *et al.*, 1989).

Tubulin in the form of dimers (rather heterodimers of α- and β-tubulins; each with 115,000 MW, see **Berns**, 1983) polymerizes into the microtubules. Thus, heterodimers of tubulins assemble to form linear "protofilaments" with the β-tubulin of one dimer in contact with the α-tubulin of the next. Since all the 13 protofilaments are aligned parallely with the same polarity, the microtubules are the **polar** structures having a **plus** or **fast growing end** and **minus** or **slow growing end**. The minus ends of cytoplasmic microtubules in cells are bound tightly to **microtubule organizing centres (MTOCs)** from which their assembly or polymerization starts. MTOCs also protect the minus ends of the microtubules from the disassembly. Generally, the plus ends of microtubules terminate near cell margins (Fig. 15.2) and are protected from disassembly by the **capping proteins** (see **Alberts** *et al.*, 1989).

Microtubule-Associated Proteins (MAPs)

Recently, a number of proteins have been identified that associate with the surface of microtubules; these proteins are called **microtubule associated proteins** or **MAPs**. The following two major classes of MAPs have been isolated from brain in association with microtubules: 1. **HMW proteins** (= high molecular weight proteins) which have molecular weights of 200,000 to 300,000 or more; 2. **tau proteins**, with molecular weights of 40,000 to 60,000. Both classes of proteins have two domains, one of which binds to microtubules; because this domain binds to several unpolymerized tubulin molecules simultaneously, these MAPs tend to speed up the nucleation (= process of grouping around a central mass) step of tubulin polymerization *in vitro*. The other domain is believed to be involved in linking the microtubule to other cell components (Fig. 15.3). Antibodies to HMW and tau proteins show that both proteins; bind along the entire length of cytoplasmic microtubules.

Microtubule Organizing Centres (MTOCs)

The microtubules are not found helter-skelter about the cell, but are organized in specific patterns designed to carry out specific function. Spontaneous nucleation, as seen *in vitro* (Fig. 15.1C), probably does not occur *in vivo*. Rather, initiation of assembly occurs at **microtubule organizing centres (MTOCs)**. Thus, MTOCs are nucleating centres that serve as templates for the polymerization of tubulin (see **Thorpe**, 1983). MTOCs exist in basal bodies (*e.g. Chlamydomonas*); in centrioles (*e.g.*, most animal cells); at the poles of mitotic spindles in dividing cells that do not have centrioles (*e.g.*, most plant cells); on chromosomes (*i.e.*, **kinetochore**); in membranes and probably many other places as well. Later studies have revealed that most cytoplasmic microtubules do not arise directly from the centrioles, but from a densely staining **pericentriolar material** that surrounds the centriole (see **King**, 1986).

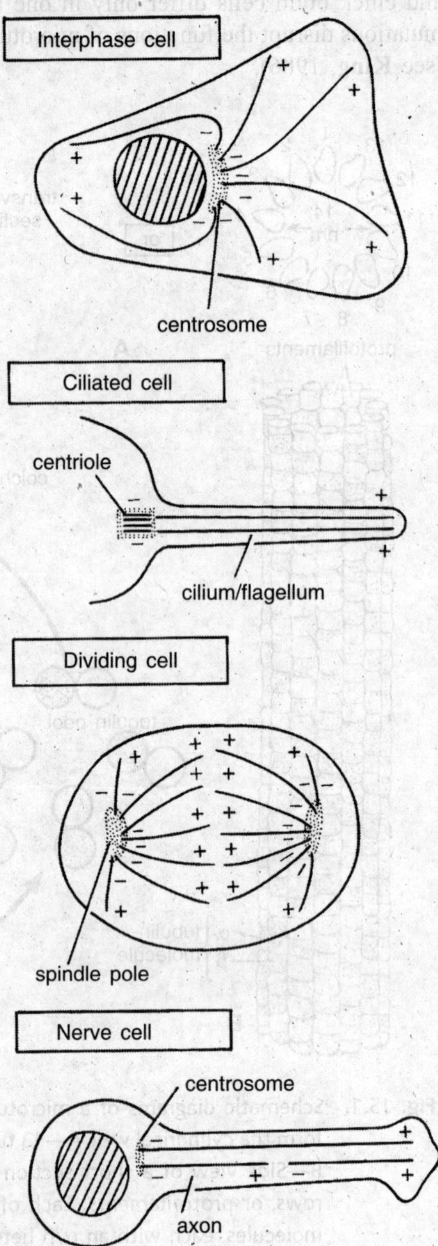

Fig. 15.2. The minus ends of microtubules in cells are generally embedded in a microtubule-organizing centre; while the plus ends are often near the plasma membrane (after Alberts *et al.*, 1989).

Turning on and off of these organizing centres for microtubule assembly at different times in the cell's life are probably regulated by one or all the following factors: changes in nucleation centres, changes in Ca^{2+} concentration and involvement of MAPs.

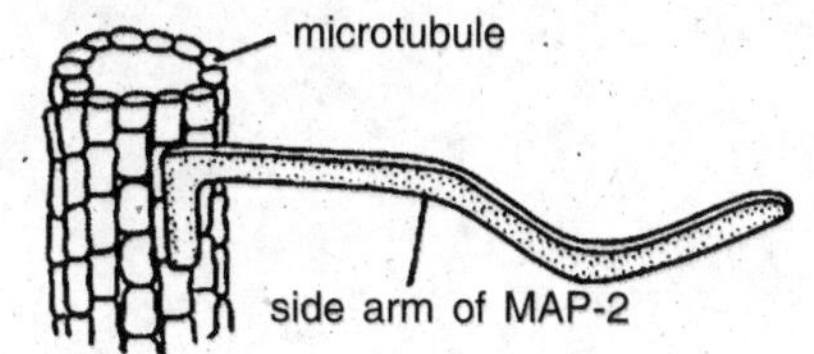

Fig. 15.3. A microtubule associated protein (known as MAP-2) showing its two domains (after Alberts *et al.*, 1989).

Assembly and Disassembly of Microtubules

Cytoplasmic microtubules are highly dynamic structures, constantly forming and disappearing depending on cell activities. They, like the microfilaments, grow by the reversible addition of subunits, accompanied by nucleotide (GTP) hydrolysis and conformational change. The process of polymerization (assembly) and depolymerization (disassembly) of the microtubules appears to be a form of **self-assembly**. The assembly of microtubules from the tubulin dimers is a specifically oriented and programmed process. In the cell, the sites of orientation are MTOCs from which the polymerization is directed. The quantity of polymerized tubulin is high at interphase (cytoplasmic microtubules) and metaphase (spindle microtubules), but low at prophase and anaphase.

Within the cell, microtubules are in equilibrium with free tubulin. Phosphorylation of the tubulin monomers by a cyclic AMP-dependent kinase favours the polymerization. A definite relationship has been found between cell shape, the number and direction of microtubules and cAMP. The assembly and disassembly of tubulin constitute a polarized phenomenon. In a microtubule, the assembly of tubulin dimers takes place at one end, while disassembly is common at the other end (Fig. 15.1C). If a cell is treated with certain drugs such as **colchicine, vincristine** or **vinblastine**, the assembly of the microtubules is inhibited, while the disassembly continues, leading to the disorganization of the microtubule. Further, the assembly is accompanied by the hydrolysis of guanosine triphosphate (GTP) to guanosine diphosphate (GDP) and lack of GTP stops the assembly. *In vivo* control of assembly and disassembly of tubulin involves Ca^{2+} and the calcium-binding protein **calmodulin**. The addition of Ca^{2+} inhibits polymerization of tubulin; this effect is also enhanced by the addition of calmodulin.

The *in vivo* mechanism involved in self-assembly of the microtubules is still little understood, however, *in vitro* studies have revealed various interesting facts about it. Thus, in a classical study using isolated bovine brain tubulin, **Weingarten** *et al.*, (1975) demonstrated that tubulin alone was not sufficient to bring about *in vitro* assembly into microtubules. Under normal conditions, if brain microtubules are isolated and caused to depolymerize into tubulin subunits, the tubulin molecules will reassemble into microtubules if Mg^{2+} and GTP (an energy source) are added to the mixture. However, according to **King** (1986), *in vitro* assembly of microtubules can occur in the presence of low calcium concentration, MAPs, GTP, and a level of free tubulin monomers above a threshold concentration.

In vitro polymerization evidently involves two distinct phases, one of **initiation** and the other of **elongation**. The initiation event seems to involve the formation of some multimeric "**nucleating**" centre, following which the addition of more subunits proceeds rapidly during elongation. Thus, during *in vitro* polymerization of microtubules, α- and β-tubulins combine to form heterodimers (Fig. 15.4). The heterodimers associate to form multimeric **rings**, **spirals** and other intermediate structures which eventually open up to form strands or protofilaments. Side-by-side assembly of the protofilaments creates sheet-like structures that curl to form a tube. Elongation of this short cylinder occurs by direct addition of new heterodimers at one end of the tubule (*i.e.*, the plus end of tubule). It is believed that during anaphase, addition of dimers to one end of a microtubule is accompanied by the loss of dimers from the other end.

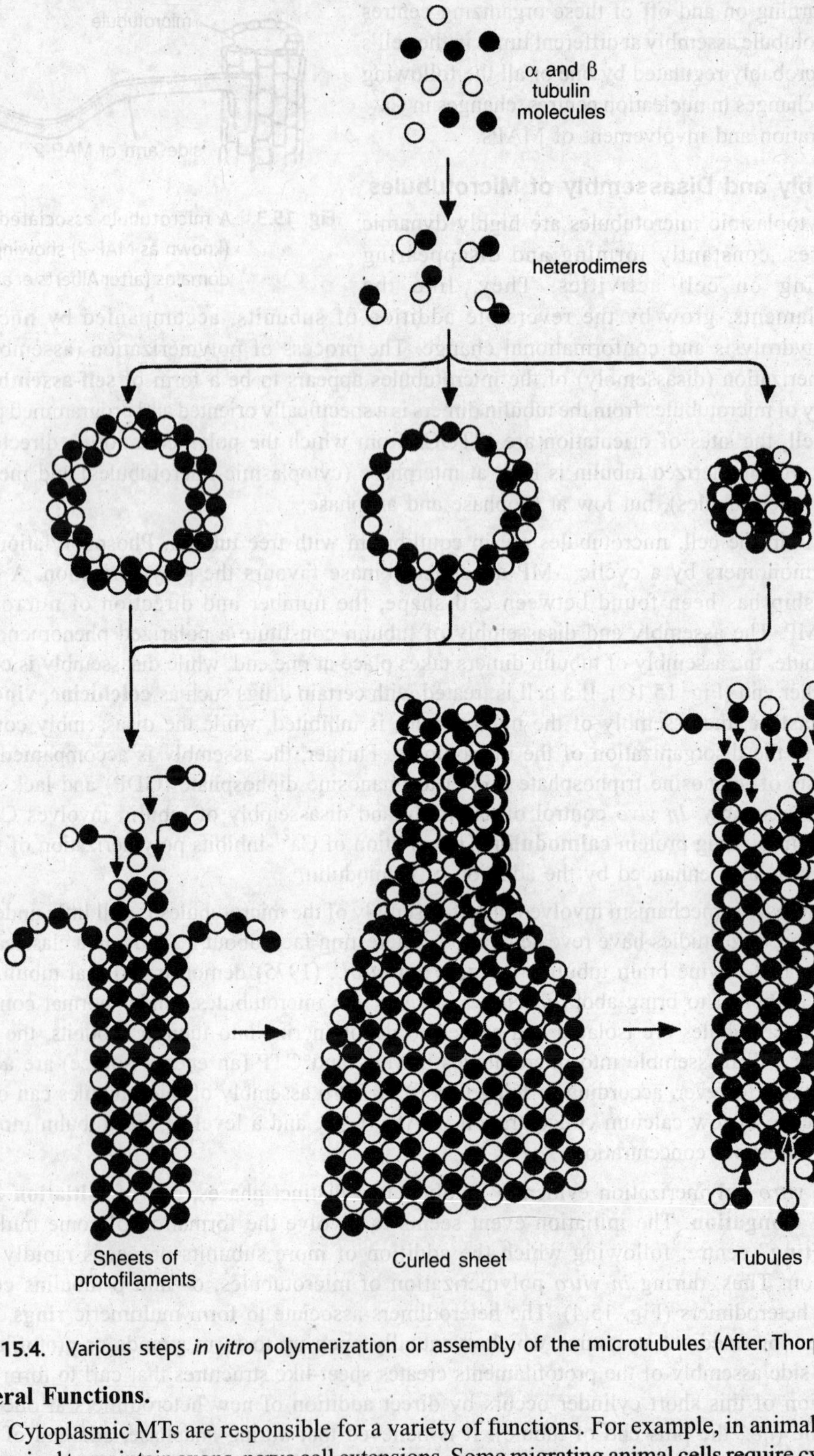

α and β tubulin molecules

heterodimers

Sheets of protofilaments

Curled sheet

Tubules

Fig. 15.4. Various steps *in vitro* polymerization or assembly of the microtubules (After Thorpe, 1984).

General Functions.

Cytoplasmic MTs are responsible for a variety of functions. For example, in animal cells they are required to maintain axons, nerve cell extensions. Some migrating animal cells require cytoplasmic

MTs to maintain their polarized shape. In plant cells, cytoplasmic MTs are thought to govern the orientation with which cellulose microfibrils are deposited during the growth of cell wall. Significantly, cytoplasmic MTs form the mitotic and meiotic spindles that are essential for the movement of chromosomes during mitosis and meiosis (see Chapter 18).

Specific Functions of Cytoplasmic Microtubules

Microtubules (MTs) have several functions in the eukaryotic cells such as follows:

1. Mechanical function. The shape of the cell (*e.g.*, red blood cells of non-mammalian vertebrates) and some cell processes or protuberances such as axons and dendrites of neurons, microvilli, etc., have been correlated to the orientation and distribution of microtubules.

Cytoplasmic microtubules also contribute to the spatial deposition and directional movement of vesicles and other organelles by providing an organized system of fibre to guide their movement. For example, cytoplasmic MTs help to govern the location of organelles, such as the Golgi apparatus and the endoplasmic reticulum, and are involved in active movement of vesicles.

2. Morphogenesis. During cell differentiation, the mechanical function of microtubules is used to determine the shape of the developing cells. For example, the enormous elongation in the nucleus of the spermatid during spermiogenesis is accompanied by the production of an orderly array of microtubules that are wrapped around the nucleus in a double helical arrangement. Likewise, the elongation of the cells during induction of the lens placode in the eye is also accompanied by the appearance of numerous microtubules.

3. Cellular polarity and motility. The determination of the intrinsic polarity of certain cells is also related to the microtubules. Directional gliding of cultured cells is found to depend on the microtubules.

4. Contraction. Microtubules play a role in the contraction of the spindle and movement of chromosomes and centrioles as well as in ciliary and flagellar motion.

5. Circulation and transport. Microtubules are involved in the transport of macromolecules, granules and vesicles within the cell. **Examples:** 1. The protozoan *Actinosphaerium* (Heliozoa) sends out long, thin pseudopodia within which cytoplasmic particles migrate back and forth. These pseudopodia contain as many as 500 microtubules disposed in a helical configuration. 2. In the protozoan *Nassula*, microtubules drive the food in the gullet. 3. In melarocytes, melanin granules move centrifugally and centripetally with different stimuli. These granules have been observed moving between channels created by the microtubules in the cytoplasmic matrix. 4. In the erythrophores found in fish scales the pigment granules may move at a speed of 25 to 30 μm per second between the microtubules. 5. They have a role in axoplasmic transport of proteins, glycoproteins and enzymes.

15.2 MICROFILAMENTS

Thin, solid **microfilaments** of actin protein, ranging between 5 to 7 nm in diameter and indeterminate length; represent the active or motile part of the cytoskeleton. They appear to play major role in cyclosis and amoeboid motion. With high voltage electron microscopy a three-dimensional view of microfilaments has been obtained (*i.e.*, an image of **microtrabecular lattice**). These microfilaments are sensitive to **cytochalasin-B**, an alkaloid that also impairs many cell activities such as beat of heart cell, cell migration, cytokinesis, endocytosis and exocytosis. It is generally assumed that the cytochalasin-B-sensitive microfilaments are the contractile machinery of non-muscle cells.

Distribution

Microfilaments are generally distributed in the cortical regions of the cell just beneath the plasma membrane. In contrast, intermediate filaments and microtubules are found in subcortical and deeper regions of the cell. Microfilaments also extend into cell processes, especially where there is movement. Thus, they are found in the microvilli of the brush border of intestinal epitheliun and in cell types where amoeboid movement and cytoplasmic streaming are prominent.

Chemical Composition

Actin is the main structural protein of microfilaments. Actin is an extremely abundant protein in virtually all eukaryotic cells, including those of plants, algae, and fungi. Actin is synthesized as a single polypeptide consisting of 375 amino acids, with a molecular weight of about 42 kDa. Once synthesized, it folds into a roughly U-shaped molecule, with a central cavity that binds ATP or ADP. Individual actin molecules are referred to as **G-actin (globular actin)**. Under the right conditions, G-actin molecule polymerize to form **microfilaments**; in this form the actin is referred to as **F-actin (filamentous actin)**. Actin in the G or F form also binds to a wide variety of other proteins. These actin-binding proteins either regulate and modify the function of actin, or are themselves regulated or organized by their association with actin.

Of the three types of cytoskeletal proteins, actin is the most highly conserved. In functional assays, all actins appear to be identical and actins from diverse organisms will copolymerize into filaments. Despite this high degree of sequence similarity, actins do differ among different organisms. Based on sequence similarity, actins can be broadly divided into two major groups: 1. The **muscle-specific actins** (*e.g.*, α-actin) and 2. **Non-muscle actins** (*e.g.*, **β-actin** and **γ-actin**).

In addition to the various types of actin, another class of protein known as **actin-related proteins (Arps)** show recognizable sequence similarity to actin Arp 2 and Arp 3 are found to be involved in nucleating assembly of new microfilaments in migrating cells.

Polymerization of G actins. Like tubulin dimers, monomers of G-actin polymerize into long filaments of F-actin with a diameter of about 7 nm (Fig. 15.5). In F-actin, a full turn of the helix occurs every 36–37 nm, with about 13.5 monomers required for a full turn. The addition of each G-actin monomer is usually followed by hydrolysis of ATP molecules so that it is tightly bound to the monomer, although the energy of ATP hydrolysis is not required to drive the polymerization reaction.

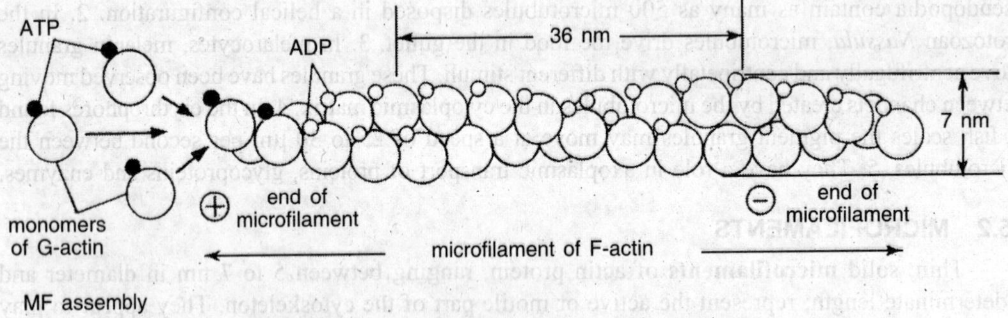

Fig. 15.5. A model for microfilament assembly in vitro (after Becker *et al.*, 2007)

Within a microfilament, all the actin monomers are oriented in the same direction, so that an MF, like a microtubule, has an inherent polarity, with one end differing chemically and structurally from the other end. This polarity can be readily demonstrated by incubating MFs with **heavy myosin (HMM)** or **myosin subfragment (S1)**, a proteolytic fragment of myosin (Fig. 15.6). HMM/S1 fragments bind to, or "decorate" the actin MFs to give a distinctive arrowhead pattern, the term **pointed end** and **barbed end** are commonly used to identify the **minus** and **plus ends** of a MF, respectively. The polarity of the microfilament is important, because it allows for independent regulation of actin assembly or disassembly of each end of the filament. Normally, plus end grows faster than the minus end.

Capping of MFs. Whether microfilament ends are available for further growth depends on whether or not the filament end is **capped**. Capping occurs when a **capping protein** binds the end of a filament and prevents further addition or loss of subunits, thereby stabilize it. One such protein that acts as a cap for the plus end of microfilament is appropriately named **Cap Z** (see **Becker** *et al.*, 2007).

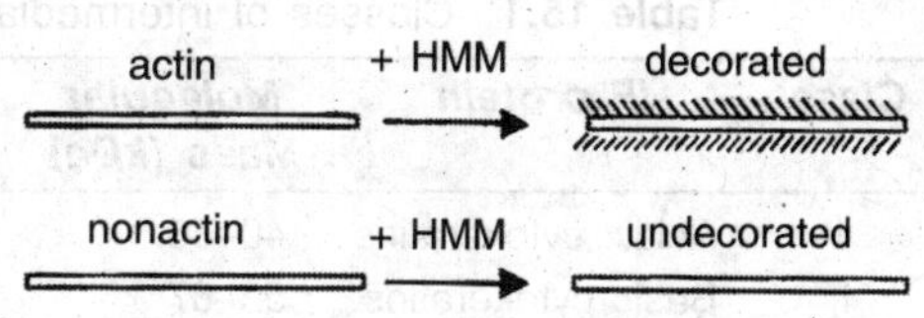

Fig. 15.6. Mode of "decoration" of actin filaments by heavy myosin or HMM. Intermediate filaments are not decorated by HMM. (after Thorpe, 1984).

Function

Microfilaments are found to be involved in movement associated with furrow formation in cell division, cytoplasmic streaming in plant cells (*e.g., Nitella* and *Chara*) and cell migration during embryonic development.

15.3 INTERMEDIATE FILAMENTS

Intermediate filaments (IFs) are tough and durable protein fibres in the cytoplasm of most higher eukaryotic cells. Constructed like woven ropes, they are typically between 8 nm to 10 nm in diameter, which is "intermediate" between the thin and thick filaments in muscle cells, where they were first described; their diameter is also between microfilaments (actin filaments) and microtubules. IFs are found resistant to colchicine and cytochalasin B and are sensitive to proteolysis.

In most animal cells IFs form a "basket" around the nucleus and extend out in gentle curving arrays to the cell periphery. IFs are particularly prominent where cells are subjected to mechanical stress, such as in epithelia, where they are linked from cell to cell at desmosomal junctions, along the length of axons, and throughout the cytoplasm of smooth muscle cells. Various names have been attached to the intermediate filaments that have a basis in the cell type in which they are observed. Thus, IFs in epidermal cells are called **tonofilaments**, in nerve cells they are referred to as **neurofilaments** and in neuroglial cells they are designated as **glial filaments**.

In cross-section, intermediate filaments have a tubular appearance. Each tubule appears to be made up of 4 or 5 protofilaments arranged in parallel fashion (**Thorpe**, 1984). IFs are composed of polypeptides of a surprisingly wide range of sizes (from about 40,000 to 240,000 daltons).

Types of Intermediate Filaments

Intermediate filaments are only found in multicellular organisms, in contrast to MTs and MFs, differ markedly in amino acid composition from tissue to tissue. Based on cell type in which they are found, IFs and their proteins can be grouped into six classes (Table 15.1).

Class I and II IFs comprise the **keratins**, proteins that make up the **tonofilaments** found in the epithelial cells that cover the body surfaces and line its cavities (the IFs visible beneath the terminal web in the intestinal mucosa cell consists of keratin). Class I keratins are **acidic keratins**, whereas class II are **basic** or **neutral keratins**.

Class III IFs include vimentin, desmin, synemin and glial fibrillarly acidic protein. **Vimentin** is present in connective tissue and other cells derived from non-epithelial cells. Vimentin-containing filaments are often prominent features, in cultured fibroblast cells. **Desmin** is found in muscle cells. **Synemin** is a protein of 230,000 daltons, which is also present in the intermediate filaments of muscle, together with desmin and vimentin. Vimentin and synemin containing IFs can be observed in the chicken erythrocytes. **Glial fibrillary acidic (GFA) protein** is characteristic of the glial cells that surround and insulate nerve cells.

Table 15.1 Classes of intermediate filaments (Source: Becker *et al.*, 2007).

Class	*IF protein*	*Molecular Mass (kDa)*	*Tissue*	*Function*
I	Acidic cytokeratins	40–56.5	Epithelial cells	Mechanical strength
II	Basic cytokeratins	53–67	Epithelial cells	Mechanical strength
III	Vimentin	54	Fibroblasts; cells of mesenchymal origin; lens of eye	Maintenance of cell shape
III	Desmin	53–54	Muscle cells, especially smooth muscle	Structure support for contractile machinery
III	GFA protein	50	Glial cells and astrocytes	Maintenance of cell shape
IV	Neurofilament protein		Central and peripheral nerves	Axon strength; determines axon size
	NF-L (major)	62		
	NF-M (minor)	102		
	NF-H (minor)	110		
V	Nuclear lamins		All cell types	Form a nuclear scaffold to give shape to nucleus
	Lamin A	70		
	Lamin B	67		
	Lamin C	60		
VI	Nestin	240	Neuronal stem cell	Unknown

Class IV IFs are the **neurofilaments** (NF) **proteins** found in the neurofilaments of nerve cells.

Class V IFs are the **nuclear lamins** A, B and C, which form a filamentous scaffold along the inner surface of the nuclear membrane of virtually all eukaryotic cells.

Neurofilaments found in cells in the embryonic nervous system are made of **nestin** protein, which constitute class VI IFs.

Each of these IF proteins tends to assemble spontaneously *in vitro* to form homopolymers and will also co-assemble with the other Types III IF proteins to form **co-polymers** and **hetero-polymers**. In fact co-polymers of vimentin and desmin, or of vimentin and glial fibrillary acidic protein, are found in some type of cells. For example, desmin remains concentrated in the Z-lines and T-tubule system of striated or skeletal system, together with vimentin, synemin and α-actinin. Since desmin links actin to plasma membrane, from this fact the name of desmin has been derived by **Lazarides** and coworkers in 1976 (in Greek desmin means link or bond).

General Structure of IFs

All six types of intermediate filaments (IFs) contain proteins which have a central rod like domain consisting of four **helical segments**. The central domain is highly conserved in size, secondary structure, and sequence, though sequence homologies are confined to the helical regions. In types I-IV, the helical segments contain a total of 276 amino acids and the linker segments are non-helical (Fig. 15.7). In type V, the helical segments contain 318 amino acids and their linker segments are also helical. The N- and C-terminal domains that flank the central section are non-helical and are much more variable in size and sequence.

Assembly of IFs

A popular model of assembly of a intermediate filament includes the following steps: 1. Two identical **monomers** pair to form a **dimer** in which the conserved helical central regions are aligned

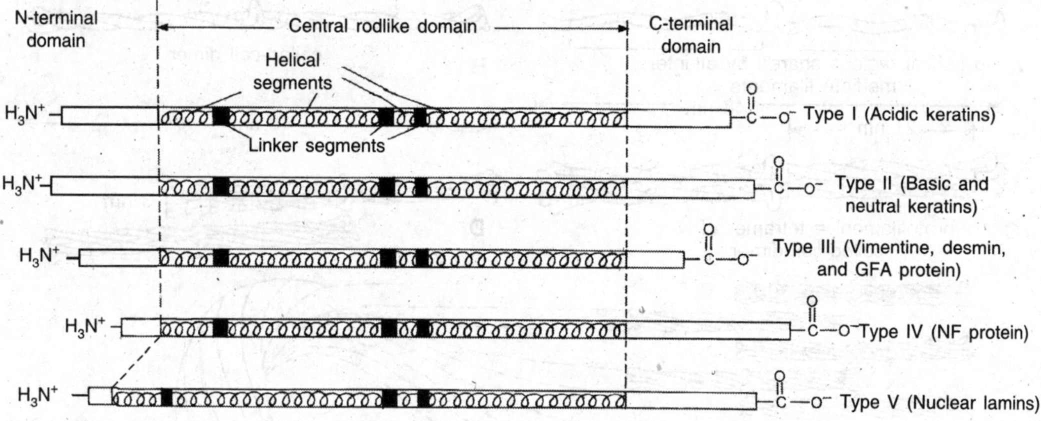

Fig. 15.7. Structural similarities of intermediate filament protein. Except VI or nestin, all the five IFs are shown (after Becker *et al.*, 2007).

in parallel and are wound together into a coiled coil. 2. Two dimers then line up side-by-side to form a 48 nm by 3 nm **protofilament** containing four polypeptide chains. 3. These protofilaments then associate in a staggered manner to form successively larger structures. 4. The final 10 nm diameter of the intermediate filament is thought to be composed of 8-protofilaments (*i.e.*, 32 polypeptide chains) joined end on end to neighbours by staggered overlap to form the long rope-like filaments (Fig. 15.8). It is still not known whether IFs are polar structures (like actin and tubulin) or non-polar (like the DNA double helix).

IFs During Mitosis

Mitosis of cultured epithelial cells shows striking changes in intermediate filaments of cytokeratin and vimentin. During prophase the 10 nm **filaments** unwind into threads of 2 to 4 nm and into spheroidal aggregates containing both types of proteins. At metaphase and anaphase most vimentin and cytokeratin appear as spheroid bodies, while at telophase the filamentous cytoskeleton become gradually reestablished. From these experimental studies, **Franke** (1982) has concluded that the living cells contain factors that promote the reversible disintegration and restoration of intermediate filaments during mitosis.

Functions of IFs

The main function of most intermediate filaments is to provide mechanical support to the cell and its nucleus. IFs in epithelia form a transcellular network that seems designed to resist external forces. The neurofilaments in the nerve cell axons probably resist stresses caused by the motion of the animal, which would otherwise break these long, thin cylinders of cytoplasm. Desmin filaments provide mechanical support for the sarcomeres in muscle cells, and vimentin filaments surround and probably support the large fat droplets in the fat cells.

Comparison of Microtubules, Intermediate Filaments and Microfilaments

The three components of the cytoskeleton, namely microtubules, intermediate filaments and microfilaments have been compared in Table 15.2.

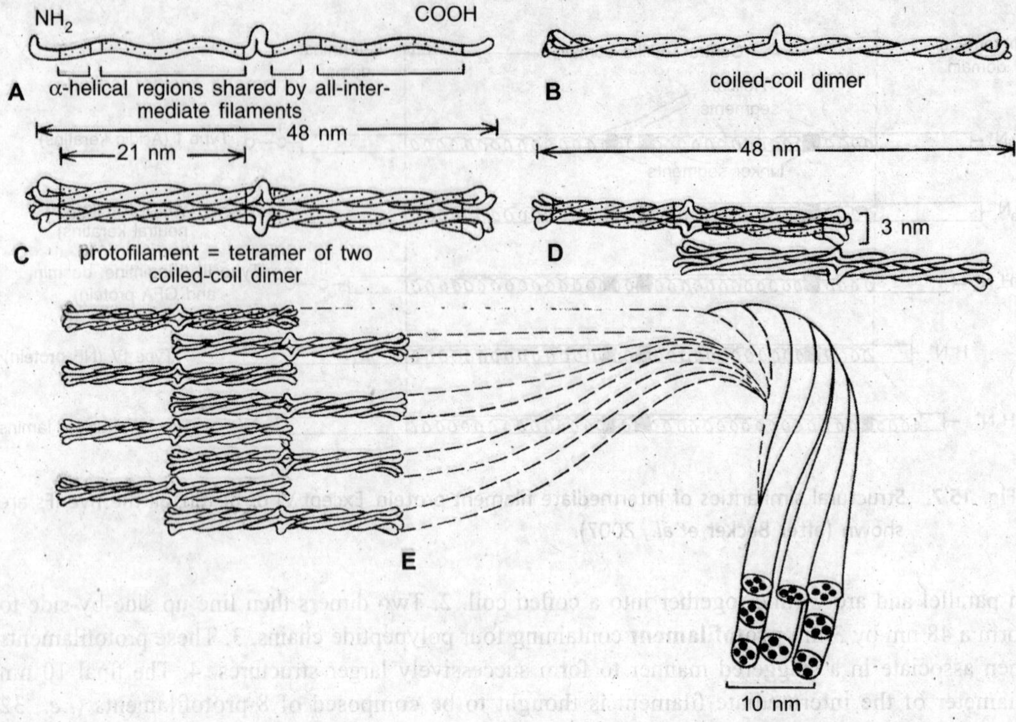

Fig. 15.8. A model to explain the mode of assembly of a 10-μm thick intermediate filament (after Albert *et al.*, 1989).

Table 15.2 Comparison of some properties of microtubules, intermediate filaments and microfilaments (Source: Thorpe, 1984).

Property	*Microtubules*	*Intermediate filaments*	*Microfilaments*
1. Structure	Hollow with walls made up of 13 protofilaments	Hollow with walls made up of 4 to 5 protofilaments	Solid made up of polymerized actin (F-actin)
2. Diameter (nm)	24–25	10	7–9
3. Monomer units	α- and β-tubulin	Six types of protein defining six major classes	G-actin
4. ATPase activity	Present in dynein arms	None	None
5. Functions	1. Motility of eukaryotes 2. Chromosome movement 3. Movements of intra-cellular materials 4. Contribute toward maintaining cell shape	1. Integrate contractile units in muscle 2. Cytoskeletal structural function in cytoplasm	1. Muscle contraction 2. Cell shape changes 3. Protoplasmic streaming 4. Cytokinesis

15.4 CENTRIOLES AND BASAL BODIES

Cytoplasm of some eukaryotic cells contains two cylindrical, rod-shaped, microtubular structures, called **centrioles**, near the nucleus. Centrioles lack limiting membrane and DNA or RNA and form a spindle of microtubules, the mitotic apparatus during mitosis or meiosis and sometimes get arranged just beneath the plasma membrane to form and bear flagella or cilia in flagellated or ciliated cells (**Fulton**, 1971). When a centriole bears a flagellum or cilium, it is called **basal body**. There are many synonyms for basal body, including kinetosome, blepharoplast, basal granule, basal corpuscle, and proximal centriole.

Due to the classic works of **Henneguy** and **Lenhossek** (1897), it has been proposed that basal bodies of cilia and flagella are homologous/with the centrioles found in mitotic spindle.

Occurrence

Centrioles occur in most algal cells (notable exception being red algae), moss cells, some fern cells and most animal cells. They are absent in prokaryotes, red algae, yeast, cone-bearing and flowering plants (conifers and angiosperms) and some non-flagellated or non-ciliated protozoans (such as amoebae). Some species of amoebae have a flagellated stage as well as an amoeboid stage; a centriole develops during the flagellated stage but disappears during the amoeboid stage.

Structure

Centrioles and basal bodies are cylindrical structures which are 0.15–0.25 μm in diameter usually 0.3–0.7 μm in length, though, some are as short as 0.16 μm and others are as long as 8 μm (see **Fulton**, 1971). Both have following ultrastructural components:

1. Cylinder wall. The most striking and regular ultrastructural feature of centrioles and basal bodies is the array of nine triplet microtubules equally spaced around the perimeter of an imaginary cylinder (Fig. 15.9). The space between and immediately around the triplet is filled with an amorphous, electron-dense material. In transverse section the triplets are arranged like vanes or blades of pinwheel or turbine. Each triplet or blade is tilted inward to the central axis at an angle of about 450 to the circumference; within each blade the tubules twist from one end to the other or have a helical course. Since centrioles have no outer membrane, the triplets are considered to form the wall of the cylinder, and arbitrarily define the inside and outside of the centriole.

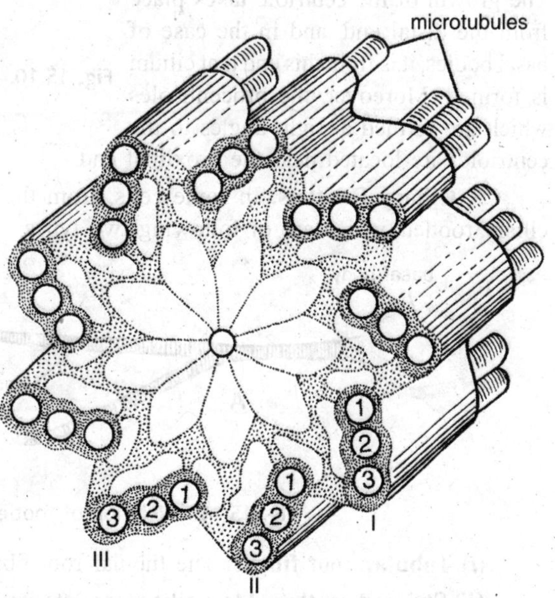

Fig. 15.9 Cross-sectional structure of a centriole. There are nine groups of three microtubules (after De Witt, 1977).

2. Triplets. The nine triplets that make up the wall are basically similar in centrioles and basal bodies. The three subunit microtubules have been designated A, B, and C, with the innermost tubule being A. Individual tubules are 200–260 Å in diameter. Only the A tubule is round; the others are incomplete, C-shaped and share their wall with the preceding tubule. At both ends the C tubule often terminates before the A and B tubules.

The substructure of A, B and C tubules, is similar to the structure of other microtubules (**Stephens**, 1 970). The A tubule has 13, 40–45 Å globular subunits around its perimeter. Three or four of these subunits are shared with the B tubules, which in turn share several of its subunits with the C tubules.

Often the triplets are thought to run parallel to one another and to the long axis of the cylinder, but this is not always the case. In the basal bodies of some organisms, the triplets get closer toward, the proximal end, so the diameter of the cylinder gets smaller. In some centrioles the triplets are parallel to one another but turn in a long-pitched helix with respect to the cylinder axis (**Fulton**, 1971).

3. Linkers. The A tubule of each triplet is linked with C tubule of neighbouring triplet by protein **linkers** at intervals along their entire length. These linkers hold the cylindrical array of the microtubules and maintain the typical radial tilt of the triplets.

4. Cartwheel. There are no central microtubules in the centrioles and no special arms. However, often faint protein spokes are radiate out to each triplet from a central core, forming a pattern like a cartwheel. Such a cartwheel configuration determines the **proximal end** of a centriole and, thus, provides a structural and functional **polarity** to it. The growth of the centriole takes place from the distal end, and in the case of basal bodies, it is from this end that cilium is formed. Moreover, the procentrioles which are formed at right angles to the centriole, are located near the proximal end.

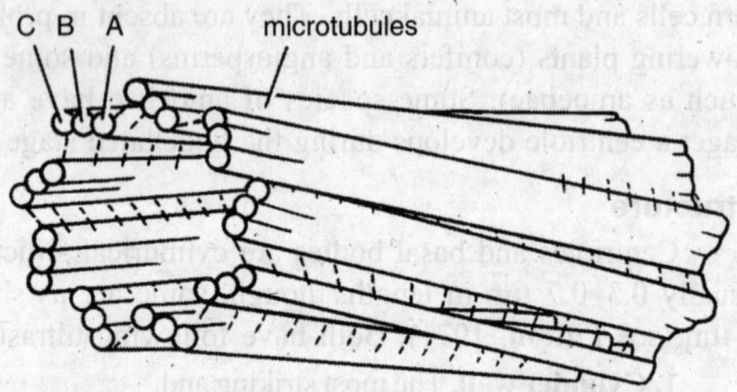

Fig. 15.10. Ultrastructure of a centriole (after Alberts *et al.*, 1989).

5. Ciliary Rootlets. In some cells, from the basal ends of the basal bodies originate the ciliary rootlets which are of following two types:

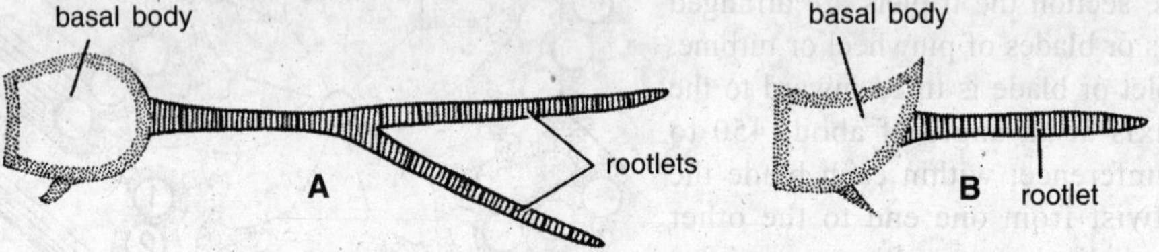

Fig. 15.11. Relation of rootlets with the basal body.

***(i)* Tubular root fibrils.** The tubular root fibrils have the diameter of 200Å.

***(ii)* Striated rootlets.** Most ciliary rootlets are striated, having a regular cross-banding with a repeating period of 55 to 70 nm. The striated fibres of rootlets are composed of parallel microfilaments; 3 to 7 nm in diameter, which in turn are formed of globular subunits. These fibres and filaments may have a structural role such as anchoring the basal body. Due to microfilaments, ciliary rootlets also have a contractile role. The rootlet may be double (*e.g.*, molluscs) or single (*e.g.*, the frog *Rana*).

6. Basal feet and satellites. The **basal feet** are dense processes that are arranged perpendicularly to the basal body. These processes impose a structural asymmetry on the basal body that is related with direction of the ciliary beat. A basal foot is composed of microfilaments that terminate in a dense bar. It may act as a focal point for the convergence of microtubules.

Satellites or **pericentriolar bodies** are electron-dense structures lying near the centriole that are probably nucleating sites for the microtubules.

Chemical Composition

The microtubules of centrioles and basal bodies contain the structural protein, **tubulin**, along with lipid molecules (**Fulton**, 1971). The centrioles and basal bodies contain a high concentration of ATPase enzyme. There exists a controversy that whether centrioles and basal bodies have DNA and RNA. **Fulton** (1971) has doubted the presence of nucleic acids in these organelles.

Origin of Centrioles and Basal Bodies

The idea prevalent years ago that new centrioles arise by the division of existing centrioles is no longer accepted. Rather it appears that new centrioles are either produced *de novo* or are synthesized using an existing centriole as some form of template (semi-autonomous replication).

1. Origin of centrioles by duplication of pre-existing centrioles. In cultured fibroblasts, centriole doubling begins at around the time that DNA synthesis begins (interphase). First the two members of a pair of centrioles separate; then a daughter centriole, called **procentriole**, is formed perpendicular to each original centriole, the two organelles being separated from each other by a distance of 50 to 100 nm. An immature centriole contains a ninefold symmetric array of single micro tubules; each microtubule then presumably acts as a template for the assembly of the triplet microtubules of mature centrioles. Each daughter centriole grows to mature size in late prophase maintaining their close proximity to and orientation at right angles to the mother centriole. As a result, when the interphase nuclei reform at the end of nuclear division, a centrosome containing two centrioles exists beside each nucleus.

Development of centriole (or basal body) has been studied in the ciliates *Paramecium* and *Tetrahymena* and in tracheal epithelium of *Xenopus* and chicks. The stages of development are virtually the same in all of them. Development of a basal body begins with the formation of a single microtubule in an amorphous mass. Microtubules are added one at a time until there is an equally spaced ring of nine. As the microtubules appear, the amorphous mass is lost, as though it were being consumed in the production of the microtubules. There is some evidence that connectives exist between the microtubules, which could act to set the distance between them. Thus, a ring of nine complete microtubules (*i.e.*, A tubules) is formed. The C-shaped B microtubule develop next and finally the C microtubules are added. The hub and cartwheel are added in the centre. The A–C links are not established until the end of development (see **Sheeler** and **Bianchi**, 1987).

2. Origin of basal bodies. In a ciliated vertebrate cell, which may contain hundreds of cilia, the centrioles of the precursor cell give rise to the many basal bodies required to nucleate the cilia in the mature cell. For example, during the differentiation of the ciliated epithelial cells that line the oviduct and the trachea, the centriole pair migrates from its normal location near the nucleus to the apical region of the cell where the cilia will form. There, instead forming a single daughter centriole in the typical manner, each centriole in the pair forms numerous electron-dense fibrogranular satellites. Many basal bodies then arise from these satellites and migrate to the membrane to initiate the formation of cilia.

3. The de novo origin of centrioles and basal bodies. There are certain cases where centrioles seem to arise *de novo*. For example, unfertilized eggs of many animals lack functional centrioles and use the sperm centriole for the first mitotic division (for cleavage), however, under certain experimental conditions—such as extreme ionic imbalance or electrical stimulation—the unfertilized egg can produce a variable number of centrioles. Each of these centrioles nucleates the formation of a small aster, one of which can be used by the egg for cleavage division, so that a haploid organism develops by a process called **parthenogenesis**. In fact, centriole precursors are stored in the cytoplasm of unfertilized eggs and can be activated to form a new centriole under special situation.

Like the centrioles, the basal bodies are found to possess some capacity for self-assembly and they appear suddenly in *Naegleria* as it changes from its amoeboid form to a typical ciliate (see **Reid** and **Leech**, 1980).

The unusual mode of duplication of centrioles and their continuity over many generations led to the earlier suggestion that centrioles might be fully **autonomous, self-replicating** organelles. Although it is now known that this is not the case and that under certain situation centrioles can arise *de novo* in the cytoplasm. So, it is possible that some information necessary for centriole formation is usually carried in the centriole itself (just as the replication of mitochondria and chloroplasts depends on extrachromosomal genes carried in the organelles). For example, in *Chlamydomonas* a set of genes that encode proteins involved in basal body structure and flagellar assembly is carried on a separate genetic element that segregate independently of the major chromosomes. The nature and location of this genetic element have still to be investigated (**Ramanis** and **Luck**, 1986).

Functions

1. Formation of basal bodies and ultimately the cilia is the specialized function of the centrioles in the cell.
2. The normal function of a pair of centrioles in most animal cells is to act as a focal point for the centrosome. The centrosome (also called the **cell centre**) organizes the array of cytoplasmic microtubules during interphase and duplicates at mitosis to nucleate the two poles of the **mitotic spindle**.
3. Sometimes centrioles can serve first one function and then another in turn: for example, prior to each division in *Chlamydomonas*, the two flagella resorb and the basal bodies leave their position to act as mitotic poles.
4. In spermatozoon one centriole give rise to the tail fibre or flagellum.
5. Centrioles and basal bodies are also found to be involved in ciliary and flagellar beat.
6. Centrioles and basal bodies have a role in the reception of optical, acoustic and olfactory signals.
7. Recently, it has been suggested that centrioles could serve as devices for locating the directions of signal sources. Such a role for them has been conceived by comparing the geometric design of centrioles (with their disposition in pairs at the right angles and their ninefold symmetry) with human-made devices such as radar scanner, that detect directional signals (**Albrecht-Buehler**, 1981).

15.5 CELLULAR MOVEMENT

Cellular movement involves motility and contractility. **Motility** occurs at the tissue, cellular and subcellular levels. The most conspicuous examples of motility, particularly in the animal world, take place at **the tissue level**. The muscle tissues common to most animals consists of cells specifically adapted to for *contraction*, and the movements produced are often obvious, whether manifested as the bending of a limb, the beating of a heart, or a uterine contraction during childbirth.

At the cellular level, motility is observed in single cells or in organisms that consist of one or only a few cells. It occurs among cell types as diverse as ciliate protozoa (*e.g.*, *Paramecium*) and motile human sperm, depending in each case on cilia or flagella, cellular appendages adapted for propulsion. Other examples of **motility at the cellular level** include cell migration during animal embryogenesis, amoeboid movement, and the invasiveness of cancer cell in malignant tumors.

Equally important is the movement of intracellular components, which might be regarded as **motility at the subcellular level**. For example, highly ordered microtubules of the mitotic spindle play a key role in the separation of chromosomes during cell division (see Chapter 18). In addition, some cells display the phenomenon of cytoplasmic streaming, in which the cytoplasm undergoes rhythmic patterns of flow (*e.g.*, cyclosis in *Paramecium*). Other examples of mechanical work at the subcellular level include the characteristic movements of molecular structures that occur during cell growth and differentiation. An example of such a process is the transport of cellulose to the growing wall of a dividing or differentiating plant cell.

The microfilaments and microtubules of the cytoskeleton provide a basic scaffolding for specialized **motor proteins**, which interact with the cytoskeleton to produce motion at the molecular level (Table 15.3). The combined effects of these molecular motions produce movement at cellular level. In cases such as muscle contraction, the combined effects of many cells moving simultaneously produce motion at the tissue level.

Table 15.3 Selected motor proteins (or mechanoenzymes) of eukaryotic cells; they produce motion at the molecular level (Source: Becker *et al.*, 2007).

Molecules	*Typical function*
A. Microtubule associated proteins	
1. Cytoplasmic matrix	Motion toward minus end of microtubule
2. Axonemal dynein	Activation of sliding in flagellar microtubule
3. Kinesins	Motion toward plus end of microtubules
B. Microfilament-associated (actin-binding) proteins	
1. Myosin I, monomer	Motion along actin filament
2. Myosin II, filament	Slides actin filaments in sarcomere of muscle

In eukaryotes, there are two motility systems. The first involves interactions between specialized motor molecules and microtubules. Such **microtubule-based movement** occurs in **fast axonal transport**, one of the processes by which a nerve cell transport materials between the central part of the cell and outlying regions. The second type of eukaryotic motility is based on interactions between the actin microfilaments and members of the myosin family of motor molecules. A familiar example of **microfilament-based movement** in muscle contraction.

Intracellular Microtubule-Based Movement

Microtubules (MTs) provide a rigid set of tracks for the transport of a variety of membrane-enclosed organelles and vesicles. The centrosome (including centrioles) provides organization and orientation to MTs, because the minus ends of most microtubules are embedded in the centrosome. The centrosome is generally located near the center of cell, so traffic towards the minus ends of MTs might be considered "inbound" traffic. Traffic directed toward the plus ends might likewise be considered "outbound", meaning that it is directed toward the periphery of the cell.

While microtubules provide an organized set of tracks along which organelles can move, they do not directly generate the force necessary for movement. The mechanical work needed for movement depend on **microtubule-associated motor proteins (motor MAPs)** which attach to vesicles or organelles and then "walk" along the microtubule using ATP to provide the needed energy. These motor MAPs recognize the polarity of the MT, with each motor MAP having a preferred direction of movement. At present, only two major families of motor MAPs namely **kinesins** and **dyneins** are known.

Role of Kinesins and Dyneins in Movement of Organelles

A historically important cell type for studying microtubule-dependent intracellular movement is the neuron. In particular, the squid (*Loligo*, a molluscan animal) giant axon has been extremely useful for the biochemical purification of components that interact with MTs. A receptor protein or neurotransmitter synthesized in cell body of the neuron must be transported over distances upto a meter between the cell body and the nerve ending. The need for such transport arises because ribosomes are present only in the cell body, so no protein synthesis occurs in the axons or synaptic knobs. Instead proteins and membranous vesicles are synthesized in the cell body and transported along the axons to the synaptic knobs. Some form of energy-dependent transport is clearly required mechanism. The process is called **fast axonal transport** and appears to involve the movement of protein containing vesicles and other organelles.

The role of microtubules in axonal transport was initially suggested because the process is inhibited by **colchicine** and other drugs that impair MT function but is insensitive to drugs such as **cytochalasins** that affect microfilaments. Since then, MTs have been visualized along the axon by the deep-etch technique and shown to be prominent feature of the axonal cytoskeleton. Moreover, axonal MTs have small membranous vesicles and mitochondria associated with them.

Further extensive investigations have revealed that in a nerve cell, **kinesin** protein mediates transport from the cell body down the axon to the nerve ending; it is called **anterograde axonal transport**. Likewise, **cytoplasmic dynein** mediates transport in the opposite direction, towards the minus ends of the MTs; it is called **retrograde axonal transport**. These two motors transport materials in opposite direction within the cytoplasm (Fig. 15.12).

Structure and Function of Kinesins

The first kinesins were originally identified in the cytoplasm of squid giant axons. Such "classic" kinesins consists of three parts (Fig. 15.13): 1. A globular head region that attaches to the microtubules and is involved in hydrolysis of ATP; 2. A coiled helical region; and 3. A light-chain region that is involved in attaching the kinesin to other proteins or organelles.

Classic kinesins move along microtubules in 8 nm steps: one of the two globular heads moves forward to make an attachment to a new β-tubulin subunit, followed by detachment of the trailing globular head which can now make an attachment to a new region of the MT (Fig. 15.13B). This movement occurs in a "hand-over-hand" fashion, as the two heads take turns serving as the front "foot". This movement is coupled to the exchange of ATP and ADP at specific sites within the heads. The result is that kinesin moves toward the plus end of a MT in an ATP-dependent fashion. A single kinesin molecule exhibits **processivity**. It can cover long distances before detaching from a MT. A single kinesin molecule can move as far as 1 μm which is a great distance relative to its size. Lastly, as a molecular motor, kinesin appears to be quite efficient, estimates of its efficiency in converting the energy of ATP hydrolysis to useful work are on the order of 60–70% (**Endow**, 2003).

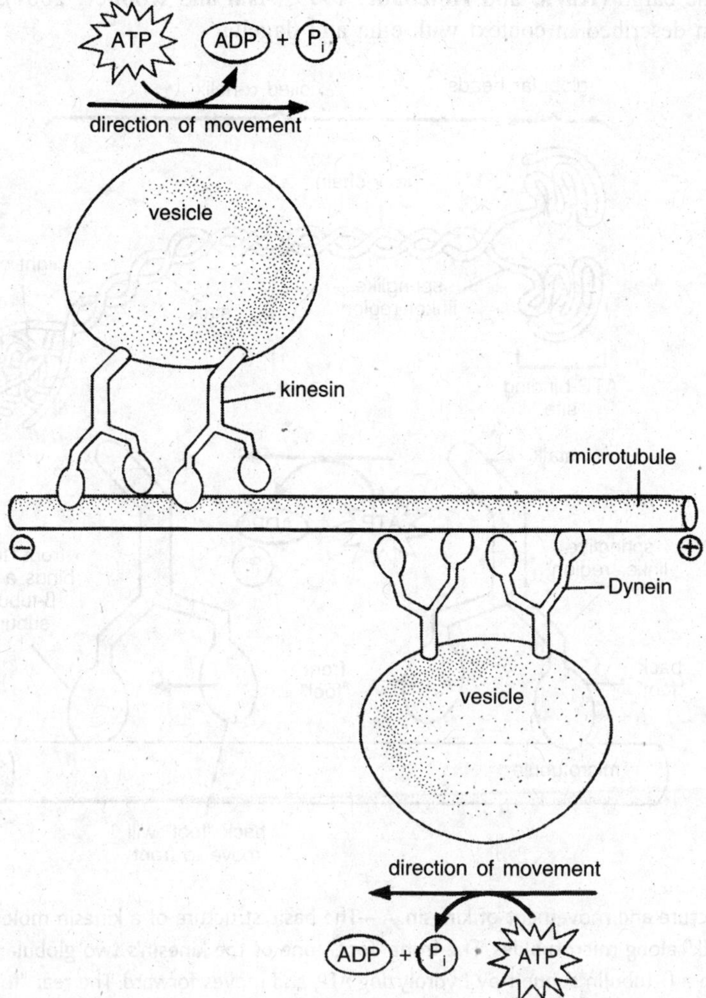

Fig. 15.12. Microtubule-based motility. Kinesins and dyneins are families of molecules that use the energy of ATP hydrolysis to "walk" along microtubules. In the process, they move intracellular structures (vesicles and organelles) along MTs. In general, members of the kinesin family move vesicles or organelles toward the plus ends of MTs-that is, from the center of the cell to the periphery (anterograde transport). Dynein move in opposite direction toward the minus ends of MTs, and therefore toward the center of cell where the MTOC (microtubule-organizing center) is located (retrograde transport)(after Becker *et al.*, 2007).

Structure and Functions of Cytoplasmic Dynein

The dynein family of motor MAPs consists of two basic types: 1. Cytoplasmic dynein and 2. Axonemal dynein. **Cytoplasmic dynein** contains two heavy chains that interacts with microtubules, three intermediate chains and four light chains. In contrast to most kinesins, cytoplasmic dynein moves toward the minus ends of MTs. Also, unlike kinesins, cytoplasmic dynein cannot efficiently bind to cytoplasmic organelles on its own. Instead, dynein is associated with the protein complex known as **dynactin** (Fig. 15.14). The dynactin complex helps to link cytoplasmic dynein to the cargo (such as membraneous vesicle) it transports along MTs, by binding to proteins such as spectrin attached to the membrane cargo (**Karki** and **Holzbaur**, 1999; **Asai** and **Koonce**, 2001). The axonemal dynein has been described in context with cilia and flagella.

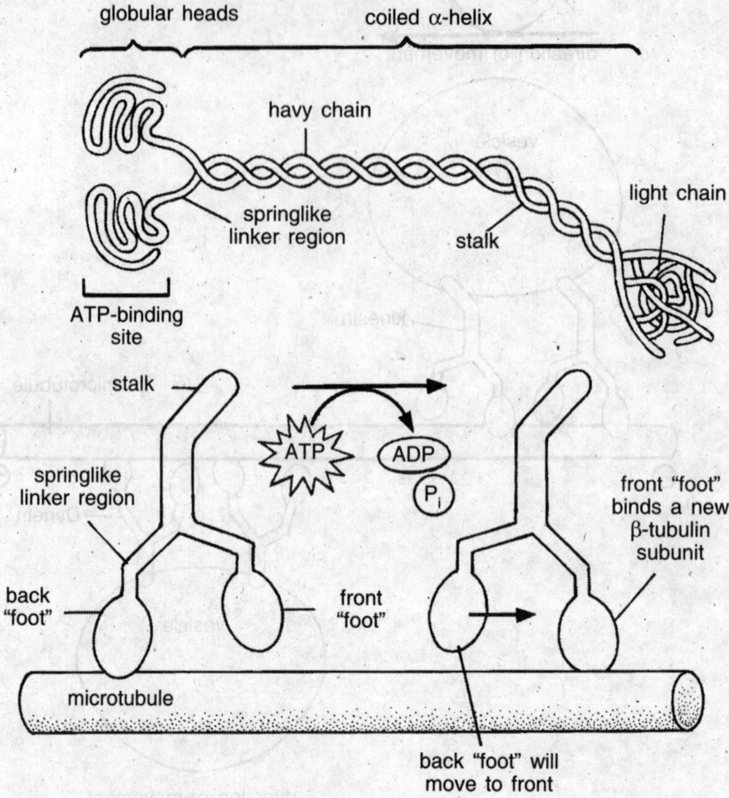

Fig. 15.13. Structure and movement of kinesin. A—The basic structure of a kinesin molecule. B—Kinesins "walk" along microtubules. The front "foot", one of the kinesin's two globular heads, detaches from a β-tubulin subunit by hydrolyzing ATP, and moves forward. The rear "foot" then releases by hydrolyzing ATP and springs forward, moving to the front in a "hand-over-hand fashion" (after Becker *et al.*, 2007).

Figure 15.15 has shown how motor MAPs are involved in transport of intracellular vesicles (**Allan, Thompson** and **McNiven**, 2002).

Microtubule-Based Motility

Microtubules are crucial not only for movement within cells, but also for the movement of **flagella** and **cilia**, the motile appendages of eukaryotic cells (cilia and flagella have been dealt in quite detail in the next section of this chapter).

Actin-Based Cell Movements: The Myosins

Movement of molecules and other cellular components also occurs along another major filament system in the cell, the actin cytoskeleton. As with microtubules, mechanoenzymes act as ATP-dependent motors that exert force on actin microfilaments within cells. These mechanoenzymes are all members of a large superfamily of proteins known as **myosins**. Currently there are at least 18 known classes of myosins. All myosins have at least one polypeptide chain, called the **heavy chain**, with a globular head group at one end attached to a tail of varying length (Fig. 15.16). The globular head binds to actin and uses the energy of ATP hydrolysis to move along an actin filament.

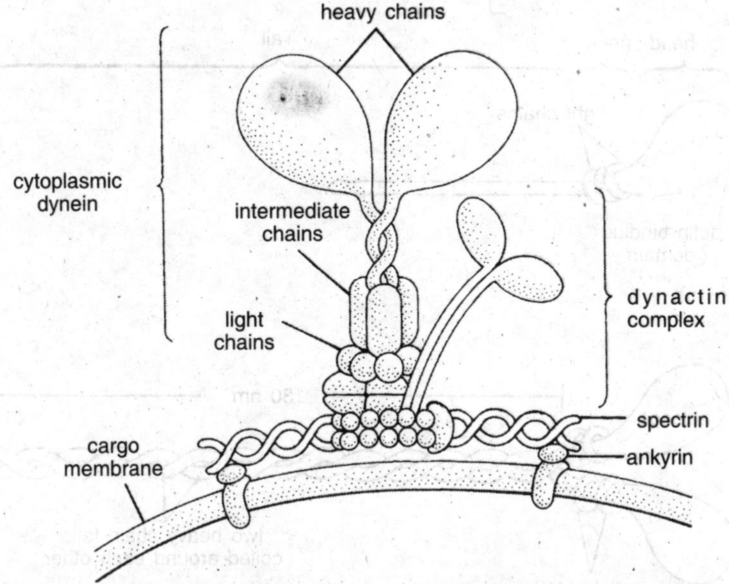

Fig. 15.14. Schematic representation of the cytoplasmic dynein/dynactin complex. Cytoplasmic dynein is linked with cargo membrane indirectly through the dynactin multiprotein complex. The dynactin complex binds to spectrin in the membrane of the cargo vesicle (after Becker *et al.*, 2007).

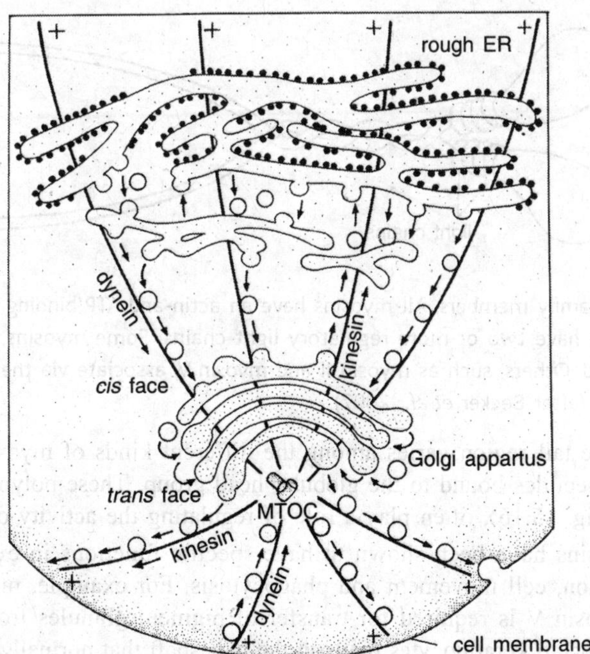

Fig. 15.15. Microtubules, Motor MAPs and the Golgi apparatus. According to this model, vesicles going to and from the Golgi apparatus are attached to microtubules and are thought to be carried by motor MAPs similar to dynein and kinesin. Dynein is a minus-end directed motor MAP, whereas kinesin is plus-end directed MAP. Thus, vesicles derived from either the ER or the plasma membrane are carried toward the Golgi apparatus and MTOC by dynein, where as vesicle derived from the Golgi apparatus a are carried toward either the ER or the cell periphery by kinesins (after Becker *et al.*, 2007).

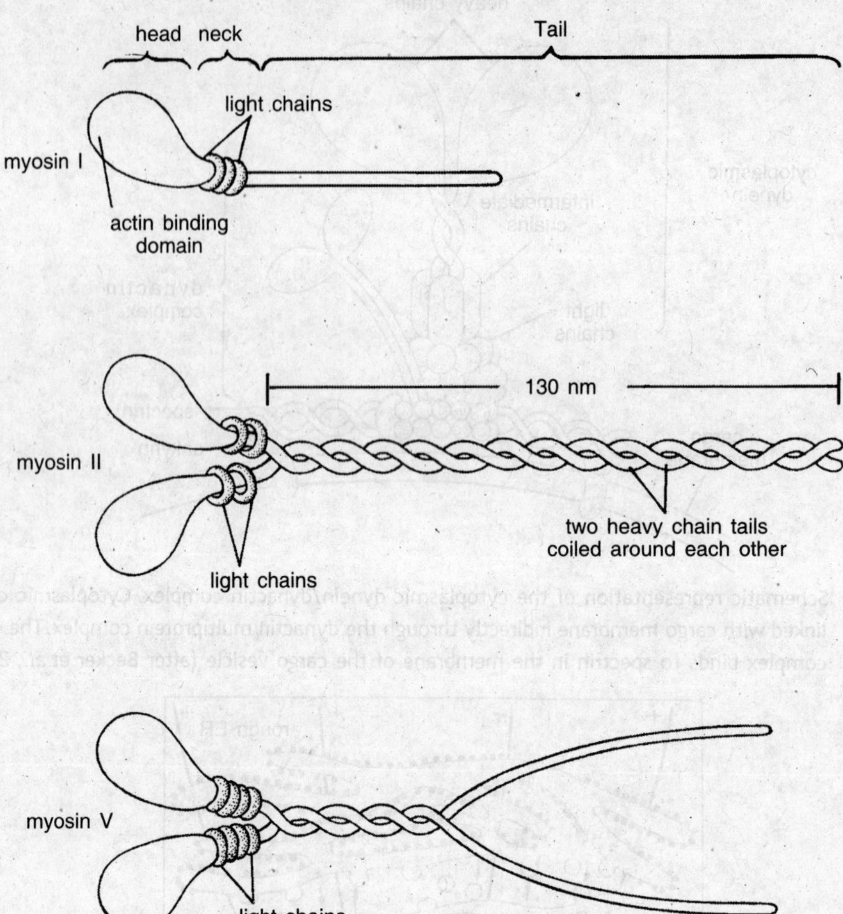

Fig. 15.16. Myosin family members. All myosins have an actin-and ATP-binding heavy chain "head" and typically have two or more regulatory light chains. Some myosins, such as myosin I, have one head. Others, such as myosin II and myosin V, associate via their tails into two headed proteins (after Becker *et al.*, 2007).

The structure of the tail region varies among the different kinds of myosin. Myosins typically contain small polypeptides bound to the globular head group. These polypeptides, referred to as the **light chains** (Fig. 15.16), often play a role of regulating the activity of the myosin ATPase.

Several myosins have been shown to have specific functions in events as wide-ranging as muscle contraction, cell movement and phagocytosis. For example, mutations in the mouse indicate that a myosin V is required for transfer of pigment granules from melanocytes (cells that produce pigment) to keratinocytes (cells in the hair shaft that normally take up the pigment).

15.6 CILIA AND FLAGELLA

The cilia (*L.*, *cili* = eye lash) and flagella (*L.*, little whip) are microscopic, contractile and filamentous processes of the cytoplasm which create food currents, act as sensory organs and perform many mechanical functions of the cell. Morphologically and physiologically, the cilia and flagella are identical structures but even then both can be distinguished from each other by their number, size and functions. Their distinguishing features are as follows:

1. The flagella are less (1 or 2) in number than the cilia which may be numerous (3000 to 14000 or more) in number.
2. The flagella occur at one end of the cell, while the cilia may occur throughout the surface of the cell.
3. The flagella are longer (up to 150 μm) processes, while the cilia are short (5 to 10 μm) appendages of the cytoplasm.
4. The flagella usually beat independently, while the cilia tend to beat in a coordinated rhythm.
5. The flagella exhibit undulatory motion, while the cilia move in a sweeping or pendular stroke.

Stereocilia and Kinocilia

The cell sometimes gives out immobile cytoplasmic extensions known as **stereocilia**. The stereocilia differ from the true cilia which are known as **kinocilia**. The stereocilia occur in most epithelial cells of the epididymis and macula and crista of the internal ear. Stereocilia of the hair cells of the inner ear are responsible for the transduction of sound. These and other stereocilia do not contain microtubules. They contain, however, about 3000 microfilaments which are disposed longitudinally but have a definite polarity and a helical symmetry, with cross-bridges around the filaments (**De Rosier**, 1980).

Distribution of the Cilia and Flagella

The flagella occur in the protozoans of the class Flagellata, choanocyte cells of the sponges, spermatozoa of the Metazoa and among plants in the algae and gamete cells. The cilia occur in the protozoans of the class Ciliata and members of other classes and ciliated epithelium of the Metazoa. The cilia may occur on external body surface and may help in the locomotion of such animals as the larvae of certain Platyhelminthes, Nemertines, Echinodermata, Mollusca and Annelida. The cilia may line the internal cavities or passages of the metazoan bodies as air passage of the respiratory system and reproductive tracts. The nematode worms and arthropods have no cilia.

Except for sperm, the cilia in mammalian systems are not organelles of locomotion. But their effect is the same, that is, to move the environment with respect to the cell surface.

Structure of the Cilia and Flagella

The ciliary apparatus is composed of following basic components: 1. The **shaft** or **cilium**, which is the slender cylindroid process that projects from the free surface of the cell; 2. The **basal body** or granule, a centriole like cellular organelle from which the cilium originates; and 3. In some cells fine fibrils—called **ciliary rootlets**. Basal body remains separated from cilium by a **ciliary** or **basal plate** which has two functions: termination of the C Tubule of each triplet of basal body; and beginning of two central microtubules. The cilia and flagella are extremely delicate, permanently formed, thread-like extension of cytoplasm and their thickness is often at the limit of the resolving power of light microscope.

Isolation and Chemical Composition of Cilia and Flagella

The first detailed chemical analysis of the protein components of the cilia of *Tetrahymena pyriformis* was conducted by **I. R. Gibbons** (1963). Ciliary movements can be analyzed easily by scraping the pharyngeal epithelium of a frog or toad with a spatula and placing the scrapings in a drop of physiological salt solution between a slide and a coverglass. By certain recent techniques, a flagellum can be severed from a cell by a laser beam and ciliary membrane can be peeled off by detergent treatments.

Axoneme of cilia has a variety of proteins such as A and B **tubulins** in the microtubules, **dynein** (the microtubule ATPase), **nexin** and others.

Ultrastructure of the Cilia and Flagella

An eukaryotic cilium or flagellum is composed of three major parts: a central axoneme or shaft, the surrounding plasma membrane and the interposed cytoplasmic matrix (Fig. 15.17).

1. Ciliary membrane. Though the ciliary membrane (9.5 nm thick) is physically continuous with plasma membrane of the cell, but it contains far less amount of proteins than the latter (*i.e.*, it is a typically protein poor; **Satir**, 1977). Further, some of the proteins present in the ciliary membrane are specific to it and have a role as the barrier against the loss of ATP and certain essential ion that are required at appropriate concentrations to provide the energy for the ciliary movement.

Ciliary necklace. An unusual feature of the membranes of all somatic cilia is the presence of multiple strands (2 to 6 and up to 11) of particles, called **ciliary necklace**, at the base of the organelle. These particles can be seen in the electron micrographs of freeze-fractured cilium (see **Fumi Suzuki**). The ciliary necklace is found at a region in the cilium where microtubules and basal bodies make contact with the membrane. According to **Thorpe** (1984), ciliary necklace may have following two functions: 1. It may position the underlying basal body from which the cilium is originated. 2. It may help in the differentiation of ciliary membrane; *i.e.*, the rings of particles may retain proteins that would otherwise diffuse out and be incorporated into ciliary membrane.

2. Matrix. The bounded space of the cilium contains a watery substance known as **matrix**. In the ciliary matrix are embedded eleven microtubules of axoneme and other interconnecting proteins.

3. Axoneme. The axial basic microtubular structure of cilia and flagella is called **axoneme**. It is the essential motile element of these organelles. The axoneme is about 0.2 to 10 μm in diameter and may range from a few microns to 1 to 2 mm in length. The cilia may be thicker at the base and may become thinner gradually along the length.

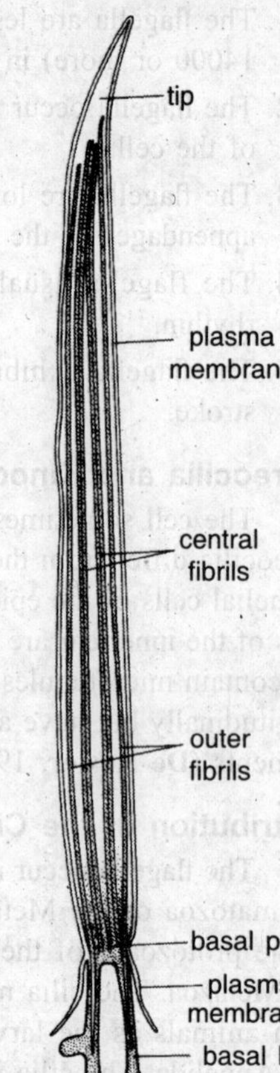

Fig. 15.17. A ciliary apparatus in L.S. showing fundamental structure of cilium or shaft, basal bodies and ciliary rootlets.

The axonemal elements of nearly all cilia and flagella (as well as the tails of sperm cells) contain the same 9 + 2 arrangement of microtubules. In the centre of the axoneme are two **singlet microtubules** or fibrils that run length of the cilium. Each of the central microtubules (25 nm in diameter) is composed of 13 protofilaments. The central fibrils, each has a wall of 6 nm thick and are located 35 nm away from each other. Both central fibrils are connected by a bridge and are enclosed in a common central **sheath**. A plane perpendicular to line joining the two central tubules divides the axoneme into a right and a left symmetrical half. It is generally accepted that the plane of the ciliary beat is perpendicular to this plane of symmetry.

Table 15.4 Major protein structures of the axoneme of the cilia and flagella (Source: Alberts *et al.*, 1989).

	Axoneme component (periodicity along axoneme)	*Function*
1.	Tubulin (8 nm)	Principal component of microtubules.
2.	Dynein (24 nm)	Project from microtubule doublets and interact with adjacent doublets to produce bending.
3.	Nexin link (86 nm)	Hold adjacent microtubule doublets together.
4.	Radial spokes (29 nm)	Extend from each of the nine outer doublets inward to the central pair.
5.	Sheath projections (14 nm)	Project as a series of side arms from the central pair of microtubules; together with the radial spokes these regulate the form of the ciliary beat.

Nine **doublet microtubules** (each 36 nm in diameter) surround the central sheath; they remain separated from each other by a distance of 20 nm and from the ciliary membrane by a distance of 25 nm. Each peripheral doublet consists of two microtubules or subfibres (18 to 25 nm in diameter), one is smaller (A) and complete, having 13 protofilaments of tubulin and lying closer to the axis; the other subfibre (B) is larger and incomplete, having only 11 protofilaments. The B subfibre lacks the wall adjacent to A subfibre and is skewed at about 10° away from the axis. Other associated structures of the doublets are the following:

1. Dynein arms. Extending from each A subfibre are two **dynein arms**—an **outer arm** and an **inner arm**, that are oriented in the same direction in all microtubules (*i.e.*, peripheral doublets). This orientation is clockwise when the axoneme is viewed from base to tip. The arms contain **dynein**, which is large protein complex (nearly 2 million daltons) composed of 9 to 10 polypeptide chains, the largest of which are about 450,000 daltons. Dynein is a Mg^{2+} and Ca^{2+}-activated ATPase enzyme which after solubilization can recombine at the same position on the A microtubule. Dynein contains two or three elliptical or globular **heads** (depending on the source) linked to a common **root**, **foot** or **base** by the thin flexible **strands** or **stalks** (Fig. 15.19). Thus, the base of the dynein molecule attaches only to A subfibre, leaving the heads free to make contact with the adjacent B tubules of neighbouring doublet. It indicates that B tubule has different structure than the A tubule, so that, the base of dynein cannot attach to it. The resulting asymmetry is required to prevent a fruitless tug-of-war between the neighbouring microtubules, which presumably explains why each of nine outer microtubules is an A–B doublet (see **Alberts** *et al.*, 1989).

2. Nexin links. Adjacent doublets are joined or linked by **peripheral**, **interdoublet** or **nexin links**; the nexin links have a periodicity of 86 nm. Nexin links extend from A tubule of one doublet to B tubule of adjacent doublet. Nexin protein has a molecular weight of about 150,000 to 160,000 daltons. Nexin links are highly elastic: their normal length is 30 nm, but they can be stretched to 250 nm without breaking (see **Darnell**, 1986). They are thought to function like the rubber bands to resist the sliding between adjacent double microtubules (*i.e.*, they maintain integrity of the axoneme during the sliding motion.

3. Radial spokes. There are 36 nm long radial bridges or links between the A subfibre and the sheath containing the central microtubules. These spokes terminate in a dense **knob** or **head**, which may have a fork-like structure. Earlier observation of **Warner** and **Satir** (1974) that the spokes are attached perpendicularly to the ciliary axis where it is straight and that they are relatively detached in bent or tilted regions of the axis has led to the hypothesis that *they may be active in the conversion of active sliding between outer doublets into local axial bending.*

Lastly, the structures of a cilium at the base and tip are slightly different from features described above (Fig. 15.18).

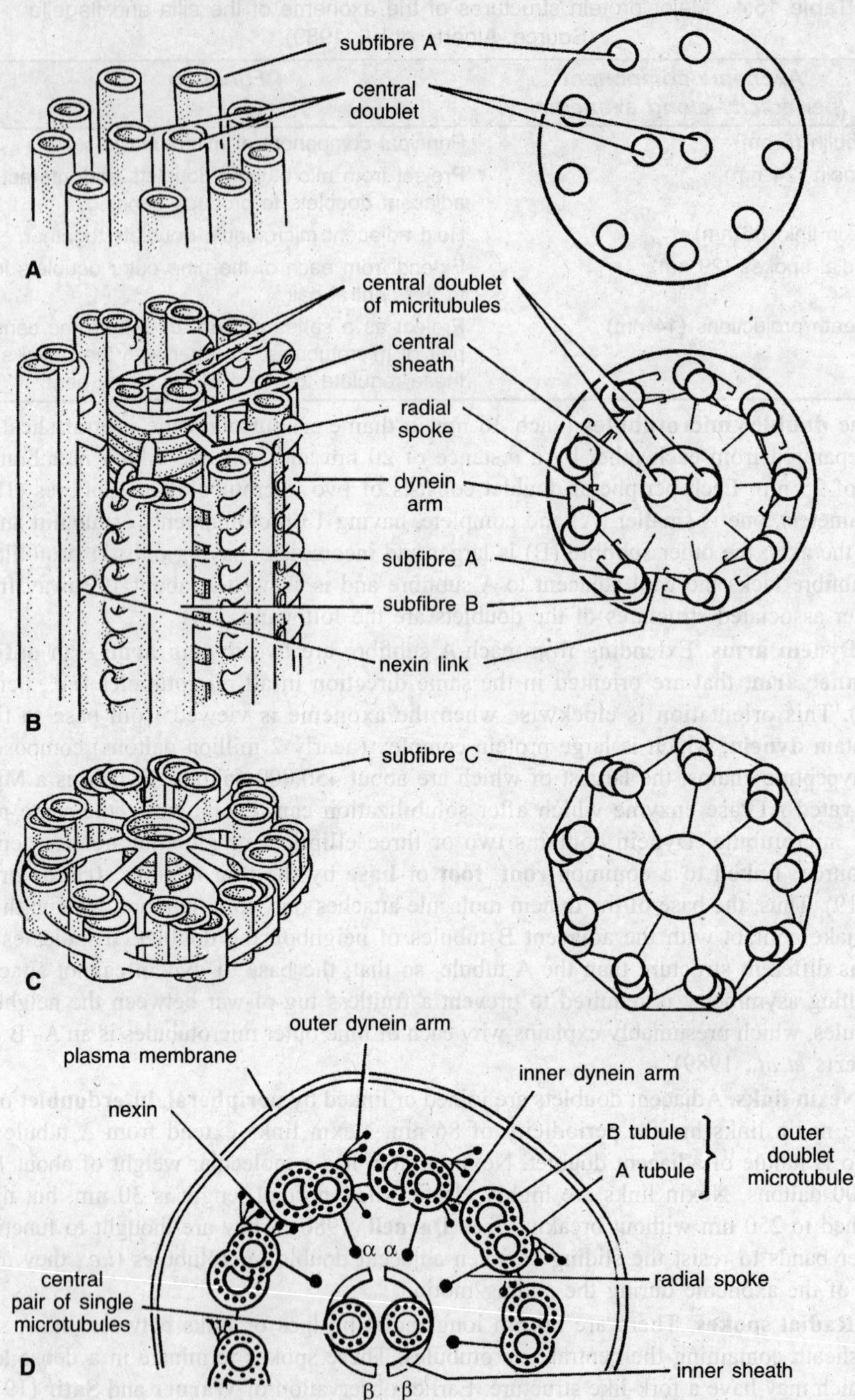

Fig. 15.18. Schematic diagram of a flagellar axoneme, showing relationships between its components at the flageller tip (A), in the middle of the flagellum (B), and at the basal body (C). More detailed picture of flagellum in the cross section (after Berns, 1983; Albert *et al.*, 1989).

Physiology of Ciliary Movement

The cilia and flagella serve many purposes and their movements propel the organism. The cilia are contractile structures and in them two types of rhythms known as **metachronic** and **isochronic** or **synchronous rhythms** produce the wave of contractions in the cilia. In the **metachronic** type of rhythm the cilia of a row beat one after the other, while in the **synchronous** or isochronic rhythm, all the cilia of a row beat simultaneously. In contrast to the cilia, the flagella exhibit undulant motion and beat independently.

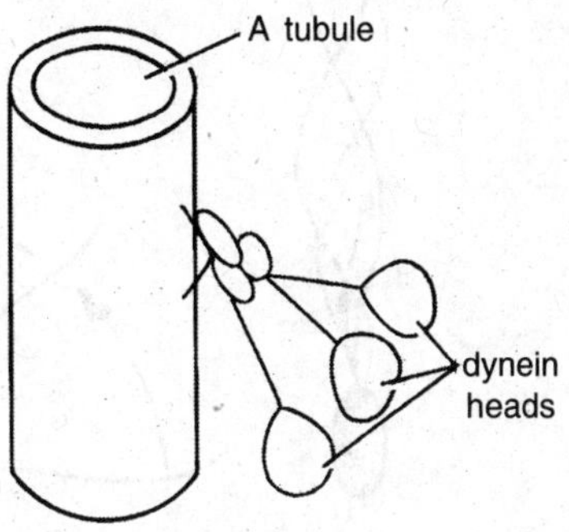

Fig. 15.19 The base of the dynein molecule binds tightly to an A tubule in an ATP-independent manner, while the large globular heads have an ATP-dependent binding site for a second microtubule (B tubule) (after Alberts *et al.*, 1989).

The beating of cilia or flagella is caused by the intraciliary excitation which is followed by the interciliary conduction. **Hayashi** (1961) has reported that the two inner filaments of a cilium transmit excitation and the nine outer filaments are the seat of ATP splitting. The movement of cilia may be under nervous or cytoplasmic control. In a few invertebrate embryos the cilia are probably under nervous control since their movement may be stopped upon stimulation of the embryo. In ciliates they are thought to be coordinated by a neuromotor centre near the mouth since destruction of the fibres connecting the centre to the cilia results in uncoordinated movements (**Taylor**, 1920). However, **Okajima** (1966) reported coordinated movements in *Euplotes* even after complete dissection of the neuromotor fibres. Recent studies have shown that cytoplasm is necessary for the ciliary movements. The ATP provides necessary amount of energy for the motion of the cilia and flagella. Four types of ciliary movements have been recognized which are as follows:

1. The pendulus ciliary movement. The pendulus type of ciliary movement is carried out in a single plane. It occurs in the ciliated protozoans which have rigid cilia. In such cases the movement of the cilia is carried out by a flexion at its base.

2. The unciform ciliary movement. The unciform (hook-like) ciliary movement occurs commonly in the metazoan cells. In such type of movement, when the cilia contract it becomes double and acquires a hook-like shape.

3. The infundibuliform ciliary movement. The infundibuliform ciliary movement occurs due to the rotary movement of the cilium and flagellum. In this case, the cilium or flagellum is passed through three mutually perpendicular planes in the space and makes conical or funnel-shaped shape.

4. The undulant movement. The undulant movement is the characteristic of the flagellum. In undulant movement the waves of the contraction proceed from the site of implantation and pass to the border.

Each beat of cilium or flagellum involves the same pattern of microtubule movement. Each cilium moves with a whip-like motion and its beat may be divided into two phases: 1. The fast **effective stroke** (or forward active stroke or power stroke) in which the cilium is fully extended and beating against the surrounding liquid (*i.e.*, it is like the action of an oar in a rowboat; Fig. 15.20B). 2. The slow **recovery stroke**, in which the cilium returns to its original position with an unrolling movement that minimizes viscous drag (Fig. 15.20C). The cycles of adjacent cilia are almost but not quite in synchrony, creating a wave-like pattern.

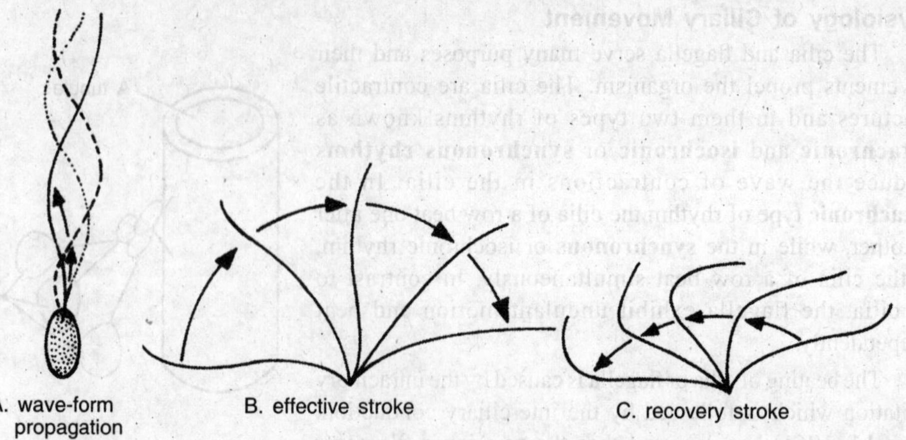

Fig. 15.20. Movement of cilia and flagella. A—Typical wave-form propagation of a flagellum, such as that found in sperm; B—Effective stroke common with cilia of all types; C—Recovery of cilium prior to the next effective stroke (after Dyson, 1978).

The flagellum instead of making whip-like movements, propagates quasi-sinusoidal waves (Fig. 15.20A), *i.e.*, successive waves move toward the tip of the flagellum, propelling the cell (*e.g.*, sperm) in the opposite direction.

The mechanism of force and movement (bending) by the flagellum has recently been studied extensively. It is well established now that the ciliary movement is generated by the microtubules and the associated structures of the flagellum. It was shown that the cell free flagella can be caused to move by adding an energy source such as ATP. Even broken pieces of cilia or isolated axoneme itself continue to beat, suggesting the role of microtubules in the movement. The contractile **axostyle** of some microorganisms such as *Metamonadida* (a dinoflagellate that lives in the gut of termite; Fig. 15.21) is another example of microtubule mediated motile process (see **Berns**, 1983). In fact, bending force is produced by the sliding of microtubules. This has been shown by exposing isolated axonemes to proteolytic enzymes, which disrupts both the nexin links and the radial spokes but leaving the dynein arms and the microtubules themselves intact. If such a partially digested structure is exposed to as little as 10 μM ATP, the axoneme elongates until it is up to nine times its original length, the component microtubules in the axoneme telescoping out of the loosened structure. It seems that the adjacent peripheral doublets can actively slide against each other once they have been freed of their lateral cross-links (such as those made of nexin). Apparently, in intact structure this sliding movement is converted to bending. Further, since the adjacent outer doublets actively slide against each other, a force must be generated between them. This force is apparently generated by dynein arms which 'walk' along the doublets, as has been suggested by **Peter Satir's** (1968) sliding filament hypothesis.

Sliding Filament Hypothesis

Recent experimental work on ciliary motion has shown notable similarities with the sliding mechanism involved in the interaction of actin and myosin in muscle. The dynein arms attached to subfibre A have been compared with the cross bridges of myosin and it has been postulated that they form intermittent attachments, by which one doublet (N_1) is able to push the adjacent one ($N_1 + 1$) toward the tip of the axoneme (Fig. 15.22). Under normal conditions, the attachment of subfibre A of N to subfibre B of N + 1 by dynein arms is not observed in an intact cilium. Only when the ciliary membrane is extracted with a detergent, the axoneme enters in a state of **rigor**

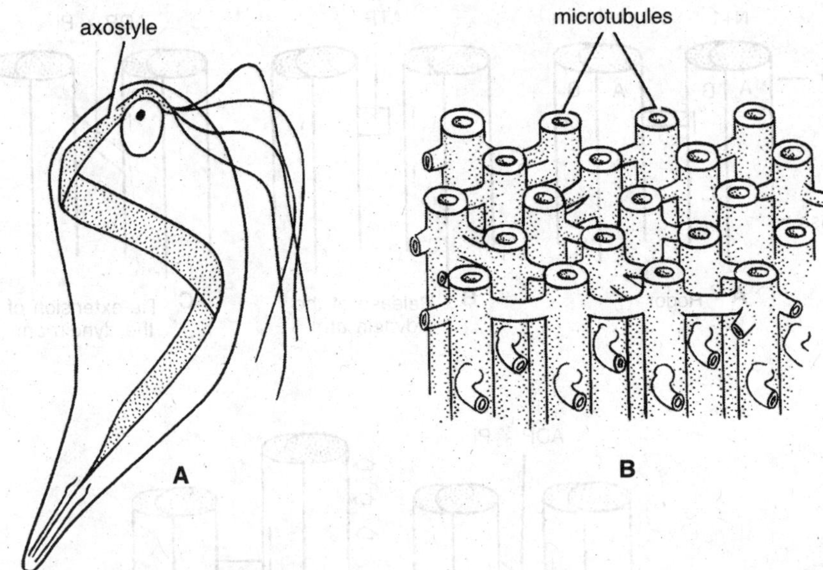

Fig. 15.21. Axostyle of *Metamonadida* flagellate. A—The relationship of the axostyle with the cell body; B—Schematic representation of the arrangement of microtubules into parallel sheets with protein cross bridges between them as seen in a T.S. of axostyle (after Berns, 1983).

in which the attachment is produced (Fig. 15.22A). Addition of ATP to axonemes in the state of rigor restores motility and causes release of the dynein arm (Fig. 15.22B). In this mechano-chemical cycle, the next step would be **re-extension** of the dynein arm (Fig. 15.22C) and its **rebinding** at an angle, with a new, more proximal site on subfibre B (Fig. 15.22D). This step involves the hydrolysis of ATP to ADP + Pi. In the last step, the arm returns to the rigor position and displacement of the doublets results (Fig. 15.22E).

Force is generated when dynein arms move. The movement of sliding is converted to bending by virtue of radial spokes that bridge each other doublet to the inner pair of microtubules (**Warner** and **Satir**, 1974; **Huang** *et al.*, 1981). The wave that is generated by sliding is propagated down the organelle from base to tip, with the cell generally moving in a direction opposite from that of wave propagation.

Immotile Cilia Syndrome (Kartagenre's Syndrome)

Ciliary motion can be affected by many deficiencies in the protein composition of the organelle. For example, in immotile cilia syndrome, a condition characterized by severe respiratory difficulty (chronic bronchitis and sinusitis) and male sterility, the underlying genetic defect is the absence of inner and outer dynein arms on the peripheral doublets of both cilia and flagella. The symptoms of this syndrome result from the immobility of cilia in the respiratory tract and of the flagella in the sperm.

In *Chlamydomonas* several mutational defects have been studied in the axoneme of flagellum which may lead to paralysis of the flagellar function (**Luck**, 1984).

Other Functions of the Cilia and Flagella

1. The ciliary or flagellar movement provides the locomotion to the cell or organism.
2. The cilia create food currents in lower aquatic animals.
3. In the respiratory tract, the ciliary movements help in the elimination of the solid particles from it.

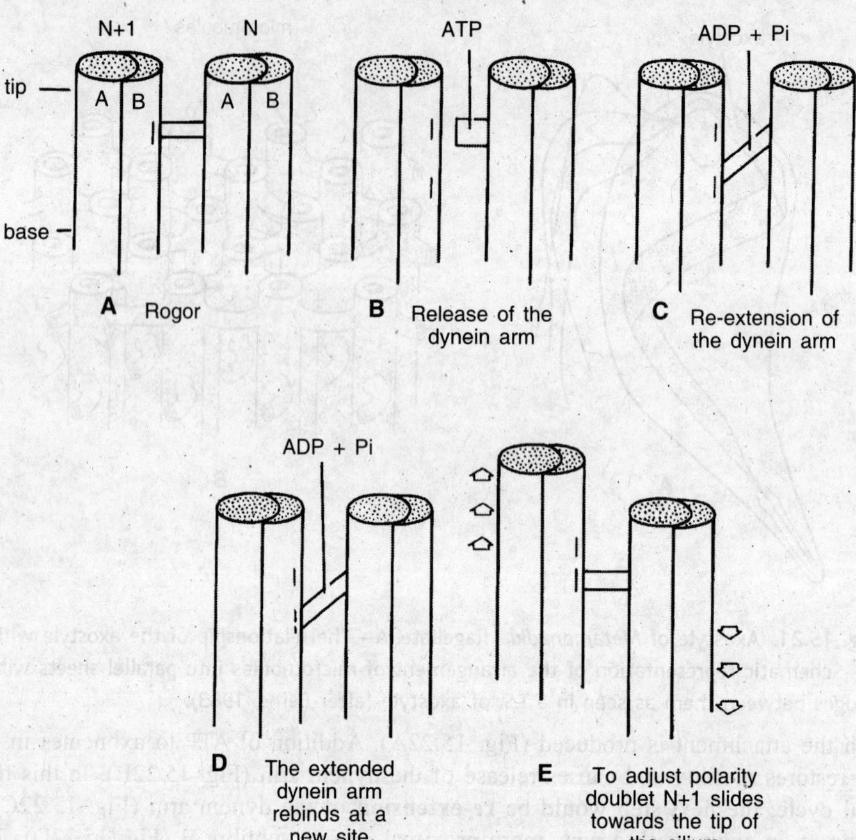

Fig. 15.22. Schematic representation of the mechano-chemical cycle involved in sliding of filament in ciliary movement (after De Robertis and De Robertis, Jr., 1987).

4. The eggs of amphibians and mammals are driven out from the oviduct by the aid of vibratile cilia of the latter.

Thus, the cilia and flagella serve many physiological processes of the cell, such as locomotion, alimentation, circulation, respiration, excretion and perception of sense.

Origin of Cilia

The newly formed basal bodies become aligned in rows beneath the apical plasma membrane and each basal body may then produce satellites from the side, a root from its base and a cilium from its apex.

The formation of the cilia from the basal bodies is started by the formation of a vesicle-like structure of the cytoplasm towards the distal end of the basal bodies. The walls of the vesicle are invaginated due to rapid growth of ciliary shaft. The walls of the vesicle are temporary and are replaced by the new and permanent ciliary sheaths.

Derivatives of Cilia

The cilia are modified into a variety of structures such as the rods and cones of the retina, crown cell of saccus vasculosus of third ventricle of fishes, primitive sensory cells of the pineal eye and cnidocil of the nematocysts of the coelenterates.

QUESTIONS

Long Answer Questions

1. What are microtubules? Describe their structure, assembly, disassembly and functions.
2. What are the intermediate filaments? Describe their types, structure and cellular functions.
3. Define the term microfilament. Describe the structure and function of microfilament in the cell.
4. Describe the ultrastructure of basal body and centrioles.
5. Differentiate between cilia and flagella. Describe the structure of the axoneme.

Short Answer Questions

1. What is cytoskeleton? Write a short note on the cytoskeleton.
2. How are the centrioles and basal bodies (kinetosomes) are originated in the cell?
3. What are the main functions of the basal bodies and the centrioles?
4. Explain sliding microtubule hypothesis of ciliary movement.
5. Flagella and cilia, though identical in structure, commonly exhibit a quite different pattern of movement. Describe the two patterns and conditions under which one or the other would be more appropriate.
6. Make a comparison of three main components of the cytoskeleton: microtubules, intermediate filaments and microfilaments.
7. Write short notes on the following:
   (*a*) axoneme; (*b*) sterocilia; (*c*) ciliary necklace; (*d*) axostyle; (*e*) sliding filament hypothesis; (*f*) Kartagenre's syndrome.
8. Describe structure and function of kinesins.
9. Give an account of structure and function cytoplasmic dyneins.

Very Short Answer Questions

1. Define the terms basal bodies and centrioles.
2. Write the names of main constituents of cytoskeleton.
3. How many subunits are found in a microtubule?
4. What is microtubule?
5. Define microfilament.
6. What is intermediate filament?
7. Define the MTOC.
8. What is cilium?
9. What is axoneme?
10. What is microtubule-based movement?
11. What is microfilament-based movement?
12. Define motor MAP.
13. What is kinesin?
14. Define the cytoplasmic dynein.
15. Define myosin.
16. What is found in the core of cilia? (*H.N.B. Garhwal 2009*)

Fill in the Blanks

1. .... filaments are .... filaments which provide mechanical support to .... of muscle cells.
2. Intermediate filaments of epidermal cells are called ....
3. .... of muscle cells depend on .... cell movement.
4. Cellular movement involves .... and contractility.
5. Microtubule-based movement occurs in fast .... by which a nerve cell transports materials between its .... and synapse.

6. Motor MAPs are of two main types: .... and cytoplasmic ....
7. .... is required in mice for transfer of pigment granules from melanocytes to ....
8. Cilia and flagella perform .... -based motility of the cells.

Multiple Choice Questions

1. Tubulin protein is absent in: (*H.N.B., Garhwal 2008*)
   (*a*) cilia (*b*) flagella
   (*c*) plasma membrane (*d*) microtubules
2. ATPase activity is associated with structural component of cilia called
   (*a*) A tubule (*b*) B tubule
   (*c*) radial spoke (*d*) dynein
3. Cytoskeleton is made up of
   (*a*) callose deposits (*b*) cellulosic microfibrils
   (*c*) proteinaceous filaments (*d*) calcium carbonate granules
4. Organelle important in spindle formation during nuclear division is
   (*a*) centriole (*b*) Golgi apparatus
   (*c*) chloroplast (*d*) mitochondrion
5. Stereocilia can be found in
   (*a*) trachea (*b*) epididymis (*c*) sperm (*d*) buccal cavity
6. Intermediate filaments are composed of
   (*a*) tubulin (*b*) actin
   (*c*) flagellin (*d*) keratin, vimentin, desmin or lamin
7. Microtubular structure of cilium is called
   (*a*) basal body (*b*) ciliary rootlet (*c*) axonem (*d*) kinetosome

ANSWERS

Very Short Answer Questions

1. *Basal body* is a microtubule containing structure located at the base of a eukaryotic flagellum or cilium that consists of nine sets of triplet microtubules; identical in appearance to a centriole.

   Centriole is microtubule containing structure embedded within the centrosome of animal cells, where centrioles lie at right angles to each other; identical in structure to the basal body of eukaryotic cilia and flagella.
2. Microtubules, microfilaments and intermediate filaments.
3. 13 subunits.
4. Microtubule (MT) is a polymer of the protein tubulin, with a diameter of about 25 nm, that is an integral part of cytoskeleton and that contributes to the support, shape and motility of eukaryotic cells.
5. Microfilament (MF) is a polymer of actin, with a diameter of about 7 nm, that is an integral part of the cytoskeleton, contributing to the support, shape and mobility of eukaryotic cells.
6. Intermediate filament (IF) is a group of protein filaments that are the most stable components of the cytoskeleton of eukaryotic cells, exhibit a diameter of 8 to 12 nm, which is intermediate between the diameter of actin microfilaments and microtubules.
7. MTOC stands for microtubule-organizing center. It is structure that initiates the assembly of microtubules, the primary example being the centrosome.
8. Cilium (plural cilia) is a membrane-bounded appendage on the surface of a eukaryotic cell, composed of a specific arrangement of microtubules and responsible for motility of the cell. It is shorter and more numerous than closely related organelles called flagella.

9. Axoneme is a group of interconnected microtubules that form the backbone of a eukaryotic cilium or flagellum, usually arranged as nine outer doublet microtubules surrounding a pair of central microtubules.

10. It is motility based on microtubules; it includes motility involving cilia, flagella and sperm tails, as well as chromosomal movements mediated by spindle microtubule.

11. It is motility based on microfilaments composed of actin and their interaction with myosin; includes muscle contraction, amoeboid movement, cytoplasmic streaming, and cytokinesis in animal cells.

12. MAP stands for microtubule-associated motor protein; it includes proteins, such as kinesin or dynein, that uses energy derived from ATP to drive the transport of vesicles and organelles along microtubules or to generate sliding forces between microtubules.

13. It is family of motor proteins that generate movement along microtubules using energy derived from ATP hydrolysis.

14. Cytoplasmic dynein is a cytoplasmic motor protein that moves along the surface of microtubules in the plus-to-minus direction driven by energy derived from ATP hydrolysis; it is associated with dynactin which links cytoplasmic dynein to cargo vesicle.

15. Myosin is a family of motor proteins that create movements by exerting force on actin microfilaments using energy derived from ATP hydrolysis; makes up thick filaments that move the actin thin filament during muscle contraction.

16. Matrix with axoneme having 9 + 2 arrangement of microtubules.

Fill in the Blanks

1. Desmin, intermediate, sarcomeres.
2. Tonofilaments.
3. Myosins, actin-based.
4. Motility.
5. Axonal transport, cell body.
6. Kinesins, dyneins.
7. Myosin V, keratinocytes.
8. Microtubule.

Multiple Choice Questions

1. (*c*) **2.** (*d*) **3.** (*c*) **4.** (*a*) **5.** (*b*) **6.** (*d*)
7. (*c*).

16

Nucleus, Nucleopore and Nucleolus

The nucleus (*L.*, *nux* = nut) is the heart of the cell. It is here that almost all of the cell's DNA is confined, replicated and transcribed. The nucleus, thus, controls different metabolic as well as hereditary activities of the cell. A synonymous term for this organelle is the Greek word **karyon**. Nucleus serves as the main distinguishing feature of eukaryotic cells, *i.e.*, this is the true nucleus as opposed to the nuclear region, prokaryon or nucleoid of the prokaryotic cells.

Nuclei were first discovered and named by **Robert Brown** in 1833 in the plant cells and were quickly recognized as a constant feature of all animal and plant cells. Nucleoli were described by **M.J. Schleiden** in 1838, although first noted by **Fontana** (1781). The term nucleolus was coined by **Bowman** in 1840. In 1879, **W. Flemming** coined the term chromatin for chromosomal meshwork.

16.1 OCCURRENCE AND POSITION

The nucleus is found in all the eukaryotic cells of the plants and animals. However, certain eukaryotic cells such as the mature sieve tubes of higher plants and mammalian erythrocytes contain no nucleus. In such cells nuclei are present during the early stages of development. Since mature mammalian red blood cells are without any nuclei, they are called red blood "corpuscles" rather than cells (*L. corpus* = body, especially dead body or corpse).

Usually the nucleus remains located in the centre. But its position may change from time to time according to the metabolic states of the cell. For example, in the embryonic cells the nucleus generally occupies the geometric centre of the cell but as the cells start to differentiate and the rate of the metabolic activities increases, the displacement in the position of the nucleus takes place. In certain cells such as the glandular cells the nucleus remains located in the basal portion of the cell.

16.2 MORPHOLOGY

Number. Usually the cells contain single nucleus but the number of the nucleus may vary from cell to cell. According to the number of the nuclei following types of cells have been recognised:

1. Mononucleate cells. Most plant and animal cells contain single nucleus, such cells are known as **mononucleate cells.**

2. Binucleate cells. The cells which contain two nuclei are known as **binucleate cells**. Such cells occur in certain protozoans such as *Paramecium* and cells of cartilage and liver.

3. Polynucleate cells. The cells which contain many (from 3 to 100) nuclei are known as **polynucleate cells**. The polynucleate cells of the animals are termed as **syncytial cells**, while the polynucleate cells of the plants are known as **coenocytes**. The most common example of the syncytial cells are the osteoblast (polykaryocytes of the bone marrow) which contain about 100 nuclei per cell and striated muscle fibres each of which contains many hundred nuclei. The siphonal algae *Vaucheria* contains hundreds of nuclei and certain fungi are the best example of the coenocytic plant cells.

Shape. The shape of the nucleus normally remains related with the shape of the cell, but certain nuclei are almost irregular in shape. The spheroid, cuboid or polyhedral cells (isodiametrical cells) contain the **spheroid** nuclei. The nuclei of the cylindrical, prismatic or fusiform cells are **ellipsoid** in shape. The cells of the squamous epithelium contain the **discoidal** nuclei. The leukocytes, certain infusoria, glandular cells of some insects and spermatozoa contain the irregular shaped nuclei. Nuclei of cells of silk glands of silk worm have finger-like extensions that greatly increase their surface area (Fig. 16.1).

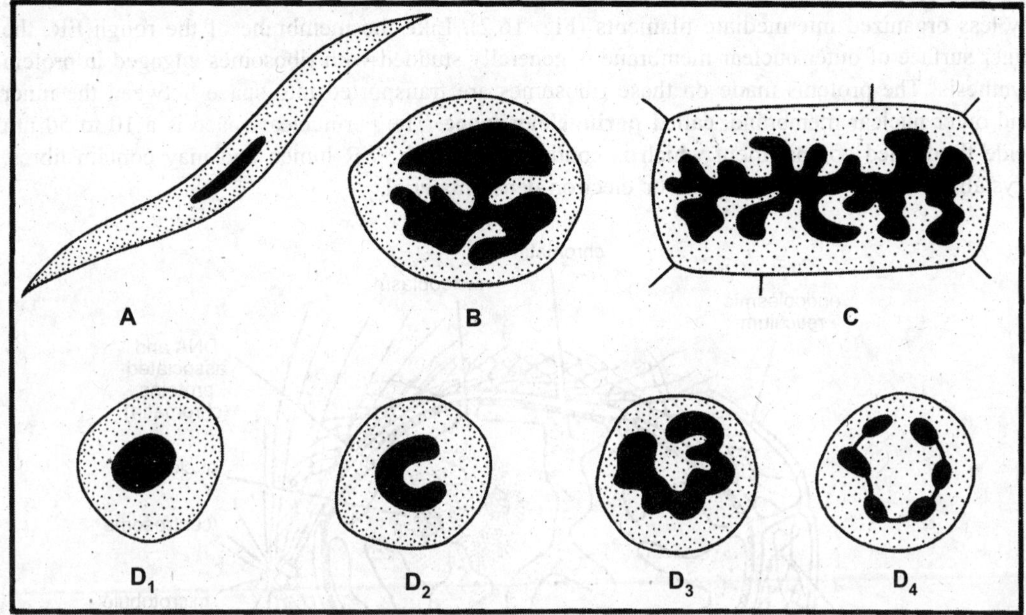

Fig. 16.1. Different shapes of the nucleus in animal cells. A—Elongated in muscle cell; B—Lobed in a human neutrophil cell; C—Branched in a silk spinning cell of an insect larva; D1 to D4—Variable shapes in leucocytes.

Size. Generally nucleus occupies about 10 per cent of the total cell volume. Nuclei vary in size from about 3 μm to 25 μm in diameter, depending on cell type and contain diploid set of chromosomes. The size of the nucleus is directly proportional to that of the cytoplasm. **R. Hertwig** has given the following formula for the deduction of the size of the nucleus of a particular cell.

$$NP = \frac{V_n}{V_c - V_n}$$

where NP is the nucleoprotein index, V_n is the volume of the nucleus and V_c is the volume of the cell.

Moreover, the size of the nucleus is related with the number of the chromosomes or ploidy. The haploid cells contain small-sized nuclei in comparison to the nuclei of the diploid cells. Likewise the polyploid cells contain larger nuclei than the diploid cells. Thus, the size of the nucleus of a cell depends on the volume of the cell, amount of the DNA and proteins and metabolic phase of the cell.

16.3 ULTRASTRUCTURE

The nucleus is composed of following structures: 1. The nuclear membrane or karyotheca or nuclear envelope; 2. The nuclear sap or nucleoplasm; 3. The chromatin fibres; and 4. The nucleolus.

1. Nuclear Envelope

The nuclear envelope (or perinuclear cisterna) encloses the DNA and defines the nuclear compartment of interphase and prophase nuclei. It is formed from two concentric unit membranes, each 5–10 nm thick. The spherical **inner nuclear membrane** contains specific proteins that act as binding sites for the supporting fibrous sheath of intermediate filaments (IF), called **nuclear lamina**. Nuclear lamina has contact with the chromatin (or chromosomes) and nuclear RNAs. The inner nuclear membrane is surrounded by the **outer nuclear membrane**, which closely resembles the membrane of the endoplasmic reticulum, that is continuous with it. It is also surrounded by less organized intermediate filaments (Fig. 16.2). Like the membrane of the rough ER, the outer surface of outer nuclear membrane is generally studded with ribosomes engaged in protein synthesis. The proteins made on these ribosomes are transported into space between the inner and outer nuclear membrane, called **perinuclear space**. The perinuclear space is a 10 to 50 nm wide fluid-filled compartment which is continuous with the ER lumen and may contain fibres, crystalline deposits, lipid droplets or electron-dense material.

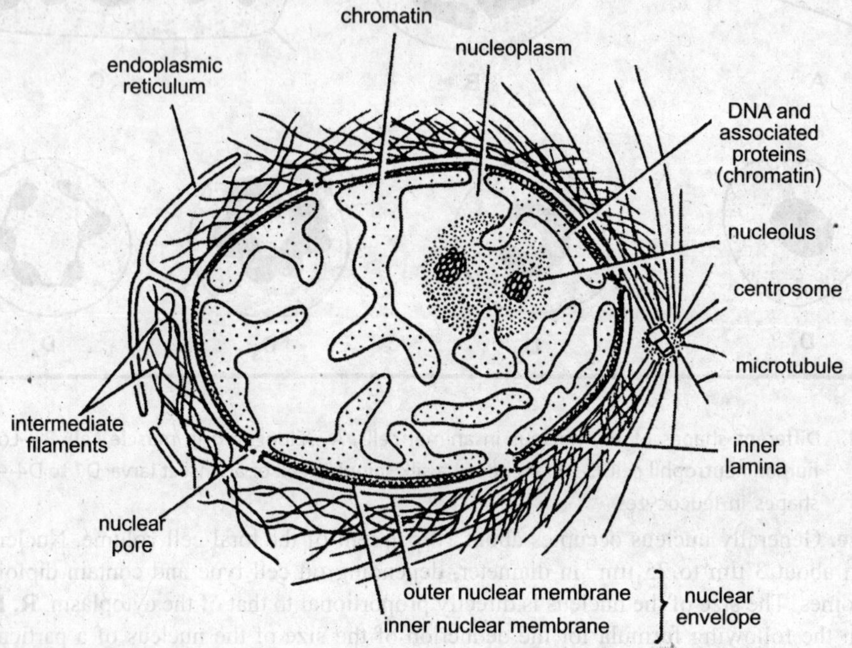

Fig. 16.2. Cross-section of a typical cell nucleus showing ultrastructure.

Nuclear lamina. It is also called **fibrous lamina**, **zonula nucleum limitans**, **internal dense lamella**, **nuclear cortex** and **lamina densa**.

The nuclear lamina is a protein meshwork which is 10 nm thick. It lines the inside surface of the inner nuclear membrane, except the areas of nucleopores, and consists of a square lattice of intermediate filaments. In mammals, these intermediate filaments are of three types: **lamins A**, **B** and **C** having M.W. 74,000, 72,000 and 62,000 daltons, respectively. The lamins form dimers that have a rod-like domain and two globular heads at one end. Under appropriate conditions of pH and ionic strength, the dimers spontaneously associate into filaments that have a diameter and repeating structure similar to those of cytoplasmic filaments.

The nuclear lamina is a very dynamic structure. In mammalian cells undergoing mitosis, the transient phosphory-lation of several serine residues on the lamins causes the lamina to reversibly disassemble into tetramers of hypo-phosphorylated lamin A and lamin C and mem-brane associated lamin B. As a result, lamin A and C become entirely soluble during mitosis, and at telophase they become dephosphory-lated again and polyme-rize around chromatin. Lamin B seems to remain associated with mem-brane vesicles during mitosis, and these vesi-cles in turn remain as a distinct subset of mem-brane components from which nuclear envelope is reassembled at telophase. Inside an interphase nucleus, chromatin binds strongly to the inner part of the nuclear lamina which is believed to interfere with chromosome condensation. In fact, during meiotic chromosome condensation, the nuclear lamina completely disappears by the pachytene stage of prophase and reappears later during diplotene in oocytes, but does not reappear at all in spermatocytes.

The lamins may play a crucial role in the assembly of interphase nuclei. For example, when cells are left for a long time in colchicine (drug which arrests cells in metaphase), the lamins assemble around individual chromosomes, which then surrounded by nuclear envelopes give rise to micronuclei containing only one chromosome. A similar phenomenon occurs during normal amphibian development. In the first few cleavages of amphibian development, the nuclear envelope initially forms around individual chromosomes, forming several vesicles that then fuse together to form a single nucleus. This suggests that chromatin is the nucleating centre for the deposition of a nuclear lamina and envelope.

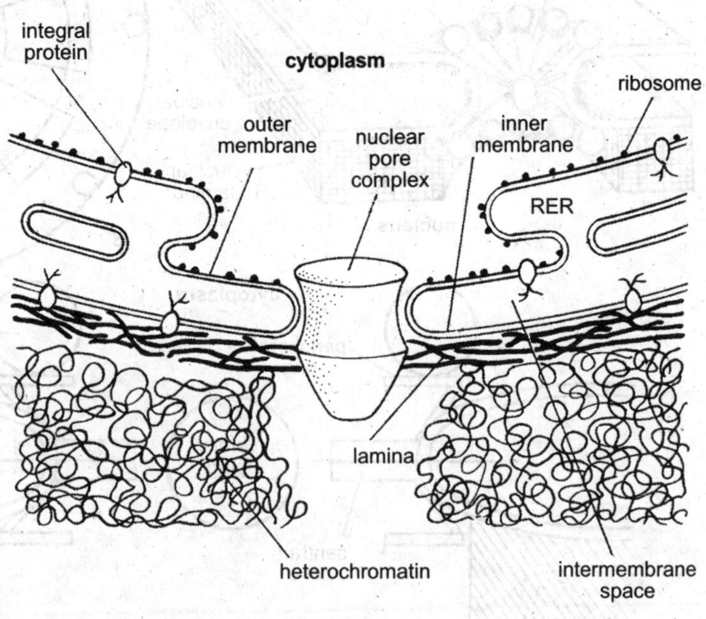

Fig. 16.3. Schematic diagram of nuclear envelope showing position of lamina and other structures.

Nuclear pores and nucleocytoplasmic traffic. The nuclear envelope in all eukaryotic forms, from yeasts to humans, is perforated by **nuclear pores** which have the following structure and function:

1. Structure of nuclear pores. Nuclear pores appear circular in surface view and have a diameter between 10 nm to 100 nm. Previously it was believed that a diaphragm made of amorphous to fibrillar material extends across each pore limiting free transfer of material. Such a diaphragm called **annulus** has been observed in animal cells, but lacking in plant cells. Recent electron microscopic studies have found that a nuclear pore has far more complex structure, so it is called **nuclear pore complex**. Each pore complex has an estimated molecular weight of 50 to 100 million daltons. Negative staining techniques have demonstrated that pore complexes have an eight-fold or octagonal symmetry. More recent computerized image-processing techniques of **Unwin** and **Mulligan** (1982) have shown that the pore complex consists of two "**rings**" (**R** or annuli) at its periphery with an inside diameter of 80 nm, a large particle that forms a **central plug (C)** and radial '**spokes**' **(S)** that extends from the plug to the rings (Figs. 16.4B and C). **Particles (P)** are anchored to cytoplasmic ring and are thought to be inactive ribosomes. The '**hole**' in the centre of the pore complex is an aqueous channel through which water-soluble molecules shuttle between the nucleus and the cytoplasm. This hole often appears to be plugged by a large central granule (**central plug**) which is believed to consist of newly made ribosomes or other particles caught in transit.

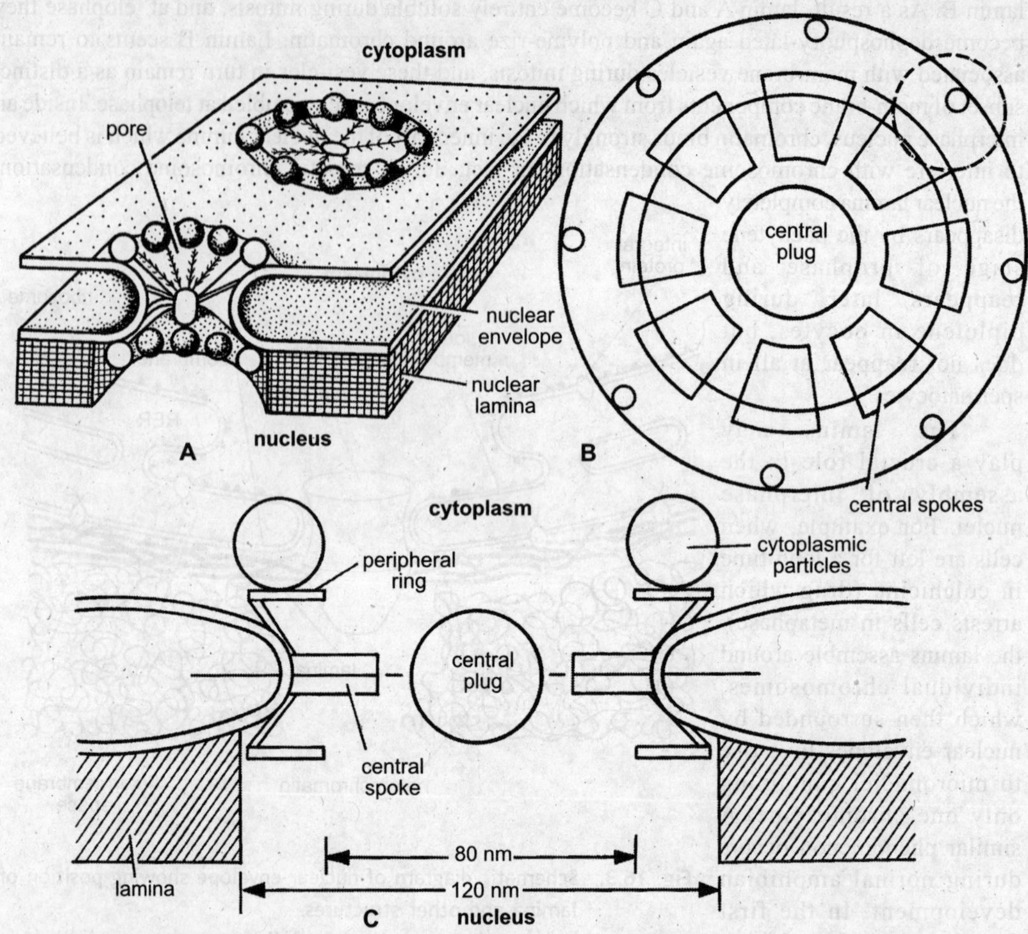

Fig. 16.4. Nuclear pore. A—A part of nuclear envelope with 80 nm pores occupied by pore complexes of octagonal radial symmetry; B—Nuclear pore in top view; C–A nuclear pore in cross section.

The pore complex perforates the nuclear envelope bringing the lipid bilayers of the inner and outer nuclear membrane together around the margins of each pore. Despite this continuity, which would seem to provide a pathway for the diffusion of membrane components between the inner and outer membranes, the two membranes remain chemically distinct.

Quite recently, following two proteins have been found to be associated to the nuclear pores : one is an integral membrane protein, a glycoprotein of 120,000 daltons that may anchor the annuli to the lipid bilayer. The second protein is a 63,000 dalton protein (that has covalently bound acetyl neuraminic acid) located on the cytoplasmic side of the electron-dense material that occludes the nuclear pores. This protein may be involved in the transport of materials through the nuclear pores.

2. Number of nuclear pores (Pore density). In nuclei of mammals it has been calculated that nuclear pores account for 5 to 15 per cent of the surface area of the nuclear membrane. In amphibian oocytes, certain plant cells and protozoa, the surface occupied by the nuclear pores may be as high as 20 to 36 per cent. The number of pores in the nuclear envelope or **pore density** seems to correlate with the transcriptional activity of the cell. Thus, pore densities as low as ~3 pores/μm^2 are seen in nucleated red blood cells and lymphocytes (which are inactive in transcription). These cells are highly differentiated but metabolically inactive and they are non- proliferating cells.

The majority of proliferating cells have pore densities between 7 and 12 pores/μm^2. Among cells of a third type, differentiated but highly active, pore densities are often 15 to 20 pores/μm^2. Liver, kidney and brain cells fall into this category. Still higher pore densities are found in specialized cells, such as salivary gland cells (~40 pores/μm^2) and the oocytes from *Xenopus laevis* (~50 pores/μm^2), both of which are very active in transcription.

3. Arrangement of nuclear pores on nuclear envelope. In somatic cells, the nuclear pores are evenly or randomly distributed over the surface of nuclear envelope. However, pore arrangement in other cell types is not random but rather range from **rows** (*e.g.,* spores of *Eqisetum*) to **clusters** (*e.g.,* oocytes of *Xenopus laevis*) to **hexagonal** (*e.g.,* Malpighian tubules of leaf hoppers) packing order.

4. Nucleo-cytoplasmic traffic. Quite evidently there is considerable trafficking across the nuclear envelope during interphase. Ions, nucleotides and structural, catalytic and regulatory proteins are imported from the cytosol (cytoplasmic matrix); mRNA, tRNA and ribosome subunits are exported to the cytosol (cytoplasmic matrix). However, one of the main functions of the nuclear envelope is to prevent the entrance of active ribosomes into the nucleus.

5. Rate of transport through the nuclear pores. As we have already described, the nuclear envelope of a typical mammalian cell contains 3000 to 4000 pores (about 11 pores/μm^2 of membrane area). If the cell is synthesizing DNA, it needs to import about 10^6 histone molecules from the cytoplasm every 3 minutes in order to package newly made DNA into chromatin, which means that on an average each pore needs to transport about 100 histone molecules per minute. Further, if the cell is growing rapidly, each nuclear pore needs to export about three newly assembled ribosomes per minute to the cytoplasm, since ribosomes are produced in nucleus but function in the cytoplasm. The export of new ribosomal subunits is particularly problematic since these particles are about 15 nm in diameter and are much too large to pass through the 9 nm channels of nuclear pores, it is believed that they are specifically exported through the nuclear pores by an active transport system. Similarly, mRNA molecules complexed with special proteins to form ribonucleoprotein particles, are thought to be actively exported from the nucleus.

Lastly, nuclear pores are not the only avenues for nucleocytoplasmic exchanges (Fig. 16.5). For example, small molecules and ions readily permeate both nuclear membranes. Larger molecules and particles may pass through the membrane by formation of small pockets and vesicles that traverse the envelope and empty on the other side.

2. Nucleoplasm.

The space between the nuclear envelope and the nucleolus is filled by a transparent, semi-solid, granular and slightly acidophilic ground substance or the matrix known as the **nuclear sap** or **nucleoplasm** or **karyolymph**. The nuclear components such as the chromatin threads and the nucleolus remain suspended in the nucleoplasm.

The nucleoplasm has a complex chemical composition. It is composed of mainly the nucleoproteins but it also contains other inorganic and organic substances, *viz.,* nucleic acids, proteins, enzymes and minerals.

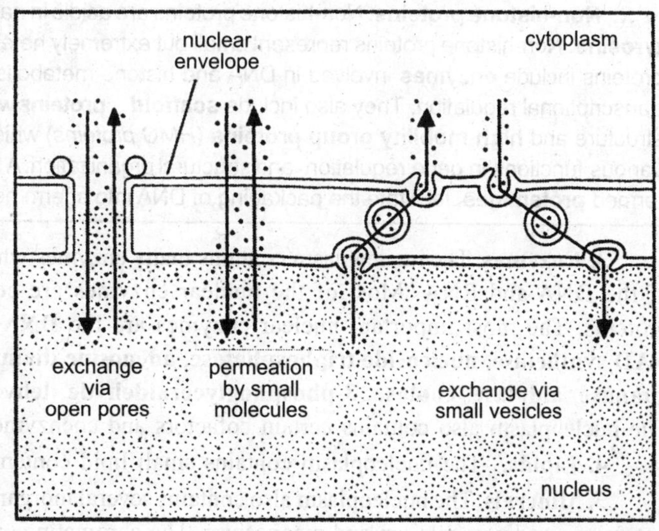

Fig. 16.5. Various avenues for transport of material from the nucleus to the cytosol (cytoplasmic matrix).

1. Nucleic acids. The most common nucleic acids of the nucleoplasm are the DNA and RNA. Both may occur in the macromolecular state or in the form of their monomer nucleotides.

2. Proteins. The nucleo-plasm contains many types of complex proteins. The nucleoproteins can be categorized into following two types:

***(i)* Basic proteins.** The proteins which take basic stain are known as the basic proteins. The most important basic proteins of the nucleus are **nucleoprotamines** and the **nucleohistones**.

The nucleoprotamines are simple and basic proteins having very low molecular weight (about 4000 daltons). The most abundant amino acid of these proteins is **arginine** (pH 10 to 11). The protamines usually remain bounded with the DNA molecules by the salt linkage. The protamines occur in the spermatozoa of the certain fishes. The nucleohistones have high molecular weight, *e.g.*, 10,000 to 18,000 daltons. The histones are composed of basic amino acids such as **arginine**, **lysine** and **histidine**. The histone proteins remain associated with the DNA by the ionic bonds and they occur in the nuclei of most organisms. According to the composition of the amino acids following types of histone proteins have been recognised, *e.g.*, histones rich in lysine, histones with arginine and histones with poor amount of the lysine.

***(ii)* Non-histone or Acidic proteins.** The acidic proteins either occur in the nucleoplasm or in the chromatin. The most abundant acidic proteins of the euchromatin (a type of chromatin) are the **phosphoproteins** (Box 16.1).

Box 16.1

Nucleoproteins are the main and the most important component of nucleus. These are compounds of nucleic acids and proteins. The proteins present in the nucleus are specific and are of following two types:

1. Histones. Histones (which are of five basic types such as H2A, H2B, H3, H4 and H1 or H5) are highly basic proteins rich in positively charged **lysine** and **arginine** residues. Histones H3 and H4 are most strongly conserved, while linker histones (H1 or H5) show most diversity.

All the histones except H4 exist as multiple isoforms (**isohistones**) whose relative predominance in chromatin varies in a cell-type-specific manner. Histones represent the bulk of protein in chromatin and are relatively homogeneous in nature.

2. Non-histone proteins. Non-histone proteins are acidic in nature and are rich in **tryptophan** and **tyrosine**. Non-histone proteins represent small but extremely heterogenous fraction. The non-histone proteins include **enzymes** involved in DNA and histone metabolism, replication, recombination, and transcriptional regulation. They also include **scaffold proteins** which organize higher order chromatin structure and **high-mobility group proteins** (*HMG proteins*) which are highly charged proteins with various functions in gene regulation and structural organization. A further class of non histone proteins, termed **protamines,** facilitate the packaging of DNA into sperm head (**Twyman** 1998).

3. Enzymes. The nucleoplasm contains many enzymes which are necessary for the synthesis of the DNA and RNA. Most of the nuclear enzymes are composed of non-histone (acidic) proteins. The most important nuclear enzymes are the **DNA polymerase, RNA polymerase, NAD synthetase, nucleoside triphosphatase, adenosine diaminase, nucleoside phosphorylase, guanase, aldolase, enolase, 3-phosphoglyceraldehyde dehydrogenase** and **pyruvate kinase**. The nucleoplasm also contains certain cofactors and coenzymes such as **ATP** and **acetyl CoA**.

4. Lipids. The nucleoplasm contains small lipid content.

5. Minerals. The nucleoplasm also contains several inorganic compounds such as phosphorus, potassium, sodium, calcium and magnesium. The chromatin comparatively contains large amount of these minerals than the nucleoplasm.

3. Chromatin Fibres.

The nucleoplasm contains many thread-like, coiled and much elongated structures which take readily the basic stains such as the basic fuchsin. These thread-like structures are known as the **chromatin** (*Gr., chrome*=colour) **substance** or **chromatin fibres**. Such chromatin fibres are observed only in the interphase nucleus. During the cell division (mitosis and meiosis) chromatin fibres become thick ribbon-like structures which are known as the **chromosomes**.

Chemically, chromatin consists of DNA and proteins. Small quantity of RNA may also be present but the RNA rarely accounts for more than about 5 per cent of the total chromatin present. Most of the protein of chromatin is histone, but "non-histone" proteins are also present. The protein: DNA weight ratio averages about 1:1. Histones are constituents of the chromatin of all eukaryotes except fungi, which, therefore, resemble prokaryotes in this respect.

The fibres of the chromatin are twisted, finely anastomosed and uniformly distributed in the nucleoplasm. Two types of chromatin material have been recognised, *e.g.*, heterochromatin and euchromatin.

A. Heterochromatin. The darkly stained, condensed region of the chromatin is known as heterochromatin. The condensed portions of the nucleus are known as **chromocenters** or **karyosomes** or **false nucleoli** (Box 16.2). The heterochromatin occurs around the nucleolus and at the periphery. It is supposed to be metabolically and genetically inert because it contains comparatively small amount of the DNA and large amount of the RNA.

Box 16.2

Chromocenters are darkly stained heterochromatic regions of chromatin of interphase nucleus. They may represent a heterochromatic region of several chromosomes or of all the chromosomes of the nucleus. Chromocenters are well marked in the nuclei of salivary gland cells of *Drosophila*. In this case, pericentromeric heterochromatin of all polytene chromosomes coalesces in a chromocenter.

B. Euchromatin. The light stained and diffused region of the chromatin is known as the euchromatin. The euchromatin contains comparatively large amount of DNA.

Heteropyknosis

Differential condensation of certain chromosomes, such as sex chromosomes, or chromosome parts is called **heteropyknosis** (*Gr.*, differential staining). During interphase and prophase stages of cell division most of the chromosomal substance is uncoiled and is in the form of light stained fine threads. But some chromosomes or their parts remain condensed and can be recognized by their dark stain. Thus, heteropyknosis can be defined as *the property of certain chromosomes or their parts to remain more condensed and stain more intensely than other chromosomes or their parts during the nuclear cycle.*

Heteropyknosis is the characteristic of sex chromosomes of many species, but may also be observed in other chromosomes as well. For example, one of two X chromosomes in females of all the mammals undergoes heteropyknosis. The heteropyknosis may involve entire chromosome, may be intercalated or localized at the extremities.

When the heteropyknotic chromosomes (*e.g.* sex chromosomes) are darkly stained, it is called **positive heteropyknosis,** however, in the toad, *Bufo arenarus,* the chromosomes exhibit **negative heteropyknosis** (under-condensation).

4. Nucleolus.

Most cells contain in their nuclei one or more prominent spherical colloidal acidophilic bodies, called **nucleoli**. However, cells of bacteria and yeast lack nucleolus. The size of the nucleolus is found to be related with the synthetic activity of the cell. Therefore, the cells with little or no synthetic activities, *e.g.*, sperm cells, blastomeres, muscle cell, etc., are found to contain smaller or no nucleoli, while the oocytes, neurons and secretory cells which synthesize the proteins or other substances contain comparatively large-sized nucleoli. The number of the nucleoli in the nucleus depends on the species and the number of the chromosomes. The number of the nucleoli in the cells may be one, two or four. The position of the nucleolus in the nucleus is eccentric.

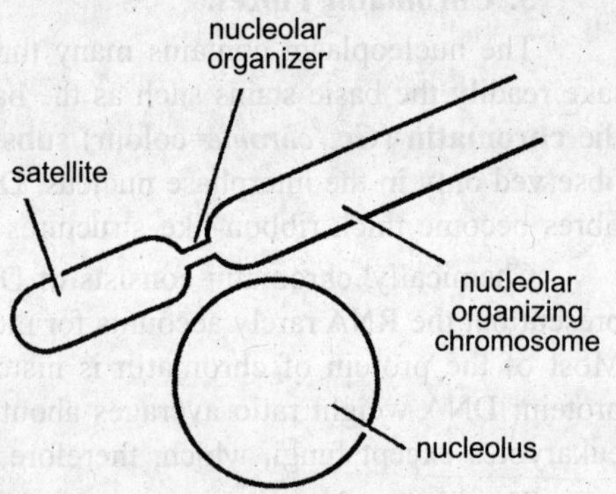

Fig. 16.6. A satellited chromosome and an attached nucleolus.

A nucleolus is often associated with the **nucleolar organizer (NO)** which represents the secondary constriction of the nucleolar organizing chromosomes, and are 10 in number in human beings (Fig. 16.6). In corn, *Zea mays* chromosome 9 and 6 contain 'darkly staining knobs' or nucleolar organizers. Nucleolar organizer consists of the genes for 18S, 5.8S and 28S rRNAs. The genes for fourth type of rRNA, *i.e.,* 5S rRNA occur outside the nucleolar organizer.

Chemical composition of nucleolus. Nucleo-lus is not bounded by any limiting membrane; calcium ions are supposed to maintain its intact organization. Chemically, nucleolus contains DNA of nucleolar organizer, four types of rRNAs, 70 types of ribosomal proteins, RNA binding proteins (*e.g.,* nucleolin) and RNA splicing nucleoproteins (U_1, U_2......U_{12}). It also contains phospholipids, orthophosphates and Ca^{2+} ions. Nucleolus also contains some enzymes such as acid phosphatase, nucleoside phosphorylase and NAD^+synthesizing enzymes for the synthesis of some coenzymes, nucleotides and ribosomal RNA. **RNA methylase** enzyme which transfers methyl groups to the nitrogen bases, occurs in the nucleolus of some cells.

Ultrastructure and function of nucleolus. Nucleolus are the sites where biogenesis of ribosomal subunits (*i.e.,* 40S and 60S) takes place. In it three types of rRNAs, namely 18S, 5.8S and 28S r RNAs, are transcribed as parts of a much longer precursor molecule (45S transcript) which undergoes processing (RNA splicing, for example) by the help of two types of proteins such as nucleolin and U3 snRNP (U3 is a 250 nucleotide containing RNA, snRNP represents small nuclear ribonucleoprotein). The 5S rRNA is transcribed on the chromosome existing outside the nucleolus and the 70 types of ribosomal proteins are synthesized in the cytoplasm. All of these components of the ribosomes migrate to the nucleolus, where they are assembled into two types of ribosomal subunits which are transported back to the cytoplasm. The smaller (40S) ribosomal subunits are formed and migrate to the cytoplasm much earlier than larger (60S) ribosomal subunits; therefore, nucleolus contains many more incomplete 60S ribosomal subunits than the 40S ribosomal subunits. Such a time lag in the migration of 60S and 40S ribosomal subunits, prevents functional ribosomes from gaining access to the incompletely processed heterogeneous RNA (hn RNA; the precursor of mRNA) molecule inside the nucleus.

Different stages of formation of ribosomes are completed in three distinct regions of the nucleolus. Thus, their **initiation**, **production** and **maturation** seem to progress from centre to periphery. Following three regions have been identified in the nucleolus (Fig. 16.7):

(*i*) Fibrillar centre. This pale-staining part represents the innermost region of nucleolus. The RNA genes of nucleolar organizer of chromosomes are located in this region. The transcription (*i.e.,* ribosomal RNA synthesis) of these genes is also initiated in this region.

(*ii*) Dense fibrillar component. This region surrounds the fibrillar centre and RNA synthesis progresses in this region. The 70 ribosomal proteins (rps) also bind to the transcripts in this region.

(*iii*) Cortical granular components. This is the outermost region of the nucleus where processing and maturation of pre-ribosomal particles occur.

Nucleolar Cycle

The appearance of nucleolus changes dramatically during the cell cycle. During meiosis as well as during mitosis the nucleolus disappears during prophase. As the cell approaches mitosis, the nucleolus first decreases in size and then disappears as the chromosomes condense and all RNA synthesis stops, so that generally there is no nucleolus in a metaphase cell. When ribosomal RNA synthesis restarts at the end of mitosis (in telophase), tiny nucleoli reappear at the chromosomal locations of the ribosomal RNA genes (NOs). For example, in humans the rRNA genes are located near the tips of each of the 5 different chromosomes (*i.e.*, paired autosomes 13, 14, 15, 21 and 22). Accordingly, 10 small nucleoli are formed after mitosis in a human diploid cell, although they are rarely seen as separate entities because they quickly grow and fuse to form the single large nucleolus typical of many interphase cells.

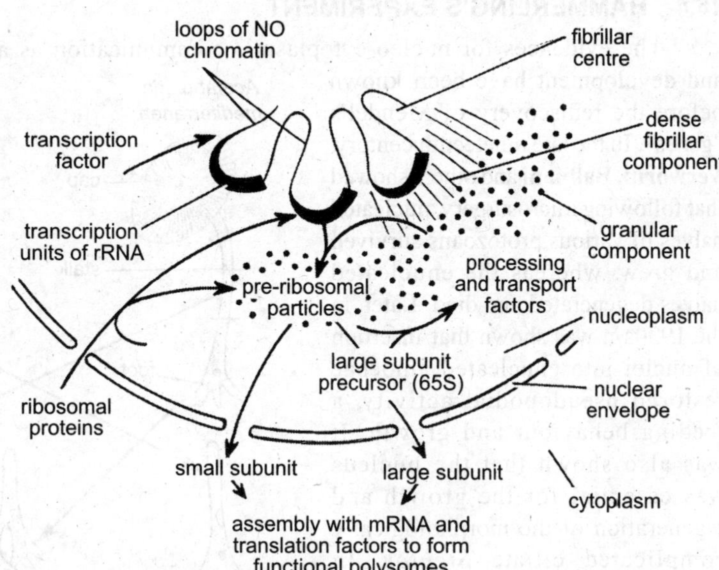

Fig. 16.7. Three different regions of the nucleolus and their involvement in ribosome assembly.

Now let us see what happens to the RNA and protein components of the disintegrated nucleolus during mitosis? It seems that at least some of them become distributed over the surface of all of the metaphase chromosomes and are carried as cargo to each of the two daughter cell nuclei. As the chromosomes decondense at telophase, these "old" nucleolar components help re-establish the newly emerging nucleoli.

Functions of Nucleus

Nucleus is the controlling centre of the cell.

(*i*) It controls all the metabolic activities of the cell by controlling the synthesis of enzymes required there in.

(*ii*) The nucleus controls the inheritance of characters from parents to offsprings.

(*iii*) It is responsible for the development of characters. Our phenotype is determined by the genes we have inherited from our parents. These genes are present in the chromosomes of nucleus.

Differences in Nucleus and Nucleolus

Table 16.1 Differences between nucleus and nucleolus.

	Nucleus		*Nucleolus*
1.	It is a large sized organelle.	1.	It is a very small part of nucleus. There occurs one nucleolus for each haploid set of chromosomes.
2.	It is bounded by a nuclear envelope of double unit membranes. Nuclear envelope bears nucleopores.	2.	It is not bounded by any membrane.
3.	It is site of replication, recombination and transcription of DNA.	3.	It is site of transcription of rDNA of nucleolar region of chromosomes. Tailoring and modifications of various types of rRNAs take place in nucleolus and they are assembled with ribosomal proteins to form ribosomal subunits.

16.4 HAMMERLING'S EXPERIMENT

The evidences for nucleo-cytoplasmic communication as a factor in cell maintenance and development have been known before the rediscovery of Mendel's "genes". In the late nineteenth century, **Verworm, Balbiani** and others showed that following microsurgery, nucleated halves of various protozoans survived and grew, whereas the enucleated halves degenerated and died. Later, in the 1930s it was shown that insertion of nuclei into enucleated amoebae restored pseudopodial activity, a feeding behaviour and growth. It was also shown that the nucleus was essential for the growth and regeneration of the morphologically complicated ciliate *Stentor*. In a classical series of experiments spanning between 1934 and 1954, on the unicellular alga *Acetabularia*, **Hammerling** demonstrated by means of interspecific nuclear transplants, that morphological features, notably the shape of cap, were determined by the nucleus. He also showed that even after removal of the nucleus, the cell was able to continue morphogenesis for a time and proposed that the cytoplasm contained a store of morphogenetic material (later on recognized as mRNA molecules) that had been produced by the nucleus. Let us closely examine the Hammerling's classical nuclear transplantation experiments:

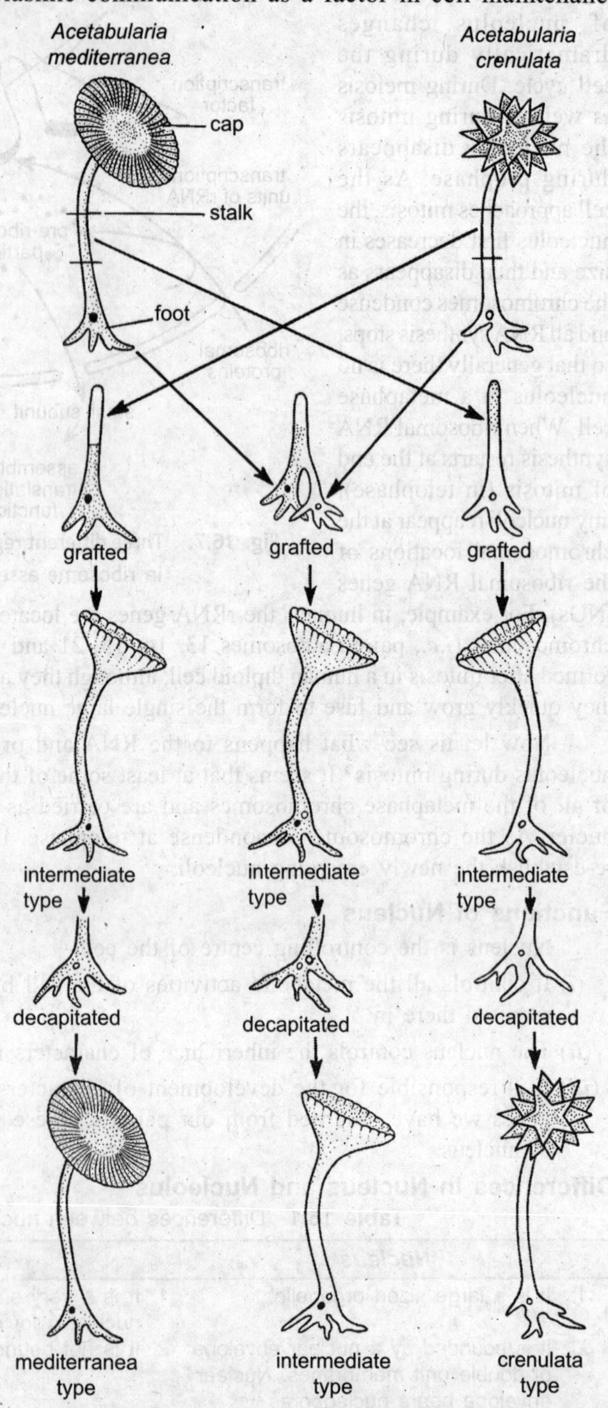

Fig. 16.8. Summary of the grafting experiments with *Acetabularia* to prove that the hereditary characters are determined by the nucleus and not by the cytoplasm.

Hammerling's experiment. The body of an alga *Acetabularia* is about six centimeters long and is differentiated into a foot, a stalk and a cap. The cap has a characteristic shape for each species and is easily regene-rated if removed. The single nucleus is situated in the rhizoid portion. *Acetabularia crenulata* has a cap, with about 31 rays, the tips of which are pointed, but *Acetabularia mediterranea* has about 81 rays with rounded tips. If the cap, stalk or even the nucleated portion of the rhizoid is removed, the remaining portion of the alga has the capacity to regenerate into a whole plant. The enucleated part loses the regeneration capacity after a few decapitations, but the nucleated portion always maintains this ability. When the stalk of one species is grafted on to the nucleated rhizoid of the other, an **intermediate type** of cap is formed. On decapitation, a second cap develops which resembles the cap of species which provides the nucleus (Fig. 16.8). When the nuclei of both the species are present in the same cytoplasm, an intermediate type of cap develops. Such experiments have clearly established that the nucleus is the storehouse for and the control tower of, all hereditary information.

QUESTIONS

Long Answer Questions

1. Describe the structure of interphase nucleus.
2. Describe the structure of nucleus. Discuss the structure and functions of different components of nucleus.
3. Discuss the structure, chemical organization and functions of nucleus. Throw light on its role in inheritance.
4. Give an account of the structure and functions of the nucleus with special reference to nucleic acids.
5. Describe the structure and functions of nucleolus.
6. 'Nucleus is the control centre of the cell'. Justify this statement with the help of two experiments.
7. With the help of Hammerling's experiment explain the importance of nucleus.
8. How can you show experimentally that nucleus is of central importance in cell heredity? Describe the nucleo-cytoplasmic inter-relationship of eukaryotic cell.

Short Answer Questions

1. Why the term nuclear membrane is replaced by nuclear envelope?
2. Describe the importance of Hammerling's experiment on *Acetabularia.*
3. Discuss the functions of nucleolus.
4. Describe the corelation between fibrillar and granular portions of nucleolus.
5. Describe in brief the nucleolar cycle.
6. Give reasons why nucleolus is small in muscle cell but large in liver cells.
7. Write short notes on the following:
   (*i*) Euchromatin
   (*ii*) Functions of nuclear envelope
   (*iii*) Functions of nucleolus
   (*iv*) Heteropyknosis
   (*v*) Nucleoproteins
   (*vi*) Chromocentres
   (*vii*) Nucleolar cycle
   (*viii*) Nucleolus
   (*ix*) Functions of nucleus
   (*x*) Nuclear pores
8. Differentiate between the following:
   1. Euchromatin and heterochromatin
   2. Nucleus and nucleolus.
9. Draw a neat and labelled diagram of nucleus.

Very Short Answer Questions

1. Why mammalian RBCs have a very short life span?
2. What will happen if nucleus in *Amoeba* is destroyed?

3. Who discovered the nucleus?

4. What is the nucleus?

5. Define the nucleolus.

6. What is nuclear envelope?

7. Define nuclear lamina.

8. What is the nuclear pore?

9. Define nuclear pore complex.

10. What is chromatin?

11. How many types of chromatin occur in the nucleus?

12. What is heterochromatin?

13. Define euchromatin.

14. What is NOR?

15. What is main function of nucleolus?

Fill in the Blanks

1. .... is site of ribosomal RNA synthesis and processing and of the assembly of .... subunits.

2. Term chromatin was coined by .... in 1879.

3. Since mature mammalian red blood cells are without any nuclei, they are called red blood .... rather than cells.

4. Polynucleate cells of the animals are termed as .... cells.

5. Nuclear .... lines the inside surface of the inner nuclear membrane and consists of a square lattice of .... filaments.

6. The number of pores in the nuclear envelope or .... seems to correlate with the .... activity of the cell.

7. High pore density occurs in the nuclei of .... gland cells and .... of *Xenopus laevis.*

8. Karyolymph is another name of ....

9. Classical .... ecperiments of Hammerling were performed on the unicellular alga, called ....

10. Nucleus is a large .... membrane enclosed cellular organelle that contains .... DNA of a .... cell.

11. Chromatin is constructed from .... spaced regularly along a .... chain.

Multiple Choice Questions

1. Site of formation of immature ribosomal subunits in a eukaryotic cell is:

(*a*) cytoplasm (*b*) nucleolus

(*c*) nucleus (*d*) nuclear pore complex

2. The role of nucleus in heredity was at first demonstrated by

(*a*) Max Hammerling (*b*) Robert Brown

(*c*) W. Fleming (*d*) Benda Hooker

3. Who coined the term nucleolus?

(*a*) Fontana (*b*) Robert Brown

(*c*) Camillo Golgi (*d*) Strasburger

4. Nucleolus is the site for the synthesis of

(*a*) DNA (*b*) mRNA

(*c*) tRNA (*d*) ribosomes

5. An example for the enucleated living plant cell is
   (*a*) xylem parenchyma (*b*) RBCs
   (*c*) sieve tubes of phloem (*d*) companion cells of phloem

ANSWERS

Very Short Answer Questions

1. Mammalian RBCs are without nucleus.
2. Such *Amoeba* will die.
3. Robert Brown in 1833.
4. Nucleus is a large, double membrane enclosed organelle that contains the chromosomal DNA of a eukaryotic cell.
5. Nucleolus is a large, spherical structure present in the nucleus of an eukaryotic cell; it is site of ribosomal RNA synthesis and processing and of the assembly of ribosomal subunits.
6. Nuclear envelope is a double membrane around the nucleus that is interrupted by numerous small pores.
7. Nuclear lamina is a thin, dense, meshwork of fibre that lines the inner surface of the inner nuclear membrane and helps support the nuclear envelope.
8. It is a small opening in the nuclear envelope through which molecules enter and exist the nucleus; it is lined by an intricate protein structure called the nuclear pore complex (NPC).
9. It is a intricate protein structure, composed of 100 or more different polypeptide subunits, that lines the nuclear pores through which molecules enter and exit the nucleus, both by simple diffusion of smaller molecules and active transport of large molecules.
10. Chromatin is DNA-protein fibre that make up chromosomes. It is constructed from nucleosomes spaced regularly along a DNA chain.
11. Euchromatin and heterochromatin.
12. Heterochromatin is highly compacted form of chromatin present during interphase; it contains DNA that is not being transcribed.
13. Euchromatin is loosely packed, uncondensed form of chromatin present during interphase; it contains DNA that is being actively transcribed.
14. NOR stands for nucleolus organizer region which is a stretch of DNA in certain chromosomes where multiple copies of the genes for ribosomal RNA are located and where nucleoli form.
15. Nucleolus is a ribosome producing organ.

Fill in the Blanks

1. Nucleolus, ribosomal.
2. Fontana.
3. Corpuscles.
4. Syncytial.
5. Lamina, intermediate.
6. Pore density, transcriptional.
7. Salivary, oocytes.
8. Nucleoplasm.
9. Nuclear, *Acetabularia*.
10. Double, chromosomal, eukaryotic.
11. Nucleosomes, DNA.

Multiple Choice Questions

1. (*b*) **2.** (*a*) **3.** (*a*) **4.** (*d*) **5.** (*c*).

17

Chromosomes

The chromosomes (Gr., *chrom* = colour, *soma* = body) are the nuclear components of special organisation, individuality and function. They are capable of self-reproduction and play a vital role in heredity, mutation, variation and evolutionary development of the species.

Chromosomes are most distinct in metaphase stage of cell division. In an interphase nucleus these form an interwoven net of fine chromatin threads. The term **chromosome** was introduced by **Waldeyer** in 1888 because of their affinity towards dyes or stains.

Types of Chromosomes

1. Viral chromosomes. In viruses, a single, linear or circular molecule of DNA or RNA represent the chromosome. Nucleic acid of the viral chromosome may be single-stranded or double-stranded.

2. Prokaryotic chromosomes (or Bacterial chromosomes). In bacteria and cyanobacteria (blue green algae), the hereditary material (DNA) is organized into a single circular chromosome. Chromosomal DNA is double stranded and lacks histone proteins. Prokaryotic chromosomes are also found in certain cellular organelles such as mitochondria and chloroplasts.

3. Eukaryotic chromosomes. In eukaryotic cells (*e.g.*, protistans plants, fungi and animals), chromosomes are formed of DNA and proteins (histones and non-histones). DNA is linear and double stranded.

In the chapter the structure and function of eukaryotic chromosomes only will be described.

Box 17.1 Distinction between viral and bacterial chromosomes

In viruses (including bacteriophages) a single molecule of DNA or RNA forms the viral chromosome. It is about 20 Å wide corresponding to the width of DNA molecule. The viral chromosome may be linear or circular.

Bacterial chromosome is a single, large circular chromosome, formed of a circular DNA molecule.

Box 17.2 Distinction between prokaryotic and eukaryotic chromosomes

Prokaryotic or bacterial chromosome is a large circular DNA molecule. It is without histones but may contain some polyamines. It lacks centromere and nucleosomes, but is folded into a series of **looped domains**.

Euchromatic chromosome is much complicated structure having centromere and nucleosomes. Each chromosome is formed by the condensation of a long chromatin thread. The chromatin thread is formed of supercoiling of a series of nuclesosomes. Each nucleosome is formed of histones and DNA.

17.1 CHEMICAL COMPOSITION OF CHROMOSOMES

Chromatin which has been isolated from rat liver contains DNA, RNA and protein. The protein of chromatin is of two types : the histones and the non-histones. Rat liver chromatin has been used as a model for chromatin. It possesses a histone to DNA ratio near 1 : 1, a non-histone protein to DNA ratio of 0.6 : 1 and a RNA/DNA ratio of 0.1 : 1.

1. DNA.

DNA is the most important chemical component of chromatin, since it plays the central role of controlling heredity. The most convenient measurement of DNA is picogram (10^{-12} gm). DNA of chromatin represent the following two phenomena:

The C-value. The DNA in nuclei was stained using the **Feulgen reactions** and the amount of stain in single nuclei was measured using a special microscope (called **cytophotometer**). Both of these techniques demonstrated that nuclei contain a constant amount of DNA. Thus, all the cells in an organism contain the same DNA content (2C) provided that they are diploid. Gametes are haploid and, therefore, have half the DNA content (1C). Some tissues such as liver, contain occasional cells that are polyploid and their nuclei have a correspondingly higher DNA content (4C or 8C) (see Table 17.1).

Table 17.1 DNA content and chromosome component

	Cells	*Mean DNA-Feulgen content*	*Presumed chromosome set*
1.	Spermatid	1.68 (1C)	Haploid (n)
2.	Liver	3.16 (2C)	Diploid (2n)
3.	Liver	6.30 (4C)	Tetraploid (4n)
4.	Liver	12.80 (8C)	Octoploid (8n)

Thus, each species has a characteristic content of DNA which is **constant** in all the individuals of that species and has, thus, been called the C-value.

The C-value paradox. Eukaryotes vary greatly in DNA content but always contain much more DNA than prokaryotes. Lower eukaryotes in general have less DNA, such as nematode *Caenorhabditis elegans* which has only 20 times more DNA than *E.coli*, or the fruit fly *D. melanogaster* which has 40 times more DNA (*i.e*, 0.18 pg or picogram per haploid genome). Vertebrates have greater DNA content (about 3 pg), in general about 700 times more then *E. coli*. One of the highest DNA content is that of the salamander *Amphiuma* which has 84 pg of DNA. Each human being has about 3 pg of DNA per haploid genome, or 3×10^9 base pairs, *i.e.*, the human genome could accommodate about 3 million average sized proteins. If all the DNA were coding (or containing structural genes) and if this was true, salamanders would have 30 times more genes than human beings. This is called **C-value paradox** (**Gall**, 1981). It was detected quite early that there was little connection between the morphological complexity of eukaryotic organisms and their DNA content. For example, *E. coli* (containing 3,400,000 base pairs in its DNA) has about 3000 genes. Although it is difficult to estimate how many different genes exist in the human genome, there are probably not more than 20,000 to 30,000 genes (Note : According to a most recent estimate, there are 100,000 genes in human genome, see **Deviah**, 1994). There is no reason to believe that salamanders should have any more. From these facts, it can be easily concluded that most of the DNA in the eukaryotic genomes must be of a non-coding nature.

2. Histones.

Histones are very basic proteins, basic because they are enriched in the amino acids arginine and lysine to a level of about 24 mole present. Arginine and lysine at physiological pH are cationic and can interact electrostatically with anionic nucleic acids. Thus, being basic, histones bind tightly to DNA which is an acid. There are five types of histones in the eukaryotic chromosomes, namely H1, H2A, H2B, H3 and H4.

One of the important discoveries that has come from chemical studies is that the primary structures of histones have been highly conserved during evolutionary history. For example, histone H4 of calf and of garden pea contains only two amino acid differences in a protein of 102 residues (**DeLange**, 1969). Likewise, the sequence of histone H3 from rat differs only in two amino acids from that of peas, out of 102 total amino acid residues. These organisms are estimated to have an evolutionary history of at least 600 million years, during which they diverged structurally. This conservation of structure suggests that over the eras, histones have had a very similar and crucial role in maintaining the structural and functional integrity of chromatin. Such an evolutionary conservation suggests that the functions of these two histones involve nearly all of their amino acids so that a change in any position is deleterious to the cell.

Histone H1 is the least rigidly conserved histone protein. It contains 210 to 220 amino acids and may be represented by a variety of forms even within a single tissue. H1 is present only once per 200 base pairs of DNA (in contrast to rest of the four types of histones each of which is present twice) and is rather loosely associated with DNA. H1 histone is absent in yeast, *Saccharomyces cerevisiae*. Histones besides determining the structure of chromatin, play a regulatory role in the repression activity of genes.

3. Non-histones. In contrast to the modest population of histones in chromatin, non-histone proteins display more diversity. In various organisms, number of non-histones can vary from 12 to 20. Heterogeneity of these proteins is not conserved in evolution as the histones. These non-histones differ even between different tissues of the same organism suggesting that they regulate the activity of specific genes.

About 50 per cent non-histones of chromatin have been found to be structural proteins and include such proteins as **actin**, and **α- and β-tubulins** and **myosin**. Although for sometime these contractile proteins were thought to be contaminants, it is now believed that they are vital ingredients of the chromosome, functioning during chromosome condensation and in the movement of chromosomes during mitosis and meiosis (see **Thorpe**, 1984). Many of the remaining 50 per cent of non-histones include all the enzymes and factors that are involved in DNA replication, in transcription and in the regulation of transcription. These proteins are not as highly conserved among organisms, although they must carry out similar enzymatic activities. Apparently they are not as important as the histones in maintaining chromosome integrity.

17.2 STRUCTURE OF CHROMOSOMES

Shape

The shape of the chromosomes is changeable from phase to phase in the continuous process of the cell growth and cell division. In the resting phase or interphase of the cell, the chromosomes occur in the form of thin, coiled, elastic and contractile, thread-like stainable structures, the chromatin threads. In the metaphase and the anaphase, the chromosomes become thick and filamentous. Each chromosome contains a clear zone, known as **centromere** or **kinetocore**, along their length. The centromere divides the chromosomes into two parts, each part is called **chromosome arm**.

Depending upon the number of centromeres, an eukaryotic chromosome may be of following types:

1. Monocentric chromosome has one centromere. Most normal chromosomes are monocentric.

2. Holocentric chromosome has a diffuse centromere, so that spindle attachment occurs over the entire chromosome, *e.g.*, Lepidoptera.

3. Dicentric chromosome is with two centromeres.

4. Polycentric chromosome is with more than two centromeres.

5. Acentric chromosome is without any centromere. Such chromosomes represent freshly broken segments of chromosomes which do not survive for long.

In a monocentric chromosome, the position of centromere varies form chromosome to chromosome and it provides different shapes to the latter which are following (Fig. 17.1):

1. Telocentric (Telomictic). It is a rod-like monocentric chromosome with a terminal centromere. It is also known as **T-chromosome** or **monobrachial chromosome**. Such chromosomes are very rare in nature.

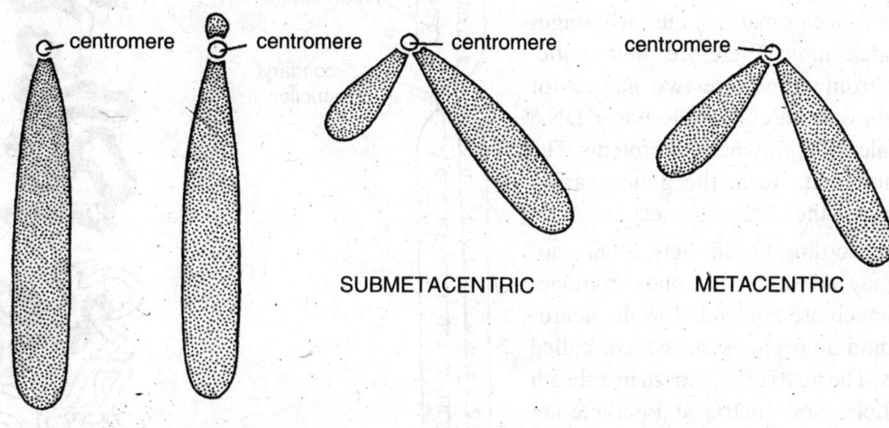

Fig. 17.1. The four morphological types of chromosomes according to the position of the centromere.

2. Atelocentric (Atelomictic). It is a monocentric chromosome with a nonterminal centromere. It is also called **dibrachial chromosome**. Atelocentric chromosomes are of following four types:

***(i)* Metacentric.** It is a monocentric V-shaped chromosome with a central or near central (Fig. 17.2) chromosome. If the centromere is exactly at the median point, it is classified as a **M-chromosome**, whilst if the centromere is not central but lies within the median region, it is classified as a **m-chromosome**. M-chromosomes are rare in nature.

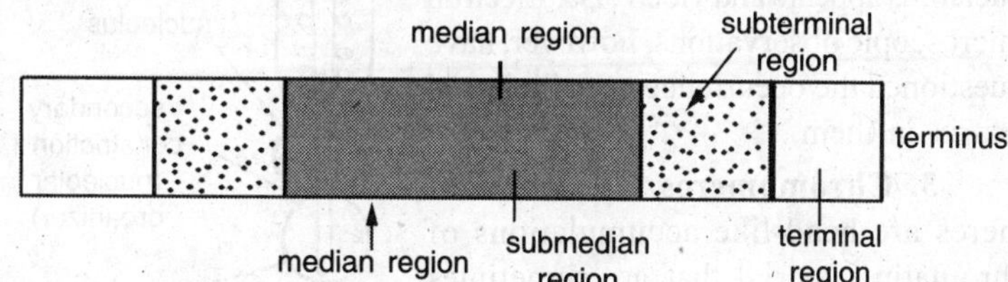

Fig. 17.2. Various regions of a chromosome used in classification of chromosomes.

***(ii)* Submetacentric.** It is a monocentric chromosome with a centromere in the **submedian region.** It is J-or L-shaped.

***(iii)* Subacrocentric.** It is a monocentric chromosome with a centromere in the **subterminal region** (Fig. 17.2).

***(iv)* Acrocentric.** It is a monocentric chromosome with its centromere in **terminal region**. It is also called **t-chromosome**, since most so-called telocentric chromosomes are actually acrocentric.

Structure (Parts of Chromosomes)

In the metaphase chromosomes following structures or parts have been recognized:

1. Chromatid. At mitotic metaphase each chromosome consists of two symmetrical structures, called **chromatids**. Each chromatid contains a single DNA molecule. Both chromatids are attached to each other only by the centromere and become separated at the beginning of anaphase, when the sister chromatids of a chromosome migrate to the opposite poles.

2. Chromonema (ta). During mitotic prophase the chromosomal material becomes visible as very thin filaments, called **chromonemata** (a term coined by **Vejdovsky** in 1912). A chromonema represents a chromatid in the early stages of condensation. Therefore, 'chromatid' and 'chromonema' are two names for the same structure: a single linear DNA molecule with its associated proteins. The chromonemata form the gene-bearing portions of the chromosomes.

According to old view, a chromosome may have more than one chromonemata which are embedded in the achromatic and amorphous substance, called **matrix**. The matrix is enclosed in a sheath or **pellicle**. Both matrix and pellicle are non-genetic materials and appear only at metaphase when the nucleolus disappears. It is believed that nucleolar material and matrix are interchangeable, *i.e.,* when chromosomal matrix disappears, the nucleolus appears and vice versa. Electron microscopic observations, however, have questioned the occurrence of pellicle and matrix in them.

3. Chromomeres. The chromomeres are bead-like accumulations of chromatin material that are sometimes visible along interphase chromosomes. The chromomere-bearing chromatin has an appearance of a necklace in which several beads occur on a string. Chromo-meres become especially clear in the polytene chromosomes, where they become aligned side by side, constituting the chromosome beads (Fig. 17.3). At metaphase the chromosomes are tightly coiled and the chromomeres are no longer visible.

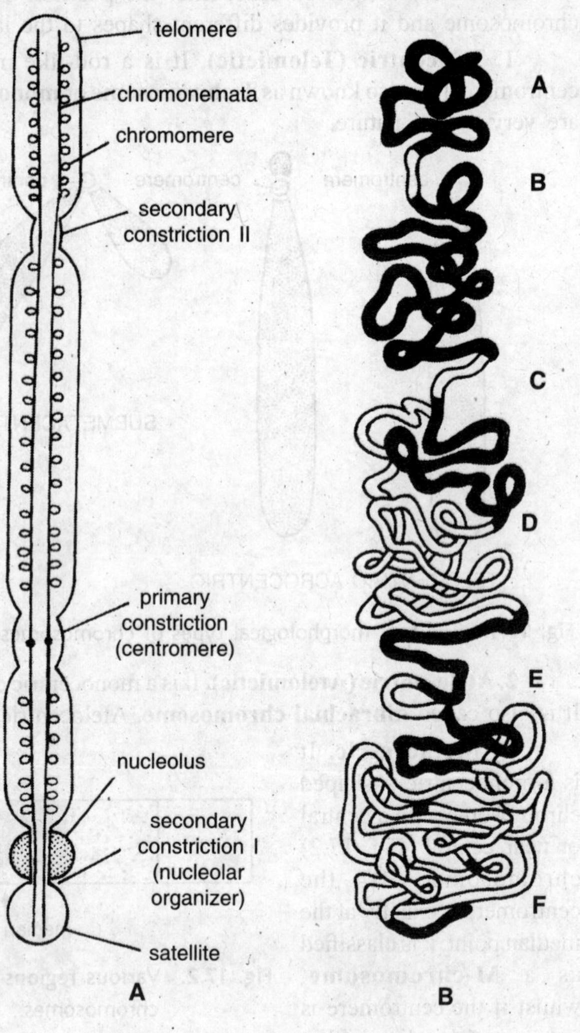

Fig. 17.3. A—Structure of a typical chromosome; B–Model of constitutive heterochromatin in a mammalian metaphase chromosome. [A—Constitutive heterochromatin; B—Secondary constriction; I or nucleolar organizer; C–Primary constriction or centromere; D—Euchromatin; E—Secondary constriction. II—possible site of 5S rRNA cistrons; F—Telomere].

Chromomeres are regions of tightly folded DNA and have great interest for the cell biologists. They are believed to correspond to the units of genetic function in the chromosomes. In fact, for long time most geneticists considered these chromomeres as genes, *i.e.*, the units of heredity.

4. Centromere and kinetochore. Originally it was considered that the centromere consists of small granules or spherules. The centromere of the chromosome of the *Trillium* has the diameter of 3μm and the spherules have the diameter of 0.2 μm. The chromonema remains connected with the spherules of the centromere. Currently it is held that centromere is the region of the chromosome to which are attached the fibres of mitotic spindle. The centromere (a term much preferred by the geneticists) lies within a thinner segment of chromosome, the **primary constriction**. The regions of chromosome flanking the centromere contain highly repetitive DNA and may stain more intensely with the basic dyes (*i.e.*, it is a constitutive heterochromatin, Fig. 17.4B). Centromeres are found to contain specific DNA sequences with special proteins bound to them, forming a disc-shaped structure, called **kinetochore** (a term that is much preferred by the cytologists). Under the EM, the kinetochore appears as a plate-or cup-like disc, 0.20 to 0.25 nm, in diameter situated upon the primary constriction or centromere. In thin electron micro-scopic sections, the kinetochore shows a trilaminar structure, *i.e.*, a 10 nm thick dense **outer proteinaceous** layer, a **middle** layer of low density and a dense **inner** layer tightly bound to the centromere (Fig. 17.4). The DNA of centromere does not exist in the form of nucleosome.

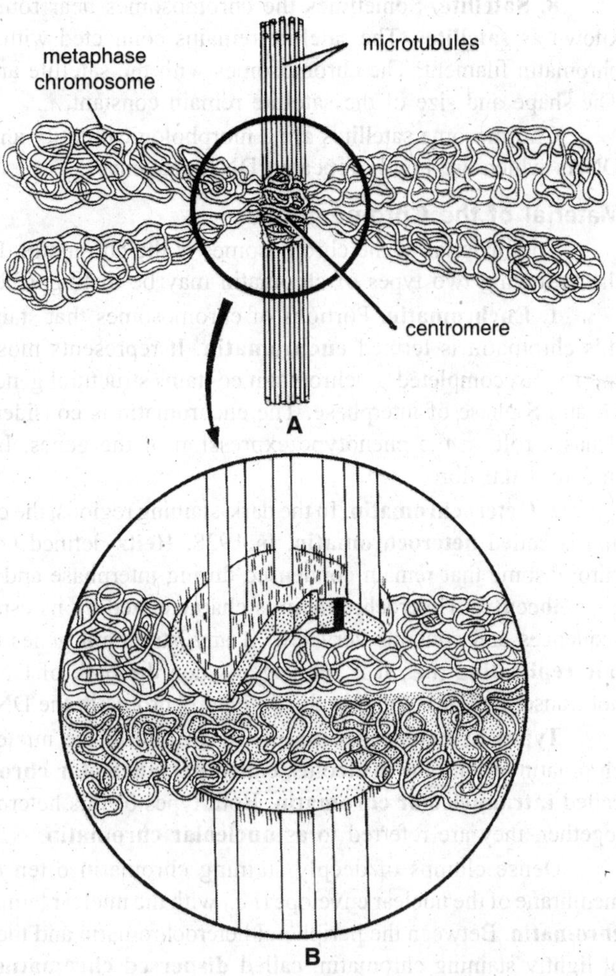

Fig. 17.4. A—A metaphase chromosome showing the folded fibre structure and the centromere with implanted microtubules; B—Inset: a higher magnification, showing the convex electron dense layer and the fibrillar material forming the "corona" of the centromere.

5. Telomere. (Gr., *telo*=for; *meros*=part). Each extremity of the chromosome has a polarity and therefore, it prevents other chromosomal segments to be fused with it. The chromosomal ends are known as the **telomeres**. If a chromosome breaks, the broken ends can fuse with each other due to lack of telomeres.

6. Secondary constriction. The chromosomes besides having the primary constriction or the centromere possess secondary constriction at any point of the chromosome. Constant in their position and extent, these constrictions are useful in identifying particular chromosomes in a set. Secondary constrictions can be distinguished from primary constriction or centromere, because chromosome bends (or exhibits angular deviation) only at the position of centromere during anaphase.

7. Nucleolar organizers. These areas are certain secondary constrictions that contain the genes coding for 5.8S, 18S and 28S ribosomal RNA and that induce the formation of nucleoli. The secondary constriction may arise because the rRNA genes are transcribed very actively and, thus, interfering with chromosomal condensation. In human beings, the nucleolar organizers are located in the secondary constrictions of chromosomes 13, 14, 15, 21 and 22, all of which are acrocentric and have satellites.

8. Satellite. Sometimes the chromosomes bear round elongated or knob-like appendages known as **satellites**. The satellite remains connected with the rest of the chromosome by a thin chromatin filament. The chromosomes with the satellite are designated as the **sat chromosomes**. The shape and size of the satellite remain constant.

Chromosome satellites are a morphological entity and should not be confused with satellite DNAs which are highly repeated DNA sequences.

Material of the Chromosomes

The material of the chromosomes is the chromatin. Depending on their staining properties, the following two types of chromatin may be distinguished in the interphase nucleus:

1. Euchromatin. Portions of chromosomes that stain lightly are only partially condensed; this chromatin is termed **euchromatin**. It represents most of the chromatin that disperse after mitosis has completed. Euchromatin contains structural genes which replicate and transcribe during G_1 and S phase of interphase. The euchromatin is considered genetically active chromatin, since it has a role in the phenotype expression of the genes. In euchromatin, DNA is found packed in 3 to 8 nm fibre.

2. Heterochromatin. In the dark-staining regions, the chromatin remains in the condensed state and is called **heterochromatin**. In 1928, **Heitz** defined heterochromatin as those regions of the chromosome that remain condensed during interphase and early prophase and form the so-called **chromocentre**. Heterochromatin is characterized by its especially high content of repititive DNA sequences and contains very few, if any, structural genes (*i.e.,* genes that encode proteins). It is **late replicating** (*i.e.,* it is replicated when the bulk of DNA has already been replicated) and is not transcribed. It is thought that in heterochromatin the DNA is tightly packed in the 30 nm fibre.

Types of heterochromatin. In an interphase nucleus, usually there is some condensed chromatin around the nucleolus, called **perinucleolar chromatin**, and some inside the nucleolus, called **intranucleolar chromatin**. Both types of this heterochromatin appear to be connected and together, they are referred to as **nucleolar chromatin**.

Dense clumps of deeply staining chromatin often occur in close contact with the inner membrane of the nuclear envelope (*i.e.,* with the nuclear lamina) and is called **condensed peripheral chromatin**. Between the peripheral heterochromatin and the nucleolar heterochromatin are regions of lightly staining chromatin, called **dispersed chromatin**. In the condensed chromosomes, the heterochromatic regions can be visualized as regions that stain more strongly or more weakly than the euchromatic regions, showing the so-called **positive** or **negative heteropyknosis** of the chromosomes (Gr., *hetero* = different + *pyknosis* = staining).

Heterochromatin has been further classified into the following types:

1. Constitutive heterochromatin. In such a heterochromatin the DNA is permanently inactive and remains in the condensed state throughout the cell cycle. This most common type of heterochromatin occurs around the centromere, in the telomeres and in the C-bands of the chromosomes. In *Drosophila virilis*, constitutive heterochromatin exists around the centromeres and such **pericentromeric heterochromatin** occupies 40 per cent of the chromosomes. In many species, entire chromosomes become heterochromatic and are called **B chromosome**, **satellite chromosomes** or **accessory chromosomes** and contain very minor biological roles. Such chromosomes comprising wholly constitutive heterochromatin occur in corn, many phytoparasitic insects and salamanders. In the fly *Sciara*, large metacentric heterochromatic chromosomes are found in the gonadal cells, but are absent in somatic cells. Entire Y chromosome of male *Drosophila* is heterochromatic, even though containing six gene loci which are necessary for male fertility.

Constitutive heterochromatin contains short repeated sequences of DNA, called **satellite DNA**. This DNA is called satellite DNA because upon ultracentrifugation, it separates from the main component of DNA. Satellite DNA may have a higher or lower G + C content than the main fraction. For example, the mouse satellite DNA is a 240 base pair sequence that is repeated about 1000,000

(10^6) times in the mouse genome, constituting 10 per cent of the total mouse DNA. The exact significance of constitutive heterochromatin is still unexplained.

2. Facultative heterochromatin. Such type of heterochromatin is not permanently maintained in the condensed state; instead it undergoes periodic dispersal and during these times is transcriptionally active. Frequently, in facultative heterochromatin one chromosome of the pair becomes either totally or partially heterochromatic. The best known case is that of the X-chromosomes in the mammalian female, one of which is active and remains euchromatic, whereas the other is inactive and forms at interphase, the **sex chromatin** or **Barr body** (Named after its discoverer, Canadian cytologist **Murray L. Barr**). Barr body contains DNA which is not transcribed and is not found in males. Indeed, the number of Barr bodies is always one less than the number of X chromosomes (*i.e.*, in humans, XXX female has two Barr bodies and XXXX female has three Barr bodies; **M.L. Barr**, 1959).

17.3 ULTRASTRUCTURE AND MOLECULAR ORGANIZATION

The structural unit of each chromosome is single DNA complexed with basic proteins (histones) in 1 : 1 ratio. This is called **single stranded hypothesis**.

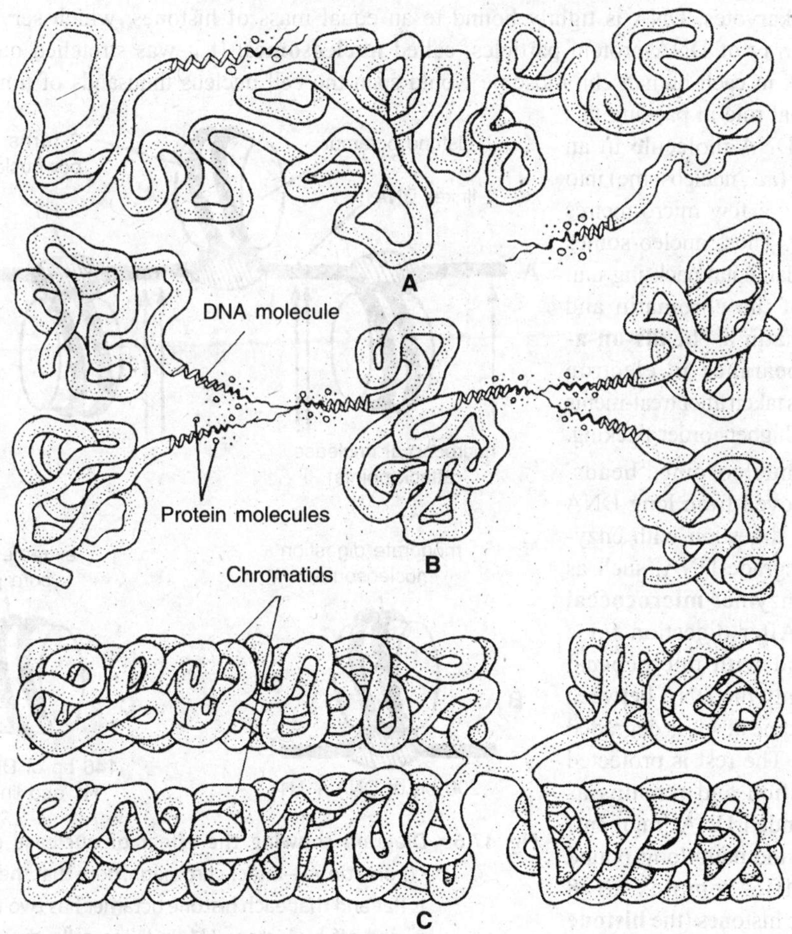

Fig. 17.5. Dupraw's folded fibre model of chromatin in interphase (A and B) and in metaphase (C).

The manner of coiling and folding of DNA was a matter of debate and dozens of models were available for this purpose; of them only two stand out and are important. A popular model is the

folded fibre model, proposed by **E.J. Dupraw** in 1965. According to it, the bulk of the chromosome is visualized to be composed of a tightly folded fibre which has a rather homogeneous diameter of 200 to 300 A°. This folded fibre is supposed to contain the DNA histone helix (of 30 A° diameter) in a supercoiled condition (Fig. 17.5). Another model is most significant and universally accepted one and is called **nucleosome model** which was proposed by **R.D. Kornberg** (1974) (Fig. 17.6) and confirmed and christened by **P. Oudet** *et al.*, (1975). Thus, while in the folded-fibre model, it was proposed that the histones were bound on the outside of the DNA coils (*i.e.*, histone shell around DNA), the nucleosome model has proposed the opposite (*i.e.*, histone particle with DNA round it). In other words, the earlier theory that basic chromatin fibre had DNA core surrounded by histones was incorrect. In fact, from a genetic perspective, a significant feature of packing mechanism through the nucleosomes lies in its topology: *at no point is the DNA buried;* instead, it is freely exposed along the entire surface of the "spool", available for genetic expression. Nucleosomes seem to be universal device for compacting the long DNA molecules of eukaryotic cells.

Nucleosomes and Solenoid Model of Chromatin

In eukaryotes, DNA is tightly bound to an equal mass of histones, which serve to form a repeating array of DNA-protein particles, called **nucleosomes**. If it was stretched out, the DNA double-helix in each human chromosome would span the cell nucleus thousands of time. His-tones play a crucial role in packing this very long DNA molecule in an orderly way (*i.e.*, nucleosome) into nucleus only a few micro-metres in diameter. Thus, nucleo-somes are the fundamental pack-ing unit particles of the chromatin and give chromatin a "**beads-on-a-string**"appearance in electron micorgraphs taken after treat-ments that unfold higher- order packing.

The nucleosome 'beads' can be removed from long DNA "string" by digestion with enzy-mes that degrade DNA, such as bacterial enzyme, **micrococcal nuclease**. After digestion for a short period with micrococcal nuclease, only the DNA between the nucleosome beads is degraded (Fig. 17.6). The rest is protected from digestion and remains as double-stranded DNA fragments 146 nucleotide pairs long bound to a specific complex of 8 nucleosome histones (the **histone octamer**). The nucleosome beads obtained in this way have been crystallized and analyzed by X-ray diffraction.

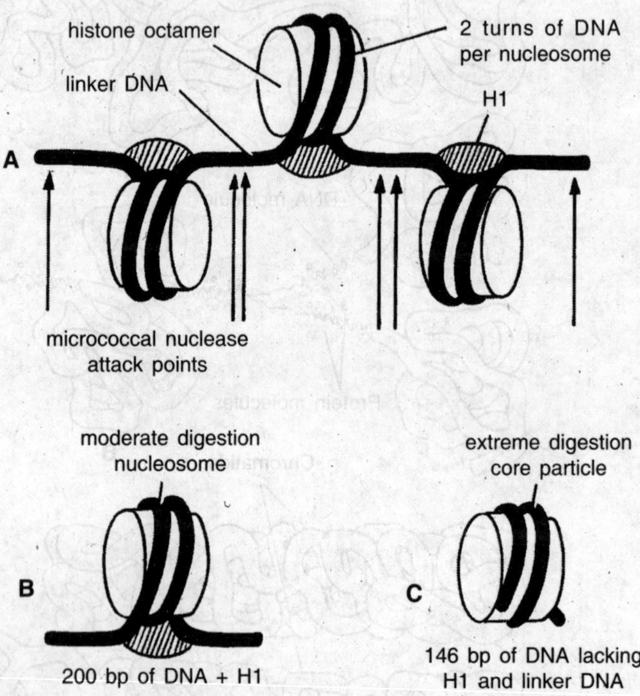

Fig. 17.6. Diagram showing the effect of nuclease enzymes on chromatin. A—Intact chromatin, note that the nucleosome is flat and that each histone octamer has two turns of DNA sealed off by histone H1. In living cells, the nucleosomes would be touching each other to form a 10 nm fibre and not stretched out (to form beads-on-a-string preparation); B—A nucleosome released by moderate digestion; C—Core particle of nucleosome obtained by extensive digestion.

Each nucleosome is a disc-shaped particle with a diameter of about 11 nm and 5.7 nm in height containing 2 copies of each 4 nucleosome histones–H2A, H2B, H3 and H4. This histone octamer forms a protein core [(*i.e.*, a core of histone tetramer $(H3, H4)_2$ and the apolar regions of 2(H2A and H2B)] around which the double-stranded DNA helix is wound 1¾ time containing 146 base pairs. In undigested chromatin the DNA extends as a continuous thread from nucleosome to nucleosome. Each nucleosome bead is separated from the next by a region of **linker DNA** which is generally 54 base pair long and contains single H1 histone protein molecule. Generally, DNA makes two complete turns around the histone octamers and these two turns (200 bp long) are sealed off by H1 molecules. [**Note** : In some organisms nucleosome DNA may vary from 162 base pairs (*e.g.*, rabbit cortical neurons) to 242 base pairs (*e.g.*, sea urchin sperm)]. Thus, on an average, nucleosomes repeat at intervals of about 200 nucleotides or base pairs. For example, an eukaryotic gene of 10,000 nucleotide pairs will be associated with 50 nucleosomes and each human cell with 6×10^9 DNA nucleotide pairs contains 3×10^7 nucleosomes.

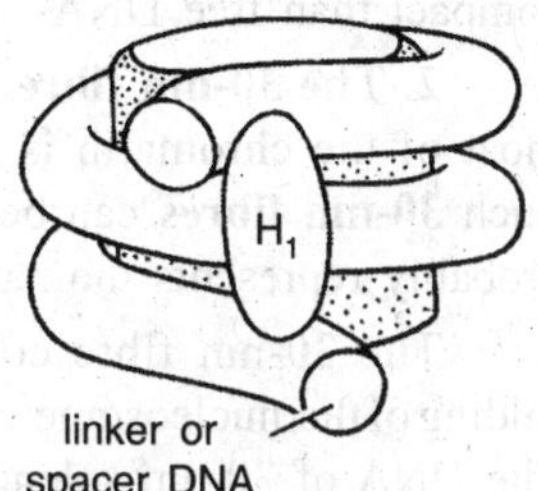

Fig. 17.7. A nucleosome.

Solenoid Models

H1 is reported to be phosphorylated just before mitotic and meiotic cell division to make possible the higher levels of coiling. During mitosis or meiosis, the prophase is the stage during which the chromosomes become shorter and thicker due to multiple coiling as proposed by **DuPraw** and others. The hypothesis of a solenoidal structure, with coils of coils had renewal since nucleosomal substructure has been discovered. Thus, due to solenoid coiling of nucleosome containing fibre, the following types of chromosomal structures can be observed during the cell cycle.

Box 17.3 Distinction between centrosome and nucleosome

Centrosome is an extranuclear structure which includes **centriole** with surrounding cytoplasm. It helps in spindle formation during cell division. Nucleosome is the unit of chromatin fibre in which 4 pairs of histones [2 (H2A, H2B, H3, H4)] jointly form a core particle and a DNA segment of about 200 base pairs long encircle the core particle. Another histone (H1 or H5) maintains this configuration.

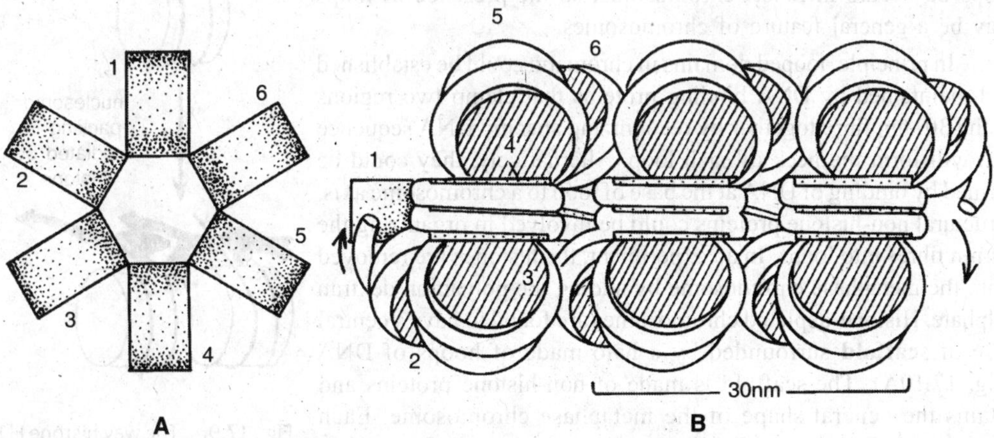

Fig. 17.8. A model suggested to explain how the "Beads-on-a-string" form of nucleosomes is packed to form the 30-nm fibre. A —Top view of 30-nm fibre; B—Side view of 30 nm fibre.

1. The 10-nm fibre. When nucleosomes are in close apposition, they form the **10-nm filaments**, in which packing of DNA is about five-to seven-fold, *i.e.*, five to seven times more compact than free DNA.

2. The 30-nm fibre. When nuclei are very gently lysed onto an electron microscopy grid, most of the chromatin is seen to be in the form of a fibre, with a diameter of about 30 nm. Such **30-nm fibres** can be observed in metaphase chromosomes and in interphase nuclei and it probably represents the natural conformation of transcriptionally inactive chromatin.

The 30-nm fibre consists of closely packed nucleosomes. It probably arises from the folding of the nucleosome chain into a **solenoid structure** having about six nucleosomes per turn. The DNA of 30-nm solenoid has a packing that is about 40-fold.

Histone H1 molecules are found responsible for packing nucleosomes into the 30-nm fibre. The H1 histone molecule has an evolutionarily conserved globular core or central region linked to extended amino-terminal and carboxyl-terminal "arms", whose amino acid sequence has evolved much more rapidly. Each H1 molecule binds through its globular portion to a unique site on a nucleosome and has arms that are thought to extend to contact with other sites on the histone cores of adjacent nucleosomes, so that the nucleosomes are pulled together into a regular repeating array. The binding of H1 molecule to chromatin tends to create a local polarity that the chromatin otherwise lacks.

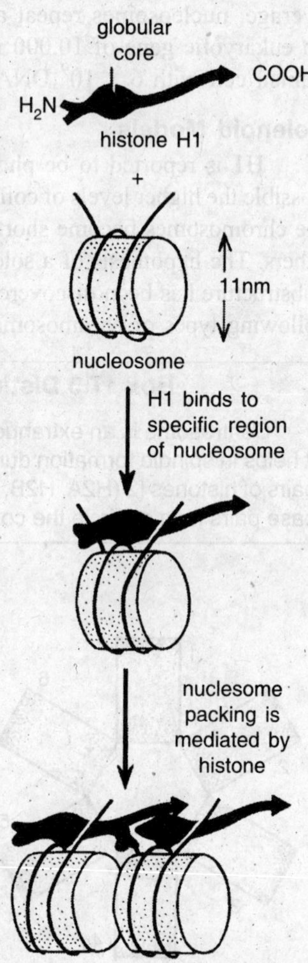

Fig. 17.9. The way histone H1 is thought to help to pack adjacent nucleosomes together.

3. Radial loops of 30-nm fibre and metaphase chromosome. The nucleus is typically about 5µm (5×10^{-4} cm) in diameter. The packaging of DNA into a 30-nm chromatin fibre leaves a human chromosome about 0.1 cm long, so there must be several higher orders of folding. The probable nature of one further level of folding was originally suggested by the appearance of specialized chromosomes—the lampbrush chromosomes and polytene chromosomes. These two types of chromosomes seem to contain a series of **looped domains**— loops of chromatin that extend at an angle from the main chromosome axis. Since such loops do occurs in *E.coli* chromosome, so the presence of loops may be a general feature of chromosomes.

In principle, looped domains in chromatin could be established and maintained by DNA binding proteins that clamp two regions of the 30- nm fibre together by recognizing specific DNA sequence that will form the neck of each loop. Alternatively, they could be formed by binding of DNA at the base of loop to a chromosome axis. Structural non-histone proteins could be involved in organizing the 30-nm fibres into loops. In an experiment, the histones are removed from the metaphase chromosome by adding the polyanion dextran sulphate. Histone-depleted chromosomes are found to have a central core of **scaffold** surrounded by a halo made of hoops of DNA (Fig. 17.10A). The scaffold is made of non-histone proteins and retains the general shape of the metaphase chromosome. Each chromosome has two scaffolds, one for each chromatid, and they are connected together at the centromere region. When the histones are removed, the DNA which has packed about 40-fold in the 30-nm chromatin fibre, becomes extended and produces loops with an average length of 25μm (75,000 base pairs). In each loop the DNA exits from the scaffold and returns to an adjacent point. On the basis of these observations a model of chromosome structure has been proposed by **Laemmli** and co-workers (1979, 1984). In Laemmli's radial loop model DNA is arranged in loops anchored to the non-histone scaffold. Because the lateral loops have 25μm DNA, after contracting 40-fold in the 30-nm fibre, they would be only about 0.6μm long, a length consistent with the diameter of metaphase chromosome (1μm). Figure 17.10B shows how the chromatin is arranged in loops which during metaphase, become arranged so that the base of the loops forms a scaffold in the centre of the chromatid. The base of the loop might be arranged on a helical coiled path (*e.g., Trillium* and *Tradescantia*).

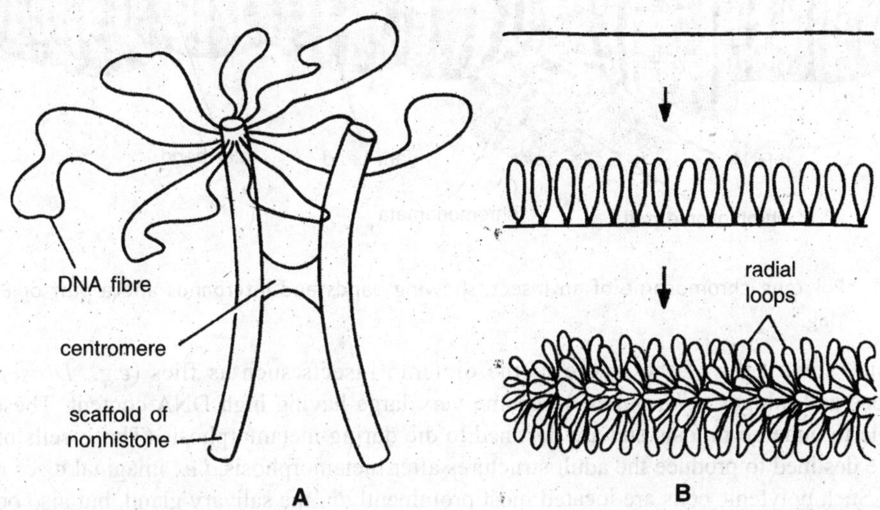

Fig. 17.10. Two methods by which DNA loops may form a metaphase chromosome. A—The non-histone proteins form two scaffolds, one per chromatid, while the naked DNA fibres form a halo around it; B—Laemmli's radial loop model of chromosome structure showing how 10-nm fibre form 30-nm fibre which further folds into radial loops.

The chromatin in mitotic chromosomes is transcriptionally inert: all RNA synthesis ceases as the chromosomes condense. Presumably the condensation prevents RNA polymerase from gaining access to the DNA, although other controlling factors might also be involved.

Functions

The function of chromosomes is to carry the genetic information from one cell generation to another. DNA being the only permanent component of chromosome structure, is the sole genetic material of eukaryotes.

17.4 GIANT CHROMOSOMES

Some cells at certain particular stages contain large nuclei with giant or large-sized chromosomes. The giant chromosomes are the **polytene** and **lampbrush** chromosomes.

1. Polytene Chromosome (Salivary Gland Chromosomes)

An Italian cytologist **E.G. Balbiani** (1881) had observed peculiar structures in the nuclei of certain secretory cells (*e.g.*, of salivary glands) of midge, *Chironomus* (Diptera). These structures were long and sausage-shaped and marked by swellings and cross striations (transverse bands). Unfortunately, he did not recognize them as chromosomes, and his report remained buried in the literature. It was not until 1933 that **Theophilus Painter**, **Ernst Heitz**, and **H. Bauer** rediscovered them in *Drosophila* and recognized them as the chromosomes. Since these chromosomes were discovered in the salivary gland cells, they were called **salivary gland chromosomes** (Fig. 17.11). The present name **polytene chromosomes** was suggested by **Kollar** due to the occurrence of many chromonemata (DNA) in them.

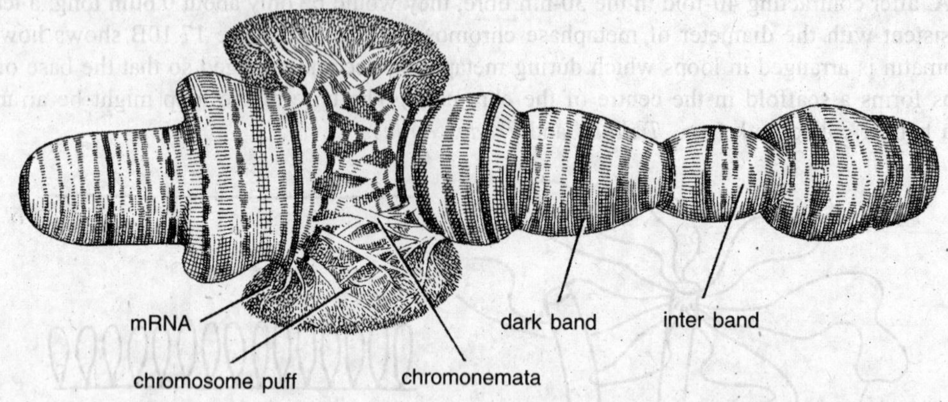

Fig. 17.11. Polytene chromosome of an insect, showing bands and interbands and a puff or Balbiani ring.

Thus, some cells of the larvae of the dipteran insects such as flies (*e.g., Drosophila*), mosquitoes and midges (*Chironomus*) become very large having high DNA content. These cells are unable to undergo mitosis and are destined to die during metamorphosis (Those cells of larva which are destined to produce the adult structures after metamorphosis, *i.e.,* imaginal discs remain diploid). Such polytenic cells are located most prominently in the salivary gland, but also occur in Malpighian tubules, rectum, gut, foot pads, fat bodies, ovarian nurse cells, etc. Polyteny of giant chromosomes is achieved by replication of the chromosomal DNA several times without nuclear division (endomitosis); and the resulting daughter chromatids do not separate but remain aligned side by side. In the process of endomitosis the nuclear envelope does not rupture and no spindle formation takes place. In fact, polyteny differs from polyploidy, in which there is also an excess DNA per nucleus, but in which the new chromoso-mes are separated from each other.

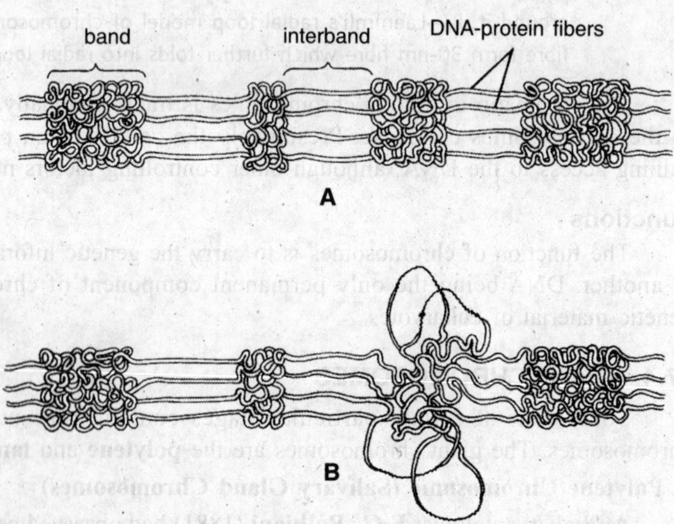

Fig. 17.12. A—Model of Dupraw and Rae showing band and interband structure in giant chromosome. B—The same model showing a possible method of "puffing".

A polytene chromosome of *Drosophila* salivary **gland** has about 1000 DNA molecules which are arranged side by side and which arise from 10 rounds of DNA replication ($2^{10} = 1024$). Other dipteran species have more

DNA, for example, *Chironomus* has 16000 DNA molecules in their each polytene chromosomes. Further, the polytene chromosomes are visible during interphase and prophase of mitosis. In them the chromomere (regions in which the chromatin is more tightly coiled) alternate with regions where the DNA fibres are folded more loosely (Fig. 17.12). The alignment of many chromomeres gives polytene chromosomes their characteristic morphology, in which a series of dark transverse **bands** alternates with clear zones called **interbands**. About 85 per cent of the DNA in polytene chromosomes is in bands and rest 15 per cent is in interbands. The crossbanding pattern of each polytene chromosome is a constant characteristic within a species and helps in chromosome mapping during cytogenetic studies. For example, in *Drosophila melanogaster* there are about 5000 bands and 5000 interbands per genome and each band and interband represent a set of 1024 identical DNA sequences arranged in file.

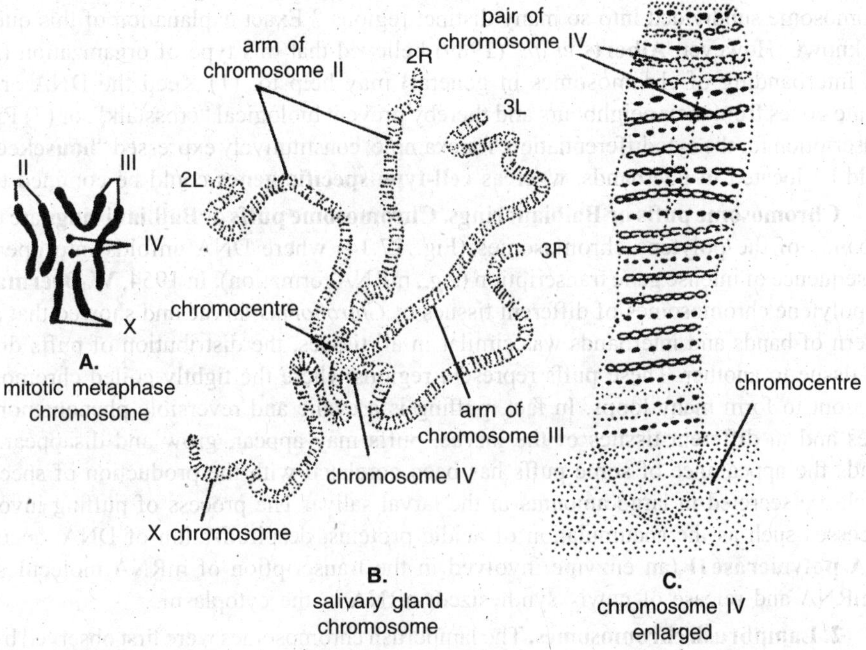

Fig. 17.13. Normal and polytene chromosomes of *Drosophila melanogaster*. A—Normal mitotic chromosomes; B—Polytene chromosomes of female; C—An enlarged IV chromosome.

Another peculiar characteristic of the polytene chromosomes is that the maternal and paternal homologous chromosomes remain associated side by side. This phenomenon is called **somatic pairing**. Consequently in the salivary gland cells the chromosome number always appear to be half of the normal somatic cells, *e.g., D. melogaster,* has only 4 polytene chromosomes. In *Drosophila*, pericentromeric heterochromatin of all polytene chromosomes also coalesces in a **chromocentre** (Fig. 17.13).

The preparation of a slide of the polytene chromosomes of dipterans for light microscopy is rather easy. The larvae are taken at the third instar stage and the salivary glands are dissected out and squashed in acetocarmine. In such preparations, these chromosomes in aggregate reach a length of as much as 2000 μm in *D. melanogaster*. In female *Drosophila,* the polytene chromosomes are found in the form of five long and one short strands radiating from a single more or less amorphous chromocentre (Fig. 17.13). One long strand corresponds to the X chromosome and remaining four long strands are the left and right arms of II and III chromosomes. The shortest strand represents the small dot-like IV chromosome. Each of these chromosomes contains maternal and paternal homologs in somatic pairing which lacks in the sex chromosomes of male fruit flies. Thus, in male *Drosophila,* X chromosome remains single and thin and Y chromosome exists indistinctly fused with the chromocentre.

One-gene, one-band hypothesis. The fixed pattern of bands and interbands in a *Drosophila* polytene chromosome suggested the early cytologists such as **Painter** (1933) and **Bridges** (1936) that each band might correspond to a single gene. Accordingly, they concluded that *Drosophila* might contain only 5000 essential genes. It was also believed that bands were the sites of genes (DNA) and interbands were relatively inert linker regions. Recent data, however, have contradicted this simple "one-band, one-gene" hypothesis, now it is held that bands as well as inter-bands contain active genes and a band may even contain more than one genes.

At this juncture a question arises, why is the single long strand of chromatin in each chromosome subdivided into so many distinct regions ? Exact explanation of this question is still not known. However, **Alberts** *et al.,* (1989) believed that this type of organization (*i.e.,* banding and interbanding of chromosomes in general) may help to: (1) Keep the DNA organized; (2) Isolate genes from their neighbours and thereby prevent biological "crosstalk", or (3) Regulate gene transcription for the cytodifferentiation, for example, constitutively expressed "**housekeeping**" **genes** could be located in interbands, whereas **cell-type-specific genes** could be confined to the bands.

Chromosome puffs or Balbiani rings. Chromosome puffs or **Balbiani rings** are the swellings of bands of the polytene chromosomes (Fig. 17.14) where DNA unfolds into open loops as a consequence of intense gene transcription (*e.g.,* mRNA formation). In 1954, **W. Beerman** compared the polytene chromosomes of different tissues of *Chironomus* larvae and showed that although the pattern of bands and interbands was similar in all tissues, the distribution of puffs differred from one tissue to another. These puffs represent regions where the tightly coiled chromosomal fibres open out to form many loops. In fact, puffing is a cyclic and reversible phenomenon; at definite times and in different tissues of the larvae, puffs may appear, grow and disappear. In salivary glands the appearance of some puffs has been corelated with the production of specific proteins which are secreted in large amounts in the larval saliva. The process of puffing involves several processes such as the accumulation of acidic proteins, despiralization of DNA, accumulation of **RNA polymerase II** (an enzyme involved in the transcription of mRNA molecules), synthesis of mRNA and release of newly synthesized mRNA in the cytoplasm.

2. Lampbrush chromosomes. The lampbrush chromosomes were first observed by **Flemming** in salamander (amphibian) oocytes in 1882. He coined the name because the chromosomes look like the brushes which were used for cleaning the glass chimneys of old-fashioned paraffin or kerosene lamps. They were described in detail in shark oocytes by **R. Ruckert** in 1892. **Thorpe** (1984) and **Burns** and **Bottino** (1989) preferred the term **test tube brush chromosomes** for them. However, due to recent investigations the structure of these exceptionally large-sized chromosomes has been interpreted in functional terms, *i.e.,* now they are merely visualized as means of *"turning on and turning off"* of the genes.

The lampbrush chromosomes occur at the diplotene stage of meiotic prophase in the primary oocytes of all animal species, both vertebrates and invertebrates. Thus, they have been described in *Sagitta* (Chaetognatha), *Sepia* (Mollusca), *Echinaster* (Echinodermata) and in several species of insects, shark, amphibians, reptiles, birds and mammals (humans). Lampbrush chromosomes are also found in spermatocytes of several species, giant nucleus of *Acetabularia* and even in plants. Generally, they are smaller and "hairy" in invertebrates than in vertebrates. Lamphrush chromosomes are best visualized in salamander oocytes because they have a high DNA content. For example, the largest chromosome having a length up to 1 mm have been observed in urodele amphibian. Thus, lampbrush chromosomes are much larger (longer) than the polytene chromosomes of insects.

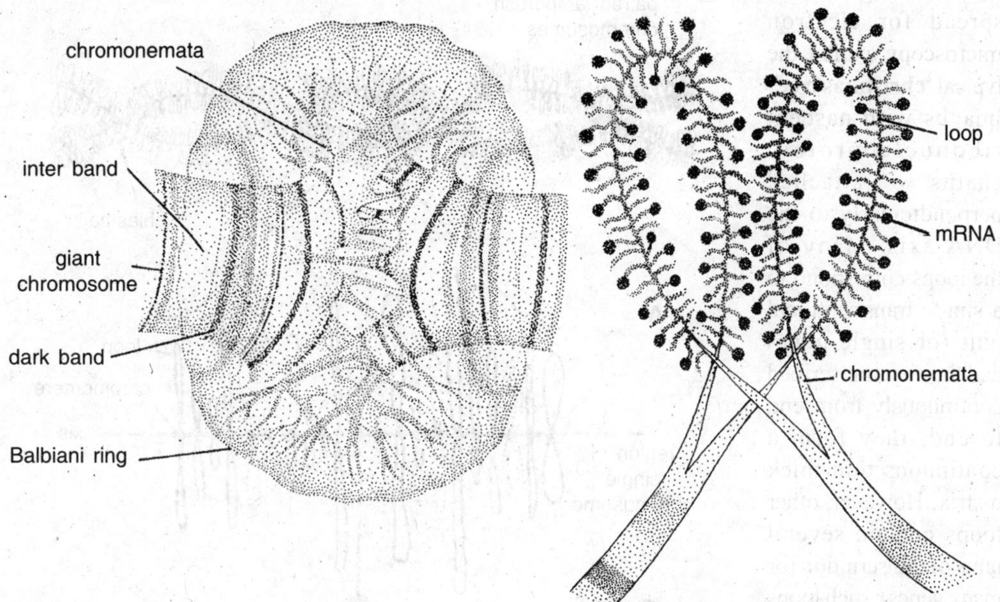

Fig. 17.14. A Balbiani ring of a polytene chromosome.

Fig. 17.15. Suggested form of the individual loops in chromosome puffs. Note the tiny fibrils of RNA and proteins attached to the loop.

Since the lampbrush chromosomes are found in the prolonged diplotene stage of meiotic prophase I, they are present in the form of **bivalents** in which the maternal and paternal chromosomes are held together by **chiasmata**, at those sites where crossing over has previously occurred (Fig. 17.16). The paired homologs are not condensed as usual chromosomes would be; instead, they are very long and stretched out. Each bivalent has four chromatids, two in each homolog. The axis of each homologs consists of a row of granules or **chromomeres** from which **lateral loops** extend. The loops are always symmetrical, each chromosome having two of them, one for each chromatid. The loops can be categorized by size, thickness and other morphological characteristics. Each loop appears at a constant position in the chromosome; this fact helps in the chromosome mapping. There are about 10,000 loops per chromosome set or haploid set (*e.g.*, oocytes of slamander *Triturus*). Each loop has an **axis** which is made of single DNA molecule that is unfolded from the chromosome for the intense RNA synthesis. Thus, about 5 to 10 per cent of the DNA exists in the lateral loops, the rest being tightly condensed in the chromomeres which are transcriptionally inactive. The centromeres of the chromosomes bear no loops.

Each loop of lampbrush chromosomes is found to perform intense transcription of **hnRNA** or **heterogeneous RNA** molecules. (*i.e.,* precursors of mRNA molecules for various ribosomal proteins and for five types of histone proteins). Electron microscopy of the loops has shown that **RNA polymerase** enzyme molecules are attached to the principal axis (DNA) of the loop from which RNA fibrils of increasing length extend. As transcription continues along the DNA strand of loop, the fibrils of RNA (*i.e.*, hnRNA) lengthen. Proteins get associated with these RNA fibrils as they are formed and ultimately ribonucleoprotein product is released.

Thus, each lateral loop is covered by a **matrix** that consists of RNA transcripts with hnRNA-binding proteins attached to them. Generally this matrix is asymmetrical, being thicker at one end of the loop than at the other. RNA synthesis starts at the thinner end and progresses toward the thicker end. Preparations spread for electron microscopy exhibit the typical '**christmas tree**" images with nascent ribonucleoprotein chains attached perpendicularly to the DNA axis. Many of the loops correspond to a single transcriptional unit (or single gene) and they are transcribed continuously from end to end; they form a continuous thin-thick matrix. However, other loops contain several units of transcription (or many genes); such loops include an extended section of chromatin that is not transcribed at all. For example, **Gall** *et al.*, (1981) found that in the American newt *Noto-phthalmus viridescens*, clusters of the five his-tone genes are tande-mly repeated in the genome but separated by about 50,000 base pairs of repeats of a 225-base pair satellite DNA.

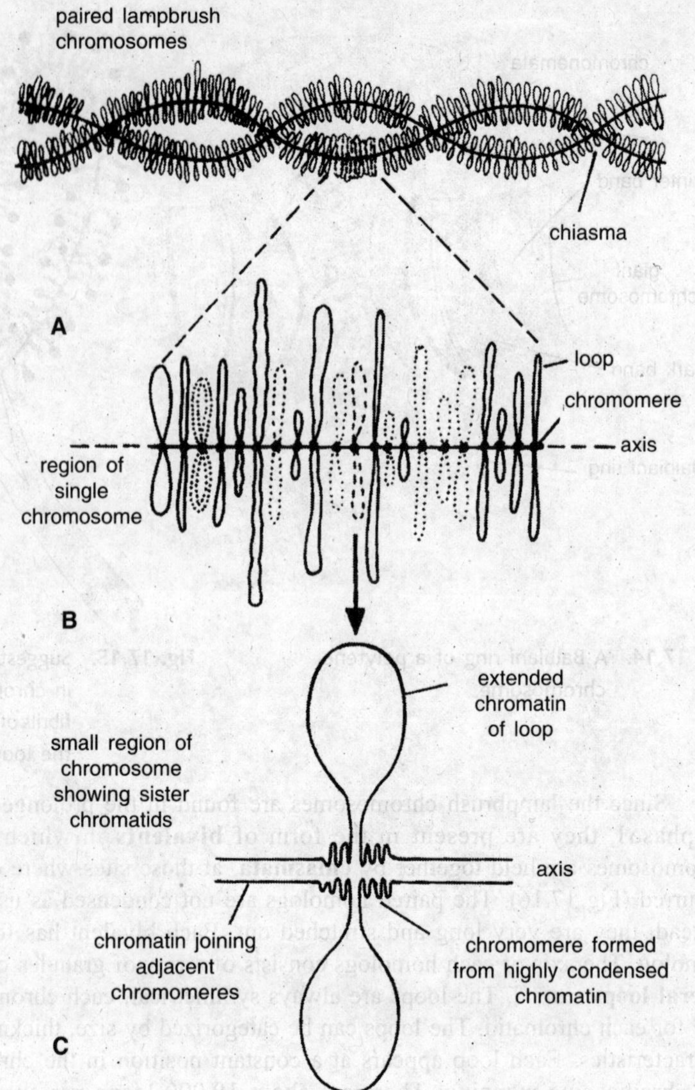

Fig. 17.16. Lampbrush chromosome structure. A—Bivalent or paired homologous chromosomes in pairing showing chiasmata, B—A part of one homologue (or homolog) showing paired loops given out by two chromatids; C—Single pair of loop.

Further, the number of pairs of loops gradually increa-ses during meiosis till it reaches maximum in diplotene. Such a lampbrush stage may persist for months or years as the primary oocyte builds up a supply of mRNA molecules and other materials required for its ultimate de-velopment into a new individual. As meiosis proceeds further, num-ber of loops gradually decreases and the loops ultimately disappear either due to disintegration or by reabsorption back into the chromosome. For example, the addition of histone proteins to the lampbrush chromosomes stops the synthesis of RNA on the loops and causes the loops to retract into the chromosomes.

Certain hypothesis regarding nature of loops. The loop of the lampbrush chromosomes can be viewed in the following two ways:

1. It may be static, unchanging in length and constantly exposing the same stretch of chromosome fibre.

2. It may be dynamic, with new loop material spinning out of one side of a chromomere and returning to a condensed state on the other side. This is called **spinning out** and **retraction hypothesis** (**Gall** and **Callan**, 1962). It means that 100 per cent of the genome can be expressed during the lampbrush stage.

Recent, DNA-RNA hybridization experiments have rejected the spinning out and retraction hypothesis.

Study of both polytene and lampbrush chromosomes provided the evidence that eukaryotic gene activity is regulated at the level of RNA synthesis (or transcription). Lampbrush chromosomes also show the possible way of gene amplification which is required during the growth phase of oocytes.

17.5 SUPERNUMERAY CHROMOSOMES

In certain plants (e.g., maize, rye) and animals (grasshopper), one or more additional chromosomes are observed in addition to the normal diploid number. Such accessory chromosomes are called **satellite** or **supernumeray chromosomes** or **B-chromosomes in plants** (to distinguish them from the invariant and essential **A-chromosomes**). They are very small and genetically inert being composed of heterochromatin. They do not take part in mitosis and do not affect the phenotype but reduce fertility and vigour of the individual. Supernumeray chromosomes may be regarded as **giant linear plasmids** (**Twyman** 1998).

QUESTIONS

Long Answer Questions

1. Describe the occurrence, ultrastructure and chemical composition of chromosome. What are its different types?
2. Define chromosome. Write the names of various types of special chromosomes.
3. Describe the structure and function of chromosome.
4. Describe the structure, chemical composition and function of the eukaryotic chromosomes.
5. Describe the structure and function of chromosomes.
6. Describe the structure of a typical chromosome.
7. Describe the structure and physiology of chromosomes.
8. Give an account of the structure of chromosome. Distinguish between chromonema and chromatid. Write a short note on the different chromosomal shapes at anaphase.
9. Give a detailed account of chromatin.
10. Describe the Oudet and other's nucleosome models of chromosomes and explain as to how this model is considered better than Dupraw's folded fibre model? Describe as to how many structure groups human chromosomes belong?
11. Give the biochemical components of the chromatin and discuss the structural organization of chromatin emphasizing the nucleosome organization.
12. What are chromosomes? Describe the chemical composition and organization of metaphase chromosome.
13. Give an account of the chemical organization of chromosomes. What are nucleosomes?
14. What do you mean by giant chromosome? Describe the physico-chemical structure of polytene chromosome.
15. What is a giant chromosome? Describe the structure of a giant chromosome and its role in heredity.
16. Give a brief description of the significance of polytene chromosomes in understanding gene regulation.
17. Describe the structure of polytene chromosome and state their significance.
18. What are lampbrush chromosomes and Balbiani rings?
19. Discuss the structure and functions of lampbrush chromosomes.
20. State the difference between the lampbrush and the salivary gland chromosomes. Write the structural and functional significance of the salivary gland chromosomes.

Short Answer Questions

1. What is the role of basic proteins which are associated with the DNA molecule?
2. What are Feulgen positive regions?
3. Why secondary constriction is known as nucleolar organizer?
4. What are supernumerary chromosomes?
5. Describe the role of centromere.
6. What is the nucleolar organizer (NO) region in the chromosome? What is its role?
7. Describe unistranded folded fibre model of chromosome structure.
8. How protein molecules are associated with DNA molecule?
9. What is nucleosome concept?
10. Describe structure of a nucleosome.
11. What do you mean by telocentric and metacentric chromosome?
12. What do you mean by polyteny? Who introduced this concept?
13. Explain significance of puffing.
14. What do you mean by gene amplification? What is its need? Explain.
15. Write short note on the following:
    (*i*) Chromosomes
    (*ii*) Chromosome shapes
    (*iii*) Types of chromosomes
    (*iv*) Centromere
    (*v*) Autosomes
    (*vi*) Barr bodies
    (*vii*) Nucleosome
    (*viii*) Heterochromatin and euchromatin
    (*ix*) Fine structure of chromosome
    (*x*) Polytene (or polytenic) chromosome
    (*xi*) Lampbrush chromosomes
    (*xii*) Balbiani chromosome
    (*xiii*) Histones
16. Prepare a well labelled diagram of each of the following:
    1. Lampbrush chromosomes
    2. Fine structure of chromosome
17. Differentiate between the following:
    1. Viral and bacterial chromosomes
    2. Prokaryotic (bacterial) and eukaryotic chromosomes
    3. Primary constriction and secondary constriction
    4. Centrosome and nucleosome
    5. Euchromatin and heterochromatin (*H.N.B., Garhwal 2009*)
    6. Lampbrush chromosome and salivary gland chromosome
    7. Heterochromatin and Barr bodies
    8. Centromere and chromocentres (*H.N.B., Garhwal 2009*)
    9. Histones and non-histones (*H.N.B., Garhwal 2009*)

Very Short Answer Questions

1. Name the scientists.
   (*i*) Who coined the term chromosome?
   (*ii*) Who described parallelism between chromosomes and transmission of hereditary factors?
2. Name the part of a chromosome where its two chromonemata are joined together.
3. Name the organism where RNA act as hereditary material.
4. What is nucleoid?
5. How many DNA double helical fibres are present in a chromosome?

6. 'Beads in the string' appearance of chromatin represents what?
7. What is the importance of telomere in chromosome? *(H.N.B., Garhwal 2009)*
8. Lampbrush chromosomes are found during which stage of cell division? *(H.N.B., Garhwal 2009)*

True/False Questions

(*i*) Histone are basic proteins rich in lysine and arginine. *(H.N.B., Garhwal 2009)*

Fill in the Blanks

1. Polytene chromosomes are also known as ....
2. Number of chromosomes in human genome is ....
3. The term chromosome was coined by .... in 1888 because of their affinity for ....
4. There are 5 types of histones in the eukaryotic chromosomes, namely H_1, ... , ... , H_3, H_4 and H_5.
5. Centromere lies within thinner segment of chromosome called ....
6. At mitotic metaphase each chromosome consists of two ....
7. The cromosomal ends are called ....
8. Heterochromatin is a .... region of chromosome where chromatin remains highly ....
9. Sex chromatin has .... heterochromatin.
10. In a nucleosome, DNA makes two complete turns around histone .... and these two turns are .... bps long.
11. Folded-fiber model of eukaryotic chromosome was proposed by .... in 1965.
12. Solenoid model of eukaryotic chromosome involves in it the ....
13. Radial loop model of metaphase chromosome was proposed by .... in 1979.
14. Lampbrush chromosomes occur at .... stage of meiotic .... in primary .... of many animal species.

Multiple Choice Questions

1. Number of chromatids present in a metaphase chromosome is
   (*a*) four (*b*) three (*c*) two (*d*) one
2. Number of chromosomes in definite nucleus
   (*a*) haploid (*b*) diploid (*c*) triploid (*d*) polyploid
3. Diploid chromosome number in humans is
   (*a*) 42 (*b*) 44 (*c*) 46 (*d*) 48
4. Two sister chromatids are attached with the
   (*a*) spindle fibre (*b*) chromatid (*c*) centromere (*d*) chromocentre
5. Autosomes in humans are
   (*a*) 11 pairs (*b*) 22 pairs (*c*) 23 pairs (*d*) 43 pairs
6. Eukaryotic chromosome is made of
   (*a*) DNA (*b*) DNA + lipid (*c*) DNA + proteins (*d*) RNA
7. The term 'nucleosome' was given by
   (*a*) Dupraw (*b*) Oudet (*c*) Flemming (*d*) Emil Heitz
8. The nucleosome
   (*a*) surrounds nuclear pores
   (*b*) has only DNA and non-histone proteins
   (*c*) is fully responsible for packing DNA into chromosomes
   (*d*) contains a core of histones with DNA wrapped around it

9. Balbiani rings occur in
   (*a*) polytene chromosomes (*b*) lampbrush chromosomes
   (*c*) polysomes (*d*) heterosomes
10. Chemical composition of chromomere is
   (*a*) DNA and proteins (*b*) DNA and lipids
   (*c*) DNA and carbohydrates (*d*) proteins and lipids
11. A chromosome in which the centromere is situated near one end is known as
   (*a*) metacentric (*b*) submetacentric
   (*c*) acrocentric (*d*) telocentric
12. The total number of genes present in a haploid cell is
   (*a*) genome (*b*) genotype (*c*) karyotype (*d*) gene pool
13. The centromere is found at the centre of the two arms in......... (*B.U. coimbatore 1997*)
   (*a*) telocentric (*b*) acrocentric (*c*) submetacentric (*d*) metacentric
14. Chromosomes can be stained with one of the following chemicals?
   (*a*) acetocarmine (*b*) safranine (*c*) light green (*d*) eosin

ANSWERS

Very Short Answer Questions

1. (*i*) Waldeyer
   (*ii*) S.W. Sutton
2. Centromere
3. Some viruses, *e.g.*, TMV.
4. The nuclear regions in bacterial cells (= prokaryotic cells) where DNA molecule of bacterial chromosome lies in a coiled state. Bacterial cells lack a nucleus.
5. One in each chromosome.
6. The 10 nm chromatin fibre (*i.e.*, nucleosomes in close apposition).
7. The chromosomal ends are called **telomeres**. Each telomere provides polarity to the chromosome and prevents other chromosomal segments to fuse with it.
8. Diplotene of prophase I of meiosis.

True/False Questions

(*i*) True.

Fill in the Blanks

1. Salivary gland chromosomes.
2. 23.
3. Waldeyer, stains.
4. H2A. H2B.
5. Primary constriction.
6. Chromatids.
7. Telomeres.
8. Dark-stained, condensed.
9. Facultative.
10. Octamer, 200.
11. E.J. Dupraw.
12. Nucleosomes.
13. Laemmli and coworkers.
14. Diplotene, prophase, oocyte.

Multiple Choice Questions

1. (*c*)	2. (*b*)	3. (*c*)	4. (*c*)	5. (*b*)	6. (*c*)
7. (*b*)	8. (*d*)	9. (*a*)	10. (*a*)	11. (*c*)	12. (*a*)
13. (*d*)	14. (*a*).				

18

Cell Cycle and Mechanics of Cell Division

18.1 CELL CYCLE

All cells are produced by divisions of pre-existing cell. Continuity of life depends on cell division. A cell born after a division, proceeds to grow by macromolecular synthesis, reaches a species-determined division size and divides. This cycle acts as a unit of biological time and defines life history of a cell. **Cell cycle** can be defined as the entire sequence of events happening from the end of one nuclear division to the beginning of the next. The cell cycle involves the following three cycles.

1. Chromosome cycle. In it **DNA synthesis** alternates with **mitosis** (or karyokinesis or nuclear division). During DNA synthesis, each double-helical DNA molecule is replicated into two identical daughter DNA molecules and during mitosis the duplicated copies of the genome are ultimately separated.

2. Cytoplasmic cycle. In it **cell growth** alternates with **cytokinesis** (or cytoplasmic division). During cell growth many other components of the cell (RNA, proteins and membranes) become double in quantity and during cytokinesis cell as a whole divides into two. Usually the karyokinesis is followed by the cytokinesis but sometimes the cytokinesis does not follow the karyokinesis and results into the multinucleate cell, *e.g.*, cleavage of egg in *Drosophila*.

3. Centrosome cycle. Both of the above cycles require that the **centrosome** be inherited reliably and duplicated precisely in order to form the two poles of the mitotic spindle; thus, centrosome cycle forms the third component of cell cycle.

Howard and **Pelc** (1953) have divided cell cycle into four phases or stages: G_1, S, G_2 and M phase. The G_1 phase, S phase and G_2 phase are combined to from the classical **interphase**.

1. G_1 Phase. After the M phase of previous cell cycle, the daughter cells begin G_1 of interphase of new cell cycle. G_1 is a resting phase. It is called **first gap phase**, since no DNA synthesis takes place during this stage; currently, G_1 is also called **first growth phase**, since it involves synthesis of RNA, proteins and membranes which leads to the growth of nucleus and cytoplasm of each daughter cell towards their mature size.

During G_1 phase, chromatin is fully extended and not distinguishable as discrete chromosomes with the light microscope. This is a time of resumption of normal cell metabolism which has slowed down during the previous cell division. Thus, G_1 involves transcription of three types of RNAs, namely rRNA, tRNA and mRNA; rRNA synthesis is indicated by the appearance of nucleolus in the interphase (G_1 phase) nucleus. Proteins synthesized during G_1 phase are (1) Regulatory proteins which control various events of mitosis; (2) Enzymes (*e.g.*, DNA polymerase) necessary for DNA synthesis of the next stage; and (3) Tubulin and other mitotic apparatus proteins.

G_1 phase is most variable as to duration (Table 18.1); it either occupies 30 to 50 per cent of the total time of the cell cycle or lacks entirely in rapidly dividing cells (*e.g.*, blastomeres of early embryo of frog and mammals). Terminally differentiated somatic cells (*i.e.*, end cells such as neurons and striated muscle cells) that no longer divide, are arrested usually in the G_1 stage; such a type of G_1 phase is called **G_0 phase.**

Table 18.1 Different stages of a mitotic cell cycle and their duration in hours.

Parts of cell cycle	*Phases*	*Description of phases*	*Duration in hours*		
			Vicia faba	*Mouse L cells*	*Human HeLa cells*
Interphase	G_1	Pre-DNA-synthesis phase	12	12	12
	S	DNA-synthesis phase	6	6–8	10
	G_2	Post-DNA synthesis phase	12	3–4	3
Mitosis	M	Mitotic phase	1	1	1

2. S phase. During the S phase or **synthetic phase** of interphase, replication of DNA and synthesis of histone proteins occur. New histones are required in massive amounts immediately at the beginning of the S period of DNA synthesis to provide the new DNA with nucleosomes. Thus, at the end of S phase, each chromosome has two DNA molecules and a duplicate set of genes. S phase occupies roughly 35 to 45 per cent of cell cycle.

3. G_2 phase. This is a **second gap** or **growth phase** or resting phase of interphase. During G_2 phase, synthesis of RNA and proteins continues which is required for cell growth. It may occupy 10 to 20 per cent time of cell cycle. As the G_2 phase draws to a close, the cell enters the M phase.

General events of interphase. The interphase is characterized by the following features:

The nuclear envelope remains intact. The chromosomes occur in the form of diffused, long, coiled and indistinctly visible chromatin fibres. The DNA amount becomes double. Due to accumulation of ribosomal RNA (rRNA) and ribosomal proteins in the nucleolus, the size of the latter is greatly increased. In animal cells, a daughter pair of centrioles originates near the already existing centrioles and, thus, an interphase cell has two pairs of centrioles.

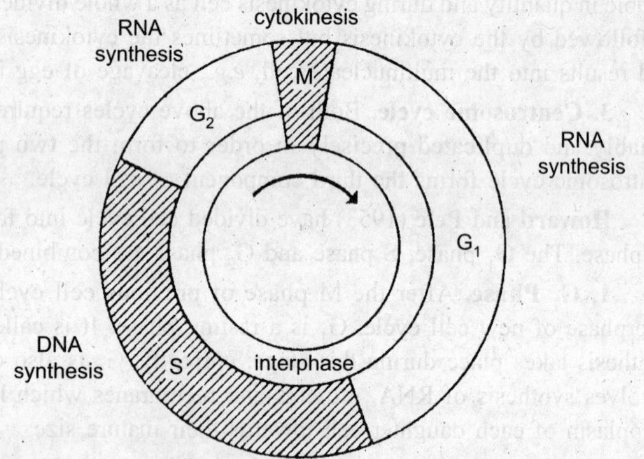

Fig. 18.1. The cell cycle or mitotic cycle, showing relative duration of phases (*e.g.*, interphase and mitotic phase) in a growing cell. S, synthesis of DNA; G_1 the first gap or growth phase; G_2, the second gap or growth phase; and M, mitotic phase.

In animal cells, net membrane biosynthesis increases just before cell division (mitosis). This extra membrane seems to be stored as **blebs** on the surface of the cells about to divide.

4. M phase or mitotic phase. The mitosis (Gr., *mitos* = thread) occurs in the somatic cells and it is meant for the multiplication of cell number during embryogenesis and blastogenesis of plants and animals. Fundamentally, it remains related with the growth of an individual from zygote to adult stage. Mitosis starts at the culmination point of interphase (*i.e.*, G_2 phase). It is a short period of chromosome condensation, segregation and cytoplasmic division. Mitosis is important for replacement of cells lost to natural friction (**attrition**), wear and tear and for wound healing.

18.2 MITOSIS

Mitosis is the process that distributes the two sets of duplicated chromosomes into two daughter nuclei. As a process, mitosis is remarkably similar in all animals and plants. It is a smoothly continuous process and is divided arbitrarily into following stages or phases for convenient reference (Fig. 18.2 and Fig. 18.3):

1. Prophase. The appearance of thin-thread like condensing chromosomes marks the first phase of mitosis, called **prophase** (Gr., *pro* = before; *phasis* = appearance). The cell becomes spheroid, more refractile and viscous.

Each prophase chromosome is composed of two coiled filaments, the **chromatids**, which are the result of the replication of DNA during the S phase. As prophase progresses, the chromatids become shorter and thicker and two sister chromatids of each chromosome are held together by a special DNA-containing region, called the **centromere** or **primary constriction**. During prophase, proteins of the trilaminar **kinetochores** (one for each chromatid) start depositing or organizing on the centromere of each chromosome. Further, during early prophase, the chromosomes are evenly distributed in the nuclear cavity; as prophase progresses, the chromosomes approach the nuclear envelope, causing the central space of the nucleus to become empty.

In the cytoplasm, the most conspicuous change is the formation of the spindle or **mitotic apparatus**. In the early prophase, there are two pairs of centrioles, each one surrounded by the so-called **aster** which is composed of microtubules radiating in all directions. The two pairs of centrioles migrate to opposite poles of the cell along with the asters and become situated in antipodal positions. Between the separating centrioles forms a spindle.

Lastly, during prophase, the nucleolus gradually disintegrates. Degeneration and disappearance of the nuclear envelope marks the end of prophase. This process is incompletely understood. However, following two factors may be involved in this process: 1. **Enzymatic action** either by some mitochondrial enzymes, cytosolic MPF kinase or nuclear RNA (or ribozyme). 2. **Physical action**, *i.e.*, physical stress exerted by microtubules which become attached to the nuclear envelope.

Prometaphase. The breakdown of nuclear envelope signals the commencement of prometaphase and enables the mitotic spindle to interact with the chromosomes. This stage is characterized by a period of frantic activity during which the spindle appears to be trying to contain and align the chromosomes at the metaphase plate.

2. Metaphase. During metaphase (Gr., *meta* = after; *phasis* = appearance) the chromosomes are shortest and thickest. Their centromeres occupy the plane of the equator of the mitotic apparatus (a region known as the **equatorial** or **metaphase plate**), although the chromosomal arms may extend in any direction. At this stage the sister chromatids are still held together by centromere and the kinetochores of the two sister chromatids face opposite poles; this would permit their proper separation in the next phase (anaphase).

Metaphase occupies a large portion of the mitotic phase (see Table 18.2), as if the cell pause until all their chromosomes are lined up correctly on the metaphase plate. During metaphase, subunits (tubulin dimers) are added to the plus end of a microtubule at the kinetochore and are removed from the minus end at the spindle pole. Thus, a poleward flux of tubulin subunits occurs, with the microtubules remaining stationary and under tension.

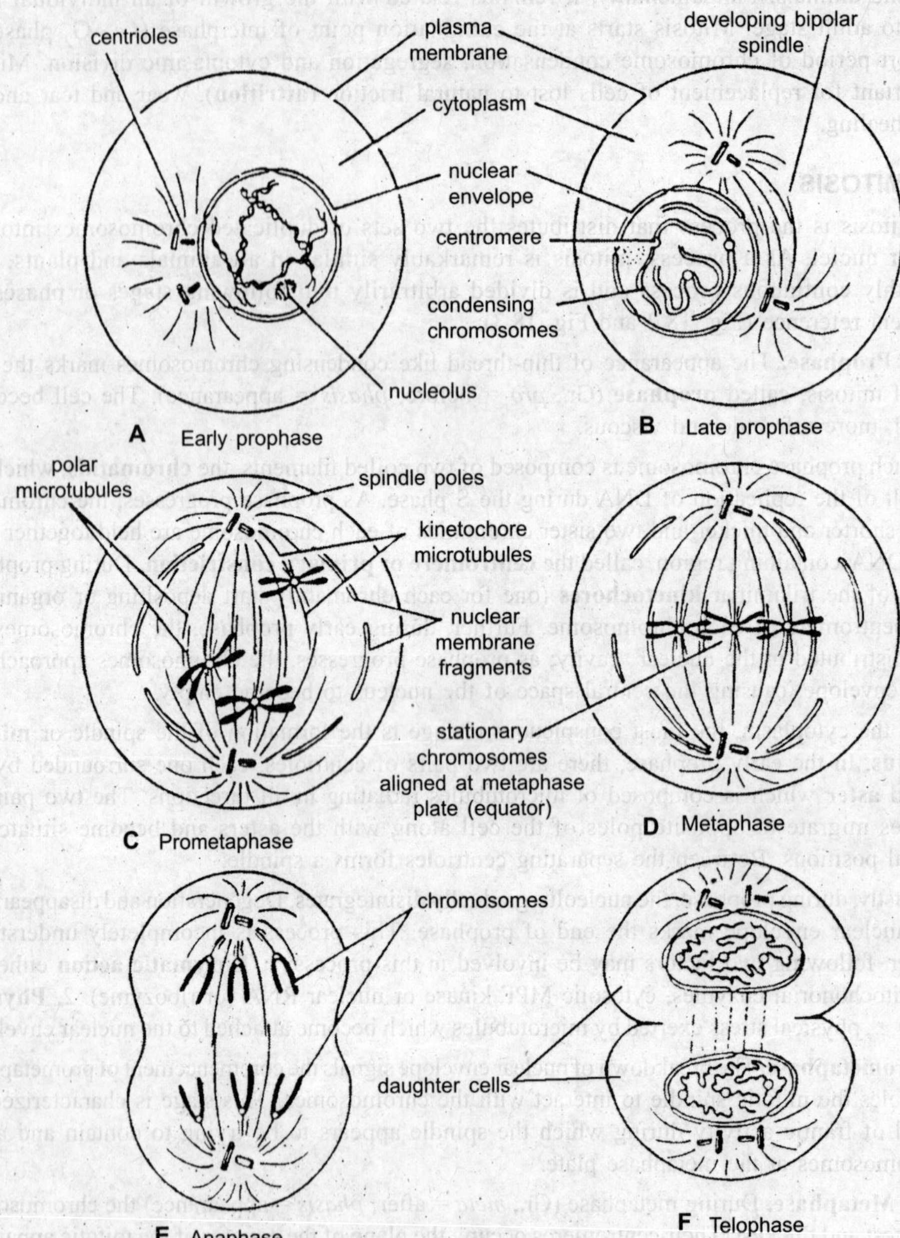

Fig. 18.2. Diagrammatic summary of mitosis in the animal cell.

3. Anaphase. The anaphase (Gr., *ana* = up; *phasis*=appearance) begins abruptly with the synchronous splitting of each chromosome into its sister chromatids, called **daughter chromosomes**, each with one kinetochore. Synchronous splitting of each centromere during prophase is evidently caused by an increase in cytosolic Ca^{2+}. In fact, Ca^{2+}–containing membrane vesicles accumulate at spindle poles and release calcium ions to initiate anaphase. Anaphase involves the following two steps:

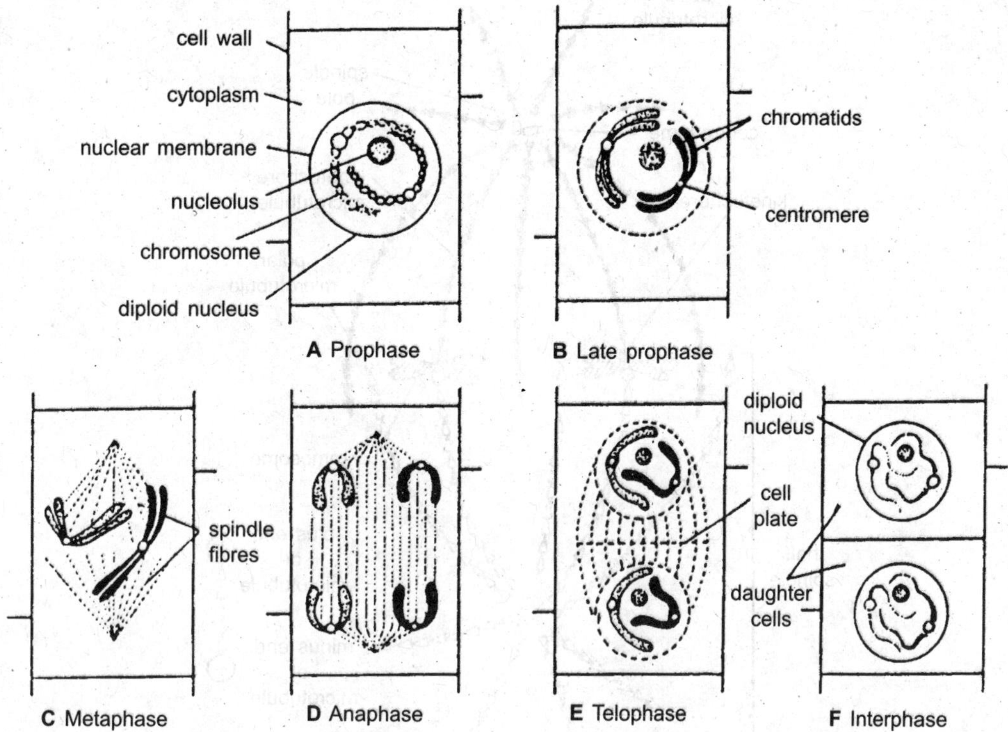

Fig. 18.3. Diagrammatic summary of mitosis in the higher plant cells.

***(i)* Anaphase A**. During it, there is poleward movement of chromatids due to shortening of the kinetochore microtubules. During their poleward migration, the centromeres (and kinetochores) remain foremost so that the chromosomes characteristically appear U, V or J-shaped.

***(ii)* Anaphase B.** It involves separation of poles themselves accompanied by the elongation of the polar microtubules. The astral microtubules also help in anaphase B by their attractive interaction with cell cortex.

Table 18.2 Duration of different phases of mitosis in certain plants and animals.

Organism	*Prophase*	*Metaphase*	*Duration in minutes*	
			Anaphase	*Telophase*
1. Mouse (spleen)	21	13	5	4
2. Grasshopper (neuroblasts)	102	13	9	57
3. Pea (root tip)	78	14.4	4.2	13.2
4. Onion (root tip)	71	6.5	2.4	3.8

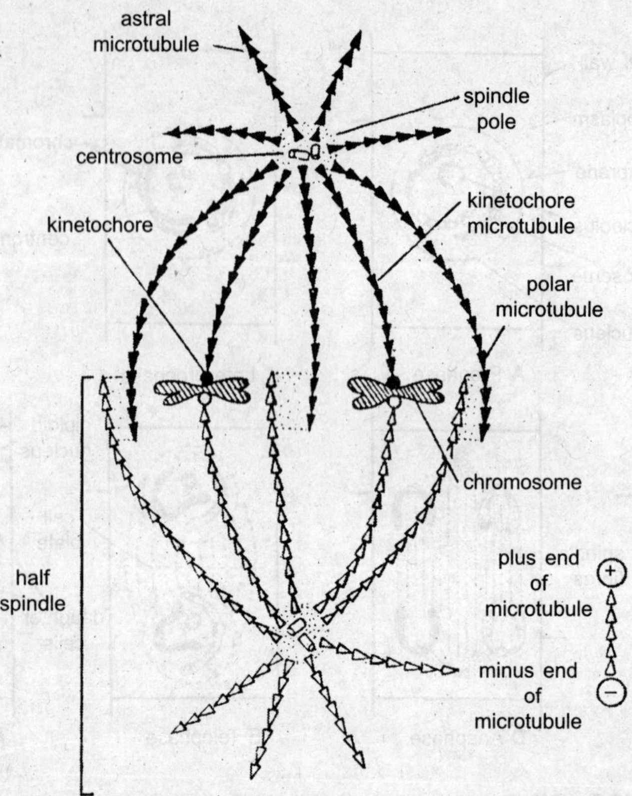

Fig. 18.4. The mitotic spindle at metaphase. The spindle is constructed from two half spindles, each composed of kinetochore, polar and astral microtubules. The polarity of the microtubules is indicated by the arrowheads.

4. Telophase. The end of the polar migration of the daughter chromosomes marks the beginning of the telophase; which in turn is terminated by the reorganization of two new nuclei and their entry into the G_1 phase of interphase. In general terms, the events of prophase occur in reverse sequence during this phase. A nuclear envelope reassembles around each group of chromosomes to form two daughter nuclei. The mitotic apparatus except the centrioles disappears; high viscosity of the cytoplasm decreases; the chromosomes resume their long, slender, extended form as their coils relax; and RNA- synthesis restarts causing the nucleolus to reappear.

Cytokinesis

Both DNA synthesis and mitosis are coupled to cytoplasmic division, or cytokinesis—the constriction of cytoplasm into two separatè cells. During cytokinesis, the cytoplasm divides by a process, called **cleavage**. The mitotic spindle plays an important role in determining where and when cleavage occurs. Cytokinesis usually begins in anaphase and continues through telophase and into interphase. The first sign of cleavage in animal cells is **puckering** and **furrowing** of the plasma membrane during anaphase. The furrowing invariably occurs in the plane of the metaphase plate, at right angles to the long axis of the mitotic spindle. A cleavage furrow tends to form midway between asters originating from two centrosomes in fertilized sand dollar eggs.

Cleavage is accomplished by the contraction of a ring composed mainly of actin filaments. This bundle of filaments, called **contractile ring**, is bound to the cytoplasmic face of the plasma membrane by unidentified attachment proteins. The contractile ring assembles in early anaphase, once assembled, it develops a force large enough to bend a fine glass needle inserted into the cell. Evidently this force is generated due to muscle-like sliding of actin and myosin filaments in the contractile ring. The actin-myosin interaction pulls the plasma membrane down into a furrow. During a normal cytokinesis, the contractile ring does not get thicker as the furrow invaginates, suggesting that it continuously reduces its volume by losing filaments. When cleavage ends, the contractile ring is finally dispensed with altogether and the plasma membrane of the cleavage furrow narrows to form the **midbody**, which remains as a tether (Tether means a rope for confining a beast within certain limits) between two daughter cells. The midbody contains the remains of the two sets of polar microtubules, packed tightly together with dense matrix material.

Cytokinesis greatly increases the total cell-surface area as two cells form from one. Therefore, the two daughter cells resulting from cytokinesis require more plasma membrane than in the plant cell. Lastly, prior to cytokinesis, in M phase large membrane-bounded organelles such as Golgi apparatus and the endoplasmic reticulum break up into smaller fragments and vesicles; this may ensure their even distribution into daughter cells during cytokinesis.

Significance of Mitosis

The mitosis has the following significance for living organisms:

1. The mitosis helps the cell in maintaining proper size.
2. It helps in the maintenance of an equilibrium in the amount of DNA and RNA in the cell.
3. The mitosis provides the opportunity for the growth and development to organs and the body of the organisms.
4. The old decaying and dead cells of body are replaced by the help of mitosis.
5. In certain organisms, the mitosis is involved in asexual reproduction.
6. The gonads and the sex cells depend on the mitosis for the increase in their number.
7. The cleavage of egg during embryogenesis and division of blastema during blastogenesis, both involve mitosis.

18.3 MECHANISM OF MITOTIC APPARATUS

The microtubules of the spindle are arranged like two cones base to base, broad at the centre or equator of the cell and narrowing to a point at either end or pole. Mitotic spindle contains three main types of fibres (Fig. 18.5): (1) **Polar fibres**, which extend from the two poles of the spindle toward the equator; (2) **Kinetochore fibres**, which attach to the kinetochores of centromeres of each mitotic chromosomes and extend toward the poles; and (3) **Astral fibres**, which radiate outward from the poles toward the periphery or cortex of cell. In cells of most higher plants, however, spindle forms without the aid of centrioles and lacks asters.

Working of mitotic spindle during anaphase. During anaphase A, a surprisingly large force acts on a chromosome as it moves from the metaphase plate to the spindle pole. By hydrodynamic analysis it has been calculated that to move a chromosome, a force about 10^{-11} dynes is needed, and that the entire displacement—from equator to the pole of a chromosome—may require the use of about 30 ATP molecules. As each chromosome moves polewards, its kinetochore microtubules disassemble, so that they have nearly disappeared at telophase. The site of subunit loss can be determined by injecting labelled tubulin into cells during metaphase. The labelled subunits are found to be added to the kinetochore end of kinetochore microtubules and then lost as anaphase A proceeds, indicating that kinetochore "eats" its way poleward along its microtubules at anaphase (Fig. 18.5). However, typically microtubule disassemble at kinetochores, poles or at both sites is probably necessary for equator-to-pole movement.

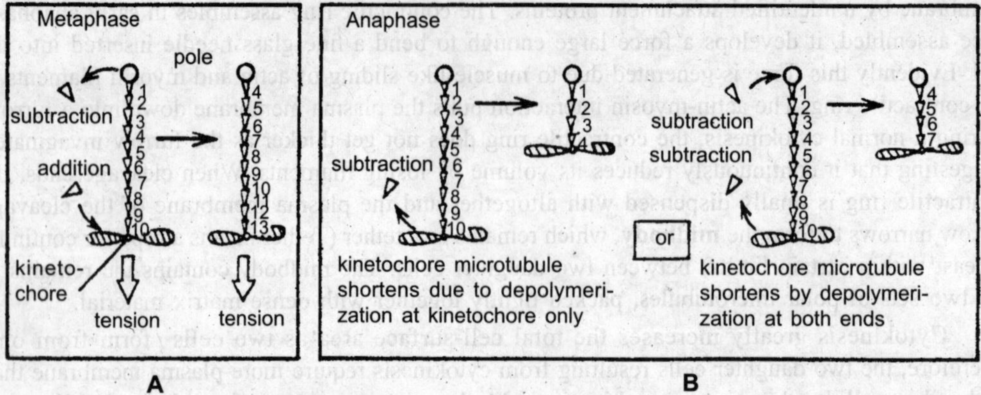

Fig. 18.5. Comparison of behaviour of kinetochore microtubules during metaphase (A) and anaphase (B). A—At metaphase, subunit are added to the plus end of a microtubule at the kinetochore and are removed from the minus end at the spindle pole. Thus, a constant poleward flux of tubulin subunits occurs, with the microtubules remaining stationary and under tension; B—At the anaphase the tension is released, and the kinetochore moves rapidly up the microtubule, removing subunits from its plus end as it goes (B I). Its attached chromatid is thereby carried to a spindle pole. In some organisms, part of the chromatid movement is due to the simultaneous shortening of the microtubules (B II).

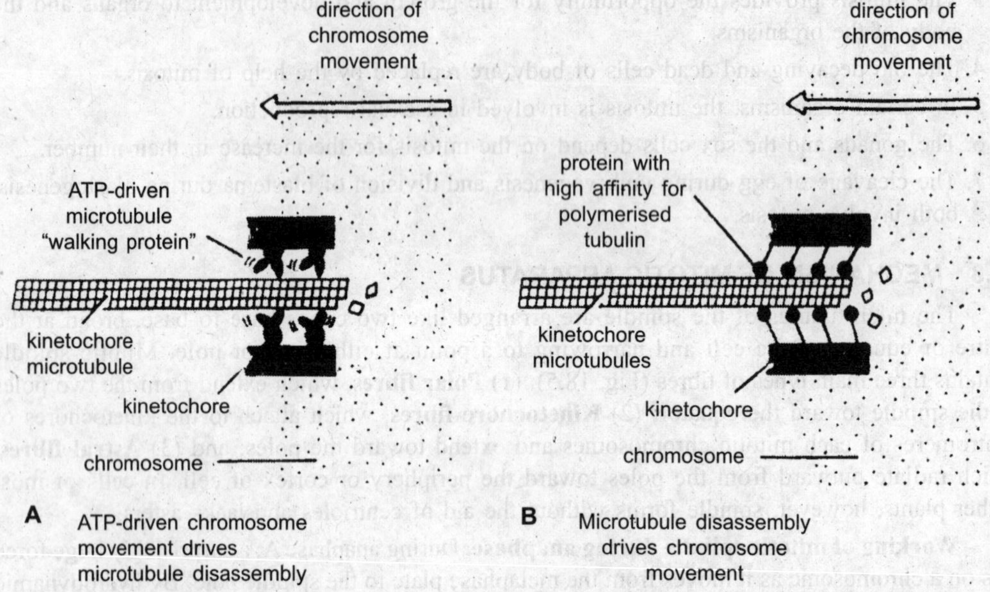

Fig. 18.6. Two alternative models to explain the generation of poleward force by the kinetochore for its chromosome during anaphase. A—Microtubule-walking proteins that resemble dynein or kinesin are part of the kinetochore, and they use the energy of ATP hydrolysis to pull the chromosome along its bound microtubules; B—Chromosome movement is driven by microtubule disassembly; as tubulin subunits dissociate, the kinetochore tends to slide poleward in order to restore its binding to the walls of the microtubule.

Further, the mechanism by which kinetochore of the chromosome moves up the spindle during anaphase A is still unknown. However, the following three models throw some light on it:

1. The kinetochore hydrolyzes ATP to move along its attached microtubule, with the plus end of the microtubule depolymerizing as it becomes exposed (Fig. 18.6A). 2. The depolymerization of the microtubule itself causes the kinetochore to move passively to optimize its binding energy on the microtubule (Fig. 18.6B). 3. A system of elastic protein filaments might connect the kinetochore to the pole and pull the kinetochore steadily poleward. Thus, in this case, microtubules merely regulate movements of the chromosomes.

In mammalian cells, anaphase B begins shortly after the chromatids have begun their voyage to the poles and stops when the spindle is about 1.5 to 2.0 times its metaphase length (15 times increase in certain protozoa). Thus, anaphase B increases the distance between two spindle poles and in contrast to anaphase A, is accompanied by the polymerization of polar microtubules at their plus ends. Further, the polar microtubules from each half-spindle overlap in a central region near the spindle equator (*e.g.*, diatoms). During anaphase B these two sets of antiparallel polar microtubules appear to slide away from each other in the region of overlap. Dynein-like force generating protein may be involved in such a directed movement of chromosomes.

18.4 REGULATION OF CELL CYCLE

Progression through the eukaryotic cell cycle is regulated by cyclin-dependent kinases (Cdks) which bind to cyclins to form Cdk-cyclin complexes (Box 18.1). At the restriction point (in yeast this point is called start), a **G1-Cdk-cyclin** catalyzes the phosphorylation of the *Rb protein* to trigger passage into S phase (Fig. 18.7). At the G_2–M boundary, a mitotic Cdk-cyclin triggers entry into mitosis by catalyzing the phosphorylation of various proteins, thereby promoting nuclear envelope breakdown, chromosome condensation and spindle formation. And at metaphase-anaphase boundary, mitotic-Cdk-cyclin contribute to activation of anaphase-promoting complex, which triggers a protein degradation pathway that initiates chromatid separation. This protein degradation pathway also targets mitotic cyclin for breakdown, leading to an inactivation of mitotic Cdk activity that in turn triggers events associated with the exit from mitosis, including chromatin decondensation and assembly of nuclear envelope.

Box 18.1

Mitosis depends upon the activation of a pre-existing protein, the **M-phase kinase**. This protein has two subunits: 1. **Cyclin dependent kinase (Cdk)** which is activated by modification at the start of M-phase and 2. **Cyclin** which is accumulated during interphase but is destroyed during mitosis.

Cdk (kinase subunit) bring about *phosphorylation*. It associates with cyclin (regulatory subunit) to form a **Cdk-cyclin complex**. This complex is also known as **cell-cycle's engine**. M-phase kinase acts as maturation promoting factor (MPF) to cause somatic cells to enter M-phase. M-phase protein (dimer) can phosphorylate a variety of protein substrates.

Rb protein is a protein whose phosphorylation controls passage through the restriction point in the cell cycle.

Lastly, cell utilises a series of **checkpoint mechanism** to monitor conditions within the cell and transiently halt the cell cycle if conditions are not suitable for continuing. The DNA replication checkpoint monitors the state of DNA replication to ensure that DNA synthesis is completed prior to permitting the cell to exit from G_2 and begin mitosis. The **p53 protein** plays a central role in a series of DNA damage check-points that halt the cell cycle at various points of DNA damage. Finally, the *anaphase promoting complex* is involved in the spindle checkpoint, which prevents anaphase chromosome movements from the beginning before the chromosomes are attached to the spindle.

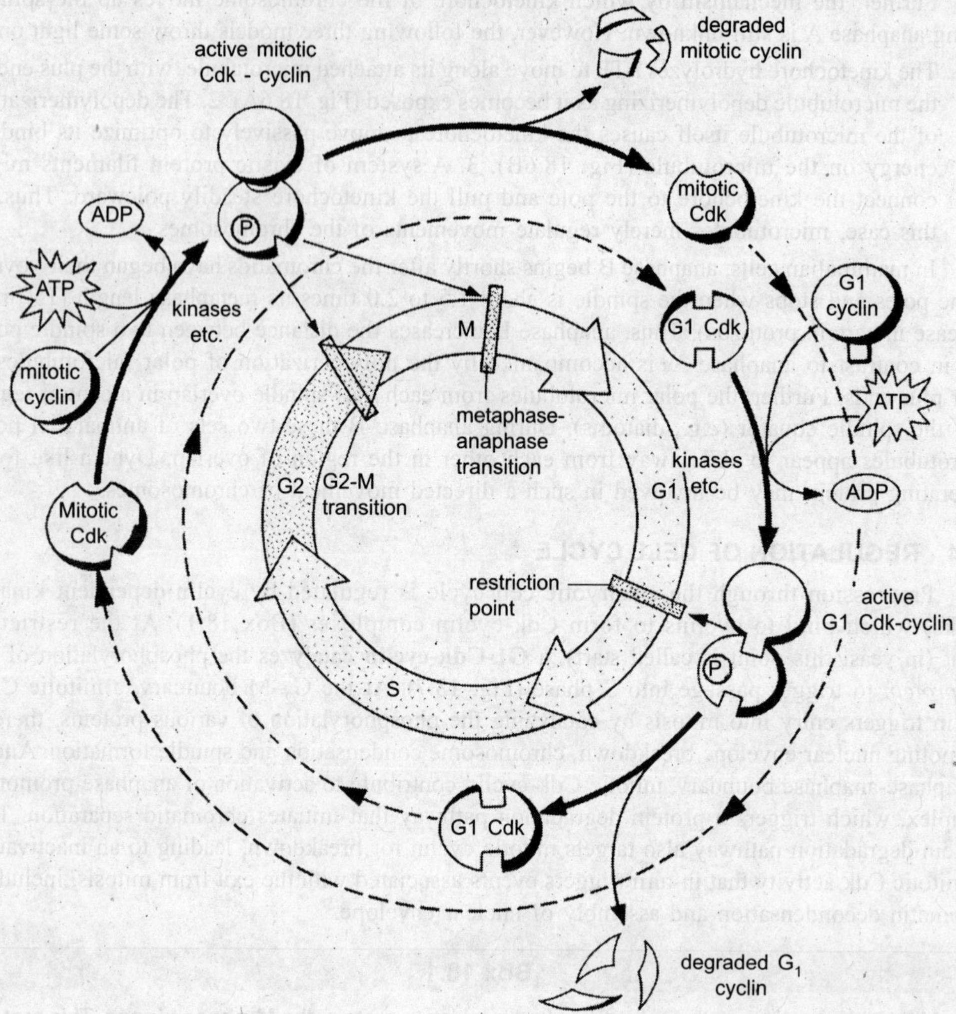

Fig. 18.7. A general model for cell cycle regulation. According to this model, passage through the three main transition points in the cell cycle is triggered by protein complexes made of cyclin and Cdk, whose phosphorylation of other proteins induces progression through the cell cycle (after Becker *et al.*, 2007).

18.5 MEIOSIS AND REPRODUCTIVE CYCLE

The term meiosis (Gr., *meioum*=to reduce or to diminish) was coined by **J.B. Farmer** in 1905. Meiosis produces a total of four haploid cells from each original diploid cell. These haploid cells either become or give rise to gametes, which through union (fertilization) support sexual reproduction and a new generation of diploid organisms. Thus, meiosis is required to run the **reproductive cycle** of eukaryotes such as microorganisms *Chlamydomonas, Neurospora*; bryophytes; plants and animals.

Kinds of Meiosis

Meiosis occurs in the germ cells of sexually reproducing organisms. In both plants and animals, germ cells are localized in the gonads. The time at which meiosis takes place varies among different organisms, and on this basis the process can be classified into: **terminal, intermediate** or **initial**.

1. Terminal meiosis. It is also called **gametic meiosis** and is found in animals and a few lower plants. In terminal meiosis, the meiotic division occurs immediately before the formation of gametes or **gametogenesis**.

2. Intermediary or **sporic meiosis.** It is the characteristic of flowering plants. This meiosis takes place at some intermediate time between fertilization and the formation of gametes. It is also involved in the production of microspores (in anthers) and megaspores (in ovary or pistil) or in **microsporogenesis** and **megasporogenesis**, respectively.

3. Initial or **zygotic meiosis.** It occurs in some algae, fungi, and diatoms. Meiotic division occurs immediately after fertilization; in this case, only the egg is diploid.

Meiocytes. The cells in which meiosis takes place are known as **meiocytes**. The meiocytes of gonads are called **gonocytes** which may be **spermatocytes** in male and **oocytes** in female. The meiocytes of the plant sporangium are called **sporocytes** (*i.e.*, **microsporocytes** and **megasporocytes**).

Process of Meiosis

Meiosis superficially resembles two mitotic divisions without an intervening period of DNA replication. The first meiotic division includes a long prophase in which the homologous chromosomes become closely associated to each other and interchange of hereditary material takes place between them. Further, in the first meiotic division the reduction of chromosome number takes place and, thus, two haploid cells are resulted by this division. The first meiotic division is also known as the **heterotypic division**. In the second meiotic division the haploid cell divides mitotically and results into four haploid cells. The second meiotic division is also known as the **homotypic division**. In the homotypic division pairing of chromosomes, exchange of the genetic material and reduction of the chromosome number do not occur.

Both the meiotic divisions occur continuously and each includes the usual stages of the meiosis, *viz.*, prophase, metaphase, anaphase and telophase. The prophase of first meiotic division is very significant phase because the most cytogenetical events such as synapsis, crossing over, etc., occur during this phase. The prophase I is the longest meiotic phase, therefore, for the sake of convenience it is divided into six substages, *viz.*, proleptonema (proleptotene), leptonema (leptotene), zygonema (zygotene), pachynema (pachytene), diplonema (diplotene) and diakinesis. The successive meiotic substages can be represented as follows:

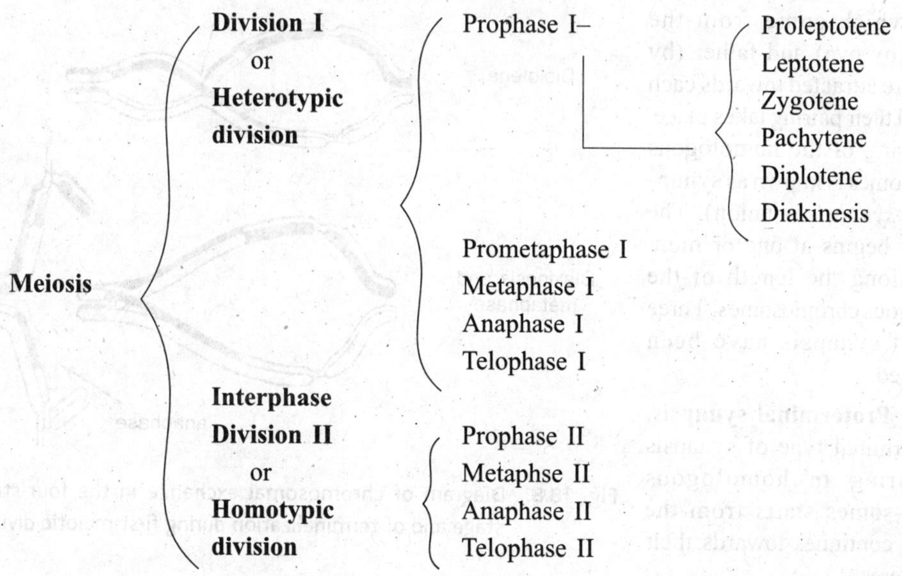

Heterotypic Division or First Meiotic Division

Meiosis starts after an **interphase** which is not very different from that of an intermitotic interphase. During the premeiotic interphase DNA duplication has occurred at the S phase. In the G_2 phase of interphase apparently there is a decisive change that directs the cell toward meiosis, instead of toward mitosis (**Stern** and **Hotta**, 1969). Further, in the beginning of the first meiotic division the nucleus of the meiocyte starts to swell up by absorbing the water from the cytoplasm and the nuclear volume increases about three folds. After these changes the cell passes to the first stage of first meiotic division which is known as prophase.

Prophase I

The first prophase is the longest stage of the meiotic division (Fig. 18.9). It includes following substages:

1. Proleptotene or **Proleptonema.** (Gr., *pro* = before; *leptas* = thin; *nema* = thread). The proleptotene stage closely resembles with the early mitotic prophase. In this stage the chromosomes are extremely thin, long, uncoiled, longitudinally single and slender thread-like structures.

2. Leptotene or **Leptonema.** In the leptotene stage the chromosomes become more uncoiled and assume a long thread-like shape. The chromosomes at this stage take up a specific orientation inside the nucleus; the ends of the chromosomes converge toward one side of the nucleus, that side where the centrosome lies (the **bouquet stage**). The centriole duplicates and each daughter centriole migrates towards the opposite poles of the cell. On reaching at the poles, each centriole duplicates and, thus, each pole of cell possesses two centrioles of a single diplosome.

3. Zygotene or **Zygonema.** (Gr., *zygon* = adjoining). In the zygotene stage, the pairing of homologous chromosomes takes place. The homologous chromosomes which come from the mother (by ova) and father (by sperm) are attracted towards each other and their pairing takes place. The pairing of the homologous chromosomes is known as **synapsis** (Gr., *synapsis* = union). The synapsis begins at one or more points along the length of the homologous chromosomes. Three types of synapsis have been recognised.

***(i)* Proterminal synapsis.** In proterminal type of synapsis the pairing in homologous chromo-somes starts from the end and continues towards their centromeres.

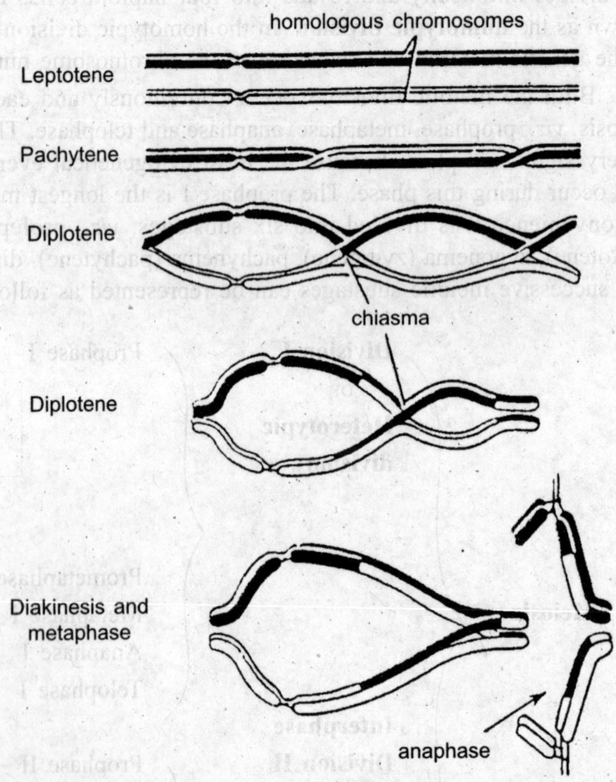

Fig. 18.8. Diagram of chromosomal exchange in the four-strand stage and of terminalization during first meiotic division.

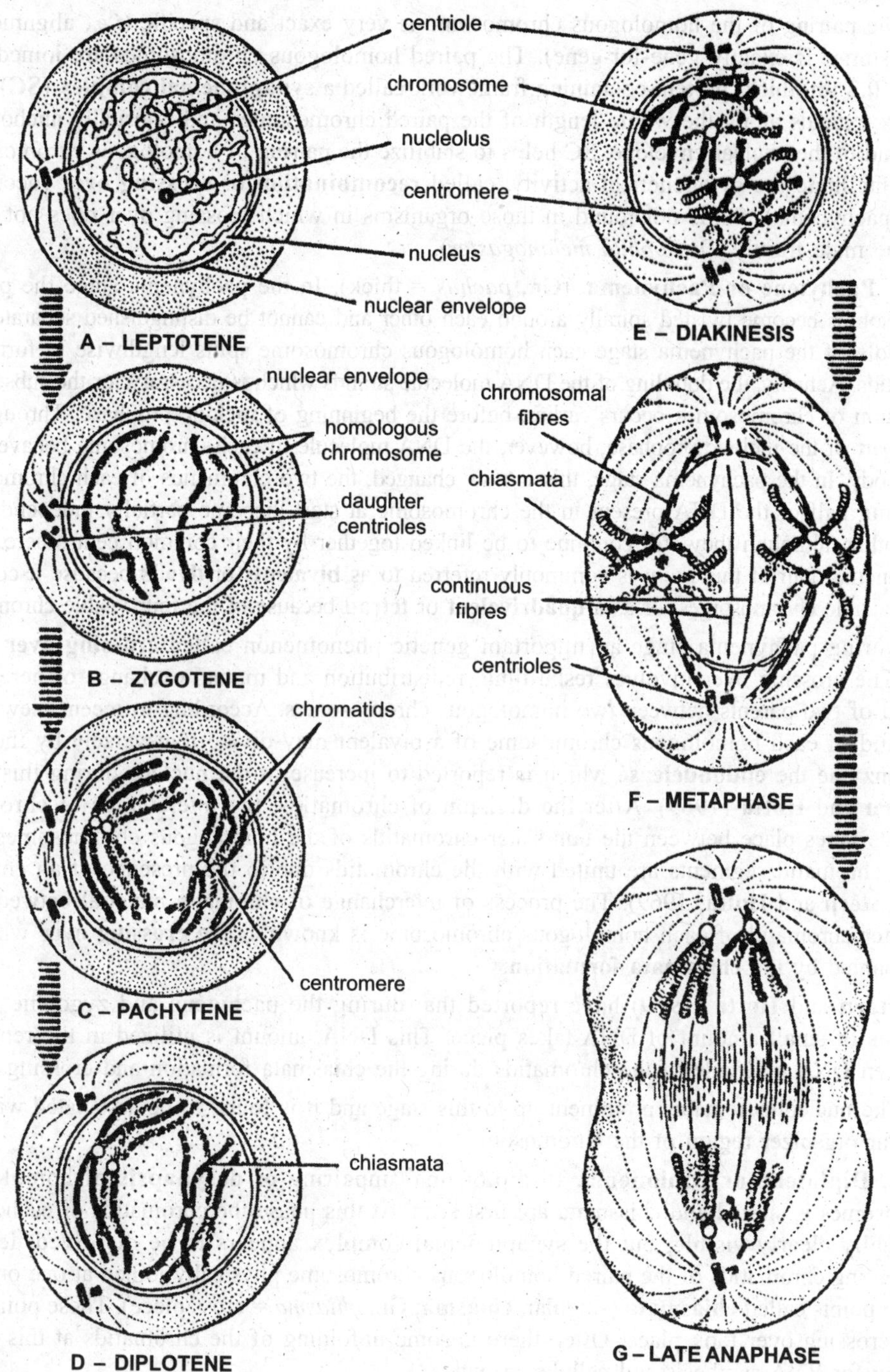

Fig. 18.9. Different stages of first meiotic division.

***(ii)* Procentric synapsis.** In procentric synapsis the homologous chromosomes start pairing from their centromeres and the pairing progresses towards the ends of the homologs.

***(iii)* Localized pairing or random synapsis.** The random type of synapsis occurs at various points of the homologous chromosomes.

The pairing of the homologous chromosome is very exact and specific (*i.e.*, alignment of chromosomes is exactly gene-for-gene). The paired homologous chromosomes are joined by a roughly 0.2 μm thick, protein-containing framework called a **synaptonemal complex (SC)**. This complex extends along the whole length of the paired chromosomes and is usually anchored at either end to the nuclear envelope. SC helps to stabilize the pairing of homologous chromosomes and to facilitate the cytogenetical activity, called **recombination** or **crossing over** (occurring during pachynema). SC is not found in those organisms in which crossing over does not occur (*e.g.*, the male fruitfly, *Drosophila melanogaster*).

4. Pachytene or Pachynema. (Gr., *pachus* = thick). In the pachynema stage the pair of chromosomes become twisted spirally around each other and cannot be distinguished separately. In the middle of the pachynema stage each homologous chromosome spilts lengthwise to form two chromatids. Actually, the doubling of the DNA molecule strands which is necessary for the subsequent duplication of chromosomes occurs earlier, before the beginning of meiotic prophase. Through the earlier part of the meiotic prophase, however, the DNA molecule in each chromosome behaves as a single body. In the pachynema stage, this is now changed, the two chromatids of each chromosome containing half of the DNA present in the chromosome at start, become partially independent of one another, although they still continue to be linked together by their common centromere. Each synaptonemal pair at this point is commonly referred to as **bivalents** or **dyads** because it consists of two visible chromosomes, or as a **quadrivalent** or **tetrad** because of the four visible chromatids.

During pachynema stage an important genetic phenomenon called **crossing over** takes place. The crossing over involves reshuffling, redistribution and mutual exchange of hereditary material of two parents between two homologous chromosomes. According to recent views, one chromatid of each homologous chromosome of a bivalent may divide transversely by the help of an enzyme the **endonuclease** which is reported to increase in the nucleus during this stage by **Stern** and **Hotta** (1969). After the division of chromatids, the interchange of chromatid segments takes place between the non-sister chromatids of the homologous chromosomes. The broken chromatid segments are united with the chromatids due to the presence of an enzyme, **ligase** (**Stern** and **Hotta**, 1969). The process of interchange of chromatin material between one non-sister chromatid of each homologous chromosome is known as the **crossing over** which is accompanied by the **chiasmata formation**.

Stern and **Hotta** (1969) have reported that during the pachytene and zygotene stage, synthesis of small amount of DNA takes place. This DNA amount is utilized in the repairing of broken DNA molecule of the chromatids during the chiasmata formation and crossing over.

The nucleolus remains prominent up to this stage and it is found to be associated with the nucleolar organizer region of the chromosome.

5. Diplotene or Diplonema. In diplonema, unpairing or **desynapsis** of homologous chromosomes is started and chiasmata are first seen. At this phase the chromatids of each tetrad are usually clearly visible, but the synaptonemal complex appears to be dissolved, leaving participating chromatids of the paired homologous chromosome physically joined at one or more discrete points called **chiasmata** (singular, **chiasma**; Gr., *chiasma* = cross piece). These points are where crossing over took place. Often there is some unfolding of the chromatids at this stage, allowing for RNA synthesis and cellular growth.

6. Diakinesis. In the diakinesis stage the bivalent chromosomes become more condensed and evenly distributed in the nucleus. The nucleolus detaches from the nucleolar organizer and ultimately disappears. The nuclear envelope breaks down. During diakinesis the chiasma moves from the centromere towards the end of the chromosomes and the intermediate chiasmata diminish. This type of movement of the chiasmata is known as **terminalization**. The chromatids still remain connected by the terminal chiasmata and these exist up to the metaphase.

Prometaphase

In the prometaphase the nuclear envelope disintegrates and the microtubules get arranged in the form of spindle in between the two centrioles which occupy the position of two opposite poles of the cell. The chromosomes become greatly coiled in the spiral manner and get arranged on the equator of the spindle.

Metaphase I

Metaphase I consists of spindle fibre attachment to chromosomes and chromosomal alignment at the equator. During metaphase I, the microtubules of the spindle are attached with the centromeres of the homologous chromosomes of each tetrad. The centromere of each chromosome is directed towards the opposite poles. The repulsive forces between the homologous chromosomes increase greatly and the chromosomes become ready to separate.

Anaphase I

At anaphase I homologous are freed from each other and due to the shortening of chromosomal fibres or microtubules each homologous chromosome with its two chromatids and undivided centromere move towards the opposite poles of the cell. The chromosomes with single or few terminal chiasma usually separate more frequently than the longer chromosomes containing many chiasmata.

The actual reduction and **disjunction** occurs at this stage. Here it should be carefully noted that the homologous chromosomes which move towards the opposite poles are the chromosomes of either paternal or maternal origin. Moreover, because during the chiasma formation out of two chromatids of a chromosome, one has changed its counterpart, therefore, the two chromatids of a chromosome do not resemble with each other in the genetical terms.

Telophase I

The arrival of a haploid set of chromosomes at each pole defines the onset of telophase I, during which nuclei are reassembled. The endoplasmic reticulum forms the nuclear envelope around the chromosomes and the chromosomes become uncoil. The nucleolus reappears and, thus, two daughter chromosomes are formed. After the karyokinesis, cytokinesis occurs and two haploid cells are formed.

Both cells pass through a short resting phase of **interkinesis**. During this phase, no DNA replication occurs, so that chromosomes at the second prophase are the same double-stranded structures that disappeared at the first telophase. In case of *Trillium* telophase I and interphase I do not occur and the anaphase I is followed by prophase II directly.

Homotypic or Second Meiotic Division

The homotypic or second meiotic division is actually the mitotic division which divides each haploid meiotic cell into two haploid cells. The second meiotic division includes following four stages.

Prophase II

In the prophase second, each centriole divides into two and, thus, two pairs of centrioles are formed. Each pair of centrioles migrates to the opposite pole. The microtubules get arranged in the form of spindle at the right angle of the spindle of first meiosis. The nuclear membrane and the nucleolus disappear. The chromosomes with two chromatids become short and thick.

Metaphase II

During metaphase II, the chromosomes get arranged on the equator of the spindle. The centromere divides into two and, thus, each chromosome produces two monads or daughter chromosomes. The microtubules of the spindle are attached with the centromere of the chromosomes.

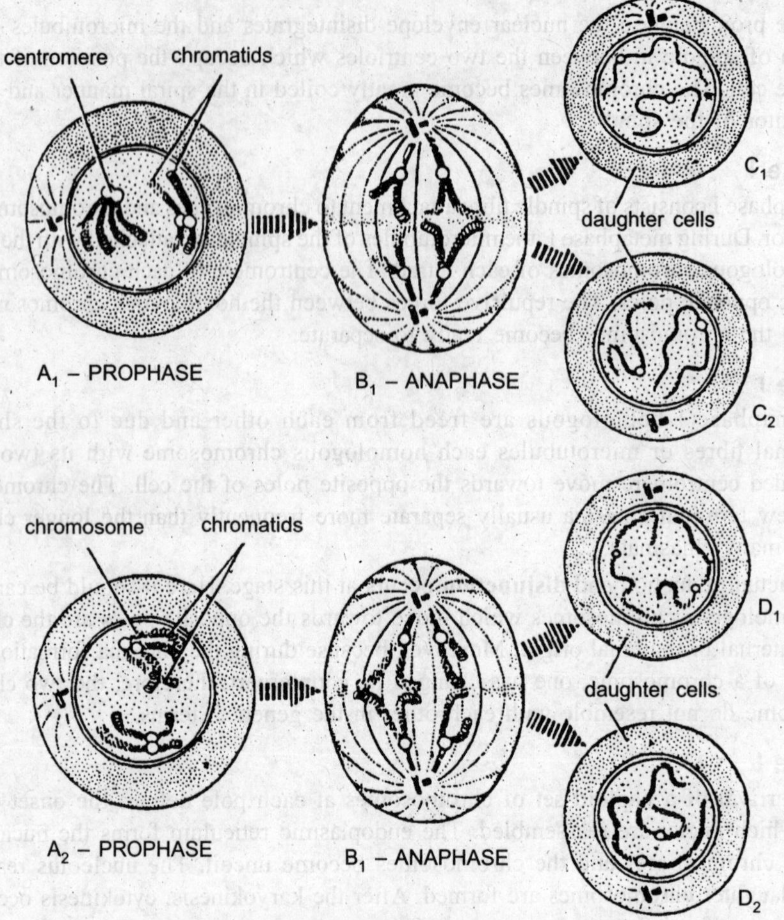

Fig. 18.10. Different stages of second meiotic division.

Anaphase II

The daughter chromosomes move towards the opposite poles due to the shortening of chromosomal microtubules and stretching of interzonal (polar) microtubules of the spindle.

Telophase II

The chromatids migrate to the opposite poles and now known as chromosomes. The endoplasmic reticulum forms the nuclear envelope around the chromosomes and the nucleolus reappears due to synthesis of ribosomal RNA (rRNA) by rDNA and also due to accumulation of ribosomal proteins.

After the karyokinesis, in each haploid meiotic cell, the cytokinesis occurs and, thus, four haploid cells are resulted. These cells have different types of chromosomes due to the crossing over in the prophase I.

Significance of Meiosis

The meiosis has the greatest significance for the biological world because of its following uses:

1. The meiosis maintains a definite and constant number of the chromosomes in the organisms.
2. By crossing over, the meiosis provides an opportunity for the exchange of the genes and, thus, causes the genetical variations among the species. The variations are the raw materials of the evolutionary process.

Thus the meiosis is a peculiar taxonomic, genetical and evolutionary process.

18.6 COMPARISON BETWEEN MITOSIS AND MEIOSIS

Table 18.3 Comparison of mitosis and meiosis.

Mitosis	*Meiosis*
1. Mitosis occurs continuously in the body or somatic cells.	1. Meiosis occurs in the germ cells (the cells of the testes or ovaries) during the process of gametogensis.
2. The whole process completes in one sequence or phase.	2. The whole process completes in two successive divisions which occur one after the other.
Prophase	
3. The prophase is of short duration and includes no substage.	3. The prophase is of longer duration and it completes in six successive stages, *viz.*, proleptotene, leptotene, zygotene, pachytene, diplotene and diakinesis.
4. The homologous chromosomes (paternal and maternal) duplicate into two chromatids. The two chromatids separate and form new chromosomes. Each daughter cell receives the daughter chromosome or chromatids of each homologous chromosome and, thus, having the chromosome number like the parental cells.	4. Out of two homologous chromosomes only one type of chromosome either maternal or paternal moves to the daughter cells. A daughter cell, thus, receives only a maternal or paternal chromosome of the homologous pair and the number of chromosomes remain half than the paternal cells.
5. No pairing or synapsis takes place between the homologous chromosomes.	5. Pairing or synapsis occurs between the homologous chromosomes.
6. Duplication of chromosomes takes place in the early prophase.	6. Duplication or splitting of chromosomes takes place in the late prophase (pachytene stage).
7. No chiasma formation or crossing over takes place.	7. Chiasma formation or crossing over takes place.
8. The exchange of the genetic material between the homologous chromosomes does not occur.	8. The exchange of the genetic material takes place between the non-sister chromatids of homologous chromosomes.
Metaphase	
9. The chromatids occur in the form of dyads.	9. The chromatids of two homologous chromosomes occurs as the tetrads.
10. The centromeres of the chromosomes remain directed towards the equator and the arms of the chromosomes remain directed towards the poles.	10. The centromeres of the chromosomes remain directed towards the poles and the chromosomal arms remain directed towards the equator.
Anaphase	
11. The chromosomes are the monads, *i.e.*, having single chromatid.	11. The chromosomes are the diads, *i.e.*, having two chromatids and single centromere.
12. The chromosomes are long and thin.	12. The chromosomes are short and thick.
Telophase	
13. The telophase always occurs.	13. The first telophase is sometimes omitted.
Significance	
14. The chromosome number in each daughter cell remains the same like the parent cell.	14. In meiotic division the chromosome number is reduced to half in the daughter cells than the parental cells.
15. A diploid cell produces two diploid cells by a mitotic division.	15. A diploid cell produces four haploid cells by a meiotic division.

QUESTIONS

Long Answer Questions

1. What do you understand by cell cycle? Give an account of the salient features of various phases of cell cycle.
2. Give an illustrated account of mitosis.
3. Give a detailed account of mitotic cell division and add a note on the significance of mitosis.
4. Describe the meiosis.
5. Describe meiosis in detail and explain its significance.
6. Describe meiotic cell division and state the significance of crossing over.
7. Describe the prophase of meiosis. Give its significance.
8. What do you mean by cell division? Describe the different stages of meiotic cell division and its significance. *(C.C.S.U., Meerut 2007)*

Short Answer Questions

1. Describe the molecular organization and functions of mitotic apparatus.
2. What are the hypotheses about the mechanism of anaphase movement of chromosomes?
3. Compare mitosis with meiosis.
4. Compare mitosis and first meiotic division with special reference to prophase.
5. Explain the difference between metaphase of mitosis and meiosis.
6. Differentiate between prophase of mitosis and meiosis. What will happen if there is no meiosis in the germinal cells?
7. Describe 'mitosis' and compare it with meiosis.
8. Write short note on the 'significance of meiosis'.
9. What are the main differences between mitosis and meiosis?
10. Differentiate between karyokinesis and cytokinesis.
11. Explain the meiotic cell division and its significance. *(C.C.S.U. Meerut 2009)*
12. Give the meiosis and its importance with diagram. *(H.N.B., Garhwal 2008)*
13. Differentiate between mitosis and meiosis cell division. *(C.C.S.U., Meerut 2008; H.N.B., Garhwal 2009)*
14. Write short notes on the following:
    (*i*) Mitotic metaphase
    (*ii*) Metaphase
    (*iii*) S-phase of cell cycle
    (*iv*) Leptotene
    (*v*) Zygotene
    (*vi*) Pachytene
    (*vii*) Diplotene
    (*viii*) Diakinesis
    (*ix*) Anaphase of meiosis
15. Draw neat and labelled diagram of the various stages of mitotic cell division (No description is required).
16. Draw neat and labelled diagram of the various stages of meiotic cell division (No description is required).

Very Short Answer Questions

1. When does synapsis occur during meiosis? *(HNB, Garhwal 2008)*
2. In which phase of meiosis chromosome number reduces to half? *(H.N.B., Garhwal 2009)*
3. In which type of cell mitosis occurs?
4. In which type of cell meiosis occurs?
5. In which phase the karyotype can be studied?
6. What is cyclin?
7. Define Cdk.

Fill in the Blanks

1. The division of cytoplasm during cell division is called ....
2. In mitotic prophase the chromatids are plectonemic but at the same phase of meiosis they are ....
3. Give one word or few words for following sentences:
   Mode of cell division involved in the formation of sperm cells in haploid male bee ....
4. Polytene chromosomes were discovered by .... *(H.N.B. Garhwal 2009)*
5. Cell cycle is divided into four phases, namely G_1, S, G_2 and M by .... and .... in 1953.
6. No DNA synthesis takes place in .... of cell cycle.
7. Mitosis is divided into four phases: .... , metaphase, ... and telophase.
8. In Anaphase A, there is .... movement of chromatids due to shortening of the .... microtubules.
9. The first sign of cytokinesis (cleavage) in animal cells is .... and .... of the plasma membrane during ....
10. The cell cycle's engine includes the complex of .... or Cdk and ....

True and False Statements

(*i*) Polytene chromosomes are in permanent interphase state.

(*ii*) Visualization of condensed chromosome during G_1 or G_2 phase of the cell cycle is possible.

(*iii*) Each chromatid is composed of single DNA duplex.

(*iv*) Crossing-over always takes place after duplication of chromonema.

(*v*) Anaphase movement is diphasic.

Multiple Choice Questions

1. Which of the following cells do not divide once it is differentiated?
   (*a*) intestine cells (*b*) nerve cells
   (*c*) blood cells (*d*) glial cells
2. Diploid somatic cells divide by
   (*a*) meiosis (*b*) mitosis only
   (*c*) both meiosis and mitosis (*d*) none of these
3. Correct sequence of stages in cell cycle is
   (*a*) G_1, S, G_2, M (*b*) G_1, G_2, S, M
   (*c*) M_1, S, G_1, G_2 (*d*) G_2, G_1, M, S
4. Condensation of chromosome with visible centromere occurs during
   (*a*) G_1 phase (*b*) G_2 phase
   (*c*) S phase (*d*) M phase

5. DNA-replication takes place during

(*a*) G_1 phase (*b*) G_2 phase
(*c*) S phase (*d*) prophase

6. Histone proteins are synthesized in

(*a*) S phase (*b*) G_1 phase
(*c*) prophase (*d*) G_2 phase

7. The term 'Meiosis' was coined by

(*a*) Knoll and Ruska (*b*) Farmer and Moore
(*c*) Beadle and Tatum (*d*) A. Flemming

8. Meiosis occurs in

(*a*) liver (*b*) kidneys
(*c*) gonads (*d*) brain

9. Pairing of homologous chromosomes occur during

(*a*) leptotene (*b*) zygotene
(*c*) pachytene (*d*) diplotene

10. Crossing-over occurs during

(*a*) leptotene (*b*) pachytene
(*c*) diakinesis (*d*) diplotene

11. Interphase is the resting stage between

(*a*) two mitotic divisions (*b*) two meiotic divisions
(*c*) mitotic and meiotic divisions

12. The structural basis for pairing of homologous chromosome during prophase I.

(*a*) kinetochore (*b*) synaptonemal complex
(*c*) chromomere (*d*) centromere

13. The spindle microtubules attach to the

(*a*) telomere (*b*) nucleosome
(*c*) kinetochore (*d*) none

14. Synapsis between homologous chromosome occurs in

(*a*) leptotene stage (*b*) pachytene
(*c*) zygotene (*d*) diplotene

15. Chromosomes become tetravalent during

(*a*) leptotene (*b*) zygotene
(*c*) pachytene (*d*) diplotene

16. The phase of cell cycle which lasts for longer duration

(*a*) G_1 (*b*) G_2
(*c*) S (*d*) M

17. Cells in Go phase of cell cycle

(*a*) exit cell cycle (*b*) enter cell cycle
(*c*) suspend cell cycle (*d*) terminate cell cycle

18. In which phase of meiosis homologous chromosome separate?

(*a*) metaphase I (*b*) metaphase II

(*c*) anaphase I (*d*) anaphase II

19. The cellular structure which always disappears during mitosis is

(*a*) mitochondria (*b*) nuclear membrane

(*c*) plastids (*d*) plasma membrane

20. Interkinesis is followed by

(*a*) prophase II (*b*) prophase I

(*c*) interphase (*d*) telophase II

ANSWERS

Very Short Answer Questions

1. Zygonema.

2. Anaphase I.

3. Somatic cell.

4. Germ cells.

5. Metaphase.

6. Cyclin is any of proteins that activate the cyclin dependent kinases (Cdks) involved in regulating progression through the eukaryotic cell cycle.

7. Cdk stands for cyclin-dependent kinase, it is any of several protein kinases that are activated by different cyclins and that control progression through the eukaryotic cell cycle by phosphorylating various proteins.

Fill in the Blanks

1. Cytokinesis. **2.** Paranemic.

3. Mitosis. **4.** E.G. Balbiani.

5. Howard, Pelc. **6.** G_1 phase.

7. Prophase, Anaphase. **8.** Poleward, kinetochore.

9. Puckering, furrowing, anaphase. **10.** Cyclin dependent kinase, cyclin.

True and False Statements

(*i*) True (*ii*) False (*iii*) True (*iv*) True

(*v*) True

Multiple Choice Questions

1. (*b*)	**2.** (*b*)	**3.** (*a*)	**4.** (*d*)	**5.** (*c*)	**6.** (*b*)
7. (*b*)	**8.** (*c*)	**9.** (*b*)	**10.** (*b*)	**11.** (*a*)	**12.** (*b*)
13. (*c*)	**14.** (*c*)	**15.** (*c*)	**16.** (*a*)	**17.** (*c*)	**18.** (*c*)
19. (*b*)	**20.** (*a*).				

19

Cell Signalling (Messengers, Receptors and Signal Transduction)

No cell lives in isolation. In eukaryotic microorganisms such as yeast, slime molds and protozoans, secreted molecules called **pheromones** coordinate the aggregation of free-living cells for sexual mating or differentiation under certain environmental conditions (see **Lodish** *et al.*, 2004). The amoeboid cells of the slime mold *Dictyostelium* secrete a compound called cyclic AMP (cAMP) at one stage in its life cycle. The compound binds to receptors on the surfaces of neighbouring cells, triggering a process whereby thousands of individual amoeboid cells aggregate and differentiate into a multicellular organism (see **Becker** *et al.*, 2007).

Factors determining the mating-types of yeast are a well understood example of pheromone-mediated cell-to-cell signalling. More important in plants and animals are extracellular **signalling molecules** that function *within* an organism to control metabolic processes within cells, the growth and differentiation of tissues, the synthesis and secretion of proteins, and the composition of intracellular and extracellular fluids. Adjacent cells often communicate by direct cell-cell contact. For example, gap junctions in the plasma membrane of adjacent cells permit them to exchange small molecules and to coordinate metabolic responses.

Extracellular signalling molecules are synthesized and released by **signalling cells** and produce a specific response only in **target cells** that have receptors for the signalling molecules. In multicellular organisms, an enormous variety of chemicals including small molecules (*e.g.,* amino acid or lipid derivatives, acetylcholine, peptides) and proteins are used in this type of cell-to-cell communication. Some signalling molecules, especially hydrophobic molecules, such as steroids, retinoids, and thyroxine, spontaneously diffuse through the plasma membrane and bind to intracellular receptors.

In this chapter we focus on signalling from a diverse group of receptor proteins located in the plasma membrane (*e.g.,* G protein-linked receptors, cytokine receptors, receptor for tyrosine kinases, TGFB receptors, etc.). The signalling molecule acts as a **ligand**, which binds to a structurally complementary site on the extracellular or membrane-spanning domains of the receptor. Binding of a ligand to its receptor causes a conformational change in the cytosolic domain (or domains) of the receptor that ultimately induces specific cellular responses. The overall process of converting signals into cellular responses, is termed **signal transduction**. As we will see, signal transduction pathways may involve relatively few or many components.

Communication by extracellular signals usually involves the following 7 step: 1. synthesis and 2. release of the signalling molecule by the signalling cell; 3. transport of the signal to the target cell; 4. binding of the signal by a specific receptor protein leading to its activation; 5. initiation of one or more intracellular signal transduction pathways by the activated receptor; 6. specific changes in cellular function, metabolism or development; and 7. removal of the signal, which often terminates the cellular response.

The vast majority of receptors are activated by binding of secreted or membrane bound molecules (*e.g.*, hormones, growth factors, neurotransmitters and pheromones). Some receptors, however, are activated by changes in the concentration of a metabolite (*e.g.*, oxygen or nutrients) or by physical stimuli (*e.g.*, light, touch, heat). In bacterium *E. coli*, for example, receptors in the cell surface membrane trigger signalling pathways that help the cell respond to changes in the external level of phosphate and other nutrients.

Signalling Pathways in *E. coli*

Bacteria (*e.g.*, *E. coli*) have two-component regulatory systems, one protein acts as a **sensor**, monitoring the level of nutrients or other components in the environment. The other protein acts as **response regulator** which in phosphorylated state binds to DNA regulatory sequences, thereby stimulating or repressing transcription of specific genes (Fig. 19.1).

For example, in *E. coli* two membrane proteins **PhoR** and **PhoB** are found to regulate transcription in response to concentration of free phosphate. PhoR is a transmembrane protein (sensor), located in the inner (*i.e.*, plasma) membrane, whose periplasmic domain binds phosphate with moderate affinity and whose cytosolic domain has protein kinase activity; PhoB (*i.e.*, response regulator) is a cytosolic protein.

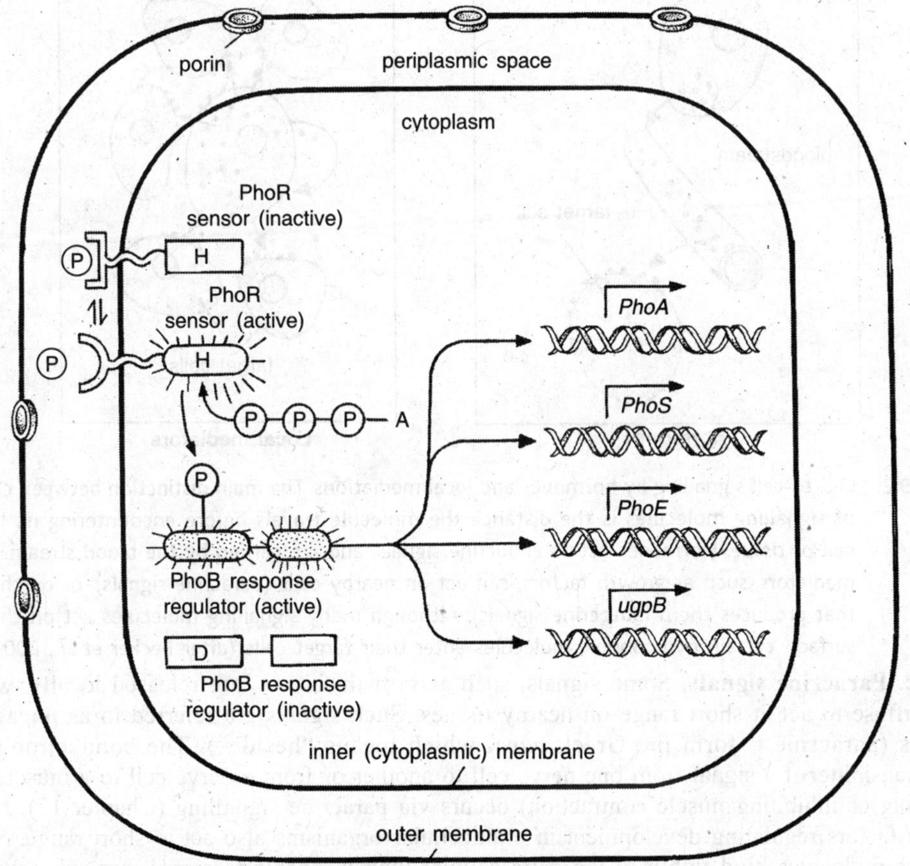

Fig. 19.1. The PhoR/PhoB two-component regulatory system in *E. coli* (see text for explanation of figure, after Lodish *et. al.*, 2004).

Large protein pores in the *E. coli* outer membrane allow ions to diffuse freely between the external environment and the periplasmic space. Consequently, when the phosphate concentration in the environment falls, it also falls in the periplasmic space, causing phosphate to dissociate from the PhoR periplasmic domain (Fig. 19.1). This causes a conformational change into the PhoR cytoplasmic domain that activates its protein kinase activity. The activated PhoR initially transfers a γ-phosphate from ATP to a histidine side chain in the PhoR kinase domain itself. The same phosphate in then transferred to a specific aspartic acid side chain in PhoB; converting it from an inactive to an active transcriptional activator. Phosphorylated, active PhoB then induces transcription from several genes (*e.g., phoA, phoS, phoE* and *ugpB*) that help the cell cope with low phosphate condition.

19.1 TYPES OF CHEMICAL SIGNALS

A variety of compounds can function as chemical messengers that relay signals between cells (Fig. 19.2). Signalling molecules are often classified based on the distance between their site of production and the target tissue(s) upon which they act.

1. Endocrine signals. These messengers are called **hormones** and they act as **endocrine signals** (endocrine = from Greek words that mean to secrete into). They are produced at great distances from their target tissues, and are carried by the circulatory system to various sites of the body.

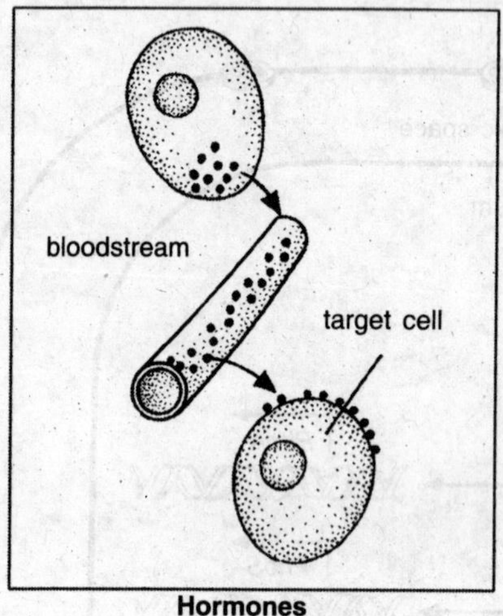

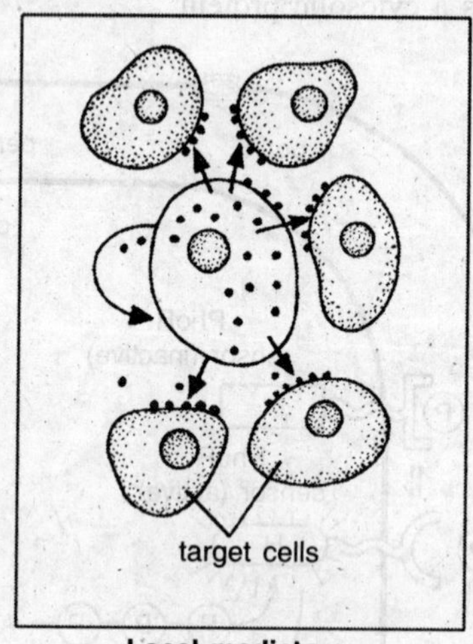

Fig. 19.2. Cell-to-cell signalling by hormones and local mediations. The main distinction between classes of signalling molecules is the distance the molecule travels before encountering its target cell or tissue. Hormones act as endocrine signals, and are carried by the blood stream. Local mediators, such as growth factors, can act on nearby cells (paracrine signals), or on the cell that produces them (autocrine signals). Although many signalling molecules act on the cell surface, certain hydrophobic molecules enter their target cells (after Becker *et al.*, 2007).

2. Paracrine signals. Some signals, such as growth factors, are released locally, where they diffuse to act at short range on nearby tissues. Such signals are referred to as **paracrine signals** (paracrine = form the Greek *para*, which means "beside"). The conduction by a neurotransmitter of a signal from one nerve cell to another or from a nerve cell to a muscle cell (inducing or inhibiting muscle contraction) occurs via paracrine signalling (Chapter 12). Many growth factors regulating development in multicellular organisms also act at short range. Some of these molecules bind tightly to the extra-cellular matrix, unable to signal, but subsequently can be released in an active form. Many developmentally important signals diffuse away from the signalling cell, forming a concentration gradient and inducing various cellular responses depending on their concentration at a particular target cell.

3. Autocrine signals. Some local mediators act on the same cell that produces them; such signals are called **autocrine signals**. Some growth factors act in this fashion. Cultured cells often secrete growth factors that stimulate their own growth and proliferation. Autocrine signalling is particularly common in tumor cells, many of which overproduce and release growth factors that stimulate inappropriate, unregulated proliferation of themselves as well as adjacent nontumor (normal) cells; this process may lead to formation of a tumor mass.

Some signalling molecules can act at both short range and long range. **Epinephrine**, for example, functions as a neurotransmitter (paracrine signalling) and as a systemic hormone (endocrine signalling). Another example is epidermal growth factor (EGF), which is synthesized as an integral plasma membrane protein. Membrane bound EGF can bind to and signal an adjacent cell by direct contact. Cleavage by an extracellular protease releases a soluble form of EGF, which can signal in either an autocrine or a paracrine manner.

Primary and Secondary Messengers

Ligands are the **first** or **primary messengers**. Once a messenger reaches its target tissue, it binds to receptors on the surface of the target cells, initiating the signalling process (Fig. 19.3). A molecule coming from either a long or a short distance functions as a ligand by binding to a receptor. A ligand often binds to a receptor embedded within the plasma membrane of the cell receiving the signal. In case of steroid hormones, the ligand binds to a receptor inside the cell.

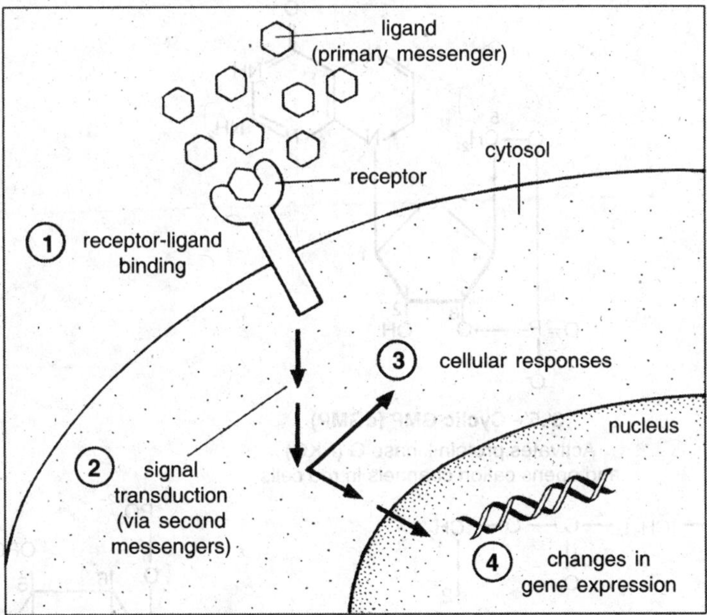

Fig. 19.3. The overall flow of information during cell signalling. Binding of ligand, the primary messenger, by a receptor activates a series of events known as signal transduction, which relays the signal to the interior of the cell, resulting in specific cellular responses and/or changes in gene expression (after Becker *et al.*, 2007).

The binding of ligand to receptor often results in a short-lived increase (or decrease) in the concentration of certain low-molecular weight intracellular signalling molecules termed **second** or **secondary messengers**. These molecules include cAMP, cGMP, DAG and IP_3 (Fig. 19.4). Other important second messengers are Ca^{2+} and various inositol phospholipids, also called **phosphoinositides**, which are embedded in cellular membranes. Second messengers relay the signals from one location in the cell, such as the plasma membrane, to the interior of the cell, initiating a cascade of changes within the receiving cell. Often these events affect the expression of specific genes within the receiving cell. The ability of a cell to translate a receptor-ligand interaction to changes in its behaviour or gene expression is the **signal transduction** pathways.

3′,5′– Cyclic AMP (cAMP)
Activates protein kinase A (PKA)

3′,5′– Cyclic GMP (cGMP)
Activates protein kinase G (PKG) and opens cation channels in rod cells

Fatty acyl groups

Glycerol

1,2-Diacylglycerol (DAG)
Activates protein kinase C (PKC)

Inositol 1,4,5-trisphosphate (IP_3)
Opens Ca^{2+} channels in the endoplasmic reticulum

Fig. 19.4. Four common intracellular second messengers. The major direct effect or effects of each compound are indicated below its structural formula (after Lodish *et al.*, 2004).

Signal transduction pathways. When a ligand (first messenger) binds to its specific (or cognate) receptor, the receptor is altered in a way that causes changes in cellular activities. In general, the binding of a ligand either induces a change in receptor conformation or causes receptors to cluster together. Once one of these changes takes place, the receptor initiates a pre-programmed sequence of events inside the cell. By *preprogammed* cell biologists mean that cells have a greater repertoire of functions than are in use at any particular time. Some of these cellular processes remain unused until particular signals are received that trigger them (see **Becker** *et al.*, 2007).

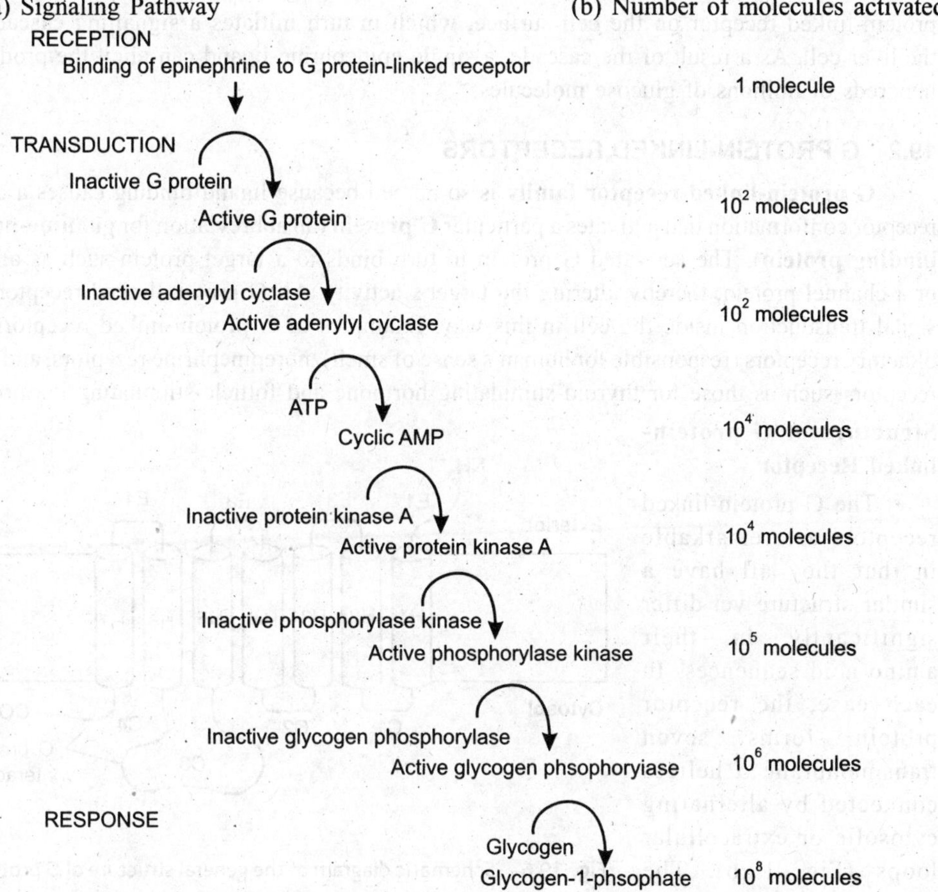

Fig. 19.5. Signal transduction pathways can amplify the cellular response to an external signal. Liver cells respond to hormone epinephrine by breaking down glycogen to liberate glucose-1-phosphate. (*a*) The epinephrine receptor is a G protein-linked receptor, which activates an enzyme known as adenylyl cyclase. Adenylyl cyclase catalyses formation of a second messenger, cAMP, which activates protein kinase (protein kinase A), which in turn activates another kinase (phosphorylase kinase). Ultimately, the enzyme glycogen phosphorylase is activated, which catalyses the breakdown of glycogen. (*b*) The approximate number of molecules produced at each step is shown on the right. One epinephrine molecule is capable of triggering the production of hundreds of millions of glucose-1-phosphate molecules (after Becker *et al.*, 2007).

Signal amplification. Signal transduction pathways allow another important aspect of a cell's response to an external signal; it is called **signal amplification**. Often exceedingly small quantities of a ligand are sufficient to elicit a response from a target cell, yet the responding cell however responds in dramatic ways. Often the strong response of the target cell results from a signalling cascade with the responding cell. At each step in the cascade, a signalling intermediate persists long enough to stimulate the production of many molecules required for the next step in the cascade, thereby multiplying the effects of a single receptor/ligand interaction on the cell surface. A well-known example of such signal amplification involves the **breakdown of glycogen** in liver cells in response to the hormone *epinephrine* (Fig. 19.5). This process involves a G protein-linked receptor on the cell surface, which in turn initiates a signalling cascade within the liver cell. As a result of the cascade, a single epinephrine ligand can elicit the production of hundreds of millions of glucose molecules.

19.2 G PROTEIN-LINKED RECEPTORS

G protein-linked receptor family is so named because ligand binding causes a change in receptor conformation that activates a particular **G protein** (an abbreviation for **guanine-nucleotide binding protein**). The activated G protein in turn binds to a target protein such as an enzyme or a channel protein, thereby altering the target's activity. All G protein-linked receptors initiate signal transduction inside the cell in this way. Examples of G protein-linked receptors include olfactory receptors (responsible for human's sense of smell), norepinephrine receptors, and hormone receptors such as those for thyroid-stimulating hormone and follicle-stimulating hormone.

Structure of G protein-linked Receptor

The G protein-linked receptors are remarkable in that they all have a similar structure yet differ significantly in their amino acid sequences. In each case, the receptor protein forms seven transmembrane α helices connected by alternating cytosolic or extracellular loops (Fig. 19.6). The N-terminus of the protein is exposed to the extracellular fluid while the C-terminus resides in the cytosol (Fig. 19.7). The extra-cellular portion of each G protein-linked receptor has a unique messenger binding site, and a cytosolic loop connecting the fifth and sixth transmembrane α helices is a specific for a particular G protein. G protein-linked receptors therefore provide a versatile method for linking different messengers to different signal transduction pathways.

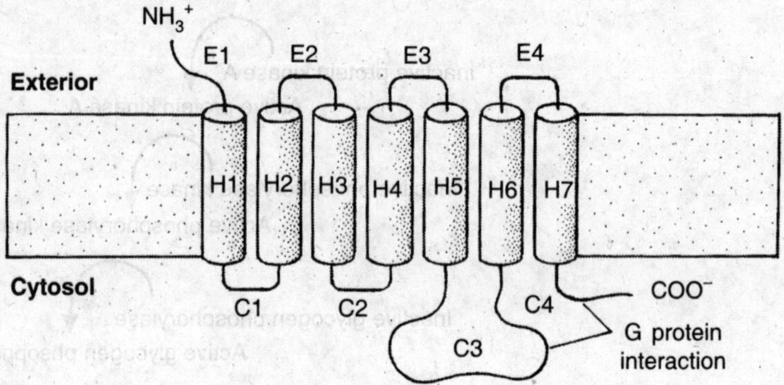

Fig. 19.6. Schematic diagram of the general structure of G protein-linked receptors. All receptors of this type have the same orientation in the membrane and contain seven transmembrane α-helical regions (H1 to H7), four extracellular segments (E1-E4), and four cytosolic segments (C1-C4). The carboxyl-terminal segment (C4), the C3 loop, and, in some receptors, also the C2 loop are involved in interactions with a linked/coupled trimeric G protein (after Lodish *et al.*, 2004).

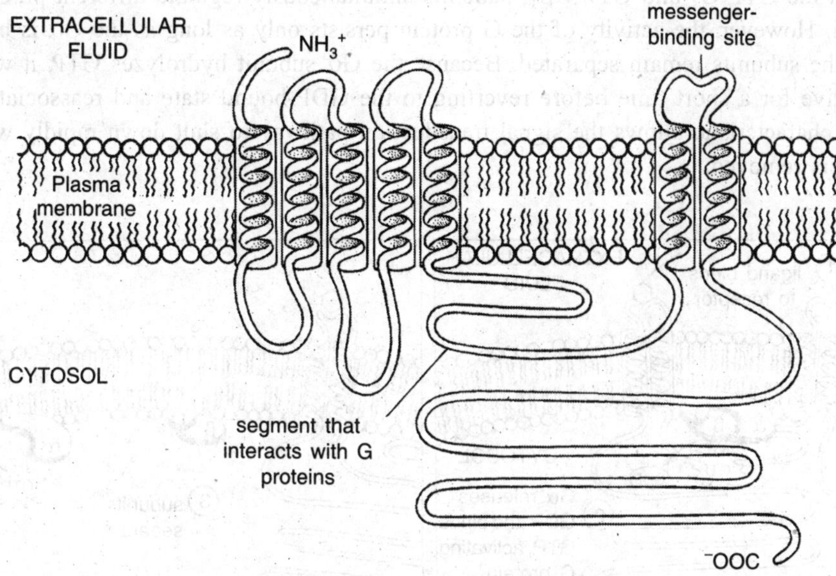

Fig. 19.7. The structure of G protein-linked receptor. Each G protein-linked receptor has seven transmembrane α helices. The primary messenger binds to the extracellular portion of the receptor. This binding causes an intracellular portion of the receptor to activate an adjacent G protein (after Becker *et al.*, 2007).

Nature of G proteins. G proteins can be described as a type of molecular switch whose "on" or "off" state depends on whether the a protein is bound to GTP or GDP. There are two distinct classes of G proteins: the **large heterotrimeric G proteins** and the **small monomeric G proteins**. The large heterotrimeric G protein contains three different subunits, called G alpha (Gα), G beta (Gβ) and G gamma (Gγ). Heterotrimeric G protein mediates signal transduction through G protein-linked receptors. The small monomeric G proteins include **Ras** protein (Box 19.1).

Box 19.1 Ras protein

The most famous member of a large family of GTP-binding proteins (called **monomeric GTPases**) that help relay signals from cell surface receptors to the nucleus. Named for the *ras* gene, first identified in viruses that cause rat sarcomas (**Alberts** *et al.*, 2002).

All G proteins have the same basic structure and mode of activation. Of the three subunits in the Gαβγ heterotrimer, Gα, the largest, binds to a guanine nucleotide (GDP or GTP). When Gα binds to GTP, it also detaches from the Gβγ complex. The Gβ and Gγ subunits, on the other hand, are permanently bound together. Some G proteins, such as **Gs,** act as stimulators of signal transduction (hence *s*, for stimulatory); others, such as G*i*, act to inhibit signal transduction (hence *i*, for inhibitory).

When a messenger (ligand) binds to a G protein-linked receptor on the surface of the cell, the change in conformation of the receptor causes the G protein to associate with the receptor, which in turn causes the *G*α subunit to release its bound GDP, acquire a GTP, and then detach from the complex (Fig. 19.8). Depending on the G-protein and the cell type, either the free GTP-Gγ subunit or the Gβγ complex can then initiate signal transduction events in the cell. Each portion of the G protein exerts its effect by binding to a particular enzyme or other protein in the cell. In some cases, both the GTP-Gγ and GTP-Gβγ, subunits simultaneously regulate different processes in the cytosol. However, the activity of the G protein persists only as long as the Gα is bound to GTP and the subunits remain separated. Because the Gα subunit hydrolyzes GTP, it will only remain active for a short time before reverting to the GDP-bound state and reassociated with Gβγ. This characteristic allows the signal transduction pathway to shut down rapidly when the messenger is removed.

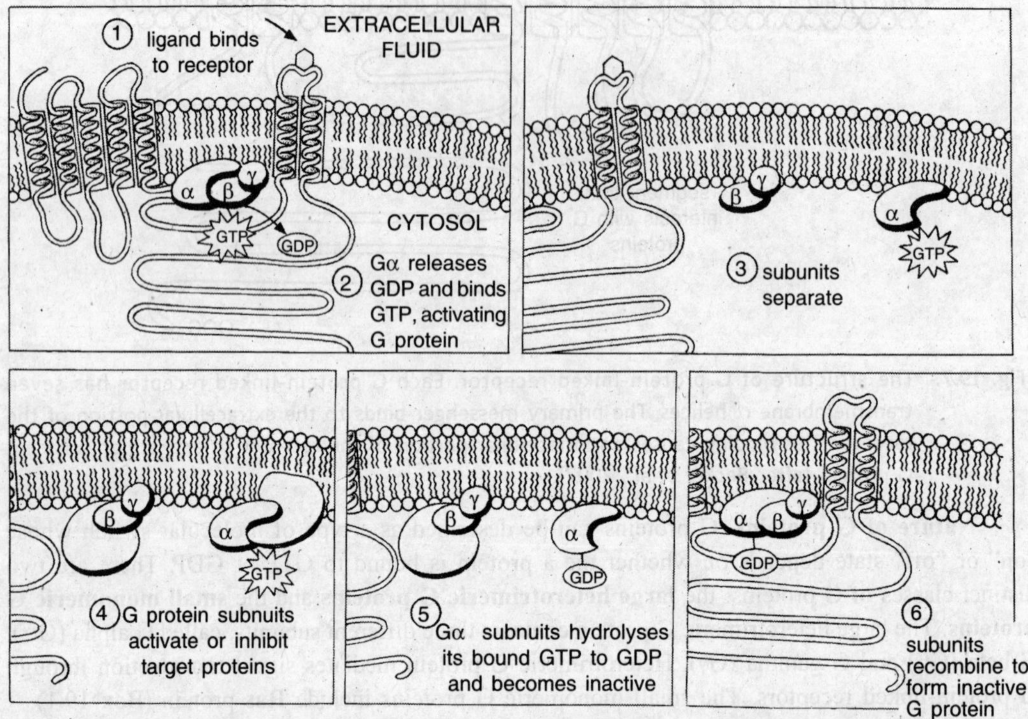

Fig. 19.8. The G protein activation and inactivation. 1. When the ligand binds, 2. The receptor activates a G protein by causing the Gα subunit to release GDP and acquire GTP, 3. The Gα and Gβγ subunits then separate and 4. Initiate signal transduction events. 5. The GTP-Gα subunit eventually hydrolyzes its bound GTP, converting the subunit back to its inactive GTP-Gα form. 6. The inactive GTP-Gα subunit then recombines with Gβγ to form the inactive heterotrimeric G protein (after Becker *et al.*, 2007).

Role of Cyclic AMP in Signal Transduction by G protein.

The second messenger, the cyclic AMP (cAMP) is formed from cytosolic ATP by the enzyme **adenylyl cyclase** (Fig. 19.9) anchored in the plasma membrane with its catalytic portion protruding into cytosol. Normally, the enzyme is inactive until it binds to the activated α subunit of a specific G protein, Gs. When the G protein-linked receptor is coupled to Gs, the binding of ligand stimulates the Gsα subunit to release GDP and acquires GTP (Fig. 19.10). This in turn causes GTP-Gsα to detach from the Gsβγ subunits and bind to adenylyl cyclase, the enzyme becomes active and converts ATP to cAMP.

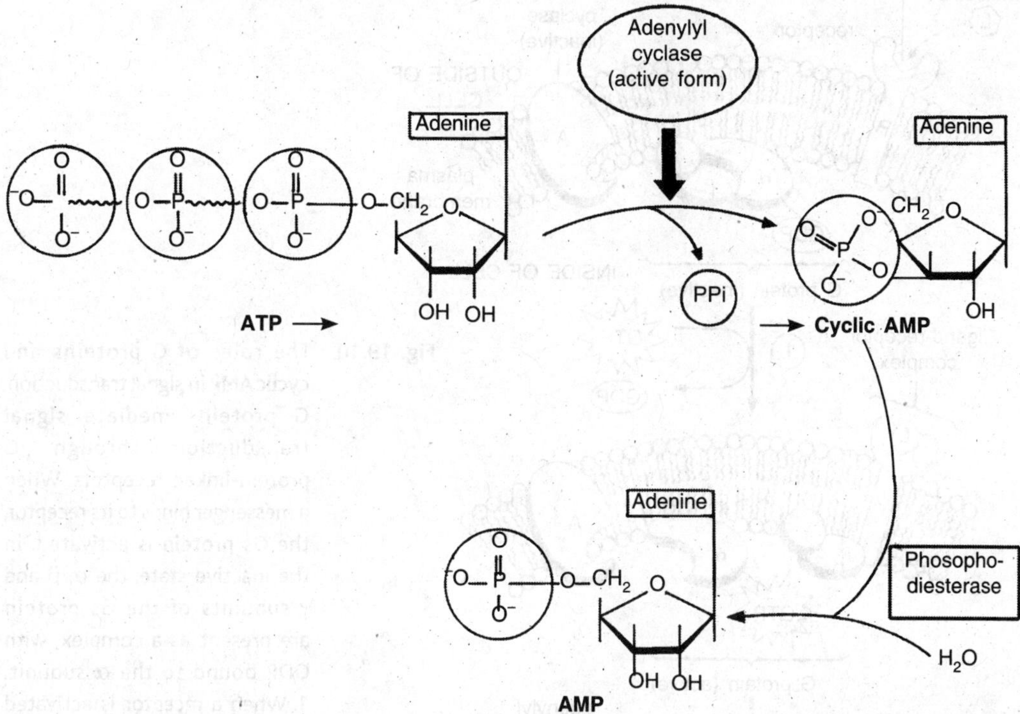

Fig. 19.9. The structure and metabolism of cAMP. Cyclic AMP (adenosine-3′, 5′-cyclic monophosphate) is generated from ATP in a reaction catalyzed by the active form of the enzyme adenylyl cyclase; it is inactivated by hydrolysis to AMP, a reaction catalyzed by the enzyme phosphodiesterase. Adenylyl cyclase is a membrane-bound enzyme, whereas phosphodiesterase is located in the cytosol (after Becker *et al.*, 2007).

The G proteins respond quickly to changes in ligand concentration because they remain active for only a short period of time before the Gα subunit hydrolyzes its bound GTP and converts to the inactive state. Once the G protein becomes inactive, the adenylyl cyclase cease to make cAMP. However, cAMP levels would still remain elevated in the cell if not for the enzyme **phosphodiesterase**, which degrades cAMP. This process further ensures that the signal transduction pathway will shut down promptly when the concentration of the ligand outside the cell falls.

Cyclic AMP has one main intracellular target, an enzyme known as cAMP-dependent kinase, or **protein kinase A (PKA)**. PKA enzyme phosphorylates a wide variety of cellular proteins by transferring phosphate from ATP to a serine or threonine found within the target protein. cAMP tends to regulate the activity of PKA.

An increase in cAMP concentration can produce many different effects in different cell types. 1. When cAMP is elevated in skeletal and liver cells, the breakdown of glycogen is stimulated. 2. In cardiac muscles, the elevation of cAMP strengthens heart contraction; 3. whereas in smooth muscle contraction is inhibited. 4. In blood platelets, the elevation of cAMP inhibits their mobilization during blood clotting, and 5. In intestinal epithelial cells, it causes the secretion of salts and water into the lumen of the gut. Each of these reactions is an example of the preprogrammed response discussed earlier.

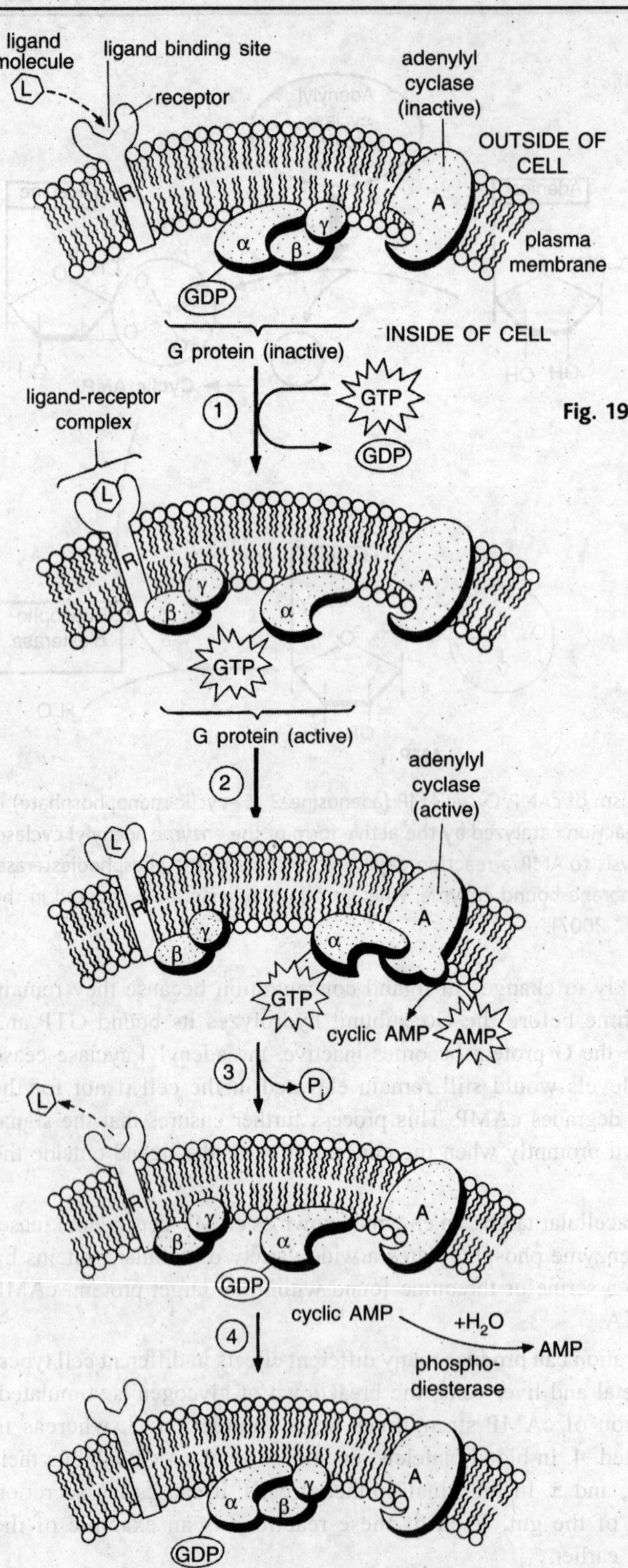

Fig. 19.10. The roles of G proteins and cyclic AMP in signal transduction. G proteins mediate signal transduction through G protein-linked receptors. When a messenger binds to its receptor, the Gs protein is activated. In the inactive state, the α, β and γ subunits of the Gs protein are present as a complex, with GDP bound to the α subunit. 1. When a receptor is activated by binding of its specific ligand on the outer surface of the plasma membrane, the receptor-messenger complex associates with Gs protein, causing the displacement of GDP by GTP and the dissociation of Gsα-GTP complex. 2. The GTP-Gsα complex then binds tightly to a molecule of membrane-bound adenylyl cyclase, activating it for synthesis of cAMP. 3. Activation ends when the ligand leaves the receptor, the GTP is hydrolyzed to GDP by the GTPase activity of the Gsα subunit, and the Gsα dissociates from the adenylyl cyclase. 4. Adenylyl cyclase then reverts to the inactive form, the Gsα reassociates with the Gsβγ complex, and cAMP molecules in the cytosol are hydrolyzed to AMP by enzyme phosphodiesterase (after Becker *et al.*, 2007).

Box 19.2

The intracellular level of cAMP can be raised without the ligands, for example, the enzyme phosphodiesterase is inhibited by **methylxanthine**, compounds such as *caffeine* and *theophylline*, found in coffee, tea and soft drinks. Theophylline is often used to treat **asthma** because it relaxes bronchial smooth muscles (see **Becker** *et al.*, 2007).

What would happen if the G protein-adenylyl cyclase system could not be shut off? Two human diseases caused by pathogenic bacteria are found to be due to certain defects in the heterotrimeric G protein (Box 19.3).

Box 19.3 Pathogenic toxins as cell biological tools

Both G*s* and G*i* proteins are targets for some medically important bacterial toxins. **Cholera toxin**, which is produced by the bacterium that causes cholera (*Vibrio cholere*) is an enzyme that catalyses the transfer of ADP ribose from intracellular NAD^+ to the α subunit of G*s* (*i.e.*, G*s*α). This ADP ribosylation alters the α subunit so than it can no longer hydrolyse its bound GTP, causing it to remain in an active state that stimulates adenylyl cyclase indefinitely. The resulting prolonged elevation in cAMP level within intestinal epithelial cells causes a large efflux of Cl^- and water into the gut, thereby causing the severe diarrhoea that characterizes cholera.

Pertussis toxin, which is made by the bacterium (*Bordetella pertussis*) that causes pertussis (whooping cough), catalyses the ADP ribosylation of the α subunit of G*i* (*i.e.*, G*i*α), preventing the subunit from interacting with receptors; as a result, this α subunit retains its bound GDP and is unable to regulate its target proteins. These two toxins are widely used as tools to determine whether a cell's response to a signal is mediated by the G*s* or by G*i* (see **Alberts** *et al.*, 2002).

ROLE OF IP_3 AND DAG AS SECOND MESSENGERS

The importance of inositol phospholipids in cell signalling was first brought to light in the pioneering studies of **Robert Michell** and **Michael Berridge**. In the early 1980s, **Berridge** noticed that when the salivary glands of certain insect larvae were stimulated to secrete, changes occurred in membrane inositol phospholipids. It is now known that **inositol 1, 4, 5-triphosphate (IP_3)**, one of the breakdown products of inositol phospholipids functions as a second messenger. IP_3 is generated from **phosphatidylinositol-4, 5-biphosphate (PIP_2)**, a relatively uncommon membrane phospholipids, when the enzyme **phospholipase C** is activated. Phospholipase C cleaves PIP_2 into two molecules: inositol triphosphate (IP_3) and **diacylglycerol (DAG)**. Later on, IP_3 and DAG were shown to be *second messengers* in a variety of regulated cell functions (Table 19.1).

Table 19.1 Examples of cell functions regulated by inositol triphosphates (IP_3) and diacylglycerol (DAG) (source Becker *et al.*, 2007).

	Regulated function	***Target tissue***	***Messenger***
1.	Platelet activation	Blood platelets	Thrombin
2.	Muscle contraction	Smooth muscle	Acetylcholine
3.	Insulin secretion	Pancreas endocrine	Acetylcholine
4.	Amylase secretion	Pancreas exocrine	Acetylcholine
5.	Glycogen degradation	Liver	Antidiuretic hormone
6.	Antibody production	B lymphocytes	Foreign antigens

The functions of IP_3 and DAG as second messengers in the inositol-phospholipid-calcium pathway are shown in Fig. 19.11. The sequence begins with the binding of a ligand to its membrane receptor, leading to the activation of a specific G protein, called Gp (the "p" stands for activation of phospholipase C). Gp then activates a form of phospholipase C known as Gβ, thereby generating both IP_3 and DAG.

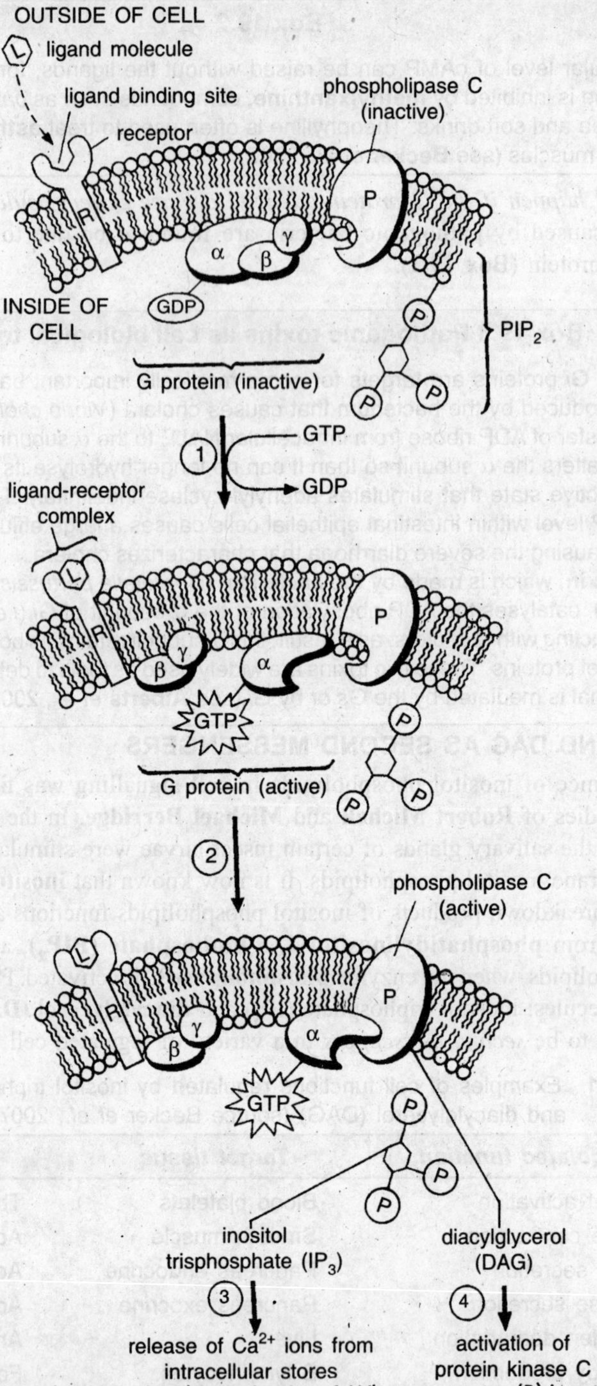

Fig. 19.11. The role of IP_3 and DAG in signal transduction. 1. When a receptor (R) is activated by the binding of its ligand (L) on the outer surface of the plasma membrane, the receptor-ligand complex associates with G protein Gp, causing the displacement of GDP by GTP and the dissociation of the GTP-Gα complex. 2. The GTP-Gα complex then binds to phospholipase (P), activating it and causing cleavage of PIP_2 into one molecule each of IP_3 and DAG. 3. The IP_3 is released into the cytosol, where it triggers the release of calcium. 4. The DAG remains in the membrane, where it activates protein kinase C (after Becker *et al.*, 2007).

IP_3 is water-soluble, so quickly diffuses through the cytosol, binding to a ligand-gated calcium channel known as the **IP_3 receptor channel** in the endoplasmic reticulum. When IP_3 binds, the channel opens, releasing calcium ions into the cytosol. Calcium then binds to a protein known as **calmodulin**, and the calcium-calmodulin complex activates the desired physiological process.

The DAG remains in the membrane, where it activates members of the **protein kinase C (PKC)** family of enzymes. PKC can then phosphorylates specific serine and threonine groups on a variety of target proteins, depending on the cell type. A variety of cellular effects have been linked to the activation of protein kinase C, including 1. The stimulation of cell growth; 2. The regulation of ion channels; 3. Changes in the cytoskeleton; 4. Increase in cellular pH, and 5. Effects on secretion of proteins and other substances.

Regulation of Calcium Ion Concentration in Cytosol

The calcium ions (Ca^{2+}) play an essential role in regulating a variety of cellular functions. An overview of the various mechanisms of calcium is normally maintained at very low levels in the cytosol because of the presence of **calcium ATPase** in the plasma membrane and the endoplasmic reticulum. Calcium ATPases in the plasma membrane transport calcium out of the cell, where as the calcium ATPase in the ER sequester (Latin for "commit for safekeeping") calcium ions in the lumen of ER. In addition, some cells have sodium-calcium exchangers that further reduce the cytosolic calcium concentration. Finally, mitochondria can transport calcium into the mitochondrial matrix. For most cells in their resting state, the action of calcium ATPase maintains the concentration of the calcium in cytosol at about 1×10^{-4} m*M*.

Cytosolic calcium concentrations can be increased by various methods. One way is by opening of calcium channels in the plasma membrane of neurons. The calcium concentration in the extracellular fluid and the blood is about 1.2 mM, more than 10,000 times as high as that of the cytosol. As a result, when calcium channels open, calcium ions rush into the cell.

The levels of calcium ions in the cytosol can also be elevated by the release of calcium from intracellular stores. Calcium ions sequestered in the ER can be released through the IP_3 receptor channel, discussed earlier, and through the **ryanodine receptor channel**. The ryanodine receptor channel, so named because it is sensitive to the plant alkaloid **ryanodine**, is particularly important for calcium release from the sarcoplasmic reticulum of cardiac and skeletal muscle, but nonmuscle cells such as neurons also have ryanodine calcium release channels.

Calcium induced calcium release. Quite interestingly, both the ryanodine receptor channel and IP_3 receptors are sensitive to calcium itself. When a neuron is depolarized, for example, calcium channels in the plasma membrane open and allow some calcium to enter the cytosol. Upon exposure to a rapid increase in calcium ions, the ryanodine receptors channel opens, allowing calcium to escape from the ER into the cytosol.

Function of calmodulin. Calcium can bind directly to some proteins and thereby alter their activity, but more often it works through the protein **calmodulin**. This protein has pivotal function in the regulation of a variety of cellular processes.

How does calmodulin mediate calcium activated events in the cell? The calmodulin molecule has been compared to a flexible "arm" with a "hand" at each end. Two calcium ions bind at each of two "hand" regions, causing the calmodulin to undergo a change in shape that form the active **calcium-calmodulin complex** (Fig. 19.12). When a protein is present that contains a calmodulin-binding site, the hands and arm bind to it by wrapping around the binding site (Fig. 19.12).

Calmodulin protein is found to have a remarkable affinity for calcium: calmodulin binds to calcium when the cytosolic calcium concentration increases to about 10^{-3} m*M*, but it releases calcium when cytosolic calcium levels decline back to the resting level of 10^{-4} m*M*. Thus calmodulin protein is uniquely suited to operate within the typical range of cytosolic calcium concentrations.

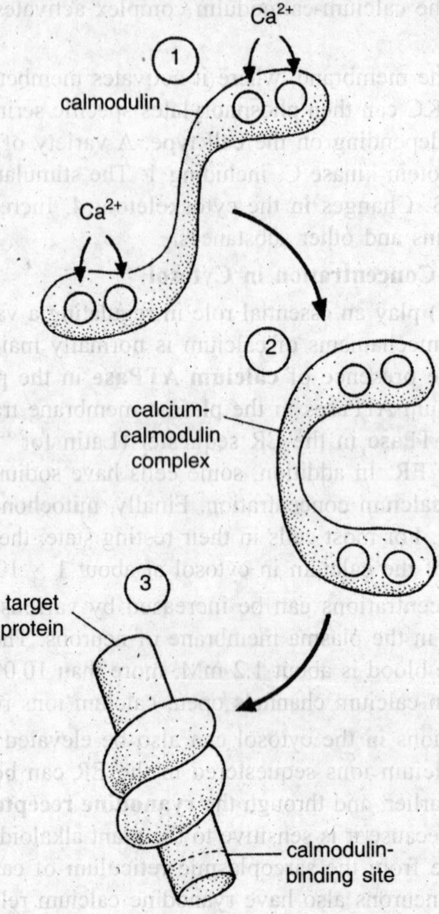

Fig. 19.12. The structure and function of the calcium-calmodulin complex. A—Calmodulin is a cytosolic calcium-binding protein. This model of molecular structure of calmodulin is based on data from X-ray crystallography. The molecule consists of two globular ends ("hands") joined by a helical region ("arm"). Each end has two calcium binding sites. B—To form a functional calcium-calmodulin complex, 1. Calmodulin binds four calcium ions. 2. Calmodulin changes conformation, resulting in an active complex. 3. The two globular "hands" of the complex wrap around a calmodulin binding site on a target protein (after Becker *et al.*, 2007).

Most calmodulin binding proteins are enzymes such as *protein kinases* and *protein phosphatases*. The response of a target cell to an increase in calcium concentration depends on the particular calmodulin-binding proteins that are present in the cell. *This means that the same change in calcium concentration can produce markedly different effects in two target cells if each possesses different calmodulin-sensitive enzyme system.*

Function of calcium in slow block to polyspermy. The process of fertilization of animal eggs (*i.e.*, ova) is an outstanding example of the importance of calcium-mediated signal transduction following a receptor-ligand interaction. In many animals, when sperm that have undergone prior steps of activation bind to the surface of mature eggs and unite with them at fertilization, a striking sequence of events arises. One of the early responses of the egg-within 30 seconds to several minutes after fertilization is the release of calcium from internal stores. Calcium release occurs initially at the site where the sperm penetrates the egg surface, and then spreads across the egg, much as a ripple on the surface of a pond spreads away from the site where a pebble strikes the water. The wavelike propagation of calcium release can be visualized using calcium indicators (*e.g.*, **aequorin**, which is a dye that when injected in fertilizing fish egg, emits light when bound to free Ca^{2+} ions in the cytosol).

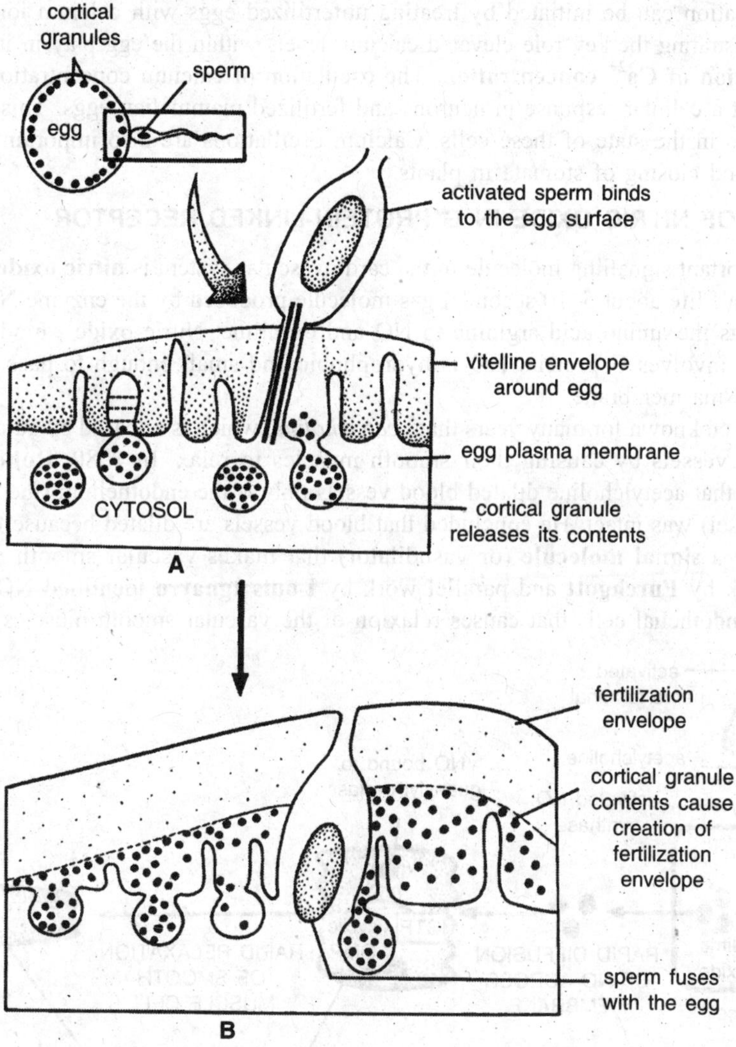

Fig. 19.13. The role of calcium in the slow block to polyspermy in sea urchins. A—A sperm cell that has been activated binds to the egg surface, resulting in local calcium release and exocytosis of cortical granule. B—The result is the creation of the fertilization envelope, which prevents additional sperm from penetrating the egg (after Becker *et al.*, 2007).

Release of calcium is necessary for two crucial events. First, it stimulates the fusion of vesicles, known as **cortical granules**, with the egg plasma membrane, resulting in the release of granule contents outside the cell (exocytosis; Fig. 19.13). Cortical granules contain several proteins and enzymes, the release of which results in alterations of the protein coat surrounding most eggs (typically known as **vitelline envelope**). These changes make the egg unable to bind additional sperm thereby preventing more than one sperm from fertilizing the egg. This process is known as **slow block to polyspermy** (An earlier process called **fast block to polyspermy** involves a transient depolarization of the egg plasma membrane).

The second major function of calcium is **egg activation**. Egg activation involves the restart of many metabolic processes, the reorganization of the internal contents of the egg, and other events that initiate the process of embryonic development. Many features of the slow block to polyspermy and egg activation can be initiated by treating unfertilized eggs with calcium ionophore in the sperm, demonstrating the key role elevated calcium levels within the egg play in its activation.

Oscillation of Ca^{2+} concentration. The oscillation of calcium concentration over time is found to elicit a cellular response in neurons and fertilized mammalian eggs. This contributes to stable changes in the state of these cells. Calcium oscillations are also important in regulating the opening and closing of stomata in plants.

FUNCTION OF NITRIC OXIDE IN G PROTEIN-LINKED RECEPTOR

An important signalling molecule in the cardiovascular system is **nitric oxide (NO)**, a toxic short-lived (half life about 5–10 seconds) gas molecule produced by the enzyme **NO synthetase**, which converts the amino acid arginine to NO and citrulline. Nitric oxide's mode of action on blood vessels involves G protein. NO is hydrophobic and small enough to pass readily across target cell plasma membrane.

It has been known for many years that *acetylcholine,* which is released by autonomic nerves, dilates blood vessels by causing their smooth muscles to relax. In 1980, **Robert Furchgott** demonstrated that acetylcholine dilated blood vessels only if the endothelium (the inner lining of the blood vessel) was intact. He concluded that blood vessels are dilated because the endothelial cells produce a **signal molecule** (or vasodilator) that makes vascular smooth muscles relax. In 1986, work by **Furchgott** and parallel work by **Louis Ignarro** identified NO as the signal released by endothelial cells that causes relaxion of the vascular smooth muscles (Fig. 19.14).

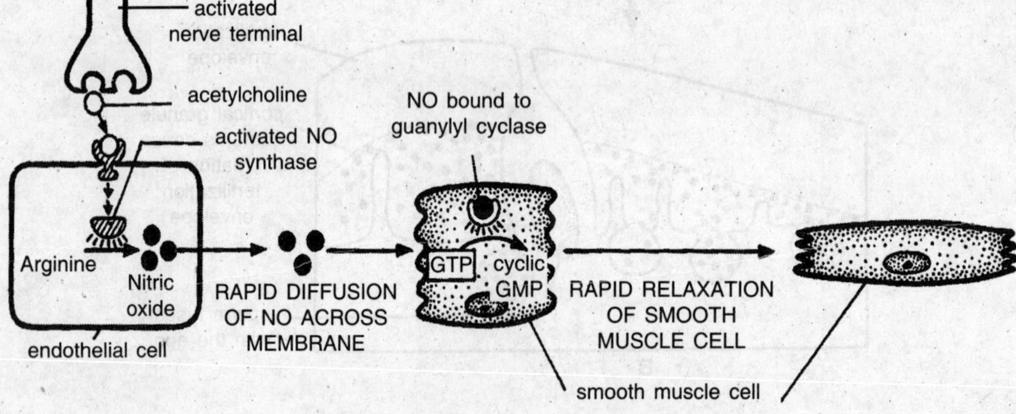

Fig. 19.14. The role of nitric oxide (NO) in smooth muscle relaxation in a blood vessel wall. Acetylcholine released by nerve terminals in the blood vessel wall activates enzyme NO synthetase in endothelial cells, lining the blood vessel, causing the endothelial cells to produce NO. The NO gas diffuses out of the endothelial cells and into the underlying smooth muscle cells, where it binds to and activates guanylyl cyclase to produce cyclic GMP. The cyclic GMP triggers a response that causes the smooth muscle cells to relax, enhancing blood flow through the blood vessel (after Alberts *et al.*, 2002).

Mode of action of NO on smooth muscle cells. In Fig. 19.15, it has been shown that how the binding of acetylcholine to the surface of vascular endothelial cells result in release of NO. This process involves the following six steps: 1. Acetylcholine binds to G protein-linked receptors that activate the phosphoinositide signalling pathway, causing IP_3 to be produced by the endothelial cells. 2. IP_3 causes the release of calcium from the endoplasmic reticulum. 3. The calcium ions bind to calmodulin, forming a complex that stimulates NO synthetase enzyme to produce nitric oxide. 4. Nitric oxide is a gas that readily diffuses through plasma membranes, allowing it to pass from the endothelial cell into the adjacent smooth muscle cell. 5. Once inside the smooth muscle cell, NO activates the enzyme **guanylyl cyclase**, which catalyzes the formation of cyclic GMP (cGMP). Cyclic GMP is derived from GTP in a manner similar to the production of cAMP from ATP, and like cAMP, cGMP concentration activates a protein known as **protein kinase G**, which induces muscle relaxation by catalyzing the phosphorylation of the appropriate muscle proteins.

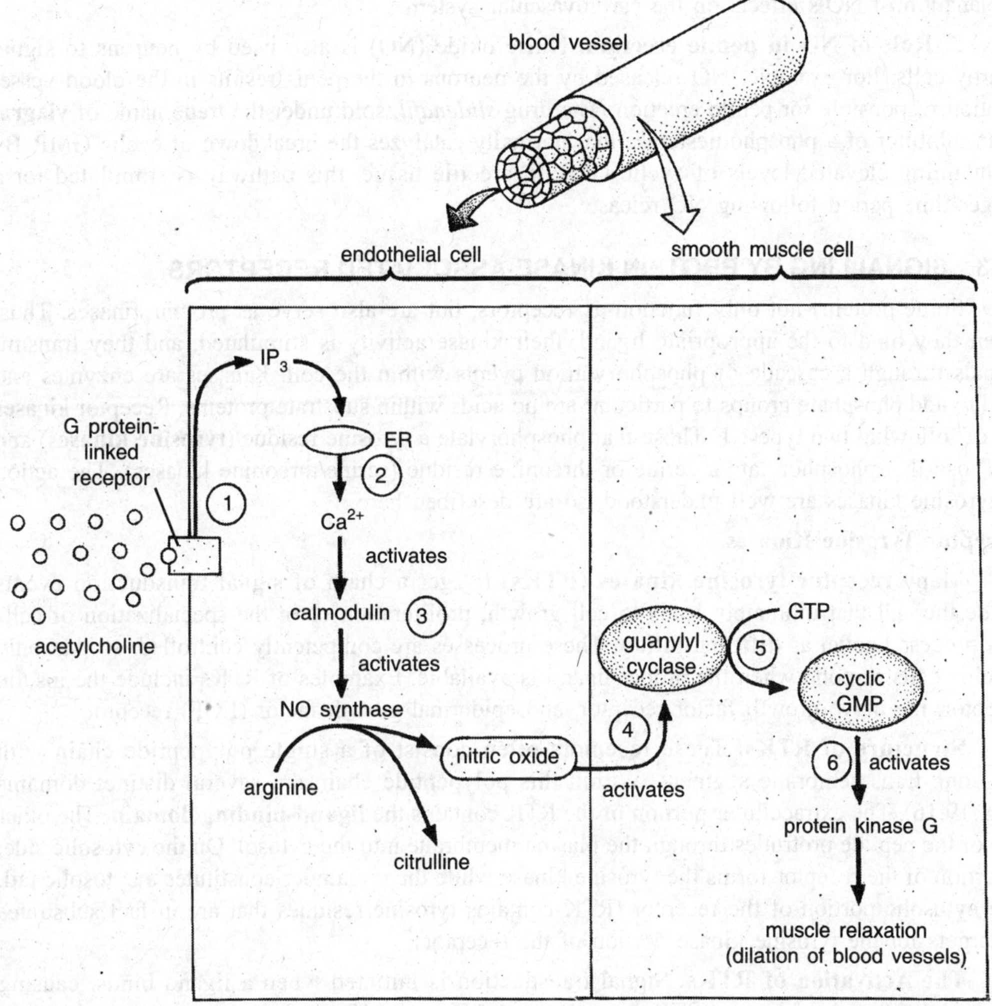

Fig. 19.15. The action of nitric oxide on blood vessels. The binding of acetylcholine to endothelial cells triggers the production of nitric oxide (NO), which diffuses into the adjacent smooth muscle cells and stimulates guanylyl cyclase, thereby leading to muscle relaxation (after Becker *et al.*, 2007).

Box 19.4

Gases such as NO (nitric oxide) and CO (carbon monoxide) are paracrine signaling molecules. They are used as intercellular signals in mammals and certain plants. For example, activated macrophages and neutrophils use NO in killing invading microorganisms. In plants, NO is involved in defensive responses to injury or infection.

Use of NO in treatment of some human ailments.

1. **Role of NO in treatment of angina.** The mechanism by which acetylcholine stimulation of the endothelial cells leads to smooth muscle relaxation also explains the mechanism of action of **nitroglycerin**. Nitroglycerin is often taken by patients with **angina** (chest pain due to inadequate blood flow to the heart) to relieve constriction of coronary arteries. In 1977, **Ferid Murad** found that nitroglycerin and similar vasodilators elicit release of nitric oxide (NO), which relaxes arterial smooth muscle cells. In 1998, **Furchogott, Ignarro** and **Murad** received a Nobel Prize for their explanation of NO's effects on the cardiovascular system.

2. **Role of NO in penile erection.** Nitric oxide (NO) is also used by neurons to signal nearby cells. For example, NO released by the neurons in the penis results in the blood vessel dilation responsible for penile erection. The drug *slidenafil,* sold under the trade name of **viagra**, is an inhibitor of a phosphodiesterase that normally catalyzes the breakdown of cyclic GMP. By maintaining elevated levels of cyclic GMP in erectile tissue, this pathway is stimulated for a longer time period following NO release.

19.3 SIGNALLING BY PROTEIN KINASE-ASSOCIATED RECEPTORS

Some proteins not only function as receptors, but are also serve as protein kinases. Thus, when they bind to the appropriate ligand, their kinase activity is stimulated, and they transmit signals through a cascade of phosphorylation events within the cell. Kinases are enzymes and tend to add phosphate groups to particular amino acids within substrate proteins. Receptor kinases are of following two types: 1. Those that phosphorylate a tyrosine residue (**tyrosine kinases**) and 2. Those that phosphorylate a serine or threonine residue (serine/threonine kinases). The action of tyrosine kinases are well understood, so are described here.

Receptor Tyrosine Kinases.

Many **receptor tyrosine kinases (RTKs)** trigger a chain of signal transduction events inside the cell that ultimately leads to cell growth, proliferation or in the specialization of cells in a process known as **differentiation**. These processes are competently controlled, so that only specific cells respond when the proper ligand is available. Examples of RTKs include the insulin receptor, the nerve growth factor receptor, and epidermal growth factor (EGF) receptor.

Structure of RTKs. These receptors often consist of a single polypeptide chain with only one transmembrane segment. Within this polypeptide chain are several distinct domains (Fig. 19.16). The extracellular portion of the RTK contains the **ligand-binding domain**. The other end of the peptide protrudes through the plasma membrane into the cytosol. On the cytosolic side, a portion of the receptor forms the tyrosine kinase while the remainder constitutes a cytosolic tail. The cytosolic portion of the receptor (RTK contains tyrosine residues that are in fact substrates or targets for the tyrosine kinase portion of the receptor).

The Activation of RTKs. Signal transduction is initiated when a ligand binds, causing the receptor tyrosine kinases (RTKs) to aggregate (Fig. 19.16B). Often two receptor molecules cluster together within the plasma membrane when they bind ligand. Once the receptors cluster in this way, the tyrosine kinase associated with each receptor phosphorylates the tyrosines of neighbouring receptors (**cross-phosphorylation**). Since the receptors phosphorylate other receptors of the same type, this process is referred to as **autophosphorylation**.

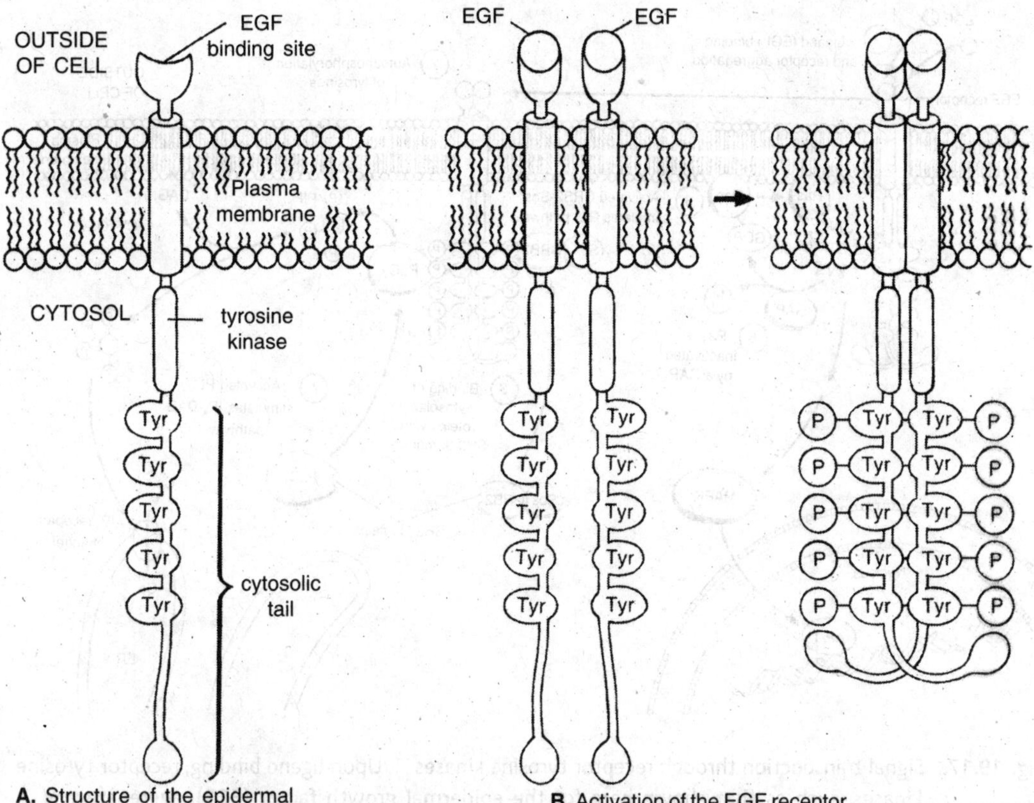

A. Structure of the epidermal growth factor (EGF) receptor

B. Activation of the EGF receptor

Fig. 19.16. The structure and activation of a receptor tyrosine kinase (RTK). A—The receptor for epidermal growth factor (EGF), shown here, is typical of many RTKs. These receptors often have only one transmembrane segment. The extracellular portion of the receptor binds to the ligand (EGF in this case). Inside the cell, a portion of the receptor has tyrosine kinase activity. The remainder of the receptor contains a series of tyrosine residues that are substrates for the tyrosine kinase. B—The activation of a receptor tyrosine kinase starts with the binding of a messenger (EGF), causing receptor aggregation or clustering. Once the receptors aggregate, they cross-phosphorylate each other at a number of tyrosine amino acid residues. The formation of tyrosine phosphate (Tyr-P) residues on the receptor creates binding sites for cytosolic proteins that contain SH_2 domains (after Becker *et al.*, 2007).

Signal transduction through RTKs. Once autophosphorylation of tyrosine residues on the cytosolic portion of the receptor occurs in response to ligand binding [(1) and (2) in Fig. 19.17], the recruits a number of cytosolic proteins to interact with itself. Each of these proteins binds to the receptor at a phosphorylated tyrosine residue. To bind to the receptor, each cytosolic protein must contain a stretch of amino acids that recognizes the phosphotyrosine and a few neighbouring amino acids on the receptor. The portion of a protein that recognizes one of these phosphorylated tyrosines is called an **SH_2 domain** [SH_2=Src homology (domain)^2].

Enrolment of different SH_2 domain-containing proteins activates different signal transduction pathways. As a result, receptor tyrosine kinases can activate several different signal transduction pathways at the same time. These include the inositol-phospholipid-calcium second messenger pathway, which we have already discussed, and the **Ras pathway**, which ultimately activates the expression of genes involved in growth or development.

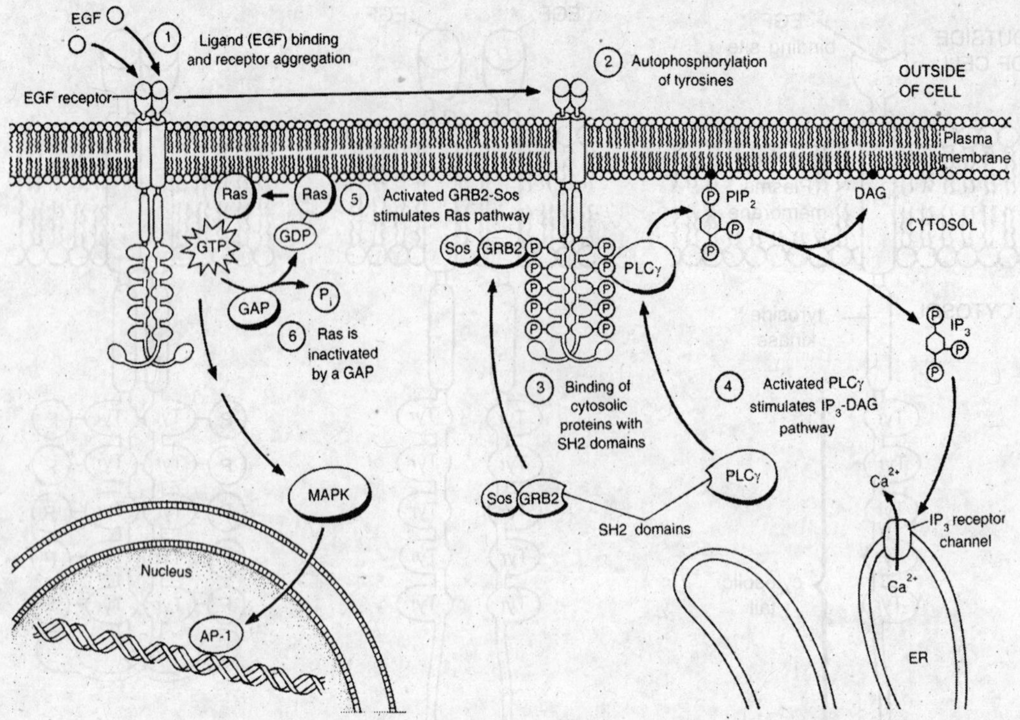

Fig. 19.17. Signal transduction through receptor tyrosine kinases. 1. Upon ligand binding, receptor tyrosine kinases, such as that shown here for the epidermal growth factor (EGF), aggregate and 2. Undergo autophosphorylation. Once a receptor is phosphorylated at residues in its cytosolic tail, 3. Proteins with SH_2 domains such as phospholipase C (PLC) and GRB2 bind to the receptor. 4. The binding of phospholipase C results in its activation and the cleavage of PIP_2 into IP_3 and DAG. 5. The binding of GRB2 causes the activation of Sos, to which it is bound. Sos (a GEF) then causes the activation of the Ras protein by helping it release GDP and acquire GTP. Activated Ras initiates a cascade of events that ultimately results in the formation of AP-1, a transcription factor in the nucleus that stimulates the expression of genes needed for growth of cell. 6. Ras is inactivated by hydrolysis of its bound GTP, a step facilitated by a GAP (after Becker *et al.*, 2007).

Characteristics of Ras pathway. Ras is important in regulating the growth of cells. It comprises a single subunit. Ras and other small monomeric G proteins are important signalling molecules. Like other types of G proteins, Ras can be bound to either GDP or GTP, but is active only when bound to GTP. In the absence of receptor stimulation, Ras is normally in the GDP-bound state. For Ras to become active, it must release GDP and acquire a molecule of GTP. For this to take place, Ras requires the help of another type of protein called a **guanine-nucleotide exchange factor (GEF)**.

The GEF that activates Ras is a Sos (so called because it was originally identified from a genetic mutation in fruit flies called *Son of sevenless* that results in the failure of cells in the compound eye to develop properly). For Sos to become active, it must bind indirectly to the receptor tyrosine kinase through another protein called **GRB2** (3. in Fig. 19.17), which contains a SH_2 domain. Thus, to activate Ras, the receptor becomes tyrosine phosphorylated, and GRB2 and Sos form a complex that binds to the receptor, activating Sos. Sos then stimulates Ras to release GDP and acquire GTP, which converts Ras to its active state.

Once Ras is active, it triggers a cascade of cellular events collectively referred to as the **Ras pathway** (5. in Fig. 19.17). One significant event in the Ras pathway is the activation of **mitogen-activated protein kinases**, or **MAP kinases (MAPKs)**. MAPKs are activated when cells receive a stimulus to grow and divide. One of the functions of MAPKs is to phosphorylate a nuclear protein called **Jun**. When Jun is phosphorylated, it assembles along with other proteins in a transcription factor called **AP-1**, which appears to stimulate the production of proteins that are needed for cells to grow and divide.

Soon Ras has to be inactivated by hydrolysis of the GTP bound to it to avoid continued stimulation of the Ras pathway. GTP hydrolysis is facilitated by a **GTPase activating protein (GAP)**. GAPs can speed up inactivation of Ras a hundredfold.

19.4 GROWTH FACTORS AS MESSENGERS

Growth factors are secreted molecules that act at short range (*i.e.*, they are mostly paracrine signals) and have specific effects on cells possessing the appropriate receptor to sense the presence of the growth factor.

Discovery of Growth Factors.

In order for a cell to grow, it must have all the nutrients needed for synthesis of its component parts, but the availability of nutrients is usually not itself sufficient for growth. Cells often also need messengers that act on specific receptors to stimulate cell growth. Biologists encountered the requirements for cell growth when they first tried to culture cells in nutrients, including the presence of blood plasma, cells would not grow. A turning point came when blood serum was used instead of plasma: *Serum was able to support the growth of cells, whereas plasma would not*. Many of the messengers present within the serum have now been purified, and they are members of various classes of proteins known as **growth factors**.

The difference between blood serum and blood plasma held an important clue about growth factors. *Plasma* is whole blood, including unreacted platelets (which contain clotting components) but without the red and white cells. *Serum* is the clear fluid remaining after blood has clotted. During clotting, platelets secrete growth factors into the blood that stimulate the growth of cells called **fibroblasts**, which form the new connective tissue that makes up a scar. After clotting, the resulting serum is full of **platelet-derived growth factor (PDGF)**. Plasma does not contain this factor because the clotting reaction has not taken place (see **Becker** *et al.*, 2007).

Examples of Growth Factors. Receptor for PDGF is found to be a receptor for tyrosine kinase. In fact, several growth factors act by stimulating receptor tyrosine kinases, including **insulin, insulin-like growth factor-I, fibroblast growth factor**, **epidermal growth factor**, and **nerve growth factor**. Many other types of growth factors have been isolated (Table 19.2).

Although collectively known as growth factors, however, these proteins have roles in many diverse events. These include not only growth and cell division, but also crucial events during the development of embryos, responses to tissues injury, and many other activities.

Table 19.2 Examples of growth factor families/superfamilies (Source: Becker *et al.*, 2007).

	Growth factor	*Target cells*	*Type of receptor complex*
1.	Epidermal growth factor (EGF)	Wide variety of epithelial and mesenchymal cells	Tyrosine kinase
2.	Transforming growth factor-α (TGFα)	Same as EGF	Tyrosine kinase
3.	Platelet-derived growth factor (PDGF)	Mesenchyme, smooth muscle, trophoblast	Tyrosine kinase
4.	Transforming growth factor-β (TGFβ)	Fibroblastic cells	Serine-threonine kinase
5.	Fibroblast growth factor (FGF)	Mesenchyme, fibroblasts, many other cell types	Tyrosine kinase
6.	Interleukin-2 (IL-2)	Cytotoxic T lymphocytes	Complex of three units
7.	Colony stimulating factor-I (CSF-1)	Macrophage precursors	Tyrosine kinase
8.	Wnts	Many types of embryonic cells	Seven-pass protein

1. Disruption of growth factor signalling through receptor tyrosine kinases. Many receptors for growth factors are tyrosine kinases, falling into families based on which growth factor (s) they can bind. One well-studied class of growth factors are the **fibroblast growth factors (FGFs)** and their receptor tyrosine kinases, the **fibroblast growth factor receptors (FGFRs)**. FGFs and FGFRs are used in signalling events in both adult animals and embryos.

For example, FGFRs have been shown to play an important role in the development of cells derived from the embryonic cell layer or germinal layer of early embryo, known as **mesoderm**. The mesoderm forms many cell types including muscle, cartilage, bone and blood cells, as well as the forerunner of the vertebral column (the notochord). When specific FGFRs fail to function properly, the development of particular mesodermal tissue is affected. In one class of FGFR defects, a **mutation** in the receptor results in *dominant effects* on the developing embryo. In other words, even though the embryo makes a substantial quantity of normal functional receptor, the presence of the mutant receptor within the embryo's cells prevents the normal receptors from functioning properly. Normal function is **inhibited** because FGFRs must act together as dimers to bind FGFs. If a normal receptor dimerizes with a mutant receptor, then the phosphorylation event that normally occur within the tyrosine kinase portion of the receptor fail to occur, blocking signal transduction (Fig. 19.18). Such a mutation that cancels the function of the normal receptor is called a **dominant negative mutation**.

Examples of dominant negative mutations. Dominant negative mutations are found to have dramatic effects on cells in developing embryos. For example, when genetically engineered dominant negative FGFRs are expressed in frog eggs and the eggs are fertilized, the embryos fail to develop mesodermal tissues in the trunk and tail, resulting in tadpoles with heads, but no bodies. In humans, dominant mutations in the transmembrane portion of the FGFR-3 gene result in the most common form of dwarfism, called **achondroplasia**. Heterozygous individuals have abnormal bone growth, in which the long bones suffer from abnormal ossification (*i.e.*, the process by which cartilage is converted to bone during childhood). A related condition known as **thanatophoric dysplasia** (a congenital dwarfism) often results from a single amino acid change in the cytosolic portion of FGFR-3 protein. In this case, more severe bone abnormalities result, and affected individuals die soon after birth.

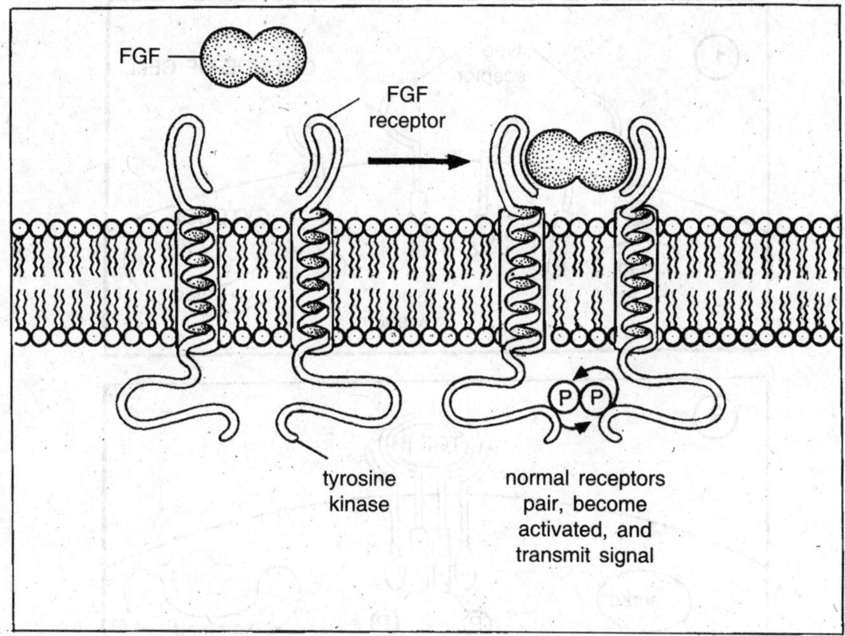

A. Normal FGFR: binds; FGFRs form a dimer

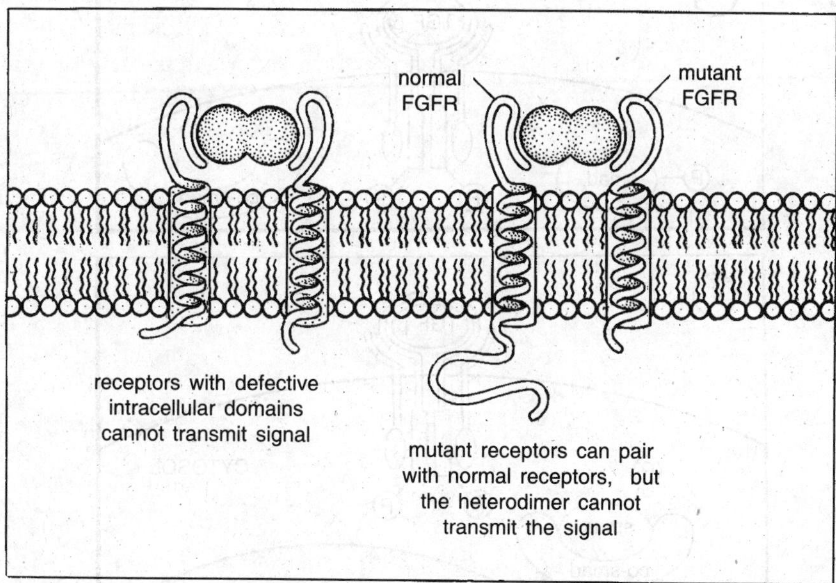

B. Dominant negative mutation in FGFR

Fig. 19.18. Dominant negative disruption (due to mutation) of FGF receptor (FGFR) function. A—normal receptors form dimers after binding FGF, and transmit the appropriate signal via Ras and MAPK. B—When a cell makes mutant FGF receptors, normal receptors can dimerize as in A, or the defective receptors can bind to FGF and dimerize with normal receptors. In this case, no signal is transmitted (after Becker *et al.*, 2007).

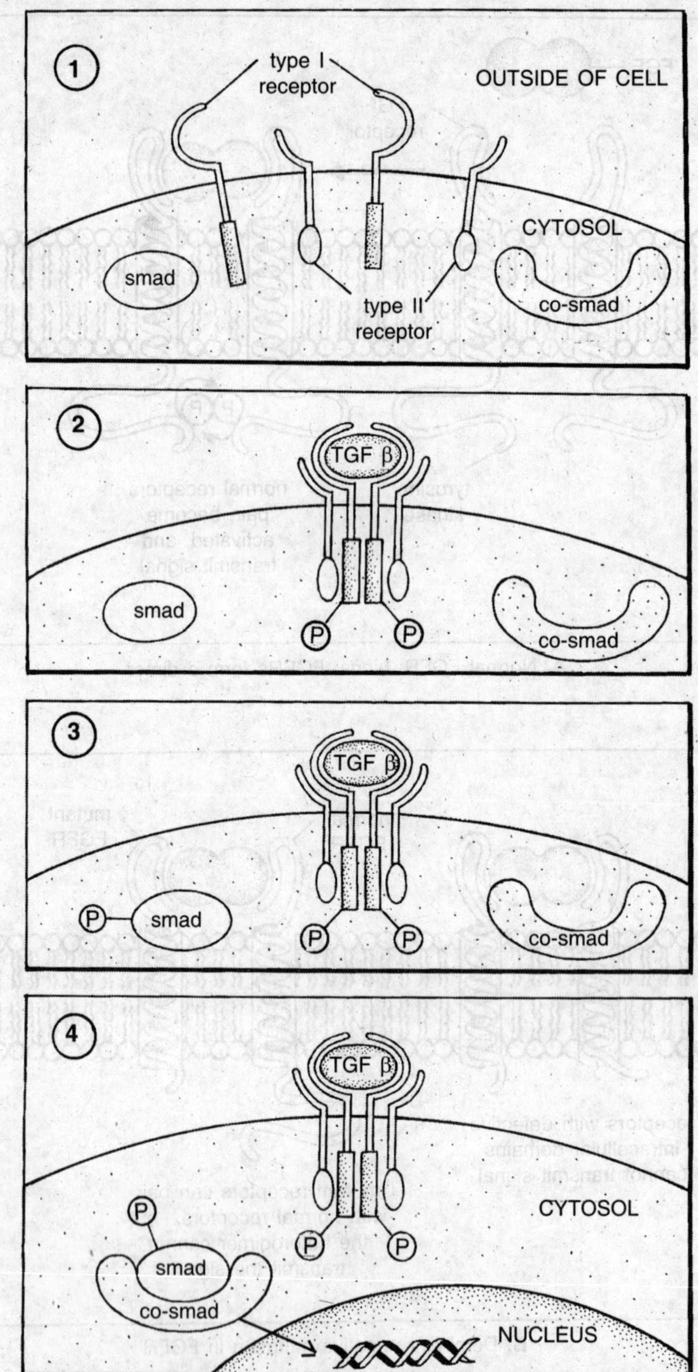

Fig. 19.19. Signal transduction by TGFβ receptor family proteins. 1. Type I and Type II receptors for TGFβ in a cell prior to binding of the growth factor. 2. Binding of growth factor results in clustering of type I and type II receptors. 3. The activated type I receptors then phosphorylate particular receptor-mediated smads. 4. These smads then bind to other smads, and together they enter the nucleus (after Becker *et al.*, 2007).

2. Disruption of growth factors signalling through receptor serine/threonine kinases. Another major class of protein kinase-associated receptors uses a very different set of signal transduction pathways to elicit changes within the cell. These receptors phosphorylate serine and threonine residues rather than tyrosines. One major class of **serine/threonine kinase receptors** comprises a family of proteins that bind members of the **transforming growth factor-β** (TGF-β) superfamily of growth factors (Fig. 19.19). TGF-β superfamily consists of a large number of structurally related, secreted, dimeric proteins. They act either as hormones or, more commonly as local mediator, to regulate a wide range of biological functions in all animals. During development, they regulate pattern formation and influence various cell behaviours, including proliferation, differentiation, extracellular matrix production, and cell death. In adults, they are involved in tissue repair and in immune regulation, as well as in many other processes. The superfamily include the TGF-βs themselves, the *activins* and the **bone morphogenetic proteins** (BMPs).

Once activated, the receptor complex uses a strategy for rapidly relaying the signal to the nucleus. The **type I receptor** directly binds and phosphorylates a latent gene regulatory protein of the **Smad family** (named after the first two identified, Sma in *C.elegans* and Mad in *Drosophila*). Activated TGF-β receptors and activin receptors phosphorylate Smad 2 or Smad 3, while activated BMP receptors phosphorylate Smad 1, Smad 5 and Smad 8. Once one of these Smads has been phosphorylated, it dissociates from the receptor and binds to Smad 4, which can form a complex with any of the above five *receptor-activated Smads*. The Smad complex then moves into the nucleus, where it associates with other gene regulatory proteins, binds to specific sites in DNA, and activates a particular set of target genes. Several mutations in Smad proteins in human cells appear to result in specific forms of cancer.

Growth factor receptors and cancer. It is well known that growth factors regulate events such as cell proliferation, cell movement, and gene expression. Embryonic development requires growth factors to tightly regulate cell biological events. If development is a prime example of the processes that are correctly regulated by growth factors, many types of cancer are exactly the opposite. It is found that a number of cancers result from loss of regula-tion of growth factor signalling. For example, one of the first cancer-causing genes to be identified was a mutant form of Ras. Mutations in Ras are often associated with cancer. In the case of receptor tyrosine kinases, for example, mutations in the epidermal growth factor receptors can result in **breast cancer, glioblastoma** (a cancer of glial cells in the brain), and **fibrosarcoma** (a type of cancer of long bones). Similar discoveries have been made

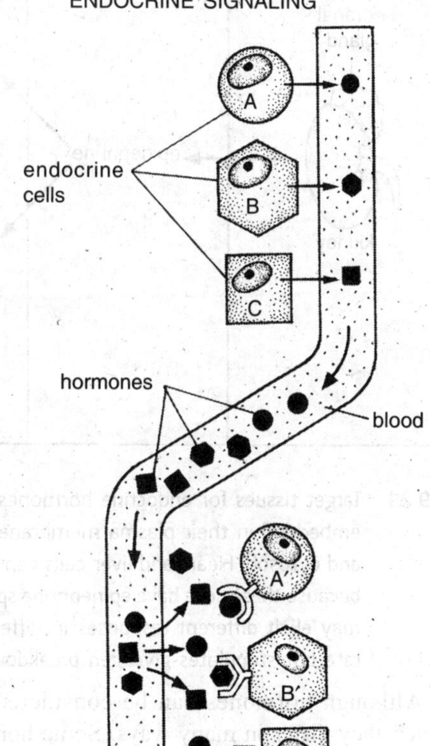

Fig. 19.20. Endocrine cells secrete hormones into the blood, which signal only the specific target cells that recognize them. These target cells have receptors for binding a specific hormone, which the cells "pull" from the extracellular fluid (after Albert *et al.*, 2002).

for receptor serine threonine kinases. Mutations in TGF-β type I receptor occur in one-third of **ovarian cancers**, and mutations in the type II receptor occur in many **colorectal cancers**. Mutations in one of the Smads (Smad 4) occur in one-half of all **pancreatic cancers**.

19.5 SIGNALLING BY HORMONES

Hormones are the chemical signals which are used by the animals to regulate the function of various cells and tissues. They are chemical messengers secreted by one tissue that regulate the function of other cells or tissues in the same organisms. In contrast to growth factors, hormones are often act over large distances. In animals, hormones are often transported via the vasculature. Hundreds of different hormones regulate a wide variety of functions, many critical for maintaining the physiological steady state of an organism.

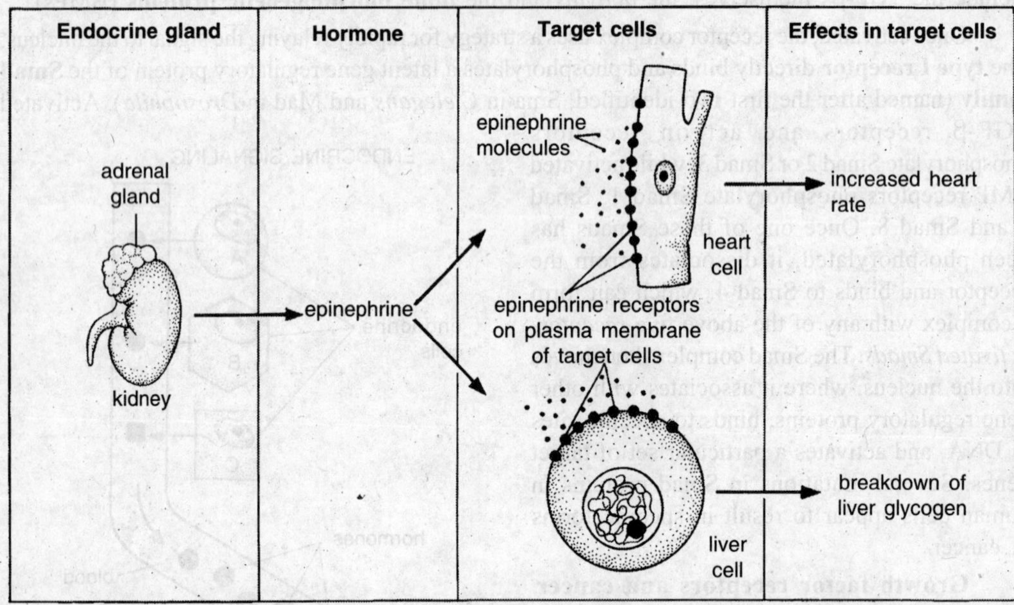

Fig. 19.21. Target tissues for endocrine hormones. Cells in a target tissue have hormone-specific receptors embedded in their plasma membrane (or in case of the steroid hormones, present in nucleus and cytosol). Heart and liver cells can respond to epinephrine synthesized by the adrenal glands because these cells have epinephrine specific receptors on their outer surfaces. A specific hormone may elicit different responses in different target cells. Epinephrine causes an increase in heart rate but stimulates glycogen breakdown in the liver (after Becker *et al.*, 2007).

Although hormones can be considered as a group based on their regulatory functions, however, they differ in many ways. Some hormones are steroids or other hydrophobic molecules that are targeted to intracellular receptors. Other hormones, such as adrenergic hormones are targeted to a wide variety of different G protein-linked receptors. Still others, such as insulin, are ligands, for receptor tyrosine kinase.

1. Classification of Hormonal Signals

Every animal hormone can be placed in one of two categories depending on the distance over which it operates. An **endocrine hormone** travels by means of the circulatory system from the cells where it is released to other cells (target cells), where it regulates one or more specific functions. A **paracrine hormone** is a more local signal that is taken up, destroyed, or immobilized so rapidly that it can act only on cells in the immediate environment.

Endocrine hormones are synthesized by the **endocrine tissue** of the body and are secreted directly into the bloodstream. Once secreted into the circulatory system endocrine hormones have a limited life span, ranging from a few seconds for *epinephrine* (a product of the adrenal gland) to many hours for *insulin* (a product of a beta (β) cells of islets of Langerhans of endocrine pancreas). As they circulate in the bloodstream, hormones come into contact with receptors in tissues throughout the body. A tissue that is specifically affected by a particular hormone is called a **target tissue** for that hormone (Fig. 19.21). For example, the heart and the liver are target tissues for epinephrine, whereas the liver and skeletal muscles are targets for insulin.

2. Functions of Hormones. Hormones regulate a wide range of physiological functions, including growth and development, rate of body processes (metabolism), concentration of substances and responses to stress and injury (Table 19.3). For example, **somatotropin** is involved in the regulation of overall growth of the body, whereas **androgens** and **estrogens**, the sex hormones, control the differentiation of tissues and the consequent attainment of secondary sex characteristics. **Thyroxine** regulates the rate at which the body makes energy available and is therefore an example of a rate-controlling hormone. Hormones that control the concentration of substances include **insulin** (control of blood glucose level), **aldosterone** (control of blood sodium and potassium levels) and **parathyroid hormones** (control of blood calcium level). The body's response to stress is regulated by **epinephrine, norepinephrine** and **cortisol** and its response to local injury is regulated by the release of **histamine** and the production of **prostaglandins**.

Table 19.3 Physiological functions of hormones (Source: Becker *et al.*, 2007).

Function under hormonal control	*Hormones*	*Source (Endocrine tissue)*
A. Growth and Development		
1. Body size	Somatotropin (growth hormones)	Anterior pituitary
2. Sexual development	Androgens (males) Estrogen (females)	Testes Ovaries
3. Reproductive cycle Follicle-stimulating hormone	Luteinizing hormone	Anterior pituitary Anterior pituitary
	Chorionic gonadotropin	Follicle
B. Rates of body processes		
1. Hormone secretion	Tropic hormones	Anterior pituitary
2. Basal metabolism	Thyroxine	Thyroid
3. Glucose uptake	Insulin	Pancreas
4. Kidney filtration	Antidiuretic hormone (vasopressin)	Posterior pituitary
5. Uterine contraction	Oxytocin	Posterior pituitary
C. Concentration of substances		
1. Blood glucose	Glucagon, insulin	Pancreas
2. Mineral balance	Corticosteroids	Adrenal cortex
3. Blood calcium	Parathyroid hormone	Parathyroid
D. Responses to stress and injury		
1. Heart rate	Epinephrine	Adrenal medulla
2. Blood pressure	Epinephrine	Adrenal medulla
3. Inflammation	Histidine Prostaglandins Corticosteroids (cortisol)	Mast cells All tissues Adrenal cortex

3. Chemical Type of Hormones

Chemically, the endocrine hormones are of four types: amino acid derivatives, peptides, proteins and lipid-like hormones such as steroids. An example of amino acid derivative is epinephrine, derived from tyrosine. **Antidiuretic hormone** (also called **vasopressin**) is an example of peptide hormone, whereas **insulin** is an example of protein hormone. **Testosterone** is an example of a steroid hormone. The steroid hormones are derivatives of cholesterol that are synthesized either in the gonads (the **sex hormones**) or in the adrenal cortex (the **cortisteroids**).

Examples of paracrine hormones are histamine and the prostaglandins. **Histamine** is produced by decarboxylation of the amino acid histidine and is responsible for local inflammatory responses. The **prostaglandins**, so named because they were first discovered in human semen as a secretion of prostate gland, are derived from arachidonic acid and are important in smooth muscle function.

Table 19.4 Chemical classification and functions of hormones (Source: Becker *et al.*, 2007).

Chemical classification	*Examples*	*Regulated functions*
A. Endocrine Hormones		
1. Amino acid derivatives	1. Epinephrine (adrenaline) and nor-epinephrine (both derived from tyrosine)	Stress response: regulation of heart rate and blood pressure; release of glucose and fatty acids from storage sites
	2. Thyroxine (derived from tyrosine)	Regulation of metabolic rate
2. Peptides	1. Antidiuretic hormones (vasopressin)	Regulation of body water and blood pressure
	2. Hypothalamic hormones (releasing factors)	Regulation of tropic hormone release from pituitary gland
3. Proteins	1. Anterior pituitary hormones	Regulation of other endocrine systems
	2. Sex hormones (androgens and estrogens)	Development and control of reproductive capacity
	3. Corticosteroids	Stress responses; control of blood electrolytes
B. Paracrine Hormones		
1. Amino acid derivatives	Histamine	Local responses to stress and injury
2. Arachidonic acid derivatives	Prostaglandins	Local responses to stress and injury

4. Modes of Endocrine Regulation of Target Cells.

Since hormones modulate the function of particular target tissues, an important aspect of studying hormones is understanding the specific functions of these target tissues. To understand how endocrine hormones act, we will look more closely at the water-soluble (nonlipid) hormones such as **adrenergic hormones** epinephrine and norepinephrine. (Epinephrine is also called adrenaline, the two words are of Greek and Latin derivation, respectively, and mean "above or near, the kidney", referring to the location in the body of the adrenal glands, which synthesize this hormone). When secreted into the blood-stream, epinephrine and norepinephrine stimulate changes in many different tissues or organs, all aimed at preparing the body for dangerous or stressful situations (the so-called "fight-or-flight response"). On the whole, the adrenergic hormones trigger increased cardiac output, shunting blood from the visceral organs to the muscles and the heart, as well as dilation of arterioles to facilitate oxygenation of the blood. In addition, these hormones stimulate the breakdown of glycogen to supply glucose to the muscles.

Adrenergic hormones bind to a family of G protein-linked receptors known as **adrenergic receptors**. The individual members of this family differ mainly in their preference for epinephrine or norepinephrine and in which G protein is linked to the receptor. They are broadly classified into α- and **β-adrenergic receptors**. The α-adrenergic receptors bind both epinephrine and norepinephrine. These receptors are located on the smooth muscles that regulate blood flow to visceral organs. The β-adrenergic receptors bind epinephrine much better than norepinephrine. These receptors are found on smooth muscles associated with arterioles that feed the heart, smooth muscles of the bronchioles in the lungs, and skeletal muscles.

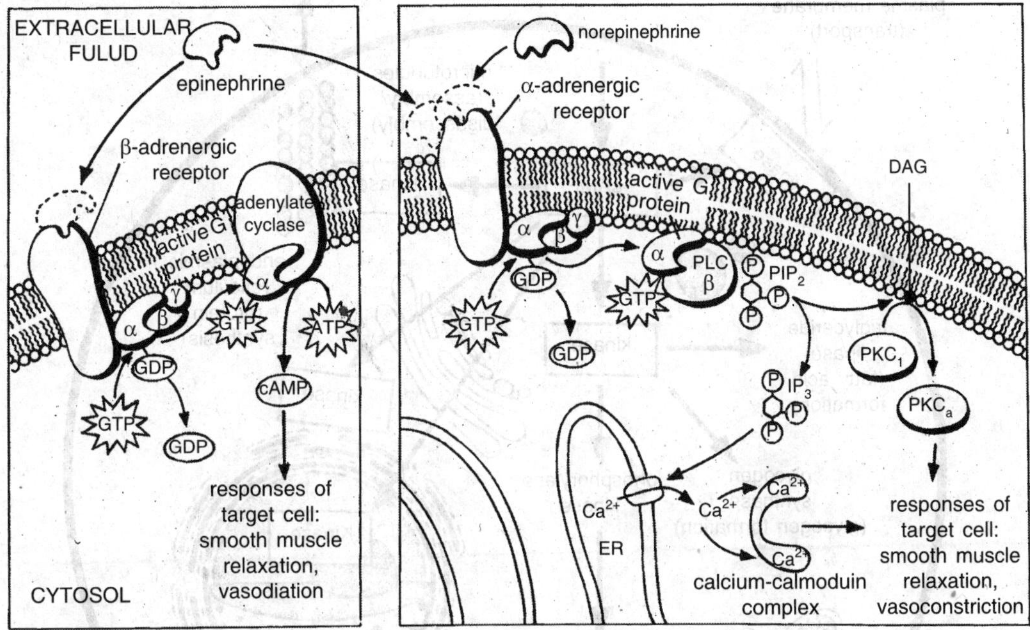

Fig. 19.22. The stimulation of G protein-linked signal transduction pathways by α- and β-adrenergic receptors (after Becker *et al.*, 2007).

The α- and β-adrenergic receptors stimulate different signal transduction pathways, because they are linked to different G proteins (Fig. 19.22). For example, the G proteins activated by one type of α-adrenergic receptor, the α_1*-adrenergic receptors*, are Gp proteins, whereas the β-adrenergic receptors activate Gs. As we discussed earlier, activation of G*s* stimulates the *cAMP signal transduction pathway*, leading to relaxation of certain smooth muscles. Activation of G*p* stimulates phospholipase C, leading to the production of IP_3 and DAG, which in turn elevates intracellular calcium levels. Table 19.5 summarises some of the main physiological functions that are regulated by cAMP. As a specific example of cAMP-mediated regulation, here will be considered the control of glycogen degradation by epinephrine in liver or muscle cells.

Table 19.5 Examples of cell functions regulated by cAMP (Source: Becker *et al.*, 2007).

	Regulated function	*Target tissue*	*Hormone*
1.	Glycogen degradation	Muscle, liver	Epinephrine
2.	Fatty acid production	Adipose	Epinephrine
3.	Heart rate, blood pressure	Cardiovascular	Epinephrine
4.	Water reabsorption	Kidney	Antidiuretic hormone
5.	Bone resorption	Bone	Parathyroid hormone

(i) Control of glycogen degradation by cAMP pathway. The breakdown of glycogen to provide muscle cells with an adequate supply of glucose is stimulated by the adrenergic hormones. The breakdown of glycogen is catalysed by the enzyme **glycogen phosphorylase**, which cleaves glucose units from glycogen as glucose-1-phosphate by the addition of inorganic phosphate (Pi). The glycogen phosphorylase system was the first cAMP-mediated regulatory sequence to be explained. The original work was published in 1956 by **Earl Sutherland**, who received a Nobel Prize in 1971 for this discovery.

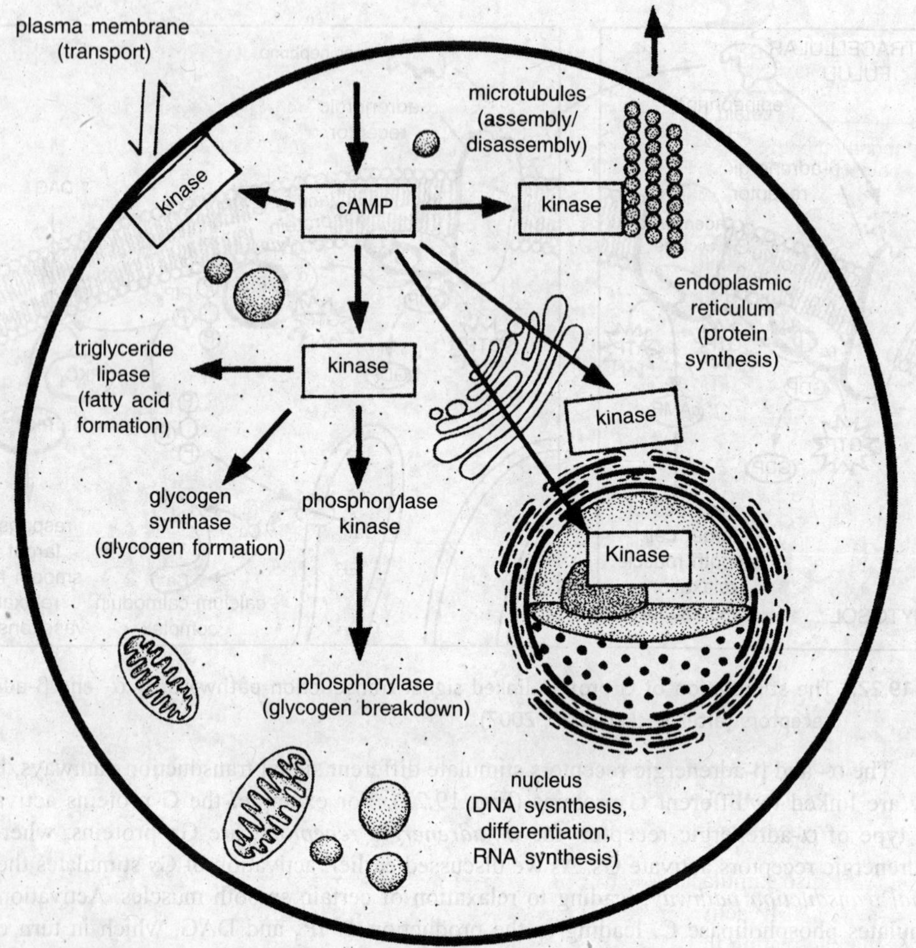

Fig. 19.23. Schematic illustration of the variety of processes that can be affected by changes in cAMP concentration. All of these effects are thought to be mediated by activation of the same basic enzyme, protein kinase A (after Karp, 2002).

Increased glycogen metabolism begins when an epinephrine (hormone) molecule binds to a β-adrenergic receptor on the plasma membrane of a liver or muscle cell. As described earlier, the receptor activates a neighboring Gs protein, and the Gs protein in turn stimulates adenylyl cyclase, the membrane-bound enzyme that generates cAMP from ATP (Fig. 19.24). The resulting transient increase in the concentration of cAMP in the cytosol activates protein kinase A (or PKA). PKA then activates another cascade of events that begins with the phosphorylation of the enzyme

phosphorylase kinase. This leads to the conversion of **glycogen phosphorylase** from phosphorylase b form and thus to an increased rate of glycogen breakdown (Fig. 19.23).

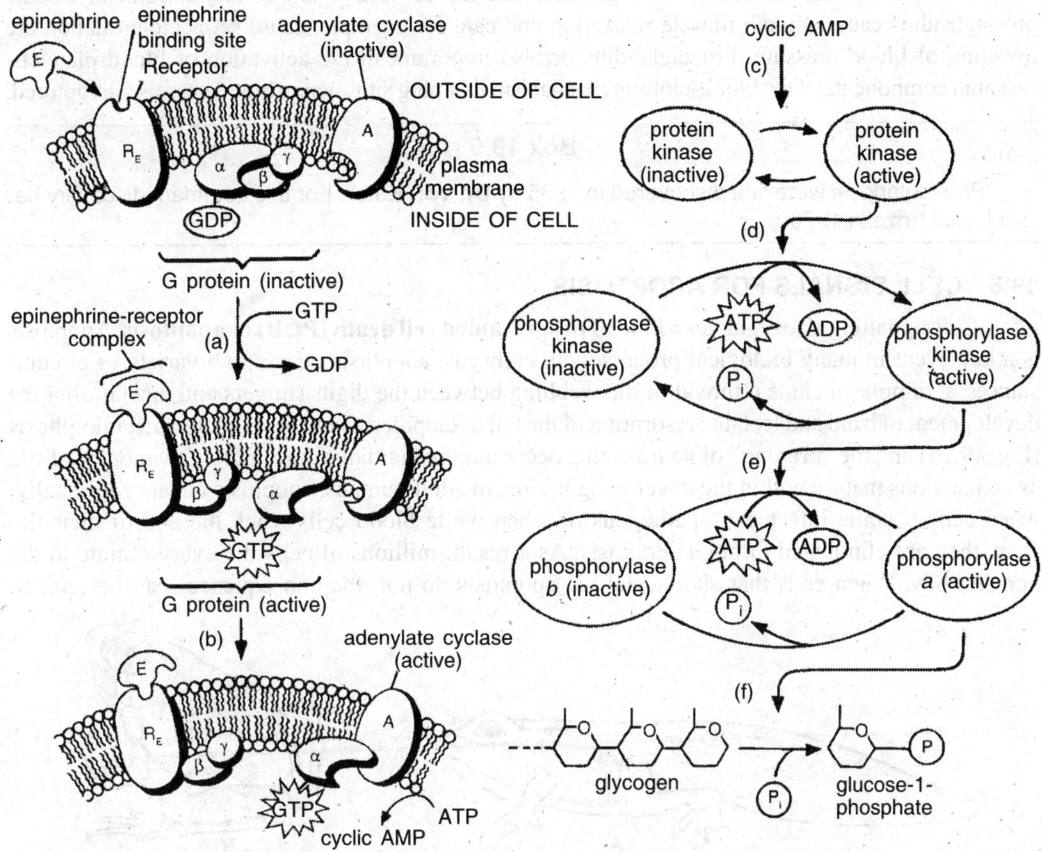

Fig. 19.24. Stimulation of glycogen breakdown by epinephrine (hormone). Muscle and liver cells respond to an increased concentration of epinephrine in the blood by increasing their rate of glycogen breakdown. The stimulatory effect of extracellular epinephrine on the intracellular glycogen metabolism is mediated by a G protein-cAMP regulatory cascade (after Becker *et al.*, 2007).

cAMP also stimulates the inactivation of the enzyme system responsible for glycogen synthesis. In this case, cAMP activates PKA which in turn phosphorylates the enzyme **glycogen synthetase**. Rather than activating this enzyme, however, phosphorylation inactivates it. Thus, the overall effect of cAMP involves both an increase in glycogen breakdown and a decrease in its synthesis.

(ii) **α-Andrenergic receptors and IP_3-DAG-Ca^{2+} pathway.** The α_1-andrenergic receptors are found mainly on smooth muscles in blood vessels, including those controlling blood flow to the intestines. When α_1-andrenergic receptors are stimulated, the formation of IP_3 causes an increase in the intercellular calcium concentration. The elevated level of calcium causes smooth muscle contraction, resulting in constriction of the blood vessels and diminished blood flow. Thus, the activation of α_1-adrenergic receptors affects smooth muscle cells in a manner opposite to that of β-adrenergic activation, which causes smooth muscles to relax.

Example of Paracrine Regulation by Prostaglandins

Typically, the prostaglandins act on G protein-linked receptors to stimulate either the cAMP or IP_3-DAG-calcium (second messenger) pathway. Prostaglandins have a variety of effects, many of which involve smooth muscle. For example, prostaglandins contained in semen stimulate uterine smooth muscle contraction, which help transport sperm to egg. Prostaglandins also help to initiate smooth muscle contraction during labour and can be used to induce labour clinically. Some prostaglandins cause smooth muscle relaxation and can, for example, cause bronchiole dilation are lowering of blood pressure. Prostaglandins or also important in the activation of blood platelets, essential components of the blood-clotting mechanism that plug sites where blood vessels are ruptured.

Box 19.5

Prostaglandins were first discovered in 1935 by **Dr. Von Euler**. For this landmark discovery he got Nobel Prize in 1970.

19.6 CELL SIGNALS FOR APOPTOSIS

Cell signalling also regulates a kind of **programmed cell death (PCD)** or **apoptosis**. Apoptosis is a key event in many biological processes. In embryos, apoptosis occurs in a variety of circumstances. Examples include removal of the webbing between the digits (fingers and toes) during the development of hand and feet, the resorption of the tail of tadpoles when they undergo metamorphosis (Fig. 19.25) and the "pruning" of neurons that occur in human infants during first few months of life as connections mature within the developing brains. In adult humans, apoptosis occurs continually; when cells become infected by pathogens or when white blood cells reach the end of their life span, they are eliminated through apoptosis. As a result, millions of cells die every minute in the human body. When cells that should die via apoptosis do not, the consequences can be critical.

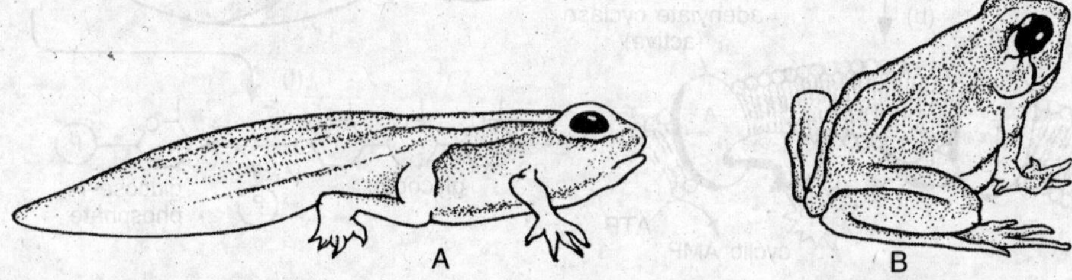

Fig. 19.25. Apoptosis during the metamorphosis of a tadpole into a frog. All the changes that occur during metamorphosis, including the induction of apoptosis of tail, are stimulated by an increase in thyroid hormone in the blood (after Albert *et al.*, 2002).

It is currently known that mutations in some of the proteins that participate in apoptosis can lead to cancer; melanoma frequently results from a mutations in Apaf-1, a protein that participates in apoptosis.

Apoptosis very different from another type of cell death, known as **necrosis**, which sometimes follows massive tissue injury. Whereas necrosis involves a specific series of events that leads to the dismantling of the internal contents of the cell (Fig. 19.26). During the early phases of apoptosis, the cell's DNA segregates near the periphery of the nucleus and volume of the cytoplasm decreases. Next, the cell begins to produce small bubble-like cytoplasmic extensions ("blebs"), and the nucleus and organelles begin to fragment. The cell's DNA is cleaved by an apoptosis-specific **DNA endonuclease** or **DNase** (an enzyme that digests DNA), at regular intervals along the DNA. As a result, the DNA fragments, which are multiples of 200 base pairs in length, form a diagnostic "ladder" of fragments. Eventually the cell is demolished into small pieces called **apoptotic bodies**. Ultimately the remnants of the affected cell are engulfed by other nearby cells (typically macrophages) via phagocytosis. The macrophages act as scavengers to remove the resulting cellular debris.

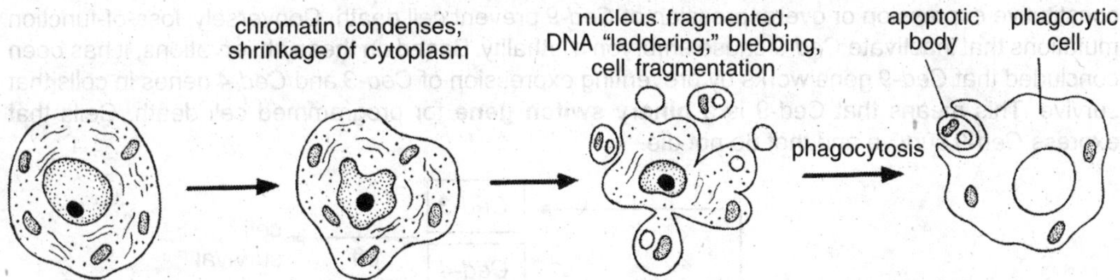

Fig. 19.26 Major steps in apoptosis. As a cell begins to undergo apoptosis, its chromatin condenses and the cytoplasm shrinks. Eventually the nucleus becomes fragmented, its DNA is digested at regular intervals ("laddering"), the cytoplasm becomes fragmented, and the cell extends numerous blebs. Ultimately the remnants of the dead cells (apoptotic bodies) are ingested by phagocytic cells (after Becker *et al.*, 2007).

Discovery of Apoptosis

The cells have a "death program" was first conclusively demonstrated in the nematode, *Caenorhabditis elegans*, where key genes that control apoptosis were first identified (Box 19.5). Subsequently research showed that many other organisms, including mammals, use similar proteins during apoptosis. As a result of these investigations many of the molecular events that underlie apoptosis are now known. For example, a key event in apoptosis of mammals is the activation of a series of enzymes called **caspases** (Caspases get their name because they contain a *c*ysteine at their active site, and they cleave proteins at sites that contain *aspar*tic acid residue followed by four amino acids that are specific to each caspase). Caspases are produced as inactive precursors known as **procaspases**, which are subsequently cleaved to create active enzymes, often by other caspases, in a proteolytic cascade. Once they are activated, caspases cleave other proteins within cells, resulting in efficient and precise killing of the cell in which they are activated.

Target proteins for caspases. Caspases are known to cleave the following target proteins of the cell (see **Karp**, 2002):

1. **More than a dozen protein kinases, including focal adhesion kinases (FAK), PKB, PKC and Raf 1**. Inactivation of FAK, for example, is presumed to disrupt cell adhesion, leading to detachment of the apoptotic cell from its neighbours.

Box 19.6

Horvitz, Sulston and **Brenner** shared the Nobel Prize in Physiology or Medicine in 2002. They worked out different aspects of apoptosis in the nematode *Caenorhabditis elegans*. **John Sulstan** performed initial cell lineage studies on *C.elegans*. He discovered occurrence of apoptosis during development of this nematode. **H.Robert Horvitz** *et al.*, provided key insight in to apoptosis and cell signalling in the formation of vulva of this nematode. **Sydney Brenner** suggested genetic basis of apoptosis in *C.elegans*.

In *C.elegans*, development relies on programmed cell death. The number of cells that die during the worm's development is always the same: 131 cells of total 1090 cells in the hermaphrodites and 147 cells of total 1178 cells in males. In addition, the point in development at which a given cell dies and the identity of the cells that die are always the same.

Genetic analysis of mutants indicates that although apoptosis occurs in cells of different developmental origins, all the cells die by using the same genetic pathway. In *C. elegans*, 15 genes are involved in cell death (apoptosis). These genes form four groups: 1. decision makers; 2. execution of the decision; 3. engulfment of dying cells, and 4. degradation of cell debris in the engulfing cells. For example, expression of *Ced-3* and *Ced-4* (= *c*ell *d*eath defective) are necessary for execution of the cell death programme; mutations that inactivates either of these genes result in survival of cells that normally die. Expression of *Ced-3* and *Ced-4* is controlled by *Ced-9*. Gain-of-function mutations that cause constitutive expression or overexpression of *Ced-9* prevent cell death. Conversely, loss-of-function mutations that inactivate *Ced-9* cause embryonic lethality. Based on these observations, it has been concluded that *Ced-9* gene works by preventing expression of *Ced-3* and *Ced-4* genes in cells that survive. This means that Ced-9 is a **binary switch gene** for programmed cell death. Cells that express *Ced-9* survive and that do not die.

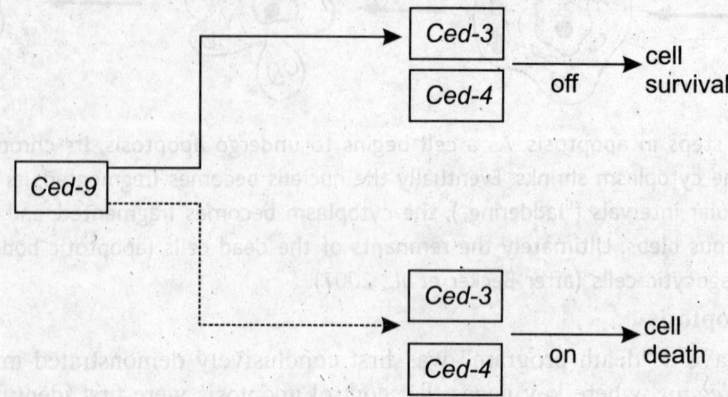

Fig. 19.27. The genetic pathway controlling cell death, the gene *Ced-9* act as a binary switch. If *Ced-9* is active, it repressed the expression of *Ced-3* and *Ced-4*, and the cell lives. If *Ced-9* is inactive, *Ced-3* and *Ced-4* are expressed and the cell die (after Klug and Cummings, 2001).

2. **Lamins**, which makes up the inner lining of the nuclear envelope. Cleavage of lamins leads to the disassembly of the nuclear lamina and shrinkage of the nucleus.

3. **Proteins required for cell structure**, such as those of intermediate filaments, actin and gelsolin. Cleavage and consequent inactivation of these proteins lead to changes in cell shape.

4. **An endonuclease called capsin activated DNase (CAD)**, which is activated following caspase cleavage of an inhibitory protein. Once activated, CAD translocates from the cytoplasm to the nucleus where it attacks DNA, serving it into fragments.

5. **Enzymes involved in DNA repair**, which are inactivated by caspase cleavage. DNA repair is a homeostatic activity that is inappropriate in an apoptotic cell in which the DNA is being dismantled.

Triggering of Apoptosis. There are two main routes by which cells can activate caspases and enter the apoptic pathway (Fig. 19.28).

1. **Cell death signals**. When cells in the human body are infected by certain viruses, a population of **killer lymphocytes** are activated and induce the infected cells to initiate apoptosis. How do lymphocytes induce cells to commit suicide via apoptosis? Typically, such activation is triggered when cells receive **cell death signals**. Two well known death signals are *tumor necrosis factor* and *CD95/Fas*. For example, **CD95** is a protein which is present on the surface of infected cells. Lymphocytes have a protein on their surfaces that binds to CD95, causing the CD95 within the infected cell aggregate (Fig. 19.28). CD95 aggregation results in the attachment of adaptor proteins to the clustered CD95, which in turn recruits a procaspase (**procaspase-8**) to sites of receptor clustering, when the procaspase is activated, it acts as an initiator of the caspase cascade. A key action of such **initiator caspases**, known as **caspase-3**. Active caspase-3 is important for activating many steps in apoptosis.

2. **Survival factors**. In other cases, apoptosis is triggered indirectly. One of the best-studied cases of this type of apoptosis involves **survival factors**. Survival factors mediate one of the many events associated with growth factors: cell survival. When such factors are withdrawn, the cell enters

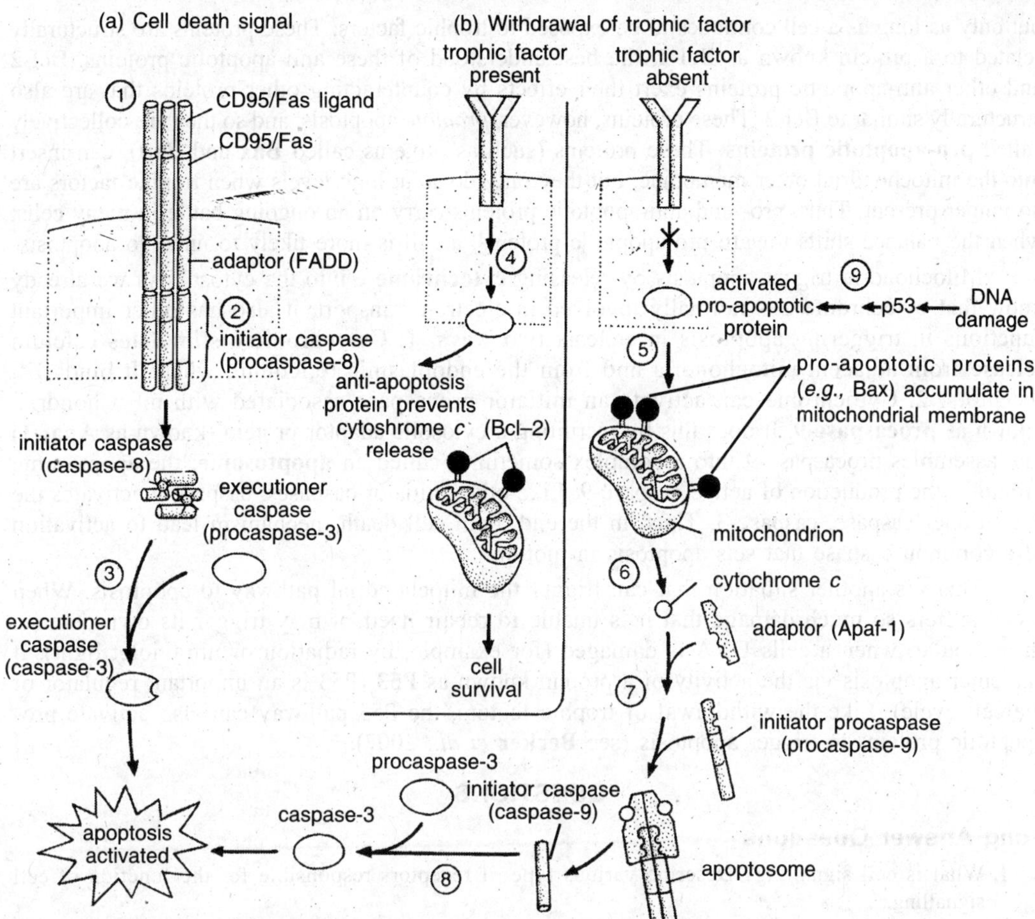

Fig. 19.28. Induction of apoptosis by cell death signals or by withdrawal of survival factors. A—Cell death signals, such as CD95/Fas ligand on the surface of a killer lymphocyte, can lead to apoptosis. 1. CD95/Fas ligand binds to CD95/Fas protein on the surface of a target cell. Binding causing clustering of receptors and recruitment of adoptor proteins in the target cell, resulting in clustering of interior procaspase (procaspase-8) protein. 2. Initiator caspases then become activated. After they are activated, 3. The initiator capases in turn activate the executioner caspase, caspase-3, a key initiator of apoptosis. B—4. When trophic factors are present, cell signalling results in inactivation of pro-apoptotic proteins and the cell does not enter apoptotis. 5. When trophic factors are withdrawn, pro-apoptotic proteins are activated, and some accumulate in the outer membrane of the mitochondrion. 6. The balance between pro-apoptotic and anti-apoptotic proteins (such as Bcl-2) at the mitochondrial outer membrane determines whether the mitochondrion releases cytochrome *C*. 7. Cytochrome C forms a complex with other proteins, resulting in activation of an initiator caspase (caspase-9).8. The initiator caspase in turn activates the executioner caspase, caspase-3, triggering apoptosis. 9. DNA damage can also lead to apoptosis through the activity of the P53 protein (after Becker *et al.*, 2007).

apoptosis. Surprisingly, a key site of action of this second pathway is the mitochondrion. The connection between mitochondria and cell death may be surprising, but it is that in addition to their role in energy production, mitochondria are important in apoptosis. If withdrawal of trophic factors is the sentence of execution, then the executioners are the mitochondria.

How does mitochondrion speed up cell death? In a healthy cell that is not committed to apoptosis, there are several **anti-apoptotic proteins** in the outer mitochondrial membrane that prevent apoptosis, but only as long as a cell continues to be exposed to trophic factors. These proteins are structurally related to a protein known as **Bcl-2**, the best understood of these anti-apoptotic proteins. Bcl-2 and other anti-apoptotic proteins exert their effects by counteracting other proteins that are also structurally similar to Bcl-2. These proteins, however, *promote* apoptosis, and so they are collectively called **pro-apoptotic proteins**. These proteins (such as proteins called **Bax** and **Bad**), can insert into the mitochondrial outer membrane, but they only do so at high levels when trophic factors are no longer present. Thus, pro- and anti-apoptotic proteins carry on an ongoing battle in many cells; when the balance shifts toward pro-apoptotic proteins, a cell is more likely to undergo apoptosis.

Mitochondria trigger apoptosis by releasing **cytochrome c** into the cytosol. As we already know that cytochrome c is normally involved in electron transport, it also has other important functions in triggering apoptosis in at least two ways. 1. Cytochrome c stimulates calcium release from adjacent mitochondria and from the endoplasmic reticulum, where it binds IP_3 receptors. 2. Cytochrome can activate an initiator procaspase associated with mitochondria, known as **procaspase-9**. It does this by recruiting a cytosolic adaptor protein (known as **Apaf-1**) that assembles procaspase-9 into a complex sometimes called an **apoptosome**; the apoptosome promotes the production of active caspase-9. Like other initiator caspases, caspase-9 activates the executioner caspase, caspase-3. Thus, in the end, both cell death mechanism lead to activation of a common caspase that sets apoptosis in motion.

There is another situation that can trigger the mitochondrial pathway to apoptosis. When a cell suffers so much damage that it is unable to repair itself, it may trigger its own demise. In particular, when a cells DNA is damaged (for example, by radiation or ultraviolet light), it can enter apoptosis via the activity of a protein known as **P53** (P53 is an important regulator of the cell cycle). Like the withdrawal of trophic factors, the P53 pathway can also activate pro-apoptotic proteins to trigger apoptosis (see **Becker** *et al.*, 2007).

QUESTIONS

Long Answer Questions

1. What is cell signalling? Describe various type of receptors responsible for the function of cell signalling.
2. What are cell surface receptors? Discuss their types and roles in cell signalling.
3. Discuss the role of second messengers in cell signalling.
4. 'Apoptosis in mediated by an intracellular proteolytic cascade'-substantiate.

Short Answer Questions

1. Explain the mechanism of cell signalling. (*V.B.S. Purvanchal, 2004*)
2. What is ligand? What is its function?
3. What are second messengers? How do they participate in cell signal transduction?
4. Explain the role of mitochondria in apoptosis.
5. Write short notes on the following:
   (*i*) Signal transduction;
   (*ii*) Apoptosis.

Very Short Answer Questions

1. What is a ligand?
2. What is the function of ligand?
3. What is the function of cell surface receptors?
4. What is the chemical nature of receptors?
5. What is receptor?
6. Define cAMP.

7. What is full form of IP_3?
8. What is the function of IP_3?
9. Define calmodulin.
10. What is a paracrine signal?
11. Define the endocrine signal.
12. What is a autocrine signal?
13. Define signal transduction.
14. What is G protein?

Fill in the Blanks

1. Communication by extracellular signals usually involves .... steps.
2. The signalling molecules act as a ...., which binds to a structurally .... site on the extracellular or membrane-spanning domains of the ....
3. Examples of secondary messenger of the cell include cAMP, ... , ... and ....
4. A well known example of signal amplification is breakdown of glycogen in .... cells in response to hormone ....
5. G-protein stand for ... .
6. The cAMP is formed from cytosolic .... by the enzyme ....
7. The full form of second messenger IP_3 is ....
8. Calcium-binding protein, ... has pivotal function in the regulation of a variety of cellular processes.
9. Nitric oxide (NO) is released by .... of blood vessel walls that causes relaxation of vascular .... muscle cells.
10. Programmed cell death or PCD is also called ....

Multiple Choice Questions

1. One of the following is a calcium-binding protein of the cell
   (*a*) actin (*b*) tubulin
   (*c*) cal modulin (*d*) dynein
2. A molecule acting as 'second messenger' in biological system is
   (*a*) cDNA (*b*) cAMP
   (*c*) tRNA (*d*) hn RNA
3. Epinephrine and nor-epinephrine together are known as
   (*a*) steroids (*b*) proteins
   (*c*) catecholamine (*d*) none of these
4. Substances which are formed at one place and express their effect at a distant place are called as
   (*a*) enzymes (*b*) hormones
   (*c*) pheromones (*d*) WBC
5. Insulin facilitates glucose uptake in
   (*a*) kidney tubules (*b*) red blood cells
   (*c*) brain (*d*) skeletal muscle
6. Cortical granules found in the egg cortex of many animals consist of
   (*a*) lipids (*b*) proteins
   (*c*) mucopoly saccharides (*d*) none of the above

ANSWERS

Very Short Answer Questions

1. Ligand is any molecule that binds to a specific site on a protein or other molecule (From Latin *ligare*, to bind).
2. Binding of ligand changes the property of the surface receptors and activate changes.
3. Communication with neighbouring cells, to bind matrix components with extracellular matrix and cell adhesion.
4. The receptors are protein molecules. These are proteoglycans.
5. A protein that contains a binding site for a specific signalling molecule.
6. cAMP is an intracellular molecule which acts as second messenger in eukaryotes and mediates the effects of various signalling molecules by activating protein.
7. Inositol-1, 4, 5-triphosphate.
8. IP_3 functions as a second messenger by triggering the release of calcium ions from storage sites within the endoplasmic reticulum.
9. Calmodulin is a calcium-binding protein involved in mediating many of the intracellular effects of calcium ions in eukaryotic cell.
10. It is a growth factor or other secreted molecule that acts on nearby cells.
11. It is a hormone or other molecule that is released into the circulatory system so that it can act on distant target cell.
12. It is a growth factor or other molecule secreted by a cell that acts on the cell that produces it.
13. Signal transduction is detection of specific signals at the cell surface and the mechanisms by which such signals are transmitted into the cell's interior, resulting in changes in cell behaviour and/or gene expression.
14. Any of numerous GTP-binding regulatory proteins located in the plasma membrane that mediate signal transduction pathways, usually by activating a specific target protein such as an enzyme or channel protein.

Fill in the Blanks

1. Seven.
2. Ligand, complementary, receptor.
3. cGMP, DAG, IP_3.
4. Liver, epinephrine.
5. Guanine-nucleotide binding protein.
6. ATP, adenylyl cyclase.
7. Ionositol, 1, 4, 5–triphosphate.
8. Calmodulin.
9. Endothelial cells, smooth.
10. Apoptosis.

Multiple Choice Questions

1. (*c*)
2. (*b*)
3. (*c*)
4. (*b*)
5. (*d*)
6. (*c*).

20

Cell Transformation and Cancer (Biology of Cancer)

The proliferation, differentiation and survival of individual cells in multicellular organisms are carefully regulated to meet the needs of the organism as a whole. This regulation is lost in cancer cells, which grow and divide in an uncontrolled manner, ultimately spreading throughout the body and interfering with the function of normal tissues and organs (**Cooper** and **Hausman**, 2007).

Cells which undergo rapid, abnormal and uncontrolled growth at the cost of remaining cells are called **neoplastic cells**. The growths resulting from the division of such cells are called **neoplastic growths** or **tumours**. Thus, any abnormal proliferation of cells is the tumour. Almost all type of differentiated cells of animals can become neoplastic or cancerous. The process of cell change in which a cell loses its ability to control its rate of division, and thus becomes a tumour cell, is called **cell transformation**.

Cancer is a genetic disease because it can be traced to alterations within specific genes, but in most cases, it is not an inherited disease. In an inherited disease, the genetic defect is present in the chromosomes of a parent and is transmitted to the zygote. In contrast, the genetic alterations that lead to most cancers arise in the DNA of a somatic cell during the lifetime of the affected individual. Because of these genetic changes, cancer cells proliferate uncontrollably, producing malignant tumours that invade surrounding healthy tissue (see **Karp**, 2002).

20.1 TYPES OF CANCER

Cancer can result from abnormal proliferation of any of the different kinds of cells in the body, so there are more than a hundred distinct types of cancer. These cancers can vary substantially in their behaviour and response to treatment. Based on differences in their growth patterns, tumours are classified as either benign or malignant. A **benign tumour** or **primary tumour**, such as a common skin wart, remains confined to its original location. A **malignant tumour** or **secondary tumour**, however, is capable of both invading surrounding normal tissue and spreading throughout the body via the circulatory or lymphatic systems (**metastasis**). Only malignant tumours are properly referred to as cancers and it is their ability to invade and metastasize that makes cancer so dangerous. The term **cancer** refers to any malignant tumour—that is, any tumour capable of spreading from its original location to other sites (Box 20.1). Whereas benign tumours can usually be removed surgically, the spread of malignant tumours to distant body sites frequently makes them resistant to such localized treatment.

Both benign and malignant tumours are classified according to the type of cell from which they arise. About two hundred distinct types of cancer have been recognized. These can be grouped into four main types: carcinomas, sarcomas, lymphomas and leukemias.

Box 20.1

The earliest recorded description of cancer is found in an Egyptian papyrus dating back to around 3000 B.C. (Papyrus refers to a manuscript written on paper prepared from papyrus plant *Cyperus papyrus*). The term cancer, which means "crab" in Latin, was coined by Hippocrates in the fifth century B.C., to describe diseases in which tissues grow and spread unrestrained throughout the body, eventually choking off life (**Becker** *et al.*, 2007); Hippocrates used the words *carcinus, carcinoma* and *cancer*, which refer to crabs in Latin, to describe the tumours (**Cooper** and **Hausman**, 2007).

1. Carcinomas. These cancers account for 90 per cent of all human cancers. Carcinomas arise from the epithelial cells that cover external and internal body surfaces. These epithelial cells may be of ectodermal or endodermal origin. Carcinomas include cervical, breast, skin, brain, lung, colon, prostate, kidney, bladder cancers. Carcinomas affect cells of ectodermal or endodermal origin.

2. Sarcomas. These cancers account for 4 per cent of all human cancers. Sarcomas are solid tumours of connective tissues such as cancers of muscle, bone, cartilage, fat cells and fibrous tissue. Sarcomas affects the cells of mesodermal origin.

3. Lymphomas. These cancers constitute about 4 per cent of human cancers. In lymphomas, there is an excessive production of lymphocytes by the lymph nodes and spleen. Hodgkin's disease is an example of lymphoma (Hodgkin's disease is a neoplastic disease that is characterized by progressive enlargement of lymph nodes, spleen, and liver and by progressive anaemia).

4. Leukemias. These cancers form about 4 per cent of human cancers. Leukemias are neoplastic growth of leucocytes (W.B.C.) and are characterized by excessive production of white blood cells (Box 20.2). Leukemias affect cells of mesodermal origin and are a class of sarcomas (**Lodish** *et al.* 2004). In addition to the four types of cancer mentioned above there may be **mixed malignant** tumours arising from ectodermal and mesodermal tissues, *e.g.*, tumours arising from ectodermal and mesodermal tissues.

Box 20.2

The leukemias grow as individual cells in the blood. The name **leukemia** is derived from the Latin for "white blood": the massive proliferation of leukemic cells can cause a patients blood to appear milky.

Box 20.3

Prior to 1900, most human deaths were due to **infectious diseases** such as pneumonia and tuberculosis, and life expectancy was less than 50 years. Cancer was rare disease that accounted for only a small percentage of deaths.

According to a recent survey conducted by American cancer society in 2005, more than 5 lac Americans die of cancer each year. The four most common cancers accounting for more than half of all cancer cases are those of the prostate, breast, lung, and colon/rectum. Lung cancer, by far the most lethal, is responsible for nearly 30% of all cancer death.

20.2 GROWTH PROPERTIES OF NORMAL AND CANCEROUS CELLS

1. Growth of cancer cells produce multilayered clumps. Information on the behaviour of human cancer cells has come largely from studies on cells grown in vitro. At the cellular level, the most important characteristic of a cancer cell—whether residing in the body or on a culture dish—is its loss of growth control. The **capacity** for growth and division is not drastically different between a cancer cell and most normal cells. When normal cells are grown in tissue culture, under conditions that promote cell proliferation, they grow and divide at a rate similar to that of their malignant counterparts. However, when the normal cells proliferate to the point where they cover the bottom of culture dish, their growth rate decreases markedly, and they tend to remain as a single layer (monolayer) of cells (Fig. 20.1A). Growth rates drop as normal cells respond to inhibitory influences from their environment. Growth-inhibiting influences may arise as the result of depletion of growth factors in the culture medium or from contact with surrounding cells on the dish (called **contact inhibition**). In contrast, when malignant cells are cultured under the same conditions, they continue to grow, piling on top of one another to form clumps (Fig. 20.1B). It is evident that malignant cells are not responsive to the types of regulatory signals that causes their normal counterparts to cease growth and division.

Stages of Tumour Development.

At the cellular level, the development of cancer is viewed as multistep process involving **mutation** and **selection** for cells with progressively increasing capacity for proliferation, survival, invasion, and metastasis (Fig. 20.2).

1. Initiation. The step of tumour initiation is thought to be the result of a genetic alteration due to mutation leading to abnormal proliferation of a single cell. Cell proliferation then leads to the outgrowth of population of clonally derived tumour cells.

2. Tumour progression. This process continues as additional mutations occur within cells of the tumour population. Some of these mutations confer a selective advantage to the cell, such as more rapid growth, and the descendants of a cell bearing such mutations will consequently become dominant within the tumour population. The process is called **clonal selection**, since a new clone of tumour cells has evolved on the basis of its increased growth rate or other properties (such as survival, invasion or metastasis) that confer a selective advantage. Clonal selection continues throughout tumour development, so tumours continuously become more rapid-growing and increasingly malignant.

Normal cells

Normal cells grow in monolyer

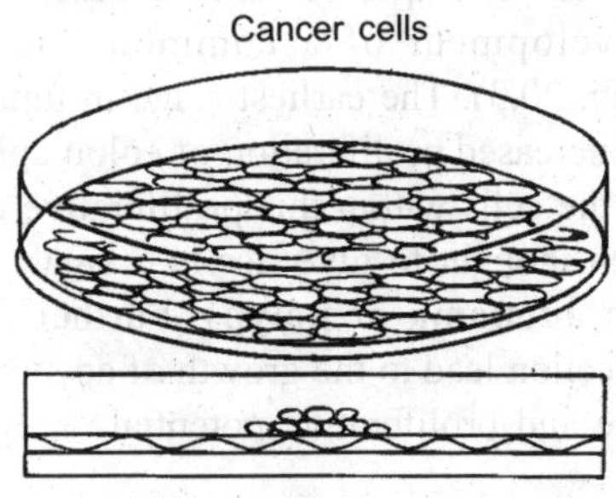

Fig. 20.1. Growth properties of normal and cancerous cells. A—Normal cells typically grow in a culture dish until they cover the surface as a monolayer. B—In contrast, cells that have been transformed by viruses or carcinogenic chemicals (or malignant cells that have been cultured from tumours) typically grow in multilayered clumps or foci (after Karp, 2002).

Angiogenesis.

Both primary and secondary tumours require *angiogenesis*, the formation of new blood vessels, in order to grow to a large mass. Most tumours induce the formation of new blood vessels that invade the tumour and nourish it, a process called **angiogenesis**. This complex process requires several distinct steps:

1. Degradation of the basal lamina that surrounds a nearby capillary,
2. Migration of endothelial cells lining the capillary into the tumour,
3. Division of these endothelial cells and
4. Formation of a new basement membrane around the newly elongated capillary.

Many tumours produce growth factors that stimulate angiogenesis; other tumours somehow induce surrounding normal cells to synthesize and secrete such factors. Basic fibroblast growth factors (bFGF), transforming growth factor α (TGFα) and vascular endothelial growth factor (VEGF), which are secreted by many tumours, all have angiogenic properties. New blood vessels nourish the growing tumour, allowing it to increase in size and thus increase the probability that additional harmful mutations will occur (see **Lodish** *et al.*, 2004).

Example of Development of a Human cancer: Colon carcinomas

Studies of colon carcinoma have provided a clear example of tumour progression during the development of a common human malignancy (Fig. 20.3). The earliest stage in tumour development is increased proliferation of colon epithelial cells. One of the cells within this proliferative cell population is then thought to give rise to a small benign neoplasm (an **adenoma** or **polyp**). Further rounds of clonal selection lead to the growth of adenomas of increasing size and proliferative potential.

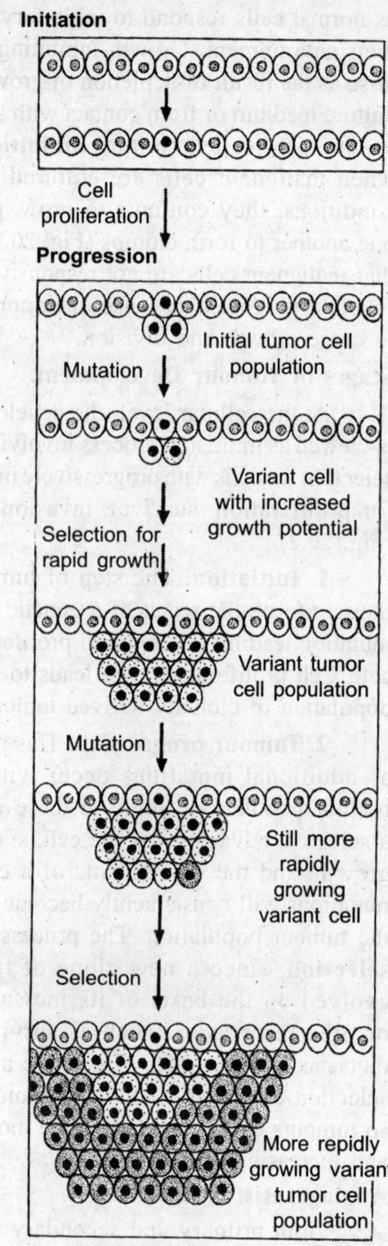

Fig. 20.2. Stages of tumour development. The development of cancer initiates when a single mutated cell begins to proliferate abnormally. Additional mutations followed by the selection for more rapid growing cells within the population then result in progression of the tumour to increasingly rapid growth and malignancy (after Cooper and Hausman, 2007).

20.3 CHARACTERISTICS OF CANCER CELLS

The cancerous cell generally retains the structural and functional characteristics of the normal cell type from which it is derived. For example, the cancerous cells of thyroid gland continue to secrete thyroxin. Neoplastic cells, however, differ from their normal counterparts in several ways.

1. Immortalization. Normal cell cultures do not survive indefinitely. For example, human cell cultures die after about 50 generations, and chicken cell cultures have a much shorter life expectancy. In contrast, transformed cell cultures are immortal and can grow indefinitely. A striking example is provided by HeLa cells, which were obtained in 1953 from a uterine cancer diagnosed in a women named **Henreitta Lacks** (hence the term "HeLa" cells). After the tumour was removed by surgeons, some of its cells were placed in culture. The cultured cells quickly began to proliferate and have continued to do so for more than 50 years, dividing more than 18,000 times with no sign of stopping (see **Becker** *et al.*, 2007). Similarly, cell cultures infected with mouse sarcoma virus can be maintained as long as nutrition is provided and overcrowding avoided.

Box 20.4

Cancer cells, like germ cells and stem cells but unlike most differentiated cells, produce **telomerase** enzyme, which prevent shortening of chromosomes during DNA replication and may contribute to their **immortalization**. The absence of telomerase is associated with resistance to generation of certain types of human tumours (*e.g.*, papillomas and colon tumours of mice) (see **Lodish** *et al.*, 2004).

2. Loss of contact inhibition. Normal cells in a culture stop growing when their plasma membranes come into contact with one another. This inhibition of growth after contact is called **contact inhibition**. Transformed cells are unable to go into a quiescent stage after division. They will grow (and divide) continuously until they kill themselves. Transformed cells apparently undergo a change in the property of their plasma membranes, which become less adhesive. This change enables the cells to dissociate from neighbouring cells and to infiltrate other organs, where they form metastatic tumours. Cancer cells thus, lack proper recognition and communication.

3. Reduced cellular adhesion. When normal cells become cancerous there is change in the **stickiness** or **adhesiveness** of their plasma membranes. If grown in a nutrient medium kept in a glass vessel, the normal cells stick to the glass rather than float in the medium. Transformed cells show a **decreased adhesiveness** and if grown in solid media, they stick to each other less than do normal cells.

Adhesiveness shows considerable specificity. For example, liver cells tend to stick to other liver cells and not to other cell types, *e.g.*, kidney cells. If the cells of the liver and the kidney are separated by the enzyme trypsin and incubated together, they aggregate to form small pieces of liver tissue and kidney tissue. Thus, kidney cells stick to kidney cells and liver cells to liver cells. Cancerous cells do show this property (*i.e.*, specificity of adhesiveness). Thus, if malignant skin cancer cells are mixed with normal kidney cells then aggregates formed contain both kidney and skin cells mixed together. This probably explains why malignant cell can invade several normal organs.

4. Invasiveness. One of the most significant characteristics of transformed cells is their invasiveness, *i.e.*, the ability of cancer cells, to invade other tissues. In contrast to normal cells, transformed cells can penetrate the chorioallantoic membrane of the hen's egg. This invasiveness could be due to changes in the plasma membrane and/or proteases released by the cancerous cells.

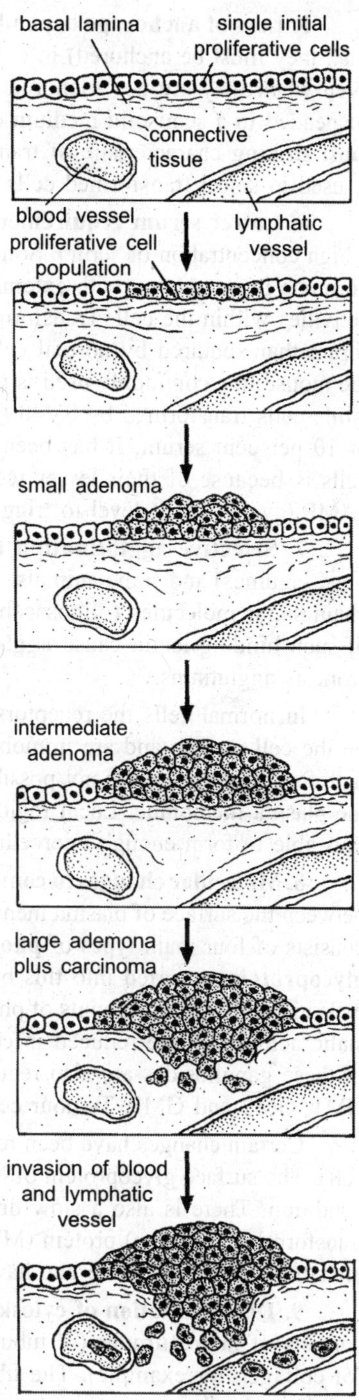

Fig. 20.3. Development of colon carcinomas. A single initially altered cell gives rise to a proliferative cell population, which progresses first to benign adenomas of increasing size and then to malignant carcinoma. The cancer cells invade the underlying connective tissue and penetrate blood and lymphatic vessels, thereby spreading throughout the body (after Cooper and Hausman, 2007).

5. Loss of anchorage dependence. Most normal cells must be attached to a rigid substratum (*i.e*., they must be anchored) in order to grow in a culture medium. Transformed cells can grow even when they are not anchored to the substratum. For example, when transformed cells are suspended in a semisolid medium containing agar or methyl cellulose. This loss of anchorage is most striking characteristic of transformed cells which form malignant tumours. This property is used to select transformed cells from a normal cell population.

6. Lower serum requirements. Growth of normal cells in tissue culture medium requires a high concentration of serum. Some serum growth factors (*e.g.*, somatomedins) resemble insulin hormone in interacting with external receptors on the plasma membrane to regulate biochemical activities within the cell. Transformed cells can grow in a culture medium containing much less serum than required by normal cells. For example, normal 3T3 cells (established fibroblasts of the mouse line commonly used is tissue culture) grow optimally in 10 per cent foetal calf-serum, while cells transformed by SV40 (simian virus number 40) can grow equally well in 1 per cent or 10 per cent serum. It has been suggested that the lower serum requirement of transformed cells is because of their lesser requirement of out side substances to lower their intracellular cAMP (cyclic AMP) level to trigger mitosis.

7. Selective agglutination by lectins. Lectins are proteins widely distributed in plants (*e.g.*, legumes) and some animals. They have the ability to bind to receptors, which are branched chain sugar molecules (oligosaccharides), on the surface of the plasma membrane. As a result of such binding lectin cause **agglutination** or **clumping** of cells. Due to this property, lectins work as agglutinins.

In normal cells the receptors or agglutin binding sites for lectins lie in a diffuse manner on the cell surface and are immobile. In such cases, lectins make few intercellular bridges, and therefore agglutination is not possible. In transformed cells, the receptors are more mobile within the plasma membrane. Local regions of high binding site concentration are formed. Lectins are thus able to form enough intercellular bridges to result in agglutination in the transformed cells.

8. Molecular changes in components of the plasma membrane. There are several differences between the surface of plasma membranes of normal and transformed cells. The plasma membrane consists of four main types of **phospholipids**, which form the lipid bilayer, with **glycolipids** and **glycoproteins** inserted into this bilayer. Cancerous cells apparently do not differ from normal cells in their relative amounts of phospholipids. However, **gangliosides** (glycolipids which contain sialic acids) become reduced in certain mouse cancer cells. Enzymes involved in biosynthesis of these gangliosides are also reduced. Normal cells possess four types of gangliosides, GM1a, GM1, GM2 and GM3. Tumour cells predominantly contain the simplest type, GM3.

Certain changes have been reported in the **glycoproteins** of plasma membrane of cancerous cells. The surface glycoprotein of MW 46,000 disappears early in transformation to the cancerous condition. There is also a slow disappearance of a major protein called **LETS** (large, external, transforming sensitive) protein (MW 240,000). Probably the most important protein to disappear after transformation is the one having MW 200,000.

9. Disorganisation of cytoskeleton. Normal cells have a well organised cytoskeleton which consists of **microtubules** (of tubulins), **microfilaments** (of actins) and **intermediate filaments** (of collagen for example). The fibre-like proteins have a regular arrangement and bring about coordinated cell movement. In transformed or cancer cells, the fibres are much fewer in number and usually much thinner. In transformed cells the cytoskeleton is found to undergo **depolymerisation**. The microtubules disaggregate. The microfilaments undergo depolymerisation and disappear, but diffuse actin persis.s. The myosin-like filaments also disappear. Thus, in transformed cells the cytoskeleton proteins become less organised than in normal cells. It has been suggested that due to this disorganisation of the cytoskeleton, there is an increased mobility of plasma membrane proteins.

The disorganisation of the cytoskeleton also affects the cell surface in another way. When cancer cells are touched, there is a constant and uncoordinated throwing out and retraction of **blebs**, **microvilli** and **ruffles** from the cell surface. Tumour cells have a more ruffled surface than normal cells, with many more surface processes.

10. Increase in negative surface charge. Comparisons of surface membrane charge by microelectrophoresis have been made between normal and malignant cells. In malignant cells anodic mobility is usually higher, indicating increase in negative surface charge.

11. Defective electrical communication. Electrical connections normally occur between individual normal cells. In some cancer cells, however, it has been reported that such electrical connections are defective.

12. Increased sugar transport. Tumour cells consume much more glucose than normal cells because they have to grow and multiply. There is a great increase in the rate of sugar transport across the surface membrane (plasma membrane) after transformation of the cell. This tends to increase in sugar intake by malignant cells.

13. Increased rate of glycolysis. In the 1920's **Warburg** pointed out that oxidative (aerobic) respiration is depressed in tumour cells and that glycolysis (anaerobic respiration) increases. This has been demonstrated by an increase in lactic acid production in cells of solid tumour.

14. Appearance of virus-specific transplantation rejection antigens. Plasma membranes of most transformed cells contain **antigens** which are not present in normal cells. Thus in cells transformed by adenoviruses and papovaviruses the **T-antigen** is always present. Similarly all cells transformed by the Epstein-Barr virus (EB virus) contain an antigen called the **EB nuclear antigen (EBNA)**. Tumour antigens can bring about an **immunity response** against themselves in genetically similar hosts. This is in contrast to normal histocompatibility antigens. The immunity response brought about by the antigens results in recognition and destruction of newly formed cancer cells and their descendants. Such a defence mechanism is called **immunological surveillance** and can lead to the elimination of transformed or cancer cells under favourable conditions. It has been suggested that only in the rare cases when this defence mechanism fails that tumours are formed.

15. Increased secretion of proteolytic enzymes. Large amount of proteolytic enzymes are secreted by all types of cancer cells, except those of blood forming tissues. The cancer cell secretes a protease called the **cell factor** (MW ~40,000). The cell factor acts on an inert serum protein **plasminogen** (MW 85,000) to form a **plasmin**, a **proteolytic enzyme** (MW 76,000). It has been suggested that plasmin removes many proteins projecting from the cell surface by enzymatic digestion and signals the cell into division.

If normal cells are treated with proteases they show many of the characteristics of transformed cells. It has been speculated that viral proteins cause the release of extracellular protease. However, there is no direct evidence for this.

16. Aldolases. In most mammalian tissues the enzyme **aldolase** exists in the form of three isozymes A, B and C. Isozymes A and C predominate in embryonic tissues while in adult differentiated tissues the B isozyme is predominant. In some tumours, especially in poorly differentiated and rapidly growing cancers such as **hepatomas**, isozyme B is replaced by isozyme A, the embryonic form.

17. Most cancers develop later in life. Because the multiple mutations that lead to formation of a tumour may require many years to accumulate, most cancers develop later in life. The occurrence of cancer after the age of reproduction may be one reason that evolutionary restraints have not done more to supress cancer. The requirement for multiple mutations also lowers the frequency of cancer compared with what it would be if tumerogenesis were triggered by a single mutation. However, huge numbers of cells are, in essence, mutagenized and tested for altered growth during our lifetime, a sort of evolutionary selection for cells that proliferate. *Fortunately the tumour itself is not inherited* (see **Lodish** *et al.*, 2004).

20.4 WHAT CAUSES CANCER?

Cancers are caused mainly by environmental agents and lifestyle factors, most of which act by triggering DNA mutations.

1. Epidemiological data. The first indication that a particular agent may cause cancer is usually provided by an approach called **epidemiology,** the branch of medical science that investigates the frequency and distribution of diseases in human populations. Epidemiological studies have revealed that cancers arise with differing frequencies in different parts of the world. For example, stomach cancer is frequent in Japan, breast cancer is prominent in the United States and liver cancer is common in Africa and Southeast Asia. To determine whether differences in heredity or environment are responsible for such differences, scientists have examined cancer rates in people who move from one country to another. For example, in Japan the incidence of stomach cancer is greater and the incidence of colon cancer is lower than in United States. When Japanese families move to the United States, their cancer rates come to resemble the rates in the United States, indicating that cancer rates are determined more by environment and lifestyle factors than by heredity.

Epidemiological data have played an important role in identifying environmental factors that can cause cancer. The most forcible statics involve lung cancer, a disease that has increased over tenfold in frequency in the United states since 1900. When the possible causes for this epidemic of lung cancer were investigated, it was discovered that virtually all lung cancer patients share one trait: a history of smoking cigarettes. As might be expected if cigarettes were responsible, heavy smokers develop lung cancer more frequently than light smokers, and long-term smokers develop lung cancer more frequently than do short-term smokers.

While the association between cigarette smoking and cancer was first detected through epidemiological studies, definitive proof of a cause-and-effect relationship requires direct experimental evidence. In the case of cigarettes and lung cancer, such evidence has come from studies showing that cigarette smoke contains several dozen chemicals that cause cancer when administered to animals. Such substances that cause cancer are called **carcinogens**.

2. Carcinogens. The idea that certain chemicals, such as those found in tobacco smoke, can cause cancer was first proposed more than 200 years ago. In 1761 a London doctor, **John Hill**, reported that people who routinely use snuff (a powdered form of tobacco that is inhaled) experience an abnormally high incidence of nasal cancer, suggesting the presence of carcinogen (cancer-causing) chemicals in tobacco. A few years later another British physician, **Percival Pott**, observed an elevated incidence of scrotum cancer among men who had served as chimney sweepers in their youth. It was common practice at the time to employ young boys to clean chimney flues because they fit into narrow spaces more readily than adults. Pott speculated that the chimney soot became dissolved in the natural oils of the scrotum, irritated the skin, and eventually triggered the development of cancer. This theory led to the discovery that scrotum cancer could be prevented among chimney sweepers through the use of protective clothing and regular bathing practices.

In forthcoming years, the list of known and suspected carcinogens has grown to include hundreds of different chemicals. Chemicals are usually labelled as carcinogens because humans or animals develop cancer when exposed to them. This does not mean, however, that each of these substances causes cancer through its own direct action. For example, consider the behaviour of **2-naphthylamine**, a potent carcinogen that causes bladder cancer in industrial workers and is present in tobacco smoke. As might be expected, feeding 2-naphthylamine to laboratory animals induces a high incidence of **bladder cancer**. But if 2-naphthylamine is implanted directly into an animal's bladder, cancer rarely develops. The explanation for the apparent difference is that

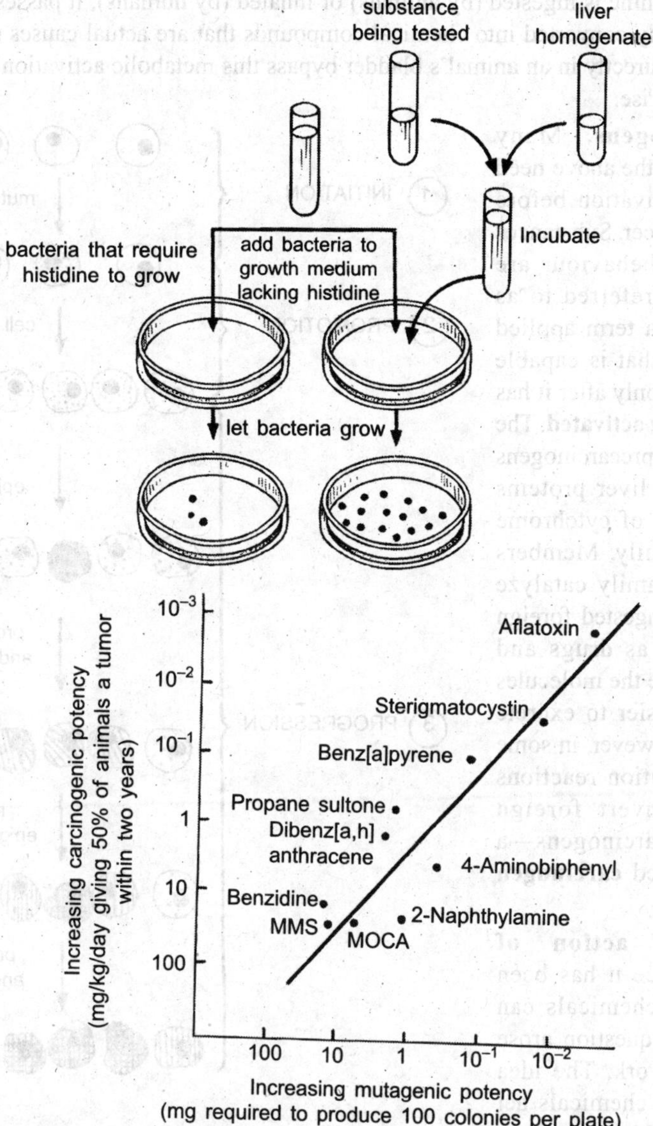

Fig. 20.4. Procedure of Ames test. Ames test is based on the grounds that most carcinogens are mutagens. A—The ability to induce mutations is measured in bacteria that lacks the ability to synthesize the amino acid histidine. When placed in a growth medium that lacks histidine, the only bacteria that can grow are those that have acquired a mutation that allows them to make histidine. The number of bacterial colonies that grow is therefore related to the mutagenic potency of the substance being tested. Chemicals being investigated with the Ames test are first incubated with a liver homogenate because many of the chemicals to which humans are exposed only become carcinogenic after they have undergone biochemical modification in the liver. B—The data in the graph reveal that substances that exhibit strong mutagenic activity in the Ames test also tend to be strong carcinogens. Here, aflatoxin is the most potent mutagen and the most potent carcinogen. (*Abbreviations:* MOCA = 4–4′ methylene-bis 2-chloroanaline, MMS = methyl methane sulphonate; after Becker *et. al.*, 2007).

2-naphthylamine is ingested (by animals) or inhaled (by humans), it passes through the liver and is metabolically converted into chemical compounds that are actual causes of cancer. Placing 2-naphthylamine directly in an animal's bladder bypass this metabolic activation and consequently cancer does not arise.

Precarcinogens. Many carcinogens share the above need for metabolic activation before they can cause cancer. Substances exhibiting such behaviour are more accurately referred to as **precarcinogens**, a term applied to any chemical that is capable of causing cancer only after it has been **metabolically activated.** The activation of most precarcinogens is carried out by liver proteins that are members of cytchrome P450 enzyme family. Members of this enzyme family catalyze the oxidation of ingested foreign chemicals, such as drugs and pollutants, to make the molecules less toxic and easier to excrete from the body. However, in some cases these oxidation reactions accidentally convert foreign chemicals into carcinogens—a phenomenon called **carcinogen activation**.

Mode of action of carcinogens. Once it has been determined that chemicals can cause cancer, the question arose as to how they work. The idea that carcinogenic chemicals act by triggering **DNA mutations** was first proposed around 1950, but there was little supporting evidence at that time because nobody had systemically compared the mutagenic potency of different chemicals with their ability to cause cancer.

Ames Test. The need for identifying potential carcinogens, inspired **Bruce Ames** (1973) to develop a simple rapid laboratory test for measuring a chemical's

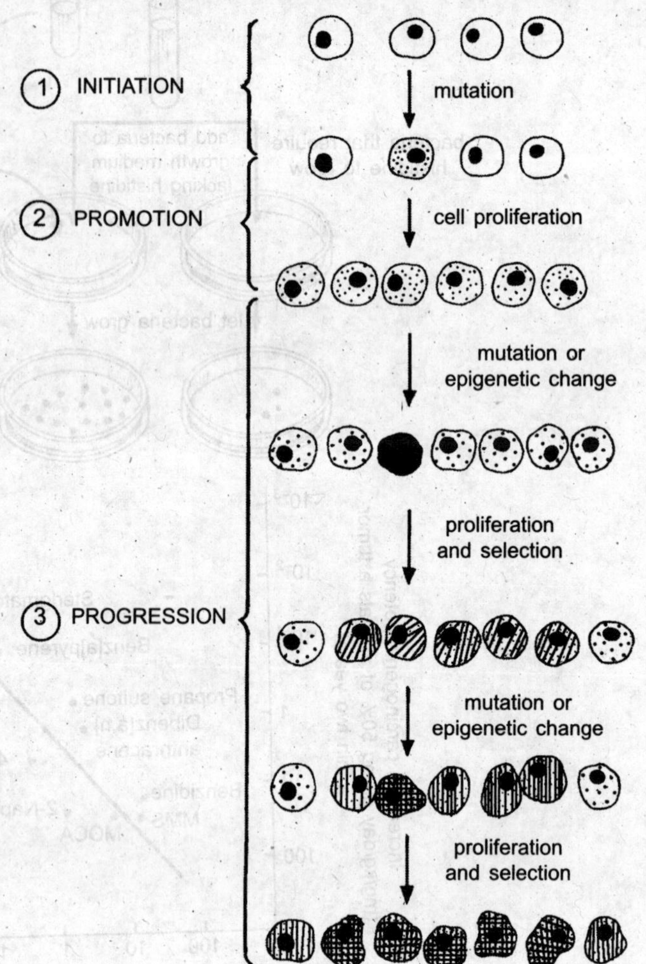

Fig. 20.5. Main stages in the development of cancer arises by a complex process involving three main stages. 1. Initiation. It is based on DNA mutation. 2. Promotion. During this stage the initiated cell is stimulated to proliferate. 3. Progression. During tumour progression, further mutations and changes in gene expression create variant cells exhibiting enhanced growth rates or other aggressive properties that give certain cells a selective advantage. Such cells tend to outgrow their companions and become the predominant cell population in the tumour. Repeated cycles of this **selection process** create a population of cells whose properties gradually change over time (after Becker *et al.*, 2007).

mutagenic activity. The procedure he developed, called the **Ames test**, utilise bacteria as a test organisms because they can be quickly grown in enormous numbers in culture. The bacteria used for the Ames test are a special strain that lack the ability to synthesize the amino acid **histidine**. As shown in Fig. 20.4, the bacteria are placed in a culture dish containing a growth medium without histidine, along with the chemicals being tested for mutagenic activity. Normally, the bacteria would not grow in the absence of histidine. However, it will trigger random mutations, some of which might restore the ability to synthesize histidine. Each bacterium acqui-ring such a mutation will grow into a visible colony, so that the total number of colonies is a measure of the **mutagenic potency** of the substance being investigated.

Since many chemicals that cause cancer only become car-cinogenic after they have been modified by liver enzymes, the Ames test includes a step in which the chemical being tested is first incubated with an extract of liver cells to mimic the reactions that normally occur in the liver. The resulting chemical mixture is then tested for its ability to cause bacterial mutations. When the Ames test is performed in this way, a strong correlation is observed between a chemical's ability to cause mutations and its ability to cause cancer.

Cancer arise through a multistep process. Chemicals induce the development of cancer through a multistep process involving **initiation, promotion**, and **tumour progression. Initiation** is based on DNA mutation, whereas **promotion** involves proliferation of the initiated cells for a prolonged period of time, a accompanied by a gradual **selection** of cells exhibiting enhanced growth properties. During **tumour progression**, cells acquire additional mutations and undergo changes in gene expression that produce cells with increasingly aberrant traits.

3. Role of ionizing and ultraviolet radiations in carcinogenesis. Shortly after the discovery of **X-rays** by **Wilhelm Roentgen** in 1895, it was noticed that people working with this type of radiation developed cancer at abnormally high rates. Animal studies subsequently confirmed that X-rays create DNA mutations and cause cancer in direct proportion to the dose administered.

X-rays type of radiations is emitted by many **radioactive elements.** An early example of the carcinogenic hazards posed by radioactivity occurred in 1920's in a New Jersey factory that produced glow-in-the-dark watch dials. A luminescent paint containing the radioactive element **radium** was used for painting the dials, and this paint was applied with a fine tipped brush that the workers frequently wetted with their tongues. As a result, tiny

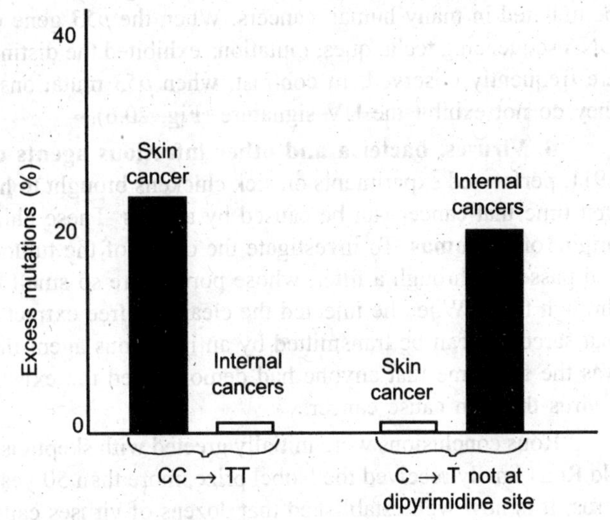

Fig. 20.6. Incidence of two types of *p53* mutations in skin cancer and internal cancers. The two bars on the left represent the frequency of CC → TT mutations, which are triggered by UV radiation. The two bars on the right represent the frequency of C → T mutations not located at dypyrimidine sites, which are not caused by UV radiation. Note that the UV-triggered type of mutation is found in the *p53* gene of skin cancers (squamous cell carcinomas), but not in cancers of internal organs. Mutations frequencies are plotted relative to what would be expected to occur randomly (after Becker *et al.*, 2007).

quantities of radium were carelessly ingested and became concentrated in their bones, leading to the development of **bone cancer**.

High rates of cancer caused by exposure to radioactivity have also been observed in people exposed to radioactive fallout from nuclear explosions. The most dramatic incidents occurred in the Japanese cities of Hiroshima and Nagasaki after atomic bombs were dropped there in 1945. Another incident occurred in the area surrounding the Chernobyl nuclear power plant in the former Soviet Union (now Ukraine), which exploded in 1986.

X-rays and related forms of radiation emitted by radioactive elements, are called **ionizing radiation** because they remove electrons from molecules, thereby generating highly reactive ions that create DNA damage. **Ultraviolet radiation (UV)** is another type of radiation that causes cancer by damaging DNA. The ability of the UV radiation in sunlight to cause cancer was first deduced from the observation that skin cancer is most prevalent in people who spend long hours in the sun, especially in tropical regions where the sunlight is very intense. UV radiation is absorbed mainly by the skin, where it imparts enough energy to trigger **pyrimidine dimer** formation—that is, the formation of covalent bonds between adjacent pyrimidine bases in DNA. If the damage is not repaired, distortion of the double helix causes improper base-pairing during DNA replication that leads to distinctive mutation patterns. For example, a CC $\rightarrow$ TT mutation (conversion of two adjacent cytosines to thymines) is a unique product of UV exposure and can therefore be used as a distinctive "signature" to identify mutations caused by sunlight.

The existence of such **signature mutations** provided a way to prove that UV-induced mutations cause skin cancer. One of the first genes studied was the *p53* gene, which is known to be mutated in many human cancers. When the *p53* gene of skin cancer cells, is examined using DNA sequencing techniques, mutations exhibited the distinctive UV signature (such as CC $\rightarrow$ TT) are frequently observed. In contrast, when *p53* mutations are detected in other types of cancer, they do not exhibit the UV signature (Fig. 20.6).

4. Viruses, bacteria and other infectious agents cause some cancers. Peyton Rous, in 1911, performed experiments on sick chickens brought to him by local farmers that showed for the first time that cancer can be caused by a virus. These chickens had cancers of connective tissue origin, or **sarcomas**. To investigate the origin of the tumours, **Rous** ground up the tumour tissue and passed it through a filter, whose pores were so small that not even bacterial cells could pass through them. When he injected the clear cell-free extract into healthy chickens, Rous concluded that sarcomas can be transmitted by an infectious agent that is smaller than a bacterial cell. This was the first time that anyone had demonstrated the existence of an **oncogenic viruses**—that is, a virus that can cause cancer.

Rous conclusions were initially greeted with skepticism. It was not until 1966 that an 87-year old Rous finally received the Nobel prize, more than 50 years after his discovery of the first cancer virus. It is now well established that dozens of viruses cause cancer in animals and that a smaller number cause cancer in humans. The first human example was discovered by **Danis Burkitt**, a British surgeon working in Africa in the late 1950's. At certain times of the year, Burkitt noted that a large number of his young patients developed massive **lymphocytic cancers** of the neck and jaw. Because this kind of cancer, now known as **Burkitt's lymphoma** occurred in periodic epidemics localized to specific geographical regions, Burkitt proposed that it was transmitted by an infections agent.

Burkitt's ideas soon attracted the attention of two virologists, named **Epstein** and **Barr**, whose electronic microscopic studies revealed virus particles in Burkitt's lymphoma cells, the virus is now called the **Epstein-Barr virus** or **EBV** in recognition of these virologists (Box 20.5).

Box 20.5 EBV and Carcinogenesis

Unlike viruses that cause cancer in animals, it is difficult to prove that a virus such as EBV causes cancers in humans because ethical considerations prevent directly testing the hypothesis by injecting the virus into healthy individuals. Nonetheless, some indirect evidence supports the conclusion that EBV causes Burkitt's lymphomas: 1. DNA sequences and proteins encoded by EBV are detected in tumour cells obtained from patients with Burkitt's lymphoma, but not in normal cells from the same individuals; 2. Adding purified EBV to normal human lymphocyte cultures cause the cells to acquire the properties of cancer cells; and 3. Injecting EBV into monkeys induces the formation of lymphomas.

Following the discovery of EBV, several additional viruses have been identified as causes of human cancer. For example, liver cancer is caused by **hepatitis B** and **hepatitis C** viruses and cervical cancer is caused by **human papillomavirus (HPV)**. In addition to viruses, the bacterium *Helicobacter pylori* (*H. pylori*) causes stomach cancer, and parasitic flatworm infections have been linked to a small number of bladder and bile duct cancers.

Infectious agents trigger the development of cancer in two fundamentally different ways. One involves those agents such as the hepatitis B and C viruses, *H. pylori*, and parasitic flatworms that cause tissue destruction and chronic inflammation. Such destruction creates chronic inflammatory conditions in which cells of the immune system infiltrate the tissue and attempt to kill the infectious agent and repair the tissue damage. Unfortunately, the mechanisms used by immune cells to fight infections often produce mutagenic chemicals, such as **oxygen free radicles** (highly reactive forms of oxygen containing an unpaired electrons). This means that proliferation of replacement cells for the injured tissue takes place under conditions in which DNA damage is likely, thereby increasing the likelihood that cancer-causing mutations will arise.

The other way in which infectious agents cause cancer is related to the ability of certain viruses to directly stimulate the proliferation of infected cells. The mechanism by which this is accomplished varies among viruses. In some cases, it involves cancer-causing viral genes, whereas in other cases a virus alters the behaviour of host cell genes. The types of genes involved in such events play a role not just in viral cancers, but in cancers caused by chemicals and radiation as well.

20.5 GENES INVOLVED IN CANCERS

Oncogenes are genes whose presence can cause cancer. While they are sometimes introduced into cells by viruses, more often they arise from normal cellular genes called **proto-oncogenes** by *point mutation, gene amplification, chromosomal translocation, local DNA rearrangements* or *insertional mutagenesis*. Most proteins produced by oncogenes are signalling pathway components, such as growth factors, receptors, plasma membrane GTP-binding proteins, non-receptor protein kinases, transcription factors, and cell-cycle or cell-death regulators. Oncogenes code for abnormal forms or excessive quantities of such proteins, thereby leading to excessive stimulation of cell proliferation.

Tumour suppressor genes are genes whose loss or inactivation can lead to cancer. Susceptibility to developing cancer is increased in people who inherit defective tumour suppressor genes. Three important tumour suppressor genes are:

1. The *RB* gene, which produces a protein that restrains passage from G1 into S phase.
2. The *p53* gene, which produces a protein that prevents cells with damaged DNA from proliferating.
3. The *APC* gene, which produces a protein that inhibits the Wnt pathway. The **Wnt pathway** plays a prominent role in controlling cell proliferation and differentiation during embryonic development. The central component of this Wnt pathway is a protein called **β-catenin**.

20.6 DIAGNOSIS, SCREENING AND TREATMENT OF CANCER

Much progress has been made in recent years in explaining the genetic and biochemical abnormalities that underlie cancer development. One of the hopes for such research is that our growing understanding of the molecular alterations exhibited by cancer cells will eventually lead to improved strategies for cancer diagnosis and treatment.

1. Microscopic Analysis of Cancer

Because cancer can arise in almost any tissue, few regular generalizations are possible regarding disease symptoms. A definitive diagnosis typically requires a **biopsy**, which involves surgical removal of a tiny tissue sample for microscopical examination. Under the microscope, cancer cells show a number of features that together indicate the presence of cancer (Table 20.1). For example, cancer cells often have large, irregularly shaped nuclei, prominent nucleoli, and a high ratio of nuclear-to-cytoplasmic volume. Cancers also tend to exhibit significant variability in cell size and shape, as well as a loss of normal tissue organisation. To varying extents, cancer cells lose the specialized structural and biochemical properties of the cells normally residing in the tissue of origin. Cancers usually have more dividing cells than normal, which means that the mitotic index (Box 20.6) will be elevated.

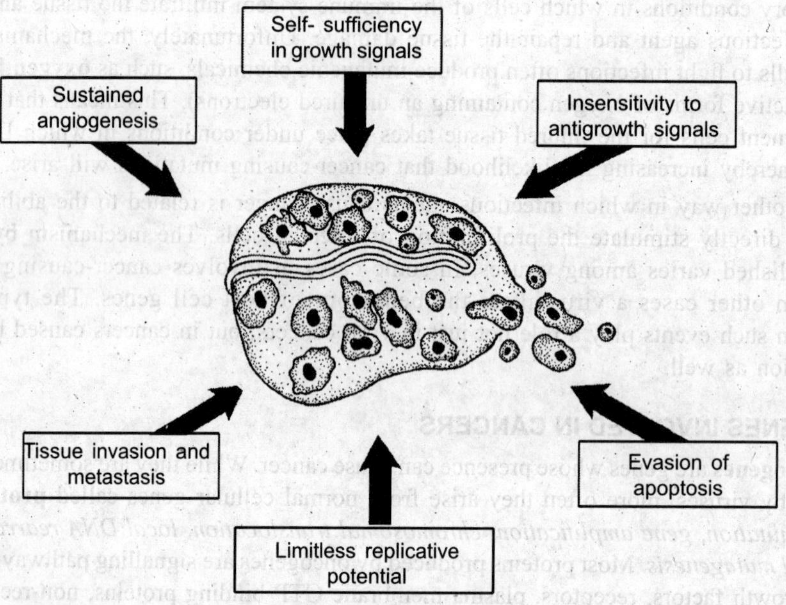

Fig. 20.7. Overview of changes in cells that cause cancer. During carcinogenesis, six fundamental cellular properties are altered, as shown here, to give rise to the complete, most destructive cancer phenotype Less dangerous tumours arise when only some of these changes occur (after Lodish *et. al.*, 2004).

Table 20.1 Some differences in the microscopic appearance of benign and malignant tumours (source: Becker *et al.*, 2006)

	Trait	*Benign*	*Malignant*
1.	Nuclear size	Small	Large
2.	N/C ratio (Ratio of nuclear to cytoplasmic volume)	Low	High
3.	Nuclear shape	Regular	Pleomorphic (irregular shape)

	Trait	Benign	Malignant
4.	Mitotic index	Low	High
5.	Tissue organisation	Normal	Disorganised
6.	Differentiation	Well-differentiated	Poorly differentiated (anaplastic)
7.	Tumour boundry	Well-defined	Poorly-defined

Box 20.6 Mitotic index

Percentage of cells in a population that are in any stage of mitosis at a certain point in time; used to estimate the relative length of the M phase of cell cycle. For example, the mitotic index for cultured mammalian cells is often about 3–5 per cent, which means that M phase lasts less than an hour (usually 30–45 minutes).

Tumour grading. If a sufficient number of these seven traits (Table 20.1) are observed upon microscopic examination of sample, it can be concluded that cancer is present. In other words, the presence of these traits indicates a tumour that, if left untreated will eventually spread by invasion and metastasis. The severity of these observed microscopic, abnormalities varies significantly among cancers, even when they arise from the same cell type and in the same organ. These variability form the basis for **tumour grading**, which is the assignment of numerical grades to tumours based on differences in their microscopic appearance.

Lower numerical grades (*e.g.*, grade 1) are assigned to tumours whose cells exhibit normal differentiated features, divide slowly, and display only modest abnormalities in the traits listed in Table 20.1. Higher numbers (*e.g.*, grade 4) are assigned to tumours containing rapidly dividing, poorly differentiated cells that bear less resemblance to normal cells and exhibit severe abnormalities in the trait, listed in Table 20.1. The highest grade cancers contain cells that are **anaplastic**, which means that they are so poorly differentiated and abnormal in appearance and organization that they bear no resemblance to the cells of the tissue in which the tumour arose. Such anaplastic, high-grade cancers tend to grow and spread more aggressively and be less responsive to therapy than lower grade cancers (see **Becker** *et. al.*, 2007).

2. Screening Techniques of Cancer

When cancer is detected before it has spread, cure rates tend to be very high. Therefore, a great need exists for screening techniques that can routinely detect cancers at an early stage. Some of these are following:

***(i)* Pap smear.** It is one of the most successful screening procedures that was developed in the early 1930s by **George Papanicolaou** (for whom it is named). The logic underlying this procedure is that the microscopic appearance of cancer cells is so distinctive that it is possible to detect the likely presence of cancer by simply examining a few isolated cells. A Pap smear is performed by taking a tiny sample of a woman's vaginal secretions and examining it with a microscope. If the cells in the fluid exhibit unusual features, such as large irregular nuclei or prominent variations in cell size and shape, it is a sign that cancer may be present and further tests need to be done. Because a Pap smear allows **cervical cancer** to be detected in its early stages before metastasis has occurred, this procedure has prevented hundreds of thousands of cancer deaths.

***(ii)* Mammography and other techniques.** The success of the Pap smear has led to development of screening techniques for other cancers. For example, **mammography** utilizes a special X-ray technique to look for early signs of **breast cancer** and **colonoscopy** uses a slender fiber-optic instrument to examine the colon (*i.e.*, the part of the large intestine that extends from the cecum to the rectum) for early signs of **colon cancer**. The ideal screening test would allow doctors to detect cancers anywhere in the body with one simple procedure, such as a blood test. **Prostate cancer** is an example of cancer that can sometimes be detected this way. Men over the age of 50 are often advised to get a **PSA test,** which measures how much **prostate-specific antigen (PSA)** is present in the bloodstream. PSA, which is a protein produced by cells of the prostate gland, normally appears in only tiny concentrations in the blood. If a PSA test reveals a high concentration of PSA, it indicates the existence of a prostate problem and further tests are performed to determine whether or not cancer is actually present.

***(iii)* Proteomic analysis.** Other cancers also release small amounts of specific proteins into the bloodstream, where there presence might be used to signal the existence of early disease. To investigate such tiny changes in blood proteins, scientists are experimenting with a general approach called **proteomic analysis** to analyse proteins present in the blood (the term proteomic refers to the complete set of proteins produced by an organism's genome). The key to most proteomic techniques is **mass spectrometry**, a high speed, extremely sensitive method for identifying proteins based on differences in mass and electrical charge. Because a blood sample contains thousands of different proteins, the data generated by proteomic analysis can be extremely complex. To deal with this problem, artificial intelligence software programs are used to compare the complex protein patterns seen in blood samples from individuals with or without cancer.

One of the first cancer to be investigated by proteomic analysis was **ovarian cancer**. When ovarian cancer is detected before it spreads, the five year survival rate exceeds 95 per cent. Using proteomic analysis, scientists have recently identified a pattern of five proteins in the blood of women with ovarian cancer that is not seen in the blood of other women.

3. Treatments of Cancer

People diagnosed with cancer have various treatment options that depend both on the type of cancer involved and how far it has spread.

***(i)* Surgery.** The most common treatment of cancer involves surgery to remove the primary tumour followed (if necessary) by radiation therapy and/or chemotherapy to destroy any remaining cancer cells.

***(ii)* Radiation therapy.** This approach employs high energy X-rays or other forms of ionizing radiation to kill cancer cells. We already know that DNA damage created by ionizing radiation can cause cancer, but paradoxically, the same type of radiation is also used in higher doses to destroy cancer cells in people who already have the disease. Ionizing radiation kills cells in two different ways. First, DNA damage caused by radiation activates the p53 signaling pathway, which then triggers cell death. However, many cancers have mutations that disable the p53 pathway, so p53-induced apoptosis plays only a modest role in the response of most cancers to radiation treatment. In the second mechanism, radiation kills cells by causing chromosomal damage that is so severe that it prevents cells from progressing through mitosis, and the cells therefore die while trying to divide.

***(iii)* Chemotherapy**. Most forms of chemotherapy use drugs that, like radiation, are intended to kill dividing cells. Such drugs can be subdividing into four major categories:

***(a)* Antimetabolites.** They inhibit metabolic pathways required for DNA synthesis by acting as competitive inhibitors that bind to enzyme active sites in place of normal substrate molecules. Examples of such antimetabolites include methotrexate, fluorouracil and mercaptopurine.

***(b)* Alkylating agents.** They inhibit DNA function by chemically cross-linking the DNA double helix, Examples of such drugs include cyclophosphamide, chlorambucil and cisplatin.

(c) Antibiotics. These are substances made by microorganisms that inhibit DNA function by either binding to DNA or inhibiting topoisomerases required for DNA replication. Examples of such antibiotics include bleomycin and doxorubicin.

(d) Plant-derived drugs. These drugs either inhibit topoisomerases or disrupt the microtubules of the mitotic spindle. Examples of such drugs include the topoisomerase-inhibitor **etoposide** and the microtubule disrupting drug, **taxol**.

One problem with such drugs (and radiation therapy) is that they are toxic to normal dividing cells as well as to cancer cells. When cancer arises in a tissue whose growth requires a specific hormone, it may be treated in a less toxic manner using drugs that block the action of that particular hormone. For example, many breast cancers require estrogen for their growth. Estrogens tend to exert their effects by binding to nuclear receptor proteins that activate the expression of specific genes. The drug **tamoxifen**, a common treatment for breast cancer, binds to estrogen receptors in place of estrogen and prevents the receptors from being activated.

Newer treatment approaches include **immunotherapies** that exploit the ability of the immune system to attack cancer cells, **molecular** targeting drugs aimed at proteins that are critical to the cancer cells, and **antiangiogenic agents** that attacks a blood supply or the tumour.

QUESTIONS

Long Answer Questions

1. Define the cancer. Describe different type of cancer. Explain various characteristics of cancer cells.
2. Explain various causes of cancer.
3. Describe diagnosis, screening and treatment of cancer.
4. Write an essay on cancer.

Short Answer Questions

1. What are preventive and control measures of cancer? *(Bangalore 2005)*
2. What is cancer? Mention any four carcinogens. *(Bangalore 2006)*
3. Write a short note on cancer. *(H.N.B, Garhwal, 2008)*
4. Explain cancer and cell transformation. *(H.N.B. Garhwal 2009)*

Very Short Answer Questions

1. Define cancer.
2. What is carcinogen?
3. What is Ames test?
4. What is a carcinoma?
5. Define the leukemia.
6. What is lymphoma?
7. Define the sarcoma.

Fill in the Blanks

1. One of the first cancers to be investigated by proteomic analysis was ....
2. The growths resulting from the division of .... are called ....
3. A .... tumour is capable of both invading surrounding normal tissues and spreading throughout the body via the circulatory or lymphatic system and is called ....
4. Four basic types of cancers include .... , ... , lymphoma and ....
5. Toxic substances that cause cancer are called ....
6. Ames test was designed by .... in 1973 to identify potential carcinogens.
7. UV radiation of .... is known to cause .... in human beings.

Multiple Choice Questions

1. Carcinoma refers to

(*a*) malignant tumours of the connective tissue

(*b*) malignant tumours of the skin or mucous membrane

(*c*) benign tumours of the connective tissue

(*d*) none of the above

2. Cancer cells are more easily damaged by radiation than normal cells because they are

(*a*) non-dividing (*b*) starved of food

(*c*) undergoing rapid division (*d*) different in structure

3. Which one of the following gives rise to cancer? (*Bihar 2003*)

(*a*) unrestricted and fast multiplication of cells by meiosis

(*b*) breakdown of mechanism regulating meiosis

(*c*) unlimited enlargement of cells by failure of formation of cleavage furrow after telophase to separate the daughter cells

(*d*) the two daughter cells after mitosis carrying different amount of DNA

4. Cancer can be defined as

(*a*) hyperplastic growth, infiltration and destruction of tissues

(*b*) non-malignant epithelial growth

(*c*) repair of damaged parts

(*d*) metabolic disorder

5. Blood cancer is called

(*a*) leukaemia (*b*) haemophilia

(*c*) thrombosis (*d*) haemolysis

6. The cells affected by leukaemia are

(*a*) plasma cells (*b*) erythrocytes

(*c*) thrombocytes (*d*) leucocytes

7. One of the faster and less expensive procedures for preliminary screening of potential carcinogens include

(*a*) Biuret (*b*) Dick

(*c*) Ames (*d*) ELISA

8. Sarcoma is the cancer of

(*a*) epithelial tissue (*b*) connective tissue

(*c*) blood (*d*) endocrine tissue

9. Cancer treatment includes

(*a*) surgery (*b*) radiotherapy

(*c*) treatment with anticancer drug (*d*) all of these

10. Which type of cancer is found in lymph nodes, gland and spleen?

(*a*) sarcoma (*b*) carcinoma

(*c*) adenoma (*d*) leukaemia

11. The spread of cancerous cells to distant sites is termed

(*a*) metamorphosis (*b*) metagenesis

(*c*) metastasis (*d*) metachrosis

ANSWERS

Very Short Answer Questions

1. Cancer is uncontrolled, growing mass of cells that is capable of invading neighboring tissues and spreading via body fluids, especially the bloodstream, to other parts of body. Cancer is also called **malignant tumour**.

2. Carcinogen is any cancer-causing agent.
3. It is screening test for potential carcinogens that assesses whether a substance causes mutations in bacteria.
4. It is a malignant tumour (*i.e.,* cancer) arising from the epithelial cells that cover external and internal body surfaces.
5. It is cancer of blood or lymphatic origin in which the cancer cells proliferate and reside mainly in the blood stream rather than growing as a solid masses of tissue.
6. It is cancer of lymphatic origin in which the cancer cells grow as solid masses of tissue.
7. Sarcoma is any cancer arising from a supporting tissue, such as bone, cartilage, fat, connective tissue and muscle.

Fill in the Blanks

1. Ovarian cancer.
2. Neoplastic, tumours.
3. Malignant, metastatis.
4. Carcinoma, sarcoma, leukemia.
5. Carcinogens.
6. Bruce Ames.
7. Sunlight, cancer.

Multiple Choice Questions

1. (*b*) **2.** (*c*) **3.** (*d*) **4.** (*a*) **5.** (*a*) **6.** (*d*) **7.** (*c*) **8.** (*b*) **9.** (*d*) **10.** (*c*) **11.** (*c*).

INDEX

D

E

I

J

K

L

Q

R

T

U

V

W

X

Y

Z